Developmental Biology

INTERNATIONAL TWELFTH EDITION

Michael J. F. Barresi

Smith College

Scott F. Gilbert

Professor Emeritus
Swarthmore College and the University of Helsinki

This version of the text has been adapted and customized. Not for sale in the USA or Canada.

 SINAUER ASSOCIATES

NEW YORK   OXFORD
OXFORD UNIVERSITY PRESS

Oxford University Press is a department of the University of Oxford. It furthers the University's objective of excellence in research, scholarship, and education by publishing worldwide.

© 2020 Oxford University Press
Sinauer Associates is an imprint of Oxford University Press.

Published in the United States of America by Oxford University Press
198 Madison Avenue, New York, NY 10016, United States of America

Oxford is a registered trade mark of Oxford University Press.

ISBN 9781605358741

Printing number: 9 8 7 6 5 4 3 2 1

Printed in the United States of America

*To those whose lives were most directly impacted
over the course of its creation.*

*My parents, Joseph and Geraldine Barresi –
Thanks for always putting your kids first.*

*My family, Heather, Samuel, Jonah, Luca, and Mateo –
This book was only accomplished with
your unwavering understanding, love and support.*

-M. J. F. B.

*To Anne, Daniel, Sarah, David, and Natalia,
whose support and humor have sustained me, and
to Alina who was born since the last edition.*

-S. F. G.

Brief Contents

I Patterns and Processes of Becoming: A Framework for Understanding Animal Development

CHAPTER 1 ● **The Making of Body and a Field**   Introduction to Developmental Biology **1**

CHAPTER 2 ● **Specifying Identity**   Mechanisms of Developmental Patterning **39**

CHAPTER 3 ● **Differential Gene Expression**   Mechanisms of Cell Differentiation **51**

CHAPTER 4 ● **Cell-to-Cell Communication**   Mechanisms of Morphogenesis **87**

CHAPTER 5 ● **Stem Cells**   Their Potential and Their Niches **127**

II Gametogenesis and Fertilization: The Circle of Sex

CHAPTER 6 ● **Sex Determination and Gametogenesis 159**

CHAPTER 7 ● **Fertilization**   Beginning a New Organism **193**

III Early Development: Cleavage, Gastrulation, and Axis Formation

CHAPTER 8 ● **Snails, Flowers, and Nematodes**   Different Mechanisms for Similar Patterns of Specification **223**

CHAPTER 9 ● **The Genetics of Axis Specification in *Drosophila* 247**

CHAPTER 10 ● **Sea Urchins and Tunicates**   Deuterostome Invertebrates **275**

CHAPTER 11 ● **Amphibians and Fish 295**

CHAPTER 12 ● **Birds and Mammals 335**

IV Building with Ectoderm: The Vertebrate Nervous System and Epidermis

CHAPTER 13 ● **Neural Tube Formation and Patterning 363**

CHAPTER 14 ● **Brain Growth 381**

CHAPTER 15 ● **Neural Crest Cells and Axonal Specificity 399**

CHAPTER 16 ● **Ectodermal Placodes and the Epidermis 441**

V Building with Mesoderm and Endoderm: Organogenesis

CHAPTER 17 ● **Paraxial Mesoderm**   The Somites and Their Derivatives **461**

CHAPTER 18 ● **Intermediate and Lateral Plate Mesoderm**   Heart, Blood, and Kidneys **491**

CHAPTER 19 ● **Development of the Tetrapod Limb 517**

CHAPTER 20 ● **The Endoderm**   Tubes and Organs for Digestion and Respiration **547**

VI Postembryonic Development

CHAPTER 21 ● **Metamorphosis**   The Hormonal Reactivation of Development **563**

CHAPTER 22 ● **Regeneration 579**

CHAPTER 23 ● **Development in Health and Disease**   Birth Defects, Endocrine Disruptors, and Cancer **619**

VII Development in Wider Contexts

CHAPTER 24 ● **Development and Evolution**   Developmental Mechanisms of Evolutionary Change **643**

Contents

PART I ● Patterns and Processes of Becoming:
A Framework for Understanding Animal Development

1

The Making of a Body and a Field
Introduction to Developmental Biology **1**

"How Are You, You?" Comparative Embryology and the Questions of Developmental Biology 2

The Cycle of Life 6
An animal's life cycle 6
A flowering plant's life cycle 7

Example 1: A Frog's Life 8
Gametogenesis and fertilization 9
Cleavage and gastrulation 9
Organogenesis 9
Metamorphosis and gametogenesis 9

Example 2: Even a Weed Can Have a Flower-Full Life 11
Reproductive and gametophytic phases 12
Embryogenesis and seed maturation 13
Vegetative phases: From sporophytic growth to inflorescence identity 13

An Overview of Early Animal Development 13
Patterns of cleavage 13
Gastrulation: "The most important time in your life" 14
The primary germ layers and early organs 17
Understanding cell behavior in the embryo 19

A Basic Approach to Watch Development 20
Approaching the bench: Find it, lose it, move it 20
Direct observation of living embryos 20
Dye marking 21
Genetic labeling 22
Transgenic DNA chimeras 22

Evolutionary Embryology 24
Understanding the tree of life to see our developmental relatedness 27
The developmental history of land plants 31

Personal Significance: Medical Embryology and Teratology 34
Genetic malformations and syndromes 35
Disruptions and teratogens 35

Coda 36

2

Specifying Identity
Mechanisms of Developmental Patterning **39**

Levels of Commitment 39
Cell differentiation 39
Cell fate maturation 40

Autonomous Specification 41
Cytoplasmic determinants and autonomous specification in the tunicate 41

Conditional Specification 44
Cell position matters: Conditional specification in the sea urchin embryo 44

Syncytial Specification 47
Opposing axial gradients define position 48

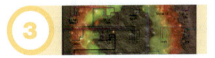

3

Differential Gene Expression
Mechanisms of Cell Differentiation **51**

Defining Differential Gene Expression 51

A Quick Primer on the Central Dogma 52

Evidence for Genomic Equivalence 53

Anatomy of the Gene 55
Chromatin composition 55
Exons and introns 56
Major parts of a eukaryotic gene 56

The transcription product and how it is processed 57

Noncoding regulatory elements: The on, off, and dimmer switches of a gene 57

Mechanisms of Differential Gene Expression:
Transcription 62

Epigenetic modification: Modulating access to genes 62

Transcription factors regulate gene transcription 65

The gene regulatory network: Defining an individual cell 69

Mechanisms of Differential Gene Expression:
Pre-messenger RNA Processing 70

Creating families of proteins through alternative pre-mRNA splicing 71

Mechanisms of Differential Gene Expression:
mRNA Translation 72

Differential mRNA longevity 72

Stored oocyte mRNAs: Selective inhibition of mRNA translation 73

Ribosomal selectivity: Selective activation of mRNA translation 73

microRNAs: Specific regulation of mRNA translation and transcription 74

Control of RNA expression by cytoplasmic localization 76

Mechanisms of Differential Gene Expression:
Posttranslational Protein Modification 77

Coda 84

Cell-to-Cell Communication
Mechanisms of Morphogenesis **87**

A Primer on Cell-to-Cell Communication 87

Adhesion and Sorting: Juxtacrine Signaling and the Physics of Morphogenesis 88

Differential cell affinity 89

The thermodynamic model of cell interactions 90

Cadherins and cell adhesion 91

The Extracellular Matrix as a Source of Developmental Signals 93

Integrins: Receptors for extracellular matrix molecules 94

The Epithelial-Mesenchymal Transition 94

Cell Signaling 95

Induction and competence 95

Paracrine Factors: Inducer Molecules 100

Morphogen gradients 100

Signal transduction cascades: The response to inducers 101

Fibroblast growth factors and the RTK pathway 102

FGFs and the JAK-STAT pathway 104

The Hedgehog family 105

The Wnt family 108

The TGF-β superfamily 111

Other paracrine factors 112

The Cell Biology of Paracrine Signaling 117

Focal membrane protrusions as signaling sources 117

Juxtacrine Signaling for Cell Identity 120

The Notch pathway: Juxtaposed ligands and receptors for pattern formation 120

Paracrine and juxtacrine signaling in coordination: Vulval induction in _C. elegans_ 121

Stem Cells
Their Potential and Their Niches **127**

The Stem Cell Concept 127

Division and self-renewal 127

Potency defines a stem cell 128

Stem Cell Regulation 130

Pluripotent Cells in the Embryo 131

Meristem cells of the _Arabidopsis thaliana_ embryo and beyond 132

Cells of the inner cell mass in the mouse embryo 134

Adult Stem Cell Niches in Animals 135

Stem cells fueling germ cell development in the _Drosophila_ ovary 136

Adult Neural Stem Cell Niche of the V-SVZ 137

The neural stem cell niche of the V-SVZ 138

The Adult Intestinal Stem Cell Niche 142

Clonal renewal in the crypt 142

Stem Cells Fueling the Diverse Cell Lineages in Adult Blood 143

The hematopoietic stem cell niche 143

The Mesenchymal Stem Cell: Supporting a Variety of Adult Tissues 146

Regulation of MSC development 146

The Human Model System to Study Development and Disease 147

Pluripotent stem cells in the lab 147

Induced pluripotent stem cells 150

Organoids: Studying human organogenesis in a culture dish 153

Stem Cells: Hope or Hype? 156

PART II ● Gametogenesis and Fertilization: The Circle of Sex

6

Sex Determination and Gametogenesis   159

Sex Determination   159

Chromosomal Sex Determination   159

The Mammalian Pattern of Sex Determination   160
Gonadal sex determination in mammals   161
Secondary sex determination in mammals:
    Hormonal regulation of the sexual phenotype   166

Chromosomal Sex Determination in *Drosophila*   169
Sex determination by dosage of X   169
The *Sex-lethal* gene   170
Doublesex: The switch gene for sex determination   172

Environmental Sex Determination   173

Gametogenesis in Animals   174
PGCs in mammals: From genital ridge to gonads   175
Meiosis: The intertwining of life cycles   176
Spermatogenesis in mammals   180
Oogenesis in mammals   182

**Sex Determination and Gametogenesis
    in Angiosperm Plants   184**

Sex Determination   184

Gametogenesis   186
Pollen   188
The ovule   188

7

Fertilization   Beginning a New Organism   193

Structure of the Gametes   193
Sperm   194
The egg   195
Recognition of egg and sperm   198

External Fertilization in Sea Urchins   198
Sperm attraction: Action at a distance   198
The acrosome reaction   200
Recognition of the egg's extracellular coat   201
Fusion of the egg and sperm cell membranes   201
Prevention of polyspermy: One egg, one sperm   202
Activation of egg metabolism in sea urchins   206
Fusion of genetic material in sea urchins   210

Internal Fertilization in Mammals   211
Getting the gametes into the oviduct:
    Translocation and capacitation   211
In the vicinity of the oocyte: Hyperactivation, directed
    sperm migration, and the acrosome reaction   212
Recognition at the zona pellucida   214
Gamete fusion and the prevention of polyspermy   214
Activation of the mammalian egg   216
Fusion of genetic material   216

Fertilization in Angiosperm Plants   217
Pollination and beyond: The progamic phase   217
Pollen germination and tube elongation   218
Pollen tube navigation   219
Double fertilization   220

Coda   221

PART III ● Early Development: Cleavage, Gastrulation, and Axis Formation

8

Snails, Flowers, and Nematodes
Different Mechanisms for Similar Patterns
of Specification   **223**

**A Reminder of the Evolutionary Context That
    Built the Strategies Governing Early
    Development   223**

The diploblastic animals: Cnidarians
    and ctenophores   224
The triploblastic animals: Protostomes
    and deuterostomes   224
What's to develop next   226

Early Development in Snails   226

Cleavage in Snail Embryos   226
Maternal regulation of snail cleavage   228
Axis determination in the snail embryo   235

Gastrulation in Snails   237

The Nematode *C. elegans*   238

Cleavage and Axis Formation in *C. elegans*   239
Rotational cleavage of the egg   240
Anterior-posterior axis formation   241
Dorsal-ventral and right-left axis formation   241
Control of blastomere identity   242

Gastrulation of 66 Cells in *C. elegans*   243

The Genetics of Axis Specification in *Drosophila*   247

Early *Drosophila* Development   249
Fertilization   249
Cleavage   250
The mid-blastula transition   252
Gastrulation   252

The Genetic Mechanisms Patterning the *Drosophila* Body   254

Segmentation and the Anterior-Posterior Body Plan   255
Maternal gradients: Polarity regulation by oocyte cytoplasm   256
The anterior organizing center: The Bicoid and Hunchback gradients   259
The terminal gene group   260
Summarizing early anterior-posterior axis specification in *Drosophila*   260

Segmentation Genes   260
Segments and parasegments   261
The gap genes   262
The pair-rule genes   263
The segment polarity genes   265

The Homeotic Selector Genes   268

Generating the Dorsal-Ventral Axis   269
Dorsal-ventral patterning in the oocyte   269
Generating the dorsal-ventral axis within the embryo   270

Axes and Organ Primordia: The Cartesian Coordinate Model   272

Sea Urchins and Tunicates
Deuterostome Invertebrates   **275**

Early Development in Sea Urchins   275
Early cleavage   276
Blastula formation   277
Fate maps and the determination of sea urchin blastomeres   278
Gene regulatory networks and skeletogenic mesenchyme specification   279
Specification of the vegetal cells   282

Sea Urchin Gastrulation   283
Ingression of the skeletogenic mesenchyme   283
Invagination of the archenteron   286

Early Development in Tunicates   288
Cleavage   288
The tunicate fate map   289
Autonomous and conditional specification of tunicate blastomeres   290

Amphibians and Fish   295

Early Amphibian Development   295

Fertilization, Cortical Rotation, and Cleavage   296
Unequal radial holoblastic cleavage   297
The mid-blastula transition: Preparing for gastrulation   298

Amphibian Gastrulation   299
Epiboly of the prospective ectoderm   300
Vegetal rotation and the invagination of the bottle cells   300
Involution at the blastopore lip   302
Convergent extension of the dorsal mesoderm   304

Progressive Determination of the Amphibian Axes   306
Specification of the germ layers   306
The dorsal-ventral and anterior-posterior axes   306

The Work of Hans Spemann and Hilde Mangold: Primary Embryonic Induction   307

Molecular Mechanisms of Amphibian Axis Formation   308
How does the organizer form?   309

Functions of the organizer   314

Induction of neural ectoderm and dorsal mesoderm: BMP inhibitors   314

Conservation of BMP signaling during dorsal-ventral patterning   318

Regional Specificity of Neural Induction along the Anterior-Posterior Axis   318

Specifying the Left-Right Axis   322

Early Zebrafish Development   322

Zebrafish Cleavages: Yolking Up the Process   324

Gastrulation and Formation of the Germ Layers   327

Progression of epiboly   327

Internalization of the hypoblast   328

The embryonic shield and the neural keel   329

Dorsal-Ventral Axis Formation   330

The fish blastopore lip   330

Teasing apart the powers of Nodal and BMP during axis determination   331

Left-Right Axis Formation   332

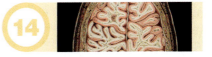

12

Birds and Mammals   335

Early Development in Birds   336

Avian Cleavage   336

Gastrulation of the avian embryo   338

Axis specification and the avian "organizer"   342

Left-right axis formation   343

Early Development in Mammals   345

Mammalian cleavage   345

Trophoblast or ICM? The first decision of the rest of your life   347

Mammalian gastrulation   348

Mammalian axis formation   351

Twins   357

Coda   359

PART IV ● Building with Ectoderm:
The Vertebrate Nervous Systyem and Epidermis

13

Neural Tube Formation and Patterning   363

Transforming the Neural Plate into a Tube: The Birth of the Central Nervous System   365

Primary neurulation   367

Secondary neurulation   374

Patterning the Central Nervous System   375

The anterior-posterior axis   375

The dorsal-ventral axis   376

Opposing morphogens   377

All Axes Come Together   379

14

Brain Growth   381

Neuroanatomy of the Developing Central Nervous System   381

The cells of the developing central nervous system   381

Tissues of the developing central nervous system   382

Developmental Mechanisms Regulating Brain Growth   386

Neural stem cell behaviors during division   386

Neurogenesis: Building from the bottom up (or from the inside out)   387

Glia as scaffold for the layering of the cerebellum and neocortex   389

Signaling mechanisms regulating development of the neocortex   390

Development of the Human Brain   392

Fetal neuronal growth rate after birth   392

Hills raise the horizon for learning   393

Genes for brain growth   396

Changes in transcript quantity   396

Teenage brains: Wired and unchained   396

15

Neural Crest Cells and Axonal Specificity   399

The Neural Crest   399

Regionalization of the Neural Crest   401

Neural Crest: Multipotent Stem Cells?   402

Specification of Neural Crest Cells   404

Neural Crest Cell Migration: Epithelial to Mesenchymal and Beyond   405
Delamination   406
The driving force of contact inhibition   408
Collective migration   409

Migration Pathways of Trunk Neural Crest Cells   410
The ventral pathway   410
The dorsolateral pathway   413

Cranial Neural Crest   414

The "Chase and Run" Model   416
An elaborate collaboration of pushes and pulls   417

Neural Crest-Derived Head Skeleton   418

Cardiac Neural Crest   419

Establishing Axonal Pathways in the Nervous System   420

The Growth Cone: Driver and Engine of Axon Pathfinding   421
Rho, Rho, Rho your actin filaments down the signaling stream   423

Axon Guidance   423

The Intrinsic Navigational Programming of Motor Neurons   424
Cell adhesion: A mechanism to grab the road   426

Local and long-range guidance molecules: The street signs of the embryo   426
Repulsion patterns: Ephrins and semaphorins   426

How Did the Axon Cross the Road?   428
…Netrin   428
Slit and Robo   428

The Travels of Retinal Ganglion Axons   431
Growth of the retinal ganglion axon to the optic nerve   431
Growth of the retinal ganglion axon through the optic chiasm   431

Target Selection: "Are We There Yet?"   432
Chemotactic proteins   432
Target selection by retinal axons: "Seeing is believing"   433

Synapse Formation   436

16

Ectodermal Placodes and the Epidermis   441

Cranial Placodes: The Senses of Our Heads   441
Cranial placode induction   443
Otic-epibranchial development: A shared experience   444
Morphogenesis of the vertebrate eye   448
Formation of the eye field: The beginnings of the retina   449
The lens-retina induction cascade   451

The Epidermis and Its Cutaneous Appendages   453
Origin of the epidermis   453
The ectodermal appendages   454
Signaling pathways you can sink your teeth into   456
Ectodermal appendage stem cells   457

PART V ● Building with Mesoderm and Endoderm: Organogenesis

17

Paraxial Mesoderm
The Somites and Their Derivatives   461

Cell Types of the Somite   464

Establishing the Paraxial Mesoderm and Cell Fates along the Anterior-Posterior Axis   465
Specification of the paraxial mesoderm   465
Spatiotemporal collinearity of Hox genes determines identity along the trunk   467

Somitogenesis   469
Axis elongation: A caudal progenitor zone and tissue-to-tissue forces   470
How a somite forms: The clock-wavefront model   473

Linking the clock-wavefront to Hox-mediated axial
identity and the end of somitogenesis   478

Sclerotome Development   479

Vertebrae formation   481

Tendon formation: The syndetome   484

Dermomyotome Development   485

Determination of the central dermomyotome   486

Determination of the myotome   487

Intermediate and Lateral Plate Mesoderm
Heart, Blood, and Kidneys   **491**

Intermediate Mesoderm: The Kidney   491

**Specification of the Intermediate Mesoderm:
Pax2, Pax8, and Lim1   493**

**Reciprocal Interactions of Developing
Kidney Tissues   494**

Mechanisms of reciprocal induction   495

**Lateral Plate Mesoderm: Heart and
Circulatory System   499**

Heart Development   501

A minimalist heart   501

Formation of the heart fields   501

Specification of the cardiogenic mesoderm   503

Migration of the cardiac precursor cells   503

Initial heart cell differentiation   505

Looping of the heart   506

Blood Vessel Formation   507

Vasculogenesis: The initial formation
of blood vessels   507

Angiogenesis: Sprouting of blood vessels
and remodeling of vascular beds   510

**Hematopoiesis: Stem Cells and Long-Lived
Progenitor Cells   510**

Sites of hematopoiesis   510

The bone marrow HSC niche   512

Coda   515

Development of the Tetrapod Limb   517

Limb Anatomy   517

The Limb Bud   518

Hox Gene Specification of Limb Skeleton Identity   519

From proximal to distal: Hox genes in the limb   519

**Determining What Kind of Limb to Form
and Where to Put It   521**

Specifying the limb fields   522

Induction of the early limb bud   523

**Outgrowth: Generating the Proximal-Distal Axis
of the Limb   527**

The apical ectodermal ridge   527

Specifying the limb mesoderm: Determining the
proximal-distal polarity   529

Turing's model: A reaction-diffusion mechanism
of proximal-distal limb development   531

Specifying the Anterior-Posterior Axis   534

Sonic hedgehog defines a zone of polarizing activity   534

Specifying digit identity by Sonic hedgehog   536

Sonic hedgehog and FGFs: Another positive
feedback loop   537

Hox genes are part of the regulatory network specifying
digit identity   538

Generating the Dorsal-Ventral Axis   539

**Cell Death and the Formation of Digits
and Joints   541**

Sculpting the autopod   541

Forming the joints   542

Evolution by Altering Limb Signaling Centers   543

The Endoderm   Tubes and Organs for
Digestion and Respiration   **547**

The Pharynx   549

The Digestive Tube and Its Derivatives   551

Specification of the gut tissue   551

Accessory organs: The liver, pancreas, and gallbladder   554

The Respiratory Tube   557

Epithelial-mesenchymal interactions and
the biomechanics of branching in the lungs   558

PART VI • Postembryonic Development

21

Metamorphosis
The Hormonal Reactivation of Development **563**

Amphibian Metamorphosis 564
Morphological changes associated
with amphibian metamorphosis 564
Hormonal control of amphibian metamorphosis 566
Regionally specific developmental programs 568

Metamorphosis in Insects 569
Imaginal discs 570
Hormonal control of insect metamorphosis 572
The molecular biology of 20-hydroxyecdysone
activity 574
Determination of the wing imaginal discs 575

22

Regeneration
The Development of Rebuilding **579**

Defining The Problem of Regeneration 579
**Regeneration, a Recapitulation of Embryonic
Development? 582**
An Evolutionary Perspective on Regeneration 583
Regenerative Mechanics 586
Plant Regeneration 586
A totipotent way of regenerating 586
A plant's meri-aculous healing abilities 587

Whole Body Animal Regeneration 591
Hydra: Stem cell-mediated regeneration, orphallaxis,
and epimorphosis 591
Stem cell-mediated regeneration in flatworms 594

Tissue-Restricted Animal Regeneration 602
Salamanders: Epimorphic limb regeneration 602

Defining the cells of the regeneration blastema 604
Luring the mechanisms of regeneration
from zebrafish organs 607

Regeneration in Mammals 613
Compensatory regeneration in the mammalian liver 613
The spiny mouse, at the tipping point between
scar and regeneration 615

23

Development in Health and Disease
Birth Defects, Endocrine Disruptors,
and Cancer **619**

The Role of Chance 619

Genetic Errors of Human Development 620
The developmental nature of human syndromes 620
Genetic and phenotypic heterogeneity 621

**Teratogenesis: Environmental Assaults
on Animal Development 621**
Alcohol as a teratogen 623
Retinoic acid as a teratogen 627

**Endocrine Disruptors: The Embryonic Origins
of Adult Disease 628**
Diethylstilbestrol (DES) 629
Bisphenol A (BPA) 632
Atrazine: Endocrine disruption through
hormone synthesis 633
Fracking: A potential new source of
endocrine disruption 634

**Transgenerational Inheritance of Developmental
Disorders 634**

Cancer as a Disease of Development 636
Development-based therapies for cancer 640

Coda 641

PART VII ● Development in Wider Contexts

Development and Evolution
Developmental Mechanisms of Evolutionary
Change **643**

The Developmental Genetic Model of Evolutionary Change 643

Preconditions for Evolution: The Developmental Structure of the Genome 643

Modularity: Divergence through dissociation 644
Molecular parsimony: Gene duplication and divergence 646

Mechanisms of Evolutionary Change 648

Heterotopy 649
Heterochrony 650
Heterometry 651
Heterotypy 653

Developmental Constraints on Evolution 654

Physical constraints 654
Morphogenetic constraints 654
Pleiotropic constraints and redundancy 655

Ecological Evolutionary Developmental Biology 656

Plasticity-First Evolution 656

Genetic assimilation in the laboratory 656
Genetic assimilation in natural environments 658

Selectable Epigenetic Variation 659

Evolution and Developmental Symbiosis 660

The evolution of multicellularity 661
The evolution of placental mammals 662

Coda 662

Appendix A-1

Glossary G-1

Index I-1

Preface: Thinking Grandly about Developmental Biology

With biology going into smaller and smaller realms, it is sometimes good to contemplate the grand scheme of things rather than the details, to "seat thyself sultanically among the moons of Saturn" (in Herman Melville's phrase). It is good, for instance, to get a perspective of developmental biology from outside the discipline rather than from inside it.

Remembering the Field's Interdisciplinary Foundations

Developmental biology, history tells us, is an interdisciplinary field that is at the foundations of biology. Indeed, before the word biology came to be used, the living world was characterized as that part of the world that was developing. The organizers of the first meeting (in 1939) of the Growth Society, which was the precursor of the Society for Developmental Biology, claimed that development must be studied by combining the insights of numerous disciplines, including genetics, endocrinology, biochemistry, physiology, embryology, cytology, biophysics, mathematics, and even philosophy. Developmental biology was to be more than embryology. It also included stem cells, which were known to generate the adult blood, and regeneration, which was seen to be the re-activation of developmental processes and which was critical for healing in vertebrates and for reproduction of hydra, flatworms, and numerous other invertebrates. The first articles published in the journal *Developmental Biology* showcased embryology, regeneration, and stem cells, and the different ways of studying them.

Throughout this new international 12th edition you will see a return to some of these founding ideas of interdisciplinary developmental biology, namely regeneration, morphomechanics, plants, and the genetic control of development.

Indeed, *regeneration* has historically been a major part of developmental biology, for it is a developmental phenomenon that can be readily studied. Experimental biology was born in the efforts of eighteenth-century naturalists to document regeneration and to examine how it was possible. The regeneration experiments of Tremblay (hydras), Réaumur (crustaceans), and Spallanzani (salamanders) set the standard for experimental biology and for the intelligent discussion of one's data.

More than two centuries later, we are beginning to find answers to the great problems of both embryology and regeneration. Indeed, the conclusions of one support the research of the other. We may soon be able to alter the human body so as to permit our own limbs, nerves, and organs to regenerate. Severed limbs could be restored, diseased organs could be removed and regrown, and nerve cells altered by age, disease, or trauma could once again function normally. The ethical issues this would exacerbate are only beginning to be appreciated. But if we are to have such abilities, we first have to understand how regeneration occurs in those species that have this ability. Our new knowledge of the roles that paracrine factors and physical factors play in embryonic organ formation, plus recent studies of stem cells and their niches, has propelled what Susan Bryant has called "a regeneration renaissance." Since "renaissance" literally means "rebirth," and since regeneration can be seen as a return to the embryonic state, the term is apt in many ways.

Notice that *biophysics* was also an early part of the mix of developmental biology. This area, too, is having a renaissance. The physical connections between cells, the strength of their bonding, and the tensile strength of the material substrates of the cells are all seen to be critical for normal development. Physical forces are necessary for sperm-egg binding, gastrulation, heart development, gut development, the branching of the kidney and lung epithelia, and even the development of tumors. Physical forces can direct the development of stem cells toward particular fates, and they can determine which part of the body is left and which is right. The patella of our kneecap doesn't form until we put pressure on it by walking. In many cases, physical forces can direct gene expression. Lev Beloussov, a pioneer in this area, has called this the "morphomechanics of development."

Another area that was prominently represented in the early programs of developmental biology was *plant development*. Plant development had much in common with regeneration, as "adult" plants could redevelop entire parts of their bodies. Whereas in animal biology the study of development diverged from the study of physiology, that separation was not evident in plant biology. Moreover, while many animals quickly set aside a germline that was to become the sperm or eggs, this was not the case in plants. Such comparisons between plants and animals are now present throughout this text, and they serve to highlight the fundamental developmental processes that are present across phyla and even kingdoms of life.

But the genes remain the center of focus in developmental biology. And the more we learn about them, the more interesting and complex these genes become. New advances in "single cell transcriptomics" have given us an amazing privilege—the ability to look at the gene expression patterns of individual cells as they develop. An individual's cells may all have the same genes, but their different positions in the embryo cause different genes to be active in each cell. It's a symphony of relationships, each cell providing the context for another. If development is the performance, then the genome is the script or score. As anyone who has gone to concerts knows, different bands perform the same score differently, and the same band will play the same song differently on two successive nights. Environment is also critical—hence, the new interest in plasticity and symbiosis in development.

Developmental biology has also taken on a new role in science. More than any other biological science, it demonstrates the critical importance of processes as opposed to entities. In many organisms, the same process can be done by different molecules. "It's the song, not the singer," say Doolittle and Booth, and we can be thankful that there are redundant pathways in development—if one pathway fails, another is often able to take over its function. The entity/process split in developmental biology mirrors the particle/wave dichotomy in physics. It is a "both, and" situation, rather than an "either/or" situation. In 1908, the Scottish physiologist J. S. Haldane said, "That a meeting point between biology and physical science may at some time be found, there is no doubting. But we may confidently predict that if that meeting-point is found, and one of the two sciences is swallowed up, that one will not be biology." Developmental biology may well solve the longstanding mysteries of physics.

New to the International Twelfth Edition

In this current volume, we have attempted to track this amazing fulfillment of the early promises of developmental biology. To this end, the book has undergone its own morphogenesis.

Plant development covered throughout

We have now incorporated plant material into the relevant chapters. Instead of segregating plant developmental biology into a single (and often unassigned)

chapter, we have integrated essential plant biology into the chapters on cell specification, gene regulation, cell communication, gamete production, fertilization,
axis determination, organ formation, and regeneration.

Upgraded and expanded chapter on regeneration

We have also expanded the chapter on regeneration, which we are proud to say
offers a unique summary of the field. It both captures the fascinating problems
of post-embryonic development that regeneration seems to solve and provides
a logical framework for the known mechanisms of regeneration, based on an
organism's degree of regenerative capacity. We feel that this chapter will be an
excellent place for anyone interested in this area to start.

Updates throughout all chapters

All of the chapters have received important updates, from the introductory chapter's broader evolutionary perspective to new material on the morphomechanics
of development during *Drosophila* gastrulation and the formation of mammalian
lungs. Special consideration was also given to the increasing use of whole-ge-
nome, transcriptomic approaches, which are dramatically shaping our understanding of cell differentiation.

A new, student-centered approach

From a pedagogical standpoint, it is also good to get an outside perspective of
how students are learning developmental biology—*the perspective of the student
experience*. For decades, it has been the responsibility of textbooks like ours to be
the most comprehensive sources for the field's foundational content. Although
this responsibility still remains, the reality is that students are inundated with an
overwhelming myriad of sources vying for their attention. If there was ever a time
a student of developmental biology needed a *guidebook* to navigate through this
dense and diverse ecosystem of texts, online resources, and infinitely expanding
scientific literature, the time is now and the guidebook this new international
volume of *Developmental Biology*.

- *Focused and streamlined coverage.* Over the years, as new knowledge has
  grown, so has our own textbook, which was reaching a size that might
  itself trigger student overload and defeat the purposes of engagement
  and deep learning. The information bombarding students is not going
  away; therefore, they need not only access to the information but also a
  clear guide that fosters movement from the essential ideas to the complex mechanisms and finally to inclusive invitations that welcome their
  research in this field. We have both reduced and reorganized the content
  in each chapter to achieve a clear and supportive lattice so that both the
  professor and the student can more easily navigate the increasing volume and complexity of developmental biology.

- *Innovative pedagogy: Empowering students to craft their own learning.* The
  first material students will encounter in each section of a chapter represents the most essential content. We have introduced a new element
  called "Further Development," which highlights content we feel represents some of the more complex ideas in the field. In addition, students
  will also come across invitations to view some Further Developments
  online. These online topics represent fantastic opportunities for students
  to *further develop* their understanding of developmental biology along
  paths of their own interest—paths of investigation that professors can
  have confidence match the standards of quality seen throughout the
  textbook (unlike some other online sources). The special features of
  previous editions—Dev Tutorials, Developing Questions, Next Step
  Investigations, and citations throughout—are still in place to play

important roles in empowering students to take that final leap to engage with the developmental biology literature. To better support students' use of the research literature, we now include a new Appendix focused on how to find and analyze research articles in developmental biology.

Thanks to this new organization of content, professors and students will now be in complete control of what level of material may be most appropriate. We are proud to introduce the international edition of *Developmental Biology* 12e, as it still provides direct access to all levels of the content but without diluting its quality and the overall learning experience.

Acknowledgments

First, the two authors gratefully acknowledge their mutual respect for one another and for the enjoyment of each other's work. Michael wants the community to know that Scott has been most accepting and welcoming to new ideas and that his enthusiasm for producing the best product has not wavered any day of any edition. Scott wants the community to know that he is thrilled with the new ideas that Michael has brought to the book and that Michael's commitment to undergraduate education is second to none.

Second, we are thrilled to acknowledge the importance of Mary Stott Tyler to this book. The winner of the Viktor Hamburger Education Award and the author of *Fly Cycle*, *Differential Expressions*, *The Developmental Biology Vade Mecum*, and *Inquiry Biology*, Mary has been a mixture of author, editor, and curator of contents for this international 12th edition, helping us decide "what to leave in/what to leave out." As we added plant studies to the book and had to remove other studies, Mary's insight and vision for the finished book was essential.

If science is like a balloon expanding into the unknown—and the larger the balloon, the more points in contact with the unknown—then developmental biology has contacted an astounding number of unknowns. The accuracy and coverage of the 12th edition owes much to the work of the many expert reviewers who took the time to provide respectful formal and informal feedback throughout the process (see list). The organization of these reviews was consistently executed by Lauren Cahillane, Nina Rodriguez-Marty and Katie Tunkavige—thank you for making this important part possible. This international 12th edition is particularly unique as it marks the new incorporation of plant developmental biology. There were numerous reviewers who offered their expertise in select chapters, thank you to all. Special thanks, however, go to Anna Edlund and Marta Laskowski for their reviews of the plant content. They were very patient with us, and any misunderstandings are those of the authors.

This edition also marks a dramatic change to the publishing of *Developmental Biology*. With the retirement of Andy Sinauer, Sinauer Associates has become an imprint of Oxford University Press. Our book overlaps these two periods, and has seen the change of managers, art directors, and our long-time editor. We thank both Sinauer Associates and Oxford University Press for their great efforts in sustaining the book during this period of metamorphosis. We wish to especially thank Dean Scudder for taking on the managerial tasks and allowing us to work on new models of science education during this transition. Moreover, half-way through production of this edition, Jason Noe of Oxford became our overseeing editor. Such a transition and short timeline for production might rattle the best of editors, but Jason helped to establish the best adaptable plans to keep things on track. Sincere thanks for your efforts, Jason. Meanwhile, in the house of Sinauer, production editors Laura Green and Kathaleen Emerson shared their expertise and their truly collaborative insights, offering us respectful considerations during key times that we will not forget. Thank you Laura for also sharing with us your most valuable plant background throughout the editorial process.

The success of this and each edition equally rests on the quality of the book's design and look, for which we sincerely appreciate the wonderful work Sinauer's art, media, and overall production team have done. The media team was headed by Suzanne Carter and supported by the creative drive of Peter Lacey. Sincere thanks to you both. Further thanks to the entire group at Dragonfly Media, who continue to do a great job taking care to represent many of Michael's original drawings with supreme accuracy. We'd also like to thank Joan Gemme, Beth Roberge, and Annette Rapier for their excellent design, layout, and production of this edition. One of the long-loved hallmarks of *Developmental Biology* has been the incorporation of actual data and images that represent the science. Special thanks to the permissions team, Mark Siddall, Tracy Marton, and Michele Beckta for their non-stop efforts in securing the rights to these essential pieces of the book. But of course, a new book can only reach the hands of the students with the help of a robust and strategic sales team. Many thanks to Susan McGlew and to all the salespersons at Oxford now helping to support this textbook.

Lastly, it needs to be acknowledged that while Scott is blissfully retired, Michael is still working his tail off doing teaching, research, committee assignments, and so forth, in addition to his strong family commitments. He would not be able to provide the time and energy to this textbook if he did not have the support of his own institution and students. Thank you, Smith College, for continuing to allow Michael to produce and disseminate his Web Conferences, Developmental Documentaries, and the Dev Tutorials freely to the community. Most sincere thanks to Michael's research students, who had to endure their principle investigator being too engrossed in all things development all the time! Know that your patience, support, and insights surely made this book possible.

—M.J.F.B.

—S.F.G.

May 24, 2019

Reviewers of the Twelfth Edition

Anna Allen, *Howard University*

William Anderson, *Harvard University*

Nicola Barber, *University of Oregon*

Madelaine Bartlett, *University of Massachusetts, Amherst*

Marianne Bronner, *California Institute of Technology*

Timothy Brush, *University of Texas, Rio Grande Valley*

Blanche Capel, *Duke University*

Jacqueline Connour, *Ohio Northern University*

D. Cornelison, *University of Missouri, Columbia*

Dr. Angus Davidson, *The University of Nottingham*

Anna Edlund, *Bethany College*

Elizabeth D. Eldon, *California State University, Long Beach*

Deborah Marie Garrity, *Colorado State University*

Bob Goldstein, *University of North Carolina*

Eric Guisbert, *Florida Institute of Technology*

Jeff Hardin, *University of Wisconsin, Madison*

Richard Harland, *University of California Berkeley*

Marcus Heisler, *The University of Sydney*

Arnold G Hyndman, *Rutgers University*

Zhi-Chun Lai, *Pennsylvania State University*

Michael Lehmann, *University of Arkansas*

Michael Levin, *Tufts University*

Yuanyuan Rose Li, *University of Alabama at Birmingham*

Barbara Mania-Farnell, *Purdue University Northwest*

Adam C. Martin, *Massachusetts Institute of Technology*

David Matus, *Stony Brook University*

Roberto Mayor, *University College London*

Dave McClay, *Duke University*

Claus Nielsen, *University of Copenhagen*

Fred Nijhout, *Duke University*

Lee Niswander, *University of Colorado, Boulder*

Julia Oxford, *Boise State University*

Mark Peifer, *University of North Carolina*

Isabelle Peter, *California Institute of Technology*

Ann Rougvie, *University of Minnesota*

Sabrina Sabatini, *Sapienza University of Rome*

Thomas F. Schilling, *University of California, Irvine*

Nick Sokol, *Indiana University*

Richard Paul Sorrentino, *Auburn University*

Ana Soto, *Tufts University*

David Stachura, *California State University, Chico*

Claudio Stern, *University College London*

Andrea Streit, *King's College London*

Keiko Sugimoto, *RIKEN*

Jonathan Sylvester, *Georgia State University*

Daniel E Wagner, *Harvard Medical School*

Zhu Wang, *University of California, Santa Cruz*

Paul M. Wassarman, *Icahn School of Medicine at Mount Sinai*

Daniel Weinstein, *Queens College, CUNY*

Jessica LaMae Whited, *Harvard University*

Jeanne Wilson-Rawls, *Arizona State University*

Colleen Winters, *Towson University*

Tracy Young-Pearse, *Harvard Medical School*

Media and Supplements

to accompany **Developmental Biology**, *International Twelfth Edition*

For the Student

Companion Website

oup.com/he/barresi12xe

Significantly enhanced for the International Twelfth Edition, and referenced throughout the textbook, the *Developmental Biology* Companion Website provides students with a range of engaging resources to help them learn the material presented in the textbook. The companion site is available free of charge and includes resources in the following categories:

- **Dev Tutorials:** Professionally produced video tutorials, presented by the textbook's authors, reinforce key concepts.
- **Watch Development:** Putting concepts into action, these informative videos show real-life developmental biology processes.
- **Further Development:** These extensive topics provide more information for advanced students, historical, philosophical, and ethical perspectives on issues in developmental biology, and links to additional online resources.
- **Scientists Speak:** In these lectures and question-and-answer interviews, developmental biology topics are explored by leading experts in the field.
- **Flashcards:** Per-chapter flashcard sets help students learn and review the many new terms and definitions introduced in the textbook.
- **Literature Cited:** Full citations are provided for all of the literature cited in the textbook (most linked to their PubMed citations).
- **Research Guide:** This illustrated and annotated guide helps students find and comprehend research articles in developmental biology.

For the Instructor

(Available to qualified adopters)

Instructor's Resource Library

The *Developmental Biology*, International Twelfth Edition Instructor's Resource Library includes the following resources:

- **Case Studies in Dev Bio:** This collection of case study problems provides instructors with ready-to-use in-class active learning exercises. The case studies foster deep learning in developmental biology by providing students an opportunity to apply course content to the critical analysis of data, to generate hypotheses, and to solve novel problems in the field. Each case study includes a PowerPoint presentation and a student handout with accompanying questions.
- **Developing Questions:** Thought-provoking questions, many with answers, references, and recommendations for further reading, are provided so that you and your students can explore questions that are posed throughout each chapter.
- **Textbook Figures & Tables:** All of the textbook's figures, photos, and tables are provided both in JPEG and PowerPoint formats. All images have been optimized for excellent legibility when projected in the classroom.

Value Options

eBook

(ISBN 978-1-60535-874-1)

Developmental Biology, International Twelfth Edition is available as an eBook, via several different eBook providers, including RedShelf and VitalSource. Please visit the Oxford University Press website at oup.com/ushe for more information.

The Making of a Body and a Field
Introduction to Developmental Biology

1

Is this plant really your cousin?

Photo credit: M. J. F. Barresi and Kathryn Lee, 2018.
Thanks to Dr. Robin Sleith for providing the charophyta algae

ONE OF THE CRITICAL DIFFERENCES between you and a machine is that a machine is never required to function until after it is built. Every multicellular organism has to function even as it builds itself. It *develops*. In the time between fertilization and birth, the organism is known as an **embryo** (**FIGURE 1.1**). The concept of an embryo is a staggering one. As an embryo, you had to build yourself from a single cell. You had to respire before you had lungs, digest before you had a gut, build bones when you were pulpy, and form orderly arrays of neurons before you knew how to think. It should thus not be surprising that most human embryos die before being born. You survived.

Multicellular organisms do not spring forth fully formed. Rather, they arise by a relatively slow process of progressive change that we call **development**. In most cases, the development of a multicellular organism begins with a single cell—an egg cell that has completed the process of fertilization and is referred to as a **zygote**. The zygote divides mitotically to produce all the cells of the body.

The study of animal development has traditionally been called **embryology**, after that phase of an organism that exists between fertilization and birth. But development does not stop at birth, or even at adulthood. Most organisms never stop developing. Each day we replace more than a gram of skin cells (the older cells being sloughed off as we move), and our bone marrow sustains the development of millions of new red blood cells every minute of our lives. Plants exhibit an astounding capacity for perpetual growth throughout their life span, a phenomenon known as **indeterminate growth** (**FIGURE 1.2A**). Plant cells even have the capacity for whole-organism regeneration

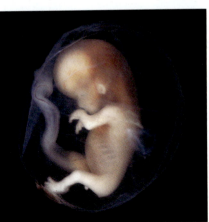

FIGURE 1.1 A 9- to 10-week-old human embryo.

lunar caustic/CC BY-SA 2.0

Original limb   After amputation   Regeneration

FIGURE 1.2 Extraordinary capacities for development. (A) "Hyperion" has been named the tallest tree in the world. It is a redwood sequoia standing over 114 meters (375 feet) tall, which is 70 feet taller than the Statue of Liberty. The two researchers shown here climbing Hyperion look like spiders hanging from its branches. (B) Axel Erlandson created the "Basket Tree" by cutting the tops off six sycamore trees and forcing engraftment of each tree's regeneration stems together. This demonstrates the remarkable plasticity and regenerative ability of plants. (C) Some animal species also exhibit a remarkable capacity for regeneration. The Mexican salamander can regrow a perfectly constructed limb following its amputation.

(**FIGURE 1.2B**). Some animals can regenerate severed parts (**FIGURE 1.2C**), and many species undergo **metamorphosis** (changing from one form into another, such as the transformation of a tadpole into a frog, or a caterpillar into a butterfly). On the most fundamental level, developmental biology seeks to elucidate the cellular and molecular mechanisms that drive changes in cells, tissues, and organs over time—a timescale that spans all of life, from fertilization through aging.[1]

"How Are You, You?" Comparative Embryology and the Questions of Developmental Biology

Aristotle, the first known embryologist, said that wonder was the source of knowledge, and animal and plant development, as Aristotle knew well, is a remarkable source of wonder. The fertilized egg has no heart. Where does the heart come from? Does it form the same way in both insects and vertebrates? Many of the questions in developmental biology are of this comparative type, and they stem from the field's embryological heritage.

The first known study of comparative developmental anatomy was undertaken by Aristotle. In his book *On the Generation of Animals* (ca. 350 BCE), he noted some of the variations on the life cycle themes: some animals are born from eggs (**oviparity**, as in birds, frogs, and most invertebrates); some by live birth (**viviparity**, as in placental mammals); and some by producing an egg that hatches inside the body (**ovoviviparity**, as in certain reptiles and sharks). Aristotle also identified the two major cell division patterns by which embryos are formed: the **holoblastic** pattern of cleavage (in which the entire egg is divided into successively smaller cells, as it is in frogs and mammals) and the

meroblastic pattern of cleavage (as in chicks, wherein only part of the egg is destined to become the embryo, while the other portion—the yolk—serves as nutrition for the embryo). And should anyone want to know who first figured out the functions of the mammalian placenta and umbilical cord, it was Aristotle.

There was remarkably little progress in embryology for the two thousand years following Aristotle. It was only in 1651 that William Harvey concluded that all animals—even mammals—originate from eggs. *Ex ovo omnia* (All from the egg) was the motto on the frontispiece of Harvey's *On the Generation of Living Creatures*, and this precluded the spontaneous generation of animals from mud or excrement.[2] Harvey also was the first to see the blastoderm of the chick embryo (the small region of the egg containing the yolk-free cytoplasm that gives rise to the embryo), and he was the first to notice that "islands" of blood tissue form before the heart does. Harvey also suggested that the amniotic fluid might function as a "shock absorber" for the embryo.

As might be expected, embryology remained little but speculation until the invention of the microscope allowed detailed observations (**FIGURE 1.3**). Marcello Malpighi published the first microscopic account of chick development in 1672. Here, for the first time, the groove of the forming neural tube, the muscle-forming somites, and the first circulation of the arteries and veins—to and from the yolk—were identified.

This development, this formation of an orderly body from relatively homogeneous material, provokes profound and fundamental questions: How does the body form with its head always above its shoulders? Why is the heart on the left side of our body? How does a simple tube become the complex structures of the brain and spinal cord that generate both thought and movement? Why can't we grow back new limbs like a salamander? How do the sexes develop their different anatomies?

Our answers to these questions must respect the complexity of the inquiry and must explain a coherent causal network from gene through functional organ. To say that mammals with two X chromosomes are usually females and those with XY chromosomes are usually males does not explain sex determination to a developmental biologist, who wants to know *how* the XX genotype produces a female and *how* the XY genotype produces a male. Similarly, a geneticist might ask how globin genes are transmitted from one generation to the next, and a physiologist might ask about the function of globin proteins in the body. But the developmental biologist asks how it is that the globin genes come to be expressed only in red blood cells and how these genes become active only at specific times in development. (We don't have all the answers yet.) The particular set of questions asked defines the field of biology, as we, too, become defined (at least in part) by the questions we ask. *Welcome to a wonderful and important set of questions!*

FIGURE 1.3 Depictions of chick developmental anatomy. (A) Dorsal view (looking "down" at what will become the back) of a 2-day chick embryo, as depicted by Marcello Malpighi in 1672. (B) Dorsal view of a late 2-day chick embryo, about 45 hours after the egg was laid. The heart starts beating during day 2. The vascular system of this embryo was revealed by injecting fluorescent beads into the circulatory system. The three-dimensionality is achieved by superimposing two separate images.

(A)

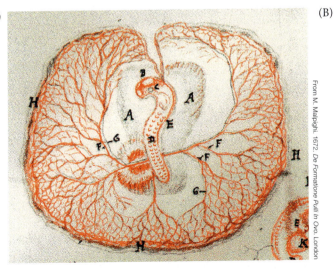

From M. Malpighi. 1672. *De Formatione Pulli In Ovo.* London

(B)

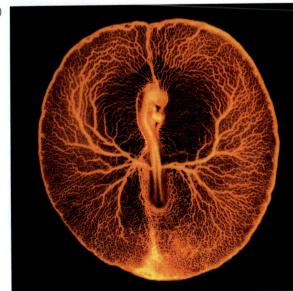

Image courtesy of Vincent Pasque

Development accomplishes two major objectives. First, it generates cellular diversity and order within the individual organism; second, it ensures the continuity of life from one generation to the next. Put another way, there are two fundamental questions in developmental biology: How does the zygote give rise to the adult body? And how does that adult body produce yet another body? These huge questions can be subdivided into several categories of questions scrutinized by developmental biologists:

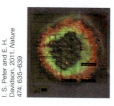

I. S. Peter and E. H. Davidson. 2011. *Nature* 474: 635–639

- **The question of differentiation**. A single cell, the fertilized egg, gives rise to hundreds of different cell types—muscle cells, epidermal cells, neurons, lens cells, lymphocytes, blood cells, fat cells, and so on. This generation of cellular diversity is called **differentiation**. Since every cell of the body (with very few exceptions) contains the same set of genes, how can this identical set of genetic instructions produce different types of cells? How can a single fertilized egg cell generate so many different cell types?[3]

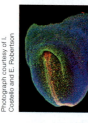

Photograph courtesy of E. M. Gorostiza

- **The question of pattern formation**. From the stripes that cover a zebra or zebrafish to the anatomical parts of our bodies, cells and tissues are stereotypically positioned in recognizable patterns. Our head is anterior, our tail posterior, and our limbs lateral to the medially positioned nervous system. Our heart is asymmetrically positioned on the left side. Indications of these patterns can be seen early in the embryo. What processes control the elaboration of cell and tissue type patterns?

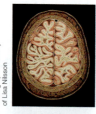

Photograph courtesy of I. Costello and E. Robertson

- **The question of morphogenesis**. How can the cells in our body organize into functional structures? Our differentiated cells are not randomly distributed. Rather, they are organized into intricate tissues and organs. During development, cells divide, migrate, and die; tissues fold and separate. The folded tubular shape of our brain and spinal cord started as a flattened plate of cells. Our digestive system functionally connects our mouth and anal openings. This creation of ordered form is called **morphogenesis**, and it involves coordinating cell growth, cell migration, and cell death.

Photograph courtesy of Lisa Nilsson

- **The question of growth**. If each cell in our face were to undergo just one more cell division, we would be considered horribly malformed. If each cell in our arms underwent just one more round of cell division, we could tie our shoelaces without bending over. How do our cells know when to stop dividing? Our arms are generally the same size on both sides of the body. How is cell division so tightly regulated?

From journal cover associated with J. Holy and G. Schatten. 1991. *Dev Biol* 147 (2), courtesy of J. Holy and G. Schatten

- **The question of reproduction**. The sperm and egg are highly specialized cells that can transmit the instructions for making an organism from one generation to the next. How are these germ cells set apart, and what are the instructions in the nucleus and cytoplasm that allow them to form the next generation?

Courtesy of Junji Morokuma and Michael Levin

- **The question of regeneration**. Some organisms can regenerate every part of their bodies. Some salamanders can regenerate their eyes and their legs, and many reptiles can regenerate their tails. While mammals are generally poor at regeneration, there are some cells in our bodies—**stem cells**—that are able to form new structures even in adults. How do stem cells retain this capacity, and can we harness it to cure debilitating diseases?

- **The question of environmental integration**. The development of many (perhaps all) organisms is influenced by cues from the environment that surrounds the embryo or larva. The sex of many species of turtles, for instance, depends on the temperature the embryo experiences while in the eggshell. The formation of the reproductive system in some insects depends on bacteria that are transmitted inside the egg. Moreover, certain chemicals in the environment can disrupt normal development, causing malformations in the adult. How is the development of an organism integrated into the larger context of its habitat?

- **The question of evolution**. Evolution involves inherited changes of development. When we say that today's one-toed horse had a five-toed ancestor, we are saying that changes in the development of cartilage and muscles occurred over many generations in the embryos of the horse's ancestors. How do changes in development create new body forms? Which heritable changes are possible, given the constraints imposed by the necessity of the organism to survive as it develops?

The questions asked by developmental biologists have become critical in molecular biology, physiology, cell biology, genetics, anatomy, cancer research, neurobiology, immunology, ecology, and evolutionary biology. Each of these disciplines has its ancestral roots in developmental biology. Yet unlike each of these descendant disciplines, which seem to continually differentiate into further sets of restricted paradigms, developmental biology remains pluripotent. In fact, it has recently been proposed that developmental biology is the "stem cell of biological disciplines" (Gilbert 2017).

CHOOSING THE ORGANISM TO STUDY THE QUESTION: THE "MODEL" SYSTEM

To answer the questions that developmental biologists ask, researchers need a tractable experimental organism best suited to their questions. What makes an organism a good "model" for addressing a given question? Just as an axe and a chain saw are suited to similar but different tasks, different animal model systems provide investigators with different advantages. Some of the common considerations in choosing a good model system are the following:

Size: A particularly practical consideration is the size of the adult organism. Is it easy to house a significant number of breeding adults in the allotted laboratory infrastructure? For example, housing 50 mice in cages requires a lot more space and expense then housing 50 flies in a vial.

Generation time: How long does it take the organism to complete its life cycle from embryo to reproductive adult? Additionally, how short is the embryonic period? The roundworm *Caenorhabditis elegans* has a full life cycle of 3 days, whereas it takes the zebrafish about 3 months to go from "egg to egg." However, early embryogenesis in a zebrafish spans only 24 hours.

Embryo accessibility: To study embryology, a researcher needs to be able to see and work with the actual embryo. Different species pose different challenges for embryo accessibility. Some embryos are dispersed in the water for easy collection, while others develop in an opaque shell, such as the avian egg, or in utero (within the womb or uterus), as with mammals.

Feasibility of genomic interrogation: Since Mendel's work with peas, developmental biologists have been driven to identify the genetic basis underlying all developmental processes, from embryology to disease. Although all life is based on the organization and use of the four nucleotide bases, no species has the same genome. Genome size, organization, and content all differ, which can affect the level of genetic interrogation that is possible. For instance, research-

FIGURE 1.4   Some of the model systems used to study developmental biology. From left to right the silhouettes represent the following model organisms: *Arabidopsis thaliana, Drosophila mela-nogaster, Hydra vulgaris, Caenorhabditis elegans, Xenopus laevis, Danio rerio, Gallus gallus, Mus musculus*, and stem cells of *Homo sapiens* (blastocyst with inner cell mass depicted).

ers studying regeneration in the Mexican salamander have to deal with the largest genome ever sequenced. Maybe the secret of regeneration lies some-where in all that DNA.

Organism type and phylogenetic position: Ideally, the research question should guide the selection of a model system. If researchers are interested in the re-markable process of metamorphosis, then clearly they are limited to a select few model species that display such transformations, such as the fruitfly or frog. If they are passionate about studying human development, they may use a mam-malian model organism, such as the mouse, or human cells in culture. If their questions are focused on deciphering the developmental changes fueling evolu-tion, they can choose species that occupy informative phylogenetic positions, such as the charophytic algae that are basal to multicellular land plants.

Ease of experimental manipulation: Last, but certainly not least, among the considerations is whether an organism is appropriate for the experimental ap-proach needed to answer the question being asked. For example, due to the long history of significant investments to develop the fruitfly and mouse model systems, a plethora of powerful molecular and genetic tools exist to manipulate gene and protein function during embryonic development of these organisms. Similarly, the extensive body of information now available on the genetics and development of the small mustard plant *Arabidopsis thaliana* has made it a wide-ly used model organism in research on flowering plants.

THE USUAL SUSPECTS   Some of the more common model systems used to study embryonic development include a flowering weed (*Arabidopsis thaliana*), sea urchin (*Strongylocentrotus purpuratus*), sea squirt (*Ciona intestinalis*), fruitfly (*Drosophila mela-nogaster*), roundworm (*Caenorhabditis elegans*), zebrafish (*Danio rerio*), African clawed frog (*Xenopus laevis*), chicken (*Gallus gallus*), and mouse (*Mus musculus*) (**FIGURE 1.4**). This short list of usual suspects is not a true representation of the diversity of organ-isms actually being used to study developmental biology, however. For instance, hydra, planarian flatworms, the *Axolotl* salamander, and the spiny mouse are among the top animals used to study regeneration. Many of the above model systems are actively being used to directly model the development of human disease. Additionally, human pluripotent stem cells are being used to study human development in a dish.

Advances in shared genomic and molecular approaches have dramatically increased the accessibility of nontraditional or nonmodel organisms for developmental research. This is one of the most exciting things about being a new student entering the field of developmental biology today. You do not have to be restricted to the conventional model systems; rather, any species could be a new model organism for you to investigate.

The Cycle of Life

Through initial studies of model organisms, descriptive embryology has brought us an understanding of the life cycles of various organisms.

An animal's life cycle

Most animals, whether earthworm or eagle, termite or beagle, pass through similar stages of development: fertilization, cleavage, gastrulation, organogenesis, hatching (or birth), metamorphosis, and gametogenesis. The stages of development between fertilization and hatching (or birth) are collectively called **embryogenesis**.

1. **Fertilization** involves the fusion of the mature sex cells, the sperm and egg, which are collectively called the **gametes**. The fusion of the gamete cells stimulates the egg to begin development and initiates a new individual. The subsequent fusion of the gamete nuclei (the male and female **pronuclei**, each of which has only half the normal number of chromosomes characteristic for the species) gives the embryo its **genome**, the collection of genes that helps instruct the embryo to develop in a manner very similar to that of its parents.

2. **Cleavage** is a series of mitotic divisions that immediately follow fertilization. During cleavage, the enormous volume of zygote cytoplasm is divided into numerous smaller cells called **blastomeres**. By the end of cleavage, the blastomeres have usually formed a sphere, known as a **blastula**.[4]

3. After the rate of mitotic division slows down, the blastomeres undergo dramatic movements and change their positions relative to one another. This series of extensive cell rearrangements is called **gastrulation**, and the embryo is said to be in the **gastrula** stage. As a result of gastrulation, the embryo contains three **germ layers** (**endoderm**, **ectoderm**, and **mesoderm**) that will interact to generate the organs of the body.

4. Once the germ layers are established, the cells interact with one another and rearrange themselves to produce tissues and organs. This process is called **organogenesis**. Chemical signals are exchanged between the cells of the germ layers, resulting in the formation of specific organs at specific sites. Certain cells will undergo long migrations from their place of origin to their final location. These migrating cells include the precursors of blood cells, lymph cells, pigment cells, and gametes (eggs and sperm).

5. In most species, the organism that hatches from the egg or is born into the world is not sexually mature. Rather, the organism needs to undergo metamorphosis to become a sexually mature adult. In most animals, the young organism is called a **larva**, and it may look significantly different from the adult. In some species, the larval stage is the one that lasts the longest, and is used for feeding or dispersal. In such species, the adult is a brief stage whose sole purpose is to reproduce. In silkworm moths, for instance, the adults do not have mouthparts and cannot feed; the larva must eat enough so that the adult has the stored energy to survive and mate. Indeed, most female moths mate as soon as they eclose from the pupa, and they fly only once—to mate and lay their eggs. Then they die.

6. In many species, a group of cells is set aside to produce the next generation (rather than forming the current embryo). These cells are the precursors of the gametes. The gametes and their precursor cells are collectively called **germ cells**, and they are set aside for reproductive function. All other cells of the body are called **somatic cells**. This separation of somatic cells (which give rise to the individual body) and germ cells (which contribute to the formation of a new generation) is often one of the first differentiations to occur during animal development. The germ cells eventually migrate to the gonads, where they differentiate into gametes. The development of gametes, called **gametogenesis**, is usually not completed until the organism has become physically mature. At maturity, the gametes may be released and participate in fertilization to begin a new embryo. The adult organism eventually undergoes senescence and dies, its nutrients often supporting the early embryogenesis of its offspring and its absence allowing less competition. Thus, the cycle of life is renewed. (See **Further Development 1.1, When Does a Human Become a Person?**, online.)

A flowering plant's life cycle

The life cycle of flowering plants (and of all other land plants) is different from that of animals in having two alternating stages, a diploid **sporophytic** (diploid spore-bearing) stage and a haploid **gametophytic** (haploid gamete-producing) stage. When you picture a beautiful rose with its flower, leaves, stem, and hidden roots, you are looking at the full-grown sporophytic stage; within its flowers are the female and male gametophytes that produce eggs and sperm. Upon fertilization, these gametes create the embryos of the next generation of sporophytes, held within the seed coats that protect them (see Figure 1.8). Under optimal environmental conditions these embryos develop, and a new cycle of life can commence.

The life cycle of a flowering plant is similar in various aspects to the general scheme of an animal's life cycle. Both male and female haploid gametes are produced, the male gamete must travel to the female gamete, and subsequent fertilization initiates mitotic cell divisions and the development of the embryo. As in animals, the embryo develops three basic cell layers, but these do not rearrange through gastrulation-like movements. In addition, the embryo, which is developing within a seed, typically pauses between completion of embryogenesis and subsequent germination and growth. This dormancy period can be exceedingly long. Unlike animals, plants have indeterminate growth. This continued growth is possible because plants retain areas of stem cells for growth called meristems, which are located at the apical and basal tips of the embryo and are maintained in the adult. (Although adult animals also retain stem cells, these are not used for indeterminate growth.) Differentiation of tissues in the developing plant results in organogenesis like in animals, but plant cells have a cell wall outside their plasma membrane that is nonexistent in animals. This plant cell wall imposes many constraints on the developmental mechanisms driving plant patterning and growth, such as inhibiting cell movement, restricting planes of cell division, requiring unique modes of molecule transport between cells, and more robust responses for regenerative repair, to name just a few.

Example 1: A Frog's Life

All animal life cycles are modifications of the generalized one described above. Here we present a concrete example, the development of the leopard frog *Rana pipiens* (**FIGURE 1.5**).

FIGURE 1.5 Developmental history of the leopard frog *Rana pipiens*. The stages from fertilization through hatching (birth) are known collectively as embryogenesis. The region set aside for producing germ cells is shown in purple. Gametogenesis, which is completed in the sexually mature adult, begins at different times during development, depending on the species. (The sizes of the varicolored wedges shown here are arbitrary and do not correspond to the proportion of the life cycle spent in each stage.)

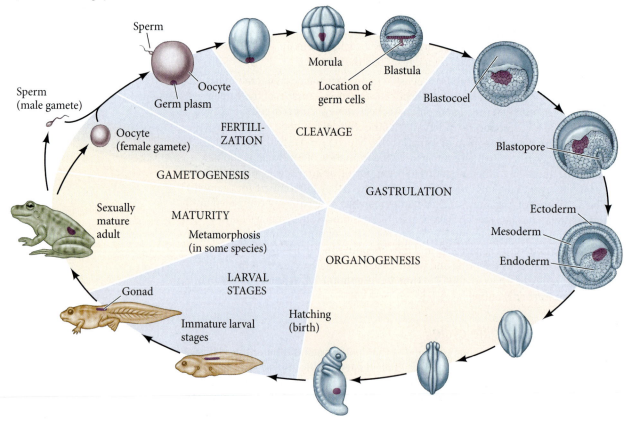

Gametogenesis and fertilization

The end of one life cycle and the beginning of the next are often intricately intertwined. Life cycles are often controlled by environmental factors (tadpoles wouldn't survive if they hatched in the fall, when their food is dying), so in most frogs, gametogenesis and fertilization are seasonal events. A combination of photoperiod (hours of daylight) and temperature informs the pituitary gland of the mature female frog that it is spring. The pituitary secretions cause the eggs and sperm to mature.

In most species of frogs, fertilization is external (**FIGURE 1.6A**). The male frog grabs the female's back and fertilizes the eggs as the female releases them (**FIGURE 1.6B**). Some species lay their eggs in pond vegetation, and the egg jelly adheres to the plants and anchors the eggs. The eggs of other species float into the center of the pond without any support. So an important thing to remember about life cycles is that they are intimately intertwined with environmental factors.

Fertilization accomplishes both sex (genetic recombination) and reproduction (the generation of a new individual). The genomes of the haploid male and female pronuclei merge and recombine to form the diploid zygote nucleus. In addition, the entry of the sperm facilitates the movement of cytoplasm inside the newly fertilized egg. This migration will be critical in determining the three body axes of the frog: anterior-posterior (head-tail), dorsal-ventral (back-belly), and right-left. And importantly, fertilization activates those molecules necessary to begin cell cleavage and gastrulation (Rugh 1950).

Cleavage and gastrulation

During cleavage, the volume of the frog egg stays the same, but it is divided into tens of thousands of cells (**FIGURE 1.6C,D**). Gastrulation in the frog begins at a point on the embryo surface roughly 180° opposite the point of sperm entry with the formation of a dimple called the **blastopore** (**FIGURE 1.6E**). The blastopore, which marks the future dorsal side of the embryo, expands to become a ring. Cells migrating through the blastopore to the embryo's interior become the mesoderm and endoderm; cells remaining outside become the ectoderm, and this outer layer expands to enclose the entire embryo. Thus, at the end of gastrulation, the ectoderm (precursor of the epidermis, brain, and nerves) is on the outside of the embryo, the endoderm (precursor of the lining of the gut and respiratory systems) is deep inside the embryo, and the mesoderm (precursor of the connective tissue, muscle, blood, heart, skeleton, gonads, and kidneys) is between them.

Organogenesis

Organogenesis in the frog begins when the cells of the most dorsal region of the mesoderm condense to form the rod of cells called the notochord.[5] These notochord cells produce chemical signals that redirect the fate of the ectodermal cells above it. Instead of forming epidermis, the cells above the notochord are instructed to become the cells of the nervous system. The cells change their shapes and rise up from the round body (**FIGURE 1.6F**). At this stage, the embryo is called a **neurula**. The neural precursor cells elongate, stretch, and fold into the embryo, forming the **neural tube**. The future epidermal cells of the back cover the neural tube.

Once the neural tube has formed, it and the notochord induce changes in the neighboring regions, and organogenesis continues. The mesodermal tissue adjacent to the neural tube and notochord becomes segmented into **somites**—the precursors of the frog's back muscles, spinal vertebrae, and dermis (the inner portion of the skin). The embryo develops a mouth and an anus, and it elongates into the familiar tadpole structure (**FIGURE 1.6G**). The neurons make connections to the muscles and to other neurons, the gills form, and the larva is ready to hatch from its egg. The hatched tadpole will feed for itself as soon as the yolk supplied by its mother is exhausted.

Metamorphosis and gametogenesis

Metamorphosis of the fully aquatic tadpole larva into an adult frog that can live on land is one of the most striking transformations in all of biology. Almost every organ is subject to modification, and the resulting changes in form are striking (**FIGURE 1.7**). The hindlimbs and forelimbs the adult will use for locomotion differentiate as the tadpole's paddle tail recedes. The cartilaginous tadpole skull is replaced by the predominantly

(A)

(B)

(C)

(D)

(E)

Dorsal blastopore lip

(F)

Open neural tube

(G)

Brain

Gill area

Expansion of
forebrain to touch
surface ectoderm
(induces eyes to form)

Stomodeum (mouth)

Somites

Tailbud

(H)

FIGURE 1.6 Early development of the frog *Xenopus laevis*. (A) Frogs
mate by amplexus, the male grasping the female around the belly and fertilizing the eggs as they are released. (B) A newly laid clutch of eggs. The
cytoplasm has rotated such that the darker pigment is where the nucleus
resides. (C) An 8-cell embryo. (D) A late blastula, containing thousands of
cells. (E) An early gastrula, showing the blastopore lip through which the
mesodermal and some endoderm cells migrate. (F) A neurula, where the
neural folds come together at the dorsal midline, creating a neural tube.
(G) A pre-hatching tadpole, as the protrusions of the forebrain begin to
induce eyes to form. (H) A mature tadpole, having swum away from the
egg mass and feeding independently.

(A)

(B)

(C)

(D)

(E)

(F)

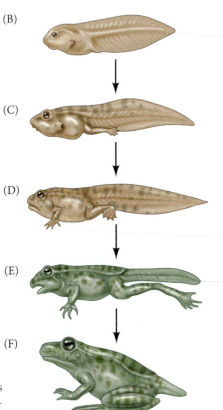

FIGURE 1.7 Metamorphosis of the frog. (A) Huge changes are obvious when one contrasts the tadpole and the adult bullfrog. Note especially the differences in jaw structure and limbs. (B) Premetamorphic tadpole. (C) Prometamorphic tadpole, showing hindlimb growth. (D) Onset of metamorphic climax as forelimbs emerge. (E,F) Climax stages.

bony skull of the young frog. The horny teeth the tadpole uses to tear up pond plants disappear as the mouth and jaw take a new shape, and the fly-catching tongue muscle of the frog develops. Meanwhile, the tadpole's lengthy intestine—a characteristic of herbivores—shortens to suit the more carnivorous diet of the adult frog. The gills regress and the lungs enlarge. Amphibian metamorphosis is initiated by hormones from the tadpole's thyroid gland; the mechanisms by which thyroid hormones accomplish these changes will be discussed in Chapter 21. The speed of metamorphosis is keyed to environmental pressures. In temperate regions, for instance, *Rana* metamorphosis must occur before ponds freeze in winter. An adult leopard frog can burrow into the mud and survive the winter; its tadpole cannot.

As metamorphosis ends, the development of the germ cells (sperm and eggs) begins. Gametogenesis can take a long time. In *Rana pipiens*, it takes 3 years for the eggs to mature in the female's ovaries. Sperm take less time; *Rana* males are often fertile soon after metamorphosis. To become mature, the germ cells must be competent to complete **meiosis**, the cell divisions that halve the number of chromosomes to produce haploid gametes. Having undergone meiosis, the mature sperm and egg nuclei can unite in fertilization, restoring the diploid chromosome number and initiating the events that lead to development and the continuation of the circle of life.

Example 2: Even a Weed Can Have a Flower-Full Life

Much of our discussions of plant development in this text will focus on research conducted on the angiosperm *Arabidopsis thaliana*. This small flowering plant, considered a weed, has all the criteria for a great laboratory model organism. Its life cycle is only 6 weeks long, its techniques for propagation are routine, and it has a comparatively small genome that has been sequenced and annotated many times over. The diversity of genetic, environmental, and other experimental approaches available to *A. thaliana* researchers has provided a wealth of understanding behind the mechanisms driving all aspects of this vascular plant's life cycle. Importantly, due to the monophyletic (descended from a single common ancestor) relationships of land plants, much of what has been learned about *A. thaliana* development is relevant to all plants (Koornneef and Meinke 2010; Provart et al. 2016). However, a flowering weed is not a sycamore tree, nor is it corn; there is diversity in the mechanisms of embryogenesis among different plants, and we will be highlighting some of these in later chapters.

Reproductive and gametophytic phases

When an adult flowering plant (angiosperm) is in the reproductive phase, the plant will have fully developed flowers with pollen-producing stamens (male reproductive organs) and ovary-containing carpels (female reproductive organs), which produce the haploid sperm and egg, respectively (**FIGURE 1.8**). These gametes are produced in the gametophytic phase. When pollen carrying the sperm delivers sperm to an egg, fertilization occurs, yielding a diploid zygote (single-celled embryo) (see Chapter 7; Huijser and Schmid 2011).

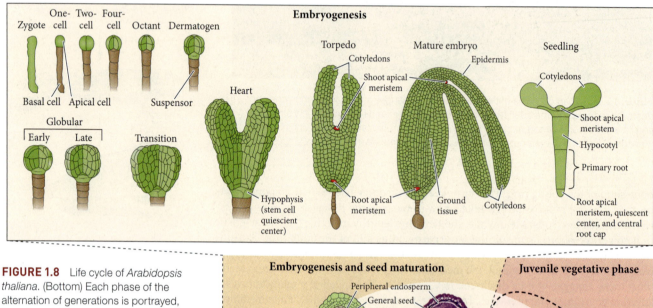

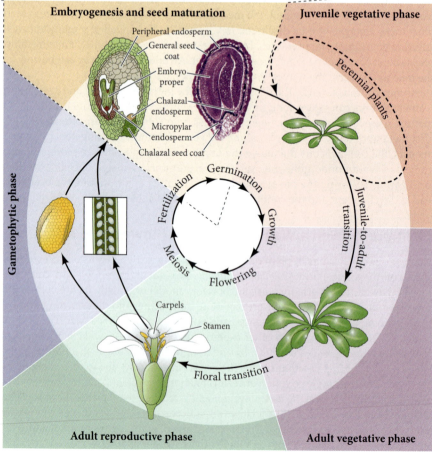

FIGURE 1.8 Life cycle of *Arabidopsis thaliana*. (Bottom) Each phase of the alternation of generations is portrayed, from the adult reproductive phase to the gametophytic phase, embryogenesis and seed maturation, and ending with the vegetative phases. Two stages of embryonic development (the torpedo and mature stages) are portrayed within the seed. (Top) Three-dimensional view of the stages of embryogenesis, from the zygote to the mature embryo. Note the shoot and root apical meristems, labeled in the torpedo and mature embryo stages. (Top, after J. Palovaara et al. 2016. *Annu Rev Cell Dev Biol* 32: 47–75; S. Yoshida et al. 2014. *Dev Cell* 29: 75–87; and courtesy of Meryl Hashimoto, Mark Belmonte, Julie Pelletier, and John Harada; bottom, after P. Huijser and M. Schmid. 2011. *Development* 138: 4117–4129.)

Embryogenesis and seed maturation

In contrast to cleavage in some animals that have large amounts of yolk in their eggs, which impedes cleavage, embryonic cleavage in seed-producing plants is not constrained by yolk, as the nutrient supply to the plant embryo comes from the surrounding endosperm in the seed (see Figure 1.8; Palovaara et al. 2016). However, of critical significance is the fact that the zygote divides, but does so asymmetrically. The first cell division yields a small (approximately one-third-size) apical cell and a much larger basal cell. The apical cell goes on to generate the embryo proper, while the basal cell becomes the suspensor, which functions to support the embryo, in part by ensuring that it develops within the lumen of the seed. This initial asymmetrical division sets up the primary apical-basal axis of the embryo, such that shoots (stem, leaves, and flowers) will grow from the apical-most cells, while roots will develop from the basal-most cells. Precisely positioned transverse and longitudinal division planes continue to build the embryo through the globular, heart, torpedo, and mature stages (see Figure 1.8). Since plant cells cannot migrate or move, there are no gastrulation movements as you would find in animal embryogenesis; instead, the different morphologies at these stages are all based on manipulation of the plane of division as well as on the directionality of cell growth.

Among the major structures that form during *A. thaliana* embryogenesis are the stem and root meristematic tissues and the embryonic leaves called **cotyledons** (see Figure 1.8). The basal-most cluster of cells of the embryo takes on stem cell behaviors and is called the **root apical meristem** (**RAM**); the cells positioned along the central axis of the embryo at the apical-most region are called the **shoot apical meristem** (**SAM**) and similarly possess self-renewing and differentiation behaviors (see Figure 1.8). Additionally, the lateral apexes producing the heart-shaped morphology give rise to the two cotyledons, which provide nutrients to support development through embryogenesis and germination of the seedling (see Figure 1.8).

Plants do not have a huge variety of different cell types, but three distinctive tissue types immediately become segregated in the embryo: **dermal**, **ground**, and **vascular tissues**. The dermal cells will produce the outer layers of the plant epidermis. Ground tissues give rise to the bulk of a plant's internal structures. The cells at the very core of the embryo will form the vascular tissues: **xylem**, which are conduits for bringing water and nutrients upward through the plant, and **phloem**, which are conduits for bringing sugars produced by photosynthesis and other metabolites, primarily from the leaves to parts of the plant that consume more than they produce.

Vegetative phases: From sporophytic growth to inflorescence identity

Upon completion of germination, the now-established sporophyte grows. This marks the beginning of the juvenile vegetative phase, which generally increases the plant's mass and overall size as it continues into the adult vegetative phase. The next phase is the adult reproductive phase, during which a change in the differentiation program of the SAM cells occurs, so that they start generating reproductive tissues instead of stems and leaves. This means the plant starts producing flowers, with their gamete-producing stamens and carpels. Once the plant is fully developed, the life cycle can repeat.

An Overview of Early Animal Development

Patterns of cleavage

E. B. Wilson, one of the pioneers in applying cell biology to embryology, noted in 1923, "To our limited intelligence, it would seem a simple task to divide a nucleus into equal parts. The cell, manifestly, entertains a very different opinion." Indeed, different organisms undergo cleavage in distinctly different ways, and the mechanisms for these differences remain at the frontier of cell and developmental biology. Cells in the cleavage stage are called blastomeres. In most animal species (mammals being the chief exception), both the initial rate of cell division and the placement of the blastomeres with respect to one another are under the control of proteins and mRNAs stored in the oocyte. Only later do the rates of cell division and the placement of cells come under the control of the newly formed organism's own genome (that is, the zygotic genome). During the initial phase of development, when cleavage rhythms are controlled by maternal factors, cytoplasmic volume does not increase. Rather, the zygote cytoplasm is divided into ever smaller cells—first in halves, then quarters, then eighths, and so forth. Cleavage occurs very rapidly in most invertebrates and many vertebrates, probably as an adaptation to generate a large number of cells quickly and to restore the somatic ratio of nuclear volume to cytoplasmic volume. The embryo often accomplishes this by abolishing the gap periods of the cell cycle (the G1 and G2 phases), when growth can occur. A frog egg, for example, can divide into 37,000 cells in just 43 hours. Mitosis in cleavage-stage *Drosophila* embryos occurs every 10 minutes for more than 2 hours, forming some 50,000 cells in just 12 hours.

The pattern of embryonic cleavage peculiar to a species is determined by two major parameters: (1) the amount and distribution of yolk protein within the cytoplasm, which determine where cleavage can occur and the relative sizes of the blastomeres; and (2) factors in the egg cytoplasm that influence the angle of the mitotic spindle and the timing of its formation.

In general, yolk impedes cleavage. When one pole of the egg is relatively yolk-free, cellular divisions occur there at a faster rate than at the opposite pole. The yolk-rich pole is referred to as the **vegetal pole**; the yolk concentration in the **animal pole** is relatively low. The zygote nucleus is frequently displaced toward the animal pole. **FIGURE 1.9** provides a classification of cleavage types and shows the influence of yolk on cleavage symmetry and pattern. At one extreme are the eggs of some sea urchins, mammals, and snails. These eggs have sparse, equally distributed yolk and are thus **isolecithal** (Greek, "equal yolk"). In these species, cleavage is holoblastic (Greek *holos*, "complete"), meaning that the cleavage furrow extends through the entire egg. With little yolk, these embryos must have some other way of obtaining food. Most will generate a voracious larval form, while placental mammals will obtain their nutrition from the maternal placenta.

At the other extreme are the eggs of insects, fish, reptiles, birds, and egg-laying mammals (monotremes). Most of their cell volumes are made up of yolk. The yolk must be sufficient to nourish these animals throughout embryonic development. Zygotes containing large accumulations of yolk undergo meroblastic cleavage (Greek *meros*, "part"), wherein only a portion of the cytoplasm is cleaved. The cleavage furrow does not penetrate the yolky portion of the cytoplasm because the yolk platelets impede membrane formation there. Insect eggs have yolk in the center (i.e., they are **centrolecithal**), and the divisions of the cytoplasm occur only in the rim of cytoplasm, around the periphery of the cell (i.e., **superficial cleavage**). The eggs of birds and fish have only one small area of the egg that is free of yolk (**telolecithal** eggs), and therefore the cell divisions occur only in this small disc of cytoplasm, giving rise to **discoidal cleavage**. These are general rules, however, and even closely related species have evolved different patterns of cleavage in different environments.

Yolk is just one factor influencing a species' pattern of cleavage. There are also inherited patterns of cell division superimposed on the constraints of the yolk. The importance of this inheritance can readily be seen in isolecithal eggs. In the absence of a large concentration of yolk, holoblastic cleavage takes place. Four major patterns of this cleavage type can be described: *radial*, *spiral*, *bilateral*, and *rotational* holoblastic cleavage (see Figure 1.9). (See **Further Development 1.2, The Cell Biology of Embryonic Cleavage**, online.)

Gastrulation: "The most important time in your life"

According to embryologist Lewis Wolpert, "It is not birth, marriage, or death, but gastrulation which is truly the most important time in your life." This is not an overstatement. Gastrulation is what makes animals animals. (Animals gastrulate; plants and

FIGURE 1.9 Summary of the main patterns of cleavage.

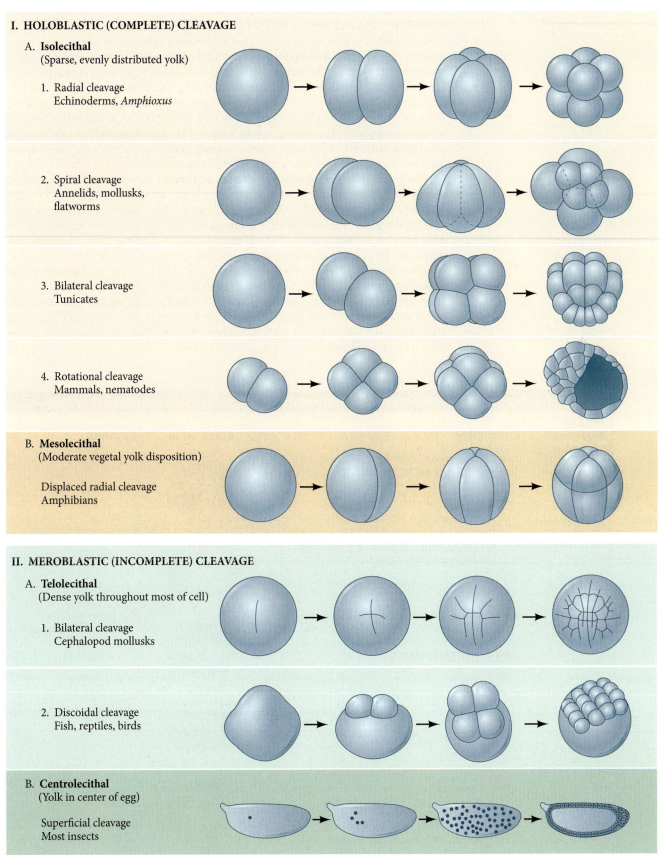

I. HOLOBLASTIC (COMPLETE) CLEAVAGE

A. **Isolecithal**
   (Sparse, evenly distributed yolk)

   1. Radial cleavage
      Echinoderms, *Amphioxus*

   2. Spiral cleavage
      Annelids, mollusks,
      flatworms

   3. Bilateral cleavage
      Tunicates

   4. Rotational cleavage
      Mammals, nematodes

B. **Mesolecithal**
   (Moderate vegetal yolk disposition)

   Displaced radial cleavage
   Amphibians

II. MEROBLASTIC (INCOMPLETE) CLEAVAGE

A. **Telolecithal**
   (Dense yolk throughout most of cell)

   1. Bilateral cleavage
      Cephalopod mollusks

   2. Discoidal cleavage
      Fish, reptiles, birds

B. **Centrolecithal**
   (Yolk in center of egg)

   Superficial cleavage
   Most insects

fungi do not.) During gastrulation, the cells of the blastula are given new positions and new neighbors, and the multilayered body plan of the organism is established. The cells that will form the endodermal and mesodermal organs are brought to the inside of the embryo, while the cells that will form the epidermis (outer layer of skin) and nervous system are spread over its outside surface. Thus, the three germ layers—outer ectoderm, inner endoderm, and, in between them, mesoderm—are first produced during gastrulation. In addition, the stage is set for the interactions of these newly positioned tissues.

Gastrulation usually proceeds by some combination of several types of movements. These movements involve the entire embryo, and cell migrations in one part of the gastrulating embryo must be intimately coordinated with other movements that are taking place simultaneously. Although patterns of gastrulation vary enormously throughout the animal kingdom, all of the patterns are different combinations of the five basic types of cell movements—**invagination**, **involution**, **ingression**, **delamination**, and **epiboly**—described in **TABLE 1.1**.

In addition to establishing which cells will be in which germ layer, embryos must develop three crucial axes that are the foundation of the body: the anterior-posterior axis, the dorsal-ventral axis, and the right-left axis (**FIGURE 1.10**). The **anterior-posterior** (**AP** or **anteroposterior**) **axis** is the line extending from head to tail (or from mouth to anus in those organisms that lack a head and tail). The **dorsal-ventral** (**DV** or **dorsoventral**) **axis** is the line extending from back (dorsum) to belly (ventrum). The **right-left axis** separates the two lateral sides of the body. Although humans (for example) may look symmetrical, recall that in most of us, the heart is in the left half of the body, while the liver is on the right. Somehow, the embryo knows that some organs belong on one side and other organs go on the other.

TABLE 1.1 Types of cell movement during gastrulation[a]

Type of movement	Description	Illustration	Example
Invagination	Infolding of a sheet (epithelium) of cells, much like the indention of a soft rubber ball when it is poked.		Sea urchin endoderm
Involution	Inward movement of an expanding outer layer so that it spreads over the internal surface of the remaining external cells.		Amphibian mesoderm
Ingression	Migration of individual cells from the surface into the embryo's interior. Individual cells become mesenchymal (i.e., separate from one another) and migrate independently.		Sea urchin mesoderm, *Drosophila* neuroblasts
Delamination	Splitting of one cellular sheet into two more or less parallel sheets. While on a cellular basis it resembles ingression, the result is the formation of a new (additional) epithelial sheet of cells.		Hypoblast formation in birds and mammals
Epiboly	Movement of epithelial sheets (usually ectodermal cells), spreading as a unit (rather than individually) to enclose deeper layers of the embryo. Can occur by cells dividing, by cells changing their shape, or by several layers of cells intercalating into fewer layers; often, all three mechanisms are used.		Ectoderm formation in sea urchins, tunicates, and amphibians

[a]The gastrulation of any particular organism is an ensemble of several of these movements.

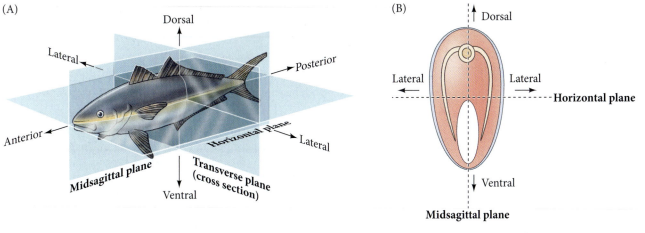

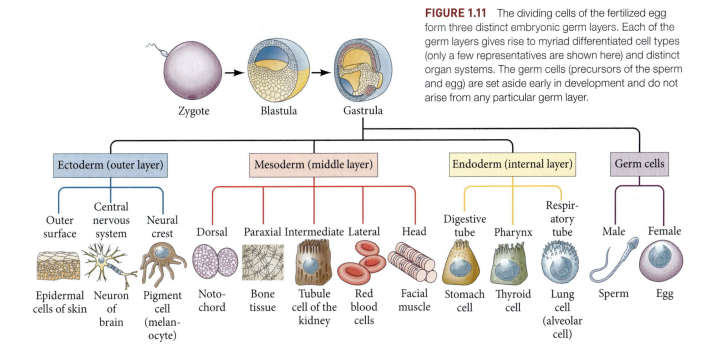

FIGURE 1.10 Axes of a bilaterally symmetrical animal. (A) A single plane, the midsagittal plane, divides the animal into left and right halves. (B) Cross sections bisecting the anterior-posterior axis.

The primary germ layers and early organs

The end of preformationism—the idea that all the organs of the adult are present in miniature in the sperm or egg (see Further Development 1.3, online)—did not come until the 1820s, when a combination of new staining techniques, improved microscopes, and institutional reforms in German universities created a revolution in descriptive embryology. The new techniques enabled microscopists to document the epigenesis of anatomical structures, and the institutional reforms provided audiences for these reports and students to carry on the work of their teachers. Among the most talented of this new group of microscopically inclined investigators were three friends, born within a year of each other, all of whom came from the Baltic region and studied in northern Germany. The work of Christian Pander, Heinrich Rathke, and Karl Ernst von Baer transformed embryology into a specialized branch of science.

Studying the chick embryo, Pander discovered that the embryo was organized into germ layers[6]—three distinct regions of the embryo that give rise through epigenesis (i.e., forming de novo, or "from scratch") to the differentiated cell types and specific organ systems (**FIGURE 1.11**). These three layers are found in the embryos of most animal phyla:

FIGURE 1.11 The dividing cells of the fertilized egg form three distinct embryonic germ layers. Each of the germ layers gives rise to myriad differentiated cell types (only a few representatives are shown here) and distinct organ systems. The germ cells (precursors of the sperm and egg) are set aside early in development and do not arise from any particular germ layer.

- The ectoderm generates the outer layer of the embryo. It produces the surface layer (epidermis) of the skin and forms the brain and nervous system.

- The endoderm becomes the innermost layer of the embryo and produces the epithelium of the digestive tube and its associated organs (including the lungs).

- The mesoderm becomes sandwiched between the ectoderm and endoderm. It generates the blood, heart, kidney, gonads, bones, muscles, and connective tissues.

Pander also demonstrated that the germ layers did not form their respective organs autonomously (Pander 1817). Rather, each germ layer "is not yet independent enough to indicate what it truly is; it still needs the help of its sister travelers, and therefore, although already designated for different ends, all three influence each other collectively until each has reached an appropriate level." Pander had discovered the tissue interactions that we now call induction. No vertebrate tissue is able to construct organs by itself; it must interact with other tissues, as we will describe in Chapter 4.

Meanwhile, Rathke followed the intricate development of the vertebrate skull, excretory systems, and respiratory systems, showing that these became increasingly complex. He also showed that their complexity took on different trajectories in different classes of vertebrates. For instance, Rathke was the first to identify the **pharyngeal arches** (**FIGURE 1.12**). He showed that these same embryonic structures became gill supports in fish, and the jaws and ears (among other things) in mammals. Interestingly, the pharyngeal arches are derived from a migrating stem cell population called **neural crest cells**. Strikingly, neural crest cells break free of the dorsal neural tube and migrate as streams of cells into a variety of peripheral parts of the head and body (see Figure 1.12A), where they give rise to such diverse cell types as cartilage and bone in the head, sensory neurons and glial cells in the body, and pigment throughout. For this extreme stem-cell-like behavior, neural crest cells are often colloquially referred to as the fourth germ layer. (See **Further Development 1.3, Epigenesis and Preformationism**, online.)

FIGURE 1.12 Evolution of pharyngeal arch structures in the vertebrate head. (A) Pharyngeal arches (also called branchial arches) in the embryo of the Mexican salamander, *Ambystoma mexicanum*. The surface ectoderm has been removed to permit visualization of the arches (highlighted in color) as they form from neural crest cells streaming down from the midline. (B) In adult fish, pharyngeal arch cells form the hyomandibular jaws and gill arches. (C) In amphibians, birds, and reptiles (a crocodile is shown here), these same cells form the quadrate bone of the upper jaw and the articular bone of the lower jaw. (D) In mammals, the quadrate has become internalized and forms the incus of the middle ear. The articular bone retains its contact with the quadrate, becoming the malleus of the middle ear. Thus, the cells that form gill supports in fish form the middle ear bones in mammals. (B, after R. Zangerl and M. E. Williams. 1975. *Paleontology* 18: 333–341.)

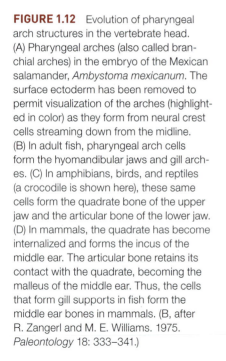

(A)

Courtesy of P. Falck and L. Olsson

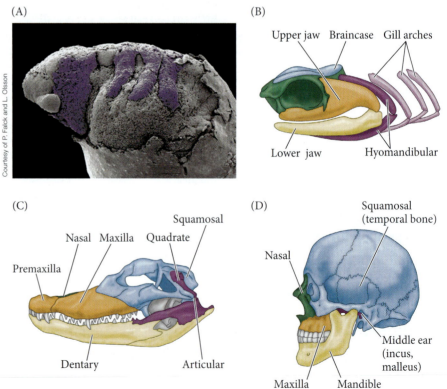

(B)

Upper jaw   Braincase   Gill arches

Lower jaw   Hyomandibular

(C)

Nasal   Maxilla   Quadrate   Squamosal

Premaxilla

Dentary   Articular

(D)

Squamosal (temporal bone)

Nasal

Middle ear (incus, malleus)

Maxilla   Mandible

Understanding cell behavior in the embryo

By the late 1800s, it had been conclusively demonstrated that the cell is the basic unit of all anatomy and physiology. Embryologists, too, began to base their field on the cell. But unlike those who studied the adult organism, developmental anatomists found that cells in the embryo do not "stay put." Indeed, one of the most important conclusions of developmental anatomists is that embryonic cells do not remain in one place, nor do they keep the same shape (Larsen and McLaughlin 1987).

There are two major types of cells in the animal embryo: **epithelial cells**, which are tightly connected to one another in sheets or tubes; and **mesenchymal cells**, which are unconnected or loosely connected to one another and can operate as independent units. Within these two types of arrangements, morphogenesis is brought about through a limited repertoire of variations in cellular processes:

- *Direction and number of cell divisions.* Think of the faces of two dog breeds—say, a German shepherd and a poodle. The faces are made from the same cell types, but the number and orientation of the cell divisions are different (Schoenebeck et al. 2012). Think also of the legs of a German shepherd compared with those of a dachshund. The skeleton-forming cells of the dachshund have undergone fewer cell divisions than those of taller dogs. And the extreme example is all plants. Their morphological diversity is highly determined by control over cell division patterns.

- *Cell shape changes.* Cell shape change is a critical feature of animal development. Changing the shapes of epithelial cells often creates tubes out of sheets (as when the neural tube forms), and a shape change from epithelial to mesenchymal is critical when individual cells migrate away from the epithelial sheet (as when muscle cells are formed). (This same type of epithelial-to-mesenchymal change operates in cancer, allowing cancer cells to migrate and spread from the primary tumor to new sites.) To be clear, mesenchymal cells are not present in plants, therefore neither are any of the behaviors they exhibit—namely migration.

- *Cell migration.* Cells have to move in order to get to their appropriate locations. For instance, the germ cells have to migrate into the developing gonad, and the primordial heart cells meet in the middle of the vertebrate neck and then migrate to the left part of the chest.

- *Cell growth.* Cells can change in size. This is most apparent in the germ cells: the sperm eliminates most of its cytoplasm and becomes smaller, whereas the developing egg conserves and adds cytoplasm, becoming comparatively huge. Many cells undergo an asymmetrical cell division that produces one big cell and one small cell, each of which may have a completely different fate. Here again, plants have monopolized this cellular mechanism of unidirectional growth to help elongate the vascular cell types, xylem and phloem.

- *Cell death.* Death is a critical part of life. The embryonic cells that constitute the webbing between our toes and fingers die before we are born. So do the cells of our tails. The orifices of our mouth, anus, and reproductive glands all form through **apoptosis**—the programmed death of certain cells at particular times and places. The sieve elements that make up the major conduits of the xylem in a plant are just skeletal remains of a cell wall following targeted apoptosis.

- *Changes in the composition of the cell membrane or secreted products.* Cell membranes and secreted cell products influence the behavior of neighboring cells. For instance, extracellular matrices secreted by one set of cells will allow the migration of their neighboring cells. Extracellular matrices made by other cell types will *prohibit* the migration of the same set of cells. In this way, "paths and guide rails" are established for migrating cells.

A Basic Approach to Watch Development

Approaching the bench: Find it, lose it, move it

As Dr. Viktor Hamburger once said, "Our real teacher has been and still is the embryo, who is, incidentally, the only teacher who is always right" (Holtfreter 1968). Hamburger was a developmental biologist who contributed to the creation of the entire chick embryo staging series used today (the Hamburger-Hamilton stages, or HH), an achievement that would have been impossible without careful observation and experimentation on the embryo.

How does the vertebrate brain develop such a precise network of connections? How are the carpels, stamens, and petals of a flower so perfectly organized in a radial distribution? Do the microbes residing in the gut influence the rate of intestinal stem cell division and differentiation, and if so, can that lead to cancer? Whatever the research question might be, developmental biologists have often approached the experimental design with a common mantra: find it, lose it, move it (Adams 2003). Admittedly, this is an oversimplification of the incredible variety of ways in which scientists have interrogated the mechanisms of developmental biology, but it is useful as an introduction to this field.

Find it: To study development, one needs to be able to see the subject in question. This could be the whole embryo, which is a different challenge depending on the species. Compare the access one has to a frog or zebrafish embryo that develops outside the mother with the access one has to a mouse that develops in utero or to a chick within the eggshell. Additionally, seeing things may mean observing select tissues or even individual cells within those tissues or, smaller still, the location of proteins and RNA transcripts. Advances in labeling techniques and innovations in microscopy are continually improving how scientists can watch development, because after all, it is a process that occurs over time. However, just seeing structures and morphogenetic events provides researchers with descriptive and correlative information about a given process. Developmental biologists need to manipulate development to get at causation.

Move it (or) Lose it: Let's ponder the fascinating phenomenon that some animals, such as the Mexican salamander, can regenerate whole limbs following their amputation. Upon amputation, one of the first structures to form is the **blastema**, which includes a small bulge of proliferating cells and the epidermal cover overlying the wound. If the blastema is removed following an amputation (lose it), regeneration does not occur. In contrast, you learn something fundamentally different when you transplant a blastema to somewhere else on the salamander (move it), say someplace weird, like its back or even its eye. As a result of this transplantation, the blastema grows a limb out of that foreign location corresponding to the handedness of the body side it originated from. Yes, you will see some crazy stuff in the pages of this book! Development is, if anything, fun. But we digress. The "lose it" experiment tells you whether that thing now lost (tissue, cells, genes, etc.) was *necessary* for a given process, whereas the "move it" experiment tells you whether that thing is *sufficient*. In this example, the blastema is both necessary (required) and sufficient for limb regeneration. When the homologous gene for Pax6 is lost in a fly or mouse, the eye does not form. However, when mouse *Pax6* is transcribed in the leg of a fly… you guessed it, a fly eye forms on the fly's leg (see Figure 24.3C)! Craziness—but craziness that can be explained, and will be, in the coming chapters.

Direct observation of living embryos

Some embryos have relatively few cells, and the cytoplasms of their early blastomeres have differently colored pigments. In such fortunate cases, it is actually possible to look through the microscope and trace the descendants of a particular cell into the organs they generate. This creates a **fate map**—a diagram that "maps" larval or adult structures onto the region of the embryo from which they arose. E. G. Conklin did this by patiently following the fates of each early cell of the tunicate *Styela partita* (**FIGURE 1.13**; Conklin 1905). The muscle-forming cells of this tunicate embryo always had a yellow color, derived from a region of cytoplasm found in one particular pair of blastomeres at the 8-cell sage. Removal of this pair of blastomeres (which according to Conklin's fate map

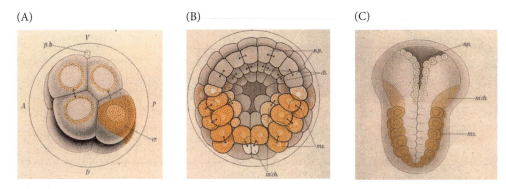

FIGURE 1.13 The fates of individual cells. Edwin Conklin mapped the fates of early cells of the tunicate *Styela partita*, using the fact that in embryos of this species, many of the cells can be identified by their different-colored cytoplasms. Yellow cytoplasm marks the cells that form the trunk muscles. (A) At the 8-cell stage, two of the eight blastomeres contain this yellow cytoplasm. (B) Early gastrula stage, showing the yellow cytoplasm in the precursors of the trunk musculature. (C) Early larval stage, showing the yellow cytoplasm in the newly formed trunk muscles. (From E. G. Conklin. 1905. *J Acad Nat Sci Phila* 13: 1–119.)

should produce the tail musculature) in fact resulted in larvae with no tail muscles, thus confirming Conklin's map (Reverberi and Minganti 1946). (See **Further Development 1.4, Conklin's Art and Science**, online.)

Dye marking

Most embryos are not so accommodating as to have cells of different colors. In the early years of the twentieth century, Vogt (1929) traced the fates of different areas of amphibian eggs by applying **vital dyes** to the region of interest. Vital dyes stain cells but do not kill them. Vogt mixed such dyes with agar and spread the agar on a microscope slide to dry. The ends of the dyed agar were very thin. Vogt cut chips from these ends and placed them on a frog embryo. After the dye stained the cells, he removed the agar chips and could follow the stained cells' movements within the embryo (**FIGURE 1.14**).

One problem with vital dyes is that they become more diluted with each cell division and thus over time become difficult to detect. One way around this is to use **fluorescent dyes** that are so intense that once injected into individual cells, they can still be detected in the progeny of these cells many divisions later. Fluorescein-conjugated dextran, for

? Developing Questions

Quiz yourself: What type of experiment did Reverberi and Minganti do? Find it, lose it, or move it? As a foreshadowing question for Chapter 2, what do their results say about those yellow blastomeres?*

*Answers to Developing Questions quiz: Lose it. The results suggest that the yellow blastomeres are the only determined cells to become muscle and that they must have some sort of muscle factor that other blastomeres lack. More will be revealed in Chapter 2.

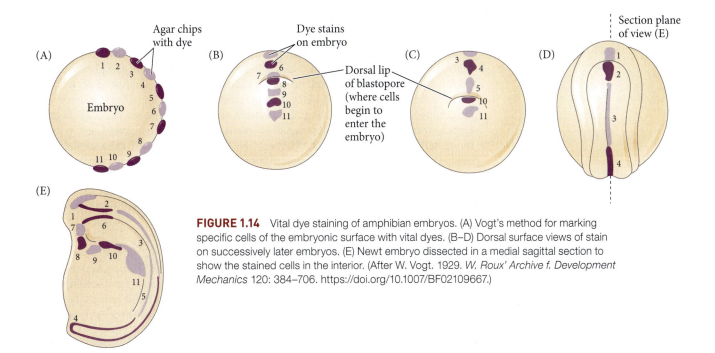

FIGURE 1.14 Vital dye staining of amphibian embryos. (A) Vogt's method for marking specific cells of the embryonic surface with vital dyes. (B–D) Dorsal surface views of stain on successively later embryos. (E) Newt embryo dissected in a medial sagittal section to show the stained cells in the interior. (After W. Vogt. 1929. *W. Roux' Archive f. Development Mechanics* 120: 384–706. https://doi.org/10.1007/BF02109667.)

(A) 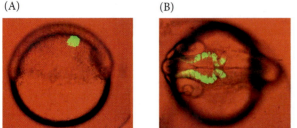 (B)

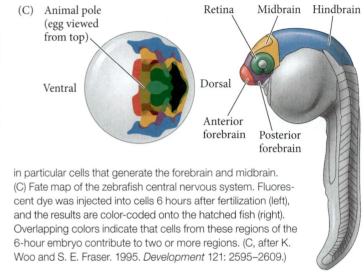

FIGURE 1.15 Fate mapping using a fluorescent dye. (A) Specific cells of a zebrafish embryo were injected with a fluorescent dye that will not diffuse from the cells. The dye was then activated by laser in a small region (about 5 cells) of the late-cleavage-stage embryo. (B) After formation of the central nervous system had begun, cells that contained the activated dye were visualized by fluorescent light. The fluorescent dye is seen in particular cells that generate the forebrain and midbrain. (C) Fate map of the zebrafish central nervous system. Fluorescent dye was injected into cells 6 hours after fertilization (left), and the results are color-coded onto the hatched fish (right). Overlapping colors indicate that cells from these regions of the 6-hour embryo contribute to two or more regions. (C, after K. Woo and S. E. Fraser. 1995. *Development* 121: 2595–2609.)

example, can be injected into a single cell of an early embryo, and the descendants of that cell can be seen by examining the embryo under ultraviolet light (**FIGURE 1.15**).

Genetic labeling

One way of permanently marking cells and following their fates is to create embryos in which the same organism contains cells with different genetic constitutions. One of the best examples of this technique is the construction of **chimeric embryos**—embryos made from tissues of more than one genetic source. Chick-quail chimeras, for example, are made by grafting embryonic quail cells inside a chick embryo while the chick is still in the egg. Chick and quail embryos develop in a similar manner (especially during the early stages), and the grafted quail cells become integrated into the chick embryo and participate in the construction of the various organs (**FIGURE 1.16A**). The chick that hatches will have quail cells in particular sites, depending on where the graft was placed. Quail cells also *differ* from chick cells in several important ways, including the species-specific proteins that form the immune system. There are quail-specific proteins that can be used to find individual quail cells, even when they are "hidden" within a large population of chick cells (**FIGURE 1.16B**). By seeing where these cells migrate, researchers have been able to produce fine-structure maps of the chick brain and skeletal system (Le Douarin 1969; Le Douarin and Teillet 1973).

Chimeras dramatically confirmed the extensive migrations of the neural crest cells during vertebrate development. Mary Rawles (1940) showed that the pigment cells (melanocytes) of the chick originate in the **neural crest**, a transient band of cells that joins the neural tube to the epidermis. When she transplanted small regions of tissue containing neural crest from a pigmented strain of chickens into a similar position in an embryo from an unpigmented strain of chickens, the migrating pigment cells entered the epidermis and late entered the feathers (**FIGURE 1.16C**). Ris (1941) used similar techniques to show that, although almost all of the external pigment of the chick embryo came from the migrating neural crest cells, the pigment of the retina formed in the retina itself and was not dependent on migrating neural crest cells. This pattern was confirmed in the chick-quail chimeras, in which the quail neural crest cells produced their own pigment and pattern in the chick feathers.

Transgenic DNA chimeras

In most animals, it is difficult to meld a chimera from two species. One way to circumvent this problem is to transplant cells from a genetically modified organism. In such a technique, the genetic modification can then be traced only to those cells that express it. One version is to infect the cells of an embryo with a virus whose genes have been

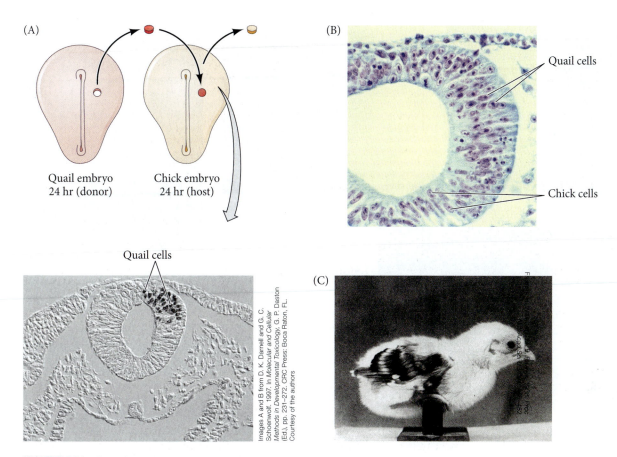

Quail cells

Images A and B from D. K. Darnell and G. C.
Schoenwolf. 1997. In *Molecular and Cellular
Methods in Developmental Toxicology*, G. P. Daston
(Ed.), pp. 231–272. CRC Press: Boca Raton, FL.
Courtesy of the authors

FIGURE 1.16 Genetic markers as cell lineage tracers. (A) Experiment in which cells from a particular region of a 1-day quail embryo have been grafted into a similar region of a 1-day chick embryo. After several days, the quail cells can be seen by using an antibody to quail-specific proteins (photograph below). This region produces cells that populate the neural tube. (B) Chick and quail cells can also be distinguished by the heterochromatin of their nuclei. Quail cells have a single large nucleus (dense purple), distinguishing them from the diffuse nuclei of the chick. (C) Chick resulting from transplantation of a trunk neural crest region from an embryo of a pigmented strain of chickens into the same region of an embryo of an unpigmented strain. The neural crest cells that gave rise to the pigment migrated into the wing epidermis and feathers.

altered such that they express the gene for a fluorescently active protein such as **green fluorescent protein** (**GFP**).[7] A gene altered in this way is called a **transgene** because it contains DNA from another species. When the infected embryonic cells are transplanted into a wild-type host, only the donor cells and their descendants express GFP; these emit a visible green glow when placed under ultraviolet light (see Affolter 2016; Papaioannou 2016).

Variations on transgenic labeling can give us a remarkably precise map of the developing body. For example, Freem and colleagues (2012) used transgenic techniques to study the migration of neural crest cells to the gut of chick embryos, where they form the neurons that coordinate peristalsis—the muscular contractions of the gut necessary to eliminate solid waste. The parents of the GFP-labeled chick embryo were infected with a replication-deficient virus that carried an active gene for GFP. This gene was inherited by the chick embryo and expressed in every cell. In this way, Freem and colleagues generated embryos in which every cell glowed green (**FIGURE 1.17A**). They then transplanted the neural tube and neural crest of a GFP-transgenic embryo into a similar region of a normal chick embryo (**FIGURE 1.17B**). A day later, they could see GFP-labeled cells migrating into the stomach region (**FIGURE 1.17C**), and 4 days after that, the entire gut glowed green up to the anterior region of the hindgut (**FIGURE 1.17D**).

(A)

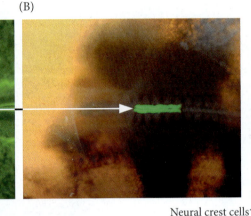

(B)

(C)

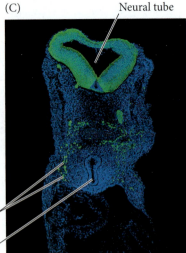

Neural tube

Neural crest cells

Foregut

All images from L. J. Freem et al. 2012. *Int J Dev Biol* 56: 245–254, courtesy of A. Burns

(D)

Stomach

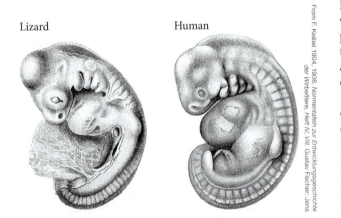

Esophagus   Midgut   Hindgut

FIGURE 1.17 Fate mapping with transgenic DNA shows that the neural crest is critical in making the gut neurons. (A) A chick embryo containing an active gene for green fluorescent protein expresses GFP in every cell. The brain is forming on the left side of the embryo, and the bulges from the forebrain (which will become the retinas) are contacting the head ectoderm to initiate eye formation. (B) The region of the neural tube and neural crest in the presumptive neck region (rectangle in A) is excised and transplanted into a similar position in an unlabeled wild-type embryo. One can see the transplanted tissue by its green fluorescence. (C) A day later, one can see the neural crest cells migrating from the neural tube to the stomach region. (D) In 4 more days, the neural crest cells have spread in the gut from the esophagus to the anterior end of the hindgut.

Evolutionary Embryology

"Community of embryonic structure reveals community of descent," Charles Darwin concluded in *On the Origin of Species* in 1859. This statement is based on Darwin's evolutionary interpretation of Karl Ernst von Baer's laws—namely, that relationships between groups can be established by finding common embryonic or larval forms. In 1828, just a few years before Darwin's voyage on the HMS *Beagle*, von Baer reported a curious observation. "I have two small embryos preserved in alcohol, that I forgot to label. At present I am unable to determine the genus to which they belong. They may be lizards, small birds, or even mammals." Drawings of such early-stage embryos allow us to appreciate his quandary (**FIGURE 1.18**).

From his detailed study of chick development and his comparison of chick embryos with the embryos of other vertebrates, von Baer derived four generalizations known as "von Baer's laws" (**TABLE 1.2**). Von Baer's laws can be summarized as describing how all vertebrates begin as simple embryos that share common characteristics, which become progressively specialized in species-specific ways. For instance, human embryos initially share characteristics in common with fish and avian embryos but diverge in form later in development, while never passing through the adult stages of lower vertebrate species. Recent research has confirmed von Baer's view that there is a **phylotypic stage** at which the embryos of the different groups of vertebrates all have a similar physical structure, such as the stage depicted in Figure 1.18. At this same stage there appears to be the least amount of difference among the *genes* expressed by the different groups within the vertebrates, indicating that this stage may be the source for the basic body plan for all the vertebrates (Irie and Kuratani 2011).[8]

Lizard   Human

From F. Keibel 1904, 1908. *Normentafeln zur Entwicklungsgeschichte der Wirbeltiere, Heft IV, VIII.* Gustav Fischer: Jena

FIGURE 1.18 The vertebrates—fish, amphibians, reptiles, birds, and mammals—all start development very differently because of the enormous differences in the size of their eggs. By the beginning of neurulation, however, all vertebrate embryos have converged on a common structure. Here, a lizard embryo is shown next to a human embryo at a similar stage. As they develop beyond the neurula stage, the embryos of the different vertebrate groups become less and less like each other.

TABLE 1.2 Von Baer's laws of vertebrate embryology

1. The general features of a large group of animals appear earlier in development than do the specialized features of a smaller group.

All developing vertebrates appear very similar right after gastrulation. All vertebrate embryos have gill arches, a notochord, a spinal cord, and primitive kidneys. It is only later in development that the distinctive features of class, order, and finally species emerge.

2. Less general characters develop from the more general, until finally the most specialized appear.

All vertebrates initially have the same type of skin. Only later does the skin develop fish scales, reptilian scales, bird feathers, or the hair, claws, and nails of mammals. Similarly, the early development of limbs is essentially the same in all vertebrates. Only later do the differences between legs, wings, and arms become apparent.

3. The embryo of a given species, instead of passing through the adult stages of lower animals, departs more and more from them.

For example, as seen in Figure 1.12, the pharyngeal arches start off the same in all vertebrates. But the arch that becomes the jaw support in fish becomes part of the skull of reptiles and becomes part of the middle ear bones of mammals. Mammals never go through a fishlike stage (Riechert 1837; Rieppel 2011).

4. Therefore, the early embryo of a higher animal is never like a lower animal, but only like its early embryo.

Human embryos never pass through a stage equivalent to an adult fish or bird. Rather, human embryos initially share characteristics in common with fish and avian embryos. Later in development, the mammalian and other embryos diverge, none of them passing through the stages of the others.

After reading Johannes Müller's summary of von Baer's laws in 1842, Darwin saw that embryonic resemblances would be a strong argument in favor of the evolutionary connectedness of different animal groups. Even before Darwin, larval forms were used in taxonomic classification. In the 1830s, for instance, J. V. Thompson demonstrated that larval barnacles were almost identical to larval shrimp, and therefore he (correctly) counted barnacles as arthropods rather than mollusks (**FIGURE 1.19**; Winsor 1969). Darwin, himself an expert on barnacle taxonomy, celebrated this finding: "Even

(A) Barnacle

(B) Shrimp

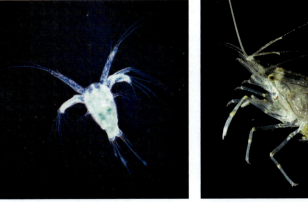

FIGURE 1.19 Larval stages reveal the common ancestry of two crustacean arthropods, barnacles (A) and shrimp (B). Barnacles and shrimp both exhibit a distinctive larval stage (the nauplius) that underscores their common ancestry as crustacean arthropods, even though adult barnacles—once classified as mollusks—are sedentary, differing in body form and lifestyle from the free-swimming adult shrimp. A larva is shown on the left in each pair of images, an adult on the right.

FIGURE 1.20 Transitional states over the course of animal evolution. (A) *Amphioxus*, or the lancelet, has a rudimentary notochord and nerve cord structures and is thus related to the common ancestor of all vertebrates. (B) A late Jurassic (~150 mya) fossil of *Archaeopteryx* showing its distinctive features of both a reptilian skeleton and avian feathered wings. (C) *Tiktaalik roseae* emerged 375 mya from the water to be the first animal hypothesized to walk on land. This fossil (upper) and reconstruction (lower) revealed characteristics of both fish fins and amphibian forelimbs, among other characteristics. (D) Scanning electron micrograph of the cnidarian, hydra. (E) A tube sponge. Dye placed at the base of the sponge is then squirted out the top, showing the pumping action of the sponge. (F) A motile larva of a sponge.

the illustrious Cuvier did not perceive that a barnacle is a crustacean, but a glance at the larva shows this in an unmistakable manner." Alexander Kowalevsky (1866) made the similar discovery that larvae of the sedentary tunicate (sea squirt) has the defining chordate structure called the notochord,[9] and that it originates from the same early embryonic tissues as the notochord does in fish and chicks. Thus, Kowalevsky reasoned, the invertebrate tunicate is related to the vertebrates, and the two great domains of the animal kingdom—invertebrates and vertebrates—are thereby united through larval structures. Darwin applauded Kowalevsky's finding, writing in *The Descent of Man* (1874) that "if we may rely on embryology, ever the safest guide in classification, it seems that we have at last gained a clue to the source whence the Vertebrata were derived." Darwin further noted that embryonic organisms sometimes form structures that are inappropriate for their adult form but that demonstrate their relatedness to other animals. He pointed out the existence of eyes in embryonic moles, pelvic bone rudiments in embryonic snakes, and teeth in baleen whale embryos.

FURTHER DEVELOPMENT

Chordates and the chord that connects us. Whether we are talking about an eagle, dinosaur, frog, or clownfish, they all have the common feature of being a vertebrate. The notochord is the most basal structure that defines an organism as a

(A)

Notochord

Oral cirri

Mouth

Segmental muscle

(B)

(C)

(D)

(E)

(F)

chordate, a group that includes the vertebrates. The notochord is a flexible rodlike structure that runs down the middle of an embryo's trunk and plays a pivotal role in organizing all surrounding tissues of the embryo. A critical moment in the transition from invertebrate to vertebrate developmental evolution is seen in *Amphioxus*, or the lancelet, a benthic, filter-feeding animal that resembles a cross between a worm and a tiny razorlike fish (**FIGURE 1.20A**). Although *Amphioxus* has no bones or even a brain of significance, it is related to the common ancestor of all chordates because it has a rudimentary notochord and nerve cord structures (Garcia-Fernàndez and Benito-Gutiérrez 2009). This discovery, made by Alexander Kowalevsky (1867), was a milestone in biology. The developmental stages of *Amphioxus* (and tunicates) united the invertebrates and vertebrates into a single "animal kingdom."

Darwin also argued that adaptations that depart from the "type" and allow an organism to survive in its particular environment develop late in the embryo.[10] He noted that the differences among species within genera become greater as development persists, as predicted by von Baer's laws. Thus, Darwin recognized two ways of looking at "descent with modification." One could emphasize *common descent* by pointing out embryonic similarities between two or more groups of organisms, or one could emphasize the *modifications* to show how development has been altered to produce structures that enable animals and plants to adapt to particular conditions.

Understanding the tree of life to see our developmental relatedness

Earth is estimated to have formed 4.56 billion years ago (bya), with evidence of the first signs of life occurring about 3.8 bya. The theory of evolution is fundamentally based on all life on Earth originating from a common ancient ancestor, so named **LUCA**, the **last universal common ancestor**. This means all forms of life are related to one another—from you, to the elephant, to the oyster toadfish,[11] to the oyster and toad alike, to the honeybee, to the horseshoe crab, to the horrible parasitic *Ascaris* roundworm, to the beautiful brain coral, to the brain puffball mushroom, to the nearly 400,000 species of flowering plants, to the 200,000 species of protists, and even to the bacteria living in your gut. If we are all related, then the mechanisms governing how a *Homo sapiens* develops are fundamentally derived from the common ancestors that connect all life along the tree—the tree of life (**FIGURE 1.21**).

FIGURE 1.21 The tree of life—an illustration of the major branches of life. A geological timescale moves radially from the bottom to the top of the diagram. All life on Earth is related. To better comprehend this reality, some of the major organismal groups are illustrated with colored branches for simplicity. The underlying layer of gray branches implies a more realistic and chaotic interconnectedness of life's lineage. The letters a–g denote the locations of common ancestors, including those of plants (b) and of multicellular organisms (a). Many of the common ancestors of acoels and flatworms, insects, vertebrates, and land animals (annelids, arthropods, mollusks, echinoderms, and vertebrates) (c–f) can be traced to the Cambrian explosion of diversity.

THE TREE OF LIFE

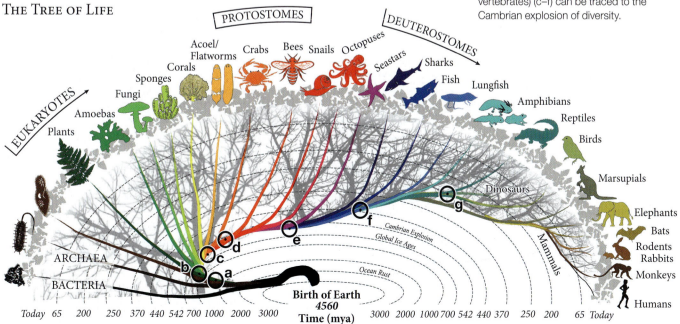

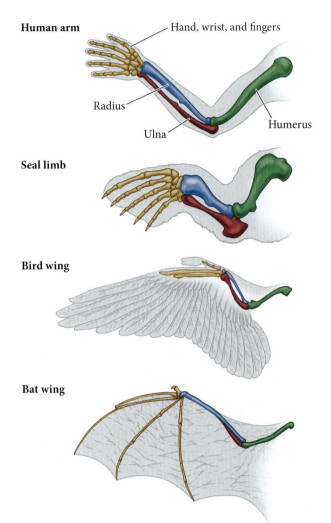

Human arm — Hand, wrist, and fingers

Radius

Ulna

Humerus

Seal limb

Bird wing

Bat wing

FIGURE 1.22 Homologies of structure among a human arm, a seal forelimb, a bird wing, and a bat wing; homologous supporting structures are shown in the same color. All four limbs were derived from a common tetrapod ancestor and are thus homologous as forelimbs. The adaptations of bird and bat forelimbs to flight, however, evolved independently of each other, long after the two lineages diverged from their common ancestor. Therefore, as wings they are not homologous, but analogous.

One of the most important distinctions made by evolutionary embryologists was the difference between *analogy* and *homology*. Both terms refer to structures that appear to be similar. **Homologous** structures are those whose underlying similarity arises from their being derived from a common ancestral structure. For example, the wing of a bird and the arm of a human are homologous, both having evolved from the forelimb bones of a common ancestor. Moreover, their respective parts are homologous (**FIGURE 1.22**).

Analogous structures are those whose similarity comes from their performing a similar function rather than their arising from a common ancestor. For example, the wing of a butterfly and the wing of a bird are analogous; the two share a common function (and thus both are called wings), but the bird wing and insect wing did not arise from a common ancestral structure that became modified through evolution into bird wings and butterfly wings. Homologies must always refer to the level of organization being compared. For instance, bird and bat wings are homologous as forelimbs but not as wings. In other words, they share an underlying structure of forelimb bones because birds and mammals share a common ancestor that possessed such bones. Bats, however, descended from a long line of non-winged mammals, whereas bird wings evolved independently, from the forelimbs of ancestral reptiles (follow the tree branches in Figure 1.21).

As we will see in Chapter 24, evolutionary change is based on developmental change. The bat wing, for example, is made in part by (1) maintaining a rapid growth rate in the cartilage that forms the fingers and (2) preventing the cell death that normally occurs in the webbing between the fingers. As seen in **FIGURE 1.23**, mice start off with webbing between their digits (as do humans and most other mammals). This webbing is important for creating the anatomical distinctions between the fingers. Once the webbing has served that function, genetic signals cause its cells to die, leaving free digits that can grasp and manipulate. Bats, however, use their fingers for flight, a feat accomplished by changing the expression of those genes in the cells of the webbing. The genes activated in embryonic bat webbing encode proteins that *prevent* cell death, as well as proteins that accelerate finger elongation (Cretekos et al. 2005; Sears et al. 2006; Weatherbee et al. 2006). Thus, homologous anatomical structures can differentiate by altering development, and such changes in development provide the variation needed for evolutionary change.

Charles Darwin observed artificial selection in pigeon and dog breeds, and these examples remain valuable resources for studying selectable variation. For instance, the short legs of dachshunds were selected by breeders who wanted to use these dogs to hunt badgers (German *Dachs*, "badger" + *Hund*, "dog") in their underground burrows. The mutation that causes the dachshund's short legs involves an extra copy of the *Fgf4* gene, which makes a protein that informs the cartilage precursor cells that they have divided enough and can start differentiating. With this extra copy of *Fgf4*, cartilage cells are told that they should stop dividing earlier than in most other dogs, so the legs stop growing (Parker et al. 2009). Similarly, long-haired dachshunds differ from their short-haired relatives in having a mutation in the *Fgf5* gene (Cadieu et al. 2009). This gene is involved in hair production and allows each follicle to make a longer hair shaft (Ota et al. 2002). It is the embryo where genotype is translated into phenotype, where

(A)

(B)

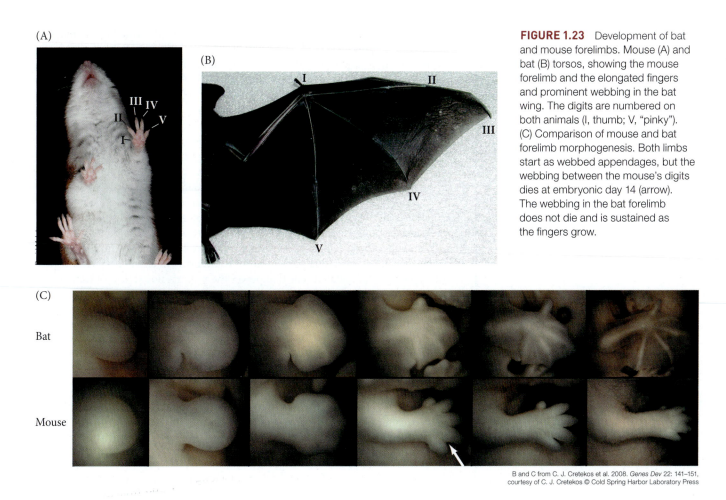

FIGURE 1.23 Development of bat and mouse forelimbs. Mouse (A) and bat (B) torsos, showing the mouse forelimb and the elongated fingers and prominent webbing in the bat wing. The digits are numbered on both animals (I, thumb; V, "pinky"). (C) Comparison of mouse and bat forelimb morphogenesis. Both limbs start as webbed appendages, but the webbing between the mouse's digits dies at embryonic day 14 (arrow). The webbing in the bat forelimb does not die and is sustained as the fingers grow.

(C)

Bat

Mouse

B and C from C. J. Cretekos et al. 2008. *Genes Dev* 22: 141–151, courtesy of C. J. Cretekos © Cold Spring Harbor Laboratory Press

inherited genes are expressed to form the adult. Thus, mutations in genes controlling developmental processes can generate selectable variation.

KEY EMBRYONIC TRANSITIONS IN ANIMALS OVER EVOLUTIONARY HISTORY   How do we know that one animal form actually preceded the evolution of another form? It's not as if we can literally see a lizard suddenly sprout feathers on its forelimbs and fly off into the sky. However, there are examples of some creatures showing traits of two closely related species, a so-called transitional morphological state. By examining such transitional organisms over the evolutionary history of metazoans (all animals), we can illuminate some important aspects of embryonic development that were altered to drive the morphological diversity we see today. For instance, the fossil record has revealed combined features of fin and leg in *Tiktaalik roseae*, suggesting it was the first aquatic species to walk on land. Similarly, fossils of the dinosaur *Archaeopteryx* possess a reptilian skeleton with feathered wings, showing the evolutionary relatedness between dinosaur and bird and the morphological transition from one to the other (**FIGURE 1.20B,C**; see also review by Stefan Rensing 2016).

FURTHER DEVELOPMENT

THE ORIGINS OF BILATERAL SYMMETRY AND OUR THREE EMBRYONIC GERM LAYERS   Bilateral symmetry found in most animal groups is thought to have evolved from organisms possessing simpler radial and spherical geometric morphologies, as we see in today's cnidarians (jellyfishes, corals, hydra, and their relatives; **FIGURE 1.20D**). As ancient organisms, cnidarians already had nervous systems, guts, and even muscles. In bilateral animals (the bilaterians), these three tissue types are derived from three separate embryonic germ layers: ectoderm, endoderm, and mesoderm. Cnidarian anatomy visibly shows only two layers, which originally were deemed to be ectoderm and endoderm, which would make the origin of the third mesodermal layer an "invention" of the earliest bilaterians. However, genetic studies have now shown the expression of mesoderm-specific genes in cnidarian embryos, leading researchers to suggest that cnidarians possess a transitional **mesendodermal** embryonic layer (Holland 2000); this would give cnidarians a transitional status in the evolution of the third germ layer, mesoderm. Interestingly, it has recently been suggested that the two layers of cnidarian construction may include regions of ectoderm that express genes typical of bilaterian endoderm, and endodermal regions that express genes typically expressed in bilaterian mesoderm—a finding that spurs questions about germ layer homologies between cnidarians and bilaterians (Steinmetz et al. 2017).

THE ORIGINS OF GASTRULATION   Some controversy surrounds the question of whether the sponge embryo undergoes the quintessential embryonic process of **gastrulation**—those cell movements in the embryo that produce the germ layers and primitive gut. Adult sponges form channels with chambers covered with choanocytes, flagellated cells that power the unidirectional flow of water through the organism (**FIGURE 1.20E**). In most cases the adult sponge is created indirectly through the metamorphosis of a free-floating larva—a physical change from a spherical embryonic and larval body type to the adult, ground-attached, filter-feeding chamber (**FIGURE 1.20F**). It is irrefutable, however, that the sponge embryo and larva have a well-delineated anterior-posterior axis with both inner and outer tissues. This is suggestive of the early origins of **epithelia** (nonmigratory tissues consisting of tightly adhering cells) with differential patterning across an axis—a developmental phenotype essential for the construction of complex tissue-layers and the formation of a primitive gut (Maldonado 2006; Nakanishi et al. 2014). It has been proposed that the larvae of some ancient sponges (homoscleromorphs) underwent sexual maturity prior to metamorphosing into the juvenile sponge. This would have freed the homoscleromorphs from maturation into the adult form, which may have opened a new door for the natural selection of tighter epithelial cell connections capable of supporting the movements of gastrulation, and ultimately the evolution of **diploblastic** (two-layered) metazoans, such as the cnidarians (Nielsen 2008).

FROM ONE TO MANY   The most fundamental evolutionary step required to build an animal was going from one cell to many different cells, or **multicellularity**. The evolution of multicellularity is estimated to have independently occurred 25–50 times over Earth's history! Nevertheless, today we have only six main groups of multicellular organisms: the brown, green, and red algae, and land plants, fungi, and animals.

Although there are many plausible ideas about how multicellularity arose, the **colonial theory** seems to be the prevailing hypothesis for the origin of metazoan multicellularity. If we consider the most basal metazoans, the sponges, then a particular flagellated cell type of the sponge comes to mind—recall the **choanocyte** we mentioned earlier. The structure of these "collar-bearing" cells, along with their water filtering functions, are considered homologous to the single-celled or colony-forming **choanoflagellates** (**FIGURE 1.24**; Nosenko et al., 2013; Nielsen, 2008). Most interesting are the types of cell-to-cell connecting proteins found in choanoflagellates, including well-conserved proteins still found in triploblastic bilaterally symmetrical animals (us). Among these proteins are cadherins that mediate cell-to-cell adhesion. In fact, loss of a leptin-like gene (known to be a bifunctional signaling and adhesion receptor that upregulates cadherin expression in some animals) in **extant** (living today) choanoflagellates prevents single cells from adhering and forming their characteristic rosette-shaped colonies (Levin et al. 2014). Thus it is hypothesized that some 3 bya, an ancient choanoflagellate formed loosely packed colonies just as they do today (see Figure 1.24). This hypothesis posits that mutations in genes encoding adhesion proteins conferred tighter junctions between neighboring choanoflagellates that fostered the sharing of nutrients between cells and a mutual interdependence for survival. This was the birth of the first multicellular organism, proposed to be a **choanoblastaea**, consisting of a

Rosetteless

Tubulin

F-actin

Composite

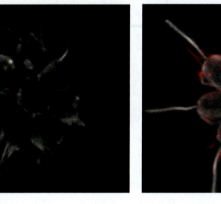

5 µm

FIGURE 1.24 Choanoflagellates were the common ancestor of all animals. Shown here are extant choanoflagellates in a rosette colony formation. These cells were immunolabeled for the proteins Rosetteless (a leptin-like protein; cyan in the composite), tubulin (marking the flagella; white in the composite), and filamentous actin (F-actin, marking the microvilli that take on a "collar-like" formation; red in the composite).

single-layered, hollow sphere of choanocytes (think 3D rosette) (Nielsen 2008). Along this metazoan branch, choanoblastea continued to adapt its epithelium for more complex functions and tissue movements, giving rise to the ancient homoscleromorphs—a special group of sponges—and the birth of the metazoan embryo (see Figure 1.25, steps 1–3). You will discover throughout this text that the same types of adhesion proteins that initially fostered multicellularity also play essential roles in likely every tissue-forming event during embryogenesis. (See **Further Development 1.5, Important Transitions in Animal Evolution**, online.)

The developmental history of land plants

Plants provide an important evolutionary and developmental comparison with animals. Did you know that all land plants undergo embryogenesis? Land plants acquired the embryonic stage as an adaptive feature when they transitioned from water to terrestrial life and thus they are aptly termed **embryophytes**. The similarities that exist between animal and plant development evolved independently yet still converged on similar mechanisms. This independent evolution of two embryo types presents a fantastically important advantage to learning about both animal and plant developmental biology (Meyerowitz 2002; Vervoort 2014; Drost et al. 2017). Being able to identify the developmental commonalities and nonconformities that drive pattern formation, morphogenesis, reproduction, and organogenesis in plants and animals can serve to highlight core principles of developmental biology.

LECA, A COMMON BEGINNING   As you surely have witnessed, embryophytes include an incredible diversity of shapes, sizes, colors, and even smells. Plants are something for the senses to marvel at, and the developmental evolution of this diversity is something for your brain to ponder. Plants are built with eukaryotic cells just as animals are, which speaks to what the last common ancestor of plants and animals might have been. Phylogenetic

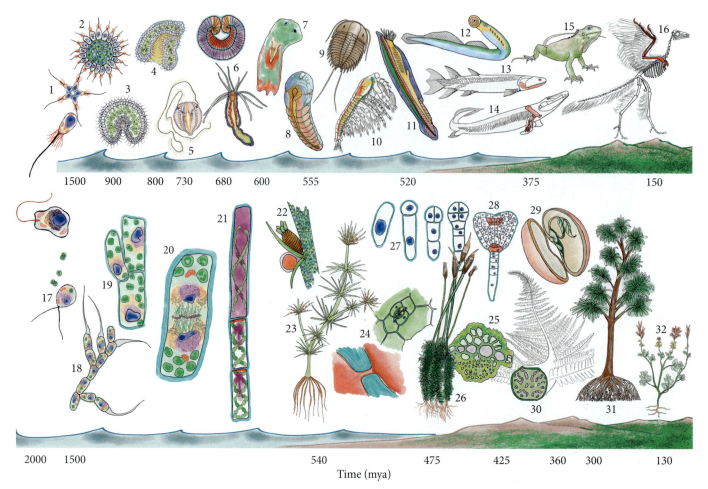

FIGURE 1.25 The developmental evolution of life. This illustration depicts key developmental adaptations that occurred over the course of evolutionary history in animals (top) and plants (bottom). The last eukaryotic common ancestor (LECA) gave rise to both plants and animals 2000 million years ago (mya). (Top) (1) Colonization of choanoflagellate cells. (2) Development of a two-layered organism with a proliferative inner layer and an epithelial filter-feeding outer layer. (3) Digestive architectures emerge with the evolution of tighter junctions and extracellular matrix (neon blue). (4) A primitive gut with aboral and oral openings appears, as in the sponge. (5) Ctenophores, such as this comb jelly, exhibit the first interconnected system of nerve-like cells. (6) Cnidarians such as the sea anemone show the first signs of gastrulation. (7) Bilateral symmetry evolves (acoels) and (8) segmentation emerges, generating (9,10) a diversity of arthropod lineages. (11) Adaptation of mesoderm produces the first axial derivative—the notochord (red)—giving rise to chordates. (12–14) From jawless fish (12, lamprey) to jawed fish (13, teleost) and from paired fins to articulating forelimbs (14, *Tiktaalik*), metazoans walk out of the water. (15,16) Among the terrestrial tetrapods, reptiles (15) further adapt their forelimbs into wings, giving rise to avian species (16). (Bottom) (17) Endosymbiosis of a cyanobacterium sets the stage for a path of photosynthesis-driven evolution. (18,19) Fixed modifications of collagen-based extracellular matrix genes foster the formation of filamentous colonies of algae (18) and a more protective cell wall (19, neon blue). (19) Integration of plastid DNA guides the biogenesis of multiplastid cells. (20) The phragmoplast builds the cell wall during cytokinesis. (21) Expansion of the phytohormone machinery opens communication across the entire plant for cell growth and morphogenesis. (22, 23) Alternation of generations is evident in the sporophytic and gametophytic phases displayed by the rhizoid-bearing charophytic algae, the common ancestor of all embryophytes. (24) Stomata and plasmodesmata provide the basis for a vascular future. (25) Hydroid cells (light purple) for nutrient transport are present in the first land plants: bryophytes (26, moss). (27) Embryonic development defines the embryophytes. (28) Pluripotent shoot and root apical meristems fuel indeterminate growth (red). (29) Seed adaptations protect and disperse embryos. (30, 31) Lignin further strengthens the cell wall for increased efficiencies of water and nutrient transport from the first vascular plants (30, ferns) to the tallest trees (31, conifers). (32) Coevolution with metazoan life helps promote an enormous diversity of angiosperms (flowering plants). To explore this figure in greater detail, go to **Further Development 1.6, The Developmental Evolution of Life**, online.

analysis suggests that the common ancestor of plants and animals (and fungi) was a unicellular protist that contained flagella and mitochondria (Niklas 2013). This hypothetical **last eukaryotic common ancestor** (**LECA**) diverged along two branches (see Figure 1.21 and **FIGURE 1.25**): one that led to the evolution of choanoflagellates and their derived multicellular animals, and another branch that acquired a second endosymbiotic relationship,[12] this time with a cyanobacterium, a photosynthesizing prokaryotic cell. This autotrophic bacterium was for some reason not digested and flourished inside the eukaryotic cell. The relationship eventually became symbiotic and gave rise to an ancestral green alga with a single photosynthesizing plastid (see Figure 1.25, step 17). **Plastids** are plant cell organelles that perform many functions, including photosynthesis.

The ability to synthesize one's own food (**autotrophy**) from the sun's energy set the primary driver for plant survival, and as such, the key factors that constrain plant evolution have always been access to sunlight, water, and carbon. The transition from freshwater algae to terrestrial land plants caused the most rapid change in atmospheric oxygen levels, which in turn progressively fueled the **Cambrian explosion**, a massive diversification of life in the water and on land (Judson 2017).

Recent sequencing of the *Chara braunii* genome strongly suggests that all embryophytes are derived from **Charophytic algae** (Nishiyama et al. 2018; see also Martin and Allen 2018). To be clear, *C. braunii* is an extant (currently living) species related to the ancestral charophycean alga (**FIGURE 1.26**). Despite being a freshwater green alga, *C. braunii* actually looks more like a land plant.[13] It possesses rudimentary roots that anchor it to solid substrates, a rigid cellulose wall, and genes orthologous to genes involved in **phytohormone** signaling systems that are essential for growth and cell differentiation in land plants (e.g., auxin and cytokinin signaling; Rensing 2018; see Figure 1.25, step 21). (See **Further Development 1.7, Transitions of the Wall, the Foot, and the Tube**, online.)

FIGURE 1.26 Transitional states over the course of plant evolution. (A) Illustration of *Chara braunii*, an extant species of charophytic alga. The reproductive organs—the oogonium (oocyte) and antheridium (sperm)—are illustrated and shown to the left. The rhizoids are illustrated to the right. (B) Transmission electron micrograph of plasmodesmata (arrow) in the charophytic alga *Chara zeylanica*. (C) Pseudocolored scanning electron micrograph of stomata (arrows) on the leaf of a coriander plant. (D) A bryophyte, the rooftop moss *Dicranoweisia cirrata*, with capsules on the tips of the setae. (E) Hydroid and leptoid cells. (F) Image of *Pteridium aquilinum*, a fern. To its right is a single transverse section of the middle stem region and two views of a high-resolution computed tomography volume rendering of xylem in this section of the stem. (G) Developmental stages of the cork oak acorn (S4–S8). Notice how the cupule retreats as the pericarp expands during seed maturation. (A, after T. Nishiyama et al. 2018. *Cell* 174: 448–464.)

(A) Thallus · Oogonium · Branchlet · Antheridium · Stipulode · Rhizoidal plate · Stem · Rhizoid · Rhizoidal cell

© Rollin Verlinde/Vildaphoto

(B) M. Cook et al. 1997. *Am J Bot* 84: 1169–1178

(C) © Steve Gschmeissner/Science Source

(D) Photo by Martin Hutten, courtesy of Bruce McCune and Kristen Whitbeck

(E) Leptoid cells · Hydroid cells
Courtesy of Dr. Jessica M. Budke

(F) From C. R. Bordersen et al. 2012. *Plant Cell Environ* 35: 1898–1911

(G) S4 · S5 · S6 · S7 · S8
From A. Miguel et al. 2015, *BMC Plant Biol* 15: 158/CC BY 4.0

A NOVEL WAY FOR A CELL TO DIVIDE AND CONQUER   Although the plant cell wall provided great protection and strength, its rigidity imposed a significant limitation—the complete inhibition of cell and tissue migration. When the cell of an embryophyte divides, it uses a unique **phragmoplast** mode of cytokinesis that ensures construction of a new cell wall between the two daughter cells, which anchors them in that position in the plant forever (see Figure 1.25, step 20). This has enormous consequences—unlike animals, plant embryos cannot use gastrulation movements to rearrange cells during their development. Instead, plants deploy targeted asymmetrical cell divisions and cell expansion to achieve morphogenesis; these mechanisms have been more than sufficient to *shape* the plant diversity seen across all terrestrial environments (Buschmann and Zachgo 2016).

A DIVISION OF LABOR   Once plants had transitioned onto land, division of the plant life cycle into sporophytic and gametophytic phases, or **alternation of generations**, supported the invasion of all terrestrial habitats (see Figure 1.25, step 22). Alternation of generations represented a significant change in developmental strategies and marked the beginning of embryogenesis in plants (Bennici 2008; Kenrick 2017). Over evolutionary time there has been a disproportionate investment in sporophyte growth, while the size of the gametophyte has been reduced.

THE EMBRYO AND ITS SEEDS OF DIVERSITY   The evolution leading to a separation of the gametophyte and sporophyte phases set the stage for the pinnacle of all innovations—the seed. Simply put, the seed holds, protects, and in part feeds the embryo.

> *Think of the fierce energy concentrated in an acorn!*
> *You bury it in the ground, and it explodes into an oak!*
> *Bury a sheep, and nothing happens but decay.*
>   George Bernard Shaw, quoted in a review by Linkies et al. 2010.

Seed plants—conifers and cousins and flowering plants—occupy a vast majority of the land on Earth, which should scream to you the rampant success of this developmental strategy. It is a strategy that was driven by the primary selection pressure that embryophytes were immediately hit with: dry land. The vital benefit of the seed is that it provides physical protection via its hardened seed coat, paired with a developmental period of **quiescence** in which the seed (embryo included) dehydrates and remains dormant until environmental conditions are optimal for germination and plant growth (Figure 1.26G). The seed strategy provides vascular plants the necessary time to wait for the best conditions. In fact, the seed can even enter a prolonged period of quiescence called **dormancy**, a mechanism that coevolved with animals to facilitate longer-range seed dispersal (see Figure 1.25, step 29).

But how? What are the mechanisms for this and all the developmental processes mentioned in this introductory chapter? The rest of this textbook will provide you with some answers and even more questions, with a final chapter that circles back to a discussion of evolution and the mechanisms of developmental biology that drove key adaptive changes leading to the life forms we know today.

Personal Significance: Medical Embryology and Teratology

While embryologists could look at embryos to describe the evolution of life and how different animals form their organs, physicians became interested in embryos for more practical reasons. Between 2% and 5% of human infants are born with a readily observable anatomical abnormality (Winter 1996; Thorogood 1997). These abnormalities may include missing limbs, missing or extra digits, cleft palates, eyes that lack certain parts, hearts that lack valves, defects in spinal cord closure, and so forth. Some birth defects are produced by mutant genes or chromosome abnormalities, and some are produced by environmental factors that impede development. The study of birth defects can tell us how the human body is normally formed. In the absence of experimental data on human embryos, nature's "experiments" sometimes offer important insights into how the human body becomes organized.

Genetic malformations and syndromes

Abnormalities caused by genetic events (gene mutations, chromosomal aneuploidies, and translocations) are called **malformations**, and a **syndrome** is a condition in which two or more malformations occur together. For instance, a hereditary disease called Holt-Oram syndrome is inherited as an autosomal dominant condition. Children born with this syndrome usually have a malformed heart (the septum separating the right and left sides fails to grow normally) and no wrist or thumb bones. Holt-Oram syndrome is caused by mutations in the *TBX5* gene (Basson et al. 1997; Li et al. 1997). The TBX5 protein is expressed in the developing heart and hand and is important for normal growth and differentiation in both locations.

Disruptions and teratogens

Developmental abnormalities caused by exogenous agents (certain chemicals or viruses, radiation, or hyperthermia) are called **disruptions**. The agents responsible for these disruptions are called **teratogens** (Greek, "monster-formers"), and the study of how environmental agents disrupt normal development is called teratology. Substances that can cause birth defects include relatively common substances, such as alcohol and retinoic acid (often used to treat acne), as well as many chemicals used in manufacturing and released into the environment. Heavy metals (e.g., mercury, lead, and selenium) can alter brain development.

Teratogens were brought to the attention of the public in the early 1960s. In 1961, Lenz and McBride independently accumulated evidence that the drug thalidomide, prescribed as a mild sedative to many pregnant women, caused an enormous increase in a previously rare syndrome of congenital anomalies. The most noticeable of these anomalies was phocomelia, a condition in which the long bones of the limbs are deficient or absent (**FIGURE 1.27A**). More than 7,000 affected infants were born to women who took thalidomide, and a woman needed only to have taken one tablet for her child to be born with all four limbs deformed (Lenz 1962, 1966; Toms 1962). Other abnormalities induced by ingesting this drug included heart defects, absence of the external ears, and malformed intestines. Nowack (1965) documented the period of susceptibility during which thalidomide caused these abnormalities (**FIGURE 1.27B**). The drug was found to be teratogenic only during days 34–50 after the last menstruation (i.e., 20–36 days postconception). From days 34 to 38, no limb abnormalities are seen, but during this period thalidomide can cause the absence or deficiency of ear components.

(A)

(B)

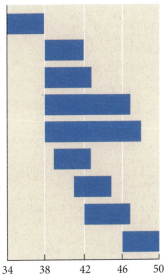

Days after last menstruation

FIGURE 1.27 A developmental anomaly caused by an environmental agent. (A) Phocomelia, the lack of proper limb development, was the most visible of the birth defects that occurred in many children born in the early 1960s whose mothers took the drug thalidomide during pregnancy. These children are now middle-aged adults; this photograph is of Grammy-nominated German singer Thomas Quasthoff. (B) Thalidomide disrupts different structures at different times of human development. (B, data from E. Nowak. 1965. *Humangenetik* 1: 516–536; graph after N. Vargesson. 2015. *Birth Defects Res C Embryo Today* 105: 140–156, and references therein.)

Malformations of the upper limbs are seen before those of the lower limbs because the developing arms form slightly before the legs.

The integration of anatomical information about congenital malformations with our new knowledge of the genes responsible for development has resulted in an ongoing restructuring of medicine. This integrated information is allowing us to discover the genes responsible for inherited malformations and to identify exactly which steps in development are disrupted by specific teratogens. We will see examples of this integration throughout this text.

Coda

It is an irrefutable fact that we are related to all the life on this planet in one way or another. What we look like, how we digest food, how we see, how we walk, even how we reproduce is evolutionarily linked to all other animals and plants. This profound relatedness can and should provide an organizing lens as you learn about developmental biology. It was direct tweaks in developmental mechanisms largely associated with embryogenesis that, in concert with natural selection, generated today's diversity of life.

Closing Thoughts on the Opening Photo

Is this plant really your cousin? This is a picture of Michael Barresi holding in his hand an extant species of characean green alga. As you learned in this chapter, *Chara braunii* is directly related to the common ancestor of all land plants, an ancestry that goes back to the last eukaryotic common ancestor (LECA) and the eukaryotic origin of animals and plants. So the answer is yes, this alga is your cousin. According to the Evogeneao interactive tree of life, this green alga would be somewhere around your 2.9 billionth cousin. So next time you see that green scum in a pond, be sure to say hello.

Photo credit: M. J. F. Barresi and Kathryn Lee, 2018.
Thanks to Dr. Robin Sleith for providing the charophyta algae

Endnotes

[1] Defining exactly what developmental biology encompasses has spurred recent debate (Pradeu et al. 2016). Some feel that the field is impossible to define, while others argue that a framework can help reduce the negative consequences associated with implicit meanings. The authors of this textbook support a more expansive definition, one that promotes inclusion of a diversity of perspectives to better support the collaborative development of the field itself.

[2] Harvey did not make this statement lightly, for he knew that it contradicted the views of Aristotle, whom Harvey venerated. Aristotle had proposed that menstrual fluid formed the substance of the embryo, while the semen gave it form and animation.

[3] More than 210 different cell types are recognized in the *adult* human, but this number tells us little about how many cell types a human body produces over the course of development.

[4] We will be using an entire "blast" vocabulary in this book. A blastomere is a cell derived from cleavage in an early embryo. A blastula is an embryonic stage composed of blastomeres; a mammalian blastula is called a blastocyst (see Chapter 12). The cavity within the blastula is the blastocoel. A blastula that lacks a blastocoel is called a stereoblastula. The invagination where gastrulation begins is the blastopore.

[5] Although adult vertebrates do not have a notochord, this embryonic organ is critical for establishing the fates of the ectodermal cells above it, as we will see in Chapter 13.

[6] From the same root as "germination," the Latin *germen* means "sprout" or "bud." The names of the three germ layers are from the Greek: ectoderm from *ektos* ("outside") plus *derma* ("skin"); mesoderm from *mesos* ("middle"); and endoderm from *endon* ("within").

[7] Green fluorescent protein occurs naturally in certain jellyfish. It emits bright green fluorescence when exposed to ultraviolet light and is widely used as a transgenic label. GFP labeling will be seen in many photographs throughout this book.

[8] Indeed, one definition of a phylum is that it is a collection of species whose gene expression at the phylotypic stage is highly conserved among them, yet different from that of other species (see Levin et al. 2016). However, controversy over what constitutes a phylum persists. For instance, some authors consider cephalochordates (*Amphioxus*), tunicates, and chordates as separate phyla, whereas others unite them in one phylum, Chordata.

[9] The notochord is a rodlike structure that runs down the middle of an embryo's trunk and functions as an organizing center for the neural and non-neural tissues that surround it. It is seen in every vertebrate embryo as well as in several invertebrate embryos, including tunicates. Thus it is a defining feature of chordates (vertebrates and their invertebrate cousins—tunicates and cephalochordates, including lancelets).

[10] As first noted by Weismann (1875), larvae must have their own adaptations. The adult viceroy butterfly mimics the monarch butterfly, but the viceroy caterpillar does not resemble the beautiful

larva of the monarch. Rather, the viceroy larva escapes detection by resembling bird droppings (Begon et al. 1986).

[11] The oyster toadfish is arguably the ugliest fish in the ocean (author opinion). So yes, due to this exemplified relationship, you could consider this a personal criticism. *Yes, we are making a joke here. It's okay to laugh (at the joke or us—both welcomed).*

[12] The first endosymbiotic event was the engulfment by a eukaryotic cell of an aerobic bacterium that evolved into the mitochondrion, and the derivation of all eukaryotic cell lineages thereafter.

[13] Why was the charophycean alga so well prepared for the conquest of land? It has recently been proposed that the common charophyte ancestor actually might have conquered land as a unicellular alga earlier than thought, during which time natural selection fostered the evolution of the traits beneficial for terrestrial life. Only after acquiring these innovations did the ancestral charophyte both fuel the terrestrial radiation of all embryophytes and return to the water, producing the land plant-like aquatic form of C. braunii present today (Harholt et al. 2016).

① Snapshot Summary
The Making of a Body and a Field: Introduction to Developmental Biology

1. The life cycle can be considered a central unit in biology; the adult form needs not be paramount. The basic animal life cycle consists of fertilization, cleavage, gastrulation, germ layer formation, organogenesis, metamorphosis, adulthood, and senescence. The basic flowering plant (angiosperm) life cycle consists of reproductive and gametophytic phases, embryogenesis and seed maturation, and juvenile and adult vegetative phases.

2. In animals, the three germ layers give rise to specific organ systems. The ectoderm gives rise to the epidermis, nervous system, and pigment cells; the mesoderm generates the kidneys, gonads, muscles, bones, heart, and blood cells; and the endoderm forms the lining of the digestive tube and the respiratory system.

3. Labeling cells with dyes or through genetic means shows that some cells differentiate where they form, whereas others migrate from their original sites and differentiate in their new locations. Migratory cells include neural crest cells and the precursors of germ cells and blood cells.

4. Plant cells do not migrate due to stiff cell walls, but plant cells use targeted cell division and growth to shape both the embryo and the adult plant.

5. Shoot and root apical meristems provide pools of stem cells to drive shoot and root genesis throughout plant life.

6. "Community of embryonic structure reveals community of descent" (Charles Darwin, *On the Origin of Species*).

7. Karl von Baer's principles state that the general features of a large group of animals appear earlier in the embryo than do the specialized features of a smaller group. As each embryo of a given species develops, it diverges from the adult forms of other species. The early embryo of a "higher" animal species is not like the adult of a "lower" animal.

8. Homologous structures in different species are those whose similarity is due to sharing a common ancestral structure. Analogous structures are those whose similarity comes from serving a similar function (but which are not derived from a common ancestral structure).

9. The evolutionary history of life on Earth tells a story of the adaptations that occurred in the developmental processes governing plant and animal traits. Key transitional morphologies in both plants and animals demonstrate their shared developmental evolution.

10. Choanoflagellates and charophytic algae are the common ancestors of metazoans (animals) and embryophytes (land plants), respectively.

11. Congenital anomalies can be caused by genetic factors (mutations, aneuploidies, translocations) or by environmental agents (certain chemicals, certain viruses, radiation).

12. Teratogens—environmental compounds that can alter development—act at specific times when certain organs are being formed.

Ⓜ **Go to oup.com/he/barresi12xe** for Further Developments, Scientists Speak interviews, Watch Development videos, Dev Tutorials, and complete bibliographic information for all literature cited in this chapter.

Specifying Identity
Mechanisms of Developmental Patterning

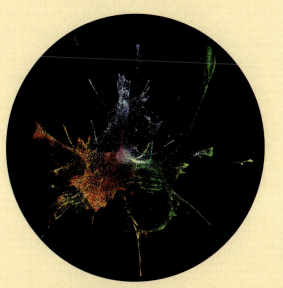

Can development be mapped?

Photograph courtesy of Dan Wagner,
Sean Megason, and Allon Klein

IN 1883, ONE OF AMERICA'S FIRST EMBRYOLOGISTS, William Keith Brooks, reflected on "the greatest of all wonders of the material universe: the existence, in a simple, unorganized egg, of a power to produce a definite adult animal." He noted that the process is so complex that "we may fairly ask what hope there is of discovering its solution, of reaching its true meaning, its hidden laws and causes." Indeed, how to get from "a simple, unorganized egg" to an exquisitely ordered body is the fundamental mystery of development. Biologists today are piecing together "its hidden laws and causes." These include how the unorganized egg becomes organized, how different cells interpret the same genome differently, and how the many modes of cell communication orchestrate the unique patterns of cell differentiation.

In this chapter, we introduce the concept of *cell specification*—how cells become specified to a specific fate—and explore how cells of different organisms use different mechanisms for determining cell fate. In Chapters 3 and 4, we will delve deeper into the genetic mechanisms underlying cell differentiation and the cell signaling involved. Chapter 5, the final chapter of this unit, will focus on the development of stem cells, which exemplifies all the principles defined in this first unit.

Levels of Commitment

To the naked eye, individual grains of sand on an expansive beach look unorganized, yet the grains can be molded together to create complex structures, as illustrated by a sand sculpture of an octopus holding children in its tentacles (**FIGURE 2.1**). How can disordered units become ordered, a pile of sand become a structured creation, or a collection of cells become a highly complex embryo? Did the sand grains contributing to the octopus's eye know they were going to become an eye as they washed up on the beach earlier that morning? Obviously, significant energy had to be applied to these inanimate and inorganic sand grains to sculpt this eye, but what about the cells of your eye? Did they know they were destined to contribute to your retina, cornea, or lens? If so, when did they know it, and how set were they in adopting this fate?

Cell differentiation

The generation of specialized cell types is called **differentiation**, a process during which a cell ceases to divide and develops specialized structural elements and distinct functional properties—a cell's unique traits. A red blood cell obviously differs radically in its protein

FIGURE 2.1 From sand grains to an organized octopus sculpture.

composition and cell structure from a neuron in the brain. These differences in cellular biochemistry and function are preceded by a process that commits each cell to a certain fate. During the course of **commitment**, a cell might not look different from its nearest or most distant neighbors in the embryo or show any visible signs of differentiation, but its developmental fate has become programmatically restricted.

Cell fate maturation

FIGURE 2.2 Cell fate determination. (A) Two differently positioned blastula cells are specified to become distinct muscle and neuronal cells when placed in isolation. (B,C) The two different blastula cells are placed together in culture. (B) In one scenario, the dark red cell was specified—but not determined—to form muscle. It adopts a neuronal fate due to its interactions with its neighbors. (C) If the dark red cell was committed and determined to become muscle at the time of culturing, it will continue to differentiate into a muscle cell despite any interactions with its neighbors.

The process of becoming committed to a specific cell identity can be divided into two stages: **specification** and then **determination** (Harrison 1933; Slack 1991). The fate of a cell or tissue is said to be **specified** when it is capable of differentiating autonomously (i.e., by itself) when placed in an environment that is neutral with respect to the developmental pathway, such as in a petri dish (**FIGURE 2.2A**). At the specification stage, commitment to cell identity is still labile (i.e., capable of being altered). If a specified cell is transplanted to a population of differently specified cells, the fate of the transplant will be altered by its interactions with its new neighbors (**FIGURE 2.2B**). It is not unlike many of you who perhaps entered your developmental biology classroom interested in chemistry but, after exposure to the awesomeness that is development, will change your interests and become a developmental biologist.

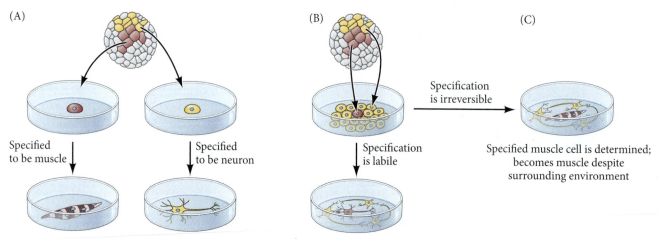

(A)

Specified to be muscle

Specified to be neuron

(B)

Specification is labile

Specified muscle cell changed to neuron

Specification is irreversible

(C)

Specified muscle cell is determined; becomes muscle despite surrounding environment

A cell or tissue is said to be **determined** when it is capable of differentiating autonomously even when placed into another region of the embryo or a cluster of differently specified cells in a petri dish (**FIGURE 2.2C**). If a cell or tissue type is able to differentiate according to its specified fate even under these circumstances, it is assumed that commitment is irreversible. This level of commitment is similar to you being unwaveringly determined to become a chemist no matter how awe-inspiring your developmental biology course might be.

In summary, during embryogenesis an undifferentiated cell matures through specific stages that cumulatively commit it to a specific fate: first specification, then determination, and finally differentiation. Embryos can exhibit three modes of specification: autonomous, conditional, and syncytial (see Further Development 2.5, online). Embryos of different species use different combinations of these modes.

Autonomous Specification

One mode of cell commitment is **autonomous specification**. Here, the blastomeres of the early embryo are apportioned a set of critical determination factors within the egg cytoplasm—so-called **cytoplasmic determinants**. In other words, the egg cytoplasm is not homogeneous; rather, different regions of the egg contain different cell fate specifying factors (molecules, often transcription factors) that regulate gene expression in a manner that directs the cell along a particular path of maturation. In autonomous specification, the cell "knows" very early what it is to become without interacting with other cells. For instance, even in the very early cleavage stages of the snail *Patella*, blastomeres that are presumptive trochoblast cells can be isolated in a petri dish. There, they develop into the same ciliated cell types that they would give rise to in the embryo and with the same temporal precision (**FIGURE 2.3**). This continued commitment to the trochoblast fate suggests that these particular early blastomeres are already specified and determined to their fate.

Cytoplasmic determinants and autonomous specification in the tunicate

Cell specification is a dynamic event that occurs over the course of embryogenesis; therefore, being able to conduct **lineage tracing** experiments—tracking the development of cell maturation over time—has become one of the most important approaches to studying cell differentiation. Groups of embryonic cells can be labeled to see what they become in the adult organism. Such studies enable the construction of a **fate map**, a diagram that "maps" the larval or adult structures onto the region of the embryo from which they arose. One of the first fate maps to be generated was based on careful observations of the tunicate (sea squirt) embryo.

Normal development of *Patella*

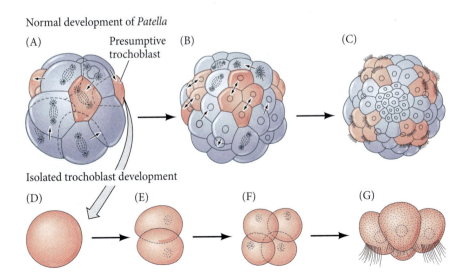

FIGURE 2.3 Autonomous specification. (A–C) Differentiation of trochoblast (ciliated) cells of the snail *Patella*. (A) 16-Cell stage seen from the side; the presumptive trochoblast cells are shown in pink. (B) 48-Cell stage. (C) Ciliated larval stage, seen from the animal pole. (D–G) Differentiation of a *Patella* trochoblast cell isolated from the 16-cell stage and cultured in vitro. Even in isolated culture, the cells divide and become ciliated at the correct time. (After E. B. Wilson. 1904. *J Exp Zool* 1: 1–72.)

(A)

Yellow crescent

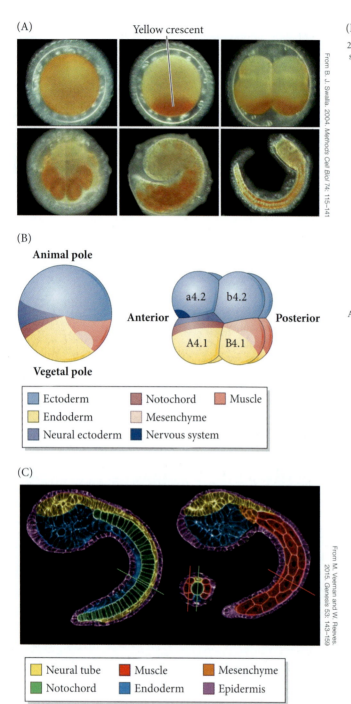

From B. J. Swalla. 2004. *Methods Cell Biol* 74: 115–141

(B)

Animal pole

a4.2 b4.2

Anterior **Posterior**

A4.1 B4.1

Vegetal pole

☐ Ectoderm	☐ Notochord	☐ Muscle
☐ Endoderm	☐ Mesenchyme	
☐ Neural ectoderm	☐ Nervous system	

(C)

From M. Veeman and W. Reeves. 2015. *Genesis* 53: 143–159

☐ Neural tube	☐ Muscle	☐ Mesenchyme
☐ Notochord	☐ Endoderm	☐ Epidermis

(D)

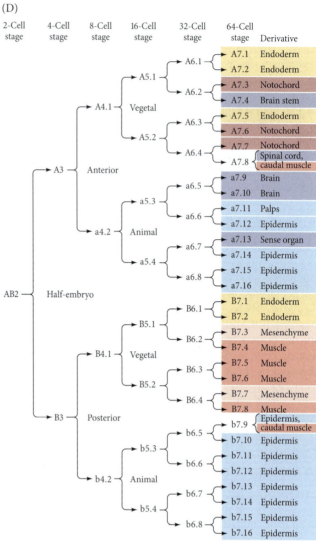

FIGURE 2.4 Autonomous specification of the tunicate. (A) The yellow crescent is seen in the tunicate from the egg to the larva (dense yellow-orange-red coloration). Original drawings by Conklin demonstrate his observations of the yellow crescent in egg and larva (golden color). (B) Schematic of a *Styela partita* zygote (left), shown shortly before the first cell division, with the fate of the cytoplasmic regions indicated. The 8-cell embryo on the right shows these regions after three cell divisions. (C) Confocal section through a larva of the tunicate *Ciona savignyi*. Different tissue types were pseudocolored. (D) A linear version of the *S. partita* fate map, showing the fates of each cell of the embryo. (B after B. I. Balinsky. 1981. *Introduction to Embryology*, 5th Ed. Saunders: Philadelphia; B, D after H. Nishida. 1987. *Dev Biol* 121: 526–541.)

In 1905, Edwin Grant Conklin, an embryologist working at the Woods Hole Marine Biological Laboratory, published a remarkable fate map of the tunicate *Styela partita*.[1] Conklin noticed a yellow coloration that was asymmetrically partitioned within the egg cytoplasm (a colored domain later dubbed the "**yellow crescent**") and ultimately segregated to muscle lineages in the larva (**FIGURE 2.4**). The yellow pigment conveniently provided Conklin a means to trace the lineages of each blastomere. Following the fates of each early cell, Conklin showed that "all the principle organs of the larva in their definitive positions and proportions are here marked out in the 2-cell stage by distinct

FIGURE 2.5 Autonomous specification in the early tunicate embryo. When the four blastomere pairs of the 8-cell embryo are dissociated, each forms the structures it would have formed had it remained in the embryo. The tunicate nervous system, however, is conditionally specified. The fate map shows that the left and right sides of the tunicate embryo produce identical cell lineages. Here the muscle-forming yellow cytoplasm is colored red to conform to its association with mesoderm. (After G. Reverberi and A. Minganti. 1946. *Pubbl Staz Zool Napoli* 20: 199–252.)

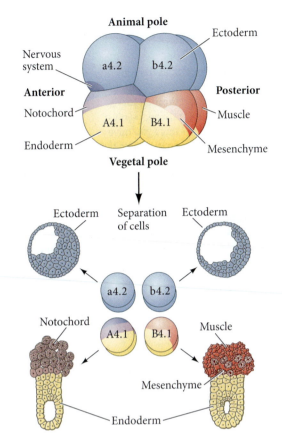

kinds of protoplasm."[2] But is each blastomere determined to its lineage? That is, *is each autonomously specified*?

The muscle-forming cells of the *Styela* embryo always retain the yellow color and are easily seen to derive from a region of cytoplasm found in the B4.1 blastomeres. Removal of the B4.1 cells resulted in a larva with no tail muscles (Reverberi and Minganti 1946). This result supports the conclusion that *only those cells derived from the early B4.1 blastomeres possess the capacity to develop into tail muscle*. Further supporting a mode of autonomous specification, each blastomere will form most of its respective cell types even when separated from the remainder of the embryo (**FIGURE 2.5**). Remarkably, if the yellow cytoplasm of the B4.1 cells is placed into other cells, those cells will form tail muscles (Whittaker 1973; Nishida and Sawada 2001). Taken together, these results suggest that critical factors that determine cell fate are present and differentially segregated in the cytoplasm of early blastomeres.

FURTHER DEVELPOMENT

WHY YELLOW IS "MACHO" In 1973, J. R. Whittaker provided dramatic biochemical confirmation of the cytoplasmic segregation of tissue determinants in early tunicate embryos. More recent studies have discovered that contained in the yellow-pigmented cytoplasm is mRNA for a muscle-specific transcription factor called Macho. Only those blastomeres that acquire this region of yellow cytoplasm (and thus Macho) give rise to muscle cells (**FIGURE 2.6A**; Nishida and Sawada 2001; reviewed by Pourquié 2001). Functionally, Macho is required for tail muscle development in *Styela*; loss of *macho* mRNA leads to a loss of muscle differentiation of the B4.1 blastomeres, whereas microinjection of *macho* mRNA into other blastomeres promotes ectopic muscle differentiation (**FIGURE 2.6B**). Thus, the tail muscles of these tunicates are formed autonomously by acquiring and retaining the *macho* mRNA from the egg cytoplasm with each round of mitosis.

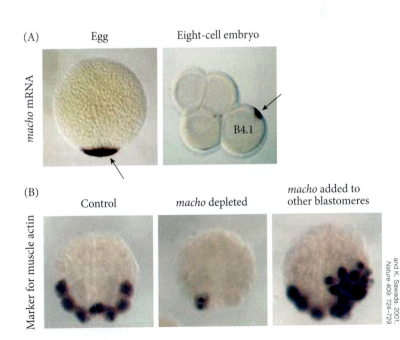

FIGURE 2.6 The *macho* gene regulates muscle development in the tunicate. (A) Like the yellow crescent, *macho* transcript is localized to the vegetal-most end of the egg and differentially expressed only in the B4.1 blastomere. (B) Knockdown of *macho* function by injection of targeting antisense oligonucleotides causes reductions in muscle differentiation, whereas ectopic misexpression of *macho* in other blastomeres results in expanded muscle differentiation.

? Developing Questions

Look closely at the localization of *macho* mRNA in the tunicate embryo (see Figure 2.6A). Is it evenly spread throughout the cell, or is it localized to only a small region? Once you have decided on its spatial distribution, contemplate whether this distribution is consistent with a mode of autonomous specification for the muscle lineage. From a cell biological perspective, how do you think this distribution of a specific mRNA is established?

Conditional Specification

We have just learned how most of the cells of an early tunicate embryo are determined by autonomous specification. However, even the tunicate embryo is not fully specified this way—its nervous system arises conditionally. **Conditional specification** is the process by which cells achieve their respective fates by interacting with other cells. This array of interactions can include cell-to-cell contacts (juxtacrine factors), secreted signals (paracrine factors), and the physical properties of the cell's local environment (mechanical stress), all of which are mechanisms we will explore in detail in Chapter 4. In conditional specification, specification depends on the conditions. For example, if cells from one region of a vertebrate blastula that have been fate mapped to give rise to dorsal tissues are transplanted into the presumptive ventral region of another embryo, the transplanted "donor" cells will change their fates and develop into ventral cell types (**FIGURE 2.7**). Moreover, the dorsal region of the donor embryo from which cells were extracted will also end up developing normally. (See **Further Development 2.1, The Germ Plasm Theory**, online.)

Cell position matters: Conditional specification in the sea urchin embryo

As far back as 1888, August Weismann proposed that each cell of the embryo developed autonomously by containing determinants not found in the other cells. This was a testable hypothesis. Based on the fate map of the frog embryo, Weismann claimed that when the first cleavage division separated the future right half of the embryo from the future left half, there would be a separation of "right" determinants from "left" determinants in the resulting blastomeres. Wilhelm Roux tested Weismann's hypothesis by using a hot needle to kill one of the cells in a 2-cell frog embryo, and only the right or left half of a larva developed (**FIGURE 2.8**). Based on this result, Roux claimed that specification was autonomous and that all the instructions for normal development were present inside each cell.

Roux's colleague Hans Driesch, however, obtained opposite results by performing isolation experiments (**FIGURE 2.9**). He separated sea urchin blastomeres from one another by vigorous shaking (or later, by placing them in calcium-free seawater). To Driesch's surprise, each of the blastomeres from a 2-cell embryo developed into a complete larva. Similarly, when Driesch separated the blastomeres of 4- and 8-cell embryos, some of the isolated cells produced complete, bilaterally symmetrical, free-swimming larvae, known as **pluteus larvae**. Here was a result drastically different from the predictions of Weismann and Roux. Rather than self-differentiating into its future embryonic part, each isolated blastomere regulated its development to produce a complete organism. These experiments provided the first experimentally observable evidence that a cell's fate depends on that of its neighbors. Furthermore, Driesch experimentally removed cells, which in turn changed the context for those cells still remaining in the embryo (they were now abutting

FIGURE 2.7 Conditional specification. (A) What a cell becomes depends on its position in the embryo. Its fate is determined by interactions with neighboring cells. (B) If cells are removed from the embryo, the remaining cells can compensate for the missing part.

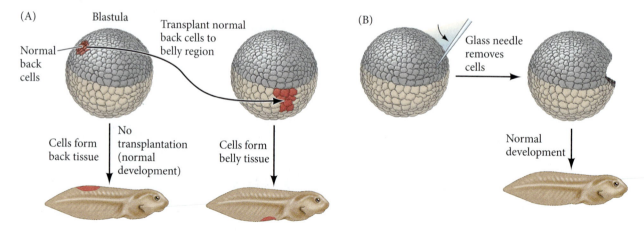

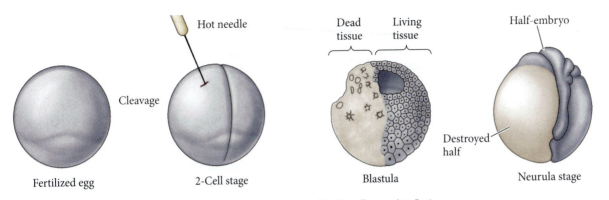

Hot needle

Cleavage

Dead tissue   Living tissue

Half-embryo

Destroyed half

Fertilized egg

2-Cell stage

Blastula

Neurula stage

FIGURE 2.8 Roux's attempt to demonstrate autonomous specification. Destroying (but not removing) one cell of a 2-cell frog embryo resulted in the development of only one half of the embryo.

new neighboring cells). As a result, all cell fates were altered, which supported complete embryonic development. In other words, the cell fates were altered to suit the *conditions*. In conditional specification, interactions between cells determine their fates rather than cell fate being specified by some cytoplasmic factor particular to that type of cell.

The consequences of these experiments were momentous, both for embryology and ultimately for Driesch, personally.[3] First, Driesch had demonstrated that the prospective potency of an isolated blastomere (i.e., those cell types that it was possible for it to form) is greater than the blastomere's actual prospective fate (those cell types that it would normally give rise to over the unaltered course of its development). According to Weismann and Roux, the prospective potency and the prospective fate of a blastomere should have been identical. Second, Driesch concluded that the sea urchin embryo is a "harmonious equipotential system" because all of its potentially independent parts interacted together to form a single organism. Driesch's experiment implies that *cell interaction is critical for normal development*. Moreover, if each early blastomere can form all the embryonic cells when isolated, it follows that in normal development the community of cells must prevent it from doing so (Hamburger 1997). Third, Driesch concluded that the fate of a nucleus depended solely on its location in the embryo (see Footnote 2).

We now know (and will see in Chapters 10 and 11) that sea urchins and frogs alike use both autonomous and conditional specification of their early embryonic cells. Moreover, both animal groups use a similar mode and even similar molecules during early development. In the 16-cell sea urchin embryo, a group of cells called the micromeres inherits a set of transcription factors from the egg cytoplasm. These transcription factors cause the micromeres to develop autonomously into the larval skeleton, but these same factors also activate genes for paracrine and juxtacrine signals that are then secreted by the micromeres and conditionally specify the cells around them.

Embryos (especially vertebrate embryos) in which most of the early blastomeres are conditionally specified have traditionally been called regulative embryos. But as we become more cognizant of the manner in which both autonomous and conditional specification are used in each embryo, the notions of "mosaic" and "regulative" embryos appear less and less tenable. Indeed, attempts to get rid of these distinctions were begun by the embryologist Edmund B. Wilson (1894, 1904) more than a century ago. (See **Further Development 2.2, Squeezing the Conditions of Specification**, online.)

FIGURE 2.9 Driesch's demonstration of conditional specification. (A) An intact 4-cell sea urchin embryo generates a normal pluteus larva. (B) When one removes the 4-cell embryo from its fertilization envelope and isolates each of the four cells, each cell can form a smaller, but normal, pluteus larva. (All larvae are drawn to the same scale.) Note that the four larvae derived in this way are not identical, despite their ability to generate all the necessary cell types. Such variation is also seen in adult sea urchins formed in this way (see Marcus 1979). (B after A. Hörstadius and A. Wolsky. 1936. *Archiv f Entwicklungsmechanik* 135: 69–113.)

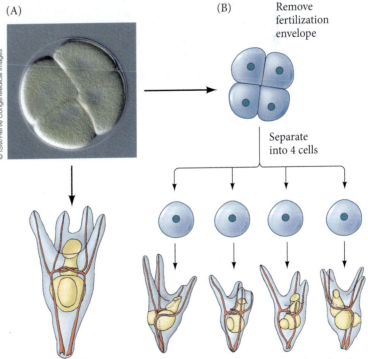

(A)

© ISM/Hervé Conge/Medical Images

(B)   Remove fertilization envelope

Separate into 4 cells

Normal pluteus larva

Plutei developed from single cells of 4-cell embryo

It depends on how you slice it: Specification in the plant embryo

Modes of cell specification in plants follow the same laws of commitment as they do in animals. Autonomous specification in plants most notably occurs as a result of the first division of the zygote. The cytoplasm of the zygote segregates asymmetrically both qualitatively and quantitatively prior to the first division, thereby establishing upon cytokinesis what is known as the **proembryo** (**FIGURE 2.10A**). This first division sets up the apical-basal axis of the embryo and of the plant it will become. The smaller, apical daughter cell of the proembryo gives rise to all parts of the plant proper except the very tip of the root. The opposing, basal daughter cell of the proembryo generates the plant's root apex and the **suspensor**, which serves to connect the plant embryo to the nutrients housed in the surrounding seed.

As a necessary criterion for autonomous specification, both the apical and the basal cells of the proembryo maintain their developmental trajectories into the embryo or suspensor, respectively, even when placed in isolation in vitro or

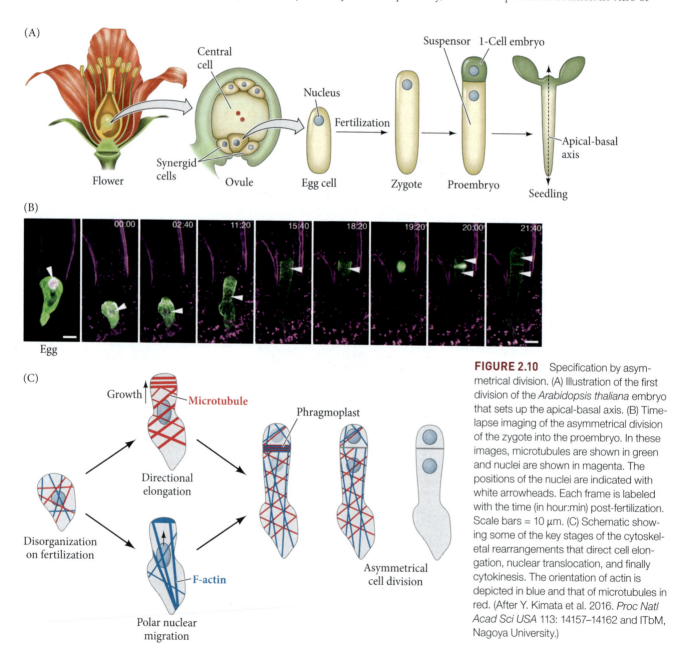

FIGURE 2.10 Specification by asymmetrical division. (A) Illustration of the first division of the *Arabidopsis thaliana* embryo that sets up the apical-basal axis. (B) Time-lapse imaging of the asymmetrical division of the zygote into the proembryo. In these images, microtubules are shown in green and nuclei are shown in magenta. The positions of the nuclei are indicated with white arrowheads. Each frame is labeled with the time (in hour:min) post-fertilization. Scale bars = 10 µm. (C) Schematic showing some of the key stages of the cytoskeletal rearrangements that direct cell elongation, nuclear translocation, and finally cytokinesis. The orientation of actin is depicted in blue and that of microtubules in red. (After Y. Kimata et al. 2016. *Proc Natl Acad Sci USA* 113: 14157–14162 and ITbM, Nagoya University.)

in vivo (Qu et al. 2017). Cytoskeletal arrays of microtubules and actin microfilaments function to first elongate the zygote, then to translocate the zygotic nucleus toward the presumptive apical side of the cell, and finally to establish the position of the preprophase band and of the phragmoplast during cytokinesis (**FIGURE 2.10B,C**; Pillitteri et al. 2016). Following this early and brief mode of autonomous specification, cell development in the plant embryo operates by conditional specification. Throughout the remainder of plant development, cell identities are highly influenced by a cell's position in the plant along the apical-basal axis. Therefore, the first asymmetrical division of the zygote sets up this governing apical-basal geography of the embryophyte, which is accomplished by putting in literal motion the signaling factors that specify cell fate. As we will discuss in Chapter 4, the polar transport of phytohormones such as auxins and cytokinins along this and other axes differentially regulates gene expression and consequently specifies different cell type identities based on a cell's position in the plant.

Syncytial Specification

A third mode of specification uses elements of both autonomous and conditional specification. A cytoplasm that contains many nuclei is called a **syncytium**,[4] and the specification of presumptive cells within a syncytium is called **syncytial specification**. Insects are notable examples of embryos that go through a syncytial stage, as illustrated by the fruit fly *Drosophila melanogaster*. During the fly's early cleavage stages, nuclei divide through 13 cycles in the absence of any cytoplasmic cleavage. This division creates an embryo of many nuclei contained within one shared cytoplasm surrounded by one common cell membrane. This embryo is called the syncytial blastoderm (**FIGURE 2.11**).

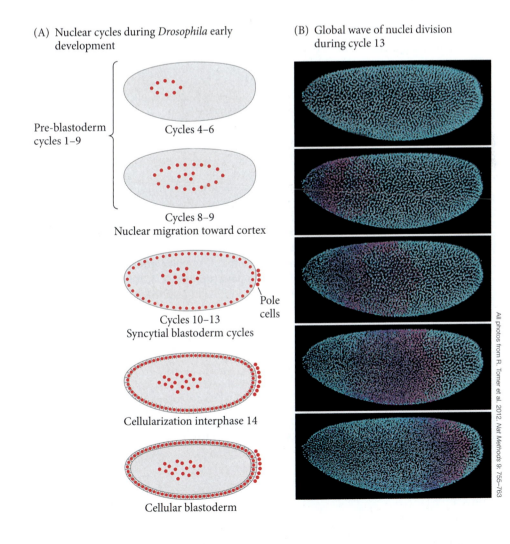

(A) Nuclear cycles during *Drosophila* early development

Pre-blastoderm cycles 1–9

Cycles 4–6

Cycles 8–9
Nuclear migration toward cortex

Cycles 10–13
Syncytial blastoderm cycles

Pole cells

Cellularization interphase 14

Cellular blastoderm

(B) Global wave of nuclei division during cycle 13

All photos from R. Tomer et al. 2012. *Nat Methods* 9: 755–763

FIGURE 2.11 The syncytial blastoderm in *Drosophila melanogaster*. (A) Schematic of the progression of blastoderm cellularization in *Drosophila* (nuclei are red). (B) Still frames from a time-lapse movie of a developing *Drosophila* embryo with nuclei that are premitotic (blue) and actively dividing in mitosis (purple). (A after A. Mazumdar and M. Mazumdar. 2002. *BioEssays* 24: 1012–1022.)

It is within the syncytial blastoderm that the *identity* of future cells is achieved simultaneously across the anterior-posterior axis of the entire embryo. Therefore, identity is established without any membranes separating nuclei into individual cells. Membranes do eventually form around each nucleus through a process called **cellularization**, which occurs after nuclear cycle 13, just prior to gastrulation (see Figure 2.11). A fascinating issue is how the cell fates—those cells determined to become the head, thorax, abdomen, and tail—are specified before cellularization. Are there determination factors segregated to discrete locations in the blastoderm to determine identity, as seen in autonomous specification? Or do nuclei in this syncytium obtain their identity from their position relative to neighboring nuclei, akin to what happens in conditional specification? The answer to both these questions is yes.

Opposing axial gradients define position

What has emerged from numerous studies is that, just as we've seen in other eggs, the cytoplasm of the *Drosophila* egg is not uniform. Instead, it contains gradients of positional information that dictate cell fate along the egg's anterior-posterior axis (reviewed in Kimelman and Martin 2012). In the syncytial blastoderm, nuclei in the anterior part of the cell are exposed to cytoplasmic determinants that are not present in the posterior part of the cell, and vice versa. It is the interaction between nuclei and the differing amounts of determination factors that specify cell fate. After fertilization, as nuclei undergo synchronous waves of division (see Figure 2.11B), each nucleus becomes positioned at specific coordinates along the anterior-posterior axis and experiences unique concentrations of determination factors.

DIFFERENTIALLY SPECIFYING ALL THE NUCLEI IN A ROOM   How do the nuclei maintain a position within the syncytial blastoderm? They do so through the action of their own cytoskeletal machinery: their centrosome, affiliated microtubules, actin filaments, and interacting proteins (Kanesaki et al. 2011; Koke et al. 2014). Specifically, when the nuclei are in between divisions (in interphase), each nucleus radiates *dynamic* microtubule extensions, organized by its centrosome, that establish an "orbit" and exert force on the orbits of other nuclei (**FIGURE 2.12**). Each time the nuclei divide, this radial microtubule array is reestablished to exert force on neighboring nuclear orbits, ensuring regular spacing of nuclei across the syncytial blastoderm. Maintaining the positional relationships between nuclei across the early embryo is essential for successful syncytial specification.

Keeping nuclear position stable during early development allows each nucleus to be exposed to different amounts of the determination factors distributed in gradients throughout the shared cytoplasmic environment. A nucleus can interpret its position (whether to become part of the anterior, midsection, or posterior part of the body) based on the concentration of cytoplasmic determinants it experiences. Each nucleus thereby becomes genetically programmed toward a particular identity. The determinants are **transcription factors**, proteins that bind DNA and regulate gene transcription. (See **Further Development 2.3, Transcription Factor Gradients Specify Fates from Head to Tail, Further Development 2.4, A Map of Cell Maturation,** and **Further Development 2.5, A Rainbow of Cell Identities**, all online.)

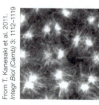

From T. Kanesaki et al. 2011. *Integr Biol (Camb)* 3: 1112–1119

Interphase

FIGURE 2.12   Positioning of nuclei during the interphase stage of nuclear cycle 13 of the *Drosophila melanogaster* syncytium. Nuclei are dynamically ordered within the syncytium of the early embryo, holding their positions using the cytoskeletal elements associated with them. (Left) EB1-GFP illuminates microtubules associated with each nucleus. The aster arrays defining nuclear orbits have some overlap with neighboring asters. (Right) An illustration of how the nuclei maintain their positions during interphase to establish orbits. This pattern of nuclei and cytoplasmic arrays was generated through computational modeling.

Closing Thoughts on the Opening Photo

"Can development be mapped?" That was the question asked about the multicolored temporal landscape of cell fate development made by Dan Wagner, Sean Megason, and Allon Klein on day 1 of zebrafish embryogenesis (Wagner et al. 2018). The philosopher Søren Kierkegaard once wrote of the truth that is inherent in the individual that can become obscured by the noise and direction of the crowd. Right now, the field of developmental biology has largely defined differentiation on the order of broad cell-type categories, and researchers are curious as to how much "truth" we may be missing on the individual cell level. Although this image is a computational reconstruction, it shows each cell with a different hue, representative of the full array of genes it expresses. It is a full-spectrum rainbow of cell identities splayed out from immature to differentiated cells, with the most similar relatives positioned closest. This approach has moved us closer to refining the differences underlying distinct individual cell identities.

Photograph courtesy of Dan Wagner, Sean Megason and Allon Klein

Endnotes

[1] Today, the most commonly researched tunicate is *Ciona intestinalis*, which has provided insight into cell lineage maturation and vertebrate evolution and development. Recent research on the species has defined the physical properties governing neural tube closure, which is similar to that in humans.

[2] Conklin fixed and stained embryos at every cleavage, and his 1905 paper records that lineage sequence. The entire study was done with a single microscope slide. He combined embryos of all stages onto that slide, mounted it in balsam, and stained the embryos. He then drew what he saw, and the resulting drawings became the lithographs of the paper. That slide is still around—in a collection at the University of North Carolina.

[3] The idea of nuclear equivalence and the ability of cells to interact eventually caused Driesch to abandon science. Driesch, who thought the embryo was like a machine, could not explain how the embryo could make its missing parts or how a cell could change its fate to become another cell type.

[4] Syncytia can be found in many organisms, from fungi to humans. Examples are the syncytium of the nematode germ cells (connected by cytoplasmic bridges), the multinucleated skeletal muscle fiber, and the giant cancer cells derived from fused immune cells.

2 Snapshot Summary
Specifying Identity

1. Cell differentiation is the process by which a cell acquires the structural and functional properties unique to a given cell type.

2. From an undifferentiated cell to a postmitotic differentiated cell type, a cell goes through a process of maturation that experiences different levels of commitment toward its end fate.

3. A cell is first specified toward a given fate, suggesting that it would develop into this cell type even in isolation.

4. A cell is committed or determined to a given fate if it maintains its developmental maturation toward this cell type even when placed in a new environment.

5. There are three different modes of cell specification: autonomous, conditional, and syncytial.

6. In autonomous specification, cells in the early embryo possess cytoplasmic determinants that commit those cells toward a specific fate. Such cells will mature only into their determined cell types even when isolated, as best exemplified by cells of the tunicate embryo.

7. Conklin first observed the yellow crescent in the tunicate embryo and showed that cells with the yellow crescent gave rise to muscle. The muscle cell fate in tunicates is dependent on the *macho* gene.

8. Conditional specification is the acquisition of a given cell identity based on the cell's position or, more specifically, on the interactions that the cell has with the other cells and molecules it comes in contact with. An extreme example of conditional specification was demonstrated by the complete normal development of sea urchin larvae from single isolated blastomeres.

9. Most species have cells that develop via autonomous specification as well as cells that develop via conditional specification. For example, many plant embryos show a first asymmetrical division that operates by autonomous specification, while remaining development of the plant follows conditional specification.

10. Syncytial specification occurs when cell fates are determined in a syncytium of nuclei, as in the *Drosophila* blastoderm.

11. Cytoskeletal arrangements maintain the positioning of nuclei in the syncytium, which enables specification of these nuclei by opposing morphogen gradients.

Go to oup.com/he/barresi12xe for Further Developments, Scientists Speak interviews, Watch Development videos, Dev Tutorials, and complete bibliographic information for all literature cited in this chapter.

Differential Gene Expression
Mechanisms of Cell Differentiation

What underlies cell differentiation?

From I. S. Peter and E. H. Davidson. 2011. *Nature* 474: 635–639

FROM ONE CELL COME MANY, and of many different types. That is the seemingly miraculous phenomenon of embryonic development. How is it possible that such a diversity of cell types within a multicellular organism can be derived from a single cell, the fertilized egg? Cytological studies done at the start of the twentieth century established that the chromosomes in each cell of an organism's body are the mitotic descendants of the chromosomes established at fertilization (Wilson 1896; Boveri 1904). In other words, each somatic cell nucleus has the same chromosomes—and therefore the same set of genes—as all other somatic cell nuclei. This fundamental concept, known as **genomic equivalence**, presented a significant conceptual dilemma. If every cell in the body contains the genes for hemoglobin and insulin, for example, why are hemoglobin proteins made only in red blood cells and insulin proteins only in certain pancreatic cells? Based on the embryological evidence for genomic equivalence (as well as on bacterial models of gene regulation), a consensus emerged in the 1960s that the answer lies in *differential gene expression*.

Defining Differential Gene Expression

Differential gene expression is the process by which cells become different from one another based on the unique combination of genes that are active, or "expressed." By expressing different genes, cells can create different proteins that lead to the differentiation of different cell types. There are three postulates of differential gene expression:

1. Every somatic cell nucleus of an organism contains the complete genome established in the fertilized egg. The DNA of all differentiated cells is identical.

2. The unused genes in differentiated cells are neither destroyed nor mutated; they retain the potential for being expressed.

3. Only a small percentage of the genome is expressed in each cell, and a portion of the RNA synthesized in each cell is specific for that cell type.

By the late 1980s, it was established that gene expression can be regulated at four levels:

1. *Level 1: Differential gene transcription* regulates which of the nuclear genes are transcribed into pre-messenger RNA.

2. *Level 2: Selective pre-messenger RNA processing* regulates which parts of the transcribed RNAs are able to enter the cytoplasm and become messenger RNAs.

3. *Level 3: Selective messenger RNA translation* regulates which of the mRNAs in the cytoplasm are translated into proteins.

4. *Level 4: Differential posttranslational protein modification* regulates which proteins are allowed to remain and/or function in the cell.

Some genes (such as those coding for the globin protein subunits of hemoglobin) are regulated at all these levels. It is because of these many different ways to regulate gene expression that a relatively small number of genes can offer an extreme diversity in the possible patterns of protein expression, which yields the enormous array of different cell types constituting both plants and animals.

A Quick Primer on the Central Dogma

To properly comprehend all the mechanisms regulating the differential expression of a gene, you must first understand the principles of the central dogma of biology. The **central dogma** pertains to the sequence of events that enables the use and transfer of information to make the proteins of a cell (**FIGURE 3.1**). *Central* to this theory is the order of deoxyribonucleotides in double-stranded DNA, which provides the informative code, or *blueprints*, for the precise combination of amino acids needed to build specific proteins. Proteins are not made directly from DNA, however; rather, the sequence of DNA bases is first copied, or *transcribed*, into a single-stranded polymer of similar molecules called heterogeneous nuclear ribonucleic acids (hnRNA), more commonly known as *pre-mRNA*. The process of copying DNA into RNA is called **transcription**, and the RNA produced from a given gene is often referred to as a *transcript*. Although the transcribed pre-mRNA includes the sequences to code for a protein, it can also hold non-protein-coding (simply called noncoding) information. The pre-mRNA strand will undergo processing to excise the noncoding domains and protect the ends of the strand to yield a **messenger RNA (mRNA)** molecule. mRNA is transported out of the nucleus into the cytoplasm where it can interact with a ribosome and present its *message* for the synthesis of a specific protein. mRNA unveils the complementary sequence of DNA three bases at a time. Each triplet, or codon, calls for a specific amino acid that will be covalently attached to its neighboring amino acid denoted by the codon next in line. In this manner, **translation** leads to the synthesis of a polypeptide chain that will undergo protein folding and potential modification by the addition of various functional moieties, such as carbohydrates, phosphates, or cholesterol groups. The completed protein is now ready to carry out its specific function serving to support the structural

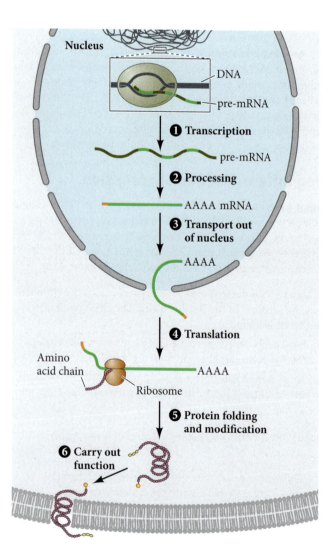

FIGURE 3.1   The central dogma of biology. A simplified schematic of the key steps in the expression of a protein-coding gene. (1) Transcription. In the nucleus, a region of the genomic DNA is seen accessible to RNA polymerase II, which transcribes an exact complementary copy of the gene in the form of a single-stranded pre-mRNA molecule. The gene is now said to be "expressed." (2) Processing. The pre-mRNA transcript undergoes processing to make a finalized messenger RNA strand, which is (3) transported out of the nucleus. (4) Translation. The mRNA complexes with a ribosome, and its information is translated into an ordered polymer of amino acids. (5) Protein folding and modification. The polypeptide adopts secondary and tertiary structures through proper folding and potential modifications (such as the addition of a carbohydrate group, as seen here). (6) The protein is now said to be "expressed" and can carry out its specific function (e.g., as a transmembrane receptor).

or functional properties of the cell. Cells that express different proteins will therefore possess different structural and functional properties, making them distinct types of cells.

Evidence for Genomic Equivalence

Until the mid-twentieth century, genomic equivalence was not so much proved as it was assumed (because every cell is the mitotic descendant of the fertilized egg). One of the first tasks of developmental genetics was to determine whether every cell of an organism does indeed have the same **genome** (set of genes) as every other cell—that is, *equivalent genomes*.

Early analysis of the chromosomes of fruit flies (*Drosophila*) provided some of the first evidence that every cell in the body has the same genome, but perhaps one used differently by different cells. The DNA of certain larval tissues in *Drosophila* undergoes numerous rounds of replication without separation, such that the structure of the chromosomes can be seen. In these **polytene** (Greek, "many strands") **chromosomes**, no structural differences were seen between cells; however, different regions of the chromosomes were "puffed up" at different times and in different cell types, which suggested that these areas were actively making RNA (**FIGURE 3.2A**; Beermann 1952). These observations were confirmed by nucleic acid in situ hybridization studies, a technique that enables the visualization of the spatial and temporal pattern of specific gene (mRNA) expression (see Figure 3.27). For instance, the mRNA of the *odd-skipped* gene is present in cells that display a segmented pattern in the *Drosophila* embryo, a pattern that changes over time (**FIGURE 3.2B**). Similarly, the mouse homolog of *odd-skipped*, called *odd-skipped related 1*, is differentially expressed in cells of the segmented pharyngeal arches, the limb buds, and the heart (**FIGURE 3.2C**).

Is the DNA in an organism's cells that is now expressing different genes truly still equivalent, however? The ultimate test of whether the nucleus of a differentiated cell has undergone irreversible functional restriction is to challenge that nucleus to generate every other type of differentiated cell in the body. If each cell's nucleus contains DNA that is identical to that of the zygote nucleus, then each cell's nucleus should also be capable of directing the entire development of the organism when transplanted into an activated enucleated egg.

Evidence for this came in 1952, when Briggs and King demonstrated that transplantation of a nucleus from a frog blastula into an enucleated egg resulted in the development of a complete embryo (Briggs and King 1952). A decade later, John Gurdon conducted the decisive experiment that would garner him a Nobel Prize in 2012: he demonstrated that a nucleus from a differentiated cell, taken from a tadpole's intestine, could direct the complete development of an enucleated egg into a *cloned* adult frog (Gurdon et al. 1958). Genomic equivalence in mammals was proved in 1997 by Ian Wilmut and colleagues when they showed that an adult mammalian somatic cell could

FIGURE 3.2 Gene expression. (A) Transmission electron micrograph of a polytene chromosome from a salivary gland cell of a midge (*Chironomus tentans*) larva showing three giant puffs indicating active transcription in these regions (arrows). (B) mRNA expression (seen here in blue, using in situ hybridization with an antisense DIG-labeled RNA probe; see "The Basic Tools of Developmental Genetics" [p. 78] and Figure 3.27) of the *odd-skipped gene* in a stage-5 and a stage-9 *Drosophila* embryo. (C) mRNA expression of the *odd-skipped related 1* gene in an 11.5-days-postconception mouse embryo (blue).

(A)
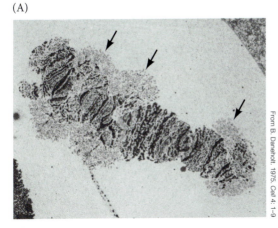
From B. Daneholt. 1975, *Cell* 4: 1–9

(B) *odd-skipped* (stage 5) *odd-skipped* (stage 9)

From R. Weiszman et al. 2009, *Nat Protoc* 4: 605

(C) *odd-skipped related 1*
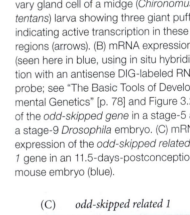

From P. L. So and P. S. Danielian. 1999, *Mech Dev* 84: 157–160

direct the development of an entire sheep; they called her Dolly (**FIGURE 3.3**; Wilmut et al. 1997). Cloning of adult mammals has been confirmed in guinea pigs, rabbits, rats, mice, dogs, cats, horses, and cows. In 2003, a cloned mule became the first sterile animal to be so reproduced (Woods et al. 2003). Thus, it appears that the nuclei of adult somatic cells in vertebrates contain all the genes needed to generate an adult organism. No genes necessary for development have been lost or mutated in the somatic cells; *the DNA of their nuclei is equivalent.*[1] (See **Further Development 3.1, Genomic Equivalence and Cloning**, online.)

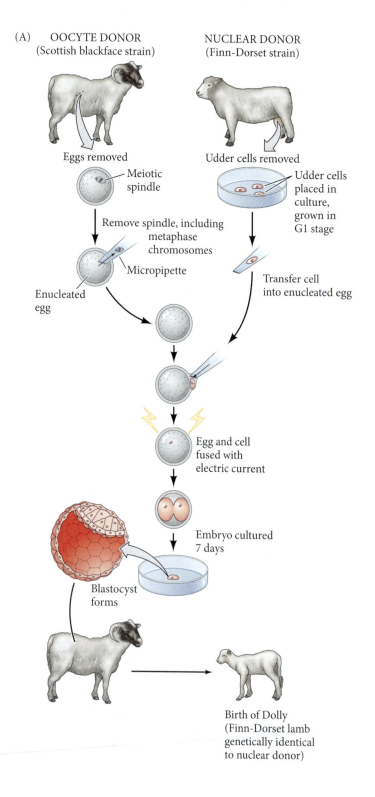

FIGURE 3.3 Cloning a mammal using nuclei from adult somatic cells. (A) Procedure used for cloning sheep. (B) Dolly, the adult sheep on the left, was derived by fusing a mammary gland cell nucleus with an enucleated oocyte, which was then implanted in a surrogate mother (of a different breed of sheep) that gave birth to Dolly. Dolly later gave birth to a lamb (Bonnie, at right) by normal reproduction. (A after I. Wilmut et al. 2000. *The Second Creation: Dolly and the Age of Biological Control.* Harvard University Press: Cambridge, MA.)

Anatomy of the Gene

So how does the same genome give rise to different cell types? To address this question, we need to first understand the anatomy of genes.

Chromatin composition

A fundamental difference distinguishing most eukaryotic genes from prokaryotic genes is that eukaryotic genes are contained within a complex of DNA and protein called **chromatin**. The protein component constitutes about half the weight of chromatin and is composed largely of **histones**. The **nucleosome** is the basic unit of chromatin structure (**FIGURE 3.4A,B**). It is composed of an octamer of histone proteins (two molecules each of histones H2A, H2B, H3, and H4) wrapped with two loops containing approximately 147 base pairs of DNA (Kornberg and Thomas 1974), with more than a dozen contacts between the DNA and the histones (Luger et al. 1997; Bartke et al. 2010). Nucleosomes result in a remarkable packaging of more than *6 feet* of DNA into the approximately 6-micrometer nucleus of each human cell (Schones and Zhao 2008).

Whereas classical geneticists have likened genes to "beads on a string," molecular geneticists liken genes to "string on the beads," an image in which the beads are nucleosomes. Much of the time, the nucleosomes appear to be wound into tight structures called **solenoids**, which are stabilized by histone H1 (**FIGURE 3.4C**). This H1-dependent conformation of nucleosomes inhibits the transcription of genes in somatic cells by packing adjacent nucleosomes together into tight arrays that prevent transcription factors and RNA polymerases from gaining access to the genes (Thoma et al. 1979; Schlissel and

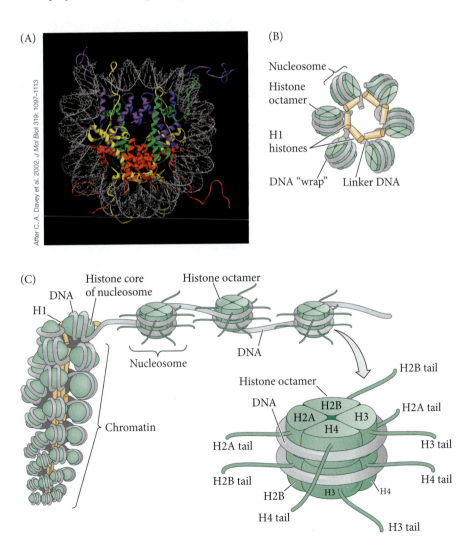

(A)

After C. A. Davey et al. 2002. *J Mol Biol* 319: 1097–1113

(B)

Nucleosome

Histone octamer

H1 histones

DNA "wrap"   Linker DNA

(C)

DNA   Histone core of nucleosome   Histone octamer

H1

Nucleosome

Chromatin

DNA

Histone octamer

DNA

H2B

H2A   H3

H4

H2A tail

H2B tail

H2B

H4 tail

H3

H4

H4 tail

H3 tail

H2B tail   H2A tail   H3 tail   H2B tail   H4 tail

FIGURE 3.4 Nucleosome and chromatin structure. (A) Model of nucleosome structure as seen by X-ray crystallography at a resolution of 1.9 Å. Histones H2A and H2B are yellow and red, respectively; H3 is purple; and H4 is green. The DNA helix (gray) winds around the protein core. The histone "tails" that extend from the core are the sites of acetylation and methylation, which may disrupt or stabilize, respectively, the formation of nucleosome assemblages. (B) Histone H1 can draw nucleosomes together into compact forms. About 147 base pairs of DNA encircle each histone octamer, and about 60–80 base pairs of DNA link the nucleosomes together. (C) Model for the arrangement of nucleosomes in the highly compacted solenoidal chromatin structure. Histone tails protruding from the nucleosome subunits allow for the attachment of chemical groups.

Brown 1984). Chromatin regions that are tightly packed are called **heterochromatin**, and regions loosely packed are called **euchromatin**. One way to achieve differential gene expression is by regulating how tightly packed a given region of chromatin may be, thereby regulating whether genes are even accessible for transcription.

Exons and introns

In addition to being contained within chromatin, another fundamental feature that distinguishes eukaryotic from prokaryotic genes is that eukaryotic genes are not co-linear with their peptide products. Rather, the single nucleic acid strand of eukaryotic mRNA that is translated into protein comes from noncontiguous regions on the chromosome. **Exons** are the regions of DNA that code for parts of a protein;[2] between exons, however, are intervening sequences called **introns**, which have nothing whatsoever to do with the amino acid sequence of the protein.

Major parts of a eukaryotic gene

To help illustrate the structural components of a typical eukaryotic gene, we highlight the anatomy of the human β-globin gene (**FIGURE 3.5**). This gene, which encodes part of the hemoglobin protein of the red blood cells, consists of the following elements:

- A **promoter**—the region where the enzyme that initiates transcription, **RNA polymerase II**, binds. The promoter of the human β-globin gene has three distinct units and extends several base pairs before ("upstream from")[3] the

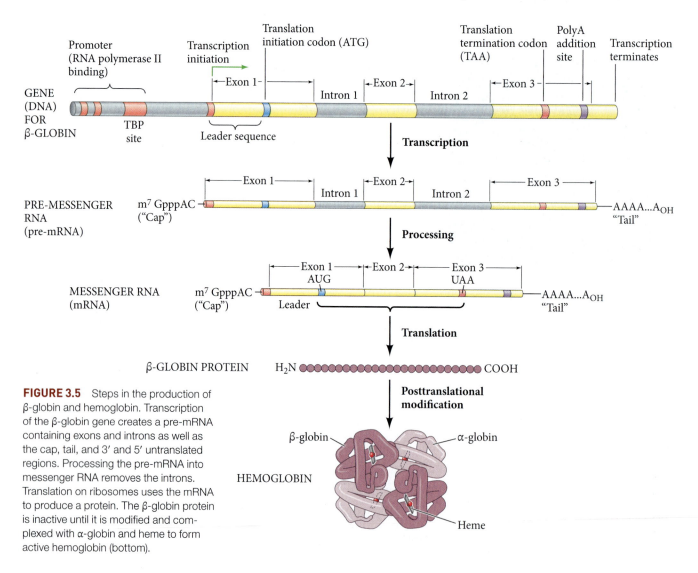

FIGURE 3.5 Steps in the production of β-globin and hemoglobin. Transcription of the β-globin gene creates a pre-mRNA containing exons and introns as well as the cap, tail, and 3′ and 5′ untranslated regions. Processing the pre-mRNA into messenger RNA removes the introns. Translation on ribosomes uses the mRNA to produce a protein. The β-globin protein is inactive until it is modified and complexed with α-globin and heme to form active hemoglobin (bottom).

transcription initiation site. Some promoters have the DNA sequence TATA (called the *TATA-box*), which binds to a TATA-binding protein (TBP) that helps anchor RNA polymerase II to the promoter.

- The **transcription initiation site**, which is often called the **cap sequence** because it is the DNA sequence that will code for the addition of a modified nucleotide "cap" at the 5′ end of the RNA soon after it is transcribed. The cap sequence begins the first exon.

- The **5′ untranslated region (5′ UTR)**, also called the **leader sequence**. This is the sequence of base pairs intervening between the initiation points of transcription and translation. The 5′ UTR can determine the rate at which translation is initiated.

- The **translation initiation site**, **ATG**. The ATG sequence, which becomes AUG in mRNA, is the same in every gene. The distance this codon is located from the transcription initiation site varies among different genes.

- The protein-coding sequences of **exons** interspersed with the noncoding sequences of **introns**; there are different numbers of exons and introns, depending on the gene.

- A **translation termination codon**, **TAA**. This codon becomes UAA in the mRNA. When a ribosome encounters this codon, the ribosome dissociates, and the protein is released. Translation termination can also be signaled by the **TAG** or **TGA** codon sequences in other genes.

- A **3′ untranslated region (3′ UTR)** that, although transcribed, is not translated into protein. This region includes the sequence AATAAA, which is needed for **polyadenylation**, the insertion of a "tail" of some 200–300 repeating adenylate residues on the RNA transcript. This polyA tail (1) confers stability on the mRNA, (2) allows the mRNA to exit the nucleus, and (3) permits the mRNA to be translated into protein.

- A **transcription termination sequence**. Transcription continues beyond the AATAAA site for about 1000 nucleotides before being terminated.

The transcription product and how it is processed

The original transcription product is called heterogeneous nuclear RNA (hnRNA) or pre-messenger RNA (pre-mRNA). Pre-mRNA contains the cap sequence, 5′ UTR, exons, introns, 3′ UTR, and a polyA tail. Both ends of these transcripts are modified before the RNAs leave the nucleus. A cap consisting of methylated guanosine is placed on the 5′ end of the RNA in opposite polarity to the RNA itself, which means there is no free 5′ phosphate group on the pre-mRNA. The 5′ cap is necessary for the binding of mRNA to the ribosome and for subsequent translation (Shatkin 1976). Both the 5′ and 3′ modifications protect the mRNA from exonucleases, enzymes that would otherwise digest it by targeting the ends of the polynucleotide chain (Sheiness and Darnell 1973; Gedamu and Dixon 1978).

Before the pre-mRNA leaves the nucleus, its introns are removed and the remaining exons are spliced together. In this way, the coding regions of the mRNA—that is, the exons—are brought together to form a single uninterrupted transcript, and this transcript is translated into a protein. The protein can be further modified to make it functional (see Figure 3.5).

Noncoding regulatory elements: The on, off, and dimmer switches of a gene

Regulatory sequences can be located on either end of the gene (or even within it). We have already mentioned one of these noncoding sequences, the promoter; additional ones include *enhancers* and *silencers*. These regulatory elements are necessary for controlling where, when, and how actively a particular gene is transcribed. When they are located on the same chromosome as the gene (and they usually are), they can be referred to as ***cis*-regulatory elements**.[4]

Promoters, as we have mentioned, are sites where RNA polymerase II binds to the DNA to initiate transcription. Promoters are typically located immediately upstream from the site where RNA polymerase II initiates transcription. Most of these promoters contain a stretch of about 1000 base pairs that is rich in the sequence CG, often referred to as CpG (a *C* and a *G* connected through the normal *p*hosphate bond). These regions are called **CpG islands** (Down and Hubbard 2002; Deaton and Bird 2011). The reason transcription is initiated near CpG islands is thought to involve proteins called **basal transcription factors**. Basal transcription factors bind to the DNA of the promoter, forming a "saddle" that can recruit RNA polymerase II and position it appropriately for the polymerase to begin transcription (**FIGURE 3.6**; Kostrewa et al. 2009).

RNA polymerase II does not bind to every promoter in the genome at the same time, however. Rather, it is recruited to and stabilized on the promoters by DNA sequences called **enhancers** that signal where and when a promoter can be used and how much gene product to make (see Figure 3.6). In other words, enhancers control the efficiency and rate of transcription from a specific promoter (see Ong and Corces 2011). In contrast, DNA sequences called **silencers** (or **repressors**) can prevent promoter use and inhibit gene transcription.

Transcription factors by definition are proteins that bind DNA with precise sequence recognition for specific promoters, enhancers, or silencers. Transcription factors function in two nonexclusive ways:

1. Transcription factors recruit nucleosome-modifying proteins to that region of the genome and make the chromatin more accessible for RNA polymerase II to carry out transcription.

2. Transcription factors can recruit **transcriptional co-regulators** to form bridges, which loop the chromatin in a conformation that brings the enhancer-bound transcription factors into the vicinity of the promoter (see Figure 3.6). Transcriptional co-regulators can function as either co-activators or co-repressors of transcription. In the activation of mammalian β-globin genes, such a bridge uniting the promoter and enhancer is formed by transcriptional co-activating proteins that bind to transcription factors on both the enhancer and the promoter sequences. These protein complexes form attractive arrangements that recruit the nucleosome-modifying enzymes and basal transcription factors, all of which together stabilize RNA polymerase II and promote transcription (see Figure 3.6; Deng et al. 2012; Noordermeer and Duboule 2013; Gurdon 2016).

To further develop your understanding of how the chromatin can be shaped to enable enhancer-promoter interactions, see **Further Development 3.2, The Mediator Complex: Linking Enhancer and Promoter**, online.

ENHANCERS Enhancers are regulatory elements that can activate and *enhance* the rate of transcription by binding to transcription factors, which build a bridge between the enhancer and the promoter (see Figure 3.6). Enhancers generally activate only *cis*-linked promoters (i.e., promoters on the same chromosome), and consequently are sometimes called *cis*-regulatory elements. Because of DNA folding, however, enhancers can regulate genes at great distances (some as far as a million bases away) from the promoter (Visel et al. 2009). Moreover, enhancers do not need to be on the 5′ (upstream) side of the gene; they can be at the 3′ end or even located in introns (Maniatis et al. 1987). An important enhancer for a gene involved in specifying the "pinky" of each of our limbs is found in an intron of *another* gene, some 1 million base pairs away from its promoter (Lettice et

FIGURE 3.6 The bridge between enhancer and promoter can be made by transcription factors. Certain transcription factors, called basal transcription factors, bind to DNA on the promoter (where RNA polymerase II will initiate transcription), whereas other transcription factors bind to the enhancer (which regulates when and where transcription can occur). Other transcriptional regulators do not bind to the DNA; rather, they link the transcription factors that have bound to the enhancer and promoter sequences. In this way, the chromatin loops to bring the enhancer to the promoter. The example shown here is the mouse β-globin gene. (A) Transcription factors assemble on the enhancer, but the promoter is not used until the Gata1 transcription factor binds to the promoter. (B) Gata1 can recruit several other factors, including Ldb1, which forms a link uniting the enhancer-bound factors to the promoter-bound factors. (After W. Deng et al. 2012. *Cell* 149: 1233–1244.)

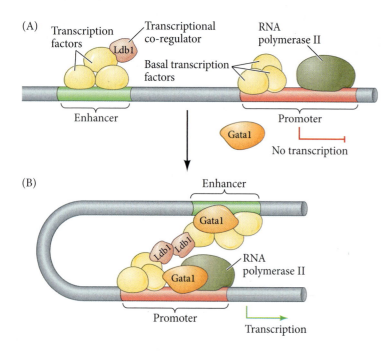

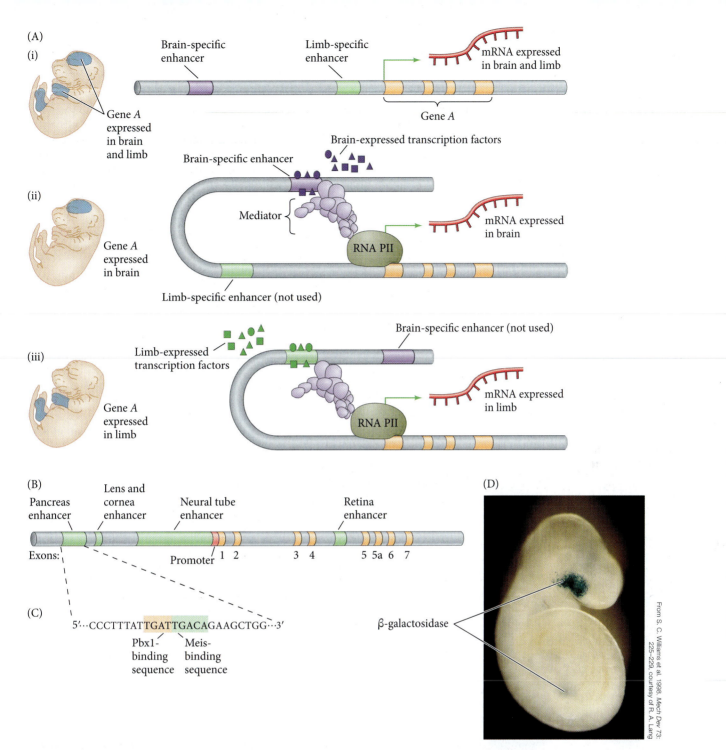

FIGURE 3.7 Enhancer region modularity. (A) Model for gene regulation by enhancers. (i) The top diagram shows the exons, introns, promoter, and enhancers of a hypothetical gene *A*, but does not show how the two enhancers are involved in the expression of the gene (see ii and iii). In situ hybridization (left) shows that gene *A* is expressed in limb and brain cells. (ii) In developing brain cells, brain-specific transcription factors bind to the brain enhancer, causing it to bind to the Mediator, stabilize RNA polymerase II (RNA PII) at the promoter, and modify the nucleosomes in the region of the promoter. The gene is transcribed in the brain cells only; the limb enhancer does not function. (iii) An analogous process allows for transcription of the same gene in the cells of the limbs. The gene is not transcribed in any cell type whose transcription factors the enhancers cannot bind. (B) The Pax6 protein is critical in the development of several widely different tissues. Enhancers direct *Pax6* gene expression (yellow exons 1–7) differentially in the pancreas, the lens and cornea of the eye, the retina, and the neural tube. (C) A portion of the DNA sequence of the pancreas-specific enhancer element. This sequence has binding sites for the Pbx1 and Meis transcription factors; both must be present to activate *Pax6* in the pancreas. (D) When the *lacZ* reporter gene (which codes for β-galactosidase) is fused to the *Pax6* enhancers for expression in the pancreas and lens/cornea, β-galactosidase enzyme activity (blue) is seen in those tissues. (A–C after A. Visel et al. 2009. *Nature* 461: 199–205.)

al. 2008; see Further Development 19.5, online). In each cell, the enhancer becomes associated with particular transcription factors, binds nucleosome regulators and the Mediator complex (see Further Development 3.6 online), and engages with the promoter to transcribe the gene in that particular type of cell (**FIGURE 3.7A**).

What are the consequences of enhancer modularity to a developing individual? To a species? How might a mutation in an enhancer affect development? For instance, what might occur in an embryo if there were a mutation in the enhancer region of the *Pax6* gene? Could such a mutation have evolutionary importance? *Hint*: It does, and it's profound!

FIGURE 3.8   The genetic elements regulating tissue-specific transcription can be identified by fusing reporter genes to suspected enhancer regions of the genes expressed in particular cell types. (A) The *GFP* gene is fused to a zebrafish gene that is active only in certain cells of the retina. The result is expression of green fluorescent protein in the larval retina (below left), specifically in the cone cells (below right). (B) The enhancer region of the gene for the muscle-specific protein Myf5 is fused to a *lacZ* reporter gene that codes for β-galactosidase and is incorporated into a mouse embryo. When stained for β-galactosidase activity (darkly stained region), the 13.5-day mouse embryo shows that the reporter gene is expressed in the muscles of the eye, face, neck, and forelimb and in the segmented myotomes (which give rise to the back musculature).

ENHANCER ACTIVATION   Even though the enhancer DNA sequences are the same in every cell type, enhancer activation can differ between different types of cells; what differs is the combination of transcription factor proteins that the enhancers experience. Once bound to enhancers, transcription factors are able to enhance or suppress the ability of RNA polymerase II to initiate transcription. Several transcription factors can bind to an enhancer, and it is the specific *combination* of transcription factors present that allows a gene to be active in a particular cell type. That is, the same transcription factor, in conjunction with different combinations of factors, will activate different promoters in different cells. Moreover, the same gene can have several enhancers, with each enhancer binding transcription factors that enable that same gene to be expressed in different cell types.

The mouse *Pax6* gene (which is expressed in the lens, cornea, and retina of the eye; in the neural tube; and in the pancreas) has several enhancers (**FIGURE 3.7B–D**; Kammandel et al. 1998; Williams et al. 1998). The enhancer sequence farthest upstream from the promoter contains the region necessary for *Pax6* expression in the pancreas, whereas a second enhancer activates *Pax6* expression in surface ectoderm (lens, cornea, and conjunctiva). A third enhancer resides in the leader sequence; it contains the sequences that direct *Pax6* expression in the neural tube. A fourth enhancer, located in an intron shortly downstream of the translation initiation site, determines the expression of *Pax6* in the retina. The *Pax6* gene illustrates the principle of enhancer modularity, wherein genes having multiple, separate enhancers allow a protein to be expressed in several different tissues but not expressed at all in others. (See **Further Development 3.3, Combinatorial Association**, online.)

FURTHER DEVELOPMENT

FINDING THE ENHANCER   Identifying enhancers has been a difficult puzzle to solve, but researchers have come up with a novel method. Researchers create constructs of possible enhancers combined with a **reporter gene** (a gene that encodes a visible marker), then insert these into embryos and monitor the spatial and temporal pattern of expression displayed by the visible protein product of the reporter gene (such as *green fluorescent protein*, or *GFP*; **FIGURE 3.8A**). If the sequence contains an enhancer, the reporter gene should become active at particular times and places based on the specificity of the enhancer. Another reporter gene often used is the *E. coli* gene for β-galactosidase (the *lacZ* gene). This has been fused to (1) a promoter that can be activated in any cell and (2) an enhancer that directs expression of a particular gene (*Myf5*) only in mouse muscles. When this construct is injected into a newly fertilized mouse egg, it becomes incorporated into

(A)

(B)

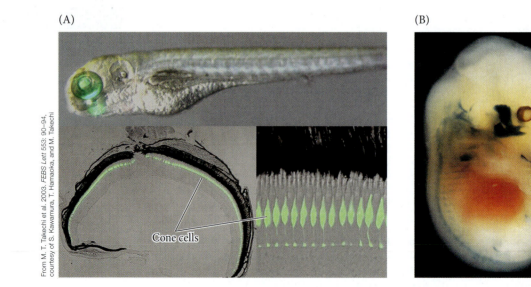

Cone cells

From M. T. Takechi et al. 2003. *FEBS Lett* 553: 90–94, courtesy of S. Kawamura, T. Hamaoka, and M. Takechi

Courtesy of A. Patapoutian and B. Wold

the mouse's genome. Once the embryo has developed muscles, the expression of β-galactosidase protein reveals the expression pattern corresponding to that of the muscle-specific *Myf5* gene (**FIGURE 3.8B**).

SILENCERS   Also called repressors, silencers are DNA regulatory elements that actively repress the transcription of a particular gene. They can be viewed as "negative enhancers," and they can silence gene expression spatially (in particular cell types) or temporally (at particular times). In the mouse, for instance, there is a DNA sequence that prevents a promoter's activation in any tissue *except* neurons. This sequence, given the name **neural restrictive silencer element** (**NRSE**), has been found in several mouse genes whose expression is limited to the nervous system: those encoding synapsin I, sodium channel type II, brain-derived neurotrophic factor, Ng-CAM, and L1. The protein that binds to NRSE is a transcription factor called **neural restrictive silencer factor** (**NRSF**). NRSF appears to be expressed in every cell that is *not* a mature neuron (Chong et al. 1995; Schoenherr and Anderson 1995). When NRSE is deleted from particular neural genes, these genes are expressed in non-neural cells (**FIGURE 3.9**; Kallunki et al. 1995, 1997).

GENE REGULATORY ELEMENTS: SUMMARY   Enhancers and silencers enable a gene to use numerous transcription factors in various combinations to control its expression. Thus, *enhancers and silencers are modular* such that, for example, the *Pax6* gene is regulated by enhancers that enable it to be expressed in the eye, pancreas, and nervous system, (see Figure 3.7B); this is the Boolean "OR" function. But *within each cis-regulatory module, transcription factors work in a combinatorial fashion* such that Pax6, L-Maf, and Sox2 proteins are all needed for the transcription of crystallin in the lens (see Figure 1A in Further Development 3.7, online); that is the Boolean "AND" function. The combinatorial association of transcription factors on enhancers leads to the spatiotemporal output of any particular gene (Zinzen et al. 2009; Peter and Davidson 2015). This "AND" function may be extremely important in activating entire groups of genes simultaneously.

(A)

L1 promoter

NRSE sequence

lacZ

(B)

L1 promoter

No NRSE sequence

lacZ

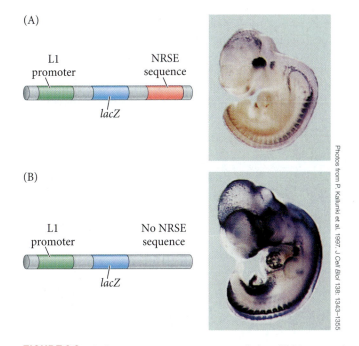

Photos from P. Kallunki et al. 1997. *J Cell Biol* 138: 1343–1355

FIGURE 3.9   A silencer represses gene transcription. (A) Mouse embryo containing a transgene composed of the L1 promoter, a portion of the neuron-specific *L1* gene, and a *lacZ* gene fused to the *L1* second exon, which contains the NRSE sequence. (B) Same-stage embryo with a similar transgene but lacking the NRSE sequence. Dark areas reveal the presence of β-galactosidase (the *lacZ* product).

Mechanisms of Differential Gene Expression: *Transcription*

Among the four levels at which gene expression can be regulated (transcription, RNA processing, protein synthesis (translation), and posttranslation), regulating how and when a gene is transcribed into pre-mRNA has by far the largest array of possible mechanisms. Two of the most prevalent mechanisms seen during embryonic development are (1) epigenetic modification of chromatin and (2) control through transcription factors.

Epigenetic modification: Modulating access to genes

In the broadest definition, **epigenetics** (*epi*, "over," "on," "in addition to") refers to phenotypic changes caused by modifying how a gene is expressed, rather than modifying the DNA sequence itself. If the DNA sequence represents a cell's genetic character, then all the factors (e.g., proteins, ions, and small molecules) acting *on* DNA represent the cell's epigenetic character. Epigenetic mechanisms most commonly involve adding or removing some of these factors, thereby altering the three-dimensional shape of DNA and its histones.

LOOSENING AND TIGHTENING CHROMATIN: HISTONES AS GATEKEEPERS   Histones are critical because they appear to be responsible for either facilitating or forbidding gene expression (**FIGURE 3.10**). Repression and activation are controlled to a large extent by modifying the "tails" of histones H3 and H4 with two small organic groups: methyl (CH_3) and acetyl ($COCH_3$) residues. In general, **histone acetylation**—the addition of negatively charged acetyl groups to histones—neutralizes the positive charge of lysine and loosens the histones, which makes transcription permissible. Enzymes known as **histone acetyltransferases** place acetyl groups on histones (especially on lysines in H3 and H4), destabilizing the tight packaging of nucleosomes so that they come apart (chromatin becomes more *euchromatic*). As might be expected, then, enzymes that *remove* acetyl groups—**histone deacetylases**—stabilize nucleosome packaging (chromatin becomes more *heterochromatic*) and prevent transcription.

   Histone methylation is the addition of methyl groups to histones by enzymes called **histone methyltransferases**. Although histone methylation more often results in heterochromatic states and transcriptional repression, it can also activate transcription depending on the amino acid being methylated and the presence of other methyl or acetyl groups in the vicinity (see Strahl and Allis 2000; Cosgrove et al. 2004). For instance, acetylation of the tails of H3 and H4 along with the addition of three methyl groups on the lysine at position four of H3 (i.e., H3K4me3; K is the abbreviation for lysine) is usually associated with actively transcribed chromatin. In

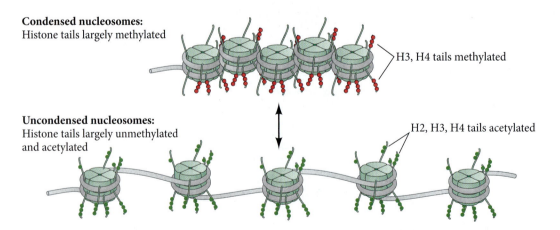

Condensed nucleosomes:
Histone tails largely methylated

H3, H4 tails methylated

Uncondensed nucleosomes:
Histone tails largely unmethylated and acetylated

H2, H3, H4 tails acetylated

FIGURE 3.10 Epigenetic regulation by histone modification. Methyl groups condense nucleosomes more tightly, preventing access to promoter sites and thus preventing gene transcription. Acetylation loosens nucleosome packing, exposing the DNA to RNA polymerase II and transcription factors that will activate the genes.

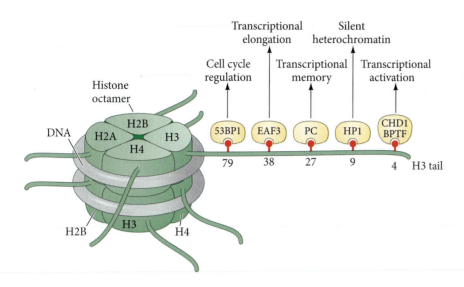

FIGURE 3.11 Histone methylations on histone H3. The tail of histone H3 (its amino-terminal sequence, at the beginning of the protein) sticks out from the nucleosome and is capable of being methylated or acetylated. Here, lysines can be methylated and recognized by particular proteins. Methylated lysine residues at positions 4, 38, and 79 are associated with gene activation, whereas methylated lysines at positions 9 and 27 are associated with repression. The proteins binding these sites (not shown to scale) are represented above the methyl group. (After T. Kouzarides and S. L. Berger. 2007. In *Epigenetics*, C. D. Allis, T. Jenuwein and D. Reinberg [Eds.], pp. 191–209. Cold Spring Harbor Press: New York.)

contrast, a combined lack of acetylation of the H3 and H4 tails and methylation of the lysine at position nine of H3 (H3K9) is usually associated with highly repressed chromatin (Norma et al. 2001). Indeed, lysine methylations at H3K9, H3K27, and H4K20 are often associated with highly repressed chromatin. **FIGURE 3.11** depicts a nucleosome with lysine residues on its H3 tail. Modifications of such residues regulate transcription.

HISTONE METHYLATION PATTERNS ARE HERITABLE Epigenetic methylation patterns can be passed down from generation to generation: they are heritable! This occurs because the modifications of histones can signal the recruitment of proteins that retain the *memory* of the transcriptional state from generation to generation as cells go through mitosis. These proteins are the proteins of the **Trithorax** and **Polycomb** families. When bound to the nucleosomes of active genes, Trithorax proteins keep these genes active, whereas Polycomb proteins, which bind to condensed nucleosomes, keep the genes in a repressed state.

The Polycomb proteins fall into two categories that act sequentially in repression. The first set serves as histone methyltransferases that methylate lysines H3K27 and H3K9 to repress gene activity. In many organisms, this repressed state is stabilized by the activity of a second set of Polycomb factors, which bind to the methylated tails of H3 and keep the methylation active and also methylate adjacent nucleosomes, thereby forming tightly packed repressive complexes (Grossniklaus and Paro 2007; Margueron et al. 2009).

The Trithorax proteins help retain the memory of activation; they act to counter the effect of the Polycomb proteins. Trithorax proteins can modify the nucleosomes or alter their positions on the chromatin, allowing transcription factors to bind to the DNA previously covered by them. Other Trithorax proteins keep the H3K4 lysine trimethylated (preventing its demethylation into a dimethylated, repressed state; Tan et al. 2008). (See **Further Development 3.4, Polycomb and Trithorax Proteins Regulate the Leaf-to-Flower Transition in Plants**, online.)

DNA METHYLATION AT PROMOTERS: CPG CONTENT MATTERS It turns out that not all promoters are the same. Rather, there are two general classes of promoters that use different methods for controlling transcription. These promoter types are catalogued as having either a relatively high or a relatively low number of CpG sequences at which DNA methylation can occur. That's right—similar to the histone protein tails, the DNA itself can also be directly methylated, and this methylation can stop transcription.

- **High CpG-content promoters** (**HCPs**) are usually found in "developmental control genes," where they regulate synthesis of the transcription factors and

(A) High CpG-content promoters (HCPs)

(B) Low CpG-content promoters (LCPs)

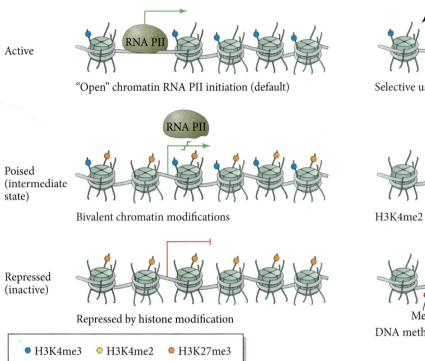

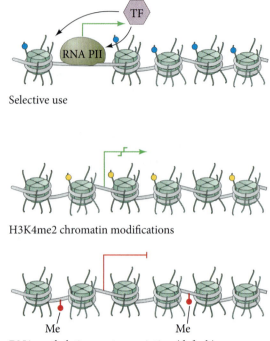

Active

"Open" chromatin RNA PII initiation (default)

Selective use

Poised
(intermediate
state)

Bivalent chromatin modifications

H3K4me2 chromatin modifications

Repressed
(inactive)

Repressed by histone modification

Me   Me
DNA methylation, no transcription (default)

● H3K4me3   ○ H3K4me2   ● H3K27me3

FIGURE 3.12   Chromatin regulation in HCPs and LCPs. Promoters with high and low CpG content have different modes of regulation. (A) HCPs are typically in an *active* state, with unmethylated DNA and nucleosomes rich in H3K4me3. The open chromatin allows RNA polymerase II (RNA PII) to bind. The *poised* state of HCPs is bivalent, having both activating (H3K4me3) and repressive (H3K27me3) modifications of the nucleosomes. RNA polymerase II can bind but not transcribe. The *repressed* state is characterized by repressive histone modification, but not by extensive DNA methylation. (B) *Active* LCPs, like active HCPs, have nucleosomes rich in H3K4me3 and low methylation but require stimulation by transcription factors (TF). *Poised* LCPs are capable of being activated by transcription factors and have relatively unmethylated DNA and nucleosomes enriched in H3K4me2. In their usual state, LCPs are *repressed* by methylated DNA nucleosomes rich in H3K27me3. (After V. W. Zhou et al. 2011. *Nat Rev Genet* 12: 7–18.)

other developmental regulatory proteins used in the *construction* of the organism (Zeitlinger and Stark 2010; Zhou et al. 2011). The default state of these promoters is "on," and they have to be actively repressed by *histone* methylation (**FIGURE 3.12A**).

- **Low CpG-content promoters** (**LCPs**) are usually found in those genes whose products characterize mature cells (e.g., the globins of red blood cells, the hormones of pancreatic cells, and the enzymes that carry out the normal maintenance functions of the cell). The default state of these promoters is "off," but they can be activated by transcription factors (**FIGURE 3.12B**). The CpG sites on these promoters are usually methylated, and this methylation is critical for preventing transcription. When these CpG sites become unmethylated, the histones associated with the promoter become modified with H3K4me3 and disperse so that RNA polymerase II can bind and transcription can occur. (See **Further Development 3.5, Mechanisms of How DNA Methylation Blocks Transcription,** and **Further Development 3.6, The Mechanisms of DNA Methylation**, both online.)

DNA METHYLATION PATTERNS ARE HERITABLE   Another enzyme recruited to the chromatin by MeCP2 is DNA methyltransferase-3 (Dnmt3). This enzyme methylates previously unmethylated cytosines on the DNA. In this way, a relatively large region can be repressed. The newly established methylation pattern is then transmitted to the next generation by DNA methyltransferase-1 (Dnmt1). This enzyme recognizes methyl cytosines on one strand of DNA and places methyl groups on the newly synthesized strand opposite it (**FIGURE 3.13**; see Bird 2002; Burdge et al. 2007). That is why it is necessary for the C to be next to a G in the sequence. Thus, in each cell division, the pattern of DNA methylation can be maintained. The newly synthesized (unmethylated) strand will become properly methylated when Dnmt1 binds to a methyl C on the old CpG sequence and methylates the cytosine of the CpG sequence on the complementary strand. In this way, once the DNA methylation pattern is established in a cell, it can be stably inherited by all the progeny of that cell.

DNA METHYLATION AND GENOMIC IMPRINTING   DNA methylation has explained at least one very puzzling phenomenon, that of genomic imprinting (Ferguson-Smith 2011). It is usually assumed that the genes one inherits from one's father and from one's mother are equivalent. In fact, the basis for Mendelian ratios (and the Punnett square analyses used to teach them) is that it does not matter whether the genes came from the sperm or from the egg. But if genes in the eggs and sperm are methylated differently, it *can* matter. In mammals, there are about 100 genes for which it matters (International Human Epigenome Consortium).[5] In these cases, the chromosomes from the male and the female are not equivalent; only the sperm-derived or only the egg-derived allele of the gene is expressed. Thus, a severe or lethal condition may arise if a mutant allele is derived from one parent, but that same mutant allele will have no deleterious effects if inherited from the other parent. In some of these cases, the nonfunctioning gene has been rendered inactive by DNA methylation. The methyl groups are placed on the DNA during spermatogenesis and oogenesis by a series of enzymes that first take the existing methyl groups off the chromatin and then place new sex-specific ones on the DNA (Ciccone et al. 2009; Gu et al. 2011). (See **Further Development 3.7, Mechanisms of DNA Methylation during Genomic Imprinting**, **Further Development 3.8, Poised Chromatin**, **Further Development 3.9, Chromatin Diminution**, and **Further Development 3.10, The Nuclear Envelope's Role in Gene Regulation**, all online.)

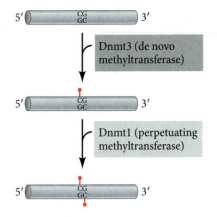

FIGURE 3.13 Two DNA methyltransferases are critically important in modifying DNA. The "de novo" methyltransferase Dnmt3 can place a methyl group on unmethylated cytosines. The "perpetuating" methyltransferase, Dnmt1, recognizes methylated Cs on one strand and methylates the C on the CG pair on the opposite strand.

Transcription factors regulate gene transcription

The science journalist Natalie Angier (1992) wrote that "a series of new discoveries suggests that DNA is more like a certain type of politician, surrounded by a flock of protein handlers and advisers that must vigorously massage it, twist it, and on occasion, reinvent it before the grand blueprint of the body can make any sense at all." These "handlers and advisers" are the transcription factors. During development, transcription factors play essential roles in every aspect of embryogenesis. When in doubt, it is usually a transcription factor's fault, a sentiment often used by politicians as well.

TRANSCRIPTION FACTOR FAMILIES AND OTHER ASSOCIATIONS   Transcription factors can be grouped together in families based on similarities in DNA-binding domains (**TABLE 3.1**). The transcription factors in each family share a common framework in their DNA-binding sites, and slight differences in the amino acids at the binding site can cause the binding site to recognize different DNA sequences. As alluded to previously, transcription factors can regulate gene expression by recruiting histone-modifying enzymes, by stabilizing RNA polymerase activity, and by coordinating the timing of RNA expression for multiple genes.

TRANSCRIPTION FACTORS RECRUIT HISTONE-MODIFYING ENZYMES   As you now know, DNA regulatory elements such as enhancers and silencers function by binding transcription factors, and each element can have binding sites for several transcription factors. Transcription factors bind to the DNA of the regulatory element using one site on the protein and other sites to interact with other transcription factors and proteins, which can serve to recruit histone-modifying enzymes. For example, when MITF (see Table 3.1), a transcription factor essential for ear development and pigment production, binds to its specific DNA sequence, it also binds to a histone acetyltransferase that facilitates the dissociation of nucleosomes, allowing for transcription (Ogryzko et al. 1996; Price et al. 1998).[6] Pax7 is yet another transcription factor that leads to the regulation of chromatin accessibility, but in muscle precursor cells (see Table 3.1). In this case, Pax7 recruits a histone methyltransferase Trithorax complex that methylates the lysine in position four

TABLE 3.1 Some major transcription factor families and subfamilies

Family	Representative transcription factors	Some functions
Homeodomain:		
Hox	Hoxa1, Hoxb2, etc.	Axis formation
POU	Pit1, Unc-86, Oct2	Pituitary development; neural fate
Lim	Lim1, Forkhead	Head development
Pax	Pax1, 2, 3, 6, 7, etc.	Neural specification; eye and muscle development
Basic helix-loop-helix (bHLH)	MyoD, MITF, daughterless	Muscle and nerve specification; *Drosophila* sex determination; pigmentation
Basic leucine zipper (bZip)	cEBP, AP1, MITF	Liver differentiation; fat cell specification
Zinc-finger:		
Standard	WT1, Krüppel, Engrailed	Kidney, gonad, and macrophage development; *Drosophila* segmentation
Nuclear hormone receptors	Glucocorticoid receptor, estrogen receptor, testosterone receptor, retinoic acid receptors	Secondary sex determination; craniofacial development; limb development
Sry-Sox	Sry, SoxD, Sox2	Bend DNA; mammalian primary sex determination; ectoderm differentiation
MADS-box	Classes A, B, C, D, E	Floral organ identity

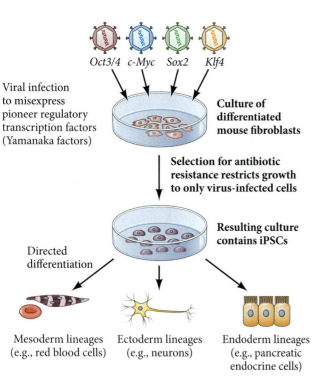

Cells from many lineages can arise from induced pluripotent stem cells

FIGURE 3.14 From differentiated fibroblast to induced pluripotent stem cell with four transcription factors. If the "Yamanaka factors" (the *Oct3/4*, *c-Myc*, *Sox2*, and *Klf4* transcription factor genes) are virally inserted into differentiated fibroblasts, these cells will dedifferentiate into *induced* pluripotent stem cells (iPSCs). Like embryonic stem cells, iPSCs can give rise to progeny of all three germ layers (mesoderm, ectoderm, and endoderm).

of histone H3 (H3K4), resulting in the *tri*methylation of this lysine and the *activation* of transcription (Adkins et al. 2004; Li et al. 2007; McKinnell et al. 2008).

TRANSCRIPTION FACTORS STABILIZE POLYMERASE In addition to recruiting histone-modifying enzymes, transcription factors can also regulate gene expression by stabilizing the transcription pre-initiation complex that enables RNA polymerase II to bind to the promoter (see Figures 3.6 and Figure 1 in Further Development 3.6, online). For instance, MyoD (see Table 3.1), a transcription factor that is critical for muscle cell development, stabilizes TFIIB, which supports RNA polymerase II at the promoter site (Heller and Bengal 1998).

SOME TRANSCRIPTION FACTORS COORDINATE THE TIMED EXPRESSION OF MULTIPLE GENES The simultaneous expression of many cell-specific genes can be explained by the binding of key transcription factors to multiple enhancer elements. For example, many different genes that are specifically activated in the lens contain an enhancer that binds Pax6. Each of these different lens genes may have all the other required transcription factors assembled at their enhancers and poised to activate gene expression, but no transcription will occur until Pax6 binds—thus coordinating the expression for many lens genes simultaneously (for other examples, see Davidson 2006). (See **Further Development 3.11, Transcription Factor Domains**, online.)

PIONEER TRANSCRIPTION FACTORS: BREAKING THE SILENCE Finding an enhancer is not easy because the DNA is usually so wound up that the enhancer sites are not accessible. Given that the enhancer might be covered by nucleosomes, how can a transcription factor find its binding site? That is the job of certain transcription factors that penetrate repressed chromatin and bind to their enhancer DNA sequences (Cirillo et al. 2002; Berkes et al. 2004). They have been called **pioneer transcription factors**, and they can uniquely bind DNA tightly wrapped around nucleosomes and heterochromatic areas (Iwafuchi-Doi 2019). As their name implies, pioneer transcription factors are the first to begin the process of making a locus available for transcription. They also appear to be critical in specifying certain cell lineages. For instance, FoxA1 is a known pioneer transcription factor that is extremely important in specifying liver cell development. FoxA1 binds to certain liver-promoting enhancers and opens up the chromatin to allow other transcription factors access to the promoter (Lupien et al. 2008; Smale 2010). Moreover, FoxA1 remains bound to the DNA even during mitosis, to provide a mechanism to reestablish normal transcription in presumptive liver cells (Zaret et al. 2008). Pax7 is also considered a pioneer transcription factor; as mentioned earlier, it supports muscle specification by recruiting a histone methyltransferase Trithorax complex that activates transcription (McKinnell et al. 2008). (See **Further Development 3.12, Insulators: Protecting Genomic Areas from Transcription Factor Binding**, online.)

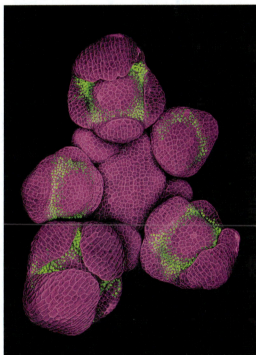

FURTHER DEVELOPMENT

PIONEER TRANSCRIPTION FACTORS WITH THE POWER TO REPROGRAM CELL IDENTITY
Recall the original cloning experiment by John Gurdon (1962), in which placing a somatic cell nucleus into an enucleated egg *reprogrammed* how the transplanted somatic cell's genome was used—in that case, it directed the full development of an adult frog! Although this experiment provided the first significant support for genomic equivalence, it did not show what proteins in the egg cytoplasm were responsible for the reprogramming. Clues came in 2006, when Shinya Yamanaka compiled a list of genes implicated in maintaining cells of the early mouse embryo in an immature state. These immature cells were from the inner cell mass of the blastocyst (the mammalian equivalent of the amphibian's blastula stage) (see Chapter 5). Yamanaka's lab experimentally expressed only four of these genes (*Oct3/4, Sox2, c-Myc,* and *Klf4*) in differentiated mouse fibroblasts and found that the fibroblasts *dedifferentiated* into inner cell mass-like cells (**FIGURE 3.14**; Takahashi and Yamanaka 2006). Such cell fate reprogramming power may reside mostly with Sox2, Oct3/4, and Klf4, as they can access and bind to closed chromatin, thus functioning as pioneer transcription factors (Soufi et al. 2012). The dedifferentiated cells have since been shown in culture to be able to give rise to any cell type of the embryo. This shows they are pluripotent, and because they were induced to this state, they are called **induced pluripotent stem cells** (**iPSCs**). Yamanaka shared the 2012 Nobel Prize in Physiology or Medicine with Gurdon for their discoveries, and iPSCs are now being used to study human development and disease in ways never before possible (further discussed in Chapter 5). (See **Further Development 3.13, Transcription Factors with the Power to Cure Diabetes**, online.)

THE ABCS OF THE TRANSCRIPTIONAL REGULATION OF FLORAL ORGAN IDENTITY GENES
In order for a plant to produce a flower, the genetic program of a shoot apical meristem (SAM) must change from producing leaves to generating the reproductive parts of the flower. Polycomb proteins repress the production of flowers, and Trithorax proteins relieve this repression. But once the repression of floral development is relieved, how are the different parts of the flower specified? *"When in doubt…"*

The genes promoting floral organ identity code for **MADS-box transcription factors**—a family of proteins found in diverse groups of eukaryotes that all share a conserved motif in their DNA-binding domains. In flowering plants, there are five classes of these genes involved in floral organ specification: A, B, C, D, and E (**FIGURE 3.15**; see Chapter 6). These floral

? Developing Questions

The precise binding of transcription factors to *cis*-regulatory elements drives differential gene expression both spatially and temporally in the developing embryo. Is a cell's identity determined by one transcription factor complex binding to one regulatory element, leading to the expression of one gene? How many genes are required to establish a specific cell's fate?

FIGURE 3.15 A fluorescence image of a living *Arabidopsis thaliana* inflorescence meristem labeled with propidium iodide for all cell membranes (magenta) and expressing GFP fused to a class B gene (*APETALA3*; green). You may recognize this particular image as it graces the cover of this book. In this image, the prominent bulges around the periphery are developing flowers, and the green fluorescence highlights the regions that will become petals and stamens (whorls 2 and 3) in each flower. The budding masses of cells within each flower are the developing floral organs.

Courtesy of Nathanaël Prunet

organ identity genes determine the identity of each organ of the flower by being expressed in different combinations.

Looking carefully at the arrangement of floral organs—the carpels, stamens, petals, and sepals—you can see that they are organized into four **whorls**, concentric rings of tissue that surround the apex of the floral meristem (**FIGURE 3.16**). The expression of class A along with class E genes induces the first whorl to differentiate as sepals; the second whorl is induced to become petals by the expression of class A, B, and E genes; the third whorl is induced to form stamens by the expression of class B, C, and E genes; and the fourth whorl is induced to form carpels by the expression of class C and E genes (see Figure 3.16). Remarkably, the wide variety of flower shapes found among the angiosperms are all generated by the overlapping and/or combinatorial interactions of this relatively small number of floral organ identity genes (Theißen et al. 2016).

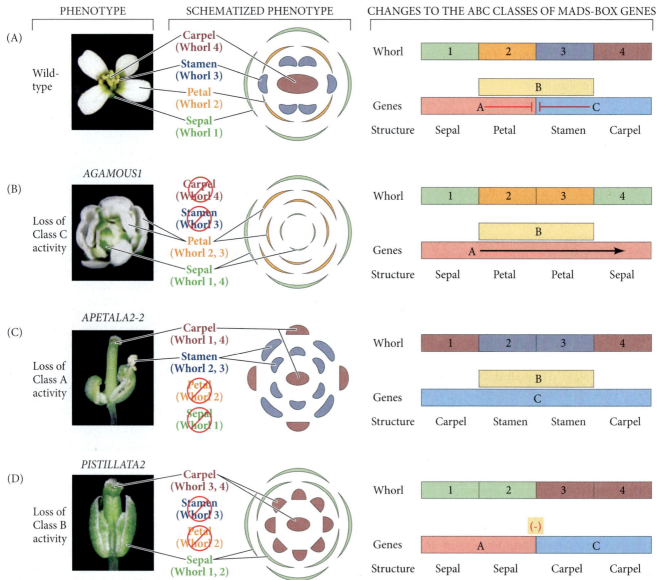

FIGURE 3.16 The ABCs of flower identity. See text for details. Note that class D genes are expressed in ovules, and that class E genes are expressed in all whorls. To simplify, these two classes of genes are not shown here. (After L. Taiz et al. *Plant Physiology and Development*, 6th Edition. Sinauer Associates: Sunderland, MA.)

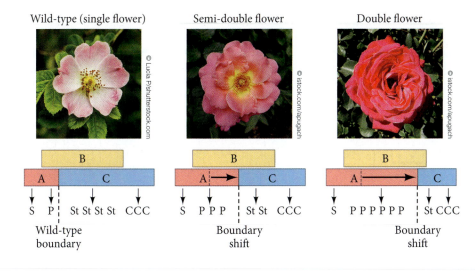

FIGURE 3.17 A + B − C = beautiful double roses. (Left) Picture of a wild-type single rose with its schematized ABC gene expression pattern. (Middle and right) Cultivated semi-double (middle) and double (right) roses were bred through rounds of selection that reduced the expression domain of the rose ortholog of *AGAMOUS1* (a class C gene), thereby enabling expansion of the zone of class A gene expression. This resulted in more petals and fewer stamens. Floral organ abbreviations: S, sepal; P, petal; St, stamen; C, carpel. (Adapted from A. Dubois et al. 2010. *PLOS One* 5: e9288/CC BY 4.0.)

Importantly, while class A and class C genes function to induce sepals and carpels, respectively, they also actively repress the expression of each other (see Figure 3.16A, right). This cross repression of genes that controls alternate cell fates reinforces the boundaries between the two tissues and is a common mechanism you will see repeated throughout plant and animal embryonic development.

The protein products of these floral organ identity genes—the MADS-box transcription factors—function as homeotic regulators of cell fate. Loss of function of class A, B, or C genes causes floral **homeotic transformations**, or the replacement of one structure by another, without affecting the initiation of flower development (see Figure 3.16B–D). For instance, loss of the class C gene *AGAMOUS1* in *Arabidopsis thaliana* results in stamens and carpels being replaced by petals and sepals. In contrast, loss of *APETALA2-2*, a class A gene, causes duplicated carpels and stamens at the expense of petals and sepals. In their pursuit of novel blossom shapes, flower breeders have unintentionally selected for mutations in the regulatory regions that control the expression domains of the different classes of MADS-box genes. The abundant petals found on many cultivated roses today are due to selection of mutations that change the expression domain of class C genes such as *AGAMOUS1* (**FIGURE 3.17**; Dubois et al. 2010). Thus, even small tweaks in the expression patterns of this small number of pioneer transcription factors in plants can contribute to the diversity of flower shapes seen today.

The gene regulatory network: Defining an individual cell

At this point in the chapter, we hope it is clear that different cell types are the result of differentially expressed genes. Although pioneer transcription factors are necessary to initiate the process, they are not sufficient for implementing an entire genomic program on their own.

Studies on sea urchin development have begun to demonstrate ways in which DNA can be regulated to specify cell type and direct morphogenesis of the developing organism. Eric Davidson's group has pioneered a network model approach in which they envision *cis*-regulatory elements (such as promoters and enhancers) in a logic circuit connected by transcription factors (**FIGURE 3.18**; see http://sugp.caltech.edu/endomes; Davidson and Levine 2008; Oliveri et al. 2008). The network receives its first inputs from maternal transcription factors in the egg cytoplasm; from then on, the network self-assembles through (1) the ability of the maternal transcription factors to recognize *cis*-regulatory elements of particular genes that encode other transcription factors and (2) the ability of this new set of transcription factors to activate paracrine signaling pathways that activate or inhibit specific transcription factors in neighboring cells (see Figure 3.18A). The studies show the regulatory logic by which the genes of the sea urchin interact to specify and generate characteristic cell types. This set of interconnections among genes specifying cell types is referred

(?) Developing Questions

How can one actually determine the GRN for a single cell? Doing so is both a conceptual and a technical challenge. The number of genes turned on and off (never mind differences in rates and amounts of expression) in a given GRN is staggeringly immense. Obtaining an accurate assessment of expressed genes is the first hurdle, techniques for which are discussed later in this chapter. These genes then have to be organized into a logic network based on experimentally determined functions. What can be learned by comparing the GRNs of cells from different regions of an embryo, or at different developmental stages, or under different conditions, or even from different species? The answer to these questions is, lots—and that represents the new frontier of developmental genetics.

(A)

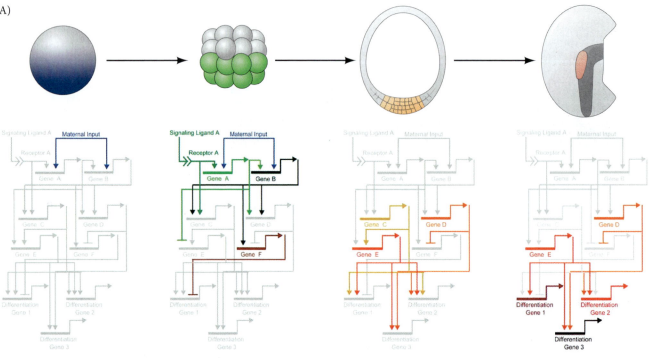

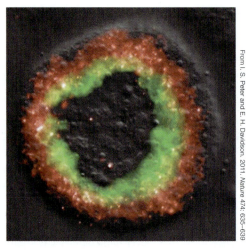

(B)

From I. S. Peter and E. H. Davidson. 2011. *Nature* 474: 635–639

FIGURE 3.18 Gene regulatory networks of endodermal lineages in the sea urchin embryo. (A) Schematics of the sea urchin embryo across four developmental stages showing the progressive specification of endodermal cell fates (top) and the corresponding gene regulatory model of this specification, from maternal contributions and signals to pioneer transcription factors leading to the final differentiation genes (bottom). (B) Double fluorescent in situ hybridization at 24 hours postfertilization showing the restricted expression of *hox11/13b* only in veg1-derived cells (red), whereas *foxa* expression is in the veg2-derived cells (green). (A after V. F. Hinman and A. M. Cheatle Jarvela. 2014. *Genesis* 52: 193–207.)

to as a **gene regulatory network** (**GRN**), a term first coined by Davidson's group. *Therefore, each cell lineage, each cell type, and likely each individual cell can be defined by the GRN that it possesses at that moment in time.*

> *Embryonic development is an enormous informational transaction, in which DNA sequence data generate and guide the system-wide deployment of specific cellular functions.*
>
> E. H. Davidson (2010)

Mechanisms of Differential Gene Expression: *Pre-messenger RNA Processing*

The regulation of gene expression is not confined to the differential transcription of DNA. Even if a particular RNA transcript is synthesized, there is no guarantee that it will create a functional protein in the cell. To become an active protein, the pre-mRNA must be (1) processed into messenger RNA by the removal of introns, (2) translocated from the nucleus to the cytoplasm, and (3) translated by the protein-synthesizing apparatus.

Differential pre-mRNA processing refers to the **splicing** (cutting, rearranging, and ligating back together) of the pre-mRNA precursor into separate messages that specify different proteins by using different combinations of potential exons. If a pre-mRNA precursor had five potential exons, one cell type might use exons 1, 2, 4, and 5; a different cell type might splice exons 1, 3, and 5 together; and yet another cell type might use all five (**FIGURE 3.19**). Thus, a single gene can produce an entire family of proteins. The different proteins encoded by the same gene are called **splicing isoforms** of the protein.

Creating families of proteins through alternative pre-mRNA splicing

Alternative pre-mRNA splicing refers to the molecular mechanism that enables the production of a wide variety of proteins from the same gene. In plants, it appears to be a means of regulating environmental fitness. For example, in *Arabidopsis thaliana* the transcripts from *FLOWERING LOCUS M* (*FLM*) can be differentially spliced in response to ambient temperature, giving rise to two different proteins that may help control the timing of flowering (Shang et al. 2017). In vertebrates, most genes make pre-mRNAs that are alternatively spliced (Wang et al. 2008; Nilsen and Graveley 2010).[7] *Recognizing* a sequence of

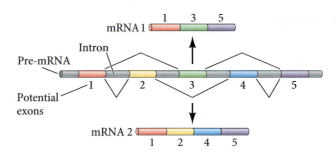

FIGURE 3.19 Differential pre-mRNA processing. By convention, splicing paths are shown by fine V-shaped lines. Differential splicing can process the same pre-mRNA into different mRNAs by selectively using different exons.

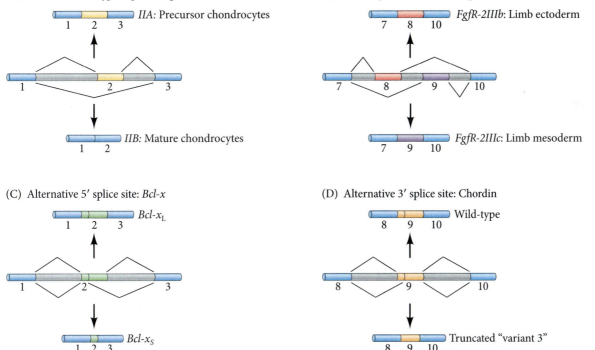

FIGURE 3.20 Some examples of alternative pre-mRNA splicing. Gray portions of the bars represent introns; all other colors represent exons. Alternative splicing patterns are shown with V-shaped lines. (A) A "cassette" (yellow) that can be used as an exon or removed as an intron distinguishes the type II collagen types of chondrocyte precursors and mature chondrocytes (cartilage cells). (B) Mutually exclusive exons distinguish fibroblast growth factor receptors found in the limb ectoderm from those found in the limb mesoderm. (C) Alternative 5' splice site selection, such as that used to create the large and small isoforms of the protein Bcl-X. (D) Alternative 3' splice sites are used to form the normal and truncated forms of Chordin. (After A. N. McAlinden et al. 2004. *Birth Def Res C* 72: 51–68.)

pre-mRNA as either an exon or an intron is a crucial first step in alternative pre-mRA splicing, which can occur in several ways. Most genes contain **consensus sequences** at the 5′ and 3′ ends of the introns. These sequences are the "splice sites" of the intron. The splicing of pre-mRNA is mediated through complexes known as **spliceosomes** that bind to the splice sites. Spliceosomes are made up of small heterogeneous nuclear RNAs and proteins called **splicing factors** that bind to splice sites or to the areas adjacent to them. By their production of specific splicing factors, cells can differ in their ability to recognize a sequence as an intron. That is to say, a sequence that is an *exon* in one cell type may be an *intron* in another (**FIGURE 3.20A,B**). In other instances, the factors in one cell might recognize different 5′ sites (at the beginning of the intron) or different 3′ sites (at the end of the intron; **FIGURE 3.20C,D**).

In some instances, alternatively spliced mRNAs yield proteins that play similar yet distinguishable roles in different cells. For instance, the WT1 isoform without the extra exon functions as a transcription factor during kidney development, whereas the isoform containing the extra exon appears to be critical in testis development (Hammes et al. 2001; Hastie 2001).

How a gene gets spliced may make the difference between the life and death of a cell. The *Bcl-x* gene undergoes alternative splicing to produce either a large or a small protein. If a particular DNA sequence is used as an exon, the "large Bcl-X protein," or Bcl-X_L, is made (see Figure 3.20C). This protein inhibits programmed cell death. If this sequence is seen as an intron and is spliced out, however, the "small Bcl-X protein" (Bcl-X_S) is made, and this protein *induces* cell death. Many tumors have a higher than normal amount of Bcl-X_L (Akgul et al. 2004; Kędzierska et al. 2017). (See **Further Development 3.14, The Dscam Gene and Its 38,016 Forms, Further Development 3.15, Control of Early Development by pre-mRNA Selection, Further Development 3.16, So You Think You Know What a Gene Is?,** and **Further Development 3.17, Splicing Enhancers and Recognition Factors**, all online.)

Mechanisms of Differential Gene Expression: mRNA Translation

The splicing of pre-mRNA is intimately connected with its export through the nuclear pores and into the cytoplasm. As the introns are removed, specific proteins bind to the spliceosome and attach the spliceosome-RNA complex to nuclear pores (Luo et al. 2001; Strässer and Hurt 2001). The proteins coating the 5′ and 3′ ends of the mRNA also change. The nuclear cap binding protein at the 5′ end is replaced by *eukaryotic initiation factor 4E* (*eIF4E*), and the polyA tail becomes bound by the cytoplasmic polyA binding protein. Although both of these changes facilitate the initiation of translation, there is no guarantee that the mRNA will be translated once it reaches the cytoplasm. The control of gene expression at the level of translation can occur by many means; some of the most important of these are described below.

Differential mRNA longevity

The longer an mRNA persists, the more protein can be translated from it. If a message with a relatively short half-life were selectively stabilized in certain cells at certain times, it would make large amounts of its particular protein only at those times and places.

The stability of a message often depends on the length of its polyA tail. The length, in turn, depends largely on sequences in the 3′ untranslated region, certain ones of which allow longer polyA tails than others. If these 3′ UTRs are experimentally traded, the half-lives of the resulting mRNAs are altered: the previously long-lived messages will now decay rapidly, whereas the normally short-lived mRNAs will remain around longer (Shaw and Kamen 1986; Wilson and Treisman 1988; Decker and Parker 1995). In some instances, mRNAs are selectively stabilized at specific times in specific cells. In the development of the nervous system, a set of RNA-binding proteins called **Hu proteins** (HuA, HuB, HuC, and HuD) stabilizes two groups of mRNAs that would otherwise perish quickly (Perrone-Bizzozero and Bird 2013). One group of target mRNAs encodes proteins that stop neuronal precursor cells from dividing, and the second group of mRNAs encodes proteins that initiate neuronal differentiation (Okano and Darnell 1997; Deschênes-Furry et al. 2006, 2007). Thus, once the Hu proteins are made, the neuronal precursor cells can become neurons.[8]

Stored oocyte mRNAs: Selective inhibition of mRNA translation

Some of the most remarkable cases of translational regulation of gene expression occur in the oocyte. Prior to meiosis, while the oocyte is still within the ovary, the oocyte often makes and stores mRNAs that will be used only after fertilization occurs. These messages stay in a dormant state until they are activated by ion signals (discussed in Chapters 6 and 7) that spread through the egg during ovulation or fertilization.

Some of these stored mRNAs encode proteins that will be needed during cleavage, when the embryo makes enormous amounts of chromatin, cell membranes, and cytoskeletal components. These maternal mRNAs include the messages for histone proteins, the transcripts for the actin and tubulin proteins of the cytoskeleton, and the mRNAs for the cyclin proteins that regulate the timing of early cell division (Raff et al. 1972; Rosenthal et al. 1980; Standart et al. 1986). The stored mRNAs and proteins are referred to as **maternal contributions** (produced from the maternal genome), and in many species (including sea urchins and zebrafish), maintenance of the normal rate and pattern of early cell divisions does not require DNA —or even a nucleus! Rather, it requires continued protein synthesis from the stored maternally contributed mRNAs (**FIGURE 3.21**; Wagenaar and Mazia 1978; Dekens et al. 2003). Stored mRNA also encodes proteins that determine the fates of cells. They include the *bicoid, caudal,* and *nanos* messages that provide information in the *Drosophila* embryo for the production of its head, thorax, and abdomen. So at some point, each of us should give a shout-out to our biological moms for giving us those transcripts early on (and to all moms for all they do for us well after oogenesis).

Most translational regulation in oocytes is negative because the "default state" of the maternal mRNA is to be *available* for translation but not actively translated. Therefore, there must be inhibitors preventing the translation of these mRNAs in the oocyte, and these inhibitors must somehow be removed at the appropriate times around fertilization. The 5′ cap and the 3′ UTR seem especially important in regulating the accessibility of mRNA to ribosomes. If the 5′ cap is not made or if the 3′ UTR lacks a polyA tail, the message will probably not be translated. The oocytes of many species have "used these ends as means" to regulate the translation of their mRNAs. For instance, the oocyte of the tobacco hornworm moth makes some of its mRNAs without their methylated 5′ caps. In this state, they cannot be efficiently translated. At fertilization, however, a methyltransferase completes the formation of the caps, and these mRNAs can then be translated (Kastern et al. 1982). (See **Further Development 3.18, Translational Regulation in Frogs and Flies**, online.)

Ribosomal selectivity: Selective activation of mRNA translation

It has long been assumed that ribosomes do not show favoritism toward translating certain mRNAs. After all, eukaryotic messages can be translated even by *E. coli* ribosomes, and ribosomes from immature red blood cells have long been used to translate mRNAs from any source. However, evidence has shown that ribosomal proteins are not

(A) Wild-type (B) *Futile cycle* mutant

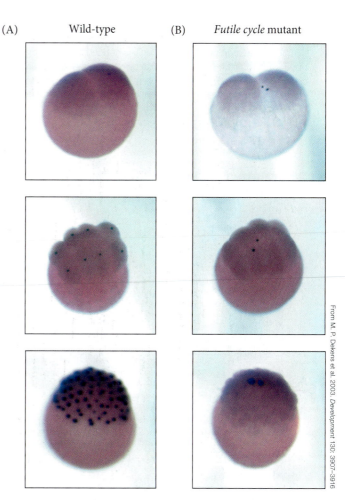

From M. P. Dekens et al. 2003. *Development* 130: 3907–3916

FIGURE 3.21 Maternal contributions to DNA replication in the zebrafish blastula. (A) Wild-type blastulae show BrdU-labeled nuclei (blue) in all cells. (B) Although the correct number of cells is present in *futile cycle* mutants, they consistently show only two labeled nuclei, indicating that these mutants fail to undergo pronuclear fusion. Even in the absence of any zygotic DNA, early cleavages progress perfectly well due to the presence of maternal contributions. However, *futile cycle* mutants arrest at the onset of gastrulation.

FIGURE 3.22 Model of ribosomal heterogeneity in mice. (A) Ribosomes have slightly different proteins depending on the tissue in which they reside. Ribosomal protein Rpl38 (i.e., protein 38 of the large ribosomal subunit) is concentrated in those ribosomes found in the somites that give rise to the vertebrae. (B) A wild-type embryo (left) has normal vertebrae and normal Hox gene translation. Mice deficient in Rpl38 have an extra pair of vertebrae, tail deformities, and reduced Hox gene translation. (After N. Kondrashov et al. 2011. *Cell* 145: 383–397.)

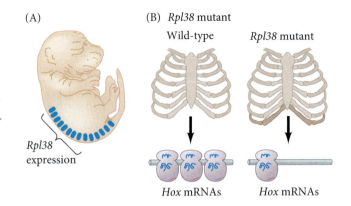

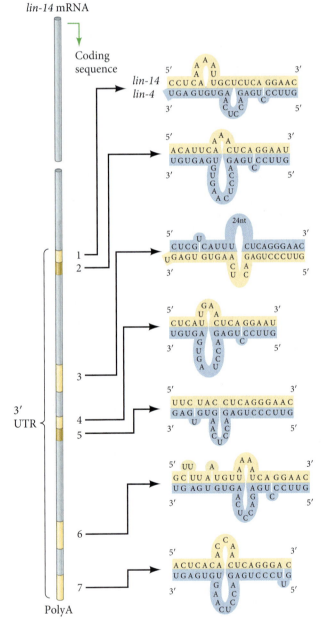

FIGURE 3.23 Model of the regulation of *lin-14* mRNA translation by *lin-4* RNAs. The *lin-4* gene does not produce an mRNA. Rather, it produces small RNAs that are complementary to a repeated sequence in the 3′ UTR of the *lin-14* mRNA, which bind to it and prevent its translation. (After M. Wickens and K. Takayama. 1995. *Nature* 367: 17–18; B. Wightman et al. 1993. *Cell* 75: 855–862.)

the same in all cells and that some ribosomal proteins are necessary for translating certain messages. When Kondrashov and colleagues (2011) mapped the gene that causes numerous axial skeleton deformities in mice, they found that the mutation was not in one of the well-known genes that control skeletal polarity. Rather, it was in ribosomal protein Rpl38. When this protein is mutated, the ribosomes can still translate most messages, but the ribosomes in the skeletal precursors cannot translate the mRNA from a specific subset of Hox genes. The Hox transcription factors, as we will see in Chapters 12 and 17, specify the type of vertebrae at each particular axial level (ribbed thoracic vertebrae, unribbed abdominal vertebrae, etc.). Without functioning Rpl38, vertebral cells are unable to form the initiation complex with mRNA from the appropriate Hox genes, and the skeleton is deformed (**FIGURE 3.22**). Mutations in other ribosomal proteins have also been found to produce deficient phenotypes (Terzian and Box 2013; Watkins-Chow et al. 2013).

microRNAs: Specific regulation of mRNA translation and transcription

If proteins can bind to specific nucleic acid sequences to block transcription or translation, you would think that RNA would do the job even better. After all, RNA can be made specifically to complement and bind a particular sequence. Indeed, one of the most efficient means of regulating the translation of a specific message is to make a small *antisense* RNA complementary to a portion of a particular transcript. Such a naturally occurring antisense RNA was first seen in the roundworm *C. elegans* (Lee et al. 1993; Wightman et al. 1993). Here, the *lin-4* gene was found to encode a 21-nucleotide RNA that bound to multiple sites in the 3′ UTR of the *lin-14* mRNA (**FIGURE 3.23**). The *lin-14* gene encodes the LIN-14 transcription factor, which is important during the first larval phase of *C. elegans* development. It is not needed afterward, and *C. elegans* is able to inhibit synthesis of LIN-14 from these messages by the small *lin-4* antisense RNA. The binding of these *lin-4* transcripts to the *lin-14* mRNA 3′ UTR causes degradation of the *lin-14* message (Bagga et al. 2005).

The *lin-4* RNA is now thought to be the "founding member" of a very large group of **microRNAs (miRNAs)**. Computer analysis of the human genome predicts that we have more than 1000 miRNA loci and that these miRNAs probably modulate 50% of the protein-coding genes in our bodies (Berezikov and Plasterk 2005; Friedman et al. 2009). These miRNAs usually contain only 22 nucleotides and are made from longer precursors. These precursors can be in independent transcription units (the *lin-4* gene is far apart from the *lin-14* gene), or they can reside in a gene's own introns (Aravin et al. 2003; Lagos-Quintana et al. 2003).

The initial RNA transcript (which may contain several repeats of the miRNA sequence) forms hairpin loops wherein the RNA finds complementary structures within its strand. Because short double-stranded RNA molecules can resemble pathogenic viral genomes, the cell has a natural mechanism to both recognize these structures and use them as guides for their eradication (Wilson and Doudna 2013). Interestingly, this protective mechanism has been co-opted to be used as yet another way that the cell can differentially regulate the expression of endogenous genes. The process by which miRNAs inhibit expression of specific genes by degrading their mRNAs is called **RNA interference** (**RNAi**) (Guo and Kemphues 1995; Sen and Blau 2006; Wilson and Doudna 2013), the characterization of which garnered Andrew Fire and Craig Mello the Nobel Prize in Physiology or Medicine in 2006 (Fire et al. 1998).

FURTHER DEVELOPMENT

THE MECHANISM OF RNAi   The miRNA double-stranded stem-loop structures are processed by a set of RNases (Drosha and Dicer) to make single-stranded microRNA (**FIGURE 3.24**). The microRNA is then packaged with a series of proteins to make an **RNA-induced silencing complex** (**RISC**). Proteins of the Argonaute family are particularly important members of this complex. These small regulatory RNAs can bind to the 3′ UTR of messages and inhibit their translation. The binding of microRNAs and their associated RISCs to the 3′ UTR can regulate translation in two ways (Filipowicz et al. 2008; see also Bartel 2004; He and Hannon 2004). First, this binding can block initiation of translation, preventing the binding of initiation factors or ribosomes. The Argonaute proteins, for instance, have been found to bind directly to

FIGURE 3.24   Model for RNA interference from double-stranded RNA (dsRNA) and miRNA. Double-stranded siRNA (short interfering RNA) or miRNA that is added to a cell or produced through transcription and processed by the Drosha RNAase (1) will interact with the RNA-induced silencing complex (RISC), made up primarily of Dicer and Argonaute, that prepares the RNA to be used as a guide for targeted mechanisms of interference. Specifically, (1) transcription of siRNA or miRNA forms several hairpin regions where the RNA finds nearby complementary bases with which to pair. The primary-miRNA (pri-miRNA) is processed into individual pre-miRNA "hairpins" by the Drosha RNase (as are the siRNAs), and they are exported from the nucleus. (2–4) Once in the cytoplasm, these double-stranded RNAs are recognized by and form the RISC complex with Argonaute and the RNase Dicer. (5) Dicer also acts as a helicase to separate the strands of the double-stranded RNA. (6) One strand (probably recognized by the placement of Dicer) is used to bind to the 3′ UTRs of target mRNAs to block translation or to induce cleavage of the target transcript, depending (at least in part) on the strength of the complementarity between the miRNA and its target. siRNA is best known for the targeting of transcript degradation. dsRBD and dsRBP are abbreviations for double-stranded RNA binding domain and protein, respectively. The gray regions of Dicer and Argonaut are other domains. (After L. He and G. J. Hannon. 2004. *Nat Rev Genet* 5: 522–531; R. C. Wilson and J. A. Doudna. 2013. *Annu Rev Biophys* 42: 217–239.)

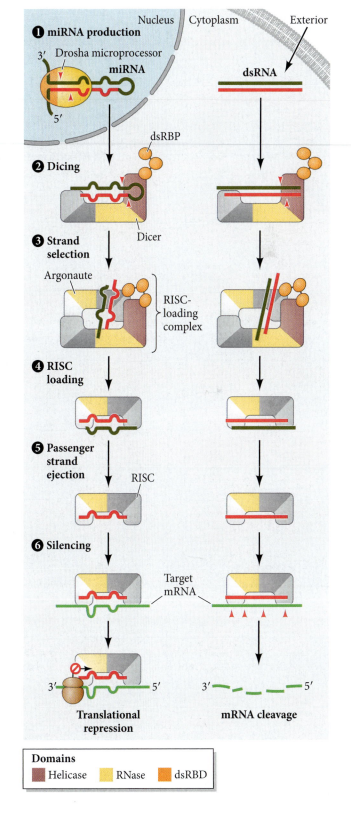

the methylated guanosine cap at the 5′ end of the mRNA (Djuranovic et al. 2010, 2011). Second, this binding can recruit endonucleases that digest the mRNA, usually starting with the polyA tail (Guo et al. 2010). The latter seems to be commonly used in mammalian cells.

MIRNAS AND THE MATERNAL-TO-ZYGOTIC TRANSITION   MicroRNAs can be used to "clean up" and fine-tune the level of gene products. We mentioned those maternal RNAs in the oocyte that allow early development to occur. How does the embryo get rid of maternal RNAs once they have been used and the embryonic cells are making their own mRNAs? In zebrafish, this cleanup operation is assigned to microRNAs such as *miR430*. That is one of the first genes transcribed by the fish embryonic cells, and there are about 90 copies of this gene in the zebrafish genome. So the level of *miR430* goes up very rapidly. This microRNA has hundreds of targets (about 40% of the maternal RNA types), and when it binds to the 3′ UTR of these target mRNAs, these mRNAs lose their polyA tails and are degraded (**FIGURE 3.25**; Giraldez et al. 2006; Giraldez 2010). In addition, *miR430* represses initiation of translation prior to promoting mRNA decay (Bazzini et al. 2012). (See **Further Development 3.19, Learn How a Mutation in a 3′ UTR Results in Bulging Biceps in Beef**, online.)

Control of RNA expression by cytoplasmic localization

Not only is the timing of mRNA translation regulated, but so is the place of RNA expression. A majority of mRNAs (about 70% in *Drosophila* embryos) are localized to specific places in the cell (Lécuyer et al. 2007). Just like the selective repression of mRNA translation, the selective localization of messages is often accomplished through their 3′ UTRs. There are three major mechanisms for the localization of an mRNA (see Palacios 2007):

1. *Diffusion and local anchoring.* Messenger RNAs such as *nanos* diffuse freely in the cytoplasm. When they diffuse to the posterior pole of the *Drosophila* oocyte, however, they are trapped there by proteins that reside particularly in these regions (**FIGURE 3.26A**).

(A)

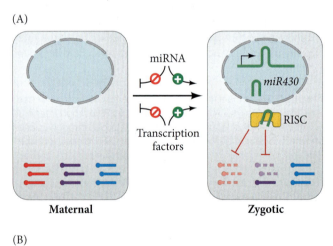

(B)

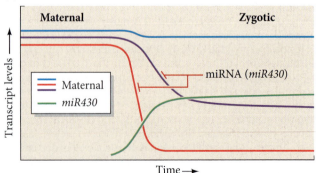

FIGURE 3.25   The role of *miR430* during the maternal-to-zygotic transition in zebrafish. (A) Numerous mRNAs derived from maternal contributions fuel development during the cleavage stages, but transitioning into the gastrula requires active transcription of the zygotic genome. miRNAs play a major role in clearing these maternally derived transcripts during this transition. (B) *miR430* has been discovered to play a major role in the interference of a majority of maternal transcripts in the zebrafish blastula as it transitions to zygotic control during gastrulation. In this graph, the different curves denote the reduction in three specific transcripts, two genes of which (purple and red) are differentially degraded by *miR430* (green). (After A. J. Giraldez. 2010. *Curr Opin Genet Dev* 20: 369–375.)

FIGURE 3.26 Localization of mRNAs. (A) Diffusion and local anchoring. *nanos* mRNA diffuses through the *Drosophila* egg and is bound (in part by the Oskar protein) at the posterior end of the oocyte. This anchoring allows the *nanos* mRNA to be translated. (B) Localized protection. The mRNA for *Drosophila* heat shock protein (Hsp83) will be degraded unless it binds to a protector protein (in this case, also at the posterior end of the oocyte). (C) Active transport on the cytoskeleton, causing the accumulation of mRNA at a particular site. Here, *bicoid* mRNA is transported along microtubules by the motor protein dynein to the anterior of the oocyte. Meanwhile, *oskar* mRNA is brought to the posterior pole by the motor protein kinesin along microtubules. (After I. M. Palacios. 2007. *Semin Cell Dev Biol* 18: 163–170.)

2. *Localized protection.* Messenger RNAs such as those encoding the *Drosophila* heat shock protein Hsp83 float freely in the cytoplasm, but like *nanos* mRNA, *hsp83* mRNA accumulates at the posterior pole. In contrast to *nanos* mRNA, *hsp83* mRNA is degraded everywhere except at the posterior pole, where localized proteins protect the *hsp83* mRNA from being destroyed (**FIGURE 3.26B**).

3. *Active transport along the cytoskeleton.* Active transport is probably the most widely used mechanism for mRNA localization. Here, the 3′ UTR of the mRNA is recognized by proteins that can bind these messages to "motor proteins" that travel along the cytoskeleton to their final destination (**FIGURE 3.26C**). We will see in Chapter 9 that this mechanism is very important for localizing transcription factor mRNAs into different regions of the *Drosophila* oocyte.

(See **Further Development 3.20, Stored Messenger RNA in Brain Cells**, online.)

Mechanisms of Differential Gene Expression: *Posttranslational Protein Modification*

The story is not over when a protein is synthesized. Once a protein is made, it becomes part of a larger level of organization. It may become part of the structural framework of the cell, for instance, or it may become involved in one of the many enzymatic pathways for the synthesis or breakdown of cellular metabolites. In any case, the individual protein is now part of a complex "ecosystem" that integrates it into a relationship with numerous other proteins. Several changes can still take place that determine whether or not the protein will be active and, if so, how it may function.

Some newly synthesized proteins remain inactive until certain inhibitory sections are cleaved away. That is what happens when insulin is made from its larger protein precursor. Some proteins must be "addressed" to their specific intracellular destinations to function. Proteins are often sequestered in certain regions of the cell, such as membranes, lysosomes, nuclei, or mitochondria. Some proteins need to assemble with other proteins to form a functional unit. The hemoglobin protein, the microtubule, and the ribosome are all examples of multiple proteins joining together to form a functional unit. In addition, some proteins are not active unless they bind an ion (such as Ca^{2+}) or are modified by the covalent addition of a phosphate or acetate group. The importance of this type of

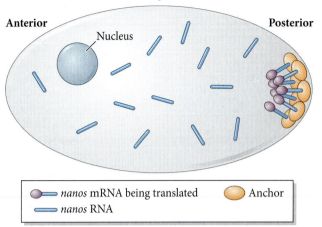

(A) Diffusion and local anchoring

Anterior   Posterior

Nucleus

- *nanos* mRNA being translated   Anchor
- *nanos* RNA

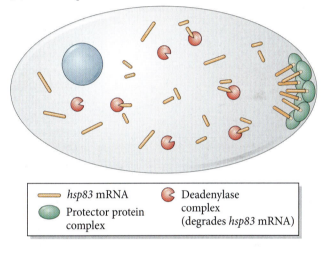

(B) Localized protection

- *hsp83* mRNA
- Protector protein complex
- Deadenylase complex (degrades *hsp83* mRNA)

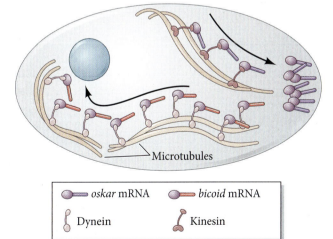

(C) Active transport along cytoskeleton

Microtubules

- *oskar* mRNA   *bicoid* mRNA
- Dynein   Kinesin

protein modification will become obvious in Chapter 4 because many of the critical proteins in embryonic cells just sit there until some signal activates them.

Finally, even when a protein may be actively translated and ready to function, the cell can immediately transport this protein to a complex called a proteasome for degradation. Why would a cell expend energy synthesizing a protein only to degrade it? If a cell needed a protein to function with rapid response at a precise moment in time, the energy expenditure might be worth it. For instance, a neuron extends a long axonal process while searching for its synaptic target in a process called axon guidance (described in Chapter 15). These pathfinding neurons synthesize certain receptor proteins only to immediately degrade them until the cell has reached an environment where a directional guidance decision is required. Signals in this location cause the cell to suspend the receptor degradation, enabling the receptors to be transported to the membrane and immediately function to guide the axon onward toward its target.

FURTHER DEVELOPMENT

The Basic Tools of Developmental Genetics

The untangling of genetic mechanisms in development, as discussed here and elsewhere in the book, depends on a sophisticated set of basic tools—tools for determining the specific time and place a gene is expressed, or the location of a particular mRNA, or where a protein is within a cell. Knowing what these tools are and how they work will greatly expand your understanding of developmental genetics and of biology in general. Techniques to detect transcripts include northern blots, RT-PCR, in situ hybridization, microarray, and next-generation sequencing technology, while western blots and immunocytochemistry enable protein detection. To ascertain the function of genes once their products are located, scientists are using new techniques such as CRISPR/Cas9-mediated knockouts, antisense, RNA interference, **morpholinos** (knockdowns), GAL4/UAS, and Cre-lox systems. Furthermore, ChIP-Seq and CUT&RUN are techniques that allow the identification of proteins that bind to specific DNA sequences. In addition, "high-throughput" RNA analysis by RNA-Seq and whole genome sequencing has enabled researchers to compare thousands of mRNAs, which when paired with computer-aided synthetic techniques can predict interactions between proteins and mRNAs. Descriptions for a majority of these procedures can be found on oup.com/he/barresi12xe. In addition, below are descriptions of some of the techniques most relevant to today's experimental methods. Enjoy using your "toolbox."

Characterizing gene expression

IN SITU HYBRIDIZATION   In **whole mount in situ hybridization**, the entire embryo (or a part thereof) can be stained for certain mRNAs. The main principle is to take advantage of the single-stranded nature of mRNA and introduce a complementary sequence to the target mRNA that enables visualization. This technique uses dyes to allow researchers to look at entire embryos (or their organs) without sectioning them, thereby observing large regions of gene expression next to regions devoid of expression. **FIGURE 3.27A** shows an in situ hybridization targeting mRNA from the *odd-skipped* gene performed on a fixed, intact *Drosophila* embryo. First an mRNA detection probe—the in situ probe—had to be created. The probe is an antisense RNA molecule about 200–2000 base pairs in length that contains uridine triphosphate (UTP) nucleosides conjugated to digoxigenin (**FIGURE 3.27B**). Digoxigenin is a compound made by particular groups of plants and not found in animal cells and, as such, is distinguishable from any other molecules in the animal cell. During the procedure, the embryo is permeabilized by lipid solvents and proteinases so that the probe can get in and out of its cells. Once in the cells, *hybridization* occurs between the probe antisense RNA and

(A)

Courtesy of Berkely *Drosophila* genome project http://insitu.
fruitfly.org/insitu_image_storage/img_dir-36/insitu36057.jpe

(B) DIG-conjugated probe

Antisense RNA complementary
to region of *odd-skipped* gene

(C)

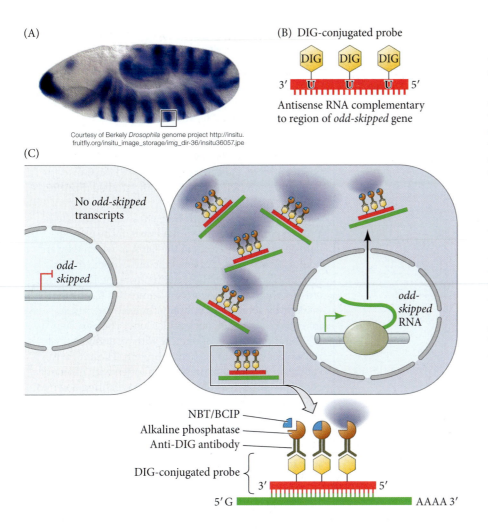

FIGURE 3.27   In situ hybridization.
(A) Whole mount in situ hybridization
for *odd-skipped* mRNA (blue) in a stage-
9 *Drosophila* embryo. (B) Antisense
RNA probe with uridine triphosphate
conjugated to digoxigenin (DIG). (C) Illustration of two cells at the border of the
odd-skipped expression pattern seen in
the box in (A). The cell on the left is not
expressing *odd-skipped*, whereas the
cell on the right is. The antisense DIG-
labeled RNA probe with complementarity to the *odd-skipped* gene becomes
hybridized to any cell expressing *odd-
skipped* transcripts. Following probe
hybridization, samples are treated with
anti-DIG antibodies conjugated to the
enzyme alkaline phosphatase. When
nitroblue tetrazolium chloride (NBT) and
5-Bromo-4-chloro-3-indolyl-phosphate
(BCIP) are then added to the sample,
alkaline phosphatase converts them
to a blue precipitate. Only those cells
expressing *odd-skipped* turn blue.

the targeted mRNA. To visualize the cells in which hybridization has occurred,
researchers apply an antibody that specifically recognizes digoxigenin. This antibody, however, has been artificially conjugated to an enzyme, such as alkaline
phosphatase. After incubation in the antibody and repeated washes to remove all
unbound antibodies, the embryo is bathed in a solution containing a substrate
for the enzyme that can be converted into a colored product by the enzyme. The
active enzyme should be present only where the digoxigenin is present—previous
steps have inactivated any endogenous enzyme in the tissue—so the digoxigenin
should be present only where the specific complementary mRNA is found. Thus,
in **FIGURE 3.27C**, the dark blue precipitate formed by the enzyme indicates the
presence of the target mRNA.

CHROMATIN IMMUNOPRECIPITATION-SEQUENCING   During the twentieth
century, we found the actors in the drama of gene transcription, but not until the
twenty-first century were their scripts discovered. How does one locate the places
on the gene where a particular transcription factor binds or where nucleosomes
with specific modifications are localized? The recent ability to identify protein-
specific DNA-binding sites using chromatin immunoprecipitation-sequencing
(ChIP-Seq), or more recently a variation of ChIP-Seq called CUT&RUN, showed
that there are different types of promoters and that they use different scripts to
transcribe their genes (Johnson et al. 2007; Jothi et al. 2008; Skene and Henikoff
2017). ChIP-Seq is based on two highly specific interactions. One is the binding
of a transcription factor or a modified nucleosome to particular sequences of DNA

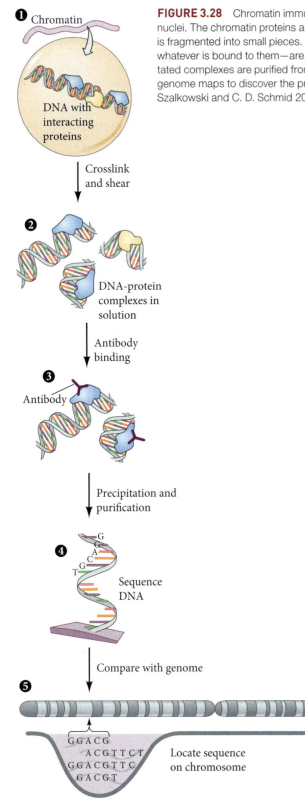

❶ Chromatin

DNA with interacting proteins

Crosslink and shear

❷ DNA-protein complexes in solution

Antibody binding

❸ Antibody

Precipitation and purification

❹ Sequence DNA

Compare with genome

❺

Locate sequence on chromosome

FIGURE 3.28 Chromatin immunoprecipitation-sequencing (ChIP-Seq). Chromatin is isolated from the cell nuclei. The chromatin proteins are crosslinked to their DNA-binding sites, and the DNA, bound to its proteins, is fragmented into small pieces. Antibodies bind to specific chromatin proteins, and the antibodies—with whatever is bound to them—are precipitated out of solution. The DNA fragments associated with the precipitated complexes are purified from the proteins and sequenced. These sequences can be compared with the genome maps to discover the precise locations of the genes these proteins may be regulating. (After A. M. Szalkowski and C. D. Schmid 2011. *Brief Bioinform* 12: 626–633 and Chris Taplin/CC BY-SA 2.0.)

(such as enhancer elements), and the other is the binding of antibody molecules specifically to the transcription factor or modified histone being studied (**FIGURE 3.28**; Liu et al. 2010).

In the first step of ChIP-Seq, chromatin is isolated, and the proteins are crosslinked (usually by glutaraldehyde or formaldehyde) to the DNA to which they are bound. This process prevents the nucleosome or transcription factors from dissociating from the DNA. After crosslinking, the DNA-protein complex is fragmented (usually by sonication, but sometimes by enzymes) into pieces about 500 nucleotides long. The next step is to bind the protein of interest with an antibody that recognizes only that particular protein. The antibody can be precipitated out of solution (often with magnetic beads that bind to antibodies), and it will bring down to the bottom of the test tube any DNA fragments bound by the protein of interest. These DNA fragments, once separated from the protein, are amplified and can be sequenced and mapped to the entire genome. In this way, the DNA sequences bound specifically by particular transcription factors or nucleosomes containing modified histones can be precisely identified.

Recently, a more efficient and sensitive variation of ChIP-Seq was developed called Cleavage Under Targets and Release Using Nuclease (CUT&RUN; Skene and Henikoff 2017). CUT&RUN is different from traditional ChIP-Seq; by avoiding fragmentation and solubilization of chromatin, it seems to yield a much higher resolution and quantitative measure of mapped targets. As you will see throughout this text, whether using ChIP-Seq or CUT&RUN, researchers are identifying important enhancer regions controlling essential and subtle variation during embryonic development. These newly identified enhancers have been extremely useful for generating transgenic reporter constructs and creating transgenic organisms, allowing us to visualize gene expression in living cells and organisms.

DEEP SEQUENCING: RNA-SEQ As emphasized in this chapter, it is the full repertoire of genes expressed by a cell that establishes the gene regulatory network controlling cell identity. Major improvements in sequencing technology have enabled whole genomes to be sequenced, but a genome does not equal the cell's transcriptome (all the RNAs expressed). To move closer to the identification of all the transcripts present in a given embryo, tissue, or even single cell, **RNA-Seq** was developed. RNA-Seq takes advantage of the high throughput capabilities of *next-generation sequencing* technology to sequence and quantify the RNA present in a cell (**FIGURE 3.29**). Specifically, RNA is isolated from samples and converted to complementary DNA (cDNA) with standard procedures using *reverse transcriptase*. This cDNA is broken up into smaller fragments, and known adaptor sequences are added to the ends. These adaptors allow immobilization and PCR-based amplification of these transcripts. Next-generation sequencing can analyze these transcripts for both nucleotide sequence and quantity (Goldman and Domschke 2014). RNA-Seq has been particularly powerful for comparing transcriptomes between identical samples differing only in select experimental parameters. One can ask, How does the array of transcripts differ between tissues located in different regions of the embryo, or the same tissue at different times of development, or the same tissue treated or untreated with a specific compound? The advent of fluorescence activated cell sorting (typically spoken as FACS for short) and microdissection has allowed for the precise isolation of tissues and individual cells, and recent advances in RNA-Seq sensitivity have permitted *transcriptomics*—the study of transcriptomes—of single cells.

A common experimental approach has been to design a targeted deep-sequencing experiment to arrive at a list of genes associated with a given condition. Researchers then use bioinformatics and an understanding of developmental biology to select gene candidates from the list to test the function of these genes in their system.

Testing gene function

Developmental biologists have used an array of methods to mutate genes to determine their functions. These methods fall into two categories: forward genetics and reverse genetics. In **forward genetics**, an organism is exposed to an agent that causes unbiased, random mutations, and the resulting phenotypes are screened for ones that affect development. Individual mutations can be maintained either in homozygotes, or in heterozygotes if the mutation seriously affects survival. The identities of the mutated loci are typically determined only after the initial phenotypic analysis. Two important forward genetics mutagenesis screens were done on *Drosophila* and zebrafish by Christiane Nüsslein-Volhard and colleagues (Nüsslein-Volhard and Wieschaus 1980). These screens have contributed immensely to the identification and functional characterization of many of the genes and pathways we know today to be important in development and disease.

In **reverse genetics**, you start with a gene in mind that you want to manipulate and then either knock down or knock out the expression of that gene. Using RNAi (RNA interference in which small RNAs are used to bind target RNAs) or morpholinos (nucleotide analogues that block transcription start sites or splice sites of target RNAs), you can target a specific gene's transcript for degradation or block its splicing or

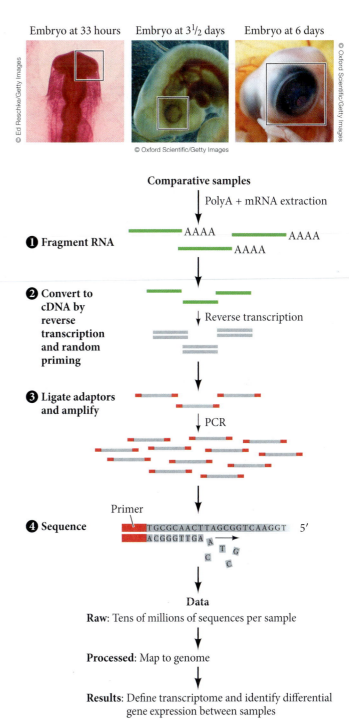

FIGURE 3.29 Deep sequencing: RNA-Seq. (Top) Researchers begin with specific sorts of tissues, often comparing different conditions, such as embryos of different ages (chick embryos, as shown here), isolated tissues (such as the eye; boxed regions) or even single cells, and samples from different genotypes or experimental manipulations. (1) RNA is isolated to obtain only those genes that are actively expressed. (2) These transcripts are then fragmented into smaller stretches and used to create cDNA with reverse transcriptase. (3) Specialized adaptors are ligated to the cDNA ends to enable PCR amplification and immobilization for (4) subsequent sequencing. (After J. H. Malone and B. Oliver. 2011. *BMC Biol* 9: 34.)

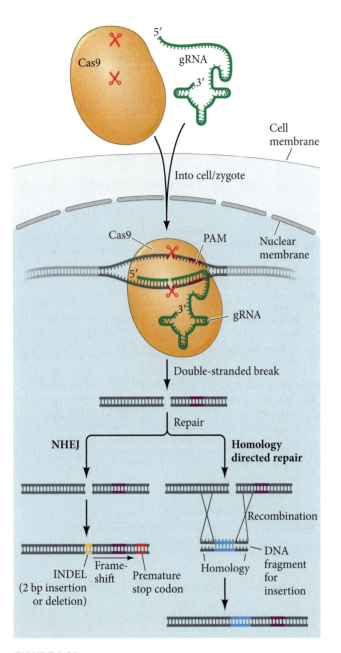

FIGURE 3.30 CRISPR/Cas9-mediated gene editing. The CRISPR/Cas9 system is used to cause targeted indel formation or insertional mutagenesis within a gene of interest. A gene-specific guide RNA (gRNA) is designed and introduced into cells together with the nuclease Cas9, for instance by co-injection into a newly fertilized zygote. The gRNA will bind to the genome with complementarity and will recruit Cas9 to this same location to induce a double-stranded break. Non-homologous end joining (NHEJ) is the cell's DNA repair mechanism that often results in small insertions or deletions (approximately 2–30 base pairs; a 2 base-pair insertion is illustrated here), which can cause the establishment of a premature stop codon and potential loss of the protein's function. In addition, plasmids carrying insertions with homology to regions surrounding the gRNA target sites are used to insert known sequences at the double-stranded break. Such methods are being explored as a way to repair mutations. PAM, protospacer adjacent motif.

translation, respectively (see Figure 3.24). These tools inhibit gene function but not always completely and only for a limited period of time because the double-stranded RNAs (dsRNA) created by RNAi or morpholinos become diluted and degraded over the course of development (hence only a gene "knockdown" and not a "knockout"). Researchers can take advantage of that and use different amounts of dsRNAs or morpholinos to achieve a dose response effect.

Targeted gene knockouts, by contrast, have been notable for completely eliminating the function of targeted genes. Such elimination has been done effectively in the mouse, where researchers have used embryonic stem cells for inserting a DNA construct called a *neomycin cassette* into a specific gene through a process of homologous recombination. This insertion both mutates the gene and, by coding for the antibiotic neomycin, provides an antibiotic selection mechanism for identifying mutated cells. These cells are injected into blastocysts, which develop into chimeric mice in which only some of the cells carry the mutation. These mice are bred to obtain homozygous mutant mice in which there is complete loss of the targeted gene's function.[9]

CRISPR/CAS9 GENOME EDITING The technique of CRISPR/Cas9 genome editing has had an enormous effect on genetic research, making gene editing faster and less expensive than ever before, and making it relatively simple in organisms from *E. coli* to primates (Jansen et al. 2002). This technique uses a system that occurs naturally in prokaryotes for defending against invading viruses (Barrangou et al. 2007). In prokaryotes, CRISPR (clustered regularly interspaced short palindromic repeats) is a stretch of DNA containing short regions that when transcribed into RNA serve as guides (guide RNAs or gRNAs) for recognizing segments of viral DNA. The RNA also binds to an endonuclease called Cas9 (CRISPR associated enzyme 9). When the gRNA binds to viral DNA, the RNA brings Cas9 with it, which catalyzes a double-stranded break in the foreign DNA, disabling the virus.

In 2012, researchers demonstrated that gRNAs and Cas9 can be efficiently used to mutate eukaryotic genes (Jinek et al. 2012). When CRISPR gRNAs specific for a gene are introduced into cells along with Cas9, the Cas9 protein is guided by the gRNA to the gene of interest and causes a double-stranded break in the DNA (**FIGURE 3.30**). Cells will naturally try to repair double-stranded breaks through a process called non-homologous end joining (NHEJ). However, NHEJ is often imperfect in its repairs, resulting in **indels** (an insertion or deletion of DNA bases). Whether the indel is an insertion or a deletion, there is a significant chance that it will cause a frameshift in the gene and consequently create a premature stop codon somewhere downstream of the mutation; hence, there will be a loss of gene function.

The CRISPR/Cas9 system has been used successfully in a variety of species, such as *Drosophila*, zebrafish, and

(A)

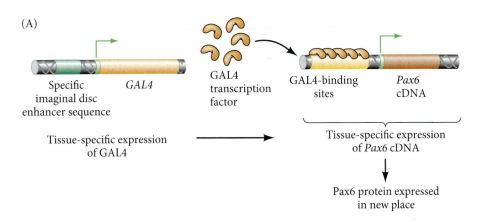

Specific imaginal disc enhancer sequence

GAL4

GAL4 transcription factor

GAL4-binding sites

Pax6 cDNA

Tissue-specific expression of GAL4

Tissue-specific expression of *Pax6* cDNA

Pax6 protein expressed in new place

(B)

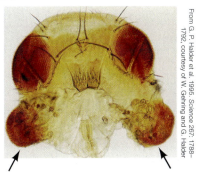

From G. P. Halder et al. 1995, *Science* 267, 1788–1792, courtesy of W. Gehring and G. Halder

FIGURE 3.31 Targeted expression of the *Pax6* gene in a *Drosophila* non-eye imaginal disc. (A) A strain of *Drosophila* was constructed wherein the gene for the yeast GAL4 transcription factor was placed downstream from an enhancer sequence that normally stimulates gene expression in the imaginal discs for mouthparts. If the embryo also contains a transgene that places GAL4-binding sites upstream of the *Pax6* gene, the *Pax6* gene will be expressed in whichever imaginal disc the GAL4 protein is made. (B) *Drosophila* ommatidia (compound eyes) emerging from the mouthparts of a fruit fly in which the *Pax6* gene was expressed in the labial (jaw) discs.

mouse, with some mutation rates exceeding 80% (Bassett et al. 2013). Researchers have been able to push CRISPR even further by using multiple gRNAs to target several genes simultaneously, yielding double and triple knockouts. Importantly, the system can be used to precisely edit a genome by including short DNA fragments as repair templates. These DNA pieces are engineered to have sequence homology on their 5′ and 3′ ends to encourage homologous recombination flanking the double-stranded breaks (see Figure 3.30). This *homology directed repair* is now being tested to repair locations of known human mutations and has potential for treating numerous genetic diseases, such as muscular dystrophy (Nelson et al. 2016). CRISPR/Cas9 is rapidly proving to be a remarkably versatile method for genome editing to further both research and therapeutic objectives across species. One of the immediate benefits is that CRISPR/Cas9 appears to be successful in all organisms. This universal utility has the potential to start a new frontier for functional gene analysis in species in which genetic approaches have previously been an insurmountable obstacle.

GAL4-UAS SYSTEM One of the most powerful uses of this genetic technology has been to activate or downregulate regulatory genes such as *Pax6* in specific tissues. For instance, using *Drosophila* embryos, Halder and colleagues (1995) placed a gene encoding the yeast GAL4 transcriptional activator protein downstream from an enhancer that was known to function in the labial imaginal discs (those parts of the *Drosophila* larva that become the adult mouthparts). In other words, the gene for the GAL4 transcription factor was placed next to an enhancer for genes normally expressed in the developing jaw. Therefore, *GAL4* should be expressed in jaw tissue. Halder and colleagues then constructed a second transgenic fly, placing the cDNA for the *Drosophila Pax6* regulatory gene downstream from a sequence composed of five GAL4-binding sites. The GAL4 protein should be made only in a particular group of cells destined to become the jaw, and when that protein is made, it should cause the transcription of *Pax6* in those particular cells (**FIGURE 3.31A**). In flies in which *Pax6* was expressed in the incipient jaw cells, part of the jaw gave rise to eyes (**FIGURE 3.31B**). In *Drosophila* and frogs (but not in mice), *Pax6* is able to turn several developing tissue types into eyes (Chow et al. 1999). It appears that in *Drosophila*, *Pax6* not only activates those genes that are necessary for the construction of eyes, but also represses those genes that are used to construct other organs.

CRE-LOX SYSTEM An important experimental use of enhancers has been the conditional elimination of gene expression in certain cell types. For example, the transcription factor Hnf4α is expressed in liver cells, but it is also expressed prior to liver formation in the visceral endoderm of the yolk sac. If this gene is deleted from mouse embryos, the embryos die before the liver can even form. So if you wanted to study the consequence of eliminating this gene's function in the liver, you would need to create a mutation that would be *conditional*; that is, you would

In most cells: No recombination

In liver cells only (expressing albumin)

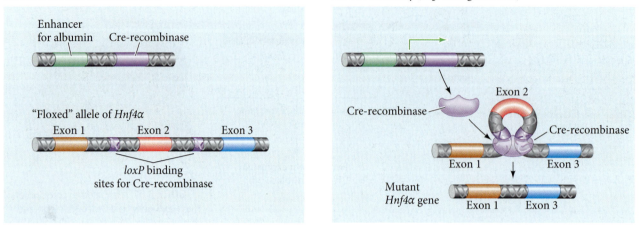

FIGURE 3.32 The Cre-lox technique for conditional mutagenesis, by which gene mutations can be generated in specific cells only. Mice are made wherein wild-type alleles (in this case, the genes encoding the Hnf4α transcription factor) have been replaced by alleles in which the second exon is flanked by *loxP* sequences. These mice are mated with mice having the gene for Cre-recombinase fused to a promoter that is active only in particular cells. In this case, the promoter is that of an albumin gene that functions early in liver development. In mice with both of these altered alleles, Cre-recombinase is made only in the cells where that promoter is activated (i.e., in the cells synthesizing albumin). The Cre-recombinase binds to the *loxP* sequences flanking exon 2 and removes that exon. Thus, in the case depicted here, only the developing liver cells lack a functional Hnf4α gene.

need a mutation that would appear only in the liver and nowhere else. How can that be done? Parviz and colleagues (2002) accomplished it using a site-specific recombinase technology called Cre-lox.

The **Cre-lox** technique uses homologous recombination to place two Cre-recombinase recognition sites (*loxP* sequences) within the gene of interest, usually flanking important exons (see Kwan 2002). Such a gene is said to be "floxed" ("*loxP*-flanked"). For example, using cultured mouse embryonic stem (ES) cells, Parviz and colleagues (2002) placed two *loxP* sequences around the second exon of the mouse *Hnf4α* gene (**FIGURE 3.32**). These ES cells were then used to generate mice that had this floxed allele. A second strain of mice was generated that had a gene encoding bacteriophage Cre-recombinase (the enzyme that recognizes the *loxP* sequence) attached to the promoter of an albumin gene that is expressed very early in liver development. Thus, during mouse development, Cre-recombinase would be made only in the liver cells. When the two strains of mice were crossed, some of their offspring carried both additions. In these double-marked mice, Cre-recombinase (made only in the liver cells) bound to its recognition sites—the *loxP* sequences—flanking the second exon of the *Hnf4α* genes. It then acted as a recombinase and deleted this second exon. The resulting DNA would encode a nonfunctional protein because the second exon has a critical function in *Hnf4α*. Thus, the *Hnf4α* gene was "knocked out" only in the liver cells.

The Cre-lox system allows for control over the spatial and temporal pattern of a gene knockout and gene misexpression. Researchers have inserted stop codons flanked with *loxP* sites to prevent transcription of a given gene until the stop codon is removed by Cre-recombinase. Moreover, Cre-recombinase expression can be controlled with greater temporal control through the use of an estrogen-responsive element sensitive to tamoxifen exposure. This control allows researchers to activate genes for specific proteins, such as reporter proteins like GFP, that are kept inactive until a timed treatment with tamoxifen. (See **Further Development 3.21, Techniques of RNA and DNA Analysis**, online.)

Coda

All the processes of differential gene expression we have discussed in this chapter are stochastic events. They depend on the concentrations of the interacting proteins (Cacace et al. 2012; Murugan and Kreiman 2012; Costa et al. 2013; Neuert et al. 2013). Each organism represents a unique "performance" coordinated by interactions that tell the individual cells which genes are to be expressed and which are to remain silent. Chapter 4 will detail the mechanisms by which cells communicate to orchestrate this differential expression of genes.

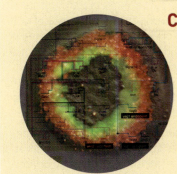

Closing Thoughts on the Opening Photo

What underlies cell differentiation? Here you see an image of a 24-hour-postfertilization sea urchin embryo differentially expressing *hox11/13b* and *foxa* in different cells. This image is overlaid on the gene regulatory network determined to "underlie" the development of endoderm. The gene regulatory network represents the combinatorial interactions that occur among genes to establish the specific array of differentially expressed genes. Networks like this one use the myriad molecular mechanisms discussed in this chapter to control gene expression and ultimately provide the most comprehensive definition of a given cell's identity.

From I. S. Peter and E. H. Davidson. 2011. *Nature* 474: 635–639

Endnotes

[1] Although all the organs were properly formed in the cloned animals, many of the clones developed debilitating diseases as they matured (Humphreys et al. 2001; Jaenisch and Wilmut 2001; Kolata 2001). As we will see shortly, this problem is due in large part to the differences in methylation between the chromatin of the zygote and the differentiated cell.

[2] The term *exon* refers to a nucleotide sequence whose RNA "exits" the nucleus. It has taken on the functional definition of a protein-coding nucleotide sequence.

[3] By convention, upstream, downstream, 5′, and 3′ directions are specified in relation to the RNA. Thus, the promoter is upstream of the region of the gene that is transcribed, near to and "before" its 5′ end.

[4] *Cis*- and *trans*-regulatory elements are so named by analogy with *E. coli* genetics and organic chemistry. Therefore, *cis*-elements are regulatory elements that reside on the same chromosome (*cis*-, "on the same side as"), whereas *trans*-elements are those that could be supplied from another chromosome (*trans*-, "on the other side of").

[5] A list of imprinted mouse genes is maintained at www.mousebook. org/all-chromosomes-imprinting-chromosome-map.

[6] MITF stands for *m*icrophthalmia-associated *t*ranscription *f*actor.

[7] Mutations can generate species-specific splicing events, and tissue-specific differences in pre-mRNA splicing among vertebrate species occur 10 to 100 times more frequently than do changes in gene transcription (Barbosa-Morais et al. 2012; Merkin et al. 2012).

[8] Interestingly, several alternatively spliced isoforms have been discovered for mouse HuD that show differential expression, different subcellular positions (posttranslational regulatory mechanism), and different functional consequences for neuronal survival and differentiation (Hayashi et al. 2015).

[9] Additional details about these and other loss-of-function methods can be found on oup.com/he/barresi12xe.

(3) Snapshot Summary
Differential Gene Expression

1. Evidence from molecular biology, cell biology, and somatic cell nuclear cloning has shown that each cell of the body (with very few exceptions) carries the same nuclear genome.

2. Differential gene expression from genetically identical nuclei creates different cell types. Differential gene expression can occur at the levels of gene transcription, pre-mRNA processing, mRNA translation, and protein modification.

3. Chromatin is made of DNA and proteins. The histone proteins form nucleosomes, and the methylation and acetylation of specific histone residues can repress or activate gene transcription, respectively.

4. Histone methylation is often used to silence gene expression. Histones can be methylated by histone methyltransferases and can be demethylated by histone demethylases.

5. Acetylated histones are often associated with active gene expression. Histone acetyltransferases add acetyl groups to histones, whereas histone deacetylases remove them.

6. Eukaryotic genes contain promoter sequences to which RNA polymerase II can bind to initiate transcription. To do so, the RNA polymerases are bound by a series of proteins called transcription factors.

7. Eukaryotic genes expressed in specific cell types contain enhancer sequences that regulate their transcription in time and space. Enhancers usually activate only genes on the same chromosome. Enhancer sequences can be upstream or downstream or within introns; they can even be millions of base pairs away from the gene they activate. Silencers act to suppress the transcription of a gene in appropriate cell types.

8. Specific transcription factors can recognize specific sequences of DNA in the promoter and enhancer regions. These proteins activate or repress transcription from the genes to which they have bound.

9. Enhancers work in a combinatorial fashion. The binding of several transcription factors can act to promote or inhibit transcription from a certain promoter. In some cases,

(Continued)

Snapshot Summary *(continued)*

transcription is activated only if both factor A and factor B are present; in other cases, transcription is activated if either factor A or factor B is present.

10. Enhancers work in a modular fashion. A gene can contain several enhancers, each directing the gene's expression in a particular cell type.

11. Transcription factors act in different ways to regulate RNA synthesis. Some transcription factors stabilize RNA polymerase II binding to the DNA, and some disrupt nucleosomes, increasing the efficiency of transcription.

12. Class A, B, C, D, and E transcription factors function as homeotic proteins for floral organ identity.

13. Even differentiated cells can be converted into another cell type by the activation of a different set of pioneer transcription factors.

14. Low CpG-content promoters are usually methylated, and their default state is "off," but they can be activated by transcription factors.

15. High CpG-content promoters have a default state that is "on," and they have to be actively repressed by histone methylation.

16. Differences in DNA methylation can account for genomic imprinting, wherein a gene transmitted through the sperm is expressed differently than the same gene transmitted through the egg. Some genes are active only if inherited from the sperm or the egg.

17. Some chromatin is "poised" to respond quickly to developmental signals. In high CpG-content promoters, RNA polymerase II binds to poised chromatin without beginning transcription, and its histones have both active and repressive marks.

18. Alternative pre-mRNA splicing can create a family of related proteins by causing different regions of the pre-mRNA to be read as exons or introns. Based on the splicing site recognition factors present in a cell, what is an exon in one set of circumstances may be an intron in another. The resulting proteins (splicing isoforms) can play different roles that lead to alternative phenotypes and disease.

19. Some messages are translated only at certain times. The oocyte, in particular, uses translational regulation to set aside certain messages that are transcribed during egg development but used only after the egg is fertilized. This activation is often accomplished either by the removal of inhibitory proteins or by the polyadenylation of the message.

20. MicroRNAs can act as translational inhibitors, binding to the 3′ UTR of the RNA. The microRNA recruits an RNA-induced silencing complex that either prevents translation or leads to the degradation of the mRNA.

21. Many mRNAs are localized to particular regions of the oocyte or other cells. This localization appears to be regulated by the 3′ UTR of the mRNA.

22. Ribosomes can differ in different cell types, and ribosomes in one cell may be more efficient at translating certain mRNAs than ribosomes in other cells.

23. A variety of molecular tools have enabled the study of differentially expressed genes, among them in situ hybridization for gene expression, ChIP-Seq to identify regulatory regions of the DNA that proteins bind to, and gene knockdown (RNA interference) and knockout (CRISPR/Cas9) to test gene function.

24. Differential gene expression is more like interpreting a musical score than decoding a code script. There are numerous events that have to take place, and each event has its own numerous interactions among component parts.

Go to oup.com/he/barresi12xe for Further Developments, Scientists Speak interviews, Watch Development videos, Dev Tutorials, and complete bibliographic information for all literature cited in this chapter.

Cell-to-Cell Communication

Mechanisms of Morphogenesis

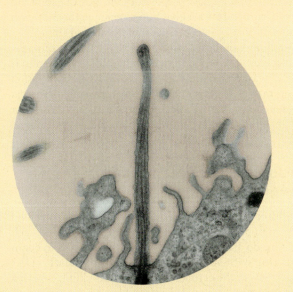

Could this be a cell's antenna? For what?

From A. Alvarez-Buylla et al.
1998. *J Neurosci* 18: 1020–1037

DEVELOPMENT IS MORE THAN JUST DIFFERENTIATION. The different cell types of an organism do not exist as random arrangements. Rather, they form organized structures such as limbs and hearts. Moreover, the types of cells that constitute our fingers—bone, cartilage, neurons, blood cells, and others—are the same cell types that make up our pelvis and legs. Somehow, the cells must be ordered to create different shapes and make different connections. This construction of organized form is called **morphogenesis**, and it has been one of the great sources of wonder for humankind.

In the mid-twentieth century, Ernest E. Just (1939) and Johannes Holtfreter (Townes and Holtfreter 1955) predicted that embryonic cells could have differences in their cell membrane components that would enable the formation of organs. In the late twentieth century, these membrane components—the molecules by which embryonic cells are able to adhere to, migrate over, and induce gene expression in neighboring cells—began to be discovered and described. Today these pathways and networks are being modeled, and we are beginning to understand how the cell integrates the information from its nucleus and from its surroundings to take its place in the community of cells in a way that fosters unique morphogenetic events.

As we discussed in Chapter 1, the cells of an embryo are either epithelial or mesenchymal (see Table 1.1). Epithelial cells adhere to one another and can form sheets and tubes, whereas mesenchymal cells can migrate individually or collectively as a group. The formation and use of diverse extracellular matrices profoundly influence epithelial and mesenchymal cell organization. There appear to be only a few processes through which cells create structured organs (Newman and Bhat 2008), and all these processes involve the cell surface and interactions between an epithelium and an underlying mesenchyme. This chapter will concentrate on three mechanisms requiring cell-to-cell communication via the cell surface: cells adhering, cells changing shape, and cells signaling.

A Primer on Cell-to-Cell Communication

An embryo at any stage is held together, organized, and formed by cell-to-cell interactions, which also define cells' methods of *communication*. Let's consider how we communicate with one another. There needs to be some initial "voice" or signal from one person that is "heard" or received by the other person, which results in a specific response (maybe a hug, a change in posture, or perhaps a sarcastic reply), much like friends conversing. Molecular communication between cells is largely carried out through highly diverse

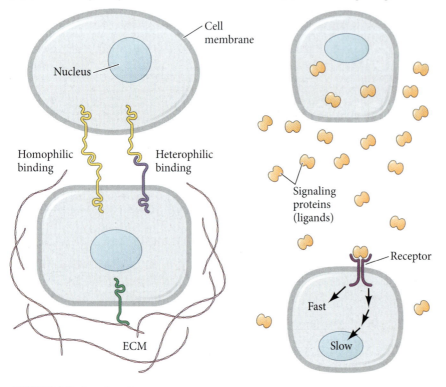

(A) Juxtacrine signaling

Cell membrane

Nucleus

Homophilic binding

Heterophilic binding

ECM

(B) Paracrine signaling

Signaling proteins (ligands)

Receptor

Fast

Slow

FIGURE 4.1   Local and long-range modes of cell-to-cell communication. (A) Local cell signaling is carried out via membrane receptors that bind to proteins in the extracellular matrix (ECM) or directly to receptors from a neighboring cell in a process called juxtacrine signaling. (B) One mechanism for signaling across multiple cell distances is through paracrine signaling, whereby one cell secretes a signaling protein (ligand) into the environment and across the distance of many cells. Only those cells expressing this ligand's corresponding receptor can respond, either rapidly through chemical reactions in the cytosol or more slowly through the process of gene and protein expression.

and specific protein-protein interactions, which have evolved to elicit an array of cellular responses, from changes in gene transcription and glucose metabolism to cell division and cell death. Interactions (or *communication*) between cells and between cells and their environment begin at the cell membrane, with proteins that are housed in, anchored to, or secreted through the membrane.

In an embryo, communication between cells can occur between neighboring cells in direct contact, called **juxtacrine signaling**, or across distances through the secretion of proteins into the extracellular matrix, called **paracrine signaling** (**FIGURE 4.1**). Proteins that are secreted from a cell and designed to communicate a response in another cell are generally referred to as signaling proteins, or **ligands**, while the proteins within a membrane that function to bind either other membrane-associated proteins or signaling proteins are called **receptors**. A receptor in the membrane of one cell that binds the same type of receptor in another cell represents a **homophilic binding**. In contrast, **heterophilic binding** occurs between different receptor types (see Figure 4.1A).

How is communication relayed to the right recipient for a specific cellular response? Protein-protein binding and protein modifications generally result in an altered shape, or *conformation*, of the proteins involved. This conformational change on the outside of the cell affects the shape of the receptor inside the cell, and this latter change can give the intracellular portion of the receptor a new property. It now has the ability to activate the enzymatic reactions that constitute a signal transduction pathway. Often the signal is relayed, or "transduced," through successive conformational changes in the molecules of the pathway—changes orchestrated through the binding of phosphate groups or other small molecules (cAMP, Ca^{2+}) that eventually lead to cellular responses. Signal transduction pathways that culminate in activating gene expression in the nucleus are typically slower than those that enzymatically activate biochemical pathways or regulate cytoskeletal proteins, thereby affecting physiological functions or movement, respectively. These signal transduction pathways are fundamental to animal and plant development.

Adhesion and Sorting: Juxtacrine Signaling and the Physics of Morphogenesis

How are separate tissues formed from populations of cells, and organs constructed from tissues? How do organs form in particular locations and migrating cells reach their destinations? For example, how do osteoblasts stick to other osteoblasts to begin bone formation rather than merging with adjacent capillary cells or muscle cells? What keeps the mesoderm separate from the ectoderm such that the skin has both a dermis and an epidermis?

Could there be a common answer to all these questions? After all, an embryo, from its molecular strands of RNA to its systemic vasculature, develops within the same physical constraints that define our world. Consider a snowman made out of sand (**FIGURE 4.2**). The thermodynamic properties governing the surface tension between water molecules and the grains of sand serve to hold the parts of "Olaf" together. Moreover, the sunlight hitting this sand sculpture will cause *differential* temperatures and associated water evaporation on the surface compared with the inner composition; consequently, the adhesion between sand grains will rapidly decline on the surface, while more centrally located grains hold tight (that is, until the tide changes). Could these same thermodynamic principles govern the connections between cells that support morphogenesis of the embryo?

Differential cell affinity

The experimental analysis of morphogenesis arguably began in 1955 when Townes and Holtfreter conducted cell recombination assays. They placed amphibian embryonic tissues into an alkaline solution that first dissociated these tissues into single cells. Townes and Holtfreter could then ask how the cells from one type of tissue might respond when recombined with cells derived from different tissues.

The results of their experiments were striking. Townes and Holtfreter found that the cells reaggregate and do so in such a way that different cell types become spatially segregated. That is, instead of two cell types remaining mixed, each type sorts out into its own region. Thus, when epidermal (ectodermal) and mesodermal cells are brought together in a mixed aggregate, the epidermal cells move to the periphery of the aggregate, and the mesodermal cells move to the inside (**FIGURE 4.3**). Importantly, the researchers found that the final positions of the reaggregated cells reflect their respective positions in the embryo. The reaggregated mesoderm migrates centrally with respect to the epidermis, adhering to the inner epidermal surface (**FIGURE 4.4A**). Surprisingly, mesoderm also migrates centrally with respect to the gut endoderm when epidermis is not present (**FIGURE 4.4B**). However, when the three germ layers are mixed together, the endoderm separates from the epidermis and mesoderm and is then enveloped by them (**FIGURE 4.4C**). In the final configuration, the epidermis is on the periphery, the endoderm is internal, and the mesoderm lies in the region between them. Holtfreter attributed these results to the cells exhibiting **selective affinity**.

Mimicry of normal embryonic structures by cell aggregates is also seen in the recombination of cell types within a given germ layer, such as the ectodermal lineages of epidermis and neural plate cells (**FIGURE 4.4D**). The presumptive epidermal cells migrate to the periphery as before; the neural plate cells migrate inward, forming a structure reminiscent of the neural tube. When axial mesoderm (notochord) cells are added to a suspension of presumptive epidermal and presumptive neural cells, cell

FIGURE 4.2 Adhesion between sand grains holds this sand sculpture of the Disney character Olaf the Snowman together.

© M. J. F. Barresi

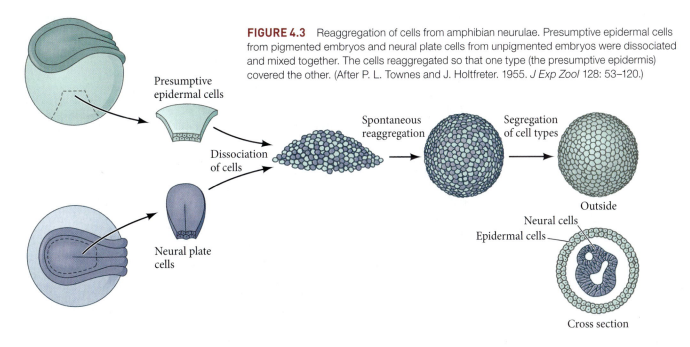

FIGURE 4.3 Reaggregation of cells from amphibian neurulae. Presumptive epidermal cells from pigmented embryos and neural plate cells from unpigmented embryos were dissociated and mixed together. The cells reaggregated so that one type (the presumptive epidermis) covered the other. (After P. L. Townes and J. Holtfreter. 1955. *J Exp Zool* 128: 53–120.)

Presumptive epidermal cells

Neural plate cells

Dissociation of cells

Spontaneous reaggregation

Segregation of cell types

Outside

Neural cells

Epidermal cells

Cross section

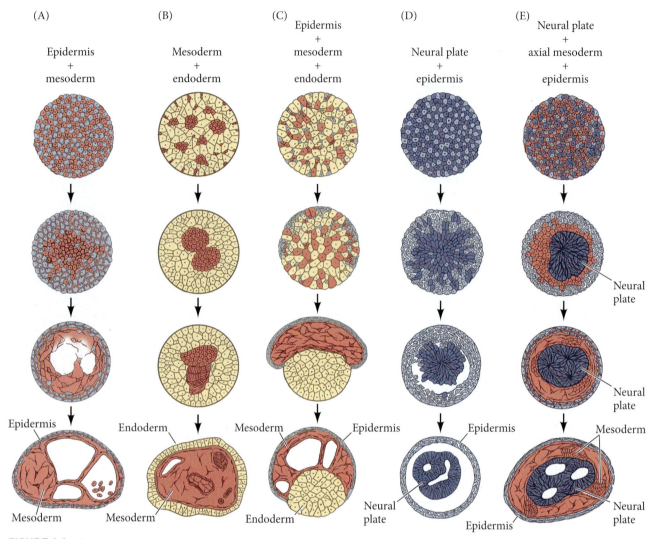

(A) Epidermis + mesoderm

(B) Mesoderm + endoderm

(C) Epidermis + mesoderm + endoderm

(D) Neural plate + epidermis

(E) Neural plate + axial mesoderm + epidermis

FIGURE 4.4 Sorting out and reconstruction of spatial relationships in aggregates of embryonic amphibian cells. (After P. L. Townes and J. Holtfreter. 1955. *J Exp Zool* 128: 53–120.)

segregation results in an external epidermal layer, a centrally located neural tissue, and a layer of mesodermal tissue between them (**FIGURE 4.4E**). *Somehow, the cells are able to sort out into their proper embryonic positions!*

Holtfreter and colleagues concluded that selective affinities change during development. For development to occur, cells must interact differently with other cell populations at specific times. Such changes in cell affinity are extremely important in the processes of morphogenesis.

The thermodynamic model of cell interactions

Cells, then, do not sort randomly, but they can actively move to create tissue organization. What forces direct cell movement during morphogenesis? In 1964, Malcolm Steinberg proposed the **differential adhesion hypothesis**, a model that sought to explain patterns of cell sorting based on thermodynamic principles. Using dissociated cells derived from trypsinized[1] embryonic tissues, Steinberg showed that certain cell types migrate centrally when combined with some cell types, but migrate peripherally when combined with others. These interactions followed a behavioral hierarchy (Steinberg 1970). If the final position of cell type A is internal to a second cell type B, and if the final position of B is internal to a third cell type C, the final position of A will always be internal to C (**FIGURE 4.5A**; Foty and Steinberg 2013). For example, pigmented retina cells migrate internally to neural retina cells, and heart cells migrate internally to pigmented retina cells. Therefore, heart cells should migrate internally to neural retina cells—and they do. This observation led Steinberg to propose that cells interact so as to form an aggregate with the smallest interfacial free energy. In other words, the cells rearrange themselves into the most thermodynamically stable pattern. If cell types A and B have different strengths of adhesion

FIGURE 4.5 Hierarchy of cell sorting of decreasing surface tensions. (A) Simple schematic demonstrating a logic statement for the properties of differential cell adhesion. (B) The equilibrium configuration reflects the strength of cell cohesion, with the cell types having the greater cell cohesion segregating inside the cells with less cohesion. These images were obtained by sectioning the aggregates and assigning colors to the cell types by computer. Black areas represent cells whose signal was edited out in the program of image optimization.

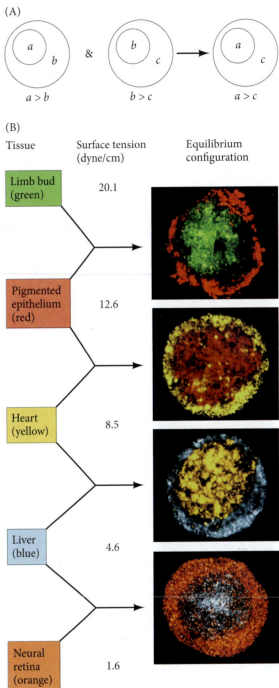

and if the strength of A-A connections is greater than the strength of A-B or B-B connections, sorting will occur, with the A cells becoming central. However, if the strength of A-A connections is less than or equal to the strength of A-B connections, the aggregate will remain as a random mix of cells. According to this hypothesis, the early embryo can be viewed as existing in an equilibrium state until there is a change in the adhesive properties of the cell membranes. The movements that result seek to restore the cells to a new equilibrium configuration. All that is required for sorting to occur is that cell types differ in the strength of their adhesions—the differential adhesion hypothesis.

In several meticulous experiments using numerous tissue types, researchers showed that those cell types that had greater surface cohesion migrated centrally when combined with cells that had less surface tension (**FIGURE 4.5B**; Foty et al. 1996; Krens and Heisenberg 2011). In the simplest form of this model, all cells could have the same type of "glue" on the cell surface. The amount of this "glue," or the cellular architecture that allows such a substance to be differentially distributed across the surface, could create a difference in the number of stable adhesions made between cell types. In a more specific version of this model, the thermodynamic differences could be caused by different types of adhesion molecules (see Moscona 1974). When Holtfreter's studies were revisited using modern techniques, Davis and colleagues (1997) found that the tissue surface tensions of the individual germ layers were precisely those required for the sorting patterns observed both in vitro and in vivo.

Cadherins and cell adhesion

Evidence shows that boundaries between tissues can indeed be created by different cell types having both different types and different amounts of cell adhesion molecules. Several classes of molecules can mediate cell adhesion, but the major cell adhesion molecules appear to be the cadherins.

As their name suggests, **cadherins** are *calcium-dependent adhesion* molecules. They are critical for establishing and maintaining intercellular connections, and they appear to be crucial to the spatial segregation of cell types and to the organization of animal form (Takeichi 1987). Cadherins are transmembrane proteins that interact with other cadherins on adjacent cells. The cadherins are anchored inside the cell by a complex of proteins called **catenins** (**FIGURE 4.6**), and the cadherin-catenin complex forms the classic adherens junctions that help hold epithelial cells together. Moreover, because the cadherins and the catenins bind to the actin (microfilament) cytoskeleton of the cell, they integrate the epithelial cells into a mechanical unit. Blocking cadherin *function* (by antibodies that bind and inactivate cadherin) or blocking cadherin *synthesis* (with antisense RNA that binds cadherin messages and prevents their translation) can prevent the formation of epithelial tissues and cause the cells to disaggregate (Takeichi et al. 1979).

Cadherins perform several related functions. First, their external domains serve to adhere cells together. Second, cadherins link to and help assemble the actin cytoskeleton, thereby providing the mechanical forces for forming sheets and tubes. Third, cadherins can initiate and transduce signals that can lead to changes in a cell's gene expression. (See **Further Development 4.1, Sorting Out the Early Embryo**, online.)

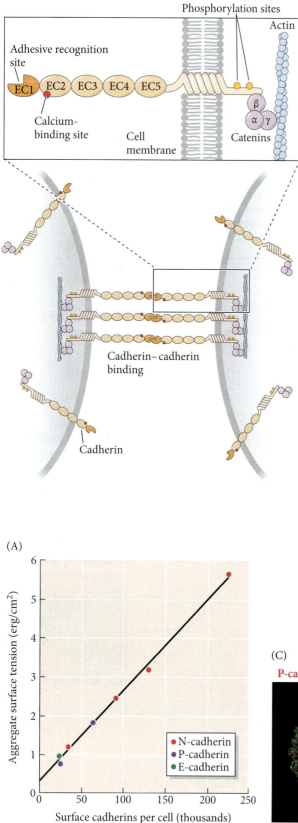

FIGURE 4.6 Simplified scheme of cadherin linkage to the cytoskeleton via catenins. (After M. Takeichi et al. 1991. *Science* 251: 1451–1455.)

QUANTITY AND COHESION The ability of cells to sort themselves based on the *amount* of cadherin expression was first shown using two cell lines that were identical except that they synthesized different amounts of P-cadherin. When these two groups of cells were mixed, the cells that expressed more P-cadherin had a higher surface cohesion and migrated internally to the cells expressing less P-cadherin (Steinberg and Takeichi 1994; Foty and Steinberg 2005). The researchers went on to demonstrate that this quantitative cadherin-dependent sorting is directly correlated with surface tension (**FIGURE 4.7A,B**). The surface tensions of these *homotypic* aggregates (meaning all cells have the same type of cadherin) are linearly related to the amount of cadherin they express on the cell surface, showing that the cell sorting hierarchy was strictly dependent on the different numbers of cadherin interactions between the cells. This thermodynamic principle also applies to *heterotypic* aggregates, in which the relative amounts of different cadherin types still predict cell sorting behavior in vitro (**FIGURE 4.7C**; Foty and Steinberg 2013). (See **Further Development 4.2, Type,**

(B)

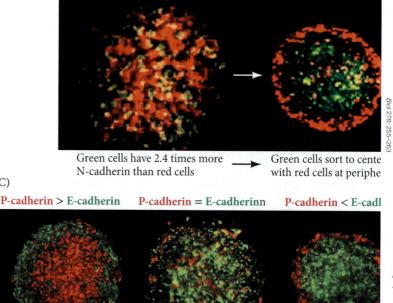

Green cells have 2.4 times more N-cadherin than red cells → Green cells sort to cente with red cells at periphe

(C)

P-cadherin > E-cadherin   P-cadherin = E-cadherinn   P-cadherin < E-cadl

From R. A. Foty and M. S. Steinberg. 2005. *Dev Biol* 278: 255–263

From D. Duguay et al. 2003. *Dev Biol* 253: 309–323

(A)

FIGURE 4.7 Importance of the amount of cadherin for correct morphogenesis. (A) Aggregate surface tension correlates with the number of cadherin molecules on the cell membranes. (B) Sorting out of two subclones having different amounts of cadherin on their cell surfaces. The green-stained cells have 2.4 times as many N-cadherin molecules in their membrane as the red cells. (These cells have no normal cadherin genes being expressed.) At 4 hours of incubation (left), the cells are randomly distributed, but after 24 hours of incubation (right), the red cells (with a surface tension of about 2.4 erg/cm²) have formed an envelope around the more tightly cohering (5.6 erg/cm²) green cells. (C) Sorting can occur based on cadherin number even if the two cells express different cadherin proteins (i.e., are heterotypic). Red indicates P-cadherin, green E-cadherin. (A,B from R. A. Foty and M. S. Steinberg. 2005. *Dev Biol* 278: 255–263.)

Timing, and Border Formation, and **Further Development 4.3, Shape Change and Epithelial Morphogenesis: "The Force Is Strong in You,"** both online.)

The Extracellular Matrix as a Source of Developmental Signals

Cell-to-cell interactions do not happen in the absence of an environment; rather, they occur in coordination with and often due to the environmental conditions surrounding the cells. This environment is called the **extracellular matrix**, which is an insoluble network consisting of macromolecules secreted by cells. These macromolecules form a region of noncellular material in the interstices between the cells. Cell adhesion, cell migration, and the formation of epithelial sheets and tubes all depend on the ability of cells to form attachments to extracellular matrices. In some cases, as in the formation of epithelia, these attachments have to be extremely strong. In other instances, as when cells migrate, attachments have to be made, broken, and made again. In some cases, the extracellular matrix merely serves as a permissive substrate for adhesion and migration; in other cases, it can hold important guidance and specification cues for directional cell movement and differentiation, respectively. Extracellular matrices are made up of the matrix protein collagen, proteoglycans, and a variety of specialized glycoprotein molecules, such as fibronectin and laminin.

Proteoglycans—large extracellular proteins with glycosaminoglycan polysaccharide (sugar) side chains—often play critical roles in the ability of matrices to present informative cues, such as paracrine factors, to cells. Two of the most widespread proteoglycans are heparan sulfate and chondroitin sulfate. Heparan sulfate can bind many members of different paracrine families, and it appears to be essential for presenting the paracrine factors in high concentrations to its receptors. In *Drosophila*, *C. elegans*, and mice, mutations that prevent the synthesis of proteoglycans block normal cell migration, morphogenesis, and differentiation (García-García and Anderson 2003; Hwang et al. 2003; Kirn-Safran et al. 2004).

The large glycoproteins contributing to the extracellular matrix are responsible for organizing the matrix and the cells associated with it into an ordered structure. **Fibronectin** is a very large (460-kDa) glycoprotein dimer that can form different quaternary structures with long fibers called fibronectin fibrils. Fibronectin generally functions as an intermediary adhesive molecule, linking cells to one another and to other substrates such as collagen and proteoglycans. Fibronectin has several distinct binding sites, and their interaction with the appropriate cell surface molecules (namely integrins) results in the proper alignment of cells with the orientation of fibril assemblies (**FIGURE 4.8A**).

? Developing Questions

Recently, a "differential interfacial tension hypothesis" proposed that cell cortex contractility governs cell sorting more than cell-to-cell adhesion does. As better in vivo tools are developed to quantitatively measure forces on the cellular and molecular levels, it will be exciting to learn how differential adhesion and differential interfacial tension cooperatively regulate morphogenesis. In the coming years, keep an eye out for a growing understanding of the role that biophysical properties play in mechanisms of morphogenesis.

(A)

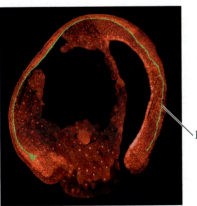

Courtesy of M. Marsden and D. W. DeSimone

Fibronectin

(B)

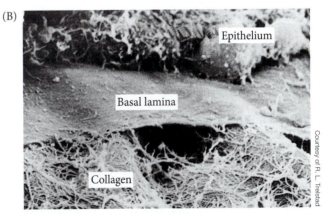

Epithelium

Basal lamina

Collagen

Courtesy of R. L. Trelstad

FIGURE 4.8 Extracellular matrices in the developing embryo. (A) Fluorescent antibodies to fibronectin show fibronectin deposition as a green band in the *Xenopus* embryo during gastrulation. The fibronectin will orient the movements of the mesodermal cells. (B) Fibronectin links together migrating cells, collagen, heparan sulfate, and other extracellular matrix proteins. This scanning electron micrograph shows the extracellular matrix at the junction of the epithelial cells (above) and mesenchymal cells (below). The epithelial cells synthesize a tight, laminin-based basal lamina, whereas the mesenchymal cells secrete a loose reticular lamina made primarily of collagen. Together, the basal lamina and reticular lamina make up the basement membrane.

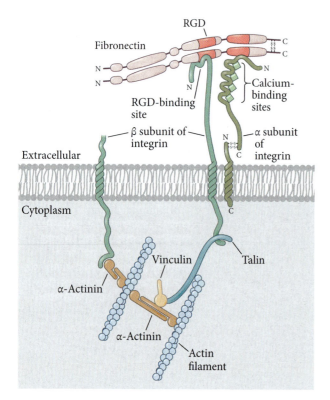

FIGURE 4.9 Simplified diagram of the fibronectin receptor complex. The integrins of the complex can form heterodimerized membrane-spanning receptor proteins that bind fibronectin on the outside of the cell while interacting with actin cytoskeleton-associated proteins such as α-actinin, vinculin, and talin on the inside of the cell. RGD, arginine-glycine-aspartate. (After E. J. Luna and A. L. Hitt. 1992. *Science* 258: 955–964.)

Fibronectin also has an important role in cell migration because the "roads" over which certain migrating cells travel are paved with this protein. Fibronectin paths lead germ cells to the gonads and heart cells to the midline of the embryo. If chick embryos are injected with antibodies that block fibronectin, the heart-forming cells fail to reach the midline, and two separate hearts develop (Heasman et al. 1981; Linask and Lash 1988).

Laminin (another large glycoprotein) and **type IV collagen** are major components of a type of extracellular matrix called the **basal lamina**. The basal lamina is characterized by closely knit sheets that underlie epithelial tissue (**FIGURE 4.8B**). The adhesion of epithelial cells to laminin (on which they sit) is much greater than the affinity of mesenchymal cells for fibronectin (to which they must bind and release if they are to migrate). Like fibronectin, laminin plays a role in assembling the extracellular matrix, promoting cell adhesion and growth, changing cell shape, and permitting cell migration (Hakamori et al. 1984; Morris et al. 2003).

Integrins: Receptors for extracellular matrix molecules

The ability of a cell to bind to adhesive glycoproteins such as laminin or fibronectin depends on its expressing membrane receptors for the cell-binding sites of these large molecules (Chen et al. 1985; Knudsen et al. 1985). The main fibronectin receptor was found to be an extremely large protein that could bind fibronectin on the outside of the cell, span the membrane, and bind cytoskeletal proteins on the inside of the cell (**FIGURE 4.9**). This family of receptor proteins is called integrins because they *integrate* the extracellular and intracellular scaffolds, allowing them to work together (Horwitz et al. 1986; Tamkun et al. 1986).

On the extracellular side, integrins bind to the amino acid sequence arginine-glycine-aspartate (RGD), found in several extracellular matrix adhesive proteins, including fibronectin and laminin (Ruoslahti and Pierschbacher 1987). On the cytoplasmic side, integrins bind to talin and α-actinin, two proteins that connect to actin filaments. This dual binding enables the cell to move by contracting the actin filaments against the fixed extracellular matrix.

Integrins can also signal from the outside of the cell to the inside of the cell, altering gene expression (Walker et al. 2002). Bissell and colleagues have shown that integrin is critical for inducing specific gene expression in developing tissues, especially those of the liver, testis, and mammary gland (Bissell et al. 1982; Martins-Green and Bissell 1995). (See **Further Development 4.4, Integrins and Cell Death**, online.)

The Epithelial-Mesenchymal Transition

One important developmental phenomenon, the **epithelial-mesenchymal transition**, or **EMT**, integrates all the processes we have discussed so far in this chapter. EMT is an orderly series of events whereby epithelial cells are transformed into mesenchymal cells. In this transition, a polarized stationary epithelial cell, which normally interacts with basal lamina through its basal surface, becomes a migratory mesenchymal cell that can invade tissues and help form organs in new places (**FIGURE 4.10A**; see Sleepman and Thiery 2011). An EMT is usually initiated when paracrine factors from neighboring cells alter gene expression in target cells that change or downregulate their expression of cadherins, releasing their attachments to other cells, or of integrins, releasing their attachments to components of the basal lamina. This is accompanied by the secretion of enzymes that break down the basal lamina, allowing the target cells to escape from the epithelium. These changes also often involve rearrangements to the target cells' actin cytoskeleton and the secretion of new extracellular matrix molecules characteristic of mesenchymal cells.

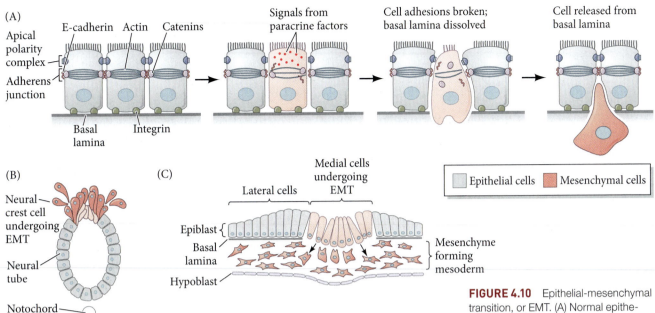

FIGURE 4.10 Epithelial-mesenchymal transition, or EMT. (A) Normal epithelial cells are attached to one another through adherens junctions containing cadherin, catenins, and actin rings. They are attached to the basal lamina through integrins. Paracrine factors can repress the expression of genes that encode these cellular components, causing the cell to lose polarity, lose attachment to the basal lamina, and lose cohesion with other epithelial cells. Cytoskeletal remodeling occurs, as well as the secretion of proteases that degrade the basal lamina and other extracellular matrix components of the basement membrane, enabling the migration of the newly formed mesenchymal cell. (B,C) EMT is seen in vertebrate embryos during the normal formation of neural crest from the dorsal region of the neural tube (B) and during the formation of the mesoderm by mesenchymal cells delaminating from the epiblast (C).

The epithelial-mesenchymal transition is critical during development (**FIGURE 4.10B,C**). Examples of developmental processes in which EMT occurs include (1) the formation of neural crest cells from the dorsalmost region of the neural tube; (2) the formation of mesoderm in chick embryos, wherein cells that had been part of an epithelial layer become mesodermal and migrate into the embryo; and (3) the formation of vertebrae precursor cells from the somites, wherein these cells detach from the somite and migrate around the developing spinal cord. EMT is also important in adults, in whom it is needed for wound healing. The most striking adult form of EMT, however, is seen in cancer metastasis, wherein cells that have been part of a solid tumor mass leave that tumor epithelium to invade other tissues as migratory mesenchymal cells that form secondary tumors elsewhere in the body. It appears that in metastasis, the processes that generated the cellular transition in the embryo are reactivated, allowing cancer cells to migrate and become invasive. Cadherins are downregulated, the actin cytoskeleton is reorganized, and the cells secrete enzymes such as metalloproteinases to degrade the basal lamina while also undergoing cell division (Acloque et al. 2009; Kalluri and Weinberg 2009).

Cell Signaling

We have just learned how cell-to-cell adhesion can influence how cells position themselves within an embryo, and in previous chapters we discussed the importance that a cell's position in the embryo can have on regulating its fate. What is so special about a given position in the embryo that it can determine a cell's fate? As you know, the experiences one has in early life greatly influence the type of person one becomes as an adult in terms of personality, career choice, or food preferences. Similarly, the experiences a cell has in its embryonic position influence the gene regulatory network under which it develops. Therefore, the real question is, In a given location, what defines the cell's experience?

Induction and competence

From the earliest stages of development through the adult, cell behaviors such as adhesion, migration, differentiation, and division are regulated by signals from one cell being received by another cell. Indeed, these interactions (which are often reciprocal, as we will describe later) are what allow organs to be constructed. The development of the vertebrate eye is a classic example used to describe the modus operandi of tissue organization via intercellular interactions.

In the vertebrate eye, light is transmitted through the transparent corneal tissue and focused by the lens tissue, eventually impinging on the neural retina. The precise arrangement of tissues in the eye cannot be disturbed without impairing its function. Such coordination in the construction of the lens and retina is accomplished by one group of cells communicating an organizing change in the behavior or developmental trajectory of an adjacent set of cells. This kind of interaction is known as an **induction**.

DEFINING INDUCTION AND COMPETENCE   There are at least two components to every inductive interaction. The first component is the **inducer**, the tissue that produces a signal (or signals) that changes the cellular behavior of the other tissue. Often this signal is a secreted protein called a paracrine factor. **Paracrine factors** are proteins made by a cell or a group of cells that alter the behavior or differentiation of adjacent cells (see Figure 4.1B). In contrast to endocrine factors (hormones), which travel through the blood and exert their effects on cells and tissues far away, paracrine factors are secreted into the extracellular space and influence their close neighbors. The second component, the **responder**, is the cell or tissue being induced. Cells of the responding tissue must have both a receptor protein for the inducing factor and the *ability* to respond to the signal. The ability to receive and respond to a specific inductive signal is called **competence** (Waddington 1940).

BUILDING THE VERTEBRATE EYE   In the initiation of vertebrate eye morphogenesis, paired regions of the brain bulge out and approach the surface ectoderm of the head. The head ectoderm is competent to respond to the paracrine factors made by these brain bulges (the **optic vesicles**), and the head ectoderm receiving these paracrine factors is induced to form the lens of the eye. Although lens specification is defined much earlier during neural plate stages (Grainger 1992; Ogina et al. 2012), lens differentiation genes are induced in the non-neural head ectodermal cells by the optic vesicle (see Chapter 16; Maddala et al. 2008). Moreover, the prospective lens cells secrete paracrine factors that instruct the optic vesicle to form the retina. Thus, the two major parts of the eye co-construct each other, and the eye forms from reciprocal paracrine interactions. Importantly, the head ectoderm is the only region competent to respond to the optic vesicle. If an optic vesicle from a *Xenopus laevis* embryo is placed underneath head ectoderm in a different part of the head from where the frog's optic vesicle normally occurs, the vesicle will induce that ectoderm to form lens tissue; trunk ectoderm, however, will not respond to the optic vesicle (**FIGURE 4.11**; Saha et al. 1989; Grainger 1992).

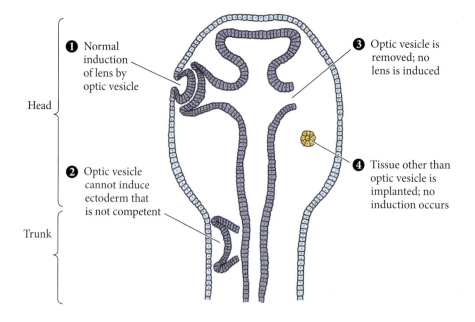

❶ Normal induction of lens by optic vesicle

❷ Optic vesicle cannot induce ectoderm that is not competent

❸ Optic vesicle is removed; no lens is induced

❹ Tissue other than optic vesicle is implanted; no induction occurs

Head

Trunk

FIGURE 4.11 Ectodermal competence and the ability to respond to the optic vesicle inducer in *Xenopus*. The optic vesicle is able to induce lens formation in the anterior portion of the ectoderm (1) but not in the presumptive trunk and abdomen (2). (3) If the optic vesicle is removed, the surface ectoderm forms either an abnormal lens or no lens at all. (4) Most other tissues are not able to substitute for the optic vesicle.

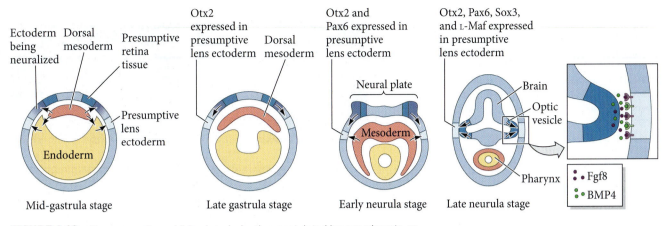

FIGURE 4.12 Sequence of amphibian lens induction postulated by experiments on embryos of the frog *Xenopus laevis*. Unidentified inducers (possibly from the foregut endoderm and cardiac mesoderm) cause the synthesis of the Otx2 transcription factor in the head ectoderm during the late gastrula stage. As the neural folds rise, inducers from the anterior neural plate (including the region that will form the retina) induce Pax6 expression in the anterior ectoderm that can form lens tissue. Expression of Pax6 protein may constitute the competence of the surface ectoderm to respond to the optic vesicle during the late neurula stage. The optic vesicle secretes BMP and FGF family paracrine factors (see signals in higher magnification of boxed area), which induce the synthesis of the Sox transcription factors and initiate observable lens formation. (After R. M. Grainger. 1992. *Trends Genet* 8: 349–356.)

Often, one induction will give a tissue the competence to respond to another inducer. Studies on amphibians suggest that one of the first inducers of the lens may be the foregut endoderm and heart-forming mesoderm that underlie the lens-forming ectoderm during the early and mid-gastrula stages (Jacobson 1963, 1966). The anterior neural plate may produce the next signals, including a signal that promotes the synthesis of the Paired box 6 (Pax6) transcription factor in the anterior ectoderm, which is required for the competence to respond to the optic vesicle's signals (**FIGURE 4.12**; Zygar et al. 1998). Thus, although the optic vesicle appears to be *the* lens inducer, the anterior ectoderm has already been induced by at least two other tissues. The optic vesicle's situation is like that of the player who kicks the "winning" goal in a soccer match, though many others helped position that ball for the final kick!

The optic vesicle appears to secrete two paracrine factors, one of which is BMP4 (Furuta and Hogan 1998), a protein that is received by the lens cells and induces the production of the Sox transcription factors (see Figure 4.12, rightmost panels). The other is Fgf8, a secreted signal that induces the appearance of the L-Maf transcription factor (Ogino and Yasuda 1998; Vogel-Höpker et al. 2000). As we saw in Chapter 3, the combination of Pax6, Sox2, and L-Maf in the ectoderm is needed for the production of the lens and the activation of lens-specific genes, such as δ-*crystallin*. Pax6 is important in providing the competence for the ectoderm to respond to the inducers from the optic cup (Fujiwara et al. 1994). If *Pax6* is lost, whether it is in fruit flies, frogs, rats, or humans, it results in a complete loss or reduction of the eyes (Quiring et al. 1994). Experiments recombining surface ectoderm with the optic vesicle from wild-type and *Pax6* mutant rat embryos demonstrated that Pax6 must be functional in the surface ectoderm for it to form a lens (**FIGURE 4.13A,B**). In humans, a spectrum of eye malformations has been associated with a variety of *Pax6* mutations. These malformations include aniridia, in which the iris is reduced or lacking (**FIGURE 4.13C**); *Pax6* mutations in *Xenopus* have revealed remarkably similar aniridia-like symptoms, enabling researchers to model and further investigate the developmental role of Pax6 in this human disorder (Nakayama et al. 2015). (See **Further Development 4.5, Reciprocal Induction**, online.)

FIGURE 4.13 The *Pax6* gene is similarly required for eye development in frogs, rats, and humans. (A) Loss of *Pax6* in rats results in the failure to form eyes as well as significant reductions in nasal structures. (B) An analysis of lens induction following recombination experiments of the optic vesicle and surface ectoderm between wild-type and *Pax6* null rat embryos. Pax6 is required only in the surface ectoderm for proper lens induction. (C) Mutations in the *Pax6* gene in *Xenopus* and humans result in similar reductions in the iris of the eye as compared with wild-type individuals. This phenotype is characteristic of aniridia.

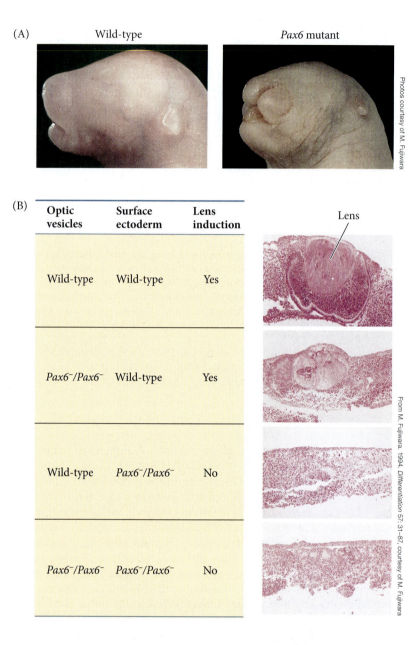

(A) Wild-type   *Pax6* mutant

Photos courtesy of M. Fujiwara

(B)

Optic vesicles	Surface ectoderm	Lens induction
Wild-type	Wild-type	Yes
Pax6⁻/Pax6⁻	Wild-type	Yes
Wild-type	*Pax6⁻/Pax6⁻*	No
Pax6⁻/Pax6⁻	*Pax6⁻/Pax6⁻*	No

Lens

From M. Fujiwara. 1994. *Differentiation* 57: 31–87, courtesy of M. Fujiwara

(C)

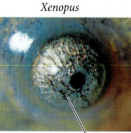
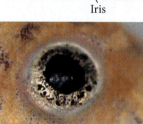

Xenopus   Human

Wild-type

Iris   Iris

Pax6-deficient iris reduced (aniridia)

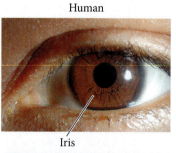

From T. Nakayama et al. 2015. *Dev Biol* 408: 328–344, courtesy of R. M. Grainger

FURTHER DEVELOPMENT

INSTRUCTIVE AND PERMISSIVE INTERACTIONS   Howard Holtzer (1968) distinguished two major modes of inductive interaction. In **instructive interaction**, a signal from the inducing cell is *necessary* for initiating new gene expression in the responding cell. Without the inducing cell, the responding cell is not capable of differentiating in that particular way. For example, one instructive interaction is when a *Xenopus* optic vesicle experimentally placed under a new region of head ectoderm causes that region of the ectoderm to form a lens.

The second type of inductive interaction is **permissive interaction**. Here, the responding tissue has already been specified and needs only an environment that allows the expression of these traits. For instance, many tissues need an extracellular matrix to develop. The extracellular matrix does not alter the type of cell that is produced, but it enables what has already been determined to be expressed.[2]

A dramatic example of permissive interaction at work comes from the regenerative medicine field, in which an extracellular matrix scaffold can promote the differentiation and rebuilding of a beating heart. Doris Taylor's research group used detergents to remove all the cells from a cadaveric rat heart, which left behind the natural extracellular matrix (**FIGURE 4.14A**; Ott et al. 2008). Proteins such as fibronectin, collagen, and laminin held together the rest of the extracellular matrix and maintained the intricate shape of the heart. The researchers then infused this extracellular matrix scaffold with cardiomyocyte progenitor cells. Surprisingly, these cells differentiated and organized into a functionally contracting "recellularized" heart (**FIGURE 4.14B**). Therefore, the environmental conditions of the decellularized extracellular matrix were equipped with instructive guidance for the development of heart muscle. (See **Further Development 4.6, From Feathers to Claws and Frogs to Newts: Further Your Understanding of Induction and Competence**, and **Further Development 4.7, The Insect Trachea: Combining Inductive Signals with Cadherin Regulation**, both online.)

(?) Developing Questions

Although rebuilding a decellularized heart is clearly an example of permissive interaction, could there be instructive interaction too? Recently, iPSC-derived cardiovascular progenitor cells successfully seeded a decellularized mouse heart and differentiated into cardiomyoctyes, smooth muscle, and endothelial cells (Lu et al. 2013). What could the extracellular matrix be providing to directly influence the differentiation of progenitor cells into these varied cell types?

(A) Decellularization

← 12 h →

(B) Recellularized beating heart

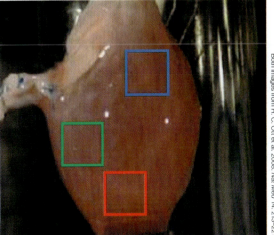

Both images from H. C. Ott et al. 2008. *Nat Med* 14: 213–221

FIGURE 4.14   Reconstructing a decellularized rat heart. (A) Whole hearts from rat cadavers were decellularized (all cells removed) over the course of 12 hours using the detergent SDS. Progression of decellularization is seen here from left to right. Ao, aorta; LA, left atrium; LV, left ventricle; RA, right atrium; RV, right ventricle. (B) A decellularized heart was mounted into a bioreactor and recellularized with neonatal cardiac cells, which developed into self-contracting cardiomyocytes and powered the beating of the heart construct. Regional ECG tracings indicate synchronous contractions of the indicated heart regions (blue, green, and red plots).

Paracrine Factors: Inducer Molecules

How are the signals between inducer and responder transmitted? While studying the mechanisms of induction that produce the kidney tubules and teeth, Grobstein (1956) and others found that some inductive events could occur despite a filter separating the epithelial and mesenchymal cells. Other inductions, however, were blocked by the filter. The researchers therefore concluded that some of the inducers were soluble molecules that could pass through the small pores of the filter and that other inductive events required physical contact between the epithelial and mesenchymal cells (Grobstein 1956; Saxén et al. 1976; Slavkin and Bringas 1976).

When membrane proteins on one cell surface interact with receptor proteins on adjacent cell surfaces (as seen with cadherins), the event is called a **juxtacrine interaction** (since the cell membranes are *juxtaposed*). When proteins synthesized by one cell can diffuse over a distance to induce changes in neighboring cells, the event is called a **paracrine interaction**. Paracrine factors are diffusible molecules that usually work in a range of about 15 cell diameters, or about 40–200 μm (Bollenbach et al. 2008; Harvey and Smith 2009).

Autocrine interactions are also possible; in this case, the same cells that secrete the paracrine factors also respond to them. In other words, the cell synthesizes a molecule for which it has its own receptor. Although autocrine regulation is not common, it is seen in placental cytotrophoblast cells; these cells synthesize and secrete platelet-derived growth factor, whose receptor is on the cytotrophoblast cell membrane (Goustin et al. 1985). The result is the explosive proliferation of that tissue.

Morphogen gradients

One of the most important mechanisms governing cell fate specification involves gradients of paracrine factors that regulate gene expression; such signaling molecules are called morphogens. A **morphogen** (Greek, "form-giver") is a diffusable biochemical molecule that can determine the fate of a cell by its concentration.[3] That is, cells exposed to high levels of a morphogen activate different genes than those cells exposed to lower levels. Morphogens can be transcription factors produced within a syncytium of nuclei, as in the *Drosophila* blastoderm (see Chapter 2). They can also be paracrine factors that are produced in one group of cells and then travel to another population of cells, specifying the target cells to have similar or different fates according to the concentration of the morphogen. Uncommitted cells exposed to high concentrations of the morphogen (nearest its source of production) are specified as one cell type. When the morphogen's concentration drops below a certain threshold, a different cell fate is specified. When the concentration falls even lower, a cell that initially was of the same uncommitted type is specified in yet a third distinct manner (**FIGURE 4.15**).

Regulation by gradients of paracrine factor concentration was elegantly demonstrated by the specification of different mesodermal cell types in the frog *Xenopus laevis* by Activin, a paracrine factor of the TGF-β superfamily (**FIGURE 4.16**; Green and Smith 1990; Gurdon et al. 1994). Activin-secreting beads were placed on unspecified cells from an early *Xenopus* embryo. The Activin then diffused from the beads. At high concentrations (~300 molecules/cell), Activin induced expression of the *goosecoid* gene, whose product is a transcription factor that specifies the frog's dorsalmost structures. At slightly lower concentrations of Activin (~100 molecules/cell), the same tissue activated the *Xbra* gene and was specified to become muscle. At still lower concentrations, these genes were not activated, and the "default" gene expression instructed the cells to become blood vessels and heart (Dyson and Gurdon 1998).

The range of a paracrine factor (and thus the shape of its morphogen gradient) depends on several aspects of that factor's synthesis, transport, and degradation. In some cases, cell surface molecules stabilize the paracrine factor and aid in its diffusion, while in other cases, cell surface moieties

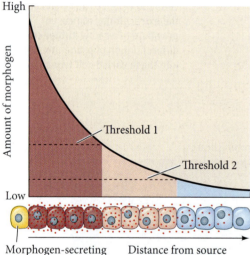

FIGURE 4.15 Specification of uniform cells into three cell types by a morphogen gradient. A morphogenetic paracrine factor (yellow dots) is secreted from a source cell (yellow) and forms a concentration gradient within the responsive tissue. Cells exposed to morphogen concentrations above threshold 1 (red) activate certain genes. Cells exposed to intermediate concentrations (between thresholds 1 and 2; pink) activate a different set of genes and also inhibit the genes induced at the higher concentrations. Those cells encountering low concentrations of morphogen (below threshold 2; blue) activate a third set of genes. (After K. W. Rogers and A. F. Schier. 2011. *Annu Rev Cell Dev Biol* 27: 377–407, based on A. Kicheva and M. González-Gaitán. 2008. *Curr Opin Cell Biol* 20: 137–143.)

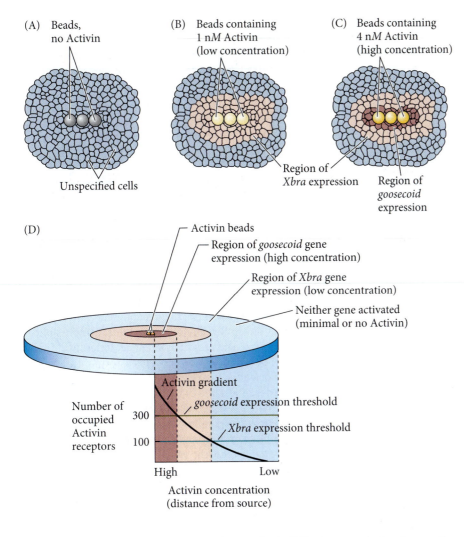

(A) Beads, no Activin

Unspecified cells

(B) Beads containing 1 nM Activin (low concentration)

(C) Beads containing 4 nM Activin (high concentration)

Region of *Xbra* expression

Region of *goosecoid* expression

(D)

Activin beads

Region of *goosecoid* gene expression (high concentration)

Region of *Xbra* gene expression (low concentration)

Neither gene activated (minimal or no Activin)

Number of occupied Activin receptors

300

100

Activin gradient

goosecoid expression threshold

Xbra expression threshold

High   Low

Activin concentration (distance from source)

FIGURE 4.16 A gradient of the paracrine factor Activin, a morphogen, causes concentration-dependent expression differences of two genes in unspecified amphibian cells. (A) Beads containing no Activin did not elicit expression (i.e., mRNA transcription) of either the *Xbra* or *goosecoid* gene. (B) Beads containing 1 nM Activin elicited *Xbra* expression in nearby cells. (C) Beads containing 4 nM Activin elicited *Xbra* expression, but only at a distance of several cell diameters from the beads. A region of *goosecoid* expression is seen near the source bead, however. Thus, it appears that *Xbra* is induced at particular concentrations of Activin and that *goosecoid* is induced at higher concentrations. (D) Interpretation of the *Xenopus* Activin gradient. High concentrations of Activin activate *goosecoid*, whereas lower concentrations activate *Xbra*. A threshold value appears to exist that determines whether a cell will express *goosecoid*, *Xbra*, or neither gene. In addition, Brachyury (the *Xbra* protein product in *Xenopus*) inhibits the expression of *goosecoid*, thereby creating a distinct boundary. This pattern correlates with the number of Activin receptors occupied on individual cells. (After J. B. Gurdon et al. 1994. *Nature* 371: 487–492 and J. B. Gurdon et al. 1998. *Cell* 95: 159–162.)

retard diffusion and enhance degradation. Such diffusion-regulating interactions between morphogens and extracellular matrix factors are very important in coordinating organ growth and shape (Ben Zvi and Barkai 2010; Ben Zvi et al. 2011).

PARACRINE FAMILIES The induction of numerous organs is effected by a relatively small set of paracrine factors that often function as morphogens. The embryo inherits a rather compact genetic "tool kit" and uses many of the same proteins to construct the heart, kidneys, teeth, eyes, and other organs. Moreover, the same proteins are used throughout the animal kingdom; for instance, the factors active in creating the *Drosophila* eye or heart are very similar to those used in generating mammalian organs. Many paracrine factors can be grouped into one of four major families on the basis of their structure:

1. The fibroblast growth factor (FGF) family
2. The Hedgehog family
3. The Wnt family
4. The TGF-β superfamily, encompassing the TGF-β family, the Activin family, the bone morphogenetic proteins (BMPs), the Nodal proteins, the Vg1 family, and several other related proteins

Signal transduction cascades: The response to inducers

For a ligand to induce a cellular response in a cell, it must bind to a receptor, which starts a cascade of events within the cell that ultimately regulates a response. This process is called a **signal transduction cascade**. Paracrine factors function by binding to a

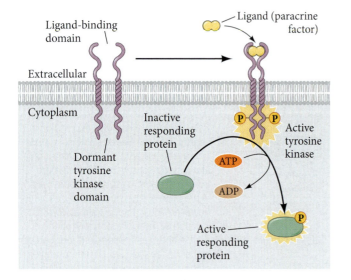

FIGURE 4.17   Structure and function of a receptor tyrosine kinase. The binding of a paracrine factor (such as Fgf8) by the extracellular portion of the receptor protein activates the dormant tyrosine kinase, whose enzyme activity phosphorylates its reciprocal receptor partner followed by specific tyrosine residues of certain intracellular proteins.

receptor that initiates a series of enzymatic reactions within the cell. These enzymatic reactions have as their end point either the regulation of transcription factors or the regulation of the cytoskeleton, which lead to changes in gene expression or cell shape and movement, respectively.

The major signal transduction pathways all appear to be variations on a common and rather elegant theme, exemplified in **FIGURE 4.17**. Each receptor spans the cell membrane and has an extracellular region, a transmembrane region, and a cytoplasmic region. When a paracrine factor binds to its receptor's extracellular domain, the paracrine factor induces a conformational change in the receptor's structure. This shape change is transmitted through the membrane and alters the shape of the receptor's cytoplasmic domain, giving that domain the ability to activate cytoplasmic proteins. Often such a conformational change confers enzymatic activity on the domain, usually a kinase activity that can use ATP to phosphorylate specific tyrosine residues of particular proteins. Thus, this type of receptor is often called a **receptor tyrosine kinase** (**RTK**). The active receptor can now catalyze reactions that phosphorylate other proteins, and this phosphorylation in turn activates their latent activities. Eventually, the *cascade* of phosphorylation activates a dormant transcription factor or a set of cytoskeletal proteins.

Below we describe some of the major characteristics of the four families of paracrine factors, their modes of secretion, gradient manipulation, and the mechanisms underlying transduction in the responding cells. As you explore the use of these pathways in various developmental events throughout the rest of this book, please consider this chapter as a great resource to reference the underlying mechanisms involved.

Fibroblast growth factors and the RTK pathway

The **fibroblast growth factor** (**FGF**) family of paracrine factors comprises nearly two dozen structurally related members, and the FGF genes can generate hundreds of protein isoforms by varying their RNA splicing or initiation codons in different tissues (Lappi 1995). Fgf1 protein is also known as acidic FGF and appears to be important during regeneration (Yang et al. 2005); Fgf2 is sometimes called basic FGF and is very important in blood vessel formation; and Fgf7 sometimes goes by the name of keratinocyte growth factor and is critical in skin development. Although FGFs can often substitute for one another, the expression patterns of the FGFs and their receptors give them separate functions.

FGF8   One member of the FGF family, Fgf8, is important for many different embryonic processes, including segmentation, limb development, and lens induction. Fgf8 is usually made by the optic vesicle that contacts the outer ectoderm of the head (**FIGURE 4.18A**; Vogel-Höpker et al. 2000). This contact by the optic vesicle induces the outer head ectoderm to form a lens. After contact with the outer ectoderm occurs, *Fgf8* gene expression becomes concentrated in the region of the presumptive neural retina (the tissue directly juxtaposed to the presumptive lens) (**FIGURE 4.18B**). Fgf8 was shown to be sufficient for inducing lens formation by placing Fgf8-containing beads adjacent to head ectoderm;[4] this ectopic Fgf8 induced the ectoderm to express the lens-associated transcription factor L-Maf and produce ectopic lenses (**FIGURE 4.18C**). FGFs often work by activating a set of receptor tyrosine kinases called the **fibroblast growth factor receptors** (**FGFRs**). (See **Further Development 4.8, Downstream Events of the FGF Signal Transduction Cascade**, online.)

When an FGFR binds an FGF ligand (and *only* when it binds an FGF ligand), the dormant kinase (part of the receptor) is activated and phosphorylates first its dimer FGFR partner and then other associated proteins within the responding cell. These proteins, once activated, can perform new functions. The RTK pathway (**FIGURE 4.19**) was one

(A)

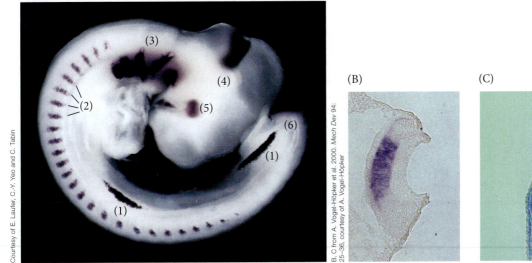

Courtesy of E. Laufer, C.-Y. Yeo and C. Tabin

(B)

(C)

B, C from A. Vogel-Höpker et al. 2000. *Mech Dev* 94: 25–36, courtesy of A. Vogel-Höpker

Contact with optic vesicle

Contact with Fgf8 bead

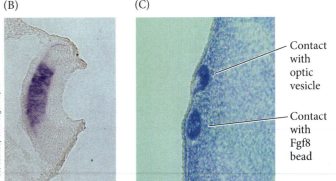

FIGURE 4.18 Fgf8 in the developing chick. (A) *Fgf8* gene expression pattern in the 3-day chick embryo, shown by in situ hybridization. Fgf8 protein (dark areas) is seen in the distalmost limb bud ectoderm (1), in the somitic mesoderm (the segmented blocks of cells along the anterior-posterior axis, 2), in the pharyngeal arches of the neck (3), at the boundary between the midbrain and hindbrain (4), in the optic vesicle of the developing eye (5), and in the tail (6). (B) In situ hybridization of *Fgf8* in the optic vesicle. The *Fgf8* mRNA (purple) is localized to the presumptive neural retina of the optic cup and is in direct contact with the outer ectodermal cells that will become the lens. (C) Ectopic expression of L-Maf in competent ectoderm can be induced by the optic vesicle (above) and by an Fgf8-containing bead (below).

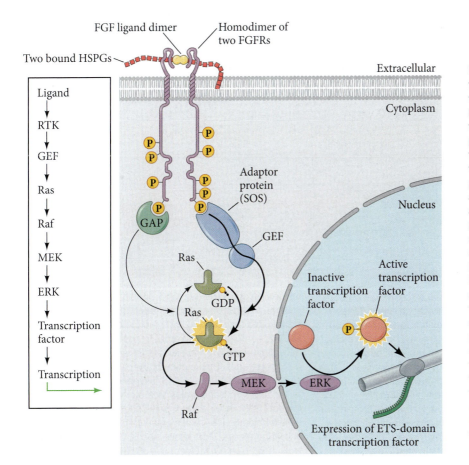

FIGURE 4.19 The widely used RTK signal transduction pathway can be activated by fibroblast growth factor. The receptor tyrosine kinase is dimerized by the ligand (a paracrine factor, such as FGF) along with heparan sulfate proteoglycans (HSPG), which together cause the dimerization and autophosphorylation of the RTKs. The adaptor protein recognizes the phosphorylated tyrosines on the RTK and activates an intermediate protein, GEF, which activates the Ras G protein by allowing phosphorylation of the GDP-bound Ras. At the same time, the GAP protein stimulates hydrolysis of this phosphate bond, returning Ras to its inactive state. The active Ras activates the Raf Protein kinase C, which in turn phosphorylates a series of kinases (such as MEK). Eventually, the activated kinase ERK alters gene expression in the nucleus of the responding cell by phosphorylating certain transcription factors (which can then enter the nucleus to change the types of genes transcribed) and certain translation factors (which alter the level of protein synthesis). In many cases, this pathway regulates the gene expression of ETS-domain (E26 transformation-specific) transcription factor. A simplified version of the pathway is shown on the left.

of the first signal transduction pathways to unite the field of developmental biology, as researchers studying *Drosophila* eyes, nematode vulvae, and human cancers found that they were all studying the same genes!

FGFs and the JAK-STAT pathway

Fibroblast growth factors can also activate the JAK-STAT cascade. This pathway is extremely important in the differentiation of blood cells, the growth of limbs, and the activation of the *Casein* gene during milk production (**FIGURE 4.20**; Briscoe et al. 1994; Groner and Gouilleux 1995). The cascade starts when a paracrine factor is bound by the extracellular domain of a receptor that spans the cell membrane, with the cytoplasmic domain of the receptor being linked to **JAK** (*Janus kinase*) proteins. The binding of paracrine factor to the receptor activates the JAK kinases and causes them to phosphorylate the **STAT** (*signal transducers and activators of transcription*) family of transcription factors (Ihle 1996, 2001). The phosphorylated STAT is a transcription factor that can now enter the nucleus and bind to its enhancers. (See **Further Development 4.9, The JAK and STAT of Bone Development**, and **Further Development 4.10, FGF Receptor Mutations**, both online.)

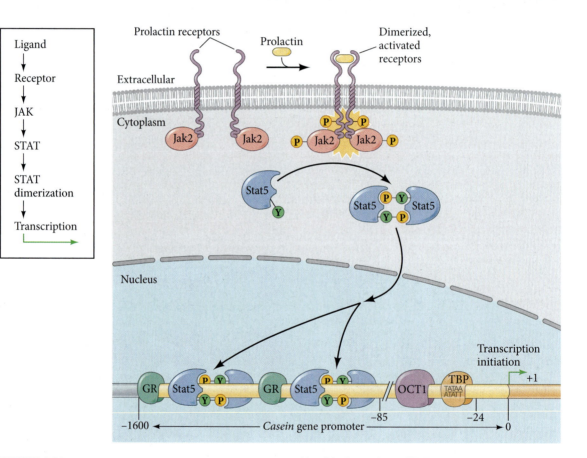

FIGURE 4.20   A JAK-STAT pathway: *casein* gene activation. The gene for casein is activated during the final (lactogenic) phase of mammary gland development, and its activating signal is the secretion of the hormone prolactin from the anterior pituitary gland. Prolactin causes the dimerization of prolactin receptors in the mammary duct epithelial cells. A particular JAK protein (Jak2) is "hitched" to the cytoplasmic domain of these receptors. When the receptors bind prolactin and dimerize, the JAK proteins phosphorylate each other and the dimerized receptors, activating the dormant kinase activity of the receptors. The activated receptors add a phosphate group to a tyrosine residue (Y) of a particular STAT protein, which in this case is Stat5. This addition allows Stat5 to dimerize, be translocated into the nucleus, and bind to particular regions of DNA. In combination with other transcription factors (which have presumably been waiting for its arrival), the Stat5 protein activates transcription of the *casein* gene. GR is the glucocorticoid receptor, OCT1 is a general transcription factor, and TBP is the major promoter-binding protein that anchors RNA polymerase II (see Chapter 3) and is responsible for binding RNA polymerase II. A simplified diagram is shown on the left. (For details, see B. Groner and F. Gouilleux. 1995. *Curr Opin Genet Dev* 5: 587–594.)

The Hedgehog family

The proteins of the **Hedgehog** family of paracrine factors are multifunctional signaling proteins that act in the embryo through signal transduction pathways to induce particular cell types and through other means to influence cell guidance. The original *hedgehog* gene was found in *Drosophila*, in which genes are named after their mutant phenotypes: the loss-of-function *hedgehog* mutation causes the fly larva to be covered with pointy denticles on its cuticle (hairlike structures), thus resembling a hedgehog. Vertebrates have at least three homologues of the *Drosophila hedgehog* gene: *sonic hedgehog* (*shh*), *desert hedgehog* (*dhh*), and *indian hedgehog* (*ihh*). The Desert hedgehog protein is found in the Sertoli cells of the testes, and mice homozygous for a null allele of *Dhh* exhibit defective spermatogenesis. Indian hedgehog is expressed in the gut and cartilage and is important in postnatal bone growth (Bitgood and McMahon 1995; Bitgood et al. 1996). Sonic hedgehog[5] has the greatest number of functions of the three vertebrate Hedgehog homologues. Among other important functions, Sonic hedgehog is responsible for assuring that motorneurons come only from the ventral portion of the neural tube (see Chapter 13), that a portion of each somite forms the vertebrae (see Chapter 17), that the feathers of the chick form in their proper places (see Figure 1 in Further Development 4.6, online), and that our pinkies are always our most posterior digits (see Chapter 19). Hedgehog signaling is capable of regulating these many developmental events because Hedgehog proteins function as morphogens; they are secreted from a cellular source, are displayed in a spatial gradient, and induce differential gene expression at different threshold concentrations that result in distinct cell identities.

HEDGEHOG SECRETION   Different modes of processing and assembly of Hedgehog proteins can significantly alter the amount secreted and the gradient that is formed (**FIGURE 4.21**). By cleaving off its carboxyl terminus and associating with both cholesterol and palmitic acid moieties, Hedgehog protein can be processed and secreted as monomers or multimers, packaged as lipoprotein assemblies, or even transported out of the cell within exovesicles.

In the mouse limb bud, it was shown that if Shh lacks the cholesterol modification, it diffuses too quickly and dissipates into the surrounding space (Li et al. 2006). These lipid modifications are also required for stable concentration gradients of Hedgehog

FIGURE 4.21 Hedgehog processing and secretion. Translation of the *hedgehog* gene in the endoplasmic reticulum produces a Hedgehog protein with autoproteolytic activity that cleaves off the carboxyl terminus (C) to reveal a signal sequence that marks the protein for secretion. The freed C-terminal segment is not involved in signaling and is often degraded, whereas the amino-terminal portion (N) of the molecule becomes the active Hedgehog protein intended for secretion. Secretion requires the addition of cholesterol and palmitic acid to the Hedgehog protein (Briscoe and Thérond 2013). Interactions between the cholesterol moiety and a transmembrane protein called Dispatched enable Hedgehog to be secreted and diffuse as monomers; both cholesterol and palmitic acid are required for multimeric assembly. In addition, Hedgehog interactions with a class of membrane-associated heparan sulfate proteoglycans (HSPGs) foster the congregation and secretion of Hedgehog molecules as lipoprotein assemblies (Breitling 2007; Guerrero and Chiang 2007). Similar clustering of Hedgehog can be used to transport Hedgehog out of the cell within exovesicles.

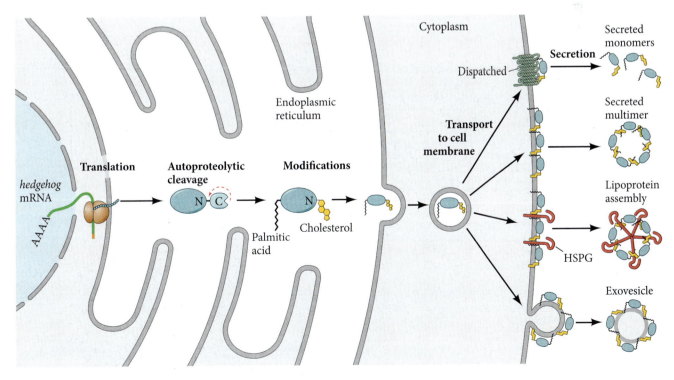

and pathway activation. Through these varied protein processing and transport mechanisms, stable gradients of Hedgehog can be established over distances of several hundred microns (about 30 cell diameters in the mouse limb, for instance).

THE HEDGEHOG PATHWAY   The cholesterol moiety on Hedgehog is not only important for modulating its extracellular transport; it is also critical for Hedgehog to anchor to its receptor on the cell membrane of the receiving cell (Grover et al. 2011). The Hedgehog binding receptor is called Patched, which is a large, 12-pass

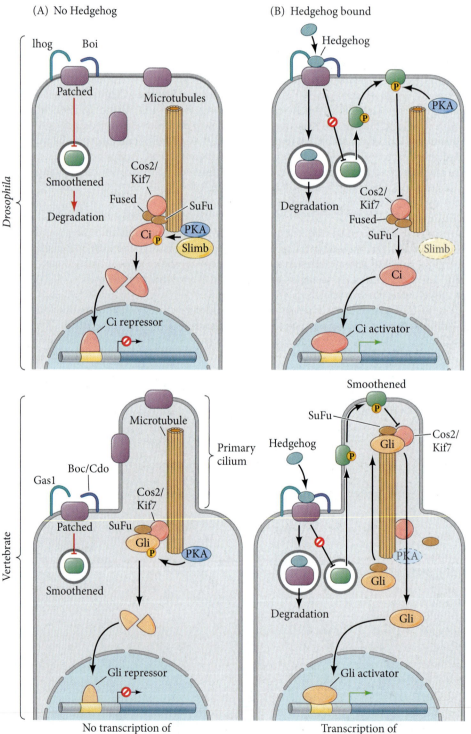

FIGURE 4.22 Hedgehog signal transduction pathway. Patched protein in the cell membrane is an inhibitor of the Smoothened protein. (A) In the absence of Hedgehog binding to Patched, Patched inhibits Smoothened, and in *Drosophila melanogaster* the Ci protein remains tethered to the microtubules by the Cos2 and Fused proteins. This tether allows the proteins PKA and Slimb to cleave Ci into a transcriptional repressor that blocks the transcription of particular genes. (B) When Hedgehog binds to Patched, its conformational changes release the inhibition of the Smoothened protein. Smoothened then releases Ci from the microtubules, inactivating the cleavage proteins PKA and Slimb. The Ci protein enters the nucleus and acts as a transcriptional activator of particular genes. In vertebrates (lower panels), the homologues of Ci are the Gli proteins, which function similarly as transcriptional activators or repressors when a Hedgehog ligand is bound to Patched or absent, respectively. Additionally in vertebrates, for Smoothened to positively regulate Gli processing into an activator form, it needs to gain access into a cellular extension called the primary cilium (see Figure 4.33)—Hedgehog ligand binding to Patched enables the transport of Smoothened into the primary cilium. Last, several co-receptors, such as Gas1 and Boc, function to enhance Hedgehog signaling. (After R. L. Johnson and M. P. Scott. 1998. *Curr Opin Genet Dev* 8: 450–456; J. Briscoe and P. P. Thérond. 2013. *Nat Rev Mol Cell Biol* 14: 416–429; E. Yao and P. T. Chuang. 2015. *J Formos Med Assoc* 114: 569–576.)

transmembrane protein (**FIGURE 4.22**). Unexpectedly, though, Patched is not a signal transducer. Rather, the Patched protein represses the function of another transmembrane receptor called Smoothened.

In the absence of Hedgehog binding to Patched, Smoothened is inactive and degraded, and a transcription factor—Cubitus interruptus (Ci) in *Drosophila* or one of its vertebrate homologues, Gli1, Gli2, or Gli3—is tethered to the microtubules of the responding cell. Although tethered to the microtubules, Ci/Gli is cleaved in such a way that a portion of it enters the nucleus and acts as a transcriptional repressor. This cleavage reaction is catalyzed by several proteins, among them Fused, Suppressor of Fused (SuFu), and Protein kinase A (PKA).

When Hedgehog is present, the responding cells express several additional co-receptors (Ihog/Cdo, Boi/Boc, and Gas1) that together foster strong Hedgehog-Patched interactions. Upon binding, the Patched protein's shape is altered such that it no longer inhibits Smoothened, and Patched enters an endocytic pathway for degradation. Smoothened releases Ci/Gli from the microtubules (probably by phosphorylation), and the full-length Ci/Gli protein can now enter the nucleus to act as a transcriptional *activator* of the same genes the cleaved Ci/Gli used to repress (see Figure 4.22; Lum and Beachy 2004; Briscoe and Thérond 2013; Yao and Chuang 2015).

FURTHER DEVELOPMENT

THE SONIC HEDGEHOG HAS A DIVERSITY OF POWERS!   The Hedgehog pathway is extremely important in vertebrate limb patterning, neural differentiation and pathfinding, retinal and pancreas development, and craniofacial morphogenesis, among many other processes (**FIGURE 4.23A**; McMahon et al. 2003). When mice were made homozygous for a mutant allele of *Sonic hedgehog*, they had major limb and facial abnormalities. The midline of the face was severely reduced, and a single eye formed in the center of the forehead, a condition known as cyclopia, after the one-eyed Cyclops of Homer's *Odyssey* (**FIGURE 4.23B**; Chiang et al. 1996). Some human cyclopia syndromes are caused by mutations in genes that encode either Sonic hedgehog or the enzymes that synthesize cholesterol (Kelley et al. 1996; Roessler et al. 1996; Opitz and Furtado 2012). Moreover, certain chemicals that induce cyclopia do so by interfering with the Hedgehog pathway (Beachy et al. 1997; Cooper et al. 1998). Two teratogens known to cause cyclopia in vertebrates are jervine and cyclopamine.[6] Both are alkaloids found in the plant *Veratrum californicum* (corn lily), and both directly bind to and inhibit Smoothened function (see Figure 4.23B; Keeler and Binns 1968).

There are other targets for Hedgehog signaling independent of Gli transcription factors, and these so-called noncanonical Hedgehog signaling mechanisms can lead to the fast remodeling of the actin cytoskeleton to cause directed cell migration. For instance, the Charron lab has shown that pathfinding axons in the neural tube can sense the presence of a gradient of Sonic hedgehog emanating from the neural tube's floor plate, which will serve to attract commissural neurons

(A)

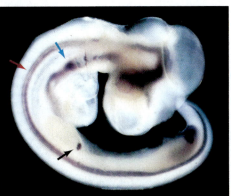

Courtesy of C. Tabin

(B)

Courtesy of L. James and USDA Poisonous Plant Laboratory

FIGURE 4.23   (A) Sonic hedgehog is shown by in situ hybridization to be expressed in the nervous system (red arrow), gut (blue arrow), and limb bud (black arrow) of a 3-day chick embryo. (B) Head of a cyclopic lamb born of a ewe that ate *Veratrum californicum* early in pregnancy. The cerebral hemispheres fused, resulting in the formation of a single central eye and no pituitary gland. The jervine alkaloid made by this plant inhibits cholesterol synthesis, which is needed for Hedgehog production and reception.

to turn toward the midline and cross to the other hemisphere of the nervous system (see Chapter 15; Yam et al. 2009; Sloan et al. 2015).

In later development, Sonic hedgehog is critical for feather formation in the chick embryo, for hair formation in mammals, and, when misregulated, for the formation of skin cancer in humans (Harris et al. 2002; Michino et al. 2003). Although mutations that inactivate the Hedgehog pathway can cause malformations, mutations that activate the pathway ectopically can have mitogenic effects and cause cancers. If the Patched protein is mutated in somatic tissues such that it can no longer inhibit Smoothened, it can cause tumors of the basal cell layer of the epidermis (basal cell carcinomas). Heritable mutations of the *PATCHED* gene cause basal cell nevus syndrome, a rare autosomal dominant condition characterized by both developmental anomalies (fused fingers, rib and facial abnormalities) and multiple malignant tumors (Hahn et al. 1996; Johnson et al. 1996). Interestingly, vismodegib, a compound that inhibits Smoothened function in a manner similar to that of cyclopamine, is currently in clinical trials as a therapy to combat basal cell carcinomas (Dreno et al. 2014; Erdem et al. 2015). (What do you think the warnings for pregnancy should be on this drug?)

The Wnt family

The Wnts are paracrine factors that make up a large family of cysteine-rich glycoproteins with at least 11 conserved Wnt members among vertebrates (Nusse and Varmus 2012); 19 separate *Wnt* genes are found in humans![7] The Wnt family was originally discovered and named *wingless* during a forward genetic screen in *Drosophila melanogaster* in 1980 by Christiane Nüsslein-Volhard and Eric Wieschaus. As you might have surmised, mutations at this locus prevent the formation of the wing. The Wnt name is a fusion of the *Drosophila* segment polarity gene *wingless* with the name of one of its vertebrate homologues, *integrated*. The enormous array of different *Wnt* genes across species speaks to their importance in an equally large number of developmental events. For example, Wnt proteins are critical in establishing the polarity of insect and vertebrate limbs, in promoting the proliferation of stem cells, in regulating cell fates along axes of various tissues, in development of the mammalian urogenital system (**FIGURE 4.24**), and in guiding the migration of mesenchymal cells and pathfinding axons. The best example of the Wnt family's relevance to all things is its evolutionary age, as *wnt* related genes have been discovered to be present in the most basal of extant metazoans (see Chapter 24). How is it that Wnt signaling is capable of mediating such diverse processes as cell division, cell fate, and cell guidance?

WNT SECRETION: PREPROCESSING   As with the building of the functional Hedgehog proteins, Wnt proteins are synthesized in the endoplasmic reticulum and modified by the addition of lipids (palmitic and palmitoleic acid). These lipid modifications are catalyzed by the *O*-acetyltransferase Porcupine. (How do you think this enzyme received this name?)[8] It is interesting that loss of the *Porcupine* gene results in reduced Wnt secretion paired with its buildup in the endoplasmic reticulum (van den Heuvel et al. 1993; Kadowaki et al. 1996), indicating that adding lipids to Wnt is important for transporting it to the cell membrane. Once at the cell membrane, Wnt can be secreted by the same mechanisms we saw for Hedgehog protein: by free diffusion, by being transported in exovesicles, or by being packaged in lipoprotein particles (Tang et al. 2012; Saito-Diaz et al. 2013; Solis et al. 2013).

WNT SECRETION: NEGATIVE FEEDBACK AT THE FRONT DOOR
Upon secretion, Wnt proteins associate with glypicans (a type of heparan sulfate proteoglycan) in the extracellular matrix, which restricts diffusion and leads to a greater accumulation of Wnt closer to the source

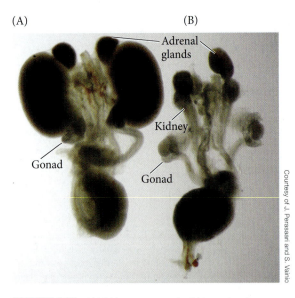

(A)   (B)

Adrenal glands

Kidney

Gonad

Gonad

Courtesy of J. Perasaari and S. Vainio

FIGURE 4.24   Wnt4 is necessary for kidney development and for female sex determination. (A) Urogenital rudiment of a wild-type newborn female mouse. (B) Urogenital rudiment of a newborn female mouse with targeted knockout of *Wnt4* shows that the kidney fails to develop. In addition, the ovary starts synthesizing testosterone and becomes surrounded by a modified male duct system.

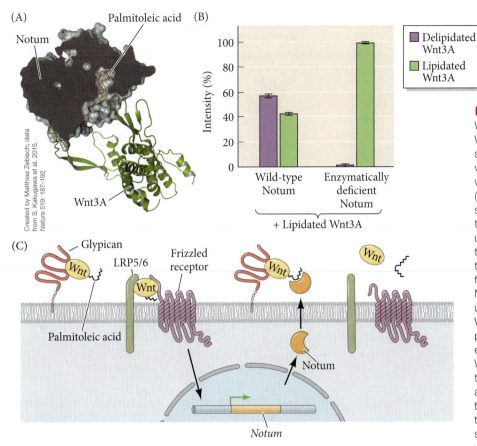

Created by Matthias Zebisch; data from S. Kakugawa et al. 2015. *Nature* 519: 187–192

FIGURE 4.25 Notum antagonism of Wnt. (A) Structures of Notum (gray) and Wnt3A (green) bound together. The active site of Notum is visualized in this cutaway view demonstrating the precise binding with the palmitoleic acid moiety of Wnt3A (orange). (B) Once bound, Notum possesses the enzymatic hydrolase activity to cleave this lipid off Wnt3A, rendering it unable to interact with the Frizzled receptor. The data shown here demonstrate the requirement of this hydrolase function for appropriate delipidation of Wnt3A. Notum lacking its enzymatic activity is unable to remove the lipid group from Wnt3A (delipidated, purple bars) as compared with wild-type Notum. (C) Model of extracellular regulation of Wnt. Lipidated Wnt can bind to its Frizzled receptor and the large transmembrane protein LRP5/6, as well as glypicans (heparan sulfate proteoglycans). Active Wnt signaling leads to the upregulation of Notum, which is secreted and interacts with glypicans, where it will also bind to and cleave off the palmitoleic acid portions of Wnt proteins. In this way, Wnt signaling leads to a Notum-mediated negative feedback mechanism. (B,C after S. Kakugawa et al. 2015. *Nature* 519: 187–192.)

of production. When Wnt attaches to the Frizzled receptor on a responding cell, the cell secretes Notum, a hydrolase that associates with glypican and then cleaves off Wnt's attached lipids in a process of *deacylation* or *delipidation* (Kakugawa et al. 2015). This process reduces Wnt signaling because the lipids are essential for Wnt to bind to Frizzled, which creates a negative feedback mechanism for preventing excessive Wnt signaling.

The Frizzled receptor possesses a unique hydrophobic cleft adapted to interact with lipidated Wnts, a binding conformation mimicked in Notum's structure as well (**FIGURE 4.25A,B**). Overexpression of Notum in the *Drosophila* imaginal wing disc causes a reduction in Wnt/Wg target gene expression; in contrast, clonal loss of *Notum* yields to expanded Wnt target gene expression. Interestingly, *Notum* gene expression is upregulated in Wnt-responsive cells, creating a mechanism of negative feedback (**FIGURE 4.25C**; Kakugawa et al. 2015; Nusse 2015). Notum is not alone in functioning to inhibit binding of Wnt to its receptor; numerous antagonists exist, including the Secreted frizzled-related protein (sFRP), Wnt inhibitory factor (Wif), and members of the Dickkopf (Dkk) family (Niehrs 2006). Together, the multiple modes of Wnt secretion, glypican-mediated restriction, secreted ligand inhibitors, and negative feedback establish both diverse and stable gradients of Wnt ligands and pathway response.

THE CANONICAL WNT PATHWAY (β-CATENIN DEPENDENT)   The first Wnt pathway to be characterized was the canonical *Wnt/β-catenin pathway*, which represents the signaling events that culminate in the activation of the β-catenin transcription factor and modulation of specific gene expression (**FIGURE 4.26A**; Chien et al. 2009; Clevers and Nusse 2012; Nusse 2012; Saito-Diaz et al. 2013). In Wnt/β-catenin signaling, lipidated Wnt family members interact with a pair of transmembrane receptor proteins: one from the Frizzled family and one large transmembrane protein called LRP5/6 (Logan and Nusse 2004; MacDonald et al. 2009).

In the absence of Wnts, the transcriptional cofactor β-catenin is constantly being degraded by a protein degradation complex containing several proteins (such as

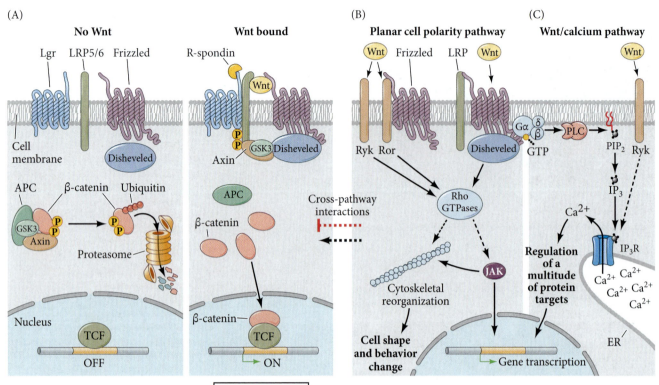

(A)

No Wnt   Wnt bound

(B) Planar cell polarity pathway

(C) Wnt/calcium pathway

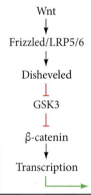

FIGURE 4.26 Wnt signal transduction pathways. (A) The canonical, or β-catenin-dependent, Wnt pathway. The Wnt protein binds to its receptor, a member of the Frizzled family, but it often does so in combination with interactions with LRP5/6 and Lgr (Leucine-rich repeat-containing G-protein coupled receptor). During periods of Wnt absence, β-catenin interacts with a complex of proteins, including GSK3, APC, and Axin, that target Wnt for protein degradation in the proteasome. The downstream transcriptional effector of Wnt signaling is the β-catenin transcription factor. In the presence of certain Wnt proteins, Frizzled then activates Disheveled, allowing Disheveled to become an inhibitor of glycogen synthase kinase 3 (GSK3). GSK3, if it were active, would prevent the dissociation of β-catenin from the APC protein. So by inhibiting GSK3, the Wnt signal frees β-catenin to associate with its co-factors (LEF or TCF) and become an active transcription factor. (B,C) Alternatively, noncanonical (β-catenin-independent) Wnt signaling pathways can regulate cell morphology, division, and movement. (B) Certain Wnt proteins can similarly signal through Frizzled to activate Disheveled but in a way that leads to the activation of Rho GTPases, such as Rac and RhoA. These GTPases coordinate changes in cytoskeleton organization and also through Janus kinase (JAK) regulate gene expression. (C) In a third pathway, certain Wnt proteins activate Frizzled and Ryk receptors in a way that releases calcium ions from the smooth endoplasmic reticulum (ER) and can result in Ca²⁺-dependent gene expression. (After B. MacDonald et al. 2009. *Dev Cell* 17: 9–26.)

Axin and APC) as well as **glycogen synthase kinase 3 (GSK3)**. GSK3 phosphorylates β-catenin so that it will be recognized and degraded by proteosomes. The result is Wnt-responsive genes being repressed by the LEF/TCF transcription factor.

When Wnts come into contact with a cell, they bring together the Frizzled and LRP5/6 receptors to form a multimeric complex. This linkage enables LRP5/6 to bind both Axin and GSK3, and enables the Frizzled protein to bind Disheveled. This complex remains bound to the cell membrane (via Disheveled), which prevents β-catenin from being phosphorylated by GSK3. Thus, this process leads to the stabilization and accumulation of β-catenin, which enters the nucleus, binds to the LEF/TCF transcription factor, and converts this former repressor into a transcriptional activator of Wnt-responsive genes (see Figure 4.26A; Cadigan and Nusse 1997; Niehrs 2012).

This model is undoubtedly an oversimplification because different cells use the pathway in different ways (see McEwen and Peifer 2001; Clevers and Nusse 2012; Nusse 2012; Saito-Diaz et al. 2013). One overriding principle already evident in both the Wnt and the Hedgehog pathways, however, is that *activation is often accomplished by inhibiting an inhibitor.*

THE NONCANONICAL WNT PATHWAYS (β-CATENIN INDEPENDENT)   In addition to sending signals to the nucleus, Wnt proteins can also cause changes within the cytoplasm that influence cell function, shape, and behavior. These alternative, or *noncanonical*, pathways can be divided into two types: the *planar cell polarity pathway* and the *Wnt/calcium pathway* (**FIGURE 4.26B,C**).

The planar cell polarity, or PCP, pathway functions to regulate the actin and microtubule cytoskeleton, thus influencing cell shape, and often results in bipolar protrusive behaviors necessary for a cell to migrate. Certain Wnts (such as Wnt5a and Wnt11) can activate Disheveled by binding to a different receptor (Frizzled paired with Ror instead of LRP5), and this Ror receptor complex phosphorylates Disheveled in a way that allows it to interact with Rho GTPases (Grumolato et al. 2010; Green et al. 2014). Rho GTPases are colloquially viewed as the "master builders" of the cell because they can activate an array of other proteins (kinases and cytoskeletal binding proteins) that remodel cytoskeletal elements to alter cell shape and movement. Wnt signaling through the PCP pathway is most notable for instructing cell behaviors along the same spatial plane within a tissue and hence is called *planar polarity*. Wnt/PCP signaling can direct cells to divide in the same plane (rather than forming upper and lower tissue compartments) and to move within that same plane (Shulman et al. 1998; Winter et al. 2001; Ciruna et al. 2006; Witte et al. 2010; Sepich et al. 2011; Ho et al. 2012; Habib et al. 2013). In vertebrates, this regulation of cell division and migration is important for establishing germ layers and for extension of the anterior-posterior axis during gastrulation.

As its name implies, the Wnt/calcium pathway leads to the release of calcium stored within cells, and this released calcium acts as an important *secondary messenger* to modulate the function of many downstream targets. In this pathway, Wnt binding to the receptor protein (possibly Ryk, alone or in concert with Frizzled) activates a phospholipase (PLC) whose enzyme activities indirectly release calcium ions from the smooth endoplasmic reticulum (see Figure 4.26C). The released calcium can activate enzymes, transcription factors, and translation factors. In zebrafish, Ryk deficiency impairs Wnt-directed calcium release from internal stores and, as a result, impairs directional cell movement (Lin et al. 2010; Green et al. 2014).

Although each of the three Wnt pathways—β-catenin, PCP, and calcium—possesses primary functions that are different from one another, mounting evidence suggests that there are significant cross interactions between these pathways (van Amerongen and Nusse 2009; Thrasivoulou et al. 2013). For instance, Wnt5-mediated calcium signaling has been shown to *antagonize* the Wnt/β-catenin pathway during vertebrate gastrulation and limb development (Ishitani et al. 2003; Topol et al. 2003; Westfall et al. 2003).

The TGF-β superfamily

There are more than 30 structurally related members of the **TGF-β superfamily**,[9] and they regulate some of the most important interactions in development (**FIGURE 4.27**). The TGF-β superfamily includes the TGF-β family, the Nodal and Activin families, the bone morphogenetic proteins (BMPs), the Vg1 family, and other proteins, including Glial-derived neurotrophic factor (GDNF; necessary for kidney and enteric neuron differentiation) and anti-Müllerian hormone (AMH), a paracrine factor involved in mammalian sex determination. Below we summarize three of these families most widely used throughout development: TGF-βs, BMPs, and Nodal/Activin.

TGF-β   Among members of the **TGF-β family**, TGF-β1, 2, 3, and 5 are important in regulating the formation of the extracellular matrix between cells and for regulating cell division (both positively and negatively). TGF-β1 increases the amount of extracellular matrix

---

(?) Developing Questions

How different are the pathways of Wnt/β-catenin, /calcium, and /PCP? Arguably the most significant challenge to understanding Wnt signaling is figuring out how the different pathways interact. Perhaps we need a more integrated comprehension of signal transduction, one that can predict interactions not only between canonical and noncanonical Wnt signaling pathways, but also among those for all the paracrine factors (Wnt, Hedgehog, FGF, BMP, etc.). What do you think? How would you go about trying to examine meaningful pathway interactions?

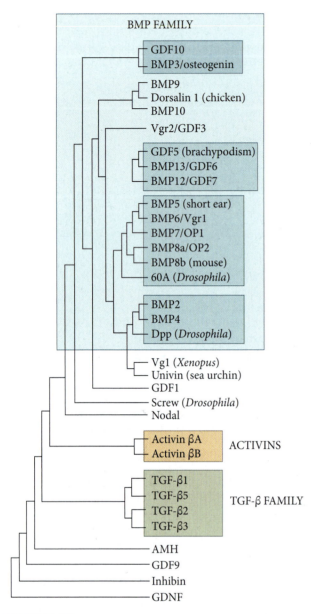

FIGURE 4.27 Relationships among members of the TGF-β superfamily. (After B. L. M. Hogan. 1996. *Genes Dev* 10: 1580–1594; BMP family organization by A. Celeste and V. Rosen, Genetics Institute, Cambridge, MA.)

that epithelial cells make (both by stimulating collagen and fibronectin synthesis and by inhibiting matrix degradation). TGF-β proteins may be critical in controlling where and when epithelia branch to form the ducts of kidneys, lungs, and salivary glands (Daniel 1989; Hardman et al. 1994; Ritvos et al. 1995). The effects of the individual TGF-β family members are difficult to sort out because members of the TGF-β family appear to function similarly and can compensate for losses of the others when expressed together.

BMP The members of the **BMP family** can be distinguished from other members of the TGF-β superfamily by having seven (rather than nine) conserved cysteines in the mature polypeptide. Because they were originally discovered by their ability to induce bone formation, they were given the name **bone morphogenetic proteins**. It turns out, though, that bone formation is only one of their many functions. BMPs have been found to regulate cell division, apoptosis (programmed cell death), cell migration, and differentiation (Hogan 1996). They include proteins such as BMP4 (which in some tissues causes bone formation, in other tissues specifies epidermis, and in other instances causes cell proliferation or cell death) and BMP7 (which is important in neural tube polarity, kidney development, and sperm formation). The BMP4 homologue in *Drosophila* is critically involved in forming appendages, including the limbs, wings, genitalia, and antennae. Indeed, the malformations of 15 such structures have given this homologue the name Decapentaplegic (Dpp). As it (rather oddly) turns out, BMP1 is not a member of the BMP family at all; rather, it is a protease. BMPs are thought to work by diffusion from the cells producing them (Ohkawara et al. 2002). Inhibitors such as Noggin and Chordin that bind directly to BMP reduce BMP-receptor interactions. We will cover this morphogenetic mechanism more directly when we discuss dorsoventral axis specification in the gastrula (see Chapter 11).

NODAL/ACTIVIN The **Nodal** and **Activin** proteins are extremely important in specifying the different regions of the mesoderm and for distinguishing the left and right sides of the vertebrate body axis. The left-right asymmetry of bilateral organisms is strongly influenced by a gradient of Nodal from right to left across the embryo. In vertebrates, this Nodal gradient appears to be created by the beating of motile cilia that promotes the graded flow of Nodal across the midline (Babu and Roy 2013; Molina et al. 2013; Blum et al. 2014; Su 2014).

THE SMAD PATHWAY Members of the TGF-β superfamily activate members of the **Smad family** of transcription factors (Heldin et al. 1997; Shi and Massagué 2003). The TGF-β ligand binds to a type II TGF-β receptor, which allows that receptor to bind to a type I TGF-β receptor. Once the two receptors are in close contact, the type II receptor phosphorylates a serine or threonine on the type I receptor, thereby activating it. The activated type I receptor can now phosphorylate the Smad[10] proteins (**FIGURE 4.28A**). Smads 1 and 5 are activated by the BMP family of TGF-β factors, whereas the receptors binding Activin, Nodal, and the TGF-β family phosphorylate Smads 2 and 3. These phosphorylated Smads bind to Smad4 and form the transcription factor complexes that will enter the nucleus to regulate gene expression (**FIGURE 4.28B**).

Other paracrine factors

Although most paracrine factors are members of one of the four major families described above (FGF, Hedgehog, and Wnt families and the TGF-β superfamily), some

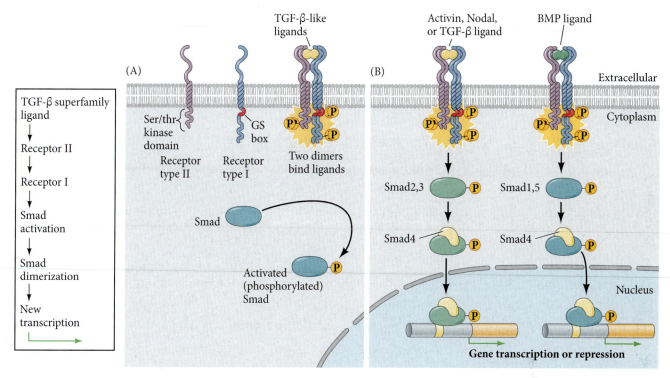

FIGURE 4.28 The Smad pathway is activated by TGF-β superfamily ligands. (A) An activation complex is formed by the binding of the ligand by the type I and type II receptors, which allows the type II receptor to phosphorylate the type I receptor on particular serine or threonine residues. The phosphorylated type I receptor protein can now phosphorylate the Smad proteins. (GS box: a domain rich in serine-glycine repeats.) (B) Those receptors that bind TGF-β family proteins or members of the Activin family phosphorylate Smads 2 and 3. Those receptors that bind to BMP family proteins phosphorylate Smads 1 and 5. These Smads can complex with Smad4 to form active transcription factors. A simplified version of the pathway is shown on the left.

paracrine factors have few or no close relatives. Epidermal growth factor, hepatocyte growth factor, neurotrophins, and stem cell factor are not included among these four groups, but each plays important roles during development. In addition, there are numerous paracrine factors involved almost exclusively with developing blood cells: erythropoietin, the cytokines, and the interleukins. Another class of paracrine factors was first characterized for its role in cell/axon guidance and includes members of the Netrin, Semaphorin, and Slit families. These classic guidance molecules (such as netrins) are now being shown to regulate gene expression as well. We will discuss all these paracrine factors in the context of their developmental relevance later in the book.

FURTHER DEVELOPMENT

Auxin: A plant morphogen

To grow, in the case of an animal or a plant, might mean that cells divide, elongate, and expand, or even differentiate into their final fates. The auxin (from the Greek *auxien*, "to grow") family of plant hormones is arguably the most studied regulator of plant growth. In fact, auxin and its collaborating machinery have a long evolutionary history that goes back to the first land plants—such as mosses and liverworts—and even to the ancestral green algae. Auxin is involved in shaping many key aspects of plant embryogenesis, from cell fate specification along the apical-basal axis to lateral root morphogenesis and the formation of the serrated edges found on some leaves. Auxin is a paracrine-like signaling molecule, satisfying most of the criteria of a morphogen, yet auxin functions a bit differently than the animal growth factors we have discussed. Here we detail a few of the key functions of auxin, how its signal transduction pathway serves these functions, and the unique mechanism used to establish auxin concentration gradients.

AUXIN-MEDIATED APICAL-BASAL POLARITY   The apical-basal organization of most adult organisms—from inside to out in animals and from shoots to roots in plants—is set up during early embryogenesis. What are the signals driving this early axis determination in plants? In his book *The Power of Movement in Plants*, Charles Darwin predicted the existence of auxin, yet the chemical structure of

auxin or **indole-3-acetic acid** (**IAA**) was discovered in 1935 by Kenneth Thimann and J. B. Koepfli. Auxin has been demonstrated to play an essential role in determining the apical-basal axis in plants (Möller and Weijers 2009; Robert et al. 2015; Smit and Weijers 2015; ten Hove et al. 2015; Weijers and Wagner 2016; Jiang et al. 2018). In *Arabidopsis thaliana*, as in most embryophytes, a morphological difference along the apical-basal axis is set up with the zygote's very first division, which results in two asymmetrically sized cells (see Figure 1.8). Recall that the apical cell develops into the embryo proper, which passes through the octant, globular, and heart stages of embryogenesis before the shoot apical and root apical meristems (SAM and RAM), hypocotyl, and cotyledons emerge. How does auxin influence the correct positional development of the meristems, hypocotyl, and cotyledons along the apical-basal axis of the plant embryo?

AUXIN SYNTHESIS AND DISTRIBUTION   A remarkable aspect of auxin signaling is how this hormone gets differentially distributed throughout the plant. To map the distribution of auxin in the plant embryo, researchers created a transgenic strain of *A. thaliana* carrying a construct (*DR5rev:GFP*) consisting of *GFP* fused to *DR5*, an auxin-responsive promoter. As a result, these plants express GFP where auxin is active, providing a convenient way to monitor auxin pathway responses over time during embryonic development (**FIGURE 4.29A**). Auxin moves from cell to cell by both diffusion and active transport, with the result that different regions accumulate different concentrations of the hormone. Regions that accumulate auxin are called auxin *sinks*, and the highest and lowest concentrations are often referred to as auxin maxima and minima, respectively. In embryos, auxin maxima can be seen in the cells that will develop into the RAM, as well as at the apex of each cotyledon (see Figure 4.29A). Enzymes such as YUCCA (YUC) and TRYPTOPHAN AMINOTRANSFERASE OF ARABIDOPSIS 1 (TAA1) catalyze the biosynthesis of auxin in the most apical cells of octant and 16-cell embryos (**FIGURES 4.29B** and **4.30A**). This suggests that the main source of auxin is at the opposite end of the embryo from the sink, in the basal-most cells showing maximal auxin responses.

Plants, because of their rigid cell walls, have evolved elaborate mechanisms for cell-to-cell communication. In the case of auxin, this communication relies heavily on the polar distribution of auxin carriers known as PIN proteins (Friml et al. 2003). PIN proteins function primarily to move auxin from inside to outside the cell—a process termed **efflux transport** (see Figure 4.29B). Specific PIN transmembrane proteins are often asymmetrically localized to one side of a cell's surface, increasing auxin efflux from that side of the cell. Thus, strategic *polar* expression of these efflux PIN carriers determines the direction of auxin movement within the embryo (and throughout tissues of the adult plant, too), and PIN-dependent auxin flows have been shown to be critical for morphogenesis. *A. thaliana pin1-1* mutants, which lack the PIN1 auxin efflux protein, fail to separate their cotyledons, and the organ forming activity of the SAM is disrupted (**FIGURE 4.29C**; Liu et al. 1993). In the octant embryo, targeted positioning of PIN carriers causes auxin to flow basally from the apical cells to the future cells of the root. As the embryo continues to develop, the flow of auxin follows a stereotypical recirculating path, which leads to apical and basal auxin maxima with a gradient of lower auxin concentrations in between (**FIGURE 4.29D**; Robert et al. 2013, 2015).

THE AUXIN PATHWAY   How does auxin signaling influence differential gene expression across the plant embryo? Auxin plays a "key" role in the regulation of cell fate-determining genes as it essentially "unlocks" the repression of auxin-dependent transcription. The downstream effector protein of the auxin pathway is AUXIN RESPONSE FACTOR (ARF), a transcription factor that activates the expression of auxin-regulated cell fate programs (**FIGURE 4.30B**; Roosjen et al. 2018).[11] *When auxin concentrations are low*, ARF function is inhibited by the binding of an auxin repressor protein named AUX/IAA. (This name

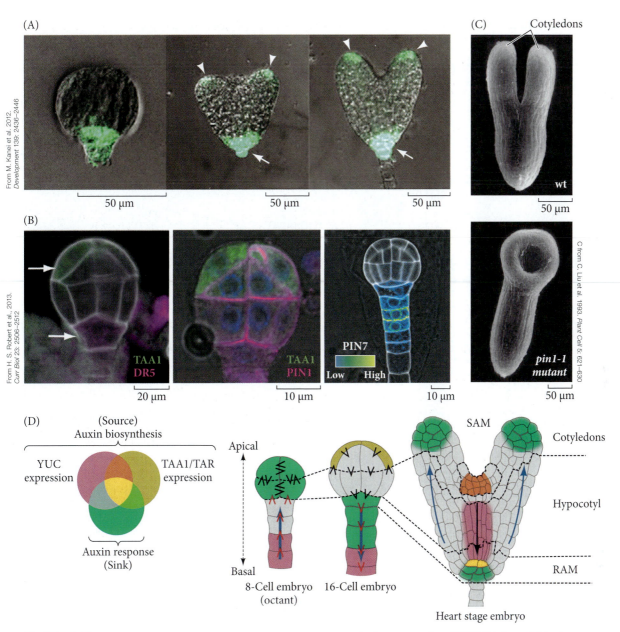

From M. Kanei et al. 2012. *Development* 139: 2436–2446

From H. S. Robert et al. 2013. *Curr Biol* 23: 2506–2512

C from C. Liu et al. 1993. *Plant Cell* 5: 621–630

FIGURE 4.29   Auxin signaling during apical-basal speci-fication in the *Arabidopsis thaliana* embryo. (A) Transgenic embryos with the *DR5rev:GFP* auxin-responsive reporter, which expresses green fluorescent protein in those cells with highest upregulation of auxin-responsive genes. Expression becomes concentrated in the apical cells of the cotyledons (arrowheads) and in the root apical meristem (RAM, arrow) over the course of early development.
(B) Enzymes such as TAA1 (first panel; green fluorescence) function to synthesize auxin in the most apical cells of the octant embryo. The auxin then travels to the precursors of the RAM (first panel; indicated by DR5-driven magenta fluo-rescence) via efflux PIN carriers such as PIN1 (second panel;

magenta fluorescence) and PIN7 (third panel; blue/green fluorescence). (C) Loss of PIN1 in the *pin1-1* mutant results in the complete failure to separate the two cotyledons. (D) Schematic model of the transport of auxin from its source to its sink. YUCs and Tryptophan Aminotransferase of Arabi-dopsis1/Tryptophan Aminotransferase-Related (TAA1/TAR) enzymes are involved in auxin synthesis. The colors indicate the embryo regions that express these enzymes and auxin-responsive genes. Black and red arrowheads indicate the direction of auxin transport by PIN1 and PIN7, respectively. Blue arrows indicate the overall direction of auxin movement. (D after H. S. Robert et al. 2015. *J Exp Bot* 66: 5029–5042.)

is confusing because it is so similar to that of auxin itself; therefore, we will unconventionally refer to this repressor as AUX-Rep for clarity.) When AUX-Rep is bound to ARF, auxin-dependent gene expression is prevented. When AUX-Rep is bound to ARF it attracts the co-repressor TOPLESS (TPL), which

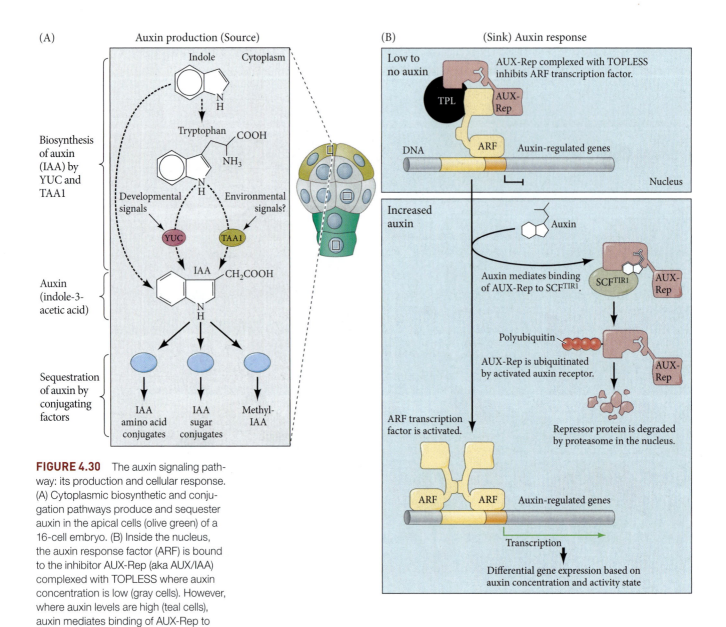

(A) Auxin production (Source)

Biosynthesis of auxin (IAA) by YUC and TAA1

Auxin (indole-3-acetic acid)

Sequestration of auxin by conjugating factors

(B) (Sink) Auxin response

FIGURE 4.30 The auxin signaling pathway: its production and cellular response. (A) Cytoplasmic biosynthetic and conjugation pathways produce and sequester auxin in the apical cells (olive green) of a 16-cell embryo. (B) Inside the nucleus, the auxin response factor (ARF) is bound to the inhibitor AUX-Rep (aka AUX/IAA) complexed with TOPLESS where auxin concentration is low (gray cells). However, where auxin levels are high (teal cells), auxin mediates binding of AUX-Rep to the ubiquitin machinery (SCF^{TIR1}), leading to ubiquitylation and proteosomal degradation of AUX-Rep and subsequent activation of ARF-mediated transcription of auxin-response genes. (A after D. Weijers and J. Friml. 2009. *Cell* 136: 1172; B after L. Taiz et al. *Plant Physiology and Development*, 6th Edition. Sinauer Associates: Sunderland, MA.)

leads to the repression of auxin-dependent gene expression. *When auxin levels are high*, the hormone leads to degradation of the AUX-Rep protein, which releases ARF from inhibition. Thus, in this double negative manner—inhibition of an inhibitor—ARF is released and ARF-dependent transcription is permitted (Weijers and Wagner 2016; Roosjen et al. 2018).

So far, we have provided a highly simplified description of auxin signaling, and it's important to understand that there are many additional ways auxin activity and distribution can be shaped. For example, auxin can be sequestered and thereby reversibly inactivated by conjugating proteins, and auxin movements can be influenced by other classes of transport proteins (see Figure 4.30B; Jiang et al. 2018). It is proposed that different concentrations of auxin result in differential gene expression that influences different cell fates across the plant, much like the classical morphogens we described earlier. Most interesting is the realization that despite the long independent evolution of plants and animals, auxin signal transduction operates through the negative regulation of a pathway inhibitor, just like Hedgehog and Wnt signaling in animals.

The Cell Biology of Paracrine Signaling

Underlying the developmental effects of paracrine signaling are the cellular mechanisms that function to shape, constrain, and otherwise support the presentation, secretion, and reception of the signaling molecule. What are those mechanisms, and how have they helped foster a wide diversity of morphological patterns?

DIFFUSION OF PARACRINE FACTORS   Paracrine factors do not flow freely through the extracellular space. Rather, factors can be bound by cell membranes and extracellular matrices of the tissues. In some cases, such binding can impede the spread of a paracrine morphogen and even target the paracrine factor for degradation (Capurro et al. 2008; Schwank et al. 2011). Wnt proteins, for instance, do not diffuse far from the cells secreting them unless helped by other proteins. Thus, the range of Wnt factors is significantly extended when the nearby cells secrete proteins that bind to the paracrine factor and prevent it from binding prematurely to the target tissue (**FIGURE 4.31**; Mulligan et al. 2012). Similarly, **heparan sulfate proteoglycans** (**HSPGs**) in the extracellular matrix often modulate the stability, reception, diffusion rate, and concentration gradient of FGF, BMP, and Wnt proteins (Akiyama et al. 2008; Yan and Lin 2009; Berendsen et al. 2011; Christian 2011; Müller and Schier 2011; Nahmad and Lander 2011). (See **Further Development 4.11, A Multitude of Ways to Shape FGF Secretion**, and **Further Development 4.12, Endosome Internalization: Morphogen Gradients Can Be Created by Literally Passing from One Cell to Another**, both online.)

Focal membrane protrusions as signaling sources

We have discussed the roles of secreted growth factors for both short- and long-range cell-to-cell communication. But is there a mechanism to present a signal without secreting it? In such a scenario, the producing cell itself *physically reaches out* and presents the signal. Here we highlight emerging ideas of how two types of dynamic membrane extensions can facilitate intercellular communication and even produce long-range gradients.

THE FILOPODIAL CYTONEME   What if the molecules we thought were diffusible paracrine factors moving through the extracellular matrix were actually transferred from one cell to another at synapse-like connections? There is now significant evidence to support the existence of specialized filopodial projections called **cytonemes**,

(A) *wingless* (Wg) expression

Wild-type (normal *swim*)

swim⁻ mutant

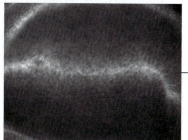

All images from K. A. Mulligan et al. 2012.
Proc Natl Acad Sci USA 109: 370–377

(B) *Distal-less* expression

Wild-type

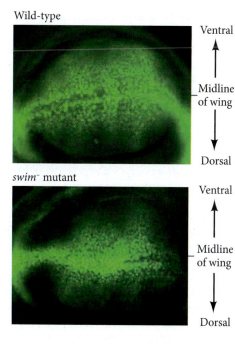

FIGURE 4.31   Wnt diffusion is affected by other proteins. (A) Diffusion of Wingless (Wg, a Wnt paracrine factor) throughout the developing wing of wild-type *Drosophila* (above) is enhanced by Swim, a protein that stabilizes Wg and that is made by some of the wing cells. When Swim is not present, as in the mutant below, Wg does not disperse and is confined to the narrow band of Wg-expressing cells. (B) Similarly, Wingless usually activates the *Distal-less* gene (green) in much of the wild-type wing (above). However, in *swim*-mutant flies (below), the range of *Distal-less* expression is confined to those areas near the band of Wg-expressing cells.

which stretch out remarkable distances (more than 100 μm) from either the target cells or the signal-producing cells, like long membrane conduits connecting the two types of cells (Roy and Kornberg 2015). Under this model, ligand-receptor binding would initially occur at the tips of cytonemes projecting from the target cells when the tips are positioned in direct apposition to the cell membrane of the producing cell. The ligand-receptor complex would then be transported down the cytoneme to the target cell body.

Cytoneme-mediated morphogen signaling was first described by Thomas Kornberg's laboratory in its study of development of the air sac and wing disc in *Drosophila* (Roy et al. 2011). A cluster of cells called the air sac primordium (ASP) develops along the basal surface of the wing disc in response to Dpp (a BMP homologue) and FGF morphogen gradients in the wing disc (**FIGURE 4.32A,B**). The Kornberg lab discovered that the ASP cells extend cytonemes toward the Dpp- and FGF-expressing cells, and that these cytonemes contain receptors for these morphogens—separate receptors in separate cytonemes. Moreover, Dpp bound to its receptor on ASP cells has been documented traveling along a cytoneme to the cell body. Anterior-posterior patterning of the wing disc by a gradient of Hedgehog (Hh) signaling also appears to be accomplished through cytonemes (**FIGURE 4.32C**). Hedgehog coming from posterior cells of the wing disc is delivered through cytonemes that extend from the basolateral surface of anterior cells of the wing disc to the Hh-producing posterior cells (**FIGURE 4.32D,E**; Bischoff et al. 2013).

CYTONEMES IN VERTEBRATES   Recent investigations have shown that vertebrates use cytonemes as well. Work in Michael Brand's lab and recent work by Steffen Scholpp's lab have shown that some gastrulating cells transport the morphogen Wnt8a along cytoneme-like extensions. In this case, the signal-producing cells are extending the cytonemes and transporting the Wnt8a morphogen to target cells (**FIGURE 4.32F**; Luz et al. 2014; Stanganello et al. 2015). Cytoneme-like interactions are also suspected in one of the classic examples of morphogen signaling, that of anterior-posterior specification in the tetrapod limb bud. Here, a posterior-to-anterior gradient of Sonic hedgehog (Shh) in the limb bud leads to the correct patterning of digits (see Chapter 19). In the chick limb bud, both the Sonic hedgehog-expressing cells and the anterior target cells extend filopodial projections toward each other and make contact at points where Sonic hedgehog receptors (Patched) are localized (**FIGURE 4.32G**; Sanders et al. 2013).

THE PRIMARY CILIUM   In many cases, the reception of paracrine factors is not uniform throughout the cell membrane; rather, receptors are often congregated asymmetrically. For instance, the reception of Hedgehog proteins in vertebrate cells occurs on the primary cilium, a focal extension of the cell membrane made by microtubules (**FIGURE 4.33A**; Huangfu et al. 2003; Goetz and Anderson 2010). The primary cilium should not be confused with motile cilia, such as those found lining the trachea or in the node of a gastrulating embryo. The primary cilium is much shorter than motile cilia and largely went unnoticed until we realized its direct role in numerous human diseases. In fact, some of these "ciliopathies," such as Bardet-Biedl syndrome, which affects numerous parts of the body, are suspected to be due to an indirect effect on Hedgehog signaling (Nachury 2014). In unstimulated cells, the Patched protein (the Hedgehog receptor; see Figure 4.22) is located in the primary cilium membrane, whereas the Smoothened protein is in the cell membrane close to the cilium or part of an endosome being targeted for degradation. Patched inhibits Smoothened function by preventing it from entering the primary cilium (Milenkovic et al. 2009; Wang et al. 2009). When Hedgehog binds to Patched, however, Smoothened is allowed to join it on the ciliary cell membrane, where it inhibits the PKA and SuFu proteins that make the repressive form of the Gli transcription factor (**FIGURE 4.33B**). The microtubules of primary cilia provide a scaffold for motor proteins to transport Patched and Smoothened as well as activated Gli proteins, and mutations that knock out cilia formation or their transport mechanism also prevent Hedgehog signaling (see Figure 4.22; Mukhopadhyay and Rohatgi 2014).

? Developing Questions

Are all the molecules that we have considered to be paracrine factors distributed solely by contact through filopodial cytoneme processes, as opposed to diffusion through the extracellular matrix? This question is increasingly coming up in debates among developmental biologists. Where do you stand? Are you a "diffusionist" or a "cytonemist"? Is there room for both mechanisms or perhaps even a developmental need for both?

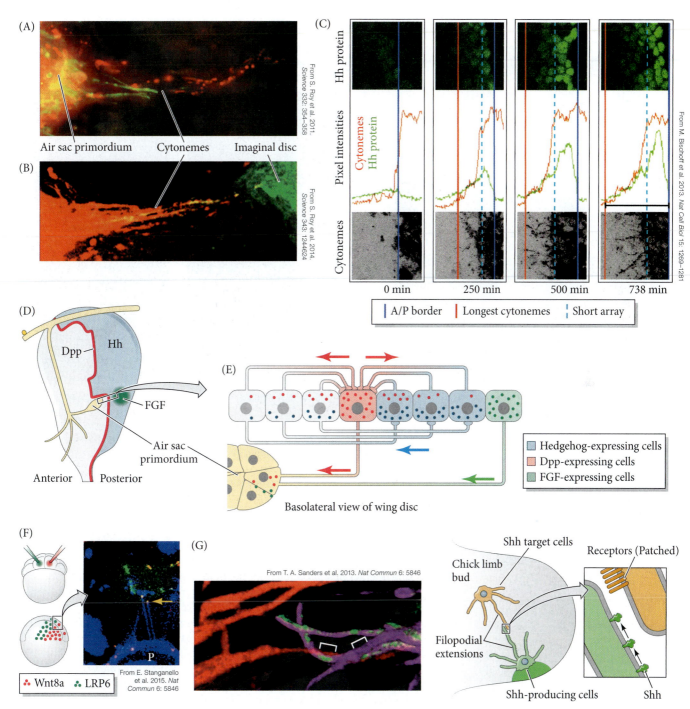

FIGURE 4.32 Filopodia-transported morphogens. (A) Cytonemes from the air sac primordium (ASP) extend toward the epithelium of the wing imaginal disc in *Drosophila* to shuttle the FGF (green) and Dpp (red) morphogens, produced by the wing disc, back to the cell bodies of the ASP. (B) Transported Dpp receptor binds Dpp produced by the wing disc cells, which gets transported back down the cytoneme to the ASP. (C) This system of cytonemes in the *Drosophila* wing disc is capable of establishing a gradient of Hedgehog (Hh) protein (green in top panels and in plot) over the course of filopodial extension (black processes in lower panels and red plot line). (D) Illustration of the *Drosophila* wing imaginal disc during its interactions with tracheal cells, namely the ASP. Hh-, Dpp-, and FGF-expressing cells are represented as blue, red, and green domains. (E) Magnified cross section of the boxed region in (D). Cytoneme extensions from the ASP as well as between cells of the wing disc are illustrated along with the morphogens produced and transported along these cytonemes (arrows). (F) Wnt8a (red) and its receptor LRP6 (green) were microinjected into two different cells of an early-stage zebrafish blastula. Live cell imaging of these cells at the gastrula stage revealed Wnt8a interactions with the LRP6 receptor at the tips of filopodial extensions from the producer cells (P, yellow arrow). (G) In the chick limb bud, long, thin filopodial protrusions have been documented extending both from Sonic hedgehog-producing cells in the posterior region (purple cell with green Shh protein in left image) and from the target cells in the anterior limb bud (red cells). These opposing filopodia directly interact (brackets, left image), and at this point of interaction it is proposed that Shh and its receptor (Patched) can bind (right illustration).

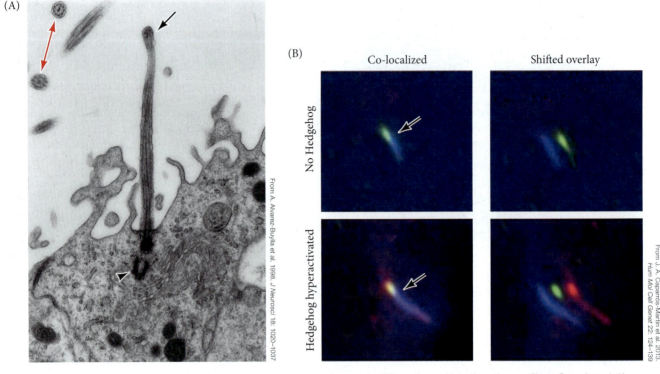

(A)

(B)

Co-localized

Shifted overlay

No Hedgehog

Hedgehog hyperactivated

From A. Alvarez-Buylla et al. 1998. *J Neurosci* 18: 1020–1037

From J. A. Caparrós-Martin et al. 2013. *Hum Mol Cell Genet* 22: 124–139

FIGURE 4.33 The primary cilium for Hedgehog reception. (A) Transmission electron micrograph showing a longitudinal section of the primary cilium (black arrow) of a "B-type cell," a neural stem cell in the adult mammalian brain (see Chapter 5). The centriole serving as a basal body at the base of this cilium is visible (arrowhead); the microtubules in this primary cilium form an 8+0 structure (other types of cilia, such as motile cilia, typically form a 9+2 arrangement; seen in upper left corner in cross section [red arrows]). (B) Activation of the Hedgehog pathway requires the transport of Smoothened into the primary cilium. Seen here is the primary cilium (arrow; immunofluorescence stained for acetylated tubulin, blue) on a fibroblast in culture. The ciliary protein Evc (stained green) co-localizes with Smoothened (red) upon hyperactivation of Hedgehog signaling by the drug SAG. Compare the co-localized labeling on the left with the overlays on the right, which have been shifted to show each individual marker. Activation of the Evc-Smoothened complex in the primary cilium leads to full-length Gli signaling.

Juxtacrine Signaling for Cell Identity

In juxtacrine interactions, proteins from the inducing cell interact with receptor proteins of adjacent responding cells without diffusing from the producing cell. Three of the most widely used families of juxtacrine factors are the **Notch proteins** (which bind to a family of ligands exemplified by the Delta protein); **cell adhesion molecules** such as cadherins; and the **Eph receptors** and their **ephrin ligands**. When an ephrin on one cell binds with the Eph receptor on an adjacent cell, signals are sent to each of the two cells (Davy et al. 2004; Davy and Soriano 2005). These signals are often those of either attraction or repulsion, and ephrins are often seen where cells are being told where to migrate or where boundaries are forming. We will see the ephrins and the Eph receptors functioning in the formation of blood vessels, neurons, and somites. We will now look more closely at the Notch proteins and their ligands.

The Notch pathway: Juxtaposed ligands and receptors for pattern formation

Although most known regulators of induction are diffusible proteins, some inducing proteins remain bound to the inducing cell surface. In one such pathway, cells expressing the Delta, Jagged, or Serrate proteins in their cell membranes activate neighboring cells that contain Notch protein in their cell membranes (see Artavanis-Tsakakonas and Muskavitch 2010). Notch extends through the cell membrane, and its external surface contacts Delta, Jagged, or Serrate proteins extending out from an adjacent cell. When complexed to one of these ligands, Notch undergoes a conformational change that enables a part of its cytoplasmic domain to be cut off by the presenilin-1 protease. The cleaved portion enters the nucleus and binds to a dormant transcription factor of the CSL family. When bound to the Notch protein, the CSL transcription factors activate their target genes (**FIGURE 4.34**; Lecourtois and Schweisguth 1998; Schroeder et al. 1998; Struhl and Adachi 1998). This activation is thought to involve the recruitment of histone acetyltransferases (Wallberg et al. 2002). Thus, Notch can be considered as a transcription factor tethered to the cell membrane. When the attachment is broken, Notch (or a piece of it) can detach from the cell membrane and enter the nucleus (Kopan 2002).

Notch proteins are involved in the formation of numerous vertebrate organs—kidney, pancreas, and heart—and they are extremely important receptors in the nervous system. In both the vertebrate and *Drosophila* nervous systems, the binding of Delta to Notch tells the receiving cell not to become neural (Chitnis et al. 1995; Wang et al. 1998). In the vertebrate eye, the interactions between Notch and its ligands regulate which cells become optic neurons and which become glial cells (Dorsky et al. 1997; Wang et al. 1998). (See **Further Development 4.13, Notch Mutations**, online.)

Paracrine and juxtacrine signaling in coordination: Vulval induction in C. elegans

Induction does indeed occur on the cell-to-cell level, and one of the best examples is the formation of the vulva in the nematode worm *C. elegans*. Remarkably, the signal transduction pathways involved turn out to be the same as those used in the formation of retinal receptors in *Drosophila*; only the targeted transcription factors are different. In both cases, an epidermal growth-factor-like inducer activates the RTK pathway, leading to the differential regulation of Notch-Delta signaling.

Most *C. elegans* individuals are hermaphrodites. In their early development, they are male, and the gonad produces sperm, which are stored for later use. As they grow older, they develop ovaries. The eggs "roll" through the region of sperm storage, are fertilized inside the nematode, and then pass out of the body through the vulva (see Chapter 8; Barkoulas et al. 2013). The formation of the vulva occurs during the larval stage from six cells called the **vulval precursor cells** (**VPCs**). The cell connecting the overlying

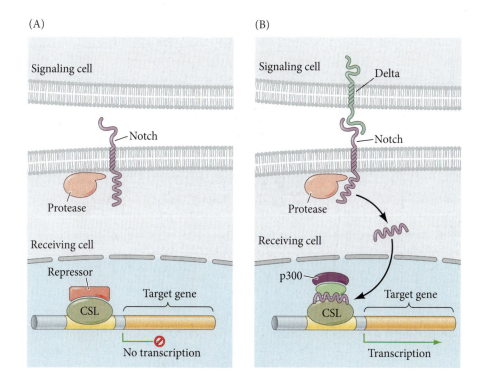

(A)

Signaling cell

Notch

Protease

Receiving cell

Repressor

CSL

Target gene

No transcription

(B)

Signaling cell

Delta

Notch

Protease

Receiving cell

p300

CSL

Target gene

Transcription

FIGURE 4.34 Mechanism of Notch activity. (A) Prior to Notch signaling, a CSL transcription factor (such as Suppressor of hairless or CBF1) is on the enhancer of Notch-regulated genes. The CSL binds repressors of transcription. (B) Model for the activation of Notch. A ligand (Delta, Jagged, or Serrate protein) on one cell binds to the extracellular domain of the Notch protein on an adjacent cell. This binding causes a shape change in the intracellular domain of Notch, which activates a protease. The protease cleaves Notch and allows the intracellular region of the Notch protein to enter the nucleus and bind the CSL transcription factor. This intracellular region of Notch displaces the repressor proteins and binds activators of transcription, including the histone acetyltransferase p300. The activated CSL can then transcribe its target genes. (A after K. Blaschuk and C. ffrench-Constant. 1998. *Curr Biol* 8: R334–R337; B after E. H. Schroeter et al. 1998. *Nature* 393: 382–386.)

gonad to the vulval precursor cells is called the **anchor cell (AC)** (**FIGURE 4.35**). The AC secretes LIN-3 protein, a paracrine factor (similar to mammalian epidermal growth factor, or EGF) that activates the RTK pathway (Hill and Sternberg 1992). If the AC is destroyed (or if the *lin-3* gene is mutated), the VPCs do not form a vulva and instead become part of the hypodermis or skin (Kimble 1981).

The six VPCs influenced by the AC form an **equivalence group**. Each member of this group is competent to become induced by the AC and can assume any of three fates, depending on its proximity to the AC. The cell directly beneath the AC divides to form the central vulval cells. The two cells flanking that central cell divide to become the lateral vulval cells, whereas the three cells farther away from the AC generate hypodermal cells. If the AC is destroyed, all six cells of the equivalence group divide once and contribute to the hypodermal tissue. If the three central VPCs are destroyed, the three outer cells, which normally form hypodermis, generate vulval cells instead.

LIN-3 secreted from the AC forms a concentration gradient, in which the VPC closest to the AC (i.e., the P6.p cell) receives the highest concentration of LIN-3 and generates the central vulval cells. The two adjacent VPCs (P5.p and P7.p) receive lower amounts of LIN-3 and become the lateral vulval cells. VPCs farther away from the AC do not receive enough LIN-3 to have an effect, so they become hypodermis (Katz et al. 1995).

NOTCH-DELTA AND LATERAL INHIBITION We have discussed the reception of the EGF-like LIN-3 signal by the cells of the equivalence group that forms the vulva. Before this induction occurs, however, an earlier interaction has formed the AC. The formation of the AC is mediated by *lin-12*, the *C. elegans* homologue of the *Notch* gene. In wild-type *C. elegans* hermaphrodites, two adjacent cells, Z1.ppp and Z4.aaa, have the potential to become the AC. They interact in a manner that causes one of them to become the AC while the other one becomes the precursor of the ventral uterine tissue. In loss-of-function *lin-12* mutants, both cells become ACs, whereas in gain-of-function mutations, both cells become ventral uterine precursors (Greenwald et al. 1983). Studies using genetic mosaics and cell ablations have shown that this decision is made in the second larval stage, and that the *lin-12* gene needs to

FIGURE 4.35 *C. elegans* vulval precursor cells (VPCs) and their descendants. (A) Location of the gonad, anchor cell, and VPCs in the second instar larva. (B,C) Relationship of the anchor cell to the six VPCs and their subsequent lineages. Primary (1°) lineages result in the central vulval cells, secondary (2°) lineages constitute the lateral vulval cells, and tertiary (3°) lineages generate hypodermal cells. (C) Outline of the vulva in the fourth instar larva. The circles represent the positions of the nuclei. (D) Model for the determination of vulval cell lineages in *C. elegans*. The LIN-3 signal from the anchor cell causes the determination of the P6.p cell to generate the central vulval lineage (dark purple). Lower concentrations of LIN-3 cause the P5.p and P7.p cells to form the lateral vulval lineages. The P6.p (central lineage) cell also secretes a short-range juxtacrine signal that induces the neighboring cells to activate the LIN-12 (Notch) protein. This signal prevents the P5.p and P7.p cells from generating the primary central vulval cell lineage. (After W. S. Katz and P. W. Sternberg. 1996. *Semin Cell Dev Biol* 7: 175–183.)

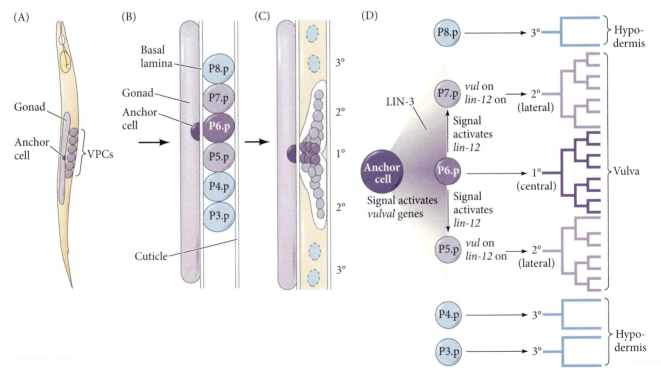

FIGURE 4.36 Model for the generation of two cell types (anchor cell and ventral uterine precursor cell) from two equivalent cells (Z1.ppp and Z4.aaa) in *C. elegans*. (A) The cells start off as equivalent, producing fluctuating amounts of signal and receptor. The *lag-2* gene is thought to encode the signal, and the *lin-12* gene is thought to encode the receptor. Reception of the signal turns down LAG-2 (Delta) production and upregulates LIN-12 (Notch). (B) A stochastic (chance) event causes one cell to produce more LAG-2 than the other cell at some particular critical time, which stimulates more LIN-12 production in the neighboring cell. (C) This difference is amplified because the cell producing more LIN-12 produces less LAG-2. Eventually, just one cell is delivering the LAG-2 signal, and the other cell is receiving it. (D) The signaling cell becomes the anchor cell, and the receiving cell becomes the ventral uterine precursor cell. (After G. Seydoux and I. Greenwald. 1989. *Cell* 57: 1237–1245.)

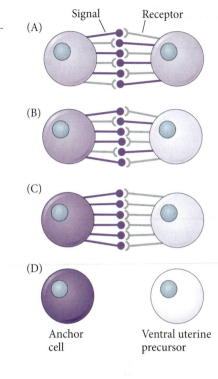

function only in that cell destined to become the ventral uterine precursor cell. The presumptive AC does not need it. As was first speculated by Seydoux and Greenwald (1989) and later shown by transgenic *lacZ* localization (Wilkinson et al. 1994), these two cells originally synthesize both the signal for uterine differentiation (the LAG-2 protein, homologous to Delta) and the receptor for this molecule (the LIN-12 protein, homologous to Notch).

During a particular time in larval development, the cell that, by chance, is secreting more LAG-2 causes its neighbor to cease its production of this differentiation signal and to increase its production of LIN-12. The cell expressing more LAG-2 becomes the AC, while the cell receiving the signal through its LIN-12 protein becomes the ventral uterine precursor cell (**FIGURE 4.36**). Thus, the two cells are thought to determine each other prior to their respective differentiation events. When LIN-12 is used again during vulva formation, it is activated by three *delta*-like genes expressed by the primary vulval lineage to stop the lateral vulval cells from forming the central vulval phenotype (Chen and Greenwald 2004; see Figure 4.35). Thus, the AC/ventral uterine precursor decision illustrates two important aspects of determination in two originally equivalent cells. First, the initial difference between the two cells is created by chance. Second, this initial difference is reinforced by feedback. This Notch-Delta mediated mechanism of restricting adjacent cell fates is called **lateral inhibition**. (See **Further Development 4.14, Hippo Signaling: An Integrator of Pathways**, online.)

Closing Thoughts on the Opening Photo

Is this a cell's antenna? If so, what is its purpose? It's for cells to communicate! This image shows a primary cilium on a neural stem cell in the brain, a structure that is in fact used like an antenna, enabling the cell to receive signals from its environment. We discussed the critical role of select signaling proteins that convey a myriad of information about position, adhesion, cell specification, and migration. New mechanisms of cell communication—such as the essential role of the primary cilium emphasized in this image; the potential reach of cytonemes, which may change our understanding of morphogen delivery; the modifying and potentially instructive roles of the extracellular matrix; and how the physical properties of cell adhesion can both sort different cells and regulate organ size—are rapidly emerging.

From A. Alvarez-Buylla et al. 1998. *J Neurosci* 18: 1020–1037

Endnotes

[1] Trypsin is the enzyme most commonly used to cleave cell surface protein connections that result in dissociated cells in culture.

[2] It is easy to distinguish permissive and instructive interactions using an analogy. This textbook is made possible by both permissive and instructive interactions. A reviewer can convince us to change the material in the chapters, which is an instructive interaction because the information expressed in the book is changed from what it would have been. However, the information in the book could not be expressed at all without permissive interactions with the publisher and printer.

[3] Although there is overlap in the terminology, a *morphogen* specifies cells in a quantitative ("more or less") manner, whereas a morphogenetic determinant specifies cells in a qualitative ("present or absent") way. Morphogens are analogue; *morphogenetic determinants* are digital.

[4] Synthetic beads can be coated with proteins and placed into the tissue of an embryo. These proteins are released from the bead slowly and then diffuse radially, creating concentration gradients.

[5] Yes, it is named after the Sega Genesis character. Riddle and colleagues (1993) discovered three genes homologous to *Drosophila hedgehog*. Two were named after existing species of hedgehogs, and the third was named after the animated character. Two other hedgehog genes, found only in fish, were originally named *echidna hedgehog* (possibly after Sonic's cartoon friend) and *tiggywinkle hedgehog*

(after Beatrix Potter's fictional hedgehog), but they are now referred to as *ihh-b* and *shh-b*, respectively.

[6] A *teratogen* is an exogenous compound capable of causing malformations in embryonic development; see Chapters 1 and 23.

[7] A comprehensive summary of all the Wnt proteins and Wnt signaling components can be found at http://web.stanford.edu/group/nusselab/cgi-bin/wnt/.

[8] In flies, the mutated porcupine gene results in segmentation defects that create denticles resembling porcupine spines in the larva (Perrimon et al. 1989). Do you recall the naming of Hedgehog? Porcupine is specific to Wnt palmitoylation, whereas Hedgehog is palmitoylated by a similar enzyme called Hhat.

[9] TGF stands for Transforming Growth Factor. The designation "superfamily" is often applied when each of the different classes of molecules constitutes a family. The members of a superfamily all have similar structures but are not as similar as the molecules within each family are to one another.

[10] Researchers named the Smad proteins by merging the names of the first identified members of this family: the *C. elegans* SMA protein and the *Drosophila* Mad protein.

[11] There are many different ARF genes, and while most upregulate gene expression as described, some also function as transcriptional repressors, providing cells a diversity of auxin responses.

(4) Snapshot Summary
Cell-to-Cell Communication

1. Cell-to-cell communication can occur between cells in direct contact with one another (juxtacrine signaling), or across a distance through cells secreting proteins into the extracellular matrix (paracrine signaling).

2. The sorting out of one cell type from another results from differences in the cell membrane.

3. The membrane structures responsible for cell sorting are often cadherin proteins, cell-cell adhesion molecules that change the surface tension properties of cells adhering to one another. Cadherins can cause cells to sort by both quantitative (different amounts of cadherin) and qualitative (different types of cadherin) differences. Cadherins appear to be critical during certain morphological changes.

4. The extracellular matrix is a source of signals and also serves to modify how such signals may be secreted across cells to influence differentiation and cell migration.

5. Components of the extracellular matrix (ECM) include proteoglycans (such as heparan sulfate and chondroitin sulfate proteoglycan), glycoproteins (such as fibronectin and laminin), and proteins (such as collagen).

6. Cells use transmembrane adhesion molecules called integrins to adhere to ECM components; on the inside of the cell, integrins are attached to the cytoskeleton; integrins, therefore, *integrate* the extracellular and intracellular scaffolds, allowing them to work together.

7. Cell migration occurs through changes in the actin cytoskeleton. These changes can be directed by internal instructions (from the nucleus) or by external instructions (such as from the extracellular matrix).

8. Cells can convert from being epithelial to being mesenchymal. The epithelial-mesenchymal transition (EMT) is a series of transformations involved in the dispersion of neural crest cells and the creation of vertebrae from somitic cells. In adults, EMT is involved in wound healing and cancer metastasis.

9. Inductive interactions involve inducing and responding tissues. The ability to respond to inductive signals depends on the competence of the responding cells.

10. Cascades of inductive events are responsible for organ formation.

11. Paracrine factors are secreted by inducing cells. These factors bind to cell membrane receptors in competent responding cells. Competence is the ability to bind and to respond to inducers, and it is often the result of a prior induction. Competent cells respond to paracrine factors through signal transduction pathways.

12. Morphogens are secreted signaling molecules that affect gene expression differently at different concentrations.

13. Many paracrine factors fit into four major families: the fibroblast growth factor (FGF) family, the Hedgehog family, the Wnt family, and the TGF-β superfamily, including Activin, BMPs, Nodal, and Vg1.

14. Signal transduction pathways begin with a paracrine or juxtacrine factor causing a conformational change in its cell membrane receptor. The new shape can result in

enzymatic activity in the cytoplasmic domain of the receptor protein. This activity allows the receptor to phosphorylate other cytoplasmic proteins. Eventually, a cascade of such reactions activates a transcription factor (or set of factors) that activates or represses specific gene activity.

15. The cell surface is intimately involved with cell signaling. Proteoglycans and other membrane components can expand or restrict the diffusion of paracrine factors.

16. The hormone auxin is a major morphogen found in plants. The essential role auxin plays in the proper morphogenesis of the embryo is conserved across plant species.

17. Auxin signal transduction operates through the negative regulation of a pathway inhibitor, just as Hedgehog and Wnt signaling does in animals.

18. The asymmetric positioning of PIN efflux transporters on select sides of plant cells functions to control the directional flow of auxin throughout the plant. The resulting graded distribution of auxin concentrations differentially regulates cell fates across the apical-basal axis.

19. Specializations of the cell surface, including the primary cilium, may concentrate receptors for paracrine and extracellular matrix proteins. Newly discovered filopodia-like extensions called cytonemes can be involved in transferring morphogens between signaling and responding cells and may be a major component of cell signaling.

20. Juxtacrine signaling involves local protein interactions between receptors. One example is Notch-Delta signaling that patterns cell fates through lateral inhibition, as seen in the generation of two different cell types (the anchor cell and ventral uterine precursor cell) from two initially equivalent cells in the *C. elegans* embryo.

Go to oup.com/he/barresi12xe for Further Developments, Scientists Speak interviews, Watch Development videos, Dev Tutorials, and complete bibliographic information for all literature cited in this chapter.

Stem Cells
Their Potential and Their Niches

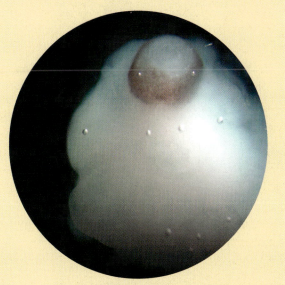

Is that really an eye and a brain in a dish?

From M. A. Lancaster et al. 2013. *Nature* 501: 373–379

WE HAVE COMPLETED AN ANALYSIS OF CELL MATURATION through the levels of cell specification, commitment, and ultimately differentiation, all of which are driven by cell-to-cell communication and the regulation of gene expression. There is no better example that encapsulates this entire process than a stem cell.

A **stem cell** retains the ability to divide and re-create itself while also having the ability to generate progeny capable of specializing into a more differentiated cell type. Stem cells are sometimes referred to as "undifferentiated" due to this maintenance of proliferative properties. There are many *different* types of stem cells, however, and their status as "undifferentiated" really pertains to the retained ability to divide and resistance to mature into a tissue's postmitotic derivatives. Because they maintain the ability to generate effectively unlimited populations of daughter cells that can go on to proliferate and differentiate, stem cells hold great potential to not only study human development but also transform modern medicine.

Currently, there are few topics in developmental biology that can rival stem cells in the pace at which new knowledge is being generated. In this chapter, we will address some of the fundamental questions regarding stem cells. What are the mechanisms governing stem cell division, self-renewal, and differentiation? Where are stem cells found, and how do they differ when in an embryo, an adult, or a culture dish? How are scientists and clinicians using stem cells to study and treat disease?

The Stem Cell Concept

A cell is a stem cell if it can divide and, in doing so, produce a replica of itself (a process called **self-renewal**) as well as a daughter cell that can undergo further development and differentiation. It thus has the power, or **potency**, to produce many different types of differentiated cells.

Division and self-renewal

Upon division, a stem cell may produce a daughter cell that can mature into a terminally differentiated cell type. Cell division can occur either symmetrically or asymmetrically. If a stem cell divides symmetrically, it could produce two self-renewing stem cells or two daughter cells that are committed to differentiate, resulting in, respectively, the expansion or reduction of the resident stem cell population. In contrast, if the stem cell divides asymmetrically, it could stabilize the stem cell pool as well as generate a daughter cell that goes on to differentiate. This strategy, in which two types of cells (a stem cell and a developmentally committed cell) are produced at each division, is called the **single stem cell asymmetry** mode and is seen in many types of stem cells (**FIGURE 5.1A**).

An alternative (but not mutually exclusive) mode of retaining cell homeostasis is the **population asymmetry** mode of stem cell division. Here, some stem cells are more prone to produce differentiated progeny, which is compensated for by another set of stem cells that divide symmetrically to maintain the stem cell pool within this population (**FIGURE 5.1B**; Watt and Hogan 2000; Simons and Clevers 2011).

Potency defines a stem cell

The diversity of cell types that a stem cell can generate in vivo defines its natural potency. A stem cell capable of producing all the cell types of a lineage is said to be **totipotent**. In organisms such as hydra, each individual cell is totipotent (see Chapter 22). In mammals, only the fertilized egg and first 4 to 8 cells are totipotent, which means that they can generate both the embryonic lineages (which form the body and germ cells) and the extraembryonic lineages (which form the placenta, amnion, allantois, and yolk sac) (**FIGURE 5.2**). Shortly after the 8-cell stage, the mammalian embryo develops an outer layer (which becomes the fetal portion of the placenta) and an inner cell mass that generates the embryo. The cells of the inner cell mass are thus said to be **pluripotent**, or capable of producing all the cells of the embryo. When these inner

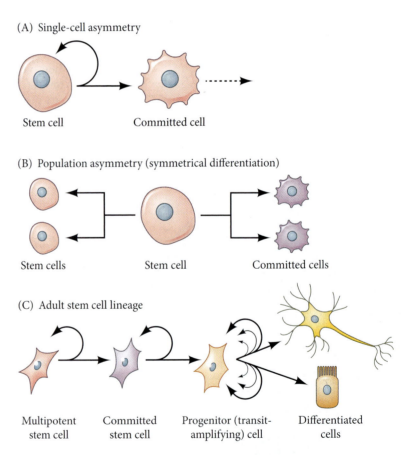

FIGURE 5.1 The stem cell concept. (A) The fundamental notion of a stem cell is that it can make more stem cells while also producing cells committed to undergoing differentiation. This process is called asymmetric cell division. (B) A population of stem cells can also be maintained through population asymmetry. Here a stem cell is shown to have the ability to divide symmetrically to produce *either* two stem cells (thus increasing the stem cell pool by one) *or* two committed cells (thus decreasing the pool by one). This is called symmetrical renewing or symmetrical differentiating. (C) In many organs, stem cell lineages pass from a multipotent stem cell (capable of forming numerous types of cells) to a committed stem cell that makes one or very few types of cells to a progenitor cell (also known as a transit-amplifying cell) that can proliferate for multiple rounds of divisions but is transient in its life and is committed to becoming a particular type of differentiated cell.

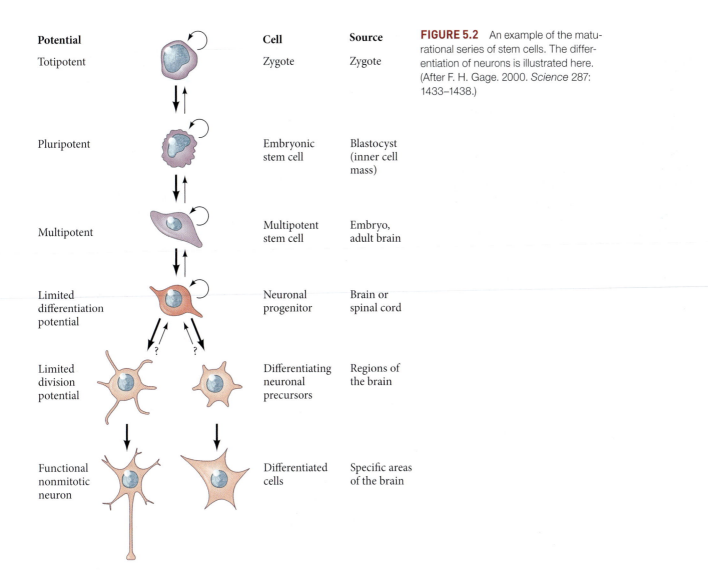

Potential		Cell	Source
Totipotent		Zygote	Zygote
Pluripotent		Embryonic stem cell	Blastocyst (inner cell mass)
Multipotent		Multipotent stem cell	Embryo, adult brain
Limited differentiation potential		Neuronal progenitor	Brain or spinal cord
Limited division potential		Differentiating neuronal precursors	Regions of the brain
Functional nonmitotic neuron		Differentiated cells	Specific areas of the brain

FIGURE 5.2 An example of the maturational series of stem cells. The differentiation of neurons is illustrated here. (After F. H. Gage. 2000. *Science* 287: 1433–1438.)

cells are removed from the embryo and cultured in vitro, they establish a population of pluripotent **embryonic stem cells** (**ESCs**).

As cell populations within each germ layer expand and differentiate, resident stem cells are maintained within these developing tissues. These stem cells are **multipotent** and function to generate cell types with restricted specificity for the tissue in which they reside (**FIGURE 5.1C** and see Figure 5.2). From the embryonic gut to the adult small intestine or from the neural tube to the adult brain, multipotent stem cells play critical roles in fueling organogenesis in the embryo and regeneration in adult tissues. Numerous adult organs possess **adult stem cells**, which in most cases are multipotent. In addition to the known hematopoietic stem cells that function to generate all the cells of the blood, biologists have discovered adult stem cells in the epidermis, brain, muscle, teeth, gut, and lungs, among other locations. Unlike pluripotent stem cells, adult or multipotent stem cells in culture have not only a restricted array of cell types that they can create, but also a finite number of generations for self-renewal. This limited renewal of adult stem cells may contribute to aging (Asumda 2013; for an in-depth discussion on mechanisms of aging, see **Further Development 5.1, A Chapter on Senescence and Longevity**, online).

When a multipotent stem cell divides asymmetrically, its maturing daughter cell often goes through a transition stage as a **progenitor** or **transit-amplifying cell**, as is seen in the formation of blood cells, sperm, and neurons (see Figures 5.1C and 5.2). Progenitor cells are not capable of unlimited self-renewal; rather, they have the capacity to divide only a few times before differentiating (Seaberg and van der Kooy 2003).

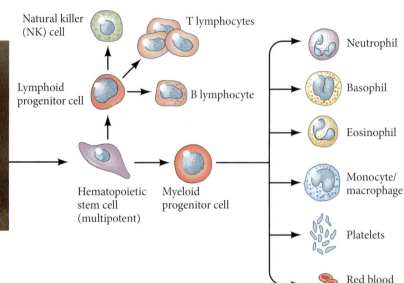

Hematopoietic stem cells

FIGURE 5.3   Blood-forming (hemato-poietic) stem cells (HSCs). These multi-potent stem cells generate blood cells throughout an individual's life. HSCs from human bone marrow (photo) can divide to produce more HSCs. Alterna-tively, HSC daughter cells are capable of becoming either lymphoid progenitor cells (which divide to form the cells of the adaptive immune system) or myeloid progenitor cells (which become the other blood cell precursors). The lineage path each cell takes is regulated by the HSC's microenvironment, or niche (see Figure 5.16). (After http://stemcells.nih.gov/; © 2001 Terese Winslow, Lydia Kibiuk.)

Although limited, this proliferation serves to *amplify* the pool of progenitors before they terminally differentiate. Cells within this progenitor pool can mature along different but related paths of specification. As an example, the hematopoietic stem cell gener-ates blood and lymphoid progenitor cells that further develop into the differentiated cell types of the blood, such as red blood cells, neutrophils, and lymphocytes (cells of the immune response), as shown in **FIGURE 5.3**. Yet another term, **precursor cell** (or simply **precursors**), is widely used to denote any ancestral cell type (either stem cell or progenitor cell) of a particular lineage; it is often used when such distinctions do not matter or are not known (see Tajbakhsh 2009). Some adult stem cells, such as spermato-gonia, are referred to as **unipotent stem cells** because they function in the organism to generate only one cell type, the sperm cell in this example. Precise control of the division and differentiation of these varied stem cell types is necessary for building the embryo as well as maintaining and regenerating tissues in the adult.

Stem Cell Regulation

As discussed above, the basic functions of stem cells revolve around self-renewal and differentiation. But how are stem cells regulated between these different states in a coordinated way to meet the patterning and morphogenetic needs of the embryo and mature tissue? Regulation is highly influenced by the microenvironment that surrounds a stem cell and is known as the **stem cell niche** (Schofield 1978). There is growing evidence that all tissue types possess a unique stem cell niche, and despite many dif-ferences among the niche architectures of different tissues, several common principles of stem cell regulation can be applied to all environments. These principles involve *extracellular* mechanisms leading to *intracellular* changes that regulate stem cell behavior (**FIGURE 5.4**). Extracellular mechanisms include:

- *Physical mechanisms* of influence, including structural and adhesion factors within the extracellular matrix that support the cellular architecture of the niche. Differences in cell-to-cell and cell-to-matrix adhesions as well as the cell density within the niche can alter the mechanical forces that influence stem cell behavior.

- *Chemical regulation* of stem cells takes the form of secreted proteins from surrounding cells that influence stem cell states and progenitor differentia-tion through endocrine, paracrine, or juxtacrine mechanisms (Moore and Lemischka 2006; Jones and Wagers 2008). In many cases, these signaling factors maintain the stem cell in an uncommitted state. Once stem cells

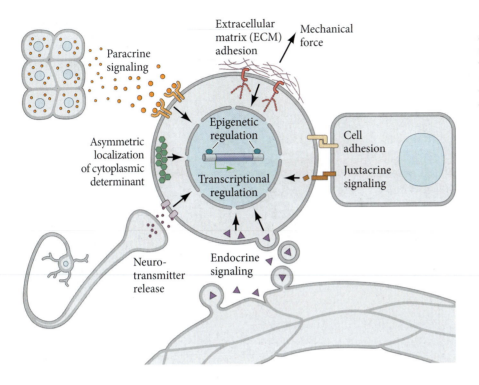

FIGURE 5.4 To divide or not to divide: an overview of stem cell regulatory mechanisms. Shown here are some of the more general external and internal molecular mechanisms that can influence the quiescent, proliferative, or differentiative behaviors of a stem cell.

become positioned farther from the niche, however, these factors cannot reach them, and differentiation commences.

Intracellular regulatory mechanisms include:

- *Regulation by cytoplasmic determinants*, the partitioning of which occurs at cytokinesis. As a stem cell divides, factors determining cell fate are either selectively partitioned to one daughter cell (asymmetric differentiating division) or shared evenly between daughter cells (symmetrical division).

- *Transcriptional regulation* occurs through a network of transcription factors that keep a stem cell in its quiescent or proliferative state and promote maturation of daughter cells toward a particular fate.

- *Epigenetic regulation* occurs at the level of chromatin. Different patterns of chromatin accessibility (e.g., by histone modifications) influence gene expression related to stem cell behavior.

The types of the intracellular mechanisms used by a given stem cell are in part the downstream net result of the extracellular stimuli in its niche. Just as important, however, is the stem cell's developmental history within its niche. Below are descriptions of some of the better-known stem cell niches, highlighting their developmental origins and the specific extracellular and intracellular mechanisms important for regulating stem cell behavior.

Pluripotent Cells in the Embryo

On a cellular basis, there are many similarities in how stem cells function in both animals and plants. In this section, we will focus on the types of stem cells that build the embryo, while later in this chapter we will explore the diversity of stem cell types residing in the adult organism. Interestingly, it is this contrast between stem cell function in the embryo versus the adult that highlights one of the key differences between plants and animals: the persistence or lack thereof of totipotent stem cells throughout life. For instance, the early mammalian blastocyst has an **inner cell mass** (**ICM**) consisting of pluripotent cells that give rise to all the cell types of the body, but these cells soon differentiate into more restricted progenitor cells within each germ layer. The early plant embryo generates two clusters of totipotent stem cells, or **initial cells**, one located at

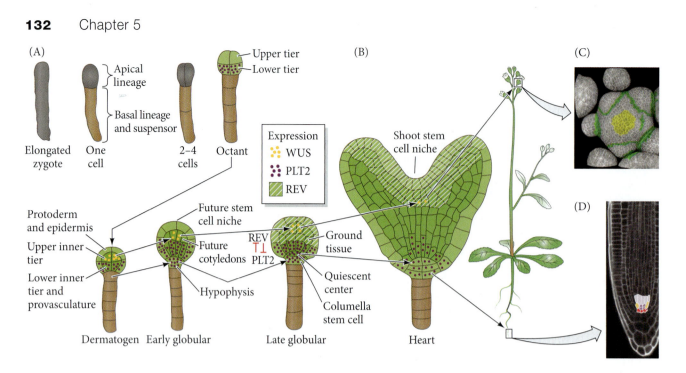

(A)

Elongated zygote | One cell | 2–4 cells | Octant

Apical lineage
Basal lineage and suspensor

Upper tier
Lower tier

Expression
WUS
PLT2
REV

Protoderm and epidermis
Upper inner tier
Lower inner tier and provasculature

Future stem cell niche
Future cotyledons
Hypophysis

Dermatogen | Early globular | Late globular | Heart

REV
PLT2

Ground tissue
Quiescent center
Columella stem cell

(B) Shoot stem cell niche

(C)

(D)

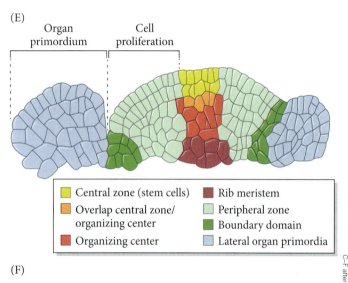

(E)
Organ primordium
Cell proliferation

Central zone (stem cells)
Overlap central zone/ organizing center
Organizing center
Rib meristem
Peripheral zone
Boundary domain
Lateral organ primordia

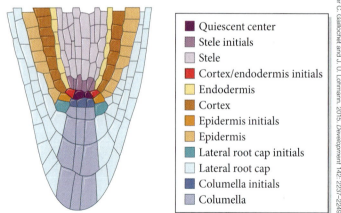

(F)
Quiescent center
Stele initials
Stele
Cortex/endodermis initials
Endodermis
Cortex
Epidermis initials
Epidermis
Lateral root cap initials
Lateral root cap
Columella initials
Columella

C–F after C. Gaillochet and J. U. Lohmann. 2015. *Development* 142: 2237–2249

the most apical (shoot) end of the embryo and the other at the most basal (root) end. These cell clusters are known, respectively, as the shoot apical meristem (SAM) and the root apical meristem (RAM). The remarkable difference from stem cells in animals is that these totipotent meristematic tissues persist throughout the life of the plant! As the embryo becomes a germinating seedling, the SAM and RAM function to generate the shoots (stems and leaves) and roots throughout vegetative growth. As we discussed in Chapter 3, the SAM can be induced to become an **inflorescence meristem** that produces reproductive tissues (flowers) instead of leaves.

How are populations of totipotent stem cells maintained throughout life in a plant but not in an animal? To answer this question, we will first examine the mechanisms driving the formation of meristems in the plant embryo, and then turn our attention to the development of the mammalian inner cell mass.

Meristem cells of the Arabidopsis thaliana embryo and beyond

All of the aerial and underground parts of a plant ultimately derive from the SAM and RAM. Amazingly, because plant cells do not move during development, the origins of these meristems can be traced back quite precisely to the early embryo (**FIGURE 5.5A–D**). Characterization of the developmental timing of meristem formation has been aided by the identification of early meristem patterning genes, such as the transcription

FIGURE 5.5 The plant meristem. (A,B) Illustration of *Arabidopsis thaliana* embryonic development from zygote to heart stages. The shoot apical meristem gives rise to leaves and flowers, and the root apical meristem gives rise to roots. (C) Top view of the shoot apical meristem. The central zone and boundary domains are pseudocolored yellow and green, respectively, which correspond to the same colored areas in the illustrated cross section of a shoot apical meristem in (E). Similarly, the stem cell parts of the root apical meristem in the micrograph (D) are pseudocolored to match the corresponding schematized cross section (F). (A,B after R. Heidstra and S. Sabatini. 2014. *Nat Rev Mol Cell Biol* 15: 301–312; C–F after C. Gaillochet and J. U. Lohmann. 2015. *Development* 142: 2237–2249.)

factors *WUSCHEL* (*WUS*), along with its paralog *WOX2,* and *REVOLUTA* (*REV*) for shoot meristems, and while members of the PLETHORA (PLT) family of transcription factors are also detected in the SAM, *PLT2* is restricted to the root meristems. Cells in the upper inner tier of the dermatogen stage *A. thaliana* embryo contribute to the *WUS*-expressing organizing center, as well as to the shoot stem cells in the apically positioned central domain (**FIGURE 5.5E**). In contrast, the lower inner tier dermatogen cells express *PLT2* and are destined for the root stem cell niche (**FIGURE 5.5F**) (Heidstra and Sabatini 2014; Zhang et al. 2017). (See **Further Development 5.2, Specifying the Shoot and Root Meristems**, online.)

MAINTAINING TOTIPOTENCY IN THE SHOOT APICAL MERISTEM *A. thaliana* development within the seed pauses when the embryo reaches the mature stage. Development recommences when environmental conditions are optimal for germination and initiation of the indeterminant growth of the vegetative phase of the plant's life cycle. One of the most amazing powers of plants is realized during this postembryonic period of growth, during which the SAM and RAM maintain totipotency. The meristem functions to balance proliferation and differentiation with maintenance of a pool of stem (initial) cells. Because of this remarkably balanced behavior, many biotechnology and pharmaceutical companies are actively investigating the mechanisms governing plant stem cell totipotency.

To best illustrate the delicate balancing act of stem cell function in the plant, we will focus our attention on the mechanisms operating in the shoot apical meristem. As do stem cells in animals, plant initial cells in the central zone of the SAM divide relatively slowly, giving rise to a self-renewed initial cell as well as a daughter progenitor cell that will progressively become displaced into the peripheral zone and differentiate. A progenitor cell located within the peripheral zone of any of the three layers will divide more rapidly (**FIGURE 5.6A**). In the two apical-most layers (L1 and L2), progenitor cells primarily undergo **anticlinal divisions**, meaning that the new cell walls are laid down perpendicular to the surface of the apical meristem. In the third, deeper layer (L3), progenitor cells divide along all planes including **periclinal** (in which the new cell wall is parallel to the surface).

As you will learn throughout this and many other chapters, the operation of negative feedback loops provides a robust mechanism to establish and maintain distinct fates and cell behaviors. Regulation of totipotency in the meristem is no exception (**FIGURE 5.6B**). Progenitor cell production by stem cells in the SAM central zone requires the expression and secretion of the signaling protein CLAVATA3 (CLV3). Just beneath the central zone is the organizing center, which expresses the pioneer regulatory transcription factor *WUS* (Laux et al. 1996; Mayer et al. 1998). WUS relays a signal to the central zone that promotes CLV3 expression and the adoption of slowly dividing stem cell behavior. With every new division, progenitor cells are gradually displaced from the central zone and enter the peripheral zone, where upregulation of the *HEC1* transcription factor intensifies the proliferative activity of progenitor cells until they reach a lateral organ primordium and differentiate. WUS also actively represses *HEC1* expression, thereby maintaining

FIGURE 5.6 Maintaining the stem cell pool in the shoot apical meristem. (A) A longitudinal section of the shoot apical meristem, with its three layers (L1–L3) pseudocolored. Illustrated on the right side are examples of the anticlinal divisions in L1 and L2 and mixed division planes within L3. (B) Illustration of the negative feedback loop operating between the organizing center and the central and peripheral zones. See text for details. (A from J. L. Bowman and Y. Eshed. 2000. *Trends Plant Sci* 5: 100–115; B after C. Gaillochet and J. U. Lohmann. 2015. *Development* 142: 2237–2249.)

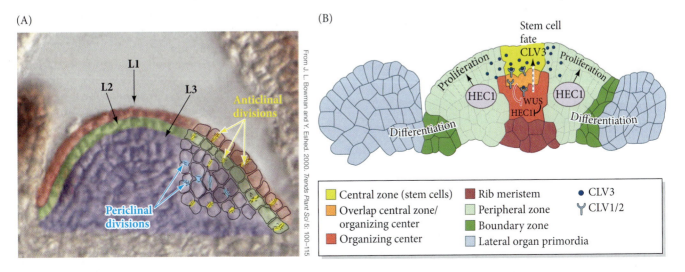

(A)   (B)

From J. L. Bowman and Y. Eshed. 2000. *Trends Plant Sci* 5: 100–115

Stem cell fate

□ Central zone (stem cells)	■ Rib meristem	● CLV3
□ Overlap central zone/ organizing center	■ Peripheral zone	Y CLV1/2
■ Organizing center	■ Boundary zone	
	□ Lateral organ primordia	

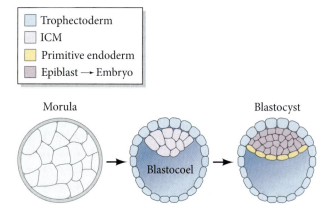

Trophectoderm
ICM
Primitive endoderm
Epiblast → Embryo

Morula

Blastocoel

Blastocyst

FIGURE 5.7 Establishment of the inner cell mass (the ICM, which will become the embryo) in the mouse blastocyst. From morula to blastocyst, the three principal cell types—trophectoderm, epiblast, and primitive endoderm—are illustrated.

(?) Developing Questions

Is there such a thing as "stem-cellness"? Is being a stem cell an intrinsic property of the cell, or is it a property acquired through interactions with the stem cell niche? Is the niche making the stem cell? What approaches might you use to determine which of these conditions exist in a particular organ?

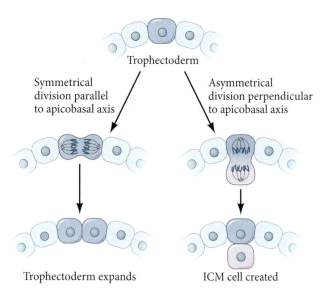

Trophectoderm

Symmetrical division parallel to apicobasal axis

Asymmetrical division perpendicular to apicobasal axis

Trophectoderm expands

ICM cell created

FIGURE 5.8 Divisions about the apicobasal axis. Depending on the axis of cell division in the trophectoderm, the trophectoderm layer can be expanded (left), or the inner cell mass (ICM) can be seeded (right).

the distinct status of the organizing center (Schuster et al. 2014). However, as CLV3 increases in concentration, it directly interacts with the receptors CLV1 and CLV2 on the cell membrane of cells in the organizing center. This interaction triggers a signal transduction pathway that suppresses WUS activity, thus creating a negative feedback loop that maintains a very consistent number of initial cells throughout vegetative, inflorescence, and floral meristem development (Somssich et al. 2016; Gaillochet and Lohmann 2015; Heidstra and Sabatini 2014).

Cells of the inner cell mass in the mouse embryo

The pluripotent stem cells of the mammalian inner cell mass (ICM) are one of the most studied types of stem cells. Following cleavages of the mammalian zygote and formation of the morula, the process of cavitation creates the blastocyst,[1] which consists of a spherical layer of **trophectoderm cells** surrounding the inner cell mass and a fluid-filled cavity called the **blastocoel** (**FIGURE 5.7**). In the early mouse blastocyst, the ICM is a cluster of approximately 12 cells adhering to one side of the trophectoderm (Handyside 1981; Fleming 1987). The ICM will subsequently develop into a cluster of cells called the epiblast and a layer of primitive endoderm (yolk sac) cells that establish a barrier between the epiblast and the blastocoel. The epiblast develops into the embryo proper, generating all the cell types (more than 200) of the adult mammalian body, including the primordial germ cells (see Shevde 2012), whereas the trophectoderm and primitive endoderm give rise to extraembryonic structures, namely the chorion, which gives rise to the embryonic side of the placenta, and the yolk sac (Stephenson et al. 2012; Artus and Chazaud 2014). Importantly, cultured cells of the ICM or epiblast produce embryonic stem cells (ESCs),[2] which retain pluripotency and can similarly generate all cell types of the body (Martin 1980; Evans and Kaufman 1981). In contrast to the in vivo behavior of ICM cells, however, ESCs can self-renew seemingly indefinitely in proper culture conditions. We discuss the properties and use of ESCs later in this chapter. Here we will focus on the mammalian blastocyst as its own stem cell niche for the development of the only cells in the vertebrate embryo that are at least transiently pluripotent.

MECHANISMS PROMOTING PLURIPOTENCY OF ICM CELLS Essential to the transient pluripotency of the ICM is expression of the transcription factors Oct4,[3] Nanog, and Sox2 (Shi and Jin 2010). These three regulatory transcription factors are necessary to maintain the uncommitted stem cell-like state and functional pluripotency of the ICM, enabling ICM cells to give rise to the epiblast and all associated derived cell types (Pardo et al. 2010; Artus and Chazaud 2014; Huang and Wang 2014). It is interesting that expression of these three transcription factors is normally lost from the ICM as the epiblast differentiates (Yeom et al. 1996; Kehler et al. 2004). In contrast, the transcription factor Cdx2 is upregulated in the *outer* cells of the morula to promote trophectoderm differentiation and repress epiblast development (Strumpf et al. 2005; Ralston et al. 2008; Ralston et al. 2010).

FIGURE 5.9 Hippo signaling and ICM development. (A) Immunolocalization of the Hippo pathway components Amot (angiomotin; green stain) and Yap (red stain)—as well as E-cadherin—from morula to blastocyst. Activated Yap is localized to the trophectoderm nuclei, while E-cadherin (purple) is restricted to the trophectoderm-ICM membrane contacts. (B) Hippo signaling in trophectoderm (top) and ICM (bottom) cells. Hippo signaling is activated through E-cadherin binding with Amot, and as a result, Yap is degraded in the ICM cell. Names in parentheses are the *Drosophila* homologues.

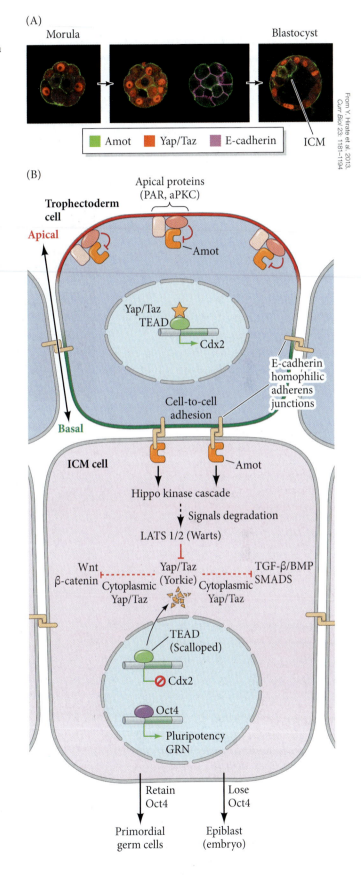

(A)

Morula → → Blastocyst

■ Amot ■ Yap/Taz ■ E-cadherin   ICM

From Y. Hirate et al. 2013. *Curr Biol* 23: 1181–1194

(B)

Apical proteins (PAR, aPKC)

Trophectoderm cell

Apical — Amot

Yap/Taz TEAD → Cdx2

E-cadherin homophilic adherens junctions

Cell-to-cell adhesion

Basal

ICM cell — Amot

Hippo kinase cascade

Signals degradation

LATS 1/2 (Warts)

Wnt β-catenin Cytoplasmic Yap/Taz ⊣ Yap/Taz (Yorkie) Cytoplasmic Yap/Taz TGF-β/BMP SMADS

TEAD (Scalloped)

Ø Cdx2

Oct4 → Pluripotency GRN

Retain Oct4 — Lose Oct4

Primordial germ cells — Epiblast (embryo)

What mechanisms are at work to control the temporal and spatial expression patterns of genes within the presumptive ICM and trophectoderm? Cell-to-cell interactions set the foundation for initial specification and architecture of these layers. First, cellular polarity along the apicobasal axis (in this case, from the outer side to the inside of the embryo) creates a mechanism by which symmetrical or asymmetrical divisions can produce two different cells. Perpendicularly positioned, asymmetrical divisions along the apicobasal axis would yield daughter cells segregated to the outside and inside of the embryo, corresponding to the development of the trophectoderm and ICM, respectively. In contrast, symmetrical divisions parallel to the apicobasal axis would distribute cytoplasmic determinants evenly to both daughter cells, further propagating cells only within either the outer trophectoderm layer or the ICM (**FIGURE 5.8**). These behaviors are remarkably similar to the periclinal and anticlinal divisions in the plant meristem (see Figure 5.6A).

At the morula stage, factors become asymmetrically localized along the apicobasal axis in the outer cells of the presumptive trophectoderm. These factors include well-known proteins in the partitioning defective (PAR) and atypical Protein kinase C (aPKC) families. One outcome of these *partitioning proteins* is the recruitment of the cell adhesion molecule E-cadherin to the basolateral membrane, where outer cells contact underlying ICM cells (**FIGURE 5.9A**; see Chapter 4; Stephenson et al. 2012; Artus and Chazaud 2014). Experimentally eliminating E-cadherin disrupts both apicobasal polarity and the specification of the ICM and trophectoderm lineages (Stephenson et al. 2010). (See **Further Development 5.3, How Does E-Cadherin Influence Blastocyst Cell Fates?**, online.)

Adult Stem Cell Niches in Animals

Many adult tissues and organs contain stem cells that undergo continual renewal. These include but are not limited to germ cells across species, and brain, epidermis, hair follicles, intestinal villi, and blood in mammals. Also, multipotent adult stem cells play major roles in organisms with high regenerative capabilities, such as hydra, axolotl, and zebrafish. Adult stem cells must maintain the long-term ability to divide, be able to produce differentiated daughter cells, and still repopulate the stem cell pool. The adult stem cell is housed in and controlled by its own **adult stem cell niche**, which regulates stem cell self-renewal, survival, and differentiation of those progeny that leave the niche. Below we describe some of the better-characterized niches, which include those for the *Drosophila* germ stem cells and mammalian neural, gut epithelial, and hematopoietic stem cells. This list is obviously not exhaustive, but it highlights some universal mechanisms that control stem cell development.

Stem cells fueling germ cell development in the *Drosophila* ovary

The *Drosophila* oocyte is derived from a germ stem cell (GSC). These GSCs are held within the ovarian stem cell niche, and positional secretion of paracrine factors influences stem cell self-renewal and oocyte differentiation in a concentration-dependent

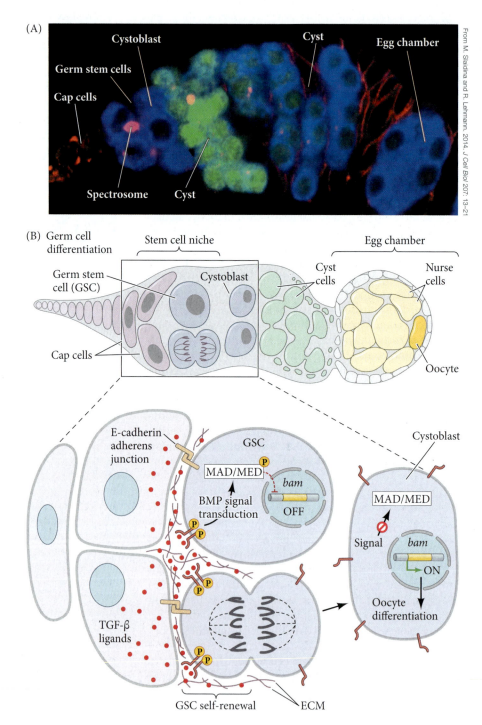

(A)

From M. Slaidina and R. Lehmann, 2014. *J Cell Biol* 207: 13–21

FIGURE 5.10 *Drosophila* ovarian stem cell niche. (A) Immunolabeling of different cell types within the *Drosophila* germarium. Germ stem cells (GSCs) are identified by the presence of spectrosomes. Differentiating germ cells (cystoblasts) are stained blue. Bam-expressing (cyst) cells are green. (B) The interactions between Cap cells and GSCs in the germarium. See text for a description of the interactions between the regulatory components. (From M. Slaidina and R. Lehmann. 2014. *J Cell Biol* 207: 13–21.)

manner. Egg production in the adult fly ovary occurs in more than 12 egg tubes or ovarioles, each one housing identical GSCs (usually two per ovariole) and several somatic cell types that construct the niche, known as the germarium (Lin and Spradling 1993). As a GSC divides, it self-renews and produces a cystoblast that (like the sperm gonialblast progenitor cell, see Further Development 5.4, online) will mature as it moves farther out of the stem cell niche—beyond the reach of the niche's regulatory signals—and becomes an oocyte surrounded by follicle cells (**FIGURE 5.10A**; Eliazer and Buszczak 2011; Slaidina and Lehmann 2014).

Although the GSCs are within the stem cell niche, they are in contact with Cap cells. Upon division of the GSC perpendicular to the Cap cells, one daughter cell remains tethered to the Cap cell by E-cadherin and maintains its self-renewal identity, whereas the displaced daughter cell begins oocyte differentiation (Song and Xie 2002). Cap cells affect GSCs by secreting TGF-β family proteins, which activate the BMP signal transduction pathway in the GSCs and, as a result, prevent GSC differentiation (**FIGURE 5.10B**). Extracellular matrix components like collagen and heparan sulfate proteoglycan restrict the diffusion of the TGF-β family proteins such that only the tethered GSCs receive significant amounts of these TGF-β signals (Akiyama et al. 2008; Wang et al. 2008; Guo and Wang 2009; Hayashi et al. 2009).[4] Activation of BMP signal transduction in the GSC prevents differentiation by repressing transcription of genes that promote differentiation, primarily that of *bag of marbles* (*bam*). When *bam* is expressed, the cell goes on to differentiate into an oocyte (see Figure 5.8).

In both the testis and ovary of *Drosophila*, coordinated cell division paired with adhesion and paracrine-mediated repression of differentiation controls GSC renewal and progeny differentiation. New insights into the epigenetic regulation of GSC development are beginning to emerge; for example, the histone methyltransferase Set1 has been discovered to play an essential role in GSC self-renewal (Yan et al. 2014). (See **Further Development 5.4, *Drosophila* Testes Stem Cell Niche**, online.)

Adult Neural Stem Cell Niche of the V-SVZ

Despite the first reports of adult neurogenesis in the postnatal rat in 1969 and in songbirds in 1983, the doctrine that "no new neurons are made in the adult brain" held for decades. At the turn of the twenty-first century, however, a flurry of investigations, primarily in the adult mammalian brain, began to mount strong support for continued neurogenesis throughout life (Gage 2002). This acceptance of **neural stem cells** (**NSCs**) in the adult central nervous system (CNS) marks an exciting time in the field of developmental neuroscience and has significant implications for both our understanding of brain development and the treatment of neurological disorders.

Whether in fish or humans, adult NSCs retain much of the cellular morphology and molecular characteristics of their embryonic progenitor cell, the radial glial cell.[5] Radial glia and adult NSCs are polarized epithelial cells spanning the full apicobasal axis of the CNS (Grandel and Brand 2013). The development of radial glia and the embryonic origins of the adult mammalian neural stem cell niche are covered in Chapter 14. In anamniotes such as teleosts (the bony fishes), radial glia function as NSCs throughout life, occurring in numerous neurogenic zones (at least 12) in the adult brain (Than-Trong and Bally-Cuif 2015).

In the adult mammalian brain, however, NSCs have been characterized only in two principal regions of the cerebrum: the **subgranular zone** (**SGZ**) of the hippocampus and the **ventricular-subventricular zone** (**V-SVZ**) of the lateral ventricles (Faigle and Song 2013; Urbán and Guillemot 2014). There are similarities and differences between these mammalian neurogenic niches such that each NSC has characteristics reminiscent of its radial glial origin, yet only the NSC of the V-SVZ maintains contact with the cerebrospinal fluid. During development of the adult V-SVZ, radial glia-like NSCs transition into **type B cells** that fuel the generation of specific types of neurons in both the olfactory bulb and striatum, as has been shown in both the mouse and the human brain (**FIGURE 5.11**; Curtis et al. 2012; Lim and Alvarez-Buylla 2014). (See **Further Development 5.5, The Subgranular Zone Niche**, online.)

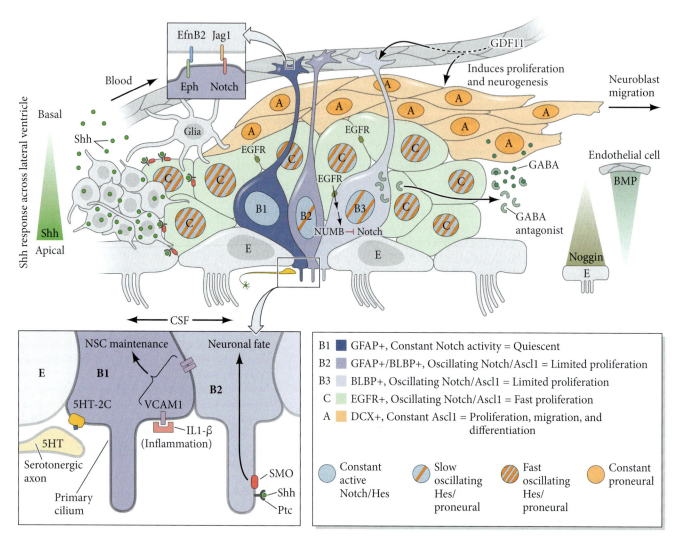

FIGURE 5.11 Schematic of the ventricular-subventricular zone (V-SVZ) stem cell niche and its regulation. Multiciliated ependymal cells (E; light gray) line the ventricle and contact the apical surface of V-SVZ NSCs (blue). Typically quiescent B1-type NSCs (dark blue) give rise to activated B2 and B3 cells (lighter shades of blue) that possess limited proliferation. The B3 cells generate the C cells (green), which, after three rounds of division, give rise to migrating neuroblasts (A cells; orange). The niche is penetrated by endothelial cell-built blood vessels that are in part enwrapped by the basal endfeet of B cells. Maintenance of the stem cell pool is regulated by VCAM1 adhesion and Notch signaling (changes in Notch pathway oscillations are depicted as color changes in the nuclei). Clusters of neurons in the ventral region of the lateral ventricle express Sonic hedgehog (Shh), which influences different neuronal cell differentiation from the niche. Antagonistic signaling between BMP and Noggin from endothelial cells and ependymal cells, respectively, balance neurogenesis along this gradient. Serotonergic (5HT) axons lace the ventricular surface, and—along with IL1-β and GDF11 from the cerebrospinal fluid (CSF) and blood, respectively—play roles as external stimuli to regulate the niche. Non-niche neurons, astrocytes, and glia can be found within the niche and influence its regulations. GFAP, glial fibrillary acidic protein; BLBP, brain lipid binding protein; DCX, double cortin. (Based on various sources, including O. Basak et al. 2012. *J Neurosci* 32: 5654–5666; C. Giachino et al. 2014. *Stem Cells* 32: 70–84; D. A. Lim and A. Alvarez-Buylla. 2014. *Trends Neurosci* 37: 563–571; and C. Ottone et al. 2014. *Nat Cell Biol* 16: 1045–1056.)

The neural stem cell niche of the V-SVZ

In the V-SVZ, B cells project a primary cilium (see Chapter 4) from their apical surface into the cerebrospinal fluid of the ventricular space, and a long basal process terminates with an endfoot tightly contacting blood vessels (akin to the astrocytic endfeet that contribute to the blood-brain barrier). The fundamental cell constituents of the V-SVZ niche include four cell types: (1) a layer of ependymal cells, E cells, along the ventricular wall; (2) the neural stem cell called the B cell; (3) progenitor (transit-amplifying) C cells; and (4) migrating neuroblast A cells (see Figure 5.11). Small clusters of B cells are surrounded by the multiciliated E cells, forming a pinwheel-like rosette structure (**FIGURE 5.12A**; Mirzadeh et al. 2008). Cell generation within the V-SVZ begins in its central core with a dividing B cell, which directly gives rise to a C cell. These type C progenitor cells proliferate and develop into type A neural precursors that stream into the olfactory bulb for final neuronal differentiation (see Figure 5.11). The B cell has been further categorized into three subtypes (B1, B2, and B3) based on differences in proliferative states that correlate with distinct radial glial gene expression patterns (Codega et al. 2014; Giachino et al. 2014). It is important to note that in the NSC niche, *type 1 B cells are quiescent or inactive*, whereas *types 2 and 3 B cells represent actively proliferating neural stem cells* (Basak et al. 2012).[6]

MAINTAINING THE NSC POOL WITH CELL-TO-CELL INTERACTIONS

Maintaining the stem cell pool is a critical responsibility of any stem cell niche because too many symmetrical differentiating and progenitor-generating divisions can deplete the stem cell pool. The V-SVZ niche is designed structurally and is equipped with signaling systems to ensure that its B cells are not lost during calls for neurogenic growth or repair in response to injury.

VCAM1 AND ADHERENCE TO THE ROSETTE NICHE The rosette or pinwheel architecture is a distinctive physical characteristic of the V-SVZ niche. It is maintained at least in part by a specific cell adhesion molecule, VCAM1 (Kokovay et al. 2012). As the mammalian brain ages, both the number of observed pinwheel structures (see Figure 5.12A) and the number of neural stem cells in those pinwheels decrease, which correlates with a reduction in neurogenic potency in later life (Mirzadeh et al. 2008; Mirzadeh et al. 2010; Sanai et al. 2011; Shook et al. 2012; Shook et al. 2014). Much like campers huddled around a fire, ependymal cells surround each type B cell; and similar to how a fire can die out or grow based on the efforts of the surrounding campers, the B cell is listening to the ependymal cells (and other niche signals) for instructions either to remain quiescent or to become active. The B cells most tightly associated with ependymal cells are the more quiescent B1 cells. The more loosely packed B cells are actively proliferating B2 and B3 cells (Doetsch et al. 1997). Experimental inhibition of VCAM1, an adhesion protein specifically localized to the apical process of B cells (see Figure 5.12A), disrupts the pinwheel pattern and causes a loss of NSC quiescence while promoting differentiation of progenitors (**FIGURE 5.12B**; Kokovay et al. 2012). *The tighter the hold, the more quiescent the stem cell.*

NOTCH, THE TIMEPIECE TO DIFFERENTIATION Notch signaling has been found to play an important role in the maintenance of the pool of B type stem cells (Pierfelice et al. 2011; Giachino and Taylor 2014). Notch family members function as transmembrane receptors, and through cell-to-cell interactions, the Notch intracellular domain (NICD) is cleaved and released to function as part of a transcription factor complex typically repressing proneural gene expression (see Figure 5.11). Higher levels of NICD activity support stem cell quiescence, whereas decreasing levels of Notch pathway activity promote progenitor proliferation and maturation toward neural fates.[7] (See **Further Development 5.6, Just Another Notch on the Clock during Neurogenesis in the V-SVZ**, online.)

PROMOTING DIFFERENTIATION IN THE V-SVZ NICHE The main purpose of a stem cell niche is to produce new progenitor cells capable of differentiating toward specific cell types. In the V-SVZ niche, a number of factors are involved, such as EGF and BMP signaling.

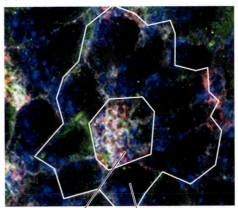

B cells   Ependymal cells

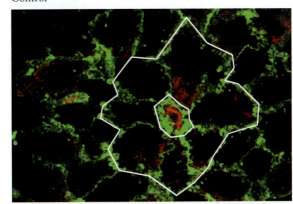

Control

VCAM1 blocked

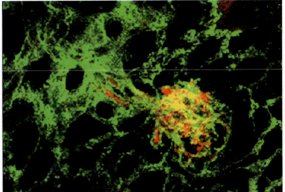

FIGURE 5.12 VCAM1 and pinwheel architecture. (A) The pinwheel arrangement of cells in the V-SVZ of the NSC niche is revealed with membrane labeling. Immunolabeling for VCAM1 (red) shows its co-localization with GFAP (green) in the B cells at the pinwheel core. The blue stain shows the presence of β-catenin; pinwheel organization is outlined in white. (B) Blocking adhesion using antibodies to VCAM1 disrupts the pinwheel organization of B cells and ependymal cells. In these photos, red visualizes GFAP; green indicates the presence of β-catenin. (After E. Kokovay et al. 2012. *Cell Stem Cell* 11: 220–230.)

? **Developing Questions**

Here, differing levels of Notch activity signify the timed emergence from quiescence to neurogenesis; however, we know quite a bit about *Notch/Delta* and *Hes* gene oscillations during embryonic development. What do oscillations of these genes in the adult neural stem cell niche actually look like? How do these oscillations result in the progression from stem cell to neuron?

EGF REPRESSES NOTCH   As discussed above, active (and constant) Notch signaling encourages quiescence and represses differentiation; therefore, one mechanism to promote neurogenesis is to attenuate (and oscillate) Notch activity. The type C progenitor cells do that by using epidermal growth factor receptor (EGFR) signaling, which upregulates NUMB, which in turn inhibits NICD (see Figure 5.11; Aguirre et al. 2010). Therefore, EGF signaling promotes the use of the stem cell pool for neurogenesis by counterbalancing Notch signaling (McGill and McGlade 2003; Kuo et al. 2006; Aguirre et al. 2010).

BONE MORPHOGENIC PROTEIN SIGNALING AND THE NSC NICHE   Further movement toward differentiation is driven by additional factors, such as BMP signaling, which promotes gliogenesis in the V-SVZ as well as other regions of the mammalian brain (Lim et al. 2000; Colak et al. 2008; Gajera et al. 2010; Morell et al. 2015). BMP signaling from endothelial cells is kept high at the basal side of the niche, whereas ependymal cells at the apical border secrete the BMP inhibitor Noggin, keeping BMP levels in this region low. Therefore, as B3 cells transition into type C progenitor cells and then move closer to the basal border of the niche, they experience increasing levels of BMP signaling, which promotes neurogenesis with a preference toward glial differentiation (see Figure 5.11).

FURTHER DEVELOPMENT

ENVIRONMENTAL INFLUENCES ON THE NSC NICHE   The adult NSC niche has to react to changes in the organism, such as injury and inflammation, exercise, and changes in circadian rhythms. How might the NSC niche respond to such changes? The cerebrospinal fluid (CSF), neural networks, and vasculature are in direct contact with the niche, and they can influence NSC behavior through paracrine release into the CSF, electrophysical activity from the brain, and endocrine signaling delivered through the circulatory system.

NEURAL ACTIVITY   Intrinsic to the niche, migrating neural precursors secrete the neurotransmitter GABA to negatively feed back upon progenitor cells and attenuate their rates of proliferation. In opposition to this action, B cells secrete a competitive inhibitor to GABA (diazepam-binding inhibitor protein) to increase proliferation in the niche (Alfonso et al. 2012). Extrinsic inputs have also been discovered from a variety of neuron types (see the references cited in Lim and Alvarez-Buylla 2014), such as synaptic connections between serotonergic neurons and both the ependymal and type B cells (Tong et al. 2014). Type B cells express serotonin receptors, and activation of the serotonin pathway in B1 cells increases proliferation in the V-SVZ, while repression of the pathway decreases proliferation (see Figure 5.11).

SONIC HEDGEHOG SIGNALING AND THE NSC NICHE   Similar to neural tube patterning in the embryo (see Chapter 13), the creation of different neuronal cell types from the V-SVZ is in part patterned by a gradient of Sonic hedgehog (Shh) signaling along the apical (high Shh) to basal (low Shh) axis of the niche (Goodrich et al. 1997; Bai et al. 2002; Ihrie et al. 2011).[8] When the Shh gene is knocked out, the loss of Shh signaling results in specific reductions in apically derived olfactory neurons (Ihrie et al. 2011). This result implies that cells derived from NSC clusters in more apical positions of the niche will adopt different neuronal fates compared to cells derived from NSCs in more basal positions due to differences in Shh signaling (see Figure 5.11).

COMMUNICATION WITH THE VASCULATURE   The vasculature heavily infiltrates the V-SVZ niche: from blood vessel cells (endothelial, smooth muscle, pericytes) to the associated extracellular matrix and substances in the blood (Licht and Keshet 2015; Ottone and Parrinello 2015). Though the apically situated cell bodies of B cells can be quite a distance from blood vessels, their basal endfeet are intimately associated with the vasculature (see Figure 5.11). As discussed earlier, Notch signaling is fundamental in controlling B1 cell quiescence. Notch receptors in the B cell's endfoot bind to the Jagged1 (Jag1) transmembrane receptor in endothelial cells, which causes Notch to be processed into its NICD transcription factor, and B1 cell quiescence is maintained as a result (Ottone et al. 2014). As the B2 and B3 cells transition into type C progenitor cells, their basal connections with endothelial cells are lost; consequently, NICD is reduced, enabling the progenitor cells to mature.

For a blood-borne substance to influence neurogenesis, it usually has to cross the tight blood-brain barrier, yet the NSC niche was found to be "leakier" than other brain regions (see Figure 5.11; Tavazoie et al. 2008). One of the most intriguing blood-borne molecules that reaches the NSC niche is growth differentiation factor 11 (GDF11, also known as BMP11), which appears to ward off some of the symptoms of aging in the brain. Similar to humans, when mice become old, they show a significantly reduced neurogenic potential. Researchers have shown that something in the circulation of young mice can prevent this decline when the circulation of a young mouse is surgically connected to that of an old mouse (heterochronic parabiosis). Doing so caused increased vasculature to develop in the brain of the heterochronic old mouse (**FIGURE 5.13A**), followed by increased NSC proliferation that restored neurogenesis and cognitive functions (Katsimpardi et al. 2014). The researchers then showed that they could similarly restore neurogenic potential in the V-SVZ of the old mouse brain using a single circulating factor, GDF11. It's important to note that there is some controversy about this result, and conflicting reports have questioned whether GDF11 levels decrease with age and whether it can similarly enhance muscle regeneration (**FIGURE 5.13B**; Loffredo et al. 2013; Poggioli et al. 2015).[9] More recently it was shown that administering GDF11 directly into the brain of a mouse after it had suffered a stroke resulted in significant neuro-regeneration (Lu et al. 2018). Additionally, systemic treatment of an aged mouse with GDF11 that stays only in the vasculature induced neurogenesis in the hippocampus of the mouse (Ozek et al. 2018). Taken together, these results suggest that communication between the NSC niches and its surrounding vasculature is a major regulatory mechanism of neurogenesis in the adult brain, and that changes in this communication over time may underlie some of the cognitive deficits associated with aging.

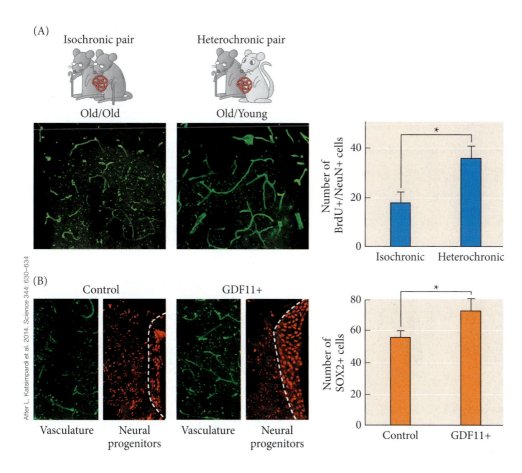

After L. Katsimpardi et al. 2014. *Science* 344: 630–634

(A) Isochronic pair — Old/Old   Heterochronic pair — Old/Young

(B) Control — Vasculature / Neural progenitors   GDF11+ — Vasculature / Neural progenitors

FIGURE 5.13 Young blood can rejuvenate an old mouse. (A) Parabiosis—fusion of the circulatory systems of two individuals—was done using mice of similar (isochronic) or different (heterochronic) ages. When an old mouse was parabiosed to a young mouse, the result was an increase in the amount of vasculature (stained green in the photographs) as well as the amount of proliferative neural progeny in the old mouse. (B) Administering GDF11 into the circulatory system of an old mouse was sufficient to similarly increase both vasculature (green in photographs) and the population of neural progenitors in the V-SVZ (outlined red population in photographs and quantified SOX2+ cells in graph). (After L. Katsimpardi et al. 2014. *Science* 344: 630–634.)

FIGURE 5.14   The ISC niche and its regulators. (A) The intestinal epithelium is composed of long, fingerlike villi that project into the lumen, and at the base of the villi, the epithelium extends into deep pits called crypts. The ISC and progenitors reside at the very bottom of the crypts (red), and cell death through anoikis occurs at the apex of the villi. (B) Along the proximodistal axis (crypt to villus), the crypt epithelium can be functionally divided into three regions: the base of the crypt houses ISC, the prolif-erative zone is made of transit amplifying cells, and the differentiation zone char-acterizes the maturation of epithelial cell types. Pericryptal stromal cells surround the basal surface of the crypt and secrete opposing morphogenic gradients of Wnt2b and Bmp4, which regulate stem-ness and differentiation, respectively. (C) Higher magnification of the cells residing in the base of the crypt. Paneth cells (P) secrete Wnt3a and D114, which stimulate proliferation of the Lrg5+ crypt base columnar cells (CBCC) in part through activation of the notch intracel-lular domain (NICD). +4 cells are the 4th cell from the Paneth cell and potentially serve as quiescent stem cells of the crypt. (LRC, label-retaining cell; PP, Paneth progenitor cell.)

The Adult Intestinal Stem Cell Niche

The neural stem cell is part of a specialized epithelium, but not all epithelial niches are created equal. The epithelial lining of the mammalian intestine projects millions of fingerlike villi into the lumen for nutrient absorption, and the base of each villus sinks into a steep valley called a **crypt** that connects with adjacent villi (**FIGURE 5.14A**). Critical to understanding the evolved function of the intestinal stem cell (ISC) niche is appreciating the rapid rate of cell turnover in the intestine.

Clonal renewal in the crypt

Cell generation occurs in the crypts, whereas cell removal largely happens at the tips of villi. Through this upward movement from cell source to cell sink, a turnover of intestinal absorptive cells occurs approximately every 2 to 3 days (Darwich et al. 2014).[10]

Several stem cells reside at the base of each crypt in the mouse small intestine; some daughter cells remain in the crypts as stem cells, whereas others become progenitor cells and divide rapidly (**FIGURE 5.14B**; Lander et al. 2012; Barker 2014; Krausova and Korinek 2014; Koo and Clevers 2014). Division of stem cells within the crypt and of the progenitor cells drives cell displacement vertically up the crypt toward the villus, and as cells become positioned farther from the crypt base, they progressively differentiate into the cells characteristic of the small intestine epithelium: enterocytes, goblet cells, and enteroendocrine cells. Upon reaching the tip of the intestinal villus, they are shed and undergo **anoikis**, a process of programmed cell death (apoptosis) caused by a loss of attachment—in this case, loss of contact with the other villus epithelial cells and extracellular matrix (see Figure 5.14A).[11]

Lineage-tracing studies (Barker et al. 2007; Snippert et al. 2010; Sato et al. 2011) have shown that intestinal stem cells (expressing the Lgr5 protein) can generate all the differentiated cells of the intestinal epithelium. Due to their specific location at the very base of the crypt, these Lgr5+ stem cells are referred to as crypt base columnar cells (CBCCs) and are found in a checkered pattern with the differentiated Paneth cells, which are also restricted to the base of the crypt (**FIGURE 5.14C**; Sato et al. 2011). One of the most convincing demonstrations that CBCCs represent "active stem cells" is that a single CBCC can completely repopulate the crypt over time (**FIGURE 5.15**; Snippert et al. 2010). After CBCC symmetrical division, one daughter cell will (by chance) be adjacent to a Paneth cell, while the other daughter cell is pushed away from the base to

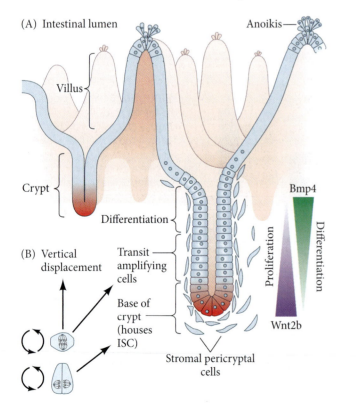

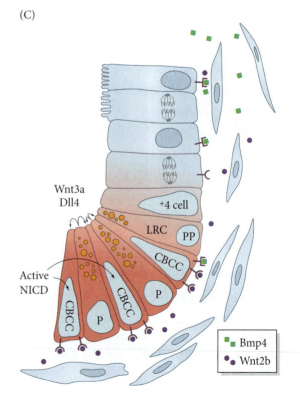

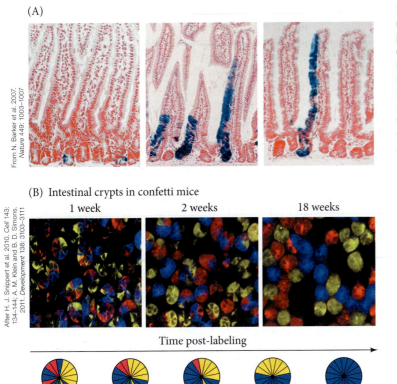

(A)

From N. Barker et al. 2007. *Nature* 449: 1003–1007

(B) Intestinal crypts in confetti mice

After H. J. Snippert et al. 2010. *Cell* 143: 134–144; A. M. Klein and B. D. Simons. 2011. *Development* 138: 3103–3111

1 week   2 weeks   18 weeks

Time post-labeling

FIGURE 5.15 Clonogenic nature of the intestinal stem cell niche. (A) Cre-responsive transgenic mice using the Lgr5 promoter and the Rosa26-LacZ reporter mark discrete clones of ISCs at the base of the crypt (blue). Retention of LacZ in cell descendants over time shows the progressive movement up the villus. (B) Mosaic labeling of ISCs in the intestinal crypt with transgenic "confetti" mice demonstrates a stochastic (predictable randomness) progression toward monoclonal (visualized as one color) crypts over time. This same progression can be mathematically modeled and simulated to produce a similar coarsening of color patterns, as seen below the photographs. (B after H. J. Snippert et al. 2010. *Cell* 143: 134–144; A. M. Klein and B. D. Simons. 2011. *Development* 138: 3103–3111.)

progress through the transit-amplifying (progenitor) fate. In this manner, the neutral competition for the Paneth cells' surfaces dictates which will remain as a stem cell and which will mature (Klein and Simons 2011). (See **Further Development 5.7, Regulatory Mechanisms in the Crypt**, online.)

Stem Cells Fueling the Diverse Cell Lineages in Adult Blood

Every day in your blood, more than 100 billion cells are replaced with new cells. Whether the needed cell type is for gas exchange or for immunity, the **hematopoietic stem cell** (**HSC**) is at the top of the hierarchical lineage powering the amazing cell generating machine that is the HSC niche (see Figure 18.24). The importance of HSCs cannot be overstated, both for its importance to the organism and its history of discovery. Since the late 1950s, stem cell therapy with HSCs has been used to treat blood-based diseases through the use of bone marrow transplantation.[12] In addition, the "niche hypothesis" of a stem cell residing in and being controlled by a specialized microenvironment was first inspired by the HSC (Schofield 1978). (See **Further Development 5.8, Were HSCs Somehow Born from Bone to Then Reside in the Marrow?**, online.)

The hematopoietic stem cell niche

In the highly vascularized tissue of the bone marrow resides the stem cell niche (**FIGURE 5.16**). HSCs are in close proximity to the bone cells (osteocytes), the endothelial cells that line the blood vessels, and the stromal cells. The hematopoietic niche can be subdivided into two regions, the endosteal niche and the perivascular niche (see Figure 5.16).[13] HSCs in the endosteal niche are often in direct contact with the osteoblasts lining the inner surface of the bone, and HSCs in the perivascular niche are in close contact with cells lining or surrounding blood vessels (endothelial cells and stromal cells). With the different physical and cellular properties of these two niches come differential regulation of the HSCs (Wilson et al. 2007). In addition, there are two subpopulations of HSCs within these niches: one population can

Developing Questions

Why might the longevity of an ISC be left to chance, and how might that actually help prevent cancer?

FIGURE 5.16 Model of adult HSC niche. Housed within the bone marrow, the HSC niche can be divided into two subniches: the endosteal and the perivascular. HSCs in the endosteal niche that are adhered to osteoblasts are long-term HSCs (purple), typically in the quiescent state, whereas short-term active HSCs (red) are associated with blood vessels (green) at oxygen-rich pores. Stromal cells—that is, the CAR cells (yellow) and mesenchymal stem cells—interact directly with mobile HSCs and progenitor cells, which can be stimulated by sympathetic connections.

divide rapidly in response to immediate needs, while a quiescent population is held in reserve and possesses the greatest potential for self-renewal (Wilson et al. 2008, 2009). Depending on physiological conditions, stem cells from one subpopulation can enter the other subpopulation.

HSCs found within the endosteal niche tend to be the most quiescent population, with long-term self-renewal serving to sustain the stem cell population for the life of the organism (Wilson et al. 2007). In contrast, more active HSCs tend to reside in the perivascular niche, exhibiting faster cycles of renewal and sustaining progenitor development for a shorter period of time (see Figure 5.16). A complex cocktail of cell adhesion molecules, paracrine factors, extracellular matrix components, hormonal signals, pressure changes from blood vessels, and sympathetic neural inputs all combine to influence the proliferative states of the HSCs (Spiegel et al. 2008; Malhotra and Kincade 2009; Cullen et al. 2014).

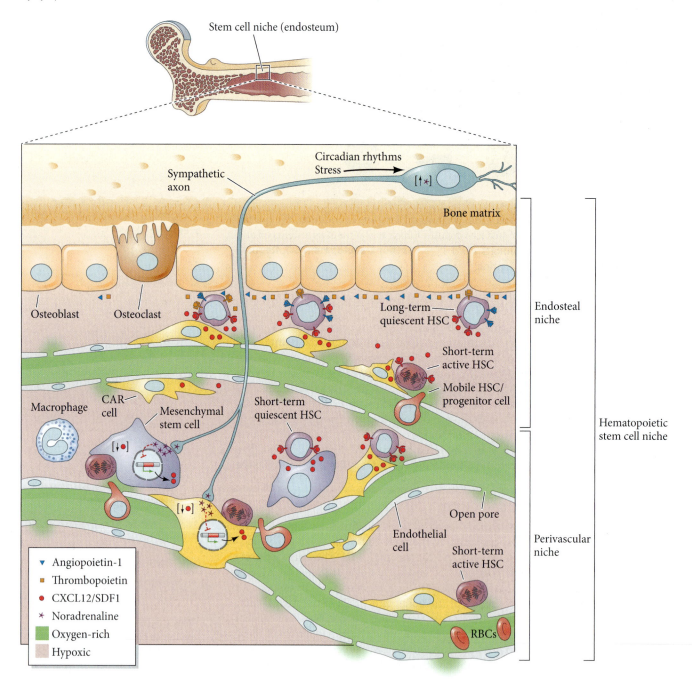

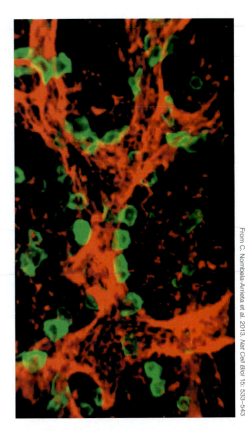

FIGURE 5.17 HSCs sit adjacent to microvasculature in the bone marrow. The c-Kit receptor (green) is a marker for HSCs and progenitors, which are seen in direct contact with the sinusoidal microvasculature in the niche (stained with anti-laminin, red). HSCs are associated with all types of vasculature in the niche.

FURTHER DEVELOPMENT

Regulatory mechanisms in the endosteal niche

In the endosteal niche, HSCs interact intimately with osteoblasts, and manipulation of osteoblast number causes similar increases or decreases in the presence of HSCs (Zhang et al. 2003; Visnjic et al. 2004; Lo Celso et al. 2009; Al-Drees et al. 2015; Boulais and Frenette 2015). Moreover, osteoblasts promote HSC quiescence by binding to the HSCs and secreting angiopoietin-1 and thrombopoietin, which keep these stem cells in reserve for long-term hematopoiesis (Arai et al. 2004; Qian et al. 2007; Yoshihara et al. 2007). The endosteal niche is permeated with sinusoidal microvessels (Nombela-Arrieta et al. 2013),[14] and some of the HSCs (*c-Kit*+) and progenitor cells are intimately associated with this highly permeable microvasculature (**FIGURE 5.17**). It has always been assumed that the endosteal niche was more hypoxic than the perivascular niche, but these microvessels undoubtedly aid in bringing oxygen to the endosteal regions, making the microlocales immediately surrounding sinusoids less hypoxic. Therefore, it is proposed that HSCs may use differences in oxygen content in the niche as a cue for locating blood vessels (Nombela-Arrieta et al. 2013).

Regulatory mechanisms in the perivascular niche

Cell-specific modulation of CXCL12 seems to be an important mechanism governing quiescence and retention of HSCs and progenitor cells in the perivascular niche. CXCL12 is secreted by several cell types, such as endothelial CXCL12-abundant reticular (CAR) cells, and the **mesenchymal stem cells** (**MSCs**) (see Figure 5.16; Sugiyama et al. 2006; Méndez-Ferrer et al. 2010). Loss of CXCL12 in CAR cells causes a significant movement of hematopoietic progenitor cells into the bloodstream, whereas selective knockout of CXCL12 in MSCs causes reductions in HSCs (Greenbaum et al. 2013).

Intriguingly, there are daily fluctuations in the rate that progenitor cells mobilize into the bloodstream: greater cell division of HSCs occurs at night, and increased migration of progenitor cells into the bloodstream happens during the day. This circadian pattern of mobilization is controlled by the release of noradrenaline from sympathetic axons infiltrating the bone marrow (see Figure 5.16; Méndez-Ferrer et al. 2008; Kollet et al. 2012). Receptors on stromal cells respond to this neurotransmitter by downregulating the expression of *CXCL12*, which temporarily reduces the hold that these stromal cells have on HSCs and progenitor cells, freeing them to circulate. Although circadian rhythms stimulate a normal round of HSC proliferation, chronic stress leads to increased release of noradrenaline (Heidt et al. 2014). This release lowers CXCL12 levels, which reduces HSC proliferation and increases their mobilization into the circulation. So, the next time you wake up, know that your sympathetic nervous system is telling your hematopoietic stem cells to wake up, too.

Additional signaling factors (Wnt, TGF-β, Notch/Jagged1, stem cell factor, and integrins; reviewed in Al-Drees et al. 2015 and Boulais and Frenette 2015) influence the production rates of different types of blood cells under different conditions; examples are the increased production of white blood cells during infections and of red blood cells when you climb to high altitudes. When this system is misregulated, it can cause diseases, such as the different types of blood cancers. Myeloproliferative disease is one such cancer that results from a failure of proper signals for blood cell differentiation (Walkley et al. 2007a,b). It stems from a failure of the osteoblasts to function properly; as a result, HSCs proliferate rapidly without differentiation (Raaijmakers et al. 2010; Raaijmakers 2012).

? Developing Questions

We discussed two distinct regions in the hematopoietic stem cell niche, but could there be more? It has been proposed that the MSCs in the bone marrow exert unique control over the HSCs and represent their own niche within a niche. What do you think? How is cell communication and HSC movement orchestrated among the endosteal, perivascular, and (potentially) MSC niches?

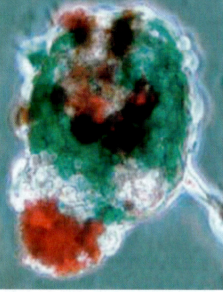

FIGURE 5.18 A mesensphere containing two derived cell types. Mesenchymal stem cells placed in culture form mesenspheres that can produce different cell types. Here, a mesensphere contains osteoblasts (bone-forming cells; teal) and adipocytes (fat-forming cells; red).

The Mesenchymal Stem Cell: Supporting a Variety of Adult Tissues

Most adult stem cells are restricted to forming only a few cell types (Wagers et al. 2002). For example, when HSCs marked with green fluorescent protein were transplanted into a mouse, their labeled descendants were found throughout the animal's blood but not in any other tissue (Alvarez-Dolado et al. 2003).[15] Some adult stem cells, however, appear to have a surprisingly large degree of plasticity. These multipotent mesenchymal stem cells (MSCs) are sometimes called **bone marrow-derived stem cells** (**BMDCs**), and their potency remains a controversial subject (Bianco 2014).

Originally found in bone marrow (Friedenstein et al. 1968; Caplan 1991), multipotent MSCs have also been found in numerous adult tissues (such as dermis of the skin, bone, fat, cartilage, tendon, muscle, thymus, cornea, and dental pulp) as well as in the umbilical cord and placenta (see Gronthos et al. 2000; Chamberlain et al. 2004; Perry et al. 2008; Traggiai et al. 2008; Kuhn and Tuan 2010; Nazarov et al. 2012; Via et al. 2012). Indeed, the finding that human umbilical cords and deciduous ("baby") teeth contain MSCs has led some physicians to propose that parents freeze cells from their child's umbilical cord or shed teeth so that these cells will be available for transplantation later in life.[16] Whether MSCs can pass the test of pluripotency—the ability to generate cells of all germ layers when inserted into a blastocyst—has not yet been shown.

Much of the controversy surrounding MSCs rests in their "split personality" as supportive stromal cells on the one hand and stem cells on the other. Morphologically, MSCs resemble fibroblasts, a cell type secreting the extracellular matrix of connective tissues (stroma). In culture, however, MSCs behave differently from fibroblasts. A single MSC in culture can self-renew to produce a clonal population of cells that can go on to form organs in vitro that contain a diversity of cell types (**FIGURE 5.18**; Sacchetti et al. 2007; Méndez-Ferrer et al. 2010; reviewed in Bianco 2014). As seen in bone marrow, MSCs in other tissues may play roles as both progenitor cells and regulators of the resident niche stem cell, possibly through paracrine signaling (Gnecchi et al. 2009; Kfoury and Scadden 2015).

Regulation of MSC development

Certain paracrine factors appear to direct development of the MSC into specific lineages. Platelet-derived growth factor (PDGF) is critical for fat formation and chondrogenesis, TGF-β signaling is also crucial for chondrogenesis, and fibroblast growth factor (FGF) signaling is necessary for the differentiation into bone cells (Pittenger et al. 1999; Dezawa et al. 2004; Ng et al. 2008; Jackson et al. 2010). Such paracrine signaling factors may underlie not only MSC differentiation but also their modulation of the resident niche stem cell. For instance, MSCs have been shown to play important dual roles as multipotent progenitor cells and stem cell niche regulators during hair follicle development and regeneration (Kfoury and Scadden 2015). The rapid turnover of epidermis and associated hair follicles in skin requires robust activation of resident stem cells (see Chapter 16). Immature adipose progenitor cells that surround the base of the growing follicle are both necessary and sufficient to trigger hair stem cell activation during growth and regeneration of the skin through a PDGF paracrine mechanism (Festa et al. 2011). (See **Further Development 5.9, Fat and Muscle, and the Role of MSCs in Aging**, online.)

The differentiation of MSCs is dependent on not only paracrine factors but also cell matrix molecules in the stem cell niche. Certain cell matrix components, especially laminin, appear to keep MSCs in a state of undifferentiated "stemness" (Kuhn and Tuan 2010). Researchers have taken advantage of the influence that the physical matrix has on MSC regulation to achieve a repertoire of derived cell types in vitro by growing stem cells on different surfaces. For example, if human MSCs are grown on soft matrices of collagen, they differentiate into neurons, a cell type that these cells do not appear to form in vivo. If instead

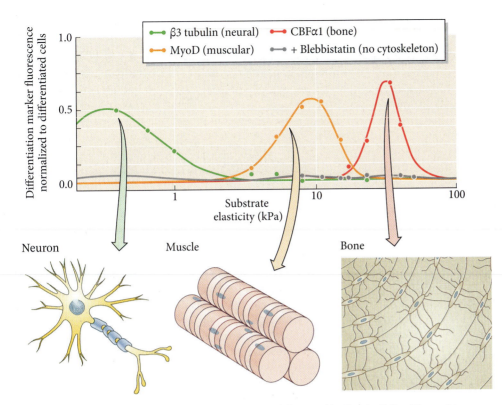

FIGURE 5.19 Mesenchymal stem cell differentiation is influenced by the elasticity of the matrices upon which the cells sit. On collagen-coated gels having elasticity similar to that of the brain (about 0.1–1 kPa), human MSCs differentiated into cells containing neural markers (such as β3-tubulin) but not into cells containing muscle cell markers (MyoD) or bone cell markers (CBFα1). As the gels became stiffer, the MSCs generated cells exhibiting muscle-specific proteins, and even stiffer matrices elicited the differentiation of cells with bone markers. Differentiation of the MSC on any matrix could be abolished with blebbistatin, which inhibits microfilament assembly at the cell membrane. (After A. J. Engler et al. 2006. *Cell* 126: 677–689.)

MSCs are grown on a moderately elastic matrix of collagen, they become muscle cells, and if grown on harder matrices, they differentiate into bone cells (**FIGURE 5.19**; Engler et al. 2006). It is not yet known whether this range of potency is found normally in the body. As technology improves, answers may come from gaining a better understanding of the properties of different MSC niches. (See **Further Development 5.10, Other Stem Cells Supporting Adult Tissue Maintenance and Regeneration**, online.)

The Human Model System to Study Development and Disease

Up to this point, we have focused on the in vivo life of stem cells. The properties of self-renewal and differentiation that define a stem cell, however, also enable their manipulation in vitro. Before we were able to culture human embryonic stem cells (Thomson et al. 1998), researchers studying human cell development used immortalized tumor cells or cells from teratocarcinomas, which are cancers that arise from germ cells (Martin 1980). The most investigated human cell has been the HeLa cell line, derived from the cervical cancer of *H*enrietta *L*acks (a cancer that took her life in 1951 and a cell line that was isolated without her or her family's knowledge or consent).[17]

Pluripotent stem cells in the lab

A major drawback to the early studies of immortalized tumor cells was that none of the cells used represented a model of normal human cells. With our present ability to grow embryonic and adult human stem cells in the lab and induce them to differentiate into different cell types, we finally have a tractable model system for studying human development and disease in vitro.

(?) Developing Questions

What molecular mechanisms may govern the change of MSCs from being a progenitor at one moment to regulating other stem cells at another?

FIGURE 5.20 Major sources of pluripotent stem cells from the early embryo. Embryonic stem cells arise from culturing the inner cell mass of the early embryo. Embryonic germ cells are derived from primordial germ cells that have not yet reached the gonads.

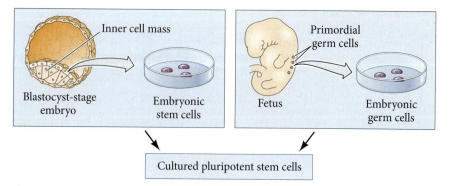

FIGURE 5.20 Major sources of pluripotent stem cells from the early embryo. Embryonic stem cells arise from culturing the inner cell mass of the early embryo. Embryonic germ cells are derived from primordial germ cells that have not yet reached the gonads.

EMBRYONIC STEM CELLS Pluripotent embryonic cells are a special case because these stem cells can generate all the cell types needed to produce the adult mammalian body (see Shevde 2012). In the laboratory, pluripotent embryonic cells are derived from two major sources: (1) the ICM of the early blastocyst, whose cells can be maintained in culture as a clonal line of ESCs (Thomson et al. 1998); and (2) primordial germ cells (PGCs) that have not yet differentiated into sperm or eggs (**FIGURE 5.20**). When PGCs are isolated from the embryo and grown in culture, they are called embryonic germ cells, or EGCs (Shamblott et al. 1998).

As in the ICM of the embryo, the pluripotency of ESCs in culture is maintained by the same core of three transcription factors: Oct4, Sox2, and Nanog. Acting in concert, these factors activate the gene regulatory network required to maintain pluripotency and repress those genes whose products would lead to differentiation (Marson et al. 2008; Young 2011). Are all pluripotent stem cells created equal, however? Although the years of experimentation with both mouse and human ESCs have demonstrated clear pluripotency (Martin 1981; Evans and Kaufman 1981; Thomson et al. 1998), they have also revealed differences in their degrees of self-renewal, the types of cells they can form, and their cellular characteristics (Martello and Smith 2014; Fonseca et al. 2015; Van der Jeught et al. 2015). It appears that these differences may be based on slight differences in the developmental stage of the original ICM cells from which the cultures were derived, which has led to recognizing two different pluripotent states of an ESC: **naïve** and **primed**.[18] The **naïve ESC** represents the most immature, undifferentiated ESC with the greatest potential for pluripotency. In contrast, the **primed ESC** represents an ICM cell with some maturation toward the epiblast lineage; hence, it is "primed," or ready for differentiation.

FURTHER DEVELOPMENT

FACTORS OF ESC DERIVATION Different methods are emerging for the maintenance of naïve human ESCs from ICM cells or even from primed ESCs (Van der Jeught et al. 2015). For example, ESCs have been cultured in the presence of leukemia inhibitory factor (LIF) in combination with at least two kinase inhibitors (called 2i) that are associated with the MAPK/ErK pathway inhibitor (MEKi) and glycogen synthase kinase 3 inhibitor (GSK3i) (see Theunissen et al. 2014). These factors, along with additional conditions, serve to prevent differentiation and maintain the ESCs in the naïve, or ground, state.

Researchers are now studying the gene networks, epigenetic modulators, paracrine factors, and adhesion molecules required for the differentiation of ESCs. These cells can respond to specific combinations and sequential application of growth factors to coax their differentiation toward specific cell fates associated with the three germ layers (**FIGURE 5.21**; Murry and Keller 2008). For instance, applying a chemically defined growth medium to a monolayer of ESCs can push their specification toward a mesodermal fate; when followed by a period of Wnt activation and then Wnt inhibition, the cells differentiate into contracting heart muscle cells (Burridge et al. 2012, 2014). In contrast, ESCs pushed toward an ectodermal fate by inhibiting Bmp4, Wnt, and activin can be subsequently induced by fibroblast growth factors (FGFs) to become neurons (see Figure 5.21; Kriks et al. 2011).

(?) Developing Questions

What is possible now that naïve human ESCs can be isolated and maintained? Proof of these cells' pluripotency was displayed when naïve human ESCs were transplanted into the mouse morula and differentiated into many cell types of an interspecies chimeric humanized mouse embryo (Gafni et al. 2013). Although federal funds cannot be used to create human-mouse chimeras in the United States, such regulations do not exist in other countries. It is theoretically plausible to create a human ICM from naïve human ESCs that is supported by a mouse trophectoderm. Minimally, doing so could enable the first direct study of human gastrulation. Should the human gastrula be studied in this way? What, if any, ethical concerns could such studies raise?

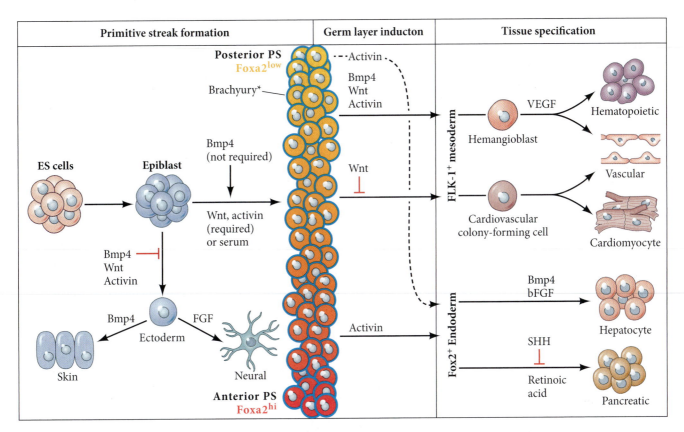

Primitive streak formation	Germ layer inducton	Tissue specification

FIGURE 5.21 Inducing stem cell differentiation from ESCs. Similar to the steps in differentiation that epiblast cells take during their maturation in the mammalian embryo, ESCs in culture can be coaxed with the same developmental factors (paracrine and transcription factors, among others) to differentiate into the cell types of each germ layer. With the inhibition of several growth factors, ESCs can make ectoderm lineages; for mesoderm or endoderm lineages, however, ESCs are first induced to become primitive streak-like cells (PS) with paracrine factors such as Wnt, Bmp4, or activin, depending on the desired differentiated cell type. (After C. E. Murry and G. Keller. 2008. *Cell* 132: 661–680.)

The physical constraints of the environment in which ESCs are cultured can also profoundly influence their differentiation. Constraining the area of cell growth to small disc shapes[19] can alone initiate a pattern of differential gene expression in the colony of cells that correlates to that of the early embryo (**FIGURE 5.22**; Warmflash et al. 2014). These results demonstrate that an incredible amount of patterning can be initiated solely by the geometry and size of the growth landscape. These discoveries are enabling further research into the structure and function of specific human cell types and their use in medical applications.

(A) Micropatterned cultures

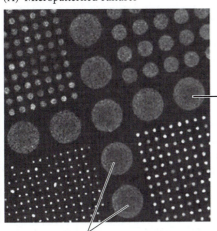

Micropatterned discs

After A. Warmflash et al. 2014. *Nat Methods* 11: 847–854

(B) Radially patterned gene expression

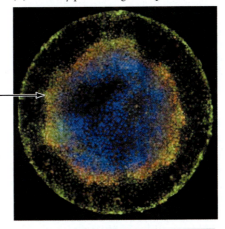

■ Ectoderm (Sox2⁺)
■ Trophectoderm (Cdx2⁺)
■ Mesoderm (Bra⁺)

FIGURE 5.22 Human ESCs cultured in confined micropatterned discs demonstrate a pattern of differential gene expression similar to that seen in the early embryo.

ESCS AND REGENERATIVE MEDICINE   A major hope for human stem cell research is that it will yield therapies for treating diseases and repairing injuries. In fact, pluripotent stem cells have opened an entirely new field of therapy called regenerative medicine (Wu and Hochelinger 2011; Robinton and Daley 2012). The therapeutic possibilities for ESCs lie in their ability to differentiate into any cell type, especially for treatment of human conditions in which adult cells degenerate (such as Alzheimer disease, Parkinson disease, diabetes, and cirrhosis of the liver). For instance, Kerr and colleagues (2003) found that human EGCs were able to cure motor neuron injuries in adult rats both by differentiating into new neurons and by producing paracrine factors (BDNF and TGF-β) that prevent the death of existing neurons. Similarly, precursor cells for dopamine-secreting neurons derived from ESCs (Kriks et al. 2011) were able to complete their differentiation into dopaminergic neurons and cure a Parkinson-like condition when engrafted into the brains of mice, rats, and even monkeys.

Although great excitement surrounds the potential of therapies using stem cells, another line of research is aimed at understanding the development of disease and assessing the effectiveness of pharmaceuticals. Such studies have already advanced our understanding of rare blood-based diseases such as Fanconi anemia, which causes bone marrow failure and consequent loss of both red and white blood cells (Zhu et al. 2011). Often, diseases like Fanconi anemia are caused by **hypomorphic mutations**—mutations that merely reduce gene function, as opposed to a "null" mutation that results in the total loss of a protein's function. Researchers used human ESCs to create a model of Fanconi anemia by using RNAi to knock *down* (not knock *out*) specific isoforms of the Fanconi anemia genes (Tulpule et al. 2010). The results gave new insights into the role of the Fanconi anemia genes during the initial steps of embryonic hematopoiesis. (See **Further Development 5.11, A Discussion of the Challenges Using ESCs**, online.)

Induced pluripotent stem cells

Although we know that the nuclei of differentiated somatic cells retain copies of an individual's entire genome, biologists have long thought that potency was like going down a steep hill with no return. Once differentiated, we believed, a cell could not be restored to an immature and more plastic state. Our newfound knowledge of the transcription factors needed to maintain pluripotency, however, has illuminated a startlingly easy way to reprogram somatic cells into embryonic stem cell-like cells.

In 2006, Kazutoshi Takahashi and Shinya Yamanaka of Kyoto University demonstrated that by inserting activated copies of four genes that encoded some of these critical transcription factors, nearly any cell in the adult mouse body could be made into an **induced pluripotent stem cell** (**iPSC**) with the pluripotency of an embryonic stem cell. These genes were *Sox2* and *Oct4* (which activated Nanog and other transcription factors that established pluripotency and blocked differentiation), c-*Myc* (which opened up chromatin and made the genes accessible to Sox2, Oct4, and Nanog), and *Klf4* (which prevents cell death; see Figure 3.14).

Within 6 months of the publication of this work (Takahashi and Yamanaka 2006), three groups of scientists reported that the same or similar transcription factors could induce pluripotency in a variety of differentiated human cells (Takahashi et al. 2007; Yu et al. 2007; Park et al. 2008). Like embryonic stem cells, iPS cell lines can be propagated indefinitely and, whether in culture or in a teratoma, can form cell types representative of all three germ layers. By 2012, modifications of the culture techniques made it possible for the gene expression of mouse iPSCs to become nearly identical to that of mouse embryonic stem cells (Stadtfeld et al. 2012). Most important was that entire mouse embryos could be generated from single iPSCs, showing complete pluripotency. Although iPSCs are functionally pluripotent, they are best at generating the cell types of the organ from which the parent somatic cell originated (Moad et al. 2013). These data suggest that, like naïve versus primed ESC, not all iPSCs are the same and that they may retain an epigenetic memory of their past home.

APPLYING iPSCS TO HUMAN DEVELOPMENT AND DISEASE   Using iPSCs provides medical researchers with the ability to experiment on diseased human tissue while avoiding the complications introduced by using human embryonic stem cells. Currently, there are four major medical uses for iPSCs: (1) making patient-specific iPSCs for studying disease pathology, (2) combining gene therapy with patient-specific iPSCs to treat disease, (3) using patient-specific iPSC-derived progenitor cells in cell transplants without the complications of immune rejection, and (4) using differentiated cells derived from patient-derived iPSCs for screening drugs.

Transplanting cells derived from mouse iPSCs back into the same donor mouse does not elicit immune rejection (Guha et al. 2013), suggesting that iPSC-based cell replacement may, in fact, be a promising therapy in the future.[20] So far, the most significant advances with iPSCs have been in modeling human diseases. Following a major study (Park et al. 2008) that created iPSCs from patients associated with 10 different diseases, numerous studies have leveraged iPSC technology to model a diverse array of diseases, including Down syndrome and diabetes among others (Singh et al. 2015).

Disease modeling is of particular importance for diseases that are not easily modeled in non-human organisms. Mice, for instance, do not get the same type of cystic fibrosis—a disease that severely compromises lung function—that humans get. After discovering what factors caused mouse iPSCs to differentiate into lung tissue (Mou et al. 2012), researchers made iPSCs from a person with cystic fibrosis and turned them into lung epithelium that showed the characteristics of human cystic fibrosis. Knowing that cystic fibrosis is often caused by mutations within a single gene (the gene for CF transmembrane conductance regulator, which encodes a chloride channel; Riordan et al. 1989; Kerem et al. 1989), researchers sought to repair the human mutation using homology directed repair approaches in these iPSCs. Crane and colleagues (2015) accomplished this task in iPSCs derived from a cystic fibrosis patient; once the cystic fibrosis mutation was corrected in these cells, they were able to make functional chloride channels when induced to differentiate in culture. The next step will be to test this approach in vivo in a non-human animal model to see if it might be used to treat cystic fibrosis in humans.

The benefits of combining the use of iPSCs and gene correction were eloquently demonstrated by Rudolf Jaenisch's lab in 2007, when the researchers cured a mouse model of sickle-cell anemia. This disease is caused by a mutation in the gene for hemoglobin. The Jaenisch lab generated iPSCs from this mouse, corrected the hemoglobin mutation (single base-pair substitution), and then differentiated the iPSCs into hematopoietic stem cells that, when implanted in the mouse, cured its sickle-cell phenotype (**FIGURE 5.23**; Hanna et al. 2007). Ongoing studies are attempting to determine if similar therapies could cure human conditions such as diabetes, macular degeneration, spinal cord injury,

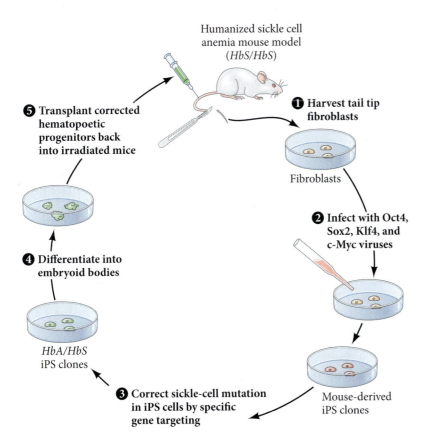

FIGURE 5.23 Protocol for curing a "human" disease in a mouse using iPS cells plus recombinant genetics. (1) Tail-tip fibroblasts are taken from a mouse whose genome contains the human alleles for sickle-cell anemia (*HbS*) and no mouse genes for this protein. (2) The cells are cultured and infected with viruses containing the four transcription factors known to induce pluripotency. (3) The iPS cells are identified by their distinctive shapes and are given DNA containing the wild-type allele of human globin (*HbA*). (4) The embryos are allowed to differentiate in culture. They form "embryoid bodies" that contain blood-forming stem cells. (5) Hematopoietic progenitor and stem cells from these embryoid bodies are injected into the original mouse, which has been irradiated to clear out its original hematopoietic cells. This treatment cures its sickle-cell anemia. (After J. Hanna et al. 2007. *Science* 318: 1920–1923.)

Parkinson disease, and Alzheimer disease, as well as liver disease and heart disease. Even sperm and oocytes have been generated from mouse iPSCs. First, skin fibroblasts were induced to form iPSCs, and these iPSCs were then induced to form primordial germ cells (PGCs). When these induced PGCs were aggregated with gonadal tissues, the cells proceeded through meiosis and became functional gametes (Hayashi et al. 2011; Hayashi et al. 2012). This work could become significant in circumventing many types of infertility as well as in allowing scientists to study the details of meiosis.

FURTHER DEVELOPMENT

MODELING MULTIGENIC HUMAN DISEASES WITH iPSCS   One challenge in studying a human disease is that individuals differ in the repertoire of genes associated with a disease as well as the timing of onset and progression of the disease. Fortunately, iPSCs have provided a new tool to help unravel this complexity. Here we highlight the use of iPSCs to study two particularly complex and multigenic diseases of the nervous system that fall at opposite ends of the developmental calendar: autism spectrum disorders and amyotrophic lateral sclerosis (ALS), also called Lou Gehrig disease.

Autism Spectrum Disorders present a range of neural dysfunctions typically affecting social and cognitive abilities that are not clearly apparent until around 3 years of age.[21] Disorders that fall within this spectrum include classic autism, Asperger syndrome, fragile-X syndrome, and Rett syndrome. Rett syndrome appears to be associated with a single gene (*methyl CpG binding protein-2*, or *MeCP2*). In contrast, autism is truly multi-allelic, with some children being non-syndromic (autism with no known cause) and likely possessing sporadic mutations (Iossifov et al. 2014; Ronemus et al. 2014; De Rubeis and Buxbaum 2015). In fact, the causative agents (genetics and environmental factors) may be unique to each autistic child, which presents significant challenges to researching autism.

One approach has been to generate iPSCs from as many children on the autism spectrum as possible to establish a more comprehensive understanding of the associated genes. This approach has been facilitated through a program called the Tooth Fairy Project, through which donations of children's deciduous (baby) teeth provide sufficient dental pulp for deriving iPSCs.[22] In using the iPSCs from a child with nonsyndromic autism, researchers created a culture of neurons and found a mutation in the TRPC6 calcium channel gene that impaired the structure and function of these neurons (Griesi-Oliveira et al. 2015). They further demonstrated improved neuronal function after exposing these cells to hyperforin, a compound found in St. John's wort and known to stimulate calcium influx. It turns out that *TRPC6* expression can be regulated by MeCP2, which confirms a direct genetic association between autism and Rett syndrome. Remarkably, the medical intervention for this child was changed to now include St. John's wort, which highlights the potential for patient-specific precision medicine in the future. This finding shows that iPSCs can play an important role in modeling a complex disease to research mechanisms that can lead to direct patient intervention.

Amyotrophic Lateral Sclerosis (ALS) is an adult-onset degenerative motor neuron disease that is multi-allelic through familial inheritance as well as sporadic mutation; unfortunately, it has no cure or treatment. Some of the first disease-specific iPSCs were derived from ALS patients in 2008 by Kevin Eggan's lab (Dimos et al. 2008). ALS-derived iPSCs can be coaxed to differentiate into motor neurons and non-neuronal cell types such as astrocytes, which are cells implicated in the ALS phenotype. More recently, motor neurons differentiated from patient-derived iPSCs harboring a known ALS familial mutation exhibited typical hallmarks of ALS cellular pathology (Egawa et al. 2012). The researchers used these differentiated motor neurons to screen for drugs that might improve motor neuron health, and they identified a histone acetyltransferase inhibitor capable of reducing the ALS cellular phenotypes. Thus, experimentation with iPSCs has revealed new insights into how ALS could be epigenetically regulated and possibly treated.

(?) Developing Questions

We have discussed modeling human disease using stem cells, but can you study vertebrate evolution in a dish? Researchers like Alysson Muotri are interested in how iPSCs generated from humans and a variety of non-human primates compare in behavior, self-renewal, and potency. By comparing the transcriptomes and physiology of the derived cell types from different species, we may gain new insights into human evolution. What specific questions would you ask, and what might your predictions be?

Organoids: Studying human organogenesis in a culture dish

We have discussed the many ways in which pluripotent stem cells (ESCs and iPSCs) can be used to better understand human development and disease at the level of the cell, but there is a vast difference between cells in culture and cells in the embryo. Human blastocysts are routinely used to research early human development and interventions for treating infertility; using human embryos for studying human organogenesis, however, has been both technically impossible and viewed as unethical by most. Through recent advances in pluripotent cell culturing techniques, however, researchers have been able to grow rudimentary organs from pluripotent stem cells. To date, the most complex structures that have been created are the optic cup of the eye, miniguts, kidney tissues, liver buds, and even brain regions (**FIGURE 5.24A**; Lancaster and Knoblich 2014).

These **organoids**, as they are called, are generally the size of a pea and can be maintained in culture for more than a year. The striking feature of organoids is that they actually mimic embryonic organogenesis. Pluripotent cells often self-organize into aggregates based on differential adhesion between cells (much like during gastrulation; see also Chapter 4), leading to cell sorting and the differentiation of cells with different fates that interact to form the tissues of an organ (**FIGURE 5.24B**). Organoids have been made from both ESCs and iPSCs derived from healthy and diseased individuals. Therefore, the same therapeutic approaches that we discussed for ESCs and iPSCs can also be applied to the organoid system. Although speculative at this point, creating organoids may prove to be a viable procedure for growing autologous structures,[23] not just for patient-specific cell replacement therapy, but also for tissue replacement. As an example, we highlight below in our "Further Development" section some of the remarkable features associated with the development of the cerebral organoid and its use in modeling a congenital brain disease.

FIGURE 5.24 Organoid derivation. (A) Schematic represents the various strategies used to promote the morphogenesis of specific tissue-type organoids. In most cases, a three-dimensional matrix (Matrigel) is used. KSR is a knockout serum replacement. (B) Early progression of organoid formation begins with differential gene expression, leading to cells with different cell adhesion molecules that confer self-organizing properties (see Chapter 4). Once sorted, cells continue to mature toward distinct lineages that interact to build a functional tissue. (After M. A. Lancaster and J. A. Knoblich. 2014. *Science* 345: 124–125.)

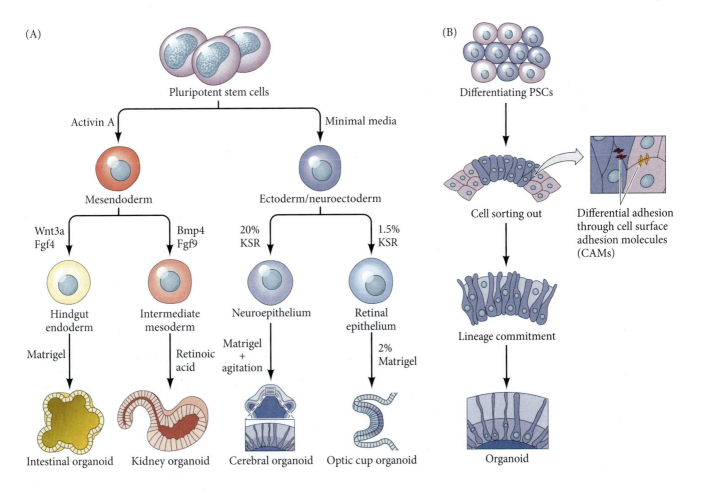

THE CEREBRAL ORGANOID The human cerebral cortex is arguably the most sophisticated tissue in the animal kingdom, so trying to build even parts of this structure may seem daunting. Ironically, neural differentiation from pluripotent cells seems to be sort of a "default state," similar to the presumptive neural-forming cells of the gastrula. Many previous studies characterizing the development of pluripotent stem cells into neural tissues have paved the way to growing multiregional brain organoids (Eiraku et al. 2008; Muguruma et al. 2010; Danjo et al. 2011; Eiraku and Sasai 2012; Mariani et al. 2012). In relatively simple growth conditions, pluripotent cells will self-organize into small spherical clusters of cells called embryoid bodies, and cells within these bodies will differentiate into a stratified neuroepithelium, similar to the neural epithelium of an embryo. The "self-organizing" ability of pluripotent cells to form three-dimensional neuroepithelial structures strongly suggests that robust intrinsic mechanisms exist that are primed for neural development (Harris et al. 2015). As seen in most adult neural stem cell niches, this neuroepithelium is polarized along the apical-basal axis and is capable of developing into brain tissue.

In a landmark study, researchers took brain tissue organoids to the next level of complexity (Lancaster et al. 2013). They placed embryoid bodies into droplets of Matrigel (a matrix made from solubilized basement membrane, the ECM

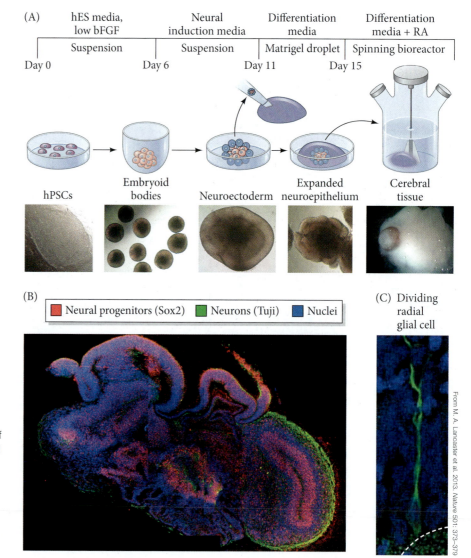

FIGURE 5.25 The cerebral organoid. (A) Schematic showing the process over time for the creation of a cerebral organoid from initial cell suspension to its growth in a bioreactor spinning at low speed. Representative light microscopic images of the developing organoid are shown below each step. (B) Section of a cerebral organoid labeled for neural progenitors (red; Sox2), neurons (green; Tuji), and nuclei (blue), which reveals the multilayered organization characteristic of the developing cerebral cortex. (C) Radial glial cell labeled with p-Vimentin (green) undergoes division and shows its characteristic morphology, with a long basal process and its apical membrane at the ventricular-like lumen (dashed white line). (After M. A. Lancaster et al. 2013. *Nature* 501: 373–379.)

(A)

hES media, low bFGF	Neural induction media	Differentiation media	Differentiation media + RA
Suspension	Suspension	Matrigel droplet	Spinning bioreactor

Day 0   Day 6   Day 11   Day 15

hPSCs   Embryoid bodies   Neuroectoderm   Expanded neuroepithelium   Cerebral tissue

(B)

Neural progenitors (Sox2)   Neurons (Tuji)   Nuclei

(C) Dividing radial glial cell

From M. A. Lancaster et al. 2013. *Nature* 501: 373–379

normally at the basal side of an epithelium) to provide a three-dimensional architecture. They next moved these neuroepithelial buds into a media-filled spinning bioreactor (**FIGURE 5.25A**; see also Lancaster and Knoblich 2014). The movement of the organoid in this three-dimensional matrix served to increase nutrient uptake, which supported the substantial growth required for multiregional cerebral organoid development. The resulting cerebral organoid showed characteristically layered tissue for a variety of brain regions, including appropriate neuronal and glial cell markers (**FIGURE 5.25B**). These cerebral organoids possessed radial glial cells adjacent to ventricular-like structures, similar to the developing neural tube and even the adult neural stem niche discussed earlier (**FIGURE 5.25C**). These human radial glial cells within the cerebral organoid displayed all patterns of mitotic behaviors: symmetrical division for stem cell expansion and asymmetrical divisions for self-renewal and differentiation (Lancaster et al. 2013).

Knoblich's group also generated iPSCs from fibroblast samples of a patient with severe microcephaly in the hope that they could study the pathologies associated with this disease (Lancaster et al. 2013). Microcephaly is a congenital disease characterized by a significant reduction in brain size (**FIGURE 5.26A**). Remarkably, cerebral organoids from this patient did show smaller developed tissues, but outer layers of the cortex-like tissues showed increased numbers of neurons compared to control organoids (**FIGURE 5.26B**). The researchers discovered that this patient had a mutation in the gene for CDK5RAP2,[24] a protein involved in mitotic spindle function. Moreover, the radial glial cells in this cerebral organoid exhibited abnormally low levels of symmetrical division (**FIGURE 5.26C**). Recall that one of the most basic functions of a stem cell is cell division. It appears that CDK5RAP2 is required for the cell division needed for expansion of the stem cell pool. Lack of symmetrical divisions leads to premature neuronal differentiation, which explains the increased number of neurons in this patient-derived organoid despite the smaller size of its tissues (Lancaster et al. 2013).

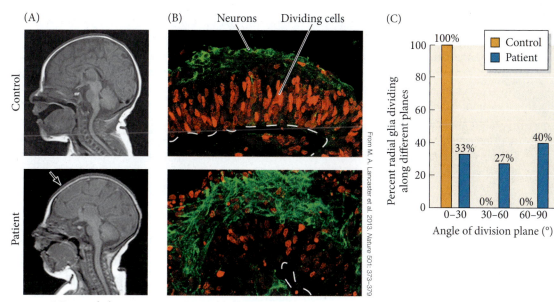

FIGURE 5.26 Modeling human microcephaly with a patient-specific cerebral organoid. (A) Sagittal views of magnetic resonance imaging scans from age-matched control (top) and patient brains at birth. The patient has a smaller brain and reduced brain folding (arrow). (B) Immunolabeling of control and patient-derived cerebral organoids. Neurons (green) and dividing cells (red) are labeled with DCX and BrdU, respectively. There is decreased proliferation and an increase in neuron numbers in the patient-derived organoid. (C) Quantification of the number of radial glial cells undergoing mitotic divisions along specific planes relative to the apical-basal axis of the organoid. Due to a loss of CDK5RAP2, patient radial glial cells divide randomly along all axes. (After M. A. Lancaster et al. 2013. *Nature* 501: 373–379.)

Stem Cells: Hope or Hype?

The ability to induce, isolate, and manipulate stem cells offers a vision of regenerative medicine wherein patients can have their diseased organs regrown and replaced using their own stem cells. Stem cells also offer fascinating avenues for the treatment of numerous diseases. Indeed, when one thinks about the mechanisms of aging, the replacement of diseased body tissues, and even the enhancement of abilities, the line between medicine and science fiction becomes thin. Developmental biologists have to consider not only the biology of stem cells, but also the ethics, economics, and justice behind their use (see Faden et al. 2003; Dresser 2010; Buchanan 2011).

Several years ago, stem cell therapy protocols were being tested in only a few human trials (Normile 2012; Cyranoski 2013). A simple search of stem cell therapies at clinicaltrials.gov will reveal a growing list of ongoing testing with stem cells (in 2018, ~7000 total). Although a majority of current clinical trials are associated with adult stem cells, progenitors derived from human ESCs (51/~7000) and iPSCs (81/~7000) are being conducted in the United States and elsewhere. Of significant concern is the increase in fraudulent stem cell therapies being offered. The International Society for Stem Cell Research (www.isscr.org) provides valuable resources to learn about stem cells and identify qualified stem cell therapies being used today.

Stem cell research may be the beginning of a revolution that will be as important for medicine (and as transformative for society) as the research on infectious microbes was a century ago. Beyond the potential for medical applications, however, stem cells can tell us a great deal about how the body is constructed and how it maintains its structure. Stem cells certainly give credence to the view that "development never stops."

Closing Thoughts on the Opening Photo

From M. A. Lancaster et al. 2013. *Nature* 501: 373–379

This chapter started out with the question "Is that really an eye and a brain in a dish?" Three-dimensional tissue construction from stem cells in a plate is a remarkable example of the "potential" that stem cells hold for the study of development and disease. Yes, that image is of a pigmented epithelium of the retina growing over the neural epithelium of a brain-like cerebral organoid. Although these organoids are certainly providing a new platform to study human organogenesis and affiliated diseases, its generated excitement must be accompanied with objectivity to understand the limitations these systems also present. What is this cerebral organoid currently lacking? Ponder these structures: blood vessels, the flow of cerebrospinal fluid, and the pituitary. Whether brain, kidney, or intestinal organoid, they are not yet complete. Perhaps in the future it will be your experiment that generates the first fully functional organ from stem cells in a dish.

Endnotes

[1] This description is a generalization; not all mammals are treated equally during early blastocyst development. For instance, marsupials do not form an inner cell mass; rather, they create a flattened layer of cells called the pluriblast that gives rise to an equivalent epiblast and hypoblast. See Kuijk et al. 2015 for further reading on the surprising divergence during early development across species.

[2] Most ESC lines begin as co-cultures of multiple cells from the ICM, after which isolated cells can be propagated as clonal lines.

[3] Oct4 is also known as Oct3, Oct3/4, and Pou5f1. Mice deficient in Oct4 fail to develop past the blastocyst stage. They lack a pluripotent ICM, and all cells differentiate into trophectoderm (Nichols et al. 1998; Le Bin et al. 2014). Oct4 expression is also necessary for the sustained pluripotency of derived primordial germ cells.

[4] Gain or loss of function of the TGF-β proteins results in tumor-like expansion of the GSC population or loss of the GSCs, respectively (Xie and Spradling 1998).

[5] Most NSCs exhibit astroglial characteristics, although there are exceptions. Self-renewing neuroepithelial-like cells persist in the zebrafish telencephalon and function as neural progenitors that lack typical astroglial gene expression. Consider the work of Michael Brand's lab for further study (Kaslin et al. 2009; Ganz et al. 2010; Ganz et al. 2012).

[6] In the mouse V-SVZ, one B cell can yield 16 to 32 A cells: each C cell that a B cell produces will divide three times, and their A cell progeny typically divide once, yielding 16 cells, but can also divide twice, yielding 32 cells (Ponti et al. 2013).

[7] Many of the roles that Notch signaling plays in neurogenesis in the adult brain are similar to its regulation of radial glia in the embryonic brain, but some important differences are beginning to emerge. For a direct comparison of Notch signaling in embryonic versus adult neurogenesis and across species, see Pierfelice et al. 2011 and Grandel and Brand 2013.

[8]The gradient of Shh in the brain is more accurately described as being oriented along the dorsal-to-ventral axis; for simplification, however, we have restricted our discussion to its presence along only the apical-to-basal axis.

[9]One recent study (Egerman et al. 2015) reported that GDF11 levels do not decline with age. In addition, despite research claiming the muscle rejuvenation capacity of GDF11 (Sinha et al. 2014), this study also states that GDF11 (like its protein cousin myostatin) inhibits muscle growth. The age-related drop in GDF11 levels has recently been confirmed (Poggioli et al. 2015), however, and GDF11's effect on neurogenesis was not being contested in the paper by Egerman and colleagues.

[10]This figure was determined through a meta-analysis of six species, including mouse and humans.

[11]This process is highly reminiscent of growth in the hydra, where each cell is formed at the animal's base, migrates to become part of the differentiated body, and is eventually shed from the tips of the tentacles (see Chapter 22).

[12]The first successful bone marrow transplantation was between two identical twins, one of whom had leukemia. It was conducted by Dr. E. Donnall Thomas, whose continued research in stem cell transplantation won him the Nobel Prize in Physiology or Medicine in 1990.

[13]*Peri* is Latin for "around." *Perivascular* refers to cells that are located on the periphery of blood vessels. The perivascular niche is also called the vascular niche, and the endosteal niche is also called the osteoblastic niche.

[14]Sinusoidal microvessels are small capillaries that are rich in open pores, enabling significant permeability between the capillary and the tissue it resides in.

[15]Initial attempts at such transplants did show incorporation of HSCs in a variety of tissues, even the brain. It turns out, however, that this finding was due to fusion events rather than actual lineage derivation from HSCs. See Alvarez-Dolado et al. 2003 and an affiliated web conference with Arturo Alvarez-Buylla in 2005 for further investigation.

[16]Another argument for saving umbilical cord cells is that they contain hematopoietic stem cells that might be transplanted into the child should he or she later develop leukemia (see Goessling et al. 2011).

[17]The story of Henrietta Lacks, HeLa cells in science, and social policy is beautifully articulated in Rebecca Skloot's 2010 book, *The Immortal Life of Henrietta Lacks*.

[18]As you examine past ESC literature, it will be important to critically consider the pluripotent state of the ESCs being depicted in each study. Are the ESCs naïve or primed, and what implications may that have on the authors' interpretations of their results? Also be aware that naïve ESCs have also been referred to as being in the "ground state."

[19]Researchers applied a micropattern of adhesive substrate to a glass plate, which restricted cell growth to a defined size and shape for systematic analysis (Warmflash et al. 2014). In a different study, lined grid substrates promoted ESC differentiation into dopamine neurons (Tan et al. 2015).

[20]At this time, the cost and scalability of iPSC-derived cell types to achieve the cell numbers required for effective cell replacement therapy are significant obstacles to the progress of this approach as a medical intervention.

[21]Although signs of some autism spectrum disorders are not overtly apparent early on, subtle early indicators—such as gazing at geometric shapes in preference to people's faces—are being identified.

[22]See Dr. Alysson Muotri describe his research and the Tooth Fairy Project in the video "Reversing Autism in the Lab with Help from Stem Cells and the Tooth Fairy," found on the Web by searching for this title. You can also access a BioWeb conference in which Dr. Muotri discusses iPSC modeling of ALS and autism.

[23]*Autologous* means derived from the same individual. In this case, cells from a patient are reprogrammed into iPSCs that are developed into a specific organoid. Cells and whole tissues from the organoid can be transplanted back into the same patient without concern of immune rejection.

[24]Cdk5 regulatory subunit-associated protein 2 (CDK5RAP2) encodes a centrosomal protein that interacts with the mitotic spindle during division.

(5) Snapshot Summary
Stem Cells

1. A stem cell maintains the ability to divide to produce a copy of itself as well as generating progenitor cells capable of maturing into different cell types.

2. Stem cell potential refers to the range of cell types a stem cell can produce. A totipotent stem cell can generate all cell types of both embryonic and extraembryonic lineages. Pluripotent and multipotent stem cells produce restricted lineages of just the embryo and of only select tissues or organs, respectively.

3. Adult stem cells reside in microenvironments called stem cell niches. Most organs and tissues possess stem cell niches, such as the germ cell, hematopoietic, gut epithelial, and ventricular-subventricular niches.

4. The niche employs a variety of mechanisms of cell-to-cell communication to regulate the quiescent, proliferative, and differentiative states of the resident stem cell.

5. The shoot apical and root apical meristems provide a continuous source of totipotent stem cells for a plant to generate a majority of its aerial and ground tissues throughout life.

6. Negative feedback mechanisms govern the ability to establish a balance between the stem cell pool and differentiation in the shoot apical meristem.

7. Inner cell mass (ICM) cells of the mouse blastocyst are maintained in a pluripotent state through E-cadherin interactions with trophectoderm cells.

8. Cadherin links the germ stem cells of the *Drosophila* oocyte to the niche, keeping them within fields of TGF-β. Asymmetric divisions push daughter cells out of this niche to promote cell differentiation of germ cells.

9. The ventricular-subventricular zone (V-SVZ) of the mammalian brain represents a complex niche architecture of B type stem cells arranged in a "pinwheel" organization,

(Continued)

Snapshot Summary (*continued*)

with a primary cilium at the apical surface and long radial processes that terminate with a basal endfoot.

10. Constant Notch activity in the V-SVZ niche keeps B cells in the quiescent state, whereas increasing oscillations of Notch activity versus proneural gene expression progressively promote maturation of B cells to transit-amplifying C cells and then into migrating neural progenitors (A cells).

11. Additional signals—from neural activity and substances like GDF11 from blood vessels to gradients of Shh, BMP4, and Noggin—all influence cell proliferation and differentiation of B cells in the V-SVZ niche.

12. The columnar cells located at the base of the intestinal crypt serve as clonogenic stem cells for the gut epithelium, which generates transit-amplifying epithelial cells that slowly differentiate as they are pushed farther up the villus.

13. Adhesion to osteoblasts keeps the hematopoietic stem cell (HSC) quiescent in the endosteal niche. Increased exposure to CXCL12 signals from CAR cells and mesenchymal stem cells can transition HSCs into proliferative behavior, yet downregulation of *CXCL12* in the perivascular niche encourages migration of short-term active HSCs into the oxygen-rich blood vessels.

14. Mesenchymal stem cells can be found in a variety of tissues, including connective tissue, muscle, cornea, dental pulp, bone, and more. They play dual roles as supportive stromal cells and multipotent stem cells.

15. Embryonic and induced pluripotent stem cells can be maintained in culture indefinitely and, when exposed to certain combinations of factors and/or constrained by the physical growth substrate, can be coaxed to differentiate into potentially any cell type of the body.

16. Embryonic stem cells (ESCs) and induced pluripotent stem cells (iPSCs) are being used to study human cell development and diseases. The use of stem cells to study patient-specific cell differentiation of the rare blood disorder Fanconi anemia or disorders of the nervous system like autism and ALS have already started to provide novel insight into disease mechanisms.

17. Pluripotent stem cells can also be used in regenerative medicine to rebuild tissues and to make structures called organoids, which seem to possess many of the multicellular hallmarks of human organs. Organoids are being used to study human organogenesis and patient-specific disease progression on the tissue level, all in vitro.

Go to oup.com/he/barresi12xe for Further Developments, Scientists Speak interviews, Watch Development videos, Dev Tutorials, and complete bibliographic information for all literature cited in this chapter.

Sex Determination and Gametogenesis

How can this northern cardinal become half male (red) and half female (light brown)?

Photo courtesy of Brian D. Peer

Sex Determination

"SEXUAL REPRODUCTION IS…THE MASTERPIECE OF NATURE," wrote Erasmus Darwin in 1791, and modern science confirms this. Different species produce male and female offspring in different ways. In mammals and flies, an individual's sex is determined by the chromosome set established at fertilization, when the **gametes**—the sperm and the egg—fuse together. As we will see, however, there are other schemes of sex determination. In certain animal species and in many plant species, an organism is both male and female (making both sperm and eggs). Moreover, there are some animals in which sex is determined not by chromosomes, but by the environment.

The gametes are the product of a **germ cell lineage** (**germ line**) that becomes separate from the somatic cell lineages that divide mitotically to generate the differentiated somatic cells of the embryo. Cells in the germ line undergo **meiosis**, a remarkable process of cell division by which the chromosomal content of a cell is halved so that the union of two gametes in fertilization restores the full chromosomal complement of the new organism. Sexual reproduction means that each new organism receives genetic material from two distinct parents, and the mechanisms of meiosis provide an incredible amount of genomic variation on which evolution can work. (See **Further Development 6.1, Sex Determination and Social Perceptions**, online.)

Chromosomal Sex Determination

There are several ways chromosomes can determine the sex of an embryo in animals. In most *mammals*, the presence of either a second X chromosome or a Y chromosome determines whether the embryo will be female (XX) or male (XY). In *birds*, the situation is reversed (Smith and Sinclair 2001): the male has the two similar sex chromosomes (ZZ) and the female has the unmatched pair (ZW). In *flies*, the Y chromosome plays no role in sex determination, but the number of X chromosomes appears to determine the sexual phenotype. In other insects (especially hymenopterans such as bees, wasps, and ants), fertilized, diploid eggs develop into females, while unfertilized, haploid eggs become males (Beukeboom 1995; Gempe et al. 2009; Ronai 2016). This chapter will discuss only two of the many chromosomal modes of sex determination in animals: sex determination in placental mammals and sex determination in fruit flies (*Drosophila*).

The Mammalian Pattern of Sex Determination

In humans and mice, the Y chromosome is critical in determining sex. XX mammals are usually females, having ovaries and producing eggs. XY mammals are usually males, having testes and making sperm. Meiosis in females produces egg cells, which have an X chromosome. Meiosis in males produces sperm, half of which have an X chromosome and half of which have a Y chromosome. If an X-bearing sperm unites with an X-bearing egg, the offspring will be XX and therefore genetically female. If a Y-bearing sperm unites with an X-bearing egg, the offspring will be XY and therefore genetically male (**FIGURE 6.1A**; Stevens 1905; Wilson 1905; see Gilbert 1978). This is the basis for the 50:50 sex ratio.

But this doesn't fully answer the question of how sex is determined in mammals. The developmental biologist wants to know how having a Y chromosome promotes testis development and sperm production, and how having two X chromosomes promotes ovary development and egg production. The importance of the Y chromosome for male sex determination in mammals was shown by analysis of people whose chromosomes differ from XX or XY. When chromosomes do not segregate properly during meiosis, fertilization can produce individuals who have an extra X chromosome. These XXY people are male (despite having two X chromosomes). Moreover, individuals having only one X chromosome (XO) are female (Ford et al. 1959; Jacobs and Strong 1959). Women with a single X chromosome begin making ovaries, but the ovarian follicles cannot be maintained without the second X chromosome. Thus, a second X chromosome is required to complete development of ovaries, whereas the presence of a Y chromosome (even when multiple X chromosomes are present) initiates the development of testes.

But what do the X and Y chromosomes do? It turns out that the gonad in early embryonic mice and humans is bipotential. This **bipotential gonad** can develop into either a testis or an ovary (**FIGURE 6.1B**). XX gonadal cells activate the Wnt pathway. This produces the transcriptional regulator β-catenin, which inhibits the development of gonadal cells into testis cell types and activates the genes that promote the development of gonadal cells into the follicle cells of the ovary.

FIGURE 6.1 Sex determination in placental mammals. (A) Mammalian chromosomal sex determination results in approximately equal numbers of male and female offspring. (B) The embryonic mammalian gonad is originally bipotential. It is neither male nor female, but has the potential to become one or the other. If the cells have an X and a Y chromosome, the bipotential gonad becomes a testis that makes sperm and the hormones that promote a male phenotype. If the cells have two X chromosomes and no Y chromosome, the bipotential gonad becomes an ovary that makes eggs and the hormones that promote a female phenotype.

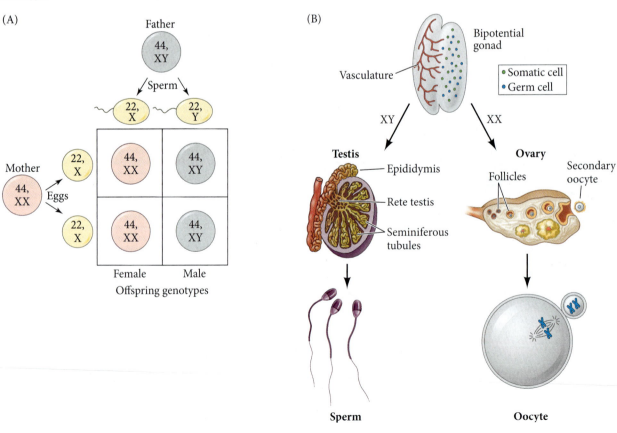

XY gonadal cells activate the gene encoding the Sry transcription factor. *Sry* is the testis-determining gene on the small arm of the Y chromosome. It is probably active for only a short duration, and it may have a single function—to activate the *Sox9* gene in these cells. The protein product of the *Sox9* gene is a transcription factor that starts the reactions that organize the bipotential gonads to become testes. The testes will form the Sertoli cells that support the sperm, and the Leydig cells that produce testosterone. Moreover, Sox9 also represses the Wnt pathway, so that gonadal cells do not form ovaries. This determination of the gonad as female or male is called **primary sex determination** (or **gonadal sex determination**), and it is accomplished by the X and Y chromosomes that control the fate of the bipotential cells of the early gonad.

Once gonadal sex determination has established the gonads, the gonads begin to produce the hormones and paracrine factors that govern **secondary sex determination**—the development of the sexual phenotype outside the gonads. This includes the male or female duct systems and the external genitalia (**FIGURE 6.2**), discussed in detail later on in the chapter.

Gonadal sex determination in mammals

The bipotential gonad of mammals represents a unique embryological situation. All other organ rudiments normally differentiate into only one type of organ—a lung rudiment can only become a lung, a liver rudiment only a liver. The gonadal rudiment, however, can develop into either an ovary or a testis—two organs with very different tissue architectures (Lillie 1917; Rey et al. 2016).

THE DEVELOPING GONADS In humans, two gonadal rudiments appear during week 4 and remain sexually indifferent until week 7. These gonadal precursors are paired regions of the mesoderm adjacent to the developing kidneys (**FIGURE 6.3A,B**; Tanaka and Nishinakamura 2014). The **germ cells**—the precursors of either sperm or eggs— migrate into the gonads during week 6 and are surrounded by the mesodermal cells.

If the fetus is XY, the mesodermal cells continue to proliferate through week 8, when a subset of these cells initiate their differentiation into **Sertoli cells**. During embryonic development, the developing Sertoli cells secrete the **anti-Müllerian hormone (AMH)** that blocks development of the female ducts. These same Sertoli epithelial cells will also form the seminiferous tubules that will support the development of sperm throughout the lifetime of the male mammal.

FIGURE 6.2 Postulated cascades leading to male and female phenotypes in placental mammals. The conversion of the genital ridge into the bipotential gonad requires the *Sf1*, *Wt1*, *Lhx9*, and *Gata4* genes; mice lacking any of these genes lack gonads. The bipotential gonad is moved into the female pathway (ovary development) by the *Wnt4* and *Rspo1* genes, which promote accumulation of β-catenin. Alternatively, the bipotential gonad can be moved into the male pathway (testis development) by the *Sry* gene (on the Y chromosome), which triggers the activity of *Sox9*. Under the influence of estrogen (first from the mother, then from the fetal ovaries), the Müllerian duct differentiates into the female reproductive tract, the internal and external genitalia develop, and the offspring develops the secondary sex characteristics of a female. The testis makes anti-Müllerian hormone (AMH), which causes the Müllerian duct to regress, and testosterone, which causes differentiation of the Wolffian duct into the male internal genitalia. In the urogenital region, testosterone is converted into dihydrotestosterone (DHT), which causes morphogenesis of the penis, prostate gland, and scrotum. (After J. Marx. 1995. *Science* 269: 1824–1825 and O. S. Birk et al. 2000. *Nature* 403: 909–913.)

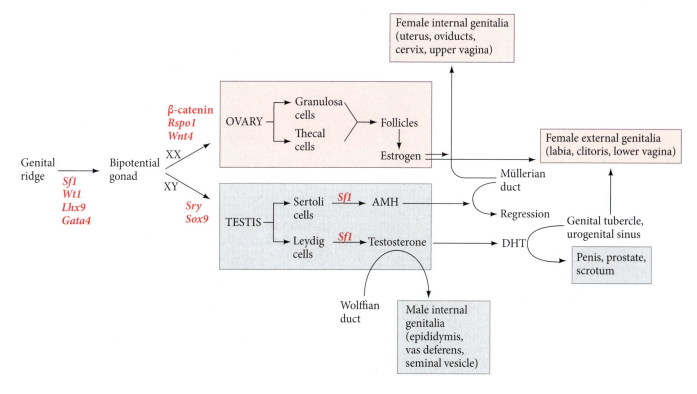

FIGURE 6.3 Differentiation of human gonads shown in transverse section. (A) Genital ridge of a 4-week embryo. (B) Genital ridge of a 6-week bipotential gonad showing expanded epithelium. (C) Testis development in week 8. The epithelial sex cords lose contact with the cortical epithelium and develop the rete testis. (D) By week 16, the testis cords are continuous with the rete testis and connect with the Wolffian duct through the efferent ducts remodeled from the mesonephric duct. (E) Ovary development in an 8-week embryo. (F) In the 20-week embryo, the ovary does not connect to the Wolffian duct, and new cortical follicle cells surround the germ cells that have migrated into the genital ridge. (After R. K. Burns. 1955. *Proc Natl Acad Sci USA* 41: 669–676.)

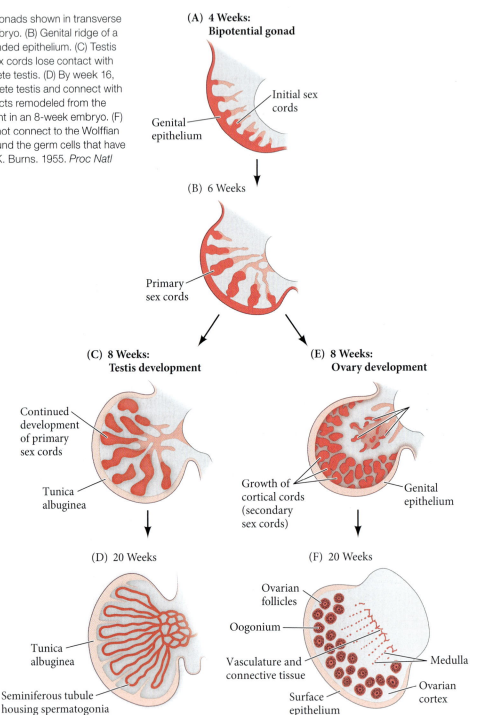

(A) **4 Weeks: Bipotential gonad**
Initial sex cords
Genital epithelium

(B) **6 Weeks**
Primary sex cords

(C) **8 Weeks: Testis development**
Continued development of primary sex cords
Tunica albuginea

(E) **8 Weeks: Ovary development**
Growth of cortical cords (secondary sex cords)
Genital epithelium

(D) **20 Weeks**
Tunica albuginea
Seminiferous tubule housing spermatogonia

(F) **20 Weeks**
Ovarian follicles
Oogonium
Vasculature and connective tissue
Surface epithelium
Medulla
Ovarian cortex

During week 8, the developing Sertoli cells surround the incoming germ cells and organize themselves into the **testis cords**. These cords form loops in the central region of the developing testis and are connected to a network of thin canals, called the **rete testis**, located near the developing kidney duct (**FIGURE 6.3C,D**). Thus, when germ cells enter the male gonads, they develop within the testis cords, *inside* the organ. They undergo several rounds of proliferation, then arrest in mitosis. Later in development (at puberty in humans; shortly after birth in mice, which procreate much faster), the testis cords mature to form the **seminiferous tubules**. The germ cells migrate to the periphery of these tubules, where they establish the **spermatogonial stem cell population** that produces sperm throughout the lifetime of the male (see Figure 6.20).

Meanwhile, the other major group of mesodermal cells (those that did not form the Sertoli epithelium) differentiate into a mesenchymal cell type, the testosterone-secreting **Leydig cells**. Thus, the fully developed testis will have epithelial tubes made of Sertoli cells that enclose the germ cells, as well as a mesenchymal cell population, the Leydig cells, that secretes **testosterone**. Each incipient testis is surrounded by a thick extracellular matrix, the tunica albuginea, which helps protect it.

If the fetus is XX, germ cells that enter the gonad are organized in clusters (cysts) surrounded by **pre-granulosa cells**. During this period of fetal life, female germ cells enter meiosis. At about the time of birth, the pre-granulosa cells in the center of the developing gonad degenerate, leaving only those at the surface (cortex) of the gonad. Each germ cell is enveloped by a separate, small cluster of pre-granulosa cells (**FIGURE 6.3E,F**). The germ cells will become developing eggs, the **oocytes**. The cells surrounding the developing eggs differentiate into **granulosa cells**. Most of the remaining mesenchymal cells differentiate into **thecal cells**. Together, the thecal and granulosa cells form **follicles** that envelop the oocytes and secrete steroid hormones such as estrogens and (during pregnancy) progesterone.

There is a reciprocal relationship between the germ cells and the somatic cells of the gonads. The germ cells are originally bipotential and can become either sperm or eggs. Once in the male or female sex cords, however, they are instructed to either (1) begin meiosis and become eggs or (2) arrest in mitosis and become spermatogonia, the sperm stem cells (McLaren 1995; Brennan and Capel 2004). In XX gonads, germ cells are essential for the maintenance of ovarian follicles. Without germ cells, the follicles degenerate. In XY gonads, the germ cells help support the differentiation of Sertoli cells but are not required for the maintenance of testis structure (McLaren 1991).

GENETIC MECHANISMS OF GONADAL SEX DETERMINATION: MAKING DECISIONS

Several human genes have been identified whose function is necessary for normal sexual differentiation. Because the phenotype of mutations in sex-determining genes is often sterility, clinical infertility studies have been useful in identifying those genes that are active in determining whether humans become male or female. Experimental manipulations to confirm the functions of these genes can then be done in mice.

The story starts in the bipotential gonad that has not yet been committed to the male or female direction. The genes for transcription factors Wt1, Lhx9, Gata4, and Sf1 are expressed, and the loss of function of any one of them will prevent the normal development of either male or female gonads. Then the decision is made. **FIGURE 6.4** shows one possible model of how gonadal sex determination can be initiated. It is a good illustration of an important rule of animal development: a pathway for cell specification often has two components, with one branch that says "Make A" and another branch that says "… and *don't* make B." In the case of the gonads, the male

FIGURE 6.4 Possible mechanism for the initiation of gonadal sex determination in mammals. While we do not know the specific interactions involved, this model attempts to organize the data into a coherent sequence. If Sry is *not present* (pink region), the interactions between transcription factors in the developing genital ridge activate *Wnt4* and *Rspo1*. *Wnt4* activates the canonical Wnt pathway, which is made more efficient by Rspo1. The Wnt pathway causes the accumulation of β-catenin, and a large accumulation of β-catenin stimulates further Wnt4 activity. This continual production of β-catenin both induces the transcription of ovary-producing genes and blocks the testis-determining pathway by interfering with *Sox9* activity. If Sry is *present* (blue region), it blocks β-catenin signaling (thus halting ovary development) and, along with Sf1, activates the *Sox9* gene. Sox9 activates Fgf9 synthesis, which stimulates testis development, blocks Wnt4, and promotes further Sox9 synthesis. Sox9 also prevents β-catenin's activation of ovary-producing genes. Sry may also activate other genes that help generate Sertoli cells. In summary, a Wnt4/β-catenin loop specifies the ovaries, whereas a Sox9/Fgf9 loop specifies the testes. One of the targets of the Wnt pathway is the *Follistatin* gene, whose product organizes the granulosa cells of the ovary. Transcription factor Foxl2, activated in the ovary, is also involved in inducing Follistatin synthesis. The XY pathway appears to have an earlier initiation; if it does not function, the XX pathway takes over. (After R. Sekido and R. Lovell-Badge. 2009. *Trends Genet* 25: 19–29 and K. McClelland et al. 2012. *Asian J Androl* 14: 164–171.)

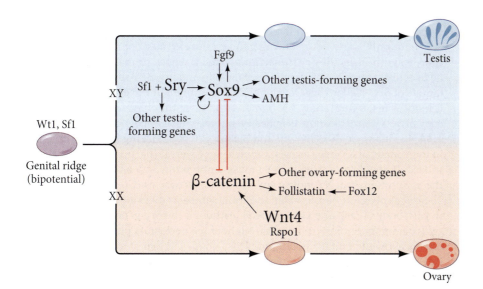

pathway says "Make testes and don't make ovaries," while the female pathway says "Make ovaries and don't make testes."

THE OVARY PATHWAY: THE IMPORTANCE OF β-CATENIN   If no Y chromosome is present, the transcription factors Wt1, Lhx9, Gata4, and Sf1 are thought to activate further expression of Wnt4 protein (already expressed at low levels in the genital epithelium) and of a small soluble protein called **R-spondin1** (**Rspo1**). Rspo1 acts in synergy with Wnt4 to produce β-catenin, which is critical both in activating further ovarian development and in blocking synthesis of the testis-promoting transcription factor Sox9 (Maatouk et al. 2008; Jameson et al. 2012). XX humans born with *RSPO1* gene mutations became phenotypic males (Parma et al. 2006; Harris et al. 2018). In XY individuals with a duplication of the region on chromosome 1 that contains both the *WNT4* and *RSPO1* genes, the pathways that make β-catenin override the male pathway, resulting in a male-to-female sex reversal. Similarly, if XY mice are induced to overexpress β-catenin in their gonadal rudiments, they form ovaries rather than testes. Indeed, β-catenin appears to be a key "pro-ovarian/anti-testis" signaling molecule in all vertebrate groups, as it is seen in the female (but not the male) gonads of birds, mammals, and turtles, three groups having very different modes of sex determination (Maatouk et al. 2008; Cool and Capel 2009; Smith et al. 2009). (See **Further Development 6.2, The Ovary Pathway: The Importance of β-catenin**, online.)

THE TESTIS PATHWAY: SRY AND SOX9   If a Y chromosome is present, the same set of transcription factors (Wt1, Lhx9, Gata4, and Sf1) in the bipotential gonad activates the *Sry* (**Sex-determining Region of the Y Chromosome**) gene on the Y chromosome (Carré et al. 2018; Kuroki and Tachibana 2018). There is extensive evidence that *SRY* is the gene that encodes the testis-determining factor. In humans, *SRY* is typically found in XY males, but it is also seen in rare XX males; it is absent from XX females and also from many XY females. Approximately 15% of human XY females have the *SRY* gene, but their copies of the gene contain point or frameshift mutations that prevent the SRY protein from binding to DNA (Pontiggia et al. 1994; Werner et al. 1995). The most impressive evidence for *Sry* being the gene for testis-determining factor comes from transgenic mice. If *Sry* induces testis formation, then inserting *Sry* DNA into the genome of a normal XX mouse zygote should cause that XX mouse to form testes. Koopman and colleagues (1991) took the 14-kilobase region of DNA that includes the *Sry* gene (and presumably its regulatory elements) and microinjected this sequence into the pronuclei of newly fertilized mouse zygotes. In several instances, XX embryos injected with this sequence developed testes, male accessory organs, and a penis (**FIGURE 6.5**).[1] Therefore, we conclude that *Sry/SRY* is the only gene on the Y chromosome required for testis determination in mammals.

For all its importance in male sex determination, the *Sry* gene is probably active for only a few hours during gonadal development in mice. During this time, it synthesizes the Sry transcription factor, whose primary role appears to be to activate an autosomal gene *Sox9* (Sekido and Lovell-Badge 2008; for other targets of Sry, see Further Development 6.5, online). In the gonadal rudiments, it induces testis formation. XX humans and mice that have an extra activated copy of *SOX9/Sox9* develop as males even if they have no *SRY/Sry* gene (**FIGURE 6.6A–C**; Huang et al. 1999; Qin and Bishop 2005). Knocking out the *Sox9* gene in the gonads of XY mice causes complete sex reversal (Barrionuevo et al. 2006). Indeed, if one deletes from an XY mouse embryo the *Sox9* enhancer that binds the Sry protein, that XY mouse embryo develops ovaries (Gonen et al. 2018).

Sox9 appears to be the older and more conserved sex determination gene in vertebrates (Pask and Graves 1999). Although the *Sry* gene is found specifically in mammals, *Sox9* is found throughout the vertebrate phyla. In mammals, *Sox9* is activated by Sry protein; in birds, frogs, and fish, it appears to be activated by the dosage of the transcription factor Dmrt1; and in those vertebrates with temperature-dependent sex determination, it is often activated (directly or

(A)

XY ♂   XX ♀   XX ♂

← *Sry*

Control ← (autosomal) gene

1   2   3

(B)

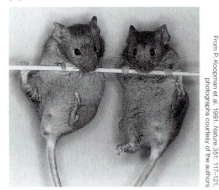

From P. Koopman et al. 1991; *Nature* 351: 117–121; photographs courtesy of the authors

FIGURE 6.5   An XX mouse transgenic for *Sry* is male. (A) Polymerase chain reaction followed by electrophoresis shows the presence of the *Sry* gene in a normal XY male and in a transgenic XX/*Sry* mouse. The gene is absent in a female XX littermate. (B) The external genitalia of the transgenic XX/*Sry* mouse are male (right) and are essentially the same as those in an XY male (left).

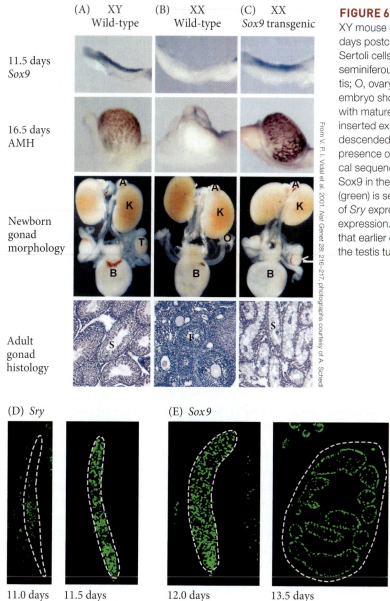

(A) XY Wild-type   (B) XX Wild-type   (C) XX *Sox9* transgenic

11.5 days *Sox9*

16.5 days AMH

Newborn gonad morphology

Adult gonad histology

From V. P. I. Vidal et al. 2001. *Nat Genet* 28: 216–217, photographs courtesy of A. Schedl

(D) *Sry*   (E) *Sox9*

11.0 days   11.5 days   12.0 days   13.5 days

From K. Kashimada and P. Koopman. 2010. *Development* 137: 3921–3930, courtesy of P. Koopman and D. Wilhelm

FIGURE 6.6 Ability of Sox9 protein to generate testes. (A) A wild-type XY mouse embryo expresses the *Sox9* gene in the genital ridge at 11.5 days postconception, anti-Müllerian hormone in the embryonic gonad Sertoli cells at 16.5 days, and eventually forms descended testes with seminiferous tubules. K, kidneys; A, adrenal glands; B, bladder; T, testis; O, ovary; S, seminiferous tubule; F, follicle cell. (B) The wild-type XX embryo shows neither *Sox9* expression nor AMH. It constructs ovaries with mature follicle cells. (C) An XX embryo with the *Sox9* transgene inserted expresses *Sox9* and has AMH in 16.5-day Sertoli cells. It has descended testes, but the seminiferous tubules lack sperm (due to the presence of two X chromosomes in the Sertoli cells). (D,E) Chronological sequence from the expression of *Sry* in the genital ridge to that of Sox9 in the Sertoli cells. (D) *Sry* expression. At day 11.0, Sry protein (green) is seen in the center of the genital ridge. At day 11.5, the domain of *Sry* expression increases and *Sox9* expression is activated. (E) *Sox9* expression. By day 12.0, Sox9 protein (green) is seen in the same cells that earlier expressed *Sry*. By day 13.5, Sox9 is seen in those cells of the testis tubule that will become Sertoli cells.

indirectly) by the male-producing temperature. In mammals, expression of the *Sox9* gene is specifically upregulated by the combined expression of Sry and Sf1 proteins in Sertoli cell precursors (**FIGURE 6.6D,E**; Sekido et al. 2004; Sekido and Lovell-Badge 2008). Thus, Sry may act merely as a "switch" operating during a very short time to activate *Sox9*, and the Sox9 protein may initiate the conserved evolutionary pathway to testis formation. So, borrowing Eric Idle's phrase, Sekido and Lovell-Badge (2009) propose that Sry initiates testis formation by "a wink and a nudge."

Once made, the Sox9 protein has several functions. First, it appears to be able to activate its own promoter, thereby allowing it to be transcribed for long periods of time (independent of Sry). Second, it blocks the ability of β-catenin to induce ovary formation, either directly or indirectly (Wilhelm et al. 2009). Third, it binds to *cis*-regulatory regions of numerous genes necessary for testis production (Bradford et al. 2009a; Rahmoun et al. 2017). These genes include those encoding anti-Müllerian hormone (which causes degeneration of the uterus-forming duct; Arango et al. 1999; de Santa Barbara et al. 2000), Dmrt1 (needed for testis maintenance), and Fgf9, a paracrine factor critical for testis development. Fgf9 is also essential for maintaining *Sox9* gene transcription, thereby establishing a positive feedback loop driving the male pathway (Kim et al. 2007). (See **Further Development 6.3, Finding the Elusive Testis-Determining Factor**; **Further Development 6.4, Fibroblast Growth Factor 9**; and **Further Development 6.5, Genes Controlled by *Sry* and *Sox9***, all online.)

Hermaphrodites are individuals in which both ovarian and testicular tissues exist; they have either ovotestes (gonads containing both ovarian and testicular tissue) or an ovary on one side and a testis on the other.[2] Experiments on the *Sry* gene in mice showed that ovotestes can be generated when the *Sry* gene is activated just a few hours later than normal, experiments that also showed that delaying activation of *Sry* by as little as 5 hours led to failure of testis development and the initiation of ovary development. Hermaphrodites can also result in those very rare instances when a Y chromosome is translocated to an X chromosome. As we will discuss later in this chapter, one of the two X chromosomes in each XX cell is inactivated. (This ensures that the X-derived products of the female aren't twice as abundant as those of the male.) In cells where the translocated Y is on the active X chromosome, the Y chromosome will be active and the *Sry* gene will be transcribed; in cells where the Y chromosome is on the inactive X chromosome, the Y chromosome will also be inactive (Berkovitz et al. 1992; Margarit et al. 2000). Such gonadal mosaicism for cells expressing *Sry* can lead to the formation of a testis, an ovary, or an ovotestis, depending on the percentage of cells expressing *Sry* in the Sertoli cell precursors (see Brennan and Capel 2004; Kashimada and Koopman 2010).

Secondary sex determination in mammals: Hormonal regulation of the sexual phenotype

Gonadal sex determination—the formation of either an ovary or a testis from the bipotential gonad—does not result in the complete sexual phenotype. In mammals, secondary sex determination is the development of the female and male phenotypes in response to hormones secreted by the ovaries and testes. Both female and male secondary sex determination have two major temporal phases. The first phase occurs within the embryo during organogenesis; the second occurs at puberty.

During embryonic development, hormones and paracrine signals coordinate the development of the gonads with the development of secondary sex organs. The reproductive duct system starts out as undifferentiated Müllerian ducts (female) and Wolffian ducts (male), both present in the embryo (**FIGURE 6.7**). In females, the Müllerian ducts persist and, through the actions of estrogen, differentiate to become the uterus, cervix, oviducts, and upper vagina (Cunha and Baskin 2018; Isaacson et al. 2018). These organs are often "bifunctional," playing important roles both in transporting sperm toward the ovary and in transporting and retaining the embryo:

- The upper vagina (not the part connected to the skin) becomes the outside entrance to a woman's reproductive system. It functions both for the entry of sperm and as the birth canal for a baby.
- The cervix, an inner muscular entrance to the uterus, secretes mucus that regulates sperm entry into the uterus. During pregnancy, it functions as a muscular band that holds the fetus in the uterus until delivery.
- The uterus can actively promote sperm movement toward the oviducts. During pregnancy, it becomes the nurturing womb where the developing embryo lodges and grows.
- The pair of oviducts (tubes) mature the sperm and guide them toward the egg. After fertilization, the oviducts guide the embryo to the uterus.

In females, the **genital tubercle** (the precursor of the external genitalia) becomes differentiated into the clitoris, and the **labioscrotal folds** become the labia majora. The Wolffian ducts require testosterone to persist, and thus they atrophy in females. In females, the portion of the **urogenital sinus** that does not become the bladder and urethra becomes Skene's glands, paired organs that make secretions similar to those of the prostate (see table in Figure 6.7).

The coordination of the male phenotype involves the secretion of two testicular factors. The first of these is anti-Müllerian hormone, a BMP-like paracrine factor made by the Sertoli cells, which causes the degeneration of the Müllerian ducts. The second is the steroid hormone testosterone, an **androgen** (masculinizing substance) secreted from the fetal Leydig cells. Testosterone is thought to inhibit the adjacent mesenchymal cells from sending a signal that instructs the Wolffian duct epithelium to undergo cell death (Zhao et al. 2017). Moreover, it causes the Wolffian ducts to differentiate into sperm-carrying tubes (the epididymis and vas deferens). Fetal testosterone also causes the genital tubercle to develop into the penis, and the labioscrotal folds to develop into the scrotum. In males, the urogenital sinus, in addition to forming the bladder and urethra, also forms the prostate gland (see table in Figure 6.7). (See **Further Development 6.6, The Origins of Genitalia**, online.)

THE GENETIC ANALYSIS OF MALE SECONDARY SEX DETERMINATION   The testes initiates two independent pathways of masculinization in mammals. Testosterone from the Leydig cells of the testes promotes the masculine body type. Anti-Müllerian hormone from the Sertoli cells causes degeneration of the duct that would otherwise give rise to the vagina, cervix, uterus, and oviducts. The independence of these two pathways is demonstrated by people with **androgen insensitivity syndrome**. These women have an XY karyotype and therefore have an *SRY* gene. Thus, they form testes that make testosterone and AMH. However, they have a mutation in the gene encoding the androgen *receptor* protein that binds testosterone to make it an active transcription factor (**FIGURE 6.8A**). Therefore, these individuals cannot respond to the testosterone made by their testes (Meyer et al. 1975; Jääskel-uäinen 2012). They can, however, respond to the estrogen made by their adrenal glands (which is normal for both XX and XY individuals), so they develop female external sex characteristics (**FIGURE 6.8B**). Despite their distinctly female appearance, these XY individuals have testes, and even though they cannot respond to testosterone, they produce and respond to AMH. Thus, their Müllerian ducts degenerate. People with androgen insensitivity syndrome develop as normal-appearing but sterile women, lacking a uterus and oviducts and having internal testes in the abdomen.

Although in most people there is an accurate correlation between the primary and secondary sexual phenotypes, about 0.5–1.7% of the population departs from the strictly dimorphic condition (about the same percentage of the population having congenitally red hair; Fausto-Sterling 2000; Hull 2003; Hughes et al.

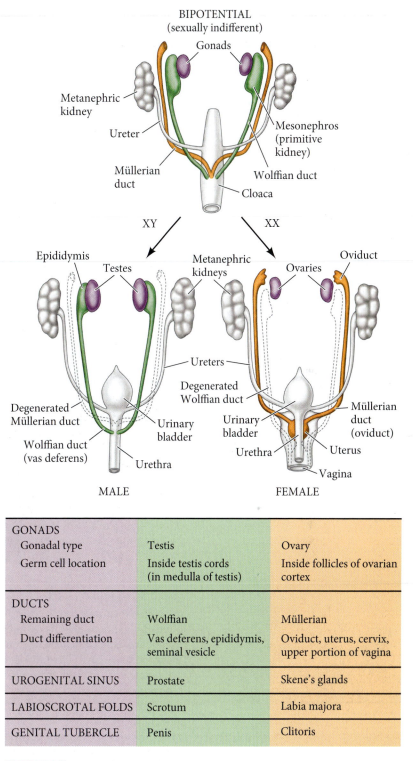

GONADS		
Gonadal type	Testis	Ovary
Germ cell location	Inside testis cords (in medulla of testis)	Inside follicles of ovarian cortex
DUCTS		
Remaining duct	Wolffian	Müllerian
Duct differentiation	Vas deferens, epididymis, seminal vesicle	Oviduct, uterus, cervix, upper portion of vagina
UROGENITAL SINUS	Prostate	Skene's glands
LABIOSCROTAL FOLDS	Scrotum	Labia majora
GENITAL TUBERCLE	Penis	Clitoris

FIGURE 6.7 Development of gonads and their ducts in mammals. Originally, a bipotential (indifferent) gonad develops, with undifferentiated Müllerian ducts (female) and Wolffian ducts (male) both present. If XY, the gonads become testes and the Wolffian duct persists. If XX, the gonads become ovaries and the Müllerian duct persists. Hormones from the gonads cause the external genitalia to develop in either the male direction (penis, scrotum) or the female direction (clitoris, labia majora) (listed in table).

FIGURE 6.8 Androgen insensitivity syndrome. (A) Mechanism of androgen receptor. Testosterone is an androgen (masculinizing) steroid hormone that can travel to cells through the blood. Inside the cytoplasm of a cell, it binds to its protein receptor (the androgen receptor, sometimes called the testosterone receptor), displacing other proteins (such as the heat shock proteins). This allows the androgen receptor to dimerize (combine with another receptor) and enter the nucleus. The bound testosterone permits the receptor protein to function as a transcription factor, binding to particular genes to produce the male phenotype. (B) A group of women with androgen insensitivity syndrome and other disorders of sexual development. Despite having an XY karyotype, individuals with androgen insensitivity syndrome develop a female phenotype. (After P. Li et al. 2017. *Cancers* 9: 20/CC BY 4.0.)

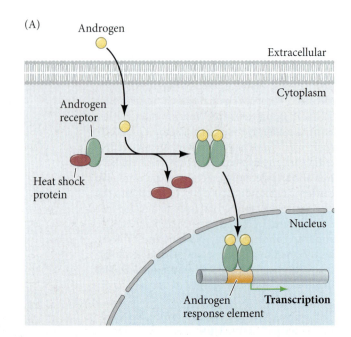

(A)

(B)

Courtesy of Kimberly Saviano/AISSG–USA

2006). Phenotypes in which male and female traits are seen in the same individual are called **intersex** conditions.[3] Androgen insensitivity syndrome is one of several intersex conditions that have traditionally been labeled **pseudohermaphroditism**. In pseudohermaphrodites, there is only one type of gonad (as contrasted with "true" hermaphroditism, in which individuals have the gonads of both sexes), but the secondary sex characteristics differ from what would be expected from the gonadal sex. Another type of pseudohermaphroditism, in which the gonadal sex is female but the person is outwardly male, can result from overproduction of androgens in the ovary or adrenal gland. The most common cause of the latter condition is **congenital adrenal hyperplasia**, in which there is a genetic deficiency of an enzyme that metabolizes cortisol steroids in the adrenal gland. In the absence of this enzyme, testosterone-like steroids accumulate and can bind to the androgen receptor, thus masculinizing the fetus (Migeon and Wisniewski 2000; Merke et al. 2002). (See **Further Development 6.7, Steroids and Secondary Sex Determination**, and **Further Development 6.8, Descent of the Testes**, both online.)

SEX CHROMOSOME DOSAGE   Another component of the mammalian female phenotype is X-chromosome inactivation. Mammals have to regulate the dosage of X-chromosome gene products. Females have two X chromosomes, while males have only one. If transcription were equal, females should have twice the amount of

(A)

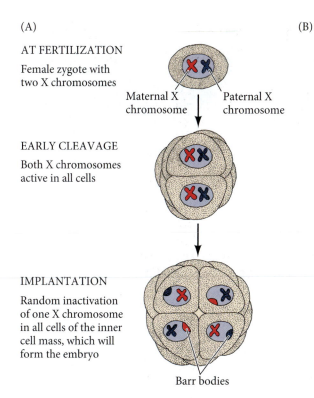

AT FERTILIZATION

Female zygote with
two X chromosomes

Maternal X
chromosome

Paternal X
chromosome

EARLY CLEAVAGE

Both X chromosomes
active in all cells

IMPLANTATION

Random inactivation
of one X chromosome
in all cells of the inner
cell mass, which will
form the embryo

Barr bodies

(B)

© iStock.com/cgbaldauf

FIGURE 6.9 X-chromosome inactivation in mammals.
(A) Schematic diagram illustrating random X-chromosome
inactivation. The DNA of the inactive X chromosome
becomes transcriptionally inert heterochromatin and often
becomes associated with the nuclear envelope. The Barr
body is the inactive X-chromosome, which stains darkly.
(B) A calico cat, with orange and black alleles of a pigment
gene on the X chromosome. Whether a splotch is orange or
black depends on which X chromosome was inactivated in
the founder cell that gave rise to the pigment in that area.

mRNA from X-chromosome genes as do males. But they don't. In general, males and
females have similar amounts of products from their X chromosomes. This is achieved
by **X-chromosome inactivation**. Randomly, one of the X chromosomes is inactivated
in each cell, and the descendants of these embryonic cells retain the same inactivated
X chromosome. This random inactivation is accomplished by making most of the DNA
of one X chromosome into transcriptionally inactive heterochromatin. Although the
details of X-chromosome inactivation differ between mice and humans, the mechanisms involve long noncoding RNAs and the placement of inhibitory histone proteins
into the nucleosomes of the inactive X chromosome (**FIGURE 6.9**; Migeon 2013, 2017).
(See **Further Development 6.9, Dosage Compensation**, online.)

In summary, gonadal sex determination in mammals is regulated by the chromosomes, which results in the production of testes in XY individuals and ovaries in XX
individuals. This type of sex determination appears to be a "digital" (either/or) phenomenon. With chromosomal sex established, the gonads then produce the hormones
that coordinate the different parts of the body to have a male or female phenotype. This
secondary sex determination is more "analogue," whereby differing levels of hormones
and responses to hormones can create different phenotypes. Secondary sex determination is thus usually, but not always, coordinated with gonadal sex determination. (See
Further Development 6.10, Brain Sex and Gender, online.)

Chromosomal Sex Determination in Drosophila

Sex determination by dosage of X

In *Drosophila*, the sex organs are specified by the number of X chromosomes in the cell
nucleus. If there is only one X chromosome in a diploid cell, the fly is male. If there are
two X chromosomes in a diploid cell, the fly is female. While XO mammals are sterile
females (no Y chromosome, thus no *Sry* gene), XO *Drosophila* are sterile males (one X
chromosome per diploid set). Since there are no hormones to mediate the phenotype,
sex determination in *Drosophila*, and in insects in general, is effectively "digital," with
each cell being a pixel. Indeed, in insects, crustaceans, and even some birds, one can
observe **gynandromorphs**—animals in which certain regions of the body are male and

(?) Developing Questions

A friend wants to bet whether a
particular calico cat is male or
female. Which do you pick? If the
cat turns out to be male, what sex
chromosomes would you expect
him to have?

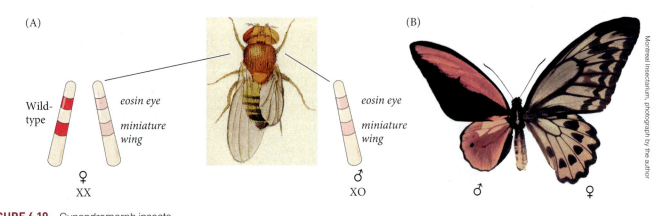

(A)

Wild-
type

eosin eye

*miniature
wing*

♀
XX

eosin eye

*miniature
wing*

♂
XO

(B)

♂   ♀

Montreal Insectarium, photograph by the author

FIGURE 6.10 Gynandromorph insects. (A) A *Drosophila melanogaster* in which the left side is female (XX) and the right side is male (XO). The male side has lost an X chromosome bearing the wild-type alleles of eye color and wing shape, thereby allowing expression of the recessive alleles *eosin eye* and *miniature wing* on the remaining X chromosome. (B) The birdwing butterfly *Ornithoptera croesus*. The male half is smaller and is red, black, and yellow, while the female half is larger and is brown and black. (Drawing by Edith Wallace from T. H. Morgan and C. B. Bridges. 1919. In *Contributions to the Genetics of* Drosophila. Publication no. 278, pp. 1–122. Carnegie Institution of Washington: Washington, DC.)

other regions are female (**FIGURE 6.10**). Gynandromorph fruit flies result when an X chromosome is lost from one embryonic nucleus. The cells descended from that cell, instead of being XX (female), are XO (male). The XO cells display male traits, whereas the XX cells display female traits, suggesting that, in *Drosophila*, each cell makes its own sexual "decision." Indeed, in their classic discussion of gynandromorphs, Morgan and Bridges (1919) concluded, "Male and female parts and their sex-linked characters are strictly self-determining, each developing according to its own aspiration," and each sexual decision is "not interfered with by the aspirations of its neighbors, nor is it overruled by the action of the gonads." Although there are organs that are exceptions to this rule (notably the external genitalia), it remains a good general principle of *Drosophila* sexual development.

The Sex-lethal *gene*

Recent molecular analyses suggest that X chromosome number alone is the primary sex determinant in normal diploid insects (Erickson and Quintero 2007; Moschall et al. 2017). The X chromosome contains genes encoding transcription factors that activate the critical gene in *Drosophila* sex determination, the X-linked locus *Sex-lethal* (*Sxl*). The Sex-lethal protein is an RNA splicing factor that initiates a cascade of RNA processing events that ultimately lead to the expression of a sexual phenotype (**FIGURE 6.11**).

ACTIVATING *SEX-LETHAL*   The number of X chromosomes is critical in activating (or not activating) the early expression of the *Sex-lethal* gene. *Sxl* encodes an RNA splicing factor that will regulate gonad development and will also regulate the amount of gene expression from the X chromosome. The gene has two promoters. The early promoter is active only in XX cells; the later promoter is active in both XX and XY cells. The X chromosome appears to encode four protein factors that activate the early promoter of *Sxl* (see Figure 6.11). If these factors accumulate so they are present in amounts above a certain threshold, the *Sxl* gene is activated through its early promoter (Erickson and Quintero 2007; Gonzáles et al. 2008; Mulvey et al. 2014). The result is the transcription of *Sxl* during the early embryonic development of XX (but not XY) embryos (**FIGURE 6.12**).

The *Sxl* pre-RNA transcribed from the *early* promoter of XX embryos lacks exon 3, which contains a stop codon. Thus, Sxl protein that is made early is spliced in a manner such that exon 3 is absent, so early XX embryos have complete and functional Sxl protein (see Figure 6.12). In XY embryos, the early promoter of *Sxl* is not active, and no functional Sxl protein is present.

However, later in development, the *late* promoter becomes active and the *Sxl* gene is transcribed in both males and females. In XX cells, Sxl protein from the early promoter is already made and can bind to the newly transcribed *Sxl* pre-mRNA (from the late promoter) and splice it in a "female" direction. In this case, Sxl binds to and blocks the splicing complex on exon 3 (Johnson et al. 2010; Salz 2011). As a result, exon 3 is skipped and is not included in the *Sxl* mRNA. Thus, early production ensures that functional full-length Sxl protein is made if the cells are XX (Bell et al. 1991; Keyes et al. 1992). In XY cells, however, the early promoter is not active (because the X-encoded transcription factors have not reached the concentration to activate the early promoter), and there is no early Sxl protein. Therefore, the *Sxl* pre-mRNA of XY cells is spliced in a manner that *includes* exon 3 and its termination codon. Protein synthesis ends at exon 3, and the Sxl is nonfunctional.

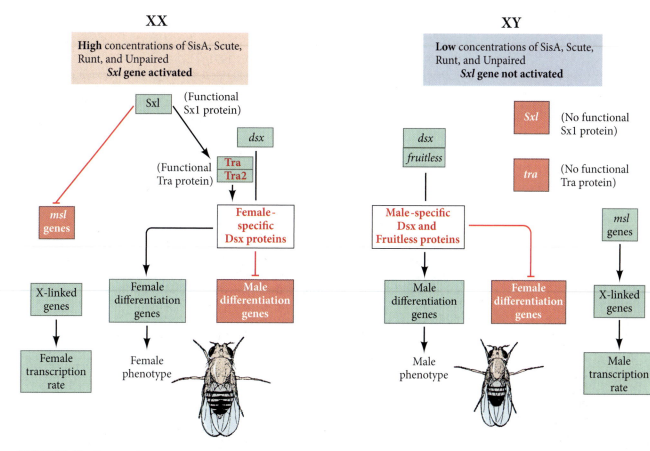

FIGURE 6.11 Proposed regulatory cascade for *Drosophila* somatic sex determination. Transcription factors from the X chromosomes activate the *Sxl* gene in females (XX) but not in males (XY). The Sex-lethal protein performs three main functions. First, it activates its own transcription, ensuring further Sxl production. Second, it represses the translation of *msl2* mRNA, a factor that facilitates transcription from the X chromosome. This equalizes the amount of transcription from the two X chromosomes in females with that from the single X chromosome in males. Third, Sxl enables the splicing of the *transformer-1* (*tra1*) pre-mRNA into functional proteins. The Tra proteins process *doublesex* (*dsx*) pre-mRNA in a female-specific manner that provides most of the female body with its sexual fate. (After B. S. Baker et al. 1987. *BioEssays* 6: 66–70.)

TARGETS OF SEX-LETHAL   The protein made by the female-specific *Sxl* transcript contains regions that are important for binding to RNA. There appear to be three major RNA targets to which the female-specific *Sxl* transcript binds. One of these, as already mentioned, is the pre-mRNA of *Sxl* itself. Another target is the *msl2* gene that controls dosage compensation (see Further Development 6.9, online). Indeed, if the *Sxl* gene is nonfunctional in a cell with two X chromosomes, the dosage compensation system will not work, and the result will be cell death (hence the gene's gory name). The third target

FIGURE 6.12 Differential RNA splicing and sex-specific expression of *Sex-lethal*. In the early blastula (syncytial blastoderm) stage of XX flies, transcription factors from the two X chromosomes are sufficient to activate the early promoter of the *Sxl* gene. This "early" transcript is spliced into an mRNA lacking exon 3 and makes a functional Sxl protein. The early promoter of XY flies is not activated, and males lack functional Sxl. Later in development (cellularization stage), the late promoter of *Sxl* is active in both XX and XY flies. In XX flies, Sxl already present in the embryo prevents the splicing of exon 3 into mRNA, and functional Sxl protein is made. Sxl then binds to its own promoter to keep it active; it also functions to splice downstream pre-mRNAs. In XY embryos, no Sxl is present and exon 3 is spliced into the mRNA. Because of the termination codon in exon 3, males do not make functional Sxl. (After H. K. Salz. 2011. *Curr Opin Genet Dev* 21: 395–400.)

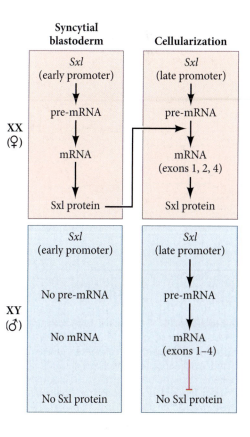

is the pre-mRNA of *transformer* (*tra*)—the next gene in the sex determination cascade (**FIGURE 6.13**; Nagoshi et al. 1988; Bell et al. 1991).

The pre-mRNA of *transformer* (so named because loss-of-function mutations turn females into males) is spliced into a functional mRNA by Sxl protein. The *tra* pre-mRNA is made in both male and female cells; however, in the presence of Sxl, the *tra* transcript is alternatively spliced to create a female-specific mRNA, as well as a nonspecific mRNA that is found in both females and males. Like the male *Sxl* message, the nonspecific *tra* mRNA message contains an early termination codon that renders the protein nonfunctional (Boggs et al. 1987). In *tra*, the second exon of the nonspecific mRNA contains the termination codon and is not used in the female-specific message (see Figures 6.11 and 6.13).

Doublesex: *The switch gene for sex determination*

The *Drosophila* **doublesex** (**dsx**) gene is active in both males and females, and it is expressed in those cells that show sexual differences in function or anatomy (Verhulst and van de Zande 2015). However, the primary transcript of *dsx* is processed in a sex-specific manner (Baker et al. 1987). This alternative RNA processing is the result of the action of the *tra* and *tra2* gene products on the *dsx* gene (see Figures 6.11 and 6.13). If the Tra2 and female Tra proteins are both present, the *dsx* transcript is processed in a female-specific manner (Ryner and Baker 1991). The female splicing pattern produces a Doublesex protein with female-specific domains that allow the protein to interact with other proteins to activate female-specific genes (such as those of the yolk proteins). If no functional Tra

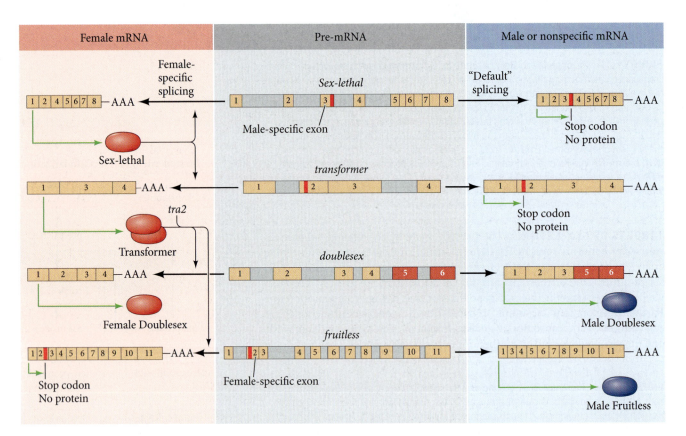

FIGURE 6.13 Sex-specific RNA splicing in four major *Drosophila* sex-determining genes. The pre-mRNAs (shown in the center of the diagram) are identical in both male and female nuclei. In each case, the female-specific transcript is shown to the left, while the default transcript (whether male or nonspecific) is shown to the right. Exons are numbered, and the positions of termination codons are marked. *Sex-lethal*, *transformer*, and *doublesex* are all part of the genetic cascade of gonadal sex determination. The transcription pattern of *fruitless* determines the secondary characteristic of courtship behavior. The transformer proteins also splice the *fruitless* pre-mRNA to make male and female forms of the Fruitless protein in the fly brain. The *fruitless* gene is discussed in Further Development 6.11, online. (After B. S. Baker. 1989. *Nature* 340: 521–524 and B. S. Baker et al. 2001. *Cell* 105: 13–24.)

is produced, the *dsx* pre-mRNA is spliced in a different manner, and a male-specific *dsx* transcript is made. This produces a protein with male-specific domains that interact with other proteins to activate those genes making male-specific traits. It appears that both the female-specific Doublesex protein (Dsx^F) and the male-specific Doublesex protein (Dsx^M) bind to the same enhancers. Dsx^F combines with other proteins to activate female-specific genes and to repress male-specific genes. Conversely, Dsx^M combines with a different set of factors to promote the expression of male-specific genes and to suppress female-specific genes.

According to this model, the result of the sex determination cascade summarized in Figure 6.11 comes down to the type of mRNA processed from the *doublesex* transcript. If there are two X chromosomes, the transcription factors activating the early promoter of *Sxl* reach a critical concentration, and *Sxl* makes a splicing factor that causes the *transformer* gene transcript to be spliced in a female-specific manner. This female-specific protein interacts with the *tra2* splicing factor, causing *dsx* pre-mRNA to be spliced in a female-specific manner.[4] If the *dsx* transcript is not acted on in this way, it is processed in a "default" manner to make the male-specific message. Interestingly, the *doublesex* gene of flies is very similar to the *Dmrt1* gene of vertebrates, and the two types of sex determination may have some common denominators. (See **Further Development 6.11, Brain Sex in *Drosophila***, online.)

Environmental Sex Determination

In many organisms, sex is determined by environmental factors such as temperature, location, and the presence of other members of the species. Here we will discuss temperature-dependent sex determination in turtles.

The sex of most turtles and of all alligators and crocodiles is determined *after* fertilization, by the embryonic environment. In these reptiles, the temperature of the eggs during a certain period of development is the critical factor in determining sex, and small changes in temperature can cause dramatic changes in the sex ratio (Bull 1980; Crews 2003). Often, eggs incubated at low temperatures produce one sex, whereas eggs incubated at higher temperatures produce the other. There is only a small range of temperatures that permits both males and females to hatch from the same brood of eggs.[5]

FIGURE 6.14 shows the abrupt temperature-induced change in sex ratios for three species of reptiles, including the red-eared slider turtle (*Trachemys scripta elegans*). If a brood of *Trachemys* eggs is incubated at a temperature below 28°C, nearly all the turtles that hatch will be male. Above 31°C, nearly every egg gives rise to a female. Temperatures in between give rise to both males and females. Variations on this theme also exist. The eggs of the snapping turtle *Macroclemys temminckii*, for instance, become female at either cool (22°C or lower) or hot (28°C or above) temperatures. Between these extremes, males predominate.

The red-eared slider has become one of the best species for studying environmental sex determination, since it is one of the few turtles that is not presently endangered by habitat destruction. (Indeed, it is the pet-store turtle that is often released into the wild, where it can take over ponds from other species.) The middle third of development appears to be the most critical for sex determination, and it is believed that the turtles cannot reverse their sex after this period.

FIGURE 6.14 Temperature-dependent sex determination in three species of reptiles: the American alligator (*Alligator mississippiensis*), red-eared slider turtle (*Trachemys scripta elegans*), and alligator snapping turtle (*Macroclemys temminckii*). (After D. A. Crain and L. J. Guillette, Jr. 1998. *Anim Reprod Sci* 53: 77–86, data from M. A. Ewert et al. 1994. *J Exp Zool* 270: 3–15 and J. W. Lang and H. V. Andrews. 1994. *J Exp Zool* 270: 28–44.)

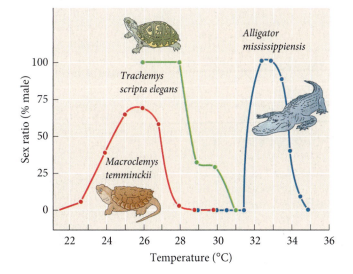

(A)

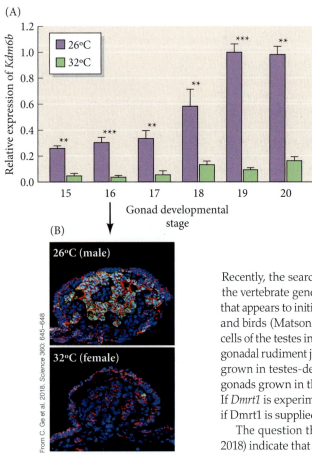

(B)

26°C (male)

32°C (female)

FIGURE 6.15 Activation of sex determination in the red-eared slider turtle. The histone demethylase KDM6B correlates with temperature-sensitive sex determination. (A) Quantitative analysis of the *Kdm6b* mRNA of gonads of turtle embryos raised in a male-producing (26°C) temperature or a female-producing (32°C) temperature. There are significantly higher levels of *Kdm6b* mRNA in embryos raised at the male-producing temperature throughout the period of sexual differentiation. **, P < 0.01; ***, P < 0.001. (B) Immunofluorescence of KDM6B protein in stage-16 gonads. The *Kdm6b* mRNA is stained green and produces a light aqua color when overlapping with the blue nuclear dye, DAPI. The red dye stains β-catenin, which is on the surface of male cells but in the nucleus and cytoplasm of female cells. (From C. Ge et al. 2018. *Science* 360: 645–648.)

For more than 50 years, scientists have tried to find the temperature-sensitive networks that generate the ovaries and testes of turtles (see Shoemaker et al. 2007; Bieser and Wibbels 2014). Recently, the search has focused on the gene encoding Dmrt1. *Dmrt1*, you may recall, is the vertebrate gene related to *doublesex* in *Drosophila*. In vertebrates, Dmrt1 is the protein that appears to initiate the testes-determining cascade in many species of fish, amphibians, and birds (Matson and Zarkower 2012). It is also responsible for maintaining the Sertoli cells of the testes in mammals (Matson et al. 2011). In *Trachemys*, *Dmrt1* is expressed in the gonadal rudiment just prior to sexual differentiation. It is expressed at high levels in gonads grown in testes-determining temperatures (26°C) and is expressed at very low levels in gonads grown in those higher temperatures (32°C) that generate ovaries (Ge et al. 2017). If *Dmrt1* is experimentally suppressed (by a virus), the gonads become ovaries. However, if Dmrt1 is supplied to these suppressed gonads, they will resume testis development.

The question then becomes: What regulates *Dmrt1*? Recent experiments (Ge et al. 2018) indicate that *Dmrt1* expression is positively regulated by the removal of a particular methyl group from nucleosomes on its promoter, a reaction catalyzed by the enzyme KDM6B. Male-producing temperatures lead to the activation of the *Kdm6b* gene during the stages when sexual specification of the gonad occurs (**FIGURE 6.15**). However, we still don't know the identity of the temperature-regulated factor(s) that activate KDM6B.

Gametogenesis in Animals

One of the most important events in sex determination is **gametogenesis**, the differentiation of the germ cells into gametes: eggs and sperm. And it is the **primordial germ cells** (**PGCs**) that are the bipotential precursors of both eggs and sperm; if they reside in the ovaries they become eggs, and if they reside in the testes they become sperm. All of these decisions are coordinated by factors produced by the developing gonads.

One of the most amazing things about germ cells is that they provide the continuity between generations. The adult body perishes, but the germ line forms the gametes that create a new body, which will also perish. The immature sperm or eggs in your body have come from a germ cell lineage that has resided in the gonads of reptiles, amphibians, fish, and invertebrates.

Second, and importantly, the cells that generate the sperm or eggs do not originally form inside the gonads. In *Drosophila* and mammals, they form in the posterior portion of the embryo and migrate into the gonads (Anderson et al. 2000; Molyneaux et al. 2001; Tanaka et al. 2005). This pattern is common throughout the animal kingdom: the germ cells are "set aside" from the rest of the embryo, and the cells' transcription and translation are shut down while they migrate from peripheral sites in the embryo and to the gonad. It is as if the germ cells were a separate entity, reserved for the next generation, and repressing gene expression makes them insensitive to the intercellular commerce going on all around them (Richardson and Lehmann 2010; Tarbashevich and Raz 2010).

Although the mechanisms used to specify the germ cells vary enormously across the animal kingdom, the proteins expressed by germ cells to suppress gene expression are remarkably conserved. These proteins, which include the Vasa, Nanos, Tudor, and Piwi family proteins, can be seen in the germ cells of cnidarians, flies, and mammals

(Ewen-Campen et al. 2010; Leclére et al. 2012). Vasa proteins, which appear to activate germ cell-specific genes, are required for germ cells in nearly all animals studied. Nanos is involved in repressing translation and preventing cell death (Kobayashi et al. 1996; Hayashi et al. 2004).

Equally remarkable is that the signal that induces the formation of PGCs also appears to be conserved throughout the animal kingdom. In mammals and in those insects that induce germ cells (e.g., crickets), BMP signals are required for the formation of PGCs (Donoughe et al. 2014; Lochab and Extravour 2017). In mammals, BMP4 from the extraembryonic cell layers induces mesenchymal cells to become PGCs by activating those genes that specify the cells to be germ line, while simultaneously blocking the expression of those genes that prevent cells from becoming part of the germ line (Fujiwara et al. 2001; Saito and Yamaji 2012; Zhang et al. 2018). This strategy of simultaneously activating one set of genes while repressing others is similar to that seen in mammalian gonadal sex determination. (See **Further Development 6.12, Theodor Boveri and the Formation of the Germ Line**, online.)

PGCs in mammals: From genital ridge to gonads

In mammals, the newly formed PGCs first enter into the hindgut and eventually migrate forward and into the bipotential gonads, multiplying as they migrate (**FIGURE 6.16**). From the time of their specification until they enter the genital ridges, the PGCs are surrounded by cells secreting stem cell factor (SCF). SCF is necessary for PGC motility and survival. Moreover, the cluster of SCF-secreting cells appears to migrate with the PGCs, forming a "traveling niche" of cells that support the persistence, division, and movement of the PGCs (Gu et al. 2009). Once they are in the

(A) Migration of PGCs to endoderm

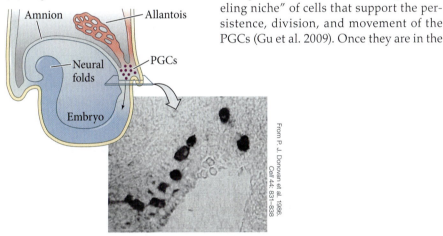

(B) Migration of PGCs into gonad   (C)

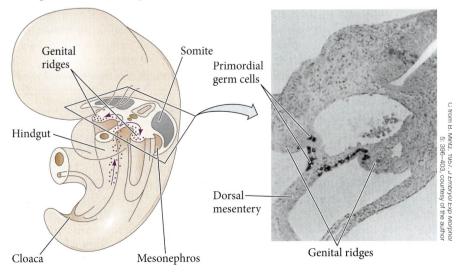

FIGURE 6.16 Primordial germ cell migration in the mouse. (A) On embryonic day 8, PGCs established in the posterior epiblast migrate into the definitive endoderm of the embryo. The photo shows large PGCs (stained for alkaline phosphatase) in the hindgut of a mouse embryo. (B) The PGCs migrate through the gut and, dorsally, into the genital ridges. (C) Alkaline phosphatase-staining cells are seen entering the genital ridges around embryonic day 11. (After J. Langman. 1981. *Medical Embryology*, 4th Ed. Williams & Wilkins: Baltimore.)

TABLE 6.1

Female oogenesis	Male spermatogenesis
Meiosis initiated once in a finite population of cells	Meiosis initiated continuously in a mitotically dividing stem cell population
One gamete produced per meiosis	Four gametes produced per meiosis
Completion of meiosis delayed for months or years	Meiosis completed in days or weeks
Meiosis arrested at first meiotic prophase and reinitiated in a smaller population of cells	Meiosis and differentiation proceed continuously without cell cycle arrest
Differentiation of gamete occurs while diploid, in first meiotic prophase	Differentiation of gamete occurs while haploid, after meiosis ends
All chromosomes exhibit equivalent transcription and recombination during meiotic prophase	Sex chromosomes excluded from recombination and transcription during first meiotic prophase

Source: After M. A. Handel and J. J. Eppig. 1998. *Curr Topics Dev Biol* 37: 333–358.

gonad, these cells are sustained by BMPs that create a niche for them in the genital ridge (Dudley et al. 2007, 2010).

The PGCs are then told by the gonad whether to initiate **oogenesis** (the formation of eggs) or to initiate **spermatogenesis** (the formation of sperm) (**TABLE 6.1**). A fundamental difference between mammalian males and females involves the timing of meiosis. In females, meiosis begins in the embryonic gonads. In males, meiosis is not initiated until puberty. The "gatekeeper" for meiosis appears to be the Stra8 transcription factor, which promotes a new round of DNA synthesis and meiotic initiation in the germ cells. In the developing ovaries, Stra8 is *upregulated* by two factors—Wnt4 and retinoic acid (RA)—coming from the adjacent kidney (Baltus et al. 2006; Bowles et al. 2006; Naillat et al. 2010; Chassot et al. 2011). In the developing testes, however, Stra8 is *downregulated* by Fgf9, and the retinoic acid produced by the developing kidney is degraded by the testes' secretion of the RA-degrading enzyme Cyp26b1 (**FIGURE 6.17**; Bowles et al. 2006; Koubova et al. 2006). During male puberty, however, retinoic acid is synthesized in the Sertoli cells and induces Stra8 in sperm stem cells. Once Stra8 is present, the sperm progenitor cells become committed to meiosis (Anderson et al. 2008; Mark et al. 2008; Nakagawa et al. 2017).

Meiosis: The intertwining of life cycles

Meiosis is perhaps the most revolutionary invention of eukaryotes, for it is the mechanism for the transmission of genes from one generation to the next and for the recombination of sperm- and egg-derived genes into new combinations of alleles. Van Beneden's 1883 observations that the divisions of germ cells caused the resulting gametes to contain half the diploid number of chromosomes "demonstrated that the chromosomes of the offspring are derived in equal numbers from the nuclei of the two conjugating germcells and hence equally from the two parents" (Wilson 1924). Meiosis is a critical starting and ending point in the cycle of life. Sexual reproduction, evolutionary variation, and the transmission of traits from one generation to the next all come down to meiosis. So to understand what germ cells do, we must first understand meiosis.

Meiosis is the means by which the gametes halve the number of their chromosomes. In the haploid condition, each cell has only one copy of each chromosome, whereas diploid cells have two copies of each chromosome. This feat is accomplished by the occurrence of a single round of DNA replication followed by two successive chromosomal divisions.

After the germ cell's final mitotic division, a period of DNA synthesis occurs, so that the cell initiating meiosis doubles the amount of DNA in its nucleus. In this state, each chromosome consists of two **sister chromatids** attached at a common kinetochore.[6] In the first of the two meiotic divisions (meiosis I), homologous chromosomes (for example, the two copies of chromosome 3 in the diploid cell) come together and are then separated into different cells. Hence the first meiotic division *splits homologous chromosomes between two daughter cells* such that each daughter cell has only one copy of each chromosome; these cells are therefore haploid. But each of

(A) Female germ cells

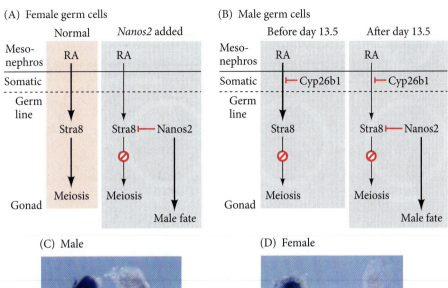

(B) Male germ cells

(C) Male

RA synthesized   RA degraded

(D) Female

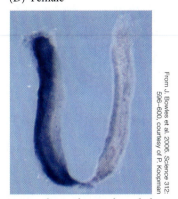

RA synthesized   RA degraded

FIGURE 6.17 Retinoic acid (RA) determines the timing of meiosis and sexual differentiation of mammalian germ cells. (A) In female mouse embryos, RA secreted from the mesonephros reaches the gonad and triggers meiotic initiation via the induction of Stra8 transcription factor in female germ cells (pink). However, if activated *Nanos2* genes are added to female germ cells, they suppress Stra8 expression, leading the germ cells into a male pathway (gray). (B) In embryonic testes, Cyp26b1 blocks RA signaling, thereby preventing male germ cells from initiating meiosis until embryonic day 13.5 (left panel). After embryonic day 13.5, when Cyp26b1 expression is decreased, Nanos2 is expressed and prevents meiotic initiation by blocking Stra8 expression. This induces male-type differentiation in the germ cells (right panel). (C,D) Day-12 mouse embryos stained for mRNAs encoding the RA-synthesizing enzyme Aldh1a2 (left gonad) and the RA-degrading enzyme Cyp26b1 (right gonad). The RA-synthesizing enzyme is seen in the mesonephros of both the male (C) and female (D); the RA-degrading enzyme is seen only in the male gonad. (A,B from Y. Saga. 2008. *Curr Opin Genet Dev* 18: 337–341.)

the chromosomes has already replicated (i.e., each has two chromatids), so the second division (meiosis II) *separates the two sister chromatids from each other*. The net result of meiosis is four cells, each with a haploid set of unreplicated chromosomes.

FURTHER DEVELOPMENT

The stages of meiosis

The first meiotic division begins with a long prophase, which is subdivided into four stages (**FIGURE 6.18A**). During the **leptotene** (Greek, "thin thread") stage, the chromatin of the chromatids is stretched out very thinly, and it is not possible to identify individual chromosomes. DNA replication has already occurred, however, and each chromosome consists of two parallel chromatids. Homologues begin to pair due to cables that pass from the cytoplasm into the nucleus and attach to the kinetochores. In this way, chromosomes can be moved by the cytoskeleton (Wynne et al. 2012; Burke 2018). The nuclear envelope also appears to be important in allowing the pairing of the homologous chromosomes (Comings 1968; Scherthan 2007; Tsai and McKee 2011). At the **zygotene** (Greek, "yoked threads") stage, the homologous chromosomes, now brought together, begin to line up side by side. This close pairing, called **synapsis**, is characteristic of meiosis, and it appears to be initiated by double-stranded DNA breaks (similar to those used for DNA repair). These breaks allow "tentacles" of single-stranded DNA to go from one chromosome to the other (Zickler and Leckner 2015). Although the mechanisms whereby each chromosome recognizes its homologue are not fully known, synapsis requires the presence of the nuclear envelope and the formation of a ladderlike proteinaceous ribbon called the **synaptonemal complex** (**FIGURE 6.19A,B**; von Wettstein 1984; Dunce et al. 2018). The configuration formed by the four chromatids and the synaptonemal complex is referred to as a **tetrad** or a **bivalent**.

FIGURE 6.18 Meiosis, emphasizing the synaptonemal complex. Before meiosis, unpaired homologous chromosomes are distributed randomly within the nucleus. (A) The four stages of meiotic prophase I. At leptotene, telomeres have attached along the nuclear envelope. The chromosomes "search" for homologous chromosomes, and synapsis—the association of homologous chromosomes—begins at zygotene, where the first evidence of the synaptonemal complex can be seen. During pachytene, homologue alignment is seen along the entire length of the chromosomes, leading to the production in diplotene of a bivalent structure. Paired homologues can recombine with each other (cross-over) during zygotene and pachytene, and even into diplotene. The synaptonemal complex dissolves at diplotene, when recombination is completed. (B) In diakinesis, chromosomes condense further and then form a metaphase plate. Segregation of the homologous chromosomes occurs at anaphase I. Only one pair of sister chromatids is shown here in meiosis II, where sister chromatids align at metaphase II and then in anaphase II segregate to opposite poles. (After J. H. Tsai and B. D. McKee. 2011. *J Cell Sci* 124: 1955–1963.)

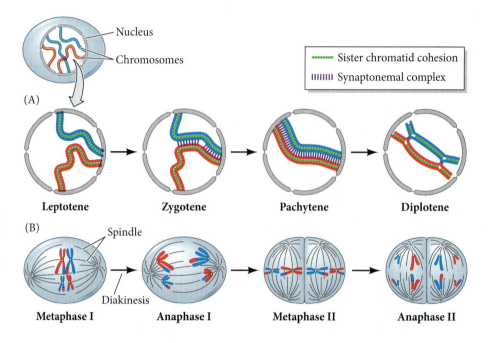

During the next stage of meiotic prophase, **pachytene** (Greek, "thick thread"), the chromatids thicken and shorten. Individual chromatids can now be distinguished under the light microscope, and crossing-over may occur. **Crossing-over** represents an exchange of genetic material whereby genes from one chromatid are exchanged with homologous genes from another chromatid. Crossing-over may continue into the next stage, **diplotene** (Greek, "double threads"). During diplotene, the synaptonemal complex breaks down and the two homologous chromosomes start to separate. Usually, however, they remain attached at various points called **chiasmata**, which are thought to represent regions where crossing-over is occurring. The diplotene stage is characterized by a high level of gene transcription.

Metaphase of the first meiotic division begins with **diakinesis** (Greek, "moving apart") of the chromosomes (**FIGURE 6.18B**). The nuclear envelope breaks down and the chromosomes migrate to form a metaphase plate. Anaphase of meiosis I does not commence until the chromosomes are properly aligned on the mitotic spindle fibers. This alignment is accomplished by proteins that prevent cyclin B from being degraded until after all the chromosomes are securely fastened to microtubules.

During anaphase I, the homologous chromosomes separate from each other in an independent fashion. This stage leads to telophase I, during which two daughter cells are formed, each cell containing one partner of each homologous chromosome pair. After a brief resting stage known as **interkinesis**, the second meiotic division takes place. During meiosis II, the kinetochore of each chromosome divides during anaphase so that each of the new cells gets one of the two chromatids, the final result being the creation of four haploid cells. Note that meiosis has also reassorted the chromosomes into new groupings. First, each of the four haploid cells has a different assortment of chromosomes. Humans have 23 different chromosome pairs; thus, 2^{23} (nearly 10 million) different haploid cells can be formed from the genome of a single person. In addition, the crossing-over that occurs during the pachytene and diplotene stages of first meiotic metaphase further increases genetic diversity and makes the number of potential different gametes incalculably large.

This organization and movement of meiotic chromosomes is choreographed by a ring of **cohesin proteins** that encircles the sister chromatids. Cohesin is found at both the kinetochore and around the chromatin arms that unite the sister chromatids (**FIGURE 6.19C**). The cohesins recruit proteins that help promote pairing between homologous chromosomes and allow recombination to occur (Pelttari et al. 2001; Villeneuve and Hillers 2001; Sakuno and Watanabe 2009). At anaphase,

(A)

(B)

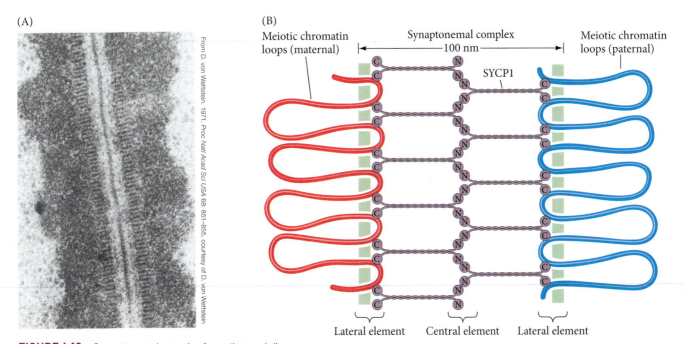

Meiotic chromatin loops (maternal)

Synaptonemal complex
100 nm

SYCP1

Meiotic chromatin loops (paternal)

Lateral element   Central element   Lateral element

FIGURE 6.19 Synaptonemal complex formation and disassembly during meiosis. During meiotic prophase I, the synaptonemal complex forms at sites nucleated by the chromatin breaks that pair the chromatids. (A) Synaptonemal complex showing the central element and the lateral elements that bind the chromatin. (B) Major structure of the synaptonemal complex, formed by interlocking molecules of the protein SYCP1. The C-terminus ends of the proteins bind to the DNA, while the N-terminus ends bind to one another in the center of the complex. (C) The complex is supported by cohesin molecules at the arms linking the two chromosomes together and at the kinetochores holding the sister chromatids together. In first meiotic metaphase, the cohesin-cleaving protease (separase) is inactive, as is the APC/C protein that can activate it. In addition, the kinetochore-bound cohesin binds the Sgo2 protein, which protects it from the separase. Anaphase begins when signals from cytoplasm activate the APC/C molecule, which activates the separase protein. Separase digests the cohesin holding the homologous chromosomes together, but it does not digest the cohesin at the kinetochores holding the sister chromatids together. As anaphase ends, the protective Sgo2 protein is lost and replaced by other proteins that allow the cohesin to be digested at the next cell division. (B from J. M. Dunce et al. 2018. *Nat Struct Mol Biol* 25: 557–569; C from A. I. Mihajlović and G. FitzHarris. 2018. *Curr Biol* 28: R671–R674.)

(C)

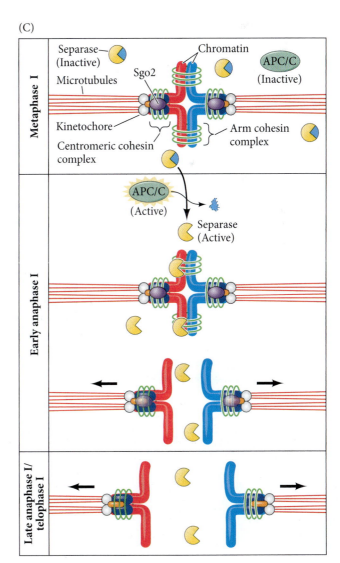

Metaphase I

Separase (Inactive)
Microtubules
Sgo2
Kinetochore
Centromeric cohesin complex
Chromatin
APC/C (Inactive)
Arm cohesin complex

Early anaphase I

APC/C (Active)
Separase (Active)

Late anaphase I / telophase I

the cohesins surrounding the chromatin are digested, allowing the two chromosomes to be pulled apart. The cohesins on the kinetochore are protected and are not digested (Argunhan et al. 2017; Mihajlović and FitzHarris 2018). These cohesin rings resist the pulling forces of the spindle microtubules, thereby keeping the sister chromatids attached during meiosis I (Haering et al. 2008; Brar et al. 2009). At the second meiotic division, these kinetochore cohesin rings are cleaved and the kinetochores of sister chromatids can separate from each other (Schöckel et al. 2011). (See **Further Development 6.13, Modifications of Meiosis**, online.)

Spermatogenesis in mammals

Spermatogenesis—the developmental pathway from germ cell to mature sperm—begins at puberty and occurs in the recesses between the Sertoli cells (**FIGURE 6.20**). Spermatogenesis is divided into three major phases (Matson et al. 2010):

1. A proliferative phase where sperm stem cells (**spermatogonia**) increase by mitosis
2. A meiotic phase, involving the two divisions that create the haploid state
3. A postmeiotic "shaping" phase called **spermiogenesis**, during which the round cells (spermatids) eject most of their cytoplasm and become the streamlined sperm

The proliferative phase begins when the mammalian PGCs arrive at the genital ridge of a male embryo. Here they are called **gonocytes** and become incorporated into the sex cords that will become the seminiferous tubules (Culty 2009). The gonocytes become undifferentiated spermatogonia residing near the basal end of the tubular cells (Yoshida et al. 2007; Yoshida 2016). These are true stem cells in that they can reestablish spermatogenesis when transferred into mice whose sperm production has been eliminated by toxic chemicals. Spermatogonia appear to take up residence in stem cell niches at the junction of the Sertoli cells (the epithelium of the seminiferous tubules), the interstitial (testosterone-producing) Leydig cells, and the testicular blood vessels. Adhesion molecules join the spermatogonia directly to the Sertoli cells, which will nourish the developing sperm (Newton et al. 1993; Pratt et al. 1993; Kanatsu-Shinohara et al. 2008).

The percentage of the gonocytes that become true stem cells probably differs greatly among groups of mammals, and the cells defining the stem cell niche may also differ (de Rooij 2017; Fayomi and Orwig 2018). This is because different groups of mammals have different strategies for sperm production. Mice have 12 amplifying divisions of progenitor cells between the sperm stem cell and the spermatocyte that undergoes meiosis; they produce 40 million sperm per gram of testis tissue per day. Humans produce more stem cells but have only five transit amplifying divisions between the stem cell and the spermatocyte; men generate 4.4 million sperm per gram of testis tissue

FIGURE 6.20 Sperm maturation. (A) Cross section of the seminiferous tubule. Spermatogonia are blue, spermatocytes are lavender, and the mature sperm appear yellow. (B) Simplified diagram of a portion of the seminiferous tubule, illustrating relationships between spermatogonia, spermatocytes, and sperm. As these germ cells mature, they progress toward the lumen of the seminiferous tubule. (See also Figure 7.1.) (B based on M. Dym. 1977. In *Histology*, 4th Ed., L. Weiss and R. O. Greep [Eds.], pp. 979–1038. McGraw-Hill: New York, courtesy of Stephane Clermont.)

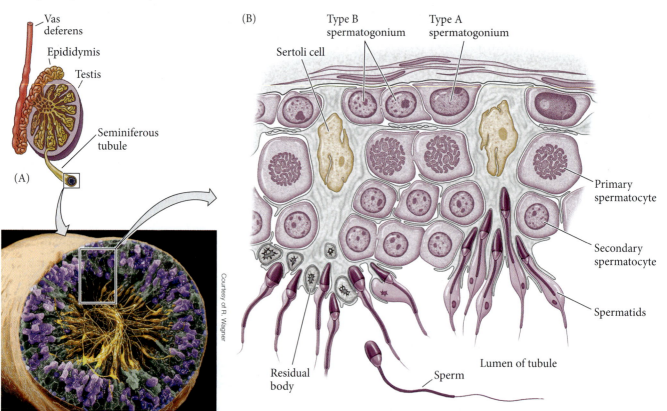

Courtesy of R. Wagner

each day. (Although this is 10-fold less efficient than mice, it means that adult human males generate more than 1000 sperm per second; Matson et al. 2010.) Each day, some 100 million sperm are made in each human testicle, and each ejaculation releases 200 million sperm. Unused sperm are either resorbed or passed out of the body in urine. During his lifetime, a human male can produce 10^{12} to 10^{13} sperm (Reijo et al. 1995).

The mitotic proliferation of the stem cells produces type A spermatogonia, which are held together by cytoplasmic bridges. However, these bridges are fragile, and a cell can split off from the group and become a stem cell again (Hara et al. 2014). Glial-derived neurotrophic factor (GDNF), secreted from the Sertoli cells, keeps the stem cells in mitosis (Chen et al. 2016a). However, BMPs and Wnts start inducing the type A spermatogonia to differentiate further to sperm (Song and Wilkinson 2014; Tokue et al. 2017).

THE MEIOTIC PHASE: GETTING TO HAPLOID SPERMATIDS   Type B spermatogonia are the precursors of the spermatocytes and contain high levels of Stra8. (**FIGURE 6.21**; de Rooij and Russell 2000; Nakagawa 2010; Griswold et al. 2012). These are the last cells of that lineage to undergo mitosis, and they divide once to generate the **primary spermatocytes**—the cells that enter meiosis. Each primary spermatocyte undergoes the first meiotic division to yield a pair of haploid **secondary spermatocytes**, which complete the second division of meiosis. The cells thus formed are called **spermatids**,

FIGURE 6.21   Overview of spermatogenesis. (A) Formation of syncytial clones (daughter cells whose cytoplasms are connected) of mammalian male germ cells. In mice, there may be 16 type B spermatogonia linked together. In humans, the cytoplasmic linkages are probably limited to four cells. (B) The principle cell types of spermatogenesis and the developmental events separating them. (C) Cells move from the basal lamina of the seminiferous tubule toward the lumen as development progresses. (A after M. Dym and D. W. Fawcett. 1971. *Biol Reprod* 4: 195–215; B,C after D. G. de Rooij. 2017. *Development* 144: 3022–3030.)

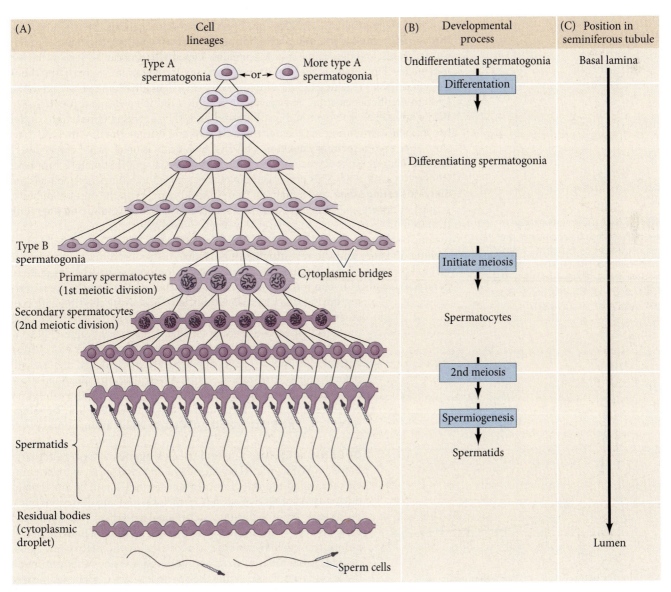

and they are still connected to one another through their cytoplasmic bridges. The spermatids that are connected in this manner have haploid nuclei but are functionally diploid, since a gene product made in one cell can readily diffuse into the cytoplasm of its neighbors (Braun et al. 1989).

During the divisions from undifferentiated spermatogonia to spermatids, the cells move farther and farther away from the basal lamina of the seminiferous tubule and closer to its lumen (see Figure 6.20; Siu and Cheng 2004).

THE POSTMEIOTIC PHASE: SPERMIOGENESIS   As the spermatids move toward the lumen, they lose their cytoplasmic connections and differentiate into spermatozoa, a process called spermiogenesis. In humans, the progression from spermatogonial stem cell to mature spermatozoa takes 65 days (Dym 1994), and the last third of it (about 21 days) is taken up by spermiogenesis. The mammalian haploid spermatid is a round, unflagellated cell that looks nothing like the mature vertebrate sperm. For fertilization to occur, the sperm has to meet and bind with an egg, and spermiogenesis prepares the sperm for these functions of motility and interaction. The process of mammalian sperm differentiation is shown in Figure 7.1 and discussed in the next chapter.

Oogenesis in mammals

Scientists who study oogenesis often write in terms of the wonder that the process generates and the huge unknown questions that remain to be solved. Mammalian oogenesis (egg production) differs greatly from spermatogenesis. The eggs mature through a symphonic coordination of hormones, paracrine factors, enzymes, chromatin structures, and tissue anatomy. Mammalian egg maturation can be seen as having four stages. First, there is the stage of proliferation. In the human embryo, the thousand or so PGCs reaching the developing ovary divide rapidly from the second to the seventh month of gestation. They generate roughly 7 million oogonia (**FIGURE 6.22**). While most of these oogonia die soon afterward, the surviving population, under the influence of retinoic acid, enter the next step and initiate the first meiotic division. They become **primary oocytes**. This first meiotic division does not proceed very far, and the primary oocytes remain in the diplotene stage of first meiotic prophase (Pinkerton et al. 1961). This prolonged diplotene stage is sometimes referred to as the **dictyate resting stage**, and it may last from 12 to 40 years. With the onset of puberty, groups of oocytes periodically resume meiosis. At that time, **luteinizing hormone** (**LH**) from the pituitary gland releases this block and permits these oocytes to resume meiotic division (Lomniczi et al. 2013; Tiwari and Chaube 2017). They complete first meiotic division, and the resulting secondary oocytes proceed to second meiotic metaphase and undergo maturation steps. This maturation involves the crosstalk of paracrine factors between the oocyte and its follicle cells, both of which are maturing during this phase. The follicle cells activate the translation of stored oocyte mRNA encoding proteins such as the sperm-binding proteins that will be used for fertilization and the cyclins that control embryonic cell division (Chen et al. 2013; Cakmak et al. 2016). After the secondary oocyte is released from the ovary, meiosis will resume only if fertilization occurs. At fertilization, calcium ions are released in the egg, and these calcium ions release the inhibitory block and allow the haploid nucleus to form. (See **Further Development 6.14, The Biochemistry of Oocyte Maturation**, online.)

OOGENIC MEIOSIS   In reviewing oogenesis, Severance and Latham (2018) write, "The oocyte is a remarkable cell with two universal roles in reproduction: correct segregation of chromosomes during two successive rounds of meiosis and sustaining viability of the early embryo until transcriptional activation." To endow the early embryo with the ability to meet its developmental demands before the nuclear genome is activated in the embryonic cells, oocytes grow to large sizes and undergo divisions that minimize the loss of cytoplasm. While the sperm loses its cytoplasm, the egg accumulates cytoplasm.

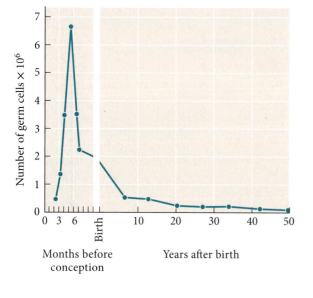

FIGURE 6.22   The number of germ cells in the human ovary changes over the life span. (After T. G. Baker. 1971. *Am J Obstet Gynecol* 110: 746–761, based on T. G. Baker and S. Zuckerman. 1963. *Proc R Soc Lond* 158: 417–433; and E. Block. 1952. *Acta Anat* 14: 108–123.)

Oogenic meiosis in mammals differs from spermatogenic meiosis in numerous ways. First, when the primary oocyte divides, its nuclear envelope breaks down, and the metaphase spindle migrates to the cortex (periphery) of the cell (see Severson et al. 2016). At the cortex, an oocyte-specific tubulin mediates the separation of chromosomes, and mutations in this tubulin have been found to cause infertility (Feng et al. 2016). At telophase, while both daughter cells now contain a haploid nucleus, one contains very little cytoplasm, while the other retains nearly the entire volume of cellular constituents (**FIGURE 6.23**). The smaller cell becomes the **first polar body**, and the larger cell is referred to as the **secondary oocyte**.

This asymmetrical cytokinesis is directed through a cytoskeletal network composed chiefly of filamentous actin that cradles the mitotic spindle and brings it to the oocyte cortex by myosin-mediated contraction (Schuh and Ellenberg 2008). A similar unequal cytokinesis takes place during the second division of meiosis. Most of the cytoplasm is retained by the mature egg (the **ovum**), and a second polar body forms but receives little more than a haploid nucleus. (In humans, the first polar body usually does not divide. It undergoes apoptosis around 20 hours after the first meiotic division.) Thus, oogenic meiosis conserves the volume of oocyte cytoplasm in a single cell rather than splitting it equally among four progeny (Longo 1997; Schmerler and Wessel 2011).

A second way in which oogenic meiosis in mammals differs from spermatogenic meiosis is in the mechanics of meiosis I. The oocyte meiotic spindle lacks centrioles. Rather, the microtubules of the meiotic spindle are organized by mRNAs and enzymes located on the chromosomes and the spindle fibers themselves (Severson et al. 2016; Severance and Latham 2018). Instead of two centrioles, numerous **microtubule organizing centers** (**MTOCs**) form around the nuclear envelope at first meiotic prophase. The MTOCs coalesce at the future spindle poles once the nuclear envelope breaks down.

OOCYTES AND AGE The retention of the oocyte in the ovary for decades has profound medical implications. Most human embryos do not survive to birth. A large proportion, perhaps even a majority, of fertilized human eggs have too many or too few chromosomes to survive. Genetic analysis has shown that such **aneuploidy** (incorrect number of chromosomes) is usually due to errors in oocyte meiosis (Hassold et al. 1984; Munné et al. 2007). Only a few aneuploidies (such as those of the sex chromosomes and chromosome 21) survive to be born, and the percentage of babies born with such aneuploidies increases greatly with maternal age. Women in their twenties have only a 2–3% chance of bearing a fetus whose cells contain an extra chromosome. This risk goes to 35% in women who become pregnant in their forties (**FIGURE 6.24A**; Hassold and Chiu 1985; Hunt and Hassold 2010). The reasons for this appear to be due to the gradual loss of cohesin proteins from the chromosomes as the cell ages (**FIGURE 6.24B,C**; Chiang et al. 2010; Lister et al. 2010; Revenkova et al. 2010), causing a less stable linkage between the kinetochore and spindle during meiotic metaphase (Holubcová et al. 2015).

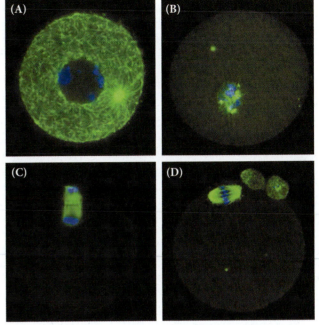

FIGURE 6.23 Meiosis in the mouse oocyte. The tubulin of the microtubules is stained green; the DNA is stained blue. (A) Mouse oocyte in meiotic prophase. The large diploid nucleus (the germinal vesicle) is still intact and actively transcribing genes whose mRNAs will be stored in the egg as maternal mRNA. (B) The nuclear envelope of the germinal vesicle breaks down as metaphase begins. (C) Meiotic anaphase I, wherein the spindle migrates to the periphery of the egg and releases a small polar body. (D) Meiotic metaphase II, wherein the second polar body is given off (the first polar body has also divided).

FIGURE 6.24 Chromosomal nondisjunction and meiosis. (A) Maternal age affects the incidence of trisomies in human pregnancy. (B,C) Reduction of chromosome-associated cohesin in aged mice. DNA (white) and cohesin (green) stained in oocyte nuclei of (B) 2-month-old (young) and (C) 14-month-old (aged, for a mouse) ovaries. A significant loss of cohesin can be seen (especially around the kinetochores) in aged mice. (A after P. Hunt and T. Hassold. 2010. *Curr Biol* 20: R699–R702.)

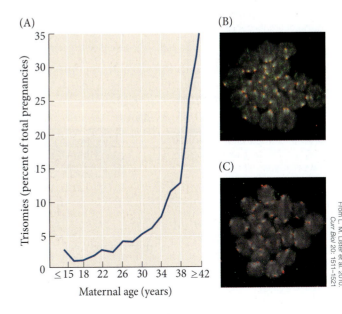

Sex Determination and Gametogenesis in Angiosperm Plants

Sex Determination

When you think about flowers, you are thinking of the sex organs of the angiosperm (flowering) plants. Remarkably, in most angiosperms, individuals are not one sex or the other. Rather, a monoecious plant can have both male and female (unisexual) flowers or can have bisexual flowers. These bisexual or "perfect" flowers have certain parts (those in the stamen) that are male, while other parts (those in the carpel) are female.[7]

The parts of the plant designated to become the flowers are determined by the expression of genes in the plant shoot apical meristem that transform it into an inflorescence meristem, which through further gene expression produces the floral meristem that gives rise to the flower. It is a cascade of gene expression with internal and environmental controls—especially photoperiod—that suppresses the genes whose expression would continue the proliferative growth of the meristem while activating a set of cells that can function for reproduction.

In the model plant species *Arabidopsis thaliana*, the flowering signal is initiated by activation of the *CONSTANS* (*CO*) gene. The transcription factor protein encoded by this gene follows a circadian rhythm that peaks in the afternoon. However, the protein is stable only in light. Thus, it reaches functional levels only on long summer days when it is still daylight until at least 12 hours after dawn (Valverde et al. 2002; Yanovsky and Kay 2002; Mizoguchi et al. 2005). The CO protein activates the *FLOWERING LOCUS T* (*FT*) gene, which results in the production of the FT protein in the leaves. This protein is transported through the phloem to the shoot apical meristem, where it complexes with the FLOWERING LOCUS D (FD) protein to become a transcription factor (Notaguchi et al. 2008). This FT/FD transcription factor activates **floral meristem identity genes** such as *APETALA1* (*AP1*), *LEAFY* (*LFY*), *AGAMOUS* (*AG*), and *PISTILLATA* (*PI*) (**FIGURE 6.25**; also see Figure 6.26; Abe et al. 2005; Wigge et al. 2005).

FIGURE 6.25 The vegetative-to-reproductive transition. (A) CO activates FT, which is transported by the phloem from leaves to the shoot apical meristem, where it forms a complex with FD. In the apical meristem, the FD/FT transcription factor activates floral meristem identity genes, such as *LEAFY* and *APETALA1*. The floral meristem identity genes activate the floral organ identity (ABCDE) genes, which encode subunits of the transcription factors that specify the parts of the flower. (B) Internal and external factors regulate whether a meristem produces vegetative or reproductive structures. Not all of the regulatory mechanisms shown are used in all species, and some species flower independently of external environmental signals.

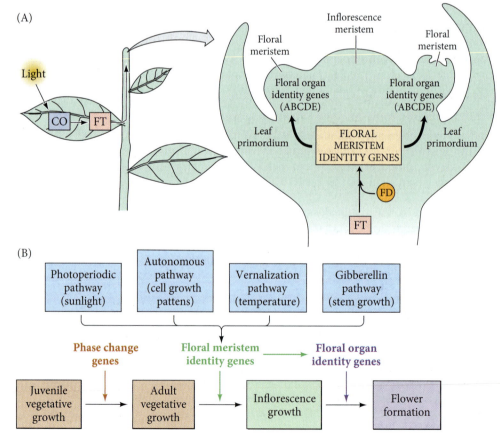

The next step is to specify individual regions of the floral meristem to produce specific parts of the flowers. In a perfect flower, four consecutive whorls develop around a central axis. The first (outer) whorl becomes the **sepals** that support and protect the flower. The second whorl becomes the **petals**, which are often brightly colored and attract pollinators. The third whorl becomes the male organs, the **stamens**, while the fourth, and central, whorl becomes the **carpels (pistil)**, the female organs, which include the style and stigma; the ovary forms at the base of the style (**FIGURE 6.26**; Meyerowitz et al. 1989; Schwarz-Sommer et al. 1990; Coen and Meyerowitz 1991; Theissen et al. 2016). Genetic and comparative studies of plant species have given rise to the ABCDE model for specifying the identity of the floral organs within these whorls (Meyerowitz et al. 1989; Theissen 2001; Theissen et al. 2016).

In the ABCDE model, the five floral organ identities are specified by five proteins that combine to form organ-specific tetramers. The proteins and **floral organ identity genes** that encode these proteins are grouped into classes: A, B, C, D, and E. A key feature of the model is that these classes are differentially expressed in the four whorls that will form the flower (**FIGURE 6.27**). Class E proteins are needed for all the proteins to function to make flower parts. It appears that these are encoded by the floral organ identity genes mentioned above. Class A proteins are needed to make sepals and petals (the nonreproductive parts of the flower), whereas the reproductive organs (stamens, carpels, and ovules) need class C proteins. Each of the other parts of the flowers need a different combination of proteins to form (see Figure 6.26). The presence of Class B gene products with the C and E gene products initiates the stamens; the presence of

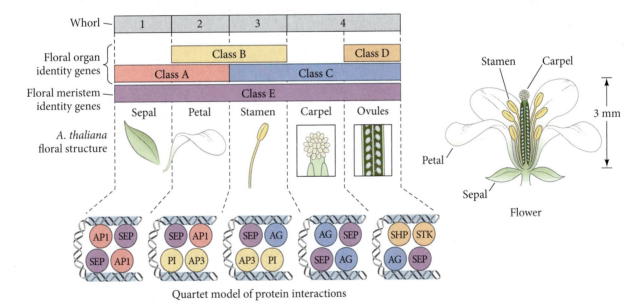

Quartet model of protein interactions

FIGURE 6.26 The floral quartet model and the underlying ABCDE model of organ identity determination in *Arabidopsis thaliana*. The bottom part of the figure depicts a version of the floral quartet model, which maintains that the five floral organ identities (sepals, petals, stamens, carpels, and ovules) are specified by the formation of floral organ-specific tetrameric complexes of MADS-domain transcription factors that bind to two nearby enhancer elements (purple), forming a DNA loop (blue) in between. A complex of two class A proteins (such as APETALA 1 [AP1]) and two class E proteins (such as SEPALLATA [SEP]) determines *sepal* identity. A complex of one class A protein, one class E protein, and one of each of the class B proteins (such as APETALA 3 [AP3] and PISTILLATA [PI]) determines *petal* identity. A complex of one class E protein, two class B proteins, and one class C protein (such as AGAMOUS [AG]) determines *stamen* identity, while a complex of two class E proteins and two class C proteins determines *carpel* identity. A complex of one class E protein, one

class C protein, and one of each of the class D proteins (SHATTER-PROOF [SHP] and SEEDSTICK [STK]) controls ovule identity. The top part of the figure illustrates the ABCDE model. In this model, flower organ specification in *A. thaliana* is controlled by five sets of floral homeotic genes providing overlapping floral homeotic functions. Class A genes are expressed in the organ primordia of the first and second whorls of the flower, class B genes in the second and third whorls, class C genes in the third and fourth whorls, class D genes in parts of the fourth whorl (ovule primordia), and class E genes throughout all four whorls. Class A and E genes specify first whorl sepals; class A, B, and E genes specify second whorl petals; class B, C, and E genes specify third whorl stamens; class C and E genes specify fourth whorl carpels; and class C, D, and E genes control development of the ovules within the fourth whorl carpels. (After Theissen et al. 2016. *Development* 143: 3259–3271 and B. A. Krizek and J. C. Fletcher. 2005. *Nat Rev Genet* 6: 688–698.)

(A)

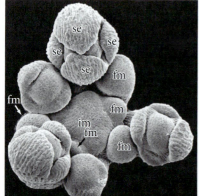

After B. A. Krizek. 2015. In *eLS*. John Wiley & Sons, Ltd: Chichester. http://doi.org/10.1002/9780470015902.a0000734.pub3

FIGURE 6.27 Specification of the floral meristem. (A) Scanning electron micrograph of an inflorescence meristem (im) and young floral meristems (fm). Four sepal primordia (se) have arisen in the older floral meristems and are indicated on one flower. (B) Expression patterns of the floral meristem identity and floral organ identity genes. *LFY*, which promotes floral meristem identity (purple), is expressed in inflorescence meristems, floral meristems, and young developing flowers. The class A gene *AP1* (red) is expressed in floral meristems, developing sepals and petals in whorls one and two of the flower and the floral pedicel. The class B genes *AP3* and *PI* (yellow) are expressed in whorls two and three, which develop into petals and stamens. The cover of the textbook shows *AP3* transcription at this stage. The class C gene *AG* (blue) is expressed in whorls three and four, which develop into stamens and carpels. The rightmost panel shows a composite: in whorl one, class A genes are expressed (red); in whorl two, class A and B genes are expressed (orange); in whorl three, class B and C genes are expressed (green); and in whorl four, class C genes are expressed (blue).

(B)

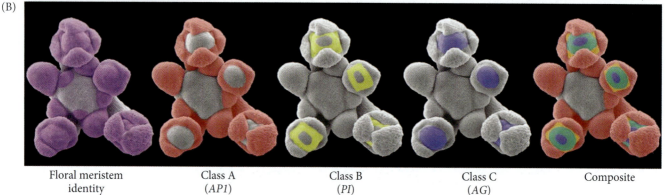

| Floral meristem identity (*LFY*) | Class A (*AP1*) | Class B (*PI*) | Class C (*AG*) | Composite |

Class D gene products with the products of C and E genes gives ovules; and when C and E genes are expressed alone, one gets carpels. The activity of Class B genes in the presence of active A and E genes gives petals.

The specific A, B, C, D, and E proteins expressed in each region come together to form tetrameric (four-member) proteins that act as transcription factors activating the genes that will form that particular organ (see Figures 6.26 and 6.27). For instance, the proteins that activate sepal-forming genes would be a class A and a class E protein. The proteins that activate the genes forming the stamens would be class B, C, and E proteins. Each of the subunits of the tetrameric protein can bind DNA, and these proteins may fold the DNA by binding to nearby enhancers. The subsequent differentiation of the flower parts involves numerous hormones as well as seasonal environmental factors (see Song et al. 2013). In congenitally dioecious plants (such as oaks and spinach, which have individuals that germinate as either males or females), this pattern is modified. In spinach plants, for instance, class B floral identity genes are expressed in high amounts in the third whorl of male plants, giving rises to anthers, while these same genes are suppressed in female plants (Pfent et al. 2005).

Gametogenesis

Unlike *Drosophila* and mammals, which have a rapid separation of their germline (gamete-producing) cells from their somatic lineages, plant germ cells are not set aside in early development.[8] Plant germ cells (like those of some invertebrates) are derived from diploid somatic cells late in development. Any meristem cell is potentially a germ cell. Remember, too, that plants have an alternation of generations. The angiosperm "plant," as we see it, comprises what would be two separate plants among the mosses and ferns (**FIGURE 6.28**). The sporophyte is the basic entity we call the plant, and it is diploid. However, within its diploid flower, a second, gametophyte generation is made. Here, some of the diploid meristem cells undergo

FLOWER OF MATURE SPOROPHYTE

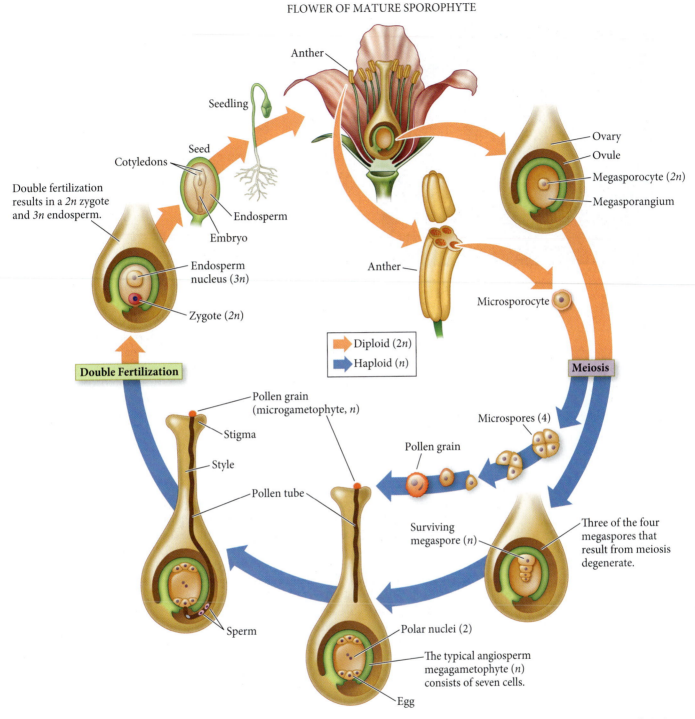

FIGURE 6.28 Life cycle of an angiosperm. The sporophyte is the dominant generation, but multicellular male and female gameto-phytes are produced within the flowers of the sporophyte. Cells of the microsporangium within the anther undergo meiosis to produce microspores. Subsequent mitotic divisions are limited, but the end result is a multicellular pollen grain. Integuments and the ovary wall protect the megasporangium. Within the megasporangium,

meiosis yields four megaspores—three small and one large. Only the large megaspore survives to produce the female gametophtye (the embryo sac). Fertilization occurs when the male gametophyte (pollen) germinates and the pollen tube grows toward the embryo sac. The sporophyte generation may be maintained in a dormant state, protected by the seed coat.

meiosis to produce haploid spores. These spores undergo mitosis to produce haploid gametophytes. Perfect flowers produce both male and female gametophytes. Some cells of the microsporangium within the anthers of the stamen undergo meiosis to produce **microspores**. These spores then undergo mitosis to produce pollen grains

(A)

© Science Photo Library/Alamy Stock Photo

(B)

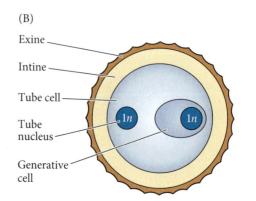

Exine

Intine

Tube cell

Tube nucleus

Generative cell

FIGURE 6.29   (A) Pollen grains have intricate surface patterns, as seen in this scanning electron micrograph of aster pollen. (B) A pollen grain consists of a cell within a cell. The generative cell will undergo division to produce two sperm cells. One will fertilize the egg, and the other will join with the polar nuclei, yielding the endosperm.

that usually contain two haploid sperm cells. Thus, in contrast to animals, plant haploid cells undergo mitosis to produce male gametes. The female gametophyte, the **embryo sac**, develops within the ovule. Here, the megasporangium undergoes meiosis to produce the haploid **megaspores**, of which one survives to produce the embryo sac. The 7-cell embryo sac contains the female gamete (the egg) as well as the central cell and accessory cells (the synergids and antipodals). The male and female gametes (the two sperm, one of which unites with the egg cell and one of which unites with the central cell) join at fertilization to form the zygote, the first cell of the next sporophyte generation.

Pollen

The stamens contain four groups of cells, called the **microsporangia** (pollen sacs), within an **anther**. The microsporangia produce the microsporocytes, which are the cells that will undergo meiosis to produce microspores, the **pollen grains** (see Figures 6.27 and 6.28). The inner wall of the pollen sac provides nourishment for the developing pollen. The pollen grain is an extremely simple multicellular structure. The outer wall of the pollen grain, the **exine**, often elaborately sculpted, is composed of resistant material provided by both the anther (sporophyte generation) and the microspore (gametophyte generation). The inner wall, the **intine**, is produced by the microspore. A mature pollen grain consists of two cells, one within the other (**FIGURE 6.29**). The **tube cell** contains a **generative cell** within it. The generative cell divides to produce two sperm. The tube cell nucleus guides pollen germination and the growth of the pollen tube after the pollen lands on the stigma of a female gametophyte. One of the two sperm will fuse with the egg cell to produce the next sporophyte generation. The second sperm will participate in the formation of the endosperm, a structure that provides nourishment for the embryo.

The ovule

The fourth whorl of organs in the flower forms the carpel, which gives rise to the female gametophyte. The carpel consists of the **stigma** (where the pollen lands), the **style**, and the **ovary** (see Figure 6.28). Following fertilization, the ovary wall will develop into the **fruit**. Thus, a fruit is the ripened ovary of the plant, a unique angiosperm structure that protects the developing embryo and often enhances seed dispersal by fruit-eating animals. Within the ovary are one or more **ovules** that contain the female gametes. Fully developed ovules are called **seeds**. The ovule has one or two outer layers of cells called the **integuments**. These enclose the **megasporangium**, which contains sporophyte cells that undergo meiosis to produce megaspores (**FIGURE 6.30**). There is a small opening in the integuments, called the **micropyle**, through which the pollen tube will grow during fertilization. The integuments develop into the **seed coat**, which protects the embryo by providing a waterproof physical barrier. Thus, when

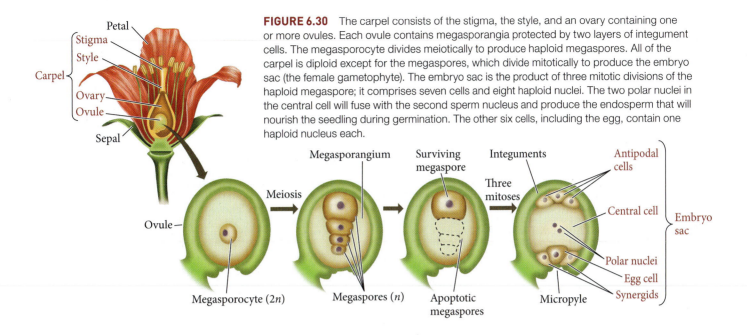

FIGURE 6.30 The carpel consists of the stigma, the style, and an ovary containing one or more ovules. Each ovule contains megasporangia protected by two layers of integument cells. The megasporocyte divides meiotically to produce haploid megaspores. All of the carpel is diploid except for the megaspores, which divide mitotically to produce the embryo sac (the female gametophyte). The embryo sac is the product of three mitotic divisions of the haploid megaspore; it comprises seven cells and eight haploid nuclei. The two polar nuclei in the central cell will fuse with the second sperm nucleus and produce the endosperm that will nourish the seedling during germination. The other six cells, including the egg, contain one haploid nucleus each.

the mature embryo disperses from the parent plant, the embryo is protected by two tissues derived from the diploid sporophyte tissue: the seed coat and the fruit.

Within the ovule, meiosis and unequal cytokinesis yield four megaspores. The largest of these megaspores undergoes three mitotic divisions to produce the female gametophyte, a 7-cell embryo sac with eight nuclei (see Figure 6.30). One of these cells is the egg, another is the central cell. Two synergid cells surround the egg, and the pollen tube enters the embryo sac by penetrating one of the synergids. The pollen tube brings with it two sperm cells. One fuses with the egg, and the other fuses with the central cell, forming the polyploid endosperm, which will nourish the embryo (and nourish you if you're eating popcorn as you study). As we will see in Chapter 7, in angiosperms this is called "double fertilization." The antipodal cells in the embryo sac will degenerate and are thought to support the endosperm.

The sex-determining mechanisms have assembled reproductive organs, and their respective gametes have been made. Plant gametes may be packaged in protective coats, but for many animals, when their gametes are released from their gonads, they are cells on the verge of death. However, if they meet, an organism with a life span of decades (and centuries, in certain plant species) can be generated. The stage is now set for one of the greatest dramas of the cycle of life—fertilization.

Closing Thoughts on the Opening Photo

This gynandromorph cardinal is split into a male half (red feathers) and a female half (light brown feathers). Half the cells are ZW (male) and half are ZZ (female; recall that birds have ZW/ZZ chromosomal sex determination), probably resulting from the egg providing too much cytoplasm to a polar body during meiosis, the subsequent fertilization of the polar body by a separate sperm, and fusion into a single mosaic embryo. In birds, each cell makes its own sexual decision. In mammals, hormones play a much larger role in making a unified phenotype, and similar male/female chimeras don't arise (see Zhao et al. 2010; Peer and Motz 2014).

Photo courtesy of Brian D. Peer

Endnotes

[1] These embryos did not form functional sperm—but they were not expected to. The presence of two X chromosomes prevents sperm formation in XXY mice and men, and the transgenic mice lacked the rest of the Y chromosome, which contains genes needed for spermatogenesis.

[2] This anatomical phenotype is named for Hermaphroditos, a young man in Greek mythology whose beauty inflamed the ardor of the water nymph Salmacis. She wished to be united with him forever, and the gods, in their literal fashion, granted her wish. Hermaphroditism is often considered to be one of the "intersex" conditions discussed later in the chapter.

[3] While the binary categories of male and female have been convenient, the biological recognition of intersexuality is not new (Fausto-Sterling 1993; Suskin 2002; Ainsworth 2015). The "intersex" language used to group these conditions is being debated. Some activists, physicians, and parents wish to eliminate the term "intersex" to avoid confusion of these anatomical conditions with identity issues such as homosexuality. They prefer to call these conditions "disorders/differences of sex development" (Kim and Kim 2012). In contrast, other activists do not want to medicalize this condition and find the "disorder" category offensive to individuals who do not feel there is anything wrong with their health. For a more detailed analysis of intersexuality, see Dreger 2000, Dreger et al. 2005, and Austin et al. 2011; also see www.isna.org/faq/conditions.

[4] Since the Tra protein is sexually distinguished (functional in females but not in males), it can also be used in sexual differentiation. This happens in larval Drosophila, in which the faster enlargement of female cells is due to Tra synthesis in the brain and fat body. This is independent of Dsx (Rideout et al. 2 Developmental Biology 12e 015; Mathews et al. 2017).

[5] Temperature-dependent sex determination puts turtle and crocodilian species at risk during the present period of global climate change (Jensen et al. 2018). The evolutionary advantages and disadvantages of temperature-dependent sex determination are discussed in Chapter 24.

[6] The terms *centromere* and *kinetochore* are often used interchangeably, but in fact the kineto chore is the complex protein structure that assembles on a sequence of DNA known as the centromere.

[7] Plants have sex. This fact was not known in Europe until Rudolph Camerarius demonstrated it in 1694. However, not only did the ancient Babylonians in the fourth century BCE know that plants had sex, but this esoteric knowledge became a secret of their agricultural success. Figs are one of the angiosperms that have male and female organs in different plants. Since only the female trees bore fruit, date farmers planted just a few male trees, then hand-pollinated the many female trees. This practice greatly increased the fruit yield per acre, and such pollination events became associated with spring fertility festivals (Roberts 1929). Erasmus Darwin (who claimed that sexual reproduction was "the masterpiece of Nature") wrote an entire plant taxonomy textbook in rhyming (and often erotic) couplets wherein he likened the stamen to men and the carpels to women.

[8] By convention, names of plant genes are fully capitalized.

[9] While the germ line in *Drosophila* and mammals is set aside rapidly, there are numerous species (possibly most invertebrates, including many insects) where the germ line arises much later, from somatic cells (Buss 1987; Ewen-Campen et al. 2013).

Snapshot Summary
Sex Determination and Gametogenesis

1. In mammals, gonadal sex determination (the determination of gonadal sex) is a function of the sex chromosomes. XX individuals are usually females, and XY individuals are usually males.

2. The mammalian Y chromosome plays a key role in male sex determination. XY and XX embryos both have a bipotential gonad. In XY embryos, Sertoli cells differentiate and enclose the germ cells within testis cords. The interstitial mesenchyme generates other testicular cell types, including the testosterone-secreting Leydig cells.

3. In XX mammals, the germ cells become surrounded by follicle cells in the cortex (outer portion) of the gonadal rudiment. The epithelium of the follicles becomes the granulosa cells; the mesenchyme generates the thecal cells.

4. In humans, the *SRY* gene encodes the testis-determining factor on the Y chromosome, a nucleic acid-binding protein that functions as a transcription factor to activate the evolutionarily conserved *SOX9* gene.

5. The *Sox9* gene product can also initiate testis formation. Functioning as a genital ridge transcription factor, it binds to the gene encoding anti-Müllerian hormone (AMH) and other genes whose products promote testis development. Fgf9 and Sox9 proteins have a positive feedback loop that activates testicular development and suppresses ovarian development.

6. Wnt4 and Rspo1 are involved in mammalian ovary formation. These proteins upregulate production of β-catenin; the functions of β-catenin include promoting the ovarian pathway of development while blocking the testicular pathway of development.

7. Secondary sex determination in mammals involves factors produced by the developing gonads. In males, the Müllerian duct is destroyed by the AMH produced by the Sertoli cells, while testosterone produced by the Leydig cells enables the Wolffian duct to differentiate into the vas deferens and seminal vesicles. In females, the Wolffian duct degenerates with the lack of testosterone, whereas the Müllerian duct persists and is differentiated by estrogen into the oviducts, uterus, cervix, and upper portion of the vagina. Individuals with mutations of these hormones or their receptors may have a discordance between their gonadal sex and secondary sex characteristics..

8. In *Drosophila*, sex is determined by the number of X chromosomes in the cell; the Y chromosome does not play a role in sex determination. There are no sex hormones, so most cells make an independent sex determination "decision."

9. The *Drosophila Sex-lethal* (*Sxl*) gene is activated in females, but the Sxl protein does not form in males because of translational termination. Sxl protein acts as an

RNA splicing factor to splice an inhibitory exon from the *transformer* (*tra*) transcript. Therefore, female flies have an active Tra protein but males do not.

10. The Tra protein also acts as an RNA splicing factor to splice exons from the *doublesex* (*dsx*) transcript. The *dsx* gene is transcribed in both XX and XY cells, but its pre-mRNA is processed to form different mRNAs, depending on whether Tra protein is present. The proteins translated from both *dsx* messages are active, and they activate or inhibit transcription of a set of genes involved in producing the sexually dimorphic traits of the fly.

11. In many invertebrates, fish, turtles, and alligators, sex is often determined by environmental agents such as temperature.

12. In animals, the precursors of the gametes are the primordial germ cells (PGCs). In most species, the PGCs form outside the gonads and migrate into the gonads during development.

13. The cytoplasm of the PGCs in many species contains inhibitors of transcription and translation, such that the PGCs are both transcriptionally and translationally silent.

14. In most animals studied, the coordination of germline sex (sperm/egg) with somatic sex (male/female) is achieved by signals coming from the gonad (testis/ovary).

15. In humans and mice, germ cells entering ovaries initiate meiosis while in the embryo; germ cells entering testes do not initiate meiosis until puberty.

16. The first division of meiosis separates homologous chromosomes, creating haploid cells. The second division of meiosis splits the kinetochore and separates sister chromatids.

17. Spermatogenic meiosis in mammals is characterized by the production of four gametes per meiosis and by the absence of meiotic arrest. Oogenic meiosis is characterized by the production of one gamete per meiosis and by a prolonged first meiotic prophase that allows the egg to grow.

18. In male mammals, the PGCs generate stem cells that last for the life of the organism. PGCs do not become stem cells in female mammals (although in many other animal groups, PGCs do become germ stem cells in the ovaries).

19. In female mammals, germ cells initiate meiosis and are retained in the first meiotic prophase (dictyate stage) until ovulation. In this stage, they synthesize mRNAs and proteins that will be used for gamete recognition and early development.

20. Certain principles of organogenesis are easily seen in gonad development: (1) gene products that promote one pathway often act to inhibit another possible pathway (think Sox9 and β-catenin); (2) a gene, once activated by one signal, can produce other signals that keep it on, allowing its activity to be independent of the original signal (think Sox9 again); and (3) an activator is often the inhibitor of an inhibitor (think oocyte meiotic spindles).

21. In angiosperm plants, the male- and female-producing gametophyte generations derive from the third and fourth whorls of the perfect flower, respectively. They are specified by a transcription factor complex made by combinations of several proteins.

22. The pollen grains contain two sperm, while the ovule contains a single female reproductive cell.

Go to oup.com/he/barresi12xe for Further Developments, Scientists Speak interviews, Watch Development videos, Dev Tutorials, and complete bibliographic information for all literature cited in this chapter.

Fertilization
Beginning a New Organism

How do the sperm and egg nuclei find each other?

From journal cover associated with J. Holy and G. Schatten. 1991. *Dev Biol* 147: 343–353

FERTILIZATION IS THE PROCESS WHEREBY THE GAMETES—sperm and egg—fuse together to begin the creation of a new organism. Fertilization accomplishes two separate ends: sex (the combining of genes derived from two parents) and reproduction (the generation of a new organism). Thus, the first function of fertilization is to transmit genes from parents to offspring, and the second is to initiate in the egg cytoplasm those reactions that permit development to proceed.

In this chapter, the term *fertilization* will include all the processes that occur after the gametes leave their respective gonads until the time that the nuclei fuse and the zygote is activated. While the actual fusion of sperm and egg has been called *amphimixis* or *syngamy* (Kondrashov 2018), the term *fertilization* emphasizes that in some species, such as humans, neither the sperm nor the egg are mature cells when they leave their respective gonads. Thus, the events that activate and mature the sperm and egg are also critical to discuss.

We will discuss only three types of fertilization in this chapter: (1) external fertilization in sea urchins, the animal group whose fertilization we know the best, (2) internal fertilization in mammals, and (3) double fertilization in angiosperm plants. These three modes of fertilization provide some hint of the truly wonderful ways in which evolution has integrated reproduction with the origin of genetic diversity. Although the details of fertilization vary from species to species, fertilization generally consists of four major events:

1. *Contact and recognition* between sperm and egg. In most cases, this ensures that the sperm and egg are of the same species.

2. *Regulation* of sperm entry into the egg. Only one sperm nucleus can ultimately unite with the egg nucleus. This is usually accomplished by allowing only one sperm to enter the egg and actively inhibiting any others from entering.

3. *Fusion* of the genetic material of sperm and egg.

4. *Activation* of egg metabolism to start development.

Structure of the Gametes

Before we investigate these aspects of fertilization, we need to consider the structures of the sperm and egg—the two cell types specialized for fertilization. The sperm and egg are in some ways very similar and in other ways very different. They both have haploid genomes, as described in Chapter 6. They also have cell membranes that have been altered to recognize and fuse with the other gamete.

However, while the sperm has eliminated most of its cytoplasm, the egg has maintained its cytoplasm and grown even larger. While the sperm is essentially a haploid nucleus with a propulsion system and an egg-recognizing membrane, the egg has equipped its haploid nucleus with a cytoplasm full of ribosomes, mitochondria, and enzymes critical for development.

Sperm

Sperm were discovered in the 1670s, but their role in fertilization was not discovered until the mid-1800s. It was only in the 1840s, after Albert von Kölliker described the formation of sperm from cells in the adult testes, that fertilization research could really begin. Even so, von Kölliker denied that there was any physical contact between sperm and egg (Farley 1982; Pinto-Correia 1997). He believed that the sperm excited the egg to develop in much the same way a magnet communicates its presence to iron. The first description of fertilization was published in 1847 by Karl Ernst von Baer, who showed the union of sperm and egg in sea urchins and tunicates (Raineri and Tammiksaar 2013). He described the fertilization envelope, the migration of the sperm nucleus to the center of the egg, and the subsequent early cell divisions of development. (See **Further Development 7.1, The Origins of Fertilization Research**, online.)

SPERM ANATOMY   Each sperm cell consists of a haploid nucleus, a propulsion system to move the nucleus, and a sac of enzymes that enable the nucleus to enter the egg. In most species, almost all of the cell's cytoplasm is eliminated during sperm maturation, leaving only certain organelles that are modified for spermatic function (**FIGURE 7.1A,B**). During the course of maturation, the sperm's haploid nucleus becomes very streamlined and its DNA becomes tightly compressed. In front or to the side of this compressed haploid nucleus lies the **acrosomal vesicle**, or **acrosome** (**FIGURE 7.1C**). The acrosome is derived from the cell's Golgi apparatus and contains enzymes that digest proteins and complex sugars. Enzymes stored in the acrosome can digest a path through the outer coverings of the egg. In many species, a region of actin proteins lies between the sperm nucleus and the acrosomal vesicle. These proteins are used to extend a fingerlike **acrosomal process** from the sperm during the early stages of fertilization. In sea urchins and numerous other species, recognition between sperm and egg involves molecules on the membrane of the acrosomal process. Together, the acrosome and nucleus constitute the **sperm head**.

The means by which sperm are propelled vary according to how the species has adapted to environmental conditions. In most species, an individual sperm is able to travel by whipping its flagellum. The major motor portion of the flagellum is the **axoneme**, a structure formed by microtubules emanating from one of the two centrioles at the base of the sperm nucleus (Avidor-Reiss and Fishman 2019). The core of the axoneme consists of two central microtubules surrounded by a row of nine doublet microtubules. These microtubules are made exclusively of the dimeric protein tubulin. The other centriole is also important, as it will enter the egg to establish the mitotic spindle of first cleavage (Fishman et al. 2018).

Although tubulin is the basis for the structure of the flagellum, other proteins are also critical for flagellar function. The force for sperm propulsion is provided by **dynein**, a protein attached to the microtubules. Dynein is an ATPase—an enzyme that hydrolyzes ATP, converting the released chemical energy into mechanical energy that propels the sperm.[1] This energy allows the active sliding of the outer doublet of microtubules, causing the flagellum to bend (Ogawa et al. 1977; Shinyoji et al. 1998). The ATP needed to move the flagellum and propel the sperm comes from rings of mitochondria located in the **midpiece** of the sperm (see Figure 7.1A). In many species (notably mammals), a layer of dense fibers has interposed itself between the mitochondrial sheath and the cell membrane. This fiber layer stiffens the sperm tail. Because the thickness of this layer decreases toward the tip, the fibers probably increase the efficiency of forward movement by preventing the sperm head from being whipped around too suddenly. Thus, the sperm cell has undergone extensive modification for the transport of its nucleus to the egg.

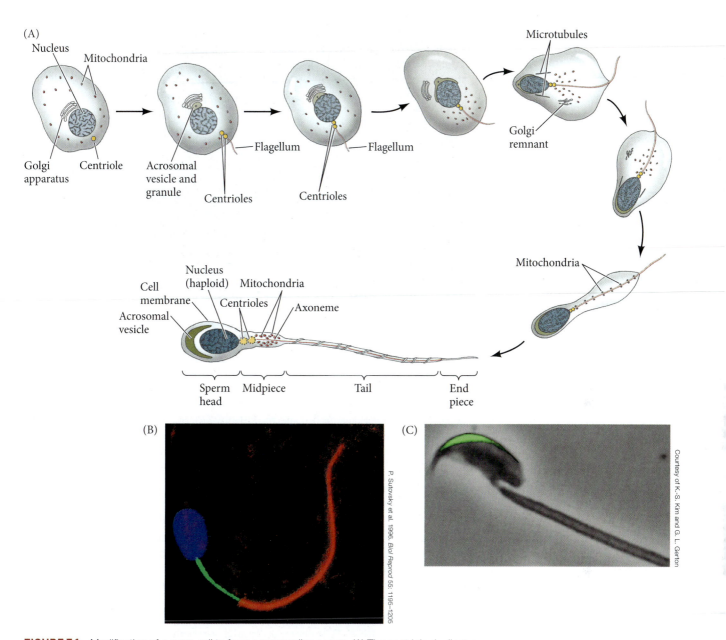

FIGURE 7.1 Modification of a germ cell to form a mammalian sperm. (A) The centriole duplicates, using one centriole to organize a long flagellum at what will be the posterior end of the sperm; the other centriole will enter the egg at fertilization. The Golgi apparatus forms the acrosomal vesicle at the future anterior end. Mitochondria collect around the flagellum near the base of the haploid nucleus and become incorporated into the midpiece ("neck") of the sperm. The remaining cytoplasm is jettisoned, and the nucleus condenses. The size of the mature sperm has been enlarged relative to the other stages. (B) Mature bull sperm. The DNA is stained blue, mitochondria are stained green, and the tubulin of the flagellum is stained red. (C) The acrosomal vesicle of this mouse sperm is stained green by the fusion of proacrosin with green fluorescent protein (GFP). (A after Y. Clermont and C. P. Leblond. 1955. *Am J Anat* 96: 229–253.)

The egg

CYTOPLASM AND NUCLEUS   All the material necessary to begin growth and development must be stored in the egg, or ovum.[2] Whereas the sperm eliminates most of its cytoplasm as it matures, the developing egg (called the oocyte before it reaches the stage of meiosis at which it is fertilized) not only conserves the material it has, but actively accumulates more. The meiotic divisions that form the oocyte conserve

FIGURE 7.2 Structure of the sea urchin egg at fertilization. Sperm can be seen in the jelly coat and attached to the vitelline envelope. The female pronucleus is apparent within the egg cytoplasm.

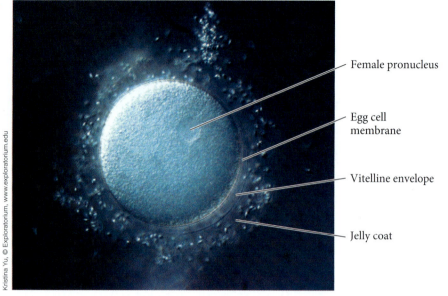

Female pronucleus

Egg cell membrane

Vitelline envelope

Jelly coat

Kristina Yu, © Exploratorium, www.exploratorium.edu

its cytoplasm rather than giving half of it away; at the same time, the oocyte either synthesizes or absorbs proteins such as yolk that act as food reservoirs for the developing embryo. Birds' eggs are enormous single cells, swollen with accumulated yolk (see Figure 12.2). Even eggs with relatively sparse yolk are large compared with sperm. The volume of a sea urchin egg is about 200 picoliters (2×10^{-4} mm^3), more than 10,000 times the volume of sea urchin sperm (**FIGURE 7.2**). So even though sperm and egg have equal haploid *nuclear* components, the egg accumulates a remarkable cytoplasmic storehouse during its maturation. This cytoplasmic trove includes the following:

- **Nutritive proteins**. The early embryonic cells must have a supply of energy and amino acids. In many species, this is accomplished by accumulating yolk proteins in the egg. Many of these yolk proteins are made in other organs (e.g., liver, fat bodies) and travel through the maternal blood to the oocyte.

- **Ribosomes and tRNA**. The early embryo must make many of its own structural proteins and enzymes, and in some species there is a burst of protein synthesis soon after fertilization. This rapid protein synthesis is accomplished by those ribosomes and tRNAs that exist in the egg *before* fertilization. The developing egg has special mechanisms for synthesizing ribosomes; certain amphibian oocytes produce as many as 10^{12} ribosomes during their meiotic prophase.

- **Messenger RNAs**. The oocyte accumulates not only proteins but also mRNAs that encode proteins for the early stages of development. It is estimated that sea urchin eggs contain thousands of different types of mRNA that remain repressed until after fertilization.

- **Morphogenetic factors**. Molecules that direct the differentiation of cells into certain cell types are present in the egg. These include transcription factors and paracrine factors. In many species, these factors are localized in different regions of the egg and become segregated into different cells during cleavage.

- **Protective chemicals**. The embryo cannot run away from predators or move to a safer environment, so it must be equipped to deal with threats. Many eggs contain ultraviolet filters and DNA repair enzymes that protect them from sunlight, and some eggs contain molecules that potential predators find distasteful. The yolk of bird eggs contains antibodies that protect the embryo against microbes.

Within the enormous volume of egg cytoplasm resides a large nucleus (see Figure 7.2). In a few species (such as sea urchins), this female pronucleus is already haploid at the time of fertilization. In other species (including many worms and insects), the egg nucleus is still diploid—the sperm enters before the egg's meiotic divisions are completed (**FIGURE 7.3**). In these species, the final stages of egg meiosis will take place

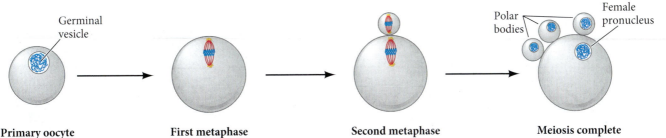

Primary oocyte	First metaphase	Second metaphase	Meiosis complete
The roundworm *Ascaris*	The nemertean worm *Cerebratulus*	The lancelet *Branchiostoma*	Cnidarians (e.g., anemones)
The mesozoan *Dicyema*	The polychaete worm *Chaetopterus*	Amphibians	Sea urchins
The sponge *Grantia*	The mollusk *Dentalium*	Most mammals	
The polychaete worm *Myzostoma*	The core worm *Pectinaria*	Fish	
The clam worm *Nereis*	Many insects		
The clam *Spisula*	Seastars		
The echiuroid worm *Urechis*			
Dogs and foxes			

FIGURE 7.3 Stages of egg maturation at the time of sperm entry in different animal species. Note that in most species, sperm entry occurs before the egg nucleus has completed meiosis. The germinal vesicle is the name given to the large diploid nucleus of the primary oocyte. The polar bodies are nonfunctional cells produced by meiosis (see Chapter 6). (After C. R. Austin. 1965. *Fertilization*. Prentice-Hall: Englewood Cliffs, NJ.)

after the sperm's nuclear material—the male pronucleus—is already inside the egg cytoplasm. (See **Further Development 7.2, The Egg and Its Environment**, online.)

CELL MEMBRANE AND EXTRACELLULAR ENVELOPE The cell membrane enclosing the egg cytoplasm regulates the flow of specific ions during fertilization and must be capable of fusing with the sperm cell membrane. Outside this egg cell membrane is an extracellular matrix that forms a fibrous mat around the egg and is often involved in sperm-egg recognition (Wassarman and Litscher 2016). In invertebrates, this structure, usually called the **vitelline envelope** (**FIGURE 7.4A**), contains several different glycoproteins. It is supplemented by extensions of membrane glycoproteins from the cell membrane and by proteinaceous "posts" that adhere the vitelline envelope to the cell membrane (Mozingo and Chandler 1991). The vitelline envelope is essential for the species-specific binding of sperm. Many types of eggs also have a layer of egg jelly outside the vitelline envelope. This glycoprotein meshwork can have numerous functions, but most commonly it is used to attract and/or activate sperm.

Lying immediately beneath the cell membrane of most eggs is a thin layer (about 5 μm) of gel-like cytoplasm called the **cortex**. It is stiffer than the internal cytoplasm and contains high concentrations of globular actin molecules. During fertilization, these actin molecules polymerize to form long cables of actin microfilaments. **Microfilaments** are necessary for cell division. They also extend the egg surface into small projections called **microvilli**, which may aid sperm entry into the cell (**FIGURE 7.4B**).

FIGURE 7.4 Sea urchin egg cell surfaces. (A) Scanning electron micrograph of an egg before fertilization. The cell membrane is exposed where the vitelline envelope has been torn. (B) Transmission electron micrograph of an unfertilized egg, showing microvilli and the cell membrane, which are closely covered by the vitelline envelope. A cortical granule lies directly beneath the cell membrane.

(A)   (B)

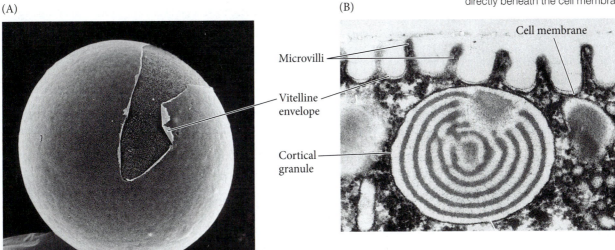

(A)

(B)

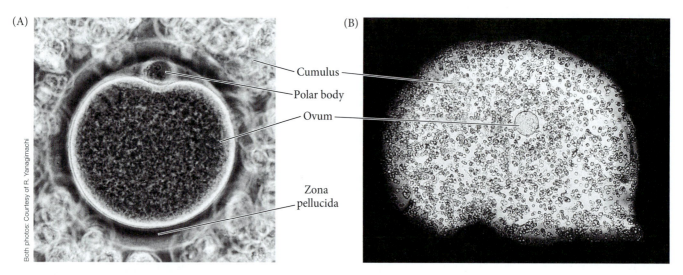

Cumulus

Polar body

Ovum

Zona pellucida

Both photos: Courtesy of R. Yanagimachi

FIGURE 7.5 Mammalian eggs immediately before fertilization. (A) The hamster egg, or ovum, is encased in the zona pellucida, which in turn is surrounded by the cells of the cumulus. A polar body cell, produced during meiosis, is visible within the zona pellucida. (B) At lower magnification, a mouse oocyte is shown surrounded by the cumulus. Colloidal carbon particles (India ink, seen here as the black background) are excluded by the hyaluronic acid matrix.

Also within the cortex are the **cortical granules**. These membrane-bound, Golgi-derived structures contain proteolytic enzymes and are thus homologous to the acrosomal vesicle of the sperm. However, whereas a sea urchin sperm contains just one acrosomal vesicle, each sea urchin egg contains approximately 15,000 cortical granules. In addition to containing digestive enzymes, the cortical granules contain glycosaminoglycans, adhesive glycoproteins, and hyalin protein. As we will soon describe, the enzymes and glycosaminoglycans help prevent polyspermy—that is, they prevent additional sperm from entering the egg after the first sperm has entered—while hyalin and the other adhesive glycoproteins that surround the early embryo provide physical support for cleavage-stage blastomeres.

In mammalian eggs, the extracellular envelope is a separate, thick matrix called the **zona pellucida**. The mammalian egg is also surrounded by a layer of cells called the **cumulus** (**FIGURE 7.5**), which is made up of the ovarian follicular cells that were nurturing the egg at the time of its release from the ovary. Mammalian sperm have to get past these cells to fertilize the egg. The innermost layer of cumulus cells, immediately adjacent to the zona pellucida, is called the **corona radiata**.

Recognition of egg and sperm

The interaction of sperm and egg generally proceeds according to five steps (Vacquier 1998):

1. *Chemoattraction* of the sperm to the egg by soluble molecules secreted by the egg
2. *Binding of the sperm* to the extracellular matrix (jelly or zona pellucida) of the egg
3. *Exocytosis* of the sperm acrosomal vesicle and release of its enzymes
4. *Passage of the sperm* through the extracellular matrix to the egg cell membrane
5. *Fusion* of the egg and sperm cell membranes

After these steps are accomplished, the haploid sperm and egg nuclei can meet and the reactions that initiate development can begin. In this chapter, we will focus on these events in three well-studied organisms: sea urchins, which undergo external fertilization; mice, which undergo internal fertilization; and *Arabidopsis thaliana*, an angiosperm plant that undergoes "double fertilization."

External Fertilization in Sea Urchins

Sperm attraction: Action at a distance

The events of sperm-egg meeting and fusing are outlined in **FIGURE 7.6**. Like many marine organisms, sea urchins release their gametes into the environment. That environment may be as small as a tide pool or as large as an ocean and is shared with other species that may shed their gametes at the same time. Such organisms are faced with two problems: How can sperm and eggs meet in such a dilute concentration, and how can sperm be prevented from attempting to fertilize eggs of another species? In addition to the production of enormous numbers of gametes, two major mechanisms have evolved to solve these problems: species-specific sperm attraction and species-specific sperm activation.

Species-specific sperm attraction has been documented in numerous species, including cnidarians, mollusks, echinoderms, amphibians, and urochordates (Miller 1985; Yoshida et al. 1993; Burnett et al. 2008). In many species, sperm are attracted toward eggs of their species by **chemotaxis**—that is, by following a gradient of a chemical secreted by the egg. These oocytes control not only the type of sperm they attract, but also the time at which they attract them, releasing the chemotactic factor only after they reach maturation (Miller 1978).

In sea urchins, the chemotactic agents are small peptides called **sperm-activating peptides (SAPs)** that diffuse away from the egg jelly into the surrounding seawater. One such SAP is **resact**, a 14-amino acid peptide that has been isolated from the egg jelly of the sea urchin *Arbacia punctulata* (Ward et al. 1985). It has a profound effect at very low concentrations (**FIGURE 7.7**), such that the binding of even a single resact molecule may be enough to provide direction for the sperm, which swim up a concentration gradient of this compound until they reach the egg (Kaupp et al. 2003; Kirkman-Brown et al. 2003). (See **Further Development 7.3, Mechanisms of Sperm Chemotaxis**, online.)

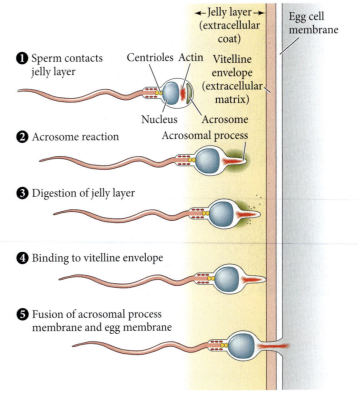

FIGURE 7.6 Summary of events leading to the fusion of egg and sperm cell membranes in sea urchin fertilization, which is external. (1) The sperm is chemotactically attracted to and activated by the egg. (2,3) Contact with the egg jelly triggers the acrosome reaction, allowing the acrosomal process to form and release proteolytic enzymes. (4) The sperm adheres to the vitelline envelope and lyses a hole in it. (5) The sperm adheres to the egg cell membrane and fuses with it. The sperm pronucleus can now enter the egg cytoplasm.

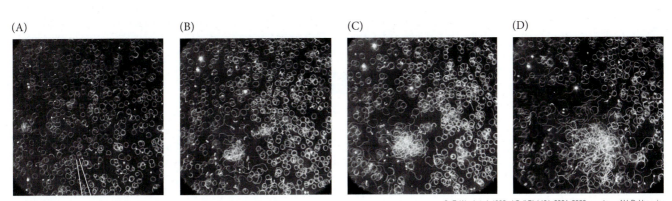

G. E. Ward et al. 1985. *J Cell Biol* 101: 2324–2329, courtesy of V. D. Vacquier

FIGURE 7.7 Sperm chemotaxis in the sea urchin *Arbacia punctulata*. One nanoliter of a 10-n*M* solution of resact is injected into a 20-μL drop of sperm suspension. (A) A 1-second photographic exposure showing sperm swimming in tight circles before the addition of resact. The position of the injection pipette is shown by the white lines. (B–D) Similar 1-second exposures showing migration of sperm to the center of the resact gradient 20, 40, and 90 seconds after injection.

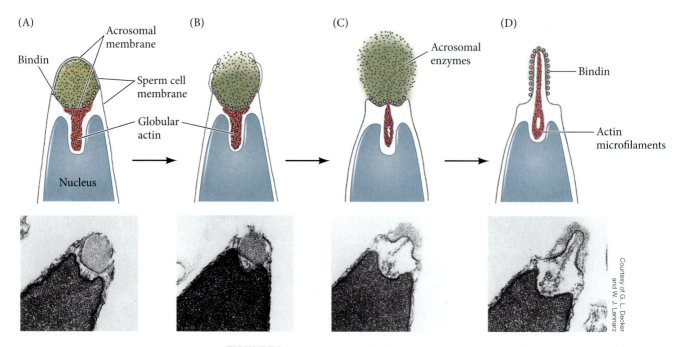

FIGURE 7.8 Acrosome reaction in sea urchin sperm. (A–C) The portion of the acrosomal membrane lying directly beneath the sperm cell membrane fuses with the cell membrane to release the contents of the acrosomal vesicle. (D) The actin molecules assemble to produce microfilaments, extending the acrosomal process outward. Photographs of the acrosome reaction in sea urchin sperm are shown below the diagrams. (After R. G. Summers and B. L. Hylander. 1974. *Cell Tissue Res* 150: 343–368.)

The acrosome reaction

A second interaction between sperm and egg jelly results in the **acrosome reaction**. In most marine invertebrates, this has two components: the fusion of the acrosomal vesicle with the sperm cell membrane (an exocytosis that results in the release of the contents of the acrosomal vesicle), and the extension of the cellular protrusion called the acrosomal process (Dan 1952; Colwin and Colwin 1963). The acrosome reaction in sea urchins is initiated by contact of the sperm with the egg jelly (**FIGURE 7.8**). The protein-digesting enzymes released from the acrosome digest a path through the jelly coat to the egg cell surface. Once the sperm reaches the egg surface, the acrosomal process adheres to the vitelline envelope and tethers the sperm to the egg. It is possible that large complexes of acrosomal protein-digesting enzymes coat the acrosomal process, allowing it to digest the vitelline envelope at the point of attachment and proceed toward the egg cell membrane (Yokota and Sawada 2007).

FURTHER DEVELOPMENT

MECHANISMS OF THE SEA URCHIN ACROSOME REACTION   In sea urchins, the acrosome reaction is initiated by sulfate-containing polysaccharides in the egg jelly that bind to specific receptors located directly above the acrosomal vesicle on the sperm cell membrane. These polysaccharides are often highly species-specific, and egg jelly factors from one species of sea urchin generally fail to activate the acrosome reaction even in closely related species (**FIGURE 7.9**; Hirohashi and Vacquier 2002; Hirohashi et al. 2002; Vilela-Silva et al. 2008). Thus, activation of the acrosome reaction serves as a barrier to interspecies (and thus unviable) fertilizations. This is important when numerous species inhabit the same habitat and when their spawning seasons overlap. (See **Further Development 7.4, Sea Urchin Acrosome Reaction and Sperm Binding**, online.)

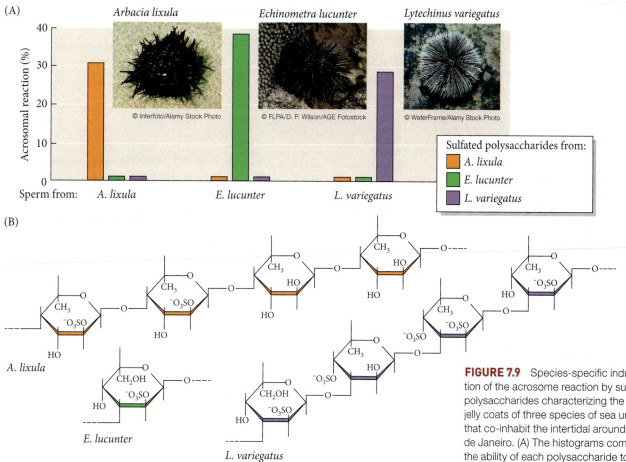

FIGURE 7.9 Species-specific induction of the acrosome reaction by sulfated polysaccharides characterizing the egg jelly coats of three species of sea urchins that co-inhabit the intertidal around Rio de Janeiro. (A) The histograms compare the ability of each polysaccharide to induce the acrosome reaction in the different species of sperm. (B) Chemical structures of the acrosome reaction-inducing sulfated polysaccharides reveal their species-specificity. (After A. P. Alves et al. 1997. *J Biol Chem* 272: 6965–6971.)

Recognition of the egg's extracellular coat

The sea urchin sperm's encounter with components of the egg's jelly coat provides the first set of species-specific recognition events (i.e., sperm attraction, activation, and acrosome reaction). Another critical species-specific binding event must occur once the sperm has penetrated the egg jelly and its acrosomal process contacts the surface of the egg (**FIGURE 7.10A**). The acrosomal protein mediating this recognition in sea urchins is an insoluble, 30,500-Da protein called **bindin**. In 1977, Vacquier and co-workers isolated bindin from the acrosome of *Strongylocentrotus purpuratus* and found it to be capable of binding to dejellied eggs of the same species (**FIGURE 7.10B**). Furthermore, sperm bindin, like egg jelly polysaccharides, is usually species-specific: bindin isolated from the acrosomes of *S. purpuratus* binds to its own dejellied eggs but not to those of *S. franciscanus* (**FIGURE 7.10C**; Glabe and Vacquier 1977; Glabe and Lennarz 1979), and the protein sequences of the bindin molecules have been shown to be species-specific (Kamei and Glabe 2003).

Fusion of the egg and sperm cell membranes

Once the sperm has traveled to the egg and undergone the acrosome reaction, the fusion of the sperm cell membrane with the egg cell membrane can begin (**FIGURE 7.11**). Sperm-egg fusion appears to cause the polymerization of actin in the egg to form a fertilization cone (Summers et al. 1975). Homology between the egg and the sperm is again demonstrated, since the sperm's acrosomal process also appears to be formed by the polymerization of actin. Actin from the gametes forms a connection that widens the cytoplasmic bridge between the egg and the sperm. The sperm nucleus, a centriole, and the tail pass through this bridge.

(A)

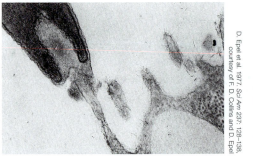

D. Epel et al. 1977. *Sci Am* 237: 128–138, courtesy of F. D. Collins and D. Epel

(B) DAB precipitate
(indicates bindin present)

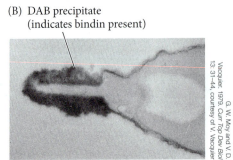

Vacquier, 1979. *Curr Top Dev Biol* 13: 31–44, courtesy of V. Vacquier

(C)

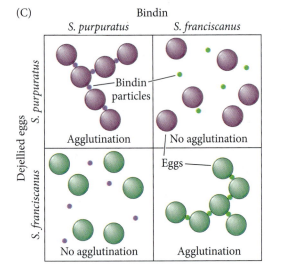

Bindin

FIGURE 7.10 Species-specific binding of the acrosomal process to the egg surface in sea urchins. (A) Contact of a sperm acrosomal process with an egg microvillus. (B) Bindin (stained black by antibodies against it) is seen to be localized to the acrosomal process after the acrosome reaction. (C) In vitro model of species-specific binding. The agglutination of dejellied eggs by bindin was measured by adding bindin particles to a plastic well containing a suspension of eggs. After 2–5 minutes of gentle shaking, the wells were photographed. Each bindin bound to and agglutinated only eggs from its own species. (C based on photographs in C. G. Glabe and V. D. Vacquier. 1977. *Nature* 267: 836–838.)

Fusion is an active process, often mediated by specific "fusogenic" proteins. In sea urchins, bindin plays a second role as a fusogenic protein. In addition to recognizing the egg, bindin contains a long stretch of hydrophobic amino acids near its amino terminus, and this region is able to fuse phospholipid vesicles in vitro (Ulrich et al. 1999; Gage et al. 2004). Under the ionic conditions present in the mature unfertilized egg, bindin can cause the sperm and egg cell membranes to fuse.

Prevention of polyspermy: One egg, one sperm

As soon as one sperm enters the egg, the fusibility of the egg membrane—which was necessary to get the sperm inside the egg—becomes a dangerous liability. In the normal case—**monospermy**—only one sperm enters the egg, and the haploid sperm nucleus combines with the haploid egg nucleus to form the diploid nucleus of the fertilized egg (zygote), thus restoring the chromosome number appropriate for the species. During cleavage, the centriole provided by the sperm divides to form the two poles of the mitotic spindle while the egg-derived centriole is degraded.

In most animals, any sperm that enters the egg can provide a haploid nucleus and a centriole. The entrance of multiple sperm—**polyspermy**—leads to disastrous consequences in most organisms. In sea urchins, fertilization by two sperm results in a triploid nucleus, in which each chromosome is represented three times rather than twice. Worse, each sperm's centriole divides to form the two poles of a mitotic apparatus, so instead of a bipolar mitotic spindle separating the chromosomes into two cells, the triploid chromosomes may be divided into as many as four cells, with some cells receiving extra copies of

FIGURE 7.11 Scanning electron micrographs (A–C) of the entry of sperm into sea urchin eggs. (A) Contact of sperm head with egg microvillus through the acrosomal process. (B) Formation of fertilization cone. (C) Internalization of sperm within the egg. (D) Transmission electron micrograph of sperm internalization through the fertilization cone.

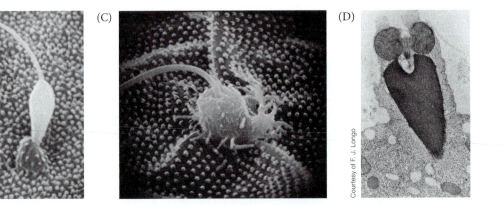

(A) Photos A–C from G. Schatten and D. Mazia. 1976. *Exp Cell Res* 98: 325–337

(B)

(C)

(D) Courtesy of F. J. Longo

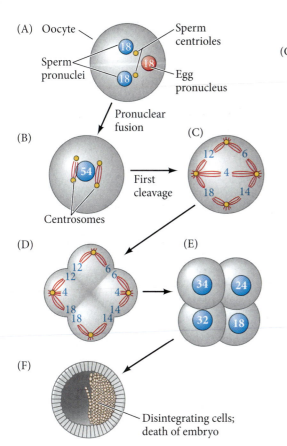

(A) Oocyte, Sperm centrioles, Sperm pronuclei, Egg pronucleus

Pronuclear fusion

(B) Centrosomes

(C)

First cleavage

(D)

(E)

(F) Disintegrating cells; death of embryo

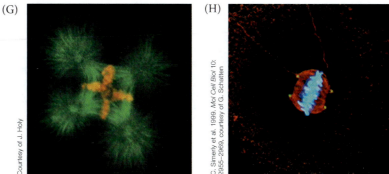

(G) Courtesy of J. Holy

(H) C. Simerly et al. 1999. *Mol Cell Biol* 10: 2955–2968, courtesy of G. Schatten

FIGURE 7.12 Aberrant development in a dispermic sea urchin egg. (A) Fusion of three haploid nuclei, each containing 18 chromosomes, and the division of the two sperm centrioles to form four centrosomes (mitotic poles). (B,C) The 54 chromosomes randomly assort on the four spindles. (D) At anaphase of the first division, the duplicated chromosomes are pulled to the four poles. (E) Four cells containing different numbers and types of chromosomes are formed, thereby causing (F) the early death of the embryo. (G) First metaphase of a dispermic sea urchin egg akin to that in (D). The microtubules are stained green; the DNA stain appears orange. The triploid DNA is being split into four chromosomally unbalanced cells instead of the normal two cells with equal chromosome complements. (H) Human dispermic egg at first mitosis. The four centrioles are stained yellow, while the microtubules of the spindle apparatus (and of the two sperm tails) are stained red. The three sets of chromosomes divided by these four poles are stained blue. (A–F after T. Boveri. 1907. *Jena Z Naturwiss* 43: 1–292.)

certain chromosomes while other cells lack them (**FIGURE 7.12**). Theodor Boveri demonstrated in 1902 that such cells either die or develop abnormally.

THE FAST BLOCK TO POLYSPERMY   The most straightforward way to prevent the union of more than two haploid nuclei is to prevent more than one sperm from entering the egg. Different mechanisms to prevent polyspermy have evolved, two of which are seen in the sea urchin egg. An initial, fast reaction, accomplished by an electric change in the egg cell membrane, is followed by a slower reaction caused by the exocytosis of the cortical granules (Just 1919).

The **fast block to polyspermy** is achieved by a change in the electric potential of the egg cell membrane that occurs immediately upon the entry of a sperm. The original resting membrane potential of the egg is generally about 70 mV, which is expressed as –70 mV because *the inside of the cell is negatively charged with respect to the exterior*. However, chemicals from the fusing sperm cytoplasm alter the sodium ion (Na^+) channels on the membrane (McCulloh and Chambers 1992; Wong and Wessel 2013). Within 1–3 seconds after the binding of the first sperm, the membrane potential shifts to a *positive* level—about +20 mV—with respect to the exterior (**FIGURE 7.13A**; Jaffe 1980; Longo et al. 1986). Sperm cannot fuse with egg cell membranes that have a positive resting potential, so the shift means that no more sperm can fuse to the egg.

The importance of Na^+ and the change in resting potential from negative to positive were demonstrated by Laurinda Jaffe and colleagues. They found that polyspermy

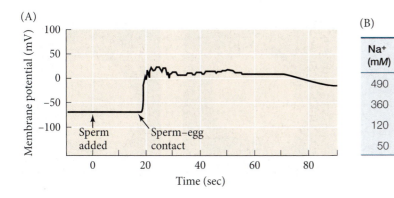

(A)

(B)

Na^+ (mM)	Polyspermic eggs (%)
490	22
360	26
120	97
50	100

FIGURE 7.13 Membrane potential of sea urchin eggs before and after fertilization. (A) Before the addition of sperm, the potential difference across the egg cell membrane is about –70 mV. Within 1–3 seconds after the fertilizing sperm contacts the egg, the potential shifts in a positive direction. (B) Table showing the rise of polyspermy with decreasing Na^+ concentration. Seawater is about 600 mM Na^+. (After L. A. Jaffe. 1980. *Dev Growth Diff* 22: 503–507.)

V. D. Vacquier and J. E. Payne. 1973. *Exp Cell Res* 82: 227–235, courtesy of V. D. Vacquier

Developing Questions

Sodium ions can readily orchestrate the fast block to polyspermy in salty seawater. But amphibians spawning in freshwater ponds also use ion channels to achieve a fast block to polyspermy. How is this achieved in an environment that lacks the ocean's high concentrations of Na⁺?

can be induced if an electric current is applied to artificially keep the sea urchin egg membrane potential negative. Conversely, fertilization can be prevented entirely by artificially keeping the membrane potential of the egg positive (Jaffe 1976). The fast block to polyspermy can also be circumvented by lowering the concentration of Na^+ in the surrounding water (**FIGURE 7.13B**). If the supply of Na^+ is not sufficient to cause the positive shift in membrane potential, polyspermy occurs (Gould-Somero et al. 1979; Jaffe 1980). An electrical block to polyspermy also occurs in frogs (Cross and Elinson 1980; Iwao et al. 2014) but probably not in most mammals (Jaffe and Cross 1983). (See **Further Development 7.5, Blocks to Polyspermy**, online.)

THE SLOW BLOCK TO POLYSPERMY The fast block to polyspermy is transient, since the membrane potential of the sea urchin egg remains positive for only about a minute. This brief potential shift is not sufficient to prevent polyspermy permanently, and polyspermy can still occur if the sperm bound to the vitelline envelope are not somehow removed (Carroll and Epel 1975). This sperm removal is accomplished by the **cortical granule reaction**, also known as the **slow block to polyspermy**. This slower, mechanical block to polyspermy becomes active about a minute after the first successful sperm-egg fusion (Just 1919). This reaction is found in many animal species, including most mammals.

Directly beneath the sea urchin egg cell membrane are about 15,000 cortical granules, each about 1 μm in diameter (see Figure 7.4B). Upon sperm entry, cortical granules fuse with the egg cell membrane and release their contents into the space between the cell membrane and the fibrous mat of vitelline envelope proteins. Several proteins are released by cortical granule exocytosis. One of these, the enzyme cortical granule serine protease, cleaves the protein posts that connect the vitelline envelope to the egg cell membrane; it also clips off the bindin receptors and any sperm attached to them (Vacquier et al. 1973; Glabe and Vacquier 1978; Haley and Wessel 1999, 2004).

The components of the cortical granules bind to the vitelline envelope to form a **fertilization envelope**. The fertilization envelope starts to form at the site of sperm entry and continues its expansion around the egg. This process starts about 20 seconds after sperm attachment and is complete by the end of the first minute of fertilization (**FIGURE 7.14**; Wong and Wessel 2004, 2008).

(A) 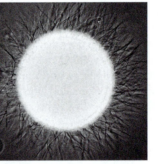 (B)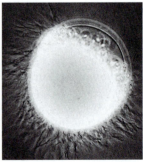

(C) 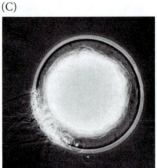 (D)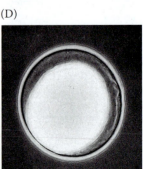

FIGURE 7.14 Formation of the fertilization envelope and removal of excess sperm. To create these photographs, sperm were added to sea urchin eggs, and the suspension was then fixed in formaldehyde to prevent further reactions. (A) At 10 seconds after sperm addition, sperm surround the egg. (B,C) At 25 (B) and 35 (C) seconds after insemination, a fertilization envelope is forming around the egg, starting at the point of sperm entry. (D) The fertilization envelope is complete, and excess sperm have been removed.

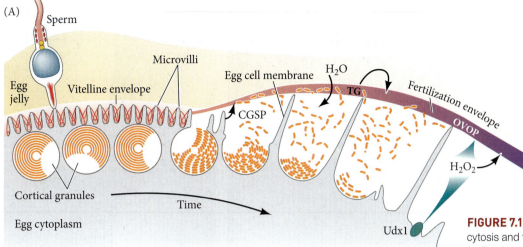

(A)

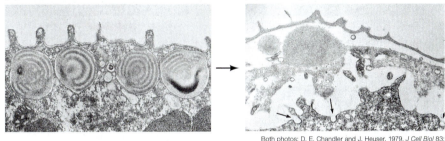

(B) Unfertilized

(C) Recently fertilized

Both photos: D. E. Chandler and J. Heuser. 1979. *J Cell Biol* 83: 91–108, courtesy of D. E. Chandler

FIGURE 7.15 Cortical granule exocytosis and formation of the sea urchin fertilization envelope. (A) Schematic diagram of events leading to the formation of the fertilization envelope. As cortical granules undergo exocytosis, they release cortical granule serine protease (CGSP), an enzyme that cleaves the proteins linking the vitelline envelope to the cell membrane. Glycosaminoglycans released by the cortical granules form an osmotic gradient, causing water to enter and swell the space between the vitelline envelope and the cell membrane. The enzyme Udx1 in the former cortical granule membrane catalyzes the formation of hydrogen peroxide (H_2O_2), the substrate for soluble ovoperoxidase (OVOP). OVOP and transglutaminases (TG) harden the vitelline envelope, now called the fertilization envelope. (B,C) Transmission electron micrographs of the cortex of an unfertilized sea urchin egg (B) and the same region of a recently fertilized egg (C), shown to the same scale. The raised fertilization envelope and the points at which the cortical granules have fused with the egg cell membrane of the egg (arrows) are visible in (C). (A after J. L. Wong and G. M. Wessel. 2008. *Development* 135: 431–440.)

The fertilization envelope is elevated from the cell membrane by glycosaminoglycans released by the cortical granules. These viscous compounds absorb water to expand the space between the cell membrane and the fertilization envelope, so that the envelope moves radially away from the egg. The fertilization envelope is then stabilized by crosslinking adjacent proteins through egg-specific peroxidase enzymes and a transglutaminase released from the cortical granules (**FIGURE 7.15**; Foerder and Shapiro 1977; Wong et al. 2004; Wong and Wessel 2009). This crosslinking allows the egg and early embryo to resist the shear forces of the ocean's intertidal waves. As this is happening, a fourth set of cortical granule proteins, including hyalin, forms a coating around the egg (Hylander and Summers 1982). The egg extends elongated microvilli, whose tips attach to this hyaline layer, which provides support for the blastomeres during cleavage.

CALCIUM IONS AS THE INITIATOR OF THE CORTICAL GRANULE REACTION The mechanism of cortical granule exocytosis is similar to that of the exocytosis of the acrosome, and it may involve many of the same molecules. Upon fertilization, the concentration of free Ca^{2+} in the egg cytoplasm increases greatly. In this high-calcium environment, the cortical granule membranes fuse with the egg cell membrane, releasing the contents of the cortical granules (see Figure 7.15A). Once the fusion of the cortical granules begins near the point of sperm entry, a wave of cortical granule exocytosis propagates around the cortex to the opposite side of the egg.

In sea urchins and mammals, the rise in Ca^{2+} concentration responsible for the cortical granule reaction is not due to an influx of Ca^{2+} into the egg, but comes from the endoplasmic reticulum within the egg itself (Eisen and Reynolds 1985; Terasaki and Sardet 1991). The release of Ca^{2+} from this intracellular storage can be monitored visually using calcium-activated luminescent dyes such as aequorin (a protein that, like GFP, is isolated from luminescent jellyfish) or fluorescent dyes such as fura-2. These dyes emit light when they bind free Ca^{2+}. When a sea urchin egg is injected with dye and then fertilized, a striking wave of Ca^{2+} release propagates across the egg and is visualized as a band of light that starts at the point of sperm entry and proceeds actively to the other end of the

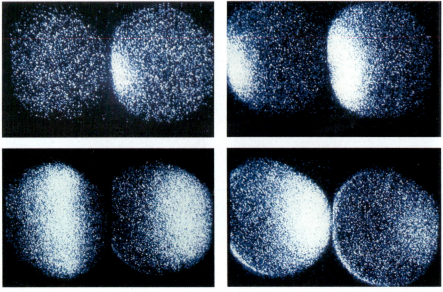

M. Hafner et al. 1988. *Cell Motil Cytoskeleton* 9: 271–277, courtesy of G. Schatten

FIGURE 7.16   Calcium release across a sea urchin egg during fertilization. The egg is pre-loaded with a dye that fluoresces when it binds Ca^{2+}. When a sperm fuses with the egg, a wave of Ca^{2+} release is seen, beginning at the site of sperm entry and propagating across the egg. Images are arranged in 3-second intervals, from left to right in the top row, and continuing from left to right in the bottom row. The wave does not simply diffuse but travels actively, taking about 30 seconds to traverse the egg.

cell (**FIGURE 7.16**; Steinhardt et al. 1977; Hafner et al. 1988). The entire release of Ca^{2+} is completed within roughly 30 seconds, and free Ca^{2+} is re-sequestered shortly after being released. (See **Further Development 7.6, The Cortical Granule Reaction**, online.)

Activation of egg metabolism in sea urchins

Although fertilization is often depicted as nothing more than the means to merge two haploid nuclei, it has an equally important role in initiating the processes that begin development. These events happen in the cytoplasm and occur without the involvement of the parental nuclei.[3] In addition to initiating the slow block to polyspermy (through cortical granule exocytosis), the release of Ca^{2+} that occurs when the sperm enters the egg is critical for activating the egg's metabolism and initiating development. Calcium ions release the inhibitors from maternally stored messages, allowing these mRNAs to be translated; they also release the inhibition of nuclear division, thereby allowing cleavage to occur. Indeed, throughout the animal kingdom, calcium ions are used to activate development during fertilization.

FURTHER DEVELOPMENT

RELEASE OF INTRACELLULAR CALCIUM IONS   The way Ca^{2+} is released varies from species to species (see Parrington et al. 2007). One way, first proposed by Jacques Loeb (1899, 1902), is that a soluble factor from the sperm is introduced into the egg at the time of cell fusion, and this substance activates the egg by changing the ionic composition of the cytoplasm (**FIGURE 7.17A**). This mechanism, as we will see later, probably works in mammals. The other mechanism, proposed by Loeb's rival Frank Lillie (1913), is that the sperm binds to receptors on the egg cell surface and changes their conformation, thus initiating reactions within the cytoplasm that activate the egg (**FIGURE 7.17B**). This is probably what happens in sea urchins.

IP_3: THE RELEASER OF CA^{2+}   If Ca^{2+} from the egg's endoplasmic reticulum is responsible for the cortical granule reaction and the reactivation of development,

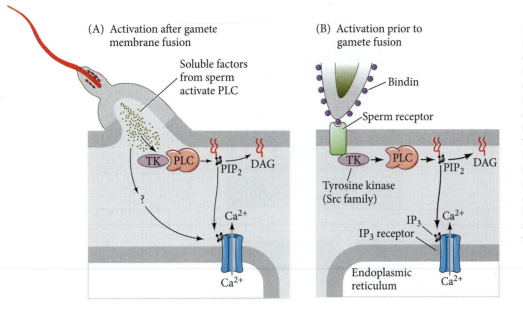

(A) Activation after gamete membrane fusion

Soluble factors from sperm activate PLC

TK PLC

PIP₂ DAG

?

Ca²⁺

Ca²⁺

(B) Activation prior to gamete fusion

Bindin

Sperm receptor

TK PLC

PIP₂ DAG

Tyrosine kinase (Src family)

IP₃

IP₃ receptor

Ca²⁺

Endoplasmic reticulum

Ca²⁺

FIGURE 7.17 Probable mechanisms of egg activation. In both cases, a phospholipase C (PLC) is activated and makes IP₃ and diacylglycerol (DAG). (A) Ca²⁺ release and egg activation by activated PLC directly from the sperm, or by a substance from the sperm that activates egg PLC. This may be the mechanism in mammals. (B) The bindin receptor (perhaps acting through a G protein) activates tyrosine kinase (TK, an Src kinase), which activates PLC. This is probably the mechanism in sea urchins.

what releases Ca²⁺? Throughout the animal kingdom, it has been found that **inositol 1,4,5-trisphosphate** (**IP₃**) is the primary agent for releasing Ca²⁺ from intracellular storage. The IP₃ pathway is shown in **FIGURE 7.18**. IP₃ is found at the point of sperm-egg contact within seconds; inhibiting IP₃ synthesis prevents Ca²⁺ release from the endoplasmic reticulum (Lee and Shen 1998; Carroll et al. 2000), and injecting IP₃ into the unfertilized egg leads to the release of Ca²⁺ and the cortical granule exocytosis (Whitaker and Irvine 1984).

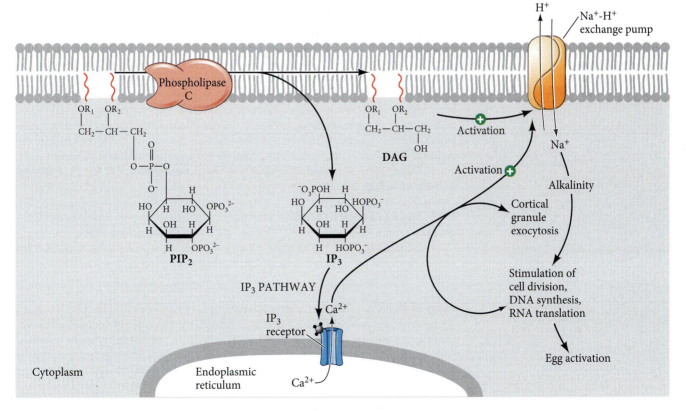

FIGURE 7.18 Roles of inositol phosphates in the release of Ca²⁺ from the endoplasmic reticulum and the initiation of development. Phospholipase C splits PIP₂ into IP₃ and DAG. IP₃ causes the release of Ca²⁺ from the endoplasmic reticulum, and DAG, with assistance from the released Ca²⁺, activates the sodium-hydrogen exchange pump in the membrane.

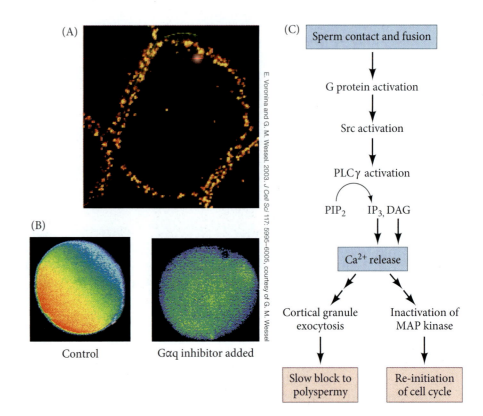

(A)

(B)

Control   Gαq inhibitor added

(C)

Sperm contact and fusion

↓

G protein activation

↓

Src activation

↓

PLCγ activation

PIP$_2$ IP$_3$, DAG

↓

Ca^{2+} release

Cortical granule exocytosis   Inactivation of MAP kinase

↓   ↓

Slow block to polyspermy   Re-initiation of cell cycle

E. Voronina and G. M. Wessel. 2003. *J Cell Sci* 117: 5995–6005, courtesy of G. M. Wessel

FIGURE 7.19 G protein involvement in Ca^{2+} entry into sea urchin eggs. (A) Mature sea urchin egg immunologically labeled for the cortical granule protein hyalin (red) and the G protein Gαq (green). The overlap of signals produces the yellow color. Gαq is localized to the cortex. (B) A wave of Ca^{2+} appears in the control egg (computer-enhanced to show relative intensities, with red being the highest) but not in the egg injected with an inhibitor of the Gαq protein. (C) Possible model for egg activation by the influx of Ca^{2+}.

PHOSPHOLIPASE C: THE GENERATOR OF IP$_3$ If IP$_3$ is necessary for Ca^{2+} release and phospholipase C is required in order to generate IP$_3$, the question then becomes, what activates PLC? Experimental results and analyses of protein phosphorylation suggest that in sea urchin eggs it involves membrane-bound kinases (Src kinases) and GTP-binding proteins that become active when the sperm contacts or fuses with the egg cell membrane (**FIGURE 7.19**; Kinsey and Shen 2000; Giusti et al. 2003; Voronina and Wessel 2004; Townley et al. 2009; Guo et al. 2015). (See **Further Development 7.7, The IP$_3$ Pathway Activates the Egg**, online.)

EFFECTS OF CALCIUM RELEASE   The flux of Ca^{2+} across the egg activates a preprogrammed set of metabolic events. The responses of the sea urchin egg to the sperm can be divided into "early" responses, which occur within seconds of the cortical granule reaction, and "late" responses, which take place several minutes after fertilization begins (**TABLE 7.1**).

EARLY RESPONSES   As we have seen, contact or fusion of a sea urchin sperm and egg activates two major blocks to polyspermy: the fast block, mediated by sodium influx into the cell; and the cortical granule reaction, or slow block, mediated by the intracellular release of Ca^{2+}. The same release of Ca^{2+} responsible for the cortical granule reaction is also responsible for the re-entry of the egg into the cell cycle and the reactivation of egg protein synthesis. Ca^{2+} levels in the egg increase from 0.05 to between 1 and 5 μM, and in almost all species this occurs as a wave or succession of waves that sweep across the egg beginning at the site of sperm-egg fusion (see Figure 7.16; Jaffe 1983; Terasaki and Sardet 1991; Stricker 1999). (See **Further Development 7.8, Rules of Evidence**, online.)

TABLE 7.1 Events of sea urchin fertilization

Event	Approximate time postinsemination[a]
EARLY RESPONSES	
Sperm-egg binding	0 sec
Fertilization potential rise (fast block to polyspermy)	within 3 sec
Sperm-egg membrane fusion	within 1 sec
Calcium increase first detected	10 sec
Cortical granule exocytosis (slow block to polyspermy)	15–60 sec
LATE RESPONSES	
Activation of NAD kinase	starts at 1 min
Increase in $NADP^+$ and NADPH	starts at 1 min
Increase in O_2 consumption	starts at 1 min
Sperm entry	1–2 min
Acid efflux	1–5 min
Increase in pH (remains high)	1–5 min
Sperm chromatin decondensation	2–12 min
Sperm nucleus migration to egg center	2–12 min
Egg nucleus migration to sperm nucleus	5–10 min
Activation of protein synthesis	starts at 5–10 min
Activation of amino acid transport	starts at 5–10 min
Initiation of DNA synthesis	20–40 min
Mitosis	60–80 min
First cleavage	85–95 min

Main sources: M. J. Whitaker and R. A. Steinhardt. 1985. *Q Rev Biophys* 15: 593–667; T. Mohri et al. 1995. *Dev Biol* 172: 139–157.

[a]Approximate times based on data from *S. purpuratus* (15–17°C), *L. pictus* (16–18°C), *A. punctulata* (18–20°C), and *L. variegatus* (22–24°C). The timing of events within the first minute is best known for *L. variegatus*, so times are listed for that species.

LATE RESPONSES: RESUMPTION OF PROTEIN AND DNA SYNTHESIS   Calcium release activates a series of metabolic reactions that initiate embryonic development (**FIGURE 7.20**). One of these is the activation of the enzyme NAD^+ kinase, which converts NAD^+ to $NADP^+$ (Epel et al. 1981). Since $NADP^+$ (but not NAD^+) can be used as a coenzyme for lipid biosynthesis, such a conversion has important consequences for lipid metabolism and thus may be important in the construction of the many new cell membranes required during cleavage. Udx1, the enzyme responsible for the reduction of oxygen to crosslink the fertilization envelope, is also NADPH-dependent (Heinecke and Shapiro 1989; Wong et al. 2004). Last, NADPH helps regenerate glutathione and ovothiols, molecules that may be crucial scavengers of free radicals that could otherwise damage the DNA of the egg and early embryo (Mead and Epel 1995).

It is thought that the Ca^{2+} elevation and the pH increase (from the replacement of H^+ by the second influx of Na^+ from seawater) act together to stimulate new DNA and protein synthesis (Winkler et al. 1980; Whitaker and Steinhardt 1982; Rees et al. 1995). If one experimentally elevates the pH of an unfertilized egg to a level similar to that of a fertilized egg, DNA synthesis and nuclear envelope breakdown ensue, just as if the egg were fertilized (Miller and Epel 1999). Calcium ions are also critical to new DNA synthesis. The wave of free Ca^{2+} inactivates the enzyme MAP kinase, converting it from a phosphorylated (active) to an unphosphorylated (inactive) form, thus removing an inhibition on DNA synthesis (Carroll et al. 2000). DNA synthesis can then resume.

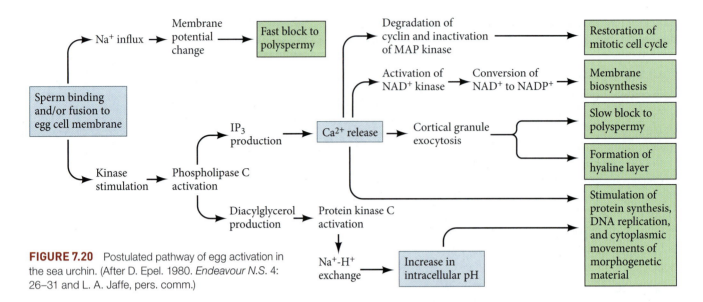

FIGURE 7.20 Postulated pathway of egg activation in the sea urchin. (After D. Epel. 1980. *Endeavour N.S.* 4: 26–31 and L. A. Jaffe, pers. comm.)

In sea urchins, a burst of protein synthesis usually occurs within several minutes after sperm entry. This protein synthesis does not depend on the synthesis of new messenger RNA, but uses mRNAs already present in the oocyte cytoplasm. These mRNAs encode proteins such as cell cycle regulators, transcription factors, histones, tubulins, and actins that are used during early development. Such a burst of protein synthesis can be induced by artificially raising the pH of the cytoplasm using ammonium ions (Sargent and Raff 1976; Winkler et al. 1980; Chassé et al. 2018).

One mechanism for this global rise in the translation of messages stored in the oocyte appears to be the release of inhibitors from the mRNA. Further Development 3.20, online, discusses maskin, an inhibitor of translation in the unfertilized amphibian oocyte. In sea urchins, a similar inhibitor acts in a similar manner to prevent several mRNAs from being translated. Upon fertilization, however, this inhibitor becomes phosphorylated and is degraded, thus allowing translation and protein synthesis from the stored sea urchin mRNAs (Cormier et al. 2001; Oulhen et al. 2007). One of the "freed" mRNAs is the message encoding cyclin B protein (Salaun et al. 2003, 2004). Cyclin B combines with Cdk1 to create mitosis-promoting factor (MPF), which is required to initiate cell division.

Fusion of genetic material in sea urchins

After the sperm and egg cell membranes fuse, the sperm nucleus and its centriole separate from the mitochondria and flagellum. The mitochondria and the flagellum disintegrate inside the egg, so very few, if any, sperm-derived mitochondria are found in developing or adult organisms. Thus, although each gamete contributes a haploid genome to the zygote, the *mitochondrial* genome is transmitted primarily by the maternal parent. Conversely, in almost all animals studied (the mouse being the major exception), the centrosome needed to produce the mitotic spindle of the subsequent divisions is derived from the sperm centriole (see Figure 7.12; Sluder et al. 1989, 1993).

Fertilization in sea urchin eggs occurs after the second meiotic division, so there is already a haploid female pronucleus present when the sperm enters the egg cytoplasm. Once inside the egg, the sperm nucleus undergoes a dramatic transformation as it decondenses to form the haploid male pronucleus. First, the nuclear envelope degenerates, exposing the compact sperm chromatin to the egg cytoplasm (Longo and Kunkle 1978; Poccia and Collas 1997). Kinases from the egg cytoplasm phosphorylate the sperm-specific histone proteins, allowing them to decondense. The decondensed histones are then replaced by egg-derived, cleavage-stage histones (Stephens et al. 2002; Morin et al. 2012). This exchange permits the decondensation of the sperm chromatin. Once decondensed, the DNA adheres to the newly formed nuclear envelope derived from precursor membranes and the endoplasmic reticulum (Poccia and Larijani 2009), and DNA polymerase initiates its replication (Infante et al. 1973).

FIGURE 7.21 Nuclear events in the fertilization of the sea urchin. (A) Sequential photographs show the migration of the egg pronucleus and the sperm pronucleus toward each other in an egg of *Clypeaster japonicus*. The sperm pronucleus is surrounded by its aster of microtubules. (B) The two pronuclei migrate toward each other on these microtubular processes. (The pronuclear DNA is stained blue by Hoechst dye.) The microtubules (stained green with fluorescent antibodies to tubulin) radiate from the centrosome associated with the (smaller) male pronucleus and reach toward the female pronucleus. (C) Fusion of pronuclei in the sea urchin egg.

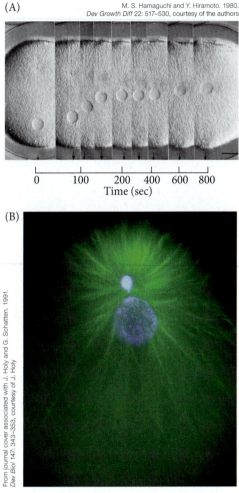

M. S. Hamaguchi and Y. Hiramoto. 1980. *Dev Growth Diff* 22: 517–530, courtesy of the authors

From journal cover associated with J. Holy and G. Schatten. 1991. *Dev Biol* 147: 343–353, courtesy of J. Holy

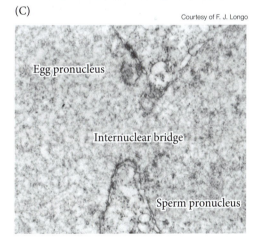

Courtesy of F. J. Longo

Egg pronucleus

Internuclear bridge

Sperm pronucleus

But how do the sperm and egg pronuclei find each other? Once the sea urchin sperm has entered the egg cytoplasm and its nucleus has separated from the tail, the sperm nucleus rotates 180° so that the sperm centriole is between the developing male pronucleus and the egg pronucleus. The sperm centriole then acts as a microtubule organizing center, extending its own microtubules and integrating them with egg microtubules to form an aster. Microtubules extend throughout the egg and contact the female pronucleus, at which point the two pronuclei migrate along the microtubules toward each other. When they make contact, the interaction of the two pronuclei activates enzymes that generate specific fusion-promoting lipids (Lete et al. 2017). This fusion forms the diploid zygote nucleus (**FIGURE 7.21**). DNA synthesis can begin either in the pronuclear stage or after the formation of the zygote nucleus, and depends on the level of Ca^{2+} released earlier in fertilization (Jaffe et al. 2001).

At this point, the diploid nucleus has formed, DNA synthesis and protein synthesis have commenced, and the inhibitions to cell division have been removed. The sea urchin can now begin to form a multicellular organism. We will describe the means by which sea urchins achieve multicellularity in Chapter 10.

Internal Fertilization in Mammals

Getting the gametes into the oviduct: Translocation and capacitation

It is very difficult to study any interactions between the mammalian sperm and egg that take place prior to these gametes making contact. One obvious reason for this is that mammalian fertilization occurs inside the oviducts of the female. Although it is relatively easy to mimic the conditions surrounding sea urchin fertilization using natural or artificial seawater, we do not yet know the components of the various natural environments that mammalian sperm encounter as they travel to the egg.

One thing known for certain is that the female reproductive tract is not a passive conduit through which sperm race, but a highly specialized set of tissues that actively regulate the transport and maturity of both gametes. Both male and female gametes use a combination of small-scale biochemical interactions and large-scale physical propulsion to get to the ampulla at the upper end of the oviduct, where fertilization takes place (see Figure 12.10).

TRANSLOCATION The meeting of sperm and egg must be facilitated by the female reproductive tract. Different mechanisms are used to position the gametes at the right place at the right time. A mammalian oocyte just released from the ovary is surrounded by a matrix containing cumulus cells. (Cumulus cells are the cells of the ovarian follicle to which the developing oocyte was attached; see Figure 7.5.) If this matrix is experimentally removed or significantly altered, the fimbriae at the upper end of the oviduct will not "pick up" the oocyte-cumulus complex (see Figure 12.10), nor will the complex be able to enter the oviduct (Talbot et al. 1999). Once the oocyte-cumulus complex is picked up, a combination of ciliary beating and muscle contractions transport it to the appropriate position for its fertilization in the oviduct.

(A)

(B)

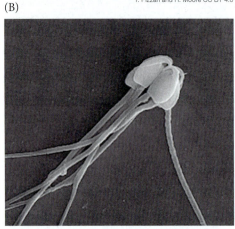

Courtesy of James Weaver, Wyss Institute, Harvard University

FIGURE 7.22 Sperm associations can occur in species in which females mate with several males in a brief time span. (A) The "sperm train" of the wood mouse (*Apodemus sylvaticus*). Sperm are joined by their acrosomal caps. (B) Close-up of the sperm heads of the field mouse *Peromyscus maniculatus*, showing hook-to-hook attachment.

The sperm must travel a longer path. In humans, about 200–300 million sperm are ejaculated into the vagina, but only one in a million enters the oviducts (fallopian tubes) (Harper 1982; Cerezales et al. 2015). And half the sperm enter the "wrong" oviduct, the one that has no oocyte. Only about 200 sperm probably reach the vicinity of the egg. The translocation of sperm from the vagina to the oviduct involves several processes that work at different times and places:

- *Sperm motility* Motility (flagellar action) is probably important in getting sperm through the cervical mucus and into the uterus. Interestingly, in those mammals where the female is promiscuous (mating with several males in rapid succession), sperm from the same male often form "trains," or aggregates, in which the combined propulsion of the flagella makes the sperm swim faster (**FIGURE 7.22**). This strategy probably evolved for competition among males. In those species without female promiscuity, the sperm usually remain as individuals (Fisher and Hoeckstra 2010; Foster and Pizzari 2010; Fisher et al. 2014).

- *Uterine muscle contractions* Sperm are found in the oviducts of mice, hamsters, guinea pigs, cows, and humans within 30 minutes of sperm deposition in the vagina—a time "too short to have been attained by even the most Olympian sperm relying on their own flagellar power" (Storey 1995). Rather, sperm appear to be transported to the oviduct by the muscular activity of the uterus.

- *Sperm rheotaxis* Sperm also receive long-distance directional cues from the flow of liquid from the oviduct to the uterus. Sperm display **rheotaxis**—that is, they migrate against the direction of the flow—using CatSper calcium channels (like sea urchin sperm) to sense calcium influx and monitor the direction of the current (Miki and Clapham 2013). Such sperm rheotaxis has been observed in mice and in humans.

CAPACITATION As mentioned in Chapter 6, newly ejaculated sperm cannot fertilize an egg. They are immature. The cells of the oviduct induce the sperm to mature, and mammalian sperm complete their development in the female oviducts (Chang 1951; Austin 1952). This maturation is called **capacitation** (the gaining of capacity). The capacities the sperm gain are those for (1) recognizing the cues that will guide them to the egg, (2) undergoing the acrosome reaction, and (3) fusing with the egg cell membrane. Sperm that are not capacitated are "held up" in the cumulus matrix and are unable to reach the egg (Austin 1960; Corselli and Talbot 1987). Here, again, we see that the oviducts are not passive tracts through which sperm race. Capacitation is possibly initiated by an efflux of cholesterol from the sperm cell membrane, an efflux caused by proteins in the female reproductive tract. The changes that follow unmask receptors on the sperm that allow binding to the zona pellucida and prepare the acrosome for the acrosome reaction (**FIGURE 7.23**). (See **Further Development 7.9, Mechanisms of Capacitation**, online.)

In the vicinity of the oocyte: Hyperactivation, directed sperm migration, and the acrosome reaction

HYPERACTIVATION Toward the end of capacitation, sperm become **hyperactivated**—they swim at higher velocities and generate greater force. Hyperactivation appears to be mediated by the opening of the sperm-specific calcium channels, the CatSper proteins, in the sperm tail (see Figure 7.23; Ren et al. 2001; Qui et al. 2007). The symmetrical beating of the flagellum is changed into a rapid asynchronous beat with a higher degree of bending. The power of the beat and the direction of sperm head movement are thought to release the

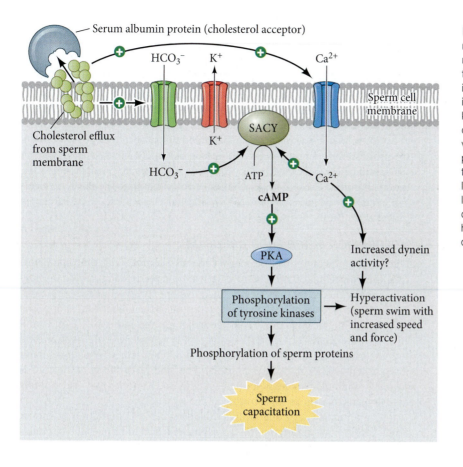

Serum albumin protein (cholesterol acceptor)

Cholesterol efflux from sperm membrane

Sperm cell membrane

SACY

HCO₃⁻ K⁺ Ca²⁺

K⁺

HCO₃⁻   ATP   Ca²⁺

cAMP

PKA

Increased dynein activity?

Phosphorylation of tyrosine kinases → Hyperactivation (sperm swim with increased speed and force)

Phosphorylation of sperm proteins

Sperm capacitation

FIGURE 7.23 Hypothetical model for mammalian sperm capacitation. The pathway is modulated by the removal of cholesterol from the sperm cell membrane, which allows the influx of bicarbonate ions (HCO_3^-) and calcium ions (Ca^{2+}). These ions activate adenylate kinase (SACY), thereby elevating cAMP concentrations. The high cAMP levels then activate Protein kinase A (PKA). Active PKA phosphorylates several tyrosine kinases, which in turn phosphorylate several sperm proteins, leading to capacitation. Increased intracellular Ca^{2+} also activates the phosphorylation of these proteins, as well as contributing to hyperactivation of the sperm. (After P. E. Visconti et al. 2011. *Asian J Androl* 13: 395–405.)

(?) Developing Questions

Sometimes the egg and sperm fail to meet and conception does not take place. What are the leading causes of infertility in humans, and what procedures are being used to circumvent these blocks?

sperm from their binding with the oviduct epithelial cells. Indeed, only hyperactivated sperm are seen to detach and continue their journey to the egg (Suarez 2008a,b). Hyperactivation may enable sperm to respond differently to the fluid current. Uncapacitated sperm move in a planar direction, allowing more time for the sperm head to attach to the oviduct epithelial cells (**FIGURE 7.24**). Capacitated sperm rotate around their long axis, probably enhancing the detachment of the sperm from the epithelium (Miki and Clapham 2013). Once a sperm has reached the oocyte-cumulus complex, hyperactivation, along with a hyaluronidase enzyme on the outside of the sperm cell membrane, enables the sperm to digest a path through the extracellular matrix of the cumulus cells (Lin et al. 1994; Kimura et al. 2009).

THERMAL AND CHEMICAL GRADIENTS   An old joke claims that the reason a man has to release so many sperm at each ejaculation is that no male gamete is willing to ask for directions. So what *does* provide the sperm with directions? Heat is one cue: there is a thermal gradient of 2 degrees Celsius between the isthmus of the oviduct and the warmer ampulla (Bahat et al. 2003, 2006). Capacitated mammalian sperm can sense thermal differences as small as 0.014 degrees Celsius over a millimeter and tend to migrate toward the higher temperature (Bahat et al. 2012). This ability to sense temperature difference and preferentially swim from cooler to warmer sites (**thermotaxis**) is found only in capacitated sperm.

The capacitated sperm are also able to detect and respond to picomolar amounts of the hormone **progesterone**, which is secreted by the cumulus cells surrounding the egg (Guidobaldi et al. 2008, 2017). Thus, as the sperm enter the ampulla, they are told where the egg can be found. However, it is not known whether the same thermal and chemical cues are used by all mammalian species.

Courtesy of S. Suarez

FIGURE 7.24   Scanning electron micrograph (artificially colored) showing a bull sperm as it adheres to the membranes of epithelial cells in the oviduct of a cow prior to entering the ampulla.

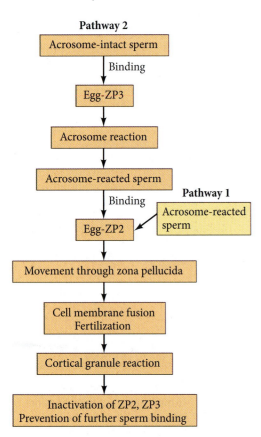

Pathway 2

Acrosome-intact sperm

↓ Binding

Egg-ZP3

↓

Acrosome reaction

↓

Acrosome-reacted sperm

↓ Binding   **Pathway 1**
   Acrosome-reacted sperm

Egg-ZP2

↓

Movement through zona pellucida

↓

Cell membrane fusion
Fertilization

↓

Cortical granule reaction

↓

Inactivation of ZP2, ZP3
Prevention of further sperm binding

FIGURE 7.25 Recent model for the recognition of sperm by the mouse zona pellucida. Sperm that have undergone the acrosome reaction bind directly to zona protein ZP2 (pathway 1) and begin making a channel toward the oocyte. Sperm with an intact acrosome bind to ZP3 (pathway 2); they undergo the acrosome reaction on the zona pellucida and then transfer their binding to ZP2. When a sperm reaches the oocyte and fuses with it, the cortical granules release proteins that digest portions of ZP2 and ZP3, making them nonfunctional. This prevents further sperm entry. (After P. M. Wassarman and E. S. Litscher. 2018. *Curr Top Dev Biol* 130: 331–356.)

ACROSOME REACTION   As they enter the ampulla of the oviduct, the capacitated sperm (but not the uncapacitated sperm) undergo the acrosome reaction. Evidence from several species of mammals (Huang et al. 1981; Yanagimachi and Phillips 1984; Jin et al. 2011) indicates that "successful" sperm (i.e., those that actually fertilize an egg) have usually undergone the acrosome reaction by the time they are seen in the cumulus. It is thought that as the sperm draw closer to the egg, the higher levels of progesterone induce the acrosome reaction (Uñates et al. 2014; Abi Nahed et al. 2016; La Spina et al. 2016). While the mechanism activated by progesterone is not known, the progesterone might trigger the release of Protein kinase A (PKA) from the sperm cell membrane, thereby allowing it to activate the sperm-specific cation channels. These CatSper channels would facilitate the transfer of calcium ions into the sperm, where they would cause exocytosis of the acrosome (Stival et al. 2018).

Recognition at the zona pellucida

Once within the cumulus, the sperm can make contact with the zona pellucida, the extracellular matrix of the egg. The zona pellucida in mammals plays a role analogous to that of the vitelline envelope in invertebrates; the zona, however, is a far thicker and denser structure than the vitelline envelope. The mouse zona pellucida is made of three major glycoproteins—ZP1, ZP2, and ZP3 (zona proteins 1, 2, and 3)—along with accessory proteins that bind to the zona's integral structure. The human zona pellucida has four major glycoproteins—ZP1, ZP2, ZP3, and ZP4. The binding of sperm to the zona is relatively, but not absolutely, species-specific, and a species may use multiple mechanisms to achieve this binding.

The mammalian egg encounters a heterogeneous population of sperm. Some sperm have undergone the acrosome reaction in or near the cumulus, whereas others have not. The egg may have pathways for accepting both types of capacitated sperm (**FIGURE 7.25**; Wassarman and Litscher 2018). In one pathway, sperm that have undergone the acrosome reaction bind directly to ZP2. In the other pathway, sperm with intact acrosomes become bound to ZP3, which causes the acrosome reaction, and the sperm binding is then transferred to ZP2 (Bleil and Wassarman 1980, 1983). (See **Further Development 7.10, Discovering How Mammalian Sperm Bind to Eggs**, online.)

Gamete fusion and the prevention of polyspermy

SPERM–EGG FUSION   The sperm and the egg now finally meet. In mammals, it is not the tip of the sperm head that makes contact with the egg (as happens in the perpendicular entry of sea urchin sperm) but the side of the sperm head (**FIGURE 7.26**). The acrosome reaction, in addition to expelling the enzymatic contents of the acrosome, also exposes the inner acrosomal membrane to the outside. The junction between the inner acrosomal

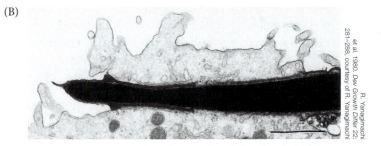

(A)

(B)

FIGURE 7.26 Entry of sperm into a golden hamster egg. (A) Scanning electron micrograph of sperm fusing with egg. The "bald" spot (without microvilli) is where the polar body has budded off. Sperm do not bind there. (B) Transmission electron micrograph of the sperm fusing parallel to the egg cell membrane.

FIGURE 7.27 Izumo protein and membrane fusion in mouse fertilization. (A) Diagram of sperm-egg cell membrane fusion. During the acrosome reaction, Izumo localized on the acrosome becomes translocated to the sperm cell membrane. There it meets the complex of Juno and other egg membrane proteins on the egg microvilli, initiating membrane fusion and the entry of the sperm into the egg. (B) Localization of Izumo to the inner and outer acrosomal membrane. Izumo is stained red, acrosomal proteins green. (After Y. Satouh et al. 2012. *J Cell Sci* 125: 4985–4990.)

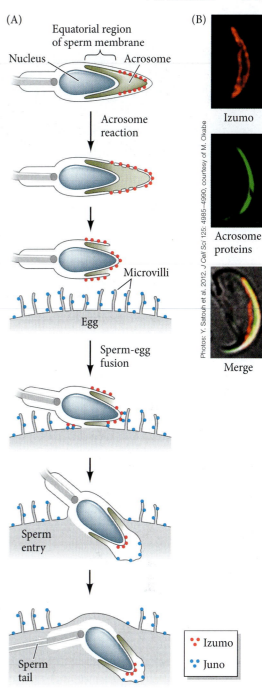

membrane and the sperm cell membrane is called the **equatorial region**, and this is where membrane fusion between sperm and egg begins. This fusion appears to involve several proteins, two of which are **Izumo**, originally in the acrosomal membrane, and **Juno**, an oocyte membrane protein (**FIGURE 7.27**; Inoue et al. 2005; Bianchi et al. 2014). These proteins bind together and appear to recruit other proteins to make an adhesion and fusion complex (Runge et al. 2006; Inoue et al. 2017). Mutations in either Juno or Izumo block fertilization. As in sea urchin gamete fusion, the sperm is bound to regions of the egg where actin polymerizes to extend microvilli to the sperm (Yanagimachi and Noda 1970).

The sperm doesn't bore or drill into the oocye. Rather, the cell membranes fuse and the two cells become one. The entire sperm is taken into the egg, including the flagellum and mitochondria. In mammals, as in sea urchins, most mitochondria brought in by the sperm are degraded in the egg cytoplasm. So, with only a few human exceptions (Luo et al. 2018), all of the mitochondria in a new individual are thought to be derived from its mother (Cummins et al. 1998; Shitara et al. 1998; Schwartz and Vissing 2002), hence our ability to trace maternal ancestry down through generations by examining mitochondrial DNA.

BLOCKS TO POLYSPERMY   Polyspermy is a problem for mammals, just as it is for sea urchins. In mammals, no electrical fast block to polyspermy has yet been detected; it may not be needed, given the limited number of sperm that reach the ovulated egg (Gardner and Evans 2006). However, there may be a *slow* block to polyspermy that, as in sea urchins, involves the cortical granule reaction. When the cortical granules fuse with the egg cell membrane, they release protein-digesting enzymes that modify the zona pellucida proteins such that they can no longer bind sperm (Bleil and Wassarman 1980). One of these cortical granule proteases is **ovastacin**. When ZP2 is cleaved by ovastacin, it loses its ability to bind sperm (**FIGURE 7.28**; Moller and Wassarman 1989). Indeed, polyspermy occurs more frequently in mouse eggs bearing mutant ZP2 that cannot be cleaved by ovastacin (Gahlay et al. 2010; Burkart et al. 2012).

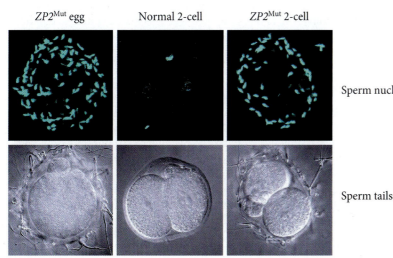

$ZP2^{Mut}$ egg   Normal 2-cell $ZP2^{Mut}$ 2-cell

Sperm nuclei

Sperm tails

G. Gahlay et al. 2010. *Science* 329: 216–219, courtesy of J. Dean

FIGURE 7.28 Cleaved ZP2 is necessary for the block to polyspermy in mammals. Eggs and embryos were visualized by fluorescence microscopy (to see sperm nuclei; top row) and brightfield microscopy (differential interference contrast, to see sperm tails; bottom row). Sperm bound normally to eggs containing a mutant ZP2 that could not be cleaved. However, the egg with normal (i.e., cleavable) ZP2 got rid of sperm by the 2-cell stage, whereas the egg with the mutant (uncleavable) ZP2 retained sperm and permitted polyspermy.

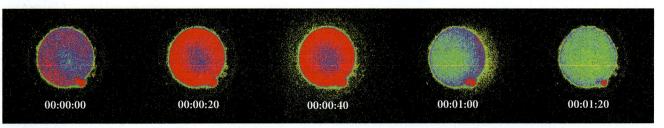

00:00:00   00:00:20   00:00:40   00:01:00   00:01:20

FIGURE 7.29 The "zinc spark" at fertilization. After the artificial activation of a human egg with a calcium channel opener, the release of zinc ions (starting with the arrowhead) can be seen in increasing and then diminishing intensity. Extracellular zinc concentrations were detected using a zinc-sensitive fluorescent dye and are represented using color to show relative intensities, with red being the highest and green the lowest.

One of the goals of modern pharmacology is to create a male contraceptive. Reviewing the steps of fertilization, what steps do you think it might be possible to block pharmacologically in order to produce a contraceptive for males?

A second slow block to polyspermy comes from the so-called zinc spark, the release of billions of zinc ions that is induced by the entry of the first sperm (**FIGURE 7.29**; Que et al. 2015, 2017; Duncan et al. 2016). These released zinc ions are seen to bind to the zona pellucida. Because two acrosomal enzymes, acrosin and MMP2, as well as the enzymes that help establish capacitation, are inhibited by zinc, the accumulation of this heavy metal on the zona pellucida and in the surrounding cumulus may form a "zinc shield" that prevents further sperm entry (Kerns et al. 2018). A third slow block to polyspermy occurs at the level of the egg cell membrane and involves Juno (Bianchi and Wright 2014). As the sperm and egg membranes fuse, Juno appears to be released from the oocyte cell membrane. Not only is this "docking site" for sperm removed, but the soluble Juno protein can bind sperm in the perivitelline space between the zona pellucida and the oocyte, thereby preventing the sperm from seeing any Juno that may still reside on the oocyte membrane.

Activation of the mammalian egg

As in every other animal studied, a transient rise in cytoplasmic Ca^{2+} is necessary for egg activation in mammals (Yeste et al. 2017; Kashir et al. 2018), and as in sea urchins, these calcium ions are released from intracellular stores. Fertilization triggers this release through the production of IP_3 by the enzyme phospholipase C (PLC) (Swann et al. 2006; Igarashi et al. 2007). Unlike in sea urchins, however, in mammals the PLC responsible for egg activation and pronucleus formation may come from the sperm rather than from the egg (Swann et al. 2006; Igarashi et al. 2007). Moreover, this PLC turns out to be a soluble sperm PLC enzyme, **PLCζ** (PLC-zeta), which is delivered to the egg by gamete fusion. (See **Further Development 7.11, PLC from Sperm Activates Mammalian Eggs**, online.)

Fusion of genetic material

As in sea urchins, the single mammalian sperm that finally enters the egg carries its genetic contribution in a haploid pronucleus. In mammals, however, the process of pronuclear migration takes about 12 hours, compared with less than 1 hour in sea urchins.

The mammalian sperm enters the oocyte while the oocyte nucleus is "arrested" in metaphase of its second meiotic division (**FIGURE 7.30A,B**; see also Figure 7.3). As described for the sea urchin, the Ca^{2+} oscillations brought about by sperm entry inactivate MAP kinase and allow DNA synthesis. But unlike in the sea urchin egg, which has already completed meiosis, the chromosomes of the mammalian oocyte are still in the middle of their second meiotic metaphase. Oscillations in the level of Ca^{2+} activate another kinase that leads to the proteolysis of cyclin (thus allowing the cell cycle to continue) and securin (the protein that holds the metaphase chromosomes together), thereby allowing the completion of meiosis and the establishment of a mature female pronucleus (Watanabe et al. 1991; Johnson et al. 1998).

DNA synthesis occurs separately in the male and female pronuclei, and in both pronuclei, the DNA is altered by the cytoplasm of the zygote (Sutovsky and Schatten 1997; Fraser and Lin 2016). The sperm DNA had epigenetic markers (mostly methyl groups) that were critical in making the cell a sperm. Similarly, the DNA of the oocyte

(A) 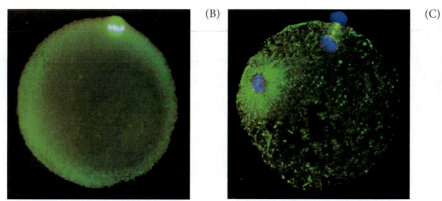 (B)

(C)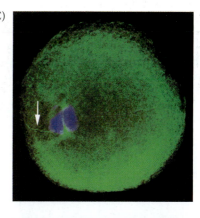

(D)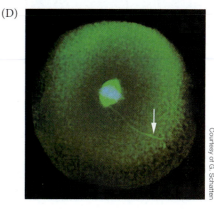

Courtesy of G. Schatten

FIGURE 7.30 Pronuclear movements during human fertilization. Microtubules are stained green, DNA is dyed blue. Arrows point to the sperm tail. (A) The mature unfertilized oocyte completes the first meiotic division, budding off a polar body. (B) As the sperm enters the oocyte (left side), microtubules condense around it as the oocyte completes its second meiotic division at the periphery. (C) By 15 hours after fertilization, the two pronuclei have come together, and the centrosome splits to organize a bipolar microtubular array. The sperm tail is still seen (arrow). (D) At prometaphase, chromosomes from the sperm and egg intermix on the metaphase plate, and a mitotic spindle initiates the first mitotic division. The sperm tail can still be seen.

pronucleus had epigenetic modifications that were crucial in making it an egg. Now, almost all these methyl groups are removed, giving the genome a "clean slate" characteristic of totipotency. The sperm DNA is also remodeled by the zygote cytoplasm, which exchanges its protamines back to histones.

The two pronuclei migrate toward each other on a microtubular belt produced from oocyte tubulin by the centrosome of the male pronucleus. Upon meeting, the two nuclear envelopes break down. However, instead of producing a common zygote nucleus (as in sea urchins), the chromatin condenses into chromosomes that orient themselves on a common mitotic spindle organized by the centriole brought in by the sperm (**FIGURE 7.30C,D**). Thus, in mammals, a true diploid nucleus is seen for the first time not in the zygote, but at the 2-cell stage. In mammals, the nucleus formed at fertilization is not active until the maternal mRNAs are used and degraded. This usually occurs in the 2-cell stage in the mouse and after two to four divisions in humans (Fraser and Lin 2016; Svoboda 2017). (See **Further Development 7.12, The Non-Equivalence of Mammalian Pronuclei**, and **Further Development 7.13, A Social Critique of Fertilization Research**, both online.)

Fertilization in Angiosperm Plants

Pollination and beyond: The progamic phase

Flowering plants present an even more baroque pattern for fertilization than echinoderms and mammals do. The process begins with pollination—the adhesion of pollen grains to the female portion (stigma) of a flower. This is followed by the **progamic phase**, the sequence of events between **pollination** and ultimate fertilization in the ovule where the egg is housed. There are marked similarities between plants and animals in these events. For example, in angiosperms, (1) pollen grains and tubes must contact and recognize first the stigma and then the ovule, respectively; (2) there must be regulation of sperm entry into the ovule; (3) there must be fusion of genetic material, both between a sperm and an egg (which will form the developing plant) and between a sperm and a central cell (which will form the endosperm that nourishes this new plant); and (4) the newly fertilized cells must become activated and begin development.

Pollen grains can be carried to the stigma of a flower by various vectors, including insects, wind, water, birds, or bats. After pollen grains land on the stigma, they adhere,

FIGURE 7.31 Pollination. (A) Pollen grains from different species of plants differ in their size, shape, color, and texture. (B) Bees and other insects (and some birds) are effective pollinators, bringing pollen from one flower to another. Note the yellow pollen on the bee's leg. (C) Gorse (*Ulex europeaus*) pollen particles carry the cells that traverse centimeters of tissue to access the ovules at the base of the style. (D) Pollen grains (green) of the opium poppy germinate pollen tubes that vie to reach the ovule.

hydrate, and germinate, which opens the pollen grain and allows the extension of a pollen tube. The pollen tube contains a vegetative cell and two sperm cells. The vegetative cell constructs the pollen tube (**FIGURE 7.31**). The timing of pollen maturation and release, pollinator activity, and stigma receptivity is closely regulated (see Bertin and Newman 1993; Edlund et al. 2004; McInnes et al. 2006). In some ways, this coordination of timing and maturity resembles that in mammals. When pollen is experimentally dusted on immature *Arabidopsis thaliana* stigmas, the immature ovules are unable to regulate sperm entry, resulting in mis-navigation of the pollen tubes and polyspermy (Nasrallah et al. 1994).

The arrival of a viable pollen grain on a receptive stigma does not guarantee fertilization. Instead, fertilization depends upon the compatibility between the plant producing the pollen and the plant containing the stigma. Among angiosperms, incompatibilities can exist between different species, between different individuals within a species, and between the pollen and stigma of the same individual plant. These recognition events are coordinated by several genes, especially the alleles of the *S* locus, which regulate several pollen-stigma interactions (Gaude and McCormick 1999; Rea and Nasrallah 2008; Tantikanjana et al. 2009).

Pollen germination and tube elongation

If the pollen and the stigma are compatible, the pollen hydrates and the pollen tube emerges. But how does the pollen tube leave the pollen grain? Most pollen grain walls contain at least one opening called an **aperture**, through which the pollen tube protrudes. In grains with multiple apertures, the pollen tube uses the opening closest to the contact point with the stigma surface. Some pollen grains have no apertures; others have them but do not depend on them for tube emergence (as is seen in the model organism *Arabidopsis thaliana*). In *A. thaliana*, pollen tube germination involves (1) an internal swelling of a pectin-rich region that strains the overlying pollen wall, and (2) a

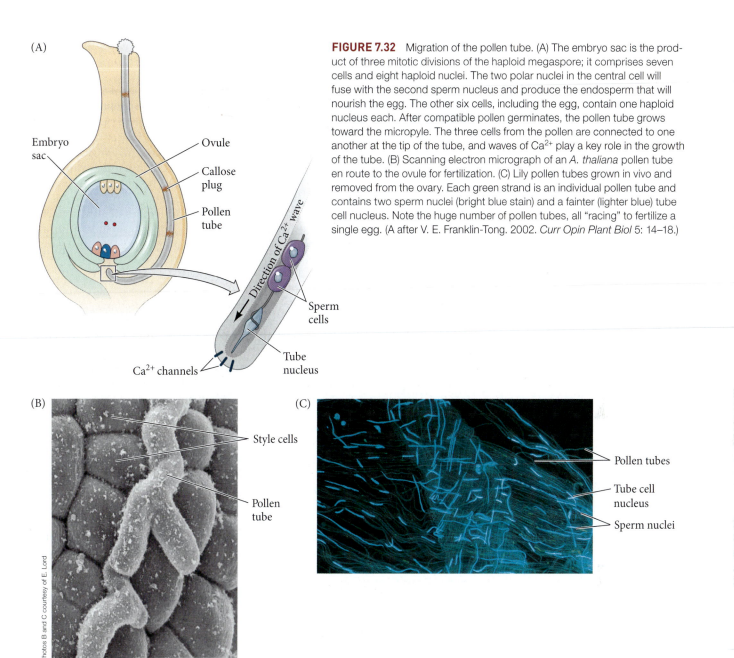

FIGURE 7.32 Migration of the pollen tube. (A) The embryo sac is the product of three mitotic divisions of the haploid megaspore; it comprises seven cells and eight haploid nuclei. The two polar nuclei in the central cell will fuse with the second sperm nucleus and produce the endosperm that will nourish the egg. The other six cells, including the egg, contain one haploid nucleus each. After compatible pollen germinates, the pollen tube grows toward the micropyle. The three cells from the pollen are connected to one another at the tip of the tube, and waves of Ca^{2+} play a key role in the growth of the tube. (B) Scanning electron micrograph of an *A. thaliana* pollen tube en route to the ovule for fertilization. (C) Lily pollen tubes grown in vivo and removed from the ovary. Each green strand is an individual pollen tube and contains two sperm nuclei (bright blue stain) and a fainter (lighter blue) tube cell nucleus. Note the huge number of pollen tubes, all "racing" to fertilize a single egg. (A after V. E. Franklin-Tong. 2002. *Curr Opin Plant Biol* 5: 14–18.)

local oxidation that weakens the pollen wall at its contact point with the stigma surface (Edlund et al. 2016, 2017). This allows the pollen to form apertures anywhere.

Once out of the grain, the pollen tube starts to enzymatically digest the stigma cell walls (Knox and Heslop-Harrison 1970) and grows down the style of the carpel toward the opening in the ovule, called the **micropyle** (**FIGURE 7.32**). The vegetative tube nucleus and the two sperm cells are kept at the growing tip by bands of callose (a complex carbohydrate). This may be an exception to the "plant cells do not move" rule, as the sperm cells appear to move ahead via adhesive molecules (Lord et al. 1996). In order to extend many centimeters from the stigma to the ovules, the sperm cells and vegetative nucleus move together as a "male germ unit" at the tip of the tube (Sprunck et al. 2014).

Pollen tube navigation

The pollen tube has a long journey ahead of it—burrowing into the stigma, growing in the maze of style tissues, and navigating into the micropyle of the ovule (Zheng et al. 2018). The successful migration of the pollen tube and its two sperm cells through the carpel depends on the crosstalk between the pollen grain's genome and the carpel. These interactions involve secreted molecules, signaling pathways, and calcium ions (Palanivelu and Tsukamoto 2012; Dresselhaus and Franklin-Tong 2013; Li et al. 2018). The tube cell grows at its tip at the prodigious speed of about 1 cm per hour, eventually entering the micropyle opening. (See **Further Development 7.14, Growth of the Pollen Tube Requires Calcium**, online.)

Two specialized cells in the embryo sac, called synergid cells, may attract the pollen tube as the final step in pollen guidance. In *Torenia fournieri*, the embryo sac protrudes from the micropyle and can be cultured in vitro, where it retains the ability to attract pollen tubes. Higashiyama and colleagues (2001) used a laser beam to destroy individual cells in the embryo sac and then tested whether pollen tubes were still attracted to the embryo sac. When both synergids were destroyed, pollen tubes were not attracted to the embryo sac, but a single synergid was sufficient to guide pollen tubes. The synergids appear to be secreting specific polypeptides that attract the pollen tube during this last stage of guidance (Okuda and Higashiyama 2010).

Double fertilization

The growing pollen tube enters the embryo sac through the micropyle and passes through one of the synergids. The two sperm cells are then released, and a **double fertilization** event occurs (reviewed by Southworth 1996; Dresselhaus et al. 2016). One sperm cell fuses with the egg cell to produce the zygote that will develop into the new embryonic plant. The second sperm cell fuses with the usually binucleate central cell, giving rise to the **endosperm**, the tissue that nourishes the developing embryo. This second event is not "true" fertilization in the sense of male and female gametes undergoing fusion. That is, it does not result in a zygote, but rather in nutritionally supportive tissue. The other accessory cells in the embryo sac degenerate after fertilization, and the cells covering the ovule harden into the seed coat—what you see on a black watermelon seed, but not on one of those soft white seeds in "seedless" watermelons. Double fertilization, first identified a century ago, is generally restricted to the angiosperms. Friedman (1998) has suggested that endosperm may have evolved from a second zygote "sacrificed" as a food supply in a gymnosperm with double fertilization.[4]

As in sea urchins and mammals, angiosperms couple fertilization with a block against polyspermy. Once the synergid cells have attracted the pollen tube into the micropyle, one of them meets the pollen tube, and catastrophic events occur involving the death of both partners (Denninger et al. 2014). First, the synergid cell degenerates. This prevents other pollen tubes from entering the micropyle (a condition half-jokingly referred to as *polytubey*) (Dresselhaus and Marton 2009; Higashiyama and Takeuchi 2015; Maruyama et al. 2015). Then the pollen tube bursts, liberating the two sperm cells.

Also, as in animals, angiosperm fertilization involves a mutual activation of gametes. When the sperm cells are released, the egg undergoes a flux of Ca^{2+}. Then, chemicals released from the egg appear to make the sperm competent for cell fusion. As one review (Dresselhaus et al. 2016) concludes, "Altogether, these findings indicate that first the egg becomes activated during or immediately after sperm release, in turn activating the sperm cells and enabling them to fuse quickly." Very little is known about the chemical processes by which one sperm fertilizes the egg and the other fuses with the central cell, but conserved gamete fusion proteins found throughout the plant kingdom are providing some interesting clues (Mori et al. 2015; Clark 2018).

Fertilization is not an absolute prerequisite for angiosperm embryonic development (Mogie 1992). Embryos can form within embryo sacs that arise from cells that did not divide meiotically. This phenomenon, common in dandelions, is called **apomixis** (Greek, "without mixing") and results in viable seeds. Indeed, viable plants can grow from cuttings, in which a ball of cells called a callus is induced to form a meristem and begin development anew (Sugimoto et al. 2011; Ikeuchi et al. 2016).

Coda

Fertilization is not a moment or an event, but a process of carefully orchestrated and coordinated events, including the contact and fusion of gametes, the fusion of nuclei, and the activation of development. It is a process whereby two cells, each at the verge of death, unite to create a new organism that will have numerous cell types and organs. It is just the beginning of a series of cell-cell interactions that characterize animal and plant development.

Closing Thoughts on the Opening Photo

When Oscar Hertwig (1877) discovered fertilization in sea urchins, he delighted in seeing what he called "the sun in the egg." This was evidence that the fertilization was going to be successful. This glorious projection turns out to be the microtubular array generated by the sperm centrosome. This set of microtubules reaches out and finds the female pronucleus, and the two pronuclei migrate toward one another on these microtubular tracks. In this micrograph, the DNA of the pronuculei is stained blue, and the female pronucleus is much larger than that derived from the sperm. The microtubules are stained green.

From journal cover associated with J. Holy and
G. Schatten. 1991. *Dev Biol* 147: 343–353

Endnotes

[1]The importance of dynein can be seen in individuals with a genetic syndrome known as Kartagener triad. These individuals lack functional dynein in all their ciliated and flagellated cells, rendering these structures immotile (Afzelius 1976). Thus, males with Kartagener triad are sterile (immotile sperm). Both men and women affected by this syndrome are susceptible to bronchial infections (immotile respiratory cilia) and have a 50% chance of having the heart on the right side of the body (a condition known as situs inversus, the result of immotile cilia in the cells necessary for gastrulation movements).

[2]Eggs over easy: the terminology used in describing the female gamete can be confusing. In general, an egg, or ovum, is a female gamete capable of binding sperm and being fertilized. An oocyte is a developing egg that cannot yet bind sperm or be fertilized (Wessel 2009). The problems in terminology come from the fact that the eggs of different species are in different stages of meiosis (see Figure 7.3). The human egg, for example, is in second meiotic metaphase when it binds sperm, whereas the sea urchin egg has completed all of its meiotic divisions when it binds sperm. The contents of the egg also vary greatly from species to species. The award for the

greatest amount of cytoplasm in a cell goes to Aepyornis, the extinct elephant bird of Madagascar, whose egg measured about 1 m in circumference and held a volume of more than 2 gallons.

[3]In certain salamanders, this function of fertilization (i.e., initiating development of the embryo) has been totally divorced from the genetic function. The silvery salamander Ambystoma platineum is a hybrid subspecies consisting solely of females. Each female produces an egg with an unreduced chromosome number. This egg, however, cannot develop on its own, so the silver salamander mates with a male Jefferson salamander (A. jeffersonianum). The sperm from the Jefferson salamander merely stimulates the egg's development; it does not contribute genetic material (Uzzell 1964). For details of this complex mechanism of procreation, see Bogart et al. 1989, 2009.

[4]Amazingly, there are some species of insects wherein the polar bodies of the egg not only survive, but also become differentiated nutritive cells (Schmerler and Wessel 2011).

Snapshot Summary
Fertilization

1. Fertilization accomplishes two separate activities: sex (the combining of genes derived from two parents) and reproduction (the creation of a new organism).

2. The events of fertilization usually include (i) contact and recognition between sperm and egg; (ii) regulation of sperm entry into the egg; (iii) fusion of genetic material from the two gametes; and (iv) activation of egg metabolism to start development.

3. In animals, the sperm head consists of a haploid nucleus and an acrosome. The acrosome is derived from the Golgi apparatus and contains enzymes needed to digest extracellular coats surrounding the egg. The midpiece of the sperm contains mitochondria and the centriole that generates the microtubules of the flagellum. Energy for flagellar motion comes from mitochondrial ATP and a dynein ATPase in the flagellum.

(Continued)

Snapshot Summary *(continued)*

4. The female gamete can be an egg (with a haploid nucleus that has finished meiosis, as in sea urchins) or an oocyte (in an earlier stage of development, as in mammals). The egg (or oocyte) has a large mass of cytoplasm storing ribosomes and nutritive proteins. Some mRNAs and proteins that will be used as morphogenetic factors are also stored in the egg. Many eggs also contain protective agents needed for survival in their particular environment.

5. Surrounding the egg cell membrane is an extracellular layer often used in sperm recognition. In most animals, this extracellular layer is the vitelline envelope. In mammals, it is the much thicker zona pellucida. Cortical granules lie beneath the egg's cell membrane.

6. Neither the egg nor the sperm is the "active" or "passive" partner; the sperm is activated by the egg, and the egg is activated by the sperm. Both activations involve calcium ions and membrane fusions.

7. In many plants and animals, eggs, or their associated cells, secrete diffusible molecules that attract and activate the sperm. These can be species-specific chemotactic molecules, as in sea urchins, providing direction toward the egg of the correct species of sperm.

8. The acrosome reaction releases enzymes exocytotically. These proteolytic enzymes digest the egg's protective coating, allowing the sperm to reach and fuse with the egg cell membrane. In sea urchins, this reaction in the sperm is initiated by compounds in the egg jelly. Globular actin polymerizes to extend the acrosomal process. Bindin on the acrosomal process is recognized by a protein complex on the sea urchin egg surface.

9. Fusion between sperm and egg is probably mediated by protein molecules whose hydrophobic groups can merge the sperm and egg cell membranes.

10. Polyspermy results when two or more sperm fertilize an egg. It is usually lethal, since it results in blastomeres with different numbers and types of chromosomes.

11. Many species have two blocks to polyspermy. The fast block is immediate and transient, and causes the egg membrane resting potential to rise. Sperm can no longer fuse with the egg. In sea urchins this is mediated by the influx of sodium ions. The slow block, or cortical granule reaction, is physical and permanent, and is mediated by calcium ions. A wave of Ca^{2+} propagates from the point of sperm entry, causing the cortical granules to fuse with the egg cell membrane. In sea urchins, the released contents of these granules cause the vitelline envelope to rise and harden into the fertilization envelope.

12. The fusion of sperm and egg causes re-initiation of the egg's cell cycle and subsequent mitotic division, and the resumption of DNA and protein synthesis.

13. In all animal species studied, free Ca^{2+}, supported by the alkalinization of the egg, activates egg metabolism, protein synthesis, and DNA synthesis. Inositol trisphosphate (IP_3) is responsible for releasing Ca^{2+} from storage in the endoplasmic reticulum. Diacylglycerol (DAG) is thought to initiate the rise in egg pH.

14. IP_3 is generated by phospholipase C. Different species may use different mechanisms to activate the phospholipases.

15. The male and female pronuclei migrate toward one another and merge to become the diploid zygote nucleus.

16. Mammalian fertilization takes place internally, within the female reproductive tract. The cells and tissues of the female reproductive tract actively regulate the transport and maturity of both the male and female gametes.

17. The translocation of sperm from the vagina to the egg is regulated by the muscular activity of the uterus, by the binding of sperm in the isthmus of the oviduct, and by directional cues from the oocyte or the cumulus cells surrounding it.

18. Mammalian sperm must be capacitated in the female reproductive tract before they are capable of fertilizing the egg. Capacitated mammalian sperm can penetrate the cumulus and bind the zona pellucida.

19. In a recent model of mammalian sperm-zona binding, the acrosome-intact sperm bind to ZP3 on the zona, and ZP3 induces the sperm to undergo the acrosome reaction on the zona pellucida; the acrosome-reacted sperm, having been induced to undergo the acrosome reaction in the cumulus, bind to ZP2.

20. In mammals, blocks to polyspermy include modification of the zona proteins by the contents of the cortical granules so that sperm can no longer bind to the zona.

21. The rise in intracellular free Ca^{2+} at fertilization in mammals causes the degradation of cyclin and the inactivation of MAP kinase, allowing the second meiotic metaphase to be completed and the formation of the mature female pronucleus.

22. In mammals, DNA replication takes place as the pronuclei are traveling toward each other. The pronuclear membranes disintegrate as the pronuclei approach each other, and their chromosomes gather around a common metaphase plate.

23. In angiosperms, fertilization is initiated in the progamic phase, when pollen is attached to the stigma. The pollen germinates, forming a long tube. The two sperm cells follow the tube nucleus at the tip of the tube, and interactions between the tube nucleus and the style allow the movement of the pollen tube to the micropyle of the ovule.

24. The pollen tube is attracted into the micropyle by synergid cells. After the tube arrives, it opens to release the sperm cells, one of which fuses with the haploid egg cell to make the diploid zygote, and one of which fuses with the binucleate central cell.

Go to oup.com/he/barresi12xe for Further Developments, Scientists Speak interviews, Watch Development videos, Dev Tutorials, and complete bibliographic information for all literature cited in this chapter.

Snails, Flowers, and Nematodes
Different Mechanisms for Similar Patterns of Specification

How are spiral patterns made?

FERTILIZATION GIVES THE ORGANISM a new genome and rearranges its cytoplasm. Once this is accomplished, the resulting zygote begins the production of a multicellular organism. During **cleavage**, rapid cell divisions divide the zygote cytoplasm into numerous cells. In the case of animals, these cells undergo dramatic displacements during **gastrulation**, a process whereby the cells move to different parts of the embryo and acquire new neighbors. We described the different patterns of cleavage and gastrulation in Chapter 1 (see Figure 1.9 and Table 1.1). Due to their rigid cell wall, plant cells are unable to migrate, and therefore the organization of their parts is based on the early patterns of cell division and growth.

During early development, the major body axes of most animals and plants are specified—a developmental process aptly called **axis determination**. In a plant, the main axis is the apical-basal axis, which we touched on in Chapters 2 and 4 (see Figures 2.10 and 4.29). But there is also a radial or spiral pattern to plant structures, such as leaves that encircle a stem and parts of flowers that radiate out from the flower's center. These patterns are controlled by a self-organizing system of cell-cell interactions. In most animals, three axes are specified in the early embryo: the anterior-posterior (head-tail), dorsal-ventral (back-belly), and left-right axes (see Figure 1.10). Different species specify these axes at different times, using different mechanisms. In some species, axis determination begins as early as oocyte formation (as in *Drosophila*). In other species, it occurs during cleavage, while in still others axis determinaton continues all the way through gastrulation (as in *Xenopus*).

The chapters in this unit will predominantly look at axis determination in several animal groups as we examine their early development from cleavage through gastrulation. We will look primarily at species that have emerged as important model organisms (including snails, nematodes, fruit flies, sea urchins, frogs, zebrafish, chickens, and mice; for a refresher on the benefits of model organisms, see Chapter 1, pp. 5 and 6, Figure 1.4). In this chapter we will also compare the development of spiral patterns in snails and plants, to gain a better understanding of how different mechanisms can achieve similar patterns.

A Reminder of the Evolutionary Context That Built the Strategies Governing Early Development

To be a **multicellular eukaryotic organism** (i.e., plant, fungus, or animal) means to be made up of multiple cells formed by mitosis, united in a functional whole. Both plants and animals evolved from a last eukaryotic common ancestor (see Figure 1.25), which suggests that molecular and cellular homologies exist between plants and animals. Importantly, though, any analogous processes or convergent mechanisms of evolution between plants and animals would help identify truly foundational principles of development.

To be a **metazoan** means to be an animal, and to be an animal means to go through gastrulation during development. All animals gastrulate, and no other organisms gastrulate. Different groups of animals undergo different patterns of development. When we say that there are 35 existing metazoan phyla, we are stating that there are 35 surviving patterns of animal development (see Davidson and Erwin 2009; Levin et al. 2016). These patterns of organization have evolved through branching pathways. We introduced these in Chapter 1 (see Figure 1.21). As we explore the mechanisms governing early embryonic development, keep in mind the four major branches of metazoans: the basal phyla, the lophotrochozoan and ecdysozoan protostomes, and the deuterostomes (**FIGURE 8.1**). Similarities and differences in early embryonic development are used in assigning animals to the different phyla. Four embryonic features are of particular importance: (1) whether an animal has two or three germ layers, (2) when and where its mouth and anus form, (3) the pattern of its early cleavages, and (4) whether it possesses the embryonic structure known as the notochord.

The diploblastic animals: Cnidarians and ctenophores

Animals that have two germ layers—ectoderm and endoderm but little or no mesoderm—are referred to as **diploblasts**, such as the cnidarians (jellyfish and hydras) and the ctenophores (comb jellies).[1] Unlike bilaterially symmetrical triploblastic animals (those with three germ layers), the diploblastic cnidarians and ctenophores have radial symmetry and have long been thought to have no mesoderm. However, such clear-cut distinctions are now being questioned, at least in regard to the cnidarians. Although cnidarians such as *Hydra* have no true mesoderm, others seem to have some mesoderm, and some display bilateral symmetry at certain stages of their life cycle (Martindale et al. 2004; Martindale 2005; Matus et al. 2006; Steinmetz et al. 2017). However, the mesoderm of cnidarians may have evolved independently from that found in the protostomes and deuterostomes.

The triploblastic animals: Protostomes and deuterostomes

The vast majority of metazoan species have three germ layers—ectoderm, endoderm, and mesoderm—and are thus **triploblasts**. The evolution of the mesoderm enabled greater mobility and larger bodies because it became the animal's musculature and circulatory system. Triploblastic animals are also called **bilaterians** because they have bilateral symmetry—that is, they have right and left sides. Bilaterians are further classified as either **protostomes** or **deuterostomes** (see Figure 8.1).

PROTOSTOMES Protostomes (Greek, "mouth first"), which include the mollusk, arthropod, and worm phyla, are so called because the mouth is formed first, at or near the opening to the gut that is produced during gastrulation. The anus forms later, at a different location. The protostome **coelom**, or body cavity, forms from the hollowing out of a previously solid cord of mesodermal cells in a process called **schizocoely**.

There are two major branches of protostomes. The **ecdysozoans** (Greek *ecdysis*, "to get out of" or "shed") are those animals that molt their exterior skeletons. The most prominent ecdysozoan group is Arthropoda, the arthropods, a well-studied phylum that includes the insects, arachnids, mites, crustaceans, and millipedes. Molecular analysis has also placed another molting group, the nematodes, in this clade. Members of the second major protostome group, the **lophotrochozoans**, are characterized by a common type of cleavage (spiral) and a common larval form (the trochophore). Lophotrochozoans include 14 of the 35 metazoan phyla, including the flatworms, annelids, and mollusks. The spiral cleavage program is so characteristic of this group that the term *spiralia* has become another way of describing this clade (Henry 2014).

DEUTEROSTOMES The major deuterostome lineages are the chordates (including the vertebrates) and the echinoderms. Although it may seem strange to classify humans,

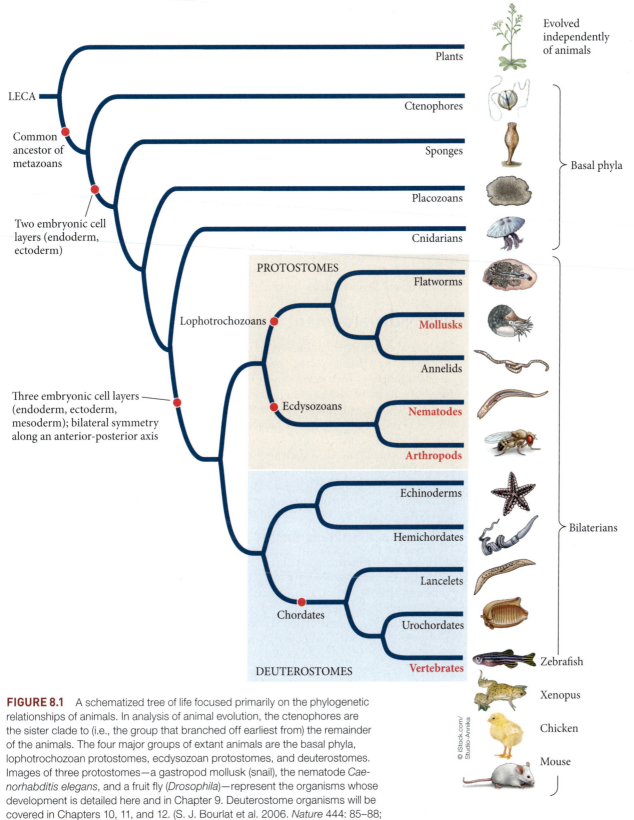

FIGURE 8.1 A schematized tree of life focused primarily on the phylogenetic relationships of animals. In analysis of animal evolution, the ctenophores are the sister clade to (i.e., the group that branched off earliest from) the remainder of the animals. The four major groups of extant animals are the basal phyla, lophotrochozoan protostomes, ecdysozoan protostomes, and deuterostomes. Images of three protostomes—a gastropod mollusk (snail), the nematode *Caenorhabditis elegans*, and a fruit fly (*Drosophila*)—represent the organisms whose development is detailed here and in Chapter 9. Deuterostome organisms will be covered in Chapters 10, 11, and 12. (S. J. Bourlat et al. 2006. *Nature* 444: 85–88; F. Delsuc et al. 2005. *Nature* 439: 965–968; B. Schierwater et al. 2009. *PLOS Biol* 7: e1000020; A. Hejnol. 2012. *Nature* 487: 181–182; J. F. Ryan et al. 2013. *Science* 342:1242592.)

fish, and frogs in the same broad group as seastars and sea urchins, certain embryological features stress this kinship. First, in most deuterostomes (Greek, "mouth second"), the oral opening is formed after the anal opening. Also, whereas protostomes generally form their body cavity by hollowing out a solid block of mesoderm (schizocoely, as mentioned earlier), most deuterostomes form their body cavity by extending mesodermal pouches from the gut (**enterocoely**). (There are many exceptions to this generalization, however; see Martín-Durán et al. 2012.)

Recall from Chapter 1 that the lancelets (Cephalochordata; *Amphioxus*) and the tunicates (Urochordata; sea squirts) are invertebrates—they have no backbone. However, the *larvae* of these organisms have a notochord and pharyngeal arches,[2] indicating that they are chordates (see Figure 1.20A). The "chord" in "chordates" refers to the **notochord**, which induces the formation of the vertebrate spinal cord.

What's to develop next

In this chapter, we will focus on two groups of protostome invertebrates, the gastropod mollusks (represented by snails) and the nematodes (represented by *Caenorhabditis elegans*). Despite the differences in early development between these invertebrate groups, they have both evolved rapid development to a larval stage (Davidson 2001). Special attention will also be paid to the events that lead to the development of spiral patterns in snails, which we will contrast with the development of similar spiral patterns in the independently evolved angiosperm *Arabidopsis thaliana*.

Early Development in Snails

Snails have a long history as model organisms in developmental biology. They are abundant along the shores of all continents, they grow well in the laboratory, and they show variations in their development that can be correlated with their environmental needs. Some snails also have large eggs and develop rapidly, specifying cell types very early in development. Although each organism uses both autonomous and conditional modes of cell specification (see Chapter 2), snails provide some of the best examples of autonomous development, in which the loss of an early blastomere causes the loss of an entire structure. Indeed, in snail embryos, the cells responsible for certain organs can be localized to a remarkable degree. The results of experimental embryology can now be extended (and explained) by molecular analyses, leading to fascinating syntheses of development and evolution (see Conklin 1897; Henry 2014).

Cleavage in Snail Embryos

"[T]he spiral is the fundamental theme of the molluscan organism. They are animals that twisted over themselves" (Flusser 2011). Indeed, the shells of snails are spirals, their larvae undergo a 180° torsion that brings the anus anteriorly above the head, and (most important) the cleavage of their early embryos is spiral. **Spiral holoblastic cleavage** (see Figure 1.9) is characteristic of several animal groups, including annelid worms, platyhelminth flatworms, and most mollusks (Hejnol 2010; Lambert 2010). The cleavage planes of spirally cleaving embryos are not parallel or perpendicular to the animal-vegetal axis of the egg; rather, cleavage is at oblique angles, forming a spiral arrangement of daughter blastomeres. The blastomeres are in intimate contact with each other, producing thermodynamically stable packing arrangements, much like clusters of soap bubbles. Moreover, spirally cleaving embryos usually undergo relatively few divisions before they begin gastrulation, making it possible to follow the fate of each cell of the blastula. When the fates of the individual blastomeres from annelid, flatworm, and mollusk embryos were compared, many of the same cells were seen in the same places, and their general fates were identical (Wilson 1898; Hejnol 2010). This represents a tight homology of blastomere development in the spiralians that is rarely seen between different phyla. Blastulae produced by spiral cleavage typically have very small or no blastocoel and are called **stereoblastulae**.

Mollusks such as the nutclam and mud snail shown in **FIGURE 8.2A** exhibit typical spiral holoblastic cleavage. As in many molluscan embryos, the first two cleavages

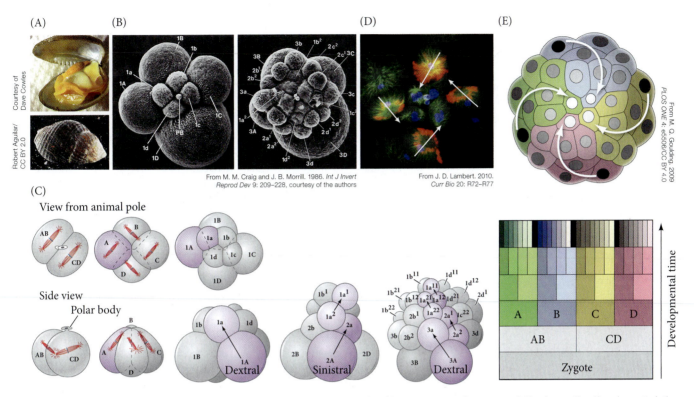

FIGURE 8.2 Spiral cleavage patterns in the molluscan embryo. (A) The nutclam (*Acila castrensis*; above) and mud snail (*Ilyanassa obsoleta*; below). (B) Scanning electron micrographs of dextral cleavage in the mud snail, showing 8-cell (left) and 32-cell (right) stages. PB, polar body (a remnant of meiosis). (C) Schematic of spiral cleavage of the snail *Trochus* viewed from the animal pole (top) and from one side (bottom). Cells derived from the A blastomere are shown in color. The mitotic spindles, sketched in the early stages, divide the cells unequally and at an angle to the vertical and horizontal axes. Each successive quartet of micromeres (lowercase letters) is displaced clockwise (dextral) or counterclockwise (sinistral) relative to its sister macromere (uppercase letters), creating the characteristic spiral pattern (arrows). (D) Animal pole view showing the segregation of an RNA (*IoLR2*, red) into the second-quartet micromeres of the mud snail embryo. The mitotic spindles of the second sinistral twisting quartet are visible (DNA, blue; microtubules, green). (E) This animal pole illustration of a spiralian embryo shows each of the four quadrant's lineages in different colors, with the different hues for each color representing first- (light) and second-quartet (dark) clones. Each quartet lineage is also represented below this illustration as a flowchart to emphasize the symmetrical derivation of each micromere clone.

are nearly **meridional** (transverse through the equatorial plane), producing four large **macromeres** (labeled A, B, C, and D; **FIGURE 8.2B,C**). In many species, these four blastomeres are different sizes (D being the largest), a characteristic that allows them to be easily identified (see Figure 8.2B). Another fascinating aspect of molluscan cleavages is that with each successive cleavage, each macromere buds off a small **micromere** at its animal pole, such that each quartet of micromeres is displaced to the right or to the left of its sister macromere. This displacement alternates from right to left through successive divisions, resulting in the characteristic spiral pattern of cells stacked on top of one another (follow the arrows in Figure 8.2C). Looking down on the embryo from the animal pole, the upper ends of the mitotic spindles are seen to alternate clockwise and counterclockwise (**FIGURE 8.2D**). This causes alternate groups of micromeres to be cleaved off to the left and then to the right of their parent cell. This pattern of cleavage gives rise to four "quadrant lineages" arranged in a spiral fashion around the animal-vegetal axis (**FIGURE 8.2E**; Goulding 2009). In normal development, the first-quartet micromeres form the head structures, the second-quartet micromeres form the statocyst (balance organ) and shell, and the third-quartet micromeres end up anterior to the blastopore on the ventral surface and form the mouth (**FIGURE 8.3**; Lambert 2010). These fates are specified both by localization of cytoplasmic factors as well as by regional inductive signals (Cather 1967; Clement 1967; Render 1991; Sweet 1998).

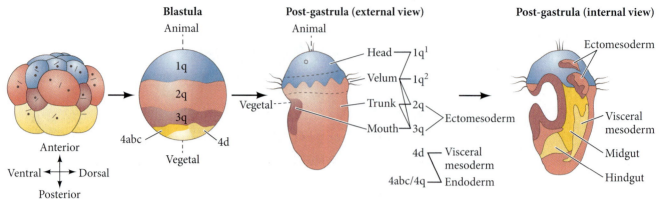

Blastula
Animal
Vegetal

Post-gastrula (external view)
Animal

Head — 1q¹
Velum — 1q²
Trunk — 2q
Mouth — 3q
Ectomesoderm

4d — Visceral mesoderm
4abc/4q — Endoderm

Post-gastrula (internal view)
Ectomesoderm
Visceral mesoderm
Midgut
Hindgut

FIGURE 8.3 The spiralian fate map through gastrulation. A clear distribution of determinants along the animal-to-vegetal axis corresponds to cell fates of ectomesoderm, visceral mesoderm, and endoderm origins. Different quartets of micromeres are indicated by q; $1q^1$, $1q^2$ represent cells derived from the 1q quartet. Importantly, the oral opening develops just anterior to the blastopore and forms at the ventral surface. (After J. D. Lambert. 2010. *Curr Biol* 20: 272–277.)

Maternal regulation of snail cleavage

The orientation of the cleavage plane to the left or to the right is controlled by cytoplasmic factors in the oocyte. This was discovered by analyzing mutations of snail coiling. In some snails, the coil curves around to the right when viewed from above and opens on the right side of the shell (**dextral coiling**), whereas the coils of other snails curve and open to the left (**sinistral coiling**). Usually the direction of coiling is the same for all members of a given species, but occasional mutants are found (i.e., in a population of right-coiling snails, a few individuals will be found with coils that open on the left). Crampton (1894) analyzed the embryos of aberrant snails and found that their early cleavage differed from the norm (**FIGURE 8.4**). The orientation of the cells after the second cleavage was different in the sinistrally coiling snails as a result of a different orientation of the mitotic apparatus. You can see in Figure 8.4 that the position of the 4d blastomere is different in the right-coiling and left-coiling snail embryos. This 4d blastomere is rather special. It is often called the **mesentoblast**, since its progeny include most of the mesodermal organs (heart, muscles, primordial germ cells) and endodermal organs (gut tube).

In snails such as *Radix* (previously known as *Lymnaea*), the direction of snail shell coiling is controlled by a single pair of genes (Sturtevant 1923; Boycott et al. 1930; Shibazaki 2004). In *Radix peregra*, mutants exhibiting sinistral coiling were found and mated with wild-type, dextrally coiling snails. These matings showed that the right-coiling allele, *D*, is dominant to the left-coiling allele, *d*. However, the direction of cleavage is determined not by the genotype of the developing snail but by the genotype of the snail's *mother*. This is called a **maternal effect**.

(A) Left-handed coiling

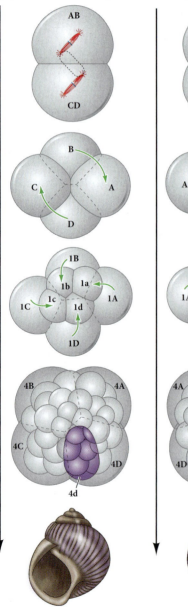

(B) Right-handed coiling

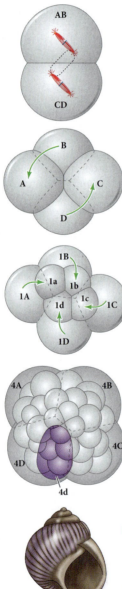

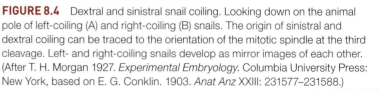

FIGURE 8.4 Dextral and sinistral snail coiling. Looking down on the animal pole of left-coiling (A) and right-coiling (B) snails. The origin of sinistral and dextral coiling can be traced to the orientation of the mitotic spindle at the third cleavage. Left- and right-coiling snails develop as mirror images of each other. (After T. H. Morgan 1927. *Experimental Embryology.* Columbia University Press: New York, based on E. G. Conklin. 1903. *Anat Anz* XXIII: 231577–231588.)

FIGURE 8.5 The *formin* gene controls left- and right-handed coiling at third cleavage. A left-coiling (sinistral) strain of the snail *Radix stagnalis* shows a complete loss of the maternally expressed *formin* mRNA (A) and protein (B) from the zygote as compared with a strain with a right-coiling (dextral) morphology. (C) Staining for actin (green) and microtubules (red) shows the helical deformation (white arrowheads) at the third cleavage in normal cleavage patterns of dextral embryos versus abnormal cleavage in sinistral embryos. White arrows denote spindle orientation; yellow arrows point in the direction of blastomere formation. pb, polar body.

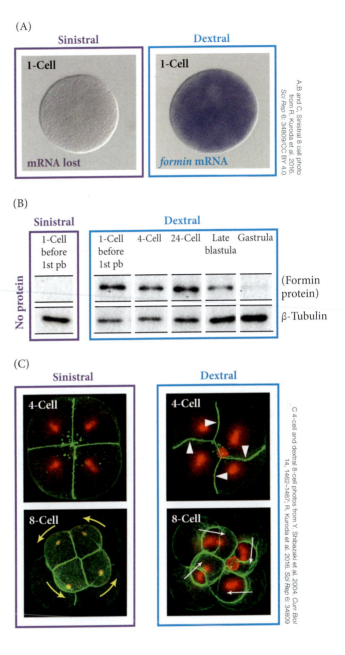

(We'll see other important maternal effect genes when we discuss *Drosophila* development.) A *dd* female snail can produce only sinistrally coiling offspring, even if the offspring's genotype is *Dd*. A *Dd* individual will coil either left or right, depending on the genotype of its mother. Such matings produce a chart like this:

Genotype		Phenotype	
DD female × *dd* male	→	*Dd*	All right-coiling
DD male × *dd* female	→	*Dd*	All left-coiling
Dd × *Dd*	→	1*DD*:2*Dd*:1*dd*	All right-coiling

Thus, it is the genotype of the *ovary* in which the oocyte develops that determines which orientation cleavage will take. The genetic factors involved in coiling are brought to the embryo in the oocyte cytoplasm. When Freeman and Lundelius (1982) injected a small amount of cytoplasm from dextrally coiling snails into the eggs of *dd* mothers, the resulting embryos coiled to the right. Cytoplasm from sinistrally coiling snails, however, did *not* affect right-coiling embryos. These findings confirmed that the wild-type mothers were placing a factor into their eggs that was absent or defective in the *dd* mothers. These experiments, among others, provided some of the first evidence for the existence of cytoplasmic determinants, and set the stage for the long journey to identify the mysterious determinants.

A major breakthrough came when two groups working with similar snail populations independently identified and mapped a gene encoding a formin protein that is active in the eggs of mothers that carry the *D* allele, but not in the eggs of *dd* mothers (**FIGURE 8.5A,B**; Liu et al. 2013; Davison et al. 2016; Kuroda et al. 2016). Thus, *DD* and *Dd* mothers produce active formin proteins. In *dd* females, however, the *formin* gene has a frameshift mutation in the coding region that renders its mRNA nonfunctional, so its message is rapidly degraded. When the egg contains functional *formin* mRNA from the mother's *D* allele, this message becomes asymmetrically positioned in the embryo as early as the 2-cell stage. The formin protein encoded by the mRNA message binds to actin and helps align the cytoskeleton. These findings are upheld by studies showing that drugs that inhibit formins cause eggs from *DD* mothers to develop into left-coiling embryos.

The first indication that the cells will divide sinistrally rather than dextrally is a helical deformation of the cell membranes at the dorsal tip of the macromeres (**FIGURE 8.5C**). Once the third cleavage takes place, the Nodal protein (a TGF-β superfamily paracrine factor) activates genes on the right side of dextrally coiling embryos and on the left side of sinistrally coiling embryos (**FIGURE 8.6A**). Using glass needles to change the direction of cleavage at the 8-cell stage

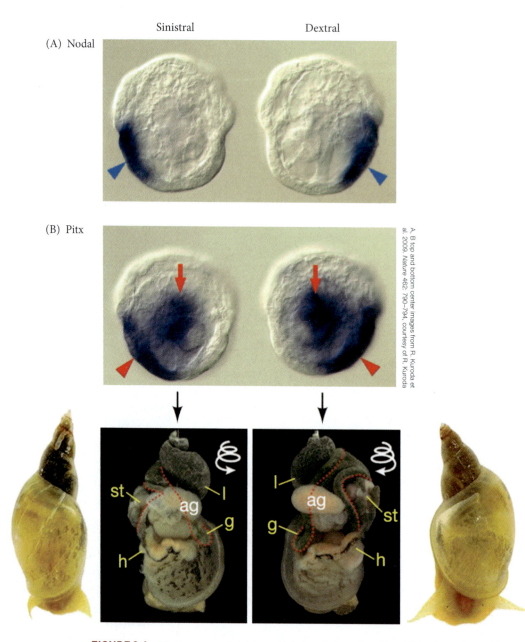

Sinistral   Dextral

(A) Nodal

(B) Pitx

FIGURE 8.6 Mechanisms of right- and left-handed snail coiling. (A) In the embryo, Nodal (blue) is activated in the shell gland (blue arrowhead) on the left side of sinistral embryos and on the right side of dextral embryos. (B) The Pitx1 transcription factor, seen expressed asymmetrically in the shell gland (red arrowheads) and visceral mass (red arrows) of the embryo (upper image; blue), is responsible for organ formation, as seen in the decapsulated ventral views of the adults (lower image). The positions of the following are indicated: ag, albumen gland; g (outlined with dotted red line), gut; h, heart; l, liver; st, stomach. The white spiraling arrow inset into the upper right corner represents the counter clockwise (left) and clockwise (right) direction of the coiling.

changes the location of *nodal* gene expression (Grande and Patel 2009; Kuroda et al. 2009; Abe et al. 2014). Nodal appears to be expressed in the C-quadrant micromere lineages (which give rise to the ectoderm) and induces asymmetrical expression of the gene for the Pitx1 transcription factor (a target of Nodal in vertebrate axis formation as well) in the neighboring D-quadrant blastomeres (**FIGURE 8.6B**). (See **Further Development 8.1, A Classic Paper Links Genes and Development**, online.)

Developing a spiral pattern: A plant's perspective

We just completed an analysis of how the snail's shell gets its particular twist, which began with the highly ordered arrangement of cell cleavages during early development. The organs of a plant, whether leaves or the reproductive floral tissues, display a similar wondrous pattern of repeated organs regularly spaced around the growing shoot or stem. One of the most recognizable patterns in nature—the spiraling arrangement of leaves on a succulent or of the yellow florets in the center of a daisy (**FIGURE 8.7A**)—follows the mathematical certainty of the Fibonacci sequence, just as does the shape of a nautilus shell. How is this spiral pattern of cell fates determined?

DEFINING PHYLLOTAXIS The regular arrangement of leaves along a plant stem is called **phyllotaxis** (Greek *phyllon*, "leaf"; *taxis*, "order") and is so predictable that it is an important feature for plant identification. A plant's phyllotaxis is determined by the positions of newly formed leaf primordia around the shoot apical meristem. The arrangement of other plant organs (e.g., flowers) is similarly determined by the pattern of primordia formation. The developmental principles underlying such phyllotactic patterns are being actively investigated in the model plant *Arabidopsis thaliana*. There are three arrangements of lateral organs in *A. thaliana*: **decussate**, in which successive opposite pairs of organs are offset by 180° (as for the cotyledons and the first pair of vegetative leaves); **spiral**, in which organs emerge sequentially around the apex in accordance with Fibonacci's

Does being a righty or a lefty matter? Being right-handed or left-handed may play a relatively minor role in one's life as a human; but if one is a snail, it has crucial implications both for individuals and for the evolution of snail populations. Among snails, lefties mate more readily with lefties and righties mate more readily with righties—it's strictly a matter of genital position and physically hooking up. Moreover, certain snake species feed on shelled snails, and the jaws of these snakes have evolved to eat right-coilers more easily than lefties. How might this evolutionary adaptation among snakes have affected the evolution of snails in regions where the species coexist? (See Hoso et al. 2010 for an interesting set of experiments.)

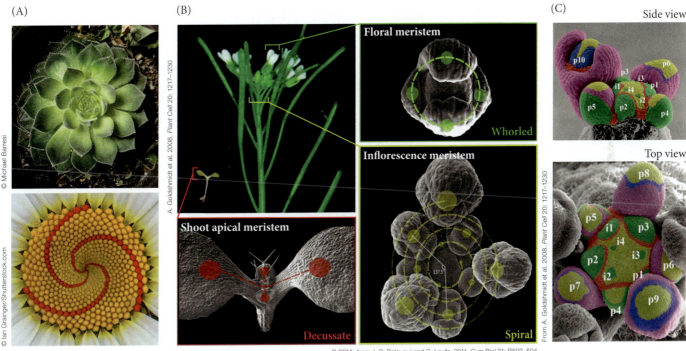

FIGURE 8.7 The shoot apical meristem generates phyllotactic patterns in plants. (A) Top views of a "candy floss" succulent plant and of the center of a daisy. In the daisy, there are 21 rows of florets; 3 equally spaced rows have been pseudocolored red to highlight the spiral arrangement of the florets. (B) The lateral organs of *Arabidopsis thaliana* fall into three types: decussate, whorled, and spiral. Only the first pair of shoot organs following the cotyledons forms in a decussate pattern (red bracket on inset of seedling); the arrangement of all subsequent leaves in *A. thaliana* follows a spiral pattern. (C) Scanning electron micrographs of an inflorescence meristem viewed from the side and from the top. These images have been pseudocolored to illustrate regions of cell specification within the meristem apex and in the developing floral buds. Primordia are labeled from youngest to oldest (p1, p2, p3, etc.). Incipient primordia (i1, i2, etc.) refer to the regions of the meristem where presumptive primordia are being specified.

"golden angle" of 137.5° between each organ (later vegetative leaves and flowers in the inflorescence); and **whorled**, in which a set of organs emerges simultaneously in a ring around the apex (flower parts) (**FIGURE 8.7B**; Palauqui and Laufs 2011). To make conceptual connections with our earlier discussion of spiral cleavages in snails, we will focus our attention on the underlying mechanisms of spiral development in the inflorescence meristem.

THE STRESSES IN THE INFLORESCENCE MERISTEM   Recall that the **inflorescence meristem** is a specialized stem cell niche through which cell division and cell expansion turn out lateral organ primordia destined to form new flowers. These lateral primordia encircle the apex of the meristem and become progressively displaced over time as new cells are produced at the apex (**FIGURE 8.7C**). The spiral pattern that emerges is controlled by biophysical properties of the meristem and the mechanisms governing transport of the hormone auxin, which together establish an amazing self-organizing system of cell-cell interactions.

To understand how the biophysical properties of the meristem play a major part in creating the spiraling pattern of flower buds, we need to realize that plant cells expand, and this expansion results in plant growth. Also, because plant cells cannot move independently of their neighbors, this expansion imposes mechanical stresses on the surrounding cells (as if you extended your arm out to the side, pushing on the student sitting next to you). Such stresses can have dramatic effects on the structural properties of a plant cell. Of particular relevance is how apical meristem cells are surprisingly adept at remodeling their cell walls in response to physical (tensile) forces (Shapiro et al. 2015), a remodeling that can determine the orientation of cell expansion. This force-dependent cell wall remodeling occurs in a two-step process:

1. First, the cortical microtubules (those just inside the cell membrane) reorient so that they are aligned perpendicular to the axis of greatest stress.

2. Then these microtubules serve as a scaffold to guide the orientation of new cellulose microfibrils added to the cell wall during remodeling.

The orientation of cellulose in the wall controls the direction in which a cell can expand (**FIGURE 8.8A**). Aligned cellulose fibers confer **mechanical anisotropy** on the wall; that is, the wall is not equally stretchable in all directions. When the cellulose is aligned parallel to the stress, cell elongation along the stress axis is resisted; when the cellulose is perpendicular, elongation is allowed (Bidhendi and Geitmann 2016). This is how the cytoskeleton and cell wall can influence the growth of a single cell, but how can this system be employed to shape a tissue-level morphogenetic event, such as the growth of a new floral primordium?

Now imagine the cells within an incipient lateral primordium. This mass of cells needs to push against and expand the overlying epidermal layer into a new floral primordium; for such outgrowth to occur, the epidermal cell walls must be able to stretch (**FIGURE 8.8B**). This is achieved by the localized disorganization of cortical microtubules in the epidermal cells (to an isotropic distribution) at the apex of the primordia, as compared with the microtubule organization observed in cells at the boundary and peripheral regions of the meristem (anisotropic distribution) (**FIGURE 8.8C,D**). Disorganized cortical microtubules mean that the cellulose microfibrils in the apical cell walls are also randomized, thus reducing resistance to cell expansion at the apex of the primordium (remember this is a tissue-level biophysical shape change). This raises the next logical question: How do these disorganized microtubules become localized to certain cells? The current hypothesis is that auxin signaling induces this localization, which then confers the positions of lateral organ primordia.

SPIRAL PATTERNS BY POSITIVE FEEDBACK   As we detailed in Chapter 4, auxin promotes cell expansion and tissue growth. Unlike animal morphogens, auxin does not move by simple diffusion, but rather is directed toward certain regions through the asymmetrical localization of PIN auxin efflux carriers (see Figures 4.29 and 4.30; Bhatia and Heisler 2018). Directed auxin flows are essential

(A)

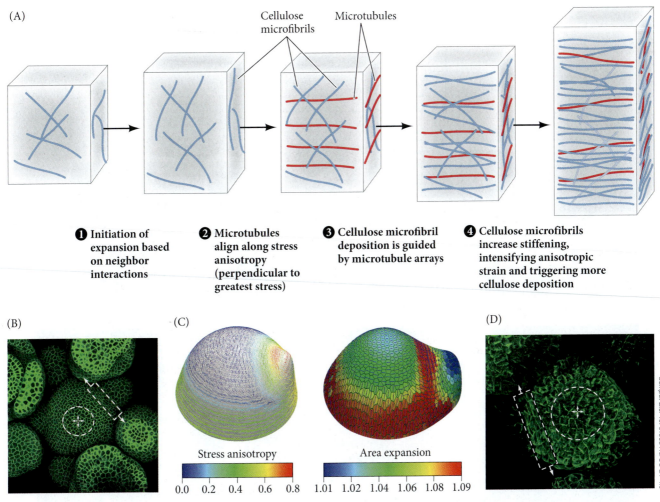

Cellulose microfibrils   Microtubules

1 Initiation of expansion based on neighbor interactions

2 Microtubules align along stress anisotropy (perpendicular to greatest stress)

3 Cellulose microfibril deposition is guided by microtubule arrays

4 Cellulose microfibrils increase stiffening, intensifying anisotropic strain and triggering more cellulose deposition

(B)

(C)

Stress anisotropy

0.0   0.2   0.4   0.6   0.8

Area expansion

1.01   1.02   1.04   1.06   1.08   1.09

(D)

FIGURE 8.8 Organization of microtubules and cellulose microfibrils in response to stress forces controls the direction of cell expansion. (A) Schematic model of how microtubule and cellulose microfibrils respond to and reinforce anisotropic growth (different amounts of growth along different axes) in plant cells. The expansion of a newly generated plant cell will be influenced by asymmetries in the rigidity of the cell wall and by forces transmitted through adherence with neighboring cells (1). The asymmetrical expansion of a plant cell along one axis over another will lead to several positively reinforcing effects that can perpetuate this anisotropic expansion. The cell will immediately respond to stress forces by building perpendicularly aligned microtubule arrays (2). These arrays then guide the deposition of cellulose microfibrils that are oriented parallel to the microtubule framework (3). This architecture creates a greater anisotropic stress that feeds back positively on the system, resulting in more cellulose deposition and further promoting cell expansion along the same axis (4). (B) Floral primordia emerging from the shoot apical meristem. Cells at the apex of the meristem (dashed circle) show uniform distribution of stress forces, while cells at the boundaries with new primordia (dashed box) are anisotropic. (C) The stress forces can be measured across the shoot apical meristem, and models such as the one shown here help visualize how anisotropic forces correlate with primordial growth (right). (D) Microtubules (green) are randomly organized at the apex of the shoot apical meristem (dashed circle), whereas microtubules in cells at the meristem periphery (and boundary with the primordium) (dashed box) are organized perpendicular to the apical-basal axis of the meristem (and primordium). (A after A. J. Bidhendi and A. Geitmann. 2016. *J Exp Bot* 67: 449–461.)

for lateral organ formation, as evidenced by *A. thaliana pin1* mutants that produce only a simple cylindrical shoot without organ primordia (**FIGURE 8.9A**). However, topical application of auxin to the apical meristem of a *pin1* mutant is sufficient to rescue primordia formation at the site of application (see Figure 8.9A; Reinhardt et al. 2003). During normal development, PIN1 efflux carriers become positioned such that several cells all direct their auxin toward a single central convergence point. This point, which accumulates auxin from all of the cells around it, becomes the center of a new primordium. What causes the cells to position their PIN1 proteins in this direction? Cells with high auxin concentrations are somehow sensed by neighboring cells, which results in the redistribution of their PIN efflux carriers to face the auxin maximum (**FIGURE 8.9B**; Heisler et al. 2005; Bhatia et al. 2016). A key component of the positive feedback loop is MONOPTEROS (also known as AUXIN RESPONSE FACTOR 5, or ARF5), an auxin-regulated auxin response transcription factor. Upregulation of MONOPTEROS precedes the polarization of PIN1 efflux carriers in cells neighboring those that already express MONOPTEROS, such that auxin will flow from a newly MONOPTEROS-expressing cell toward the convergence point (**FIGURE 8.9C**). This creates a positive feedback loop that increases the auxin concentration in the convergence point (Shapiro et al. 2015).

(A)

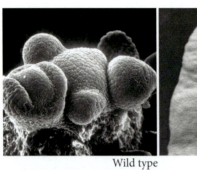

From N. Bhatia and M. G. Heisler. 2018. *Development* 145: dev149336; E. M. Meyerowitz et al. 1991. *Dev Suppl* 1: 157–167

From D. Reinhardt et al. 2000. *Plant Cell* 12: 507–518

Wild type   *pin1-1*

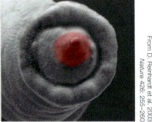

From D. Reinhardt et al. 2003. *Nature* 426: 255–260

pin1-1 + auxin at apex

(B)

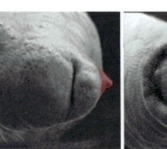

Left and center from M. G. Heisler et al. 2005. *Curr Biol* 15: 1899–1911

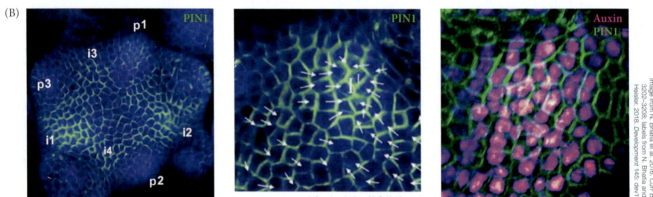

Image from N. Bhatia et al. 2016. *Curr Biol* 26: 3202–3208; labels from N. Bhatia and M. G. Heisler. 2018. *Development* 145: dev149336

(C)

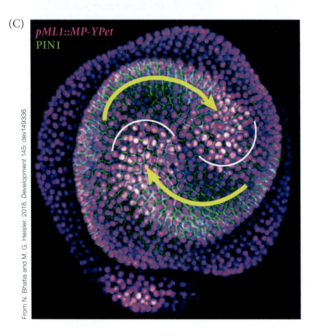

pML1::MP-YPet
PIN1

From N. Bhatia and M. G. Heisler. 2018. *Development* 145: dev149336

FIGURE 8.9 PIN-mediated auxin transport is required for organogenesis by the inflorescence meristem. (A) Loss of *PIN1* arrests lateral organ formation by the inflorescence meristem (leftmost two SEMs). However, local application of auxin (red shading) to the side or apex of the meristem of *pin1-1* mutants can induce primordium formation at the auxin source (rightmost two SEMs). (B) PIN1 auxin efflux carriers are polarized to distribute auxin to incipient primordia and promote their growth. The inflorescence meristem in these images is expressing a *PIN1:GFP* fusion. Left image: In a top-down view of the entire meristem, PIN1 can be seen as green fluorescence polarized to select sides of a cell's membrane. p1–p3 are primordia, i1–i4 are incipient primordia. Middle image: A magnified view of position i1. PIN1 polarity (marked by white arrows) is oriented toward the center of the incipient primordium (I). Right image: Another magnified view of position i1. The predicted activity of auxin is visualized using a ratiometric auxin-specific sensor (R2D2). PIN1 (green) is oriented away from the edges of the primordium and toward its center at the very tip, which is where the greatest activity of auxin is detected. Magenta shows low auxin activity; white shows high auxin activity. (C) In a *monopteros* mutant, restricted expression of MONOPTEROS (MP) only in the epidermis (*pML1::MP-YPet*) results in a single continuous spiral-shaped primordium. The shoot apical meristem in this image is from a mutant plant lacking MP but engineered to express MP only in the epidermal layer of cells. This image from a time-lapse video shows the localization of transgenic reporters for MP (magenta) and PIN1 (green) in the epidermis. Maximum MP expression spirals over time (white curves), while PIN1 follows behind (yellow arrows).

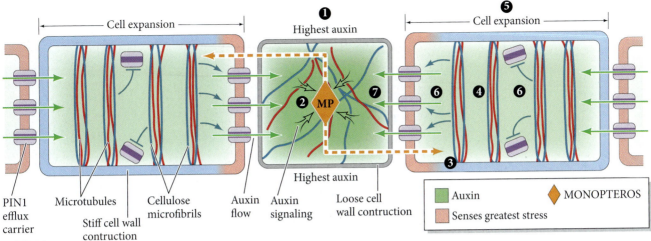

PIN1 efflux carrier · Microtubules · Stiff cell wall contruction · Cellulose microfibrils · Auxin flow · Auxin signaling · Loose cell wall contruction

Cell expansion · ❶ Highest auxin · ❺ Cell expansion · ❷ MP · ❼ · ❻ ❹ ❻ · ❸ · Highest auxin

Auxin · MONOPTEROS · Senses greatest stress

FIGURE 8.10 Model of phyllotaxis by a mechanism of morphogen-mechanical stress feedback. (1) The cell with maximum auxin has both the loosest wall construction and the highest activation of the auxin response factor, MONOPTEROS (MP, 2). Through unknown mechanisms, MP organizes microtubule assembly in neighboring cells perpendicular to the auxin maximum cell source (3). These microtubules are used as a scaffold for the guidance of cellulose deposition and microfibril assembly (4). This cell wall configuration creates anisotropic cell expansion perpendicular to the cellulose microfibrils (5). The loose cell wall of the auxin maximum cell imposes high stresses on the cell walls of cells directly bordering it (pink cell wall). These anisotropic stresses feed back on the system by reinforcing the polar distribution of PIN1 efflux carrier proteins within bordering cells to the end closest to the auxin maximum (6), which leads to the continued transport of auxin toward the auxin maximum (7).

How is the auxin maximum translated into a directed cell expansion? The current hypothesis of lateral organ induction is based on the auxin maximum in the convergence point loosening the walls of those cells (**FIGURE 8.10**). Because all cells of the plant are attached to one another, when one set of cell walls relaxes, this local mechanical anisotropy (stress) is felt by the other cells around it. The localization of PIN1 correlates with cell boundaries of highest stress; PIN1 is localized in cells next to the convergence point, consequently directing auxin flow toward the auxin maximum (Heisler et al. 2010). A fascinating finding supporting this model is that the independent loosening of cell walls (mimicking the proposed role of auxin) in a *pin1* mutant was capable of inducing primordium outgrowth on its own (Pien et al. 2001; Peaucelle et al. 2008; Sassi et al. 2014)!

A morphogen-mechanical stress model is starting to emerge to explain the reciprocal patterning of successive lateral organ formation in plants. As a given primordium grows, so will the mechanical stress caused by auxin-induced cell expansion at the apex, which will eventually feed back to negatively regulate auxin signaling. This occurs by triggering a reorientation of microtubules and cellulose microfibrils that stabilizes the cell walls. This stabilization is sensed by neighboring cells, which respond by repolarizing PIN1 toward other auxin maxima in the meristem, thereby stimulating the outgrowth of new primordia. The reciprocal interaction between positive and negative feedback loops—involving auxin morphogenetic signaling and the biophysical properties of the meristem—determines the positions of new primordia emerging from the shoot apex. This positioning in relation to the upward growth of the shoot is what determines the spiral phyllotactic pattern. Although the resulting spiral pattern of cell and organ specification in the *A. thaliana* inflorescence resembles the spiraling of a snail's shell, it is achieved through a different mechanism involving self-organizing cell-cell interactions.

Axis determination in the snail embryo

Mollusks provide some of the most impressive examples of autonomous development, in which the blastomeres are specified by cell fate specifying cytoplasmic determinants located in specific regions of the oocyte (see Chapter 2). Autonomous specification of early blastomeres is especially prominent in those groups of animals having spiral cleavage, all of which initiate gastrulation at the vegetal pole when only a few dozen cells have formed (Lyons et al. 2015). In mollusks, the mRNAs for some transcription factors and paracrine factors are placed in particular cells by associating with certain centrosomes (**FIGURE 8.11**; Lambert and Nagy 2002; Kingsley et al. 2007; Henry et al. 2010a,b). This association allows the mRNA to enter specifically into one of the two daughter cells. In many instances, the mRNAs that get transported together into a particular tier of blastomeres have 3' tails that form very similar shapes, thus suggesting that the identity of the micromere tiers may be controlled largely by the 3' untranslated regions (UTRs) of the mRNAs that attach to the centrosomes at each division (**FIGURE 8.12**; Rabinowitz and Lambert 2010). In other cases, the patterning

(?) Developing Questions

How are the anisotropic stress forces of the epidermis across the meristem interpreted to result in changes in PIN polarity? What similarities do you see between spiral pattern development in the plant and in the snail? Could you imagine the snail blastocyst operating under self-organizing principles?

(A) (B) (C)

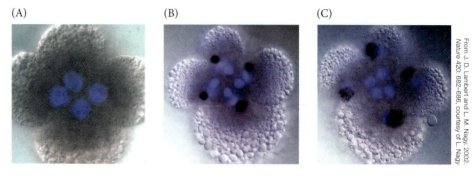

From J. D. Lambert and L. M. Nagy. 2002. *Nature* 420: 682–686, courtesy of L. Nagy

FIGURE 8.11 Association of *decapentaplegic* (*dpp*) mRNA with specific centrosomes of *Ilyanassa*. (A) In situ hybridization of the mRNA for Dpp in the 4-cell snail embryo shows no Dpp accumulation. (B) At prophase of the 4- to 8-cell stage, *dpp* mRNA (black) accumulates at one centrosome of the pair forming the mitotic spindle. (The DNA is light blue.) (C) As mitosis continues, *dpp* mRNA is seen to attend the centrosome in the macromere rather than the centrosome in the micromere of each cell. The BMP-like paracrine factor encoded by *dpp* is critical to molluscan development.

molecules (of still unknown identities) appear to be bound to a certain region of the egg that will form a unique structure called the polar lobe. The **polar lobe** is a protrusion that forms and then is absorbed at the vegetal pole of the embryo. On the first cleavage, the cytoplasm that flows into the polar lobe is absorbed into the CD blastomere; on the second cleavage, the cytoplasm flowing into the lobe is absorbed into the D blastomere. (See **Further Development 8.2, The Snail Fate Map**, and **Further Development 8.3, The Role of the Polar Lobe in Cell Specification**, both online.)

FIGURE 8.12 Importance of the 3′ UTR for association of mRNAs with specific centrosomes. In *Ilyanassa*, the *R5LE* message is usually segregated into the first tier of micromeres. The message binds to one side of the centrosome complex (the side that will be in the small micromere). (A) Normal *R5LE* mRNA distribution from the 2-cell through the 24-cell stage. The mRNA (green) associates with the centrosomic region (blue) that will generate the micromere tier and becomes localized to particular blastomeres by the 24-cell stage. (B) Hairpin loop of the 3′ UTR of the *R5LE* message. (After J. S. Rabinowitz and J. D. Lambert. 2010. *Development* 137: 4039–4049.)

THE D BLASTOMERE The development of the D blastomere can be traced in Figure 8.2B–C. This macromere, having received the contents of the polar lobe, is larger than the other three (Clement 1962). When one removes the D blastomere or its first or second macromere derivatives (i.e., 1D or 2D), one obtains an incomplete larva lacking heart, intestine, velum, shell gland, eyes, and foot. This is essentially the same phenotype seen when one removes the polar lobe. Since the D blastomeres do not directly contribute cells to many of these structures, it appears that the D-quadrant macromeres are involved in inducing other cells to have these fates.

When one removes the 3D blastomere shortly after the division of the 2D cell to form the 3D and 3d blastomeres, the larva produced looks similar to those formed by the removal of the D, 1D, or 2D macromeres. However, ablation of the 3D blastomere at a later time produces an almost normal larva, with eyes, foot, velum, and some shell gland, but no heart or intestine. After the 4d cell is given off (by the division of the 3D blastomere), removal of the D derivative (the 4D cell) produces no qualitative difference

(A) 2-Cell interphase   4-Cell interphase   4-Cell prophase   4-Cell metaphase   4-Cell anaphase   8-Cell interphase   16-Cell   24-Cell

(B)

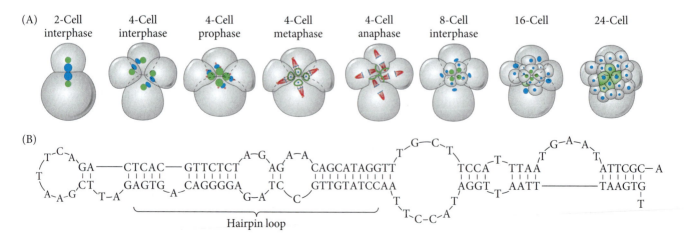

Hairpin loop

FIGURE 8.13 Cell fate specifying determinants in the 4d snail blastomere. (A) β-Catenin expression in ML and MR, the two cells (left and right) produced by division of the 4d blastomere of *Crepidula*. (B) Localization of *nanos* mRNA (dark purple) in the dividing 4d blastomere and in its right and left progeny, 4dR and 4dL, of *Ilyanassa*. (Nuclei are light blue.)

(A)

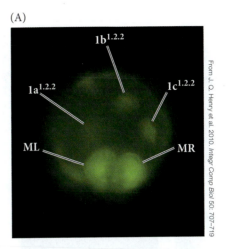

From J. Q. Henry et al. 2010. *Integr Comp Biol* 50: 707–719

in development. In fact, all the essential determinants for heart and intestine formation are now in the 4d blastomere (also called the mesentoblast, as mentioned earlier), and removal of that cell results in a heartless and gutless larva (Clement 1986). The 4d blastomere is responsible for forming (at its next division) the two bilaterally paired blastomeres that give rise to both the mesodermal (heart) and endodermal (intestine) organs (Lyons et al. 2012; Chan and Lambert 2014).

The mesodermal and endodermal determinants of the 3D macromere, therefore, are transferred to the 4d blastomere. At least two cell fate specifying determinants are involved in regulating the development of 4d. First, the cell appears to be specified by the presence of the transcription factor **β-catenin**, which enters into the nucleus of the 4d mesentoblast and its immediate progeny (**FIGURE 8.13A**; Henry et al. 2008; Rabinowitz et al. 2008). When translational inhibitors suppressed β-catenin protein synthesis, the 4d cell underwent a normal pattern of early cell divisions, but these cells failed to differentiate into heart, muscles, or hindgut; gastrulation also failed to occur in those embryos (Henry et al. 2010a). Indeed, β-catenin may have an evolutionarily conserved role in mediating autonomous specification and specifying endomesodermal fates throughout the animal kingdom; in subsequent chapters we will see a similar role for this protein in both sea urchin and frog embryos.

(B)

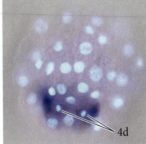

From J. S. Rabinowitz et al. 2008. *Curr Biol* 18: 331–336

The 4d mesentoblast also contains the protein and mRNA for the translation suppressor *nanos* (**FIGURE 8.13B**). As with β-catenin, blocking translation of *nanos* mRNA prevents formation of the larval muscles, heart, and intestine from the 4d blastomere (Rabinowitz et al. 2008). In addition, the germline cells (sperm and egg progenitors) do not form. As we will see throughout the book, the Nanos protein is often involved in specification of germ cell progenitors. (See **Further Development 8.4, 4D and the Role of Notch**, and **Further Development 8.5, Altering Evolution by Altering Cleavage Patterns: An Example from a Bivalve Mollusk**, both online.)

Gastrulation in Snails

The snail stereoblastula is relatively small, and its cell fates have already been determined by the D series of macromeres. Gastrulation is accomplished by a combination of processes, including the invagination of the endoderm to form the primitive gut, and the epiboly of the animal cap micromeres that multiply and "overgrow" the vegetal macromeres (Collier 1997; van den Biggelaar and Dictus 2004; Lyons and Henry 2014). Eventually, the micromeres cover the entire embryo, leaving a small blastopore slit at the vegetal pole (**FIGURE 8.14A**). The first- through third-quartet micromeres form an epithelial animal cap that expands to cover vegetal endomesodermal precursors. As the blastopore narrows, cells derived from $3a^2$ and $3b^2$ undergo epithelial-mesenchymal transition and move into the archenteron. Posteriorly, cells derived from $3c^2$ and $3d^2$ undergo convergence and extension that involves a zipper-like mechanism and their intercalation across the ventral midline (**FIGURE 8.14B**; Lyons et al. 2015).

The mouth of the snail forms from cells around the circumference of the blastopore. The anus arises from the $2d^2$ cells, which are briefly part of the blastopore lip, but whose progeny later form a separate hole, not related to the blastopore, that becomes the anus. Thus, these animals are protostomes, forming their mouths in the area where the blastopore is first seen.

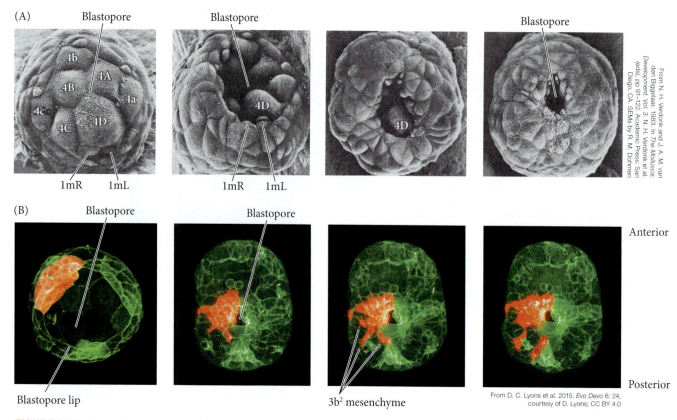

From N. H. Verdonk and J. A. M. van den Biggelaar. 1983. In *The Mollusca: Development, Vol. 3*. N. H. Verdonk et al. (eds), pp 91–122, Academic Press: San Diego, CA. SEMs by R. M. Dohmen

From D. C. Lyons et al. 2015. *Evo Devo* 6: 24, courtesy of D. Lyons; CC BY 4.0

FIGURE 8.14 Gastrulation in the snail *Crepidula*. (A) Scanning electron micrographs focusing on the blastopore region show internalization of the endoderm, which is derived from the macromeres plus the fourth tier of micromeres. The 1mR and 1mL (right and left mesendoderm cells, respectively) are in the 4d cell lineage. The ectoderm undergoes epiboly from the animal pole and envelops the other cells of the embryo. (B) Live cell labeling of *Crepidula* embryos shows gastrulation occurring by epiboly. Cells derived from the 3b micromere are stained orange.

The Nematode *C. elegans*

Unlike the snail, with its long embryological pedigree, the nematode *Caenorhabditis elegans* (usually referred to as *C. elegans*) is a thoroughly modern model system, uniting developmental biology with molecular genetics. In the 1970s, Sydney Brenner sought an organism wherein it might be possible to identify each gene involved in development as well as to trace the lineage of each and every cell (Brenner 1974). Nematode roundworms seemed like a good group to start with because embryologists such as Richard Goldschmidt and Theodor Boveri had already shown that several nematode species have a relatively small number of chromosomes and a small number of cells with invariant cell lineages.

Brenner and his colleagues eventually settled on *C. elegans*, a small (1 mm long) free-living (i.e., nonparasitic) soil nematode with relatively few cell types. *C. elegans* has a rapid period of embryogenesis—about 16 hours—which it can accomplish in a petri dish (**FIGURE 8.15A**). Moreover, its predominant adult form is hermaphroditic, with each individual producing both eggs and sperm. These roundworms can reproduce either by self-fertilization or by cross-fertilization with infrequently occurring males.

The body of an adult *C. elegans* hermaphrodite contains exactly 959 somatic cells, and the entire cell lineage has been traced through its transparent cuticle (**FIGURE 8.15B**; Sulston and Horvitz 1977; Kimble and Hirsh 1979). It has what is called an **invariant cell lineage**, which means that each cell gives rise to the same number and type of cells in every embryo. This allows one to know which cells have the same precursor cells. Thus, for each cell in the embryo, we can say where it came from (i.e., which cells in earlier embryonic stages were its progenitors) and which tissues it will contribute to forming. Furthermore, unlike vertebrate cell lineages, the *C. elegans* lineage is almost entirely invariant from one individual to the next; there is little room for randomness (Sulston et al. 1983). It also has a very compact genome. The *C. elegans* genome was the first complete sequence ever obtained for a multicellular organism (*C. elegans* Sequencing Consortium 1998). Although it has about the same number of genes as humans (*C. elegans* has 19,000–20,000 genes; *Homo sapiens* has 20,000–25,000), the nematode has only about 3% the number of nucleotides in its genome (Hodgkin et al. 1998; Hodgkin 2001).

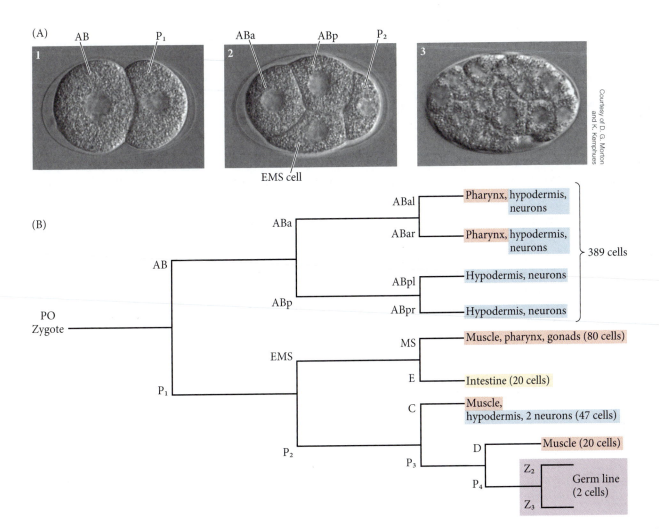

(A)

1 AB   P₁

2 ABa   ABp   P₂

EMS cell

3

Courtesy of D. G. Morton and K. Kemphues

(B)

PO
Zygote

AB

ABa

ABal — Pharynx, hypodermis, neurons

ABar — Pharynx, hypodermis, neurons

ABp

ABpl — Hypodermis, neurons

ABpr — Hypodermis, neurons

} 389 cells

P₁

EMS

MS — Muscle, pharynx, gonads (80 cells)

E — Intestine (20 cells)

C — Muscle, hypodermis, 2 neurons (47 cells)

P₂

P₃

D — Muscle (20 cells)

P₄

Z₂

Z₃

} Germ line (2 cells)

FIGURE 8.15 Development in the nematode *Caenorhabditis elegans* is rapid and results in an adult with exactly 959 somatic cells. Individual cell lineages have been traced through the course of the animal's development. AB, E, MS, C, and D are founder cells. (A) Differential interference micrographs of the cleaving embryo. (1) The AB cell (left) and the P₁ cell (right) are the result of the first asymmetrical division. Each will give rise to a different cell lineage. (2) The 4-cell embryo shows ABa, ABp, P₂, and EMS cells. (3) Gastrulation is initiated by the movement of E-derived cells toward the center of the embryo. (B) Abbreviated cell lineage chart. The germ line segregates into the posterior portion of the most posterior (P) cell. The first three cell divisions produce the AB, C, MS, and E lineages. The number of derived cells (in parentheses) refers to the 558 cells present in the newly hatched larva. Some of these continue to divide to produce the 959 somatic cells of the adult. (B after M. Pines. 1992. *From Egg to Adult: What Worms, Flies, and Other Creatures Can Teach Us about the Switches That Control Human Development—A Report from the Howard Hughes Medical Institute*. Howard Hughes Medical Institute: Bethesda, MD, based on J. E. Sulston and H. R. Horvitz. 1977. *Dev Biol* 56: 110–156 and J. E. Sulston et al. 1983. *Dev Biol* 100: 64–119.)

C. elegans displays the rudiments of nearly all the major bodily systems (feeding, nervous, reproductive, etc., although it has no skeleton), and it exhibits an aging phenotype before it dies. Neurobiologists celebrate its minimal nervous system (302 neurons), and each one of its 7,600 synapses (neuronal connections) has been identified (White et al. 1986; Seifert et al. 2006). In addition, *C. elegans* is particularly friendly to molecular biologists. DNA injected into *C. elegans* cells is readily incorporated into their nuclei, and *C. elegans* can take up double-stranded RNA from its culture medium. Last, with the versatility of gene editing techniques such as the CRISPR/Cas9 system (see Chapter 3), researchers have taken full advantage of making targeted gene knockouts and knockins in *C. elegans* (Dickinson and Goldstein 2016).

Cleavage and Axis Formation in *C. elegans*

Fertilization in *C. elegans* is a not your typical sperm-meets-egg story. Most *C. elegans* individuals are hermaphrodites, producing both sperm and eggs, and fertilization occurs within a single adult individual. The egg becomes fertilized by rolling through

(?) Developing Questions

Humans have trillions of cells, a regionalized brain, and intricate limbs. The nematode has 959 cells and can fit under a fingernail. However, humans and *C. elegans* have nearly the same number of genes, leaving Jonathan Hodgkin, curator of the *C. elegans* gene map, to ask, "What does a worm want with 20,000 genes?" Any suggestions?

a region of the adult worm (the spermatheca) containing mature sperm (**FIGURE 8.16A,B**). The sperm are not the typical long-tailed, streamlined cells, but are small, round, unflagellated cells that travel slowly by amoeboid motion. When a sperm fuses with the egg cell membrane, polyspermy is prevented by the rapid synthesis of chitin (the protein comprising the cuticle) by the newly fertilized egg (Johnston et al. 2010). The fertilized egg undergoes early divisions and is extruded through the vulva.

Rotational cleavage of the egg

The zygote of *C. elegans* exhibits rotational holoblastic cleavage (**FIGURE 8.16C**). During early cleavage, each asymmetrical division produces one founder cell (denoted AB, E, MS, C, and D) that produces differentiated descendants, and one stem cell (the P_1–P_4 lineage). The anterior-posterior axis is determined before the first cell division, and the first cleavage furrow is located asymmetrically along this axis of the egg, closer to what will be the posterior pole. The first cleavage forms an anterior founder cell (AB) and a posterior stem cell (P_1). The dorsal-ventral axis is determined during the second division. The founder cell (AB) divides equatorially (longitudinally, 90° to the anterior-posterior axis), while the P_1 cell divides meridionally (transversely) to produce another founder cell (EMS) and a posterior stem cell (P_2). The EMS cell marks the ventral region of the developing embryo. The stem cell lineage always undergoes meridional division to produce (1) an anterior founder cell and (2) a posterior cell that will continue the stem cell lineage. The first left-right asymmetry is seen between the 4- and 8-cell stages, as two of the "granddaughters" shift anteriorly as they form. Here, the locations of two granddaughters of the AB cell (ABal and ABpl) are on the left side, while two others (ABar and ABpr) are on the right (see Figure 8.16C).

FIGURE 8.16 Fertilization and early cleavages in *C. elegans*. (A) Side view of adult hermaphrodite. Sperm are stored such that a mature egg must pass through the sperm on its way to the vulva. (B) The germ cells undergo mitosis near the distal tip of the gonad. As they move farther from the distal tip, they enter meiosis. Early meioses form sperm, which are stored in the spermatheca. Later meioses form eggs, which are fertilized as they roll through the spermatheca. (C) Early development occurs as the egg is fertilized and moves toward the vulva. The P lineage consists of stem cells that will eventually form the germ cells. (After M. Pines. 1992. *From Egg to Adult: What Worms, Flies, and Other Creatures Can Teach Us about the Switches that Control Human Development—A report from the Howard Hughes Medical Institute*. Howard Hughes Medical Institute: Bethesda, MD, based on J. E. Sulston and H. R. Horvitz. 1977. *Dev Biol* 56: 110–156 and J. E. Sulston et al. 1983. *Dev Biol* 100: 64–119.)

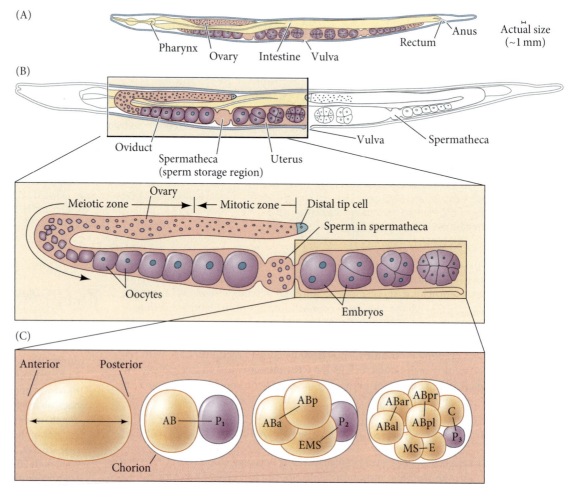

Anterior-posterior axis formation

The decision as to which end of the egg will become the anterior and which the posterior seems to reside with the position of the sperm pronucleus at fertilization (**FIGURE 8.17**). When the sperm pronucleus along with its centrosome enter the oocyte cytoplasm, the oocyte has no polarity. Alterations in the site of sperm entry change the orientation of the anterior-posterior axis. This suggests that the sperm provides a mechanism for specifying the anterior-posterior axis of the zygote (Goldstein and Hird 1996). However, components of the oocyte also play a role. The oocyte has a specific arrangement of **PAR proteins** in its cytoplasm (Motegi and Seydoux 2013), and mutations in the *par* genes lead to defects in the ability to asymmetrically partition cytoplasmic determinants. PAR-3 and PAR-6, interacting with the protein kinase PKC-3, are uniformly distributed in the cortical cytoplasm.[3] PKC-3 restricts PAR-1 and PAR-2 to the internal cytoplasm by phosphorylating them (see Figure 8.17A). Following fertilization, the sperm centrosome organizes microtubules to contact the oocyte's cortical cytoplasm, and this initiates cytoplasmic movements that push the male pronucleus to the nearest end of the oblong oocyte. That end becomes the posterior pole (Goldstein and Hird 1996). The microtubules organized by the sperm centrosome locally protect PAR-2 from phosphorylation, thereby allowing PAR-2 (and its binding partner, PAR-1) into the cortex nearest the centrosome. Once PAR-1 is in the cortical cytoplasm, it phosphorylates PAR-3, causing PAR-3 (and its binding partner, PKC-3) to leave the cortex. At the same time, the microtubules organized by the sperm centrosome induce the contraction of the actin-myosin cytoskeleton toward the anterior, thereby clearing PAR-3, PAR-6, and PKC-3 from the posterior of the 1-cell embryo. During first cleavage, the metaphase plate starts forming centrally and moves posteriorly, and the fertilized egg is divided into two cells, one having the anterior PARs (PAR-6 and PAR-3) and one having the posterior PARs (PAR-2 and PAR-1) (see Figure 8.17D–G; Goehring et al. 2011; Motegi et al. 2011; Rose and Gönczy 2014).

Dorsal-ventral and right-left axis formation

The dorsal-ventral axis of *C. elegans* is established in the division of the AB cell. As the cell divides, it becomes longer than the eggshell is wide. This squeezing causes the daughter cells to slide, one becoming anterior and one posterior (hence their respective names, ABa and ABp; see

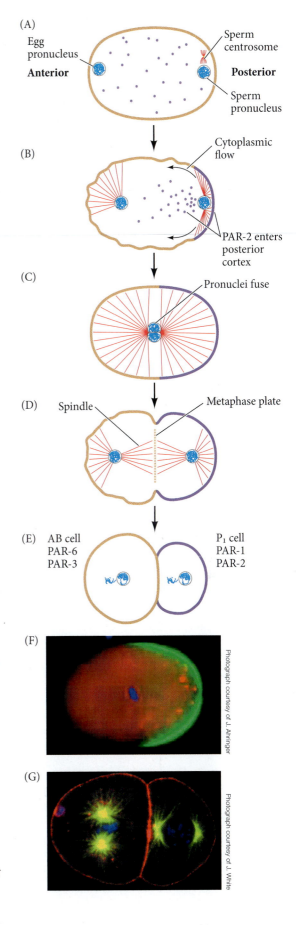

FIGURE 8.17 PAR proteins and the establishment of polarity. (A) When sperm enters the egg, the egg nucleus is undergoing meiosis (left). The cortical cytoplasm (orange) contains PAR-3, PAR-6, and PKC-3, and the internal cytoplasm contains PAR-2 and PAR-1 (purple dots). (B,C) Microtubules organized by the sperm centrosome initiate contraction of the actin-based cytoskeleton toward the future anterior side of the embryo. These microtubules also protect PAR-2 protein from phosphorylation, allowing it to enter the cortex along with its binding partner, PAR-1. PAR-1 phosphorylates PAR-3, causing PAR-3 and its binding partners PAR-6 and PKC-3 to leave the cortex. (D) The posterior of the cell becomes defined by PAR-2 and PAR-1, while the anterior of the cell becomes defined by PAR-6 and PAR-3. The metaphase plate is asymmetrical, as the microtubules of the spindle apparatus are closer to the posterior pole. (E) The metaphase plate separates the zygote into two cells, one having the anterior PARs and one the posterior PARs. (F) In this dividing *C. elegans* zygote, PAR-2 protein is stained green; DNA is stained blue. (G) In second division, the AB cell and the P_1 cell divide perpendicularly (90° differently from each other). (A–E after R. Bastock and D. St. Johnston. 2011. *Dev Cell* 21: 981–982.)

Labels in figure:
(A) Egg pronucleus; **Anterior**; Sperm centrosome; **Posterior**; Sperm pronucleus
(B) Cytoplasmic flow; PAR-2 enters posterior cortex
(C) Pronuclei fuse
(D) Spindle; Metaphase plate
(E) AB cell PAR-6 PAR-3; P_1 cell PAR-1 PAR-2
(F) Photograph courtesy of J. Ahringer
(G) Photograph courtesy of J. White

Figure 8.16C). The squeezing also causes the ABp cell to take a position above the EMS cell, which results from the division of the P_1 blastomere. The ABp cell thus defines the future dorsal side of the embryo, while the EMS cell—the precursor of the muscle and gut cells—marks the future ventral surface of the embryo.

The left-right axis is not readily seen until the 12-cell stage, when the MS blastomere (from the division of the EMS cell) contacts half the "granddaughters" of the ABa cell, distinguishing the right side of the body from the left side (Evans et al. 1994). This asymmetrical signaling sets the stage for several other inductive events that make the right side of the larva differ from the left (Hutter and Schnabel 1995). Indeed, even the different neuronal fates seen on the left and right sides of the *C. elegans* brain can be traced back to that single change at the 12-cell stage (Poole and Hobert 2006). Although left-right asymmetry is readily seen at the 12-cell stage, the first indication of it probably occurs at the zygote stage. Just prior to first cleavage, the embryo rotates 120° inside its vitelline envelope. This rotation is always in the same direction relative to the already established anterior-posterior axis, indicating that the embryo already has a left-right chirality, or a mirror-image asymmetry. If cytoskeleton proteins or the PAR proteins are inhibited, the direction of the rotation and subsequent chirality become random (Wood and Schonegg 2005; Pohl 2011).

Control of blastomere identity

C. elegans demonstrates both conditional and autonomous modes of cell specification. Both modes can be seen if the first two blastomeres are experimentally separated (Priess and Thomson 1987). The P_1 cell develops autonomously without the presence of AB, generating all the cells it would normally make, and the result is the posterior half of an embryo. However, the AB cell in isolation makes only a small fraction of the cell types it would normally make. For instance, the resulting ABa blastomere fails to make the anterior pharyngeal muscles that it would have made in an intact embryo. Therefore, specification of the AB blastomere is conditional, and it needs to interact with the descendants of the P_1 cell in order to develop normally.

AUTONOMOUS SPECIFICATION The determination of the P_1 lineages appears to be autonomous, with cell fates determined by internal cytoplasmic factors rather than by interactions with neighboring cells (see Maduro 2006). The SKN-1, PAL-1, and PIE-1 proteins encode transcription factors that act intrinsically to determine the fates of cells derived from the four P_1-derived somatic founder cells (MS, E, C, and D). (See **Further Development 8.6, Defining the Role of SKN-1 and PAL-1 in Early Cell Specification in *C. elegans***, online.)

CONDITIONAL SPECIFICATION As mentioned earlier, the *C. elegans* embryo uses both autonomous and conditional modes of specification. Conditional specification can be seen in the development of the endoderm cell lineage. At the 4-cell stage, the EMS cell requires a signal from its neighbor (and sister cell), the P_2 blastomere. Usually, the EMS cell divides into an MS cell (which produces mesodermal muscles) and an E cell (which produces the intestinal endoderm). If the P_2 cell is removed at the early 4-cell stage, the EMS cell will divide into two MS cells, and no endoderm will be produced. This instructive interaction with P_2 is further confirmed by experiments that move P_2 to the other side of the presumptive EMS blastomeres. This results in the two sides of EMS (E and MS) swapping fates. These results, taken together with others, show that it is this interaction with the P_2 blastomere that specifies the differences between E and MS cell fates (Goldstein 1993, 1995).

Specification of the MS cell begins with maternal SKN-1 activating the genes encoding transcription factors such as MED-1 and MED-2. The POP-1 signal (which encodes the TCF protein that binds β-catenin to the DNA) blocks the pathway to the E (endodermal) fate in the prospective MS cell by blocking the ability of MED-1 and MED-2 to activate the *tbx-35* gene (**FIGURE 8.18**; Broitman-Maduro et al. 2006; Maduro 2009). Throughout the animal kingdom, TBX proteins are known to be active in mesoderm formation; TBX-35 acts to activate the mesodermal genes, such as *pha-4* in the pharynx and the *myoD* homolog *hlh-1* in the muscles of *C. elegans*.

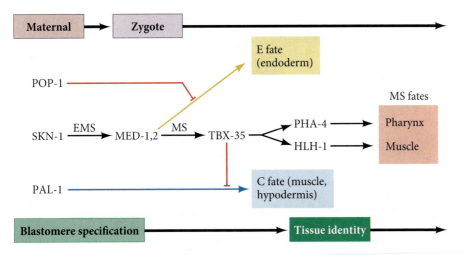

FIGURE 8.18   Model for specification of the MS blastomere. Maternal SKN-1 activates Gata transcription factors MED-1 and MED-2 in the EMS cell. The POP-1 signal prevents these proteins from activating the endodermal transcription factors (such as END-1) and instead activates the *tbx-35* gene. The TBX-35 transcription factor activates mesodermal genes in the MS cell, including *pha-4* in the pharynx lineage and *hlh-1* (which encodes a myogenic transcription factor) in muscles. TBX-35 also inhibits *pal-1* gene expression, thereby preventing the MS cell from acquiring the C-blastomere fates. (After G. Broitman-Maduro et al. 2006. *Development* 133: 3097–3106.)

The P_2 cell produces a signal that interacts with the EMS cell and instructs the EMS daughter next to it to become the E cell. This message is transmitted through the Wnt signaling cascade (**FIGURE 8.19**; Rocheleau et al. 1997; Thorpe et al. 1997; Walston et al. 2004). The P_2 cell produces the MOM-2 protein, a *C. elegans* Wnt protein. MOM-2 is received in the EMS cell by the MOM-5 protein, a *C. elegans* version of the Wnt receptor protein Frizzled. When the EMS cell divides, this signaling cascade is confined to the posterior daughter cell and downregulates the expression of the *pop-1* gene. This induces the posterior daughter cell to become an E cell. The expression of the *pop-1* gene in the anterior daughter cell results in it becoming an MS cell. In *pop-1*-deficient embryos, both EMS daughter cells become E cells (Lin et al. 1995; Park et al. 2004). Thus, the Wnt pathway induces cell fates along the anterior-posterior axis. Remarkably, as we will see, Wnt signaling appears to induce fates along the anterior-posterior axis throughout the animal kingdom. (See **Further Development 8.7, Cell-Cell Interactions in the Early *C. elegans* Embryo,** and **Further Development 8.8, Integration of Autonomous and Conditional Specification: Differentiation of the *C. elegans* Pharynx**, both online.)

Gastrulation of 66 Cells in *C. elegans*

Gastrulation in *C. elegans* starts extremely early, just after the generation of the P_4 cell in the 26-cell embryo (**FIGURE 8.20A**; Skiba and Schierenberg 1992). At this time, the two daughters of the E cell (Ea and Ep) move from the ventral side into the center of the embryo. This internalization is initiated by the common mechanism of cell shape changes known as **apical constriction**, during which actinomyosin contraction on the apical side reduces its surface area relative to the basal side. As seen during the invagination events in *Drosophila*, *Xenopus*, zebrafish, chick, and mouse gastrulation, this polarized shape creates the site for inward invagination. Once in the center, the E cell divides to form a gut consisting of 20 cells. There is a very small and transient blastocoel prior to the movement of the Ea and Ep cells. Sixty-four other cells internalize, and each cell's identity is known and has been mapped onto the *C. elegans* cell lineage (**FIGURE 8.20B**). The next cells to internalize are some from the AB neural lineage, followed by the $P_{4'}$ precursor

FIGURE 8.19   Cell-cell signaling in the 4-cell embryo of *C. elegans*. The P_2 cell produces two signals: (1) the juxtacrine protein APX-1 (a Delta homologue), which is bound by GLP-1 (Notch) on the ABp cell, and (2) the paracrine protein MOM-2 (Wnt), which is bound by the MOM-5 (Frizzled) protein on the EMS cell. (After M. Han. 1998. *Cell* 90: 581–584.)

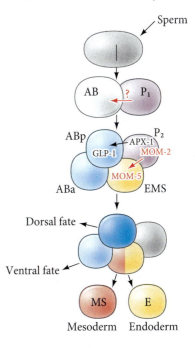

(A)

Lateral view

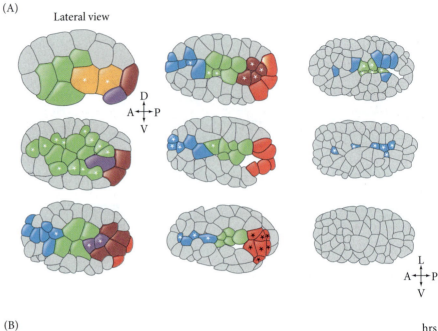

(B)

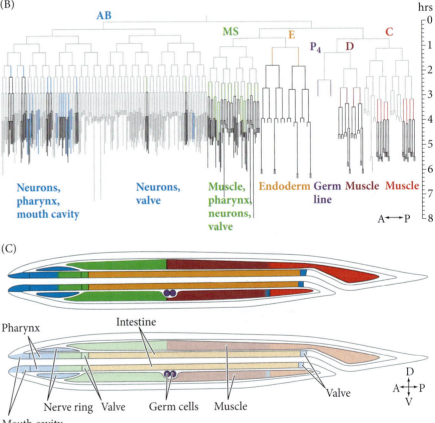

(C)

Pharynx

Intestine

Nerve ring Valve   Germ cells   Muscle   Valve

Mouth cavity

FIGURE 8.20 Gastrulation in *C. elegans*. (A) Time series of the 66 gastrulating cells color-coded to represent the lineages of E, MS, P₄, D, and all of their descendants, as well as those cells of AB and C lineages that gastrulate (see B for color key). The top left diagram is a lateral view; the rest are ventral. Asterisks denote cells that are going to internalize. (B) Cell lineage map for all 66 gastrulating cells of *C. elegans*. Each horizontal line represents a cell division; the vertical lengths of the lines are proportional to the time between cell divisions (see axis on right; hrs, hours.) (C) Final positions of lineages in the larval worm. (A–C from J. R. Harrell and B. Goldstein. 2011. *Dev Biol* 350: 1–12; C based on J. E. Sulston et al. 1983. *Dev Biol* 100: 64–119.)

of the germ cells. The P_4 cell moves to a position beneath the gut primordium. The mesodermal cells move in next: the descendants of the MS cell internalize, and the C- and D-derived muscle precursors follow. These cells flank the gut tube on the left and right sides (Schierenberg 1997). Finally, about 6 hours after fertilization, additional AB-derived cells that contribute to the pharynx are brought inside, while the hypoblast cells (precursors of the hypodermal skin cells) move ventrally by epiboly, eventually closing the blastopore. The two sides of the hypodermis are sealed by E-cadherin on the tips of the leading cells that meet at the ventral midline (Raich et al. 1999; Harrell and Goldstein 2011).

During the next 6 hours, the cells move and develop into organs, while the ball-shaped embryo stretches out to become a worm with 556 somatic cells and 2 germ-line stem cells (**FIGURE 8.20C**; see Priess and Hirsh 1987; Schierenberg 1997; Harrell and Goldstein 2011). Other modeling takes place as well; an additional 115 cells undergo apoptosis (programmed cell death). After four molts, the worm is a sexually mature, hermaphroditic adult, containing exactly 959 somatic cells as well as numerous sperm and eggs.

FURTHER DEVELOPMENT

Cell fusion in the C. elegans embryo

One characteristic that distinguishes *C. elegans* development from that of most other well-studied organisms is the prevalence of cell fusion. During *C. elegans* gastrulation, about one-third of all the cells fuse together to form syncytial cells containing many nuclei. The 186 cells that comprise the hypodermis (skin) of the nematode fuse into 8 syncytial cells, and cell fusion is also seen in the vulva, uterus, and pharynx. The functions of these fusion events can be determined by observing mutations that prevent syncytia from forming (Shemer and Podbilewicz 2000, 2003). It seems that fusion prevents individual cells from migrating beyond their normal borders. In the vulva (see Chapter 4), fusion prevents hypodermis cells from adopting a vulval fate and making an ectopic (and nonfunctional) vulva.

Even in an organism as "simple" as *C. elegans*, with a small genome and a small number of cell types, the right side of the body is made in a different manner from the left. The identification of the genes mentioned above is just a starting point as we continue to explore the complex interactions of development. (See **Further Development 8.9, Heterochronic Genes and the Control of Larval Stages**, online.)

Closing Thoughts on the Opening Photo

In 1923, Alfred Sturtevant identified left-coiling of snail shells as one of the first developmental mutations known. He was able to link the genetics of *Radix* snails with their coiling patterns, establishing that the left-coiling (sinistral) phenotype was a maternal effect. His work demonstrated in a highly visible manner the profound effect of genes on development. In 2016, the genetic basis of snail coiling may have been identified and the pathway leading to right-left asymmetry outlined (see Davison et al. 2016; Kuroda et al. 2016). In comparison, the highly stereotypical spiral pattern of floral organs in plants provides a unique perspective on how a combined morphogenetic and mechanical mechanism can yield a self-regulating process to attain pattern. What new insights can be gleaned by further direct comparisons between plants and animals such as this daisy and this snail?

Endnotes

[1]The hypothesis that ctenophores rather than sponges represent the outgroup or sister clade to all other animal phyla (potentially the most ancient extant metazoan lineage) remains controversial (Borowiec et al. 2015; Chang et al. 2015; Pisani et al. 2015).

[2]Pharyngeal arches give rise to head structures in tetrapods, to gills in fish, and to filter-feeding structures in Amphioxus and tunicates.

[3]Although originally discovered in *C. elegans,* PAR proteins are used by many species in establishing cell polarity. They are critical for forming the anterior and posterior regions of *Drosophila* oocytes, and they distinguish the basal and apical ends of *Drosophila* epithelial cells. *Drosophila* par proteins are also important in distinguishing which product of a neural stem cell division becomes the neuron and which remains a stem cell. PAR-1 homologues in mammals also appear to be critical in neural polarity (Goldstein and Macara 2007; Nance and Zallen 2011).

8 Snapshot Summary
Snails, Flowers, and Nematodes

1. Body axes are established in different ways in different species. In some species the axes are established at fertilization through determinants in the egg cytoplasm. In others, the axes are established by cell interactions later in development.

2. Both snails and nematodes have holoblastic cleavage. In snails, cleavage is spiral; in nematodes, it is rotational.

3. In snails and *C. elegans*, gastrulation begins when there are relatively few cells.

4. Spiral cleavage in snails results in stereoblastulae (i.e., blastulae with no blastocoels). The direction of the cleavage spirals is regulated by a factor encoded by the mother and placed in the oocyte.

5. The phyllotactic spiral arrangement of floral organs in *Arabidopsis thaliana* is achieved through a self-regulating mechanism involving the interaction between auxin hormone signaling and mechanical strain in the meristem.

6. The polar lobe of certain mollusks contains the cell fate specifying determinants for mesoderm and endoderm. These determinants enter the D blastomere.

7. The soil nematode *Caenorhabditis elegans* was chosen as a model organism because it has a small number of cells, has a small genome, is easily bred and maintained, has a short life span, can be genetically manipulated, and has a cuticle through which one can see cell movements.

8. In the early divisions of the *C. elegans* zygote, one daughter cell becomes a founder cell (producing differentiated descendants), and the other becomes a stem cell (producing other founder cells and the germ line).

9. Blastomere identity in *C. elegans* is regulated by both autonomous and conditional specification.

Go to oup.com/he/barresi12xe for Further Developments, Scientists Speak interviews, Watch Development videos, Dev Tutorials, and complete bibliographic information for all literature cited in this chapter.

The Genetics of Axis Specification in *Drosophila*

Four wings! Is the extra pair additional, or did it replace something else now lost?

Courtesy of Nipam Patel

THANKS LARGELY TO STUDIES spearheaded by Thomas Hunt Morgan's laboratory during the first two decades of the twentieth century, we know more about the genetics of *Drosophila melanogaster* than that of any other multicellular organism (**FIGURE 9.1**). The reasons have to do with both the flies themselves and with the people who first studied them. *Drosophila* is easy to breed, hardy, prolific, and tolerant of diverse conditions. Moreover, in many larval cells, the DNA replicates multiple times without separating. This leaves hundreds of strands of DNA adjacent to each other, forming polytene (Greek, "many strands") chromosomes (**FIGURE 9.2**). The unused DNA is more condensed and stains darker than the regions of active DNA. The banding patterns were used to indicate the physical location of the genes on the chromosomes. Morgan's laboratory established a database of mutant strains (see Figure 9.1), as well as an exchange network whereby any laboratory could obtain them.

Jack Schultz (originally in Morgan's laboratory) and others attempted to relate the burgeoning supply of data on the genetics of *Drosophila* to its development. But *Drosophila* was a difficult organism in which to study embryology. Fly embryos proved complex and intractable, being neither large enough to manipulate experimentally nor transparent enough to observe microscopically. It was not until the techniques of molecular biology allowed researchers to identify and manipulate the insect's genes and RNA that its genetics could be related to its development. And when that happened, a revolution occurred in the field of biology. This revolution is continuing, in large part because of the availability of the complete *Drosophila* genome sequence and our ability to generate transgenic flies at high frequency (Pfeiffer et al. 2010; del Valle Rodríguez et al. 2011). Researchers are now able to identify developmental interactions taking place in very small regions of the embryo, to identify enhancers and their transcription factors, and to mathematically model the interactions to a remarkable degree of precision (Hengenius et al. 2014; Markow 2015; Kaufman 2017).

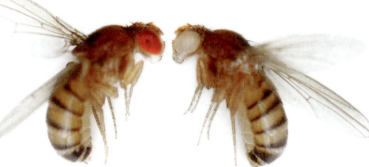

FIGURE 9.1 The T. H. Morgan Lab and development of a powerful genetic model system. Many researchers in Thomas Hunt Morgan's lab helped to define *Drosophila melanogaster* as one of the preeminent model systems for genetics and development. The picture to the left is of Lilian Vaughan Morgan in the "Fly Room" of Thomas Hunt Morgan's laboratory. Lilian was married to Thomas and worked independently on the developmental genetics of sex-linked traits (Keenan 1983). To the right is a photograph of a wild type (left, red eye) and a mutant (right, white eye) fly. The *white* gene was one of the first sex-linked traits discovered by the Morgan lab.

(A)

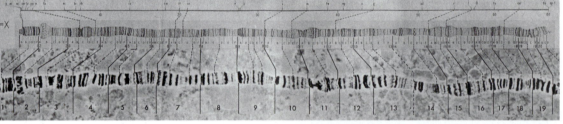

FIGURE 9.2 Polytene chromosomes of *Drosophila*. DNA in the larval salivary glands and other larval tissue replicates without separating. (A) Photograph of the *D. melanogaster* X chromosome. The chart above it was made by Morgan's student Calvin Bridges in 1935. (B) Chromosomes from salivary gland cells of a third instar *D. melanogaster* male. Each polytene chromosome has 1024 strands of DNA (blue stain). Here, an antibody (red) directed against the MSL transcription factor binds only to genes on the X chromosome. MSL accelerates gene expression in the single male X chromosome so that it can match the amount of gene expression by females, with their two X chromosomes.

(B)

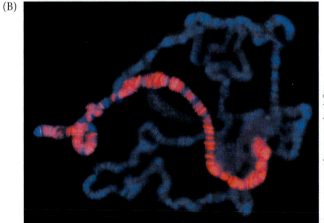

Early *Drosophila* Development

We have already discussed the specification of early embryonic cells by cytoplasmic determinants stored in the oocyte. The cell membranes that form during cleavage establish the region of cytoplasm incorporated into each new blastomere, and the morphogenetic determinants in the incorporated cytoplasm then direct differential gene expression in each cell (see Further Development 2.3, online; **FIGURE 9.3**). But in *Drosophila* development, cell membranes do not compartmentalize nuclei until after the thirteenth nuclear division. Prior to this time, the dividing nuclei all share a common cytoplasm, and material can diffuse throughout the whole embryo. This is **syncytial specification**, and different cell types along the anterior-posterior and dorsal-ventral axes are specified by the interactions of components *within* the single multinucleated cell. Whereas the sperm entry site may fix the axes in nematodes and tunicates, the fly's anterior-posterior and dorsal-ventral axes are specified by interactions between the egg and its surrounding follicle cells prior to fertilization.

Fertilization

Drosophila fertilization is a remarkable series of events and is quite different from fertilizations we've described previously.

- *The sperm enters an egg that is already activated.* Egg activation in *Drosophila* is accomplished at ovulation, a few minutes *before* fertilization begins. As the *Drosophila* oocyte squeezes through a narrow orifice, calcium channels open and Ca²⁺ flows in. The oocyte nucleus then resumes its meiotic divisions and the cytoplasmic mRNAs become translated without fertilization (Mahowald et al. 1983; Fitch and Wakimoto 1998; Heifetz et al. 2001; Horner and Wolfner 2008).

- *There is only one site where the sperm can enter the egg.* This is the **micropyle**, a tunnel in the chorion (eggshell) located at the future dorsal anterior region of the embryo. The micropyle allows sperm to pass through it one at a time and

FIGURE 9.3 Life cycle of *Drosophila melanogaster*. (A) Following fertilization, embryogenesis begins with the division of nuclei (cleavage) and their subsequent cellularization, which is then followed by the cell and tissue movements of gastrulation and organ formation. The embryo hatches out as a first instar larva that grows, going through two molts to become a third instar larva. The third instar larva becomes a pupa, which metamorphoses into the adult fly. (B) The anterior-to-posterior patterning of the embryonic segments are visualized with a fluorescent histone reporter in the live fly embryo imaged with state-of-the-art light-sheet microscopy.

(A)   (B)

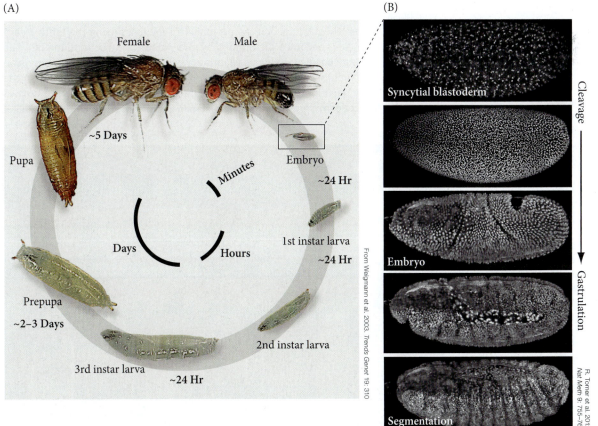

From Weigmann et al. 2003. *Trends Genet* 19: 310

R. Tomer et al. 2012. *Nat Meth* 9: 755–763

probably prevents polyspermy in *Drosophila*. There are no cortical granules to block polyspermy, although cortical changes are seen.

- *By the time the sperm enters the egg, the egg has already begun to specify the body axes;* thus, the sperm enters an egg that is already organizing itself as an embryo.

- *The sperm and egg cell membranes do not fuse. Rather, the sperm enters the egg intact.* The DNA of the male and female pronuclei replicate before the pronuclei have fused, and after the pronuclei fuse, the maternal and paternal chromosomes remain separate until the end of the first mitosis (Loppin et al. 2015).

(See **Further Development 9.1, *Drosophila* Fertilization**, online.)

Cleavage

Most insect eggs undergo **superficial cleavage**, wherein a large mass of centrally located yolk confines cleavage to the cytoplasmic rim of the egg (see Figure 1.9). One of the fascinating features of this cleavage pattern is that cells do not form until after the nuclei have divided several times. In the *Drosophila* egg, karyokinesis (nuclear division) occurs without cytokinesis (cell division) so as to create a **syncytium**, a single cell with many nuclei residing in a common cytoplasm (**FIGURE 9.4**; see also Figure 2.11). The zygote nucleus undergoes several nuclear divisions within the central portion of the egg; 256 nuclei are produced by a series of eight nuclear divisions averaging 8 minutes each (**FIGURE 9.5A,B**). This rapid rate of division is accomplished by repeated rounds of alternating S (DNA synthesis) and M (mitosis) phases in the absence of the gap (G) phases of the cell cycle. During the ninth division cycle, approximately five nuclei reach the surface of the posterior pole of the embryo. These nuclei become enclosed by cell membranes and generate the **pole cells** that give rise to the gametes of the adult. At cycle 10, the other nuclei migrate to the cortex (periphery) of the egg and the mitoses continue, albeit at a progressively slower rate (**FIGURE 9.5C,D**; Foe et al. 2000). During these stages of nuclear division, the embryo is called a **syncytial blastoderm**, since no cell membranes exist other than that of the egg itself.

Although the nuclei divide within a common cytoplasm, the cytoplasm itself is far from uniform. Karr and Alberts (1986) have shown that each nucleus within the syncytial blastoderm is contained within its own little territory of cytoskeletal proteins (see Figure 2.12). When the nuclei reach the periphery of the egg during the tenth cleavage cycle, each nucleus becomes surrounded by microtubules and actin filaments. The nuclei and their associated cytoplasmic islands are called **energids**. Following division cycle 13, the cell membrane (which had covered the egg) folds inward between the nuclei, eventually partitioning off each energid into a single cell. This process creates the **cellular blastoderm**, in which all the cells are arranged in a single-layered jacket

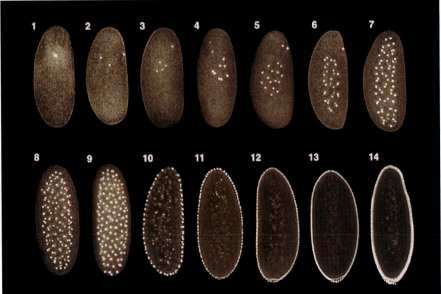

FIGURE 9.4 Laser confocal micrographs of stained chromatin showing syncytial nuclear divisions and superficial cleavage in a series of *Drosophila* embryos. The future anterior end is positioned upward; numbers refer to the nuclear division cycle. The early nuclear divisions occur centrally within a syncytium. Later, the nuclei and their cytoplasmic islands (energids) migrate to the periphery of the cell. This creates the syncytial blastoderm. After cycle 13, the cellular blastoderm forms by ingression of cell membranes between nuclei. The pole cells (germ cell precursors) form in the posterior.

(A)

(B)

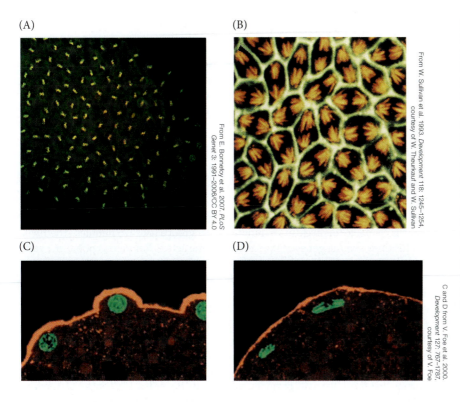

(C)

(D)

FIGURE 9.5 Nuclear and cell division in *Drosophila* embryos. (A) Nuclear division (but not cell division) can be seen in a syncytial *Drosophila* embryo using a dye that stains DNA. The first region to cellularize, the pole region, can be seen forming the cells in the posterior region of the embryo that will eventually become the germ cells (sperm or eggs) of the fly. (B) Chromosomes dividing at the cortex of a syncytial blastoderm. Although there are no cell boundaries, actin (green) can be seen forming regions within which each nucleus divides. The microtubules of the mitotic apparatus are stained red with antibodies to tubulin. (C,D) Cross section of a part of a cycle 10 *Drosophila* embryo showing nuclei (green) at the cortex of the syncytial cell, adjacent to a layer of actin filaments (red). (C) Interphase nuclei. (D) Nuclei in anaphase, dividing parallel to the cortex and enabling the nuclei to stay in the cell periphery.

around the yolky core of the egg (Turner and Mahowald 1977; Foe and Alberts 1983; Mavrakis et al. 2009).

As with all cell formation, the formation of the cellular blastoderm involves a delicate interplay between microtubules and actin filaments (**FIGURE 9.6**). The membrane movements, nuclear elongation, and actin polymerization all appear to be coordinated by the microtubules (Riparbelli et al. 2007). The first phase of blastoderm cellularization is characterized by the invagination of cell membranes between the nuclei to form furrow

(A)

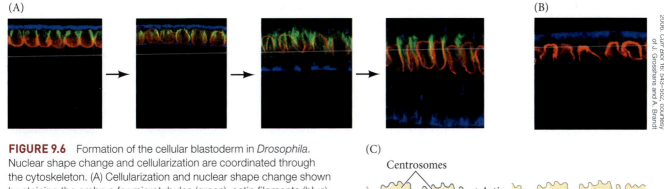

(B)

FIGURE 9.6 Formation of the cellular blastoderm in *Drosophila*. Nuclear shape change and cellularization are coordinated through the cytoskeleton. (A) Cellularization and nuclear shape change shown by staining the embryo for microtubules (green), actin filaments (blue), and nuclei (red). The red stain in the nuclei is due to the presence of the Kugelkern protein, one of the earliest proteins made from the zygotic nuclei. It is essential for nuclear elongation. (B) This embryo was treated with nocadozole to disrupt microtubules. The nuclei fail to elongate, and cellularization is prevented. (C) Diagrammatic representation of cell formation and nuclear elongation. (After A. Brandt et al. 2006. *Curr Biol* 16: 543–552.)

(C)

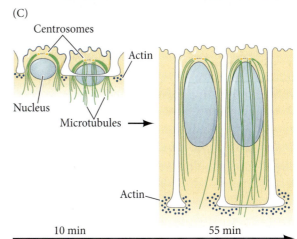

Centrosomes

Actin

Nucleus

Microtubules

Actin

10 min

55 min

Cellularization

canals. This process can be predictably inhibited by drugs that block microtubules. After the furrow canals have passed the level of the nuclei, the second phase of cellularization occurs. The rate of invagination increases and the actin-membrane complex begins to constrict at what will be the basal end of the cell (Foe et al. 1993; Schejter and Wieschaus 1993; Mazumdar and Mazumdar 2002). In *Drosophila*, the cellular blastoderm consists of approximately 6000 cells and is formed within 3 hours of fertilization.

The mid-blastula transition

After the nuclei reach the periphery, the time required to complete each of the next four divisions becomes progressively longer. Whereas cycles 1–10 average 8 minutes each, cycle 13—the last cycle in the syncytial blastoderm—takes 25 minutes to complete. Cycle 14, in which the *Drosophila* embryo forms cells (i.e., after 13 divisions), is asynchronous. Some groups of cells complete this cycle in 75 minutes; other groups take 175 minutes (Foe 1989).

It is at this time that the genes of the nuclei become active. Before this point, the early development of *Drosophila* is directed by proteins and mRNAs placed into the egg during oogenesis. These are the products of the *mother's* genes, not the genes of the embryo's own nuclei. Such genes that are active in the mother to make products for the early development of the offspring are called **maternal effect genes**, and the mRNAs in the oocyte are often referred to as **maternal messages**. Zygotic gene transcription (i.e., the activation of the embryo's own genes) begins around cycle 11 and is greatly enhanced at cycle 14. This slowdown of nuclear division, cellularization, and concomitant increase in new RNA transcription is often referred to as the **mid-blastula transition** (Yuan et al. 2016). It is at this stage that the maternally provided mRNAs are degraded and control of development is handed over to the zygote's own genome (Brandt et al. 2006; De Renzis et al. 2007; Benoit et al. 2009; Laver et al. 2015). Such a **maternal-to-zygotic transition** is seen in the embryos of numerous vertebrate and invertebrate phyla. (See **Further Development 9.2, Mechanisms of the *Drosophila* Mid-Blastula Transition**, online.)

Gastrulation

The general body plan of *Drosophila* is the same in the embryo, the larva, and the adult, each of which has a distinct head end and a distinct tail end, between which are repeating segmental units. Three of these segments form the thorax, while another eight segments form the abdomen. Each segment of the adult fly has its own identity. The first thoracic segment, for example, has only legs; the second thoracic segment has legs and wings; and the third thoracic segment has legs and halteres (flight balancing organs).

Gastrulation begins shortly after the mid-blastula transition. The first movements of *Drosophila* gastrulation segregate the presumptive mesoderm, endoderm, and ectoderm. The prospective mesoderm—about 1000 cells constituting the ventral midline of the embryo—folds inward to produce the **ventral furrow** (**FIGURE 9.7A**). This furrow eventually pinches off from the surface to become a ventral tube within the embryo. The prospective endoderm invaginates to form two pockets at the anterior and posterior ends of the ventral furrow. The pole cells are internalized along with the endoderm (**FIGURE 9.7B,C**). At this time, the outer tissue layer (ectoderm) bends to form the **cephalic furrow**.

The ectodermal cells on the surface and the mesoderm undergo convergence and extension, migrating toward the ventral midline to form the **germ band**, a collection of cells along the ventral midline that includes all the cells that will form the trunk of the embryo. The germ band extends posteriorly and, perhaps because of the egg case, wraps around the top (dorsal) surface of the embryo (**FIGURE 9.7D**). Thus, at the end of germ band formation, the cells destined to form the most posterior larval structures are located immediately behind the future head region (**FIGURE 9.7E**). Although this largely marks the end of gastrulation, with all three germ layers now formed, there are several important morphogenesis events that still need to transpire. For one, the body segments begin to appear, dividing the ectoderm and mesoderm. The germ band then retracts, placing the presumptive posterior segments at the posterior tip of the embryo (**FIGURE 9.7F,G**). At the dorsal surface, the two sides of the epidermis are brought together in a process called **dorsal closure**. The amnioserosa (the extraembryonic layer that surrounds the embryo), which had been the most dorsal structure, interacts with the epidermal cells to stimulate their migration (reviewed in Panfilio 2008; Heisenberg 2009). (See **Further Development 9.3, It Takes Strength to Bend a Fly**, online.)

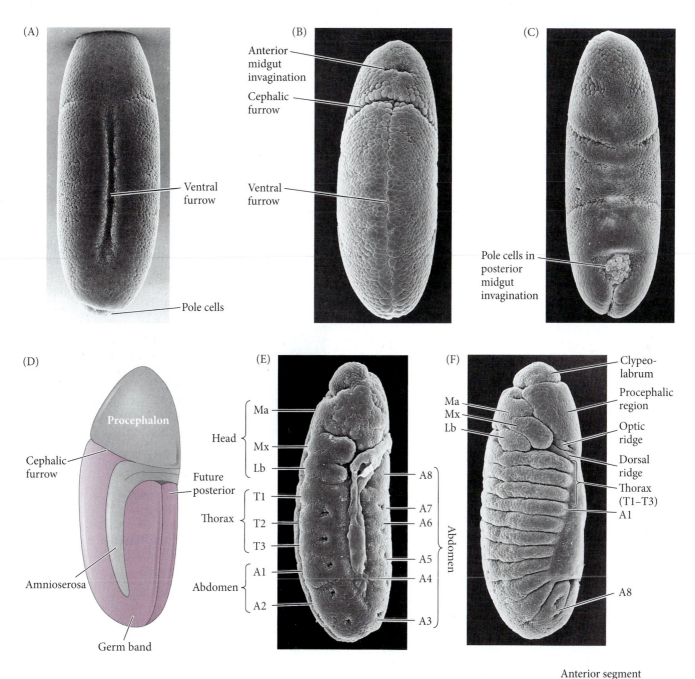

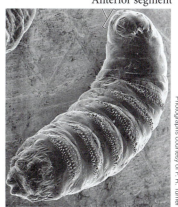

FIGURE 9.7 Gastrulation in *Drosophila*. The anterior of each gastrulating embryo points upward in this series of scanning electron micrographs. (A) Ventral furrow beginning to form as cells flanking the ventral midline invaginate. (B) Closing of ventral furrow, with mesodermal cells placed internally and surface ectoderm flanking the ventral midline. (C) Dorsal view of a slightly older embryo, showing the pole cells and posterior endoderm sinking into the embryo. (D) Schematic representation showing dorsolateral view of an embryo at fullest germ band extension, just prior to segmentation. The cephalic furrow separates the future head region (procephalon) from the germ band, which will form the thorax and abdomen. (E) Lateral view, showing fullest extension of the germ band and the beginnings of segmentation. Subtle indentations mark the incipient segments along the germ band. Ma, Mx, and Lb correspond to the mandibular, maxillary, and labial head segments; T1–T3 are the thoracic segments; and A1–A8 are the abdominal segments. (F) Germ band reversing direction. The true segments are now visible, as well as the other territories of the dorsal head, such as the clypeolabrum, procephalic region, optic ridge, and dorsal ridge. (G) Newly hatched first instar larva. (D after J. A. Campos-Ortega and V. Hartenstein. 1985. *The Embryonic Development of* Drosophila melanogaster. Springer-Verlag: New York.)

FIGURE 9.8   Axis formation in *Drosophila*. (A) Comparison of larval (left) and adult (right) segmentation. In the adult, the three thoracic segments can be distinguished by their appendages: T1 (prothorax) has legs only; T2 (mesothorax) has wings and legs; T3 (metathorax) has halteres (not visible) and legs. (B) During gastrulation, the meso-dermal cells in the most ventral region enter the embryo, and the neurogenic cells expressing *Short gastrulation* (*Sog*) become the ventralmost cells of the embryo. *Sog*, blue; *ventral nervous system defective*, green; *intermediate neuroblast defective*, red. (Larva, after A. Martinez-Arias and P. A. Lawrence. 1985. *Nature* 313: 639–642; adult after M. Peifer et al. 1987. *Genes & Dev* 1: 891–898.)

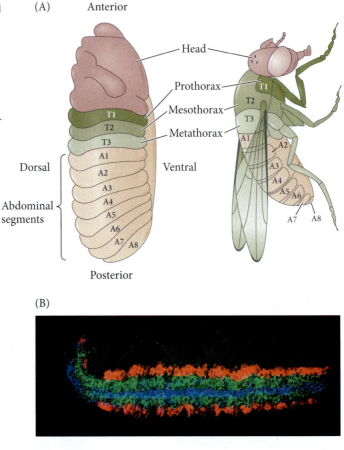

While the germ band is in its extended position, several key morphogenetic processes occur: organogenesis, segmentation (**FIGURE 9.8A**), and segregation of the imaginal discs.[1] The nervous system forms from two regions of ventral ectoderm. Neuroblasts (i.e., the neural progenitor cells) differentiate from this neurogenic ectoderm and migrate inward within each segment (and also from the nonsegmented region of the head ecto-derm). Therefore, in insects such as *Drosophila*, the nervous system is located ventrally rather than being derived from a dorsal neural tube as it is in vertebrates (**FIGURE 9.8B**; see also Figure 9.28).

The Genetic Mechanisms Patterning the *Drosophila* Body

Most of the genes involved in shaping the larval and adult forms of *Drosophila* were iden-tified in the early 1980s using a powerful forward genetics approach (i.e., identifying the genes responsible for a particular phenotype). The basic strategy was to randomly muta-genize flies and then screen for mutations that disrupted the normal formation of the body plan. Some of these mutations were quite fantastic, including embryos and adult flies in which specific body structures were either missing or in the wrong place. These mutant collections were distributed to many different laboratories. The genes involved in the mutant phenotypes were sequenced and then characterized with respect to their expression patterns and their functions. This combined effort has led to a molecular understanding of body plan development in *Drosophila* that is unparalleled in all of biol-ogy, and in 1995 the work resulted in the Nobel Prize in Physiology or Medicine being awarded to Edward Lewis, Christiane Nüsslein-Volhard, and Eric Wieschaus.

The rest of this chapter details the genetics of *Drosophila* development as we have come to understand it over the past three decades. First we will examine how the ante-rior-posterior axis of the embryo is established by interactions between the developing oocyte and its surrounding follicle cells. Next we will see how dorsal-ventral patterning gradients are formed within the embryo, and how these gradients specify different tissue types. Finally we will briefly show how the positioning of embryonic tissues along the two primary axes specifies these tissues to become particular organs.

Segmentation and the Anterior-Posterior Body Plan

The processes of embryogenesis may officially begin at fertilization, but many of the molecular events critical for *Drosophila* embryogenesis actually occur during oogenesis. Each oocyte is descended from a single female germ cell—the **oogonium**. Before oogenesis begins, the oogonium divides four times with incomplete cytokinesis, giving rise to 16 interconnected cells. These 16 germline cells, along with a surrounding epithelial layer of somatic follicle cells, constitute the **egg chamber** in which the oocyte will develop. These germline cells include 15 metabolically active **nurse cells** that make mRNAs and proteins that are transported into the single cell that will become the oocyte. As the oocyte precursor develops at the posterior end of the egg chamber, numerous mRNAs made in the nurse cells are transported along microtubules through the cellular interconnections into the enlarging oocyte.

The genetic screens pioneered by Nüsslein-Volhard and Wieschaus identified a hierarchy of genes that (1) establish anterior-posterior polarity and (2) divide the embryo into a specific number of segments, each with an established polarity and a different identity (**FIGURE 9.9**). This hierarchy is initiated by maternal effect genes that produce messenger RNAs localized to different regions of the egg. These mRNAs encode transcriptional and translational regulatory proteins that diffuse through the syncytial blastoderm and act as morphogens to activate or repress the expression of certain zygotic genes.

The first such zygotic genes to be expressed are called **gap genes** because mutations in them cause gaps in the segmentation pattern. These genes are expressed in certain broad

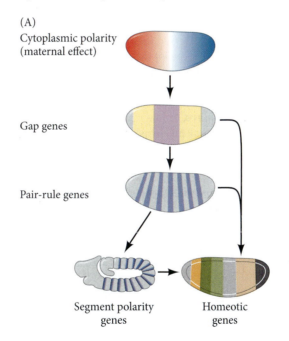

(A)
Cytoplasmic polarity
(maternal effect)

Gap genes

Pair-rule genes

Segment polarity genes

Homeotic genes

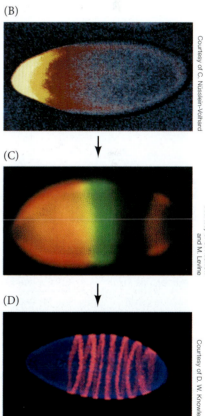

(B)

(C)

(D)

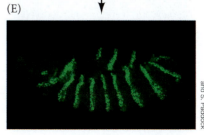

(E)

FIGURE 9.9 *Generalized model of* Drosophila *anterior-posterior pattern formation. Anterior is to the left; the dorsal surface faces upward.* (A) The pattern is established by maternal effect genes that form gradients and regions of morphogenetic proteins. These proteins are transcription factors that activate the gap genes, which define broad territories of the embryo. The gap genes enable the expression of the pair-rule genes, each of which divides the embryo into regions about two segments wide. The segment polarity genes then divide the embryo into segment-sized units along the anterior-posterior axis. Together, the actions of these genes define the spatial domains of the homeotic genes that define the identities of each of the segments. In this way, periodicity is generated from nonperiodicity, and each segment is given a unique identity. (B) Maternal effect genes. The anterior axis is specified by the gradient of Bicoid protein (yellow through red; yellow being the highest concentration). (C) Gap gene protein expression and overlap. The domain of Hunchback protein (orange) and the domain of Krüppel protein (green) overlap to form a region containing both transcription factors (yellow). (D) Products of the *fushi tarazu* pair-rule gene form seven bands across the blastoderm of the embryo. (E) Products of the segment polarity gene *engrailed*, seen here at the extended germ band stage.

(about three segments wide), partially overlapping domains. Gap genes encode transcription factors, and differing combinations and concentrations of gap gene proteins regulate the transcription of **pair-rule genes**, which divide the embryo into periodic units. The transcription of the different pair-rule genes results in a striped pattern of seven transverse bands perpendicular to the anterior-posterior axis. The transcription factors encoded by the pair-rule genes activate the **segment polarity genes**, whose mRNA and protein products divide the embryo into 14-segment-wide units, establishing the periodicity of the embryo. At the same time, the protein products of the gap, pair-rule, and segment polarity genes interact to regulate another class of genes, the **homeotic selector genes**, whose transcription determines the developmental fate of each segment (see Figure 9.9A). (See **Further Development 9.4, Anterior-Posterior Polarity in the Oocyte**, online.)

Maternal gradients: Polarity regulation by oocyte cytoplasm

A series of ligation experiments (see Further Development 9.4 online) showed that two organizing centers control insect development: a head-forming center anteriorly and a posterior-forming center in the rear of the embryo. These centers appeared to secrete substances that generated a head-forming gradient and a tail-forming gradient. In the late 1980s, this gradient hypothesis was united with a genetic approach to the study of *Drosophila* embryogenesis. If there were gradients, what were the morphogens whose concentrations changed over space? Recall that a morphogen is a secreted signaling molecule capable of regulating the expression of different genes in a temporal and concentration-dependent manner. What were the genes that shaped these morphogen gradients? And did these morphogens function by activating or inhibiting certain genes in the areas where they were concentrated? Christiane Nüsslein-Volhard led a research program that addressed these questions. The researchers found that one set of genes encoded morphogens for the anterior part of the embryo, another set of genes encoded morphogens responsible for organizing the posterior region of the embryo, and a third set of genes encoded proteins that produced the terminal regions at both ends of the embryo: the acron and the tail (**TABLE 9.1**). (See **Further Development 9.5, Insect Signaling Centers**, online.)

Two maternal messenger RNAs, *bicoid* and *nanos*, were found to correspond to the anterior and posterior signaling centers and to initiate the formation of the

TABLE 9.1 Maternal effect genes that establish the anterior-posterior polarity of the *Drosophila* embryo

Gene	Mutant phenotype	Proposed function
ANTERIOR GROUP		
bicoid (bcd)	Head and thorax deleted, replaced by inverted telson	Graded anterior morphogen; contains homeodomain; represses *caudal* mRNA
exuperantia (exu)	Anterior head structures deleted	Anchors *bicoid* mRNA
swallow (swa)	Anterior head structures deleted	Anchors *bicoid* mRNA
POSTERIOR GROUP		
nanos (nos)	No abdomen	Posterior morphogen; represses *hunchback* mRNA
tudor (tud)	No abdomen, no pole cells	Localization of *nanos* mRNA
oskar (osk)	No abdomen, no pole cells	Localization of *nanos* mRNA
vasa (vas)	No abdomen, no pole cells; oogenesis defective	Localization of *nanos* mRNA
valois (val)	No abdomen, no pole cells; cellularization defective	Stabilizes Nanos localization complex
pumilio (pum)	No abdomen	Helps Nanos protein bind *hunchback* message
caudal (cad)	No abdomen	Activates posterior terminal genes
TERMINAL GROUP		
torsolike	No termini	Possible morphogen for termini
trunk (trk)	No termini	Transmits Torsolike signal to Torso
fs(1)Nasrat[fs(1)N]	No termini; collapsed eggs	Transmits Torsolike signal to Torso
fs(1)polehole[fs(1)ph]	No termini; collapsed eggs	Transmits Torsolike signal to Torso

Source: K. V. Anderson. 1989. In *Genes and Embryos (Frontiers in Molecular Biology series)*, D. M. Glover and B. D. Hames (Eds.) pp. 1–37. IRL: New York.

FIGURE 9.10 Syncytial specification in *Drosophila*. Anterior-posterior specification originates from morphogen gradients in the egg cytoplasm. *bicoid* mRNA is stabilized in the most anterior portion of the egg, while *nanos* mRNA is tethered to the posterior end. (The anterior can be recognized by the micropyle on the shell; this structure permits sperm to enter.) When the egg is laid and fertilized, these two mRNAs are translated into proteins. The Bicoid protein forms a gradient that is highest at the anterior end, and the Nanos protein forms a gradient that is highest at the posterior end. These two proteins form a coordinate system based on their ratios. Each position along the axis is thus distinguished from any other position. When the nuclei divide, each nucleus is given its positional information by the ratio of these proteins. The proteins forming these gradients activate the transcription of the genes specifying the segmental identities of the larva and the adult fly.

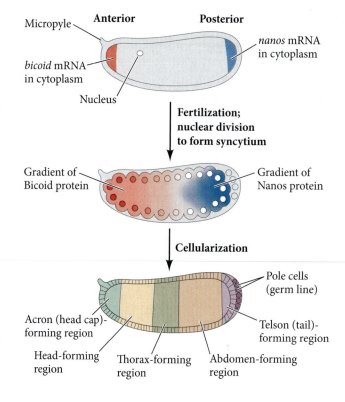

anterior-posterior axis. The *bicoid* mRNAs are located near the anterior tip of the unfertilized egg, and *nanos* messages are located at the posterior tip. These distributions occur as a result of the dramatic polarization of the microtubule networks in the developing oocyte (see Further Development 9.4, Anterior-Posterior Polarity in the Oocyte, online). After ovulation and fertilization, the *bicoid* and *nanos* mRNAs are translated into proteins that can diffuse in the syncytial blastoderm, forming gradients that are critical for anterior-posterior patterning (**FIGURE 9.10**; see also Figure 9.9B).

FURTHER DEVELOPMENT

BICOID AS THE ANTERIOR MORPHOGEN   That Bicoid was the head morphogen of *Drosophila* was demonstrated by a "find it, lose it, move it" experimentation scheme (see Chapter 1, p. 20). Christiane Nüsslein-Volhard, Wolfgang Driever, and their colleagues (Driever and Nüsslein-Volhard 1988a,b; Driever et al. 1990) showed that (1) Bicoid protein was found in a gradient, highest in the anterior (head-forming) region; (2) embryos lacking Bicoid could not form a head; and (3) when *bicoid* mRNA was added to Bicoid-deficient embryos in different places, the place where *bicoid* mRNA was injected became the head (**FIGURE 9.11**). Moreover, the areas around the site of Bicoid injection became the thorax, as expected from a concentration-dependent signal. When injected

FIGURE 9.11 Schematic representation of experiments demonstrating that the *bicoid* gene encodes the morphogen responsible for head structures in *Drosophila*. The phenotypes of *bicoid*-deficient and wild-type embryos are shown at left. When *bicoid*-deficient embryos are injected with *bicoid* mRNA, the point of injection forms the head structures. When the posterior pole of an early-cleavage wild-type embryo is injected with *bicoid* mRNA, head structures form at both poles. (After W. Driever et al. 1990. *Development* 109: 811–820.)

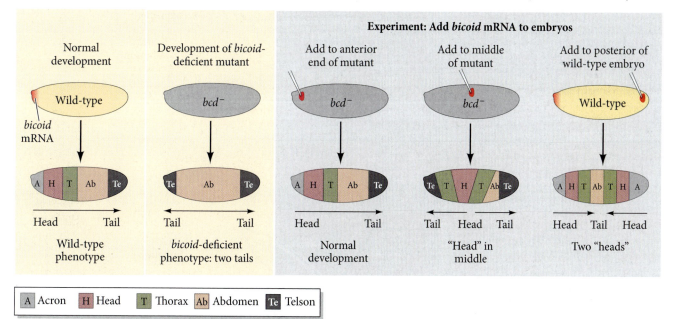

Anterior   Posterior

FIGURE 9.12   Caudal protein gradient of a wild-type *Drosophila* embryo at the syncytial blastoderm stage. Anterior is to the left. The protein (stained darkly) enters the nuclei and helps specify posterior fates. Compare with the complementary gradient of Bicoid protein in Figure 1 in Further Development 9.6, Bicoid mRNA Localization in the Anterior Pole of the Oocyte, online.

into the anterior of *bicoid*-deficient embryos (whose mothers lacked *bicoid* genes), the *bicoid* mRNA "rescued" the embryos and they developed normal anterior-posterior polarity. If *bicoid* mRNA was injected into the center of an embryo, then that middle region became the head, with the regions on either side of it becoming thorax structures. If a large amount of *bicoid* mRNA was injected into the posterior end of a wild-type embryo (with its own endogenous *bicoid* message in its anterior pole), two heads emerged, one at either end (Driever et al. 1990).

At the completion of oogenesis, the *bicoid* message is anchored at the anterior end of the oocyte, and the *nanos* message is tethered to the posterior end (Frigerio et al. 1986; Berleth et al. 1988; Gavis and Lehmann 1992; Little et al. 2011). These two mRNAs are dormant until ovulation and fertilization, at which time they are translated. Since the Bicoid and Nanos *protein products* are not bound to the cytoskeleton, they diffuse toward the middle regions of the early embryo, creating the two opposing gradients that establish the anterior-posterior polarity of the embryo. Mathematical models indicate that these gradients are established by protein diffusion as well as by the active degradation of the proteins (Little et al. 2011; Liu and Ma 2011). (See **Further Development 9.6, Bicoid mRNA Localization in the Anterior Pole of the Oocyte**, online.)

GRADIENTS OF SPECIFIC TRANSLATIONAL INHIBITORS   Two other maternally provided mRNAs—*hunchback, hb*; and *caudal, cad*—are critical for patterning the anterior and posterior regions of the body plan, respectively (Lehmann and Nüsslein-Volhard 1987; Wu and Lengyel 1998). These two mRNAs are synthesized by the nurse cells of the ovary and transported to the oocyte, where they are distributed ubiquitously throughout the syncytial blastoderm. But if they are not localized, how do they mediate their localized patterning activities? It turns out that translation of the *hb* and *cad* mRNAs is repressed by the diffusion gradients of Nanos and Bicoid proteins, respectively.

In the anterior region, Bicoid protein prevents translation of the *caudal* message. Bicoid binds to a specific region of *caudal*'s 3′UTR. Here, it binds Bin3, a protein that stabilizes an inhibitory complex that prevents the binding of the mRNA 5′ cap to the ribosome. By recruiting this translational inhibitor, Bicoid prevents translation of *caudal* in the anterior of the embryo (**FIGURE 9.12**; Rivera-Pomar et al. 1996; Cho et al. 2006; Singh et al. 2011). This suppression is necessary; if Caudal protein is made in the embryo's anterior, the head

FIGURE 9.13   Model of anterior-posterior pattern generation by *Drosophila* maternal effect genes. (A) The *bicoid, nanos, hunchback*, and *caudal* mRNAs are deposited in the oocyte by the ovarian nurse cells. The *bicoid* message is sequestered anteriorly; the *nanos* message is localized to the posterior pole. (B) Upon translation, the Bicoid protein gradient extends from anterior to posterior, while the Nanos protein gradient extends from posterior to anterior. Nanos inhibits the translation of the *hunchback* message (in the posterior), while Bicoid prevents the translation of the *caudal* message (in the anterior). This inhibition results in opposing Caudal and Hunchback gradients. The Hunchback gradient is secondarily strengthened by transcription of the *hunchback* gene in the anterior nuclei (since Bicoid acts as a transcription factor to activate *hunchback* transcription). (C) Parallel interactions whereby translational gene regulation establishes the anterior-posterior patterning of the *Drosophila* embryo. (C after P. M. Macdonald and C. A. Smibert. 1996. *Curr Opin Genet Dev* 6: 403–407.)

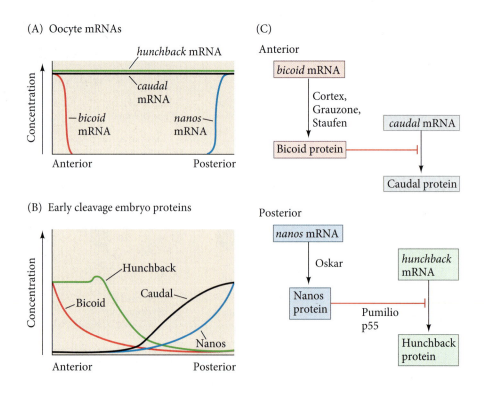

and thorax do not form properly. Caudal activates the genes responsible for the invagination of the hindgut and thus is critical in specifying the posterior domains of the embryo.

In the posterior region, Nanos protein prevents translation of the *hunchback* message. Nanos in the posterior of the embryo forms a complex with several other ubiquitous proteins, including Pumilio and Brat. This complex binds to the 3′UTR of the *hunchback* message, where it recruits d4EHP and prevents the *hunchback* message from attaching to ribosomes (Tautz 1988; Cho et al. 2006).

The result of these interactions is the creation of four maternal protein gradients in the early embryo (**FIGURE 9.13**):

- An anterior-to-posterior gradient of Bicoid protein
- An anterior-to-posterior gradient of Hunchback protein
- A posterior-to-anterior gradient of Nanos protein
- A posterior-to-anterior gradient of Caudal protein

The stage is now set for the activation of zygotic genes in the insect's nuclei, which were busy dividing while these four protein gradients were being established.

The anterior organizing center: The Bicoid and Hunchback gradients

In *Drosophila*, the phenotype of *bicoid* mutants provides valuable information about the function of morphogenetic gradients (**FIGURE 9.14A–C**). Instead of having anterior structures (acron, head, and thorax) followed by abdominal structures and a telson, the structure of a *bicoid* mutant is telson-abdomen-abdomen-telson (**FIGURE 9.14D**). It would appear that these embryos lack whatever substances are needed for the formation of the anterior structures. Moreover, one could hypothesize that the substance

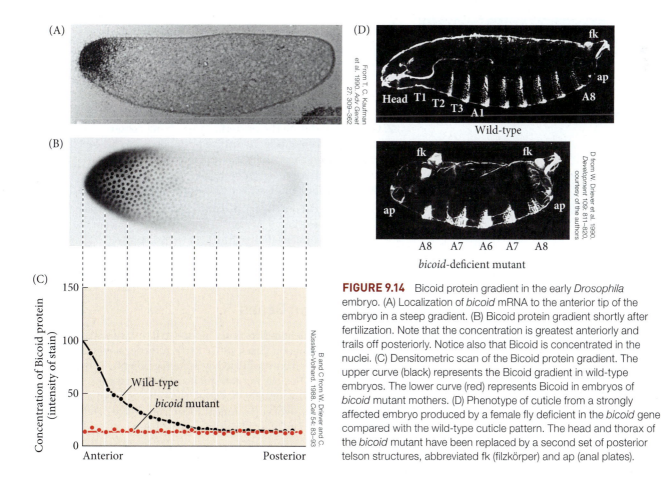

FIGURE 9.14 Bicoid protein gradient in the early *Drosophila* embryo. (A) Localization of *bicoid* mRNA to the anterior tip of the embryo in a steep gradient. (B) Bicoid protein gradient shortly after fertilization. Note that the concentration is greatest anteriorly and trails off posteriorly. Notice also that Bicoid is concentrated in the nuclei. (C) Densitometric scan of the Bicoid protein gradient. The upper curve (black) represents the Bicoid gradient in wild-type embryos. The lower curve (red) represents Bicoid in embryos of *bicoid* mutant mothers. (D) Phenotype of cuticle from a strongly affected embryo produced by a female fly deficient in the *bicoid* gene compared with the wild-type cuticle pattern. The head and thorax of the *bicoid* mutant have been replaced by a second set of posterior telson structures, abbreviated fk (filzkörper) and ap (anal plates).

From T. C. Kaufman et al. 1990, *Adv Genet* 27: 309–362

D from W. Driever et al. 1990, *Development* 109: 811–820, courtesy of the authors

B and C from W. Driever and C. Nüsslein-Volhard. 1988, *Cell* 54: 83–93

these mutants lack is the one postulated by Sander and Kalthoff to turn on genes for the anterior structures and turn off genes for the telson structures.

Bicoid protein appears to act as a morphogen (i.e., a substance that differentially specifies the fates of cells by different concentrations; see Chapter 4). High concentrations of Bicoid produce anterior head structures. Slightly less Bicoid tells the cells to become mouthparts. A moderate concentration of Bicoid is responsible for instructing cells to become the thorax, whereas the abdomen is characterized as lacking Bicoid. How might a gradient of Bicoid protein control the determination of the anterior-posterior axis? Bicoid's primary function is to act as a transcription factor that activates the expression of target genes in the anterior part of the embryo.[2] The first target of Bicoid to be discovered was the *hunchback* (*hb*) gene. In the late 1980s, two laboratories independently demonstrated that Bicoid binds to and activates *hb* (Driever and Nüsslein-Volhard 1989; Struhl et al. 1989; Wieschaus 2016). Bicoid-dependent transcription of *hb* is seen only in the anterior half of the embryo—the region where Bicoid is found. In fact, Bicoid and Hunchback function synergistically to upregulate the transcription of head-specific genes. (See **Further Development 9.7, What Genes Make a Fly's Head? Bicoid Plus Hunchback Equal a Buttonhead**, online.)

The terminal gene group

In addition to the anterior and posterior morphogens, there is a third set of maternal genes whose proteins generate the unsegmented extremities of the anterior-posterior axis: the **acron** (the terminal portion of the head that includes the brain) and the **telson** (tail). Mutations in these terminal genes result in the loss of both the acron and most anterior head segments *and* the telson and most posterior abdominal segments (Degelmann et al. 1986; Klingler et al. 1988). (See **Further Development 9.8, The Terminal Gene Group**, online.)

Summarizing early anterior-posterior axis specification in Drosophila

The anterior-posterior axis of the *Drosophila* embryo is specified by three sets of genes:

1. **Genes that define the anterior organizing center**. Located at the anterior end of the embryo, the anterior organizing center acts through a gradient of Bicoid protein. Bicoid functions both as a *transcription factor* to activate anterior-specific gap genes and as a *translational repressor* to suppress posterior-specific gap genes.

2. **Genes that define the posterior organizing center**. The posterior organizing center is located at the posterior pole. This center acts *translationally* through the Nanos protein to inhibit anterior formation, and *transcriptionally* through the Caudal protein to activate those genes that form the abdomen.

3. **Genes that define the terminal boundary regions**. The boundaries of the acron and telson are defined by the product of the *torso* gene, which is activated at the tips of the embryo.

The next step in development will be to use these gradients of transcription factors to activate specific genes along the anterior-posterior axis.

Segmentation Genes

Cell fate commitment in *Drosophila* appears to have two steps: specification and determination (Slack 1983). Early in fly development, the fate of a cell depends on cues provided by protein gradients. This specification of cell fate is flexible and can still be altered in response to signals from other cells. Eventually, however, the cells undergo a transition from this loose type of commitment to an irreversible determination. At this point, the fate of a cell becomes cell-intrinsic.[3]

TABLE 9.2 Major genes affecting segmentation pattern in *Drosophila*	
Category	**Gene name**
Gap genes	*Krüppel* (*Kr*)
	knirps (*kni*)
	hunchback (*hb*)
	giant (*gt*)
	tailless (*tll*)
	huckebein (*hkb*)
	buttonhead (*btd*)
	empty spiracles (*ems*)
	orthodenticle (*otd*)
Pair-rule genes (primary)	*hairy* (*h*)
	even-skipped (*eve*)
	runt (*run*)
Pair-rule genes (secondary)	*fushi tarazu* (*ftz*)
	odd-paired (*opa*)
	odd-skipped (*odd*)
	sloppy-paired (*slp*)
	paired (*prd*)
Segment polarity genes	*engrailed* (*en*)
	wingless (*wg*)
	cubitus interruptus (*ci*)
	hedgehog (*hh*)
	fused (*fu*)
	armadillo (*arm*)
	patched (*ptc*)
	gooseberry (*gsb*)
	pangolin (*pan*)

interactions between the different gap gene products themselves. (These interactions are facilitated by the fact that they occur within a syncytium, in which the cell membranes have not yet formed.) These boundary-forming inhibitions are thought to be directly mediated by the gap gene products, because all four major gap genes (*hunchback, giant, Krüppel,* and *knirps*) encode DNA-binding proteins (Knipple et al. 1985; Gaul and Jäckle 1990; Capovilla et al. 1992). One such model—established by genetic experiments, biochemical analyses, and mathematical modeling—is presented in **FIGURE 9.17A** (Papatsenko and Levine 2011). The model depicts a network with three major toggle switches (**FIGURE 9.17B–D**). Two of these switches are the strong mutual inhibition between Hunchback and Knirps, and the strong mutual inhibition between Giant and Krüppel (Jaeger et al. 2004). The third is the concentration-dependent interaction between Hunchback and Krüppel. At high doses, Hunchback inhibits the production of Krüppel protein, but at moderate doses (at about 50% of the embryo length), Hunchback promotes Krüppel formation (see Figure 9.17C).

The end result of these repressive interactions is the creation of a precise system of overlapping mRNA expression patterns. Each domain serves as a source for diffusion of gap proteins into adjacent embryonic regions. This creates a significant overlap (at least eight nuclei, which accounts for about two segment primordia) between adjacent gap protein domains. This was demonstrated in a striking manner by Stanojevíc and co-workers (1989). They fixed cellularizing blastoderms (see Figure 9.4), stained Hunchback protein with an antibody carrying a red dye, and simultaneously stained Krüppel protein with an antibody carrying a green dye. Cellularizing regions that contained both proteins bound both antibodies and stained bright yellow (see Figure 9.9C). Krüppel overlaps with Knirps in a similar manner in the posterior region of the embryo (Pankratz et al. 1990). The precision of these patterns is maintained by having redundant enhancers; if one of these enhancers fails to work, there is a high probability that the other will still function (Perry et al. 2011).

The pair-rule genes

The first indication of segmentation in the fly embryo comes when the pair-rule genes are expressed during nuclear division cycle 13, as the cells begin to form at the periphery of the embryo. The transcription patterns of these genes divide the embryo into regions that are precursors of the segmental body plan. As can be seen in **FIGURE 9.18** (and in Figure 9.9D), one vertical band of nuclei (the cells are just beginning to form) expresses a pair-rule gene, the next band of nuclei does not express it, and then the next band expresses it again. The result is a "zebra stripe" pattern along the anterior-posterior axis, dividing the embryo into 15 subunits (Hafen et al. 1984). Eight genes are currently known to be capable of dividing the early embryo in this fashion, and they overlap one another so as to give each cell in the parasegment a specific set of transcription factors (see Table 9.2).

The primary pair-rule genes include *hairy, even-skipped,* and *runt,* each of which is expressed in seven stripes. All three build their striped patterns from scratch, using distinct enhancers and regulatory mechanisms for each stripe. These enhancers are often modular: control over expression in each stripe is located in a discrete region of the DNA, and these DNA regions often contain binding sites recognized by the transcription factor family of gap proteins. Thus, it is thought that the different concentrations of gap proteins determine whether or not a pair-rule gene is transcribed.

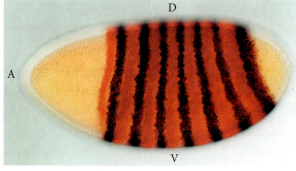

FIGURE 9.18 Messenger RNA expression patterns of two pair-rule genes, *even-skipped* (red) and *fushi tarazu* (black), in the *Drosophila* blastoderm. Each gene is expressed as a series of seven stripes. Anterior is to the left, dorsal is up.

FURTHER DEVELOPMENT

Don't "Even-skip" this segment!

One of the best-studied primary pair-rule genes is *even-skipped* (**FIGURE 9.19**). Its enhancer region is composed of modular units arranged such that each enhancer regulates a separate stripe or a pair of stripes. For instance, *even-skipped* stripe 2 is controlled by a 500-bp region that is activated by Bicoid and Hunchback and repressed by both Giant and Krüppel proteins (**FIGURE 9.20**; Small et al. 1991, 1992; Stanojević et al. 1991; Janssens et al. 2006). The anterior border is maintained by repressive influences from Giant, while the posterior border is maintained by Krüppel. DNase I footprinting showed that the minimal enhancer region for this stripe contains five binding sites for Bicoid, one for Hunchback, three for Krüppel, and three for Giant. Thus, this region is thought to act as a switch that can directly sense the concentrations of these proteins and make on/off transcriptional decisions.

The importance of these enhancer elements can be shown by both genetic and biochemical means. First, a mutation in a particular enhancer can delete its particular stripe and no other. Second, if a reporter gene (such as *lacZ*, which encodes β-galactosidase) is fused to one of the enhancers, the reporter gene is expressed only in that particular stripe (see Figure 9.19; Fujioka et al. 1999). Third, placement of the stripes can be altered by deleting the gap genes that regulate them. Thus, stripe placement is a result of (1) the modular *cis*-regulatory enhancer elements of the pair-rule genes and (2) the *trans*-regulatory gap gene and maternal gene proteins that bind to these enhancer sites.

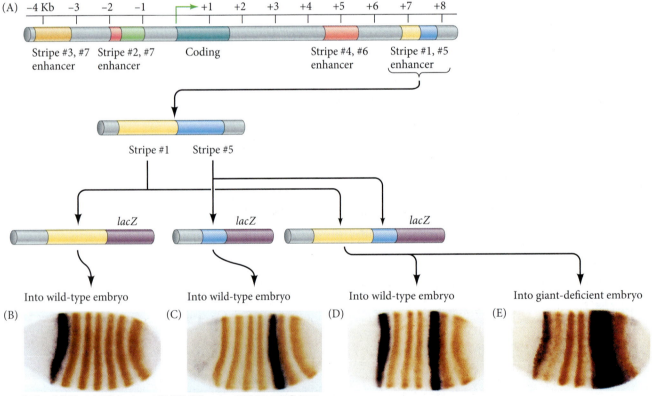

B–E from M. Fujioka et al. 1999. *Development* 126: 2527–2538, courtesy of M. Fujioka and J. B. Jaynes

FIGURE 9.19 Specific promoter regions of the *even-skipped* (*eve*) gene control specific transcription bands in the embryo. (A) Partial map of the *eve* promoter, showing the regions responsible for the various stripes. (B–E) A reporter β-galactosidase gene (*lacZ*) was fused to different regions of the *eve* promoter and injected into fly embryos. The resulting embryos were stained (orange bands) for the presence of Even-skipped protein. (B–D) Wild-type embryos that were injected with *lacZ* transgenes containing the enhancer region specific for stripe 1 (B), stripe 5 (C), or both regions (D). (E) The enhancer region for stripes 1 and 5 was injected into an embryo deficient in *giant*. Here, the posterior border of stripe 5 is missing. (A after C. Sackerson et al. 1999. *Dev Biol* 211: 39–52.)

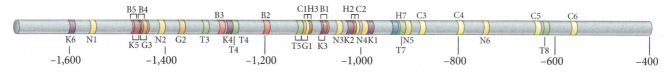

Once initiated by the gap gene proteins, the transcription pattern of the primary pair-rule genes becomes stabilized by interactions among their products (Levine and Harding 1989). The primary pair-rule genes also form the context that allows or inhibits expression of the later-acting secondary pair-rule genes, such as *fushi tarazu* (*ftz*; **FIGURE 9.21**). The eight known pair-rule genes are all expressed in striped patterns, but the patterns are not coincident with each other. Rather, each row of nuclei within a parasegment has its own array of pair-rule products that distinguishes it from any other row. These products activate the next level of segmentation genes, the segment polarity genes.

The segment polarity genes

So far our discussion has described interactions between molecules within the syncytial embryo. But once cells form, interactions take place between the cells. These interactions are mediated by the segment polarity genes, and they accomplish two important tasks. First, they reinforce the parasegmental periodicity established by the earlier transcription factors. Second, through this cell-to-cell signaling, cell fates are established within each parasegment.

The segment polarity genes encode proteins that are constituents of the Wnt and Hedgehog signaling pathways (Ingham 2016; see Chapter 4, Figures 4.22 and 4.26). Mutations in these genes lead to defects in segmentation and altered gene expression patterns across each parasegment. The development of the normal pattern relies on the fact that only one row of cells in each parasegment is permitted to express the Hedgehog protein, and only one row of cells in each parasegment is permitted to express the Wingless protein. (Wingless is the *Drosophila* Wnt protein.) The key to this pattern is the activation of the *engrailed* (*en*) gene in those cells that are going to express Hedgehog.

FIGURE 9.20 Model for formation of the second stripe of transcription from the *even-skipped* gene. The enhancer element for stripe 2 regulation contains binding sequences for several maternal and gap gene proteins. Activators (e.g., Bicoid and Hunchback) are noted above the line; repressors (e.g., Krüppel and Giant) are shown below. Note that nearly every activator site is closely linked to a repressor site, suggesting competitive interactions at these positions. (Moreover, a protein that is a repressor for stripe 2 may be an activator for stripe 5; it depends on which proteins bind next to them.) B, Bicoid; C, Caudal; G, Giant; H, Hunchback; K, Krüppel; N, Knirps; T, Tailless. (After H. Janssens et al. 2006. *Nat Genet* 38: 1159–1165.)

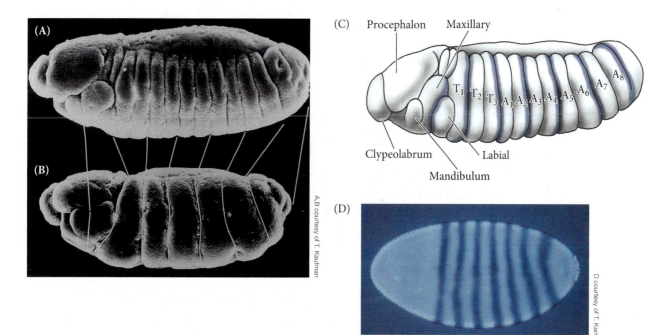

FIGURE 9.21 Defects seen in the *fushi tarazu* mutant. Anterior is to the left; dorsal surface faces upward. (A) Scanning electron micrograph of a wild-type embryo, seen in lateral view. (B) A *fushi tarazu*-mutant embryo at the same stage. The white lines connect the homologous portions of the segmented germ band. (C) Diagram of wild-type embryonic segmentation. The areas shaded in purple show the parasegments of the germ band that are missing in the mutant embryo. (D) Transcription pattern of the *fushi tarazu* gene. (C after T. C. Kaufman et al. 1990. *Adv Genet* 27: 309–362.)

FIGURE 9.22 Model for transcription of the segment polarity genes *engrailed* (*en*) and *wingless* (*wg*). (A) Expression of *wg* and *en* is initiated by pair-rule genes. The *en* gene is expressed in cells that contain high concentrations of either Even-skipped or Fushi tarazu proteins. The *wg* gene is transcribed when neither *eve* nor *ftz* genes are active, but when a third gene (probably *sloppy-paired*) is expressed. (B) The continued expression of *wg* and *en* is maintained by interactions between the Engrailed- and Wingless-expressing cells. Wingless protein is secreted and diffuses to the surrounding cells. In those cells competent to express Engrailed (i.e., those having Eve or Ftz proteins), Wingless protein is bound by the Frizzled and Lrp6 receptor proteins, which enables the activation of the *en* gene via the Wnt signal transduction pathway. (Armadillo is the *Drosophila* name for β-catenin.) Engrailed protein activates the transcription of the *hedgehog* gene and also activates its own (*en*) gene transcription. Hedgehog protein diffuses from these cells and binds to the Patched receptor protein on neighboring cells. The Hedgehog signal enables the transcription of the *wg* gene and the subsequent secretion of the Wingless protein. For a more complex view, see Sánchez et al. 2008. (After M. S. Levine and K. W. Harding. 1989. In D. M. Glover and B. D. Hames [Eds.], *Genes and Embryos*. IRL, New York, pp. 39–94; M. Peifer and A. Bejsovec. 1992. *Trends Genet* 8: 243–249; E. Siegfried et al. 1994. *Nature* 367: 76–80.)

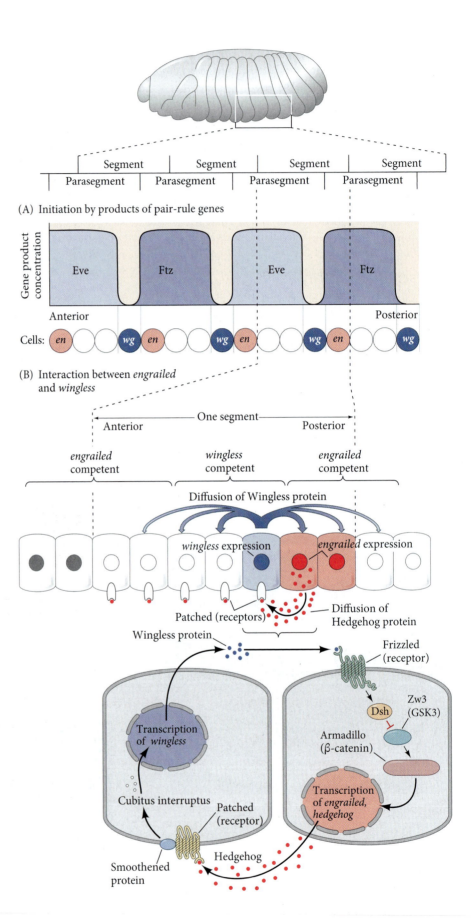

The *engrailed* gene is activated in cells that have high levels of the Even-skipped, Fushi tarazu, or Paired transcription factors; *engrailed* is repressed in those cells with high levels of Odd-skipped, Runt, or Sloppy-paired proteins. As a result, the Engrailed protein is found in 14 stripes across the anterior-posterior axis of the embryo (see Figure 9.9E). (Indeed, in *ftz*-deficient embryos, only seven bands of *engrailed* are expressed.)

These stripes of *engrailed* transcription mark the anterior compartment of each parasegment (and the posterior compartment of each segment). The *wingless* (*wg*) gene is activated in those bands of cells that receive little or no Even-skipped or Fushi tarazu protein, but that do contain Sloppy-paired. This pattern causes *wingless* to be transcribed solely in the column of cells directly anterior to the cells where *engrailed* is transcribed (**FIGURE 9.22A**).

Once *wingless* and *engrailed* expression patterns are established in adjacent cells, this pattern must be maintained to retain the parasegmental periodicity of the body plan. It should be remembered that the mRNAs and proteins involved in initiating these patterns are short-lived, and that the patterns must be maintained after their initiators are no longer being synthesized. The maintenance of these patterns is regulated by reciprocal interaction between neighboring cells: cells secreting Hedgehog protein activate *wingless* expression in their neighbors, and the Wingless protein signal, which is received by the cells that secreted Hedgehog, serves to maintain *hedgehog* (*hh*) expression (**FIGURE 9.22B**). Wingless protein also acts in an autocrine fashion, maintaining its own expression (Sánchez et al. 2008). (See **Further Development 9.9, Flying "Wingless"**, online.)

The diffusion of these proteins is thought to provide the gradients by which the cells of the parasegment acquire their identities. This process can be seen in the dorsal epidermis, where the rows of larval cells produce different cuticular structures depending on their position in the segment. The 1° row of cells consists of large, pigmented spikes called denticles. Posterior to these cells, the 2° row produces a smooth epidermal cuticle. The next two cell rows have a 3° fate, making small, thick hairs; they are followed by several rows of cells that adopt the 4° fate, producing fine hairs (**FIGURE 9.23**).

FIGURE 9.23 Cell specification by the Wingless/Hedgehog signaling center. (A) Dark-field photograph of wild-type *Drosophila* embryo, showing the position of the third abdominal segment. Anterior is to the left; the dorsal surface faces upward. (B) Close-up of the dorsal area of the A3 segment, showing the different cuticular structures made by the 1°, 2°, 3°, and 4° rows of cells. (C) A model for the roles of Wingless and Hedgehog. Each signal is responsible for roughly half the pattern. Either each signal acts in a graded manner (shown here as gradients decreasing with distance from their respective sources) to specify the fates of cells at a distance from these sources, or each signal acts locally on the neighboring cells to initiate a cascade of inductions (shown here as sequential arrows). (After J. Heemskerk and S. DiNardo. 1994. *Cell* 76: 449–460.)

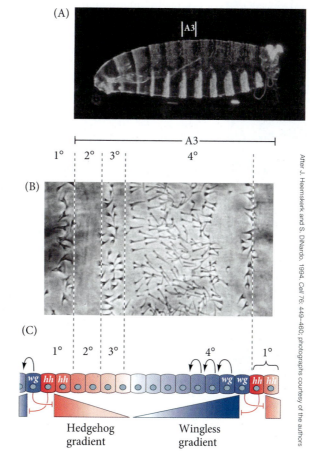

After J. Heemskerk and S. DiNardo. 1994. *Cell* 76: 449–460; photographs courtesy of the authors

(A)

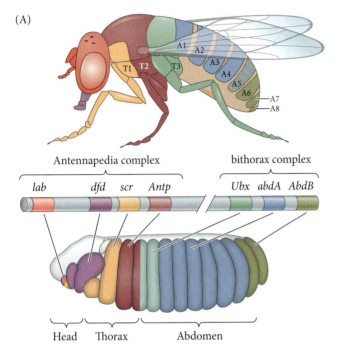

Antennapedia complex   bithorax complex

lab   dfd   scr   Antp   Ubx abdA AbdB

Head   Thorax   Abdomen

(B)

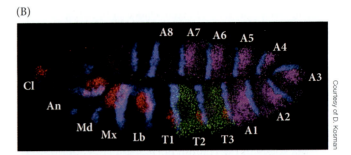

FIGURE 9.24 Homeotic gene expression in *Drosophila*. (A) Expression map of the homeotic genes. In the center are the genes of the Antennapedia and bithorax complexes and their functional domains. Below and above the gene map, the regions of homeotic gene expression (both mRNA and protein) in the blastoderm of the *Drosophila* embryo and the regions that form from them in the adult fly are shown. (B) In situ hybridization for four genes at a slightly later stage (the extended germ band). The *engrailed* (blue) expression pattern separates the body into segments; *Antennapedia* (green) and *Ultrabithorax* (purple) separate the thoracic and abdominal regions; *Distal-less* (red) shows the placement of jaws and the beginnings of limbs. (A top, after M. Peifer et. al 1987. Genes Dev 1: 891–898; bottom, after S. D. Hueber et al. 2010. *PLOS ONE* 5: e10820/CC BY 4.0. doi:10.1371/journal.pone.0010820.)

The Homeotic Selector Genes

After the segmental boundaries are set, the pair-rule and gap genes interact to regulate the homeotic selector genes, which specify the characteristic structures of each segment (Lewis 1978). By the end of the cellular blastoderm stage, each segment primordium has been given an individual identity by its unique constellation of gap, pair-rule, and homeotic gene products (Levine and Harding 1989). Two regions of *Drosophila* chromosome III contain most of these homeotic genes (**FIGURE 9.24**). The first region, known as the **Antennapedia complex**, contains the homeotic genes *labial* (*lab*), *Antennapedia* (*Antp*), *sex combs reduced* (*scr*), *deformed* (*dfd*), and *proboscipedia* (*pb*). The *labial* and *deformed* genes specify the head segments, while *sex combs reduced* and *Antennapedia* contribute to giving the thoracic segments their identities. The *proboscipedia* gene appears to act only in adults, but in its absence, the labial palps of the mouth are transformed into legs (Wakimoto et al. 1984; Kaufman et al. 1990; Maeda and Karch 2009).

The second region of homeotic genes is the **bithorax complex** (Lewis 1978; Maeda and Karch 2009). Three protein-coding genes are found in this complex: *Ultrabithorax* (*Ubx*), which is required for the identity of the third thoracic segment; and the *Abdominal A* (*AbdA*) and *Abdominal B* (*AbdB*) genes, which are responsible for the segmental identities of the abdominal segments (Sánchez-Herrero et al. 1985). The chromosome region containing both the Antennapedia complex and the bithorax complex is often referred to as the **homeotic complex**, or **Hom-C**.

Because the homeotic selector genes are responsible for the specification of fly body parts, mutations in them lead to bizarre phenotypes. In 1894, William Bateson called these organisms **homeotic mutants**, and they have fascinated developmental biologists for decades.[5] For example, the body of the normal adult fly contains three thoracic segments, each of which produces a pair of legs. The first thoracic segment does not produce any other appendages, but the second thoracic segment produces a pair of wings in addition to its legs. The third thoracic segment produces a pair

(A)

Second thoracic segment

(B)

Second thoracic segment   Third thoracic segment

FIGURE 9.25 (A) Wings of the wild-type fruit fly emerge from the second thoracic segment. (B) A four-winged fruit fly constructed by putting together three mutations in *cis*-regulators of the *Ultrabithorax* gene. These mutations effectively transform the third thoracic segment into another second thoracic segment (i.e., transform halteres into wings).

(A)

Antenna

(B)

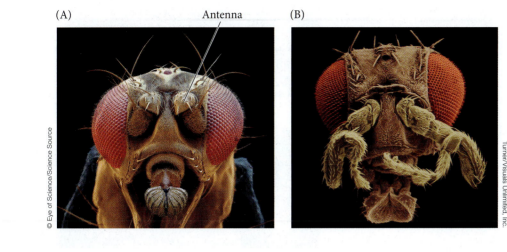

FIGURE 9.26 (A) Head of a wild-type fruit fly. (B) Head of a fly containing the *Antennapedia* mutation that converts antennae into legs.

of legs and a pair of balancers known as **halteres**. In homeotic mutants, these specific segmental identities can be changed. When the *Ultrabithorax* gene is deleted, the third thoracic segment (characterized by halteres) is transformed into another second thoracic segment. The result is a fly with four wings (**FIGURE 9.25**)—an embarrassing situation for a classic dipteran (two-winged organism).[6]

Similarly, Antennapedia protein usually specifies the second thoracic segment of the fly. But when flies have a mutation wherein the *Antennapedia* gene is expressed in the head (as well as in the thorax), legs rather than antennae grow out of the head sockets (**FIGURE 9.26**). This is partly because in addition to promoting the formation of thoracic structures, the Antennapedia protein binds to and represses the enhancers of at least two genes, *homothorax* and *eyeless*, which encode transcription factors that are critical for antenna and eye formation, respectively (Casares and Mann 1998; Plaza et al. 2001). Therefore, one of Antennapedia's functions is to repress the genes that would trigger antenna and eye development. In the recessive mutant of *Antennapedia*, the gene fails to be expressed in the second thoracic segment, and antennae sprout in the leg positions (Struhl 1981; Frischer et al. 1986; Schneuwly et al. 1987).

The major homeotic selector genes have been cloned, their expression analyzed by in situ hybridization and shown to encode the homeobox-containing transcription factors (Harding et al. 1985; Akam 1987). Transcripts from each gene can be detected in specific regions of the embryo (see Figure 9.24B) and are especially prominent in the central nervous system. (See **Further Development 9.10, Initiation and Maintenance of Homeotic Gene Expression**, online.)

Generating the Dorsal-Ventral Axis

Dorsal-ventral patterning in the oocyte

As oocyte volume increases, the oocyte nucleus is pushed by the growing microtubules to a position that becomes the dorsal anterior corner of the oocyte—a critical symmetry-breaking event (Zhao et al. 2012). Here the *gurken* message, which had been critical in establishing the anterior-posterior axis, initiates the formation of the dorsal-ventral axis. The *gurken* mRNA becomes localized in a crescent between the oocyte nucleus and the oocyte cell membrane, and its protein product forms an anterior-posterior gradient along the dorsal surface of the oocyte (**FIGURE 9.27**; Neuman-Silberberg and Schüpbach 1993). Since it can diffuse only a short distance, Gurken protein reaches only those follicle cells closest to the oocyte nucleus, and it signals through the Torpedo receptor to those cells to become the more columnar dorsal follicle cells (Montell et al. 1991; Schüpbach et al. 1991). This establishes the dorsal-ventral polarity in the follicle cell layer that surrounds the growing oocyte.

Maternal deficiencies of either the *gurken* or the *torpedo* gene cause ventralization of the embryo. However, *gurken* is active only in the oocyte, whereas *torpedo* is active only in the somatic follicle cells (Schüpbach 1987). The Gurken-Torpedo signal that specifies

FIGURE 9.27 Expression of Gurken between the oocyte nucleus and the dorsal anterior cell membrane. (A) The *gurken* mRNA is localized between the oocyte nucleus and the dorsal follicle cells of the ovary. Anterior is to the left; dorsal faces upward. (B) A more mature oocyte shows Gurken protein (yellow) across the dorsal region. Actin is stained red, showing cell boundaries. As the oocyte grows, follicle cells migrate across the top of the oocyte, where they become exposed to Gurken.

(A)

From R. P. Ray and T. Schüpbach. 1996. *Genes Dev* 10: 1711–1723, courtesy of T. Schüpbach

(B)

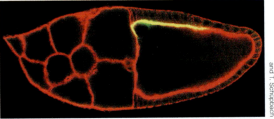

Courtesy of C. van Buskirk and T. Schüpbach

dorsalized follicle cells initiates a cascade of gene activity that creates the dorsal-ventral axis of the embryo. (See **Further Development 9.11, Torpedos Away: The Downstream Signaling Events**, online.)

Generating the dorsal-ventral axis within the embryo

The protein that distinguishes dorsum (back) from ventrum (belly) in the fly embryo is the product of the *dorsal* gene. The Dorsal protein is a transcription factor that activates the genes that generate the ventrum. (Note that this is another *Drosophila* gene named after its mutant phenotype: the *dorsal* gene product is a morphogen that ventralizes the region in which it is present.) The mRNA transcript of the mother's *dorsal* gene is deposited in the oocyte by the nurse cells. However, Dorsal protein is not synthesized from this maternal message until about 90 minutes after fertilization. When Dorsal is translated, it is found throughout the embryo, not just on the ventral or dorsal side. How can this protein act as a morphogen if it is located everywhere in the embryo?

The answer to this question was unexpected (Roth et al. 1989; Rushlow et al. 1989; Steward 1989). Although Dorsal protein is found throughout the syncytial blastoderm of the early *Drosophila* embryo, it is translocated into nuclei only in the ventral part of the embryo. In the nucleus, Dorsal acts as a transcription factor, binding to certain genes to activate or repress their transcription. If Dorsal does not enter the nucleus, the genes responsible for specifying ventral cell types are not transcribed, the genes responsible for specifying dorsal cell types are not repressed, and all the cells of the embryo become specified as dorsal cells.

This model of dorsal-ventral axis formation in *Drosophila* is supported by analyses of maternal effect mutations that give rise to an entirely dorsalized or an entirely ventralized phenotype, where there is no "back" to the larval embryo, which soon dies (Anderson and Nüsslein-Volhard 1984). In mutants in which all the cells are dorsalized (evident from their dorsal-specific exoskeleton), Dorsal does not enter the nucleus of any cell. Conversely, in mutants in which all cells have a ventral phenotype, Dorsal protein is found in every cell nucleus (**FIGURE 9.28A**).

FURTHER DEVELOPMENT

Establishing a nuclear Dorsal gradient

So how does Dorsal protein enter into the nuclei only of the ventral cells? When Dorsal is first produced, it is complexed with a protein called Cactus in the cytoplasm of the syncytial blastoderm. As long as Cactus is bound to it, Dorsal remains in the cytoplasm. Dorsal enters ventral nuclei in response to a

(A)

(B)

From S. Roth et al. 1989. *Cell* 59: 1189–1202, courtesy of the authors

Dorsal

Amnioserosa

Dorsal ectoderm

Lateral ectoderm

Neurogenic ectoderm

Mesoderm

Ventral

Lateral section   Transverse section

(C)

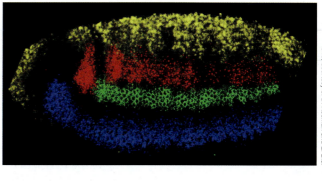

From D. Kosman et al. 2004. *Science* 305: 846, courtesy of D. Kosman and E. Bier

FIGURE 9.28 Specification of cell fate by the Dorsal protein. (A) Transverse sections of embryos stained with antibody to show the presence of Dorsal protein (dark area). The wild-type embryo (left) has Dorsal protein only in the ventralmost nuclei. A dorsalized mutant (center) has no localization of Dorsal protein in any nucleus. In the ventralized mutant (right), Dorsal protein has entered the nucleus of every cell. (B) Fate maps of cross sections through the *Drosophila* embryo at division cycle 14. The most ventral part becomes the mesoderm; the next higher portion becomes the neurogenic (ventral) ectoderm. The lateral and dorsal ectoderm can be distinguished in the cuticle, and the dorsalmost region becomes the amnioserosa, the extraembryonic layer that surrounds the embryo. The translocation of Dorsal protein into ventral, but not lateral or dorsal, nuclei produces a gradient whereby the ventral cells with the most Dorsal protein become mesoderm precursors. (C) Dorsal-ventral patterning in *Drosophila*. Following invagination of the mesoderm, the readout of the Dorsal gradient can be seen in the trunk region of this whole-mount stained embryo. The expression of the most ventral gene, *ventral nervous system defective* (blue), is from the neurogenic ectoderm. The *intermediate neuroblast defective* gene (green) is expressed in lateral ectoderm. Red represents the *muscle-specific homeobox* gene, expressed in the mesoderm above the intermediate neuroblasts. The dorsalmost tissue expresses *decapentaplegic* (yellow). (B after C. A. Rushlow et al. 1989. *Cell* 59: 1165–1177.)

signaling pathway that frees it from Cactus (see Figure 1B in Further Development 9.11, online). This separation of Dorsal from Cactus is initiated by the ventral activation of the Toll receptor. When Spätzle binds to and activates the Toll protein, Toll activates a protein kinase called Pelle. Another protein, Tube, is probably necessary for bringing Pelle to the cell membrane, where it can be activated (Galindo et al. 1995). The activated Pelle protein kinase (probably through an intermediate) can phosphorylate Cactus. Once phosphorylated, Cactus is degraded and Dorsal can enter the nucleus (Kidd 1992; Shelton and Wasserman 1993; Whalen and Steward 1993; Reach et al. 1996). Since Toll is activated by a gradient of Spätzle protein that is highest in the most ventral region, there is a corresponding gradient of Dorsal translocation in the ventral cells of the embryo, with the highest concentrations of Dorsal in the most ventral cell nuclei, which become the mesoderm (**FIGURE 9.28B**).

The Dorsal protein signals the first morphogenetic event of *Drosophila* gastrulation. The 16 ventralmost cells of the embryo—those cells containing the highest amount of Dorsal in their nuclei—invaginate into the body and form the mesoderm (**FIGURE 9.29**). All of the body muscles, fat bodies, and gonads derive from these mesodermal cells (Foe 1989). The cells that will take their place at the ventral midline will become the nerves and glia (**FIGURE 9.28C**; see also **Further Development 9.12, Effects of the Dorsal Protein Gradient**, online).

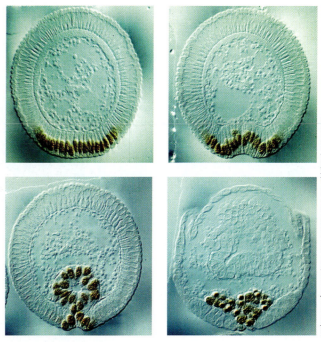

From M. Leptin. 1991. In *Gastrulation: Movements, Patterns, and Molecules*, R. Keller et al. (Eds.), pp. 199–212. Plenum: New York, courtesy of M. Leptin

FIGURE 9.29 Gastrulation in *Drosophila*. In this cross section, the mesodermal cells at the ventral portion of the embryo buckle inward, forming the ventral furrow (see Figure 9.7A,B). This furrow becomes a tube that invaginates into the embryo and then flattens and generates the mesodermal organs. The nuclei are stained with antibody to the Twist protein, a marker for the mesoderm.

Axes and Organ Primordia: The Cartesian Coordinate Model

The anterior-posterior and dorsal-ventral axes of *Drosophila* embryos form a coordinate system that can be used to specify positions within the embryo (**FIGURE 9.30A**). Theoretically, cells that are initially equivalent in developmental potential can respond to their position by expressing different sets of genes. This type of specification has been demonstrated in the formation of the salivary gland rudiments (Panzer et al. 1992; Bradley et al. 2001; Zhou et al. 2001).

Drosophila salivary glands form only in the strip of cells defined by the activity of the *sex combs reduced* (*scr*) gene along the anterior-posterior axis (parasegment 2). No salivary glands form in *scr*-deficient mutants. Moreover, if *scr* is experimentally expressed throughout the embryo, salivary gland primordia form in a ventrolateral stripe along most of the length of the embryo. The formation of salivary glands along the dorsal-ventral axis is repressed by both Decapentaplegic and Dorsal proteins, which inhibit salivary gland formation dorsally and ventrally, respectively. Thus, the salivary glands form at the intersection of the vertical *scr* expression band (parasegment 2) and the horizontal region in the middle of the embryo's circumference that has neither Decapentaplegic nor Dorsal (**FIGURE 9.30B**). The cells that form the salivary glands are directed to do so by the intersecting gene activities along the anterior-posterior and dorsal-ventral axes.

A similar situation is seen in neural precursor cells found in every segment of the fly. Neuroblasts arise from 10 clusters of 4 to 6 cells each that form on each side in every segment in the strip of neural ectoderm at the midline of the embryo (Skeath and Carroll 1992). The cells in each cluster interact (via the Notch pathway discussed in Chapter 4) to generate a single neural cell from each cluster. Skeath and colleagues (1992) have shown that the pattern of neural gene transcription is imposed by a coordinate system. Their expression is repressed along the dorsal-ventral axis by the Decapentaplegic and Snail proteins, while positive enhancement by pair-rule genes along the anterior-posterior axis causes neural gene repetition in each half-segment. It is very likely, then, that the positions of organ primordia in the fly are specified via a two-dimensional coordinate system based on the intersection of the anterior-posterior and dorsal-ventral axes. (See **Further Development 9.13, The Right-Left Axes**, and **Further Development 9.14, Early Development of Other Insects**, both online.)

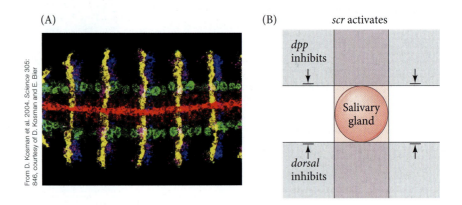

(A)

From D. Kosman et al. 2004. *Science* 305: 846, courtesy of D. Kosman and E. Bier

(B)   *scr* activates

dpp inhibits

Salivary gland

dorsal inhibits

FIGURE 9.30   Cartesian coordinate system mapped out by gene expression patterns. (A) A grid (ventral view, looking "up" at the embryo) formed by the expression of *short-gastrulation* (red), *intermediate neuroblast defective* (green), and *muscle segment homeobox* (magenta) along the dorsal-ventral axis, and by the expression of *wingless* (yellow) and *engrailed* (purple) transcripts along the anterior-posterior axis. (B) Coordinates for the expression of genes giving rise to *Drosophila* salivary glands. These genes are activated by the protein product of the *sex combs reduced* (*scr*) homeotic gene in a narrow band along the anterior-posterior axis, and they are inhibited in the regions marked by *decapentaplegic* (*dpp*) and *dorsal* gene products along the dorsal-ventral axis. This pattern allows salivary glands to form in the midline of the embryo in the second parasegment. (B after S. Panzer et al. 1992. *Development* 114: 49–57.)

Closing Thoughts on the Opening Photo

Courtesy of Nipam Patel

In the fruit fly, inherited genes produce proteins that interact to specify the normal orientation of the body, with the head at one end and the tail at the other. As you studied this chapter, you should have observed how these interactions result in the specification of entire blocks of the fly's body as modular units. A patterned array of homeotic proteins specifies the structures to be formed in each segment of the adult fly. Mutations in the genes for these proteins, called homeotic mutations, *can change the structure specified, resulting in wings where there should have been halteres*, or legs where there should have been antennae (see pp. 268–269). Remarkably, the proximal-distal orientation of the mutant appendages corresponds to the original appendage's proximal-distal axis, indicating that the appendages follow similar rules for their extension. We now know that many mutations affecting segmentation of the adult fly in fact work on the embryonic modular unit, the parasegment. You should keep in mind that, in both invertebrates and vertebrates, the units of embryonic construction are often not the same units we see in the adult organism.

Endnotes

[1] Imaginal discs are cells set aside to produce the adult structures. Imaginal disc differentiation will be discussed as a part of metamorphosis in Chapter 21.

[2] *bicoid* appears to be a relatively "new" gene that evolved in the Dipteran lineage (two-winged insects such as flies); it has not been found in other insect lineages. The anterior determinant in other insect groups includes the Orthodenticle and Hunchback proteins, both of which can be induced in the anterior of the *Drosophila* embryo by Bicoid (Wilson and Dearden 2011).

[3] Aficionados of information theory will recognize that the process by which the anterior-posterior information in morphogenetic gradients is transferred to discrete and different parasegments represents a transition from analog to digital specification. Specification is analog, determination digital. This process enables the transient information of the gradients in the syncytial blastoderm to be stabilized so that it can be used much later in development (Baumgartner and Noll 1990).

[4] The two modes of segmentation may be required for the coordination of movement in the adult fly. In arthropods, the ganglia of the ventral nerve cord are organized by parasegments, but the cuticle grooves and musculature are segmental. This shift in frame by one compartment allows the muscles on both sides of any particular

epidermal segment to be coordinated by the same ganglion. This, in turn, allows rapid and coordinated muscle contractions for locomotion (Deutsch 2004). A similar situation occurs in vertebrates, where the posterior portion of the anterior somite combines with the anterior portion of the next somite.

[5] *Homeo*, from the Greek, means "similar." *Homeotic mutants* are mutants in which one structure is replaced by another (as where an antenna is replaced by a leg). *Homeotic genes* are those genes whose mutation can cause such transformations; thus, homeotic genes are genes that specify the identity of a particular body segment. The *homeobox* is a conserved DNA sequence of about 180 base pairs that is shared by many homeotic genes. This sequence encodes the 60-amino acid *homeodomain*, which recognizes specific DNA sequences. The homeodomain is an important region of the transcription factors encoded by homeotic genes. However, not all genes containing homeoboxes are homeotic genes.

[6] Dipterans—two-winged insects such as flies—are thought to have evolved from four-winged insects, and it is possible that this change arose via alterations in the bithorax complex. Chapter 24 includes further speculation on the relationship between the homeotic complex and evolution.

⑨ Snapshot Summary
The Genetics of Axis Specification in *Drosophila*

1. *Drosophila* cleavage is superficial. The nuclei divide 13 times before being compartmentalized. Before cell formation, the nuclei reside in a syncytial blastoderm. Each nucleus is surrounded by actin-filled cytoplasm.

2. When the cell membranes form around the nuclei, the *Drosophila* embryo undergoes a mid-blastula transition, wherein the cleavages become asynchronous and new mRNA is made. At this time, there is a transfer from maternal to zygotic control of development.

3. Gastrulation begins with the invagination of the most ventral region (the presumptive mesoderm), which involves

formation of a ventral furrow. The germ band expands such that the future posterior segments curl just behind the presumptive head.

4. The genes regulating pattern formation in *Drosophila* operate according to certain principles:

   - There are *morphogens*—such as Bicoid and Dorsal—whose gradients determine the specification of different cell types. In syncytial embryos, these morphogens can be transcription factors.

   - *Boundaries* of gene expression can be created by the interaction between transcription factors and their

(Continued)

Snapshot Summary   *(continued)*

gene targets. Here, the transcription factors transcribed earlier regulate the expression of the next set of genes.

- *Translational control* is extremely important in the early embryo, and localized mRNAs are critical in patterning the embryo.

- *Individual cell fates* are not defined immediately. Rather, there is a stepwise specification wherein a given field is divided and subdivided, eventually regulating individual cell fates.

5. There is a temporal order wherein different classes of genes are transcribed, and the products of one gene often regulate the expression of another gene.

6. Maternal effect genes are responsible for the initiation of anterior-posterior polarity. *bicoid* mRNA is bound by its 3′UTR to the cytoskeleton in the future anterior pole; *nanos* mRNA is sequestered by its 3′UTR in the future posterior pole; *hunchback* and *caudal* messages are seen throughout the embryo.

7. Bicoid and Hunchback proteins activate the genes responsible for the anterior portion of the fly; Caudal activates genes responsible for posterior development.

8. The unsegmented anterior and posterior extremities are regulated by the activation of Torso protein at the anterior and posterior poles of the egg.

9. The gap genes respond to concentrations of the maternal effect gene proteins. Their protein products interact with each other such that each gap gene protein defines specific regions of the embryo.

10. The gap gene proteins activate and repress the pair-rule genes. The pair-rule genes have modular enhancers such that they become activated in seven "stripes." Their boundaries of transcription are defined by the gap genes. The pair-rule genes form seven bands of transcription along the anterior-posterior axis, each one comprising two parasegments.

11. The pair-rule gene products activate segment polarity genes *engrailed* and *wingless* expression in adjacent cells. The *engrailed*-expressing cells form the anterior boundary of each parasegment. These cells form a signaling center that organizes the cuticle formation and segmental structure of the embryo.

12. Homeotic selector genes are found in two complexes on chromosome III of *Drosophila*. Together, these regions are called Hom-C, the homeotic gene complex. The genes are arranged in the same order as their transcriptional expression. Genes of the Hom-C specify the individual segments, and mutations in these genes are capable of transforming one segment into another.

13. Dorsal-ventral polarity is initiated when the nucleus moves to the dorsal-anterior of the oocyte and sequesters the *gurken* message, enabling it to synthesize proteins in the dorsal side of the egg.

14. Dorsal protein is activated in a gradient as it enters the various nuclei. Those nuclei at the most ventral surface incorporate the most Dorsal protein and become mesoderm; those more lateral become neurogenic ectoderm.

15. Organs form at the intersection of dorsal-ventral and anterior-posterior regions of gene expression.

Go to oup.com/he/barresi12xe for Further Developments, Scientists Speak interviews, Watch Development videos, Dev Tutorials, and complete bibliographic information for all literature cited in this chapter.

Sea Urchins and Tunicates
Deuterostome Invertebrates

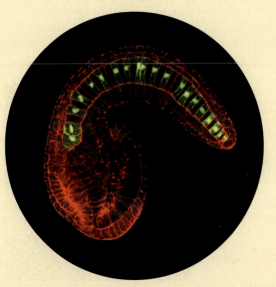

How do the fluorescing cells of this tunicate embryo proclaim its kinship with you?

Photograph courtesy of Anna Di Gregorio

HAVING DESCRIBED THE PROCESSES of early development in representative species from three protostome groups—mollusks, nematodes, and insects—we turn now to the deuterostomes. Although there are far fewer species of deuterostomes than there are of protostomes, the deuterostomes include the members of all the vertebrate groups—fish, amphibians, reptiles, birds, and mammals. Several invertebrate groups also follow the deuterostome pattern of development (in which the blastopore becomes the anus during gastrulation). These include the hemichordates (acorn worms), cephalochordates (*Amphioxus*), echinoderms (sea urchins, seastars, sea cucumbers, and others), and urochordates (tunicates, also called sea squirts) (**FIGURE 10.1**). This chapter covers the early development of echinoderms (notably sea urchins) and tunicates, both of which have been the subjects of critically important studies in developmental biology.

Indeed, conditional specification (historically, but less today, also referred to as "regulative development") was first discovered in sea urchins, while tunicates provided the first evidence for autonomous specification ("mosaic development"; see Chapter 2). As we will see, it turns out that both groups use both modes of specification.

Early Development in Sea Urchins

Sea urchins have been exceptionally important organisms in studying how genes regulate the formation of the body. Hans Driesch discovered regulative development when he was studying sea urchins. He found that the early stages of sea urchin development had a regulative mode, since a single blastomere isolated from the 4-cell stage could form an entire sea urchin pluteus larva (Driesch 1891; also see Chapter 2). However, cells isolated from later stages could not become all the cells of the larval body, indicating that at some point along their maturation, cells become committed to their fate.

Sea urchin embryos also provided the first evidence that chromosomes were needed for development, that DNA and RNA were present in every animal cell, that messenger RNAs directed protein synthesis, that stored messenger RNAs provided the proteins for early embryonic development, that cyclins controlled cell division, and that enhancers were modular (Ernst 2011; McClay 2011). The first cloned eukaryotic gene encoded a sea urchin histone protein (Kedes et al. 1975), and the first evidence for chromatin remodeling concerned histone alterations during sea urchin development (Newrock et al. 1978). With the advent of new genetic techniques, sea urchin embryos continue to be critically important organisms for delineating the mechanisms by which genetic interactions specify different cell fates.

FIGURE 10.1 The echinoderms and tunicates represent deuterostome invertebrates. The tunicates are also classified as chordates because their larvae possess a notochord, dorsal neural tube, and pharyngeal arches. The tunicates are called urochordates (tail + chord) because their notochord is located only in their larval tail. The green sea urchin (*Lytechinus variegatus*) and the tunicate *Ciona intestinalis* are two widely studied model organisms (red text and accompanying images).

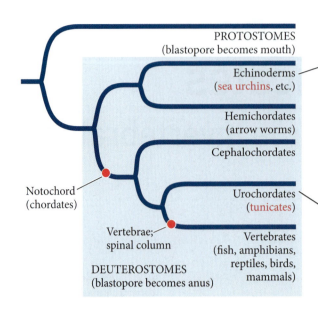

PROTOSTOMES
(blastopore becomes mouth)

Echinoderms
(sea urchins, etc.)

Hemichordates
(arrow worms)

Cephalochordates

Urochordates
(tunicates)

Vertebrates
(fish, amphibians, reptiles, birds, mammals)

Notochord
(chordates)

Vertebrae;
spinal column

DEUTEROSTOMES
(blastopore becomes anus)

Lytechinus variegatus

Ciona intestinalis

Early cleavage

Sea urchins exhibit **radial holoblastic cleavage** (**FIGURES 10.2** and **10.3**). Recall from Chapter 1 that this type of cleavage occurs in eggs with sparse yolk, and that holoblastic cleavage furrows extend through the entire egg (see Figure 1.9). In sea urchins, the first seven cleavage divisions are stereotypic in that the same pattern is followed in every individual of the same species. The first and second cleavages are both meridional and are perpendicular to each other (that is, the cleavage furrows pass through the animal and vegetal poles). The third cleavage is equatorial, perpendicular to the first two cleavage planes, and separates the animal and vegetal hemispheres from each other (see Figure 10.2A, top row, and Figure 10.3A–C). The fourth cleavage, however, is very different. The four cells of the animal tier divide meridionally into eight blastomeres, each with the same volume. These eight cells are called **mesomeres**. The vegetal tier, however, undergoes an unequal equatorial cleavage (see Figure 10.2B) to produce four large cells—the **macromeres**—and four smaller **micromeres** at the vegetal pole.

In *Lytechinus variegatus*, a species often used for experimentation, the ratio of cytoplasm retained in the macromeres and micromeres is 95:5. As the 16-cell embryo

FIGURE 10.2 Cleavage in the sea urchin. (A) Planes of cleavage in the first three divisions, and the formation of tiers of cells in divisions 3–6. (B) Confocal fluorescence micrograph of the unequal cell division that initiates the 16-cell stage (asterisk in A), highlighting the unequal equatorial cleavage of the vegetal blastomeres to produce the micromeres and macromeres. (A after J. W. Saunders, Jr. 1982. *Developmental Biology: Patterns, Problems, and Principles*. Macmillan: New York.)

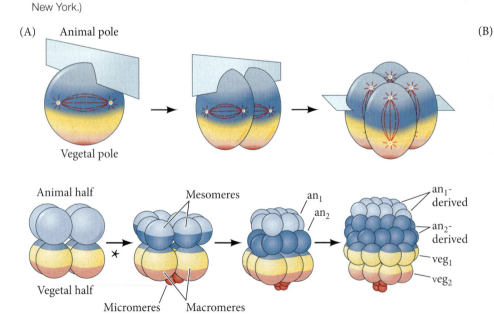

(A) Animal pole

Vegetal pole

Animal half

Mesomeres

an₁
an₂

an₁-
derived

an₂-
derived

veg₁

veg₂

Vegetal half

Micromeres   Macromeres

(B)

(A) (B) Fertilization envelope (C)

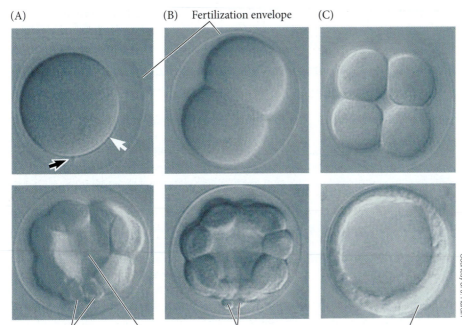

(D) Micromeres Blastocoel (E) Micromeres (F) Future vegetal plate

Courtesy of J. Hardin

FIGURE 10.3 Micrographs of cleavage in live embryos of the sea urchin *Lytechinus variegatus*, seen from the side. (A) The 1-cell embryo (zygote). The site of sperm entry is marked with a black arrow; a white arrow marks the vegetal pole. The fertilization envelope surrounding the embryo is clearly visible. (B) 2-Cell stage. (C) 8-Cell stage. (D) 16-Cell stage. Micromeres have formed at the vegetal pole. (E) 32-Cell stage. (F) The blastula has hatched from the fertilization envelope. The vegetal plate is beginning to thicken.

cleaves, the eight "animal" mesomeres divide equatorially to produce two tiers—an_1 and an_2, one staggered above the other. The macromeres divide meridionally, forming a tier of eight cells below an_2 (see Figure 10.2A, bottom row). Somewhat later, the micromeres divide unequally, producing a cluster of four small micromeres at the vegetal pole, beneath a tier of four large micromeres. The small micromeres divide once more, then stop dividing until the larval stage. At the sixth division, the animal hemisphere cells divide meridionally while the vegetal cells divide equatorially (see Figure 10.2A, bottom row); this pattern is reversed in the seventh division. At that time, the embryo is a 120-cell blastula,[1] in which the cells form a hollow sphere surrounding a central cavity called the **blastocoel** (see Figure 10.3D). From here on, the pattern of divisions becomes less regular.

As we will discuss later, these repeated asymmetrical divisions provide a strategy to make the decisions of cell specification through the unequal partitioning of maternally contributed cytoplasmic determinants with every mitosis.

Blastula formation

By the blastula stage, all the cells of the developing sea urchin are the same size, the micromeres having slowed down their cell divisions. Every cell is in contact with the proteinaceous fluid of the blastocoel on the inside and with the hyaline layer on the outside. Tight junctions unite the once loosely connected blastomeres into a seamless epithelial sheet that completely encircles the blastocoel. As the cells continue to divide, the blastula remains one cell layer thick, thinning out as it expands. This is accomplished by the adhesion of the blastomeres to the hyaline layer and by an influx of water that expands the blastocoel (Dan 1960; Wolpert and Gustafson 1961; Ettensohn and Ingersoll 1992).

These rapid and invariant cell cleavages last through the ninth or tenth division, depending on the species. By this time, the fates of the cells have become specified (discussed in the next section), and each cell becomes ciliated on the region of the cell membrane farthest from the blastocoel. Thus, there is apical-basal (outside-inside) polarity in each embryonic cell, and there is evidence that PAR proteins (like those of the nematode) are involved in distinguishing the basal cell membranes (Alford et al. 2009). The ciliated blastula begins to rotate within the fertilization envelope. Soon afterward, differences are seen in the cells. The cells at the vegetal pole of the blastula begin to thicken, forming a **vegetal plate** (see Figure 10.3F). The cells of the animal hemisphere synthesize and secrete a hatching enzyme that digests the fertilization envelope (Lepage et al. 1992). The embryo is now a free-swimming **hatched blastula**. (See **Further Development 10.1, Urchins in the Lab**, online.)

Fate maps and the determination of sea urchin blastomeres

Early fate maps of the sea urchin embryo followed the descendants of each of the 16-cell-stage blastomeres. More recent investigations have refined these maps by using injectable fluorescent dyes to track a cell and its progeny. Such studies have shown that by the 60-cell stage, most of the embryonic cell fates are restricted to subsets of cells, but their specification is not irreversibly committed (**FIGURE 10.4**). In other words, particular blastomeres consistently produce the same cell types in each embryo, but these cells remain pluripotent and can give rise to other cell types if experimentally placed in a different part of the embryo.

The animal half of the embryo consistently gives rise to the ectoderm—the larval skin and its neurons (see Figure 10.4). The veg_1 layer produces cells that can enter into either the ectodermal or the endodermal organs of the larva. The veg_2 layer gives rise to cells that can populate three different structures—the endoderm, the coelom (internal mesodermal body wall), and the **non-skeletogenic mesenchyme** (sometimes called **secondary mesenchyme**), which generates pigment cells, immunocytes, and muscle cells. The upper tier of micromeres (the large micromeres) produces the **skeletogenic mesenchyme** (also called **primary mesenchyme**), which forms the larval skeleton. The lower-tier micromeres (the small micromeres) contribute cells to the larval coelom, from which the tissues of the adult are derived during metamorphosis (Logan and McClay 1997, 1999; Wray 1999). The small micromeres also contribute to producing the germline cells (Yajima and Wessel 2011).

The fates of the different cell layers are determined in a two-step process:

1. Unlike most cells of the 16-cell embryo, the large micromeres are *autonomously* specified. They inherit maternal determinants that were deposited at the vegetal pole of the egg; these become incorporated into the large micromeres at the fourth cleavage. As a result, these four micromeres become skeletogenic mesenchyme cells that will leave the blastula epithelium, enter the blastocoel, migrate to particular positions along the blastocoel wall, and then differentiate into the larval skeleton. Even if these micromeres are isolated from the 16-cell embryo and placed in a petri dish, they will divide the appropriate number of times and produce the skeletal spicules, thereby showing that they do not need any external signals to generate their skeletal fates (Okazaki 1975). This demonstrates how cleavage asymmetries can contribute to molecular asymmetries and differential fate outcomes.

2. The autonomously specified large micromeres are now able to produce paracrine and juxtacrine factors that *conditionally* specify the fates of their neighbors. These factors signal the cells above the micromeres to become endomesoderm (the endoderm and the non-skeletogenic secondary mesenchyme cells) and to invaginate

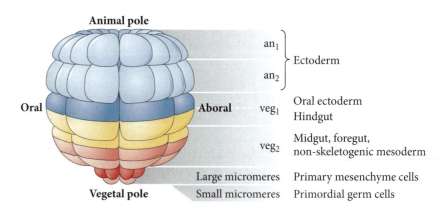

FIGURE 10.4 Fate map and cell lineage of the sea urchin *Strongylocentrotus purpuratus*. The 60-cell embryo is shown, with the left side facing the viewer. Blastomere fates are segregated along the animal-vegetal axis of the egg.

FIGURE 10.5 Ability of micromeres to induce presumptive ecto-dermal cells to acquire other fates. (A) Normal development of the 60-cell sea urchin embryo, showing the fates of the different layers. (B) An isolated animal hemisphere becomes a ciliated ball of undif-ferentiated ectodermal cells called a *Dauerblastula* (permanent blastula). (C) When an isolated animal hemisphere is combined with isolated micromeres, a recognizable pluteus larva is formed, with all the endoderm derived from the animal hemisphere. (After S. Hörstadius. 1939. *Biol Rev* 14: 132–179.)

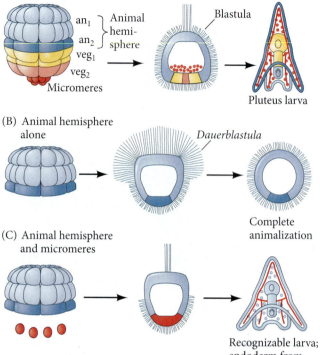

into the embryo. This inducing ability is so pronounced that if micromeres are removed from the embryo and placed in contact with an isolated **animal cap**—cells that normally become ectoderm—the animal cap cells gener-ate endoderm and a more or less normal larva develops (**FIGURE 10.5**; Hörstadius 1939). Moreover, if skeletogenic micromeres are transplanted into the animal region of the blastula,[2] in addition to forming skeletal spicules, the transplanted micromeres also alter the fates of nearby ectoderm cells, inducing them to become respecified as endoderm, form a secondary site of gastrulation, and produce a secondary gut (**FIGURE 10.6**; Hörstadius 1973; Ransick and Davidson 1993).

Gene regulatory networks and skeletogenic mesenchyme specification

According to the embryologist E. B. Wilson, heredity is the transmis-sion from generation to generation of a particular pattern of develop-ment, and evolution is the hereditary alteration of such a plan. As far back as 1895 in his analysis of sea urchin development, Wilson wrote that the instructions for development were somehow stored in chro-mosomal DNA and were transmitted by the chromosomes at fertil-ization. However, he had no way of knowing how the chromosomal information was translated into instructions for forming an embryo.

Studies from the sea urchin developmental biology community are now unraveling how DNA is involved in directing sea urchin mor-phogenesis (McClay 2016). Eric Davidson's group envisions a network involving *cis*-regulatory elements (such as promoters and enhancers) in a logic circuit connected to one another by transcription factors (see Figure 3.7; Davidson and Levine 2008; Oliveri et al. 2008; Peter and Davidson 2015). The network receives its first inputs from transcrip-tion factors in the egg cytoplasm. From then on, the network self-assembles from (1) the ability of the maternal transcription factors to recognize *cis*-regulatory elements of particular genes that encode other transcription factors, and (2) the ability of this new set of tran-scription factors to activate paracrine signaling pathways that activate specific transcription factors in neighboring cells. The researchers

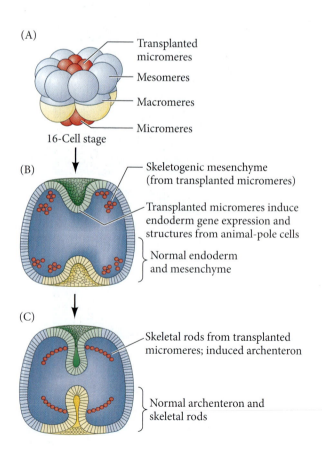

FIGURE 10.6 Ability of micromeres to induce a secondary axis in sea urchin embryos. (A) Micromeres are transplanted from the vegetal pole of a 16-cell embryo into the animal pole of a host 16-cell embryo. (B) The transplanted micromeres invaginate into the blastocoel to create a new set of skeletogenic mesenchyme cells, and they induce the animal-pole cells next to them to become vegetal plate endoderm cells. (C) The transplanted micromeres form skeletal rods while the induced animal cap cells form a secondary archenteron. Meanwhile, gastrulation proceeds normally from the original vegetal plate of the host. (After A. Ransick and E. H. Davidson. 1993. *Science* 259: 1134–1138.)

refer to this network of interconnections among genes that specify cell types as a **gene regulatory network**, or **GRN**. (See **Further Development 10.2, The Echinobase of Sea Urchin Development**, online.)

Here we will focus on one such GRN: the regulatory network involved in skeletogenic mesenchyme cells receiving their developmental fate and inductive properties.

DISHEVELED AND β-CATENIN: SPECIFYING THE MICROMERES The specification of the micromere lineage (and hence of the rest of the embryo) begins inside the undivided egg. Two transcription regulators, Disheveled and β-catenin, both of which are found in the egg cytoplasm, are inherited by the micromeres as soon as they are formed (i.e., at the fourth cleavage). Disheveled is located in the vegetal cortex of the egg (**FIGURE 10.7A**; Weitzel et al. 2004; Leonard and Ettensohn 2007), where it prevents the degradation of β-catenin in the micromere and veg$_2$-tier macromere cells. The β-catenin then enters the nuclei of these

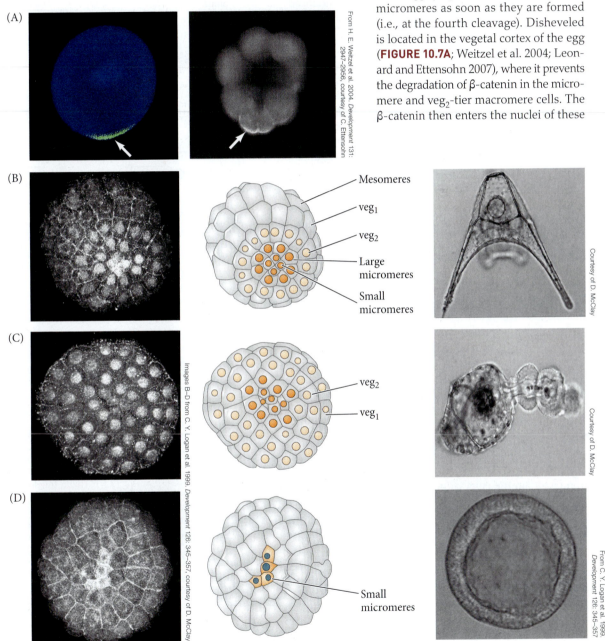

FIGURE 10.7 Role of the Disheveled and β-catenin proteins in specifying the vegetal cells of the sea urchin embryo. (A) Localization of Disheveled (arrows) in the vegetal cortex of the sea urchin oocyte before fertilization (left) and in the region of a 16-cell embryo about to become the micromeres (right). (B) During normal development, β-catenin accumulates predominantly in the micromeres and somewhat less in the veg$_2$ tier cells. (C) In embryos treated with lithium chloride, β-catenin accumulates in the nuclei of all blastula cells (probably by LiCl blocking the GSK3 enzyme of the Wnt pathway), and the cells of the animal pole become specified as endoderm and mesoderm. (D) When β-catenin is prevented from entering the nuclei (i.e., it remains in the cytoplasm), the vegetal cell fates are not specified, and the entire embryo develops as a ciliated ectodermal ball.

cells, where it combines with the TCF transcription factor to activate gene expression from specific promoters.

β-Catenin appears to be one of the earliest specifiers of the micromeres and of the endomesoderm of the macromeres. β-Catenin accumulates in the nuclei of micromeres at the 16-cell stage, then in the endomesoderm nuclei at the 32-cell stage (**FIGURE 10.7B**). This accumulation is autonomous and can occur even if the micromere precursors are separated from the rest of the embryo. The different timing of this nuclear accumulation is important. In the micromeres, the early activity of β-catenin represses endomesoderm development. However, in the endomesoderm, the later activity of β-catenin is too late to repress the same targets repressed in the micromeres (such as HesC; see Further Development below, *Pmar1* and *HesC*: A double-negative gate), and so the genes that promote endomesoderm specification have already started to be expressed.

Nuclear β-catenin accumulation may also help determine the mesodermal and endodermal fates of the vegetal cells (Kenny et al. 2003). Treating sea urchin embryos with lithium chloride allows β-catenin to accumulate in every cell and transforms presumptive ectoderm into endoderm (**FIGURE 10.7C**). Conversely, experimental procedures that inhibit β-catenin accumulation in the vegetal cell nuclei prevent the formation of endoderm and mesoderm (**FIGURE 10.7D**; Logan et al. 1998; Wikramanayake et al. 1998).

FURTHER DEVELOPMENT

***PMAR1* AND *HESC*: A DOUBLE-NEGATIVE GATE**   How does the timing of β-catenin activity support micromere versus macromere development? The Otx transcription factor may be involved. It also is enriched in the micromere cytoplasm and it interacts with the β-catenin/TCF complex to activate *Pmar1* transcription in the micromeres shortly after their formation (16-cell stage) (**FIGURE 10.8A**; Oliveri et al. 2008). The Pmar1 protein represses *HesC*; HesC inhibits micromere identity, and it is expressed in every cell of the sea urchin embryo *except* the micromeres. Recall that β-catenin functions early in the micromeres and therefore upregulates Pmar1, leading to the prevention of *HesC* expression and specification of the micromere identity.

(A)

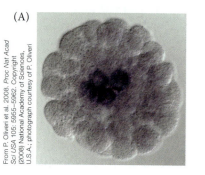

FIGURE 10.8   Simplified illustration of the double-negative gated "circuit" for micromere specification. (A) In situ hybridization reveals the accumulation of *Pmar1* mRNA (dark purple) in the micromeres. (B) Otx, a general transcription factor, and β-catenin from the maternal cytoplasm are concentrated at the vegetal pole of the egg. These transcriptional regulators are segregated to the micromeres and activate the *Pmar1* gene. *Pmar1* encodes a repressor of *HesC*, which in turn encodes a repressor (hence the "double-negative") of several genes involved in micromere specification (e.g., *Alx1*, *Tbr*, and *Ets*). Genes encoding signaling proteins (e.g., *Delta*) are also under the control of HesC. In the micromeres, where activated Pmar1 protein represses the *HesC* repressor, the micromere specification and signaling genes are active. In the veg₂ cells, *Pmar1* is not activated and HesC shuts down the skeletogenic genes; however, those cells containing Notch can respond to the Delta signal from the skeletogenic mesenchyme. The gene expression patterns are seen below. U represents ubiquitous activating transcription factors. (B after Oliveri et al. 2008. *Proc Nat Acad Sci USA* 105: 5955–5962. © 2008 National Academy of Sciences, U.S.A.)

(B)

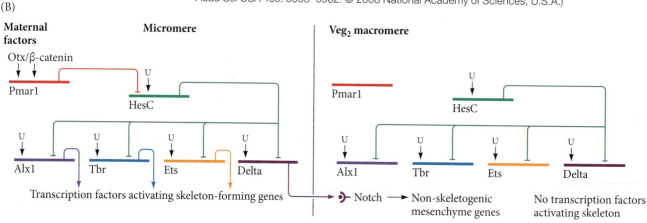

FIGURE 10.9 "Logic circuits" for gene expression. (A) In a double-negative gate, a single gene encodes a repressor of an entire battery of genes. When this repressor gene is repressed, the battery of genes is expressed. (B) In a feedforward circuit, gene product A activates both gene B and gene C, and gene B also activates gene C. Feedforward circuits provide an efficient way to amplify a signal in one direction. (After P. Oliveri et al. 2008. *Proc Nat Acad Sci USA* 105: 5955–5962. Copyright [2008] National Academy of Sciences, U.S.A.)

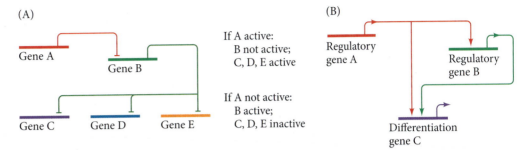

(A)

Gene A

Gene B

Gene C   Gene D   Gene E

If A active:
  B not active;
  C, D, E active

If A not active:
  B active;
  C, D, E inactive

(B)

Regulatory gene A

Regulatory gene B

Differentiation gene C

Developing Questions

Evolution is accomplished by changes in development. Such developmental changes, in turn, can be accomplished by changes in GRNs. In considering the evolution of two closely related species, how might the GRN of the mesenchyme cells of the sea star, whose larva lacks skeletal elements, differ from that of the skeletogenic mesenchyme cells of the sea urchin? (See also Chapter 24.)

This mechanism—whereby a repressor locks the genes of specification and these genes can be unlocked by the repressor of that repressor (in other words, when activation occurs by the repression of a repressor)—is called a **double-negative gate** (**FIGURES 10.8B** and **10.9A**). Such a gate allows for tight regulation of fate specification: it *promotes* the expression of specification genes where the input occurs, and it *represses* the same genes in every other cell type (Oliveri et al. 2008).

HesC represses a number of micromere specification genes when bound to their enhancers; these include *Alx1*, *Ets1*, *Tbr*, *Tel*, and *SoxC*. However, when the Pmar1 protein is present, as it is in micromeres, Pmar1 represses *HesC*, and all these genes become active, moving micromeres toward their skeletogenic cell fates (Revilla-i-Domingo et al. 2007; see also Peter and Davidson 2016).

In contrast to the double-negative gate regulatory module supporting the micromere fate, another GRN specifying skeletogenic cells involves a feedforward process (**FIGURE 10.9B**). Here, gene A regulates the expression of gene B, and gene B regulates the expression of gene C. Gene C feeds back to regulate the expression of gene A to ensure that the network is on and relatively stable and irreversible. (See **Further Development 10.3, How to Specify Yourself**, and **Further Development 10.4, Evolution by Subroutine Co-option**, both online.)

Specification of the vegetal cells

The skeletogenic micromeres also produce signals that can induce changes in adjacent cells. One of these signals is the TGF-β superfamily paracrine factor activin. Expression of the gene for activin is also under the control of the *Pmar1-HesC* double-negative gate, and activin secretion appears to be critical for endoderm formation (Sethi et al. 2009). Indeed, if *Pmar1* mRNA is injected into a cell from the animal hemisphere, that *Pmar1* overexpressing cell will develop into a skeletogenic mesenchyme cell, and the cells adjacent to it will start developing like macromeres (Oliveri et al. 2002). If the activin signal is blocked, the adjacent cells do not become endoderm (Ransick and Davidson 1995; Sherwood and McClay 1999; Sweet et al. 1999).[3]

Another cell-specifying signal from the micromeres is the juxtacrine protein Delta, also a factor that is controlled by the double-negative gate. Delta functions by activating Notch proteins on the adjacent veg$_2$ cells and will later act on the adjacent small micromeres. Delta causes the veg$_2$ cells to become the non-skeletogenic mesenchyme cells by activating the Gcm transcription factor and repressing the Foxa transcription factor (which activates the endoderm-specific genes). The upper veg$_2$ cells, since they do not receive the Delta signal, retain Foxa expression, and this pushes them in the direction of becoming endodermal cells (Croce and McClay 2010).

In sum, gene expression in the sea urchin micromeres specifies their cell fates autonomously and also specifies the fates of their neighbors conditionally. The original inputs come from the maternal cytoplasm and activate genes that unlock repressors of a specific cell fate. Once the maternal cytoplasmic factors accomplish their functions, the embryonic genome takes over.

Sea Urchin Gastrulation

Architect Frank Lloyd Wright wrote in 1905 that "form and function should be one, joined in a spiritual union." While Wright never used sea urchin skeletons as inspiration, other architects (such as Antoni Gaudí) may have; the characteristic sea urchin **pluteus larva** is a feeding structure in which form and function are remarkably well integrated.

The late blastula of the sea urchin is like a hollow ball formed by about 750 epithelial cells. By this time, ectoderm, mesoderm, and endoderm cells are already at least partly specified toward their eventual fates. The next stage of development is responsible for moving these cells into new positions in the morphogenetic process known as **gastrulation**. This process is responsible for moving mesoderm cells beneath the outer epithelium and for invagination of the gut tube. As you will learn in the next two chapters, during gastrulation in most other embryos these two major movements occur simultaneously. In the sea urchin, however, they are sequential, such that mesoderm cells enter the blastocoel first, and later the **archenteron**, or primitive gut, invaginates.

Ingression of the skeletogenic mesenchyme

FIGURE 10.10 illustrates development of the blastula through gastrulation to the pluteus larva stage (hour 24). Shortly after the blastula hatches from its fertilization envelope, the descendants of the large micromeres undergo an epithelial-mesenchymal transition. The epithelial cells change their shape, lose their adhesions to their neighboring cells, and break away from the epithelium to enter the blastocoel as skeletogenic mesenchyme cells (see Figure 10.10, 9–10 hours). The skeletogenic mesenchyme cells then begin extending and contracting long, thin (250 nm in diameter and 25 μm long) processes called **filopodia**. At first the cells appear to move randomly along the inner blastocoel surface, actively making and breaking filopodial connections to the wall of the blastocoel. Eventually, however, they become localized within the prospective ventrolateral region of the blastocoel. Here they fuse into syncytial cables that will form the axis of the calcium carbonate spicules of the larval skeletal rods. This is coordinated through the same GRN that specified the skeletogenic mesenchyme cells.

EPITHELIAL-MESENCHYMAL TRANSITION The ingression of the large micromere descendants into the blastocoel is a result of their losing their affinity for their neighbors and for the hyaline membrane; instead these cells acquire a strong affinity for a group of proteins that line the blastocoel. Initially, all the cells of the blastula are connected on their outer surface to the hyaline layer, and on their inner surface to a basal lamina secreted by the cells. On their lateral surfaces, each cell has another cell for a neighbor. Fink and McClay found that the prospective ectoderm and endoderm cells (descendants of the mesomeres and macromeres, respectively) bind tightly to one another and to the hyaline layer, but adhere only loosely to the basal lamina. The micromeres initially display a similar pattern of binding. However, the micromere pattern changes at

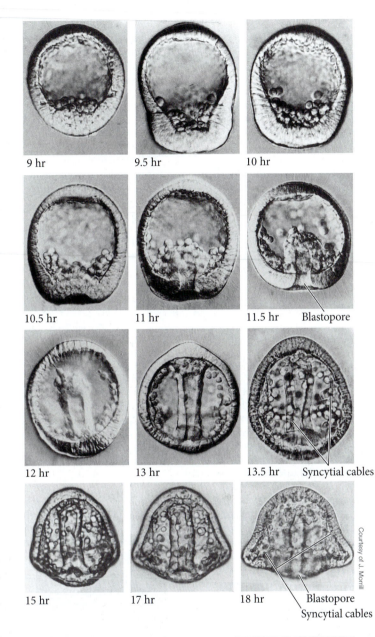

9 hr   9.5 hr   10 hr

10.5 hr   11 hr   11.5 hr   Blastopore

12 hr   13 hr   13.5 hr   Syncytial cables

15 hr   17 hr   18 hr   Blastopore / Syncytial cables

Courtesy of J. Morrill

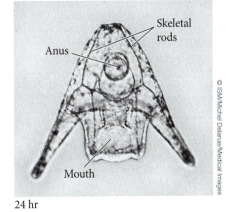

Skeletal rods

Anus

Mouth

24 hr

© ISM/Michel Delarue/Medical Images

FIGURE 10.10 Entire sequence of gastrulation in *Lytechinus variegatus*. Times show the length of development at 25°C.

gastrulation. Whereas the other cells retain their tight binding to the hyaline layer and to their neighbors, the skeletogenic mesenchyme precursors lose their affinities for these structures (which drop to about 2% of their original value), while their affinity for components of the basal lamina and extracellular matrix increases 100-fold. This accomplishes an **epithelial-mesenchymal transition** (**EMT**), whereby cells that had formerly been part of an epithelium lose their attachments and become individual, migrating cells (**FIGURE 10.11A**; see also Chapter 4).

EMTs are important events throughout animal development, and the pathways to EMT are revisited in cancer cells, where the EMT is often necessary for the formation of secondary tumor sites. There appear to be five distinct processes in the EMT, and all of these events are regulated by the same micromere GRN that specifies and forms the skeletogenic mesenchyme. However, each of these processes is controlled by a different subset of transcription factors. Even more surprising, none of the transcription factors function as pioneering regulators of the EMT (Saunders and McClay 2014). These five events are:

1. *Apical-basal polarity.* The vegetal cells of the blastula elongate to form a thickened vegetal plate epithelium (see Figure 10.10, 9 hours).

FIGURE 10.11 Ingression of skeletogenic mesenchyme cells. (A) Depiction of changes in the adhesive affinities of the skeletogenic mesenchyme cells (pink). These cells lose their affinities for hyalin and for their neighboring blastomeres while gaining an affinity for the proteins of the basal lamina. Nonmesenchymal blastomeres retain their original high affinities for the hyaline layer and neighboring cells. (B–D) Skeletogenic mesenchyme cells breaking through extracellular matrix. The matrix laminin is stained pink, the mesenchyme cells are green, and cell nuclei are blue. (B) Laminin matrix is uniformly spread throughout the lining of the blastocoel. (C) A hole is made in blastocoel laminin above the vegetal cells, and the mesenchyme begins to pass through it into the blastocoel. (D) Within an hour, cells are in the blastocoel. (E) Scanning electron micrograph of skeletogenic mesenchyme cells enmeshed in the extracellular matrix of an early *Strongylocentrotus* gastrula. (F) Gastrula-stage mesenchyme cell migration. The extracellular matrix fibrils of the blastocoel lie parallel to the animal-vegetal axis and are intimately associated with the skeletogenic mesenchyme cells. (A after H. Katow and M. Solursh. 1980. *J Exp Zool* 213: 231–246.)

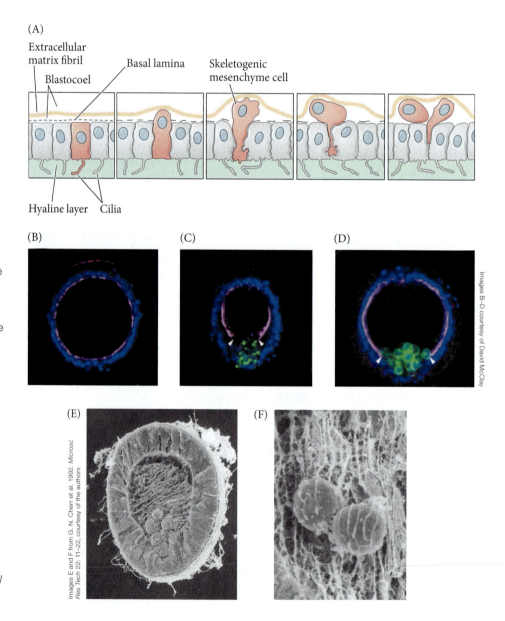

(A)

Extracellular matrix fibril
Blastocoel
Basal lamina
Skeletogenic mesenchyme cell

Hyaline layer   Cilia

(B)   (C)   (D)

Images B–D courtesy of David McClay

(E)   (F)

Images E and F from G. N. Cherr et al. 1992. *Microsc Res Tech* 22: 11–22, courtesy of the authors

2. *Apical constriction of the micromeres.* The cells alter their shape, wherein the apical end (away from the blastocoel) becomes constricted. Apical constriction is seen during gastrulation and neurulation in both vertebrates and invertebrates, and is one of the most important cell shape changes associated with morphogenesis (Sawyer et al. 2010).

3. *Basal lamina remodeling.* The cells must pass through the laminin-containing basal lamina. Originally, this membrane is uniform around the blastocoel. However, the micromere cells secrete proteases (protein-digesting enzymes) that digest a hole in this membrane, shortly before the first mesenchymal cells are seen inside the blastocoel (**FIGURE 10.11B–D**).

4. *De-adhesion.* The cadherins that couple epithelial cells together are degraded, thereby allowing the cells to become free from their neighbors. Downregulation of cadherins is controlled by the transcription factor Snail. The *snail* gene is activated by the Alx1 transcription factor, which in turn is regulated by the double-negative gate of the GRN (Wu et al. 2007). The Snail transcription factor is involved in de-adhesion throughout the animal kingdom (including in cancers).

5. *Cell motility.* The transcription factors of the GRN activate those proteins causing the active migration of the cells out of the epithelium and into the blastocoel. One of the most critical of these is Foxn2/3. This transcription factor is also seen in regulation of the motility of neural crest cells after their EMT (to form the face in vertebrates). The cells bind to and travel on extracellular matrix proteins within the blastocoel (**FIGURE 10.11E,F**).

FURTHER DEVELOPMENT

SKELETOGENIC DEVELOPMENT AFTER EMT   Once inside the blastocoel, the mesenchyme continues to be influenced by the same autonomous GRN that enabled its earlier EMT behavior, but now this GRN controls differentiation toward skeletogenic fates. Additionally, a new nonautonomous feature of skeletogenesis emerges: paracrine signals provide positional guidance and promote skeletogenic differentiation.

At two sites near the future ventral side of the larva, many skeletogenic mesenchyme cells cluster together, fuse with one another, and initiate spicule formation (Hodor and Ettensohn 1998; Lyons et al. 2014). If a labeled micromere from another embryo is injected into the blastocoel of a gastrulating sea urchin embryo, it migrates to the correct location and contributes to the formation of the embryonic spicules (Ettensohn 1990; Peterson and McClay 2003). It is thought that the necessary positional information is provided by the prospective ectodermal cells and their basal laminae (**FIGURE 10.12A**; Harkey and Whiteley 1980; Armstrong et al. 1993; Malinda and Ettensohn 1994). Only the skeletogenic mesenchyme cells (and not other cell types or latex beads) are capable of responding to these patterning cues (Ettensohn and McClay 1986). The extremely fine filopodia on the skeletogenic mesenchyme cells explore and sense the blastocoel

(A)

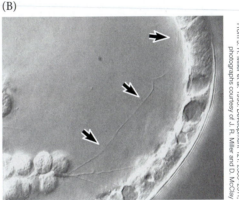

Courtesy of J. R. Miller and D. McClay

(B)

From J. R. Miller et al. 1995. *Development* 121: 2505–2511, photographs courtesy of J. R. Miller and D. McClay

(C)

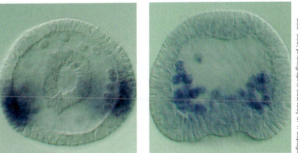

From E. Röttinger et al. 2008. *Development* 135: 353–365, photographs courtesy of T. Lepage

FIGURE 10.12   Positioning of skeletogenic mesenchyme cells in the sea urchin. (A) Positioning of the micromeres to form the calcium carbonate skeleton is determined by the ectodermal cells. Skeletogenic mesenchyme cells are stained green; β-catenin is red; skeletogenic mesenchyme cells appear to accumulate in those regions characterized by high β-catenin concentrations. (B) Nomarski videomicrograph showing a long, thin filopodium extending from a skeletogenic mesenchyme cell to the ectodermal wall of the gastrula (arrows), as well as a shorter filopodium extending inward from the ectoderm. Mesenchymal filopodia extend through the extracellular matrix and directly contact the cell membrane of the ectodermal cells. (C) Seen in cross section through the archenteron (top), the surface ectoderm expresses FGF in the particular locations where skeletogenic micromeres congregate. Moreover, the ingressing skeletal micromeres (bottom; longitudinal section) express the FGF receptor. When FGF signaling is suppressed, the skeleton does not form properly.

FIGURE 10.13 Formation of syncytial cables by skeletogenic mesenchyme cells of the sea urchin. (A) Skeletogenic mesenchyme cells in the early gastrula align and fuse to lay down the matrix of the calcium carbonate spicule (arrows). (B) Scanning electron micrograph of the syncytial cables formed by the fusing of skeletogenic mesenchyme cells.

(A)

(B)

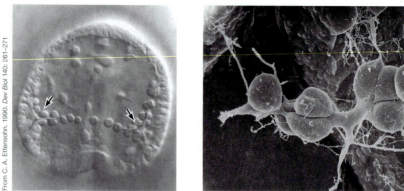

From J. B. Morrill and L. L. Santos. 1985. In *The Cellular and Molecular Biology of Invertebrate Development*, R. H. Sawyer and R. M. Showman (Eds.), pp. 3–33. University of South Carolina Press: Columbia

From C. A. Ettensohn. 1990. *Dev Biol* 140: 261–271

FIGURE 10.14 Invagination of the vegetal plate. (A) Vegetal plate invagination in *Lytechinus variegatus*, seen by scanning electron microscopy of the external surface of the early gastrula. The blastopore is clearly visible. (B) Fate map of the vegetal plate of the sea urchin embryo, looking "upward" at the vegetal surface. The central portion becomes the non-skeletogenic mesenchyme cells. The concentric layers around it become the foregut, midgut, and hindgut. The boundary where the endoderm meets the ectoderm marks the anus. The non-skeletogenic mesenchyme and foregut come from the veg$_2$ layer; the midgut comes from veg$_1$ and veg$_2$ cells; the hindgut and the ectoderm surrounding it come from the veg$_1$ layer. (B after C. Y. Logan and D. R. McClay. 1999. In *Cell Lineage and Determination*, S. A. Moody [Ed.], pp. 41–58. Academic Press: New York, and S. W. Ruffins and C. A. Ettensohn. 1996. *Development* 122: 253–263.)

wall and appear to be sensing dorsal-ventral and animal-vegetal patterning cues from the ectoderm (**FIGURE 10.12B**; Malinda et al. 1995; Miller et al. 1995).

To initiate skeletal production, skeletogenic cells immediately beneath two locations of ectoderm receive VEGF and FGF. The VEGF paracrine factors are emitted from two small regions of the ectoderm (Duloquin et al. 2007), and an FGF paracrine factor is made in the equatorial belt between endoderm and ectoderm (**FIGURE 10.12C**; Röttinger et al. 2008; McIntyre et al. 2014). The skeletogenic mesenchyme cells migrate to these points of VEGF and FGF synthesis and arrange themselves in a ring along the animal-vegetal axis (**FIGURE 10.13**). The receptors for these paracrine factors appear to be specified by the double-negative gate (Peterson and McClay 2003). As the syncytial cables of mesenchyme cells begin to extend, other signals contribute further positional information from the ectoderm, so that the skeleton grows in the correct shape in response to these non-cell autonomous patterning signals. (See **Further Development 10.5, Axis Specification in Sea Urchin Embryos**, online.)

Invagination of the archenteron

FIRST STAGE OF ARCHENTERON INVAGINATION As the skeletogenic mesenchyme cells leave the vegetal region of the spherical embryo, important changes are occurring in the cells that remain there. These cells thicken and flatten to form a vegetal plate, changing the shape of the blastula (see Figure 10.10, 9 hours). The vegetal plate cells remain bound to one another and to the hyaline layer of the egg, and they move to fill the gaps caused by the ingression of the skeletogenic mesenchyme. The vegetal plate involutes inward by altering its cell shape, then invaginates about one-fourth to one-half of the way into the blastocoel before invagination suddenly ceases. The invaginated region is called the archenteron (primitive gut), and the opening of the archenteron at the vegetal pole is the **blastopore** (**FIGURE 10.14A**; see also Figure 10.10, 10.5–11.5 hours).

(A)

Courtesy of J. B. Morrill

(B)

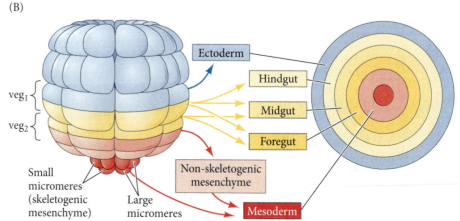

The movement of the vegetal plate into the blastocoel appears to be initiated by shape changes in the vegetal plate cells and in the extracellular matrix underlying them (see Kominami and Takata 2004). Actin microfilaments collect in the apical ends of the vegetal cells, causing these ends to constrict, forming bottle-shaped vegetal cells that pucker inward (Kimberly and Hardin 1998; Beane et al. 2006). Destroying these cells with lasers retards gastrulation. In addition, the hyaline layer at the vegetal plate buckles inward due to changes in its composition, directed by the vegetal plate cells (Lane et al. 1993).

At the stage when the skeletogenic mesenchyme cells begin ingressing into the blastocoel, the fates of the vegetal plate cells have already been specified (Ruffins and Ettensohn 1996). The non-skeletogenic mesenchyme is the first group of cells to invaginate, forming the tip of the archenteron and leading the way into the blastocoel. The non-skeletogenic mesenchyme will form the pigment cells and the musculature around the gut, and will contribute to the coelomic pouches. The endodermal cells adjacent to the macromere-derived non-skeletogenic mesenchyme become foregut, migrating the farthest into the blastocoel. The next layer of endodermal cells becomes midgut, and the last circumferential row to invaginate forms the hindgut and anus (**FIGURE 10.14B**).

SECOND AND THIRD STAGES OF ARCHENTERON INVAGINATION After a brief pause following initial invagination, the second stage of archenteron formation begins. The archenteron elongates dramatically, sometimes tripling in length. In this process of extension, the wide, short gut rudiment is transformed into a long, thin tube (**FIGURE 10.15A**; see also Figure 10.10, 12 hours). To accomplish this extension, numerous cellular phenomena work together. First, the endoderm cells proliferate as they enter into the embryo. Second, the clones derived from these cells slide past one another, like the extension of a telescope. And last, the cells rearrange themselves by intercalating

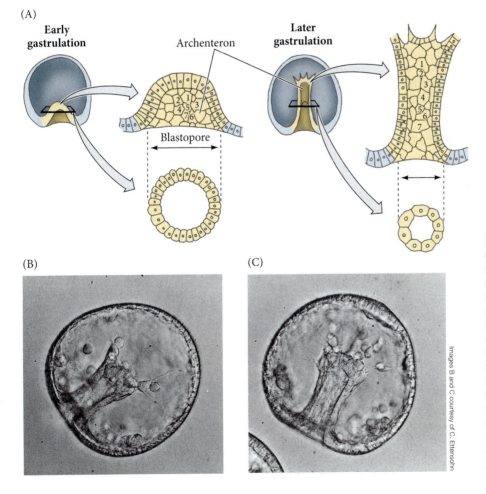

(A)

Early gastrulation

Archenteron

Later gastrulation

Blastopore

FIGURE 10.15 Extension of the archenteron in sea urchin embryos. (A) Cell rearrangement during extension of the archenteron in sea urchin embryos, showing the early archenteron with 20–30 cells around its circumference. Later in gastrulation, the archenteron has a circumference made by only 6–8 cells. (B) At the mid-gastrula stage of *Lytechinus pictus*, filopodial processes extend from non-skeletogenic mesenchyme located at the tip of the archenteron. (C) Syncytial cables connect the blastocoel wall to the archenteron tip. The tension of the cables can be seen as they pull on the blastocoel wall at the points of attachment. (A after J. D. Hardin. 1990. *Semin Dev Biol* 1: 335–345.)

(B)

(C)

Images B and C courtesy of C. Ettensohn

between one another, like lanes of traffic merging (Ettensohn 1985; Hardin and Cheng 1986; Martins et al. 1998; Martik and McClay 2012). This phenomenon, whereby cells intercalate to narrow the tissue and at the same time lengthen it, is called **convergent extension**.

The third and final stage of archenteron elongation is initiated by the tension provided by non-skeletogenic mesenchyme cells at the tip of the archenteron. These cells extend filopodia through the blastocoel fluid to contact the inner surface of the blastocoel wall (Dan and Okazaki 1956; Schroeder 1981). The filopodia attach to the wall at the junctions between the blastomeres and then shorten, pulling up the archenteron (**FIGURE 10.15 B,C**; see also Figure 10.10, 12 and 13 hours). Hardin (1988) ablated non-skeletogenic mesenchyme cells of *Lytechinus pictus* gastrulae with a laser, with the result that the archenteron could elongate only to about two-thirds of the normal length. If a few non-skeletogenic mesenchyme cells were left, elongation continued, although at a slower rate. Thus, in this species the non-skeletogenic mesenchyme cells play an essential role in pulling the archenteron upward to the blastocoel wall during the last stage of invagination. (See **Further Development 10.6, Filopodia Feel for Their Target Tissue**, online.)

As the top of the archenteron meets the blastocoel wall in the target region, many of the non-skeletogenic mesenchyme cells disperse into the blastocoel, where they proliferate to form the mesodermal organs (see Figure 10.10, 13.5 hours). Where the archenteron contacts the wall, a mouth eventually forms. The mouth fuses with the archenteron to create the continuous digestive tube of the pluteus larva. The remarkable metamorphosis from pluteus larva to adult sea urchin is described in Further Development 21.12, Metamorphosis of the Pluteus Larva, online.

Early Development in Tunicates

Tunicates (also known as ascidians or sea squirts) are fascinating animals for several reasons, but the foremost is that they are the closest evolutionary relatives of the vertebrates. As Lemaire (2009) has written, "Looking at an adult ascidian, it is difficult, and slightly degrading, to imagine that we are close cousins to these creatures." Even though tunicates lack vertebrae at all stages of their life cycles, the free-swimming tunicate larva, or "tadpole," has a notochord and a dorsal nerve cord, making these animals invertebrate chordates (see Figure 10.1). When the tadpole undergoes metamorphosis, its nerve cord and notochord degenerate, and it secretes a cellulose tunic that is the source of the tunicates' name.

Cleavage

Tunicates have **bilateral holoblastic cleavage** (**FIGURE 10.16**). The most striking feature of this type of cleavage is that the first cleavage plane establishes the earliest axis of symmetry in the embryo, separating the embryo into its future right and left sides. Each successive division is oriented to this plane of symmetry, and the half-embryo formed on one side of the first cleavage plane is the mirror image of the half-embryo on the other side.[4] The second cleavage is meridional like the first, but unlike the first division, it does not pass through the center of the egg. Rather, it creates two large anterior cells

FIGURE 10.16 Bilateral symmetry in the egg of the tunicate *Styela partita*. (The cell lineages of *Styela* are shown in Figure 1.13.) (A) Uncleaved egg. The regions of cytoplasm destined to form particular organs are labeled here and coded by color throughout the diagrams. (B) 8-Cell embryo, showing the blastomeres and the fates of various cells. The embryo can be viewed as left and right 4-cell halves; from here on, each division on the right side of the embryo has a mirror-image division on the left. (C,D) Views of later embryos from the vegetal pole. The dashed line shows the plane of bilateral symmetry. (After B. I. Balinsky. 1981. *Introduction to Embryology*, 5th Ed. Saunders: Philadelphia.)

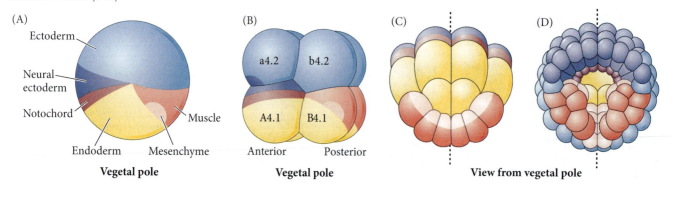

(the A and a blastomeres) and two smaller posterior cells (the B and b blastomeres). Each side now has a large and a small blastomere.

Indeed, from the 8- through the 64-cell stages, every cell division is asymmetrical, such that the posterior blastomeres are always smaller than the anterior blastomeres (Nishida 2005; Sardet et al. 2007). Prior to each of these unequal cleavages, the posterior centrosome in the blastomere migrates toward the **centrosome-attracting body** (**CAB**), a macroscopic subcellular structure composed of endoplasmic reticulum. The CAB connects to the cell membrane through a network of PAR proteins that position the centrosomes asymmetrically in the cell (as in *C. elegans*; see Figure 8.17), resulting in one large and one small cell at each of these three divisions. The CAB also attracts particular mRNAs in such a way that these messengers are placed in the posteriormost (i.e., smaller) cell of each division (Hibino et al. 1998; Nishikata et al. 1999; Patalano et al. 2006). In this way, the CAB integrates cell patterning with cell determination.[5] At the 64-cell stage, a small blastocoel is formed, and gastrulation begins from the vegetal pole.

The tunicate fate map

Most early tunicate blastomeres are specified autonomously, each cell acquiring a specific type of cytoplasm that will determine its fate. In many tunicate species, the different regions of cytoplasm have distinct pigmentation, so the cell fates can easily be seen to correspond to the type of cytoplasm taken up by each cell. These cytoplasmic regions are apportioned to the egg during fertilization.

Figure 2.4 shows the fate map and cell lineages of the tunicate *Styela partita*. In the unfertilized egg, a central gray cytoplasm is enveloped by a cortical layer containing yellow lipid inclusions (**FIGURE 10.17A**). During meiosis, the breakdown of the nucleus releases a clear substance that accumulates in the animal hemisphere of the egg. Within 5 minutes of sperm entry, the inner clear and cortical yellow cytoplasms contract into the vegetal (lower) hemisphere of the egg (**FIGURE 10.17B**; Prodon et al. 2005, 2008; Sardet et al. 2005). As the male pronucleus migrates from the vegetal pole to the equator of the cell along the future posterior side of the embryo, the yellow lipid inclusions migrate with it. This migration forms the **yellow crescent**, extending from the vegetal pole to the equator (**FIGURE 10.17C,D**); this region will produce most of the tail muscles of the tunicate larva. The movement of these cytoplasmic regions depends on microtubules that are organized by the sperm centriole and on a wave of calcium ions that contracts the animal-pole cytoplasm (Sawada and Schatten 1989; Speksnijder et al. 1990; Roegiers et al. 1995).

As we noted in Chapter 2, Edwin Conklin (1905) took advantage of the differing coloration of these regions of cytoplasm to follow each of the cells of the tunicate embryo to its fate in the larva (see Figure 2.4A and Further Development 1.4, online). Conklin found that cells receiving clear cytoplasm become ectoderm; those containing yellow cytoplasm give rise to mesodermal cells; those incorporating slate gray inclusions become endoderm; and light gray cells become the neural tube and notochord. The cytoplasmic regions are arrayed bilaterally on both sides of the plane of symmetry, so they are bisected by the first cleavage furrow into the right and left halves of the

FIGURE 10.17 Cytoplasmic rearrangement in the fertilized egg of *Styela partita*. (A) Before fertilization, yellow cortical cytoplasm surrounds the gray (yolky) inner cytoplasm. (B) After sperm entry into the vegetal hemisphere of the oocyte, the yellow cortical cytoplasm and the clear cytoplasm derived from the breakdown of the oocyte nucleus contract vegetally toward the sperm. (C) As the sperm pronucleus migrates animally toward the newly formed egg pronucleus, the yellow and clear cytoplasms move with it. (D) The final position of the yellow cytoplasm marks the location where cells will give rise to tail muscles. (After E. G. Conklin. 1905. *J Acad Nat Sci Phila* 13: 5–119.)

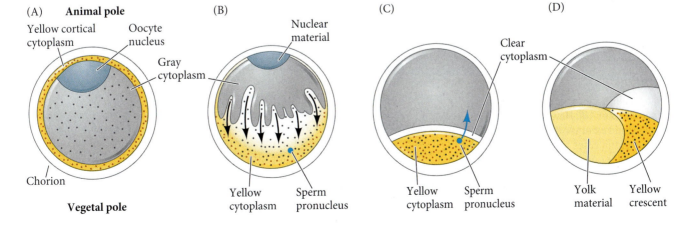

(A) **Animal pole** — Yellow cortical cytoplasm; Oocyte nucleus; Gray cytoplasm; Chorion; **Vegetal pole**

(B) Nuclear material; Yellow cytoplasm; Sperm pronucleus

(C) Yellow cytoplasm; Sperm pronucleus

(D) Clear cytoplasm; Yolk material; Yellow crescent

embryo. The second cleavage causes the prospective mesoderm to lie in the two posterior cells, while the prospective neural ectoderm and chordamesoderm (notochord) will be formed from the two anterior cells (see Figure 10.16). The third division further partitions these cytoplasmic regions such that the mesoderm-forming cells are confined to the two vegetal posterior blastomeres, while the chordamesoderm cells are restricted to the two vegetal anterior cells.

Autonomous and conditional specification of tunicate blastomeres

The autonomous specification of tunicate blastomeres was one of the first observations in the field of experimental embryology (Chabry 1888). Cohen and Berrill (1936) confirmed Chabry's and Conklin's results, and by counting the number of notochord and muscle cells, they demonstrated that larvae derived from only one of the first two blastomeres (either the right or the left) had half the expected number of cells.[6] When the 8-cell embryo is separated into its four doublets (the right and left sides being equivalent), both autonomous and conditional specification are seen (Reverberi and Minganti 1946). Autonomous specification is seen in the gut endoderm, muscle mesoderm, and skin ectoderm (see Lemaire 2009). Conditional specification (by induction) is seen in the formation of the brain, notochord, heart, and mesenchyme cells. Indeed, a majority of tunicate cell lineages involve some inductions.

AUTONOMOUS SPECIFICATION OF THE MYOPLASM: THE YELLOW CRESCENT AND MACHO-1 As mentioned in Chapter 2, when yellow crescent cytoplasm is experimentally transferred from the B4.1 (muscle-forming) blastomere to the b4.2 or a4.2 (ectoderm-forming) blastomeres of an 8-cell tunicate embryo, the ectoderm-forming blastomeres generate muscle cells as well as their normal ectodermal progeny. Nishida and Sawada (2001) showed that this muscle-forming determinant was an mRNA that encoded a transcription factor they named Macho-1. They demonstrated that *macho-1* mRNA was present at the right place and time, and was both necessary and sufficient for cells to make muscles (see Figure 2.6).

The Macho-1 protein is a transcription factor that is required for the activation of several mesodermal genes, including *tbx6*, *snail*, and the genes for muscle actin and myosin (Yagi et al. 2004; Sawada et al. 2005). Of these gene products, only the Tbx6 protein phenocopied (mimicked) *macho-1* function by inducing ectopic muscle differentiation. Macho-1 thus appears to directly activate a set of *tbx6* genes, and Tbx6 proteins activate the rest of muscle development (Yagi et al. 2005; Kugler et al. 2010). Macho-1 and Tbx6 also appear to activate (possibly in a feedforward manner) the muscle-specific gene *snail*. Snail protein is important in preventing *Brachyury* expression in presumptive muscle cells, and thereby prevents the muscle precursors from becoming notochord cells.[7] It appears, then, that Macho-1 is a critical transcription factor of the tunicate yellow crescent, muscle-forming cytoplasm. Macho-1 activates a transcription factor cascade that promotes muscle differentiation while at the same time inhibiting notochord specification. (See **Further Development 10.7, The Search for the Myogenic Factor**, online.)

AUTONOMOUS SPECIFICATION OF ENDODERM: β-CATENIN Presumptive endoderm originates from the vegetal A4.1 and B4.1 blastomeres. The specification of these cells coincides with the localization of β-catenin (recall that β-catenin is also involved in endoderm specification in the sea urchin embryo). Inhibition of β-catenin results in the loss of endoderm and its replacement by ectoderm in the tunicate embryo (**FIGURE 10.18**; Imai et al. 2000). Conversely, increasing β-catenin synthesis causes an increase in the endoderm at the expense of the ectoderm (just as in sea urchins). The β-catenin transcription factor appears to function by activating the synthesis of the homeobox transcription factor Lhx3. Inhibition of the *lhx3* message prohibits differentiation of the endoderm (Satou et al. 2001). (See **Further Development 10.8, Specification of the Larval Axes in Tunicate Embryos**, online.)

(?) Developing Questions

The tunicate nervous system is present during the larval stage but degenerates during metamorphosis. Consider the neural tube of a vertebrate such as a fish and how it becomes divided into forebrain, midbrain, hindbrain, and spinal cord portions (see Chapter 13). How would one determine whether the neural tubes of tunicates and vertebrates are similar? Are they homologous or analogous (see Chapter 24)?

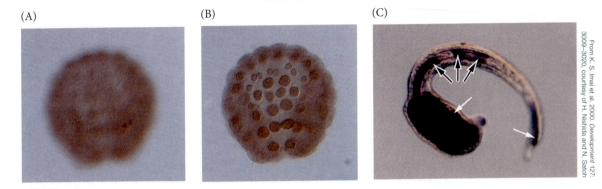

(A) (B) (C)

From K. S. Imai et al. 2000. *Development* 127: 3009–3020, courtesy of H. Nishida and N. Satoh

FIGURE 10.18 Antibody staining of β-catenin protein shows its involvement with endoderm formation. (A) No β-catenin is seen in the animal-pole nuclei of a 110-cell *Ciona* embryo. (B) In contrast, β-catenin is readily seen in the nuclei of the vegetal endoderm precursors at the 110-cell stage. (C) When β-catenin is expressed in notochordal precursor cells, those cells will become endoderm and express endodermal markers such as alkaline phosphatase. The white arrows show normal endoderm; the black arrows show notochordal cells that are expressing endodermal enzymes.

CONDITIONAL SPECIFICATION OF MESENCHYME AND NOTOCHORD BY THE ENDODERM Although most tunicate muscles are specified autonomously from the yellow crescent cytoplasm, the most posterior muscle cells form through conditional specification by interactions with the descendants of the A4.1 and b4.2 blastomeres (Nishida 1987, 1992a,b). Moreover, the notochord, brain, heart, and mesenchyme also form through inductive interactions. In fact, the notochord and mesenchyme appear to be induced by FGFs secreted by the endoderm cells (Nakatani et al. 1996; Kim et al. 2000; Imai et al. 2002). These FGF proteins induce expression of Brachyury, which binds to the *cis*-regulatory elements that specify notochord development (**FIGURE 10.19**; Davidson and Christiaen 2006).

Interestingly, those genes that are turned on early by the Brachyury transcription factor have multiple binding sites for this protein and need all these sites occupied for maximal effect. Those genes that are turned on a bit later (also by Brachyury) have only one site, and this site may not bind as well as those on genes that are activated early. Notochord genes that are activated even later are activated indirectly. In the last instance, Brachyury activates a second transcription factor, which then will bind to activate these later genes (Katikala et al. 2013; José-Edwards et al. 2015). In this way, the timing of gene expression in the notochord can be carefully regulated. (See **Further Development 10.9, How Being Macho Can Give You a "Spine"**, and **Further Development 10.10, Gastrulation in Tunicates**, both online.)

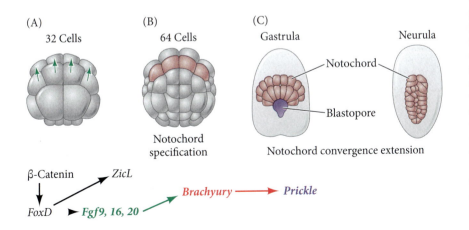

(A) 32 Cells

(B) 64 Cells

Notochord specification

(C) Gastrula — Neurula

Notochord

Blastopore

Notochord convergence extension

β-Catenin → ZicL
β-Catenin → FoxD
FoxD → Fgf9, 16, 20
Fgf9, 16, 20 → *Brachyury* → *Prickle*

FIGURE 10.19 Simplified version of the gene network leading to notochord development in the early tunicate embryo. (A,B) Vegetal views of 32- and 64-cell *Ciona* embryos. (A) β-Catenin accumulation leads to expression of the *foxd* gene. Foxd protein helps specify the cells to become endoderm and to secrete FGFs. (B) FGFs induce *Brachyury* expression in neighboring cells; these are the cells that will become the notochord. (C) Dorsal views. Brachyury activates regulators of cellular activity such as Prickle, which regulates cell polarity, leading to convergent extension of the notochord in the gastrula and neurula stages. (After B. Davidson and L. Christiaen. 2006. *Cell* 124: 247–250.)

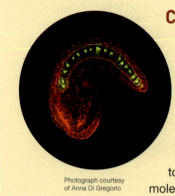

Closing Thoughts on the Opening Photo

The cells of this tunicate larva are fluorescing because they express a labeled target gene for the Brachyury protein, thus identifying them as the cells that will form the notochord. In tunicate and vertebrate embryos, the notochord is a primitive backbone that instructs the ectodermal cells above it to become the neural tube. Thus, the tunicate is an invertebrate (it has no spinal cord) chordate (it does have a notochord). When Alexander Kowalevsky discovered this in 1866, Charles Darwin was thrilled, realizing that tunicates are an evolutionary link between the invertebrate and vertebrate phyla. Today the Brachyury transcription factor is known to be important in both tunicate and vertebrate notochord formation, thus providing molecular support for Kowalevsky's finding.

Photograph courtesy of Anna Di Gregorio

Endnotes

[1] You might have been expecting a 128-cell blastula, but remember that the small micromeres stopped dividing.

[2] Micromeres can be placed into a "hole" that forms between the mesomeres and that closes shut and holds the transplanted micromeres in place. This intimate contact positions neighboring cells within range of the micromeres' inductive signals, to which the cells comply by changing their fate (D. McClay, Pers. Comm.).

[3] Recall the experiments in Figure 10.6, which demonstrated that the micromeres are able to induce a second embryonic axis when transplanted to the animal hemisphere. However, micromeres in which β-catenin is prevented from entering the nucleus are unable to induce the animal-hemisphere cells to form endoderm, and no second axis forms (Logan et al. 1998).

[4] This conclusion turns out to be a good first approximation—indeed, it has lasted over a century. However, new labeling techniques have shown that there is some left-right asymmetry in later tunicate embryos (B. Davidson, Pers. Comm.).

[5] This description should remind you of the discussion of the posterior cytoplasm of *Drosophila* eggs in Chapter 9. Indeed, mRNAs are localized in the CAB by their 3′ UTRs, the CAB is enriched with vesicles, and some of the mRNAs of the CAB become partitioned into the germ cells while others help construct the anterior-posterior axis (Makabe and Nishida 2012).

[6] Chabry and Driesch each seem to have gotten the results the other desired (see Fischer 1991). Driesch, who saw the embryo as a machine, expected autonomous specification but showed conditional specification. Chabry, a Socialist who believed everyone starts off equally endowed, expected to find conditional specification but instead discovered autonomous specification. Recent research on gene regulatory networks has begun to provide a molecular basis for this regulation (Peter et al. 2012).

[7] We will see the importance of *Brachyury* in vertebrate notochord formation as well. Indeed, the notochord is the "cord" that links the tunicates with the vertebrates, and *Brachyury* appears to be the gene that specifies the notochord (Satoh et al. 2012). As we will also see, *tbx6* (which is closely related to *Brachyury*) is important in forming vertebrate musculature.

10 Snapshot Summary
Sea Urchins and Tunicates

1. In both sea urchins and tunicates, the blastopore becomes the anus and the mouth is formed elsewhere; this deuterostome mode of gastrulation is also characteristic of chordates (including vertebrates).

2. Sea urchin cleavage is radial and holoblastic. At fourth cleavage, however, the vegetal tier divides into large macromeres and small micromeres. The animal hemisphere forms the mesomeres.

3. Sea urchin cell fates are determined by both autonomous and conditional modes of specification. The micromeres are specified autonomously and become a major signaling center for conditional specification of other lineages. Maternal β-catenin is important for the autonomous specification of the micromeres.

4. Differential cell adhesion is important in sea urchin gastrulation. First the micromeres detach from the vegetal plate and move into the blastocoel. They then form the skeletogenic mesenchyme, which forms the skeletal rods of the pluteus larva. The vegetal plate invaginates to form the endodermal archenteron, with a tip of non-skeletogenic mesenchyme cells. The archenteron elongates by convergent extension and is guided to the future mouth region by the non-skeletogenic mesenchyme.

5. The large micromeres form the skeleton of the larva; the small micromeres contribute to the coelomic pouches and to the germ cells.

6. The micromeres regulate the fates of their neighboring cells through juxtacrine and paracrine pathways. They can convert animal-hemisphere cells into endoderm.

7. Gene regulatory networks coordinate differentiation. The micromeres integrate maternal components such that the placement of Disheveled at the vegetal pole enables the

formation of β-catenin to help activate the *Pmar1* gene, whose products inhibit the *HesC* gene. HesC inhibits skeletogenic genes. Thus, by locally inhibiting the inhibitor, the most vegetal cells become committed to skeleton production. This is called a double-negative gate.

8. The ingression of the skeletogenic mesenchyme is accomplished through an epithelial-mesenchymal transition in which these cells lose cadherins and gain affinity to adhere to the matrix within the blastocoel.

9. Archenteron invagination and growth are coordinated by cell shape changes, cell proliferation, and convergent extension. In the final stage in invagination, the tip of the archenteron is actively pulled to the blastocoel roof by the filopodia of the non-skeletogenic mesenchyme cells.

10. The tunicate embryo divides holoblastically and bilaterally.

11. The yellow cytoplasm contains muscle-forming determinants that act autonomously. The heart and nervous system are formed conditionally by signaling interactions between blastomeres.

12. Macho-1 is the tunicate muscle determinant (a transcription factor that is sufficient to activate muscle-specifying genes). The notochord and mesenchyme are generated conditionally by paracrine factors such as FGFs.

13. FGFs induce expression of *Brachyury* in neighboring cells, inducing these cells to become the notochord.

Go to oup.com/he/barresi12xe for Further Developments, Scientists Speak interviews, Watch Development videos, Dev Tutorials, and complete bibliographic information for all literature cited in this chapter.

Amphibians and Fish

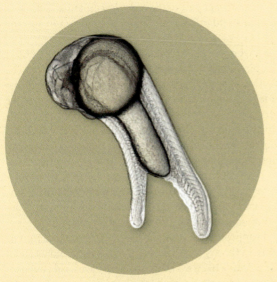

Two opposing bodies, made by two opposing...

Courtesy of Christine Thisse

DESPITE VAST DIFFERENCES IN ADULT MORPHOLOGY, early development in each of the vertebrate groups is very similar. Fish and amphibians are among the most easily studied vertebrates. In both cases, hundreds of eggs are laid, complete fertilization, and develop externally. Fish and amphibians are **anamniotic** vertebrates (**FIGURE 11.1**), meaning that they do not form the amnion that permits embryonic development to take place on dry land. However, developing amphibians and fish employ many of the same processes and genes used by other vertebrates (including humans) to generate body axes and organs.

Early Amphibian Development

Amphibian embryos once dominated the field of experimental embryology. With their large cells and rapid development, salamander and frog embryos were excellently suited for transplantation experiments. However, amphibian embryos fell out of favor during the early days of developmental genetics, in part because these animals undergo a long period of growth before they become fertile and because their chromosomes are often found in duplicated copies, precluding easy mutagenesis. But with the advent of molecular techniques such as in situ hybridization, chromatin immunoprecipitation, and the use of antisense oligonucleotides and dominant-negative proteins, researchers have returned to the study of amphibian embryos and have been able to integrate their molecular analyses with earlier experimental findings. The results have been spectacular, revealing new vistas of how vertebrate bodies are patterned and structured. As Jean Rostand wrote in 1960, "Theories come and theories go. The frog remains."

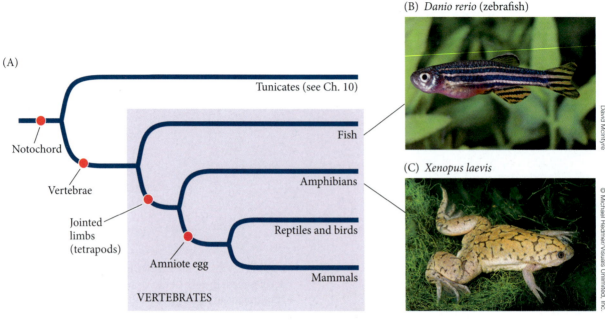

(A)

Notochord

Vertebrae

Jointed limbs (tetrapods)

Amniote egg

VERTEBRATES

Tunicates (see Ch. 10)

Fish

Amphibians

Reptiles and birds

Mammals

(B) *Danio rerio* (zebrafish)

David McIntyre

(C) *Xenopus laevis*

© Michael Redmer/Visuals Unlimited, Inc.

FIGURE 11.1   (A) Phylogenetic tree of the chordates showing the relationship of the vertebrate groups. The embryonic development of fish and amphibians must be carried out in moist environments. The evolution of the shelled amniote egg permitted development to proceed on dry land for the reptiles and their descendants, as we will see in Chapter 12. (B) The zebrafish (*Danio rerio*) has become a popular model organism for the study of development. It is the first vertebrate species to be subjected to mutagenesis studies similar to those that have been carried out on *Drosophila*. (C) *Xenopus laevis*, the African clawed frog, is one of the most studied amphibians because it has the rare property of not having a breeding season and thus can generate embryos year-round.

Fertilization, Cortical Rotation, and Cleavage

Most frogs have external fertilization, with the male fertilizing the eggs as the female is laying them. Even before fertilization, the frog egg has polarity, in that the dense yolk is at the vegetal (bottom) end, whereas the animal part of the egg (the upper half) has less yolk. As we will also see, certain proteins and mRNAs are already localized in specific regions of the unfertilized egg.

Fertilization most often occurs anywhere in the animal hemisphere of the amphibian egg. This tendency for sperm-egg binding to occur in the animal hemisphere may be due to the spatially restricted interaction of surface glycoproteins on the sperm with the vitelline envelope and cell membrane in the egg's animal hemisphere (Nagai et al. 2009; Kubo et al. 2010). The point of sperm entry is important because it influences dorsal-ventral polarity. The point of sperm entry marks the ventral (belly) side of the embryo, while the site 180° opposite the point of sperm entry will mark the dorsal (spinal) side.

The sperm centriole, which enters the egg with the sperm nucleus, organizes the microtubules of the egg into parallel tracks in the vegetal cytoplasm, separating the outer cortical cytoplasm from the yolky internal cytoplasm (**FIGURE 11.2A,B**). These microtubular tracks allow the cortical cytoplasm to rotate with respect to the inner cytoplasm. Indeed, these arrays are first seen immediately before rotation starts, become progressively aligned during rotation, and disappear when rotation ceases (Elinson and Rowning 1988; Houliston and Elinson 1991). In the zygote, the cortical cytoplasm rotates about 30° with respect to the internal cytoplasm (**FIGURE 11.2C**). In some cases, this exposes a region of gray-colored inner cytoplasm directly opposite the sperm entry point (**FIGURE 11.2D**; Roux 1887; Ancel and Vintenberger 1948). This region, the **gray crescent**, is where gastrulation will begin (Manes and Elinson 1980; Vincent et al. 1986).

Gastrulation begins on the side of the egg opposite the point of sperm entry, and this region will become the dorsal portion of the embryo. The microtubular arrays organized by the sperm centriole at fertilization support the specification of dorsal-to-ventral cell

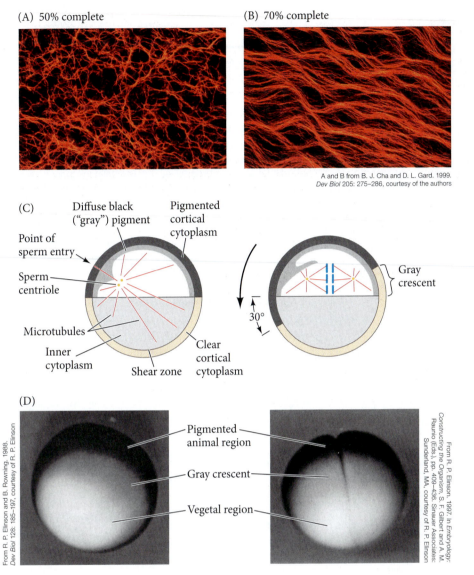

(A) 50% complete

(B) 70% complete

A and B from B. J. Cha and D. L. Gard. 1999.
Dev Biol 205: 275–286, courtesy of the authors

(C)

Diffuse black ("gray") pigment

Pigmented cortical cytoplasm

Point of sperm entry

Sperm centriole

Gray crescent

Microtubules

Inner cytoplasm

Clear cortical cytoplasm

Shear zone

30°

(D)

From R. P. Elinson and B. Rowning. 1988.
Dev Biol 128: 185–197, courtesy of R. P. Elinson

Pigmented animal region

Gray crescent

Vegetal region

From R. P. Elinson. 1997. In *Embryology:
Constructing the Organism*, S. F. Gilbert and A. M.
Raunio (Eds.), pp. 409–436. Sinauer Associates:
Sunderland, MA, courtesy of R. P. Elinson

FIGURE 11.2 Reorganization of the cytoplasm and cortical rotation produce the gray crescent in frog eggs. (A,B) Parallel arrays of microtubules (visualized here using fluorescent antibodies to tubulin) form in the vegetal hemisphere of the egg, along the future dorsal-ventral axis. (A) With the first cell cycle 50% complete, microtubules are present, but they lack polarity. (B) By 70% completion, the vegetal shear zone is characterized by a parallel array of microtubules; cortical rotation begins at this time. At the end of rotation, the microtubules will depolymerize. (C) Schematic cross section of cortical rotation. At left, the egg is shown midway through the first cell cycle. It has radial symmetry around the animal-vegetal axis. The sperm nucleus has entered at one side and is migrating inward. At right, 80% into first cleavage, the cortical cytoplasm has rotated 30° relative to the internal cytoplasm. Gastrulation will begin in the gray crescent—the region opposite the point of sperm entry, where the greatest displacement of cytoplasm occurs. (D) Gray crescent of *Rana pipiens*. Immediately after cortical rotation (left), lighter gray pigmentation is exposed beneath the heavily pigmented cortical cytoplasm. The first cleavage furrow (right) bisects this gray crescent. (C after J. C. Gerhart et al. 1989. *Development Suppl* 107: 37–51.)

fate, and thereby influence these movements. Therefore, the dorsal-ventral axis of the larva can be traced back to the point of sperm entry! Said another way, the frog zygote already "knows" its head from its tail.

Unequal radial holoblastic cleavage

Cleavage in most frog and salamander embryos is radially symmetrical and holoblastic, like echinoderm cleavage. The amphibian egg, however, is much larger than echinoderm eggs and contains much more yolk. As cleavages proceed, this yolk is disproportionally apportioned to the cells in the vegetal hemisphere. Recall from Chapter 1 that heavy concentrations of yolk act as an impediment to cell division; therefore, cleavage furrows in the vegetal hemisphere are slowed down by this yolk. Consequently, the first division begins at the animal pole and slowly extends down into the vegetal region (**FIGURE 11.3A,D**). Moreover, while the first cleavage furrow is still cleaving the yolky cytoplasm of the vegetal hemisphere, the second cleavage has already started near the animal pole.

In those species having a gray crescent (especially salamanders and frogs of the genus *Rana*), the first cleavage usually bisects the gray crescent (see Figure 11.2D). The second cleavage is at right angles to the first one and is also meridional. The third cleavage is equatorial but displaced asymmetrically toward the animal pole, again due

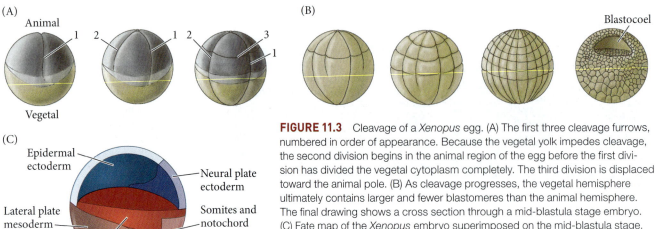

(A) Animal / Vegetal — 1 2 1 2 3 1

(B) Blastocoel

(C)

Epidermal ectoderm

Neural plate ectoderm

Lateral plate mesoderm

Somites and notochord

Blastopore

Somites

Endoderm

Heart

FIGURE 11.3 Cleavage of a *Xenopus* egg. (A) The first three cleavage furrows, numbered in order of appearance. Because the vegetal yolk impedes cleavage, the second division begins in the animal region of the egg before the first division has divided the vegetal cytoplasm completely. The third division is displaced toward the animal pole. (B) As cleavage progresses, the vegetal hemisphere ultimately contains larger and fewer blastomeres than the animal hemisphere. The final drawing shows a cross section through a mid-blastula stage embryo. (C) Fate map of the *Xenopus* embryo superimposed on the mid-blastula stage. (D) Scanning electron micrographs of the first, second, and fourth cleavages. Note the size discrepancies of the animal and vegetal cells after third cleavage. (A,B after B. M. Carlson. 1981. *Patten's Foundations of Embryology*. McGraw-Hill, New York; C after M. C. Lane and W. C. Smith. 1999. *Development* 126: 423–434 and C. S. Newman and P. A. Kreig. 1999. In *Cell Lineage and Fate Determination*, S. A. Moody [Ed.], pp. 341–351. Academic Press: New York.)

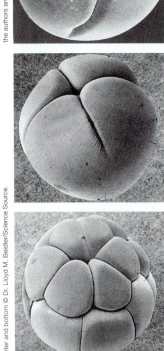

From H. W. Beams and R. G. Kessel. 1976. *Am Sci* 64: 279–290, courtesy of the authors and L. Biedler

Cleavage furrow

Center and bottom © Dr. Lloyd M. Beidler/Science Source.

to the influence of the vegetally placed yolk (Valles et al. 2002). It divides the amphibian embryo into four small animal blastomeres (micromeres) and four large blastomeres (macromeres) in the vegetal region. Despite their unequal sizes, continued cleavage of these blastomeres occurs at the same rate until the twelfth cell cycle. As cleavage progresses, the animal region becomes packed with numerous small cells, while the vegetal region contains a smaller number of large, yolk-laden macromeres. An amphibian embryo containing 16 to 64 cells is commonly called a **morula** (plural *morulae*; Latin, "mulberry," whose shape it vaguely resembles). At the 128-cell stage, the blastocoel becomes apparent, and the embryo is considered a **blastula** (**FIGURE 11.3B**).

The amphibian blastocoel serves several functions, one of which is to prevent the cells beneath it from interacting prematurely with the cells above it. When Nieuwkoop (1973) took embryonic newt cells from the roof of the blastocoel in the animal hemisphere (a region often called the **animal cap**) and placed them next to the yolky vegetal cells from the base of the blastocoel, the animal cap cells differentiated into mesodermal tissue instead of ectoderm. Thus, the blastocoel prevents premature contact of the vegetal cells with the animal cap cells, and keeps the animal cap cells undifferentiated.

Although amphibian development differs from species to species (see Hurtado and De Robertis 2007; Elinson and del Pino 2012), in general the animal hemisphere cells will give rise to the ectoderm, the vegetal cells will give rise to the endoderm, and the cells beneath the blastocoel cavity will become mesoderm (**FIGURE 11.3C**). The cells opposite the point of sperm entry will become the neural ectoderm, the notochord mesoderm, and the pharyngeal (head) endoderm (Keller 1975, 1976; Landström and Løvtrup 1979).

The mid-blastula transition: Preparing for gastrulation

An important precondition for gastrulation is the activation of the zygotic genome (that is, the genes within each nucleus of the embryo). In *Xenopus laevis*, only a few genes appear to be transcribed during early cleavage. For the most part, nuclear genes are not activated until late in the twelfth cell cycle (Newport and Kirschner 1982a,b; Yang et al. 2002). At that time, the embryo experiences a **mid-blastula transition**, or **MBT**, as different genes begin to be transcribed in different cells, the cell cycle acquires gap phases, and the blastomeres acquire the capacity to become motile. It is thought that some factor in the egg is being absorbed by the newly made chromatin because as in *Drosophila* (and, as you will learn, in zebrafish too), the time of this transition can be changed experimentally by altering the ratio of chromatin to cytoplasm in the cell (Newport and Kirschner 1982a,b).

It is thought that once the chromatin has been remodeled into euchromatic states (more open), various transcription factors (such as the VegT protein, formed in the vegetal cytoplasm from localized maternal mRNA) bind to the promoters and initiate new transcription. For instance, the vegetal cells (under the direction of the VegT protein) become the endoderm and begin secreting factors that induce the cells above them to become the mesoderm (see Figure 11.12). (See **Further Development 11.1, Mechanisms of the MBT by Chromatin Modifications**, online.)

Amphibian Gastrulation

The study of amphibian gastrulation is both one of the oldest and one of the newest areas of experimental embryology (see Beetschen 2001; Braukmann and Gilbert 2005). Most of our theories concerning the mechanisms of gastrulation and axis specification have been revised over the past two decades. The study of the developmental movements associated with gastrulation has also been complicated by the fact that there is no single way amphibians gastrulate; different species employ different means to achieve the same goal. In recent years, the most intensive investigations have focused on *Xenopus laevis* (pronounced "*Zeno-puss lay-vis*"), so we will concentrate on the mechanisms of gastrulation in that species.

Amphibian blastulae are faced with the same tasks as the invertebrate blastulae we followed in Chapters 8 through 10—namely, to bring inside the embryo those areas destined to form the endodermal organs; to surround the embryo with cells capable of forming the ectoderm; and to place the mesodermal cells in the proper positions between the ectoderm and the endoderm. Unlike in sea urchin gastrulation, many of the cell and tissue movements in *Xenopus* gastrulation happen simultaneously. We can, however, still identify the different types of behaviors that occur during *Xenopus* gastrulation (**FIGURE 11.4**):

1. *Epiboly:* The thinning and spreading of the animal cap cells over the vegetal hemisphere, powered by proliferation and radial intercalation.

2. *Vegetal rotation:* Vegetal cells asymmetrically press up against the inner blastocoel roof on the dorsal side.

3. *Bottle cell formation and invagination:* Localized apical constriction at the dorsal blastopore lip creates anisotropic forces that foster invagination.

4. *Involution and cell migration:* The leading edge of invaginating cells crawls up onto the blastocoel roof.

5. *Convergence and extension:* The targeted medial-to-lateral intercalation of cells on the midline (convergence) drives anterior-posterior axis elongation (extension).

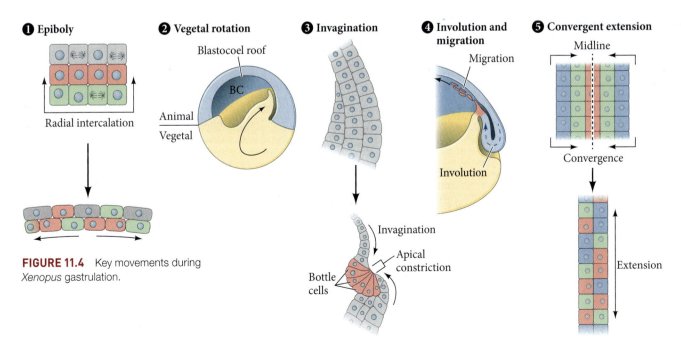

FIGURE 11.4 Key movements during *Xenopus* gastrulation.

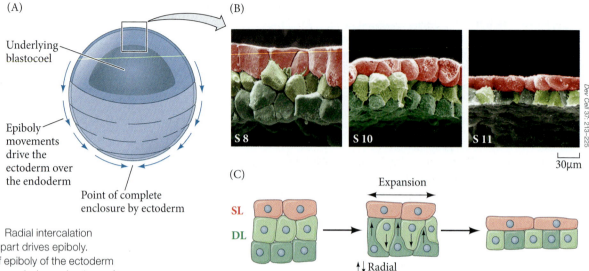

(A)

Underlying blastocoel

Epiboly movements drive the ectoderm over the endoderm

Point of complete enclosure by ectoderm

(B)

S 8   S 10   S 11

From A. Szabó et al. 2016. *Dev Cell* 37: 213–225

30µm

(C)

Expansion

SL

DL

Radial intercalation

FIGURE 11.5 Radial intercalation of ectoderm in part drives epiboly. (A) Depiction of epiboly of the ectoderm layer (blue) progressively moving toward the vegetal pole to completely enclose the endoderm. (B) Scanning electron micrographs of the *Xenopus* blastocoel roof (black box in A), showing the changes in cell shape and arrangement during radial intercalation. Stage 8 (S 8) is a blastula; stages 10 and 11 (S 10 and S 11) represent progressively later gastrulae. (C) Diagrammatic representations of the blastocoel roof at the same stages shown in (B). SL, superficial layer; DL, deep cell layer. (A and C after A. Szabó et al. 2016. *Dev Cell* 37: 213–225.)

Epiboly of the prospective ectoderm

The epidermis—the outer layer of your skin—stretched over your entire body, is derived from the ectoderm that initially resided solely in the animal hemisphere. It is the gastrulation movements of epiboly that pushed the ectoderm over the rest of the embryo, with one outcome being the establishment of an epidermal covering (**FIGURE 11.5A**). Two essential mechanisms powering epiboly are an increase in cell number (through division) and a concurrent integration of several deep cell layers into upper cell layers (through radial intercalation) (Keller and Schoenwolf 1977; Keller and Danilchik 1988; Saka and Smith 2001; Szabó et al. 2016). The surface ectoderm at the blastocoel roof can be seen to go from three to four cell layers thick to just two layers of flattened cells (**FIGURE 11.5B**). Cells from the deeper layer send protrusive processes *radially* toward the surface layer of cells, and then physically *intercalate* between more superficial cells as the ectoderm expands outward (**FIGURE 11.5C**). It is this radial intercalation of deeper cells that translates into a tissue-wide pushing force that reverberates through the entire population of ectodermal cells, expanding this layer vegetally over the endoderm. (See **Further Development 11.2, How Do Deeper Ectodermal Cells "Know" to Intercalate toward the Surface Ectoderm?**, online.)

Vegetal rotation and the invagination of the bottle cells

As epiboly extends the ectoderm over the vegetal hemisphere, it is doing so while the mesoderm and endoderm are actively moving inside. These actions correlate with the induction of mesoderm and also lead to the infolding of the endoderm to form the embryonic or primitive gut (known as the **archenteron**). This infolding requires the buckling of the outer epithelium at an invagination point. Most important, this invagination is initiated on the future *dorsal* side of the embryo, just below the equator, in the region of the gray crescent (i.e., the region opposite the point of sperm entry; see Figure 11.2C). Here the cells invaginate to form the slitlike **blastopore**, and thus on the dorsal side this region undergoing invagination is termed the **dorsal blastopore lip**. The formation of the dorsal blastopore lip is facilitated by a change in the shape of epithelial cells to resemble bottles (**FIGURE 11.6A,B**). The main body of each cell is displaced toward the inside of the embryo while maintaining contact with the outside surface by way of a slender neck (see Figure 11.4, step 3). As in sea urchins, these bottle cells will initiate the formation of the archenteron. However, unlike in sea urchins, gastrulation in the frog begins not in the most vegetal region but in the **marginal zone**—the region surrounding the equator of the blastula, where the animal and vegetal hemispheres meet (see Figure 11.6A).

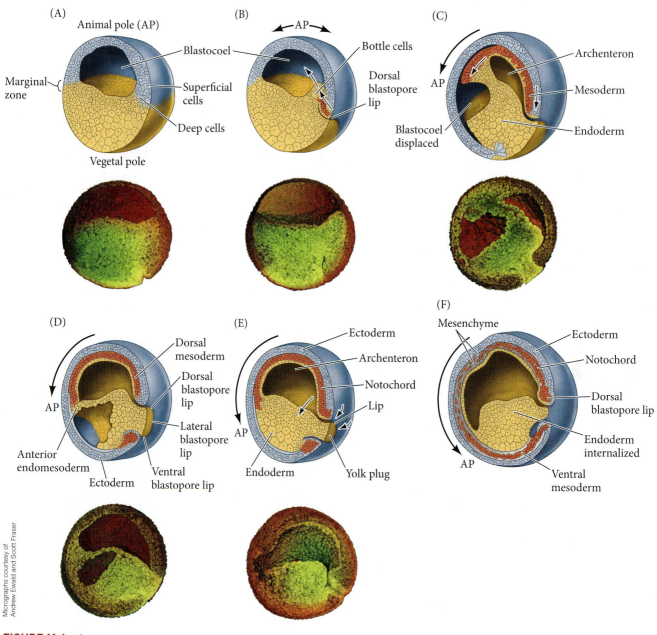

FIGURE 11.6 Cell movements during frog gastrulation. The drawings show meridional sections cut through the middle of the embryo and positioned so that the vegetal pole is tilted toward the observer and slightly to the left. The major cell movements are indicated by arrows, and the superficial animal hemisphere cells are colored so that their movements can be followed. Below the drawings are corresponding micrographs imaged with a surface imaging microscope (see Ewald et al. 2002). (A,B) Early gastrulation. The bottle cells of the margin move inward to form the dorsal lip of the blastopore, and the mesodermal precursors involute under the roof of the blastocoel. AP marks the position of the animal pole, which will change as gastrulation continues. (C,D) Mid-gastrulation. The archenteron forms and displaces the blastocoel, and cells migrate from the lateral and ventral lips of the blastopore into the embryo. The cells of the animal hemisphere migrate down toward the vegetal region, moving the blastopore to the region near the vegetal pole. (E,F) Toward the end of gastrulation, the blastocoel is obliterated, the embryo becomes surrounded by ectoderm, the endoderm has been internalized, and the mesodermal cells have been positioned between the ectoderm and endoderm. (Drawings after R. E. Keller. 1986. In *Developmental Biology: A Comprehensive Synthesis*, Vol. 2, L. Browder [Ed.], pp. 241–327. Plenum: New York.)

The formation of the dorsal blastopore lip isn't the first event of gastrulation, however. At least 2 hours before the bottle cells form, the cells of the *floor* of the blastocoel move upward toward the animal cap, but preferentially spiraling in the direction of the presumptive dorsal side of the early gastrula. This movement is called **vegetal rotation** (see Figure 11.4), and it places the prospective pharyngeal endoderm cells adjacent to

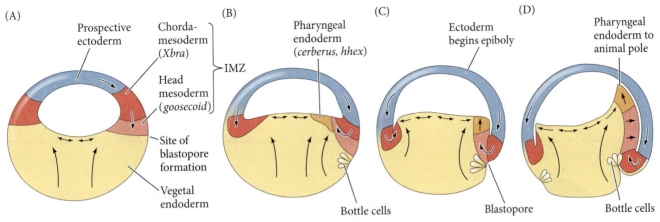

FIGURE 11.7 Early movements of *Xenopus* gastrulation. (A) At the beginning of gastrulation, the involuting marginal zone (IMZ) forms. Pink represents the prospective head mesoderm (*goosecoid* expression). Chordamesoderm (*Xbra* expression) is red. (B) Vegetal rotation (arrows) pushes the prospective pharyngeal endoderm (orange; specified by *hhex* and *cerberus* expression) to the side of the blastocoel. (C,D) The vegetal endoderm (yellow) movements push the pharyngeal endoderm forward, driving the mesoderm passively into the embryo and toward the animal pole. The ectoderm (blue) begins epiboly. (After R. Winklbauer and M. Schürfeld. 1999. *Development* 126: 3703–3713.)

the blastocoel and immediately above the involuting cells that will give rise to mesoderm (**FIGURE 11.7**). This combination of endodermal and mesodermal cells (termed **endomesoderm**) then migrates along the basal surface of the blastocoel roof, traveling toward the future anterior of the embryo (**FIGURE 11.6C–F**; Nieuwkoop and Florschütz 1950; Winklbauer and Schürfeld 1999; Ibrahim and Winklbauer 2001). (See **Further Development 11.3, A Fountain of Vegetal Movement**, online.)

Involution at the blastopore lip

As the migrating marginal cells reach the lip of the blastopore, they turn inward and travel along the inner surface of the outer animal hemisphere of ectodermal cells (i.e., the blastocoel roof). The behavior of a tissue turning inward and spreading over an internal surface (often of itself) is known as **involution**, and the cells in the amphibian gastrula that exhibit this behavior are often referred to as the **involuting marginal zone** (**IMZ**) (see Figure 11.6D–F). Importantly, the layers of ectoderm and underlying endomesoderm (a.k.a. mesendoderm) stay separated by a tight space of extracellular matrix called the **cleft of Brachet** (Gorny and Steinbeisser 2012). The first IMZ cells to compose the dorsal blastopore lip and *involute* into the embryo are the cells of the prospective pharyngeal endoderm of the foregut. This involuting mass of cells is led by cells of the deep endoderm that were repositioned by the movements of vegetal rotation (see Figure 11.7). These cells collectively migrate anteriorly beneath the surface ectoderm of the blastocoel (Papan et al. 2007a,b; Winklbauer and Damm 2012; Moosmann et al. 2013). These anterior endoderm cells transcribe the *hhex* gene, which encodes a transcription factor that is critical for forming the head and heart (Rankin et al. 2011). As these first cells pass into the interior of the embryo, the dorsal blastopore lip becomes composed

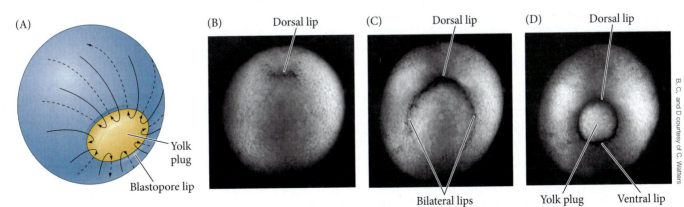

FIGURE 11.8 Formation of the blastopore lips in *Xenopus laevis*. (A) Illustration of epiboly of the ectoderm (blue) as the yolky cells of the endoderm (yellow) and mesoderm fold into the embryo. (B–D) Embryo seen from the vegetal surface. (B) The site of the dorsal blastopore lip is evident by the pigmented cells at its rim, coming from the animal cap. (C) This region of involution later spreads to form the lateral lips. (D) The blastopore eventually encircles a small yolk plug, with cells involuting along each side. This entire sequence takes about 7 hours.

B, C, and D courtesy of C. Watters

of cells that involute into the embryo to become the **prechordal plate**, the precursor of the head mesoderm. Prechordal plate cells transcribe the *goosecoid* gene, whose product is a transcription factor that activates numerous genes controlling head formation. It achieves this activation indirectly by repressing those genes (e.g., *Wnt8*) that repress head development. This phenomenon—the activation of genes by repressing their repressors—is a major feature of animal development, as we saw in the double-negative gate that specifies sea urchin micromeres (see Chapter 10).

Next to involute through the dorsal blastopore lip are the cells of the **chordamesoderm**. These cells will form the **notochord**, the transient mesodermal rod that plays an important role in inducing and patterning the nervous system. Chordamesoderm cells express the *tbxt* (*brachyury*) gene, whose product (as we saw in Chapter 10) is a transcription factor critical for notochord formation. Thus, the cells constituting the dorsal blastopore lip are constantly changing as the original cells migrate into the embryo and are replaced by cells migrating downward, inward, and upward.

As the new cells enter the embryo, the blastocoel is displaced to the side opposite the dorsal lip. Meanwhile, the lip expands laterally and ventrally as bottle cell formation and involution continue around the blastopore. The widening blastopore "crescent" develops lateral lips and, finally, a ventral lip over which additional mesodermal and endodermal precursor cells pass (**FIGURE 11.8**). These cells include the precursors of the heart and kidney. With the formation of the ventral lip, the blastopore has formed a ring around the large endodermal cells that remain exposed on the vegetal surface. This remaining patch of endoderm is called the **yolk plug**; it, too, is eventually internalized (at the site of the anus). At that point, all the endodermal precursors have been brought into the interior of the embryo, the ectoderm has enveloped the surface, and the mesoderm has been brought between them.

FURTHER DEVELOPMENT

A fibronectin road for the collective migration of involuting mesendoderm

Just as you need a firm surface to apply force as you walk, the cells of the embryo require a stable substrate to accomplish the many movements of gastrulation. One such road in the *Xenopus* gastrula is established through the assembly of fibronectin fibrils in the extracellular matrix underlying the blastocoel roof. In *Xenopus* and many other amphibians, it appears that the involuting mesendodermal precursors migrate toward the animal pole by traveling on an extracellular lattice of **fibronectin (FN)** secreted by the presumptive ectoderm cells of the blastocoel roof (**FIGURE 11.9A**). Loss of FN or of its fibril assemblies results in the arrest of involution and a failure to complete epiboly (Boucaut et al. 1984; Rozario et al. 2009).

If fibronectin is the road, then how do cells as part of a layer move along this road? Have you ever witnessed a road race? Whether it's a 5K or a marathon, the top runners are positioned at the leading edge of the pack. Those lead runners have earned their spot at the front and are presumed to be the most prepared for that race. Now imagine if everyone in the long crowd of runners held on to the shirts of the runners in front of them, which continued all the way to the leaders at the front of the pack. This would *force* the runners to move together as a single collective. In this analogy, where the runners are cells, this type of group behavior is known as **collective cell migration** (**FIGURE 11.9B**). Can you fathom the amount of force being applied to those cells at the very front of this pack?

Recently, the laboratory of Doug DeSimone characterized the dynamics of IMZ cells during *Xenopus* gastrulation as that of a population of collectively migrating cells (Sonavane et al. 2017). It appears that the involuting endomesoderm cells adhere tightly to one another, with the leading row of cells behaviorally distinct from the many rows of following cells that together are pulling on those leaders. Leading row cells uniquely display polarized Rac activity concurrent with lamellipodia formation along their leading edge. These leading-edge lamellipodia are necessary for migration (see Figure 11.9B). Using traction force microscopy, the researchers demonstrated that the leading row of cells at the front of the pack

? Developing Questions

The following IMZ cells send long cytonemes into the cleft of Brachet during their collective migration. Imagine you are gliding along the water in a canoe, and you reach out your arms and fingers to touch the water. This sends ripples throughout the water in three dimensions, transmitting a physical signal of sorts. Could the IMZ cell cytonemes be like your fingers, touching the surface of the overlying ectoderm as the collective mass moves across the blastocoel roof? What sort of signal could they be relaying? Answers to these questions are currently unknown but may profoundly change how we think axis determination is achieved.

(A)

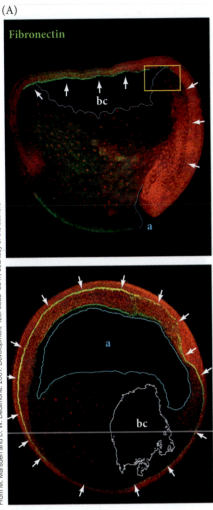

(B)

Lateral view of collective cell migration

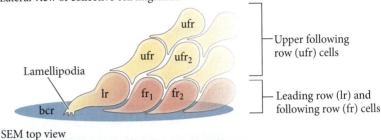

SEM top view

ICC top view; **fibronectin**, actin, **β-catenin**

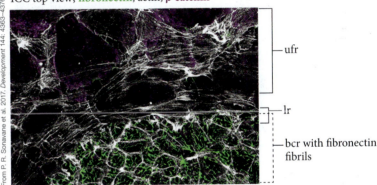

FIGURE 11.9 Fibronectin and amphibian gastrulation. (A) Sagittal section of *Xenopus* embryos at early (top) and late (bottom) gastrulation. The fibronectin lattice on the blastocoel roof is identified by fluorescent antibody labeling (green and yellow, arrows), while the embryonic cells are counterstained red. The blastocoel (bc) is delineated with a white outline, and the archenteron (a) with a teal line. (B) High magnification of the IMZ cells (yellow boxed area in A) from a lateral schematized view (top) and from top views by scanning electron microscopy (SEM; image has been colorized; middle) and immunocytochemistry (ICC; bottom). The leading row of cells exerts a tractional force on the fibronectin matrix of the blastocoel roof, while being pulled on by following mesendodermal cells. The leading cells are the only cells in this collective migration to consistently show lamellipodia formation (in bottom figure actin microfilaments are shown in white).

are the only migrating cells that exhibit "traction stresses" (**FIGURE 11.10A**). This means that only the leading cells are actually attached to the extracellular matrix (FN fibrils) of the blastocoel roof and applying true traction-type pulling forces. As you surmised from our earlier analogy, the amount of force exerted on this leading row of cells is tremendous and even results in an enlargement of the cleft of Brachet just behind the leaders (**FIGURE 11.10B**). Consequently, specialized cadherin-keratin intermediate filament complexes are found in the leading cells that serve to withstand the pulling stress from the following rows of cells. DeSimone's lab has also raised awareness of the presence of cytoneme-like filopodial protrusions (see Chapter 4) extending into the cleft of Brachet (see Figure 11.10B; Sonavane et al. 2017). What role these cytonemes play in gastrulation or axis determination is currently unknown.

Convergent extension of the dorsal mesoderm

Figure 11.11 depicts the behavior of the involuting marginal zone cells at successive stages of *Xenopus* gastrulation (Keller and Schoenwolf 1977; Hardin and Keller 1988). The IMZ is originally several layers thick. Shortly before their involution through the blastopore lip, the several layers of deep IMZ cells intercalate radially to form one thin, broad layer. This intercalation further extends the IMZ vegetally (**FIGURE 11.11A**). At the same time, the superficial cells spread out by dividing and flattening. When the deep cells reach the blastopore lip, they involute into the embryo and initiate a second type of intercalation. This intercalation causes a convergent extension along the mediolateral axis that integrates several mesodermal streams to form a long, narrow band (**FIGURE 11.11B**). The anterior part of this band migrates toward the animal cap. Thus,

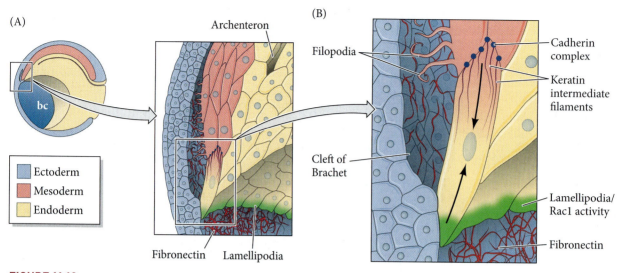

FIGURE 11.10 Collective cell migration of the involuting mesendoderm. See text for details. (After P. R. Sonavane et al. 2017. *Development* 144: 4363–4376.)

the mesodermal stream continues to migrate toward the animal pole, and the overlying layer of superficial cells (including the bottle cells) is passively pulled toward the animal pole, thereby forming the endodermal roof of the archenteron (see Figures 11.6 and 11.11A). The radial and mediolateral intercalations of the deep layer of cells appear to be responsible for the continued movement of mesoderm into the embryo. Several forces appear to drive convergent extension, including, but not limited to, polarized cell cohesion through the Wnt/planar cell polarity (PCP) pathway, cadherin-mediated differential adhesion, and midline waves of Ca^{2+} signaling (see Shindo and Wallingford 2014; Shindo 2017; Shindo et al. 2019).

Those mesodermal cells entering through the dorsal lip of the blastopore give rise to the central dorsal mesoderm (notochord and somites), while the remainder of the body mesoderm (which forms the heart, kidneys, bones, and parts of several other organs) enters through the ventral and lateral blastopore lips to create the **mesodermal mantle**. The endoderm is derived from the superficial cells of the involuting marginal zone that form the lining of the archenteron roof and from the subblastoporal vegetal cells that become the archenteron floor (Keller 1986). The remnant of the blastopore—where the endoderm meets the ectoderm—now becomes the anus. As gastrulation expert Ray Keller famously remarked, "Gastrulation is the time when a vertebrate takes its head out of its anus." (See **Further Development 11.4, The Forces of Convergent Extension,** and **Further Development 11.5, Migration of the Mesodermal Mantle**, both online.)

FIGURE 11.11 *Xenopus* gastrulation continues. (A) The deep marginal cells flatten, and the formerly superficial cells form the wall of the archenteron. (B) Radial intercalation, looking down at the dorsal blastopore lip from the dorsal surface. In the noninvoluting marginal zone (NIMZ) and the upper portion of the IMZ, deep (mesodermal) cells are intercalating radially to make a thin band of flattened cells. This thinning of several layers into a few causes convergent extension (white arrows) toward the blastopore lip. Just above the lip, mediolateral intercalation of the cells produces stresses that pull the IMZ over the lip. After involuting over the lip, mediolateral intercalation continues, elongating and narrowing the axial mesoderm. (After P. Wilson and R. Keller. 1991. *Development* 112: 289–300; R. Winklbauer and M. Schürfeld. 1999. *Development* 126: 3703–3713.)

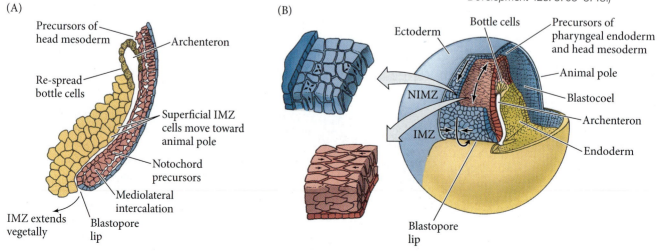

Progressive Determination of the Amphibian Axes

As we have seen, the unfertilized amphibian egg has polarity along the animal-vegetal axis, and the germ layers can be mapped onto the oocyte even before fertilization. The animal hemisphere blastomeres become the cells of the ectoderm (epidermis and nerves); the vegetal hemisphere cells become the inner lining of the gut and associated organs (endoderm); and the equatorial cells form the mesoderm (bone, muscle, heart, blood, kidneys, etc.).

Specification of the germ layers

The general fate of these different regions is thought to be imposed on the embryo by the vegetal cells, which have two major functions: (1) to differentiate into endoderm and (2) to induce the cells immediately above them to become mesoderm.

The mechanism for this "bottom-up" specification of the frog embryo resides in a set of mRNAs that are tethered to the vegetal cortex. This includes the mRNA for the transcription factor VegT, which becomes apportioned to the vegetal cells during cleavage. VegT is critical in generating both endodermal and mesodermal lineages. When VegT transcripts are destroyed by antisense oligonucleotides, the entire embryo becomes epidermis, with no mesodermal or endodermal components (Zhang et al. 1998; Taverner et al. 2005). *VegT* mRNA is translated shortly after fertilization. Its product activates a set of genes prior to the mid-blastula transition. One of the genes activated by this VegT protein encodes the Sox17 transcription factor. Sox17, in turn, is critical for activating the genes that specify cells to be endoderm. Thus, the fate of the vegetal cells is to become endodermal.

Another set of early genes activated by VegT encodes Nodal paracrine factors that instruct the cell layers *above* them to become mesoderm (Skirkanich et al. 2011). Nodal secreted from the vegetal cells in the nascent endoderm signals the cells above them to accumulate phosphorylated Smad2. Phosphorylated Smad2 helps upregulate the *eomesodermin* and *Brachyury* (*tbxt*) genes in those cells, the protein products of which cause the cells to become specified as mesoderm. The Eomesodermin and Smad2 proteins working together can activate the zygotic genes for the VegT proteins, thus creating a positive feedforward loop that is critical in sustaining the mesoderm (**FIGURE 11.12**). In the absence of such induction, cells become ectoderm (Fukuda et al. 2010).

In addition, the *Vg1* mRNA that has been stored in the vegetal cytoplasm is also translated. The production of Vg1 (another Nodal-like protein) is needed to activate other genes in the dorsal mesoderm. If either Nodal or Vg1 signaling is blocked, there is little or no mesoderm induction (Kofron et al. 1999; Agius et al. 2000; Birsoy et al. 2006). Thus, by the late blastula stage, the fundamental germ layers are becoming specified. The vegetal cells are specified as endoderm through transcription factors such as Sox17. The equatorial cells are specified as mesoderm by transcription factors such as Eomesodermin. And the animal cap—which has not begun receiving signals yet—becomes specified as ectoderm (see Figure 11.12).

The dorsal-ventral and anterior-posterior axes

Although animal-vegetal polarity initiates specification of the germ layers, the anterior-posterior, dorsal-ventral, and left-right axes are specified by events triggered at

FIGURE 11.12 Model for the specification of the mesoderm. The vegetal region of the oocyte has accumulated mRNA for the transcription factor VegT and (in the future dorsal region) mRNA for the Nodal paracrine factor Vg1. At the late blastula stage, the *Vg1* mRNA is translated and Vg1 induces the future dorsal mesoderm to transcribe the genes for several Wnt antagonists (such as Dickkopf). The *VegT* message is also translated, and VegT activates nuclear genes encoding Nodal proteins. These TGF-β superfamily members activate the expression of the transcription factor Eomesodermin in the presumptive mesoderm. Eomesodermin, with the help of activated Smad2 from the Nodal proteins, activates nuclear genes encoding VegT. In this way, VegT expression has gone from maternal mRNAs in the presumptive endoderm to nuclear expression in the presumptive mesoderm. (After M. Fukuda et al. 2010. *Int J Dev Biol* 54: 81–92.)

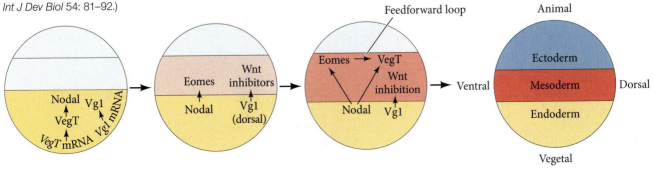

fertilization but not realized until gastrulation. In *Xenopus* (and in other amphibians), the formation of the anterior-posterior axis is inextricably linked to the formation of the dorsal-ventral axis. This, as we will see, is predicated on fertilization events that will place the transcription factor β-catenin in the region of the egg opposite the point of sperm entry and will specify that region of the egg to be the dorsal region of the embryo.

Once β-catenin is localized in this region of the egg, the cells containing β-catenin will induce expression of certain genes and thus initiate the movement of the involuting mesoderm. This movement will establish the anterior-posterior axis of the embryo. The first mesodermal cells to migrate over the dorsal blastopore lip will induce the ectoderm above them to produce anterior structures such as the forebrain; mesoderm that involutes later will signal the ectoderm to form more posterior structures, such as the hindbrain and spinal cord. This process, whereby the central nervous system forms through interactions with the underlying mesoderm, has been called **primary embryonic induction** and is one of the principal ways in which the vertebrate embryo becomes organized. Indeed, its discoverers called the dorsal blastopore lip and its descendants the **organizer** and found that this region is different from all the other parts of the embryo. In the early twentieth century, experiments by Hans Spemann and his students at the University of Freiburg, Germany, framed the questions that experimental embryologists would continue to ask for most of the rest of the century and resulted in a Nobel Prize for Spemann in 1935 (see Hamburger 1988; De Robertis and Aréchaga 2001; Sander and Fässler 2001).

The Work of Hans Spemann and Hilde Mangold: Primary Embryonic Induction

The most spectacular transplantation experiments published by Spemann and his doctoral student Hilde Mangold in 1924[1] showed that, of all the tissues in the early gastrula, only one has its fate autonomously determined. This self-determining tissue is the dorsal lip of the blastopore—the tissue derived from the gray crescent cytoplasm opposite the point of sperm entry. When this tissue was transplanted into the presumptive belly skin region of another gastrula, it not only continued to be dorsal blastopore lip but also initiated gastrulation and embryogenesis in the surrounding tissue!

In these experiments, Spemann and Mangold used the differently pigmented embryos of two species of newts, *Triturus taeniatus* (darkly pigmented) and *T. cristatus* (nonpigmented), so that they could identify host and donor tissues on the basis of color. When the dorsal lip of an early *T. taeniatus* gastrula was removed and implanted into the region of an early *T. cristatus* gastrula fated to become ventral epidermis (the outer layer of belly skin), the dorsal lip tissue invaginated just as it normally would have done (showing self-determination) and disappeared beneath the vegetal cells (**FIGURE 11.13A**). The pigmented donor tissue then continued to self-differentiate into the chordamesoderm (notochord) and other mesodermal structures that normally form from the dorsal lip (**FIGURE 11.13B**). As the donor-derived mesodermal cells moved forward, host cells began to participate in the production of a new embryo, becoming organs that they normally would have never formed. In this secondary embryo, a somite could be seen containing both pigmented (donor) and nonpigmented (host) tissue. Even more spectacularly, the dorsal lip cells were able to interact with the host tissues to form a complete neural plate from host ectoderm. Eventually, a secondary embryo formed, conjoined face to face with its host (**FIGURE 11.13C**). The results of these technically difficult experiments have been confirmed many times and in many amphibian species, including *Xenopus* (**FIGURE 11.13D,E**; Capuron 1968; Gimlich and Cooke 1983; Smith and Slack 1983; Recanzone and Harris 1985).

Spemann referred to the dorsal lip cells and their derivatives (notochord and head endomesoderm) as the organizer because (1) they induced the host's ventral tissues to change their fates to form a neural tube and dorsal mesodermal tissue (such as somites), and (2) they organized host and donor tissues into a secondary embryo with clear anterior-posterior and dorsal-ventral axes. He proposed that during normal development, these cells "organize" the dorsal ectoderm into a neural tube and

FIGURE 11.13 Organization of a secondary axis by dorsal blastopore lip tissue. (A–C) Spemann and Mangold's 1924 experiments visualized the process by using differently pigmented newt embryos. (A) Dorsal lip tissue from an early *Triturus taeniatus* gastrula is transplanted into a *T. cristatus* gastrula in the region that normally becomes ventral epidermis. (B) The donor tissue invaginates and forms a second archenteron, and then a second embryonic axis. Both donor and host tissues are seen in the new neural tube, notochord, and somites. (C) Eventually, a second embryo forms, joined to the host. (D) Live twinned *Xenopus* larvae generated by transplanting a dorsal blastopore lip into the ventral region of an early-gastrula host embryo. (E) Similar twinned larvae are seen from below and stained for notochord; the original and secondary notochords can be seen. (A,C after J. Holtfreter and V. Hamburger. 1955. In *Analysis of Development*, B. H. Willier, P. Weiss, and V. Hamburger [Eds.], pp. 230–296. W. B. Saunders Company: Philadelphia and London.)

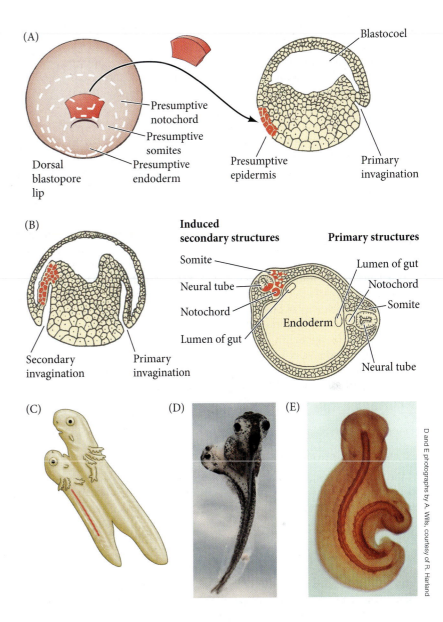

D and E photographs by A. Wills, courtesy of R. Harland

transform the flanking mesoderm into the anterior-posterior body axis (Spemann 1938). It is now known (thanks largely to Spemann and his students) that the interaction of the chordamesoderm and ectoderm is not sufficient to organize the entire embryo. Rather, it initiates a series of sequential inductive events. Because there are numerous inductions during embryonic development, this key induction—in which the progeny of dorsal lip cells induce the dorsal axis and the neural tube—is traditionally called the primary embryonic induction. This classic term has been a source of confusion, however, because the induction of the neural tube by the notochord is no longer considered to be the first inductive process in the embryo. As we have already seen, there are inductive events that precede this "primary" induction (see Figure 11.12). (See **Further Development 11.6, Autonomous Specification versus Inductive Interactions**, online.)

Molecular Mechanisms of Amphibian Axis Formation

The experiments of Spemann and Mangold showed that the dorsal lip of the blastopore, along with the dorsal mesoderm and pharyngeal endoderm that form from it, constituted an "organizer" able to instruct the formation of embryonic axes. But the mechanisms

by which the organizer itself was constructed and through which it operated remained a mystery. Indeed, it is said that Spemann and Mangold's landmark paper posed more questions than it answered. Among those questions were:

- How did the organizer get its properties? What caused the dorsal blastopore lip to differ from any other region of the embryo?

- What factors were being secreted from the organizer to cause the formation of the neural tube and to create the anterior-posterior, dorsal-ventral, and left-right axes?

- How did the different parts of the neural tube become established, with the most anterior becoming the sensory organs and forebrain and the most posterior becoming spinal cord?

Spemann and Mangold's description of the organizer was the starting point for one of the first truly international scientific research programs (see Gilbert and Saxén 1993; Armon 2012). Researchers from Britain, Germany, France, the United States, Belgium, Finland, Japan, and the Soviet Union all joined in the search for the remarkable substances responsible for the organizer's ability. R. G. Harrison referred to the amphibian gastrula as the "new Yukon to which eager miners were now rushing to dig for gold around the blastopore" (see Twitty 1966, p. 39). Unfortunately, their early picks and shovels proved too blunt to uncover the molecules involved. The proteins responsible for induction were present in concentrations too small for biochemical analyses, and the large quantity of yolk and lipids in the amphibian egg further interfered with protein purification (Grunz 1997). The analysis of organizer molecules had to wait until recombinant DNA technologies enabled investigators to make cDNA clones from blastopore lip mRNA, thus allowing them to see which of these clones encoded factors that could dorsalize the embryo (Carron and Shi 2016). We are now able to take up each of the above questions in turn.

How does the organizer form?

Why are the dozen or so initial cells of the organizer positioned opposite the point of sperm entry, and what determines their fate so early? Recent evidence provides an unexpected answer: these cells are in the right place at the right time, at a point where two signals converge. The first signal tells the cells that they are dorsal. The second signal says that these cells are mesoderm. These signals interact to create a polarity within the mesoderm that is the basis for specifying the organizer and for creating dorsal-ventral polarity.

THE DORSAL SIGNAL, PART 1: THE NIEUWKOOP CENTER Experiments by Pieter Nieuwkoop and Osamu Nakamura showed that the organizer receives its special properties from signals coming from the prospective endoderm beneath it. Nakamura and Takasaki (1970) showed that the mesoderm arises from the marginal (equatorial) cells at the border between the animal and vegetal poles. The Nakamura and Nieuwkoop laboratories then demonstrated that the properties of this newly formed mesoderm can be induced by the vegetal (presumptive endoderm) cells underlying them. Nieuwkoop (1969, 1973, 1977) removed the equatorial cells (i.e., presumptive mesoderm) from a blastula and showed that neither the animal cap (presumptive ectoderm) nor the vegetal cap (presumptive endoderm) produced any mesodermal tissue. However, when the two caps were recombined, the animal cap cells were induced to form mesodermal structures, such as notochord, muscles, kidney cells, and blood cells. The polarity of this induction (i.e., whether the animal cap cells formed dorsal mesoderm or ventral mesoderm) depended on whether the endodermal (vegetal) fragment was taken from the dorsal or the ventral side: ventral and lateral vegetal cells (those closer to the site of sperm entry) induced ventral (mesenchyme, blood) and intermediate (kidney) mesoderm, while the dorsalmost vegetal cells specified dorsal mesoderm components (somites, notochord)—including those having the properties of the organizer. These dorsalmost vegetal cells of the blastula, which are capable of inducing the organizer, have been called the **Nieuwkoop center** (Gerhart et al. 1989). (See **Further Development 11.7, Play Mix and Match with Vegetal Blastomeres to Prove the Inductive Power of the Nieuwkoop Center**, online.)

(A)

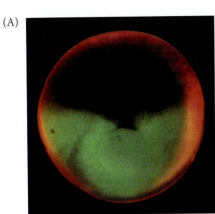

(D)

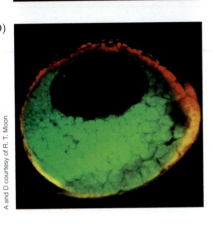

(B) (C)

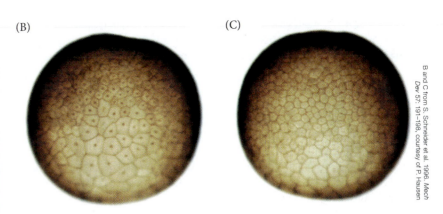

FIGURE 11.14 Role of β-catenin in dorsal-ventral axis specification. (A–D) Differential translocation of β-catenin into *Xenopus* blastomere nuclei. (A) Early 2-cell stage, showing β-catenin (orange) predominantly at the dorsal surface. (B) Presumptive dorsal side of a blastula stained for β-catenin shows nuclear localization. (C) Such nuclear localization is not seen on the ventral side of the same embryo. (D) Dorsal localization of β-catenin persists through the gastrula stage.

THE DORSAL SIGNAL, PART 2: β-CATENIN Once the special nature of the Nieuwkoop center was discovered, an important question became, What gives these dorsalmost vegetal cells their special properties? The major candidate for the factor that forms the Nieuwkoop center in these vegetal cells was β-catenin. We saw in Chapter 10 that β-catenin is responsible for specifying the micromeres of the sea urchin embryo. This multifunctional protein also proved to be a key player in the formation of the dorsal amphibian tissues. Experimental depletion of this molecule results in the lack of dorsal structures, while injection of exogenous β-catenin into the *ventral* side of an embryo produces a secondary axis (McMahon and Moon 1989; Smith and Harland 1991; Sokol et al. 1991; Heasman et al. 1994a; Funayama et al. 1995; Guger and Gumbiner 1995).

In *Xenopus* embryos, β-catenin is initially synthesized throughout the embryo from maternal mRNA (Yost et al. 1996; Larabell et al. 1997). It begins to accumulate in the dorsal region of the egg during the cytoplasmic movements of fertilization and continues to accumulate preferentially at the dorsal side throughout early cleavage. This accumulation is seen in the nuclei of the dorsal cells and appears to cover both the Nieuwkoop center and organizer regions (**FIGURE 11.14**; Schneider et al. 1996; Larabell et al. 1997).

If β-catenin is originally found throughout the embryo, how does it become localized specifically to the side opposite sperm entry? The answer appears to reside in the localizations of three proteins in the egg cortical cytoplasm. The proteins Wnt11, GSK3-binding protein (GBP), and Disheveled (Dsh) all are translocated from the vegetal pole of the egg to the future dorsal side of the embryo after fertilization. From research on the Wnt pathway, we have learned that β-catenin is targeted for destruction by glycogen synthase kinase 3 (GSK3; see Chapter 4). Indeed, activated GSK3 stimulates degradation of β-catenin and blocks axis formation when added to the egg, and if endogenous GSK3 is knocked out by a dominant-negative form of GSK3 in the ventral cells of the early embryo, a second axis forms (**FIGURE 11.15A**; He et al. 1995; Pierce and Kimelman 1995; Yost et al. 1996).

GSK3 can be inactivated by GBP and Disheveled. These two proteins release GSK3 from the degradation complex and prevent it from binding β-catenin and targeting it for destruction. During the first cell cycle, when the microtubules form parallel tracts in the vegetal portion of the egg, GBP travels along the microtubules by binding to kinesin, an ATPase motor protein that travels on microtubules. Kinesin always migrates toward the growing end of the microtubules, and in this case, that means moving to the point opposite sperm entry—that is, the future dorsal side (**FIGURE 11.15B–D**). Disheveled, which is originally found in the vegetal

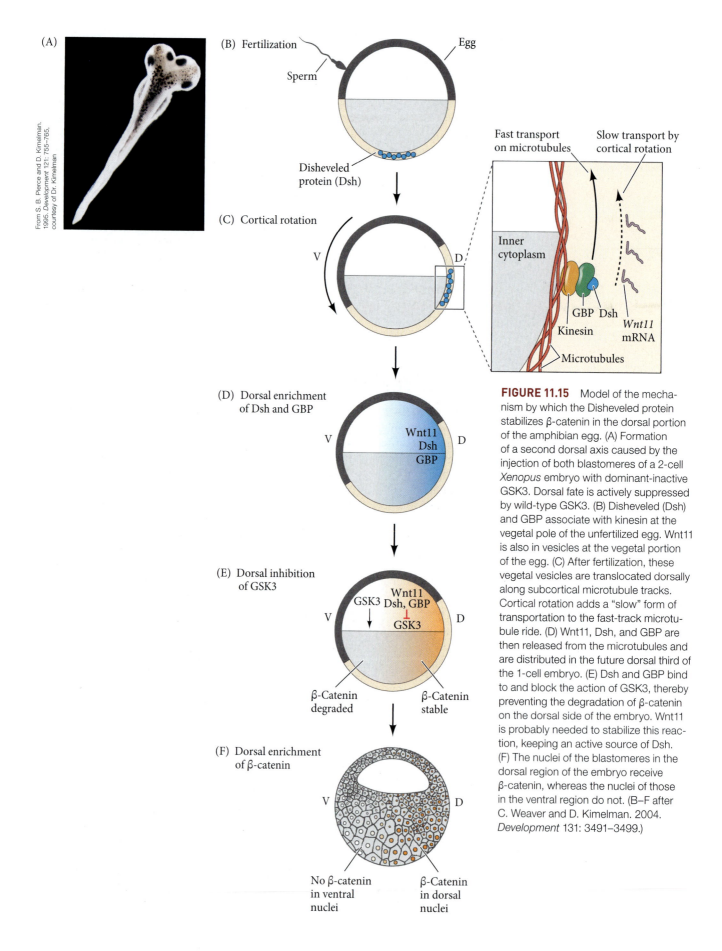

FIGURE 11.15 Model of the mechanism by which the Disheveled protein stabilizes β-catenin in the dorsal portion of the amphibian egg. (A) Formation of a second dorsal axis caused by the injection of both blastomeres of a 2-cell *Xenopus* embryo with dominant-inactive GSK3. Dorsal fate is actively suppressed by wild-type GSK3. (B) Disheveled (Dsh) and GBP associate with kinesin at the vegetal pole of the unfertilized egg. Wnt11 is also in vesicles at the vegetal portion of the egg. (C) After fertilization, these vegetal vesicles are translocated dorsally along subcortical microtubule tracks. Cortical rotation adds a "slow" form of transportation to the fast-track microtubule ride. (D) Wnt11, Dsh, and GBP are then released from the microtubules and are distributed in the future dorsal third of the 1-cell embryo. (E) Dsh and GBP bind to and block the action of GSK3, thereby preventing the degradation of β-catenin on the dorsal side of the embryo. Wnt11 is probably needed to stabilize this reaction, keeping an active source of Dsh. (F) The nuclei of the blastomeres in the dorsal region of the embryo receive β-catenin, whereas the nuclei of those in the ventral region do not. (B–F after C. Weaver and D. Kimelman. 2004. *Development* 131: 3491–3499.)

pole cortex, grabs onto the GBP, and it too becomes translocated along the microtubular monorail (Miller et al. 1999; Weaver et al. 2003). Once at the site opposite the point of sperm entry, GBP and Dsh are released from the microtubules. Here, they inactivate GSK3, allowing β-catenin to accumulate and start the gene regulatory network for dorsal cell specification while β-catenin on the opposite side (future ventral fates) is degraded (**FIGURE 11.15E,F**; Weaver and Kimelman 2004).

But the mere translocation of these proteins to the dorsal side of the embryo does not seem to be sufficient for protecting β-catenin. It appears that a Wnt paracrine factor has to be secreted there to activate the β-catenin protection pathway; this is accomplished by Wnt11. If Wnt11 synthesis is suppressed (by the injection of antisense Wnt11 oligonucleotides into the oocytes), the organizer fails to form. Furthermore, *Wnt11* mRNA is localized to the vegetal cortex during oogenesis and is translocated to the future dorsal portion of the embryo by the cortical rotation of the egg cytoplasm (Tao et al. 2005; Cuykendall and Houston 2009). Here it is translated into a protein that becomes concentrated in and secreted on the dorsal side of the embryo (Ku and Melton 1993; Schroeder et al. 1999; White and Heasman 2008). Thus, during first cleavage, GBP, Dsh, and Wnt11 are brought into the future dorsal section of the embryo, where GBP and Dsh can *initiate* the inactivation of GSK3 and the consequent protection of β-catenin. The signal from Wnt11 amplifies the signal and *stabilizes* GBP and Dsh and organizes them to protect β-catenin; β-catenin can associate with other transcription factors, giving these factors new properties.

FURTHER DEVELOPMENT

THE DOWNSTREAM TARGETS OF β-CATENIN *Xenopus* β-catenin is most well-known to combine with Tcf3, a ubiquitous transcription factor, resulting in the conversion of Tcf3 from a repressor into an activator of transcription. Expression of a mutant form of Tcf3 that lacks the β-catenin binding domain results in embryos without dorsal structures (Molenaar et al. 1996). The β-catenin/Tcf3 complex binds to the promoters of several genes critical for axis formation. For example, *twin* and *siamois* encode homeodomain transcription factors and are expressed in the organizer region immediately following the mid-blastula transition. If these genes are ectopically expressed in the ventral cells, a secondary axis emerges on the former ventral side of the embryo; and if cortical microtubular polymerization is prevented, *siamois* expression is eliminated (Lemaire et al. 1995; Brannon and Kimelman 1996). The Tcf3 protein is thought to inhibit *siamois* and *twin* transcription when it binds to those genes' promoters in the absence of β-catenin. However, when β-catenin binds to Tcf3, the repressor is converted into an activator, and *twin* and *siamois* are transcribed (**FIGURE 11.16**).

The proteins Siamois and Twin bind to the enhancers of several genes involved in organizer function (Fan and Sokol 1997; Bae et al. 2011). These include genes encoding the transcription factors Goosecoid and Xlim1 (which are critical in specifying the dorsal mesoderm) and the paracrine factor antagonists Noggin, Chordin, Frzb, and Cerberus (which specify the ectoderm to become neural; Laurent et al. 1997; Engleka and Kessler 2001). In the vegetal cells, Siamois and Twin appear to combine with vegetal transcription factors to help activate endodermal genes (Lemaire et al. 1995). Thus, one could expect that if the dorsal side of the embryo contained β-catenin, this β-catenin would allow the region to express Twin and Siamois, which in turn would initiate formation of the organizer.

THE DORSAL SIGNAL, PART 3: SYNERGIZING WITH VEGETAL SIGNALS The phosphorylated Smad2 transcription factor is not only essential for forming different mesoderm cell fates along the dorsal-ventral axis (as discussed earlier), but also critical for activating the genes that characterize the organizer itself. Smad2 is activated in the mesodermal cells when it becomes phosphorylated in response to Nodal-related paracrine factors secreted by the vegetal cells beneath the mesoderm (Brannon and Kimelman 1996; Engleka and Kessler 2001). Importantly, Nodal-related proteins are shaped

---

(?) Developing Questions

It is clear that β-catenin has many transcriptional targets of its own, but what if β-catenin could influence an entire "dorsalizing" network of genes? What if it had a grander chromatin organizing power? If you are intrigued, consider checking out the work by Blythe et al. 2010.

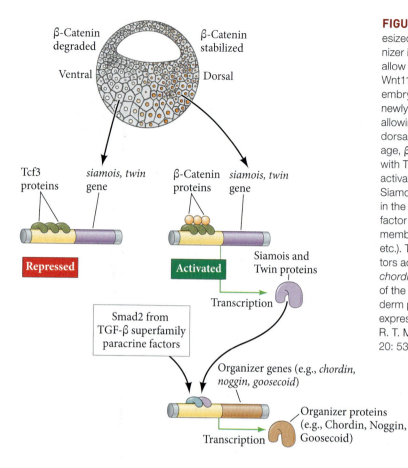

FIGURE 11.16 Summary of events hypothesized to bring about induction of the organizer in the dorsal mesoderm. Microtubules allow the translocation of Disheveled and Wnt11 proteins to the dorsal side of the embryo. Dsh (from the vegetal cortex and newly made by Wnt11) binds GSK3, thereby allowing β-catenin to accumulate in the future dorsal portion of the embryo. During cleavage, β-catenin enters the nuclei and binds with Tcf3 to form a transcription factor that activates genes encoding proteins such as Siamois and Twin. Siamois and Twin interact in the organizer with the Smad2 transcription factor activated by vegetal TGF-β superfamily members (Nodal-related proteins, Vg1, activin, etc.). Together, these three transcription factors activate the "organizer" genes, such as *chordin, noggin,* and *goosecoid*. The presence of the VegT transcription factor in the endoderm prevents the organizer genes from being expressed outside the organizer area. (After R. T. Moon and D. Kimelman. 1998. *BioEssays* 20: 536–545.)

into a gradient across the endoderm, with low concentrations ventrally and high concentrations dorsally (Onuma et al. 2002; Rex et al. 2002; Chea et al. 2005). Vg1 and the Nodal-related proteins act together to produce an additive activation of Smad2, and the graded distribution of Nodal-related proteins can yield differential transcriptional responses (Agius et al. 2000). The question then becomes, How is this dorsal-to-ventral gradient of Nodal-related proteins established?

The Nodal-related gradient is produced in large part by β-catenin. High β-catenin levels activate Nodal-related gene expression (**FIGURE 11.17**). In the dorsalmost (Nieuwkoop center) blastomeres, β-catenin cooperates with the VegT transcription factor to activate the *Xenopus nodal-related 1, 5,* and *6 (Xnr1, 5,* and *6)* genes even before the midblastula transition. The more ventral blastomeres in the endoderm lack the expression of these Nodal-related genes. In the region that will become the anteriormost portion of the organizer—the pharyngeal endoderm—higher levels of Nodal-related proteins produce higher concentrations of activated Smad2. Smad2 can bind to the promoter of the *hhex* gene, and in concert with Twin and Siamois (induced by β-catenin), Hhex activates genes that specify pharyngeal endoderm cells to become foregut endoderm and to induce anterior brain development (Smithers and Jones 2002; Rankin et al. 2011). Slightly lower levels of Smad2 are believed to activate *goosecoid* expression in the cells that will become the prechordal mesoderm and notochord. Even lower amounts of Smad2 result in the formation of lateral and ventral mesoderm.

In summary, then, the formation of the dorsal mesoderm and the organizer originates through the activation of critical transcription factors by intersecting and synergizing pathways. The first pathway is the Wnt/β-catenin pathway, which activates genes encoding the Siamois and Twin transcription factors. The second pathway is the vegetal pathway, which activates the expression of Nodal-related paracrine factors, which in turn activate the Smad2 transcription factor in the overlying mesodermal cells. The high levels of Smad2 and Siamois/Twin transcription factor proteins work within

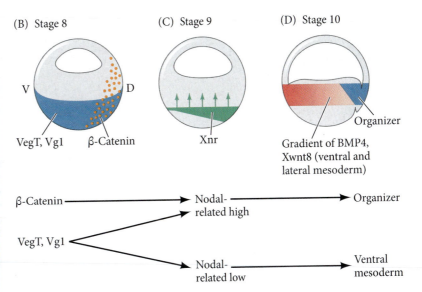

FIGURE 11.17 Vegetal induction of mesoderm. (A) The maternal RNA encoding Vg1 (bright white crescent) is tethered to the vegetal cortex of a *Xenopus* oocyte. The message (along with the maternal VegT message) will be translated at fertilization. Both proteins appear to be crucial for the ability of vegetal cells to induce cells above them to become mesoderm. (B–D) Model for mesoderm induction and organizer formation by the interaction of β-catenin and TGF-β proteins. (B) At late blastula stages, Vg1 and VegT are found in the vegetal hemisphere; β-catenin is located in the dorsal region. (C) β-Catenin acts synergistically with Vg1 and VegT to activate the *Xenopus nodal-related* (*Xnr*) genes. This creates a gradient of Xnr proteins across the endoderm, highest in the dorsal region. (D) The mesoderm is specified by the Xnr gradient. Mesodermal regions with little or no Xnr have high levels of BMP4 and Xwnt8; they become ventral mesoderm. Those having intermediate concentrations of Xnr become lateral mesoderm. Where there is a high concentration of Xnr, *goosecoid* and other dorsal mesodermal genes are activated and the mesodermal tissue becomes the organizer. (B–D after E. Agius et al. 2000. *Development* 127: 1173–1183.)

the dorsal mesoderm cells and activate the genes that give these cells their "organizer" properties (Germain et al. 2000; Cho 2012; review Figures 11.15–11.17).

Functions of the organizer

While the Nieuwkoop center cells remain endodermal, the cells of the organizer become the dorsal mesoderm and migrate inward to take up a position underneath the dorsal ectoderm. The cells of the organizer ultimately contribute to four cell types: pharyngeal endoderm, head mesoderm (prechordal plate), dorsal mesoderm (primarily the notochord), and the dorsal blastopore lip (Keller 1976; Gont et al. 1993). The pharyngeal endoderm and prechordal plate lead the migration of the organizer tissue and induce the forebrain and midbrain. The dorsal mesoderm induces the hindbrain and trunk. The dorsal blastopore lip remaining at the end of gastrulation eventually becomes the chordaneural hinge that induces the tip of the tail. The properties of the organizer tissue can be divided into four major functions:

1. The ability to self-differentiate into dorsal mesoderm (prechordal plate, chordamesoderm, etc.)

2. The ability to dorsalize the surrounding mesoderm into paraxial (somite-forming) mesoderm when it would otherwise form ventral mesoderm

3. The ability to dorsalize the ectoderm and induce formation of the neural tube

4. The ability to initiate the movements of gastrulation

Induction of neural ectoderm and dorsal mesoderm: BMP inhibitors

Evidence from experimental embryology showed that one of the most critical properties of the organizer was its production of soluble factors. The evidence for such diffusible signals from the organizer came from several sources. First, Hans Holtfreter (1933) showed that if the notochord fails to migrate beneath the ectoderm, the ectoderm will not become neural tissue (and will become epidermis). More definitive evidence for the importance of soluble factors came later from the transfilter studies of Finnish investigators (Saxén 1961; Toivonen et al. 1975; Toivonen and Wartiovaara 1976). Here, newt dorsal lip tissue was placed on one side of a filter fine enough that no cellular processes could fit through the pores, and competent gastrula ectoderm was placed on the other side. After several hours, neural structures were observed in the ectodermal tissue. The identities of the factors diffusing from the organizer, however, took another quarter of a century to find.

It turned out that the mechanism by which the organizer "induces" neural development was not intuitive. Scientists were searching for a molecule secreted

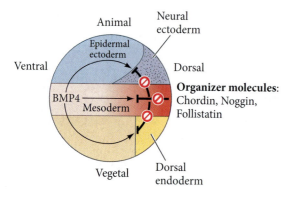

FIGURE 11.18 Model for the action of the organizer. BMP4 (along with certain other molecules) is a powerful ventralizing factor. Organizer proteins such as Chordin, Noggin, and Follistatin block the action of BMP4; their inhibitory effects can be seen in all three germ layers. (After E. De Robertis and Y. Sasai. 1996. *Nature* 380: 37–40 and Y. Sasai. et al. 1996. *EMBO J* 15: 4547–4555.)

by the organizer and received by the ectoderm that would then induce the ectoderm to become neural tissue. However, molecular studies led to a remarkable and not-so-obvious conclusion: *it is the epidermis (and ventral mesoderm) that is induced to form, not the neural tissue.* The ectoderm is induced to become epidermal tissue by binding **bone morphogenetic proteins** (**BMPs**), whereas the nervous system forms from that region of the ectoderm that is *protected* from epidermal induction by BMP-inhibiting molecules (Hemmati-Brivanlou and Melton 1994, 1997). In other words, (1) the "default fate" of the ectoderm is to become neural tissue; (2) certain parts of the embryo induce the ectoderm to become epidermal tissue by secreting BMPs; and (3) the organizer tissue acts by secreting molecules that block BMPs, thereby allowing the ectoderm "protected" by these BMP inhibitors to become neural tissue.

Thus, BMPs induce naïve ectodermal cells to become epidermal, while the organizer produces substances that block this induction (Wilson and Hemmati-Brivanlou 1995; Piccolo et al. 1996; Zimmerman et al. 1996; Iemura et al. 1998). In *Xenopus*, the major epidermal inducers are BMP4 and its close relatives BMP2, BMP7, and ADMP. Initially, BMPs such as BMP4 are expressed throughout the ectodermal and mesodermal regions of the late blastula. However, during gastrulation, transcription factors (such as Goosecoid) induced by Siamois and Twin prevent the transcription of *bmp4* in the dorsal region of the embryo, restricting their expression to the ventrolateral marginal zone (Blitz and Cho 1995; Hemmati-Brivanlou and Thomsen 1995; Northrop et al. 1995; Steinbeisser et al. 1995; Yao and Kessler 2001). In the ectoderm, BMPs repress the genes (such as *sox3* [Rogers et al. 2008, 2009a,b], *foxD4*, and *neurogenin*) involved in forming neural tissue, while activating other genes involved in epidermal specification (Lee et al. 1995). In the mesoderm, it appears that graded levels of BMP4 activate different sets of mesodermal genes: an absence of BMP4 specifies the dorsal mesoderm; a low amount specifies the intermediate mesoderm; and a high amount specifies the ventral mesoderm (**FIGURE 11.18**; Gawantka et al. 1995; Hemmati-Brivanlou and Thomsen 1995; Dosch et al. 1997).

The organizer acts by blocking the BMPs. Three of the major BMP inhibitors secreted by the organizer are Noggin, Chordin, and Follistatin. The genes encoding these proteins are some of the most critical genes activated by Smad2 and Siamois/Twin (Carnac et al. 1996; Fan and Sokol 1997; Kessler 1997). A fourth BMP inhibitor, Norrin, appears to be stored in the animal hemisphere of the oocyte and functions to block BMPs in the dorsal ectoderm (Xu et al. 2012).

FURTHER DEVELOPMENT

The organizer, a whole noggin of BMP antagonists!

In 1992, Smith and Harland constructed a cDNA plasmid library from dorsalized (lithium chloride-treated) *Xenopus* gastrulae. Messenger RNAs synthesized from sets of these plasmids were injected into ventralized embryos (having no neural tube) that were produced by irradiating early embryos with ultraviolet light. Those plasmid sets whose mRNAs rescued dorsal structures in these embryos were split into smaller sets, and so on, until single-plasmid clones were isolated whose mRNAs were able to restore the dorsal tissue in such embryos.

(A)

UV ventralized

Concentration of Noggin

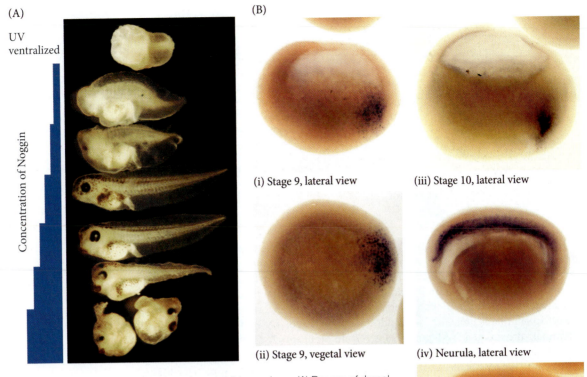

(B)

(i) Stage 9, lateral view

(iii) Stage 10, lateral view

(ii) Stage 9, vegetal view

(iv) Neurula, lateral view

(v) Neurula, dorsal view

All images courtesy of R. M. Harland

FIGURE 11.19   The soluble protein Noggin dorsalizes the amphibian embryo. (A) Rescue of dorsal structures by Noggin protein. When *Xenopus* eggs are exposed to ultraviolet radiation, cortical rotation fails to occur, and the embryos lack dorsal structures (top). If such an embryo is injected with *noggin* mRNA, it develops dorsal structures in a dosage-related fashion (top to bottom). If too much *noggin* message is injected, the embryo produces dorsal and anterior tissue at the expense of ventral and posterior tissue, becoming little more than a head (bottom). (B) Localization of *noggin* mRNA in the organizer tissue, shown by in situ hybridization. At gastrulation (i and ii, stage 9), *noggin* mRNA (dark areas) accumulates in the dorsal marginal zone. When cells involute (i and iii, stages 9 and 10), *noggin* mRNA is seen in the dorsal blastopore lip. During convergent extension (iii, stage 10), *noggin* is expressed in the precursors of the notochord, prechordal plate, and pharyngeal endoderm, which in the neurula (iv, v) extend beneath the ectoderm in the center of the embryo.

NOGGIN   One of these clones contained the gene for the protein Noggin (**FIGURE 11.19A**). Injection of *noggin* mRNA into UV-irradiated, 1-cell embryos completely rescued dorsal development and allowed the formation of a complete embryo (Lamb et al. 1993; Smith et al. 1993). Noggin is a secreted protein that is able to accomplish two of the major functions of the organizer: it induces dorsal ectoderm to form neural tissue, and it dorsalizes mesoderm cells that would otherwise contribute to the ventral mesoderm (Smith et al. 1993). Smith and Harland showed that newly transcribed *noggin* mRNA is first localized in the dorsal blastopore lip region and then becomes expressed in the notochord (**FIGURE 11.19B**). Noggin binds to BMP4 and BMP2 and inhibits their binding to receptors (Zimmerman et al. 1996).

CHORDIN   Chordin protein was isolated from clones of cDNA whose mRNAs were present in dorsalized, but not in ventralized, embryos (Sasai et al. 1994). These cDNA clones were tested by injecting them into ventral blastomeres and seeing whether they induced secondary axes. One of the clones capable of inducing a secondary neural tube contained the *chordin* gene; *chordin* mRNA was found to be localized in the dorsal blastopore lip and later in the notochord (**FIGURE 11.20**). Morpholino antisense oligomers directed against the *chordin* message blocked the ability of an organizer graft to induce a secondary central nervous system (Oelgeschläger et al. 2003). Of all organizer genes observed, *chordin* is the one most acutely activated by β-catenin (Wessely et al. 2004). Like Noggin, Chordin binds directly to BMP4 and BMP2 and prevents their complexing with their receptors (Piccolo et al. 1996).

(A)

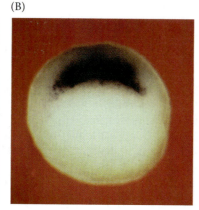
(B)

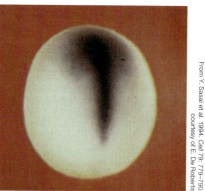
(C)

From Y. Sasai et al. 1994, *Cell* 79: 779–790, courtesy of E. De Robertis

FIGURE 11.20 Localization of *chordin* mRNA. (A) Whole mount in situ hybridization shows that just prior to gastrulation, *chordin* mRNA (dark area) is expressed in the region that will become the dorsal blastopore lip. (B) As gastrulation begins, *chordin* is expressed at the dorsal blastopore lip. (C) In later stages of gastrulation, the *chordin* message is seen in the organizer tissues.

FOLLISTATIN The mRNA for a third organizer-secreted protein, Follistatin, is also transcribed in the dorsal blastopore lip and notochord. Follistatin was found in the organizer as an unexpected result of an experiment that was looking for something else. Ali Hemmati-Brivanlou and Douglas Melton (1992, 1994) wanted to see whether the protein activin was required for mesoderm induction. In searching for the mesoderm inducer, they found that Follistatin, an inhibitor of both activin and BMPs, caused ectoderm to become neural tissue. They then proposed that under normal conditions, ectoderm becomes neural unless induced to become epidermal by the BMPs. This model was supported by, and explained, certain cell dissociation experiments that had also produced odd results. Three 1989 studies—by Grunz and Tacke, Sato and Sargent, and Godsave and Slack—showed that when whole embryos or their animal caps were dissociated, they formed neural tissue. This result would be explainable if the "default state" of the ectoderm was not epidermal but neural, so that tissue had to be induced to have an epidermal phenotype. Thus, we conclude that the organizer blocks epidermalizing induction by inactivating BMPs.

ECTODERMAL BIAS The ectoderm above the notochord also appears to have been biased to become neural ectoderm by β-catenin that extended into the periphery of the egg. This causes the expression of Siamois and Twin proteins in the cells that will become the neural ectoderm. Here, these transcription factors perform two critical functions. First, they activate those genes (such as *foxD4* and *sox11*) that will permit these cells to become neural ectoderm (**FIGURE 11.21**). However, these genes can be suppressed by BMPs, so as a second step during gastrulation, the organizer mesoderm produces proteins that block the BMP signal from reaching the ectoderm (Klein and Moody 2015). (See **Further Development 11.8, The Experiments That Confirmed Ectodermal Bias**, online.)

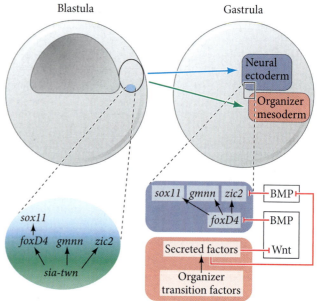

FIGURE 11.21 Schematic diagram of Siamois (Sia) and Twin (Twn) inducing the activation of the neuroepithelium genes. During the blastula stage, cells expected to give rise to both organizer mesoderm and the neural ectoderm express both *sia* and *twn*. The products of these genes activate the neuroectoderm genes *foxD4*, *gmnn*, and *zic2*. These genes encode transcription factors that will activate other neural genes, such as *sox11*. During the gastrula stage, the descendants of these cells have become the organizer mesoderm and the neural ectoderm. Here, the neuroepithelial genes are upregulated by the factors secreted by the organizer that inhibit the BMP and Wnt pathways. If BMPs and Wnts are not blocked, *sox11*, *gmnn*, *foxD4*, and *zic2* transcription declines. (After S. L. Klein and S. A. Moody. 2015. *Genesis* 53: 308–320.)

(?) Developing Questions

Chordin and BMPs seem to be homologous between flies and vertebrates—but are the *processing pathways* of Chordin and BMPs homologous as well? How might the Chordin-BMP axis allow the regulation of the embryonic axes such that the same pattern always occurs, even if the embryo is much smaller or larger?

Conservation of BMP signaling during dorsal-ventral patterning

Remarkably, the ability of BMPs to induce the skin ectoderm and of BMP antagonists to specify the neural ectoderm has been seen across the animal kingdom. In *Drosophila*, the BMP homologue Decapentaplegic (Dpp) specifies the hypodermis (skin), while the BMP antagonist Short gastrulation (Sog) blocks the actions of Dpp and specifies the neural system. Sog protein is a homologue of Chordin. These insect homologues not only appear to be similar to their vertebrate counterparts, but can actually substitute for each other. When *sog* mRNA is injected into ventral regions of *Xenopus* embryos, it induces the amphibian notochord and neural tube. Injecting *chordin* mRNA into *Drosophila* embryos produces ventral nervous tissue. Although Chordin dorsalizes the *Xenopus* embryo, it ventralizes *Drosophila*. In *Drosophila*, Dpp is made dorsally; in *Xenopus*, BMP4 is made ventrally. In both cases, Sog/Chordin helps specify neural tissue by blocking the effects of Dpp/BMP4 (Hawley et al. 1995; Holley et al. 1995; De Robertis et al. 2000; Bier and De Robertis 2015). Thus, arthropods appear to be upside-down vertebrates—an observation French anatomist Geoffroy Saint-Hilaire pointed out in his attempts to convince other anatomists of the unity of the animal kingdom in the 1840s (see Appel 1987; Genikhovich et al. 2015; De Robertis and Moriyama 2016).

Regional Specificity of Neural Induction along the Anterior-Posterior Axis

Just as the dorsal-ventral axis across the animal kingdom is predicated on BMP and its inhibitors (the neural region being the area of lowest BMPs), the specification of the anterior-posterior axis is predicated on a gradient of Wnt proteins, with the head being characterized by the lowest concentrations of Wnts (Petersen and Reddien 2009). There are a few exceptions (such as *Drosophila*) where the Wnt gradient does not provide major patterning cues; but even in these cases, vestigial patterns of Wnt are still seen (Vorwald-Denholtz and De Robertis 2011).

In vertebrates, one of the most important phenomena along the anterior-posterior axis is the regional specificity of the neural structures that are produced. Forebrain, hindbrain, and spinocaudal regions of the neural tube must be properly organized in an anterior-to-posterior direction. The organizer tissue not only induces the neural tube, it also specifies the regions of the neural tube. This region-specific induction was demonstrated by Hilde Mangold's husband, Otto Mangold, in 1933. He removed four successive regions of the archenteron roof of late-gastrula newt embryos—the archenteron roof is where the organizer tissue is located in these embryos—and transplanted the individual pieces into the blastocoels of early-gastrula embryos. The anteriormost portion of the archenteron roof (containing head mesoderm) induced balancers and portions of the oral apparatus; the next most anterior section induced the formation of various head structures, including nose, eyes, balancers, and otic vesicles; the third section (including the notochord) induced the hindbrain structures; and the posteriormost section induced the formation of dorsal trunk and tail mesoderm[2] (**FIGURE 11.22A–D**).

In further experiments, Mangold demonstrated that when dorsal blastopore lips from early salamander gastrulae were transplanted into other *early* salamander gastrulae, the transplant induced the formation of secondary heads. When dorsal lips from *later* gastrulae were transplanted into early salamander gastrulae, however, the transplant induced the formation of secondary tails (**FIGURE 11.22E,F**; Mangold 1933). These results show that the first cells of the organizer to enter the embryo induce the formation of brains and heads, while those cells that form the dorsal lip of later-stage embryos induce the cells above them to become spinal cords and tails.

The question then became, What are the molecules being secreted by the organizer in a regional fashion such that the first cells involuting through the blastopore lip (the endomesoderm) induce head structures, whereas the next portion of involuting mesoderm (notochord) produces trunk and tail structures? **FIGURE 11.23** shows a possible model for these inductions, the elements of which we describe in the following "Further Development."

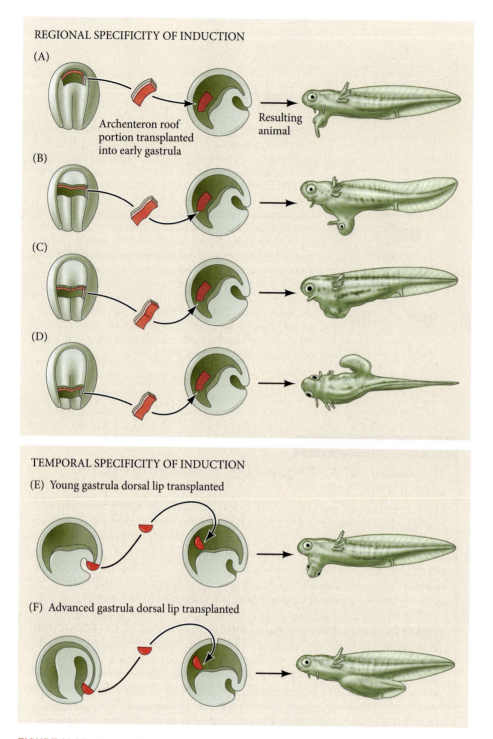

REGIONAL SPECIFICITY OF INDUCTION

(A)

Archenteron roof portion transplanted into early gastrula

Resulting animal

(B)

(C)

(D)

TEMPORAL SPECIFICITY OF INDUCTION

(E) Young gastrula dorsal lip transplanted

(F) Advanced gastrula dorsal lip transplanted

FIGURE 11.22   Regional and temporal specificity of induction. (A–D) Regional specificity of structural induction can be demonstrated by implanting different regions (color) of the archenteron roof (where the organizer tissue is located) into early *Triturus* gastrulae. The resulting embryos develop secondary dorsal structures. (A) Head with balancers. (B) Head with balancers, eyes, and forebrain. (C) Posterior part of head, diencephalon, and otic vesicles. (D) Trunk-tail segment. (E,F) Temporal specificity of inducing ability. (E) Young dorsal lips (which will form the anterior portion of the organizer) induce anterior dorsal structures when transplanted into early newt gastrulae. (F) Older dorsal lips transplanted into early newt gastrulae produce more posterior dorsal structures. (A–D after O. Mangold. 1933. *Naturwissenschaften* 21: 761–766; E,F after L. Saxén and S. Toivonen. 1962. *Primary Embryonic Induction*. Prentice-Hall, Englewood Cliffs, NJ.)

(A)

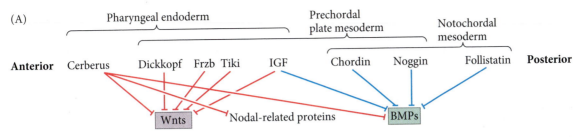

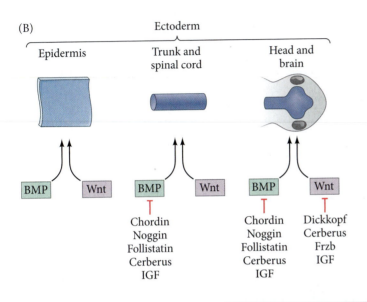

(B)

FIGURE 11.23 Paracrine factor antagonists from the organizer are able to block specific paracrine factors to distinguish head from tail. (A) The pharyngeal endoderm that underlies the head secretes Dickkopf, Frzb, and Cerberus, among others. Dickkopf and Frzb block Wnt proteins; Cerberus blocks Wnts, Nodal-related proteins, and BMPs. The prechordal plate secretes the Wnt blockers Dickkopf and Frzb, as well as BMP blockers Chordin and Noggin. The notochord contains the BMP blockers Chordin, Noggin, and Follistatin but does not secrete Wnt blockers. Insulin-like growth factor (IGF) from the head endomesoderm probably acts at the junction of the notochord and prechordal mesoderm. (B) Summary of paracrine antagonist function in the ectoderm. Brain formation requires inhibiting both the Wnt and the BMP pathways. Spinal cord neurons are produced when Wnt functions without the presence of BMPs. Epidermis is formed when both the Wnt and the BMP pathways are operating.

FURTHER DEVELOPMENT

The head inducer: Wnts and BMPs wage mythic battles of heads and tails

The anteriormost regions of the head and brain are underlain not by notochord but by pharyngeal endoderm and head (prechordal) mesoderm (see Figures 11.6C,D and 11.23A). This endomesodermal tissue constitutes the leading edge of the dorsal blastopore lip. Remarkably, these cells not only induce the anteriormost head structures, but do so by blocking the Wnt pathway as well as by blocking BMPs. The Wnt antagonists appear to be induced by the high levels of phosphorylated Smad2 in response to Nodal and Vg1 secreted by the vegetal cells (Agius et al. 2000; Birsoy et al. 2006).

CERBERUS: AN ALL-PURPOSE PARACRINE INHIBITOR FOR HEAD PRODUCTION
The induction of trunk structures may be caused by the blockade of BMP signaling from the notochord, while Wnt signals are allowed to proceed. However, to produce a head, both the BMP signal and the Wnt signal must be blocked. This dual functioning blockade comes from the endomesoderm, the anteriormost portion of the organizer (Glinka et al. 1997). In 1996, Bouwmeester and colleagues showed that the induction of the anteriormost head structures could be accomplished by a secreted protein called Cerberus (named after the three-headed dog that guarded the entrance to Hades in Greek mythology). When *cerberus* mRNA was injected into a vegetal ventral *Xenopus* blastomere at the 32-cell stage, ectopic head structures were formed. These head structures arose from the injected cell as well as from neighboring cells.

The *cerberus* gene is expressed in the pharyngeal endomesoderm cells that arise from the deep cells of the early dorsal lip. Cerberus protein can bind BMPs, Nodal-related proteins, and Xwnt8 (see Figure 11.23A and 11.25; Piccolo et al. 1999). When Cerberus synthesis is blocked, the levels of BMP, Nodal-related proteins, and Wnts all rise in the anterior of the embryo, and the ability of the anterior endomesoderm to induce a head is severely diminished (Silva et al. 2003). (See **Further Development 11.9, Frzb, Dickkopf, Notum, and Tiki: More Ways to Block Wnts**, online.)

INSULIN-LIKE GROWTH FACTORS AND FIBROBLAST GROWTH FACTORS

All the Wnt inhibitors we have discussed so far have been extracellular. In addition to these, the head region contains yet another set of proteins that prevent BMP and Wnt signals from reaching the nucleus. **Fibroblast growth factors (FGFs)** and **insulin-like growth factors (IGFs)** are also required for inducing the brain and sensory placodes (Pera et al. 2001, 2014). IGFs and FGFs are especially prominent in the anterior region of the embryo, and both initiate the receptor tyrosine kinase (RTK) signal transduction cascade (see Chapter 4). These tyrosine kinases interfere with the signal transduction pathways of both BMPs and Wnts (Richard-Parpaillon et al. 2002; Pera et al. 2014). When injected into ventral mesodermal blastomeres, mRNA for IGFs causes the formation of ectopic heads, while blocking IGF receptors in the anterior results in the lack of head formation. Last, as you will learn in Chapters 17, 18, and 19, FGFs also play direct roles in the induction and maintenance of genes important for mesodermal lineages and limb outgrowth, often working similarly in concert with Wnt and BMP signals.

TRUNK PATTERNING: WNT SIGNALS AND RETINOIC ACID
Toivonen and Saxén provided evidence for a gradient of a posteriorizing factor that would act to specify the trunk and tail tissues of the amphibian embryo (Toivonen and Saxén 1955, 1968; reviewed in Saxén 2001).[3] This factor's activity would be highest in the posterior of the embryo and would weaken anteriorly. Recent studies have extended this model and have proposed that Wnt proteins, especially Wnt8, are the posteriorizing molecules (Domingos et al. 2001; Kiecker and Niehrs 2001). In the *Xenopus* gastrula, an endogenous gradient of Wnt signaling and β-catenin is highest in the posterior and absent in the anterior (**FIGURE 11.24A**). Moreover, if Xwnt8 is added to developing embryos, spinal cord-like neurons are seen more anteriorly in the embryo, and the anteriormost markers of the forebrain are absent. Conversely, suppressing Wnt signaling (by adding Frzb or Dickkopf to the developing embryo) leads to the expression of the anteriormost markers in more posterior neural cells. While the Wnt proteins play a major role in specifying the anterior-posterior axis, they are probably not the only agents involved. In fact, posterior-to-anterior Wnt, FGF, and retinoic acid (RA) gradients function to determine the boundaries of the Hox genes along the anterior-posterior axis (Wacker et al. 2004; Durston et al. 2010a,b). We will discuss these mechanisms further in Chapter 17.

In summary, there appear to be two major gradients in the amphibian gastrula—a BMP gradient that specifies the dorsal-ventral axis and a Wnt gradient specifying the anterior-posterior axis (**FIGURE 11.24B**). It must be remembered, too, that both of these axes are established by the initial axes of Nodal-like TGF-β paracrine factors and β-catenin across the vegetal cells. The basic model of neural induction, then, looks like the diagram in **FIGURE 11.25**. (See **Further Development 11.10, Gradients and Hox Gene Expression in *Xenopus***, online.)

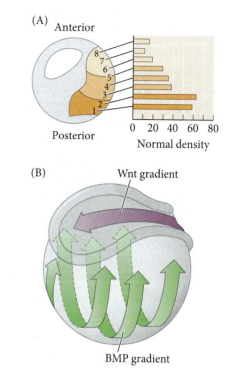

FIGURE 11.24 Signaling gradients and axis specification. (A) A Wnt signaling pathway posteriorizes the neural tube. Gastrulating embryos were stained for β-catenin and the density of the stain compared between regions of the ectodermal cells, revealing a gradient of β-catenin in the presumptive neural plate. (B) The Wnt gradient specifies posterior-anterior polarity, and the BMP gradient specifies dorsal-ventral polarity. This double-gradient interaction, first discovered in amphibians, has now been shown to be characteristic of animal development. (A after C. Kiecker and C. Niehrs. 2001. *Development* 128: 4189–4201; B after C. Niehrs. 2004. *Nat Rev Genet* 5: 425–434.)

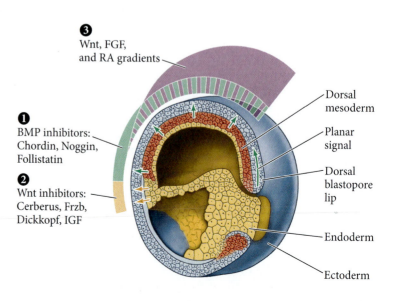

FIGURE 11.25 Model of organizer function and axis specification in the *Xenopus* gastrula. (1) BMP inhibitors from organizer tissue (dorsal mesoderm and pharyngeal mesendoderm) block the formation of epidermis, ventrolateral mesoderm, and ventrolateral endoderm. (2) Wnt inhibitors in the anterior of the organizer (pharyngeal endomesoderm) allow the induction of head structures. (3) A gradient of caudalizing factors (Wnts, FGFs, and retinoic acid) results in the regional expression of Hox genes, which specify the regions of the neural tube. (After R. E. Keller. 1986. In *Developmental Biology: A Comprehensive Synthesis*, Vol. 2, L. Browder (Ed.), pp. 241–327. Plenum: New York.)

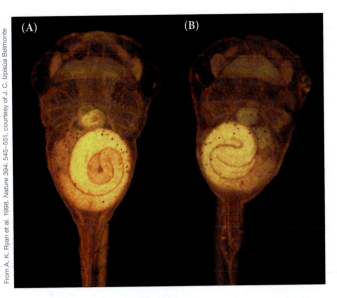

FIGURE 11.26 Pitx2 determines the direction of heart looping and gut coiling. (A) A wild-type *Xenopus* tadpole viewed from the ventral side, showing rightward heart looping and counterclockwise gut coiling. (B) If an embryo is injected with Pitx2 so that this protein is present in the mesoderm of both the right and left sides (instead of just the left side), heart looping and gut coiling are random with respect to each other. Sometimes this treatment results in complete reversals, as in this embryo, in which the heart loops to the left and the gut coils in a clockwise manner.

(?) Developing Questions

How are the cilia of the left-right organizer determined? A recent study suggests that this mechanism of morphogen dispersal is itself created by a differential mechanical force across the gastrulating embryo's anterior-posterior axis. This new information may just strain your brain until it spins. See Chien et al. 2018.

Specifying the Left-Right Axis

Although the developing tadpole outwardly appears symmetrical, several internal organs, such as the heart and the gut tube, are not evenly balanced on the right and left sides. In other words, in addition to its dorsal-ventral and anterior-posterior axes, the embryo has a left-right axis. In all vertebrates studied so far, the crucial event in left-right axis formation is the expression of a Nodal gene in the lateral plate mesoderm on the left side of the embryo. In *Xenopus*, this gene is *Xnr1* (*Xenopus nodal-related 1*). If *Xnr1* expression is reversed (so as to be solely on the right-hand side), the position of the heart (normally found on the left) is reversed, as is the coiling of the gut. If *Xnr1* is expressed on both sides, coiling and heart placement are random.

But what limits *Xnr1* expression to the left-hand side? In Chapter 8, we saw that left-right patterning in snails is controlled by Pitx2 and Nodal and is regulated by cytoskeletal proteins that are active during the first cleavage cycles. The case is similar in frogs, where during the first cleavages maternal deposition of Pitx2 and Nodal appears to be under the influence of the cytoskeleton. Early disruptions in tubulin-associated proteins cause left-right defects, such as randomizing the placement of the heart and coiling of the gut (Lobikin et al. 2012).

Although there are indications of left-right laterality in the early embryo, it is evident that left-right patterning requires a mechanical mechanism: cilia driving a leftward flow of extracellular factors, namely Nodal (Blum et al. 2014). In *Xenopus*, these specific cilia are formed at the dorsal blastopore lip during the later stages of gastrulation (i.e., after the original specification of the mesoderm); this specific ciliated region has been termed the left-right organizer (LRO; Schweickert et al. 2007; Blum et al. 2009). These cilia are located in the posterior region of the embryo, at the site where the archenteron is still forming, and perform a clockwise rotation that has been proposed to create a leftward current across the midline of the extending axis. If rotation of these cilia is blocked, *Xnr1* expression fails to occur in the mesoderm, and laterality defects result (Walentek et al. 2013).

One of the key genes activated by Xnr1 protein appears to encode the transcription factor Pitx2, which is normally expressed only on the left side of the embryo. Pitx2 persists on the left side as the heart and gut develop, controlling their respective positions. If *pitx2* is injected into the right side of an embryo, causing the protein product to be present on both left and right sides of the embryo, heart placement and gut coiling are randomized (**FIGURE 11.26**; Ryan et al. 1998). As we will see, the pathway through which Nodal protein establishes left-right polarity by activating Pitx2 on the left side is conserved throughout all vertebrate lineages.

Early Zebrafish Development

In recent years, the teleost (bony) fish *Danio rerio*, commonly known as the zebrafish, has joined *Xenopus* as a widely studied model of vertebrate development (see Figure 11.1B). Despite differences in their cleavage patterns (the *Xenopus* egg is holoblastic, dividing the entire egg, whereas the yolky zebrafish egg is meroblastic—only a small portion of its cytoplasm forms cells), *Xenopus* and *Danio* form their body axes and specify their cells in very similar ways.

Why zebrafish? The zebrafish satisfies many of the criteria for a great model system to study developmental biology (see Chapter 1, pp. 5–6). The adult fish is small, permitting thousands to be maintained in a small laboratory. Zebrafish breed all year and produce hundreds of externally fertilized embryos a week. As with *Xenopus*, developing outside the mother enables total accessibility for manipulation, yet unlike frog embryos, zebrafish embryos are transparent, facilitating the use of remarkable microscopic imaging techniques. In addition, these fish develop rapidly. By 24 hours after fertilization, the embryo has already formed most of its organ primordia and displays a characteristic tadpole-like form (**FIGURE 11.27**; see Granato and Nüsslein-Volhard 1996; Langeland

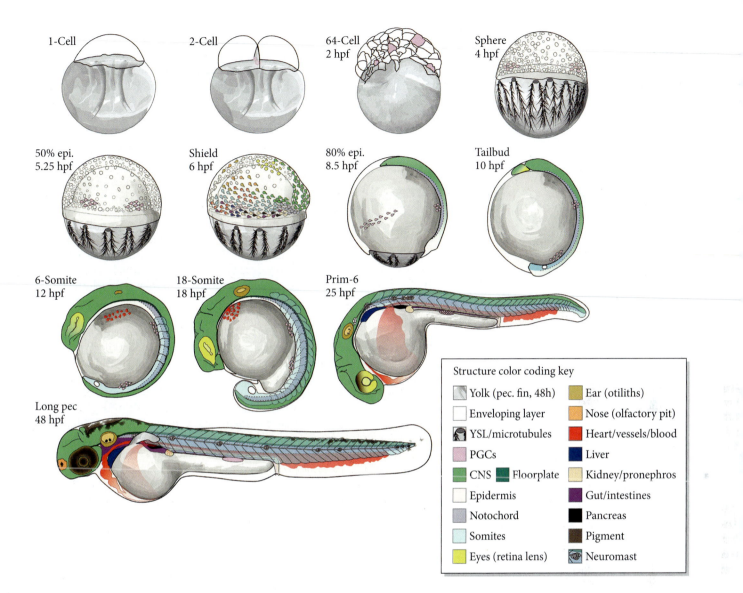

Structure color coding key

Yolk (pec. fin, 48h)		Ear (otiliths)	
Enveloping layer		Nose (olfactory pit)	
YSL/microtubules		Heart/vessels/blood	
PGCs		Liver	
CNS	Floorplate	Kidney/pronephros	
Epidermis		Gut/intestines	
Notochord		Pancreas	
Somites		Pigment	
Eyes (retina lens)		Neuromast	

FIGURE 11.27 Illustration of zebrafish embryogenesis over 48 hours. (hpf, hours postfertilization.) Illustrations in the top row are representative of the cleavage period during which cells get smaller with every division. Primordial germ plasm (pink) is evident at the 2-cell stage along the division plane, while four discrete primordial germ cells (PGCs) can be seen at the 64-cell stage. These four PGCs establish four multi-cell clusters (two of which are shown) by the sphere stage. At the sphere stage, the enveloping layer covers the animal hemisphere, stretching down to the yolk syncytial layer (YSL), where yolk syncytial nuclei and microtubules extend toward the vegetal pole. The second row of images shows gastrulation, with the vegetal microtubule arrays helping draw down the embryonic cells from 50% epiboly to 80%, and finally full closure of the blastopore at the tailbud stage. The dorsal-ventral axis is first visible at the shield stage, where a convergence of cells on the midline produces a visible budge of cells. Dorsal is to the right of each embryo. Individual cells and their trajectory of movement are mapped at the shield stage in locations and colors indicative of their future fates (see key). Kupffer's vesicle (white circular structure in tail) is an early sign of posterior growth and left-right patterning. Mesoderm induction during gastrulation gives rise to the notochord (light mauve, delineated with dotted lines) and somites through subdivision of the paraxial mesoderm (light blue block/chevron-shaped superficially positioned structures [transparently illustrated]). The PGC migration into the gonad is completed about 24 hpf. Placodal structures in the ectoderm form the eye (yellow), ear (honey color), and nose (orange). Development of the heart and circulatory system begins with bilaterally positioned cells that migrate over to the midline and converge to form the heart (only one population of cells is shown, in red), after which an elaborate vasculature system is created (only the dorsal aorta and vena cava with posterior blood island are shown). Development of the gut, kidney, liver, and pancreas are all illustrated in their approximate positions. The lateral line primordium emerges from the brain about 24 hpf and migrates to the tail atop the middle portion of the somites, depositing neuromasts along the way. 48 hpf is known as the long pec stage due to the presence of the extended pectoral fin (shown above the yolk, gray pattern). Although pigment can first be seen at 24.5 hpf in the dorsal region of the retina (eye, brown), body pigment increasingly appears between days 1 and 2 (only the dorsalmost melanocytes are shown, at 48 hpf). (Illustration by Michael J. F. Barresi © 2018.)

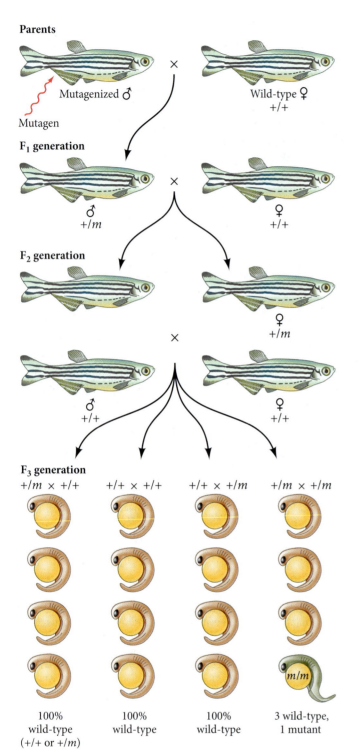

Parents

Mutagenized ♂

Mutagen

Wild-type ♀
+/+

F₁ generation

♂
+/m

♀
+/+

F₂ generation

♀
+/m

♂
+/+

♀
+/+

F₃ generation

+/m × +/+   +/+ × +/+   +/+ × +/m   +/m × +/m

100%
wild-type
(+/+ or +/m)

100%
wild-type

100%
wild-type

3 wild-type,
1 mutant

m/m

FIGURE 11.28 Screening protocol for identifying mutations of zebrafish development. The male parent is mutagenized and mated with a wild-type (+/+) female. If some of the male's sperm carry a recessive mutant allele (m), then some of the F₁ progeny of the mating will inherit that allele. F₁ individuals (here shown as a male carrying the mutant allele m) are then mated with wild-type partners. This creates an F₂ generation wherein some males and some females carry the recessive mutant allele. When the F₂ fish are mated, approximately 25% of their progeny will show the mutant phenotype. (After P. Haffter et al. 1996. *Development* 123: 1–36.)

and Kimmel 1997). Furthermore, the ability to microinject fluorescent dyes and nucleic acids into single blastomeres has enabled powerful fate mapping and gene manipulation to test gene function and to generate transgenic fish lines that display fluorescence in specific cell types.

Due to the large numbers of embryos produced in a single mating, the zebrafish was the first vertebrate to be studied through intensive mutagenesis screens. By treating parent fish with mutagens and selectively breeding the progeny, scientists have found thousands of mutations in genes whose normal functioning is critical for development. The traditional method of genetic screening (modeled after large-scale screens in *Drosophila*) begins with treating male parental fish with a chemical mutagen that will cause random mutations in their germ cells. Each mutagenized male is then mated with a wild-type female fish to generate F₁ fish. Individuals in the F₁ generation carry the mutations inherited from their father. If the mutation is dominant, it will be expressed in the F₁ generation. If the mutation is recessive, the F₁ fish will not show a mutant phenotype, since the wild-type dominant allele will mask the mutation. The F₁ fish are then mated with wild-type fish to produce an F₂ generation that includes both males and females that carry the mutant allele. When two F₂ parents carry the same recessive mutation, there is a 25% chance that their offspring will show the mutant phenotype (**FIGURE 11.28**). Since zebrafish development occurs in the open (as opposed to within an opaque shell or inside the mother's body), abnormal developmental stages can be readily observed, and the defects in development can often be traced to changes in a particular group of cells (Driever et al. 1996; Haffter et al. 1996). Recently, high-throughput methods of gene analysis and the CRISPR genome editing system have accelerated the analysis of zebrafish development, enabling mutations in particular genes to be rapidly generated, identified, and bred (see Gonzales and Yeh 2014; Varshney et al. 2015).

Zebrafish embryos are also permeable to small molecules placed in the water—a property that allows us to test drugs that may be deleterious to vertebrate development. For instance, zebrafish development can be altered by the addition of ethanol or retinoic acid, both of which produce malformations in the fish that resemble human developmental syndromes known to be caused by these molecules (Blader and Strähle 1998). This attribute has enabled both big pharmaceutical companies and small laboratories alike to conduct chemical screens of large libraries of drug compounds to determine their effects on embryonic and larval zebrafish. These efforts have already identified new therapies for human conditions. As one zebrafish researcher joked, "Fish really are just little people with fins" (Bradbury 2004).

Zebrafish Cleavages: Yolking Up the Process

The eggs of most bony fish are **telolecithal**, meaning that most of the cytoplasm is occupied by yolk. Cleavage can take place only in the **blastodisc**, a thin region of yolk-free cytoplasm at the animal pole. The cell divisions do not completely divide the egg, so this type of cleavage is called **meroblastic**

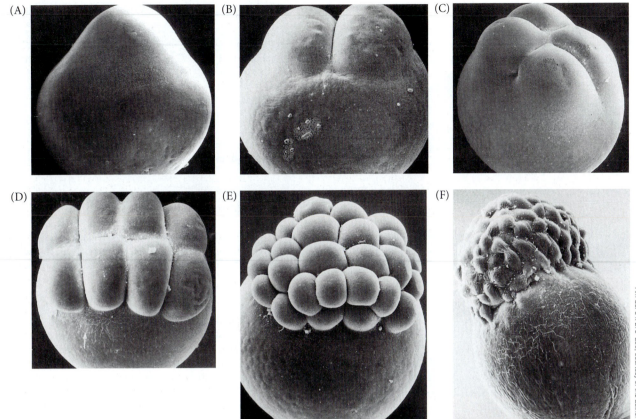

(A)   (B)   (C)

(D)   (E)   (F)

From H. W. Beams and R. G. Kessel. 1976. Am Sci 64: 279–290, courtesy of the authors

FIGURE 11.29 Discoidal meroblastic cleavage in a zebrafish egg. (A) 1-Cell embryo. The mound atop the cytoplasm is the blastodisc. (B) 2-Cell embryo. (C) 4-Cell embryo. (D) 8-Cell embryo, wherein two rows of four cells are formed. (E) 32-Cell embryo. (F) 64-Cell embryo, wherein the blastodisc can be seen atop the yolk cell.

(Greek *meros*, "part"). Since only the blastodisc becomes the embryo, this type of mero-blastic cleavage is referred to as **discoidal**.

Scanning electron micrographs show beautifully the incomplete nature of discoi-dal meroblastic cleavage in fish eggs (**FIGURE 11.29**). The calcium waves initiated at fertilization stimulate the contraction of the actin cytoskeleton to squeeze nonyolky cytoplasm into the animal pole of the egg. This process converts the spherical egg into a pear-shaped structure with an apical blastodisc (Leung et al. 1998, 2000). In fish, there are many waves of calcium release, and they orchestrate the processes of cell division. The calcium ions are critical for coordinating mitosis. They integrate the movements of the mitotic spindle with those of the actin cytoskeleton, deepen the cleavage furrow, and heal the membrane after the separation of the blastomeres (Lee et al. 2003).

The first cell divisions follow a highly reproducible pattern of meridional and equa-torial cleavages. These divisions are rapid, taking only about 20 minutes each. The first 10 divisions occur synchronously, forming a mound of cells that sits at the animal pole of a large **yolk cell**. This mound of cells constitutes the **blastoderm**. Initially, all the cells maintain some open connection with one another and with the underlying yolk cell, so that moderately sized (17 kDa) molecules can pass freely from one blastomere to the next (Kimmel and Law 1985; Kane and Kimmel 1993). Remarkably, as the daughter cells migrate away from one another, they often retain these bridges through long tunnels connecting the cells (Caneparo et al. 2011).

Maternal effect mutations have shown the importance of oocyte proteins and mRNAs in embryonic polarity, cell division, and axis formation (Dosch et al. 2004; Langdon and Mullins 2011). As in frogs, the microtubules are important roads along which cell fate specifying cytoplasmic determinants travel, and maternal mutants affecting the formation of the microtubule cytoskeleton prevent the normal position-ing of the cleavage furrow and of mRNAs in the early embryo (Kishimoto et al. 2004).

Fish embryos, like many other embryos, undergo a mid-blastula transition (seen around the tenth cell division in zebrafish) when zygotic gene transcription begins,

(A)

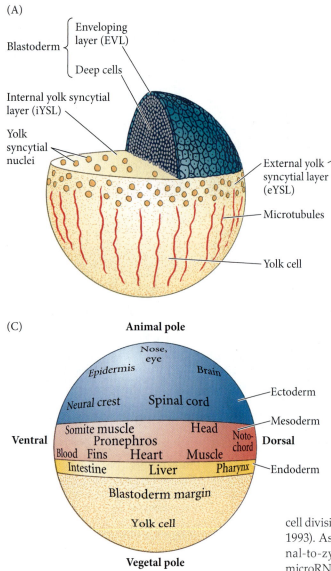

Blastoderm
{ Enveloping layer (EVL)
  Deep cells

Internal yolk syncytial layer (iYSL)

Yolk syncytial nuclei

External yolk syncytial layer (eYSL)

Microtubules

Yolk cell

(B)

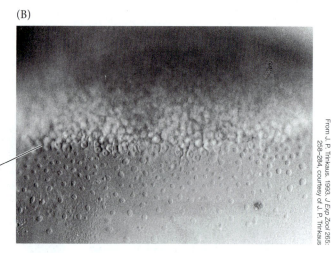

From J. P. Trinkaus. 1993. *J Exp Zool* 265: 258–284, courtesy of J. P. Trinkaus

FIGURE 11.30 Fish blastula. (A) Prior to gastrulation, the deep cells are surrounded by the enveloping layer (EVL). The animal surface of the yolk cell is flat and contains the nuclei of the yolk syncytial layer (YSL). Microtubules extend through the yolky cytoplasm and the external YSL. (B) Late-blastula-stage embryo of the minnow *Fundulus*, showing the external YSL. The nuclei of these cells were derived from cells at the margin of the blastoderm, which released their nuclei into the yolky cytoplasm. (C) Fate map of the deep cells after cell mixing has stopped. This is a lateral view; for the sake of clarity, not all organ fates are labeled. (After L. Solnica-Krezel and W. Driever. 1994. *Development* 120: 2443–2455; B from From J. P. Trinkaus. 1993. *J Exp Zool* 265: 258–284, courtesy of J. P. Trinkaus; C after C. B. Kimmel et al. 1995. *Dev Dyn* 203: 253–310; based on C. B. Kimmel et al. 1990. *Development* 108: 581–594.)

(C)

Animal pole

Nose, eye

Epidermis

Brain

Neural crest

Spinal cord — Ectoderm

Somite muscle   Head — Mesoderm

Ventral   Pronephros   Noto-chord   **Dorsal**

Blood Fins Heart   Muscle

Intestine   Liver   Pharynx — Endoderm

Blastoderm margin

Yolk cell

Vegetal pole

cell divisions slow, and cell movements become evident (Kane and Kimmel 1993). As discussed in Chapter 3 (see Figures 3.21 and 3.25), the maternal-to-zygotic transition requires the function of maternally deposited microRNAs (miR430, for instance), which progressively degrade maternal messages and thereby foster a fast transfer of transcriptional responsibilities to the zygotic genome. At this same time, three distinct cell populations can be distinguished. The first of these is the **yolk syncytial layer**, or **YSL** (Agassiz and Whitman 1884; Carvalho and Heisenberg 2010). The YSL will not contribute cells or nuclei to the embryo, but it is critical for generating the fish organizer, patterning the mesoderm, and leading the epiboly of the ectoderm over the embryo (Chu et al. 2012). The YSL is formed at the tenth cell cycle, when the cells at the vegetal edge of the blastoderm fuse with the underlying yolk cell. This fusion produces a ring of nuclei in the part of the yolk cell cytoplasm that sits just beneath the blastoderm. Later, as the blastoderm expands vegetally to surround the yolk cell, some of the yolk syncytial nuclei will move under the blastoderm to form the **internal YSL** (**iYSL**), and others will move vegetally, staying ahead of the blastoderm margin, to form the **external YSL** (**eYSL**; **FIGURE 11.30A,B**).

The second cell population distinguished at the mid-blastula transition is the **enveloping layer** (**EVL**). It is made up of the most superficial cells from the blastoderm, which form an epithelial sheet a single cell layer thick. The EVL is a protective covering that is sloughed off after 2 weeks. It allows the embryo to develop in a hypotonic solution (such as fresh water) that would otherwise burst the cells (Fukazawa et al. 2010). Between the EVL and the YSL is the third set of blastomeres, the deep cells, that give rise to the embryo proper.

The fates of the early blastoderm cells are not determined, and cell lineage studies (in which a nondiffusible fluorescent dye is injected into a cell so that its descendants can be followed) show that there is much cell mixing during cleavage. Moreover, any one of these early blastomeres can give rise to an unpredictable variety of tissue descendants

(Kimmel and Warga 1987; Helde et al. 1994). A fate map of the blastoderm cells can be made shortly before gastrulation begins. At this time, cells in specific regions of the embryo give rise to certain tissues in a highly predictable manner (**FIGURE 11.30C**), although they remain plastic, and cell fates can change if tissue is grafted to a new site.

Gastrulation and Formation of the Germ Layers

All three layers of the zebrafish blastoderm undergo epiboly. The first cell movement of fish gastrulation is the epiboly of the blastoderm cells over the yolk, and this is thought to be controlled both by maternal proteins (such as Eomesodermin) and by new proteins transcribed from the YSL nuclei (Du et al. 2012). Similar to the drivers of *Xenopus* epiboly, intercalation of the deeper blastula cells of the epiblast into the more superficially positioned layers forces the epiblast layer to spread radially outward (Warga and Kimmel 1990). As we discussed in Chapter 4, differential expression of E-cadherin is responsible for these radial intercalation movements (Kane et al. 2005; McFarland et al. 2005). This intercalation of cells causes a flattening of the "dome" of the blastoderm cells (**FIGURE 11.31A**).

Progression of epiboly

When about half the yolk is covered, a new set of movements is initiated by the internal and external YSLs (Bruce 2016). The YSL nuclei divide such that some nuclei (constituting the external YSL, or eYSL) remain in the upper cortex of the yolk cell, while the iYSL (internal YSL) lies beneath the blastoderm. The enveloping layer is tightly joined to the iYSL by E-cadherins and tight junctions (Shimizu et al. 2005a; Siddiqui et al. 2010) and is dragged ventrally as the iYSL nuclei migrate "downward." That the vegetal migration of the blastoderm margin is dependent on the epiboly of the YSL can be demonstrated by severing the attachments between the YSL and the EVL. When this is done, the EVL and the deep cells spring back to the top of the yolk, while the YSL continues its expansion around the yolk cell (Trinkaus 1984, 1992).

The migration of the YSL vegetally depends partially on the expansion of this layer by cell division and intercalation, and partly on the cytoskeletal network within the yolk cell (see Lepage and Bruce 2010; Bruce 2016). An actomyosin band forms in the eYSL, at the boundary between the YSL and the EVL. This pulls down the YSL/EVL at its vegetal connection by means of contraction and friction (Behrndt et al. 2012).

FIGURE 11.31 Cell movements during zebrafish gastrulation. (A) The blastoderm at 30% completion of epiboly (~4.2 hours). (B) Formation of the hypoblast, either by involution of cells at the margin of the epibolizing blastoderm or by delamination and ingression of cells from the epiblast (~6 hr). A close-up of the marginal region is at the right. (C) As ectodermal epiboly nears completion, the hypoblast, carrying the mesoderm and endoderm precursors, begins to cover the yolk (~9 hr). (D) Completion of gastrulation (~10.3 hr). The germ layers (yellow endoderm, blue ectoderm, red mesoderm) are present. (After W. Driever. 1995. *Curr Opin Genet Dev* 5: 610–618; L. Carvalho and C. P. Heisenberg. 2010. *Trends Cell Biol* 20: 586–592.)

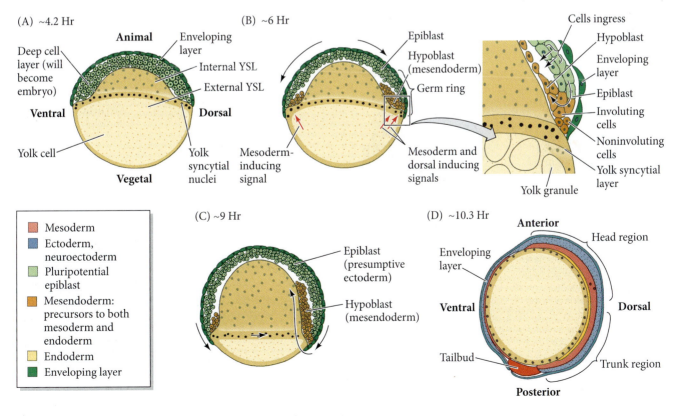

Meanwhile, the eYSL nuclei appear to migrate along the microtubles aligned along the animal-vegetal axis of the yolk cell, presumably pulling the iYSL and its accompanying EVL over the yolk cell. (Radiation or drugs that block the polymerization of tubulin slow epiboly; Strähle and Jesuthasan 1993; Solnica-Krezel and Driever 1994.) At the end of gastrulation, the entire yolk cell is covered by the blastoderm.

Internalization of the hypoblast

After the blastoderm cells have covered about half the zebrafish yolk cell, a thickening occurs throughout the margin of the deep cells. This thickening, called the **germ ring**, is composed of a superficial layer, the epiblast (which will become the ectoderm), and an inner layer, the **hypoblast** (which will become endoderm and mesoderm). The hypoblast forms in a synchronous "wave" of internalization (Keller et al. 2008) that has some characteristics of ingression (especially in the dorsal region; see Carmany-Rampey and Schier 2001) and some elements of involution (especially in the future ventral regions). Thus, as the cells of the blastoderm undergo epiboly around the yolk, they are also internalizing cells at the blastoderm margin to form the hypoblast. The epiblast cells (presumptive ectoderm) do not involute, whereas the deep cells—the future mesoderm and endoderm—do (**FIGURE 11.31B–D**). As the hypoblast cells internalize, the future mesoderm cells (the majority of the hypoblast cells) initially migrate vegetally while proliferating to make new mesoderm cells. Later, they alter direction and proceed toward the animal pole. The endodermal precursors, however, appear to move randomly over the yolk (Pézeron et al. 2008). The coordination of migration and cell specification is accomplished by physical forces rather than by chemicals. When the cortical cytoskeleton is disrupted by drugs, the cells fail to turn and the mesodermal genes are not activated. However, if the cells are injected with magnetic particles before being hit with the drugs, they can be mechanically towed around the embryo. The cells don't involute, but the mesodermal genes do turn on. Thus, during normal development, epiboly and cell specification may be coordinated by the mechanical stress of involution (**FIGURE 11.32**; Brunet et al. 2013).

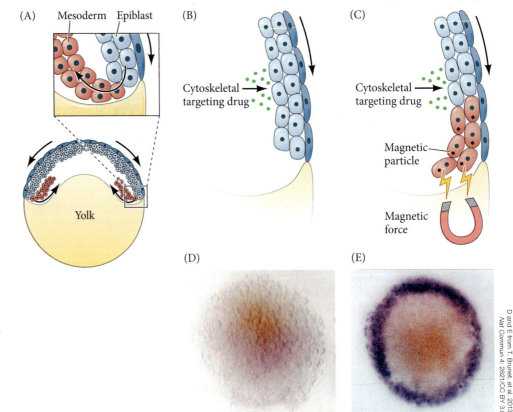

FIGURE 11.32 Stretching the zebrafish epiblast cells generates mesoderm. (A) During epiboly, cells at the border undergo structural changes and involute. As they do, mesodermal genes (red) are activated. (B) When the cortical cytoskeleton is prevented from contracting, the animal cap cells remain ectodermal and do not involute. (C) However, if these cells are pulled by a magnetic field, the mesodermal genes become expressed. (D,E) Circumpolar views of B and C, respectively, visualize expression of the mesodermal gene *no tail* (the zebrafish homologue of the *brachyury* gene). (D) *no tail* expression is blocked by the lack of involution. (E) *no tail* expression induced by stretching and subsequent involution. (A–C after S. Piccolo. 2013. *Nature* 504: 223–225.)

D and E from T. Brunet, et al. 2013. *Nat Commun* 4: 2821/CC BY 3.0

The embryonic shield and the neural keel

Once the hypoblast has formed, cells of the epiblast and hypoblast intercalate on the future dorsal side of the embryo to form a localized thickening, the **embryonic shield** (Schmitz and Campos-Ortega 1994). Here, the cells converge and extend anteriorly, eventually narrowing along the dorsal midline (**FIGURE 11.33A**). This convergent extension in the hypoblast forms the chordamesoderm, the precursor of the notochord (**FIGURE 11.33B,C**; Trinkaus 1992). This convergent extension is similar to that discussed in *Xenopus*, and is similarly accomplished by the Wnt-mediated planar cell polarity pathway (see Vervenne et al. 2008).

As we will see, the embryonic shield is functionally equivalent to the dorsal blastopore lip of amphibians, since it can organize a secondary embryonic axis when transplanted to a host embryo (Oppenheimer 1936; Ho 1992). The cells adjacent to the chordamesoderm—the paraxial mesoderm cells—are the precursors of the mesodermal somites (see Chapter 17). Concomitant convergence and extension in the epiblast bring presumptive neural cells from the epiblast into the dorsal midline, where they form

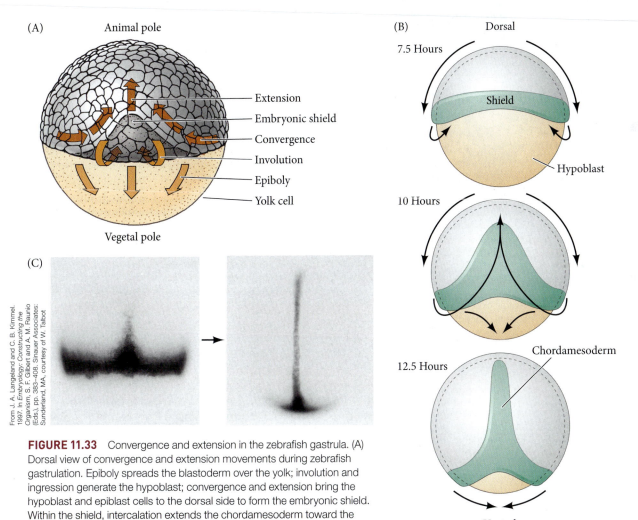

FIGURE 11.33 Convergence and extension in the zebrafish gastrula. (A) Dorsal view of convergence and extension movements during zebrafish gastrulation. Epiboly spreads the blastoderm over the yolk; involution and ingression generate the hypoblast; convergence and extension bring the hypoblast and epiblast cells to the dorsal side to form the embryonic shield. Within the shield, intercalation extends the chordamesoderm toward the animal pole. (B) Model of mesendoderm (hypoblast) formation. Numbers indicate hours after fertilization. On the future dorsal side, the internalized cells undergo convergent extension to form the chordamesoderm (notochord) and the paraxial (somitic) mesoderm adjacent to it. On the ventral side, the hypoblast cells migrate with the epibolizing epiblast toward the vegetal pole, eventually converging there. (C) Convergent extension of the chordamesoderm of the hypoblast cells. These cells are marked by their expression of the *no tail* gene (dark areas) encoding a T-box transcription factor. (B after P. J. Keller et al. 2008. *Science* 322: 1065–1069.)

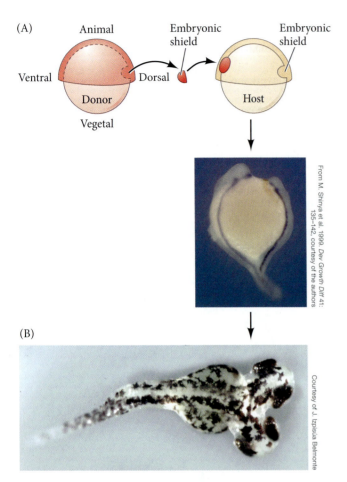

(A)

Animal

Embryonic shield

Embryonic shield

Ventral

Dorsal

Donor

Host

Vegetal

From M. Shinya et al. 1999. *Dev Growth Diff* 41: 135–142, courtesy of the authors

(B)

Courtesy of J. Izpisúa Belmonte

FIGURE 11.34   The embryonic shield as organizer in the fish embryo. (A) A donor embryonic shield (about 100 cells from a stained embryo) is transplanted into a host embryo at the same early-gastrula stage. The result is two embryonic axes joined to the host's yolk cell. In the photograph, both axes have been stained for *sonic hedgehog* mRNA, which is expressed in the ventral midline. (The embryo to the right is the secondary axis.) (B) The same effect can be achieved by activating nuclear β-catenin in embryos at sites opposite from where the embryonic shield will form. (A illustration after M. Shinya et al. 1999. *Dev Growth Diff* 41: 135–142.)

the **neural keel**. The neural keel, a band of neural precursors that extends over the axial and paraxial mesoderm, eventually develops a slitlike lumen to become the neural tube and to enter into the embryo.[4] Those cells remaining in the epiblast become the epidermis. On the ventral side (see Figure 11.33B), the hypoblast ring moves toward the vegetal pole, migrating directly beneath the epiblast that is epibolizing itself over the yolk cell. Eventually, the ring closes at the vegetal pole, completing the internalization of those cells that will become mesoderm and endoderm (Keller et al. 2008).

By different mechanisms, the *Xenopus* egg and zebrafish egg have reached the same state: they have become multicellular; they have undergone gastrulation; and they have positioned their germ layers such that the ectoderm is on the outside, the endoderm is on the inside, and the mesoderm lies between them. We will now see that zebrafish form their body axes in ways very similar to *Xenopus*, using very similar molecules.

Dorsal-Ventral Axis Formation

As mentioned above, the embryonic shield of fish is homologous to the dorsal blastopore lip of amphibians, and it is critical in establishing the dorsal-ventral axis. Shield tissue can convert lateral and ventral mesoderm (blood and connective tissue precursors) into dorsal mesoderm (notochord and somites), and it can cause the ectoderm to become neural rather than epidermal. This transformative capacity was shown by transplantation experiments in which the embryonic shield of an early-gastrula embryo was transplanted to the ventral side of another (**FIGURE 11.34**; Oppenheimer 1936; Koshida et al. 1998). Two axes formed, sharing a common yolk cell. Although the prechordal plate and notochord were derived from the donor embryonic shield, the other organs of the secondary axis came from host tissues that would normally form ventral structures. The new axis had been induced by the donor cells.

Like the amphibian blastopore lip, the embryonic shield forms the prechordal plate and the notochord of the developing embryo. The prechordal plate cells are the first to involute, and they migrate toward the animal pole (Dumortier et al. 2012). The presumptive prechordal plate and notochord are responsible for inducing ectoderm to become neural ectoderm, and they appear to do this in a manner very much like the homologous structures in amphibians.[5] Furthermore, the signals secreted from the zebrafish shield/organizer are the same as those we discussed for *Xenopus*. TGF-β superfamily members Nodal and BMP function to induce differential fates of the germ layers across the dorsal-ventral axis, and the shield/organizer-secreted signals Chordin, Noggin, and Follistatin all serve to block ventral induction to promote neural development (**FIGURE 11.35**). (See **Further Development 11.11, Fish Signals Are Like Amphibian Signals**, online.)

The fish blastopore lip

Perhaps unique to fish, the entire blastopore lip appears to be another important source of organization. Recall that the blastopore of a fish extends around the entire yolk cell. The dorsal lip (the shield) will induce head structures when placed into the ventral region of the blastopore margin. However, it will not induce any structures from the neighboring tissue when placed onto the animal cap of a blastula, which contains thoroughly undifferentiated cells. When a graft from the *ventral* blastopore lip is placed on animal cap cells, a well-organized tail structure is formed, having epidermis, somites, and neural tube, but no dorsal mesoderm (Agathon et al. 2003). Much of this structure is induced from the host tissue. So the ventral blastopore lip in

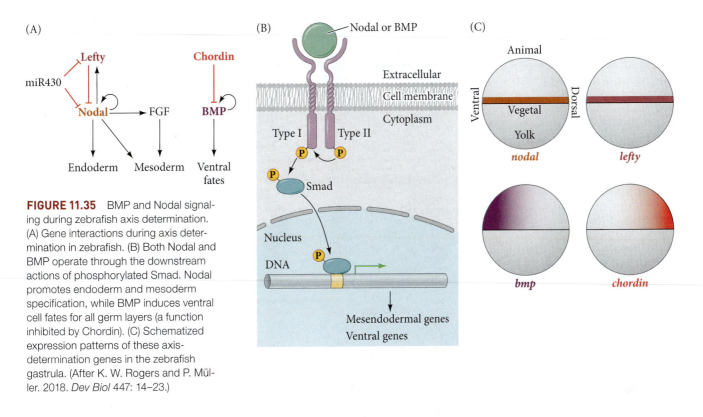

FIGURE 11.35 BMP and Nodal signaling during zebrafish axis determination. (A) Gene interactions during axis determination in zebrafish. (B) Both Nodal and BMP operate through the downstream actions of phosphorylated Smad. Nodal promotes endoderm and mesoderm specification, while BMP induces ventral cell fates for all germ layers (a function inhibited by Chordin). (C) Schematized expression patterns of these axis-determination genes in the zebrafish gastrula. (After K. W. Rogers and P. Müller. 2018. *Dev Biol* 447: 14–23.)

zebrafish is a "tail organizer." Cells from the *lateral* blastopore lips will induce trunk and posterior head structures, having notochord tissue. Moreover, these transplanted tissues do not express BMPs, Wnts, or their antagonists.

Teasing apart the powers of Nodal and BMP during axis determination

It appears that in addition to the classic shield organizer (see Figure 11.34), the entire blastopore lip is involved in forming posterior head, trunk, and tail, through another means. This second set of axis-determining factors seems to be a dual gradient of Nodal and BMP proteins (Fauny et al. 2009; Thisse and Thisse 2015). Along the blastopore lip, from the ventral to the dorsal margin, there forms a continuous gradation in the ratio of BMP to Nodal activity. BMP is highest at the ventral margin, low dorsolaterally, and approaches 0 in the dorsalmost domain, where only Nodal is active. Thus, each region of the blastopore lip is characterized by a specific BMP:Nodal ratio of activity. Remarkably, an entire ectopic axis can be made by injecting one of the animal cap blastomeres with *nodal* mRNA and another animal cap cell with *bmp* mRNA (Xu et al. 2014). A gradient is formed between them, and the neighboring cells respond by constructing a new axis (**FIGURE 11.36**).

Moreover, by injecting different amounts of *bmp* and *nodal* mRNAs into a single blastula animal cap cell, one can mimic the effect of the blastopore lip. Injections of mRNAs with a high *bmp* and *nodal* ratio induce the formation of new tails growing from the animal pole of the embryo. Wnt8, a posterior morphogen, is produced in these cells. Injection of mRNAs with decreasing *bmp* and *nodal* ratios induces the formation of secondary trunks from these animal pole cells. When *bmp* and *nodal* are injected in the same amounts, a posterior head is induced (Thisse et al. 2000). As previously mentioned, in *Xenopus*, Nodal proteins are critical for the formation of the organizer; and in zebrafish, ectopic expression of Nodal in the ventral blastopore margin will convert the ventral blastopore lip into a shield, inducing an entire secondary axis. The shield in zebrafish may be the "head organizer," while the blastopore lip cells 180° away from the shield become the "tail organizer."

The engine for integrating the BMP-Chordin and BMP-Nodal axes appears to be β-catenin. As in *Xenopus*, β-catenin activates the *nodal* genes. In addition, β-catenin

(?) Developing Questions

All axes are developing simultaneously. How is the timing of dorsal versus ventral cell fates coordinated with the timing of development of the anterior-posterior and left-right axes? Consider also that morphological changes are occurring amidst these patterning events. Throughout this chapter we have described gastrulation and axis determination separately, only because this makes them easier to comprehend. But in reality, the two processes have evolved together as part of a whole. So could the physical nature of morphogenetic movements somehow help support the timing of axis development? If you're interested, visit the work by Mary Mullins on this topic (e.g., Tuazon and Mullins 2015 and others).

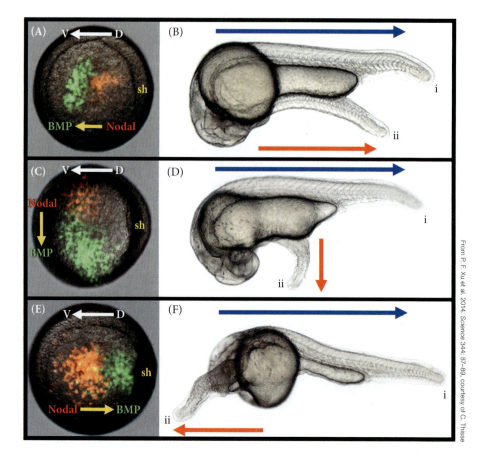

From P. F. Xu et al. 2014. *Science* 344: 87–89, courtesy of C. Thisse

FIGURE 11.36 Correlation between the relative position of BMP- and Nodal-secreting clones and the orientation of the secondary embryonic axis induced at the animal cap. (A,B) When the Nodal-BMP vector (yellow arrow in A) is parallel to the dorsal-ventral axis (white arrow) of the embryonic margin (where Nodal is strong dorsally and BMP strong ventrally), the original axis (blue arrow in B) and the secondary axis (red arrow) are parallel. (C,D) When the Nodal-BMP vector is perpendicular to the original dorsal-ventral axis, the secondary embryonic axis grows perpendicular to the primary axis. (E,F) When the Nodal-BMP vector is against that of the original dorsal-ventral axis, the primary and secondary axes grow in opposite directions. sh, embryonic shield; i, primary embryo; ii, induced secondary embryo. A, C, and E are animal pole views at the shield stage; B, D, and F are lateral views at 30 hours after fertilization.

activates the genes encoding FGFs and other factors that repress BMP and Wnt expression on the dorsal side of the embryo while activating the genes for *goosecoid*, *noggin*, and *dickkopf* there (Sampath et al. 1998; Gritsman et al. 2000; Schier and Talbot 2001; Solnica-Krezel and Driever 2001; Fürthauer et al. 2004; Tsang et al. 2004). As in *Xenopus*, β-catenin accumulates specifically in the nuclei destined to become the dorsal cells (Langdon and Mullins 2011). And as in *Xenopus*, this appears to be regulated by a maternal Wnt protein, in this case Wnt8a (Lu et al. 2011). The presence of β-catenin distinguishes dorsal YSL from the lateral and ventral YSL regions[6] (**FIGURE 11.37**; Schneider et al. 1996), and inducing β-catenin accumulation on the *ventral* side of the egg results in its dorsalization and a second embryonic axis (Kelly et al. 1995). (See **Further Development 11.12, Anterior-Posterior Axis Formation in Zebrafish**, online.)

Left-Right Axis Formation

In all vertebrates studied, the right and left sides differ both anatomically and developmentally. In fish, the heart is on the left side, and there are different structures in the left and right regions of the brain. Moreover, as in other vertebrates, the cells on the left side of the body are given that information by Nodal signaling and by the Pitx2 transcription factor. The ways that the different vertebrate classes accomplish this asymmetry differ, but recent evidence suggests that the currents produced by motile cilia in the node may be responsible for left-right axis formation in all the vertebrate classes (Okada et al. 2005).

In zebrafish, the Nodal structure housing the cilia that control left-right asymmetry is a transient fluid-filled organ called **Kupffer's vesicle**. As shown in Figure 11.27, Kupffer's vesicle arises from a group of dorsal cells near the embryonic shield shortly after gastrulation. Essner and colleagues (2002, 2005) were able to inject small beads into Kupffer's vesicle and see their translocation from one side of the vesicle to the other. Blocking ciliary function by preventing the synthesis of dynein or by ablating the

Courtesy of S. Schneider

FIGURE 11.37 β-Catenin activates organizer genes in the zebrafish. Nuclear localization of β-catenin marks the dorsal side of the *Xenopus* blastula (larger image) and helps form its Nieuwkoop center beneath the organizer. In the zebrafish late blastula (smaller image), nuclear localization of β-catenin is seen in the yolk syncytial layer nuclei beneath the future embryonic shield.

precursors of the ciliated cells resulted in abnormal left-right axis formation. Cilia are responsible for the left-side specific activation of the Nodal signaling cascade. Nodal target genes are critically important in instructing asymmetrical organ migration and morphogenesis in the body (Rebagliati et al. 1998; Long et al. 2003).

Closing Thoughts on the Opening Photo

The zebrafish embryo in the photo has two body axes, a normal one and a second axis (arrow) that was induced by adding a region of an embryo containing high amounts of Nodal (see also Figure 11.36). New hypotheses concerning conjoined twinning propose that during gastrulation, ectopic expression of signaling molecules such as Nodal might lead to new axis formation. Human conjoined twins will be discussed in more detail in Chapter 12.

Courtesy of Christine Thisse

Endnotes

[1] Hilde Proescholdt Mangold died in a tragic accident in 1924, when her kitchen's gasoline heater exploded. She was 26 years old, and her paper was just about to be published. Hers is one of the very few doctoral theses in biology that has directly resulted in the awarding of a Nobel Prize. For more information about Hilde Mangold, her times, and the experiments that identified the organizer, see Hamburger 1984, 1988; Fässler and Sander 1996.

[2] The induction of dorsal mesoderm—rather than the dorsal ectoderm of the nervous system—by the posterior end of the notochord was confirmed by Bijtel (1931) and Spofford (1945), who showed that the posterior fifth of the neural plate gives rise to tail somites and the posterior portions of the pronephric kidney duct.

[3] The tail inducer was initially thought to be part of the trunk inducer, since transplantation of the late dorsal blastopore lip into the blastocoel often produced larvae with extra tails. However, it appears that tails are normally formed by interactions between the neural plate and the posterior mesoderm during the neurula stage (and thus are generated outside the organizer). Here, Wnt, BMPs, and Nodal signaling all seem to be required (Tucker and Slack 1995;

Niehrs 2004). Interestingly, all three of these signaling pathways have to be inactivated if the head is to form.

[4] This is different from the formation of the neural tube in frog embryos and is probably equivalent to "secondary" neural tube formation in the posterior of mammalian embryos (see Chapter 13).

[5] Another similarity between amphibian and fish organizers is that they can be duplicated by rotating the egg and changing the orientation of the microtubules (Fluck et al. 1998). One difference in the axial development of these groups is that in amphibians, the prechordal plate is necessary for inducing the anterior brain to form. In zebrafish, although the prechordal plate appears to be necessary for forming ventral neural structures, the anterior regions of the brain can form in its absence (Schier et al. 1997; Schier and Talbot 1998).

[6] Some of the endodermal cells that accumulate β-catenin will become the precursors of the ciliated cells of Kupffer's vesicle (Cooper and D'Amico 1996). As we will discuss in the final section of this chapter, these cells are critical in determining the left-right axis of the embryo.

(11) Snapshot Summary
Amphibians and Fish

1. Amphibian cleavage is holoblastic, but it is unequal because of the presence of yolk in the vegetal hemisphere.

2. Amphibian gastrulation begins with epiboly of the ectoderm followed by the invagination of the bottle cells and the coordinated involution of the mesoderm. Vegetal rotation plays a significant role in directing the involution.

3. The driving forces for ectodermal epiboly and the convergent extension of the mesoderm are the intercalation events in which several tissue layers merge. Fibronectin plays a critical role in enabling the mesodermal cells to migrate into the embryo.

4. The dorsal lip of the blastopore forms the organizer tissue of the amphibian gastrula. This tissue dorsalizes the ectoderm, transforming it into neural tissue, and transforms ventral mesoderm into lateral and dorsal mesoderm.

5. The organizer consists of pharyngeal endoderm, head mesoderm, notochord, and dorsal blastopore lip tissues. The organizer functions by secreting proteins (Noggin, Chordin, and Follistatin) that block the BMP signal that would otherwise ventralize the mesoderm and activate the epidermal genes in the ectoderm.

6. Dorsal-ventral specification begins with maternal messages and proteins stored in the vegetal cytoplasm. These include

(Continued)

Snapshot Summary *(continued)*

Nodal-like paracrine factors, transcription factors (such as VegT), and agents that protect β-catenin from degradation.

7. The organizer is itself induced by the Nieuwkoop center, located in the dorsalmost vegetal cells. This center is formed by the translocation of the Disheveled protein and Wnt11 to the dorsal side of the egg to stabilize β-catenin in the dorsal cells of the embryo.

8. The Nieuwkoop center is formed by the accumulation of β-catenin, which combines with Tcf3 to form a transcription factor complex that can activate transcription of the *siamois* and *twin* genes on the dorsal side of the embryo.

9. The Siamois and Twin proteins collaborate with activated Smad2 transcription factors generated by the TGF-β pathway (Nodal, Vg1) to activate genes encoding BMP inhibitors. These inhibitors include the secreted factors Noggin, Chordin, and Follistatin, as well as the transcription factor Goosecoid.

10. In the presence of BMP inhibitors, ectodermal cells form neural tissue. The action of BMP on ectodermal cells causes them to become epidermis.

11. In the head region, an additional set of proteins (Cerberus, Frzb, Dickkopf, Tiki) blocks the Wnt signal from the ventral and lateral mesoderm.

12. Wnt signaling causes a gradient of β-catenin along the anterior-posterior axis of the neural plate, which appears to specify the regionalization of the neural tube.

13. Insulin-like growth factors (IGFs) help transform the neural tube into anterior (forebrain) tissue.

14. The left-right axis appears to be initiated by the activation of a Nodal protein solely on the left side of the embryo. In *Xenopus*, as in other vertebrates, Nodal protein activates expression of *pitx2*, which is critical in distinguishing left-sidedness from right-sidedness.

15. Cleavage in fish is meroblastic. The deep cells of the blastoderm form between the yolk syncytial layer and the enveloping layer. These deep cells migrate over the top of the yolk, forming the hypoblast and epiblast.

16. On the future dorsal side, the hypoblast and epiblast intercalate to form the embryonic shield, a structure homologous to the amphibian organizer. Transplantation of the embryonic shield into the ventral side of another embryo will cause a second embryonic axis to form.

17. In both amphibians and fish, neural ectoderm is permitted to form where the BMP-mediated induction of epidermal tissue is prevented. The fish embryonic shield, like the amphibian dorsal blastopore lip, secretes the BMP antagonists. Like the amphibian organizer, the shield receives its abilities by being induced by β-catenin and by underlying endodermal cells expressing Nodal-related paracrine factors.

Go to oup.com/he/barresi12xe for Further Developments, Scientists Speak interviews, Watch Development videos, Dev Tutorials, and complete bibliographic information for all literature cited in this chapter.

Birds and Mammals

Can you see the head and tail? How does this mammalian embryo determine them?

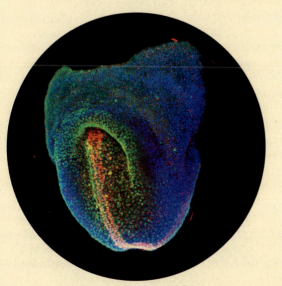

Photograph courtesy of I. Costello and E. Robertson

THIS FINAL CHAPTER ON THE PROCESSES OF EARLY DEVELOPMENT extends our survey of vertebrate development to include the **amniotes** —those vertebrates whose embryos form an amnion, or water sac (i.e., the reptiles, birds, and mammals). Birds and reptiles follow a very similar pattern of development (Gilland and Burke 2004; Coolen et al. 2008), and birds are considered by modern taxonomists to be a reptilian clade (**FIGURE 12.1A**).

The **amniote egg** is characterized by a set of membranes that together enable the embryo to survive on land (**FIGURE 12.1B**). First, the **amnion**, for which the amniote egg is named, is formed early in embryonic development and enables the embryo to float in a fluid environment that protects it from desiccation. Another cell layer derived from the embryo, the **yolk sac**, enables nutrient uptake and the development of the circulatory system. The **allantois**, developing at the posterior end of the embryo, stores waste products. The **chorion** contains blood vessels that exchange gases with the outside environment. In birds, most reptiles, and egg-laying mammals (monotremes), the embryo and its membranes are enclosed in a hard or leathery shell within which the embryo develops outside the mother's body. Cleavage in bird and reptile eggs, like that of the bony fishes described in the last chapter, is meroblastic, with only a small portion of the egg's volume being used to make the cells of the embryo. The vast majority of the large egg is composed of yolk that will nourish the growing embryo.

In most mammals, holoblastic cleavage is modified to accommodate the formation of a **placenta**, an organ containing tissues and blood vessels from both the embryo and the mother. Gas exchange, nutrient uptake, and waste elimination take place through the placenta, enabling the embryo to develop inside another organism. (See **Further Development 12.1, The Extraembryonic Membranes**, online.)

Ever since Aristotle first observed and recorded the details of its 3-week-long development, the domestic chicken (*Gallus gallus*) has been a favorite organism for embryological studies. It is accessible year-round and is easily maintained. Moreover, at any particular incubation temperature, its developmental stage can be accurately predicted, so large numbers of same-stage embryos can be obtained and manipulated. Chick organ formation is accomplished by molecular components and cell movements similar to those of mammalian organ formation, and the chick is one of the few organisms whose embryos are amenable to both surgical and genetic manipulations (Stern 2005a). Thus, the chick embryo has often served as a model for human embryos, as has the ubiquitous laboratory mouse.

(A)

AMNIOTE
VERTEBRATES

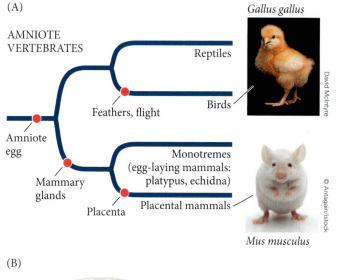

Gallus gallus

Mus musculus

FIGURE 12.1 The membranes of the amniote egg characterize reptiles, birds, and mammals. (A) Phylogenetic relationships of the amniotes. Note that birds are considered reptiles by most modern taxonomists, but for physiological studies they are often treated as separate taxa. (Other flying and feathered reptile groups have not survived to the present day.) The domestic chicken (*Gallus gallus*) is the most widely studied bird species. Among mammals, the development of the laboratory mouse *Mus musculus* is the most widely studied. Both avian and mouse studies contribute to our understanding of human development. (B) The shelled amniote egg (as exemplified by the chicken egg on the left) permitted animals to develop away from bodies of water. The amnion provides a "water sac" in which the embryo develops, the allantois stores wastes, and the blood vessels of the chorion exchange gases and nutrients from the yolk sac. In mammals (right), this arrangement is modified such that the blood vessels acquire nutrients and exchange gases via a placenta joined to the mother's uterus rather than from the yolk sac.

(B)

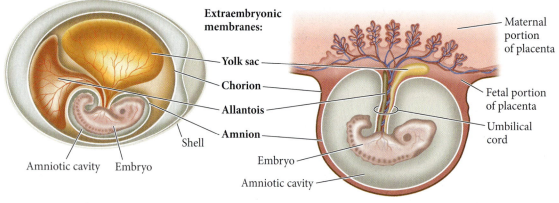

Extraembryonic membranes:
- **Yolk sac**
- **Chorion**
- **Allantois**
- **Amnion**

Shell
Amniotic cavity Embryo

Maternal portion of placenta
Fetal portion of placenta
Umbilical cord
Embryo
Amniotic cavity

The mouse is the most commonly used mammalian model organism and is the subject of many studies involving genetic and surgical manipulation.[1] In addition, the mouse was the first mammalian genome to be sequenced, and when its sequence was first published, many scientists felt it was more valuable than knowing the human genome sequence. Their reasoning was that "working on mouse models allows the manipulation of each and every gene to determine their functions" (Gunter and Dhand 2002). We cannot do that with humans. Human development is a subject of medical as well as general scientific interest, however, and the latter sections of this chapter will cover early human development, illustrating the application of many of the principles we have described in model organisms.

Early Development in Birds

Fertilization of the chick egg occurs in the hen's oviduct, before the albumin ("egg white") and shell are secreted to cover it. Cleavage occurs during the first day of development, while the egg is still inside the hen, during which time the embryo progresses from a zygote through late blastula stages (Sheng 2014). Like the egg of the zebrafish, the chick egg is telolecithal, with a small disc of yolk-free cytoplasm—the **blastodisc**—sitting on top of a large mass of yolky cytoplasm (**FIGURE 12.2A**).

Avian Cleavage

Like fish eggs, the yolky eggs of birds undergo **discoidal meroblastic cleavage**. Cleavage occurs only in the blastodisc, which is about 2–3 mm in diameter and is located at the animal pole of the egg. The first cleavage furrow appears centrally in the blastodisc; other cleavages follow to create a **blastoderm** (**FIGURE 12.2B,C**).

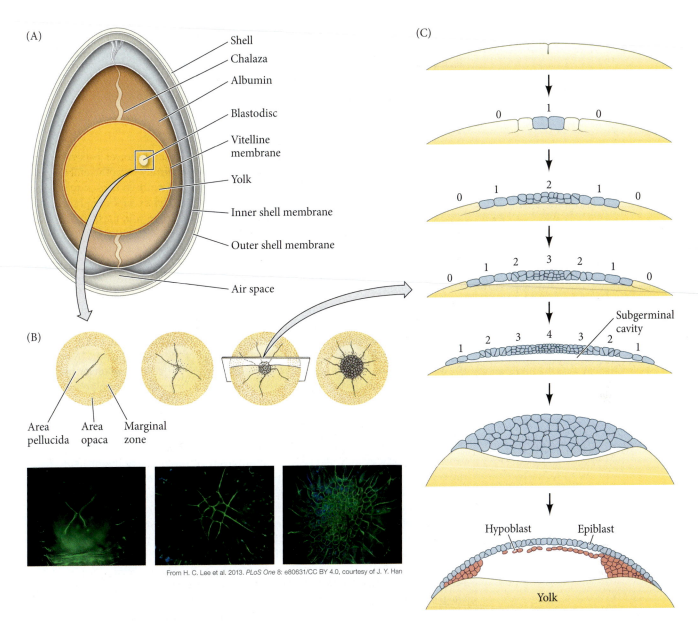

From H. C. Lee et al. 2013. *PLoS One* 8: e80631/CC BY 4.0, courtesy of J. Y. Han

FIGURE 12.2 Discoidal meroblastic cleavage in a chick egg. (A) Avian eggs include some of the largest cells known (inches across), but cleavage takes place in only a small region. The yolk fills up the entire cytoplasm of the egg cell, with the exception of a small blasto-disc, in which cleavage and development will take place. The chalaza are protein strings that keep the yolky egg cell centered in the shell. The albumin (egg white) is secreted onto the egg in its passage out of the oviduct. (B) Early cleavage stages viewed from the animal pole (the future dorsal side of the embryo). In the micrographs, the tightly apposed cell membranes have been stained with phalloidin (green). (C) Schematic view of cellularization in the chick egg during the day it is fertilized and still inside the hen. The numbers refer to the layers of cells. (B after R. Bellairs et al. 1978. *J Embryol Exp Morphol* 43: 55–69; C after Nagai et al. 2015. *Development* 142: 1279–1286.)

As in the fish embryo, the cleavages do not extend into the yolky cytoplasm, so the early-cleavage cells are continuous with one another and with the yolk at their bases. Thereafter, equatorial and vertical cleavages divide the blastoderm into a tissue about 4 cell layers thick, with the cells linked together by tight junctions (see Figure 12.2C; Bellairs et al. 1978; Eyal-Giladi 1991; Nagai et al. 2015). The switch from maternal to zygotic gene expression occurs at about the seventh or eighth division, when there are around 128 cells (Nagai et al. 2015).

Between the blastoderm and the yolk of avian eggs is a space called the **subgerminal cavity**, which is created when the blastoderm cells absorb water from the albumin and secrete the fluid between themselves and the yolk (New 1956). At this stage, the deep cells in the center of the blastoderm appear to be shed and die, leaving behind a 1-cell-thick **area pellucida**; this part of the blastoderm forms most of the actual embryo. The peripheral ring of blastoderm cells that have not shed their deep cells constitutes the **area opaca**. Between the area pellucida and the area opaca is a thin layer of cells called the **marginal zone** (Eyal-Giladi 1997; Arendt and Nübler-Jung 1999). Some marginal zone cells become very important in determining cell fate during early chick development.

Gastrulation of the avian embryo

The blastoderm will form an upper layer called the epiblast and a lower layer called the hypoblast (see Figure 12.2B). The avian embryo comes entirely from the epiblast; the hypoblast does not contribute any cells to the developing embryo (Rosenquist 1966, 1972). Rather, the hypoblast cells form portions of the extraembryonic membranes (see Figure 12.1B), especially the yolk sac and the stalk linking the yolk mass to the endodermal digestive tube. Hypoblast cells also provide chemical signals that specify the migration of epiblast cells. However, the three germ layers of the embryo proper (plus the amnion, chorion, and allantois extraembryonic membranes) are formed solely from the epiblast (Schoenwolf 1991).

THE HYPOBLAST   By the time a hen has laid an egg, its blastoderm contains some 50,000 cells. At this time, most of the cells of the area pellucida remain at the surface, forming the **epiblast**. Shortly after the egg is laid, a local thickening of the epiblast, called **Koller's sickle**, is formed at the posterior edge of the area pellucida. In between the area opaca and Koller's sickle is a belt-like region called the **posterior marginal zone** (**PMZ**). A sheet of cells at the posterior boundary between the area pellucida and marginal zone migrates anteriorly beneath the surface. Meanwhile, cells in more anterior regions of the epiblast have delaminated and stay attached to the epiblast to form hypoblast "islands," an archipelago of disconnected clusters of 5–20 cells each that migrate and become the **primary hypoblast** (**FIGURE 12.3A,B**). The sheet of cells that grows anteriorly from Koller's sickle combines with the primary hypoblast to form the complete hypoblast layer, also called the **secondary hypoblast** or **endoblast** (**FIGURE 12.3C–E**; Eyal-Giladi et al. 1992; Bertocchini and Stern 2002; Khaner 2007a,b). The resulting two-layered blastoderm (epiblast and hypoblast) is joined together at the marginal zone, and the space between the layers forms a blastocoel-like cavity. Thus, although the shape and formation of the avian blastodisc differ from those of the blastula of the amphibian, fish, or echinoderm, the overall spatial relationships are retained.

FORMATION OF THE PRIMITIVE STREAK   Although many reptile groups may initiate gastrulation by migration through an amphibian-like blastopore, avian and mammalian gastrulation takes place through the **primitive streak**. This can be considered the equivalent of an elongated blastopore lip of amphibian embryos (Alev et al. 2013; Bertocchini et al. 2013; Stower et al. 2015). Dye-marking experiments and time-lapse cinemicrography indicate that the primitive streak first arises from Koller's sickle and the epiblast above it (Bachvarova et al. 1998; Lawson and Schoenwolf 2001a,b; Voiculescu et al. 2007). As cells converge to form the primitive streak, a depression called the **primitive groove** forms within the streak. Most migrating cells pass through the primitive groove, which serves as a gateway into the deep layers of the embryo (**FIGURE 12.4**; Voiculescu et al. 2014). Thus, the primitive groove is homologous to the amphibian blastopore, and the primitive streak is homologous to the blastopore lip.

At the anterior end of the primitive streak is a regional thickening of cells called **Hensen's node** (also known as the **primitive knot**; see Figure 12.4C). The center of Hensen's node contains a funnel-shaped depression (sometimes called the **primitive pit**) through which cells can enter the embryo to form the notochord and prechordal plate. Hensen's node is the functional equivalent of the dorsal lip of the amphibian blastopore (i.e., the organizer)[2] and the fish embryonic shield (Boettger et al. 2001).

The primitive streak defines the major body axes of the avian embryo. It extends from posterior to anterior; migrating cells enter through its dorsal side and move to its ventral side; and it separates the left portion of the embryo from the right. The axis of the streak is equivalent to the dorsal-ventral axis of amphibians. The anterior end of

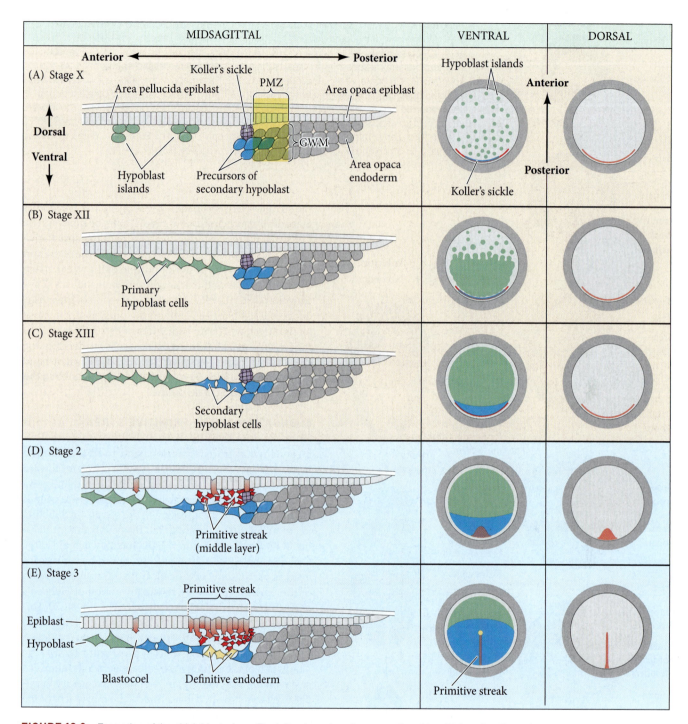

FIGURE 12.3 Formation of the chick blastoderm. The left column is a diagrammatic midsagittal section through part of the blastoderm. The middle column depicts the entire embryo viewed from the ventral side, showing the migration of the primary hypoblast and the secondary hypoblast (endoblast) cells. The right column shows the entire embryo seen from the dorsal side. (A–C) Events prior to primitive streak formation. (A) Stage X embryo, where islands of hypoblast cells can be seen, as well as a congregation of hypoblast cells around Koller's sickle. Posterior marginal zone (PMZ) of the epiblast is bracketed with a yellow overlay as well as the deep cells of this region called the posterior germ wall margin (GWM). (B) By stage XII, a sheet of cells that grows anteriorly from Koller's sickle combines with the hypoblast islands to form the complete hypoblast layer. (C) By stage XIII, just prior to primitive streak formation, the formation of the hypoblast has just been completed. (D) By stage 2 (12–14 hours after the egg is laid), the primitive streak cells form a third layer that lies between the hypoblast and the epiblast cells. (E) By stage 3 (15–17 hours post laying), the primitive streak has become a definitive region of the epiblast, with cells migrating through it to become the mesoderm and endoderm. (After C. D. Stern. 2004. In *Gastrulation: From Cells to Embryo*, C. D. Stern [Ed.] pp. 219–232. Cold Spring Harbor Laboratory Press: Cold Spring Harbor, NY.)

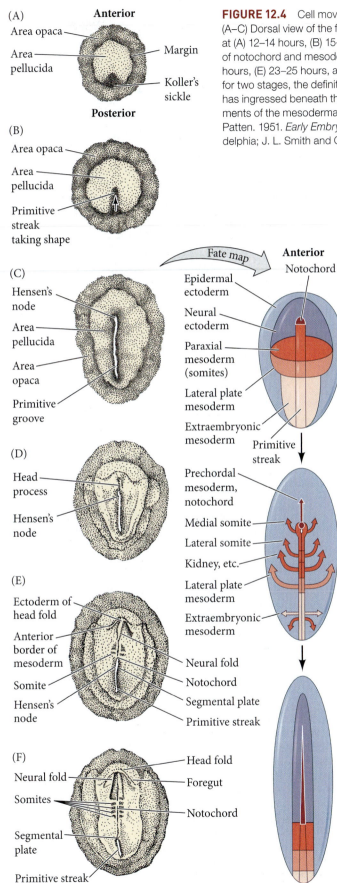

(A)

Anterior

Area opaca
Area pellucida
Margin
Koller's sickle

Posterior

(B)

Area opaca
Area pellucida
Primitive streak taking shape

(C)

Hensen's node
Area pellucida
Area opaca
Primitive groove

Fate map → **Anterior**

Notochord

Epidermal ectoderm
Neural ectoderm
Paraxial mesoderm (somites)
Lateral plate mesoderm
Extraembryonic mesoderm
Primitive streak

(D)

Head process
Hensen's node

Prechordal mesoderm, notochord
Medial somite
Lateral somite
Kidney, etc.
Lateral plate mesoderm
Extraembryonic mesoderm

(E)

Ectoderm of head fold
Anterior border of mesoderm
Somite
Hensen's node

Neural fold
Notochord
Segmental plate
Primitive streak

(F)

Neural fold
Somites
Segmental plate
Primitive streak

Head fold
Foregut
Notochord

FIGURE 12.4 Cell movements of the primitive streak and fate map of the chick embryo. (A–C) Dorsal view of the formation and elongation of the primitive streak. The blastoderm is seen at (A) 12–14 hours, (B) 15–17 hours, and (C) 18–20 hours after the egg is laid. (D–F) Formation of notochord and mesodermal somites as the primitive streak regresses, shown at (D) 20–22 hours, (E) 23–25 hours, and (F) the four-somite stage. Fate maps of the chick epiblast are shown for two stages, the definitive primitive streak stage (C) and neurulation (F). In (F), the endoderm has ingressed beneath the epiblast, and convergent extension is seen in the midline. The movements of the mesodermal precursors through the primitive streak at (C) are shown. (After B. M. Patten. 1951. *Early Embryology of the Chick*, 4th ed., pp. 70–85: The Blakiston Company: Philadelphia; J. L. Smith and G. C. Schoenwolf. 1998. *Curr Top Dev Biol* 40: 79–110.)

the streak—Hensen's node—gives rise to the prechordal mesoderm, notochord, and medial part of the somites. Cells that ingress through the middle of the streak give rise to the lateral part of the somites and to the heart and kidneys. Cells in the posterior portion of the streak make the lateral plate and extraembryonic mesoderm (Psychoyos and Stern 1996). After the ingression of the mesoderm cells, epiblast cells remaining outside of, but close to, the streak will form medial (dorsal) structures such as the neural plate, while those epiblast cells farther from the streak will become epidermis (see Figure 12.4, right-hand panels). (See **Further Development 12.2, Organizing the Chick Primitive Streak**, online.)

ELONGATION OF THE PRIMITIVE STREAK As cells enter the primitive streak, they undergo an epithelial-mesenchymal transformation, and the basal lamina beneath them breaks down. The streak elongates toward the future head region as more anterior cells migrate toward the center of the embryo. Convergent extension is responsible for the progression of the streak—a doubling in streak length is accompanied by a concomitant halving of its width (see Figure 12.4B; Voiculescu et al. 2007). Cell division adds to the length produced by convergent extension, and some of the cells from the anterior portion of the epiblast contribute to the formation of Hensen's node (Streit et al. 2000; Lawson and Schoenwolf 2001b).

At the same time, the secondary hypoblast (endoblast) cells continue to migrate anteriorly from the posterior marginal zone of the blastoderm (see Figure 12.3E). The elongation of the primitive streak appears to be coextensive with the anterior migration of these secondary hypoblast cells, and the hypoblast directs the movement of the primitive streak (Waddington 1933; Foley et al. 2000; Voiculescu et al. 2007, 2014). The streak eventually extends to 60%–75% of the length of the area pellucida.

FORMATION OF ENDODERM AND MESODERM The basic rule of amniote cell specification is that germ layer identity (ectoderm, mesoderm, or endoderm) is established before gastrulation starts (see Chapman et al. 2007), but the specification of cell type is controlled by inductive influences during and after migration through the primitive streak. As soon as the primitive streak has formed, epiblast cells begin to migrate through it and into the space between epiblast and hypoblast (reminiscent

of the amphibian blastocoel). The streak thus has a continually changing cell population. Cells migrating through the anterior end pass down into the embryonic space and migrate anteriorly, forming the endoderm, head mesoderm, and notochord; cells passing through the more posterior portions of the primitive streak give rise to the majority of mesodermal tissues (**FIGURE 12.5**; Rosenquist 1966; Schoenwolf et al. 1992).

The first cells to migrate through Hensen's node are those destined to become the pharyngeal endoderm of the foregut. Once deep within the embryo, these endodermal cells migrate anteriorly and eventually displace the hypoblast cells, causing the hypoblast cells to be confined to a region in the anterior portion of the area pellucida. This anterior region, the **germinal crescent**, does not form any embryonic structures, but it does contain the precursors of the germ cells, which later migrate through the blood vessels to the gonads.

The next cells entering through Hensen's node also move anteriorly, but they do not travel as far ventrally as the presumptive foregut endodermal cells. Rather, they remain between the endoderm and the epiblast to form the **prechordal plate mesoderm** (Psychoyos and Stern 1996). Thus, the head of the avian embryo forms anterior (rostral) to Hensen's node.

The next cells passing throug Hensen's node become the **chordamesoderm**. The chordamesoderm has two components: the head process and the notochord. The most anterior part, the **head process**, is formed by central mesoderm cells migrating anteriorly, behind the prechordal plate mesoderm and toward the rostral tip of the embryo (see Figures 12.4 and 12.5). The head process underlies those cells that will form the forebrain and midbrain. As the primitive streak regresses, the cells deposited by the regressing Hensen's node will become the notochord. In the ectoderm, most of the initial neural plate corresponds to the future brain region (from forebrain to the level of the future ear vesicle, which lies adjacent to Hensen's node at full primitive streak stage). A small region of neural ectoderm just lateral and posterior to Hensen's node (sometimes called the caudal lateral epiblast) will give rise to the rest of the nervous system, including the posterior hindbrain and all of the spinal cord. As the primitive streak regresses, this latter region regresses with Hensen's node and adds cells to the caudal end of the elongating neural plate. It appears that FGF signaling in the streak and paraxial (future somite) mesoderm keeps this region "young" and undifferentiated as it

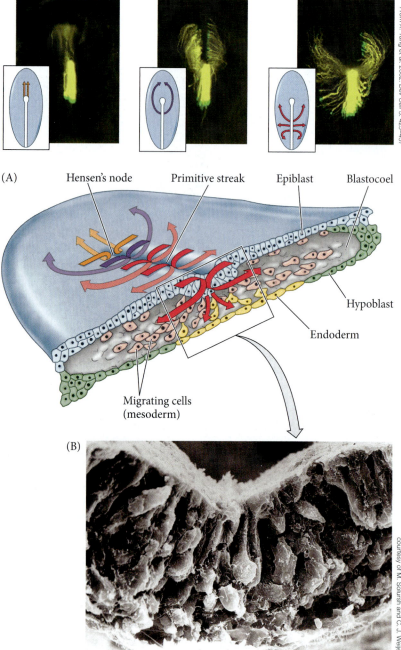

From X. Yang et al. 2002. *Dev Cell* 3: 425–437

(A) Hensen's node   Primitive streak   Epiblast   Blastocoel

Hypoblast

Endoderm

Migrating cells (mesoderm)

(B)

From M. Solursh and J. P. Revel. 1978. *Differentiation* 11: 185–190, courtesy of M. Solursh and C. J. Weijer

FIGURE 12.5 Migration of endodermal and mesodermal cells through the primitive streak. (A) Stereogram of a gastrulating chick embryo, showing the relationship of the primitive streak, the migrating cells, and the hypoblast and epiblast of the blastoderm. The lower layer becomes a mosaic of hypoblast and endodermal cells; the hypoblast cells eventually sort out to form a layer beneath the endoderm and contribute to the yolk sac. Above each region of the stereogram are micrographs showing the tracks of GFP-labeled cells at that position in the primitive streak. Cells migrating through Hensen's node travel anteriorly to form the prechordal plate and notochord; those migrating through the next anterior region of the streak travel laterally but converge near the midline to make notochord and somites; those from the middle of the streak form intermediate mesoderm and lateral plate mesoderm (see the fate maps in Figure 12.4). Farther posterior, the cells migrating through the primitive streak make the extraembryonic mesoderm (not shown). (B) This scanning electron micrograph shows epiblast cells passing into the blastocoel and extending their apical ends to become bottle cells. (A after B. I. Balinsky. 1975. *Introduction to Embryology*, 4th Ed. Saunders: Philadelphia.)

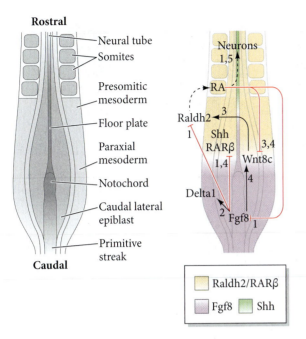

Rostral

- Neural tube
- Somites
- Presomitic mesoderm
- Floor plate
- Paraxial mesoderm
- Notochord
- Caudal lateral epiblast
- Primitive streak

Caudal

Neurons
1,5

RA

Raldh2
3
1
Shh
RARβ
1,4
3,4
Wnt8c
4
Delta1
2
Fgf8
1

☐ Raldh2/RARβ
☐ Fgf8 ☐ Shh

FIGURE 12.6 Signals that regulate axis extension in chick embryos. In the stage 10 chick embryo, Fgf8 inhibits expression of the retinoic acid (RA) synthesizing enzyme Raldh2 in the presomitic mesoderm (1) and the expression of the RA receptor RARβ in the neural ectoderm (4), thus preventing RA from triggering differentiation in the caudal-lateral epiblast cells (those cells adjacent to the node-streak border and which give rise to lateral and dorsal neural tube) and the caudalmost paraxial mesoderm (1,5). In addition, Fgf8 inhibits Sonic hedgehog (Shh) expression in the neural tube floor plate, controlling the onset of ventral patterning genes (1). FGF signaling is also required for expression of Delta1 in the medial portion of the caudal-lateral epiblast cells (2) and promotes expression of Wnt8c (4). As Fgf8 decays in the caudal paraxial mesoderm, Wnt signaling, most likely provided by Wnt8c, now acts to promote Raldh2 in the adjacent paraxial mesoderm (4). RA produced by Raldh2 activity represses Fgf8 (1) and Wnt8c (3,4). (After V. I. Wilson et al. 2009. *Development* 136: 1591–1604.)

regresses, and that this is antagonized by retinoic acid (RA) activity as cells leave this zone (**FIGURE 12.6**; see also Chapter 17; Diez del Corral et al. 2003). (See **Further Development 12.3, Molecular Mechanisms of Migration through the Primitive Streak**, online.)

REGRESSION OF THE PRIMITIVE STREAK AND EPIBOLY OF THE ECTODERM Now a new phase of development begins. As mesodermal ingression continues, the primitive streak starts to regress, moving Hensen's node from near the center of the area pellucida to a more posterior position (**FIGURE 12.7**). The regressing streak leaves in its wake the dorsal axis of the embryo, including the notochord. The notochord is laid down in a head-to-tail direction, starting at the level where the ears and hindbrain form and extending caudally to the tailbud. As in the frog, the pharyngeal endoderm and head mesendoderm will largely induce the anterior parts of the brain, while the notochord will induce the hindbrain and spinal cord. By this time, all the presumptive endodermal and mesodermal cells have entered the embryo and the epiblast is composed entirely of presumptive ectodermal cells.

While the presumptive mesodermal and endodermal cells are moving inward, the ectodermal precursors proliferate and migrate to surround the yolk by epiboly. The enclosure of the yolk by the ectoderm (again, reminiscent of the epiboly of the amphibian ectoderm) is a Herculean task that takes the greater part of 4 days to complete. It involves the continuous production of new cellular material and the migration of the presumptive ectodermal cells along the underside of the vitelline envelope (New 1959; Spratt 1963). Interestingly, only the cells of the outer margin of the area opaca attach firmly to the vitelline envelope. These cells are inherently different from the other blastoderm cells, as they can extend enormous (500 μm) cytoplasmic processes onto the vitelline envelope. These elongated filopodia are believed to be the locomotor apparatus of the marginal cells, by which the marginal cells pull other ectodermal cells around the yolk (Schlesinger 1958). The filopodia bind to fibronectin, a laminar protein that is a component of the chick vitelline envelope. If the contact between the marginal cells and the fibronectin is experimentally broken by adding a soluble polypeptide similar to fibronectin, the filopodia retract and ectodermal migration ceases (Lash et al. 1990).

Thus, as avian gastrulation draws to a close, the ectoderm has surrounded the embryo, the endoderm has replaced the hypoblast, and the mesoderm has positioned itself between these two regions. (See **Further Development 12.4, Epiblast Cell Heterogeneity**, online.)

Axis specification and the avian "organizer"

As a consequence of the sequence in which the head endomesoderm and notochord are established, gastrulating avian (and mammalian) embryos exhibit a distinct anterior-to-posterior gradient. While cells of the posterior portions of the embryo are still part of a primitive streak and entering the inside of the embryo, cells at the anterior end are already starting to form organs (see Darnell et al. 1999). For the next several days, the anterior end of the embryo is more advanced in its development (having had a "head start," if you will) than the posterior end.

THE ROLE OF GRAVITY AND THE PMZ The conversion of the radially symmetrical blastoderm into a bilaterally symmetrical structure appears to be determined by gravity. As the ovum passes through the hen's reproductive tract, it is rotated for about 20 hours in the shell gland. This spinning, at a rate of 15 revolutions per hour, shifts the yolk such that its lighter components (probably containing stored maternal determinants for development) lie beneath one side of the blastoderm. This imbalance tips up one end of the blastoderm, and that end becomes the posterior marginal zone (PMZ), adjacent to where primitive streak formation begins (**FIGURE 12.8**; Kochav and Eyal-Giladi 1971; Bachvarova et al. 1998; Callebaut et al. 2004).

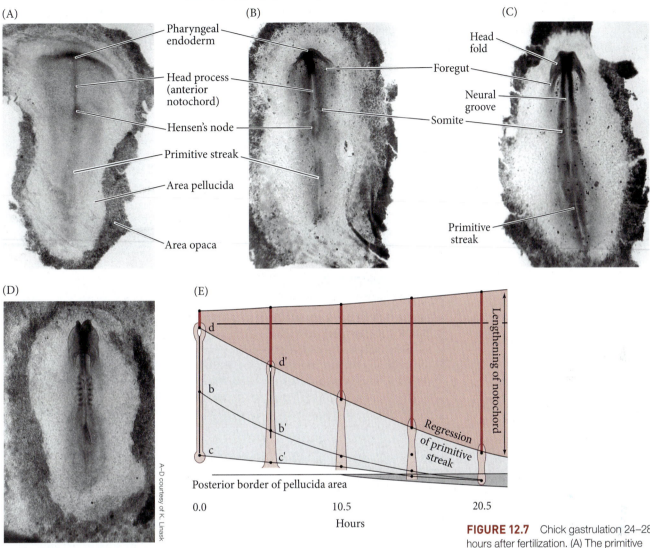

(A)
Pharyngeal endoderm
Head process (anterior notochord)
Hensen's node
Primitive streak
Area pellucida
Area opaca

(B)
Pharyngeal endoderm
Foregut
Somite
Primitive streak

(C)
Head fold
Foregut
Neural groove
Somite
Primitive streak

(D)

A–D courtesy of K. Linask

(E)
d
d'
b
b'
c
c'
Posterior border of pellucida area
Regression of primitive streak
Lengthening of notochord
0.0   10.5   20.5
Hours

FIGURE 12.7 Chick gastrulation 24–28 hours after fertilization. (A) The primitive streak at full extension (24 hours). The head process (anterior notochord) can be seen extending from Hensen's node. (B) Two-somite stage (25 hours). Pharyngeal endoderm is seen anteriorly, while the anterior notochord pushes up the head process beneath it. The primitive streak is regressing. (C) Four-somite stage (27 hours). (D) At 28 hours, the primitive streak has regressed to the caudal portion of the embryo. (E) Regression of the primitive streak, leaving the notochord in its wake. Various points of the streak (represented by letters) were followed after it achieved its maximum length. The *x* axis (time) represents hours after achieving maximum length (the reference line is about 18 hours of incubation). (E after N. T. Spratt Jr. 1947. *J Exp Zool* 104: 69–100.)

It is not known what interactions cause this specific portion of the blastoderm to become the PMZ. Early on, the ability to initiate a primitive streak is found throughout the marginal zone; if the blastoderm is separated into parts, each with its own marginal zone, each part will form its own primitive streak (Spratt and Haas 1960; Bertocchini and Stern 2012; Torlopp et al. 2014). However, once the PMZ has formed, it controls the other regions of the margin. Not only do the cells of the PMZ initiate gastrulation, but they also prevent other regions of the margin from forming their own primitive streaks (Khaner and Eyal-Giladi 1989; Eyal-Giladi et al. 1992; Bertocchini et al. 2004). (See **Further Development 12.5, The PMZ is Equivalent to the Amphibian Nieuwkoop Center**, online.)

Left-right axis formation

The vertebrate body has distinct right and left sides. The heart and spleen, for instance, are generally on the left side of the body, whereas the liver is usually on the right. The distinction between the sides is primarily regulated by the left-sided expression of two proteins: the paracrine factor Nodal and the transcription factor Pitx2. However, the mechanism by which *Nodal* gene expression is activated in the left side of the body differs among the vertebrate classes. The ease with which chick embryos can be manipulated has allowed scientists to elucidate the pathways of left-right axis determination in birds more readily than in other vertebrates.

(A) Surface view of egg   (B) Cross section through egg   (C) Surface view of yolk

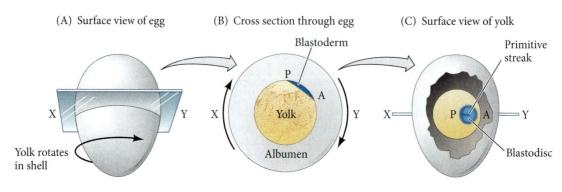

Yolk rotates in shell

FIGURE 12.8 Specification of the chick anterior-posterior axis by gravity. (A) Rotation in the shell gland results in (B) the lighter components of the yolk pushing up one side of the blastoderm. (C) This more elevated region becomes the posterior of the embryo. (After L. Wolpert et al. 1998. *Principles of Development*. Current Biology Ltd.: London.)

FURTHER DEVELOPMENT

THE MOLECULAR MECHANISM OF LEFT-RIGHT AXIS SPECIFICATION IN THE CHICK As the primitive streak reaches its maximum length, transcription of the *Sonic hedgehog* gene (*Shh*) becomes restricted to the left side of the embryo, controlled by activin and its receptor (**FIGURE 12.9A**). Activin signaling, along with BMP4, appears to block the expression of Sonic hedgehog protein and to activate expression of Fgf8 protein on the right side of the embryo. Fgf8 blocks expression of the paracrine factor Cerberus on the right-hand side; it may also

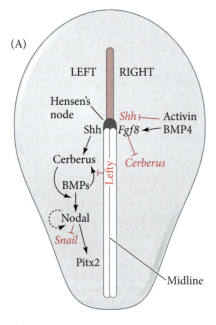

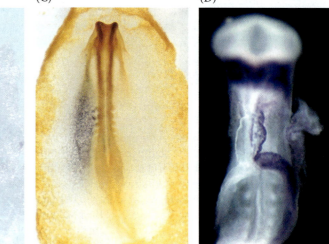

From C. Rodríguez Esteban et al. 1999. *Nature* 401: 243–251, courtesy of J. Izpisúa Belmonte

Courtesy of C. Stern

From M. Logan et al. 1998. *Cell* 94: 307–331, courtesy of C. Tabin

FIGURE 12.9 Model for generating left-right asymmetry in the chick embryo. (A) On the left side of Hensen's node, Sonic hedgehog (Shh) activates Cerberus, which stimulates BMPs to induce the expression of Nodal. In the presence of Nodal, the *Pitx2* gene is activated. Pitx2 protein is active in the various organ primordia and specifies which side will be the left. On the right side of the embryo, activin is expressed, along with activin receptor IIa. This activates Fgf8, a protein that blocks expression of the gene for Cerberus. In the absence of Cerberus, Nodal is not activated and thus Pitx2 is not expressed. (B) Whole-mount in situ hybridization of *Cerberus* mRNA. This view is from the ventral surface ("from below," so the expression seems to be on the right). Viewed from the dorsal surface, the expression pattern would be on the left. (C) Whole-mount in situ hybridization using probes for the chick *Nodal* message (stained purple) shows its expression in the lateral plate mesoderm only on the left side of the embryo. This view is from the dorsal side. (D) Similar in situ hybridization, using the probe for *Pitx2* at a later stage of development. The embryo is seen from its ventral surface. At this stage, the heart is forming, and *Pitx2* expression can be seen on the left side of the heart tube (as well as symmetrically in more anterior tissues). (A after A. Raya and J. C. Izpisúa Belmonte. 2004. *Mech Dev* 121: 1043–1054.)

activate a signaling cascade that instructs the mesoderm to have right-sided capacities (Schlueter and Brand 2009).

Meanwhile, on the left side of the body, Shh protein activates Cerberus (**FIGURE 12.9B**), which in this case acts with BMP to stimulate the synthesis of Nodal protein (Yu et al. 2008). Nodal activates the *Pitx2* gene while repressing *Snail*. In addition, Lefty1 in the ventral midline prevents the Cerberus signal from passing to the right side of the embryo (**FIGURE 12.9C,D**). As in *Xenopus*, Pitx2 is crucial in directing the asymmetry of the embryonic structures. Experimentally induced expression of either Nodal or Pitx2 on the right side of the chick embryo reverses the asymmetry or causes randomization of asymmetry on the right or left sides (Levin et al. 1995; Logan et al. 1998; Ryan et al. 1998).[3]

Early Development in Mammals

Mammalian eggs are among the smallest in the animal kingdom, making them difficult to manipulate experimentally. The human zygote, for instance, is only 100 µm in diameter—barely visible to the eye and less than one-thousandth the volume of a *Xenopus laevis* egg. Also, mammalian zygotes are not produced in numbers comparable to sea urchin or frog zygotes; a female mammal usually ovulates fewer than 10 eggs at a given time, so it is difficult to obtain enough material for biochemical studies. As a final hurdle, the development of placental mammalian embryos is accomplished inside another organism rather than in the external environment (although early embryos prior to implantation can be cultured and observed in vitro). Most research on mammalian development has focused on the mouse, since mice are relatively easy to breed, have multiple progeny in their litters, and are easily housed in laboratories.

Mammalian cleavage

Prior to fertilization, the mammalian oocyte, wrapped in cumulus cells, is released from the ovary and swept by the fimbriae into the oviduct (**FIGURE 12.10**). Fertilization occurs in the **ampulla** of the oviduct, a region close to the ovary. Meiosis is completed after sperm entry, and the first cleavage begins about a day later (see Figure 7.32). The positioning of the first cleavage plane may depend on the point of sperm entry (Piotrowska and Zernicka-Goetz 2001), and in mice, a sperm-borne microRNA (miRNA-34c) is required to initiate this first cell division. This miRNA appears to bind and inhibit Bcl-2, a protein that prevents the cell from entering the S phase of the cell cycle (Liu et al. 2012). The two nuclei produced by this cleavage are the first nuclei to contain the entire genome, since the haploid pronuclei enter cell division upon meeting (see Chapter 7).

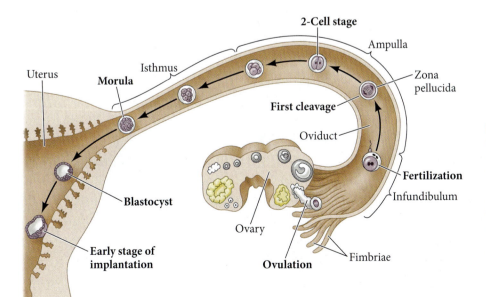

(?) Developing Questions

The real mystery is, what processes create the original asymmetry of Shh and Fgf8? Curiously, the first asymmetry seen during the formation of Hensen's node in chicks involves Fgf8- and Shh-expressing cells rearranging themselves to converge on the right-hand side of Hensen's node (Cui et al. 2009; Gros et al. 2009). What establishes this initial asymmetry is still unknown. Perhaps it's a chemoattractant for cell migration, or it may involve a physical displacement of cells around Hensen's node (Gros et al., 2009; Tsikolia et al. 2012; Otto et al. 2014). Pick a *side* and check it out.

FIGURE 12.10 Development of a human embryo from fertilization to implantation. Compaction of the human embryo occurs on day 4, at the 10-cell stage. The embryo "hatches" from the zona pellucida upon reaching the uterus. During its migration to the uterus, the zona prevents the embryo from prematurely adhering to the oviduct rather than traveling to the uterus. (After H. Tuchmann-Duplessis et al. 1971. *Embryogenesis. Illustrated Human Embryology*, vol 1. Springer: New York.)

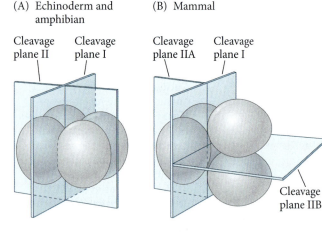

(A) Echinoderm and amphibian

Cleavage plane II   Cleavage plane I

(B) Mammal

Cleavage plane IIA   Cleavage plane I

Cleavage plane IIB

FIGURE 12.11 Comparison of early cleavage in (A) echinoderms and amphibians (radial cleavage) and (B) mammals (rotational cleavage). Nematodes also have a rotational form of cleavage, but they do not form the blastocyst structure characteristic of mammals. (After B. J. Gulyas. 1975. *J Exp Zool* 193: 235–248.)

THE UNIQUE NATURE OF MAMMALIAN CLEAVAGE Cleavages in mammalian eggs are among the slowest in the animal kingdom, taking place some 12–24 hours apart. The cilia in the oviduct push the embryo toward the uterus, and the first cleavages occur along this journey. In addition to the slowness of cell division, several other features distinguish mammalian cleavage, including the unique orientation of mammalian blastomeres relative to one another. In many but not all mammalian embryos, the first cleavage is a normal meridional division; however, in the second cleavage, one of the two blastomeres divides meridionally and the other divides equatorially (**FIGURE 12.11**). This is called **rotational cleavage** (Gulyas 1975).

Another major difference between mammalian cleavage and that of anamniotes is the marked asynchrony of early cell division. Mammalian blastomeres do not all divide at the same time. Thus, mammalian embryos do not increase exponentially from 2 to 4 to 8 cells, but frequently contain odd numbers of cells. Furthermore, the mammalian genome, unlike the genomes of rapidly developing animals, is activated during early cleavage, and zygotically transcribed proteins are necessary for cleavage and development. Maternally encoded proteins can persist through most of the cleavage stages and play important roles in early development. In the mouse and goat, the activation of zygotic (i.e., nuclear) genes begins in the late zygote and continues through the 2-cell stage (Zeng and Schultz 2005; Rother et al. 2011). In humans, the zygotic genes are activated slightly later, around the 8-cell stage (Pikó and Clegg 1982; Braude et al. 1988; Dobson et al. 2004). (See **Further Development 12.6, Epigenetic Regulation of Histone States Is Required for the Maternal to Zygotic Transition in the Mouse**, online.)

COMPACTION One of the most crucial events of mammalian cleavage is **compaction**. Mouse blastomeres through the 8-cell stage form a loose arrangement (**FIGURE 12.12A–C**). Following the third cleavage, however, the blastomeres undergo a spectacular change in their behavior. Cell adhesion proteins such as E-cadherin become expressed, and the blastomeres gradually huddle together and form a compact ball of cells (**FIGURE 12.12D**; Peyrieras et al. 1983; Fleming et al. 2001). This tightly packed arrangement is stabilized by tight junctions that form between the outside cells of the ball, sealing off the inside of the sphere. The cells within the sphere form gap junctions, thereby enabling small molecules and ions to pass between them (much like in the fish early blastula).

FIGURE 12.12 Cleavage of a single mouse embryo in vitro. (A) 2-Cell stage. (B) 4-Cell stage. (C) Early 8-cell stage. (D) Compacted 8-cell stage. (E) Morula. (F) Blastocyst. (G) Electron micrograph through the center of a mouse blastocyst.

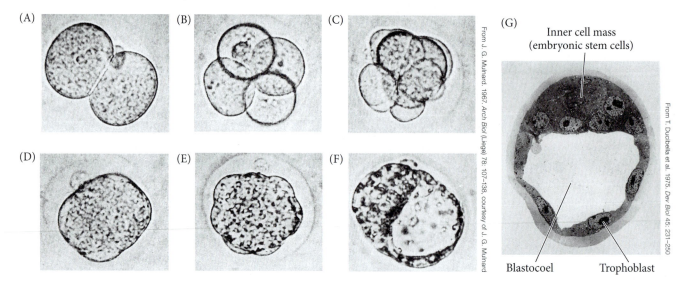

(A)   (B)   (C)

(D)   (E)   (F)

(G)

Inner cell mass (embryonic stem cells)

Blastocoel   Trophoblast

From J. G. Mulnard. 1967. *Arch Biol* (Liege) 78: 107–138, courtesy of J. G. Mulnard

From T. Ducibella et al. 1975. *Dev Biol* 45: 231–250

The cells of the compacted 8-cell embryo divide to produce a 16-cell **morula** (**FIGURE 12.12E**). The morula consists of a small group of internal cells surrounded by a larger group of external cells (Barlow et al. 1972). Most of the descendants of the external cells become **trophoblast** (trophectoderm) cells, whereas the internal cells give rise to the **inner cell mass** (**ICM**). The inner cell mass, which will give rise to the embryo, becomes positioned on one side of the ring of trophoblast cells; the resulting **blastocyst** is another hallmark of mammalian cleavage (**FIGURE 12.12F,G**; see also Figure 5.7).

The trophoblast cells produce no embryonic structures, but rather form the tissues of the chorion, the extraembryonic membrane and portion of the placenta that enables the fetus to get oxygen and nourishment from the mother. The chorion also secretes hormones that cause the mother's uterus to retain the fetus, and it produces regulators of the immune response so that the mother will not reject the embryo.

It is important to remember that a crucial outcome of these first divisions is the generation of cells that attach the embryo to the uterus. Thus, formation of the trophectoderm is the first differentiation event in mammalian development. The earliest blastomeres (such as each blastomere of a 2-cell embryo) can form both trophoblast cells and the embryo precursor cells of the ICM. These very early cells are said to be **totipotent** (Latin, "capable of everything"). The cells of the inner cell mass are said to be **pluripotent** (Latin, "capable of many things"). That is, each cell of the ICM can generate any cell type in the body but is no longer able to form the trophoblast. These pluripotent cells of the inner cell mass can be isolated and placed in culture, where under appropriate conditions they will self-renew indefinitely, maintain their pluripotency, and thus acquire the characteristic properties of embryonic stem cells (see Chapter 5).

Trophoblast or ICM? The first decision of the rest of your life

The philosopher and theologian Søren Kierkegaard wrote that we define ourselves by the choices we make. It seems that the embryo already knows this. The decision to become either trophoblast or inner cell mass is the first binary decision in mammalian life. As we described in Chapter 5, embryonic cells must eventually lose their pluripotency and decide on what they are going to grow up to be (see Figure 5.8). In the first decision, Oct4 and Cdx2 reciprocally repress each other's gene transcription, enabling some cells to be trophoblast and other cells to become the pluripotent cells of the ICM. In the second decision, each of the cells of the ICM expresses (among other proteins) either Nanog or Gata6, thereby retaining its pluripotency (Nanog) or becoming primitive endoderm (Gata6) (Ralston and Rossant 2005; Rossant 2016).

Prior to blastocyst formation, each embryonic blastomere expresses both Cdx2 and Oct4 transcription factors (Niwa et al. 2005; Dietrich and Hiiragi 2007; Ralston and Rossant 2008) and appears to be capable of becoming either ICM or trophoblast (Hiiragi and Solter 2004; Motosugi et al. 2005; Kurotaki et al. 2007). However, once the decision to become either trophoblast or ICM is made, the cell expresses a set of genes specific to each region. As mentioned, the pluripotency of the ICM is maintained by a core of three transcription factors: Oct4, Sox2, and Nanog. These proteins bind to the enhancers of their own genes to maintain their expression while at the same time activating one another's enhancers (**FIGURE 12.13**). Thus, when one of these genes is activated, the other ones are too. Acting in concert, Sox2 and Oct4 form a dimer and often reside on enhancers adjacent to Nanog, activating those genes required to maintain pluripotency in embryonic stem (ES) cells and repressing those genes whose products would lead to differentiation (Marson et al. 2008; Kagey et al. 2010; Adamo et al. 2011; Young 2011). (See **Further Development 12.7, The Role of Hippo Signaling during Trophoblast-ICM Determination**, and **Further Development 12.8, Mechanisms of Compaction and Formation of the Inner Cell Mass**, both online. Additionally, consult Chapter 5, Figure 5.9.)

(?) Developing Questions

We have been discussing eutherian mammals, those organisms such as mice and humans that retain the fetus during its development. But what about monotreme mammals (such as the platypus) that lay eggs, or marsupial mammals (such as kangaroos) that have extremely short pregnancies? Do their embryos have blastocysts?

(A) (B)

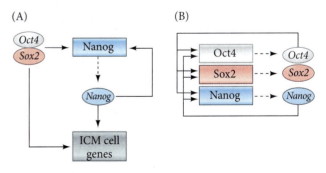

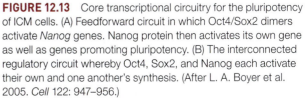

FIGURE 12.13 Core transcriptional circuitry for the pluripotency of ICM cells. (A) Feedforward circuit in which Oct4/Sox2 dimers activate *Nanog* genes. Nanog protein then activates its own gene as well as genes promoting pluripotency. (B) The interconnected regulatory circuit whereby Oct4, Sox2, and Nanog each activate their own and one another's synthesis. (After L. A. Boyer et al. 2005. *Cell* 122: 947–956.)

CAVITATION In mice, the embryo proper is derived from the inner cell mass of the 16-cell stage, supplemented by cells dividing from the outer cells of the morula during the transition to the 32-cell stage (Pedersen et al. 1986; Fleming 1987; McDole et al. 2011). The cells of the ICM give rise to the embryo and its associated yolk sac, allantois, and amnion. By the 64-cell stage, the ICM (comprising approximately 13 cells at that stage) and the trophoblast cells have become separate cell layers, neither of which contributes cells to the other group (Dyce et al. 1987; Fleming 1987). The ICM actively supports the trophoblast, secreting proteins that stimulate the trophoblast cells to divide (Tanaka et al. 1998).

Initially, the morula does not have an internal cavity. However, during a process called **cavitation**, the trophoblast cells secrete fluid into the morula to create a blastocoel. The membranes of trophoblast cells contain sodium pumps (Na^+-K^+ ATPase and the Na^+-H^+ exchanger) that pump Na^+ into the central cavity. The subsequent accumulation of Na^+ draws in water osmotically, creating and enlarging the blastocoel (Borland 1977; Ekkert et al. 2004; Kawagishi et al. 2004). Interestingly, this sodium-pumping activity appears to be stimulated by the oviduct cells on which the embryo is traveling toward the uterus (Xu et al. 2004). As the blastocoel expands, the inner cell mass becomes positioned on one side of the ring of trophoblast cells, resulting in the distinctive mammalian blastocyst.[4] (See **Further Development 12.9, Escape from the Zona Pellucida and Implantation**, online.)

Mammalian gastrulation

Birds and mammals are both descendants of reptilian species. It is not surprising, therefore, that mammalian development parallels that of reptiles and birds. What *is* surprising is that the gastrulation movements of reptilian and avian embryos, which evolved as an adaptation to yolky eggs, are retained in the mammalian embryo even in the absence of large amounts of yolk. The mammalian inner cell mass can be envisioned as sitting atop an imaginary ball of yolk, following instructions that seem more appropriate to its reptilian ancestors.

MODIFICATIONS FOR DEVELOPMENT INSIDE ANOTHER ORGANISM The mammalian embryo obtains nutrients directly from its mother and does not rely on stored yolk. This adaptation has entailed a dramatic restructuring of the maternal anatomy (such as expansion of the lower oviduct to form the uterus) as well as the development of a combined embryonic-maternal organ (the placenta) capable of absorbing maternal nutrients. The origins of early mammalian tissues are summarized in **FIGURE 12.14**. As we saw above, the first distinction is between the inner cell mass and the trophoblast. The trophoblast develops through several stages, eventually becoming the chorion, the embryonically derived portion of the placenta. Trophoblast cells also induce the mother's uterine cells to form the maternal portion of the placenta, the **decidua**. The decidua becomes rich in the blood vessels that will provide oxygen and nutrients to the embryo. The inner cell mass gives rise to the epiblast and the hypoblast (**primitive endoderm**). The hypoblast will generate yolk sac cells, while the epiblast will generate the embryo, the amnion, and the allantois.

THE PRIMITIVE ENDODERM: THE MAMMALIAN HYPOBLAST A significant amount of differentiation happens in the blastocyst prior to its implanting into the wall of the uterus. This marks a period of mammalian development called the **peri-implantation** period, which pertains to the time from when the blastocyst is free in the uterus through its initial interactions with the uterine endometrium. First cells within the inner cell mass segregate to form two layers (**FIGURE 12.15**). The lower layer, in contact with the blastocoel, is called the primitive endoderm (PrE), and it is homologous to the hypoblast of the chick embryo. The upper layer is called the epiblast. The primitive endoderm will form the yolk sac of the embryo and, like the chick hypoblast, will be used for positioning the site of gastrulation, regulating cell movements in the epiblast, and promoting the maturation of blood cells. Moreover, the primitive endoderm, like the chick hypoblast, is largely an extraembryonic layer and does not provide many cells to the actual embryo (see Stern and Downs 2012).

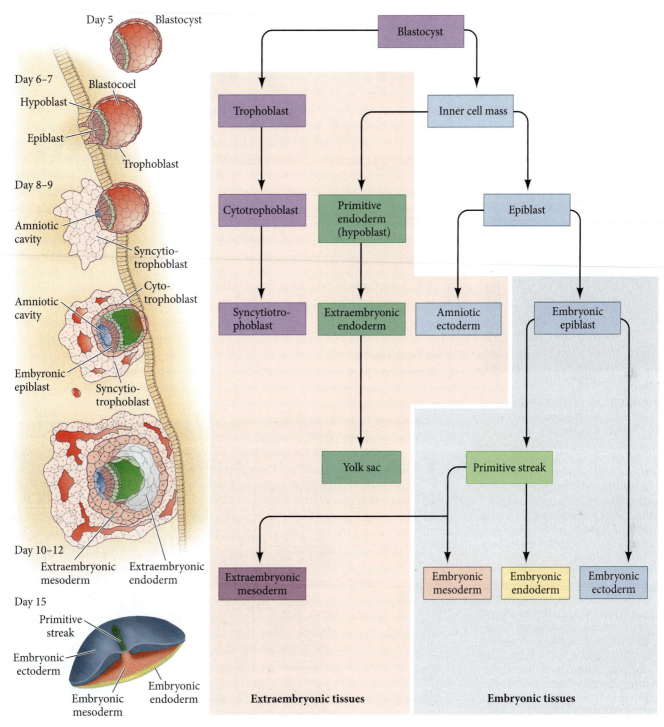

FIGURE 12.14 Tissue and germ layer formation in the early human embryo. Days 5–9: Implantation of the blastocyst. The inner cell mass delaminates hypoblast cells that line the blastocoel, forming the extraembryonic endoderm of the primitive yolk sac and a bilayered (epiblast and hypoblast) blastodisc. Days 10–12: The trophoblast divides into the cytotrophoblast, which will form the villi, and the syncytiotrophoblast, which will ingress into the uterine tissue to form the chorion. Days 12–15: Gastrulation and formation of primitive streak. Meanwhile, the epiblast splits into the amniotic ectoderm (which encircles the amniotic cavity) and the embryonic epiblast. The adult mammal (ectoderm, endoderm, mesoderm, and germ cells) forms from the cells of the embryonic epiblast. The extraembryonic endoderm forms the yolk sac. The actual size of the embryo at this stage is about that of the period at the end of this sentence. (After W. P. Luckett. 1978. *Am J Anat* 152: 59–97.)

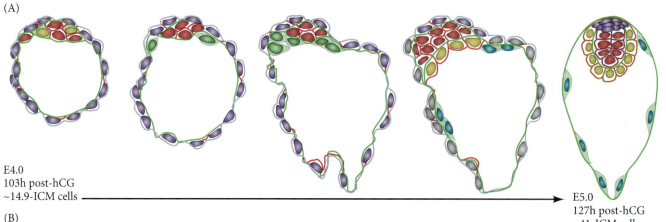

(A)

E4.0
103h post-hCG
~14.9-ICM cells

E5.0
127h post-hCG
~41-ICM cells

(B)

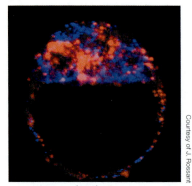

Courtesy of J. Rossant

Nanog (EPI); Gata6

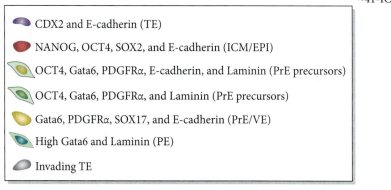

- CDX2 and E-cadherin (TE)
- NANOG, OCT4, SOX2, and E-cadherin (ICM/EPI)
- OCT4, Gata6, PDGFRα, E-cadherin, and Laminin (PrE precursors)
- OCT4, Gata6, PDGFRα, and Laminin (PrE precursors)
- Gata6, PDGFRα, SOX17, and E-cadherin (PrE/VE)
- High Gata6 and Laminin (PE)
- Invading TE

FIGURE 12.15 Stages of cell differentiation in the peri-implantation embryo. (A) Changes in morphology and gene expression during peri-implantation development of the mouse embryo between 3.75 and 4.75 days after fertilization. The trophectoderm (TE; purple) is the outermost layer. The cells of the inner cell mass begin as a mosaic pattern of fates, but sort themselves out into an upper epiblast, where the body of the embryo will form (red), and a lower primitive endoderm, which then forms the visceral endoderm (yellow) below the epiblast and the parietal endoderm (blue) spreading out along the inner side of the trophectoderm. (B) Mouse embryo at day 3.5 (early blastocyst), showing the random expression of Nanog (blue, for the epiblast) and Gata6 (red, for the primitive endoderm, among other potential fates) in the inner cell mass. C) Immunofluorescence of transverse sections (a–d) of a single embryo at 5 days of development; levels of these sections are shown by black lines labeled a–d on the diagram on the left. The differential expression of Gata6 is evident in a variety of developing cell types in the peri-implantation blastocyst. Arrows indicate the parietal endoderm. ICM/EPI, inner cell mass/epiblast; TE, trophectoderm; PrE, primitive endoderm; PPrE, peripheral primitive endoderm; PE, parietal endoderm; VE, visceral endoderm; ExE, extraembryonic endoderm; CDH1, E-cadherin.

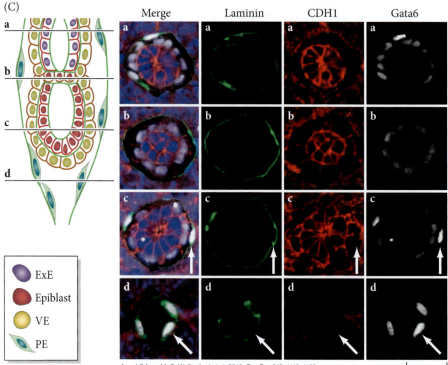

(C)

| | Merge | Laminin | CDH1 | Gata6 |

- ExE
- Epiblast
- VE
- PE

A and C from M. C. Wallingford et al. 2013. *Dev Dyn* 242: 1110–1120

50 μm

Whether a mouse ICM cell becomes epiblast or primitive endoderm may depend on *when* the cell became part of the ICM (Bruce and Zernicka-Goetz 2010; Morris et al. 2010). Cells that become internalized in the division from 8 to 16 cells appear biased to become pluripotent epiblast cells, while the future primitive endoderm may be generated by cells entering the ICM during the division from 16 to 32 cells (see Figure 12.15A). At that stage, the blastomeres of the ICM are a mosaic of future epiblast cells (expressing the Nanog transcription factor, which promotes pluripotency) and primitive endoderm cells (expressing Gata6 transcription factor) a full day before the layers segregate at day 4.5 (see Figure 12.15B; Chazaud et al. 2006).

The epiblast and primitive endoderm form the **bilaminar germ disc** (**FIGURE 12.16A**). The primitive endoderm cells expand to line the blastocoel cavity, where they give rise to the yolk sac. The primitive endoderm cells contacting the epiblast are the **visceral endoderm**, while those yolk sac cells contacting the trophoblast are the **parietal endoderm** (see Figure 12.15). The epiblast cell layer is split by small clefts that eventually coalesce to separate the embryonic epiblast from the other epiblast cells that form the amnion. Once the amnion is completed, the amniotic cavity fills with **amniotic fluid**, a secretion that serves as a shock absorber and prevents the developing embryo from drying out. The embryonic epiblast is thought to contain all the cells that will generate the actual embryo and is similar in many ways to the avian epiblast.

Gastrulation begins at the posterior end of the embryo, and this is where the primitive streak arises (**FIGURE 12.16B,C**). Like the chick epiblast cells, mammalian mesoderm and endoderm cells originate in the epiblast, undergo an epithelial-mesenchymal transition, lose E-cadherin, and migrate through the primitive streak as individual mesenchymal cells (**FIGURE 12.17**; Burdsal et al. 1993; Williams et al. 2012). Eventually, a **node** forms as a thickened bulb at the anterior end of the primitive streak.[5] Those cells arising from the node give rise to the notochord. However, in contrast to notochord formation in the chick, the cells that form the mouse notochord are thought to become integrated into the endoderm of the primitive gut (Jurand 1974; Sulik et al. 1994). These cells can be seen as a band of small, ciliated cells extending rostrally from the node. They form the notochord by converging medially and "budding" off in a dorsal direction from the roof of the gut. The timing of these developmental events varies enormously in mammals. In humans, the migration of cells forming the mesoderm doesn't start until day 16—around the time that a mouse embryo is almost ready to be born (see Figure 12.16C; Larsen 1993). (See **Further Development 12.10, The Role of FGF in Early Mouse Gastrulation**, and **Further Development 12.11, Placental Formation and Function**, both online.)

Mammalian axis formation

Biologist and poet Miroslav Holub (1990) remarked:

> *Between the fifth and tenth days the lump of stem cells differentiates into the overall building plan of the [mouse] embryo and its organs. It is a bit like a lump of iron turning into the space shuttle. In fact it is the profoundest wonder we can still imagine and accept, and at the same time so usual that we have to force ourselves to wonder about the wondrousness of this wonder.*

It is, indeed, wonderful, and we are just beginning to find out how really amazing it is.

FIGURE 12.16 Amnion structure and cell movements during human gastrulation. (A,B) Human embryo and uterine connections at day 15 of gestation. (A) Sagittal section through the midline. (B) View looking down on the dorsal surface of the embryo. Movements of the epiblast cells through the primitive streak and the node and underneath the epiblast are superimposed on the dorsal surface view. (C) At days 14 and 15 the ingressing epiblast cells are thought to replace the hypoblast cells (which contribute to the yolk sac lining), and at day 16 the ingressing cells fan out to form the mesodermal layer. (After W. J. Larsen. 1993. *Human Embryology*. Churchill Livingstone: New York.)

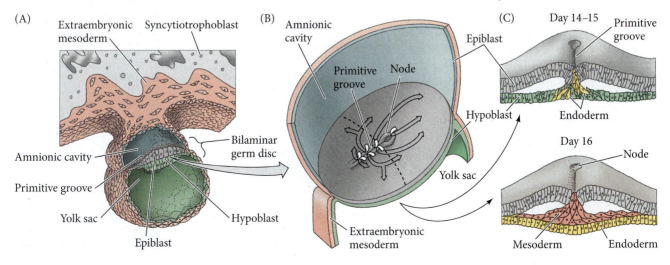

FIGURE 12.17 Epithelial-mesenchymal transitions at the primitive streak precede ingression of the mesoderm. (A) Epiblast cells in the mouse embryo express E-cadherin and apical cell markers such as apical polarity protein (aPKC; green arrows). The localized breakdown of the basal lamina (red fluorescence, Collagen IV) reveals the position of the primitive streak (brackets), at which epiblast cells undergo an epithelial-mesenchymal transition and ingress to form the underlying mesoderm, which lacks both E-cadherin and aPKC expression (white arrowhead). (B) Using live 4-dimensional imaging of mouse embryos in culture, the morphology of epiblast cells can be followed as they change from epithelial (blue asterisks), to bottle-shaped cells during apical constriction (yellow asterisks), and finally into a rounded mesenchymal cell entering into the mesoderm layer (red asterisks). (C) Tracking cells (colored lines) over time during primitive streak formation reveals cell intercalation behaviors that converge on the primitive streak at the midline, resulting in the extension of the anterior-posterior axis (double arrow).

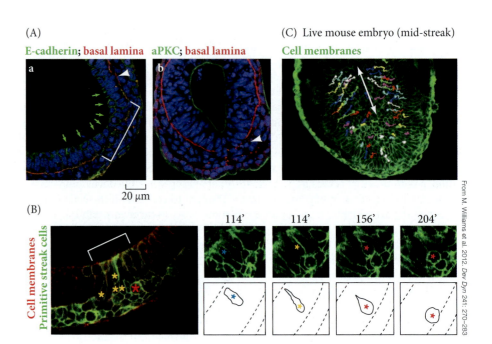

(A)

E-cadherin; basal lamina   aPKC; basal lamina

(C) Live mouse embryo (mid-streak)

Cell membranes

20 µm

(B)

Cell membranes
Primitive streak cells

114'   114'   156'   204'

From M. Williams et al. 2012. *Dev Dyn* 241: 270–283

THE ANTERIOR-POSTERIOR AXIS: TWO SIGNALING CENTERS

The formation of the mammalian anterior-posterior axis has been studied most extensively in mice. The structure of the mouse epiblast, however, differs from that of humans in that it is cup-shaped rather than disc-shaped. Whereas the human embryo looks very much like the chick embryo, the mouse embryo "drops" such that it looks like a droplet enclosed by the primitive endoderm (**FIGURE 12.18A**).

The mammalian embryo appears to have two signaling centers: one in the node (equivalent to Hensen's node and the trunk portion of the amphibian organizer), and one in the **anterior visceral endoderm** (**AVE**; Beddington and Robertson 1999; Foley et al. 2000). Although the node and the AVE have some overlapping functions, the node appears to be primarily responsible for neural induction and for the patterning of most of the anterior-posterior axis, while the AVE is most critical for positioning the primitive streak (see Bachiller et al. 2000).

FIGURE 12.18 Axis and notochord formation in the mouse. (A) In the 7-day mouse embryo, the dorsal surface of the epiblast (embryonic ectoderm) is in contact with the amniotic cavity. The ventral surface of the epiblast contacts the newly formed mesoderm. In this cuplike arrangement, the endoderm covers the surface of the embryo. The node is at the bottom of the cup, and it has generated chordamesoderm. The two signaling centers, the node and the anterior visceral endoderm (AVE), are located on opposite sides of the cup. Eventually, the notochord will link them. The caudal side of the embryo is marked by the presence of the allantois. (B) Confocal fluorescence image of *Cerberus* gene expression, with the *Cerberus* gene fused to a gene for GFP. At this stage, the Cerberus-synthesizing cells (green) are migrating to the most anterior region of the visceral endoderm.

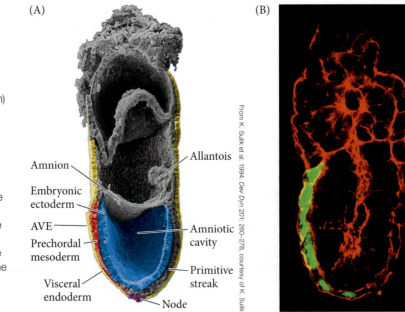

(A)

Amnion
Embryonic ectoderm
AVE
Prechordal mesoderm
Visceral endoderm
Allantois
Amniotic cavity
Primitive streak
Node

(B)

From K. Sulik et al. 1994. *Dev Dyn* 201: 260–278, courtesy of K. Sulik

Courtesy of J. Belo

The signals that initiate the primitive streak appear to come from interactions between the trophoblast-derived extraembryonic ectoderm and the epiblast. BMP4 originating from the extraembryonic ectoderm instructs the adjacent epiblast cells to make Wnt3a and Nodal. However, the AVE prevents Wnt3a and Nodal from having an effect on the anterior side of the embryo, by secreting antagonists of these paracrine factors, Lefty-1, Dickkopf, and Cerberus (**FIGURE 12.18B**; Brennan et al. 2001; Perea-Gomez et al. 2001; Yamamoto et al. 2004). As in amphibian embryos, the anterior region is protected from Wnt signals. Thus, Wnt3a activates the *Brachyury* gene in cells of the *posterior* but not the *anterior* epiblast, generating mesoderm cells (Bertocchini and Stern 2002; Perea-Gomez et al. 2002). Once formed, the node secretes Chordin; the head process and notochord will later add Noggin. Mice missing *both* genes lack a forebrain, nose, and other facial structures.

But how does the mouse AVE form? The answer was unexpected. The AVE, and thus the anterior-posterior axis of the mammal, appears to be generated by an environmental force—the shape of the uterus. The uterus constrains the embryo such that growth occurs only in one direction. This stretching "downward" breaks the extracellular matrix and induces new gene expression in the distalmost epiblast cells (**FIGURE 12.19**). The products of these newly expressed genes cause the cells to migrate anteriorly and become the AVE. Hiramatsu and colleagues (2013) found that if an embryo grows in nonconstricted chambers, the A-P axis doesn't form. Thus, the mechanical force generated by the uterus is critical in instructing normal development.

The restriction of Nodal signaling to the posterior part of the embryo is related to the formation of the primitive streak in this region, and the inhibition of Nodal anteriorly is related to the inhibition of streak formation in this region. Later in development, however, Nodal *is* expressed anteriorly and is important in head-tail patterning: one of its key roles is in left/right specification—it is expressed in the left lateral mesoderm in the early hindbrain region, for example. (See **Further Development 12.12, Anterior-Posterior Patterning by FGF and RA Gradients**, online.)

ANTERIOR-POSTERIOR PATTERNING: THE HOX CODE HYPOTHESIS   In all vertebrates, anterior-posterior polarity becomes specified by the expression of Hox genes. If you reviewed Further Development 12.10 online, then you would have learned that the posterior gradient of FGF along with an opposing anterior gradient of retinoic acid in the region of the embryo where the primitive streak is located leads to the differential activation of *Hox* gene expression along the anterior-to-posterior axis. Vertebrate Hox genes are homologous to the homeotic selector genes (Hom-C genes) of the fruit fly (see Chapter 9). The *Drosophila* homeotic gene complex on chromosome 3 contains the *Antennapedia* and *bithorax* clusters (see Figure 9.24) and can be seen as a single functional unit. (Indeed, in some other insects, such as the flour beetle *Tribolium*, it *is* a single physical unit.) All of the known mammalian genomes contain four copies of the Hox complex per haploid set, located on four different chromosomes (*Hoxa* through *Hoxd* in

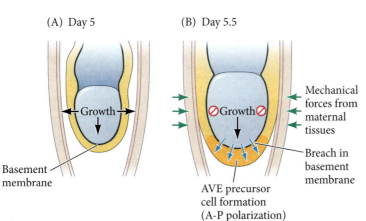

(A) Day 5   (B) Day 5.5

FIGURE 12.19 Formation of the AVE precursor cells by mechanical stress in the mouse embryo. (A) At day 5 of embryogenesis, growth is not restricted by the shape of the uterus, and the embryo grows in several directions. (B) About 12 hours later, embryonic growth becomes restricted, and the embryo grows only in the proximal-distal direction. The basement membrane at the distal region breaks, and epiblast cells enter the visceral endoderm layer (blue arrows), forming the precursors of the AVE. (After R. Hiramatsu et al. 2013. *Dev Cell* 27: 131–144.)

the mouse, *HOXA* through *HOXD* in humans; see Boncinelli et al. 1988; McGinnis and Krumlauf 1992; Scott 1992).

The order of these genes on their respective chromosomes is remarkably similar in insects and humans, as is the expression pattern of these genes. Those mammalian genes homologous to the *Drosophila labial*, *proboscipedia*, and *deformed* genes are expressed anteriorly and early, whereas those genes homologous to the *Drosophila AbdB* gene are expressed posteriorly and later. As in *Drosophila*, a separate set of genes in mice encodes the transcription factors that regulate head formation. In *Drosophila*, these are the *orthodenticle* and *empty spiracles* genes. In mice, the midbrain and forebrain are made through the expression of genes homologous to these—*Otx2* and *Emx* (see Kurokawa et al. 2004; Simeone 2004).

Mammalian Hox/HOX genes are numbered from 1 to 13, starting from the end of each complex that is expressed most anteriorly. **FIGURE 12.20** shows the relationships between the *Drosophila* and mouse homeotic gene sets. The equivalent genes in each mouse complex (such as *Hoxa4*, *b4*, *c4*, and *d4*) are **paralogues**—that is, it is thought that the four mammalian Hox complexes were formed by chromosome duplications. Because the correspondence between the *Drosophila* Hom-C genes and mouse Hox genes is not one-to-one, it is likely that independent gene duplications and deletions have occurred since these two animal groups diverged (Hunt and Krumlauf 1992). Indeed, the most posterior mouse Hox gene (equivalent to *Drosophila AbdB*) underwent its own set of duplications in some mammalian chromosomes.

Hox gene expression can be seen along the mammalian body axis (in the neural tube, neural crest, paraxial mesoderm, and surface ectoderm) from the anterior boundary of the hindbrain through the tail. The regions of expression are not in register, but the 3′ Hox genes (homologous to *labial*, *proboscipedia*, and *deformed* of the fly) are expressed more anteriorly than the 5′ Hox genes (homologous to *Ubx*, *AbdA*, and *AbdB*). Thus, one

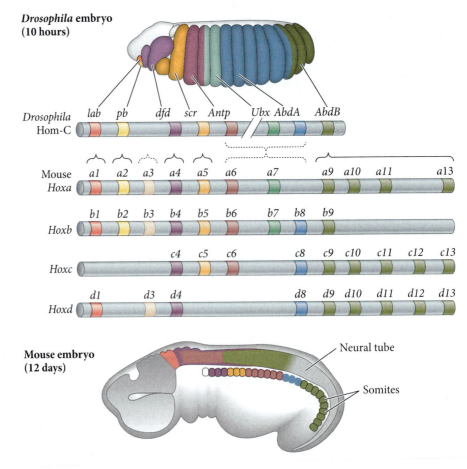

FIGURE 12.20 Evolutionary conservation of homeotic gene organization and transcriptional expression in fruit flies and mice is seen in the similarity between the Hom-C cluster on *Drosophila* chromosome 3 and the four Hox gene clusters in the mouse genome. Genes with similar structures occupy the same relative positions on each of the four chromosomes, and paralogous gene groups display similar expression patterns. The mouse genes in the higher-numbered groups are expressed later in development and more posteriorly. The comparison of the transcription patterns of the Hom-C and *Hoxb* genes of *Drosophila* and mice are shown above and below the chromosomes, respectively. (After S. D. Huber et al. 2010. *PLOS ONE* 5(5): e10820. doi:10.1371/journal.pone.0010820; S. B. Carroll. 1995. *Nature* 376: 479–485.)

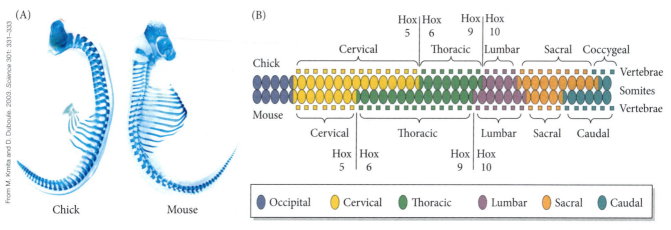

(A) Chick   Mouse

From M. Kmita and D. Duboule. 2003. *Science* 301: 331–333

(B) Chick / Mouse

Occipital   Cervical   Thoracic   Lumbar   Sacral   Caudal

FIGURE 12.21   Schematic representation of the chick and mouse vertebral pattern along the anterior-posterior axis. (A) Axial skeletons stained with alcian blue at comparable stages of development. The chick has twice as many cervical vertebrae as the mouse. (B) The boundaries of expression of certain Hox gene paralogous groups (*Hox5/6* and *Hox9/10*) have been mapped onto the vertebral type domains. (B after A. C. Burke et al. 1995. *Development* 121: 333–346.)

generally finds the genes of paralogous group 4 expressed anteriorly to those of paralogous group 5, and so forth (see Figure 12.20; Wilkinson et al. 1989; Keynes and Lumsden 1990). Mutations in the Hox genes suggest that the regional identity along the anterior-posterior axis is determined primarily by the most posterior Hox gene expressed in that region. (See **Further Development 12.13, Experimental Analysis of the Hox Code**, online.)

COMPARATIVE ANATOMY AND HOX GENE EXPRESSION   A new type of comparative embryology has emerged based on the comparison of gene expression patterns that produce the characteristics of different species. Gaunt (1994) and Burke and her collaborators (1995) have compared the vertebrae of the mouse and the chick (**FIGURE 12.21A**). Although the mouse and the chick have a similar number of vertebrae, they apportion them differently. Mice (like all mammals, be they giraffes or whales) have 7 cervical (neck) vertebrae. These are followed by 13 thoracic (rib) vertebrae, 6 lumbar (abdominal) vertebrae, 4 sacral (hip) vertebrae, and a variable (20+) number of caudal (tail) vertebrae. The chick, by contrast, has 14 cervical vertebrae, 7 thoracic vertebrae, 12 or 13 (depending on the strain) lumbosacral vertebrae, and 5 coccygeal (fused tail) vertebrae. The researchers asked, Does the constellation of Hox gene expression correlate with the type of vertebra formed (e.g., cervical or thoracic) or with the relative position of the vertebrae (e.g., number 8 or 9)?

The answer is that the constellation of Hox gene expression predicts the type of vertebra formed. In the mouse, the transition between cervical and thoracic vertebrae is between vertebrae 7 and 8; in the chick, it is between vertebrae 14 and 15 (**FIGURE 12.21B**). In both cases, the *Hox5* paralogues are expressed in the last cervical vertebra, while the anterior boundary of the *Hox6* paralogues extends to the first thoracic vertebra. Similarly, in both animals, the thoracic-lumbar transition is seen at the boundary between the *Hox9* and *Hox10* paralogous groups. It appears there is a code of differing Hox gene expression along the anterior-posterior axis, and that code determines the type of vertebra formed.

FURTHER DEVELOPMENT

HOMEOTIC TRANSFORMATIONS IN THE MOUSE   As noted above, there is a specific pattern to the number and type of vertebrae in mice, and the Hox gene expression pattern dictates which vertebral type will form (**FIGURE 12.22A**). This was demonstrated when all six copies of the *Hox10* paralogous group (i.e., *Hoxa10*, *c10*, and *d10* in Figure 12.23) were knocked out and no lumbar vertebrae developed. Instead, the presumptive lumbar vertebrae formed ribs and other characteristics similar to those of thoracic vertebrae (**FIGURE 12.22B**). This was a homeotic transformation comparable to those seen in insects; however, the redundancy of genes in the mouse made it much more difficult to produce, because the existence of even one copy of the *Hox10* group genes prevented the transformation (Wellik and Capecchi 2003; Wellik 2009). Similarly, when all six copies of the *Hox11* group were knocked out, the thoracic and lumbar vertebrae were normal, but the sacral vertebrae failed to form and were replaced

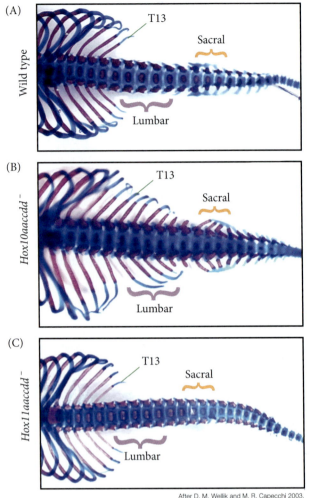

(A) Wild type

T13
Sacral
Lumbar

(B) *Hox10aaccdd⁻*

T13
Sacral
Lumbar

(C) *Hox11aaccdd⁻*

T13
Sacral
Lumbar

After D. M. Wellik and M. R. Capecchi 2003.
Science 301: 363–367, courtesy of M. Capecchi

FIGURE 12.22 Axial skeletons of mice in gene knockout experiments. Each photograph is of an 18.5-day embryo, looking upward at the ventral region from the middle of the thorax toward the tail. (A) Wild-type mouse. (B) Complete knockout of *Hox10* paralogues (*Hox10aaccdd*) converts lumbar vertebrae (after the thirteenth thoracic vertebrae) into ribbed thoracic vertebrae. (C) Complete knockout of *Hox11* paralogues (*Hox11aaccdd*) transforms the sacral vertebrae into copies of lumbar vertebrae. (After D. M. Wellik and M. R. Capecchi. 2003. *Science* 301: 363–367, courtesy of M. Capecchi.)

by lumbar vertebrae (**FIGURE 12.22C**). More recently, a *Hoxb6* gene was placed on a Delta enhancer, causing it to be expressed in every somite. The result was a "snakelike" mouse in which each somite had formed a rib-bearing thoracic vertebra (see Figure 17.7C; Guerreiro et al. 2013).

THE LEFT-RIGHT AXIS The internal organs of the mammalian body are not symmetric, as seen by the position of the spleen, heart, and liver along either the left or the right side of the body (**FIGURE 12.23A–C**). As in the chick embryo, the left-right axis appears to be due to the activation of Nodal proteins and the Pitx2 transcription factor on the left side of the lateral plate mesoderm, while Cerberus, an inhibitor of Nodal signaling, is expressed on the right (see Figure 12.9; Collignon et al. 1996; Lowe et al. 1996; Meno et al. 1996). However, each amniote group may have different ways of initiating this pathway (Vanderberg and Levin 2013). In mammals, the distinction between left and right sides can be seen in the ciliary cells of the node (**FIGURE 12.23D**). The cilia cause fluid in the node to flow from right to left (clockwise when viewed from the ventral side). When Nonaka and colleagues (1998) knocked out a mouse gene encoding the ciliary motor protein dynein (see Chapter 7), the nodal cilia did not move, and the lateral position of each asymmetrical organ was randomized. (This helped explain the clinical observation that humans with a dynein deficiency have immotile cilia and a random chance of having their heart on the left or right side of the body; see Afzelius 1976). Moreover, when Nonaka and colleagues (2002) cultured early mouse embryos under an artificial flow of medium from left to right, they obtained a reversal of the left-right axis.

The mechanism for this rotation appears to be the placement of the basal body of the cilium on each of the 200 or so monociliated node cells. The basal body giving rise to each cilium is at the posterior side of each cell, and the cilium extends out the ventral surface (see Figure 12.23D). Thus, the placement of cilia integrates information concerning the anterior-posterior and dorsal-ventral axes to construct the right-left axis (Guirao et al. 2010; Hashimoto et al. 2010). The placement of the cilia is governed by the planar cell polarity (PCP) pathway, possibly directed by a Wnt. Mutations in PCP pathway signaling molecules can randomize localization of cilia in these cells, also causing randomization of the left-right axis.

But how does rotation generate a body axis? It appears that the cells neighboring the node, the **crown cells**, are responsible for sensing the flow. Crown cells have immobile cilia, and these cilia are affected by the movement of fluids. The fluid movement activates the Pkd2 protein on their cilia. A cascade initiated by Pdk2 (in a manner still unknown) appears to suppress the synthesis of Cerberus and thereby activate the expression of Nodal (Kawasumi et al. 2011; Yoshiba et al. 2012). Nodal is thought to bind in an autocrine manner to crown cells to maintain its own transcription, so Cerberus (which is made by the crown cells on the right side) would inhibit the maintenance of Nodal expression. In this way, Nodal expression is maintained on the left side, where it can activate *Pitx*, which determines the left- and right-sidedness of the tissue.

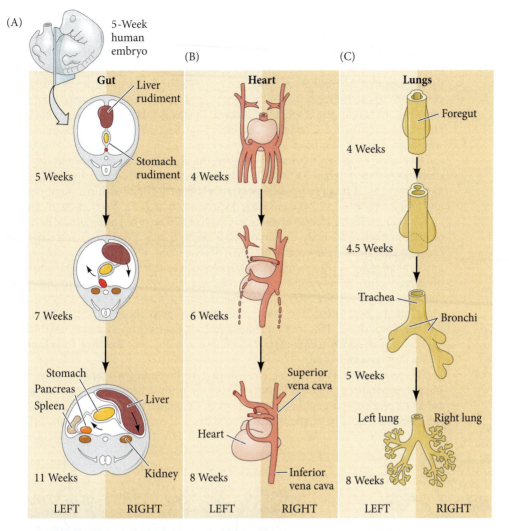

(A) 5-Week human embryo

Gut
Liver rudiment
5 Weeks
Stomach rudiment
7 Weeks
Stomach
Pancreas
Spleen
Liver
11 Weeks
Kidney
LEFT　RIGHT

(B) **Heart**
4 Weeks
6 Weeks
Superior vena cava
Heart
8 Weeks
Inferior vena cava
LEFT　RIGHT

(C) **Lungs**
Foregut
4 Weeks
4.5 Weeks
Trachea
Bronchi
5 Weeks
Left lung　Right lung
8 Weeks
LEFT　RIGHT

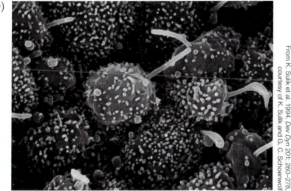

(D)

From K. Sulik et al. 1994. *Dev Dyn* 201: 260–278, courtesy of K. Sulik and G. C. Schoenwolf

FIGURE 12.23 Left-right asymmetry in the developing human. (A) Abdominal cross sections show that the originally symmetrical organ rudiments acquire asymmetric positions by week 11. The liver moves to the right and the spleen moves to the left. (B) Not only does the heart move to the left side of the body, but the originally symmetrical veins of the heart regress differentially to form the superior and inferior venae cavae, which connect only to the right side of the heart. (C) The right lung branches into three lobes, while the left lung (near the heart) forms only two lobes. (In human males, the scrotum also forms asymmetrically.) (D) Ciliated cells of the mouse node, each with a cilium extending from the posterior ventral region of the cell. (A–C after K. Kosaki and B. Casey. 1998. *Semin Cell Dev* 9: 89–99.)

Twins

The early cells of the mammalian embryo can replace each other and compensate for a missing cell. This regulative ability was first demonstrated in 1952, when Seidel destroyed one cell of a 2-cell rabbit embryo and the remaining cell produced an entire embryo. The regulative capacity of the early embryo is also seen in humans. Human twins are classified into two major groups: monozygotic (one-egg; identical) twins and dizygotic (two-egg; fraternal) twins. Fraternal twins are the result of two separate fertilization events, whereas identical twins are formed from a single embryo whose cells somehow become dissociated from one another.

Identical twins, which occur in roughly 1 in 400 human births, may be produced by the separation of early blastomeres, or even by the separation of the inner cell mass into two regions within the same blastocyst. About 33% of identical twins have two complete and separate chorions, indicating that separation may have occurred before the formation of the trophoblast tissue at day 5 (**FIGURE 12.24A**). Other identical twins share a common chorion, often interpreted to mean that the split occurred within the inner cell mass after the trophoblast formed. By day 9, the human embryo has completed the construction of another extraembryonic layer, the lining of the amnion. If separation of the embryo comes after the formation of the chorion on day 5 but before the formation of the amnion on day 9, then the resulting embryos should have one chorion and two amnions (**FIGURE 12.24B**). This happens in about two-thirds of human identical twins. A small percentage of identical twins are born within a single chorion and amnion (**FIGURE 12.24C**), meaning the division of the embryo came after day 12.

According to these studies of twins, each cell of the inner cell mass should be able to produce any cell of the body. This hypothesis has been confirmed, and it has important consequences for the study of mammalian development. When ICM cells are isolated and grown under certain conditions, they remain undifferentiated and continue to divide in culture (Evans and Kaufman 1981; Martin 1981). Such cells are embryonic stem cells (ES cells). When ES cells are injected into a mouse blastocyst, they can integrate into the host inner cell mass. The resulting embryo has cells from both the host and the donor tissue. This technique has become extremely important in determining the function of genes during mammalian development. (See **Further Development 12.14, Chimerism**, online.)

Whereas fraternal twins are created during fertilization (two separate eggs meeting two separate sperm) and identical twins are formed during cleavage, **conjoined twins** are probably created during gastrulation. Conjoined twins are most often identical (there have only been a few documented cases of conjoined twins being different

FIGURE 12.24 The timing of human monozygotic twinning with relation to extraembryonic membranes. (A) Splitting occurs before formation of the trophoblast, so each twin has its own chorion and amnion. (B) Splitting occurs after trophoblast formation but before amnion formation, resulting in twins having individual amniotic sacs but sharing one chorion. (C) Splitting after amnion formation leads to twins in one amniotic sac and a single chorion. (After T. W. Sadler. 2018. *Langman's Medical Embryology*. Wolters Kluwer Health: Philadelphia.)

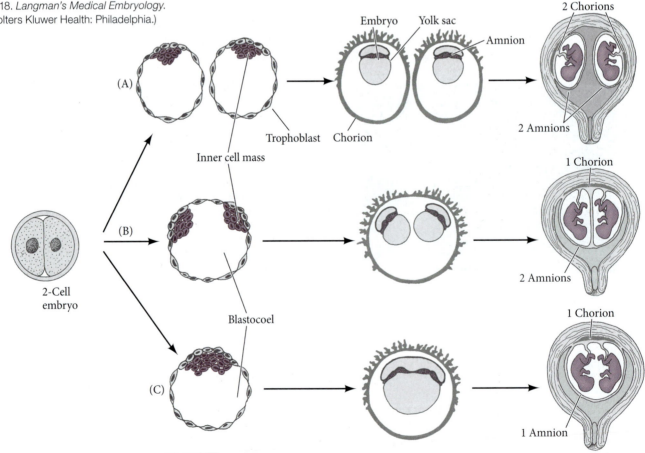

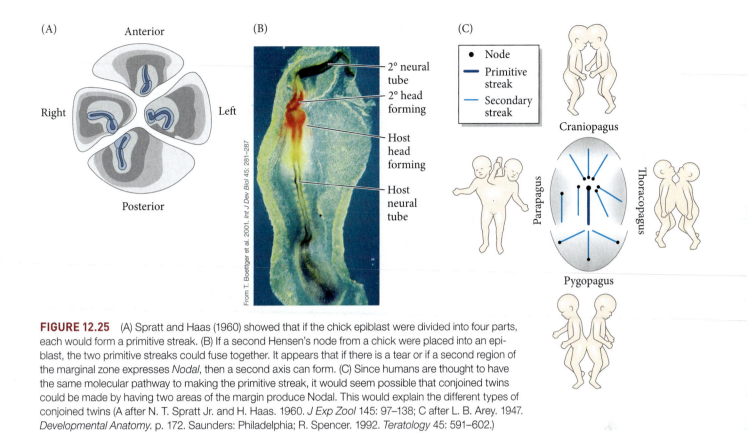

FIGURE 12.25 (A) Spratt and Haas (1960) showed that if the chick epiblast were divided into four parts, each would form a primitive streak. (B) If a second Hensen's node from a chick were placed into an epiblast, the two primitive streaks could fuse together. It appears that if there is a tear or if a second region of the marginal zone expresses *Nodal*, then a second axis can form. (C) Since humans are thought to have the same molecular pathway to making the primitive streak, it would seem possible that conjoined twins could be made by having two areas of the margin produce Nodal. This would explain the different types of conjoined twins (A after N. T. Spratt Jr. and H. Haas. 1960. *J Exp Zool* 145: 97–138; C after L. B. Arey. 1947. *Developmental Anatomy*. p. 172. Saunders: Philadelphia; R. Spencer. 1992. *Teratology* 45: 591–602.)

sexes; see Martínez-Frías 2009) and occur approximately once in every 200,000 live births.[6] Spratt and Haas (1960) showed that if the chick epiblast is divided into four parts, each forms a primitive streak (**FIGURE 12.25A**). Moreover, if a second Hensen's node from a chick is placed into an epiblast, the two primitive streaks can fuse together. It appears that if there is a tear in the marginal zone (allowing a new center of Nodal expression to form), or if a second region of the marginal zone should express *Nodal*, then a second axis can form (**FIGURE 12.25B**; Bertocchini and Stern 2002; Perea-Gomez et al. 2002; Torlopp et al. 2014). Since humans are thought to have the same molecular pathway to making the primitive streak as chicks do, it seems possible that monozygotic twins could be the result of two areas of the margin producing Nodal. Levin (1999) has shown that this would explain the different types of conjoined twins (**FIGURE 12.25C**). Though we do not know how conjoined twins form, the generation of multiple axes during gastrulation might begin to explain this phenomenon.

Coda

Variations on the important themes of development have evolved in the different vertebrate groups (**FIGURE 12.26**). The major themes of vertebrate gastrulation include the following:

- Internalization of the endoderm and mesoderm
- Epiboly of the ectoderm around the entire embryo
- Convergence of the internal cells to the midline
- Extension of the body along the anterior-posterior axis

Although fish, amphibian, avian, and mammalian embryos have different patterns of cleavage and gastrulation, they use many of the same molecules to accomplish the same goals. Each group uses gradients of Nodal and Wnt proteins to establish polarity along

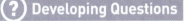

? Developing Questions

In an episode of the U.S. television series *CSI: Crime Scene Investigation*, the DNA of the obvious perpetrator did not match the DNA of the cells at the crime scene. The episode was based on actual rare instances in which a mammal has been found to have two different sets of DNA. How do you think such a situation might come about?

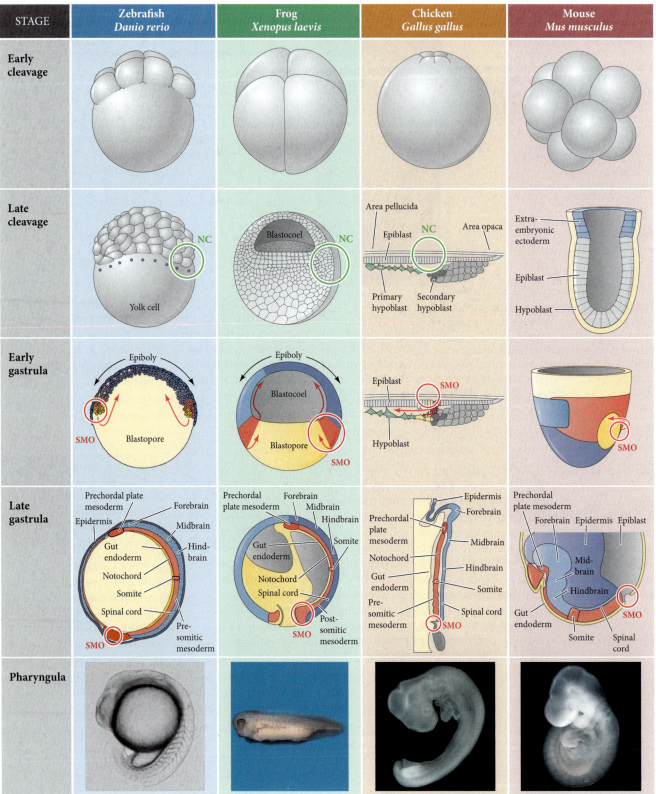

STAGE	Zebrafish *Danio rerio*	Frog *Xenopus laevis*	Chicken *Gallus gallus*	Mouse *Mus musculus*
Early cleavage				
Late cleavage	NC / Yolk cell	Blastocoel / NC	Area pellucida / Epiblast / NC / Area opaca / Primary hypoblast / Secondary hypoblast	Extra-embryonic ectoderm / Epiblast / Hypoblast
Early gastrula	Epiboly / SMO / Blastopore	Epiboly / Blastocoel / Blastopore / SMO	Epiblast / SMO / Hypoblast	SMO
Late gastrula	Prechordal plate mesoderm / Forebrain / Epidermis / Midbrain / Gut endoderm / Hind-brain / Notochord / Somite / Spinal cord / SMO / Pre-somitic mesoderm	Prechordal plate mesoderm / Forebrain / Midbrain / Hindbrain / Gut endoderm / Somite / Notochord / Spinal cord / SMO / Post-somitic mesoderm	Epidermis / Forebrain / Prechordal plate mesoderm / Midbrain / Notochord / Hindbrain / Gut endoderm / Somite / Pre-somitic mesoderm / Spinal cord / SMO	Prechordal plate mesoderm / Forebrain / Epidermis / Epiblast / Mid-brain / Hindbrain / Gut endoderm / Somite / SMO / Spinal cord
Pharyngula				

From L. Solnica-Krezel, 2005. *Curr Biol* 15: R213–R228

◀ **FIGURE 12.26** Early development in four vertebrates. Cleavage differs greatly among the four groups. Zebrafish and chicks have meroblastic discoidal cleavage, frogs have unequal holoblastic cleavage, and mammals have equal holoblastic cleavage. These cleavage patterns form different structures, but there are many conserved features, such as the Nieuwkoop center (NC; green circles). As gastrulation begins, each of the groups has cells equivalent to the Spemann-Mangold organizer (SMO; red circles). The SMO marks the beginning of the blastopore region, and the remainder of the blastopore is indicated by the red arrows extending from the organizer. By the late gastrula stage, the endoderm (yellow) is inside the embryo, the ectoderm (blue, purple) surrounds the embryo, and the mesoderm (red) is between the endoderm and ectoderm. The regionalization of the mesoderm has also begun. The bottom row shows the pharyngula stage that immediately follows gastrulation. This stage—with a pharynx, a central neural tube and notochord flanked by somites, and a sensory cephalic (head) region—characterizes the vertebrates. (After L. Solnica-Krezel. 2005. *Curr Biol* 15: R213–R228.)

the anterior-posterior axis. In *Xenopus* and zebrafish, maternal factors induce Nodal proteins in the vegetal hemisphere or marginal zone. In the chick, Nodal expression is induced by Wnt and Vg1 emanating from the posterior marginal zone, while elsewhere Nodal activity is suppressed by the hypoblast. In the mouse, the hypoblast similarly restricts Nodal activity—the chick embryo uses Cerberus to do this, while the mammalian embryo uses Cerberus and Lefty1.

Each of these vertebrate groups uses BMP inhibitors to specify the dorsal axis. Similarly, Wnt inhibition and Otx2 expression are important in specifying the anterior regions of the embryo, but different groups of cells may express these proteins. In all cases, the region of the body from the hindbrain to the tail is specified by Hox genes. Finally, the left-right axis is established through the expression of Nodal on the left-hand side of the embryo. Nodal activates *Pitx2*, leading to the differences between the left and right sides of the embryo. How Nodal becomes expressed on the left side appears to differ among the vertebrate groups. But overall, despite their initial differences in cleavage and gastrulation, the different vertebrate groups have maintained very similar ways of establishing the three body axes.

Closing Thoughts on the Opening Photo

This photograph of the primitive streak and head endoderm of the 7.5-day mouse embryo shows the beginnings of anterior specification. This one photograph illustrates many of the signaling systems at play to put the head and tail in the right places. The nuclei of all cells are stained blue. The transcription factors Lhx1, Foxa2 (green), and Otx2 interact to regulate differentiation of the anterior mesendoderm. Brachyury (red) stains the central mesoderm. Foxa2 and Brachyury are co-expressed (yellow) in the anterior midline structure and node. This pattern is regulated by Nodal and Wnt proteins and establishes the regions of the embryo that will form the anterior (head) and posterior (tail) structures of the mammalian body.

Photograph courtesy of I. Costello and E. Robertson

Endnotes

[1] Although most research has centered around rodents (mouse and rats), it is important to know that many other mammals present a genesis of different morphological events during early development. As new doors open to nonmodel systems it will be exciting to learn about the foundational mechanisms driving gastrulation and axis determination in other mammalian species.

[2] Francis M. Balfour proposed the homology of the amphibian blastopore and the chick primitive streak in 1873, while he was still an undergraduate (Hall 2003). August Rauber (1876) provided further evidence for their homology.

[3] In humans, homozygous loss of *PITX2* causes Rieger's syndrome, a condition characterized by asymmetry anomalies. A similar condition is caused by knocking out the *Pitx2* gene in mice (Fu et al. 1998; Lin et al. 1999).

[4] Although the mammalian blastocyst was discovered by Rauber in 1881, its first public display was probably in Gustav Klimt's 1907 painting *Danae*, in which blastocyst-like patterns are featured on the heroine's robe as she becomes impregnated by Zeus (Gilbert and Braukmann 2011).

[5] In mouse development, Hensen's node is usually just called "the node," despite the fact that Viktor Hensen discovered this structure in rabbit and guinea pig embryos.

[6] See also the online Hernandez article in *International Business Times*, June 6, 2016, "Conjoined male and female twins born in India share every vital organ."

Snapshot Summary
Birds and Mammals

1. Reptiles and birds, like fish, undergo discoidal meroblastic cleavage, wherein the early cell divisions do not cut through the yolk of the egg. These early cells form a blastoderm.

2. In chick embryos, early cleavage forms an area opaca and an area pellucida. The region between them is the marginal zone. Gastrulation begins in the area pellucida next to the posterior marginal zone, as the hypoblast and primitive streak both start there.

3. The primitive streak is derived from epiblast cells and the central cells of Koller's sickle. As the primitive streak extends rostrally, Hensen's node is formed. Cells migrating out of Hensen's node become prechordal mesendoderm and are followed by the head process and notochord cells.

4. The prechordal plate helps induce formation of the forebrain; the chordamesoderm induces formation of the midbrain, hindbrain, and spinal cord. The first cells migrating laterally through the primitive streak become endoderm, displacing the hypoblast. The mesoderm cells then migrate through the primitive streak. Meanwhile, the surface ectoderm undergoes epiboly around the yolk.

5. In birds, gravity helps determine the position of the primitive streak, which points in a posterior-to-anterior direction and whose differentiation establishes the dorsal-ventral axis. The left-right axis is formed by the expression of Nodal protein on the left side of the embryo, which signals Pitx2 expression on the left side of developing organs.

6. The hypoblast helps determine the body axes of the embryo, and its migration determines the cell movements that accompany formation of the primitive streak and thus its orientation.

7. Mammals undergo a variation of holoblastic rotational cleavage that is characterized by a slow rate of cell division, a unique cleavage orientation, lack of divisional synchrony, and formation of a blastocyst.

8. The blastocyst forms after the blastomeres undergo compaction. It contains outer cells—the trophoblast cells—that become the chorion, and an inner cell mass that becomes the amnion and the embryo.

9. The inner cell mass cells are pluripotent and can be cultured as embryonic stem cells. They give rise to the epiblast and to the visceral endoderm (hypoblast).

10. The chorion forms the fetal portion of the placenta, which functions to provide oxygen and nutrition to the embryo, to provide hormones for the maintenance of pregnancy, and to block the potential immune response of the mother to the developing fetus.

11. Mammalian gastrulation is not unlike that of birds. There appear to be two signaling centers, one in the node and one in the anterior visceral endoderm. The latter center is critical for establishing the body axes, while the former is critical in inducing the nervous system and in patterning axial structures caudally from the midbrain.

12. Hox genes pattern the anterior-posterior axis and help specify positions along that axis. If Hox genes are knocked out, segment-specific malformations can arise. Similarly, causing the ectopic expression of Hox genes can alter the body axis.

13. The homology of gene structure and the similarity of expression patterns between *Drosophila* and mammalian Hox genes suggest that this patterning mechanism is extremely ancient.

14. The mammalian left-right axis is specified similarly to that of the chick, but with some significant differences in the roles of certain genes.

15. In amniote gastrulation, the pluripotent epithelium, or epiblast, produces the mesoderm and endoderm, which migrate through the primitive streak, and the precursors of the ectoderm, which remain on the surface. By the end of gastrulation, the head and anterior trunk structures are formed. Elongation of the embryo continues through precursor cells in the caudal epiblast surrounding the posteriorized Hensen's node.

16. In each class of vertebrates, neural ectoderm is permitted to form where the BMP-mediated induction of epidermal tissue is prevented.

17. Fraternal twins arise from two separate fertilization events. Identical twins result from the splitting of the embryo into two cellular groups during stages where there are still pluripotent cells in the embryo. Experimental evidence suggests that conjoined twins may occur through the formation of two organizers within a common blastodisc.

Go to oup.com/he/barresi12xe for Further Developments, Scientists Speak interviews, Watch Development videos, Dev Tutorials, and complete bibliographic information for all literature cited in this chapter.

Neural Tube Formation and Patterning

13

"**LIKE THE ENTOMOLOGIST IN SEARCH** of brightly colored butterflies, my attention hunted, in the garden of the gray matter, cells with delicate and elegant forms, the mysterious butterflies of the soul." Thus reflected Santiago Ramón y Cajal, often referred to as the father of neuroscience, on his study of the brain. His 1937 quotation masterfully captures the fascination and mystery of the brain as part of a larger system that controls communication, consciousness, memory, emotion, motor control, digestion, sensory perceptions, and so much more. How the development of this central organ is coordinated with the development of the rest of the organism for integrated connectivity will remain one of the most fundamental questions in developmental biology for the next century. The first pivotal event is the transformation of an epithelial sheet into a tube. This initial structure will provide the foundation for the regionalization and diversification of brain structures along the anterior-to-posterior axis; then, through strategic mechanisms of cell growth and differentiation, the elaborate and highly connected structure of the vertebrate central nervous system can be realized. Over the next three chapters, we will study the development of the nervous system, beginning in this chapter with the formation of the neural tube and the specification of cell fates within it (**FIGURE 13.1**). In Chapter 14, we will delve into the mechanisms governing cell fate patterning and neurogenesis along the dorsal-ventral axis of the central nervous system. Then, in Chapter 15, we will navigate the molecular guidance mechanisms underlying the wiring of the nervous system and the development of neural crest lineages.

The vertebrate ectoderm, the outer germ layer covering the late-stage gastrula, has three major responsibilities (**FIGURE 13.2**):

1. One part of the ectoderm will become the **neural plate**, the presumptive neural tissue induced by the prechordal plate and notochord during gastrulation. The neural plate moves into the body to form the **neural tube**, the precursor of the **central nervous system** (**CNS**)—the brain and spinal cord.

2. Another part of this germ layer will become the **epidermis**, the outer layer of the skin (which is the largest organ of the vertebrate body). The epidermis forms an elastic, waterproof, and constantly regenerating barrier between the organism and the outside world.

3. Between the compartments forming the epidermis and the central nervous system lies the presumptive **neural crest**. The cells of the neural crest delaminate from these epithelia at the dorsal midline and migrate away (between the neural tube and epidermis) to generate, among other things, the **peripheral nervous system** (all the nerves and neurons lying outside the CNS) and pigment cells (e.g., melanocytes).

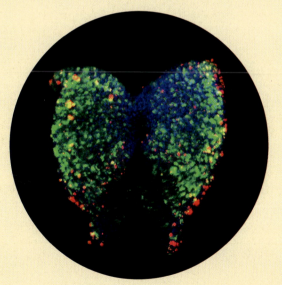

What's preventing this brain from closing?

Courtesy of Lee Niswander and Huili Li

FIGURE 13.1 The major questions to be addressed in Chapters 13, 14, and 15. Questions of neurulation and cell fate specification (A,B) will be answered in this chapter. How the neural tube (NT) is expanded into the elaborate structures of the brain (C) will be covered in Chapter 14. How the peripheral nervous system (D) is largely derived from neural crest cells (NCC) migrating out of the dorsal neural tube, and how the newly born neurons extend long processes to find their synaptic partners (E) and thus connect up the nervous system will be covered in Chapter 15.

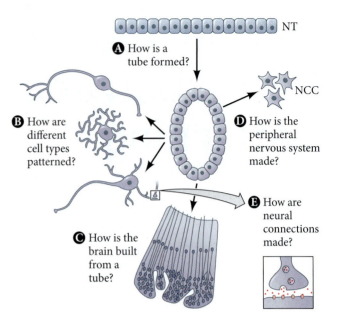

Ⓐ How is a tube formed?

NT

NCC

Ⓑ How are different cell types patterned?

Ⓓ How is the peripheral nervous system made?

Ⓔ How are neural connections made?

Ⓒ How is the brain built from a tube?

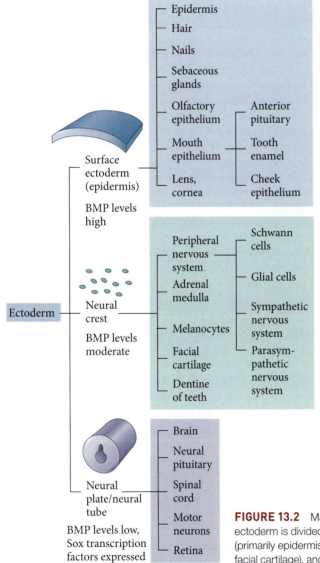

Ectoderm

Surface ectoderm (epidermis)
BMP levels high
- Epidermis
- Hair
- Nails
- Sebaceous glands
- Olfactory epithelium
- Mouth epithelium
  - Anterior pituitary
  - Tooth enamel
- Lens, cornea
  - Cheek epithelium

Neural crest
BMP levels moderate
- Peripheral nervous system
  - Schwann cells
  - Glial cells
- Adrenal medulla
  - Sympathetic nervous system
- Melanocytes
- Facial cartilage
  - Parasympathetic nervous system
- Dentine of teeth

Neural plate/neural tube
BMP levels low, Sox transcription factors expressed
- Brain
- Neural pituitary
- Spinal cord
- Motor neurons
- Retina

FIGURE 13.2 Major derivatives of the ectoderm germ layer. The ectoderm is divided into three major domains: the surface ectoderm (primarily epidermis), the neural crest (peripheral neurons, pigment, facial cartilage), and the neural tube (brain and spinal cord).

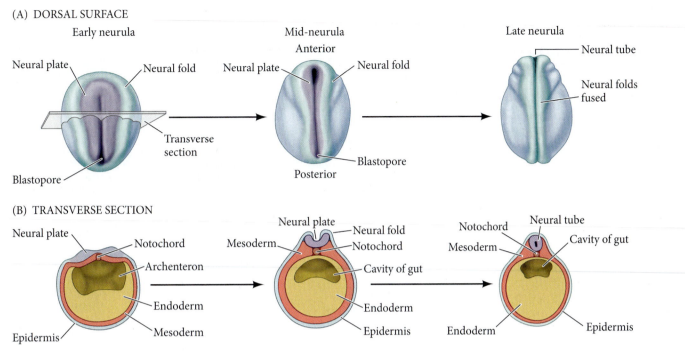

(A) DORSAL SURFACE

FIGURE 13.3 Two views of primary neurulation in an amphibian embryo, showing early (left), middle (center), and late (right) neurulae in each case. (A) Looking down on the dorsal surface of the whole embryo. (B) Transverse section through the center of the embryo. (After B. I. Balinsky. 1981. *Introduction to Embryology*, 5th Ed. Saunders: Philadelphia.)

The processes by which the three ectodermal regions are made physically and functionally distinct from one another is called **neurulation**, and an embryo undergoing these processes is called a **neurula** (**FIGURE 13.3**; Gallera 1971). As we saw in the preceding chapters, the specification of the ectoderm is accomplished during gastrulation, primarily by regulating the levels of BMP experienced by the ectodermal cells. High levels of BMP specify the cells to become epidermis. Very low levels specify the cells to become neural plate. Intermediate levels effect the formation of the neural crest cells. Neurulation directly follows gastrulation.

Transforming the Neural Plate into a Tube: The Birth of the Central Nervous System

The cells of the neural plate are characterized by expression of the Sox family of transcription factors (Sox1, 2, and 3). These factors (1) activate the genes that specify cells to be neural plate and (2) inhibit the formation of epidermis and neural crest by blocking the transcription and signaling of BMPs (Archer et al. 2011). In this process, we see once again an important principle of development: *often the signals promoting the specification of one cell type also block the specification of an alternative cell type*. The expression of Sox transcription factors establishes the neural plate cells as neural precursors that can form all the cell types of the central nervous system (Wilson and Edlund 2001).

Although the neural plate lies on the surface of the embryo, the nervous system will not lie on the outside of the mature body. Somehow, the neural plate has to move inside the embryo and form a neural tube. This process is accomplished through neurulation, which occurs with some diversity across vertebrates (Harrington et al. 2009). There are two principal modes of neurulation. In **primary neurulation**, the cells surrounding the neural plate direct the neural plate cells to proliferate, invaginate into the body, and separate from the surface ectoderm to form an underlying hollow tube. In **secondary neurulation**, the neural tube arises from the aggregation of mesenchyme cells into a solid cord that subsequently forms cavities that coalesce to create a hollow tube. In many vertebrates, primary and secondary neurulation are

FIGURE 13.4 Primary and secondary neurulation and the transition zone between them. The bottom image is a lateral view of the neural tube surface. The illustrations above the neural tube correspond to transverse sections through the axial level indicated as the neural tube forms in a rostral-to-caudal direction. Different cell types are represented in different colors, as indicated in the key. (After A. Dady et al. 2014. *J Neurosci* 34: 13208–13221.)

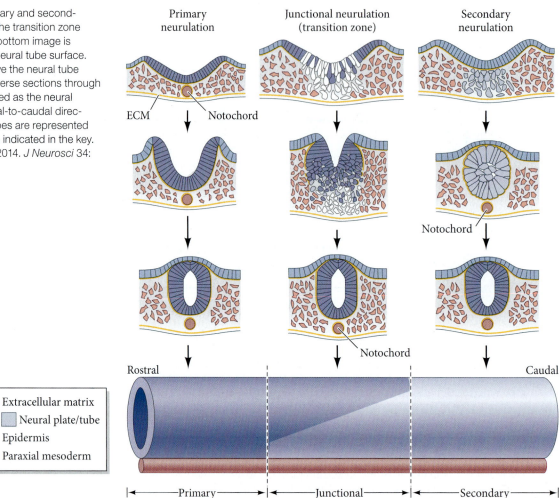

divided spatially in the embryo such that primary neurulation forms the *anterior* portion of the neural tube and the *posterior* portion of the neural tube is the product of secondary neurulation (**FIGURE 13.4**).

In birds, primary neurulation generates the neural tube anterior to the hindlimbs (Pasteels 1937; Catala et al. 1996). In mammals, secondary neurulation begins at the level of the sacral vertebrae of the tail (Schoenwolf 1984; Nievelstein et al. 1993). In fish and amphibians (e.g., zebrafish and *Xenopus*), only the tail neural tube is derived from secondary neurulation (Gont et al. 1993; Lowery and Sive 2004). More basal chordates, such as *Amphioxus* and *Ciona*, only exhibit mechanisms of primary neurulation, suggesting that primary neurulation was the ancestral condition and that secondary neurulation evolved much the way limbs did—that is, as a vertebrate novelty—and in the case of secondary neurulation, one that was associated with tail elongation (Handrigan 2003).

The neural tube is finally complete when these two separately formed tubes join together (Harrington et al. 2009). The size of the **transition zone** between the primary and secondary neural tubes varies among species, from relatively abrupt in the mouse, to a region spanning the thoracic vertebrae in the chick, to the thoracolumbar region in humans (Dady et al. 2014). Formation of the neural tube in this transition zone has been named **junctional neurulation** (Dady et al. 2014) because it involves a combination of mechanisms involved in both primary and secondary neurulation (see Figure 13.4).

Primary neurulation

Although some species differences exist, the process of primary neurulation is relatively similar in all vertebrates.[1] To explore the mechanisms of neural plate folding, we will largely focus on the process of primary neurulation in amniotes. Shortly after the neural plate has formed in the chick, its edges thicken and move upward to form the **neural folds**, and a U-shaped **neural groove** appears in the center of the plate, dividing the future right and left sides of the embryo (**FIGURE 13.5**). The neural folds on the lateral sides of the neural plate migrate toward the midline of the embryo, eventually fusing to form the neural tube beneath the overlying ectoderm.

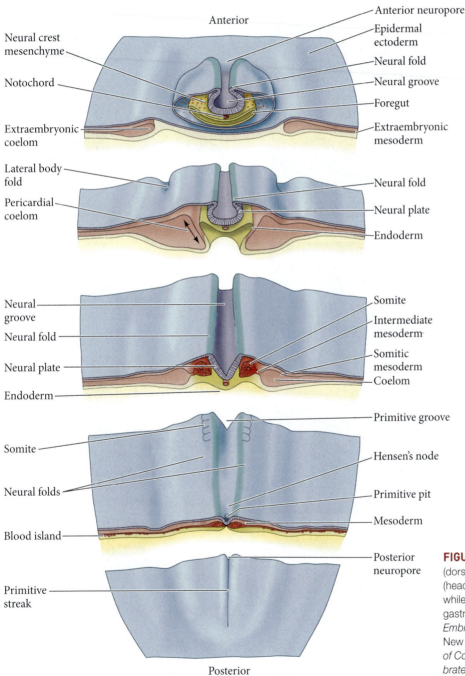

FIGURE 13.5 The neurulating chick embryo (dorsal view) at about 24 hours. The cephalic (head) region has undergone neurulation, while the caudal (tail) region is still undergoing gastrulation. (After B. M. Patten. 1971. *Early Embryology of the Chick*, 5th Ed. McGraw-Hill: New York; A. F. Huettner. 1943. *Fundamentals of Comparative Embryology of the Vertebrates*. The Macmillan Company: New York.)

(?) Developing Questions

Why is there a need for two separate mechanisms to complete the neural tube? What were the evolutionary pressures that forced the adoption of secondary neurulation as opposed to a posterior extension of primary neurulation? As you ponder these questions, consider the embryo's first morphogenetic mechanism, that of gastrulation. Does the timing of the end of gastrulation and its capacity for axis elongation (or lack thereof) influence your ideas? It is surprising that we still do not fully understand the evolutionary history of the nervous system.

Primary neurulation can be divided into four distinct but spatially and temporally overlapping stages:

1. Elongation and folding of the neural plate. Cell divisions within the neural plate are preferentially in the anterior-posterior direction (often referred to as the **rostral-caudal**, or beak-to-tail, direction), which fuels continued axial elongation associated with gastrulation. These events occur even if the neural plate tissue is isolated from the rest of the embryo. To roll into a neural tube, however, the presumptive epidermis is also needed (**FIGURE 13.6A,B**; Jacobson and Moury 1995; Moury and Schoenwolf 1995; Sausedo et al. 1997).

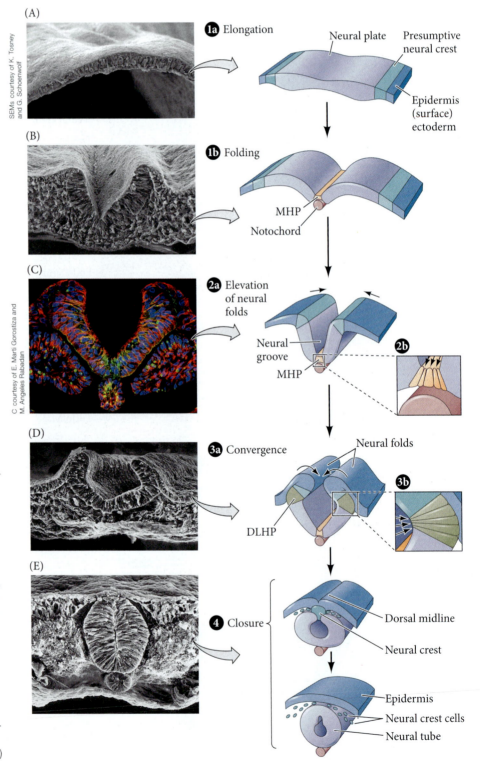

FIGURE 13.6 Primary neurulation: neural tube formation in the chick embryo. (A, 1a) Cells of the neural plate can be distinguished as elongated cells in the dorsal region of the ectoderm. (B, 1b) Folding begins as the medial hinge point (MHP) cells anchor to the notochord and change their shape while the presumptive epidermal cells move toward the dorsal midline. (C, 2a) The neural folds are elevated as the presumptive epidermis continues to move toward the dorsal midline. Asymmetric constriction of actin on the apical side changes cell shapes to promote MHP bending (B, C, 2b). (C) Elevated neural folds stained to show the extracellular matrix (green) and the actin cytoskeleton (red) concentrated in the apical portions of the neural plate cells. (D, 3a) Convergence of the neural folds occurs as the cells at the dorsolateral hinge point (DLHP) become wedge-shaped and the epidermal cells push toward the center. (D, 3b) Similar apical constriction occurs at the DLHP. (E, 4) The neural folds are brought into contact with one another. The neural crest cells disperse, leaving the neural tube separate from the epidermis. (Drawings after J. L. Smith and G. C. Schoenwolf. 1997. *Trends Neurosci* 20: 510–517.)

2. Bending of the neural plate. The bending of the neural plate involves the formation of hinge regions where the neural plate contacts surrounding tissues. In birds and mammals, the cells at the midline of the neural plate form the **medial hinge point**, or **MHP** (Schoenwolf 1991a,b; Catala et al. 1996). MHP cells are reported to be firmly anchored to the notochord beneath them and form a hinge, which enables the creation of a furrow, or **neural groove**, at the dorsal midline (**FIGURE 13.6C**).

3. Convergence of the neural folds. Shortly thereafter, two **dorsolateral hinge points (DLHPs)** are induced by and anchored to the surface (epidermal) ectoderm. After the initial furrowing of the neural plate, the plate bends around the hinge regions. Each hinge acts as a pivot that directs the rotation of the cells around it (Smith and Schoenwolf 1991). Continued convergence of the surface ectoderm pushes toward the midline of the embryo, providing another motive force for bending the neural plate, causing the neural folds to converge (**FIGURE 13.6D**; Alvarez and Schoenwolf 1992; Lawson et al. 2001). This movement of the presumptive epidermis and the anchoring of the neural plate to the underlying mesoderm may also be important for ensuring that the neural tube invaginates, folding inward, into the embryo and not outward (Schoenwolf 1991a).

4. Closure of the neural tube. The neural tube closes as the paired neural folds are brought in contact with one another at the dorsal midline. The folds adhere to each other, and the neural and surface ectoderm cells from one side fuse with their respective counterparts from the other side. During this fusion event, cells at the apex of the neural folds delaminate and become neural crest cells (**FIGURE 13.6E**).

REGULATION OF HINGE POINTS   To fold the neural plate means to bend a sheet of epithelial cells. How can a row of attached boxlike epithelial cells be bent? While in the shape of a rectangular box (i.e., epithelial), they cannot; however, if in a region of boxes the surface area of one side of each box were reduced relative to its apposing side (creating the shape of a truncated pyramid), each of these cells should introduce a displacing angle with its neighboring cells and cause the row of boxes to bend. The MHP and two DLHPs are three regions of the neural plate where such cell shape changes occur (see Figure 13.6B–D). The epithelial cells in these locations adopt a "wedge-shaped" (or truncated pyramid) morphology along the apicobasal axis, one that is wider basally than apically (Schoenwolf and Franks 1984; Schoenwolf and Smith 1990). Similar to the bottle cells that initiate invagination during gastrulation (see Figure 11.4), localized contraction of actinomyosin complexes at the apical border reduces the size of the apical half of the cell relative to the basal compartment, a process known as **apical constriction**. This apical constriction pairs with the basal location of nuclei to yield the wedge-shaped hinge point cells (see Figure 13.6C,D; Smith and Schoenwolf 1987, 1988). In addition, recent findings suggest that the division rates in the dorsolateral domains of the neural plate are significantly faster than those in the ventral regions; this increases the cell density in the neural folds and adds a force that is hypothesized to promote buckling at the DLHP (McShane et al. 2015). The physical forces exerted by different regions of the neural plate have yet to be quantified, but at the cellular level, hinge points are formed by (1) apical constriction; (2) basal thickening, with retention of the nucleus within the basal portion of cells; and (3) cell packing in the neural folds.

What regulates these cellular changes in the correct locations of the neural plate? *Here's the short answer:* Hinge point formation appears to center around the

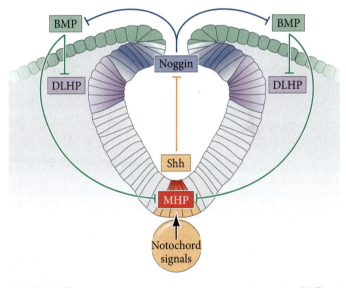

FIGURE 13.7   Morphogen regulation of hinge point formation. BMPs are expressed by the surface ectoderm (green), *Noggin* is expressed in the dorsal neural folds (blue), and *Shh* is expressed ventrally in the noto-chord and floor plate (orange). The regulation of hinge points revolves around BMP as an antagonist to both DLHP and MHP formation. Shh is required for the specification of floor plate, while additional signals from the notochord induce MHP morphology. Noggin directly inhibits BMP ligands, thus alleviating BMP repression of the hinge points. The DLHPs, however, form only at the correct size and dorsal-ventral position, which is based on Noggin's distance from inhibitory Shh gradients ascending from the floor plate. Therefore, apical constriction occurs only in those cells experiencing low enough concentrations of both BMP (MHP and DLHP) and Shh (DLHP) morphogens.

(?) Developing Questions

What induces medial hinge point formation? Two findings—that (1) an extra notochord can induce ectopic hinge point formation and (2) Sonic hedgehog represses the DLHP—suggest that factor(s) beyond the precise control of BMP signaling may be responsible. Could it be the early expression of Noggin in the notochord (and thus still be all about repressing BMPs)? Here is an additional fact to bear in mind: in the anteriormost region of the neural plate only an MHP forms, whereas in the posteriormost region of the neural plate only DLHPs form. Only in the central regions of the neural plate are both types of hinge points present. Why are these hinge points located in different positions along the anterior-to-posterior axis, and how is this difference regulated?

precise control of BMP signaling. BMP inhibits MHP and DLHP formation, whereas repression of BMP by Noggin enables DLHPs to form, and Shh from the notochord and floor plate prevent precocious and ectopic hinges from forming in the neural plate (**FIGURE 13.7**). To *further* understand the signaling network in control of hinge point formation and the data to support it, see **Further Development 13.1, Molecular Regulation of Hinge Point Formation**, online.

EVENTS OF NEURAL TUBE CLOSURE   Closure of the neural tube does not occur simultaneously throughout the neural ectoderm. This phenomenon is best seen in amniote vertebrates (reptiles, birds, and mammals), whose body axis is elongated prior to neurulation. In amniotes, induction occurs in an anterior-to-posterior

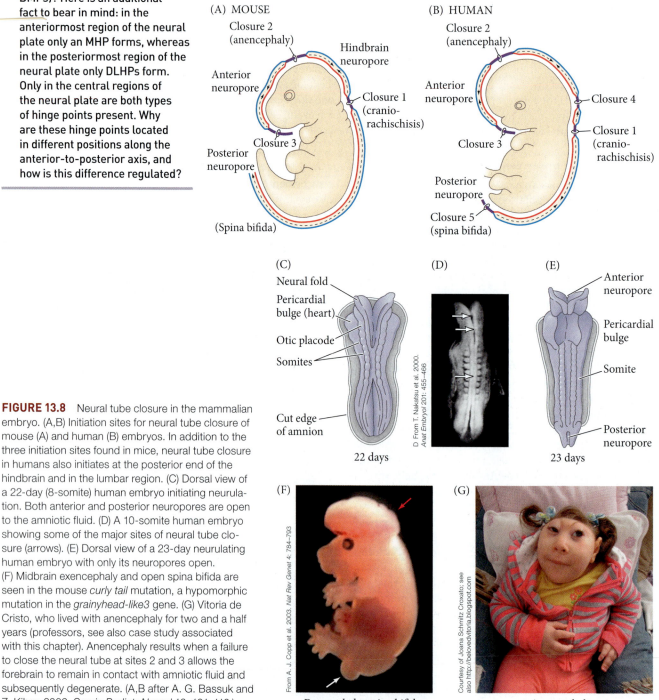

FIGURE 13.8   Neural tube closure in the mammalian embryo. (A,B) Initiation sites for neural tube closure of mouse (A) and human (B) embryos. In addition to the three initiation sites found in mice, neural tube closure in humans also initiates at the posterior end of the hindbrain and in the lumbar region. (C) Dorsal view of a 22-day (8-somite) human embryo initiating neurulation. Both anterior and posterior neuropores are open to the amniotic fluid. (D) A 10-somite human embryo showing some of the major sites of neural tube closure (arrows). (E) Dorsal view of a 23-day neurulating human embryo with only its neuropores open. (F) Midbrain exencephaly and open spina bifida are seen in the mouse *curly tail* mutation, a hypomorphic mutation in the *grainyhead-like3* gene. (G) Vitoria de Cristo, who lived with anencephaly for two and a half years (professors, see also case study associated with this chapter). Anencephaly results when a failure to close the neural tube at sites 2 and 3 allows the forebrain to remain in contact with amniotic fluid and subsequently degenerate. (A,B after A. G. Bassuk and Z. Kibar. 2009. *Semin Pediatr Neurol* 16: 101–110.)

fashion. So, in the 24-hour chick embryo, neurulation in the **cephalic** (head) region is well advanced, but the **caudal** (tail) region of the embryo is still undergoing gastrulation (see Figure 13.5). The two open ends of the neural tube are called the **anterior neuropore** and the **posterior neuropore**.

In chicks, neural tube closure is initiated at the level of the future midbrain and "zips up" in both directions. By contrast, in mammals, neural tube closure is initiated at several places along the anterior-posterior axis (**FIGURE 13.8**). In humans, there are probably five sites of neural tube closure (see Figure 13.5B; Nakatsu et al. 2000; O'Rahilly and Muller 2002; Bassuk and Kibar 2009), and the closure mechanism may differ at each site (Rifat et al. 2010). The rostral closure site (closure site 1) is located at the junction of the spinal cord and hindbrain and appears to close, as does the chick neural tube, by zipping up the neural folds. Similarly, at closure site 2, located at the midbrain/forebrain boundary, a directional zipper-like mechanism paired with dynamic cell extension appears to be at work. At closure site 3 (the rostral forebrain), the dorsolateral hinge points appear to be fully responsible for the neural tube closure.

How do the apices of the neural folds zip up? Are there interlocking cell membranes and some mysterious force that sequentially puts them together one at a time along the anterior-to-posterior axis? One way to better understand a process as complex as neural tube closure is to simply watch it. Rather remarkable in toto live-cell imaging has been conducted on mouse embryos in culture (Pyrgaki et al. 2010; Massarwa and Niswander 2013). During DLHP bending, dynamic cell processes extend from the juxtaposed tips of the neural folds (**FIGURE 13.9**). This cellular behavior is being displayed by non-neural surface ectoderm cells, which ultimately extend long filopodial processes toward the opposing fold. These filopodial extensions establish temporary "cellular bridges," whose functions are currently unknown. (See **Further Development 13.2, The Biomechanics of Neural Fold Zippering Revealed by the Ancestral Chordate**, online.)

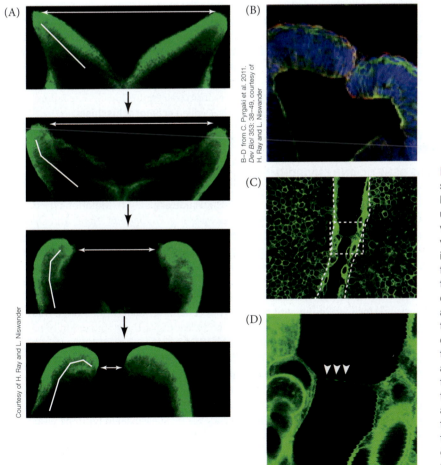

Courtesy of H. Ray and L. Niswander

B–D from C. Pyrgaki et al. 2011. *Dev Biol* 353: 38–49, courtesy of H. Ray and L. Niswander

FIGURE 13.9   Neural tube closure at mouse site 2 (midbrain region; see Figure 13.8A). (A) Live imaging of a 15-somite stage embryo using a transgenic CAG:Venus^{myr} mouse to visualize all cell membranes. Optical dorso-ventral (cross) sections seen from the top image to the bottom image show DLHP formation (curving of white line on left fold) to the point of near closure (decreasing size of double arrow). (B) Optical section through a mouse embryo as the neural folds are touching but not yet closed. The single layer of non-neural surface ectoderm (large, flattened cells; stained green) has wrapped itself around the neural ectoderm (stained blue) at the edge of the closing neural folds. (C) Dotted lines show the border between neural and non-neural ectoderm. Cellular bridges from the non-neural ectoderm connect the two juxtaposed neural folds. (D) A close-up of one of these bridges (from boxed area in C) is marked by arrowheads.

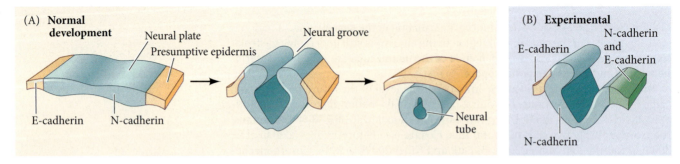

FIGURE 13.10 Expression of N- and E-cadherin adhesion proteins during neurulation in *Xenopus*. (A) Normal development. In the neural plate stage, N-cadherin is seen in the neural plate, whereas E-cadherin is seen on the presumptive epidermis. Eventually, the N-cadherin-bearing neural cells separate from the E-cadherin-containing epidermal cells. (Neural crest cells express neither N- nor E-cadherin, and they disperse.) (B) No separation of the neural tube occurs when one side of the frog embryo is injected with N-cadherin mRNA so that N-cadherin is expressed in the epidermal cells as well as in the presumptive neural tube.

FUSION AND SEPARATION The neural tube eventually forms a closed cylinder that separates from the surface ectoderm. This separation appears to be mediated by the expression of different cell adhesion molecules. Although the cells that will become the neural tube originally express E-cadherin, they stop producing this protein as the neural tube forms and instead synthesize N-cadherin (**FIGURE 13.10A**). As a result, the surface ectoderm and neural tube tissues no longer adhere to each other. If the surface ectoderm is experimentally made to express N-cadherin (by injecting N-cadherin mRNA into one cell of a two-cell *Xenopus* embryo), the separation of the neural tube from the presumptive epidermis is dramatically impeded (**FIGURE 13.10B**; Detrick et al. 1990; Fujimori et al. 1990). Loss of the gene for N-cadherin in zebrafish also results in failure to form a neural tube (Lele et al. 2002). The Grainyhead transcription factors are especially important in this process (Rifat et al. 2010; Werth et al. 2010; Pyrgaki et al. 2011). Grainyhead-like2, for instance, controls a battery of cell adhesion molecules and downregulates E-cadherin synthesis in the neural folds. Mice with mutations in *Grainyhead-like2* or *Grainyhead-like3* genes have severe neural tube defects, which include a split face, exencephaly, and spina bifida (see Figure 13.8F; Pyrgaki et al. 2011).

NEURAL TUBE CLOSURE DEFECTS In humans, neural tube closure defects occur in about 1 in every 1000 live births. Failure to close the posterior neuropore (closure site 5; see Figure 13.8B) around day 27 of development results in a condition called **spina bifida**, the severity of which depends on how much of the spinal cord remains exposed. Failure to close site 2 or site 3 in the rostral neural tube keeps the anterior neuropore open, resulting in a usually lethal condition called **anencephaly**, in which the forebrain remains in contact with the amniotic fluid and subsequently degenerates. The fetal forebrain ceases development, and the vault of the skull fails to form (see Figure 13.8G). The failure of the entire neural tube to close over the body axis is called **craniorachischisis**.

FURTHER DEVELOPMENT

THE GENETIC AND ENVIRONMENTAL CAUSES OF NTDs Failure to close the neural tube can result from both genetic and environmental causes (Fournier-Thibault et al. 2009; Harris and Juriloff 2010; Wilde et al. 2014). Mutations (first found in mice) in genes such as *Pax3*, *Sonic hedgehog*, *Grainyhead*, *Tfap2*, and *Open-brain* show that these genes are essential for the formation of the mammalian neural tube; in fact, more than 300 genes appear to be involved. Environmental factors including drugs, maternal dietary factors (such as deficiencies in cholesterol, zinc and folate, also known as folic acid or vitamin B_9), diabetes, obesity, and toxins can all influence human neural tube closure. How these factors lead to neural tube defects is largely unknown. For instance, a recent report has demonstrated that zinc deficiency disrupts neural tube closure by leading to the stabilization of p53 and consequently increased apoptosis (**FIGURE 13.11A**; Li et al. 2018). Folic acid deficiencies, however, have been one of the leading causes of neural tube defects. An emerging idea posits that a major outcome of environmental perturbations is the modification of the embryo's epigenome, which in turn causes transcription variability leading to neural tube defects (**FIGURE 13.11B**; Feil et al. 2012; Shyama-sundar et al. 2013; Wilde et al. 2014). This idea is most associated with the potential downstream consequences of folic acid metabolism.

Although the exact role of folate remains unknown, the early use of drugs that are folic acid antagonists and were given to women led to fetuses with neural tube defects. Since then, many large-scale human studies have demonstrated clear

(?) Developing Questions

What initiates the directionality of neural tube closure? Zipping proceeds in a posterior-to-anterior direction in the tunicate *Ciona*, as well as in certain closure points in mammals, yet it proceeds in opposite directions to close other regions of the mammalian brain. Moreover, are the cell forces that seem to advance the zipper in the primitive chordate *Ciona* conserved throughout vertebrates?

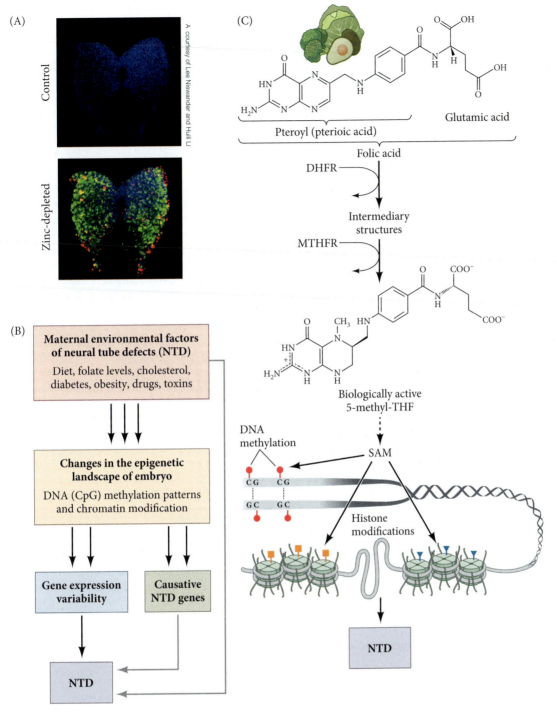

FIGURE 13.11 Environmental influences on neural tube defects and the role of folic acid. (A) Dorsal view of the anterior neural tube (developing brain) of control mice and of mice treated with the zinc chelator TPEN. Zinc depletion causes a dramatic increase in apoptosis, as indicated by DNA fragmentation (TUNEL labeling, green) and cleaved Caspase3 (red). Nucleic acids are stained with Hoechst dye (blue). (B) Overview of the connection that environmental factors are proposed to have with neural tube defects (NTD). Black arrows represent the main proposal for how environmental factors may lead to neural tube defects. Gray arrows represent other possible modes leading to NTD. (C) Simplified biochemical pathway for the metabolism of folic acid leading to epigenetic regulation through DNA methylation or histone modification. DHFR, dihydrofolate reductase; MTHFR, methylenetetrahydrofolate reductase; 5-methyl-THF, 5-methyltetrahydrofolate; SAM, S-adenosylmethionine.

correlations of neural tube disorders with folic acid deficiency, which is the reason dietary folic acid is not only recommended for pregnant women but also systematically fortified in foods (reviewed in Wilde et al. 2014). How folic acid deficiency leads to neural tube disorders is currently an active area of research. Folic acid is an important nutrient used for regulating DNA synthesis during cell division in the brain (Anderson et al. 2012) and is also critical in regulating DNA methylation (**FIGURE 13.11C**). Further evidence that epigenetic mechanisms are essential for proper neural tube development are the findings that functional manipulation of histone-modifying enzymes (acetyltransferases, deacetylases, demethylases) cause neural tube defects (Artama et al. 2005; Bu et al. 2007; Shpargel et al. 2012; Welstead et al. 2012; Murko et al. 2013). Whatever the mechanisms, it has been estimated that 25%–30% of human neural tube birth defects can be prevented if pregnant women take supplemental folate. Therefore, the U.S. Public Health Service recommends that women of childbearing age take 0.4 milligram of folate daily (Milunsky et al. 1989; Centers for Disease Control 1992; Czeizel and Dudás 1992).

Secondary neurulation

Secondary neurulation, which takes place in the most posterior region of the embryo during tailbud elongation, produces a neural tube through a very different process than primary neurulation (see Figure 13.4). Secondary neurulation involves the production of mesenchyme cells from the prospective ectoderm and mesoderm, followed by the condensation of these cells into a **medullary cord** beneath the surface ectoderm (**FIGURE 13.12A,B**). After this mesenchymal-epithelial transition, the central portion of this cord undergoes **cavitation** to form several hollow spaces, or **lumens** (**FIGURE 13.12C**); the lumens then coalesce into a single central cavity (**FIGURE 13.12D**; Schoenwolf and Delongo 1980).

We have seen that after Hensen's node has migrated to the posterior end of the embryo, the caudal region of the epiblast contains a precursor cell population that gives rise to both neural ectoderm and paraxial (somite) mesoderm as the embryo's trunk elongates (Tzouanacou et al. 2009). The ectodermal cells that will form the posterior (secondary) neural tube express the *Sox2* gene, whereas the ingressing mesodermal cells (which no longer encounter high levels of BMPs as they migrate beneath the epiblast) do not express *Sox2*. Rather, the ingressing mesodermal cells express *Tbx6* and form somites (see Chapter 17; Shimokita et al. 2010; Takemoto et al. 2011). The ability of the Tbx6 transcription factor to repress neural-inducing *Sox2* expression explains the bizarre phenotype of homozygous *Tbx6* mouse mutants, which have three neural tubes, posteriorly (see Figure 17.4; Chapman and Papaioannou 1998; Takemoto et al. 2011). In these mutants, the two rods of paraxial mesoderm have become neural tubes that even express regionally appropriate genes (such as *Pax6*). Thus, the epiblast surrounding the rostral primitive streak (the caudal lateral epiblast; see Chapter 12) contains a common precursor pool for paraxial mesoderm and for the neural plate that forms the spinal cord (Cambray and Wilson 2007; Wilson et al. 2009). This distinction emphasizes another fundamental difference between primary and secondary neurulation. During primary neurulation, the surface ectoderm and neural ectoderm are intimately connected through the process of neural tube closure and fusion, whereas during secondary neurulation, these two tissues are essentially uncoupled and develop independently of each other.[2] (See **Further Development 13.3, Closure at the Junction: A Human-Avian Connection**, online.)

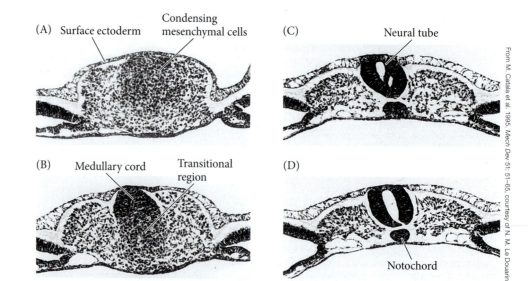

(A) Surface ectoderm   Condensing mesenchymal cells   (C)   Neural tube

(B) Medullary cord   Transitional region   (D)

Notochord

FIGURE 13.12 Secondary neurulation in the caudal region of a chick embryo. (A–D) A 25-somite chick embryo. (A) Mesenchymal cells condense to form the medullary cord at the most caudal end of the chick tailbud. (B) The medullary cord at a slightly more anterior position in the tailbud. (C) The neural tube is cavitating and the notochord is forming; note the presence of separate lumens. (D) The lumens coalesce to form the central canal of the neural tube.

From M. Catala et al. 1996. *Mech Dev* 51: 51–65, courtesy of N. M. Le Douarin

Patterning the Central Nervous System

The early development of most vertebrate brains is similar (**FIGURE 13.13A–D**). Because the human brain may be the most organized piece of matter in the solar system and is arguably the most interesting organ in the animal kingdom, we will concentrate on the development that is supposed to make *Homo* sapient.[3]

The anterior-posterior axis

The early mammalian neural tube is a straight structure, but even before the posterior portion of the tube has formed, the most anterior portion of the tube is undergoing

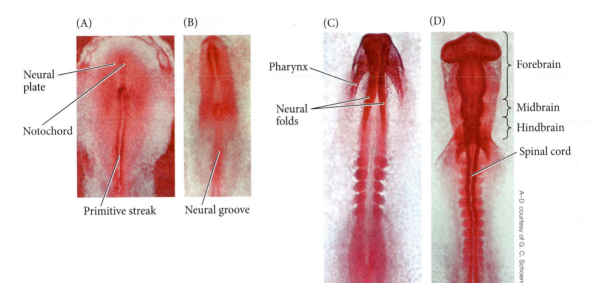

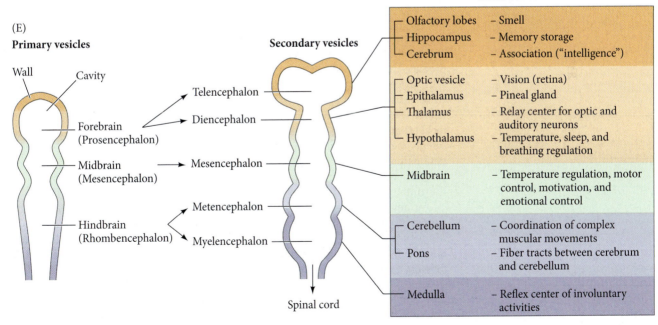

Adult derivatives

Olfactory lobes	– Smell
Hippocampus	– Memory storage
Cerebrum	– Association ("intelligence")
Optic vesicle	– Vision (retina)
Epithalamus	– Pineal gland
Thalamus	– Relay center for optic and auditory neurons
Hypothalamus	– Temperature, sleep, and breathing regulation
Midbrain	– Temperature regulation, motor control, motivation, and emotional control
Cerebellum	– Coordination of complex muscular movements
Pons	– Fiber tracts between cerebrum and cerebellum
Medulla	– Reflex center of involuntary activities

FIGURE 13.13 Early brain development and formation of the first brain chambers. (A–D) Chick brain development. (A) Flat neural plate with underlying notochord (head process). (B) Neural groove. (C) Neural folds begin closing at the dorsalmost region, forming the incipient neural tube. (D) Neural tube, showing the three brain regions and the spinal cord. The neural tube remains open at the anterior end, and the optic bulges (which become the retinas) have extended to the lateral margins of the head. (E) As in humans, the three primary brain vesicles become further subdivided as development continues. At the right is a list of the adult derivatives formed by the walls and cavities of the brain along with some of their functions. (E after K. L. Moore and T. V. M. Persaud. 1977. *The Developing Human: Clinically Oriented Embryology* 2e, p. 337. W. B. Saunders: Philadelphia.)

FIGURE 13.14   Rhombomeres of the chick hindbrain. (A) Hindbrain of a 3-day chick embryo. The roof plate has been removed so that the segmented morphology of the neural epithelium can be seen. The r1/r2 boundary is at the upper arrow, and the r6/r7 boundary is at the lower arrow. (B) A chick hindbrain at a similar stage stained with antibody to a neurofilament subunit. The rhombomere boundaries are emphasized because they serve as channels for neurons crossing from one side of the brain to the other.

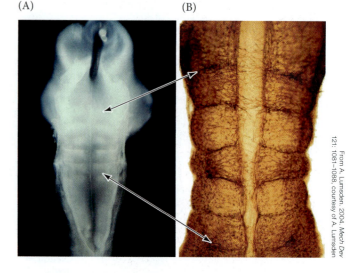

(A)   (B)

From A. Lumsden, 2004. *Mech Dev* 121: 1081–1088, courtesy of A. Lumsden

drastic changes. In the anterior region, the neural tube balloons into the three primary vesicles: the forebrain (**prosencephalon**), which forms the cerebral hemispheres; the midbrain (**mesencephalon**), whose neurons are involved in motivation, movement, and depression (Niwa et al. 2013; Tye et al. 2013); and the hindbrain (**rhombencephalon**), which becomes the cerebellum, pons, and the medulla oblongata (the most primitive area of the brain and the center of involuntary activities such as breathing; **FIGURE 13.13E**). By the time the posterior end of the neural tube closes, secondary vesicles have formed. The forebrain becomes the telencephalon (which forms the cerebral hemispheres) and the diencephalon (which will form the optic vesicle that initiates eye development).

The rhombencephalon develops a segmental pattern that specifies the places where certain nerves originate. Periodic swellings called **rhombomeres** divide the rhombencephalon into smaller compartments. The rhombomeres represent separate "territories," in that the cells within each rhombomere mix freely within it but do not mix with cells from adjacent rhombomeres (Guthrie and Lumsden 1991; Lumsden 2004). Each rhombomere expresses a unique combination of transcription factors, thereby generating rhombomere-specific patterns of neuronal differentiation. Thus, each rhombomere produces neurons with different fates. As we will see in Chapter 15, the neural crest cells derived from the rhombomeres will form **ganglia**, clusters of neuronal cell bodies whose axons form a nerve. Each rhombomeric ganglion produces a different type of nerve. The generation of the cranial nerves from the rhombomeres has been studied most extensively in the chick, in which the first neurons appear in the even-numbered rhombomeres r2, r4, and r6 (**FIGURE 13.14**; Lumsden and Keynes 1989). Neurons originating from r2 ganglia form the fifth (trigeminal) cranial nerve, those from r4 form the seventh (facial) and eighth (vestibuloacoustic) cranial nerves, and those from r6 form the ninth (glossopharyngeal) cranial nerve.

The anterior-to-posterior patterning of the hindbrain and spinal cord is controlled by a series of genes that include the Hox gene complexes. For more details on the mechanisms that pattern cell fates along the anterior-to-posterior axis, see **Further Development 13.4, Dividing the Central Nervous System**, and **Further Development 13.5, Specifying the Brain Boundaries**, both online.

The dorsal-ventral axis

The neural tube is polarized along its dorsal-ventral axis. In the spinal cord, for instance, the dorsal region is the place where the spinal neurons receive input from sensory neurons, whereas the ventral region is where the motor neurons reside. In the middle are numerous interneurons that relay information between the sensory and motor neurons (**FIGURE 13.15**). These differentiated cell types organized along the dorsoventral axis arose from progenitor cell populations located adjacent to the cavities (ventricles) of the brain and spinal cord that run along the

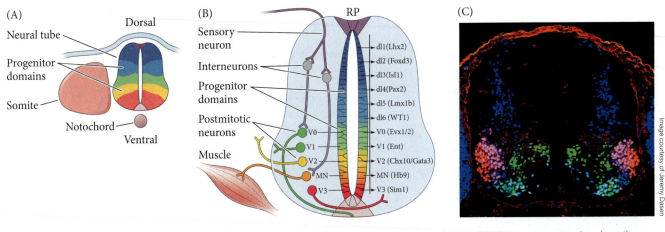

Image courtesy of Jeremy Dasen

FIGURE 13.15 Differential expression of transcription factors define progenitor domains and derived cell types along the dorso-ventral axis. (A) The early neural tube is made up of neuroepithelial progenitor cells that can be divided into discrete domains based on their unique repertoire of transcription factor expression. Pax3 and Pax7 define the most dorsal domain (dark blue), Nkx6-1 is expressed ventrally (red), and Pax6 is located in the central region of the neural tube (green). Overlap in expression of these different transcription factors creates further subdomains (yellow and light blue). (B) As the neural tube develops, these progenitor zones expand and continue to diversify as their gene regulatory networks mature, until the full differentiative program is adopted and derived cell types emerge (such as the different neuronal cell types illustrated). (C) Immunolabeling for some of the transcription factors expressed by different regions shown in B: the Isl1 (blue), Foxp1 (red), and Lhx3 (green) transcription factors in an embryonic (day 12.5) mouse spinal cord at the cervical level. (A,B after C. Catela et al. 2015. *Annu Rev Cell Dev Biol* 31: 669–698.)

anterior-to-posterior axis (i.e., in the ventricular zone). Each progenitor domain can be defined by its expression of specific transcription factors (such as the products of homeobox genes like Pax7), which specify progeny to differentiate into the specific classes of neuronal and glial cells that make up the CNS (Catela et al. 2015). This presents a logical question: How does a cell sense its position within the neural tube such that it develops into the progenitor cell population that generates the precise type and properly positioned neurons and glia? Said another way, how is pattern created in the neural tube?

Opposing morphogens

The dorsal-ventral polarity of the neural tube is induced by morphogenetic signals coming from its immediate environment. In the spinal region, the ventral pattern is imposed by the notochord, whereas the dorsal pattern is induced by the overlying epidermis (**FIGURE 13.16A–D**). Specification of the axis is initiated by two major paracrine factors: Sonic hedgehog protein (Shh) originating from the notochord, and TGF-β proteins originating from the dorsal ectoderm (**FIGURE 13.16E**). In both cases, these factors induce a second signaling center within the neural tube itself.

Sonic hedgehog secreted from the notochord induces the medial hinge point cells to become the **floor plate** of the neural tube. The floor plate cells also secrete Sonic hedgehog, which forms a gradient that is highest at the most ventral portion of the neural tube (see Figure 13.16B,C,E). Cells experiencing the highest concentrations of Shh develop into the progenitor cells for motor neurons and a class of interneurons called V3 neurons, whereas moderate and lower levels of Shh induce increasingly more dorsal progenitor populations (see Figure 13.16D and **FIGURE 13.16F,G**; Roelink et al. 1995; Briscoe et al. 1999). (See **Further Development 3.6, Determining the Sonic Signal**, online.)

The dorsal fates of the neural tube are established by proteins of the TGF-β super-family, especially BMPs 4 and 7, Dorsalin, and Activin (Liem et al. 1995, 1997, 2000). Initially, BMP4 and BMP7 are found in the epidermis. Just as the notochord establishes a secondary signaling center—the floor plate cells—on the ventral side of the neural tube, the epidermis establishes a secondary signaling center by inducing BMP4 expression in the **roof plate** cells of the neural tube. The BMP4 protein from the roof plate induces a cascade of TGF-β proteins in adjacent cells (see Figure 13.16). Dorsal sets of cells are thus exposed to TGF-β proteins at higher concentrations and at earlier times when compared with adjacent neural cells. The importance of TGF-β superfamily

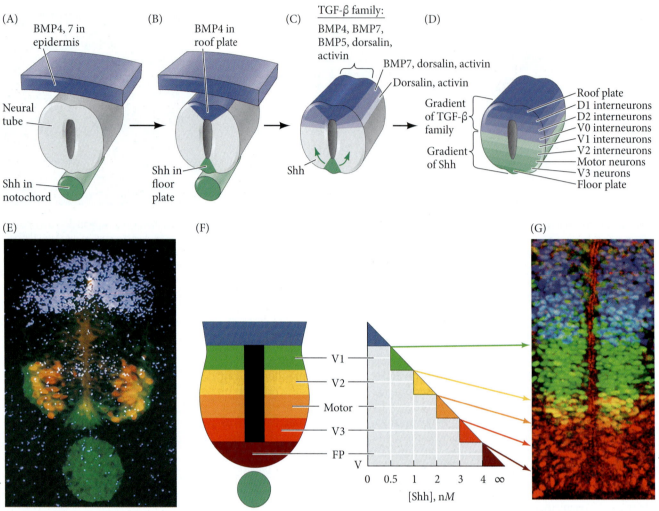

FIGURE 13.16 Dorsal-ventral specification of the neural tube. (A) The newly formed neural tube is influenced by two signaling centers. The roof of the neural tube is exposed to BMP4 and BMP7 from the epidermis, while the floor of the neural tube is exposed to Sonic hedgehog (Shh) from the notochord. (B) Secondary signaling centers are established in the neural tube. BMP4 is expressed and secreted from the roof plate cells; Shh is expressed and secreted from the floor plate cells. (C) BMP4 establishes a nested cascade of TGF-β factors spreading ventrally into the neural tube from the roof plate. Sonic hedgehog diffuses dorsally as a gradient from the floor plate cells. (D) The neurons of the spinal cord are given their identities by their exposure to these gradients of paracrine factors. The amounts and types of paracrine factors present cause different transcription factors to be activated in the nuclei of these cells, depending on their position in the neural tube. (E) Chick neural tube showing areas of Shh (green) and the expression domain of the protein Dorsalin (blue; Dorsalin is a member of the TGF-β superfamily). Motor neurons induced by a particular concentration of Shh are stained orange/yellow. (F) Relationship between Sonic hedgehog concentrations, the generation of particular neuronal types in vitro, and distance from the notochord. Cells closest to the notochord become the floor plate neurons; motor neurons and V3 interneurons emerge on the ventro-lateral sides. (G) In situ hybridization for three other transcription factors: Pax7 (blue; characteristic of the dorsal neural tube cells), Pax6 (green), and Nkx6-1 (red). Where Nkx6-1 and Pax6 overlap (yellow), motor neurons become specified. (F after J. Briscoe et al. 1999. *Nature* 398: 622–627.)

factors in patterning the dorsal portion of the neural tube was demonstrated by the phenotypes of zebrafish mutants. Those mutants deficient in certain BMPs lacked dorsal and intermediate types of neurons (Nguyen et al. 2000). (See **Further Development 13.7, How Much and How Long?, Further Development 13.8, Transcriptional Cross-Repression by the Downstream Shh and TGF-ß Effector Proteins,** and **Further Development 13.9, Gli Activation**, all online.)

All Axes Come Together

The model of dorsal-ventral patterning by TGF-β and Shh morphogens pertains to cell fates throughout the CNS along the rostral-caudal axis. Remember, though, that there are differences in how the anterior regions of the neural tube form and how the posteriormost region forms—differences that are defined by primary and secondary neurulation. Progenitor cells in the *anterior* regions of the neural tube (which become the brain and most of the spinal cord) adopt a proneural fate directly from the epiblast (Harland 2000; Stern 2005). Cells in the *posterior* begin as bipotential **neuromesodermal progenitors** (**NMPs**) that undergo a transition to become neural or somitic cell types, with the neural cells forming the caudal end of the neural tube (**FIGURE 13.17**). NMPs are born in the caudal lateral epiblast during tailbud elongation and are positively maintained by Fgf8 and Wnt signals (see Chapter 17 for details on axis elongation). In opposition to the caudal Fgf/Wnt signals, retinoic acid is expressed by somitic mesoderm and inhibits Fgf8 signaling. It is the antagonistic gradients of retinoic acid and Fgf/Wnt along the rostral-caudal axis that establish a road to NMP maturation. Cells that enter the neural mesenchyme become preneural progenitor cells and are initially competent to respond to either Shh or BMP signals by differentiating into either floor plate or roof plate, respectively. As the tailbud elongates, these preneural NMP cells become positioned farther from Fgf/Wnt and closer to retinoic acid; this repositioning triggers a switch in their competency to respond to Shh/TGF-β signals, thus allowing patterning of the proneural progenitors along the dorsoventral axis of the maturing neural tube (Sasai et al. 2014; Gouti et al. 2015; Henrique et al. 2015; Steventon and Martinez Arias 2017).

(?) Developing Questions

We have described an elaborate morphogenic system that relays positional information for the development of cell identity. What if the responding cells were not static but moving? How would this change the dynamics of gradient interpretation? Work in the Megason lab has shown that specified progenitors in the zebrafish neural tube do in fact move about the epithelium, sorting themselves into discrete domains (Xiong et al. 2013). Thus, a new dynamic of cell movement among the morphogenetic signals needs to be factored in to the model of pattern formation in the neural tube.

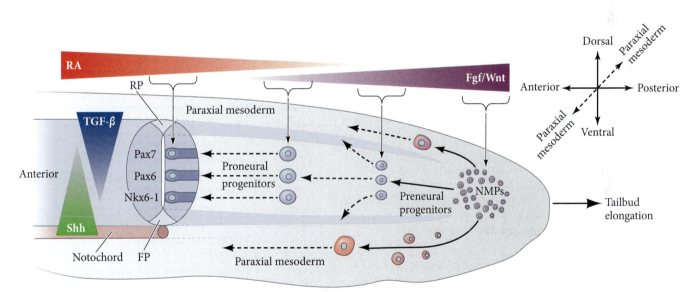

FIGURE 13.17   Model for converging signals for maturation and specification of neural progenitors in the developing caudal region of the spinal cord. During spinal cord development, the tail is undergoing elongation, in part fueled by the caudal lateral epiblast, which houses the proliferative and motile neuromesodermal progenitors (NMPs). NMPs leave the tailbud and either progress into the neural mesenchyme/neural plate or the mesenchyme of the paraxial mesoderm, where they will give rise to the neural tube or somites, respectively. (Dashed arrows indicate that those cells will contribute to those regions of the neural tube, but the cells are not actively migrating into those regions.) Opposing antagonistic morphogens of retinoic acid (RA) and Fgf/Wnt establish inverse gradients along the rostral-caudal axis, setting up graded positional instructions along this axis. High Fgf/Wnt maintains the NMP pool. Moderate Fgf/Wnt and low RA promote early preneural progenitors to be competent to respond to TGF-β and Shh dorsoventral signals and develop into the roof plate (RP) and floor plate (FP), respectively. As the tailbud continues to elongate, preneural progenitors will experience low Fgf/Wnt and moderate RA, which broadens their competency to initiate gene regulatory programs specific for proneural progenitor populations. In this way, morphogens along all axes pattern the cell fates in the neural tube.

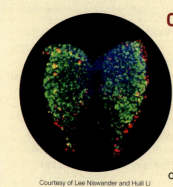

Closing Thoughts on the Opening Photo

You came across this beautiful fluorescent image by the Niswander lab earlier in the chapter (see Figure 13.11), which shows that deficiencies in maternal dietary zinc can cause neural tube defects in a fetus. This further emphasizes the importance of maternal diet for fetal development. Working to understand the many ways in which the environment affects development is a new frontier in developmental biology. Here, a zinc deficiency caused increased cell death and a failure of the neural tube to close. As mentioned above, a folate deficiency may cause an epigenetic change that disturbs neural tube formation. We are only at the beginning of this new frontier, and it will be up to *you*, the new generation of scientists, to explore this new field of environment-embryo interactions.

Courtesy of Lee Niswander and Huili Li

Endnotes

[1] In teleost (bony) fish such as zebrafish, the neural plate does not fold; rather, convergence at the midline generates the neural keel, and the lumen of the neural tube is formed through a process of cavitation (Lowery and Sive 2004; see also Harrington et al. 2009).

[2] Failure to form the nerve (medullary) cord in zebrafish does not prevent surface ectoderm from spanning the dorsal midline (Harrington et al. 2009).

[3] Our species name comes from the Latin *sapio*, meaning "to be capable of discerning."

Snapshot Summary
Neural Tube Formation and Patterning

1. The neural tube forms from the shaping and folding of the neural plate. In primary neurulation, the surface ectoderm folds into a tube that separates from the surface ectoderm. In secondary neurulation, ectoderm and mesoderm cells coalesce as a mesenchyme to form first a cord and then a cavity (lumen) within the cord.

2. Primary neurulation is regulated by both intrinsic and extrinsic forces. Intrinsic forces causing changes in cell shape from boxlike to a wedge shape occur within cells of the hinge regions, bending the neural plate. Extrinsic forces include the movement of the surface ectoderm toward the center of the embryo.

3. Neural tube closure is also the result of extrinsic and intrinsic forces. In humans, congenital anomalies can result if the neural tube fails to close. Folate provided to the fetus through the maternal diet is of great importance for proper neural tube closure in the fetus.

4. After the node has reached the posterior end of the epiblast, certain cells contribute to both the paraxial mesoderm and the neural tube.

5. The neural crest cells arise at the borders of the neural tube and surface ectoderm. They become located between the neural tube and surface ectoderm, and they migrate away from this region to become a number of cell types, including peripheral neural, glial, and pigment cells.

6. There is a gradient of maturity in many embryos (especially those of amniotes) such that the anterior develops earlier than the posterior.

7. The vertebrate brain forms three primary vesicles: the prosencephalon (forebrain), the mesencephalon (midbrain), and the rhombencephalon (hindbrain). The prosencephalon and rhombencephalon become further subdivided.

8. Dorsal-ventral patterning of the neural tube is accomplished by proteins of the TGF-β superfamily secreted from the surface ectoderm and roof plate of the neural tube, and by Sonic hedgehog protein secreted from the notochord and floor plate cells. Temporal and spatial gradients of Shh trigger the synthesis of particular transcription factors that pattern the neuroepithelium.

9. Bipotential neuromesodermal progenitor cells (NMPs) in the caudal end of the developing embryo can become either neural or somitic cells. Exposure of the preneural progenitors to opposing gradients of Fgf/Wnt and retinoic acid leads to the patterning of the caudal neural tube along its dorsal-ventral axis.

Go to oup.com/he/barresi12xe for Further Developments, Scientists Speak interviews, Watch Development videos, Dev Tutorials, and complete bibliographic information for all literature cited in this chapter.

Brain Growth

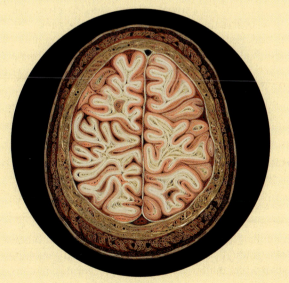

The complexities of being human: How deep do they fold?

Courtesy of Lisa Nilsson

"WHAT IS PERHAPS THE MOST INTRIGUING QUESTION OF ALL is whether the brain is powerful enough to solve the problem of its own creation," declared Gregor Eichele in 1992. Determining how the brain—an organ that perceives, thinks, loves, hates, remembers, changes, deceives itself, and coordinates all our conscious and unconscious bodily processes—is constructed is undoubtedly the most challenging of all developmental enigmas. A combination of genetic, cellular, and systems level approaches is now giving us a very preliminary understanding of how the basic anatomy of the brain becomes ordered.

Differentiation of the neural tube into the various regions of the brain and spinal cord occurs simultaneously in three different ways. On the gross anatomical level, the neural tube and its lumen bulge and constrict to form the vesicles of the brain and spinal cord. At the tissue level, the cell populations in the wall of the neural tube arrange themselves into the different functional regions of the brain and spinal cord. Finally, on the cellular level, the neuroepithelial cells differentiate into the numerous types of nerve cells (**neurons**) and associated cells (**glia**) present in the body. In this chapter, we will concentrate on the development of the mammalian brain in general, as well as the human brain in particular, as we consider what makes us human.

Neuroanatomy of the Developing Central Nervous System

Your brain contains approximately 130–200 billion cells, of which half (about 80 billion) are neurons, about 60 billion are glial cells, and about 20 billion are thought to be endothelial cells (Azevedo et al. 2009; Andrade-Moraes et al. 2013; von Bartheld et al. 2016; von Bartheld 2018). There is a wide variety of neuronal and glial cell *types*, however, from the relatively small (e.g., granule cells) to the comparatively enormous (e.g., Purkinje neurons). All this diversity begins with the multipotent neuroepithelial cells of the neural tube.

The cells of the developing central nervous system

Neuroepithelial cells make up the neural plate and early neural tube, and as epithelial cells, they are polarized along their apical-basal axis (**FIGURE 14.1**). Once the plate closes into a neural tube, the apical surface of the neuroepithelium borders the internal cavity of the tube, which will become filled with cerebrospinal fluid. The basal surface of each cell terminates with an **endfoot**, or swelling of its basal membrane at the outer surface of the neural tube. The surface of the CNS is also referred to as the **pial surface**, after the pia mater that represents the fibrous membranes that surround nervous tissues.

(A) Basal   Pial surface   Endfeet

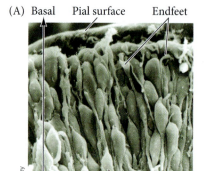

Courtesy of K. Tosney

Apical   Soma

FIGURE 14.1   Cell polarity in the CNS. Scanning electron micrograph of a newly formed chick neural tube, showing neuroepithelial cells at different stages of their cell cycles spanning the full width of the epithelium.

NEURAL STEM CELLS OF THE EMBRYO   Neuroepithelial cells are the first multipotent neural stem cells of the embryo. As stem cells, they are highly proliferative, generating progenitor cells for the first neuronal and glial cell types of the neural tube (Turner and Cepko 1987).

Neuroepithelial cells are only present in the early embryo and eventually transform into **ventricular (ependymal) cells** and **radial glial cells**, or **radial glia**. Ependymal cells remain an integral component of the neural tube lining and secrete the cerebrospinal fluid. Radial glia maintain a polarized morphology spanning the apicobasal axis of the central nervous system (CNS) and carry out two primary functions. First, they serve as the major neural stem cell throughout embryonic and fetal development, demonstrating self-renewal and the multipotent generation of both neurons and glial cells (Doetsch et al. 1999; Kriegstein and Alvarez-Buylla 2009); and second, they serve as a scaffold for the migration of other progenitor cells and newborn neurons (Bentivoglio and Mazzarello 1999). These two functions provide the foundational mechanism driving brain growth. (See **Further Development 14.1, A Primer on the Basic Anatomy and Function of Neurons and Glia**, online.)

Tissues of the developing central nervous system

The neurons of the brain are organized into layers (**laminae**) and clusters (**nuclei**),[1] each having different functions and connections. The original neural tube is composed of a **germinal neuroepithelium**, a layer of rapidly dividing neural stem cells one cell layer thick. Over the course of evolution, adaptations have led to the germinal neuroepithelium producing a diversity of highly complex regions within the CNS. All of these regions, however, are elaborations of the same basic three-zone pattern of layers: ventricular (next to the lumen), mantle (intermediate), and marginal (outer) (**FIGURE 14.2**).

The formation of these three layers begins as the stem cells in the **ventricular zone** continue to divide, the migrating cells forming a second layer around the original neural tube. This layer becomes progressively thicker as more cells are added to it from the germinal neuroepithelium. This new layer is the **mantle**, or **intermediate**, **zone**. The mantle zone cells differentiate into both neurons and glia. The neurons make connections among themselves and send forth axons away from the lumen, thereby creating a **marginal zone** poor in neuronal cell bodies. Eventually, oligodendrocytes cover many of the axons in the marginal zone in myelin sheaths, giving them a whitish appearance. Hence, the axonal marginal layer is often called **white matter**, while the mantle zone, containing the neuronal cell bodies, is referred to as **gray matter** (see Figure 14.2). The germinal epithelium of the ventricular zone will later shrink to become the **ependyma** that lines the brain cavity.

Here we will focus our investigation of CNS structure on the architecture associated with the spinal cord and medulla, cerebellum, and cerebrum.

SPINAL CORD AND MEDULLA ORGANIZATION   The basic three-zone pattern of ventricular (ependymal), mantle, and marginal layers is retained throughout development of the spinal cord and the medulla (the posterior region of the hindbrain). When viewed in cross section, the mantle gradually becomes a butterfly-shaped structure surrounded by the marginal zone, or white matter, and both become encased in connective tissue. As the neural tube matures, a longitudinal groove—the **sulcus limitans**—divides it into dorsal and ventral halves. The dorsal portion receives input from sensory neurons, whereas the ventral portion is involved in effecting various motor functions (**FIGURE 14.3**). This developmental anatomy generates the basis of medullary and spinal cord physiology (such as the reflex arc).

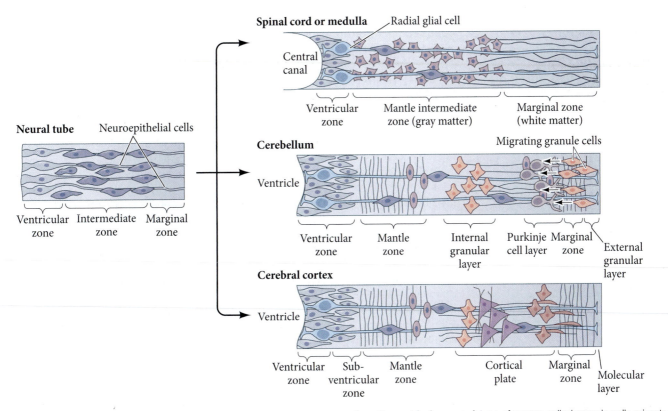

FIGURE 14.2 Differentiation of the walls of the neural tube. A section of a 5-week human neural tube (left) reveals three zones: ventricular (ependymal), intermediate (mantle), and marginal. In the spinal cord and medulla (top right), the ventricular zone remains the sole source of neurons and glial cells. In the cerebellum (middle right), a second mitotic layer, the external granular layer, forms at the region farthest removed from the ventricular zone. A type of neuron called granule cells migrate from this layer back into the intermediate zone to form the internal granular layer. In the cerebral cortex (bottom right), the migrating neurons and glioblasts form a cortical plate containing six layers. (After M. Jacobson. 1978. *Developmental Neurobiology*. Springer: Boston, MA.)

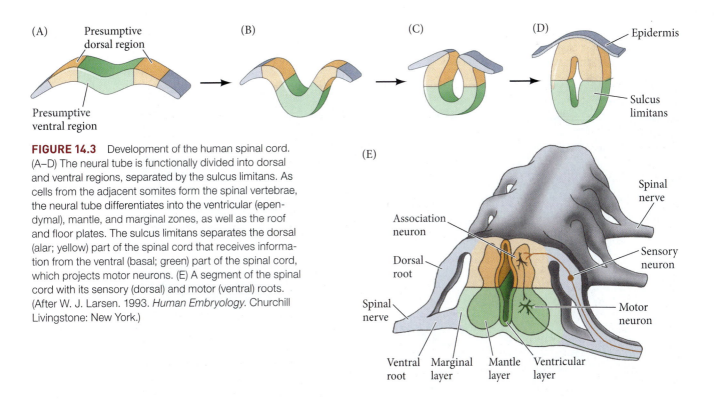

FIGURE 14.3 Development of the human spinal cord. (A–D) The neural tube is functionally divided into dorsal and ventral regions, separated by the sulcus limitans. As cells from the adjacent somites form the spinal vertebrae, the neural tube differentiates into the ventricular (ependymal), mantle, and marginal zones, as well as the roof and floor plates. The sulcus limitans separates the dorsal (alar; yellow) part of the spinal cord that receives information from the ventral (basal; green) part of the spinal cord, which projects motor neurons. (E) A segment of the spinal cord with its sensory (dorsal) and motor (ventral) roots. (After W. J. Larsen. 1993. *Human Embryology*. Churchill Livingstone: New York.)

CEREBELLAR ORGANIZATION In the cerebellum, cell migration and selective proliferation and cell death produce modifications of the three-zone pattern shown in Figure 14.2. Cerebellar development results in a highly folded cortex (outer region) composed of Purkinje neurons and granule neurons integrated into nuclei that control balance functions and relay information from the cerebellar cortex to other brain regions. In the development of the cerebellum, the critical event appears to be the migration of neural progenitor cells to the outer surface of the developing cerebellum. Here they form a new germinal zone—the **external granular layer**—near the outer boundary of the neural tube.

At the outer boundary of the external granular layer, which is 1–2 cell bodies thick, neural progenitor cells proliferate and come into contact with cells that secrete bone morphogenetic proteins (BMPs). The BMPs specify the postmitotic cells derived from neural progenitor divisions to become a type of neuron called **granule cells** (Alder et al. 1999). Granule cells migrate back toward the ventricular (ependymal) zone, where they form a region called the **internal granular layer** (see Figure 14.2). Meanwhile, the original ventricular zone of the cerebellum generates a wide variety of neurons and glial cells, including the distinctive and large **Purkinje neurons**, the major cell type of the cerebellum (**FIGURE 14.4**). Purkinje neurons secrete Sonic hedgehog, which sustains the division of granule cell precursors in the external granular layer (Wallace 1999). Each Purkinje neuron has an enormous **dendritic arbor**, which spreads like a tree above a bulb-like cell body (see Figure 2A in Further Development 14.1, online). A typical Purkinje neuron may form as many as 100,000 synapses with other neurons—more connections than any other type of neuron studied. Each Purkinje neuron also sends out a slender axon, which connects to neurons in the deep cerebellar nuclei.

Purkinje neurons are critical in the electrical pathway of the cerebellum. All electric impulses eventually regulate their activity because Purkinje neurons are the only output neurons of the cerebellar cortex. Such regulation requires the proper cells to differentiate at the appropriate places and times. *How is this complicated series of events accomplished?* Contemplate a few ideas about what mechanisms might control the pattern of neuronal differentiation, as we will return to this overarching question later in the chapter.

CEREBRAL ORGANIZATION The three-zone arrangement of the neural tube is also seen, although modified, in the cerebrum. The cerebrum (also commonly referred to as the cerebral cortex) is organized in two distinct ways. First, like the cerebellum, it is

FIGURE 14.4 Cerebellar organization. (A) Sagittal section of a fluorescently labeled rat cerebellum photographed using dual-photon confocal microscopy. (B) Enlargement of the boxed area in (A) illustrates the highly structured organization of neurons and glial cells. Purkinje neurons are light blue with bright green processes, Bergmann glia are red, and granule cells are dark blue.

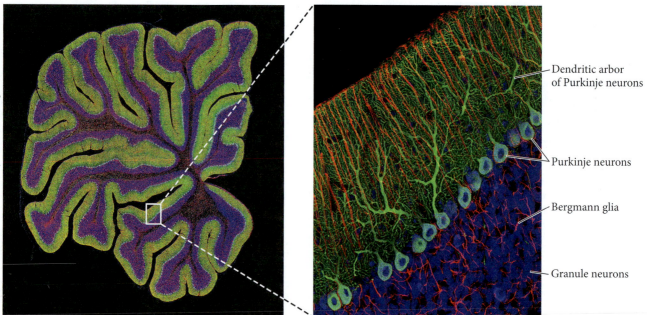

(A)

(B)

Courtesy of T. Deerinck and M. Ellisman, University of California, San Diego

Dendritic arbor of Purkinje neurons

Purkinje neurons

Bergmann glia

Granule neurons

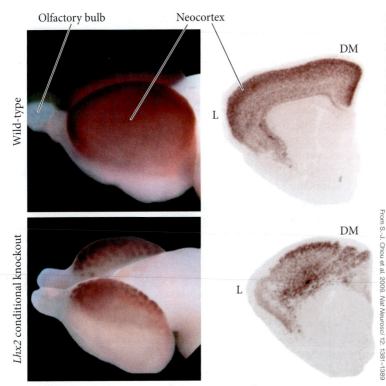

Olfactory bulb   Neocortex

Wild-type

Lhx2 conditional knockout

DM

L

DM

L

From S.-J. Chou et al. 2009, *Nat Neurosci* 12: 1381–1389

organized radially into layers that interact with one another. Certain neural precursor cells from the mantle zone migrate on radial glial processes toward the outer surface of the brain and accumulate in a new layer, the cortical plate (see Figure 14.2). This new layer of gray matter will become the **neocortex**,[2] a distinguishing feature of the mammalian brain. Specification of the neocortex involves the Lhx2 transcription factor, which activates numerous other cerebral genes. In *Lhx2*-deficient mice, the cerebral cortex fails to form (**FIGURE 14.5**; Mangale et al. 2008; Chou et al. 2009).

The neocortex eventually stratifies into six layers of neuronal cell bodies; the adult forms of these layers are not fully mature until the middle of childhood. Each layer of the neocortex differs from the others in its functional properties, the types of neurons found there, and the sets of connections they make (**FIGURE 14.6**). For instance, neurons

(A)   Pial surface   (B)

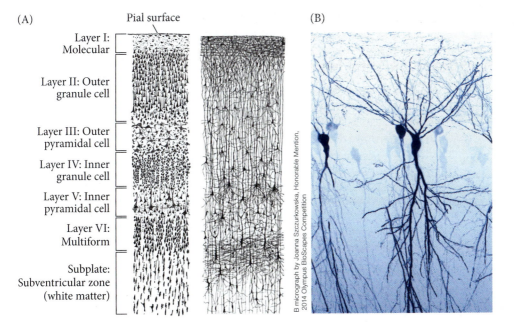

Layer I: Molecular

Layer II: Outer granule cell

Layer III: Outer pyramidal cell

Layer IV: Inner granule cell

Layer V: Inner pyramidal cell

Layer VI: Multiform

Subplate: Subventricular zone (white matter)

B micrograph by Joanna Szczurkowska, Honorable Mention, 2014 Olympus BioScapes Competition

FIGURE 14.6 Different neuronal cell types are organized into the six layers of the neocortex. (A) Different cellular stains reveal neocortical layering in these exquisite drawings by Santiago Ramón y Cajal from his 1899 work "Comparative study of the sensory areas of the human cortex." (B) Pyramidal neurons of mouse hippocampus (postnatal day 7).

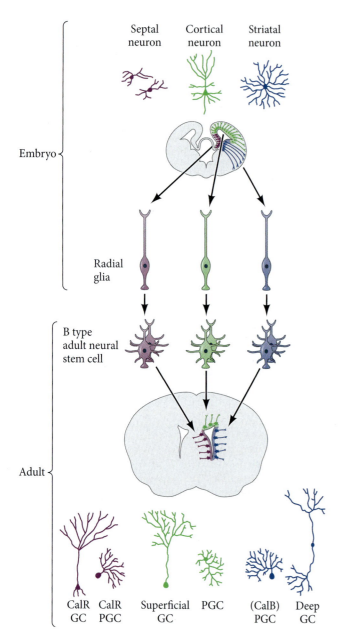

FIGURE 14.7 Regional specification of embryonic radial glia translates into restricted progenitor derivation. This schematic shows the positions of ventricular radial glia in the embryonic brain (above), and the clonal derivation of B type stem cells and their related differentiated neurons in the adult brain (below). (GC, granule cells; PGC, periglomerular cell; CalB, calbindin; CalR, calretinin.) (After L. C. Fuentealba et al. 2015. *Cell* 161: 1644–1655.)

in cortical layer 4 *receive* their major input from the thalamus (a region that forms from the diencephalon), whereas neurons in layer 6 *send* their major output to the thalamus.

In addition to the six vertical layers, the cerebral cortex is organized horizontally into more than 40 regions that regulate anatomically and functionally distinct processes. For instance, neurons of the visual cortex in layer 6 project axons to the lateral geniculate nucleus of the thalamus, which is involved in vision, whereas neurons of the auditory cortex of layer 6 (located more anteriorly than the visual cortex) project axons to the medial geniculate nucleus of the thalamus, which functions in hearing.

FURTHER DEVELOPMENT

ADULT ORGANIZATION IS SPECIFIED IN THE EMBRYONIC BRAIN   One of the major questions in developmental neurobiology is whether the different functional regions of the cerebral cortex are already specified in the ventricular region, or if specification is accomplished much later by synaptic connections between the regions. Evidence that specification is early (and that there might be some "protomap" of the cerebral cortex) is suggested by certain human mutations that destroy the layering and functional abilities in only one part of the cortex, leaving the other regions intact (Piao et al. 2004). More direct evidence for the existence of a protomap in the embryonic cortex recently emerged when Fuentealba and colleagues (2015) followed ventricular radial glial cells from different regions of the embryonic mouse brain using retroviral barcoding until the cells' direct clonal descendants could be identified in the adult cortex (**FIGURE 14.7**). They discovered that the differentiated neurons of the cortex were descended from stem cells that resided in comparable areas in the embryo (which were themselves derived from radial glia from comparable ventricular zone areas). These results support a model in which ventricular zone radial glia are regionally specified in the embryo and give rise to similarly specified adult stem cells, which propagate regionally restricted progeny.

Developmental Mechanisms Regulating Brain Growth

Growing the vertebrate brain is much like constructing a multilevel, multicolored brick building. First, the bricks need to be made and an appropriate number of the correctly colored bricks supplied to the right locations. Second, scaffolding is used throughout the structure to transport the necessary bricks and supplies to their destined locations. The building is constructed from bottom to top, building outward in the various dimensions to create increasingly complex architecture. In the developing brain, precisely controlled cell division of stem and progenitor cells generates the necessary numbers and types of cells ("the bricks"). Radial glial cells not only serve as stem cells, but also provide the scaffolding required for movement of progenitor cells and newborn neurons to increasingly more superficial layers in a manner that effectively builds the brain from the inside outward.

Neural stem cell behaviors during division

INTERKINETIC NUCLEAR MIGRATION DURING DIVISION   Sauer and colleagues' 1935 study of the germinal neuroepithelium not only indicated that the cells spanned

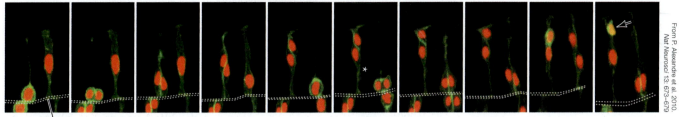

0 h, 00 min   0 h, 07 min   0 h, 24 min   0 h, 59 min   1 h, 17 min   1 h, 24 min   1 h, 41 min   2 h, 55 min   5 h, 15 min   7 h, 00 min

Apical surface

From P. Alexandre et al. 2010.
Nat Neurosci 13: 673–679

the width of the epithelium, but also showed that the *cell* nuclei are at different heights in this tissue (see Figure 14.1), and that the nuclei move as the cell goes through the cell cycle. During DNA synthesis (S phase of the cell cycle), the nucleus is near the basal end of the cell, near the outside edge of the neural tube, and translocates toward the apical end of the cell as the cycle proceeds. By mitosis (M phase), the nucleus is at the cell's apical end, near the ventricular surface. Following mitosis (G1 phase), the nucleus slowly migrates basally again (**FIGURE 14.8**). This process, called **interkinetic nuclear migration**, is also seen in the radial glial cells and occurs in a broad range of vertebrates (Alexandre et al. 2010; Meyer et al. 2011; Spear and Erickson 2012). The mechanisms involved are not fully understood, but microtubules and motor proteins appear to be involved. When a gene for a motor protein that is important for mitotic spindle pole separation is mutated in zebrafish, radial glial cells can successfully initiate interkinetic nuclear migration but fail to progress through mitosis, and the somas of these radial glia accumulate at the luminal (apical) surface over time (Johnson et al. 2016).

SYMMETRY OF DIVISION   When neuroepithelial cells or radial glial cells divide, what options do they have? Recall from Chapter 5 our descriptions of division in other stem cells (see Figure 5.1). A stem cell can divide symmetrically to produce two copies of itself, thus increasing the pool of stem cells. Alternatively, symmetrical division can produce two differentiating daughter cells, which depletes the stem cell pool. A stem cell can also divide asymmetrically to self-renew and yield a differentiating daughter cell. How might you investigate which of these divisions occurs in the neuroepithelium? Labeling the cells with a tracer, such as radioactive thymidine, that is incorporated only into dividing cells would allow you to trace cell lineages. When mammalian neuroepithelial cells are labeled in this way during early development, 100% of them incorporate the radioactive thymidine into their DNA, indicating that they are all undergoing some form of division (Fujita 1964). Shortly thereafter, however, certain cells stop incorporating this thymidine analog, indicating that they are no longer dividing. These cells can then be seen to migrate away from the lumen of the neural tube and differentiate into neuronal and glial cells (Fujita 1966; Jacobson 1968). When a cell of the germinal neuroepithelium is ready to generate neurons (instead of more neural stem cells), the division plane often shifts to create an asymmetrical division (the arrow in Figure 14.8). Instead of both daughter cells remaining attached to the luminal surface, one of them becomes detached (the asterisk in Figure 14.8). The cell remaining connected to the luminal surface usually remains a stem cell, while the other cell migrates and differentiates into a neuron or another type of progenitor (Chenn and McConnell 1995; Hollyday 2001).

Neurogenesis: Building from the bottom up (or from the inside out)

In a 2008 paper, Nicholas Gaiano summarized neurogenesis:

> *The construction of the mammalian neocortex is perhaps the most complex biological process that occurs in nature. A pool of seemingly homogeneous stem cells first undergoes proliferative expansion and diversification and then initiates the production of successive waves of neurons. As these neurons are generated, they take up residence in the nascent cortical plate where they integrate into the developing neocortical*

FIGURE 14.8   Live imaging of neuro-epithelial cell interkinetic nuclear migration and division of neural stem cells in the zebrafish embryonic hindbrain. Two nearly adjacent progenitor cells in the germinal epithelium were recorded over 7 hours. Cells are labeled to show cell membranes (green) and nuclei (red). A reporter gene specifically marks neurons (yellow). The progenitor cell on the left underwent an asymmetrical division, generating a neuron (arrow at 7 h) and another progenitor (below the neuron). The cell on the right underwent symmetrical division, giving rise to two progenitor cells. The asterisk at 1 h, 24 min, indicates the point at which the neuronal daughter cell detached from the apical surface (white dotted double lines). Notice the translocation of the nucleus in a progenitor cell as it proceeds through the cell cycle. The cell is undergoing DNA synthesis (S phase) when its nucleus is toward the basal end of the cell (away from the dotted white lines) and is in mitosis (M phase) when its nucleus is near the apical end of the cell.

(?) Developing Questions

It has been said that a picture is worth a thousand words, in which case a movie must be worth a million. As you watch movies of interkinetic nuclear migration, we challenge you to avoid seeing something new each time. Hidden in those movies are answers to many questions, including: Why does cytokinesis need to happen at the apical surface in the neuroepithelium? Do these cells maintain their basal process? What role do centrosomes—key organizing structures for microtubules and mitosis—play in this nuclear migration?

circuitry. The spatial and temporal coordination of neuronal generation, migration, and differentiation is tightly regulated and of paramount importance to the creation of a mature brain capable of processing and reacting to sensory input from the environment and of conscious thought.

As the neural tube matures, the progeny of the neuroepithelial stem cells become radial glial cells. Only recently have cell lineage studies demonstrated that radial glia are neural stem cells that undergo symmetrical and asymmetrical divisions (Malatesta et al. 2000, 2003; Miyata et al. 2001; Noctor et al. 2001; Anthony et al. 2004; Casper and McCarthy 2006; Johnson et al. 2016). The divisions of the radial glia take place in the ventricular zone (the zone lining the ventricle and therefore in contact with the cerebrospinal fluid). In the cerebrum, as the progenitor cells delaminate from the ventricular zone, they form a **subventricular zone** basal to it. Together, these zones form the germinal strata that generate the neurons that migrate into the cortical plate and form the layers of neurons of the neocortex (**FIGURE 14.9A,B**; Frantz et al. 1994; for reviews, see Kriegstein and Alvarez-Buylla 2009; Lui et al. 2011; Kwan et al. 2012; Paridaen and Huttner 2014).

A single stem cell in the ventricular layer can give rise to both neurons and glial cells in any of the cortical layers (Walsh and Cepko 1988). There are three major progenitor cell types in the human germinal strata: **ventricular radial glia (vRG)**, **outer radial glia (oRG)**, and **intermediate progenitor (IP)** cells. During the early stages of CNS development, neuroepithelial cells transform into ventricular radial glia that, as their name suggests, maintain contact with the luminal surface. The vRG serve as the parental stem cell type and, in addition to directly generating neurons, will give rise to both the oRG and the IP cells (**FIGURE 14.9C–E**). Self-renewing, symmetrical divisions dominate early in neurogenesis to expand the progenitor pool, and then more asymmetrical divisions govern progenitor differentiation. (See **Further Development 14.2, Defining Ventricular and Outer Radial Glia and Their Immediate Progenitors**, online.)

FIGURE 14.9 Summary model of neurogenesis in the cerebral cortex. (SVZ, subventricular zone; VZ, ventricular zone.) (Model based on A. Kriegstein and A. Alvarez-Buylla. 2009. *Annu Rev Neurosci* 32: 149–184; K. Y. Kwan et al. 2012. *Development* 139: 1535–1546; J. T. Paridaen and W. B. Huttner 2014. *EMBO Rep* 15: 351–364.)

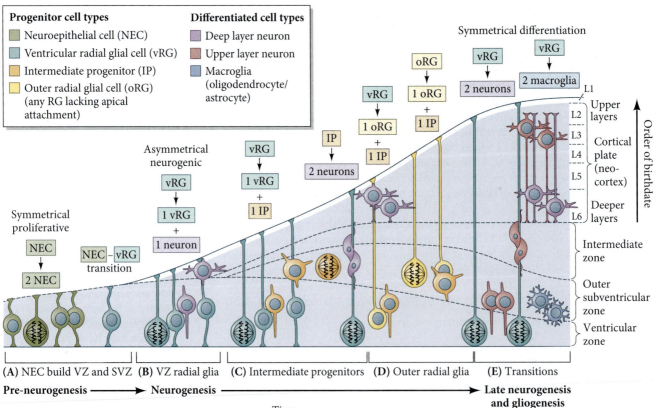

Glia as scaffold for the layering of the cerebellum and neocortex

Different types of neurons and glial cells are born at different times. Labeling cells at different times during development of the cerebrum shows that the cells with the earliest birthdays migrate the shortest distances; those with later birthdays migrate farther to form the more superficial regions of the brain cortex. Subsequent differentiation depends on the positions the neurons occupy once outside the germinal neuroepithelium (Letourneau 1977; Jacobson 1991). What are the developmental mechanisms governing this pairing of neuronal birth with differentiation along the apicobasal axis of the brain?

It has been known for decades that radial glial cells guide neural progenitor cell migration from the inner (luminal) region to the outer zones throughout the CNS (Rakic 1971). Thus, the progenitor cells formed as the progeny of radial glia also use their "sister" stem cell's connection between luminal and outer surfaces to migrate to their appropriate positions. We will explore the mechanisms of radial glia-enabled migration in the cerebellum and cerebrum. (See **Further Development 14.3, The Scaffolding of Bergmann Glia in the Cerebellum**, online.)

RADIAL GLIA IN THE NEOCORTEX   In the developing cerebrum, most of the neurons generated in the ventricular zone migrate outward along radial glial processes to form the **cortical plate** near the outer surface of the brain, where they set up the six layers of the neocortex. As in the rest of the brain, those neurons with the earliest birthdays form the layer closest to the ventricle (**FIGURE 14.10A,B**). Subsequent neurons travel greater distances to form the more superficial layers of the cortex. This process forms an "inside-out" gradient of development (Rakic 1974). McConnell and Kaznowski (1991)

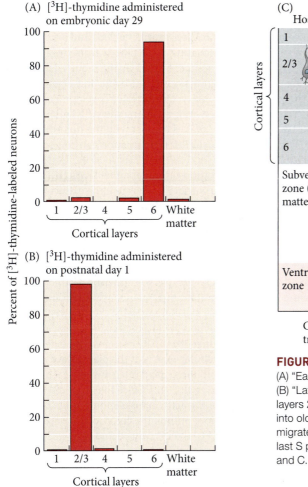

(A) [³H]-thymidine administered on embryonic day 29

(B) [³H]-thymidine administered on postnatal day 1

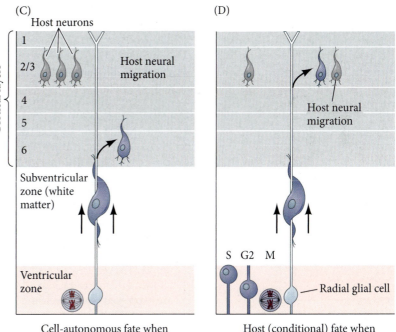

(C) Host neurons

Cell-autonomous fate when transplanted after last S phase

(D) Host neural migration

Host (conditional) fate when transplanted in S phase

FIGURE 14.10 Determination of cortical laminar identity in the ferret cerebrum. (A) "Early" neuronal precursors (birthdays on embryonic day 29) migrate to layer 6. (B) "Late" neuronal precursors (birthdays on postnatal day 1) migrate farther, into layers 2 and 3. (C) When early neuronal precursors (dark blue) are transplanted into older ventricular zones after their last mitotic S phase, the neurons they form migrate to layer 6. (D) If these precursors are transplanted before or during their last S phase, they migrate with the host neurons to layer 2. (After S. K. McConnell and C. E. Kaznowski. 1991. *Science* 254: 282–285.)

have shown that the determination of laminar identity (i.e., which layer a cell migrates to) is made during the final cell division. Newly generated neuronal precursors transplanted after this last division from young brains (where they would have formed layer 6) into older brains, whose migratory neurons are forming layer 2, are committed to their fate and migrate only to layer 6. However, if these cells are transplanted prior to their final division (i.e., during mid-S phase), they are uncommitted and can migrate to layer 2 (**FIGURE 14.10C,D**). The fates of neuronal precursors from older brains are more fixed. The neuronal precursor cells formed early in development have the potential to become any neuron (at layer 2 or 6, for instance); later precursor cells give rise only to upper level (layer 2) neurons (Frantz and McConnell 1996). Once the cells arrive at their final destination, it is thought that they express specific adhesion molecules that organize them into brain nuclei (Matsunami and Takeichi 1995).

Signaling mechanisms regulating development of the neocortex

How do migrating neural progenitors become segregated to the correct layer? As mentioned above, earlier-born neurons establish the deeper layers, and later-born neurons form the more superficial layers. *Think about this.* It means that the cerebrum is growing from inside to outside. One outcome of such growth is that with each new expanding layer, the outer pial surface moves farther away from the ventricular surface. Therefore, the pial surface is an ever-expanding outer limit, and the neurons embarking on their outward trek have farther to travel than their predecessors. This important dynamic ultimately influences the layering of the brain (see Frotscher 2010).

FURTHER DEVELOPMENT

CAJAL-RETZIUS CELLS: "MOVING TARGETS" IN THE NEOCORTEX   When the luminal and pial surfaces are relatively close during early development of the neocortex, a newly born neuron extends basal filopodia toward the pial surface, establishes adhesive contact, and then simply displaces its nucleus and associated cytoplasm toward the pial surface, translocating the cell body from the apical to the basal region of the cell. The basal attachment provides the required physical resistance and tension that enables this translocation (Miyata and Ogawa 2007). Thus, no actual cell migration is necessary. During later development, however, each precursor cell needs to actively migrate along the radial glial cell's basal process until its own basal membrane makes contact with the outermost region of the cortical plate, at which point a similar translocation can complete the journey (**FIGURE 14.11A**).

The cells influencing this outward migration of precursor cells are the **Cajal-Retzius cells**, which lie under the pial surface and secrete the extracellular protein Reelin, which regulates layering in the cerebellum (D'Arcangelo et al. 1995, 1997). Translocating precursor cells express transmembrane receptors for Reelin (Trommsdorff et al. 1999), and when these receptors bind to Reelin, they set off a series of signal transduction pathways mediated by Disabled-1 (see Figure 14.11A, cell 1). As a result, the cells increase their expression of N-cadherin, allowing them to attach to other cells that also express N-cadherins. Cadherins are expressed with increasing intensity from the ventricular zone to their highest levels in the marginal zone, overlapping Cajal-Retzius cells; thus, newly born neurons expressing N-cadherin become oriented toward regions of increasing adhesion (Franco et al. 2011; Jossin and Cooper 2011). The neurons also extend filopodia toward the fibronectin-rich extracellular matrix at the pial surface (Chai et al. 2009) and use transmembrane proteins called integrins to attach the filopodia to this extracellular matrix (Sekine et al. 2012). Once the filopodia are attached, Disabled-1-mediated regulation of the actin cytoskeleton powers contraction of the filopodia in a springlike motion, pulling the cell body forward as the cell's apical end detaches (see Figure 14.11A, cell 2; Miyata and Ogawa 2007).

The same Reelin signal that initiates this migration also triggers a negative feedback such that at the highest levels of Reelin (near the marginal zone), the neurons lose their cell adhesion molecules and integrate into the layers of the cortical plate

(A)

2 Reelin-Disabled-1 signaling:
Promotes integrin association
with filopodial tip

**High levels of Reelin
inhibit Disabled-1
(negative feedback):**
Leads to destabilized F-actin

Result:
Destabilized F-actin and final
translocation

**1 Moderate levels of
Reelin activate
Disabled-1:**
Activates N-cadherin
expression and stabilizes
F-actin

Result:
Promotes filopodial
extensions and cell
translocation toward region
of highest N-cadherin

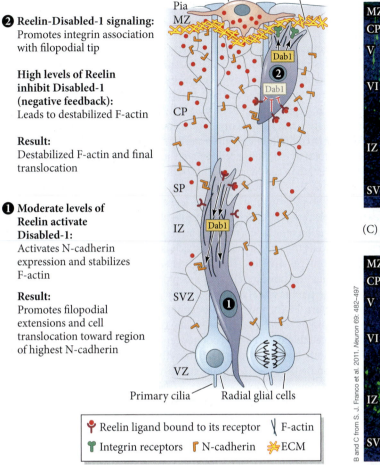

Cajal-Retzius cell   ECM (fibronectin)
Pia
MZ

CP

SP

IZ

SVZ

VZ

Primary cilia   Radial glial cells

🧍 Reelin ligand bound to its receptor   \ F-actin
🧍 Integrin receptors   ∫ N-cadherin   ✳ ECM

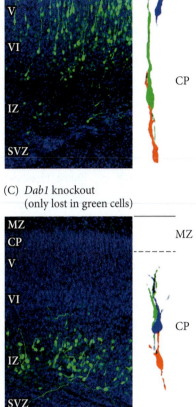

(B) Wild-type

MZ
CP
V

VI

IZ

SVZ

MZ

CP

(C) *Dab1* knockout
(only lost in green cells)

MZ
CP
V

VI

IZ

SVZ

MZ

CP

B and C from S. J. Franco et al. 2011. *Neuron* 69: 482–497

FIGURE 14.11   Model of Reelin regulation of directed neuronal migration. (A) Secreted from Cajal-Retzius cells, Reelin (red circles) is distributed in a gradient in the extracellular matrix. Reelin instructs the newly born migrating neurons (labeled 1 and 2) to extend filopodia from their basal membrane toward the pial surface. Disabled-1 (Dab1) is activated by Reelin. The product of the *Dab1* gene stabilizes filamentous actin (F-actin) as well as upregulating *N-cadherin* expression. N-cadherin is also localized to the membranes of radial glial fibers and other cells throughout the epithelium, increasing to its highest concentrations closest to the marginal zone. Initial Reelin-Dab1 signaling results in marginal zone-directed extension of the filopodium and translocation of neuron 1. In a migrating neuron approaching the marginal zone (cell 2), Dab1 upregulates integrin expression in the tip of the filopodium to anchor this cell to the fibronectin-rich extracellular matrix. However, at the highest Reelin concentrations, a negative feedback mechanism is triggered that inhibits Dab1 by protein degradation (cell 2), ending migration and enabling cell differentiation within the specified cortical layer. (B,C) Conditional inactivation of Dab1 in newly born neurons and migrating progenitor cells. Two types of mice were used, wild type and a strain carrying a conditional *Dab1* mutation that is activated only when combined with a second gene (*CRE*). *CRE* has no effect on wild-type mice. A plasmid carrying *CRE* and *GFP* was introduced into progenitor cells of the two mouse strains. Cells that received the plasmid could be identified by their GFP expression (green). (B) The wild-type control shows that the treated progenitor cells successfully reached the layer of the cortical plate. (C) In the *Dab1* conditional mutant, *Dab1* is knocked out in the green (CRE-containing, GFP-expressing) cells. These cells were retained in the intermediate zone. Time-lapse imaging of a single cell demonstrates that a typical progenitor cell will initiate migratory cell elongation (red), then extend its basal process to the marginal zone (green), and finally translocate the apical compartment to the outer layers (blue) (B, drawing on right). Similar imaging shows migration is initiated in *Dab1* knockout cells, but they fail to advance productive basal extensions, nor do they show translocation (C, drawing on right).

in a progressive inside-out manner (see Figure 14.11A, cell 2; Feng et al. 2007). Loss of *Reelin*, its receptors, or *Disabled-1* results in an inversion of cortical layering; neurons typically found within the inner layers (layers 4 and 5) are positioned near the marginal zone (layer 1), and cells of the external layers (layers 2 and 3) are found near the subplate when these genes are lost (**FIGURE 14.11B,C**; Olson et al. 2006; Franco et al. 2011; Sekine et al. 2011). (See **Further Development 14.4, Horizontal and Vertical Specification of the Cerebrum**, online.)

N-cadherin was shown to be important for neural progenitor migration in the neocortex, but how? Cells along the apicobasal axis have increasing levels of cadherin expression that appear to provide a road to the next layer, but cadherins are typically thought of as playing a role in differential adhesion and cell sorting. Could the properties of cell sorting be driving the movement of neural progenitors toward the marginal zone, much like E-cadherin functions in the zebrafish gastrula (see Chapter 4)? In one interesting study, whole-transcriptome analysis of the different proliferative zones of the mouse and human neocortex revealed a high level of expression of cell adhesion and extracellular matrix genes (Fietz et al. 2012), which suggests that the use of cell adhesion is likely both a complex and foundational mechanism for layering the cortex.

TO BE OR NOT TO BE … A STEM, A PROGENITOR, OR A NEURON?   Whether a radial glial cell undergoes symmetrical versus asymmetrical division depends on the plane of division (which in turn depends on the orientation of the mitotic spindle) and is correlated with the type of progeny generated. Cytokinesis that separates the radial glial cell perfectly perpendicular (planar) to the luminal surface (i.e., the mitotic spindle is parallel to the lumen) can result in two radial glial stem cells (Xie et al. 2013). Although such perpendicular cleavages can sometimes yield different progeny—a radial glial cell and a neuron—more often it is oblique division planes that give rise to these two distinct progeny. When the mitotic spindle is altered in such a way that cytokinesis occurs along random axes, it increases early asymmetrical divisions and triggers premature neurogenesis (Xie et al. 2013).

The fate of a daughter cell following cytokinesis has been linked to the centriole it inherits. The two centrioles in a dividing cell are not the same in regard to their age: the parental centriole is "older" than the daughter centriole it creates when it replicates. At each division, the cell receiving the "old" centriole will stay in the ventricular zone as a stem cell, while the cell receiving the "young" centriole leaves and differentiates (Wang et al. 2009). These two centrioles are bound to different proteins and structures, which results in an asymmetrical localization of those factors that influence gene expression and cell fate. Of particular significance is the primary cilium, which is connected to the older centriole and remains with it during cell division. The daughter cell that inherits this older centriole along with the primary cilium can quickly present the primary cilium to the lumen and, consequently, to the cerebrospinal fluid. Cerebrospinal fluid contains factors such as insulin-like growth factors, FGFs, and Sonic hedgehog, which induce proliferation and signal the cell to maintain its radial glial stem cell fate (Lehtinen et al. 2011; Paridaen et al. 2013). The daughter cell that inherits the younger centriole will eventually form a new primary cilium. This cilium, however, will extend from the cell's basal process rather than from its apical surface and will experience a different array of signals that will influence its development into a progenitor cell or a neuron (Wilsch-Brauninger et al. 2012). (See **Further Development 14.5, Asymmetric Divisions: vRG Are a Notch above Par**, online.)

This separation of high and low Notch is directly due to the co-transport of the Notch-inhibitor Numb with Par-3. It may seem counterintuitive to recruit this inhibitor into the stem cell that requires high Notch, but Par-3 actually sequesters and deactivates Numb function. Daughter cells lacking Par-3 exhibit freely active Numb, which functions to reduce Notch and thus enable an alternative (Delta-mediated) cell fate (Gaiano et al. 2000; Rasin et al. 2007; Bultje et al. 2009).

Development of the Human Brain

There are many differences between humans and our closest relatives, chimpanzees and bonobos (Prüfer et al. 2012). These differences include our hairless, sweaty skin and our striding, bipedal posture. The most striking and significant differences, however, occur in brain development. The enormous growth and asymmetry of the human neocortex and our advanced ability to reason, remember, plan for the future, and learn language and cultural skills make humans unique among the animals (Varki et al. 2008). The development of the human neocortex is strikingly plastic and is an almost constant work in progress. Several developmental phenomena, some of which are shared with other primates, distinguish the development of the human brain from that of other species. These include:

- Cerebral cortical folding and brain size
- Activity of human-specific genes
- Differences in the levels of gene transcription
- Human-specific alleles of developmental regulatory genes
- Prolonged brain maturation into adulthood

Fetal neuronal growth rate after birth

If there is one developmental trait that distinguishes humans from the rest of the animal kingdom, it is our retention of the fetal neuronal growth rate. Both human and ape brains have a high growth rate before birth. After birth, however, this rate slows greatly

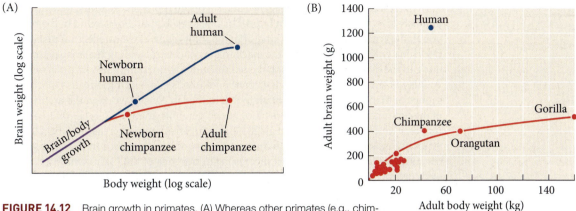

FIGURE 14.12 Brain growth in primates. (A) Whereas other primates (e.g., chimpanzees) attenuate neurogenesis around the time of birth, the generation of neurons in newborn humans occurs at the same rate as in the fetal brain. (B) The brain:body weight ratio (encephalization index) of humans is about 3.5 times higher than that of apes. (After B. Bogin. 1997. *Yrbk Phys Anthropol* 40: 63–89; see also the more recent quantification done by S. Herculano-Houzel. 2012. *Proc Natl Acad Sci USA* 109: 10661–10668 and S. Herculano-Houzel et al. 2015. *Brain Behav Evol* 86: 145–163.)

in the apes, whereas human brain growth continues at a rapid rate for about 2 years (**FIGURE 14.12A**; Martin 1990; see Leigh 2004). Portmann (1941), Montagu (1962), and Gould (1977) have each made the claim that we are essentially "extrauterine fetuses" for the first year of life.

It has been estimated that during early postnatal development, we add approximately 250,000 neurons per minute (Purves and Lichtman 1985). The ratio of brain weight to body weight at birth is similar for great apes and humans, but by adulthood, the ratio for humans is literally off the chart when compared with that of other primates (**FIGURE 14.12B**; Bogin 1997). Indeed, if one follows the charts of ape maturity, human gestation should be 21 months. Our "premature" birth is an evolutionary compromise based on maternal pelvic width, fetal head circumference, and fetal lung maturity. The mechanism for retaining the fetal neuronal growth rate beyond birth has been called **hypermorphosis**, the extension of development beyond its ancestral state (Vrba 1996; Vinicius and Lahr 2003).

In addition to the neurons made after birth, the number of synapses increases by an astronomical number. At the cellular level, no fewer than 30,000 synapses per cm³ of cortex are formed *every second* during the first few years of human life (Rose 1998; Barinaga 2003). It is speculated that these new neurons and rapidly proliferating neural connections enable plasticity and learning, create an enormous storage potential for memories, and enable us to develop skills such as language, humor, and music—that is, they enable those things that help make us human. (See **Further Development 14.6, Neuronal Growth and the Invention of Childhood**, online.)

Hills raise the horizon for learning

A particularly important feature of the cerebral cortex that is associated with human brain evolution is the number and intricacy of the hills and valleys of the brain—that is, its gyri and sulci (Hofman 1985). There is diversity in the number and complexity of cortical convolutions among mammalian species; for example, the cerebral cortex is highly folded (**gyrencephalic**) in humans and elephants, is only moderately gyrencephalic in ferrets, and completely lacks folds (is **lissencephalic**) in mice (**FIGURE 14.13**). The amount and complexity of gyrification are usually

(A) Human

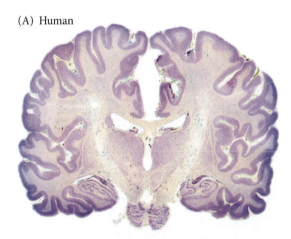

(B) Mouse

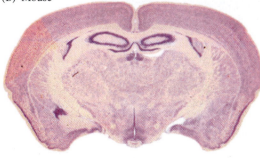

From J. H. Lui et al. 2011. *Cell* 146: 18–36

FIGURE 14.13 Transverse sections of the human and mouse brain. Nissl staining marks the nuclei of the gyrencephalic human brain (A) and lissencephalic mouse brain (B).

associated with the level of intelligence and therefore represent a significant adaptation uniquely leveraged in the human brain.[3] In fact, specific mutations found in the human population cause lissencephaly, and affected individuals have intellectual disabilities (Mochida et al. 2009). What are the mechanisms of cortical folding that may be contributing to the diversity of gyrencephalic brains seen within mammals?

It is not surprising that studying cortical folding is challenging, because it occurs in groups of mammals that are difficult to test in the lab. Recent work on the architecture of the mammalian cortex, however, along with genomic analyses, has begun to unravel the story (reviewed in Lewitus et al. 2013). It is surprising that increased cortical folding is not necessarily associated with an increased number of neurons in the cerebral cortex, although it is correlated with increased surface area of the brain. One study modeled cortical folding in comparison to crumpled paper and demonstrated that when total surface area expands at a faster rate than the thickness of the cortex (or sheets of paper), gyrencephaly will follow (Mota and Herculano-Houzel 2015). In agreement with this finding, larger cerebrums tend to contain more folds than smaller ones. Moreover, in the human disorder pachygyria, the cerebrum has reduced folding and reduced surface area, despite a normal number of neurons (Ross and Walsh 2001).

Cells that could be candidates for providing the mechanical force needed to create cerebral folds are the radial glial cells. Remember that in addition to functioning as stem cells, radial glia span the width of the cerebral cortex and provide a structural scaffolding that may generate mechanical forces. Interestingly, there is a greater percentage of proliferative radial glial cells (particularly outer radial glia) in gyrencephalic than in lissencephalic brains. Moreover, in gyrencephalic brains, the distribution and organization of the radial glial cells relative to gyri and sulci are appropriate for providing the tension necessary for folding (**FIGURE 14.14**; Hansen et al. 2010; Shitamukai et al. 2011; Wang et al. 2011; Pollen et al. 2015). Taken together, the increase of oRG cells in gyrencephalic brains and the biomechanics of their radial fiber arrays provide strong support for the direct involvement of radial glial cells in the evolved mechanisms of cortical folding.

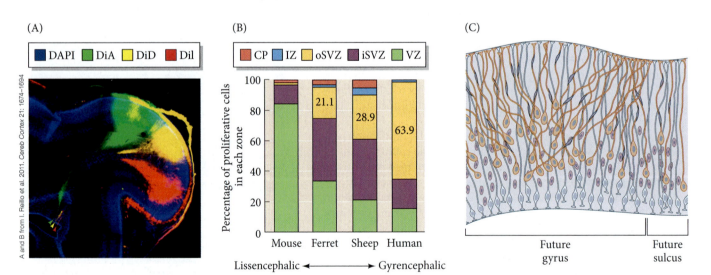

A and B from I. Reillo et al. 2011. *Cereb Cortex* 21: 1674–1694

FIGURE 14.14 Characterization of radial glial fanning during cortical folding in the ferret neocortex. (A) Retrograde tracing of radial glial cells in the ferret neocortex over the course of neurogenesis and gyrification. Note the triangle-shaped distribution of the dye-filled cells, indicating a progressive fanning out of the radial fibers along the apicobasal axis. (Arrowheads at the lower left show tight clusters at the luminal surface compared to the extreme width of each dye fill at the pial surface.) (DAPI, a dye that stains nuclei; DiA, DiD, and DiI, fluorescent dyes that were used to label different populations of radial glial cells different colors. (B) Quantification of mitotic cells in the different regions of the cerebrum of lissencephalic and gyrencephalic species. Brains that are more gyrencephalic show higher percentages of proliferative cells in the outer subventricular zone (oSVZ), which is the same region that houses more outer radial glial cells compared to lissencephalic species. (CP, cortical plate; IZ, intermediate zone; iSVZ, inner subventricular zones; VZ, ventricular zone) (C) The orientation of ventricular (light blue) and outer radial glial (orange) fibers in regions that will form a gyrus and sulcus. Outer radial glia are depicted to show more obliquely oriented fibers—a structural organization proposed to support gyrus formation. (C from E. Lewitus et al. 2013. *Front Hum Neurosci* 7: 424/CC BY.)

FURTHER DEVELOPMENT

IDENTIFYING THE GENES THAT MAKE THE HUMAN BRAIN HUMAN The ability to compare whole transcriptomes (total mRNAs expressed by genes in an organism) from different brain types has led to the discovery of an additional correlation between radial glial cells and cortical folding in humans (Florio et al. 2015; Johnson et al. 2015; Pollen et al. 2015). For instance, in looking at the transcriptomes of different radial glia in humans and rodents, Walsh and colleagues (see Johnson et al. 2015) discovered that compared to the radial glial cells of the mouse, the outer radial glial cells of humans show transcriptional heterogeneity relevant to cortical development, including differential expression of genes involved in calcium signaling, epithelial-mesenchymal transitions, cell migration, and specific activation of the proneural transcriptional regulator neurogenin.

The identification of the *ARHGAP11B* gene focused further attention on the unique role outer radial glial cells may play in human cortex development. This gene is found *only* in humans and is expressed *specifically* in radial glial cells (and not in cortical neurons). When Huttner and colleagues inserted the *ARHGAP11B* gene by electroporation into the developing cortex of the mouse brain (which is normally lissencephalic), the mouse cortex developed folds that resembled gyri (**FIGURE 14.15**). The exact mechanism whereby *ARHGAP11B* expression results in the formation of cortical folds is not clear, but it appears to be connected to a specific and significant increase in the number of outer radial glial cells being produced (Florio et al. 2015). This discovery is of particular significance to our understanding of the evolution of the human brain. The *ARHGAP11B* gene arose in humans from a partial duplication of *ARHGAP11A* (a gene found in animals in general) and arose in the human lineage after early hominids diverged from the chimpanzee lineage (see Figure 14.15A). (See **Further Development 14.7, Speech, Language, and the Foxp2 Gene**, online.)

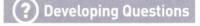

Developing Questions

If oRG cells are able to apply tension to the pial surface and promote folding of the cerebral cortex, perhaps something is holding their apical end within the subventricular zone to provide the necessary resistance for tissue folding. What's tethering these oRG cells, given that the hallmark difference between these stem cells and the vRG is the lack of an attachment to the luminal surface? How much pulling force is necessary to fold the cortex?

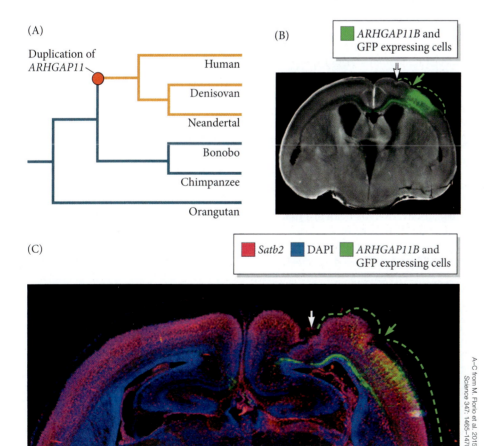

FIGURE 14.15 *ARHGAP11B* is an evolutionarily novel human gene that can induce the formation of gyri in the mouse neocortex. (A) Phylogenetic tree of primates showing the point in the human lineage where the *ARHGAP11B* gene arose through a partial duplication of the *ARHGAP11A* gene. (B) Transverse section through the mouse brain showing the expression of GFP (green) in cells that were electroporated in utero with a construct encoding both GFP and ARHGAP11B. (C) Immunolabeling using a marker for the neocortex (*Satb2*; red) in a mouse electroporated with ARHGAP11B (green). Nuclei are stained with DAPI (blue). The dashed lines in (B) and (C) denote induced gyri; the arrows denote sulci.

Genes for brain growth

What other genes besides *ARHGAP11B* distinguish us from our closest relatives, the chimpanzees and bonobos? Humans and these two non-human primates have remarkably similar genomes. When protein-encoding DNAs are compared, the three genomes are around 99% identical. Protein-coding regions, however, comprise only around 2% of these genomes. When the total genomes are compared, humans and chimpanzees differ at about 4% of their nucleotide sequences, most of the differences occurring in noncoding regions (see Varki et al. 2008). King and Wilson (1975) concluded from their studies of human and chimpanzee proteins that "the organismal differences between chimpanzees and humans would then result chiefly from genetic changes in a few regulatory systems, while amino acid substitutions in general would rarely be a key factor in major adaptive shifts." Theirs was one of the first suggestions that evolution can occur through changes in developmental regulatory genes.

Although there are some brain growth genes (e.g., *ASPM*, also called *microcephalin-5* and *microcephalin-1*) whose DNA sequences differ between humans and apes, these differences have not been correlated with the huge growth of human brains. Rather, the critical differences appear to reside in the sequences that control these genes. These sequences could be in enhancer regions of the DNA or in the DNA that produces noncoding RNAs. Noncoding RNAs are highly expressed in the developing brain, and although not producing any protein products themselves, they may regulate the transcription or translation of neuronal transcription factors. Computer analyses comparing different mammalian genomes suggest that such noncoding RNAs are important factors in human brain evolution (Pollard et al. 2006a,b; Prabhakar et al. 2006). First, these studies identified a relatively small group of noncoding DNA regions where the sequences were conserved among the non-human mammals studied. This group represents about 2% of the genome, and it was assumed that if these regions have been conserved throughout mammalian evolution, they must be important.

The studies then compared these sequences to their human homologues to see if any of these regions differ between humans and other mammals. About 50 regions were found where the sequence is highly conserved among mammals but has diverged rapidly between humans and chimpanzees. The most rapid divergence is seen in the sequence *HAR1* (*human accelerated region-1*), where 18 sequence changes were seen between chimpanzees and humans. *HAR1* is expressed in the developing brains of humans and apes, especially in the *Reelin*-expressing Cajal-Retzius neurons that are known to be responsible for directing neuronal migration during the formation of the six-layered neocortex (see Figure 14.11). Research is ongoing to discover the function of *HAR1* and the other HAR genes that are in the conserved noncoding region of the genome. (See **Further Development 14.8, Human Enhancement with the Loss of an Enhancer**, online.)

Changes in transcript quantity

In the 1970s, A. C. Wilson suggested that the difference between humans and chimpanzees might reside in the *amount* of proteins made from their genes (see Gibbons 1998). Today, there is evidence that supports this hypothesis. Using microarrays to study global patterns of gene expression, several recent investigations have found that although the quantities and types of genes expressed in human and chimpanzee liver and blood are indeed extremely similar, human *brains* produce over 5 times more mRNA than do chimpanzee brains (Enard et al. 2002a; Preuss et al. 2004). In humans, transcription of some genes (such as *SPTLC1*, a gene whose defect causes sensory nerve damage) is elevated 18-fold over the same genes' expression in the chimpanzee cortex. These data are not meant to suggest that all human genes in the brain are at a higher transcriptional level. There are many genes whose activity is lower in humans than it is in non-human primates. For example, the gene *DDX17*, whose product is involved in RNA processing, is expressed 10 times *less* in human than in chimpanzee cortices.

Teenage brains: Wired and unchained

Until recently, most scientists thought that in humans, after the initial growth of neurons during fetal development and early childhood, rapid brain growth ceased. However, magnetic resonance imaging (MRI) studies have shown that the brain keeps developing

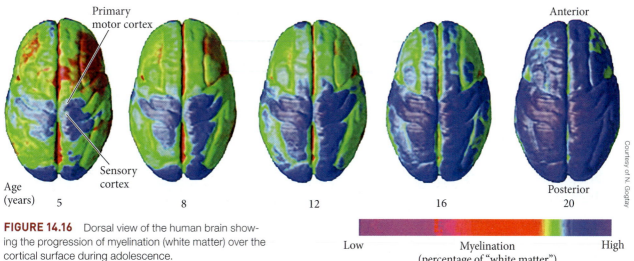

FIGURE 14.16 Dorsal view of the human brain showing the progression of myelination (white matter) over the cortical surface during adolescence.

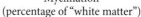

until around puberty and that not all areas of the brain mature simultaneously (Giedd et al. 1999; Sowell et al. 1999). Soon after puberty, brain growth ceases, and pruning of some neuronal synapses continues to occur. The time of this pruning correlates with the time when new language acquisition becomes difficult (which may be why children learn language more readily than adults). There is also a wave of myelin production ("white matter" from the glial cells that surround neuronal axons) in certain areas of the brain at this time. Myelination is critical for proper neural functioning, and although myelination continues throughout adulthood (Lebel and Beaulieu 2011), the greatest differences between brains in early puberty and those in early adulthood involve the frontal cortex (**FIGURE 14.16**; Sowell et al. 1999; Gogtay et al. 2004). The teenage brain is still actively developing, which could explain why teenagers demonstrate different responses to certain stimuli, as well as their ability (or inability) to learn certain tasks.

In tests using functional MRI to scan subjects' brains while emotion-charged pictures flashed on a computer screen, the brains of young teenagers showed activity in the amygdala, which mediates fear and strong emotions. When older teens were shown the same pictures, most of their brain activity was centered in the frontal lobe, an area involved in more reasoned perceptions (Baird et al. 1999; Luna et al. 2001). These data come primarily from studies comparing different groups of individuals. Improvements in technology, however, are beginning to allow assessment of brain maturation of a single individual over time (Dosenbach et al. 2010). The teenage brain is a complicated and dynamic entity that is not easily understood (as any parent knows). Once through the teen years, however, the resulting adult brain is usually capable of making reasoned decisions, even in the onslaught of emotional situations.

? Developing Questions

The onset of certain neuropsychiatric symptoms is often first observed during the teenage years, and these can lead to schizophrenia, bipolar disorder, and depression. What are the important neurodevelopmental events occurring during this period of teenage brain development—the dysregulation of which could lead to these neuropsychiatric symptoms?

Closing Thoughts on the Opening Photo

The complexities of being human: How deep do they fold? Our big, gyrified brains are part of what makes us human. This photograph is of a sculpture made by Lisa Nilsson, who intricately *folds* colored paper to make anatomical cross sections of human anatomy—in this case, the *folds* of the brain (http://lisanilssonart.com/home.html). Aside from its beauty, this is an artistic representation of the study that modeled cortical folding in comparison to crumpled paper that was discussed in this chapter. Moreover, you have learned that the increase in the number of stem cell types in the cortex and the changes in unique gene expression (e.g., *HAR1* and *ARHGAP11B*) are some of the major contributors to the complexity of the human brain and underlie its evolution. Lastly, throughout adulthood, the human brain continues to grow and develop into a highly myelinated structure that is so remarkable it can create its own structure through art.

Courtesy of Lisa Nilsson

Endnotes

[1]In neuroanatomy, the term *nucleus* refers to an anatomically discrete collection of neurons within the brain that typically serves a specific function. Note that it is a distinct structure from the cell nucleus.

[2]The neocortex, also referred to as the isocortex, is the largest of the three major cortical types, which includes the 3- to 6-layered mesocortex and 3-layered allocortex. The neocortex is positioned to the outermost basal regions of the brain, the allocortex at the deepest apical portions, and the mesocortex in between the two.

[3]This criterion is not exact; dolphins and whales, for example, have a greater amount of folding in the cortex than do humans.

Snapshot Summary
Brain Growth

1. Dendrites receive signals from other neurons, and axons transmit signals to other neurons. The gap between cells where signals are transferred from one neuron to another (through the release of neurotransmitters) is called a synapse. There is an enormous array of different neuronal morphologies throughout the CNS.

2. Macroglial cell types in the CNS are the astroglia (astrocytes) and oligodendrocytes (myelinating cells). Microglial cells also play important roles as the "immune" cells of the nervous system.

3. Radial glial cells serve as the neural stem cells in the embryonic and fetal brain. Humans continue making neurons throughout life in certain brain regions, although at nowhere near the fetal rate.

4. The neurons of the brain are organized into laminae (layers) and nuclei (clusters).

5. New neurons are formed by the division of neural stem cells (neuroepithelial cells, radial glial cells) in the wall of the neural tube (called the ventricular zone). The resulting newborn neurons can migrate away from the ventricular zone and form a new layer, called the mantle zone (gray matter). Neurons forming later have to migrate through the existing layers. This process forms the cortical layers.

6. Newborn neurons and progenitor cells migrate out of the ventricular zone on the processes of radial glial cells.

7. In the cerebellum, migrating neurons form a second germinal zone, called the external granular layer.

8. The newest evolved part of the cerebral cortex in mammals, called the neocortex, has six layers. Each layer differs in function and in the type of neurons located there.

9. Ventricular radial glia can give rise to outer radial glial cells that populate the subventricular zone. Both stem cells can also generate intermediate progenitors that are themselves capable of further symmetrical and asymmetrical divisions.

10. In both the cerebellum and the cerebrum, secreted Reelin guides migrating neurons to the correct superficial layer in an "inside-out" growth progression; it does so through the regulation of N-cadherin and integrin expression.

11. The number and complexity of gyri and sulci (folds) of the neocortex are correlated with level of intelligence. Humans have a highly folded (gyrencephalic) neocortex. Radial glial cells are likely to play a major role in the development of these folds.

12. Human brains appear to differ from those of other primates by their retention of the fetal neuronal growth rate during early childhood, the altered transcriptional activity of certain genes, and the presence of human-specific alleles of developmental regulatory genes.

Go to oup.com/he/barresi12xe for Further Developments, Scientists Speak interviews, Watch Development videos, Dev Tutorials, and complete bibliographic information for all literature cited in this chapter.

Neural Crest Cells and Axonal Specificity

Destined to be a face?

CONTINUING THE DISCUSSION OF ECTODERMAL DEVELOPMENT, this chapter focuses on two remarkable entities: (1) the neural crest, whose cells generate the facial skeleton, pigment cells, and peripheral nervous system; and (2) nerve axons, whose growth cones guide them to their destinations. Neural crest cells and axon growth cones share at least two core features: both are motile, and both invade tissues that are external to the nervous system.

The Neural Crest

Although it is derived from the ectoderm, the neural crest is so important that it has sometimes been called the "fourth germ layer" (see Hall 2009). It has even been said—somewhat hyperbolically—that "the only interesting thing about vertebrates is the neural crest" (Thorogood 1989). Certainly the emergence of the neural crest is one of the pivotal events of animal evolution, as it led to the jaws, face, skull, and bilateral sensory ganglia of the vertebrates (Northcutt and Gans 1983).

The neural crest is a transient structure. Adults do not have a neural crest, nor do late-stage vertebrate embryos. Rather, the cells of the neural crest undergo an epithelial-mesenchymal transition from the dorsal neural tube, after which they migrate extensively along the anterior-posterior axis and generate a prodigious number of differentiated cell types (**FIGURE 15.1**; **TABLE 15.1**).

FIGURE 15.1 Neural crest cell migration. (A) The neural crest is a transient structure dorsal to the neural tube. Neural crest cells (stained blue in this micrograph) undergo an epithelial-mesenchymal transition from the dorsalmost portion of the neural tube (the top of this micrograph). (B) When the epidermis has been removed from the dorsal surface of a vertebrate embryo, the neural crest cells (here computer-colored gold against the purple somites) can be seen as a collection of mesenchymal cells above the neural tube. (C) Illustration of the sequential steps of neural crest development, starting with the cells' specification at the border of the neural plate (1) and subsequent location at the apex of neural folds (2), followed by their delamination at the point of neural tube closure (3), and final migration out of ectodermal tissues (4). (C after M. Simões-Costa and M. E. Bronner. 2015. *Development* 142: 242–27.)

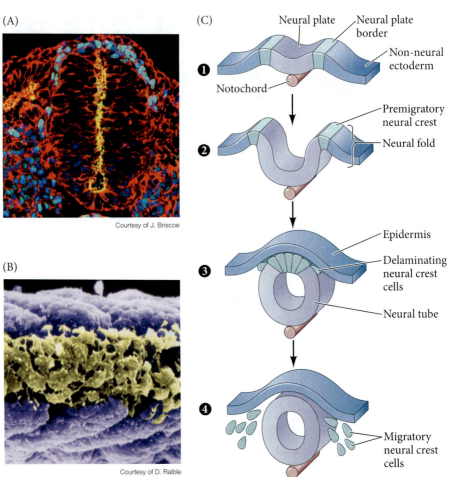

(A)

Courtesy of J. Briscoe

(B)

Courtesy of D. Raible

(C)

Neural plate
Neural plate border
Non-neural ectoderm
Notochord
Premigratory neural crest
Neural fold
Epidermis
Delaminating neural crest cells
Neural tube
Migratory neural crest cells

TABLE 15.1 Some derivatives of the neural crest	
Derivative	**Cell type or structure derived**
Peripheral nervous system (PNS)	Neurons, including sensory ganglia, sympathetic and parasympathetic ganglia, and plexuses Neuroglial cells Schwann cells and other glial cells
Endocrine and paraendocrine derivatives	Adrenal medulla Calcitonin-secreting cells Carotid body type I cells
Pigment cells	Epidermal pigment cells
Facial cartilage and bones	Facial and anterior ventral skull cartilage and bones
Connective tissue	Corneal endothelium and stroma Tooth papillae Dermis, smooth muscle, and adipose tissue of skin, head, and neck Connective tissue of salivary, lachrymal, thymus, thyroid, and pituitary glands Connective tissue and smooth muscle in arteries of aortic arch origin

Source: After M. Jacobson. 1991. *Developmental Neurobiology*, 2nd ed., Plenum: New York, based on multiple sources.

Regionalization of the Neural Crest

The **neural crest** is a population of cells that can produce tissues as diverse as (1) the neurons and glial cells of the sensory, sympathetic, and parasympathetic nervous systems; (2) the epinephrine-producing (medulla) cells of the adrenal gland; (3) the pigment-containing cells of the epidermis; and (4) many of the skeletal and connective tissue components of the head. The crest can be divided into four main (but overlapping) anatomical regions, each with characteristic derivatives and functions (**FIGURE 15.2**):

1. **Cranial**, or **cephalic**, **neural crest** cells migrate to produce the craniofacial mesenchyme, which differentiates into the cartilage, bone, cranial neurons, glia, pigment cells, and connective tissues of the face. These cells also enter the pharyngeal arches[1] and pouches to give rise to thymic cells, the odontoblasts of the tooth primordia, and the bones of the middle ear and jaw.

2. The **cardiac neural crest** is a subregion of the cranial neural crest and extends from the otic (ear) placodes to the third somites (Kirby 1987; Kirby and Waldo 1990). Cardiac neural crest cells develop into melanocytes, neurons, cartilage, and connective tissue (of the third, fourth, and sixth pharyngeal arches). This region of the neural crest also produces the entire muscular-connective tissue wall of the large arteries (the "outflow tracts") as they arise from the heart; it also contributes to the septum that separates pulmonary circulation from the aorta (Le Lièvre and Le Douarin 1975; Sizarov et al. 2012).

3. **Trunk neural crest** cells migrate along either the ventrolateral or the dorsolateral pathways. Ventrolaterally migrating crest cells go through the anterior half of each somitic sclerotome in the chick (see Chapter 17),[2] where they differentiate into the sensory neurons of the **dorsal root ganglia**[3]. Cells that continue traveling more ventrally form the sympathetic ganglia, the adrenal medulla, and the nerve clusters surrounding the aorta. Trunk neural crest cells traveling along the dorsolateral pathway move through the dermis of the skin from the dorsum to the belly and develop into pigment cells (e.g., melanocytes) (Harris and Erickson 2007).

4. The **vagal** and **sacral neural crest** cells generate the **parasympathetic** (**enteric**) **ganglia** of the gut (Le Douarin and Teillet 1973; Pomeranz et al. 1991). The vagal (neck) neural crest overlaps the cranial/trunk crest boundary, lying opposite chick somites 1–7, while the sacral neural crest lies posterior to somite 28.

Trunk neural crest cells and cranial neural crest cells are not equivalent. Cranial crest cells can form cartilage, muscle, and bone as well connective tissue of the cornea, whereas trunk crest cells cannot. When trunk neural crest cells are transplanted into the head region, they can migrate to the sites of cartilage and cornea formation, but they make neither cartilage nor cornea (Noden 1978; Nakamura and Ayer-Le Lievre 1982; Lwigale et al. 2004). However, both cranial and trunk neural crest cells can generate neurons, melanocytes, and glia (Noden 1978; Schweizer et al. 1983).

FURTHER DEVELOPMENT

Hox gene control over neural crest potency

The inability of the trunk neural crest to form skeleton is most likely due to the expression of Hox genes in the trunk neural crest. If Hox genes are expressed in the cranial neural crest, these cells fail to make skeletal

FIGURE 15.2 Regions of the chick neural crest. The cranial neural crest migrates into the pharyngeal arches and the face to form the bones and cartilage of the face and neck. It also contributes to forming the cranial nerves. The vagal neural crest (near somites 1–7) and the sacral neural crest (posterior to somite 28) form the parasympathetic nerves of the gut. The cardiac neural crest cells arise near somites 1 through 3; they are critical in making the division between the aorta and the pulmonary artery. Neural crest cells of the trunk (from about somite 6 through the tail) make sympathetic neurons and pigment cells (melanocytes), and a subset of them (at the level of somites 18–24) forms the medulla portion of the adrenal gland. (After N. M. Le Douarin. 2004. *Mech Dev* 121: 1089–1102, based on N. M. LeDouarin and M.-A. Teillet. 1973. *Development* 30: 31–48.)

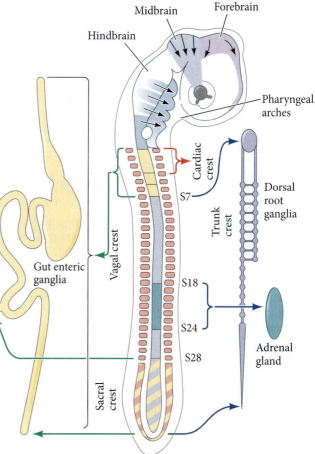

tissue; if trunk crest cells lose Hox gene expression, they can form skeleton. Moreover, if transplanted into the trunk region, cranial crest cells participate in forming trunk cartilage that normally does not arise from neural crest components. This ability to form skeletal tissue may have been a primitive property of the neural crest and may have been critical for forming the bony armor found in several extinct fish species (Smith and Hall 1993). In other words, rather than the cranial crest *acquiring* the ability to form bone, the trunk crest has apparently *lost* this ability.

So even though the cells of the cranial neural crest and trunk neural crest are multipotent (a cranial crest cell can form neurons, cartilage, bone, and muscles; a trunk neural crest cell can form glia, pigment cells, and neurons), they have different repertoires of cell types that they can generate under normal conditions.[4] Interestingly, under experimental conditions, transfection of three late cranial neural crest-specific genes (*sox8, tfap2b,* and *ets1*) into trunk neural crest cells of the chick is sufficient to reprogram them into cells that express known cranial neural crest genes in vivo. More remarkable is that transplantation of these reprogrammed trunk neural crest cells into cranial regions led to their differentiation into ectopic cartilage within the jaw; non-transfected trunk neural crest similarly grafted gave rise only to neurons and melanocytes (Simoes-Costa and Bronner 2016). These results suggest that trunk and cranial neural crest cells are under the control of distinct regulatory circuits that can reprogram at least trunk neural crest cells into having cranial neural crest cell fates.

Neural Crest: Multipotent Stem Cells?

There has long been controversy as to whether the majority of the individual cells leaving the neural crest are multipotent or whether most are already restricted to certain fates. Bronner-Fraser and Fraser (1988, 1989) provided the first evidence that many individual trunk neural crest cells are multipotent as they leave the crest. They injected fluorescent dextran molecules into individual chick neural crest cells while the cells were still within the neural tube, then recorded what types of cells their descendants became after migration. The progeny of a single neural crest cell could become sensory neurons, melanocytes (pigment-forming cells), glia (including Schwann cells), and cells of the adrenal medulla. In contrast, previous studies suggested that the initial avian trunk neural crest population was a heterogeneous mixture of precursor cells, half of which generate only a single cell type (Henion and Weston 1997; Harris and Erickson 2007).

With the advent of better lineage-tracing methods, this controversy may now be over. Researchers in the Sommer's lab have used the "confetti" mouse model[5] to trace the journey of individual trunk neural crest cells and their progeny from both the premigratory and migratory stages (Baggiolini et al. 2015). Tracings of nearly 100 cell clones of premigratory and migratory neural crest cells showed that approximately 75% of them proliferated, and their progeny showed multiple types of lineages that differentiated into different cell types: dorsal root ganglia, sympathetic ganglia, Schwann cells wrapping the ventral root, and melanocytes (**FIGURE 15.3**). Although a small population of these mapped neural crest cells appeared to be unipotent, the large majority displayed multipotency throughout their migration—a finding strongly suggesting that mouse embryonic trunk neural crest cells are multipotent stem cells, an important step forward in the field of neural crest research.

Nicole Le Douarin and others proposed a model of neural crest development that likely still holds true, even in light of the new information about multipotency. In this model, an original multipotent neural crest cell divides and progressively refines its developmental potentials (**FIGURE 15.4**; see Creuzet et al. 2004; Martinez-Morales et al. 2007; Le Douarin et al. 2008). To test this model directly, an individual neural crest cell would need to be exposed to different environments to determine the array of different cell types it could generate.

(?) Developing Questions

It seems we can now be confident that trunk neural crest cells are multipotent stem cells. Is it possible that some small population of these cells, dispersed widely like novel developmental seeds, could be retained as adult stem cells in each of their final destinations? The ontogeny of adult stem cells is unknown for many tissues. The migratory and multipotent nature of neural crest cells suggests the hypothesis that they may be capable of seeding these adult tissues.

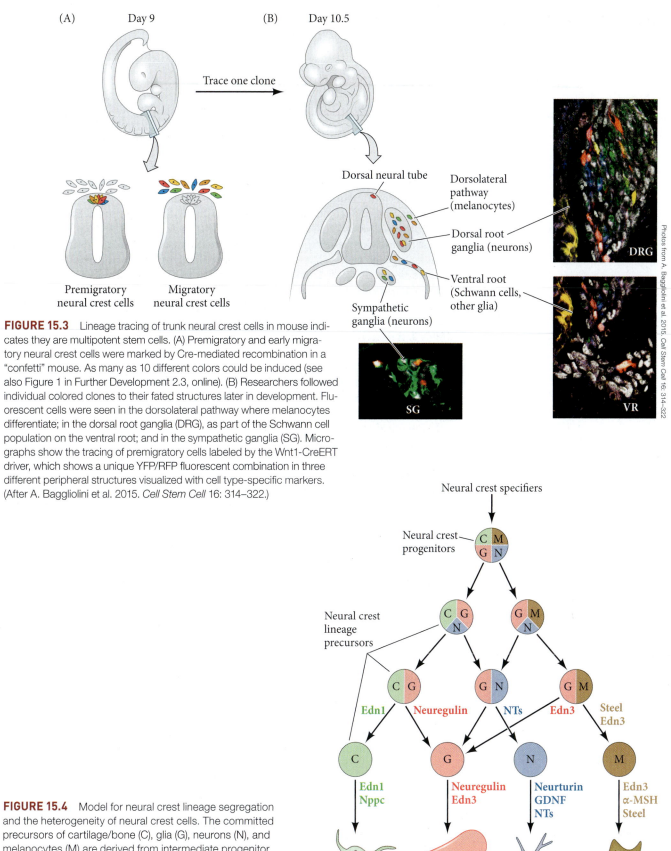

FIGURE 15.3 Lineage tracing of trunk neural crest cells in mouse indicates they are multipotent stem cells. (A) Premigratory and early migratory neural crest cells were marked by Cre-mediated recombination in a "confetti" mouse. As many as 10 different colors could be induced (see also Figure 1 in Further Development 2.3, online). (B) Researchers followed individual colored clones to their fated structures later in development. Fluorescent cells were seen in the dorsolateral pathway where melanocytes differentiate; in the dorsal root ganglia (DRG), as part of the Schwann cell population on the ventral root; and in the sympathetic ganglia (SG). Micrographs show the tracing of premigratory cells labeled by the Wnt1-CreERT driver, which shows a unique YFP/RFP fluorescent combination in three different peripheral structures visualized with cell type-specific markers. (After A. Baggliolini et al. 2015. *Cell Stem Cell* 16: 314–322.)

FIGURE 15.4 Model for neural crest lineage segregation and the heterogeneity of neural crest cells. The committed precursors of cartilage/bone (C), glia (G), neurons (N), and melanocytes (M) are derived from intermediate progenitor cells, some of which could act as stem cells. The paracrine factors regulating these steps are shown in colored type. α-MSH, alpha-melanocyte-stimulating hormone; Edn, endothelin; GDNF, glial-derived neurotropic factor; NTs, neurotropins; Nppc, natriuretic peptide precursor. (After J. R. Martinez-Morales et al. 2007. *Genome Biol* 8: R36/CC BY 2.0.)

Specification of Neural Crest Cells

Induction of neural crest cells first occurs well before their migration from the neural tube during early gastrulation, at the border between the presumptive epidermis and the presumptive neural plate (see Huang and Saint-Jeannet 2004; Meulemans and Bronner-Fraser 2004). In the anterior region, this same border location will give rise to the **placodes**—thickenings in the surface ectoderm that will generate the eye lens, inner ear, olfactory epithelium, and other sensory structures (see Chapter 16). This border between the neural plate and the epidermis appears to be specified by the interplay between a number of **inductive signals**, including BMPs, Wnts, and FGFs (Basch et al. 2006; Schmidt et al. 2007; Ezin et al. 2009).

In the anterior region, the timing of BMP and Wnt expression is critical for discriminating between neural plate, epidermis, placode, and neural crest tissues (**FIGURE 15.5**). As we saw in Chapters 11 and 12, if both BMP and Wnt signaling are continuous, the fate of the ectoderm is epidermal; but if BMP antagonists (e.g., Noggin or FGFs) block BMP signaling, the ectoderm becomes neural. Studies by Patthey and colleagues (2008, 2009) have shown that if Wnts induce BMPs and then Wnt signaling is turned *off*, the cells become committed to be anterior placodes, whereas if the Wnt signaling induces BMPs but stays *on*, the cells become capable of becoming neural crest. The output of these morphogen interactions *in part* serves to yield the intermediate concentration of BMP signaling that is necessary for neural crest specification. Additionally, it was recently shown that an intracellular mechanism of SMAD degradation (the downstream effectors of BMP signaling) is also required to properly attenuate BMP signaling, keeping it within the intermediate levels required for neural crest induction (Piacentino and Bronner 2018).

The last few decades of research on neural crest specification has helped to assemble the gene regulatory network (GRN) involved in the maturation of neural crest cells (**FIGURE 15.6**). This GRN begins with Wnt and BMP inducing the expression of a set of transcription factors in the ectoderm (including Gbx2, Zic1, Msx1, and Tfap2), which in turn regulate the **neural plate border specifiers**. These specifiers, including Pax3/7 and dlx5/6, collectively confer upon the border region the ability to form neural crest as well as dorsal neural tube cell types. The border-specifying transcription factors then induce a second set of more specific transcription factors, the **neural crest specifiers**, in those cells that are to become the neural crest. These neural crest specifiers include genes encoding the transcription factors FoxD3, Sox9, Snail (premigratory), and Sox10 (migratory) (Simões-Costa and Bronner 2015). (See **Further Development 15.1, Learn How These Four Pioneer Transcription Factors Differentiate Neural Crest Cell Fates**, and **Further Development 15.2, Induced Neural Crest Cells**, both online.)

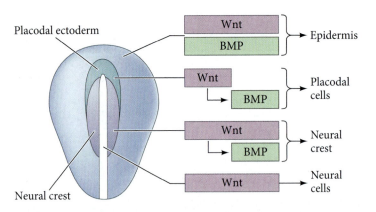

FIGURE 15.5 Specification of neural crest cells. The neural plate is bordered by neural crest anteriorly and caudally and by placodal ectoderm anteriorly. If the ectodermal cells receive both BMP and Wnt for an extended period of time, they become epidermis. If Wnt induces BMPs and is then downregulated, the cells become placodal cells (expressing the placode specifier genes *Six1*, *Six4*, and *Eya2*). If Wnt induces BMP but remains active, these border cells between the neural plate and the epidermis become neural crest (expressing neural crest specifier genes *Pax7*, *Snail2*, and *Sox9*). If they receive Wnt only (because the BMP signal is blocked by Noggin or FGF), the ectodermal cells become neural cells. (After C. L. Patthey et al. 2009. *PLOS ONE* 3: e1625.)

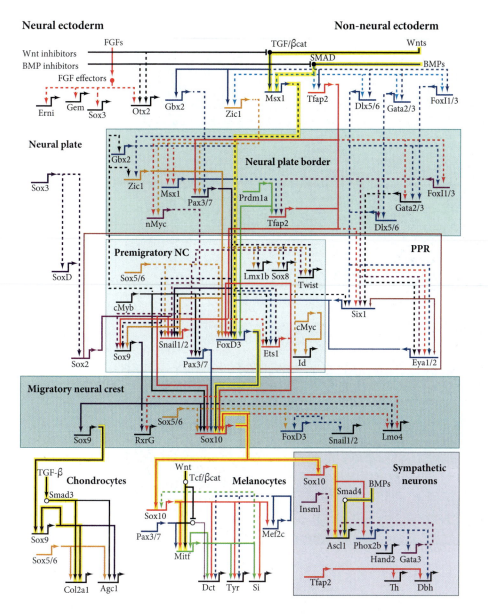

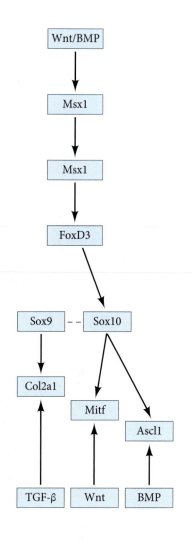

FIGURE 15.6 The gene regulatory network for neural crest development. This GRN is a compilation of data from a variety of vertebrate organisms. One of the most significant circuits (highlighted in yellow) shows gene expression in cells from early ectoderm (top) to the derived cell types (bottom). This circuit is replicated in a simpler linear flow chart to the right. (Note: Not all derived cell types are illustrated.) (After M. Simões-Costa and M. E. Bronner. 2015. *Development* 142: 242–257.)

Neural Crest Cell Migration: Epithelial to Mesenchymal and Beyond

The environment through which neural crest cells migrate differs along the anterior-posterior axis, which offers different experiential journeys for neural crest cells from different regions. Like cars in traffic, these cells must navigate their routes using environmental cues, and their passage is affected by the cells around them (**FIGURE 15.7**). Like cars using their engines and wheels to move on the road, the cells use the protrusive forces of their cytoskeletons to extend lamellipodia, reaching out to grab hold of the extracellular matrix in front of them while releasing the brakes from behind. A car can scoot freely down an open road or may have to travel collectively in traffic, being responsive to the pace and distance of neighboring cars. Similarly, neural crest cells can migrate individually as well as move as a collective

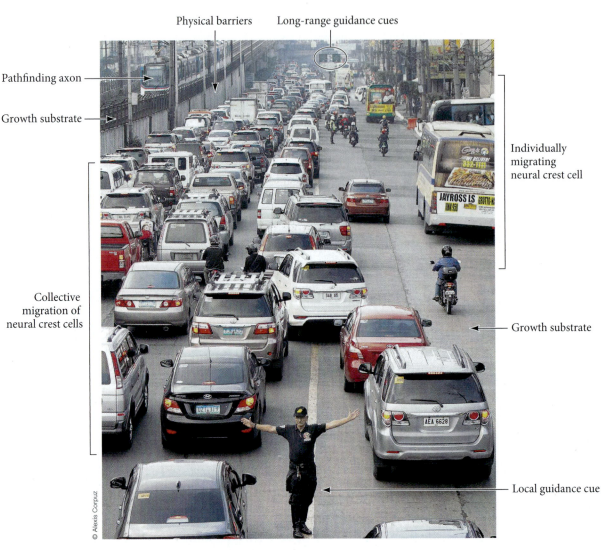

Physical barriers   Long-range guidance cues

Pathfinding axon

Growth substrate

Individually migrating neural crest cell

Collective migration of neural crest cells

Growth substrate

Local guidance cue

© Alexis Corpuz

FIGURE 15.7 Analogy of neural crest and axonal migration to the guidance and movement of traffic. See text for narrative.

cluster of cells responsive to the distance of neighboring cells. Just as traffic officers, structured barriers, and street signs guide car traffic, local adhesive cues and long-range secreted factors displayed in gradients guide migrating cells through an embryonic environment. And, like drivers who cannot see ahead to the final turn into their parking spot, migrating cells must make decisions in an incremental way, moving from one turn to the next to reach their final destination. Consider this analogy as you continue reading about neural crest cell migration.

Delamination

After cell specification, the first visible indication of neural crest cells is their epithelial-mesenchymal transition (EMT) in preparation for leaving the neural tube. Neural crest cells lose their adhesive junctions and separate from the epithelium in a process known as **delamination** (**FIGURE 15.8**). The timing of neural crest delamination is controlled by the neural tube's environment. The trigger for the EMT appears to be the activation of the Wnt genes by BMPs. The BMPs (which

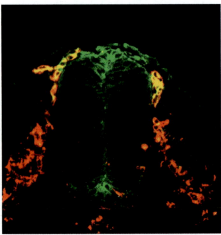

From J.-P. Liu and T. M. Jessell. 1998. *Development* 125: 5055–5067, courtesy of T. M. Jessell

FIGURE 15.8 Migrating neural crest cells are stained red by HNK-1 antibody, which recognizes a cell surface carbohydrate involved in neural crest cell migration. RhoB protein (green stain) is expressed in cells as they delaminate. Cells expressing both HNK-1 and RhoB appear yellow.

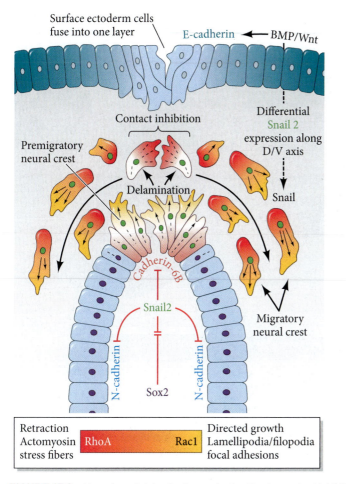

FIGURE 15.9 Neural crest delamination and migration by contact inhibition. The process of neural crest delamination is shown here at the time when the neural and surface ectoderms have separated and are both in the process of fusing at the midline into the neural tube and epidermis, respectively. BMP and Wnt signals specify the three major regions of the neuroepithelium, which are distinguished by their expression of unique adhesion proteins: surface ectoderm (E-cadherin), neural tube (N-cadherin), and the premigratory neural crest (cadherin-6B). In the premigratory domain, BMP levels are the highest, with Wnt at intermediate amounts; this situation supports the upregulation of *Snail-2* (and *Zeb-2*) in these cells. Snail-2 proteins repress N-cadherin and E-cadherin in this domain. Cadherin-6B is upregulated only in the apical half of premigratory neural crest cells and functions to activate RhoA and actomyosin contractile fibers for apical constriction and the initiation of delamination. Noncanonical Wnt signaling (not shown) establishes the polar activity of RhoA (red) and Rac1 (yellow) along the migratory axis of migrating neural crest cells. When neural crest cells contact one another, they experience contact inhibition, during which they will stop, turn, and migrate away in the opposite direction.

can be produced by the dorsal region of the neural tube; see Chapter 13) are held in check by Noggin produced by the notochord and somites. When Noggin expression is reduced, BMPs can function and activate EMT in the premigratory neural crest cells (Burstyn-Cohen et al. 2004).

Prior to delamination, the different regions of ectoderm in the area of the neural crest can be identified by their expression of different cell-cell adhesion molecules: surface ectoderm expresses E-cadherin, premigratory neural crest expresses cadherin-6B, and the neural tube expresses N-cadherin (**FIGURE 15.9**). The Wnt and BMP signals lead to the expression of the "core EMT regulatory factors" (e.g., Snail-2, Zeb-2, Foxd3, and Twist) in the delaminating premigratory neural crest. Sox2 is expressed by the cells of the neural ectoderm (plate/fold/tube) and functions in part to transcriptionally repress *Snail-2* expression, whereas the more dorsally expressed Snail-2 in the premigratory neural crest region cross-represses *Sox2* expression (reviewed in Duband et al.

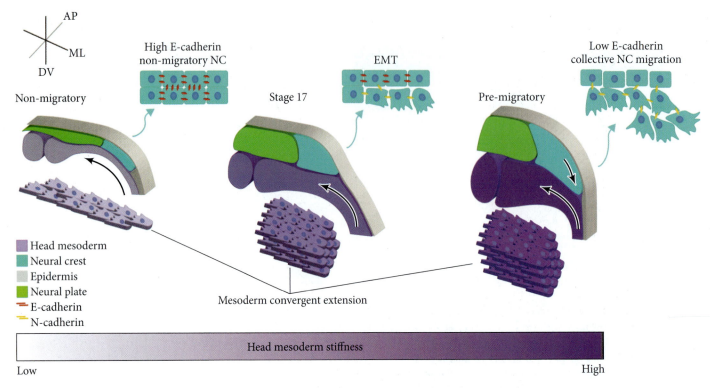

FIGURE 15.10 Mesodermal compaction by convergent extension provides a tissue stiffening force underlying the neural plate that is required for epithelial-mesenchymal transitions (EMTs) by cephalic neural crest in the embryonic *Xenopus* head. As E-cadherin expression shifts to N-cadherin in non-migratory to pre-migratory neural crest cells, the underlying mesoderm is progressively compacting toward the midline, which increases the stiffness of the mesodermal layer over time. This stiffening provides both a greater force upon the neural plate to promote EMTs, as well as a more stable substrate for neural crest migration. (After E. H. Barriga et al. 2018. *Nature* 554: 523–527.)

2015). As seen in the patterning of the ventral neural tube (see Chapter 13), this cross-transcriptional repression helps to refine the boundaries between the neural tube epithelium (N-cadherin), premigratory neural crest (cadherin-6B), and surface ectoderm (E-cadherin); see Figure 15.9.

As the molecular regulatory networks are priming pre-migratory neural crest cells for an EMT, what biomechanical factors might help trigger this event? Recent studies of the head region in the *Xenopus* embryo have shown that the underlying mesodermal environment is dramatically changing into a much more rigid substrate, and these biomechanical changes appear to be required for initiation of cephalic neural crest EMT. Similar to how walking on loosely packed sand is more difficult than walking on a paved road, the early specified, non-migratory neural crest cells appear to be less likely to delaminate and migrate through the loosely packed population of underlying early mesodermal cells (**FIGURE 15.10**). However, over the course of gastrulation, convergent extension continues to pack these mesodermal cells toward the midline into an increasingly stiffened layer. It appears that this mesodermal stiffening is both sufficient and necessary to trigger EMTs of cephalic neural crest cells and support their migration in *Xenopus* (Barriga et al. 2018). (See **Further Development 15.3, Rho GTPases Are the Link between Wnt and Cytoskeletal-Mediated Cell Migration**, online.)

The driving force of contact inhibition

The pushing out of neural crest cells from the dorsal neural tube appears to be facilitated by fellow neural crest cells (Abercrombie 1970; Carmona-Fontaine et al. 2008). The phenomenon known as **contact inhibition of locomotion** occurs when two migrating cells make contact. Resulting depolymerizing changes in each cell's cytoskeleton halt the protrusive activity along the contacting cell surfaces, and new protrusive extensions form away from the point of contact (**FIGURE 15.11**;

(?) Developing Questions

Several different embryonic progenitor cells, such as border cells in the *Drosophila* ovary and the lateral line primordia in zebrafish, display collective migration, but do any cell types in the adult exhibit similar collective migratory behaviors that result in "invasion" of tissues? If they did, would the cells go through an epithelial-mesenchymal transition as the neural crest cells do?

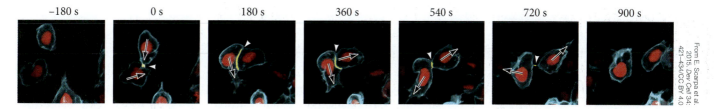

From E. Scarpa et al. 2015, Dev Cell 34: 421–434/CC BY 4.0

Carmona-Fontaine et al. 2008; Scarpa et al. 2015). As you might imagine, this behavior can cause cells to disperse. In fact, wherever neural crest cells are in close contact with each other, contact inhibition will repress protrusive activity on all sides of the cells except for those at the leading edge of the stream (Roycroft and Mayor 2016). (See **Further Development 15.4, Rho GTPases Are the Link between Wnt and Cytoskeletal-Mediated Contact Inhibition**, online.)

Collective migration

Traveling as part of a group of cars destined for a similar location is a different experience than traveling alone on the open road. Being part of the group requires cooperation and sticking together during the journey. A similar pattern of cell migration in the embryo is called **collective migration** (**FIGURE 15.12**).

Both epithelial and mesenchymal cells can migrate collectively, with the cells on the leading edge guiding and driving the movement of the cluster (Scarpa and Mayor 2016). In neural crest cell migration, cranial neural crest cells typically undergo collective migration and can do so even in culture (Alfandari et al. 2003; Theveneau et al. 2010). This ability suggests that external factors such as chemotaxis are not required for collective migration, but that properties intrinsic to the cells are sufficient to maintain cluster integrity and directional movement. Simulations modeling collective migration predict that both contact inhibition of locomotion and co-attraction between cells are necessary for efficient collective migration, which matches what is seen in vivo and in vitro (Carmona-Fontaine et al. 2011; Woods et al. 2014). Indeed, cranial neural crest cells in *Xenopus* show not only a Wnt/PCP-mediated RhoA mechanism of contact inhibition of locomotion but also secretion of Complement 3a (C3a), which attracts neural crest cells expressing the C3a receptor. These same cranial neural crest cells also express low levels of N-cadherin (see Figure 15.12). Experimentally increasing N-cadherin results in a more tightly adherent population of neural crest unable to invade spaces with the same migratory speed, suggesting that optimal levels of N-cadherin may be necessary for normal collective migration of these cells and for their invasion of tissues (Theveneau et al. 2010; Kuriyama et al. 2014).

FIGURE 15.11 Migrating neural crest cells demonstrate contact inhibition of locomotion in a live zebrafish embryo. A time series of neural cells that have been made to express *mCherry* in the nucleus (red) and green fluorescent protein in the cell membrane (blue). After membrane contact (yellow; arrowhead), the two cells send protrusions away from the location of contact. Arrows denote the direction of cell movement.

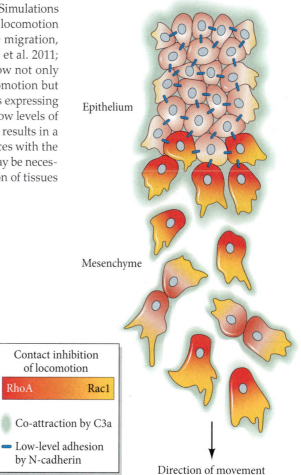

FIGURE 15.12 Model of collective migration of neural crest cells. Some populations of neural crest cells are known to migrate collectively as a large group of cells. This "collective migration" requires some amount of cell-to-cell adhesion, which is mediated by a low level of N-cadherin expression (blue receptors). Additionally, collectively migrating neural crest cells will also secrete an attractive signal (Complement 3a, C3a) to ensure that neural crest cells continually move toward each other instead of dispersing. The pattern of migration by the group has a *collective direction* due to ongoing contact inhibition among the cells at the leading edge. Contact of inhibition is represented by the differential activation of Rho GTPases (red to yellow). (After E. Scarpa and R. Mayor. 2016. *J Cell Biol* 212: 143–155.)

Migration Pathways of Trunk Neural Crest Cells

Following early specification and delamination, neural crest cells migrate along different paths to their specific locations for final differentiation. How do these cells "know" where to go?

Neural crest cells migrating from the neural tube at the axial level of the trunk follow one of two major pathways (**FIGURE 15.13**). Many cells that leave early follow a **ventral** or **ventrolateral pathway** away from the neural tube. Fate-mapping experiments show that these cells become sensory (dorsal root) and autonomic neurons, adrenomedullary cells, and Schwann cells and other glial cells (Weston 1963; Le Douarin and Teillet 1974). In birds and mammals (but not in fish and amphibians), these cells migrate ventrally through the anterior, but not the posterior, section of the sclerotomes (Rickmann et al. 1985; Bronner-Fraser 1986; Loring and Erickson 1987; Teillet et al. 1987).[6]

Trunk crest cells that emigrate via the second pathway—the **dorsolateral pathway**—become melanocytes, the melanin-forming pigment cells. These cells travel between the epidermis and the dermis, entering the ectoderm through minute holes in the basal lamina (which they themselves may create). Once in the ectoderm, they colonize the skin and hair follicles (Mayer 1973; Erickson et al. 1992). The dorsolateral pathway was demonstrated in a series of classic experiments by Mary Rawles (1948), who transplanted a region of neural tube with its overlying neural crest from a pigmented strain of chickens into the neural tube of an albino chick embryo and saw pigmented feathers on otherwise white wings (see Figure 1.14). These studies, along with other notable chick-quail chimeras (Teillet et al. 1987), showed that neural crest cells initially located opposite the posterior region of a somite migrate anteriorly or posteriorly along the neural tube so that they can specifically enter the anterior region of a somite. These cells join with neighboring streams of neural crest cells entering the same anterior portion of a somite and will contribute to the same structures. Thus, each dorsal root ganglion comprises neural crest cell populations from three adjacent somites: one from the neural crest opposite the anterior portion of the somite, and one each from the two neural crest regions opposite the posterior portions of its own and the neighboring somites (see Figure 15.13).

The ventral pathway

The choice between the dorsolateral versus the ventral trunk pathway is made at the dorsal neural tube shortly after neural crest cell specification (Harris and Erickson

FIGURE 15.13 Neural crest cell migration in the trunk of the chick embryo. Schematic diagram of trunk neural crest cell migration. Cells taking the ventral pathway (1) travel through the anterior of the sclerotome (that portion of the somite that generates vertebral cartilage). Those cells initially opposite the posterior portion of a sclerotome migrate along the neural tube until they come to an anterior region. These cells contribute to the sympathetic and parasympathetic ganglia as well as to the adrenomedullary cells and dorsal root ganglia. Somewhat later, other trunk neural crest cells enter the dorsolateral pathway (2) at all axial positions of the somite. These cells travel beneath the ectoderm and become pigment-producing melanocytes. (Migration pathways are shown on only one side of the embryo.) (After N. M. LeDouarin et al. 1984. In *The Role of Extracellular Matrix in Development* (*42nd Symp Soc Dev Biol*), R. L. Trelstad (ed.), pp. 373–398. Alan R. Liss: New York.)

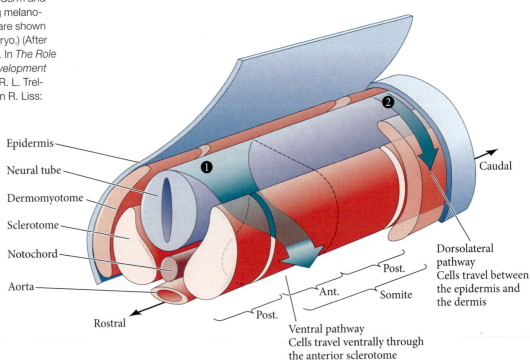

2007). The earliest migrating cells are inhibited from entering the dorsolateral pathway by chondroitin sulfate proteoglycans, ephrins, Slit proteins, and probably several other molecules. Because they are so inhibited, these cells turn around and migrate ventrally, and there they give rise to the neurons and glial cells of the peripheral nervous system.

The next choice concerns whether these ventrally migrating cells migrate *between* the somites (to form the sympathetic ganglia of the aorta) or *through* the somites (Schwarz et al. 2009). In the mouse embryo, the first few neural crest cells that form go between the somites, but this pathway is soon blocked by **semaphorin-3F**, a protein that repels neural crest cells; thus, most neural crest cells traveling ventrally migrate through the somites. These cells migrate through the *anterior* portion of each sclerotome (**FIGURE 15.14A**) and associate with proteins of the extracellular matrix, such as fibronectin and laminin, that are permissive for migration (Newgreen and Gooday 1985; Newgreen et al. 1986).

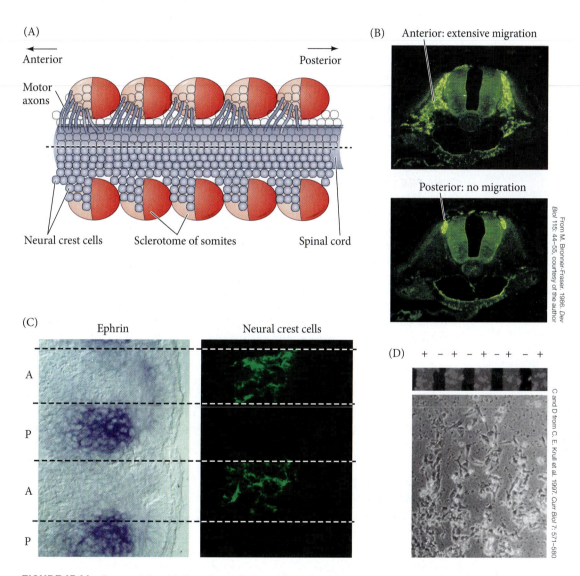

FIGURE 15.14 Segmental restriction of neural crest cells and motor neurons by the ephrin proteins of the sclerotome. (A) Composite scheme showing the migration of spinal cord neural crest cells and motor neurons through the ephrin-deficient anterior regions of the sclerotomes. (For clarity, the neural crest cells and motor neurons are each depicted on only one side of the spinal cord.) (B) Cross sections through these areas in a chick embryo, showing extensive migration through the anterior portion of the sclerotome (top), but no migration through the posterior portion (bottom). Antibodies to HNK-1 are stained green. (C) Negative correlation between regions of ephrin in the sclerotome (dark blue stain, left) and the presence of neural crest cells (green HNK-1 stain, right) in the chick embryo. (D) When neural crest cells from a quail are plated on fibronectin-containing matrices with alternating stripes of ephrin, they bind to those regions lacking ephrin. (A after D. D. M. O'Leary and D. G. Wilkinson. 1999. *Curr Opin Neurobiol* 9: 65–73.)

DEFINING EPHRIN AS A REPELLENT TO NEURAL CREST MIGRATION
The extracellular matrices of the sclerotome differ in the anterior and posterior regions of each somite, and only the extracellular matrix of the *anterior* sclerotome allows neural crest cell migration (**FIGURE 15.14B**). Like the extracellular matrix molecules that prevented neural crest cells from migrating dorsolaterally, the extracellular matrix of the *posterior* portion of each sclerotome contains proteins that actively exclude neural crest cells (**FIGURE 15.14C**). Besides semaphorin-3F, these proteins include the **ephrins**. The ephrin on the posterior sclerotome is recognized by its receptor, Eph, on the neural crest cells. Similarly, semaphorin-3F on the posterior sclerotome cells is recognized by its receptor, neuropilin-2, on the migrating neural crest cells. When neural crest cells are plated on a culture dish containing stripes of immobilized cell membrane proteins alternately with and without ephrins, the cells leave the ephrin-containing stripes and move along the stripes that lack ephrin (**FIGURE 15.14D**; Krull et al. 1997; Wang and Anderson 1997; Davy and Soriano 2007).

This patterning of neural crest cell migration generates the overall segmental character of the peripheral nervous system, reflected in the positioning of the dorsal root ganglia and other neural crest-derived structures (see Figures 15.2 and 15.14A). (See **Further Development 15.5, Cell Differentiation in the Ventral Pathway**, online.)

GOING FOR THE GUT   Which neural crest cells colonize the gut, and which do not? This distinction involves both extracellular matrix components and soluble paracrine factors. Neural crest cells from the vagal and sacral regions form the enteric ganglia of the gut tube and control intestinal peristalsis. Cells from the vagal neural crest, once past the somites, enter into the foregut and spread to most of the digestive tube, while the sacral neural crest cells colonize the hindgut (see Figure 1B in Further Development 15.5, Cell Differentiation in the Ventral Pathway, online). Various inhibitory extracellular matrix proteins (including the Slit proteins) block the more ventral migration of trunk neural crest cells into the gut, but these inhibitory proteins are absent around the vagal and sacral crest, allowing these neural crest cells to reach the gut tissue. Once in the vicinity of the developing gut, these crest cells are attracted to the digestive tube by **glial-derived neurotrophic factor** (**GDNF**), a paracrine factor produced by the gut mesenchyme (Young et al. 2001; Natarajan et al. 2002). GDNF from the gut mesenchyme binds to its receptor, Ret, on the neural crest cells. The vagal neural crest cells have more Ret in their cell membranes than do the sacral cells, which makes the vagal cells more invasive (Delalande et al. 2008). (See **Further Development 15.6, GDNF and Hirschsprung Disease: How Loss of Ventral Migrating Neural Crest Can Cause a Syndrome Wherein the Intestine Cannot Properly Void Solid Wastes**, online.)

A MUTUALISTIC RELATIONSHIP BETWEEN NEURAL CREST AND PATHFINDING AXON TO WIRE UP THE GUT   The enteric neural crest cells perform one of the longest migratory journeys because they are chasing a moving target—the caudalmost, or distal extent, of the growing gut. This caudal migration of enteric neural crest cells has been likened to a wave that has a leading (caudal) "crest" of cells (Druckenbrod and Epstein 2007). As the wave progresses down the developing gut, neural crest cells must spread evenly throughout the tissue to ensure complete innervation and function. The process by which enteric neural crest cells are deposited in the gut has been called "directional dispersal" (Theveneau and Mayor 2012), but little has been discovered about the cell behaviors mediating this process. Enteric neural crest cells do not migrate collectively as clusters but rather in long chains (Corpening et al. 2011; Zhang et al. 2012). Moreover, as the wave progresses, enteric neural crest cells migrate in seemingly random directions and explore domains along all axes; overall dispersal, however, preferentially occurs in the posterior direction

(?) Developing Questions

Enteric neural crest cells are positioned at the right time and place to influence the trajectory of pathfinding enteric neurons. Does that in fact occur? Are these leading neural crest cells expressing membrane receptors or secreting diffusible proteins that instruct the growing neuron to extend its axon in the direction of the neural crest cell? If the enteric neural crest cells are leading the way for the axons, what is guiding the neural crest cells toward the caudal end of the gut while also establishing a homogeneously dispersed array of neurons?

(A)   Dispersal of ENCC in the gut over time   (B)   ENCC lead and migrate on growing neurites

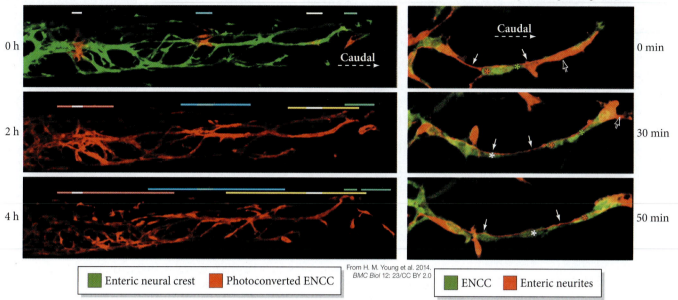

0 h

2 h

4 h

Caudal

0 min

30 min

50 min

■ Enteric neural crest   ■ Photoconverted ENCC   ■ ENCC   ■ Enteric neurites

FIGURE 15.15 Following the movement of individual enteric neural crest cells (ENCCs) in the developing gut. Transgenic *Ednrb-hKikGR* mice were used to fluorescently label enteric neural crest (green). KikGR is a photoconvertable protein that changes its emission when exposed to ultraviolet light from green to red. (A) Four separate foci were photoconverted in the developing gut (red; bars at top). The caudalmost tip of the moving ENCC wave is located to the far right of the 0-hour time point. Early chains of ENCCs can be seen sparsely spread throughout the gut (green, 0 h). Photoconverted cells actively migrate and disperse with a caudal preference over time (2 h, 4 h; width of bar at top). (B) ENCCs (green) can be seen at the growth tip of differentiating neurites (red, arrows). ENCCs are also seen to use neurites as a substrate for migration (note movement of asterisks over time; different-colored asterisks denote different cells).

(**FIGURE 15.15A**; Young et al. 2014). Enteric neural crest cells are differentiating into neurons along this journey, with both the soma (cell body) and the projecting axon still quite mobile during gut development. It is interesting that enteric neural crest cells can be found migrating all along the growing axon as well as just ahead of the leading tip of the growing axon (**FIGURE 15.15B**). These findings suggest a mutualistic relationship between migrating enteric neural crest cells and pathfinding enteric neurons such that the neural crest cells use the nerve axons as a substrate for migration and the neurons' axons follow "trailblazing" neural crest cells (Young et al. 2014).

The dorsolateral pathway

In vertebrates, all pigment cells except those of the pigmented retina are derived from the neural crest. It appears that the cells that take the dorsolateral pathway have already become specified as melanoblasts—pigment cell progenitors—and that they are led along the dorsolateral route by chemotactic factors and cell matrix glycoproteins (**FIGURE 15.16**). In the chick (but not in the mouse), the first neural crest cells to migrate enter the ventral pathway, whereas cells that migrate later enter the dorsolateral pathway (see Harris and Erickson 2007). These late-migrating cells remain above the neural tube in what is often called the "staging area," and it is these cells that become specified as melanoblasts (Weston and Butler 1966; Tosney 2004).

FIGURE 15.16 Neural crest cell migration in the dorsolateral pathway through the skin. (A) Whole mount in situ hybridization of day-11 mouse embryo stained for neural crest-derived melanoblasts (purple). (B) Stage-18 chick embryo seen in cross section through the trunk. Melanoblasts (arrows) can be seen moving through the dermis, from the neural crest region toward the periphery.

(A)

(B)

(A)   (B)   (C)   (D)

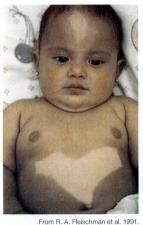

© anetapics/Shutterstock.com

© Mark J. Barrett/Alamy

From R. A. Fleischman et al. 1991.
Proc Natl Acad Sci USA 88: 10885,
courtesy of R. A. Fleischman

Courtesy of R. A. Fleischman

FIGURE 15.17 Variable melanoblast migration caused by different mutations. (A,B) In several animals, the random death of melanoblasts provides spotted pigmentation. Migrating melanoblasts induce the blood vessels to form in the inner ear, and without these vessels, the cochlea degenerates, and the animal cannot hear in that ear. That is often the case with Dalmation dogs (A), which are heterozygous for *Mitf*, and American Paint horses (B), which are thought to be heterozygous for endothelin receptor B. (C) Piebaldism in a human infant. Pigment fails to form in regions of the body, the result of a mutation in the *KIT* gene. Kit protein is essential for the proliferation and migration of neural crest cells, germ cell precursors, and blood cell precursors. (D) Mice can also have a *Kit* mutation, and they provide important models for piebaldism and melanoblast migration.

CELL DIFFERENTIATION IN THE DORSOLATERAL PATHWAY The switch between glial/neural precursor and melanoblast precursor seems to be controlled by the Foxd3 transcription factor. If Foxd3 is present, it represses expression of the gene for **MITF**,[7] a transcription factor necessary for melanoblast specification and pigment production (see Figure 15.6). If Foxd3 expression is downregulated, MITF is expressed, and the cells become melanoblasts. MITF is involved in three signaling cascades. The first cascade activates those genes responsible for pigment production; the second allows these neural crest cells to travel along the dorsolateral pathway into the skin; and the third prevents apoptosis in the migrating cells (Kos et al. 2001; McGill et al. 2002; Thomas and Erickson 2009). In humans heterozygous for *MITF*, fewer pigment cells reach the center of the body, resulting in a hypopigmented (white) streak through the hair. In some animals, including certain breeds of dogs and horses, heterozygosity for *Mitf* causes a random death of melanoblasts (**FIGURE 15.17**).

CELL GUIDANCE IN THE DORSOLATERAL PATHWAY Once specified, the melanoblasts in the staging area upregulate the ephrin receptor (EphB2) and the endothelin receptor (EDNRB2). Doing so allows melanoblasts to migrate along extracellular matrices that contain ephrin and endothelin-3 (see Figure 15.16B; Harris et al. 2008). Indeed, the melanocyte lineage migrates on exactly those same molecules that *repelled* the glial/neural lineage of crest cells. Ephrin expressed along the dorsolateral migration pathway stimulates the migration of melanocyte precursors. Ephrin activates its own receptor, EphB2, on the neural crest cell membrane, and this Eph signaling appears to be critical for promoting neural crest migration to sites of melanocyte differentiation. Disruption of Eph signaling in late-migrating neural crest cells prevents their dorsolateral migration (Santiago and Erickson 2002; Harris et al. 2008). An interesting recent discovery is that certain breeds of chickens displaying white plumage are the result of naturally occurring mutations in the *Ednrb2* gene (Kinoshita et al. 2014). (See **Further Development 15.7, Kit-SCF Guides and Determines Melanoblasts to Their Fate**, online.)

Thus, the differentiation of the trunk neural crest is accomplished by (1) autonomous factors (such as the Hox genes distinguishing trunk and cranial neural crest cells, or MITF committing cells to a melanocyte lineage), (2) specific conditions of the environment (such as the adrenal cortex inducing adjacent neural crest cells into adrenomedullary cells), or (3) a combination of the two (as when cells migrating through the dorsolateral pathway respond to Kit for both fate and guidance). The fate of an individual neural crest cell is determined both by its starting position (anterior-posterior along the neural tube) and by its migratory path.

Cranial Neural Crest

The head, comprising the face and the skull, is the most anatomically sophisticated portion of the vertebrate body (Northcutt and Gans 1983; Wilkie and Morriss-Kay 2001). The head is largely the product of the cranial neural crest, and the evolution of jaws, teeth, and facial cartilage occurs through changes in the placement of these cells (see Chapter 24).

Like the trunk neural crest, the cranial crest can form pigment cells, glial cells, and peripheral neurons; but in addition, it can generate bone, cartilage, and connective tissue. The cranial neural crest is a mixed population of cells in different stages of commitment, and about 10% of the population is made up of multipotent progenitor cells that can differentiate to become neurons, glia, melanocytes, muscle cells, cartilage, and bone (Calloni et al. 2009). In mice and humans, the cranial neural crest cells migrate from the neural folds even before they have fused together (Nichols 1981; Betters et al. 2010). Subsequent migration of these cells is directed by an underlying segmentation of the hindbrain. As mentioned in Chapter 13, the hindbrain is segmented along the anterior-posterior axis into compartments called rhombomeres. The cranial neural crest cells migrate ventrally from those regions anterior to rhombomere 8 into the pharyngeal arches and the frontonasal process that forms the face (**FIGURE 15.18**). The final destination of these crest cells determines their eventual fate (**TABLE 15.2**).

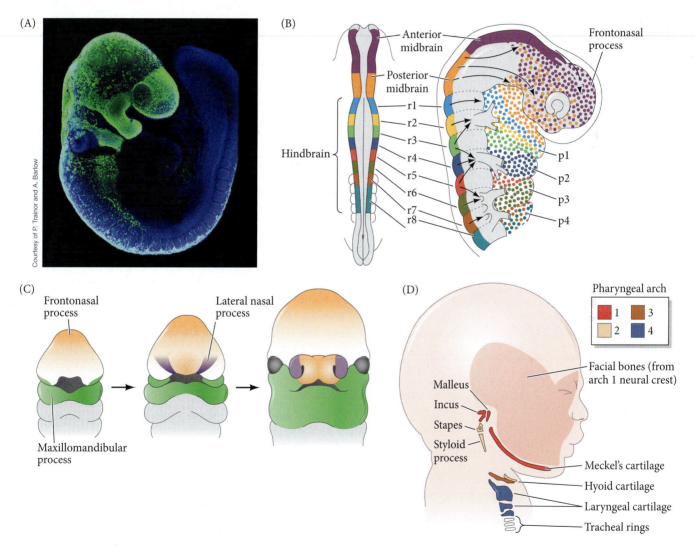

FIGURE 15.18 Cranial neural crest cell migration in the mammalian head. (A) Migration of GFP-labeled neural crest cells in a 9.5-day mouse embryo, emphasizing the colonization of the pharyngeal arches and frontonasal process. (B) Migrational pathways of the cranial neural crest into the pharyngeal arches (p1–p4) and frontonasal process (r, rhombomere). (C) Continued migration of the cranial neural crest to produce the human face. The frontonasal process contributes to the forehead, nose, philtrum of the upper lip (the area between the lip and the nose), and primary palate. The lateral nasal process generates the sides of the nose. The maxillomandibular process gives rise to the lower jaw, much of the upper jaw, and the sides of the middle and lower regions of the face. (D) Structures formed in the human face by the mesenchymal cells of the neural crest. The cartilaginous elements of the pharyngeal arches are indicated by colors, and the darker pink region indicates the facial skeleton produced by anterior regions of the cranial neural crest. (B after N. M. Le Douarin. 2004. *Mech Dev* 121: 1089–1102; C after J. A. Helms et al. 2005. *Development* 132: 851–861; D after B. M. Carlson. 1999. *Human Embryology and Developmental Biology*, 2nd Ed. Mosby: St. Louis.)

TABLE 15.2   Some derivatives of the pharyngeal arches in humans

Pharyngeal arch	Skeletal elements (neural crest plus mesoderm)	Arches, arteries (mesoderm)	Muscles (mesoderm)	Cranial nerves (neural tube)
1	Incus and malleus (from neural crest); mandible, maxilla, and temporal bone regions (from neural crest)	Maxillary branch of the carotid artery (to the ear, nose, and jaw)	Jaw muscles; floor of mouth; muscles of the ear and soft palate	Maxillary and mandibular divisions of trigeminal nerve (V)
2	Stapes bone of the middle ear; styloid process of temporal bone; part of hyoid bone of neck (all from neural crest cartilage)	Arteries to the ear region: corticotympanic artery (adult); stapedial artery (embryo)	Muscles of facial expression; jaw and upper neck muscles	Facial nerve (VII)
3	Lower rim and greater horns of hyoid bone (from neural crest)	Common carotid artery; root of internal carotid	Stylopharyngeus (to elevate the pharynx)	Glossopharyngeal nerve (IX)
4	Laryngeal cartilages (from lateral plate mesoderm)	Arch of aorta; right subclavian artery; original spouts of pulmonary arteries	Constrictors of pharynx and vocal cords	Superior laryngeal branch of vagus nerve (X)
6[a]	Laryngeal cartilages (from lateral plate mesoderm)	Ductus arteriosus; roots of definitive pulmonary arteries	Intrinsic muscles of larynx	Recurrent laryngeal branch of vagus nerve (X)

Source: W. J. Larsen. 1993. *Human Embryology*. Churchill Livingstone, New York.

[a] The fifth arch degenerates in humans.

The cranial crest cells follow one of three major streams:

1. Neural crest cells from the midbrain and rhombomeres 1 and 2 of the hindbrain migrate to the first pharyngeal arch (the mandibular arch), forming the jawbones as well as the incus and malleus bones of the middle ear. These cells will also differentiate into neurons of the trigeminal ganglion—the cranial nerve that innervates the teeth and jaw—and will contribute to the ciliary ganglion that innervates the ciliary muscle of the eye. These neural crest cells are also pulled by the expanding epidermis to generate the **frontonasal process**, the bone-forming region that becomes the forehead, the middle of the nose, and the primary palate. Thus, the cranial neural crest cells generate much of the facial skeleton (see Figure 15.18B,C; Le Douarin and Kalcheim 1999; Wada et al. 2011).

2. Neural crest cells from rhombomere 4 populate the second pharyngeal arch, forming the upper portion of the hyoid cartilage of the neck as well as the stapes bone of the middle ear (see Figure 15.18B,D). These cells will also contribute neurons of the facial nerve. The hyoid cartilage eventually ossifies to provide the bone in the neck that attaches the muscles of the larynx and tongue.

3. Neural crest cells from rhombomeres 6–8 migrate into the third and fourth pharyngeal arches and pouches to form the lower portion of the hyoid cartilage as well as contribute cells to the thymus, parathyroid, and thyroid glands (see Figure 15.18B; Serbedzija et al. 1992; Creuzet et al. 2005). These neural crest cells also go to the region of the developing heart, where they help construct the outflow tracts (i.e., the aorta and pulmonary artery). If the neural crest is removed from those regions, these structures fail to form (Bockman and Kirby 1984). Some of these cells migrate caudally to the clavicle (collarbone), where they settle at the sites that will be used for the attachment of certain neck muscles (McGonnell et al. 2001).

The "Chase and Run" Model

Due to the many cell types that the cranial neural crest generates in the anterior embryo, much importance has been placed upon understanding the molecular and cellular mechanisms governing cranial neural crest migration in this region. Recall that streams of cranial neural crest cells collectively migrate through the autonomous mechanisms of contact inhibition of locomotion, co-attraction, and low-level adhesion (see Figure 15.12). But how is each stream kept separate to move collectively in the correct direction? The three streams of cranial neural crest cells are kept from dispersing through interactions of the cells with their environment and with one another. Observations of the migration patterns produced by rhombomeres 3 and 5 in the chick hindbrain revealed that they do not migrate laterally, but rather join the even-numbered streams anterior and posterior to these odd-numbered rhombomeres. The migration of individually marked cranial neural crest cells in these regions can be monitored with cameras focused through a Teflon membrane window in the egg, and such experiments have shown that the migrating cells were "kept in line" not only by restrictions provided by neighboring cells, but also by the lead cells passing material to those cells behind them. It appears that cranial neural crest cells extend long, slender bridges that temporarily connect cells and influence the migration of the later cells to "follow the leader" (Kulesa and Fraser 2000; McKinney et al. 2011).[8]

An elaborate collaboration of pushes and pulls

Recently, analysis of frog and fish cranial neural crest migratory streams revealed that the separate streams seem to be kept apart by the chemorepellent properties of ephrins and semaphorins and the proteoglycan versican (**FIGURE 15.19A**; Szabó et al. 2016).

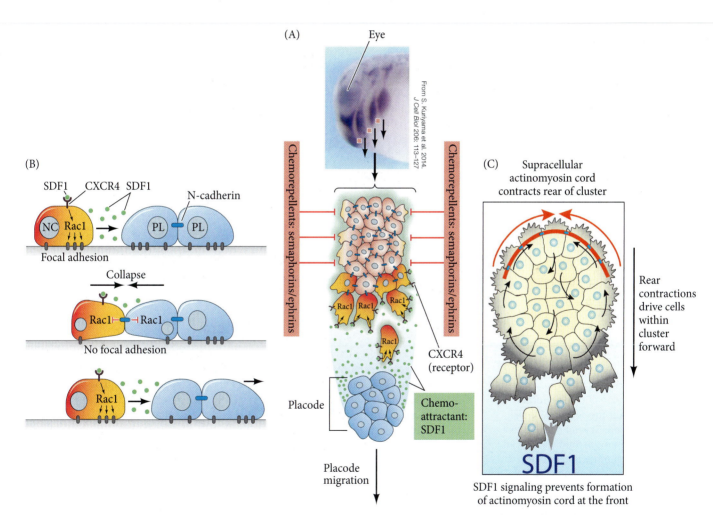

FIGURE 15.19 "Chase and run" model for chemotactic cell migration. (A) The micrograph is a lateral view of streams of cranial neural crest cells in the *Xenopus* head, visualized by in situ hybridization for *FoxD3* and *Dlx2* for both premigratory and migratory cranial neural crest populations (dark purple). The illustration depicts the collective migration of cranial neural crest cells using the autonomous mechanisms described in Figure 15.12. Here, a ventrally positioned placode (blue cells) attracts the leading edge of the cranial neural crest stream via SDF1-CXCR4 signaling (the "chase"). The cranial neural crest contacts the placode, triggering contact inhibition that pushes the placode forward (the "run"). Chemorepellents restrict crest cells from lateral wandering outside the cluster. (B) Internal molecular events and resulting cellular behaviors that yield forward migratory movement of both placodal (blue) and cranial neural crest (red/yellow) cells. (C) Illustration of the cell movements of cranial neural crest cells within the cluster. Along the side of the cluster sensing the highest concentrations of SDF1, actinomyosin contractile cords fail to form. A supracellular actinomyosin cord does form along the rear periphery of the neural crest cluster and functions to pull cells along the periphery toward the rear while driving cells within the cluster forward. The contractile forces power the forward collective migration of cranial neural crest cells. (A,B after E. Theveneau et al. 2013. *Nat Cell Biol* 15: 763–772; E. Scarpa and R. Mayor. 2016. *J Cell Biol* 212: 143–155. C generously provided by Roberto Mayor.)

Developing Questions

You have learned that Shh is in part responsible for attracting commissural axons to the floor plate. What about other classical morphogen signals? Can BMPs from the dorsal neural tube affect dorsoventral pathfinding, or can Wnts expressed in an anterior-posterior gradient in the neural tube influence longitudinal axon guidance?

Blocking the activity of the Eph receptors causes cells from the different streams to mix (Smith et al. 1997; Helbling et al. 1998; reviewed in Scarpa and Mayor 2016). In addition, it appears that the ventrally directed stream is guided by placode cells (Theveneau et al. 2013). If an explant of cranial neural crest cells is placed next to an explant of a placode, the neural crest cells appear to "chase" the placodal explant, a behavior that can be abolished by knocking down CXCR4. These and other data led Theveneau and colleagues (2013) to propose the "chase and run" model to explain how this relationship results in directed collective migration.

It was found that placodal cells secrete the chemoattractant **stromal-derived factor-1** (**SDF1**), thereby setting up a gradient of the factor that is highest at the placode. The cranial neural crest cells express the receptor for this ligand, CXCR4, which allows them to sense the attraction of the SDF1 gradient and directs the migration of neural crest cells up the gradient toward the placode (the "chase"). Once the neural crest cells reach the placode, however, contact inhibition of locomotion between the neural crest and placode cells causes the placode cells to migrate away from the site of contact (the "run"). The chemoattractive force of SDF1, however, will start the chase again in the ventral direction toward the "running" placode (**FIGURE 15.19B**; Theveneau et al. 2013; Scarpa and Mayor 2016).

FURTHER DEVELOPMENT

NEURAL CREST MIGRATION WORKS BY REAR-WHEEL DRIVE Although it is clear that chemoattractants like SDF1 are essential for promoting directional migration of the neural crest, the mechanism of how such chemoattractants work has remained elusive until recently. It has been discovered that each neural crest cluster exhibits a supracellular ring of actomyosin at its rear edge that contracts uniformly at the periphery of the cluster. The chemoattractant SDF1 inhibits this contraction at the front of the cluster, leaving intact the contractions at the back of the cluster (**FIGURE 15.19C**). This asymmetric contraction triggered by SDF1 is necessary and sufficient to drive the movement of the cluster toward SDF1. Thus, collective chemotaxis of neural crest is driven by contraction of the rear cells, equivalent to squeezing the back of a tube of toothpaste to allow it to flow to the front (Shellard et al. 2018; see also comment by Adameyko 2018). (See **Further Development 15.8, Intramembranous Bone and the Role of Neural Crest in Building the Head Skeleton**, online.)

Neural Crest-Derived Head Skeleton

The vertebrate skull, or **cranium**, is composed of the **neurocranium** (skull vault and base) and the **viscerocranium** (jaws and other pharyngeal arch derivatives). The bones of the cranium are called **intramembranous bones** because they are created by laying down calcified spicules directly in connective tissue without a cartilaginous precursor. Skull bones are derived from both the neural crest and the head mesoderm (Le Lièvre 1978; Noden 1978; Evans and Noden 2006). Although the neural crest origin of the viscerocranium is well documented, the contributions of cranial neural crest cells to the skull vault are more controversial. In 2002, Jiang and colleagues constructed transgenic mice that expressed β-galactosidase only in their cranial neural crest cells.[9] When the embryonic mice were stained for β-galactosidase, the cells forming the anterior portion of the head—the nasal, frontal, alisphenoid, and squamosal bones—stained blue; the parietal bone of the skull did not (**FIGURE 15.20A,B**). The boundary between neural crest-derived head bone and mesoderm-derived head bone is between the frontal and parietal bones (**FIGURE 15.20C**; Yoshida et al. 2008). Although the specifics may vary among the vertebrate groups, in general the front of the head is derived from the neural crest, while the back of the skull is derived from a combination of neural crest-derived and mesodermal bones. The neural crest contribution to facial muscle mixes with the cells of the cranial mesoderm such that facial muscles probably also have dual origins (Grenier et al. 2009).

Given that the neural crest forms our facial skeleton, it follows that even small variations in the rate and direction of cranial neural crest cell divisions will determine what

Developing Questions

We discussed the progressive maturation of trunk neural crest cells over the course of their migration. Cranial neural crest cells are more tightly adherent to one another and undergo collective migration, implying that a different kind of mechanism might regulate the progressive differentiation of cranial neural crest cells. Do the cells located more centrally within the collective stream get patterned differently from those at the periphery, or is the spatiotemporal positioning of the cells along their migration route correlated with their specification?

(A) *Wnt1-Cre*: Neural crest-derived bone

(B) *Mesp-Cre*: Mesoderm-derived bone

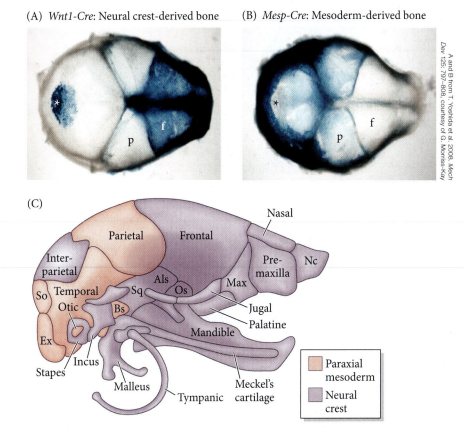

(C)

A and B from T. Yoshida et al. 2008. *Mech Dev* 125, 797–808, courtesy of G. Morriss-Kay

FIGURE 15.20 Cranial neural crest cells in embryonic mice, stained for β-galactosidase expression. (A) In the *Wnt1-Cre* strain, β-galactosidase is expressed wherever Wnt1 (a neural crest marker) would be expressed. This dorsal view of a 17.5-day embryonic mouse shows staining in the frontal bone (f) and interparietal bone (asterisk) but not in the parietal bone (p). (B) The *Mesp-Cre* strain of mice expresses β-galactosidease in those cells derived from the mesoderm. Here, a reciprocal pattern of staining is seen, and the parietal bone is blue. (C) Summary diagram of results from mapping with Sox9 and Wnt1 markers (Als, alisphenoid; Bs, basisphenoid; Ex, exoccipital; Max, maxilla; Nc, nasal capsule; Os, orbitosphenoid; So, supraoccipital; Sq, squamosal). (C from several sources, including D. M. Noden and R. A. Schneider. 2006. *Adv Exp Med Biol* 589: 1–23; P. Francis-West et al. 1998. *Mech Dev* 75(1-2): 3–28.)

we look like. Moreover, because we resemble our biological parents, such small variations must be hereditary. The regulation of our facial features is probably coordinated in large part by numerous paracrine growth factors. BMPs (especially BMP3) and Wnt signaling cause the protrusion of the frontonasal and maxillary processes, giving shape to the face (Brugmann et al. 2006; Schoenebeck et al. 2012). FGFs from the pharyngeal endoderm are responsible for the attraction of the cranial neural crest cells into the arches as well as for patterning the skeletal elements within the arches. Fgf8 is both a survival factor for the cranial crest cells and is critical for the proliferation of cells forming the facial skeleton (Trocovic et al. 2003, 2005; Creuzet et al. 2004, 2005). The FGFs work in concert with BMPs, sometimes activating them and sometimes repressing them (Lee et al. 2001; Holleville et al. 2003; Das and Crump 2012). (See **Further Development 15.9, Coordination of Face and Brain Growth**, and **Further Development 15.10, Why Birds Don't Have Teeth**, both online.)

Cardiac Neural Crest

The heart originally forms in the neck region, directly beneath the pharyngeal arches, so it should not be surprising that it acquires cells from the neural crest. The pharyngeal ectoderm and endoderm both secrete Fgf8, which acts as a chemotactic factor to draw neural crest cells into the area. Indeed, if beads containing large amounts of Fgf8 are placed dorsal to the chick pharynx, the cardiac neural crest cells will migrate there instead (Sato et al. 2011). The caudal region of the cranial neural crest is called the cardiac neural crest because its cells (and only those particular neural crest cells) generate the endothelium of the aortic arch arteries and the septum between the aorta and the pulmonary artery (**FIGURE 15.21**; Kirby 1989; Waldo et al. 1998). Cardiac crest cells also enter pharyngeal arches 3, 4, and 6 to become portions of other neck structures such as the thyroid, parathyroid, and thymus glands. These cells are often referred to as the circumpharyngeal crest (Kuratani and Kirby 1991, 1992). In the thymus, neural crest-derived cells are especially important in one of the most critical functions of adaptive

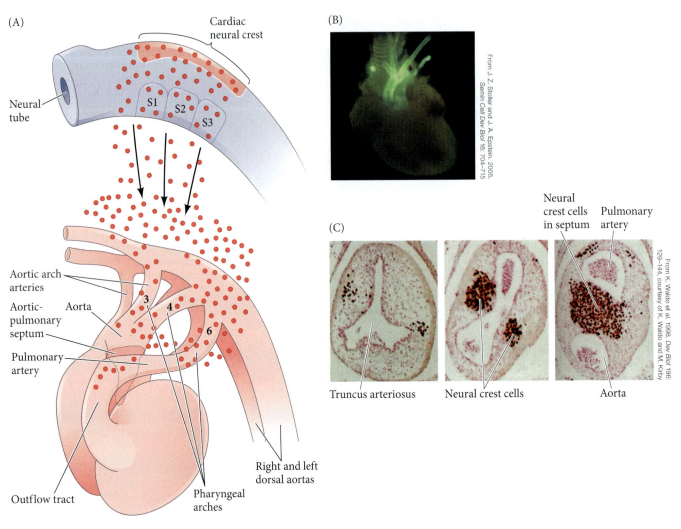

(A)

Cardiac neural crest

Neural tube

S1 S2 S3

Aortic arch arteries

Aortic-pulmonary septum

Aorta

3 4

6

Pulmonary artery

Outflow tract

Pharyngeal arches

Right and left dorsal aortas

(B)

From J. Z. Stoller and J. A. Epstein. 2005. *Semin Cell Dev Biol* 16: 704–715

Neural crest cells in septum

Pulmonary artery

(C)

Truncus arteriosus

Neural crest cells

Aorta

From K. Waldo et al. 1998. *Dev Biol* 196: 129–144, courtesy of K. Waldo and M. Kirby

FIGURE 15.21 The septum of the heart (which separates the truncus arteriosus into the pulmonary artery and the aorta) forms from cells of the cardiac neural crest. (A) During the fifth week of human gestation, cardiac neural crest cells migrate to pharyngeal arches 3, 4, and 6 and enter the truncus arteriosus to generate the septum (S1, S2, S3, somites 1, 2, and 3). (B) In a transgenic mouse where the fluorescent green protein is expressed only in cells having the Pax3 cardiac neural crest marker, the outflow regions of the heart become labeled. (C) Quail cardiac neural crest cells were transplanted into the analogous region of a chick embryo, and the embryos were allowed to develop. Quail cardiac neural crest cells are visualized by a quail-specific antibody (dark stain). In the heart, these cells can be seen separating the truncus arteriosus into the pulmonary artery and the aorta. (A after M. R. Hutson and M. L. Kirby. 2007. *Cell Dev Biol* 18: 101–110.)

immunity: regulating the exit of mature T cells from the thymus and into the circulation (Zachariah and Cyster 2010). It is also likely that the carotid body, which monitors oxygen in the blood and regulates respiration accordingly, is derived from the cardiac neural crest (see Pardal et al. 2007). (See **Further Development 15.11, Mice, Humans, and How CNCC May Connect Disparate Diseases**, online.)

Establishing Axonal Pathways in the Nervous System

At the beginning of the twentieth century, there were many competing theories on how axons formed. Theodor Schwann (yes, he discovered Schwann cells) believed that numerous neural cells linked themselves together in a chain to form an axon. Viktor Hensen, the discoverer of the embryonic node in birds, thought that axons formed around preexisting cytoplasmic threads between the cells. Wilhelm His (1886) and Santiago Ramón y Cajal (1890) postulated that the axon was an outgrowth (albeit an extremely large one) of the neuron's soma.

In 1907, Ross Granville Harrison demonstrated the validity of the outgrowth theory in an elegant experiment that birthed the science of developmental neurobiology and the technique of tissue culture. Harrison isolated a portion of the neural tube from a 3-mm frog tadpole. (At this stage, shortly after the closure of the neural tube, there is no visible differentiation of axons.) He placed this neuroblast-containing tissue in a drop of frog lymph on a coverslip and inverted the coverslip over a depression slide so

FIGURE 15.22   A starry sky of synaptic connections contact Purkinje dendrites in this mouse cerebellum. This fluorescence image shows the elaborate array of dendrites emerging from each Purkinje neuron (red), which receives information from other neurons through hundreds of synapses (blue). DNA is stained green.

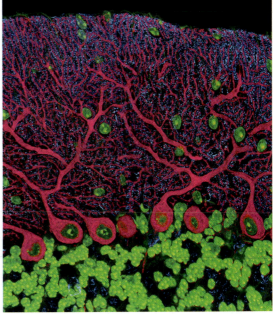

that he could watch what was happening within this "hanging drop." What Harrison saw was the emergence of axons as outgrowths from the neuroblasts, elongating at about 56 mm per hour.

Unlike most cells, neurons are not confined to their immediate space; instead, they can produce axons that may extend for meters. Each of the 86 billion neurons in the human brain has the potential to interact in specific ways with thousands of other neurons (**FIGURE 15.22**; Azevedo et al. 2009). A large neuron (such as a Purkinje cell or motor neuron) can receive input from more than 10^5 other neurons (Gershon et al. 1985). Understanding the generation of this stunningly ordered complexity is one of the greatest challenges of modern science. How is this complex circuitry established? We will explore this question as we discuss how neurons extend their axons, how axons are guided to their target cells, how synapses are formed, and what determines whether a neuron lives or dies. (See **Further Development 15.12, The Evolution of Developmental Neurobiology**, online.)

The Growth Cone: Driver and Engine of Axon Pathfinding

Earlier in this chapter, we presented an analogy comparing cell migration to the navigation of cars in traffic (see Figure 15.7). A similar analogy can be used for axon pathfinding by comparing it to a moving train. A neuron needs to build an axonal connection with a target cell that may lie a great distance away. Like the engine of a train, the *locomotory* apparatus of an axon—the **growth cone**—is at the front, and like new carriages added behind the engine, the axon grows through polymerization of microtubules (**FIGURE 15.23A**). The growth cone has been called a "neural crest cell on a leash" because, like neural crest cells, it migrates and senses the environment. Moreover, it can respond to the same types of signals that migrating cells sense.

The growth cone does not move forward in a straight line but rather "feels" its way along the substrate. The growth cone moves by the elongation and contraction of pointed filopodia called **microspikes** (**FIGURE 15.23B**). These microspikes contain microfilaments, which are oriented parallel to the long axis of the axon. (This mechanism is similar to that seen in the filopodial cytoskeleton of secondary (skeletogenic) mesenchyme cells in echinoderms; see Chapter 10). Within the axon itself, structural support is provided by microtubules (Yamada et al. 1971; Forscher and Smith 1988).

As in most migrating cells, the exploratory microspikes of the growth cone attach to the substrate and exert a force that pulls the rest of the cell forward. Axons will not grow if the growth cone fails to advance (Lamoureux et al. 1989). In addition to their structural role in axonal migration, microspikes have a sensory function. Fanning out in front of the growth cone, each microspike samples the microenvironment and sends signals back to the cell body (Davenport et al. 1993). The navigation of axons to their appropriate targets depends on guidance molecules in the extracellular environment, and it is the growth cone that turns or does not turn in response to guidance cues as the axon seeks to make appropriate synaptic connections. Such differential responsiveness is due to disparities in the expression of receptors on the growth cone cell membrane. Growth cones have the ability to sense the environment and translate the extracellular signals into a directed movement (**FIGURE 15.23C**). This use of directional cues to facilitate specific migration is accomplished by altering the cytoskeleton, changing membrane growth, and coordinating cell adhesion and cell movement (Vitriol and Zheng 2012). (See **Further Development 15.13, Primer on the Molecular Anatomy of the Growth Cone**, and **Further Development 15.14, "Plus Tips" and Actin-Microtubule Interactions during Growth Cone Guidance**, both online.)

(A)

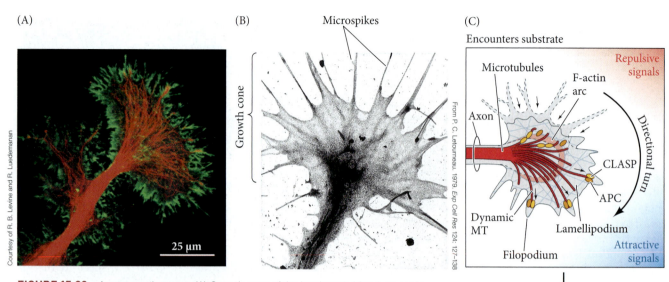

Courtesy of R. B. Levine and R. Luedemanan

25 µm

(B)

Microspikes

Growth cone

From P. C. Letourneau. 1979. *Exp Cell Res* 124: 127–138

(C)

Encounters substrate

Microtubules

F-actin arc

Axon

Repulsive signals

CLASP

APC

Directional turn

Dynamic MT

Lamellipodium

Filopodium

Attractive signals

Protrusion

Filopodium and lamellipodium extend forward

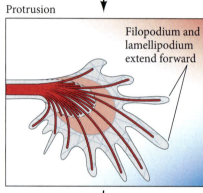

Engorgement

C-domain moves forward

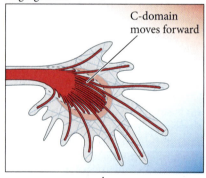

Consolidation

New axon shaft forms

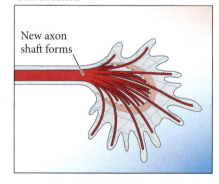

FIGURE 15.23 Axon growth cones. (A) Growth cone of the hawkmoth *Manduca sexta* during axon extension and pathfinding. The actin in the filopodia is stained green with fluorescent phalloidin, and the microtubules are stained red with a fluorescent antibody to tubulin. (B) Actin microspikes in an axon growth cone, seen by transmission electron microscopy. (C) The periphery of the growth cone contains lamellipodia and filopodia. The lamellipodia are the major motile apparatus and are seen in the regions that are turning toward a stimulus; the filopodia are sensory. Both structures contain actin filaments. There is also a central region of microtubules, some of which extend outward into the filopodia. The microtubules (MT) entering the peripheral area can be lengthened or shortened by proteins activated by attractive or repulsive stimuli. During attraction, regulatory proteins (such as CLASP and APC) bind to the plus ends of the microtubules, stabilizing and lengthening them. On the side opposite the attractive cue, microtubules are removed from the periphery. (C-domain, central domain of the growth cone, contains microtubules that extend the axon shaft.) (C after L. A. Lowery and D. Van Vactor. 2009. *Nat Rev Mol Cell Biol* 10: 332–343.)

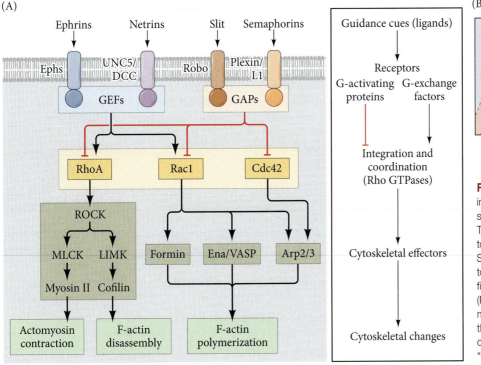

(A)

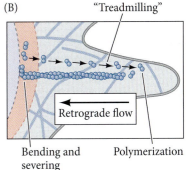

(B)

FIGURE 15.24 Rho GTPases interpret and relay external guidance signals to the actin cytoskeleton. (A) The four major ligands providing cues to the growth cone (ephrins, netrins, Slit, and semaphorins) bind to receptors that stabilize or destabilize actin filaments. The Rho family of GTPases (RhoA, Rac1, and Cdc42) act as mediators between the receptors and the agents carrying out the cytoskeletal changes. (B) Diagram depicting actin "treadmilling," during which the position of actin monomers flows from the plus end to the minus end, powered by polarized polymerization and depolymerization. (A after L. A. Lowery and D. Van Vactor. 2009. *Nat Rev Mol Cell Biol* 10: 332–343; B after G. M. Cammarata et al. 2016. *Cytoskeleton* 73: 461–476.)

Rho, Rho, Rho your actin filaments down the signaling stream

The regulation of actin polymerization drives growth cone movement and thus is the target of many molecular guidance pathways. **Rho GTPases** regulate the growth of actin filaments. These GTPases can be activated or repressed by receptors binding ephrins, netrins, Slit proteins, or semaphorins (**FIGURE 15.24A**). Similarly, the regulation of tubulin polymerization into microtubules is important because tubulin is encouraged to polymerize on the side of the growth cone receiving attractant stimuli, and it is inhibited from polymerizing (indeed, the tubulin is depolymerized and recycled) on the side opposite the attractive stimuli (Vitriol and Zheng 2012).

Adhesion is thought to provide the "clutch" for directional movement. Visualize actin as being linked to the cell membrane. Now contemplate that the dynamics of actin assembly and disassembly cause a retrograde flow of actin—that is, picture actin moving *away from* the tip of the growth cone and *toward* the cell body ("treadmilling"; **FIGURE 15.24B**). If the cell membrane is anchored to an external adhesion molecule (through its integrins or cadherins), however, the membrane will be propelled forward by actin treadmilling (Bard et al. 2008; Chan and Odde 2008). If there is no such anchoring adhesion, then there won't be any net movement. Importantly, though, if the adhesion is too stable, the growth cone will stop moving altogether. Thus, adhesions have to be made and broken for the growth cone to progress dynamically. These transitory adhesive complexes are referred to as **focal adhesions**, and they bind actin internally and the extracellular environment externally. Focal adhesions may have as many as 100 different protein components (Geiger and Yamada 2011), and one such component is focal adhesion kinase (FAK) which is critical for the assembly, stabilization, and degradation of the focal adhesions (Mitra et al. 2005; Chacon and Fazzari 2011). (See **Further Development 15.15, Turning the Growth Cone Requires Membrane Endocytosis**, online.)

Axon Guidance

How does the growth cone "know" to traverse numerous potential target cells and make a specific connection? Harrison (1910) first suggested that the specificity of axonal growth is due to **pioneer nerve fibers**, axons that go ahead of other axons and serve as guides for them.[10] This observation simplified but did not solve the problem of how neurons form appropriate patterns of interconnection. Harrison also noted, however, that axons must grow on a solid substrate, and he speculated that differences among embryonic surfaces might allow axons to travel in certain specified directions. The final connections would occur by complementary interactions on the target cell surface:

> *That it must be a sort of a surface reaction between each kind of nerve fiber and the particular structure to be innervated seems clear from the fact that sensory and motor fibers, though running close together in the same bundle, nevertheless form proper peripheral connections, the one with the epidermis and the other with the muscle. . . . The foregoing facts suggest that there may be a certain analogy here with the union of egg and sperm cell.*

Research on the specificity of neuronal connections has focused on three major systems: (1) motor neurons, whose axons travel from the spinal cord to a specific muscle; (2) commissural neurons, whose axons must cross the midline plane of the embryo to innervate targets on the opposite side of the central nervous system; and (3) the optic system, where axons originating in the retina must find their way back into the brain. In all cases, the specificity of axonal connections unfolds in three steps (Goodman and Shatz 1993):

1. *Pathway selection.* The axons travel along a route that leads them to a particular region of the embryo.

2. *Target selection.* The axons, once they reach the correct area, recognize and bind to a set of cells with which they may form stable connections.

3. *Address selection.* The initial patterns are refined such that each axon binds to a small subset (sometimes only one) of its possible targets.

The first two processes are independent of neuronal activity. The third process involves interactions between several active neurons and converts the overlapping projections into a fine-tuned pattern of connections.

How are the neurons' axons instructed where to go?

The Intrinsic Navigational Programming of Motor Neurons

Neurons at the ventrolateral margin of the vertebrate neural tube become motor neurons, and one of their first steps toward maturation involves target specificity (Dasen et al. 2008). The cell bodies of the motor neurons projecting to a single muscle are pooled in a longitudinal column of the spinal cord (**FIGURE 15.25A**; Landmesser 1978; Hollyday 1980; Price et al. 2002). The pools are grouped into the **columns of Terni** and the lateral and medial motor columns (LMC and MMC, respectively), and neurons in similar places have similar targets (see Figure 13.15C). For instance, in the chick hindlimb, LMC motor neurons innervate the dorsal musculature, whereas the motor neurons of the MMC innervate ventral limb musculature (Tosney et al. 1995; Polleux et al. 2007). This arrangement of motor neurons is consistent throughout the vertebrates.

FURTHER DEVELOPMENT

The guidance mechanisms directing Lim-specific motor neurons to their muscle cell targets

The targets of motor neurons are specified before their axons extend into the periphery. This was shown by Lance-Jones and Landmesser (1980), who reversed segments of the chick spinal cord so that the motor neurons were placed in new locations. The axons went to their original targets, not to the ones expected from their new positions (**FIGURE 15.25B–D**). The molecular basis for this target specificity resides in the members of the Hox and Lim protein families that are induced during neuronal specification (Tsushida et al. 1994; Sharma et al. 2000; Price and Briscoe 2004; Bonanomi and Pfaff 2010). For instance, all motor neurons express the Lim proteins Islet1 and (slightly later) Islet2. If no other Lim protein is expressed, the neurons project to the ventral limb muscles (**FIGURE 15.26**) because the axons (just like the trunk neural crest cells) synthesize neuropilin-2, the receptor for the chemorepellant semaphorin-3F, which is made in the dorsal part of the

(A)

(B) 2.5 Days

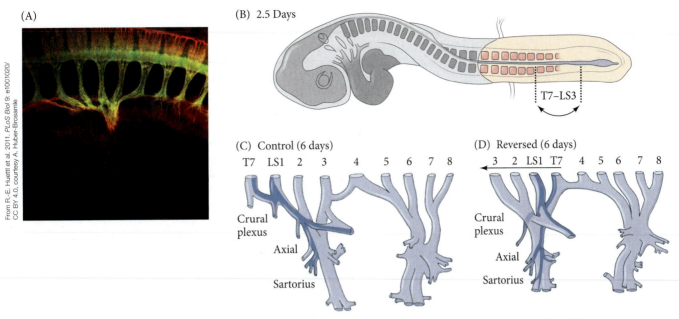

T7–LS3

(C) Control (6 days)

T7 LS1 2 3 4 5 6 7 8

Crural plexus

Axial

Sartorius

(D) Reversed (6 days)

3 2 LS1 T7 4 5 6 7 8

Crural plexus

Axial

Sartorius

FIGURE 15.25 Compensation for small dislocations of axonal initiation position in the chick embryo. (A) Axons from motor neurons and sensory neurons group together (fasciculate) before finding their muscle targets. Here, motor nerves (stained green with GFP) and sensory neurons (stained red with antibodies) fasciculate before entering the limb bud of a 10.5-day mouse embryo. (B) A length of spinal cord comprising segments T7–LS3 (seventh thoracic through third lumbo-sacral segments) is reversed in a 2.5-day embryo. (C) Normal pattern of axon projection to the forelimb muscles at 6 days. (D) Projection of axons from the reversed segment at 6 days. The ectopically placed neurons eventually found their proper neural pathways and innervated the appropriate muscles. (B–D after C. Lance-Jones and L. Landmesser. 1980. *J Physiol* 302: 581–602.)

limb bud. If Lim1 protein is also synthesized, however, the motor neurons project dorsally to the dorsal limb muscles. This axonal growth toward the dorsal muscle is because Lim1 induces the expression of EphA4, which is the receptor for the chemorepellent protein ephrin A5 that is made in the ventral part of the limb bud. Thus, the innervation of the limb by motor neurons depends on repulsive signals. The motor neurons entering the axial muscles of the body wall, however, are brought there by chemoattraction—indeed, these axons make an abrupt turn to get to the developing musculature—because these motor neurons express Lhx3, which induces the expression of a receptor for FGFs, such as those secreted by the dermomyotome (the somitic region that contains muscle precursor cells, see Chapter 17). This process is an example of classic morphogens (FGFs) also functioning to directly guide axons in a manner that is classically attributed to guidance molecules. We will revisit this point again later in this chapter. In summary, motor neurons seek their targets through intrinsic "programs" that assign different motor neurons different cell surface molecules, which determine the responsiveness of the axon growth cones to guidance cues in their path and on their targets.

FIGURE 15.26 Motor neuron organization and Lim specification in the spinal cord innervating the chick hindlimb. Neurons in each of three different columns express specific sets of Lim family genes (including *Isl1* and *Isl2*), and neurons within each column make similar pathfinding decisions. Neurons of the medial motor column are attracted to the axial muscles by FGFs secreted by the dermomyotome. Neurons of the lateral motor column send axons to the limb musculature. Where these columns are subdivided, medial subdivisions project to ventral positions because they are repelled by semaphorin-3F in the dorsal limb bud, and lateral subdivisions send axons to dorsal regions of the limb bud because they are repelled by the ephrin A5 synthesized in the ventral half. (After F. Polleux et al. 2007. *Nat Rev Neurosci* 8: 331–340.)

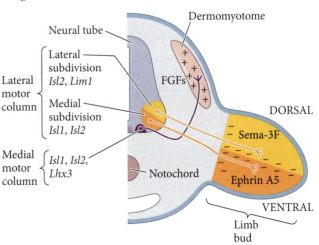

Cell adhesion: A mechanism to grab the road

The initial pathway an axonal growth cone follows is determined by the environment the growth cone experiences. The polarity of a neuron—that is, which part of the cell will extend the axon—is determined largely by the neuron's response to cell adhesive cues in its immediate environment. Integrins and N-cadherins serve as receptors to orient the neuron in accordance with cues from the extracellular matrices (ECM) and membranes of surrounding cells (Myers et al. 2011; Randlett et al. 2011; Gärtner et al. 2012; Ligon et al. 2001). Once an axon begins to form, its growth cone encounters different substrates. The growth cone adheres to certain substrates and moves in their direction. Other substrates cause the growth cone to retract, preventing its axon from growing in that direction. Growth cones prefer to migrate on surfaces that are more adhesive than their surroundings, and a track of adhesive molecules (such as laminin) can direct them to their targets (Letourneau 1979; Akers et al. 1981; Gundersen 1987). In addition to cell-to-ECM adhesion, there are important cell-to-cell adhesion contacts that can provide permissive substrates for growth cone walking. (See **Further Development 15.16, Mechanisms of Cell-to-Cell Adhesion**, online.)

Local and long-range guidance molecules: The street signs of the embryo

Navigation through the embryonic environment is literally guided by molecules that function much like the traffic signs, lights, and other directional cues that we use to find our way around our environment. Scientists have interpreted many of the decisions a growth cone makes during its journey as equivalent to choices between being attracted to or repelled from a particular region of the embryo (see Figure 15.23). The signals that elicit attractive and repulsive responses in growing axons fall into four protein families: ephrins, semaphorins, netrins, and the Slit proteins; they are some of the same proteins that we saw regulating neural crest cell migration (see Kolodkin and Tessier-Lavigne 2011). Whether a guidance signal is attractive or repulsive can depend on (1) the type of cell receiving that signal and (2) the time when a cell receives the signal. Most intriguing is that neural development has employed dynamic mechanisms to alter the responsiveness of a growth cone, allowing growth cones to become repelled by cues they either ignored or were actively attracted to previously.

Repulsion patterns: Ephrins and semaphorins

Members of two membrane protein families, the ephrins and the semaphorins, are most well-known (but not exclusively) for their role as repellent guidance cues during the patterning of axonal anatomy. Just as neural crest cells are inhibited from migrating across the posterior portion of a sclerotome, the axons from the dorsal root ganglia and motor neurons also pass only through the anterior portion of each sclerotome and avoid migrating through the posterior portion (**FIGURE 15.27A**; see also Figure 15.14). Davies and colleagues (1990) showed that membranes isolated from the posterior portion of a somite cause the growth cones of these neurons to collapse (**FIGURE 15.27B,C**). These growth cones contain Eph receptors (which bind ephrins) and neuropilin receptors (which bind semaphorins) and are thus responsive to ephrins and semaphorins on the posterior sclerotome cells (Wang and Anderson 1997; Krull et al. 1999; Kuan et al. 2004). In this way, the same signals that pattern neural crest cell migration also pattern the spinal neuronal outgrowths. (See **Further Development 15.17, The Classic Example of Semaphorin-Mediated Repulsion of the Grasshopper Sensory Axon**, online.)

FURTHER DEVELOPMENT

SEMAPHORINS, MAKING THE GROWTH CONE COLLAPSE   The proteins of the semaphorin-3 family, also known as collapsins, are found in mammals and birds. These secreted proteins collapse the growth cones of axons originating in the dorsal root ganglia (Luo et al. 1993). There are several types of neurons in the dorsal root ganglia whose axons enter the dorsal spinal cord. Most of these axons are

FIGURE 15.27 Repulsion of dorsal root ganglion growth cones. (A) Motor axons migrating through the rostral (anterior), but not the caudal (posterior), compartments of each sclerotome. (B) In vitro assay, wherein ephrin stripes (+) were placed on a background surface of laminin. Motor axons grew only where the ephrin was absent (–). (C) Inhibition of growth cones by ephrin after 10 minutes of incubation. The left-hand photograph shows a control axon exposed to a similar (but not inhibitory) compound; the axon on the right was exposed to an ephrin found in the posterior somite.

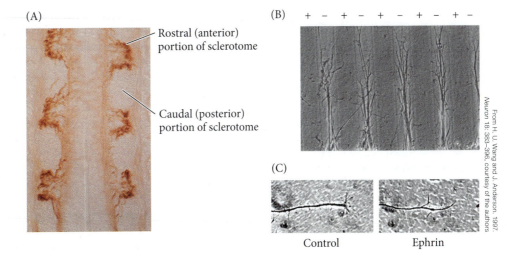

(A)

Rostral (anterior) portion of sclerotome

Caudal (posterior) portion of sclerotome

(B)

(C)

Control   Ephrin

From H. U. Wang and J. Anderson. 1997. *Neuron* 18: 383–396, courtesy of the authors

prevented from traveling farther and entering the ventral spinal cord; however, a subset of them does travel ventrally through the other neural cells (**FIGURE 15.28**). These particular axons are not inhibited by semaphorin-3, whereas those of the other neurons are (Messersmith et al. 1995). This finding suggests that semaphorin-3 patterns sensory projections from the dorsal root ganglia by selectively repelling certain axons so that they terminate dorsally. A similar scheme is seen in the brain, where semaphorin made in one region of the brain prevents the entry of neurons that originated in another region (Marín et al. 2001).

(A)

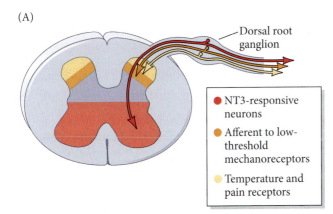

Dorsal root ganglion

● NT3-responsive neurons

● Afferent to low-threshold mechanoreceptors

● Temperature and pain receptors

FIGURE 15.28 Semaphorin-3 as a selective inhibitor of axonal projections into the ventral spinal cord. (A) Trajectory of axons in relation to semaphorin-3 expression in the spinal cord of a 14-day embryonic rat. Neurons that are responsive to neurotrophin 3 (NT3) can travel to the ventral region of the spinal cord, but the afferent axons for the mechanoreceptors and for temperature and pain receptor neurons terminate dorsally. (B) Transgenic chick fibroblast cells that secrete semaphorin-3 inhibit the outgrowth of mechanoreceptor axons. These axons are growing in medium treated with nerve growth factor (NGF), which stimulates their growth, but they are still inhibited from growing toward the source of semaphorin-3. (C) Neurons that are responsive to NT3 for growth are not inhibited from extending toward the source of semaphorin-3 when grown with NT3. (A after E. K. Messersmith et al. 1995. *Neuron* 14: 949–959; J. Marx. 1995. *Science* 268: 971–973.)

(B)

(C)

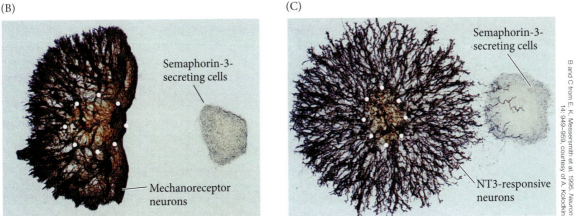

Semaphorin-3-secreting cells

Mechanoreceptor neurons

Semaphorin-3-secreting cells

NT3-responsive neurons

B and C from E. K. Messersmith et al. 1995. *Neuron* 14: 949–959, courtesy of A. Kolodkin

How Did the Axon Cross the Road?

The idea that chemotactic cues guide axons in the developing nervous system was first proposed by Santiago Ramón y Cajal (1892). He suggested that diffusible molecules might signal the commissural neurons of the spinal cord to send axons from their dorsal positions in the neural tube to the ventral floor plate. Commissural neurons coordinate right and left motor activities. To accomplish this, they somehow must migrate to (and through) the ventral midline. The axons of commissural neurons begin growing ventrally down the side of the neural tube. About two-thirds of the way down, however, the axons change direction and project through the ventrolateral (motor) neuron area of the neural tube toward the floor plate (**FIGURE 15.29**).

There appear to be two systems involved in attracting the axons of dorsal commissural neurons to the ventral midline. The first involves the Sonic hedgehog protein (Shh), which starts the attraction of commissural axons ventrally, followed by the additional attraction of Netrin to cross the ventral midline. (Recall from Chapter 13 the importance of Shh as a morphogen for patterning cell fate; see Figures 13.16 and 4.22). Shh is made in and secreted from the floor plate and is distributed in a concentration gradient that is high ventrally and low dorsally. Loss of Shh signal transduction through pharmacological inhibition or conditional knockout of Smoothened results in the reduction of commissural axons reaching the ventral midline (Charron et al. 2003). Interestingly, commissural axon guidance by Shh signaling operates in a *noncanonical* manner through the alternative receptor, Brother of Cdo (Boc), and is independent of Gli-mediated transcriptional regulation (see Figure 4.22; Okada et al. 2006; Yam et al. 2009). Curiously, loss of the Shh reception does not eliminate all midline crossing of commissural axons, which suggests that some other factor is also involved. That other factor is…

…Netrin

In 1994, Serafini and colleagues developed an assay that allowed them to screen for the presence of a presumptive diffusible molecule that might be guiding the axons of commissural neurons. When dorsal spinal cord explants from chick embryos were cultured on collagen gels in the presence of a piece of floor plate tissue, these conditions promoted the outgrowth of commissural axons from the dorsal spinal cord explant. Serafini and colleagues took fractions of embryonic chick brain homogenate and tested them to see if any of the proteins therein elicited the same explant response. This research resulted in the identification of two proteins, **netrin-1** and **netrin-2**. Like Shh, netrin-1 is made by, and secreted from, the floor plate cells, whereas netrin-2 is synthesized in the lower region of the spinal cord but not in the floor plate (see Figure 15.29B). The current model proposes that growth cones of commissural neurons first encounter a gradient of netrin-2 and Shh, which brings them into the domain of the steeper netrin-1 gradient. The netrins are recognized by the receptors DCC and DSCAM, which are expressed by commissural axon growth cones (Liu et al. 2009). (See **Further Development 15.18, Netrin, the Bifunctional Guidance Cue**, online.)

Slit and Robo

It seems that for a commissural axon to cross the midline and grow away from it on the *contralateral side* (opposite to the side of the CNS in which the soma is residing), repulsive cues are needed as a driving force. One important chemorepulsive group of molecules is the **Slit** proteins, which are expressed and secreted by midline cells (reviewed in Neuhaus-Follini and Bashaw 2015; Martinez and Tran 2015). In *Drosophila*, Slit is secreted by the glial cells at the midline of the nerve cord, and it acts to prevent most axons from crossing the midline from either side. **Roundabout (Robo)** proteins (Robo1,[11] Robo2, and Robo3) are the receptors for Slit (Rothberg et al. 1990; Kidd et al. 1998; Kidd et al. 1999). Expression of these Robo receptors in the growth cones of pathfinding neurons function to interpret Slit signals as repulsion, which prevents migration across the midline and, depending on the repertoire of Robo receptors expressed, delineates the lateral positioning of longitudinal tracts[12] from the midline (Rajagopalan et al. 2000;

FIGURE 15.29 Trajectory of commissural axons in the rat spinal cord. (A) Schematic drawing of a model wherein commissural neurons first experience a gradient of Sonic hedgehog and netrin-2 and then a steeper gradient of netrin-1. The commissural axons are chemotactically guided ventrally down the lateral margin of the spinal cord toward the floor plate. Upon reaching the floor plate, contact guidance from the floor plate cells causes the axons to change direction. (B) Autoradiographic localization of *netrin-1* mRNA by in situ hybridization of antisense RNA to the hindbrain of a young rat embryo. *netrin-1* mRNA (dark area) is concentrated in the floor plate neurons.

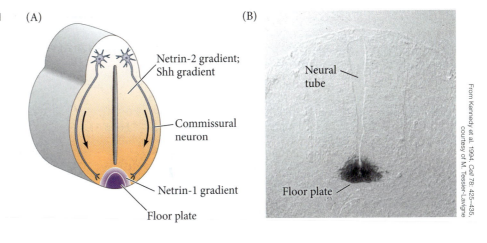

(A)

Netrin-2 gradient; Shh gradient

Commissural neuron

Netrin-1 gradient

Floor plate

(B)

Neural tube

Floor plate

From Kennedy et al. 1994, *Cell* 78: 425–435, courtesy of M. Tessier-Lavigne

Simpson et al. 2000; Bhat 2005; Spitzweck et al. 2010). Loss of Slit or the combined loss of Robo1 and Robo2 results in precocious crossing and the consolidated longitudinal growth of axons along the midline itself (**FIGURE 15.30**). These and other results have led to a model in which the axons of commissural neurons that cross from one side of the midline to the other temporarily avoid this repulsion by downregulating Robo1/2 proteins as they approach the midline. Once the growth cone is across the middle of the embryo, the neurons re-express Robos at the growth cone to once again become sensitive to the midline repulsive actions of Slit (**FIGURE 15.31**; Brose et al. 1999; Kidd et al. 1999; Orgogozo et al. 2004). (See **Further Development 15.19, How a Growth Cone Can Rapidly Change Its Responsiveness to Guidance Cues,** and **Further Development 15.20, The Early Evidence for Chemotaxis**, both online.)

(A) Slit protein

(B) Robo protein

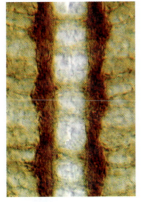

(C) Wild-type

(D) *Slit*⁻/⁻

All images from T. Kidd et al. 1999, *Cell* 96: 785–794, courtesy of C. S. Goodman

FIGURE 15.30 Robo/Slit regulation of midline crossing by axons of neurons. Robo and Slit in the *Drosophila* central nervous system. (A) Antibody staining reveals Slit protein in the midline glial cells. (B) Robo protein appears along the neurons of the longitudinal tracts of the CNS axon scaffold. (C) Wild-type CNS axon scaffold shows the ladderlike arrangement of axons crossing the midline. (D) Staining of the CNS axon scaffold with antibodies to all CNS neurons in a *Slit* loss-of-function mutant shows axons entering but failing to leave the midline (instead of running alongside it).

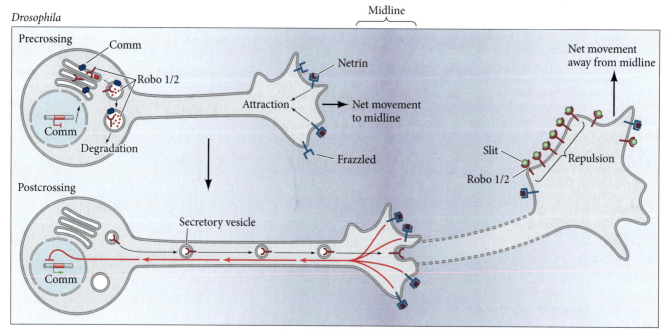

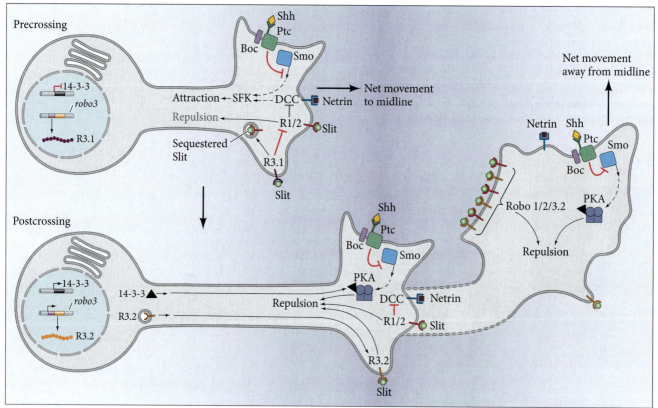

FIGURE 15.31 Model of axon guidance of commissural neurons crossing the midline in the fly and the mouse. Illustration shows a single commissural neuron residing in the left hemisphere of the fly ventral nerve cord (top) or mouse neural tube (bottom). Slit-Robo signaling mediates repulsion. In flies, precrossing axons are essentially blind to Slit repulsive cues due to the redirecting of Robo receptors to the lysosome by Commissureless (Comm) for degradation. Once at the midline, however, increased signaling through the Netrin receptor Frazzled triggers the downregulation of Comm. Robo is then returned to the growth cone, and Slit-mediated repulsion ensues so that the axon will not recross. In mouse, Netrin and Shh elicit attraction to the midline in precrossing commissural axons through their Frazzled/DCC and the Ptc-Boc-Smo receptor complex, respectively. In postcrossing axons, expression of 14-3-3 protein becomes upregulated, which shifts the responsiveness to Shh by influencing PKA downstream of Shh signaling. In vertebrates, Robo1/2 (R1/2) receptors are capable of inhibiting Netrin-DCC binding; therefore, in precrossing axons, this repression and Slit-Robo repulsion in general needs to be attenuated. The Robo3.1 (R3.1) isoform *may* function to inhibit Robo1/2, thus permitting Netrin-mediated attraction. Moreover, Robo3.1 may also competitively sequester Slit, with no direct downstream guidance outcome, which would serve to reduce available Slit to bind Robo1/2. In postcrossing axons, however, the Robo3.2 (R3.2) isoform is upregulated, and the 3.1 isoform is lost. Robo3.2 seems to function similarly as a canonical Slit repellent.

The Travels of Retinal Ganglion Axons

Nearly all the mechanisms for neuronal specification and axon specificity mentioned in this chapter can be seen in the ways individual retinal neurons send axons to the visual processing areas of the brain. Despite some differences, retinal development and axon guidance are largely conserved across vertebrates. For instance, as with motor neurons, the LIM family of transcription factors (Islet-2) is differentially expressed in the developing retinal ganglion layer specifying cell fate, which ultimately determines the repertoire of growth cone receptors to guide axon pathfinding to the tectum (reviewed in Bejarano-Escobar et al. 2015). As you read over the mechanisms that connect the retina to the brain, try to *see* their *connections* to already familiar developmental strategies, now seen in a new context.

Growth of the retinal ganglion axon to the optic nerve

The first steps in getting retinal ganglion cell (RGC) axons to their specific regions of the optic tectum take place within the retina (the neural retina of the optic cup). As the RGCs differentiate, their position in the inner margin of the retina is determined by cadherin molecules (N-cadherin and retina-specific R-cadherin) in their cell membranes (Matsunaga et al. 1988; van Horck et al. 2004). The RGC axons grow along the inner surface of the retina toward the optic disc (the head of the optic nerve). The mature human optic nerve will contain more than a million retinal ganglion axons. (See **Further Development 15.21, Intraretinal Guidance: How Do RGC Axons Even Leave the Eye in the Right Place?** online.)

Growth of the retinal ganglion axon through the optic chiasm

In non-mammalian vertebrates, the final destination for RGC axons is a portion of the brain called the optic tectum, while mammalian RGC axons go to the lateral geniculate nuclei. At many points, the journey of RGC axons within the brain occurs on an astroglial substrate (Bovolenta et al. 1987; Marcus and Easter 1995; Barresi et al. 2005). Laminin appears to promote crossing of the optic chiasm. Both across the midline (optic chiasm) and on the "optic tract" to the optic tectum, RGC axons of non-mammalian vertebrates travel over glial cells whose surfaces are coated with laminin. Very few areas of the brain contain laminin, and the laminin in this pathway exists only when the optic nerve fibers are growing on it (Cohen et al. 1987). Therefore, glial cells present a transient laminin road for RGC axons to preferentially grow along from the eye to the tectum. However, simply having a road to drive upon does not mean you actually know which direction to drive in. How are RGC growth cones guided across the midline and then on to the tectum or lateral geniculate nuclei?

RGC axons are guided out of the eye and through the optic nerve by the pathway expression of netrin, and kept on this path by the surrounding repulsion of semaphorins (see Harada et al. 2007). Upon reaching the midline, RGC axons have to "decide" if they are to continue straight across the midline and form the optic chiasm, or if they are to turn 90° and remain on the ipsilateral side of the brain. The mammalian and fish optic chiasm area seems to be surrounded by a corridor of Slit expression (perpendicular to the midline) through which RGC axons are channeled. This Slit repulsion serves to prevent inappropriate exploration outside the chiasm location. Thus, although the optic chiasm occurs at the midline, it is using a very different strategy of Slit-mediated repulsion from that used in the midline of the ventral spinal cord (see Figure 15.31). As in the retina, Robo2 appears to be the major mediator of RGC guidance at the ventral forebrain where the optic chiasm is forming (**FIGURE 15.32**; Erskine et al. 2000; Hutson and Chien 2002; Plump et al. 2002; Barresi et al. 2005).

In fish and frogs, all RGC axons cross at the optic chiasm to the contralateral side, but in mammals, some portion of RGC axons remain on the ipsilateral side. It appears that those axons not destined to cross to the opposite side of the brain are repulsed from doing so when they enter the optic chiasm (Godement et al. 1990). This repulsion appears to be influenced by the synthesis of ephrin and Shh in the cells occupying the chiasm. These midline cues are interpreted by the receptors Eph and Boc that are uniquely expressed on ipsilaterally projecting RGCs (Cheng et al. 1995; Marcus et al. 2000; Fabre et al. 2010). (See **Further Development 15.22, How Does an Ipsilateral RGC Axon Interpret Ephrin and Shh as a Decision Not to Cross?** online.)

(?) Developing Questions

It appears that RGC axons contact astroglia throughout much of their journey. What is the significance of this neuron-glial interaction, and what specific molecules are important in facilitating this interaction?

(A)

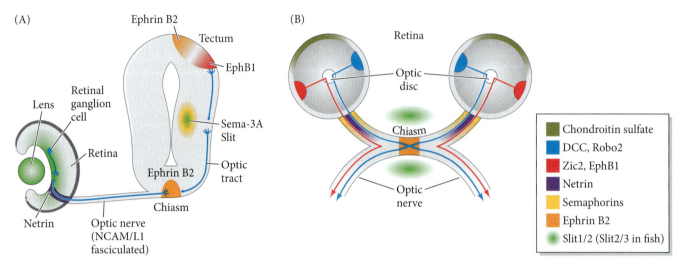

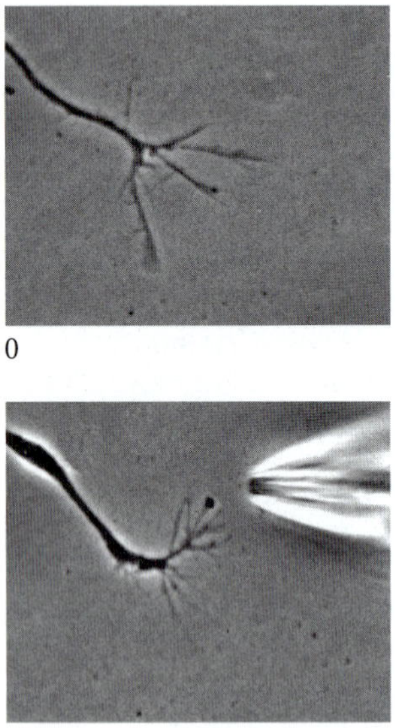

From H. Paves and M. Saarma, 1997, *Cell Tissue Res* 290: 285–297, courtesy of M. Saarma

FIGURE 15.32 Multiple guidance cues direct the movement of retinal ganglion cell (RGC) axons to the optic tectum. Guidance molecules belonging to the netrin, Slit, semaphorin, and ephrin families are expressed in discrete regions at several sites along the pathway to direct the RGC growth cones. RGC axons are repelled from the retinal periphery, probably by chondroitin sulfate. At the optic disc, the axons exit the retina and enter the optic nerve, guided by netrin/DCC-mediated attraction. Once in the optic nerve, the axons are kept within the pathway by inhibitory interactions. Slit proteins in the optic chiasm create zones of inhibition. Zic2-expressing ganglia in the ventrotemporal retina project EphB1-expressing axons, which are repelled at the chiasm by ephrin B2, thus terminating at ipsilateral (same-side) targets. Neurons from the medial portions of the retina do not express EphB1 and proceed to the opposite (contralateral) side. (A) Cross section. (B) Dorsal view. Not all cues are shown. (A after F. P. van Horck et al. 2004. *Curr Opin Neurobiol* 14: 61–66; B after T. C. Harada et al. 2007. *Genes Dev* 21: 367–378.)

Target Selection: "Are We There Yet?"

In some cases, nerves in the same ganglion can have several different targets.[13] How does a neuron know which cell to form a synapse with? The general mechanisms of ligand-receptor specificity that lead a growth cone to its target tissue can also take on a refining role at the final destination. What are the important signals that direct axons to the right addresses?

Chemotactic proteins

ENDOTHELINS Some neurons in the superior cervical ganglia (the largest ganglia in the neck) go toward the carotid artery, whereas other neurons from these same ganglia do not. It appears that those axons extending from the superior cervical ganglia to the carotid artery follow blood vessels that also lead there. These blood vessels secrete small peptides called **endothelins**. In addition to their adult roles constricting blood vessels, endothelins appear to have an embryonic role, as they are able to direct the migration of certain neural crest cells (such as those entering the gut) and of certain sympathetic axons that have endothelin receptors on their membranes (Makita et al. 2008). (See **Further Development 15.23, Bmp4 and Trigeminal Ganglion Neurons**, online.)

NEUROTROPHINS Some target cells produce a set of chemotactic proteins collectively called **neurotrophins**.[14] The neurotrophins include nerve growth factor (NGF), brain-derived neurotrophic factor (BDNF), conserved dopamine neurotrophic factor (CDNF), and neurotrophins 3 and 4/5 (NT3, NT4/5). These proteins are released from potential target tissues and work at short ranges as either chemotactic factors or chemorepulsive factors (Paves and Saarma 1997). Each can promote and attract the growth of some axons to its source while inhibiting other axons. For instance, sensory neurons from the rat dorsal root ganglia are attracted to sources of NT3 (**FIGURE 15.33**) but are inhibited by BDNF.

CHEMOTROPHINS: QUALITY AND QUANTITY The attachment of an axon to its target can be either "digital" or "analogue." In "analogue" mode, different axons recognize the same molecule on the target, but the *amount* of the molecule on the

FIGURE 15.33 Embryonic axon from a rat dorsal root ganglion turning in response to a source of NT3. The photographs document the growth cone's turn over a 10-minute period. The same growth cone was insensitive to other neurotrophins.

target appears to be critical to the connections that form; this may be the case in the attachment of retinal neurons to the tectum in the fish brain (Gosse et al. 2008). In other situations, there may be extremely molecule-specific qualitative ("digital") binding such that certain connections are neuron-specific. This may be the case for retinal neurons in *Drosophila*. Dscam protein has several thousand splicing isoforms (see Chapter 3), and this variety may enable highly specific recognition of a given neuron with its target neurons (Millard et al. 2010; Zipursky and Sanes 2010). Given the complexity of neural connections, it is probable that both qualitative and quantitative cues are used (Winberg et al. 1998).

Target selection by retinal axons: "Seeing is believing"

When the retinal axons come to the end of the laminin-lined optic tract, they spread out and find their specific targets in the optic tectum. Studies on frogs and fish (in which retinal neurons from each eye project to the opposite side of the brain) have indicated that each retinal ganglion axon sends its impulse to one specific site (a cell or small group of cells) within the optic tectum (**FIGURE 15.34A**; Sperry 1951). There are two optic tecta in the frog brain. The axons from the right eye form synapses in the left optic tectum, and those from the left eye form synapses in the right optic tectum.

The map of retinal connections to the frog optic tectum (the **retinotectal projection**) was detailed by Marcus Jacobson (1967). Jacobson created this map by shining a narrow beam of light on a small, limited region of the retina and noting, by means of a recording electrode in the tectum, which tectal cells were being stimulated. The retinotectal projection of *Xenopus laevis* is shown in **FIGURE 15.34B**. Light illuminating the ventral part of the retina stimulates cells on the lateral surface of the tectum. Similarly, light focused on the temporal (posterior) part of the retina stimulates cells in the caudal portion of the tectum. These studies demonstrate a point-for-point correspondence between the cells of the retina and the cells of the tectum. When a group of retinal cells is activated, a

(A)

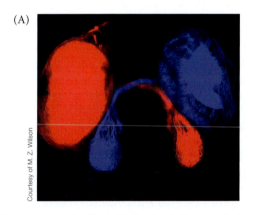

(B)

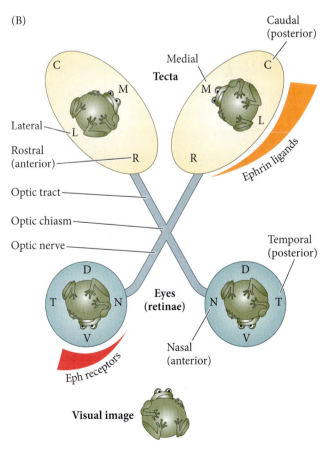

FIGURE 15.34 Retinotectal projections. (A) Confocal micrograph of axons entering the tecta of a 5-day zebrafish embryo. Fluorescent dyes were injected into the eyes of zebrafish embryos mounted in agarose. The dyes diffused down the axons and into each tectum, showing the retinal axons from the right eye going to the left tectum and vice versa. (B) Map of the normal retinotectal projection in adult *Xenopus*. The right eye innervates the left tectum, and the left eye innervates the right tectum. The dorsal (D) portion of the retina innervates the lateral (L) regions of the tectum. The nasal (anterior) region of the retina projects to the caudal (C) region of the tectum. (After K. G. Johnson et al. 2001. In *eLS*. doi:10.1038/npg.els.0000789.)

very small and specific group of tectal cells is stimulated. Furthermore, the points form a continuum; in other words, adjacent points on the retina project onto adjacent points on the tectum. This arrangement enables the frog to see an unbroken image. This intricate specificity caused Sperry (1965) to put forward the **chemoaffinity hypothesis**:

> *The complicated nerve fiber circuits of the brain grow, assemble, and organize themselves through the use of intricate chemical codes under genetic control. Early in development, the nerve cells, numbering in the millions, acquire and retain thereafter, individual identification tags, chemical in nature, by which they can be distinguished and recognized from one another.*

Current theories do not propose a point-for-point specificity between each axon and the neuron that it contacts. Rather, evidence now demonstrates that gradients of adhesivity (especially those involving repulsion) play a role in defining the territories that the axons enter, and that activity-driven competition between these neurons determines the final connection of each axon.

FURTHER DEVELOPMENT

Adhesive specificities in different regions of the optic tectum: Ephrins and Ephs

There is good evidence that retinal ganglion cells can distinguish between regions of the optic tectum. Cells taken from the ventral half of the chick neural retina preferentially adhere to dorsal (medial) halves of the tectum and vice versa (Gottlieb et al. 1976; Roth and Marchase 1976; Halfter et al. 1981). Retinal ganglion cells are specified along the dorsal-ventral axis by a gradient of transcription factors. Dorsal retinal cells are characterized by high concentrations of Tbx5 expression, whereas ventral cells have high levels of Pax2 (Koshiba-Takeuchi et al. 2000). Thus, retinal ganglion cells are specified according to their location.

One gradient that has been identified and functionally characterized is a gradient of repulsion, which is highest in the posterior tectum and weakest in the anterior tectum. Tectal "carpet" experiments by Bonhoeffer and colleagues with alternating "stripes" of membrane derived from the posterior and the anterior tecta demonstrated that the nasal ganglion cells extended axons equally well on both the anterior and the posterior tectal membranes. The neurons from the temporal side of the retina, however, extended axons only on the anterior tectal membranes (Walter et al. 1987; Baier and Bonhoeffer 1992). When the growth cone of a temporal retinal ganglion axon contacted a posterior tectal cell membrane, the growth cone's filopodia withdrew, and the cone collapsed and retracted (Cox et al. 1990).

The basis for this axial specificity by cells in the retina for the spatial recognition of tectal domains appears to be controlled by two sets of gradients along the tectum and retina. The first gradient set consists of ephrin proteins and their receptors. In the optic tectum, ephrin proteins (especially ephrins A2 and A5) are found in gradients that are highest in the posterior (caudal) tectum and decline anteriorly (rostrally) (**FIGURE 15.35A**). Moreover, cloned ephrin proteins have the ability to repulse axons, and ectopically expressed ephrin will prohibit axons from the temporal (but not from the nasal) regions of the retina from projecting to where it is expressed (Drescher et al. 1995; Nakamoto et al. 1996). The complementary Eph receptors have been found on chick retinal ganglion cells, expressed in a temporal-to-nasal gradient along the retinal ganglion axons (see Figure 15.35A; Cheng et al. 1995).

Ephrins appear to be remarkably pliable molecules. Concentration differences in ephrin A in the tectum can account for the smooth topographic map (wherein the position of neurons in the retina maps continuously onto the targets). Hansen and colleagues (2004) have shown that ephrin A can be an attractive as well as

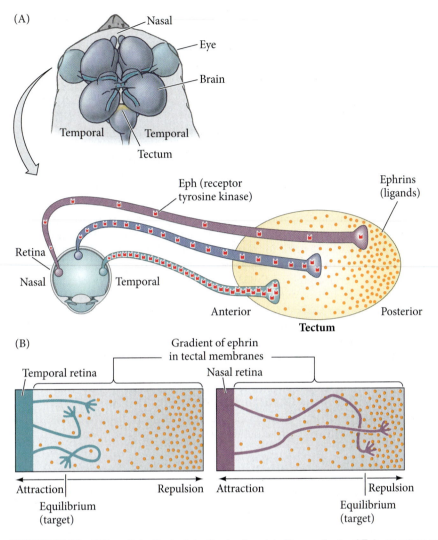

FIGURE 15.35 Differential retinotectal adhesion is guided by gradients of Eph receptors and their ligands. (A) Representation of the dual gradients of Eph receptor tyrosine kinase in the retina and its ligands (ephrin A2 and A5) in the tectum. (B) Experiment showing that temporal, but not nasal, retinal ganglion axons respond to a gradient of ephrin ligand in tectal membranes by turning away or slowing down. An equilibrium of attractive and repulsive forces inherent in the gradient may lead specific axons to their targets. (After M. Barinaga. 1995. *Science* 269: 1668–1670; M. J. Hansen et al. 2004. *Neuron* 42: 717–730.)

a repulsive signal for retinal axons. Moreover, their quantitative assay for axon growth showed that the origin of the axon determined whether it was attracted or repulsed by ephrins. Axon growth is promoted by low ephrin A concentrations anterior to the proper target and is inhibited by higher concentrations posterior to the correct target (**FIGURE 15.35B**). Each axon is thus led to the appropriate place and then told to go no farther. At that equilibrium point, there would be no growth and no inhibition, and the synapses with the target tectal neurons could be made.

To further refine this retinotectal mapping, a second set of gradients exist that parallel the ephrins and Ephs. The tectum has a gradient of Wnt3 that is highest at the medial region and lowest laterally (like the ephrin gradient). In the retina, a gradient of Wnt receptor is highest ventrally (like the Eph proteins). The two sets of gradients are both required to specify the coordinates of axon-to-tectum targets (Schmitt et al. 2006).

? Developing Questions

There is significant synaptic plasticity in the brain over one's life, and problems with forming new synapses may underlie many disorders, such as autism spectrum disorder. What role do the guidance and target-specifying cues play in synapse remodeling later in life?

Synapse Formation

When an axon contacts its target (usually either a muscle cell or another neuron), it forms a specialized junction called a **synapse**. The axon terminal of the **presynaptic neuron** (i.e., the neuron transmitting the signal) releases chemical neurotransmitters that depolarize or hyperpolarize the membrane of the target cell (the **postsynaptic cell**). The neurotransmitters are released into the synaptic cleft between the two cells, where they bind to receptors in the target cell.

The construction of a synapse involves several steps (Burden 1998). When motor neurons in the spinal cord extend axons to muscles, the growth cones that contact newly formed muscle cells migrate over their surfaces. When a growth cone first adheres to the cell membrane of a muscle fiber, no specializations can be seen in either membrane. However, the axon terminal soon begins to accumulate neurotransmitter-containing synaptic vesicles, the membranes of both cells thicken at the region of contact, and the synaptic cleft between the cells fills with an extracellular matrix that includes a specific form of laminin (**FIGURE 15.36A–C**). This muscle-derived laminin specifically binds the growth cones of motor neurons and may act as a "stop signal" for axonal growth (Martin et al. 1995; Noakes et al. 1995). In at least some neuron-to-neuron synapses, the synapse is stabilized by N-cadherin. The activity of the synapse releases N-cadherin from storage vesicles in the growth cone (Tanaka et al. 2000).

In muscles, after the first axon makes contact, growth cones from other axons converge at the site to form additional synapses. During mammalian development, all muscle cells that have been studied are innervated by at least two axons. This *polyneuronal innervation* is transient, however; soon after birth, all but one of the axon branches are retracted (**FIGURE 15.36D–F**). This "address selection" is based on competition among the axons (Purves and Lichtman 1980; Thompson 1983; Colman et al. 1997). When one of the motor neurons is active, it suppresses the synapses of the other neurons, possibly through a nitric oxide-dependent mechanism (Dan and Poo 1992; Wang et al. 1995). Eventually, the less active synapses are eliminated. The remaining axon terminal expands and is ensheathed by a Schwann cell (see Figure 15.36E). (See **Further Development 15.24, A Program of Cell Death**, and **Further Development 15.25, The Uses of Apoptosis**, both online.)

FURTHER DEVELOPMENT

Activity-dependent neuronal survival

The apoptotic death of a neuron is not caused by any obvious defect in the neuron itself. Indeed, before dying, these neurons have differentiated and successfully extended axons to their targets. Rather, it appears that the target tissue regulates the number of axons innervating it by selectively supporting the survival of certain synapses. A recent study of motor neuron pathfinding in the limb exemplifies how neuronal survival is dependent on neuron activity (Hua et al. 2013). When manipulations of guidance systems cause motor axons to be misrouted to the incorrect muscle cells in the limb, the motor neurons survive despite the errors in target selection because they form successful synapses. In contrast, when motor neurons are unable to find their target muscle cells in the limb of mice in which the gene for Frizzled-3 (a Wnt/PCP receptor) is knocked out, the neurons never form synapses and consequently undergo apoptosis (**FIGURE 15.37**; Hua et al. 2013). These results strongly support that neuronal survival is dependent upon successful synapse formation and is also dependent upon the target cell (muscle in this case) supplying a signal to the presynaptic cell that promotes the neuron's survival. *What is this survival signal?* (See **Further Development 15.26, Differential Survival after Innervation: The Role of Neurotrophins**, and **Further Development 15.27, The Development of Behaviors: Constancy and Plasticity**, both online.)

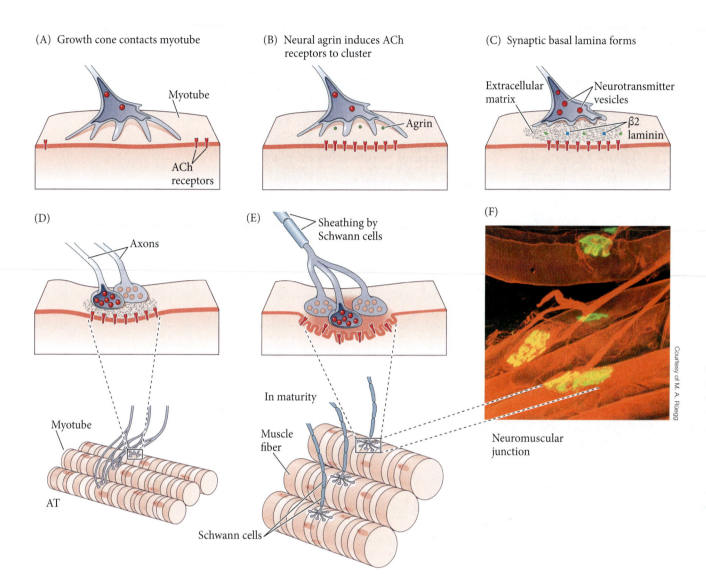

(A) Growth cone contacts myotube

(B) Neural agrin induces ACh receptors to cluster

(C) Synaptic basal lamina forms

Myotube

ACh receptors

Agrin

Extracellular matrix

Neurotransmitter vesicles

β2 laminin

(D)

Axons

Myotube

AT

(E)

Sheathing by Schwann cells

In maturity

Muscle fiber

Schwann cells

(F)

Neuromuscular junction

Courtesy of M. A. Ruegg

FIGURE 15.36 Differentiation of a motor neuron synapse with a muscle in mammals. (A) A growth cone approaches a developing muscle cell. (B) The axon stops and forms an unspecialized contact on the muscle surface. Agrin, a protein released by the neuron, causes acetylcholine (ACh) receptors to cluster near the axon. (C) Neurotransmitter vesicles enter the axon terminal (AT), and an extracellular matrix connects the axon terminal to the muscle cell as the synapse widens. This matrix contains a nerve-specific laminin. (D) Other axons converge on the same synaptic site. The wider view (below) shows single muscle cells innervated by several axons (seen in mammals at birth). (E) All but one of the axons are eliminated. The remaining axon can branch to form a complex neuromuscular junction with the muscle fiber. Each axon terminal is sheathed by a Schwann cell process, and folds form in the muscle cell membrane. The overview shows muscle innervation several weeks after birth. (F) Whole mount view of mature neuromuscular junction in a mouse. (A–E after Z. W. Hall. 1995. *Science* 269: 362–363, based on Z. W. Hall and J. R. Sanes. 1993. *Cell* 72 [Suppl.]: 99–121; D. Purves 1994. *Neural Activity and the Growth of the Brain.* Cambridge University Press: New York.)

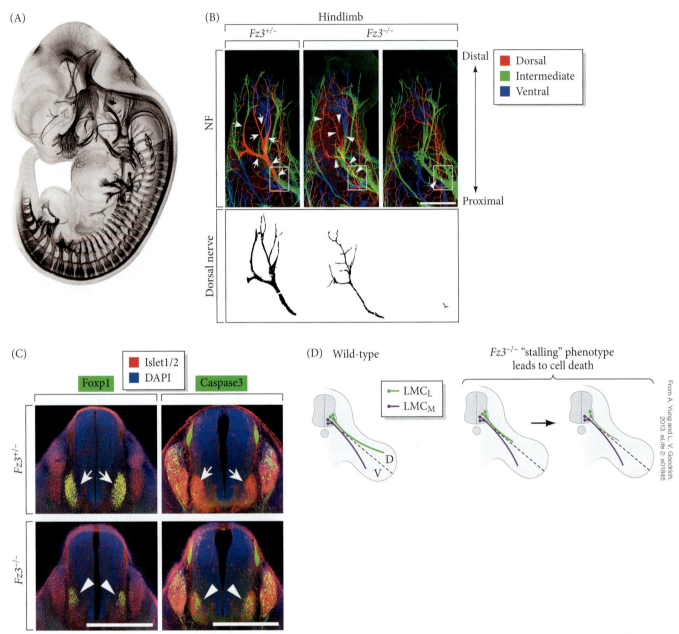

(A)

(B) Hindlimb

$Fz3^{+/-}$ $Fz3^{-/-}$

NF

Dorsal nerve

Distal

■ Dorsal
■ Intermediate
■ Ventral

Proximal

(C)

■ Islet1/2
■ DAPI

Foxp1   Caspase3

$Fz3^{+/-}$

$Fz3^{-/-}$

A–C from Z. L. Hua et al. 2013. *eLife* 2: e01482/CC BY 3.0

(D) Wild-type $Fz3^{-/-}$ "stalling" phenotype leads to cell death

● LMC$_L$
● LMC$_M$

D
V

From A. Yung and L. V. Goodrich. 2013. *eLife* 2: e01845

FIGURE 15.37 Analysis of motor neuron axon stalling and cell death in *Frizzled-3* knockout mice. (A) Whole mount immunocytochemistry for the protein Neurofilament, which labels all axons in the embryo. (B) Close examination of the axonal projections into the hindlimb are visualized with the neurofilament antibody (NF), and axons at different depths along the dorsal-ventral axis of this limb have been pseudocolored differently. *Fz3* heterozygous mice show normal dorsal nerve projections, whereas *Fz3* homozygous knockout mice exhibit varying degrees of dorsal nerve loss (two examples shown here). The axon trajectories of the LMC$_L$ motor neurons (dorsal nerve) are extracted and shown below each image to emphasize the reduction of axonal projections distal to the plexus. (C) Transverse sections of the spinal cord labeled for discrete populations of motor neurons and any cells undergoing apoptosis, as indicated by Caspase3 labeling (green, in right-hand photos). *Frizzled-3* knockout mice (lower images) show reductions in the motor neuron specification markers Islet1/2 (red) and Foxp1 (green, in left-hand photos), along with increased cell death specifically in the motor columns. (D) Schematics describe the phenotypes associated with a loss of *Frizzled-3/PCP* signaling. A stalling defect in motor neurons destined for the dorsal limb is followed by cell death.

Closing Thoughts on the Opening Photo

This image shows the cranial neural crest cells populating the pharyngeal arches of a 42-hour zebrafish embryo expressing GFP driven by the *fli1a* promoter. It is a lateral view with anterior to the left. The first two major streams contribute neural crest cells to the major jaw cartilages, and the more posterior streams contribute to the "branchial arches" and gill structures. Cranial neural crest cells play a major role in building the craniofacial skeleton—the "face"—in fish and humans alike. We have only recently come to understand that cranial neural crest cells operate through the collective migration mechanism described in this chapter. The pattern of cranial neural crest-built pharyngeal arches may appear complicated, yet there is a way you can grasp these structures—literally. This image was collected with a laser-scanning confocal microscope and, as such, possesses three-dimensional data. You can go to the National Institutes of Health three-dimensional print exchange website (http://3dprint.nih.gov/discover/3dpx-001506), download a file of the pharyngeal arches of this 42-hour-postfertilization transgenic zebrafish embryo [*tg(fli1a:EGFP)*], and use this file to print a three-dimensional model that you can hold in your hand. (Image and three-dimensional modeling generated and provided by the Barresi lab; Barresi et al. 2015.)

© Barresi et al. 2015

Endnotes

[1] The pharyngeal (branchial) arches (see Figure 1.12A) are outpocketings of the head and neck region into which cranial neural crest cells migrate. The pharyngeal pouches form between these arches and become the thyroid, parathyroid, and thymus.

[2] In contrast to crest migration in the chick anterior somite, trunk neural crest cells migrate through the medial region of each somite in zebrafish and the posterior region in *Xenopus*.

[3] Recall from Chapter 13 that ganglia are clusters of neurons whose axons form a nerve.

[4] To learn the role that neural crest plays in tooth, hair, and cranial nerve development, seek out Chapter 16.

[5] We introduced Brainbow-based fate mapping in Chapter 2 and described the utility of the "confetti" mouse model for lineage tracing in Chapter 5.

[6] The sclerotome is the portion of the somite that gives rise to the cartilage of the spine. In the migration of fish neural crest cells, the sclerotome is less important; rather, the myotome appears to guide the migration of the crest cells ventrally (Morin-Kensicki and Eisen 1997).

[7] *MITF* stands for *microphthalmia-associated transcription factor*, so named because one result of mutations of the gene, as described in mice, is small eyes (i.e., microphthalmia). The effects of MITF are widespread, however, as we will see in this chapter.

[8] We have seen similar phenomena before, as in the cells of the chick neural folds (some of which probably become neural crest cells) and limb buds, early zebrafish blastomeres, and the extensions of sea urchin micromeres.

[9] These experiments were done using the Cre-lox technique. The mice were heterozygous for both (1) a β-galactosidase allele that could be expressed only when Cre-recombinase was activated in that cell, and (2) a Cre-recombinase allele fused to a *Wnt1* gene promoter. Thus, the gene for β-galactosidase was activated (blue stain) only in those cells expressing Wnt1, a protein that is activated in the cranial neural crest and certain brain cells.

[10] The growth cones of pioneer neurons migrate to their target tissue while embryonic distances are still short and the intervening embryonic tissue is still relatively uncomplicated. Later in development, other neurons bind to pioneer neurons and thereby enter the target tissue. Klose and Bentley (1989) have shown that in some cases, pioneer neurons die after the "follow-up" neurons reach their destination. Yet if the pioneer neurons are prevented from differentiating, the other axons do not reach their target tissue.

[11] Although historically Roundabout1 in *Drosophila* has been referred to simply as Robo (without the number 1), to avoid confusion we call it Robo1 throughout this text.

[12] Axon pathways in the central nervous system are called *tracts*, whereas axon pathways in the peripheral nervous system are called *nerves*.

[13] As is seen throughout developmental biology, the metaphor of a "target" is problematic. Here, the target is not a passive entity, but an importantly active one.

[14] There is some confusion between the terms *neurotropic* and *neurotrophic*. Neurotropic (Latin, *tropicus*, "a turning movement") means that something attracts the neuron. Neurotrophic (Greek, *trophikos*, "nursing" or "nourishing") refers to a factor's ability to keep the neuron alive, usually by supplying growth factors. Because many agents have both properties, they are alternatively called *neurotropins* and *neurotrophins*. In the recent literature, *neurotrophin* appears to be more widely used.

(15) ## Snapshot Summary
Neural Crest Cells and Axonal Specificity

1. The neural crest is a transitory structure. Its cells migrate to become numerous different cell types. The path a neural crest cell takes depends on the extracellular environment it meets.

2. Trunk neural crest cells can migrate dorsolaterally to become melanocytes and dorsal root ganglia cells. They can also migrate ventrally to become sympathetic and parasympathetic neurons and adrenomedullary cells.

(Continued)

Snapshot Summary *(continued)*

3. Cranial neural crest cells enter the pharyngeal arches to become the cartilage of the jaw and the bones of the middle ear. They also form the bones of the frontonasal process, the papillae of the teeth (including odontoblasts, dentin forming cells), and cranial nerves.

4. Cardiac neural crest cells enter the heart and form the septum (separating wall) between the pulmonary artery and the aorta.

5. The formation of the neural crest depends on interactions between the prospective epidermis and the neural plate. Paracrine factors from these regions induce the formation of transcription factors that enable neural crest cells to emigrate.

6. Collective migration of neural crest cells is powered by contact inhibition of locomotion and a co-attraction to leading cells.

7. Trunk neural crest cells will migrate through the anterior portion of each sclerotome, but not through the posterior portion of a sclerotome. Semaphorin and ephrin proteins expressed in the posterior portion of each sclerotome can prevent neural crest cell migration.

8. Some neural crest cells appear to be capable of forming a large repertoire of cell types. Other neural crest cells may be restricted even before they migrate. The final destination of the neural crest cell can sometimes change its specification.

9. The fates of the cranial neural crest cells are influenced by Hox genes. They can acquire their Hox gene expression pattern through interaction with neighboring cells.

10. Motor neurons are specified according to their position in the neural tube. The Lim family of transcription factors plays an important role in this specification before their axons have even extended into the periphery.

11. The growth cone is the locomotory organelle of the neuron and rearranges its cytoskeletal architecture in response to environmental cues. Axons can find their targets without neuronal activity.

12. Some proteins are generally permissive to neuron adhesion and provide substrates on which axons can migrate. Other substances prohibit migration.

13. Some growth cones recognize molecules that are present in very specific areas and are guided by these molecules to their respective targets.

14. Some neurons are "kept in line" by repulsive molecules. If the neurons wander off the path to their target, these molecules send them back. Some molecules, such as the semaphorins and Slits, are selectively repulsive to particular sets of neurons.

15. Some neurons sense gradients of a protein and are brought to their target by following these gradients. The netrins and Shh may work in this fashion.

16. Changes in growth cone responsiveness to the attractive and repulsive cues secreted from the midline enable commissural axons to cross the midline and connect to two sides of the central nervous system.

17. Target selection can be brought about by neurotrophins, proteins that are made by the target tissue and that stimulate the particular set of axons able to innervate it. In some cases, the target makes only enough of these factors to support a single axon.

18. Retinal ganglion cells in frogs and chicks send axons that bind to specific regions of the optic tectum. This process is mediated by numerous interactions, and target selection appears to be mediated through ephrins.

19. Synapse formation has an activity-dependent component. An active neuron can suppress synapse formation by other neurons on the same target.

20. Lack of synapse formation and neuronal activity can lead to the induction of programmed cell death, or apoptosis, which unleashes a cascade of caspase enzymes that result in cell death.

Go to oup.com/he/barresi12xe for Further Developments, Scientists Speak interviews, Watch Development videos, Dev Tutorials, and complete bibliographic information for all literature cited in this chapter.

Ectodermal Placodes and the Epidermis

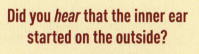

Did you *hear* that the inner ear started on the outside?

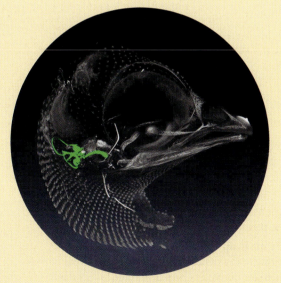

A. Kumar et al. 2018. *Birth Defects Research* 110: 1194–1204, courtesy of Nobue Itasaki

IRONICALLY, THE BUILDING OF SOME OF OUR SENSES depends on the way different ectodermal cells "sense" one another, using touch and other molecular means of communication to instruct the development of complex and highly integrated sensory organ systems. Our organs of sight, hearing, and smell, and even the hairs on our skin started as simple thickenings in the surface ectoderm called **placodes**. These placodes, as well as the epidermis of the skin, are formed from the non-neural ectoderm that remains external after the neural ectoderm has moved inward to form the central and peripheral nervous systems. Non-neural ectodermal placodes are born through interactions with surrounding cells and tissues of the endoderm, mesoderm, and neural plate, and the types of interactions differ depending on the placode. In the head, **sensory placodes** give rise to the olfactory epithelium of the nasal organs, the entire inner ear responsible for sound and balance perception, the lens of the eye, and the distal parts of the cranial sensory ganglia (Streit 2007, 2008; Steventon et al. 2014; Saint-Jeannet and Moody 2014; Schlosser 2010, 2014; Moody and LaMantia 2015). Non-sensory placodes produce the oral epithelium for tooth development and cutaneous structures throughout the body, such as hair, feathers, and mammary and sweat glands (Pispa and Thesleff 2003). In this chapter, we will be highlighting the importance of reciprocal interactions in placode differentiation and the similar signaling pathways used to coordinate placode development, from head to turtle shell.

Cranial Placodes: The Senses of Our Heads

The vertebrate head has a unique concentration of sensory organs—organs that are more than just the neurons they contain. The eyes, nose, ears, and taste buds are all in the head. The head also has its own highly integrated nervous system for sensing pain (think of the trigeminal nerve innervating your teeth) and temperature (think of the receptors on your lips and tongue). The elements of this nervous system arise from the **cranial sensory placodes**—local and transient thickenings of the ectoderm in the head and neck regions of the embryo. The precursors to the cranial sensory placodes originate in the early blastoderm as an intermingled population of progenitor cells for the central nervous system (CNS), placodes, neural crest, and epidermis. These progenitor cells become progressively specified and sorted out over the course of gastrulation (**FIGURE 16.1**). With some contributions from the cranial neural crest, the cranial placodes generate most of the sensory neurons of the head associated with

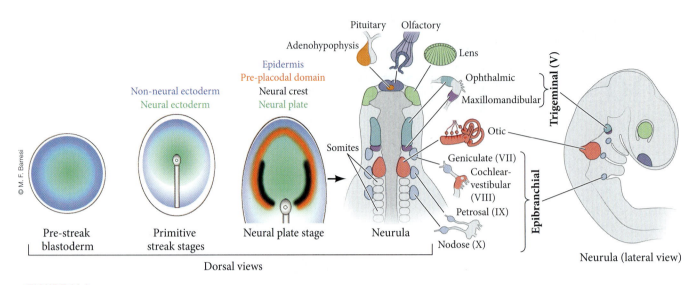

Dorsal views

Neurula (lateral view)

FIGURE 16.1 Cranial sensory placodes contribute to the sense organs and cranial ganglia. Diagrams show a generalized amniote embryo from blastoderm to neural tube stages. At the blastula and early gastrula stages, placode progenitors are mixed with future neural plate and neural crest cells at the border of the presumptive neural plate and surface ectoderm, where neural (blue) and non-neural (light blue) ectoderm marker gene expression initially overlap (green-to-blue region; blastoderm) and then become distinct (primitive streak). Placodes are first specified during the neural plate stage in the pre-placodal region (green), which is distinct from the premigratory neural crest domain (red), and then separate into individual placodes by the neural tube stage. The derived systems are designated in various colors in the illustrated dorsal view of the neural tube stage. Lateral view of a later neural tube stage embryo further illustrates the positions of these placodes. (After A. Streit. 2004. *Dev Biol* 276: 1–15, based on A. D'amico-Martel and D. M. Noden. 1983. *Am J Anat* 166: 445–468; S. Singh and A. K. Groves. 2016. *WIREs Dev Biol* 5: 363–376.)

hearing, balance, smell, taste, pain, temperature and touch sensation, and even blood pressure; the cranial neural crest contributes all of the glia and proximal portions of the sensory ganglia (see Singh and Groves 2016).

The anterior cranial placodes are:

- The adenohypophyseal placode first develops into a structure known as Rathke's pouch before differentiating into the anterior lobe of the pituitary gland.

- The **lens placode** invaginates to form the lens vesicle, which later forms the lens of the eye. The adenohypophyseal and lens placodes are the only cranial placodes that do not generate sensory neurons—all other placodes in the head do.

- The **olfactory placode** gives rise to the sensory neurons involved in smell (1st cranial nerve—the olfactory nerve), as well as to a variety of migratory neurons that will travel into the brain, including, but not limited to, those that will secrete gonadotropin-releasing hormone.

The intermediate cranial placodes are:

- The **trigeminal placode** is subdivided into the ophthalmic and maxillomandibular placodes and generates the distal neurons of the trigeminal ganglion, whereas the proximal neurons of this ganglion are formed from neural crest cells (Baker and Bronner-Fraser 2001; Hamburger 1961). The trigeminal (5th cranial) nerve functions to support the sensations of touch, temperature, and pain.

The posterior cranial placodes are:

- The **otic placode** gives rise to the sensory epithelium of the inner ear and to neurons that form the cochlear-vestibular ganglion (8th cranial nerve), which are important for transmitting sound and balance information to the brain.

- In anamniotes (fish and amphibians), the **lateral line placodes** generate neuromasts containing mechanosensing hair cells and their innervating neurons; they lie superficially along the trunk to detect water flow.

- The **epibranchial placodes** are made up of the **geniculate**, **petrosal**, and **nodose placodes**, which give rise to the distal parts of different cranial nerves (7th, 9th, and 10th). The geniculate-derived nerves innervate taste buds, tonsils, and ear lobes; the petrosal-derived nerves innervate the tongue and carotid sinus and body; the nodose-derived vagus nerve innervates many other organs of the body, such as the heart, lungs, and gastrointestinal tract.

(See **Further Development 16.1, The Human Cranial Nerves**, and **Further Development 16.2, Kallmann Syndrome**, both online.)

Cranial placode induction

Detailed fate mapping studies during the neurula stages have shown that all the cranial placodal precursors are located in a horseshoe-shaped domain (the pre-placodal region) that surrounds the anterior neural plate and cranial neural folds (see Figure 16.1; Kozlowski et al. 1997; Streit 2002; Bhattacharyya et al. 2004; Xu et al. 2008; Pieper et al. 2011). These progenitors are induced by inductive signals arising from the head mesoderm and endoderm, with some contribution of neural plate signals (**FIGURE 16.2**; Platt 1896; Brugmann et al. 2004; Schlosser and Ahrens 2004; Litsiou et al. 2005; Schlosser 2005; Streit 2018). Jacobson (1963) showed that the presumptive placodal cells adjacent to the anterior neural plate are competent to give rise to any placode in amphibians. More recently these studies have been extended to amniotes, showing that at neural plate stages, the pre-placodal region is the only territory competent to respond to the placode-inducing signal. Induction of the pre-placodal region is largely influenced by signaling coming from the underlying tissues. This involves Wnt and Bmp and repression of these signals by Fgf and antagonists to Wnt and Bmp (possibly Cerberus) (see Figure 16.2A,B). This signaling system first leads to the gradual separation of the neural plate from the more laterally positioned non-neural epidermis. Fgf and Wnt/Bmp antagonists from the underlying head mesoderm induce pre-placodal specification by repressing Wnt and Bmp activity from the surrounding ectodermal regions (Streit 2007; Nakajima 2015; Singh and Groves 2016; Schlosser 2017). In addition, Sonic hedgehog and neuropeptide signaling, as well as retinoic acid, influence the anterior-posterior and lateral boundaries of the pre-placodal fields, respectively (Kondoh et al. 2000; Janesick et al. 2012; Hintze et al. 2017; Streit 2018).

FIGURE 16.2 Specification of the pre-placodal region. (A) A simplified relationship of the key paracrine signals that specify neural, neural crest (NC), pre-placodal region (PPR), and future epidermal (non-neural) lineages. Generally, an antagonism between Wnt/Bmp and FgfGF signaling specifies different ectodermal fates. (B) Dorsal and cross-sectional view of an early neural plate stage embryo. From most medial to most lateral, the neural plate (NP), neural crest (NC), pre-placodal region (PPR), and epidermal ectoderm (Epi) are shown. Wnt and Bmp signals from both the dorsal neural plate and lateralmost ectoderm induce neural crest cells while repressing the PPR fate. In contrast, Fgfs and Wnt/Bmp antagonists from the underlying mesoderm inhibit Bmp/Wnt signaling locally, and the overlying ectoderm is specified as PPR. (C) Placodes along the anterior-posterior axis develop under the influence of local signaling activities, including Bmps, Wnts, and Fgfs, as well as other pathways, such as Shh and Pdgf. These signals induce a unique combination of transcription factors that is characteristic for each placode (key transcription factors are noted). (A and C, after S. Singh and A. K. Groves. 2016. *WIREs Dev Biol* 5: 363–376; B after Y. Nakajima. 2015. *Congenit Anom* 55: 17–25.)

(A)

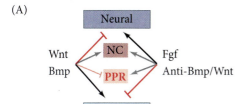

(B)

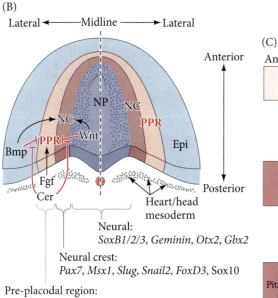

(C)

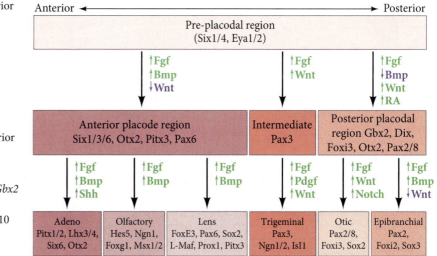

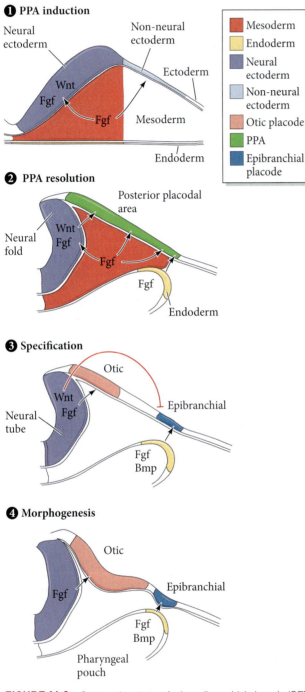

❶ PPA induction

Neural ectoderm

Non-neural ectoderm

Ectoderm

Wnt
Fgf

Fgf

Mesoderm

Endoderm

■	Mesoderm
■	Endoderm
■	Neural ectoderm
■	Non-neural ectoderm
■	Otic placode
■	PPA
■	Epibranchial placode

❷ PPA resolution

Posterior placodal area

Neural fold

Wnt
Fgf

Fgf

Fgf

Endoderm

❸ Specification

Otic

Neural tube

Wnt
Fgf

Epibranchial

Fgf
Bmp

❹ Morphogenesis

Otic

Fgf

Epibranchial

Fgf
Bmp

Pharyngeal pouch

FIGURE 16.3 Successive steps of otic-epibranchial placode (OEP) induction. (1,2) The Fgf pathway activates Wnt and Fgfs in the neural ectoderm (light blue). (1) First, cells in the posterior pre-placodal region of the non-neural ectoderm are induced to become otic-epibranchial progenitors by mesoderm-derived Fgf signaling (PPA, posterior placodal area). (2) Meanwhile, additional Fgfs are supplied by the pharyngeal endoderm (yellow). (3,4) OEPs become subdivided into separate placodes. (3) Wnt signaling promotes otic (pink) but represses epibranchial (dark blue) fate. (4) Meanwhile, continued Fgf signaling activates the epibranchial program, and Bmps from the lateral endoderm initiate epibranchial neurogenesis. (After R. K. Ladher et al. 2010. *Development* 137: 1777–1785.)

The above paracrine signaling systems lead to the induction of specific gene regulatory networks throughout the pre-placodal region, which begins with the expression of the Six1/4, and Eya1/2 transcription factors (see Figure 16.2C). These proteins are maintained in all the placodes and are downregulated in the interplacodal regions as the pre-placodal region separates into discrete placodes (Streit 2002; Bhattacharyya et al. 2004; Schlosser and Ahrens 2004; Xu et al. 2006; Breau and Schneider-Maunoury 2014; Singh and Groves 2016). Different sets of paracrine factors now induce each discrete placode toward its respective fate, such that each placode expresses its own unique set of transcription factors (Groves and LaBonne 2014; Moody and LaMantia 2015; Chen et al. 2017).

To exemplify the tissue interactions and gene regulatory networks at play during placode development, we will focus on (1) the shared emergence and separation of the otic and epibranchial placodes, and (2) the morphogenesis of lens formation.

Otic-epibranchial development: A shared experience

The sense of hearing and balance is accomplished through the transformation of mechanical information into electrical stimuli by specialized cells in the inner ear, the sensory hair cells. These are located in the sensory patches at the base of the semicircular canals responsible for the perception of balance and acceleration, and in the cochlea or basilar papilla to mediate sound perception. In the adult amniote, sound waves in the air are first caught by the unique folds of the outer ear and channeled to the tympanic membrane, or ear drum, causing it to vibrate. These vibrations are amplified by the movement of the three bones of the middle ear (the smallest bones of the body) and transferred as waves to the fluid in the cochlea of the inner ear. The mammalian **cochlea** is a remarkable coiled tube that possesses three separate chambers.[1] The middle fluid-filled chamber contains the **Organ of Corti** (known as the basilar papilla in birds), which houses the sensory **hair cells** that transform the movement of the fluid into electrical signals, which are then transmitted to the brain through the auditory nerve. Along the length of the cochlea, hair cells differ in their morphology and physiological properties and are tuned to different frequencies to allow sensing of a full range of frequencies—a range that varies among different species. This tonotopic organization is maintained by the neurons innervating the hair cells and by the entire auditory pathway in the central nervous system.

The inner ear is a most remarkable hearing machine, with a long history of evolutionary adaptions. How do you build an organ with such complex morphology—specifying fifty different cell types and coordinating their morphogenesis—all from a simple placode?

OTIC AND EPIBRANCHIAL INDUCTION The same signals that specify placode progenitors are again at play in initiating otic and epibranchial placode development. First, Fgf signaling from the underlying head mesoderm induces otic-epibranchial precursors to form from the posterior pre-placodal region. This is soon reinforced by Fgf coming from the pharyngeal endoderm and the neural plate (**FIGURE 16.3**). Wnt signaling, also emanating from the neural plate, functions to promote otic identity while repressing epibranchial development. In the otic placode, the Notch pathway potentiates Wnt activity, thus promoting otic fate. Later, upregulation of Bmp signaling from the pharyngeal endoderm supports the specification of epibranchial neurons. Thus, Fgf signaling induces progenitors for both placodes, while Wnt signaling is critical to delineate otic from epibranchial placode identities (Ladher et al. 2010; Nakajima 2015; Sai and Ladher 2015; Ladher 2017).

OTIC MORPHOGENESIS Soon after the otic placode begins to thicken, changes in cell shape along the apical-basal axis initiate the placode's invagination (**FIGURE 16.4A,B**). First, the basal surfaces of the placode cells expand their area relative to their apical surfaces, which causes an indentation—this denotes the **otic pit** stage. This initial invagination is then amplified by the added force of apical constriction, which drives this epithelium toward the **otic cup** stage. Finally, the otic cup forms the otic vesicle by bringing its edges together and fusing them, and finally separating as a vesicle from the surface ectoderm (Sai and Ladher 2008; Sai and Ladher 2015; Ladher 2016). The mechanisms behind the closure of the otic vesicle are largely unknown, but they are hypothesized to be akin to the cellular processes governing neural tube closure (see Chapter 13).

SENSORY GANGLION BY DELAMINATION Most cranial placodes give rise to sensory neurons that converge into clusters of neuronal cell bodies known as **ganglia** (sing.,

> **? Developing Questions**
>
> Many fundamental questions regarding otic development exist. How do you build different hair cells, sensitive for different frequencies and with physiologically different properties? How are the different hair cell morphologies created and organized tonotopically?

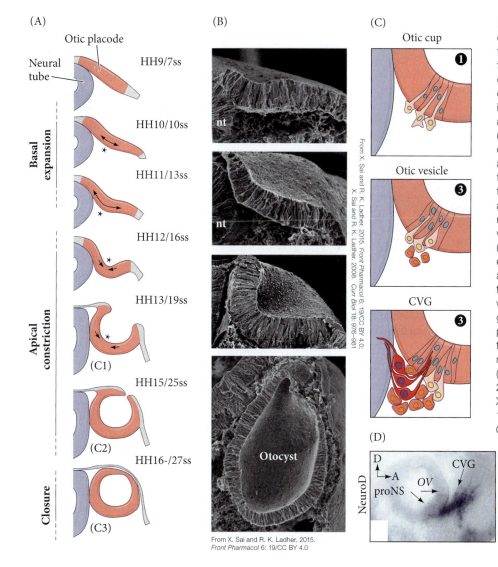

From X. Sai and R. K. Ladher. 2015. *Front Pharmacol* 6: 19/CC BY 4.0.

FIGURE 16.4 Morphogenesis of the otic placode and vesicle. (A,B) Invagination of the pre-placodal ectoderm next to the future hindbrain in the developing chick otic placode region is shown in diagrams (A; ss, somite stage; C1–C3 indicate corresponding diagrams in C) and scanning electron micrographs (B; nt, neural tube; otocyst = otic vesicle). The invagination is first powered by an expansion of the basal surfaces of the epithelial cells, followed by constriction of the apical surfaces (asterisks, arrows). (C) Neural progenitor cells (light red cells with orange nuclei) along the ventromedial portion of the otic cup (1) and vesicle (2) delaminate from the epithelium (pink cells with yellow nuclei) and differentiate into the cochleovestibular ganglion (3, CVG, red neurons with purple nuclei). (D) NeuroD is a gene expression marker (blue) for the proneural-sensory domain (proNS) of the otic vesicle (OV) and for delaminated neuroblast cells migrating into the CVG. (A–C after X. Sai and R. K. Ladher. 2015. *Front Pharmacol* 6: 19/CC BY 4.0; X. Sai and R. K. Ladher. 2008. *Curr Biol* 18: 976–981; R. K. Ladher. 2017. *Semin Cell Dev Biol* 65: 39–46.)

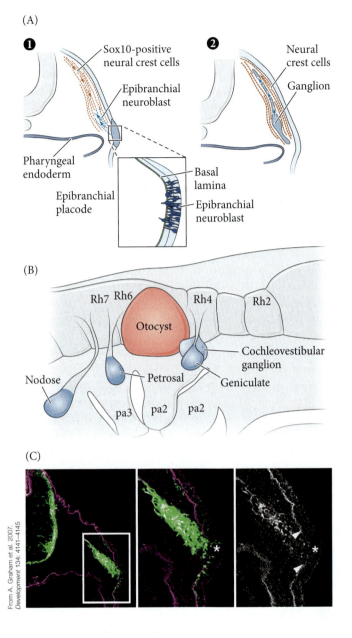

(A)

❶ Sox10-positive neural crest cells

Epibranchial neuroblast

Pharyngeal endoderm

Epibranchial placode

Basal lamina

Epibranchial neuroblast

❷ Neural crest cells

Ganglion

(B)

Rh7 Rh6   Rh4   Rh2

Otocyst

Cochleovestibular ganglion

Nodose   Petrosal   Geniculate

pa3   pa2   pa2

(C)

From A. Graham et al. 2007. *Development* 134: 4141–4145

FIGURE 16.5 Epibranchial sensory neurons delaminate from the placode as neuroblasts. (A) The neural progenitors for epibranchial-derived sensory neurons delaminate from the basal surface of the placode (1). These "epibranchial neuroblasts" migrate dorsally through ventrally streaming neural crest cells. (B) Lateral view of final positions of the otic vesicle (otocyst), cochleovestibular ganglion, and the epibranchial placodes (geniculate, petrosal, and nodose) that give rise to the distal parts of the 7th, 9th, and 10th cranial nerves. (pa, pharyngeal arch; Rh, rhombomere of the neural tube.) (C) Cross section of a 2½-day chick embryo at the epibranchial placode position, immunolabeled for precursor neuroblasts (green) of the cochleovestibular ganglion and for the basal lamina (magenta) with antibodies to neurofilament medium chain and laminin, respectively. The two images on the right are magnified views of the boxed area in the left image. The basal lamina is reduced only at the location where neuroblast delamination is occurring (asterisks). The extent of the breach in the basal lamina is indicated by arrowheads (last image). (A,B after R. K. Ladher et al. 2010. *Development* 137: 1777–1785.)

ganglion), which in turn grow fibers to generate the cranial nerves. Ganglia are generated from both the otic and the epibranchial placodes by cell **delamination**, a process by which epithelial cells lose their tight adhesions and migrate away from their layered epithelium. Such epithelial cells that are located at the ventromedial side of the otic cup and vesicle generate neural progenitors, which delaminate as neuroblasts over a protracted period by a mechanism that does not involve an epithelial-mesenchymal transition (**FIGURE 16.4C,D**). These neural progenitor cells differentiate into the **cochleovestibular ganglion**, which sits adjacent to the otic vesicle (otocyst) and will form the major neural connection between the brain and the inner ear structures derived from the otocyst (Hemond and Morest 1991; Magariños et al. 2012). Similarly, neural progenitor cells also delaminate as neuroblasts from the epibranchial placodes, yet these cells continue to migrate dorsally, guided by neural crest cell tunnels (**FIGURE 16.5**). Manipulations that disrupt neural crest migration result in misplaced epibranchial ganglia, suggesting that neural crest cells provide guidance cues and/or organization to the ganglion (Begbie and Graham 2001; Golding et al. 2004; Osborne et al. 2005; Schwarz et al. 2008; Ladher et al. 2010; Freter et al. 2013; Ladher 2017). Together, both the epibranchial neuroblasts and neural crest cell populations form the epibranchial ganglia and nerves.

FURTHER DEVELOPMENT

FROM OTIC VESICLE TO THE ORGAN OF CORTI   Induction and morphogenesis of the otic placode into the otocyst are merely the initial events that set the stage for the development of the inner ear. To achieve the vestibular (balance and acceleration) and auditory (hearing) functions of the ear, the cells of the otocyst need to be patterned across all axes of the otocyst to give rise to one of the more cell-diverse and complex sensory organs that vertebrates possess. The dorsal portions of the otocyst will generate the three semicircular canals of the vestibular systems, whereas the ventral portions will give rise to the sensory and supportive cells within the cochlear auditory system (**FIGURE 16.6A**). We will focus specifically on the development of the mammalian cochlea to exemplify how some of the core morphogenetic signals that are well known to pattern the overall embryonic axes have been co-opted to pattern the different cell types along the different axes of the inner ear.

ANATOMY OF THE INNER EAR Although the coiled shape of the mammalian cochlea may seem strikingly different from the homologous straight or curved cochlea of lizards and birds, and although the organ of Corti in mammals may seem especially different from the basilar papilla of our aquatic tetrapod ancestors, the development and function of these structures are evolutionarily conserved (Manley 2012; Basch et al. 2016). As mentioned above, the cochlea plays a pivotal role in transforming sound waves into electrical signals due to the presence of sensory hair cells in the organ of Corti (**FIGURE 16.6B**). Three rows of outer and one row of inner sensory hair cells run down the length of the organ of Corti and extend their mechanosensitive microvilli into the fluid-filled chamber of the cochlear duct. These hair cells are surrounded by strategically arranged supportive cells: the inner and outer hair cell groups are separated by the pillar cells, which form the "tunnel of Corti"; the inner hair cells are surrounded by phalangeal and border cells; the outer hair cells sit directly upon Deiters' cells, which extend a specialized phalangeal process to the surface of the cochlear duct (see Figure 16.6B; Basch et al. 2016). How are these different cell types stereotypically organized throughout the cochlea?

AXIS DETERMINATION OF THE OTOCYST Before the hair cells of the cochlea can develop, the axes of the otocyst need to be determined. As is the case with many organogenesis events, the specification of cell fate along the axes of the otic vesicle is determined by the same morphogenetic signals that regulated the cardinal axes of the embryo (**FIGURE 16.6C**). As early as the otic cup stage, it appears that specification of the medial-lateral axis has been influenced by the combined signaling of Wnt and Fgf from the developing hindbrain. This signaling is required for the proper formation of the endolymphatic duct and semicircular canals of the vestibular portions of the inner ear (see Figure 16.6A; Lin et al. 2005; Riccomagno et al. 2005; Brown et al. 2015). The opposing morphogen gradients of Sonic hedgehog from the ventral neural tube (and notochord) and Wnt from the dorsal neural tube are necessary for patterning the otocyst along its dorsal-ventral axis.

FIGURE 16.6 Axis determination in the otic vesicle. (A) Diagrams show the developmental progression of the inner ear in the embryonic mouse from 11.5 days (E11.5) through 17.5 days (E17.5) of development. Areas where hair cells develop are shown in turquoise. The endolymphatic duct (green) is a non-sensory component of the inner ear. (ASC, PSC, and LSC, anterior, posterior, and lateral semicircular canals; SM, saccular macula; UM, utricular macula.) (B) Drawing of a cross-section of the mammalian organ of Corti, showing the cell types that make up this structure. (C) Drawing of the different tissue-derived signals that influence axis determination in the otocyst. The different signals for the three axes of the otocyst are represented by the colors of the arrows within the otocyst. The colors correspond to the wedges above, below, and at left, which indicate the gradients of the signals required for patterning of these axes. (A,B after M. L. Basch et al. 2016. *J Anat* 228: 233–254.)

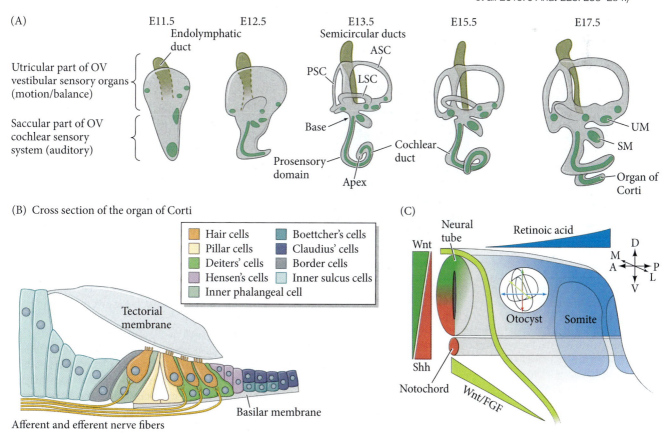

Indeed, loss of sonic hedgehog signaling within the inner ear leads to a loss of the ventrally derived cochlear cell types (Brown and Epstein 2011). Lastly, retinoic acid is important for patterning cell fates along the anterior-posterior axis. Indeed, if retinoic acid is overexpressed or misexpressed anteriorly, posterior cell fates of the otocyst are expanded at the expense of cochlear development (Bok et al. 2011). You will see in Chapter 17 that retinoic acid has a similar role in patterning the anterior-posterior axis of the embryonic trunk. Thus, the early determination of cell fates within the otocyst relies heavily on co-opting existing axial morphogens (see Basch et al. 2016). However, further differentiation of the supportive and prosensory cell types in the organ of Corti within the cochlea seems to use less common and more organ-specific mechanisms—these involve an unusual uncoupling of cell differentiation from the process of exiting the cell cycle and an intricate spatial and temporal regulation of signaling pathways within the developing otocyst. (See **Further Development 16.3, Cell Fate Determination in the Organ of Corti of the Mouse Ear**, online.)

Morphogenesis of the vertebrate eye

The most basic parts of the embryonic vertebrate eye include the retinal pigmented epithelium, the neural retina, and the lens. While the retinal pigmented epithelium and the neural retina arise from the central nervous system, the lens, which starts out as a lens placode, is derived from the non-neural ectoderm. Unlike most other sensory placodes, the lens placode does not form neurons. Rather, it forms the transparent lens that focuses incoming light onto the neural retina. The retina develops from the optic vesicle that forms as a lateral bulge of the diencephalon, a region of the forebrain. The interactions between the cells of the lens placode and the presumptive retina structure the eye via a cascade of reciprocal signaling events that enable the construction of an intricately complex organ.

During gastrulation, the involuting prechordal plate and foregut endoderm interact with the overlying preplacodal region, inducing it to have an anterior character and conferring upon it a lens-forming bias (Saha et al. 1989; Dutta et al. 2005; Hintze et al. 2017). These tissues induce a number of anterior genes, including *Pax6*, which codes for a transcription factor that is critical for the ectoderm to respond to subsequent signals. Interestingly, the entire pre-placodal region is initially specified as lens (Bailey et al. 2006), even cells that normally form the ear or other placodes. Since not all parts of the pre-placodal region eventually form lenses, lens fate must be repressed so that the lens only forms with a precise spatial relationship to the retina. The repression of lens potential is achieved by migrating neural crest cells (von Woellworth 1961; Bailey et al. 2006). These neural crest cells are blocked from making contact with the pre-lens region by the outgrowing optic vesicle. This, along with signals from the **optic vesicle**, positions the lens in relation to the retina.

FIGURE 16.7 shows the development of the vertebrate eye. The optic vesicle, where it contacts the head ectoderm, causes changes in cell shape that create a thickened lens placode. The optic vesicle then bends inward, invaginating to form the two-layered **optic cup**, and, in so doing, draws the developing lens into the forming eye. This invagination is accomplished by three changes: (1) The cells of the lens

(A) 4-mm embryo

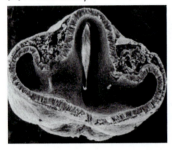

(B) 4.5-mm embryo

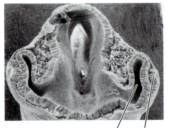

Optic   Lens
vesicle placode

(C) 5-mm embryo   (D)   (E) 7-mm embryo
Lens vesicle   Retina   Lens

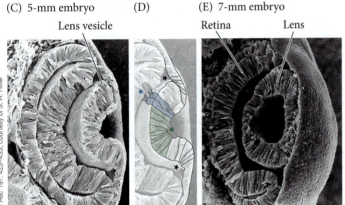

Optic cup   Cornea

A–D from S. R. Hilfer and J.-J. W. Yang, 1980. *Anat Rec* 197: 423–433, courtesy of S. R. Hilfer

Courtesy of K. Tosney

FIGURE 16.7 Development and reciprocal signaling in the vertebrate eye. (A) The optic vesicle evaginates from the brain and contacts the overlying ectoderm, inducing cells of the lens placode to become columnar. (B,C) The lens placode differentiates into lens cells as the optic vesicle folds in on itself, and the lens placode becomes the lens vesicle. (C) The optic vesicle becomes the neural retina and retinal pigmented epithelium as the lens is internalized. (D) The three principle changes in cell shape are superimposed on the SEM of the forming lens: basal constriction (black asterisks), apical constriction (green asterisk), and filopodial protrusions to the retina (blue asterisk). (E) The lens vesicle induces the overlying ectoderm to become the cornea.

placode extend adhesive filopodia to contact the optic vesicle (Chauhan et al. 2009). (2) The cells at the edge of the invaginating layer undergo basal constriction. (3) Meanwhile, the cells at the center of the invaginating layer undergo apical constriction.

As the optic vesicle becomes the optic cup, its two layers differentiate. The cells of the outer layer produce melanin (being one of the few tissues other than the neural crest cells that can form this pigment) and ultimately become the retinal **pigmented epithelium**. The cells of the inner layer proliferate rapidly and generate a variety of glial cells, ganglion cells, interneurons, and light-sensitive photoreceptor neurons that collectively constitute the **neural retina** (see Figure 1 in Further Development 16.6, online). While the photoreceptors are responsible for light perception, the retinal ganglion cells are neurons that transmit this information to the brain. Their axons meet at the base of the eye and travel down the optic stalk, which is then called the **optic nerve**. Cross talk between the inner cells of the optic cup (which will become the neural retina) and the lens placode is required for retina differentiation, lens vesicle formation, and lens epithelial and fiber cell differentiation.

Formation of the eye field: The beginnings of the retina

The precise organization of the eye is the result of multiple inductive events involving different signals and changes in gene expression in time and place. Development of the retina begins with formation of the **eye field** in the anterior neural plate at early somite stages. The anterior neural plate, where both Bmp and Wnt pathways are inhibited, is specified by *Otx2* expression. Noggin is especially important, as it not only blocks Bmps (thus allowing *Otx2* expression), but also inhibits expression of the transcription factor *ET*, one of the first genes expressed in the eye field. However, the differential expression of *Otx2* across the dorsal-ventral axis of the forebrain causes a differential repression of Noggin's ability to inhibit *ET* expression, thus allowing ET protein to be produced.

One of the genes controlled by ET is *Rx* (for *retinal homeobox*), which in turn is required to specify the retina. Rx is a transcription factor with dual functions: it acts first by inhibiting *Otx2* and second by activating *Pax6*, the major gene in forming the eye field in the anterior neural plate (**FIGURE 16.8A–C**; Zuber et al. 2003; Zuber 2010). Pax6 protein is

FIGURE 16.8 Dynamic formation of the eye field in the anterior neural plate. (A) Formation of the eye field in *Xenopus* embryos. Light blue represents the neural plate; moderate blue indicates the area of *Otx2* expression (forebrain); and dark blue indicates the region of the eye field as it forms in the forebrain. (B) Dynamic expression of transcription factors leading to specification of the eye field. Prior to stage 10 in the *Xenopus* embryo, Noggin inhibits *ET* expression but promotes *Otx2* expression. Otx2 protein then blocks the inhibition of *ET* by Noggin signaling. The resulting ET transcription factor activates the *Rx* gene, which encodes a transcription factor that blocks *Otx2* and promotes *Pax6* expression. Pax6 protein initiates the cascade of gene expression constituting the eye field (at right). (C) Location of the transcription factors in the nascent eye field of stage 12.5 (early neurula) and stage 15 (mid-neurula) *Xenopus* embryos, showing a concentric organization of transcription factors having domains of decreasing size: Six3 > Pax6 > Rx > Lhx2 > ET. (D) Eye development in a normal mouse embryo (left) and lack of eyes in a mouse embryo whose *Rx* gene has been knocked out (right). (E) Expression pattern of the *Xenopus Xrx1* gene in the single eye field of the early neurula (left) and in the two developing retinas (as well as in the pineal gland, an organ that has a presumptive retina-like set of photoreceptors) of a newly hatched tadpole (right). (A–C after M. E. Zuber et al. 2003. *Development* 130: 5155–5167.)

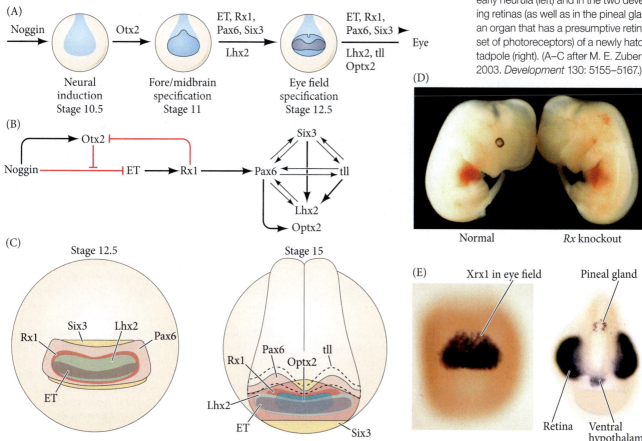

D and E from T. J. Bailey et al. 2004. *Int J Dev Biol* 48: 761–770

especially important in specifying the lens and retina; indeed, it appears to be a common denominator for specifying photoreceptive cells in most phyla, vertebrate and invertebrate (Halder et al. 1995).

Humans and mice heterozygous for loss-of-function mutations in *Pax6* have small eyes, whereas homozygous mutants of mice and humans (and *Drosophila*) lack eyes altogether (see Figure 4.13), as do *Rx* mutant mice and zebrafish (**FIGURE 16.8D**; Jordan et al. 1992; Glaser et al. 1994; Quiring et al. 1994; Rojas-Muñoz et al. 2005; Stigloher et al. 2006). In both flies and vertebrates, Pax6 protein initiates a cascade of transcription factors (such as Six3, Rx, and Sox2) with overlapping functions. These factors mutually activate one another to generate a single eye-forming field in the center of the ventral forebrain (**FIGURE 16.8E**; Tétreault et al. 2009; Fuhrmann 2010). The final result, however, is two eyes that are more lateral in the head. The main player in separating the single vertebrate eye field into two bilateral fields is our old friend Sonic hedgehog (Shh).

Shh from the prechordal plate suppresses *Pax6* expression in the center of the neural tube, dividing the field in two (**FIGURE 16.9**). If the mouse *Shh* gene is mutated, or if the processing of this protein is inhibited, the single median eye field does not split. The result is **cyclopia**—a single eye in the center of the face, usually below the nose (see Figure 16.6C; Chiang et al. 1996; Kelley et al. 1996; Roessler et al. 1996). Conversely, if too much Shh is synthesized by the prechordal plate, the *Pax6* gene is suppressed in too large an area, and the eyes fail to form at all. This phenomenon may explain why cave-dwelling fish are blind. Yamamoto and colleagues (2004) demonstrated that the difference between surface populations of the Mexican tetra fish (*Astyanax mexicanus*) and eyeless cave-dwelling populations of the same species is the amount of Shh secreted from the prechordal plate. Elevated Shh was probably selected in cave-dwelling species because it resulted in heightened oral sensing, including larger olfactory placodes and larger jaws (Yamamoto et al. 2009). However, Shh also downregulates *Pax6*, resulting in the disruption of optic cup development, apoptosis of lens cells, and arrested eye development (**FIGURE 16.10**). Alternations in Shh expression are not alone in accounting for eye and forebrain differences in cave fish; recent studies suggest that changes to Fgf heterochrony (timing of expression) and expression of Lhx transcription factors are also critically involved (Rétaux et al. 2008; Pottin et al. 2011; Alié et al. 2018).

(A)

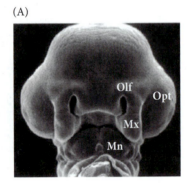

(B)

(C)

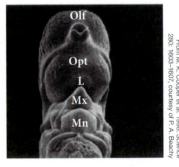

(D)

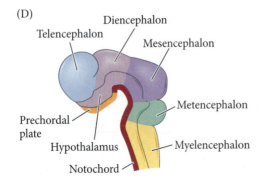

FIGURE 16.9 Sonic hedgehog separates the eye field into bilateral fields. Jervine, an alkaloid found in certain plants, inhibits endogenous Shh signaling. (A) Scanning electron micrograph showing the external facial features of a normal mouse embryo. (B) Mouse embryos exposed to 10 μM jervine had variable loss of midline tissue and resulting fusion of the paired, lateral olfactory processes (Olf), optic vesicles (Opt), and maxillary (Mx) and mandibular (Mn) processes of the jaw. (C) Complete fusion of the mouse optic vesicles and lenses (L) resulted in cyclopia. (D) Drawing showing the location of the prechordal plate (the source of Shh) in the 12-day mouse embryo.

The lens-retina induction cascade

Once the eye field split is accomplished, how do the two fields form the eyes? Modern studies of vertebrate eye formation were initiated by Hans Spemann (1901), who found that when he destroyed the presumptive optic vesicle region of the anterior neural plate on one side of an amphibian embryo, no lens formed on the affected side (see Figure 4.11). Something in the neural plate was necessary for the lens to form. Soon afterward, Warren Lewis (1904) continued these experiments on older amphibian embryos. He placed the optic vesicle underneath trunk epidermal ectoderm, and a lens formed in this region. For many years, these types of experiments were used as evidence that the optic vesicle alone could induce epidermal ectoderm to form a lens. More recent studies that traced the lineage of the tissues, however, showed that transplanted optic vesicles that seemed to induce lens formation in ectopic sites were actually contaminated with presumptive lens cells (see Grainger 1992; Ogino et al. 2012); this indicated that the optic vesicle alone cannot be considered sufficient for inducing a lens. In fact, lens induction takes multiple steps that begin long before the optic vesicle meets the epidermal ectoderm. The story, still unfolding, reveals that the steps in lens and retina development are a beautiful example of reciprocal embryonic signaling (**FIGURE 16.11**).

As mentioned earlier, the non-neural ectoderm has to become competent to respond to lens induction. Jacobson (1963, 1966) showed that the ectoderm is first conditioned by the prechordal mesoderm and anterior endoderm toward the end of gastrulation. These tissues supply the area with antagonists that block the Bmp and Wnt pathways and with FGFs and neuropeptides. These tissues and signals are critical for inducing *Pax6* and other genes specific for the anterior ectoderm (Donner et al. 2006; Lleras-Forero et al. 2013).

Meanwhile, in the brain, the bilateral forebrain eye fields evaginate as the Rx protein activates *Nlcam*, a gene whose cell-surface product regulates these bilateral evaginations (Brown et al. 2010; Bazin-Lopez et al. 2015). These evaginations become the optic vesicles.

When the cells of the optic vesicles touch surface ectoderm, both tissues are changed. The optic vesicle cells flatten against the surface ectoderm and produce Bmp4, Fgf8, and Delta (Furuta and Hogan 1998; Faber et al. 2002; Plageman et al. 2010; Ogino et al. 2012). These signals instruct the cells of the surface ectoderm to elongate and acquire placode morphology. As the lens placode forms, it secretes Fgfs that instruct the adjacent cells of the optic vesicle to activate the *Vsx2* gene that characterizes the neural retina.

The neural crest-derived mesenchyme surrounding the optic vesicle instructs most of the outer optic vesicle cells to activate the *Mitf* gene, which will activate the production of melanin pigment (Burmeister et al. 1996; Nguyen and Arnheiter 2000). Thus, the most distal part of the optic vesicle (those cells touching the surface ectoderm) is instructed to become neural retina, while the cells adjacent to this region are instructed to become retinal pigmented epithelium (see Fuhrmann 2010).

The presumptive neural retina now adheres to the lens placode and draws it inward as the optic vesicle changes its shape to form the optic cup. Fgfs from the optic cup activate a new set of genes in the lens placode, transforming the placode into the lens vesicle, which will form the cells of the lens. (See **Further Development 16.4, The Autonomous Development of the Optic Cup**, **Further Development 16.5, Lens and Cornea Differentiation**, **Further Development 16.6, Neural Retina Differentiation**, and **Further Development 16.7, Why Babies Don't See Well**, all online.)

(A) Surface-dwelling populations   (B) Cave-dwelling populations

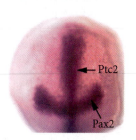
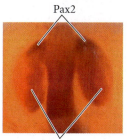

Ptc2

Pax2

Pax2

Pax6   Pax2

From Y. Yamamoto et al. 2004. *Nature* 431: 844–847, photographs courtesy of W. Jeffery

FIGURE 16.10 Surface-dwelling (A) and cave-dwelling (B) Mexican tetras (*Astyanax mexicanus*). The eye fails to form in the population that has lived in caves for more than 10,000 years (top right). Two genes, *Ptc2* and *Pax2*, respond to Shh and are expressed in broader domains in the cave fish embryos than in those of surface dwellers (center). The embryonic optic vesicles (bottom) of surface-dwelling fish are a normal size and have small domains of *Pax2* expression (specifying the optic stalk). The optic vesicles of the cave-dwelling fishes' embryos (where *Pax6* is usually expressed) are much smaller, and the *Pax2*-expressing region has grown at the expense of the *Pax6* region.

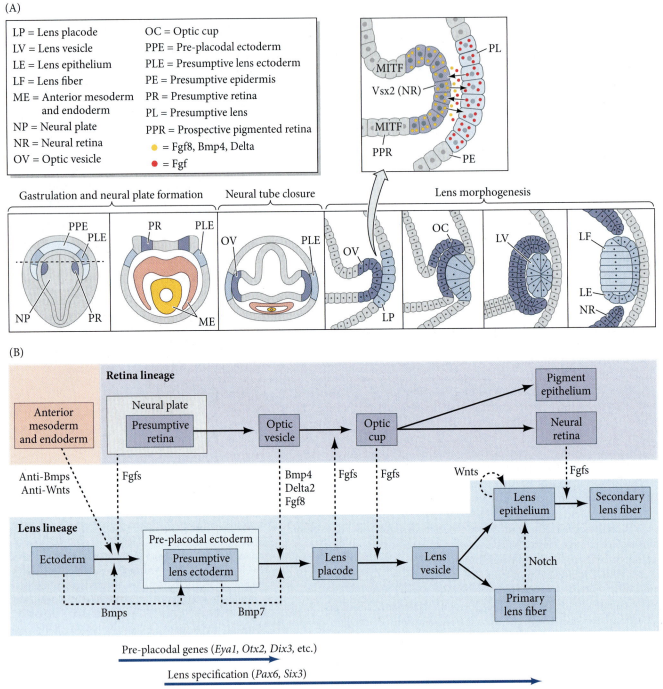

FIGURE 16.11 Reciprocal embryonic interactions between the developing lens placode and the optic vesicle from the brain. (A) Diagrammatic rendering of the major anatomical changes from gastrulation through lens morphogenesis. These interactions start with the entire pre-placodal region, including the eventual lens ectoderm, being specified as lens, influenced by the neural plate, cardiac mesoderm, and pharyngeal mesoderm. Later, the optic vesicle—a bulge from the diencephalon—touches the presumptive lens ectoderm, physically preventing lens-repressing signals from the neural crest from reaching the lens ectoderm and triggering a series of interactions that turns the optic vesicle into a two-layered optic cup, converts the inner layer of the optic cup into the neural retina, and causes the lens placode to involute and form the lens vesicle. The expanded inset (upper right) shows the critical interactions between the optic vesicle and the presumptive lens cells of the lens placode. (B) Some of the paracrine factors involved in lens development. At different stages and in different tissues, the Fgfs and their receptors may be different. The three arrows below show some of the genes that become expressed in the presumptive lens during the indicated time-frames. (After H. Ogino et al. 2012. *Dev Biol* 363: 333–347.)

The Epidermis and Its Cutaneous Appendages

The skin—a tough, elastic, water-impermeable membrane—is the largest organ in our bodies. Mammalian skin has three major components: (1) a stratified epidermis; (2) an underlying dermis composed of loosely packed fibroblasts; and (3) neural crest-derived melanocytes that reside in the basal epidermis and hair follicles. It is the melanocytes (discussed in Chapter 15) that provide the skin's pigmentation. In addition, a subcutaneous (below the skin) fat layer is present beneath the dermis. The epidermis of our skin is constantly being renewed. The average adult human replaces enough epidermal cells to cover five king-sized beds every year! This regenerative ability is possible due to a population of epidermal stem cells that last a lifetime.

Origin of the epidermis

The epidermis originates from the ectodermal cells covering the embryo after neurulation. As detailed in Chapter 13, this surface ectoderm forms epidermis rather than neural tissue because of the actions of Bmps. Bmps promote epidermal specification and at the same time induce transcription factors that block the neural pathway (see Bakkers et al. 2002). Once again, we see the principle that the specification of one tissue also involves blocking the specification of an alternative tissue.

The epidermis is only one cell layer thick to start with, but in most vertebrates, it soon becomes a two-layered structure. The outer layer gives rise to the **periderm**, a temporary covering that is shed once the inner layer differentiates to form a true epidermis. The inner layer, called the **basal layer** or **stratum germinativum**, contains epidermal stem cells attached to a basal lamina that the stem cells themselves help to make (**FIGURE 16.12**). Just as in neural stem cells, this differentiation is positively regulated by the Notch pathway (Nguyen et al. 2006; Aguirre et al. 2010). In the absence of Notch signaling, there is hyperproliferation of the dividing cells (Ezratty et al. 2011). The Notch signal promotes the synthesis of the keratins characteristic of epidermis and promotes their formation into dense intermediate filaments (Lechler and Fuchs 2005; Williams et al. 2011). There is some evidence that, like the neural stem cells of the ependymal layer, epidermal stem cells divide asymmetrically. The daughter cell that remains attached to the basal lamina remains a stem cell, while the cell that leaves the basal layer migrates outward and starts differentiating. However, it is also possible that asymmetrical and symmetrical divisions both play important roles in forming and sustaining the epidermis (Hsu et al. 2014; Yang et al. 2015). Moreover, it is still not determined whether there is a discrete population of long-lived stem cells in the basal layer (Mascré et al. 2012), or whether all basal cells have stem cell-like properties (Clevers 2015).

Cell division from the basal layer produces younger cells and pushes the older cells to the surface of the skin. This is much like the development of the gut epithelium, where cells are generated from the stem cell crypt and are pushed out to the tip of the villus (see Figure 5.12), but is unlike the "inside-out" patterning of the neural tube, where newly generated neurons migrate through layers of older cells on their way to the periphery. After the synthesis of the differentiated products, the cells cease transcriptional and metabolic activities. These differentiated epidermal cells, the **keratinocytes**, are bound tightly together and produce a water-impermeable seal of lipid and protein.

As they reach the surface, keratinocytes are dead, flattened sacs of keratin protein, and their nuclei are pushed to one edge of the cell. These cells constitute the **cornified layer**, or **stratum corneum**. Throughout life, the dead keratinocytes of the cornified layer are shed and are replaced by new cells.[2] In mouse epidermis, the journey from the basal layer to the sloughed cell takes about two weeks. Human epidermis turns over a bit more slowly; the proliferative ability

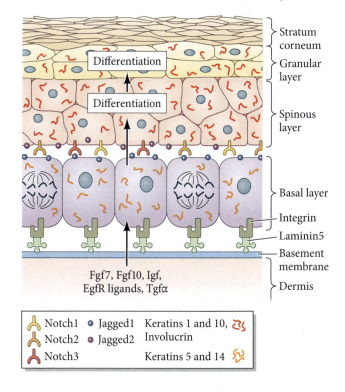

FIGURE 16.12 Layers of the human epidermis and the signals that enable the continued regeneration of the epidermis of mammalian skin. The basal cells are mitotically active, whereas the fully keratinized cells characteristic of external skin are dead and are shed continually. Self-renewing stem cells reside within the basal layer, which adheres through integrins to an underlying laminin-rich basement membrane that separates the epidermis from the underlying dermis. The dermal fibroblasts secrete factors such as Fgf7, Fgf10, Igf, EgfR ligands, and Tgf-α to promote the proliferation of basal epidermal cells. Proliferative basal progenitor cells generate columns of Notch-activated terminally differentiating cells that pass through three stages, each expressing particular keratins: spinous layers, granular layers, and finally dead stratum corneum layers, which are shed from the surface. (After Y. C. Hsu et al. 2014. *Nat Med* 20: 847–856.)

Differentiation

Differentiation

Stratum corneum
Granular layer
Spinous layer
Basal layer
Integrin
Laminin5
Basement membrane
Dermis

Fgf7, Fgf10, Igf, EgfR ligands, Tgfα

Notch1	Jagged1	Keratins 1 and 10,
Notch2	Jagged2	Involucrin
Notch3		Keratins 5 and 14

of the basal layer is remarkable in that it can supply the cellular material to continuously replace 1–2 m² of epidermis about every 30 days throughout adult life. (See **Further Development 16.8, Epidermal Factors**, online.)

The ectodermal appendages

The ectodermal epidermis and the mesenchymal dermis interact inductively at specific sites to create the **ectodermal appendages**: hairs, scales, scutes (e.g., the coverings of turtle shells), teeth, sweat glands, mammary glands, or feathers, depending on the species and type of mesenchyme. The formation of these appendages requires a series of reciprocal inductive interactions between the mesenchyme and the ectodermal epithelium, resulting in the formation of **epidermal placodes** that are the precursors of the epithelia of these structures. Remarkably, the early development of structures as different as hair, teeth, and mammary glands follows the same patterns and appears to be governed by reciprocal induction using the same paracrine factors.

In all these ectodermal appendages, the first obvious sign of morphogenesis is a local epithelial thickening, the placode. If we look at a developing mammal, we see that in many regions of the trunk and abdominal ectoderm, thousands of individual hair placodes develop independently. In each jaw, there is a broad epidermal thickening called the **dental lamina**, which (like the pre-placodal stage of cranial sensory placodes) later resolves into separate placodes, each of which forms the enamel portion of a tooth (see Figure 16.14). In the ventral ectoderm, two mammary ridges (or "milk lines") extend from the forelimbs to the hindlimbs. In the mouse, five pairs of mammary placodes typically survive on each side, each becoming a mammary gland. In humans, usually only one pair survives, although sometimes a third or fourth placode remains, forming supernumerary nipples. Mammary placodes form in both sexes but are fully developed only in females (Biggs and Mikkola 2014).

After the placode stage, there is a bud stage during which the ectoderm grows into the mesenchyme. In the placode-forming regions, superficial cells of the placode contract and intercalate toward the common center. This causes them to pucker inwardly and drive the epithelium into the underlying mesenchyme. Such contractile forces have been seen in tooth, mammary, and hair buds (Panousopoulou and Green 2016). The initial bud looks similar for each of these epidermal appendages. However, as the bud continues to interact with the underlying mesenchyme, differences begin to be observed. The hair follicle elongates, grows inward, and grows around the condensed inductive mesenchyme. The tooth epithelium likewise grows into the mesenchyme and, in the center of the epithelium, generates an **enamel knot**. This signaling center will control the proliferation and differentiation of the surrounding cells (Jernvall et al. 1998). The mammary epithelium grows through the inductive mesenchymal cells and into the developing fat pad, where it undergoes extensive branching (**FIGURE 16.13**).

(?) Developing Questions

How is the avian feather formed? What patterns the placement of the feather placodes?

FIGURE 16.13 Ectodermal appendages arise through shared placode and bud stages prior to diversification and epithelial morphogenesis. The developmental schema of hair (top row), tooth (center), and mammary gland (bottom row) are shown for the three stages. (After L. C. Biggs and M. L. Mikkola. 2014. *Semin Cell Dev Biol* 26: 11–21.)

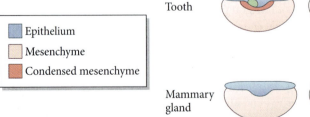

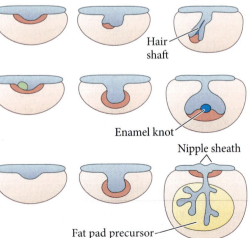

FURTHER DEVELOPMENT

Recombination experiments: The roles of epithelium and mesenchyme

The inductive interactions between epithelium and mesenchyme are very specific. By separating epithelial and mesenchymal components and then recombining them, twentieth-century developmental biologists were able to discern which part held the specificity. For instance, dental epithelium from the jaw of a day-10 mouse embryo caused tooth formation when combined with non-dental jaw mesenchyme of the same stage embryo. By embryonic day 12, however, the dental epithelium lost this ability, while the newly condensed dental mesenchyme gained it (Mina and Kollar 1987). The expression pattern of *bmp4* shifts from the epithelium to the mesenchyme concomitant with this switch in odontogenic ability (Vainio et al. 1993). Moreover, after this transition, the site of the epidermis became less important. Dental mesenchyme could interact with foot epidermis to generate teeth (Kollar and Baird 1970). The reverse combination did not form teeth. In the mouse jaw, the mesenchyme is given tooth-forming ability through Fgf8 and is prevented from tooth formation by Bmps (**FIGURE 16.14**; Neubüser et al. 1997).

Similarly, condensed dermal cells can induce hair follicles even in epithelia (such as the soles of the feet) that usually do not produce hair (Kollar 1970). The hair placode epithelium cannot induce new hairs. Early mouse mammary gland mesenchyme (but not epithelium) will induce the early stages of mammary formation in epidermis derived from head and neck (Propper and Gomot 1967; Kratochwil 1985). Thus, in the developing hair, mammary gland, and tooth, once the condensed mesenchyme has formed, it appears to have the inductive capacity that specifies the type of epidermal appendage that forms. Interestingly, this capacity to induce placodes is absent in certain skin mesenchymes, namely those mesenchymes associated with palms, soles, and external genitalia. In these locations, *HoxA13* gene expression appears to promote the synthesis of paracrine factors that induce the ectodermal epithelium to express specific keratin proteins rather than hair (Rinn et al. 2008; Johansson and Headon 2014). (See **Further Development 16.9, Recombination of Mammary Gland Epithelium Reveals Sex-Based Differences in Its Development**, online.)

(?) Developing Questions

Why don't chickens have teeth? Can chick embryos be manipulated to make them? The mesenchyme of the tooth comes from neural crest cells, and in 2003 Mitsiadis and colleagues created teeth in a chick embryo by transplanting neural crest cells from a mouse embryo into the chick embryo. Can you suggest a probable explanation for how, evolutionarily, birds lost the ability to make teeth?

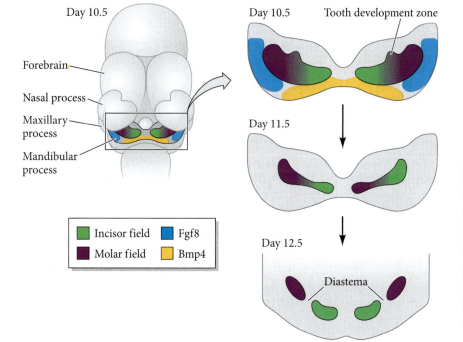

FIGURE 16.14 Division of the dental lamina into incisor and molar fields. (A) Schematic diagram showing the mandible of a 10.5-day mouse embryo. A mutually antagonistic interaction between Bmp4 and Fgf8 is thought to define the tooth field in the oral ectoderm. Each tooth field is continuous until day 11, but subsequently divides into the incisor and molar fields in the distal and proximal areas of the jaw, respectively. The two fields are separated by a toothless region called a diastema. (After Y. Ahn. 2015. *Curr Top Dev Biol* 111: 421–452.)

Signaling pathways you can sink your teeth into

Just about all the major signaling pathways are involved in the formation of the ecto-dermal appendages (Biggs and Mikkola 2014; Ahn 2015). In some instances, such as the enamel knot of mammalian teeth, the same signaling center sends out paracrine factors belonging to nearly every family (**FIGURE 16.15**). The canonical Wnt/β-catenin pathway was implicated by the loss of properly formed hair, teeth, and mammary glands in mice deficient in components of this pathway (van Genderen et al. 1994). Manipulating embryos to have β-catenin expression throughout the epidermis eventually transforms the entire epidermis to a hair follicle fate (Närhi et al. 2008; Zhang et al. 2008), and it creates supernumerary teeth in the jaw (Järvinen et al. 2006; Liu et al. 2008). Indeed, Wnt signaling may help induce the enamel knot to form and may be critical in the abil-ity (lost in mammals) to regenerate teeth (Järvinen et al. 2006). Mouse mutants lacking negative regulators of the Wnt pathway display more and larger mammary placodes (Närhi et al. 2012; Ahn et al. 2013).

Fgfs probably play multiple roles in ectodermal appendage development. One of these roles is to regulate the migration of mesenchymal cells to become the conden-sates beneath the placode. In teeth, Fgf8 from the placode appears to attract mesen-chyme cells to the tooth placode and sustain them there (Trumpp et al. 1999; Mam-moto et al. 2011). In hair formation, the placodal Wnt activates the secretion of Fgf20, which may stimulate mesenchymal cell migration to the placode (Huh et al. 2015). In the mammary gland, Fgf10 from the somites (and possibly the limb buds) is thought to induce placode formation (Mailleux et al. 2002; Veltmaat et al. 2006). Mice lack-ing the genes for Fgf10 or its receptor lack mammary placodes 1, 2, 3, and 5. Some Tgf-β family members, especially the Bmps, also play important roles in ectodermal appendage formation. Indeed, the shifting expression pattern of Bmp4 from the epi-thelium to the mesenchyme coordinates the shifting of tooth-generating potential and is critical for the bud-to-cap transition. Bmps are known to induce several genes involved in tooth development (Vainio et al. 1993; Jussila and Thesleff 2012), and Bmps and Wnts most likely regulate each other to control tooth shape (Munne et al. 2009; O'Connell et al. 2012). While Bmp expression is necessary for tooth formation, Bmp activity must be *suppressed* for the induction of hair placodes (Jussila and Thesleff 2012; Sennett and Rendl 2012).

Other signaling pathways, such as those initiated by hedgehog and ectodyspla-sin, are also important to varying degrees (Ahn 2015; Biggs and Mikkola 2014). The ectodysplasin pathway (activating the NF-κβ transcription factor) is active in every cutaneous appendage. People (and other animals) with anhidrotic ectodermal dys-plasia thus have defective growth of hair, teeth, and sweat glands (Mikkola 2015). (See **Further Development 16.10, The Ectodysplasin Pathway and Mutations of Hair Development**, online.)

(?) Developing Questions

Turtle scutes (the keratinous external plates covering the back and belly) all display the same pattern, whether they are on marine turtles or desert tortoises. The mathematically inclined might want to ask: How is this pattern generated?

(A)

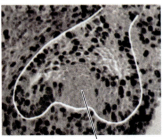

Enamel knot

(B)   (C)   (D)

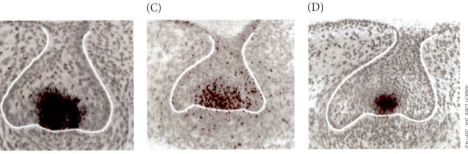

From A. Vaahtokari et al. 1996. Mech Dev 54: 39–43

FIGURE 16.15 Tooth generation in mammals. The enamel knot is the signaling center direct-ing tooth morphogenesis. These photographs show the cap stage of development where the epithelium is growing into the mesenchyme. (A) Staining for cell division with radioactive BrdU indicates a region of non-dividing cells, the enamel knot. (B–D) In situ hybridization reveals that the enamel knot transcribes the genes for paracrine factors initiating several signaling cas-cades. These genes include *Sonic hedgehog* (B), *Bmp7* (C), and *Fgf4* (D).

Ectodermal appendage stem cells

In many cases, the epidermal appendages generate or retain adult stem cells that allow the regrowth of these structures at particular times. Whether or not such stem cells are present differs among species. Fish and reptiles can regenerate their teeth, but mammals cannot. Most mammals have two sets of teeth, one for children ("milk" teeth) and a "permanent" set for adults. Both sets of teeth started developing before birth. Once we grow adult teeth, the dental lamina decays, and we cannot regenerate lost or damaged teeth. While humans cannot grow new teeth if they are lost in old age (realize, though, that throughout much of human history, most people died before the age of 40), other mammals, including rodents and elephants, have teeth that grow continually (Thesleff and Tummers 2009). In the ever-growing mouse incisors, there is a stem cell niche that retains epithelial stem cells for generating enamel-forming ameloblasts as well as mesenchymal stem cells for generating dentin-forming odontoblasts and pulp cells (An et al. 2018). In some reptiles, such as alligators, part of the dental lamina is retained, and it contains epithelial stem cells capable of regenerating lost teeth. When teeth are lost, β-catenin accumulates in these cells, while Wnt inhibitors are lost (Wu et al. 2013).

The mammary gland (after which our vertebrate class was named) contains stem cells for the reactivation of its growth at puberty and pregnancy (**FIGURE 16.16A**). During puberty, estrogens cause extensive branching of the ducts and elongation of the **terminal end buds**. During pregnancy, the ducts, stimulated by progesterone and prolactin, form tertiary side branches that differentiate into the milk-producing alveoli (Oakes et al. 2006; Sternlicht et al. 2006). The mammalian mammary gland probably contains a stem cell that is able to generate all the lineages of the gland. Data from genetically labeled mammary cells (Rios et al. 2014; Wang et al. 2015) suggest that there may be a single stem cell type that can make the two major progenitor cells of the breasts: one that generates the ducts and alveoli, and one that generates the myoepithelial cells that contract to push the milk out of the alveoli and toward the nipple (**FIGURE 16.16B**).

FIGURE 16.16 Stem cells and breast development. (A) The stages of mammalian breast development in mice. Breast development begins around embryonic day 11. The gland remains in the newborn condition until puberty, when the ducts expand. During pregnancy and lactation, the alveoli differentiate and produce milk. After pregnancy, the alveoli undergo apoptosis, but they can be regenerated during subsequent pregnancy. (B) Possible stem and progenitor cells of the breast. The mammary stem cell can generate two types of progenitor cells (in addition to making another mammary stem cell). One of these progenitors gives rise to the contractile myoepithelial cells that line the alveoli and ducts, while the other progenitor generates the ducts and alveoli. (After J. E. Visvader and J. Stingl. 2014. *Genes Dev* 28: 1143–1158.)

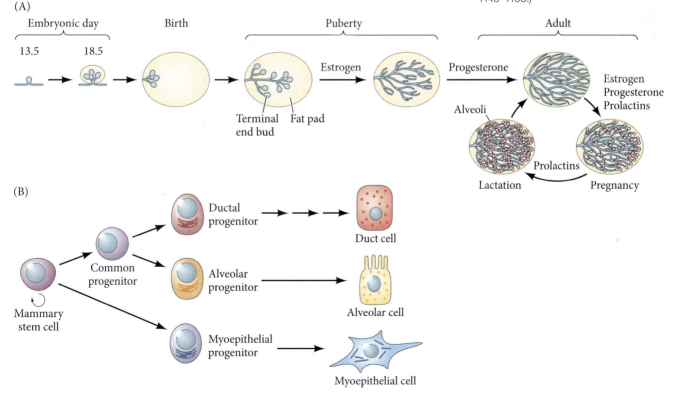

(?) Developing Questions

Humans have sweat glands throughout their skin, whereas most mammals have them only in their palms and soles. With increasing brain size came the necessity for a very efficient cooling system—thus hairless skin with its many sweat glands. How might sweat glands arise as an evolutionary modification of ectodermal appendages? What is the evidence that sweat glands may be a modification of hair that develop at the expense of hair follicles, thereby explaining both our large number of sweat glands and our bodies' relative hairlessness?

The best studied of the epidermal appendage stem cells is the hair stem cell. There appear to be three stem cell populations involved in producing epidermal structures. One, discussed earlier, is found in the germinal layer of the epidermis and generates the keratinocytes that characterize the interfollicular epidermis. A second group of stem cells is critical for forming the sebaceous gland of each hair shaft, and a third group is critical for regenerating the hair shaft itself. Interestingly, it seems that there is also a primitive stem cell that can form all the others (Snippert et al. 2010), and members of each stem cell group can be recruited to any of the other pools if needed, as when the skin repairs itself when wounded (Levy et al. 2007; Fuchs and Horsley 2008).

Hair is one structure that mammals are able to regenerate. Throughout life, hair follicles undergo cycles of growth (**anagen**), regression (**catagen**), rest (**telogen**), and regrowth. Hair length is determined by the amount of time the hair follicle spends in the anagen phase. Human scalp hair can spend several years in anagen, whereas arm hair grows for only 6–12 weeks in each cycle. The ability of hair follicles to regenerate depends on the existence of a population of epithelial stem cells that forms in the permanent **bulge** region of the follicle late in embryogenesis. When Philipp Stöhr drew the histology of the human hair for his 1903 textbook, he showed this bulge ("Wulst") as the attachment site for the arrector pili muscles (which give a person "goosebumps" when they contract). Research carried out during the 1990s suggested that the bulge houses populations of at least two adult stem cell types: the **hair follicle stem cells** (HFSCs), which give rise to the hair shaft and sheath (Cotsarelis et al. 1990; Morris and Potten 1999; Taylor et al. 2000); and the **melanocyte stem cells**, which give rise to the pigment of the skin and hair (Nishimura et al. 2002). The bulge appears to be an important niche that allows adult cells to retain the quality of "stemness." Follicle stem cells in the bulge can regenerate all the epithelial cell types of the hair, and without the stem cells, there is no new follicle. However, if stem cells are selectively ablated by a laser, certain bulge epithelial cells (which are normally not used for hair growth) repopulate the stem cell population and can sustain hair follicle regeneration (Rompolas et al. 2013).

There appear to be two populations of HFSCs: a quiescent population in the bulge, and a population primed for cell division just below the bulge. The entire skin organ seems to be involved in hair cycling (**FIGURE 16.17**). The HFSCs reside in the outer layer of the bulge. The inner bulge cells are the progeny of the HFSCs, and they secrete Bmp6 and Fgf18, two repressors of HFSC proliferation. In addition, the dermal fibroblasts and the subcutaneous fat cells also make growth-suppressive Bmps. The HFSCs are activated at the beginning of the anagen phase by signals from the dermal papilla of condensed mesenchyme. These signals are Fgfs and Wnts, as well as Bmp antagonists, and they direct the epidermal stem cells to migrate out of the bulge. There, the epidermal stem cells produce progenitor cells that proliferate downward and generate the seven concentric columns of cells that form the outer root sheath from bulge to matrix.

This activation of the dermal papilla is regulated by the microenvironment of the dermis. The underlying dermis appears to make more Wnts and less Bmps, while the adipocyte precursor cells make more of the paracrine factor Pdgf, which stimulates the dermal papilla (Plikus et al. 2008; Rendl et al. 2008; Hsu and Fuchs 2012). As the dermal papilla is moved farther away by the downward growth of the epithelial cells, its signals are not received by the stem cells, and the bulge returns to quiescence. During the latter part of anagen, prostaglandin PGD_2 appears to prevent the production of the progenitor cells. During catagen, most of the basal (outer root sheath) epithelial cells undergo apoptosis. Upper follicular stem cells remain, however. The outer root sheath cells close to the old bulge contain HFSCs and become the outer layer, while those nearer to the matrix differentiate to become the inner layer of the bulge. Apoptosis causes the outer cells to be in contact with the dermal papilla in readiness for the next cycle (Hsu et al. 2011; Mesa et al. 2015). (See **Further Development 16.11, The Progenitor Cell Is a Runway Signal**, **Further Development 16.12, Normal Variation in Human Hair Production**, and **Further Development 16.13, The Hair Follicle Niche: Its Role in Baldness and Long Lashes**, all online.)

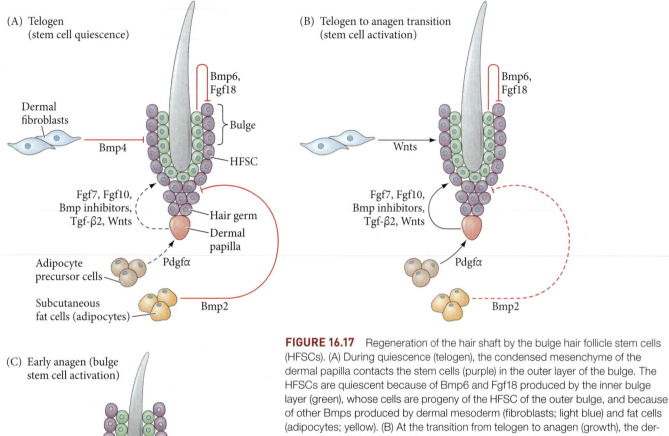

(A) Telogen (stem cell quiescence)

(B) Telogen to anagen transition (stem cell activation)

(C) Early anagen (bulge stem cell activation)

FIGURE 16.17 Regeneration of the hair shaft by the bulge hair follicle stem cells (HFSCs). (A) During quiescence (telogen), the condensed mesenchyme of the dermal papilla contacts the stem cells (purple) in the outer layer of the bulge. The HFSCs are quiescent because of Bmp6 and Fgf18 produced by the inner bulge layer (green), whose cells are progeny of the HFSC of the outer bulge, and because of other Bmps produced by dermal mesoderm (fibroblasts; light blue) and fat cells (adipocytes; yellow). (B) At the transition from telogen to anagen (growth), the dermal papilla (pink) is induced by the mesenchyme cells to produce activators of hair growth (Fgfs and Wnts) as well as antagonists of Bmp signals. This causes the proliferation and differentiation of the hair follicle. (C) At anagen, the cells contacting the dermal papilla at the base divide rapidly to generate the hair shaft and its channel. The cells close to the earlier bulge become the outer layer of the bulge and have stem cell properties. An inner layer, several cell layers thick, is also derived from the HFSCs; this layer eventually inhibits HFSC proliferation. At catagen (not shown), most cells undergo apoptosis, but the remaining stem cells survive in the bulge region. An epithelial strand then brings the dermal papilla up to the region of the bulge, and the interaction between them appears to generate the hair germ of the next generation of hair. (After Y. C. Hsu and E. Fuchs. 2012. *Nat Rev Mol Cell Biol* 23: 103–114; Y. C. Hsu et al. 2014. *Nat Med* 20: 847–856.)

Closing Thoughts on the Opening Photo

Did you *hear* that the inner ear started on the outside? Dr. Nobue Itasaki won the Runner Up Best Image Prize at the Anatomical Society in 2017 for this micro-CT image of a 10-day-old chick embryo. This image shows the inner ear structures in green, namely the semicircular canals, vestibule, and cochlea in the context of the chick embryo's head. In this chapter, we described how an elaborate network of changing signals communicates with the early ectoderm to first specify otic fate and then turn surface ectoderm (outside) into a placode, followed by differentiation of the otic vesicle (moves inside)—a *melody* that *resonates* with the patterning of hair cells in the cochlea.

Endnotes

[1]There is great diversity in the anatomy of the cochlea across vertebrates. For instance, only the mammalian cochlea has a coiled morphology. For a review, see Manley 2017.

[2]Humans lose about 1.5 grams of these cells every day. Most of this skin becomes "house dust." Deficient Notch signaling has been implicated in psoriasis (Kim et al. 2016).

Snapshot Summary
Ectodermal Placodes and the Epidermis

1. Ectodermal placodes are areas of columnar-shaped cells. In the head, sensory placodes contribute to the sense organs forming the olfactory epithelium, the inner ear, and the lens of the eye, and to the cranial sensory ganglia. Non-sensory placodes form the epithelial portions of hair, teeth, feather, scutes, and scales that cover the epidermis.

2. The sensory placodes form in a band of ectoderm surrounding the anterior of the neural plate called the pre-placodal region. A combination of Fgf signaling and inhibition of Wnt and Bmp pathways induces this territory and distinguishes it from neural crest cells.

3. The pre-placodal region then separates into individual placodes, a process controlled by local signals from the neural tube and underlying mesoderm or endoderm. For example, Fgf signaling promotes the specification of the otic-epibranchial progenitors, while Wnt and Bmp differentiate these two placode types.

4. The otic vesicle, which forms the inner ear, is subsequently patterned by a dynamic series of morphogens, which also set up different cell types in the outgrowing cochlea.

5. Eye development starts with specification of the eye field in the ventral diencephalon. Pax6 plays a major role in eye formation, and its downregulation by Sonic hedgehog signaling from the underlying mesoderm splits the eye field so that two separate optic vesicles grow out from the brain. The optic vesicles develop into the retina and the retinal pigmented epithelial cells, while the lens of the eye forms from the lens placode.

6. Reciprocal signaling is critical in differentiation of the retina and lens. The cells that form the organs have two "lives." In the embryonic life, they construct the organs; in the adult life, they function as part of an organ.

7. The basal layer of the surface ectoderm becomes the germinal layer of the skin. Epidermal stem cells divide to produce differentiated keratinocytes and more stem cells.

8. The enamel knot is the signaling center for tooth shape and development.

9. The hair follicular stem cells, which regenerate hair follicles during periods of cyclical growth, reside in the bulge of the hair follicle.

Go to oup.com/he/barresi12xe for Further Developments, Scientists Speak interviews, Watch Development videos, Dev Tutorials, and complete bibliographic information for all literature cited in this chapter.

Paraxial Mesoderm

The Somites and Their Derivatives

17

What, when, where, and how many?

Photograph courtesy of Anne C. Burke

SEGMENTATION OF THE BODY PLAN is a highly conserved physical feature across all vertebrate species. Repetition of form through segmentation has provided a developmental mechanism for the evolution of increasingly sophisticated functions. For instance, although humans and giraffes have the same number of cervical vertebrae, the sizes of these segments are profoundly different and adapted to their environmental pressures. Thoracic vertebrae are the only segments to have ribs, which function in part to provide protection to organs. The number of thoracic vertebrae differs wildly among a human, mouse, and snake. The number and size of segments and their bone and muscle derivatives are decided by modifications in the fission of mesoderm along the anterior-posterior axis. How can a tissue be developmentally cut up into precisely sized segments? How can snakes have some 300 pairs of segments while humans have only about 38?

One of the major tasks of gastrulation is to create a mesodermal layer between the endoderm and the ectoderm. As seen in **FIGURE 17.1**, the formation of mesodermal tissues in the vertebrate embryo is not subsequent to neural tube formation, but occurs synchronously. The notochord extends beneath the neural tube, from the posterior region of the forebrain into the tail. On either side of the neural tube lie thick bands of mesodermal cells divided into the paraxial mesoderm, intermediate mesoderm, and lateral plate mesoderm. The early paraxial mesoderm directly adjacent to the notochord lacks somites; it takes the form of bilateral streaks of continuous mesenchymal cells, referred to as the **presomitic mesoderm**, or **PSM** (also known as the **segmental plate**). As has been studied extensively in amniotes, when the primitive streak is regressing and the neural folds begin to gather at the center of the embryo, the cells of the presomitic mesoderm will form somites. **Somites** are epithelial, "blocklike" clusters of cells that are bilaterally positioned adjacent to the neural tube.

The regions of trunk and head mesoderm and their derivatives can be summarized as follows (see Figure 17.1):

1. The central region of trunk mesoderm is the **chordamesoderm** (generally referred to as the axial mesoderm). This tissue forms the notochord, a transient tissue whose major functions include inducing and patterning the neural tube and establishing the anterior-posterior body axis. Notochord cells are hydrostatically pressurized with large vacuoles to provide a rigid rodlike structure for the developing embryo. Despite many notochord cells succumbing to apoptotic clearance, the jelly-like core of the

FIGURE 17.1 Major lineages of the mesoderm are shown in this schematic of the mesodermal compartments of the amniote embryo.

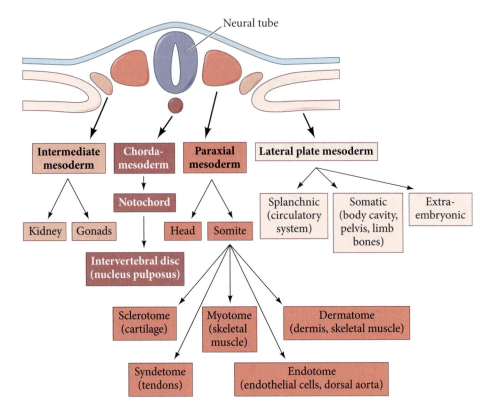

intervertebral disc, called the nucleus pulposus, is derived from notochord cells (**FIGURE 17.2**; Choi et al. 2008; McCann et al. 2011).

2. Flanking the notochord on both sides is the **paraxial**, or **somitic**, **mesoderm**. The tissues developing from this region will be located in the back of the embryo, surrounding the spinal cord, and for some muscle descendants, in the limb and ventral (abdominal wall) regions. Before those regions can be populated, the cells of the paraxial mesoderm will form somites—transitory epithelial blocks of mesodermal cells on either side of the neural tube—which will produce muscle and many of the connective tissues of the back (dermis, muscle, and skeletal elements such as the vertebrae and ribs; see Figure 17.2E,F). The anteriormost paraxial mesoderm does not segment; it becomes the **head mesoderm**, which (along with the neural crest) forms the skeleton, muscles, and connective tissue of the face and skull.

3. The **intermediate mesoderm** is positioned directly lateral to the paraxial mesoderm and forms the urogenital system, consisting of the kidneys, the gonads, and their associated ducts. The outer (cortical) portion of the adrenal gland also derives from this region (see Figure 17.2C).

4. Farthest away from the notochord, the **lateral plate mesoderm** gives rise to the heart, blood vessels, and blood cells of the circulatory system as well as to the lining of the body cavities. It also gives rise to the pelvic and limb skeleton (but not the limb muscles, which are somitic in origin). The lateral plate mesoderm also helps form a series of extraembryonic membranes that are important for transporting nutrients to the embryo (see Figure 17.2B,C).

5. Anterior to the trunk mesoderm is the head mesoderm, consisting of the unsegmented paraxial mesoderm and the prechordal mesoderm. This region provides the head mesenchyme that forms much of the connective tissues and musculature of the head (Evans and Noden 2006). The muscles derived from the head mesoderm form differently than those formed from the somites. Not only do they have their own set of transcription factors, but the head and trunk muscles are affected by different types of muscular dystrophies (Emery 2002; Bothe and Dietrich 2006; Harel et al. 2009).

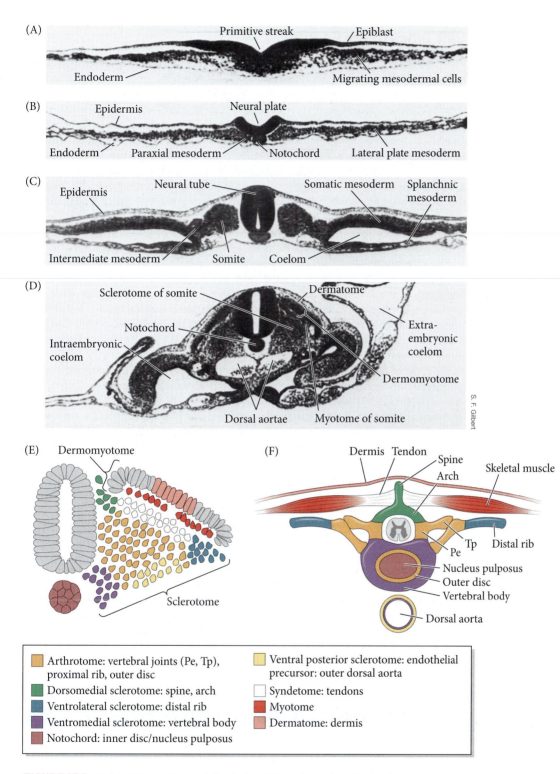

FIGURE 17.2 Gastrulation and neurulation in the chick embryo, focusing on the mesodermal component. (A–C) show 24-hour embryos, (D–F) show 48-hour embryos. (A) Primitive streak region, showing migrating mesodermal and endodermal precursors. (B) Formation of the notochord and paraxial mesoderm. (C,D) Differentiation of the somites, the coelom, and the two dorsal aortae (which will eventually fuse). (E,F) Color-coded schematic of one half of a somite from a 48-hour embryo in cross section (E), with the derivative structures those somitic cells contribute to in the adult (F). Gray cells of the dermamyotome represent the myogenic progenitor zones (E). Pe, pedicle of the vertebal arch; Tp, transverse process (F). (E,F adapted from M. Scaal. 2015. *Semin Cell Dev Biol* 49: 83–91.)

Cell Types of the Somite

When a somite first forms, it is shaped by epithelial cell contacts that create a blocklike mass with mesenchymal cells at its core. The cells within a somite become committed toward a particular cell fate relatively late, after the somite has already formed. When a somite is first separated from the presomitic mesoderm, all its cells are capable of becoming any of the somite-derived structures. As the somite matures, portions of it undergo an epithelial-mesenchymal transition (EMT), and its various regions become committed to forming only certain cell types (see Figure 17.2A–D). These mature somites contain two major compartments: the sclerotome and the dermomyotome (see Figure 17.2D,E). The sclerotome is formed from the ventromedial cells of the somite (those cells closest to the neural tube and notochord), where they undergo mitosis, lose their round epithelial characteristics, and become mesenchymal cells again. The dermomyotome is formed from the remaining epithelial portion of the somite, which gives rise to the muscle-forming myotome and the dermis-forming dermatome (see Figure 17.2E,F).

The **sclerotome** gives rise to the vertebrae, with associated tendons and rib cartilage (see Figure 17.2E). It can be further subdivided into the progenitor zones for specific cell lineages. The **syndetome** arises from the most dorsal sclerotome cells and generates the tendons, while the most internal cells of the sclerotome (sometimes called the **arthrotome**) become the vertebral joints, the outer portion of the intervertebral discs, and the proximal portion of the ribs (Mittapalli et al. 2005; Christ et al. 2007). The lateralmost mesenchyme will build the distalmost aspects of the ribs. Cells occupying the ventromedial sclerotome will migrate to the notochord and form the vertebral body, while additional cells from the arthrotome will combine with the notochord cells to form the intervertebral discs. The most dorsomedial sclerotome cells will establish the spine and arch of the vertebrae. Finally, *endothelial precursor cells* in a region recently named the endotome, in the ventral-posterior sclerotome, generate differentiated vascular cells of the dorsal aorta and intervertebral blood vessels (**TABLE 17.1**; Pardanaud et al. 1996; Sato et al. 2008; Ohata et al. 2009; Nguyen et al. 2014).

The **dermomyotome** contains progenitor cells for making skeletal muscle and the dermis of the back (see Figure 17.2E). The ventralmost portion of the dermomyotome is partitioned off as the **myotome**, which forms the musculature of the back, rib cage, and ventral body wall. Additional muscle progenitors are provided by cells that detach from the lateral edge of the dermomyotome and migrate into the limbs to generate the musculature of the forelimbs and hindlimbs. The dorsalmost surface of the dermomyotome develops into the **dermatome**, which gives rise to the dermis of the back.

Thus, the somite contains a population of multipotent cells whose specification is correlated with (and depends on) their location within the somite. What is it about their position that influences their differentiation? Consider the structures neighboring each of the progenitor domains within the somite. How might they influence the development of sclerotomal or dermomyotomal cells? At the midline are the notochord and neural tube, laterally are the other mesodermal derivatives, and above them all is the epidermis. Later in this chapter, we will discuss how paracrine signals from these surrounding

TABLE 17.1 Derivatives of the somite	
Traditional view	**Current view**
DERMOMYOTOME	
Myotome forms skeletal muscles	Lateral edges generate primary myotome that forms muscle
Dermatome forms back dermis	Central region forms muscle, muscle stem cells, dermis, brown fat cells
SCLEROTOME	
Forms vertebral and rib cartilage	Forms vertebral and rib cartilage
	Dorsal region forms tendons (syndetome)
	Medial region forms blood vessels and meninges
	Central mesenchymal region forms joints (arthrotome)
	Forms smooth muscle cells of dorsal aorta

tissues pattern the cell fates of the somite. First, however, we need to understand how the paraxial mesoderm and somites are specified.

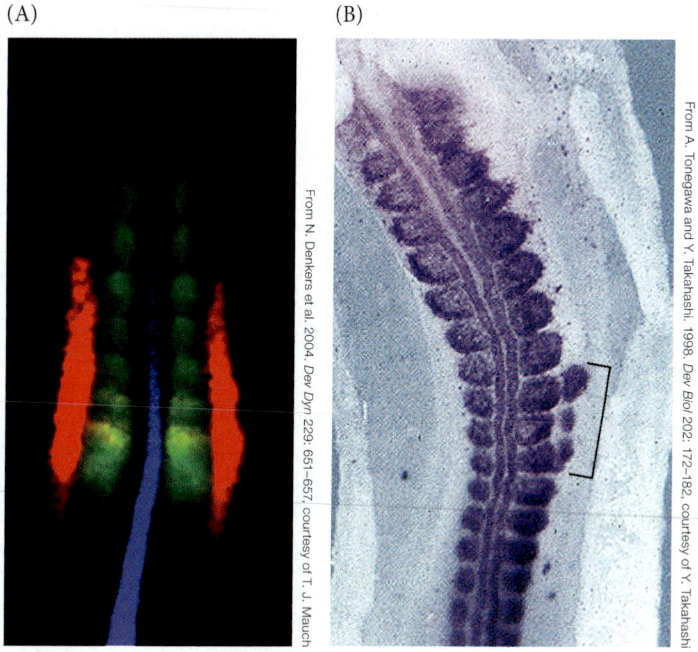

Establishing the Paraxial Mesoderm and Cell Fates along the Anterior-Posterior Axis

Specification of the paraxial mesoderm

The mesodermal subtypes (chordamesoderm, paraxial, intermediate, and lateral plate) are specified along the mediolateral (center-to-side) axis by increasing amounts of BMPs (**FIGURE 17.3A**; Pourquié et al. 1996; Tonegawa et al. 1997). The more lateral mesoderm of the chick embryo expresses higher levels of BMP4 than do the midline areas, and one can change the identity of the mesodermal tissue by altering BMP expression. One mechanism to shape the BMP gradient uses Noggin, an inhibitor of BMP, which is expressed first in the notochord and then in the somitic mesoderm (Tonegawa and Takahashi 1998). If Noggin-expressing cells are placed into presumptive lateral plate mesoderm, the lateral plate tissue will be respecified into somite-forming paraxial mesoderm (**FIGURE 17.3B**; Tonegawa and Takahashi 1998; Gerhart et al. 2011). (See **Further Development 17.1, BMP to Fox**, online.)

Brachyury (T), Tbx6, and Mesogenin have been identified as the pioneer transcriptional regulators for the early specification of the presomitic mesoderm (Van Eeden et al. 1998; Nikaido et al. 2002; Windner et al. 2012). In zebrafish, presomitic mesoderm development requires both *Tbx6* and *Tbx16* (*spadetail*); in the mouse, however, *Tbx6* appears to serve the function of both those genes. Loss of *Tbx6* in mouse converts the presumptive PSM into neural tissue. These *Tbx6* knockout mice express the neural progenitor determination factor *Sox2* (among other neural genes) in the presumptive PSM, and strikingly, this generates ectopic neural tubes in place of the PSM (Chapman and Papaioannou 1998; Takemoto et al. 2011; Nowotschin et al. 2012). These embryos actually have three neural tubes (**FIGURE 17.4**)! These results suggest that Tbx6 normally promotes PSM fates, in part by repressing *Sox2* and neural fates.

From N. Denkers et al. 2004. *Dev Dyn* 229: 651–657, courtesy of T. J. Mauch

From A. Tonegawa and Y. Takahashi. 1998. *Dev Biol* 202: 172–182, courtesy of Y. Takahashi

FIGURE 17.3 (A) Staining for the medial mesodermal compartments in the trunk of a 12-somite chick embryo (~33 hours). In situ hybridization was performed with probes binding to *Chordin* mRNA (blue) in the notochord, *Paraxis* mRNA (green) in the somites, and *Pax2* mRNA (red) in the intermediate mesoderm. (B) Specification of somites. Placing Noggin-secreting cells into a prospective region of chick lateral plate mesoderm will respecify that mesoderm into somite-forming paraxial mesoderm. Induced somites (bracket) were detected by in situ hybridization with a *Pax3* probe.

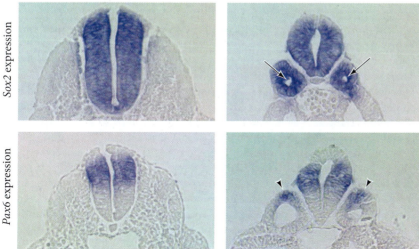

FIGURE 17.4 Three neural tubes: loss of the *Tbx6* gene transforms the paraxial mesoderm into neural tubes. In situ hybridization for the mRNA expression (blue) of the neural specification markers *Sox2* and *Pax6* in wild-type mice (A) and in *Tbx6* knockout mice (B). *Sox2* is ectopically expressed throughout the presumptive paraxial mesoderm in the *Tbx6⁻/⁻* embryo, which has also taken on a neural tube-like morphology, even displaying a central lumen (arrows). Similarly, the dorsal neural tube marker *Pax6* shows regional cell specification within these ectopic neural tubes (arrowheads).

From T. Takemoto et al. 2011. *Nature* 470: 394–398

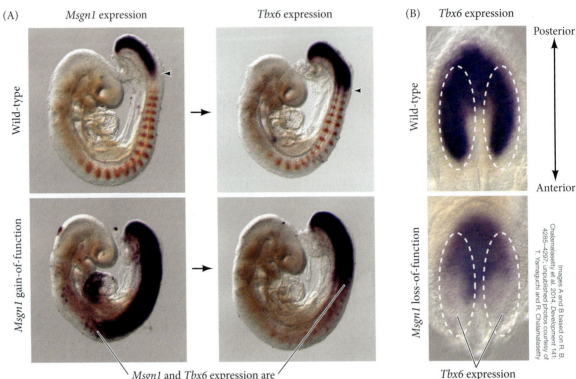

(A)

Msgn1 expression

Tbx6 expression

(B) *Tbx6* expression

Wild-type

Msgn1 gain-of-function

Wild-type

Msgn1 loss-of-function

Posterior

Anterior

Msgn1 and *Tbx6* expression are expanded into the anterior PSM

Tbx6 expression reduced in PSM

Images A and B based on R. B. Chalamalasetty et al. 2014. *Development* 141: 4285–4297; unpublished photos courtesy of T. Yamaguchi and R. Chalamalasetty

FIGURE 17.5 *Mesogenin 1* (*Msgn1*) is both sufficient and necessary for *Tbx6* expression. (A) *Msgn1* gain-of-function. As visualized by in situ hybridization, misexpression of *Msgn1* in the presomitic mesoderm (lower two photos) is expanded throughout the trunk (left), which causes an expansion of *Tbx6* expression (right). (B) In contrast, loss of *Msgn1* function results in the reduction of *Tbx6* expression (blue, circled domains). (A, lateral views; B, dorsal views.)

Tbx6 is not the only PSM determination factor. Another transcriptional regulator of PSM fates, Mesogenin 1, may function as a pioneer transcription factor by acting upstream of Tbx6 (Yabe and Takada 2012; Chalamalasetty et al. 2014). Gain- and loss-of-function analyses of mouse *Mesogenin 1* have demonstrated that it is both sufficient and necessary for *Tbx6* expression in the PSM (**FIGURE 17.5**).

Taken together, these findings posit that a bipotential stem cell population is retained in the posterior region of the embryo, which maintains the extreme plasticity needed to give rise to cell fates spanning the mesodermal and ectodermal lineages (reviewed in Kimelman and Martin 2012; Neijts et al. 2014; Beck 2015; Carron and Shi 2015; Henrique et al. 2015). Although we have identified the transcriptional regulators of these stem cells, what signaling systems are in place to induce cell-type maturation from this **caudal (posterior) progenitor zone**?

Several opposing morphogens are displayed along the anterior-posterior axis of the paraxial mesoderm. Specifically, Fgf8 and Wnt3a are highly expressed in the vertebrate tailbud, whereas an anterior originating gradient of retinoic acid (RA) is produced from the somites and neural plate. RA directly represses *Fgf8* and *Tbx6* expression enough to foster neural cell fate maturation through upregulation of *Sox2* (**FIGURE 17.6A**; Kumar and Duester 2014; Cunningham et al. 2015; Garriock et al. 2015). Importantly, Fgf8 activates Cyp26b (a direct inhibitor of RA synthesis), which promotes mesodermal cell fate development (**FIGURE 17.6B**). Therefore, a balance of signaling between these opposing and antagonistic morphogens patterns cell migration, proliferation, and differentiation into the appropriate neural or mesodermal cell fates (Cunningham and Duester 2015; Henrique et al. 2015). However, this model must still enable the development of distinct somite identities along the anterior-posterior axis.

(A)

Wild-type   *Raldh2*^{-/-}

(B)

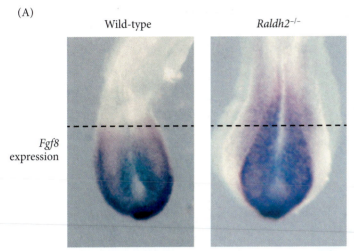

Fgf8
expression

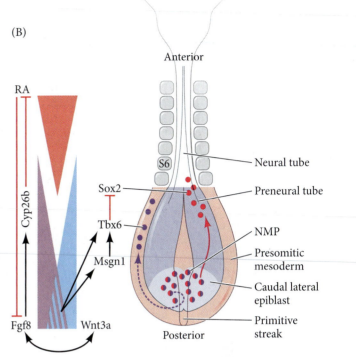

FIGURE 17.6 Antagonistic signals along the anterior-posterior axis pattern neuromesodermal progenitors (NMPs) during paraxial meso-derm development. (A) Loss of retinoic acid synthesis in the *Raldh2* knockout mouse shows an anterior expansion of *Fgf8* expression (blue past the dashed line). (B) Model of signaling systems regulating the NMP from the caudal progenitor zone (the caudal lateral epiblast) into the pre-neural tube or presomitic mesoderm. Posterior signals of FGF and Wnt antagonize anterior retinoic acid signaling. Fgf8 and Wnt3a upregulate *Mesogenin 1 (Msgn1)* and *Tbx6* expression to promote presomitic pro-genitor specification and repress neural cell fate specification (Sox2). (B after D. Henrique et al. 2015. *Development* 142: 2864–2875.)

Spatiotemporal collinearity of Hox genes determines identity along the trunk

All somites may resemble one another, but they form different structures along the rostral-caudal (anterior-posterior) axis. For instance, the somites that form the cervical vertebrae of the neck and lumbar vertebrae of the abdomen are not capable of form-ing ribs; ribs are generated only by the somites that form the thoracic vertebrae, and this specification of thoracic vertebrae occurs very early in development. The different regions of presomitic mesoderm are determined by their position along the anterior-posterior axis before somite formation. If presomitic mesoderm from the thoracic region of a chick embryo is transplanted to the cervical (neck) region of a younger embryo, the host embryo will develop ribs in its neck on the side of the transplant (**FIGURE 17.7A**; Kieny et al. 1972; Nowicki and Burke 2000).

The anterior-posterior specification of the somites is determined by Hox genes (see Chapter 12). Hox gene expression is spatially presented in a collinear fashion. Thus, the Hox genes organized in more 3′ positions of their cluster in the genome are expressed in more anterior regions of the paraxial mesoderm; conversely, the Hox genes at more 5′ positions are expressed in more posterior regions of the embryo (see Figure 12.20; Wellik and Capecchi 2003). If this pattern of Hox gene expression is altered, so is the specification of the mesoderm. For instance, if the entire pre-somitic mesoderm ectopically expresses *Hoxa10*, ribs are completely lost due to the replacement of thoracic vertebrae with lumbar vertebrae (**FIGURE 17.7B**). If *Hoxb6* is misexpressed throughout the PSM, however, all vertebrae form ribs (**FIGURE 17.7C**; Carapuço et al. 2005; Guerreiro et al. 2013). In both cases, these transformations of vertebral identity were induced by the ectopic expression within the PSM, not in the somites, which suggests that cells of the PSM receive instructions for axial-level specification, and these instructions are then implemented later during somite

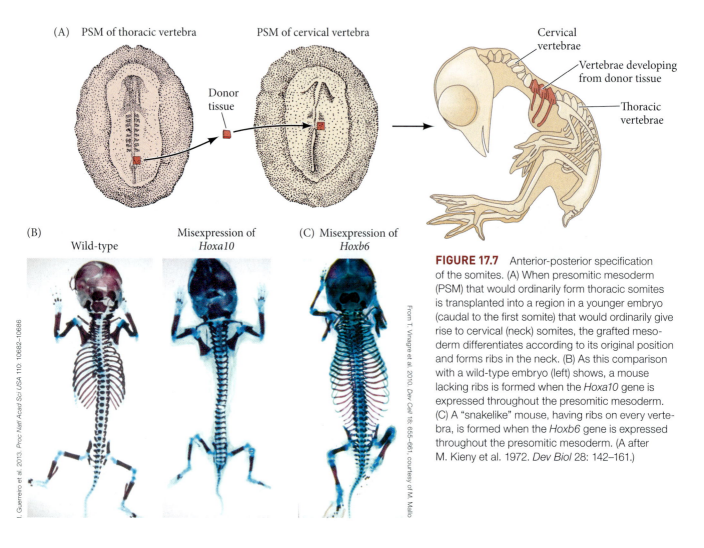

(A) PSM of thoracic vertebra   PSM of cervical vertebra   Cervical vertebrae

Donor tissue

Vertebrae developing from donor tissue

Thoracic vertebrae

(B) Wild-type   Misexpression of *Hoxa10*   (C) Misexpression of *Hoxb6*

I. Guerreiro et al. 2013. *Proc Natl Acad Sci USA* 110: 10682–10686

From T. Vinagre et al. 2010. *Dev Cell* 18: 655–661, courtesy of M. Mallo

FIGURE 17.7 Anterior-posterior specification of the somites. (A) When presomitic mesoderm (PSM) that would ordinarily form thoracic somites is transplanted into a region in a younger embryo (caudal to the first somite) that would ordinarily give rise to cervical (neck) somites, the grafted meso-derm differentiates according to its original position and forms ribs in the neck. (B) As this comparison with a wild-type embryo (left) shows, a mouse lacking ribs is formed when the *Hoxa10* gene is expressed throughout the presomitic mesoderm. (C) A "snakelike" mouse, having ribs on every verte-bra, is formed when the *Hoxb6* gene is expressed throughout the presomitic mesoderm. (A after M. Kieny et al. 1972. *Dev Biol* 28: 142–161.)

differentiation. In the chick, prior to cell migration through the primitive streak, Hox gene expression can be labial; once the paraxial mesoderm cells have taken up posi-tion in the presomitic mesoderm, however, Hox gene expression appears more fixed. Indeed, once established, each somite retains its pattern of Hox gene expression, even if that somite is transplanted into another region of the embryo (Nowicki and Burke 2000; Iimura and Pourquié 2006; McGrew et al. 2008).

FURTHER DEVELOPMENT

Hox genes and their temporal collinearity

The **temporal collinearity** of Hox genes refers to a temporal control mechanism of Hox gene activation that sets up the spatial collinearity of Hox gene expression along the anterior-posterior axis of the embryo, such that this expression corresponds to the genomic organization of Hox genes along the 3′-to-5′ orientation. In other words, the earlier expressed Hox genes are both expressed by cells positioned more anteriorly in the embryo and found in more 3′ chromosomal locations. In fact, it is the dynamic temporal pairing of 3′-to-5′ Hox gene activation with the timing of cell ingression/migration into the paraxial mesoderm that then lays out the spatial pattern of Hox gene expression along the trunk (**FIGURE 17.8A**; Izpisúa-Belmonte et al. 1991a,b). Initially, a progress zone model was proposed in which the progres-sive activation of 3′ to 5′ Hox genes occurs, respectively, in earlier to later cells of the PSM (Kondo and Duboule 1999; Kmita and Duboule 2003). More recent investiga-tions on chick embryos, however, have favored a model in which Hox genes control the timing of cell ingression through the primitive streak; hence, cells that express

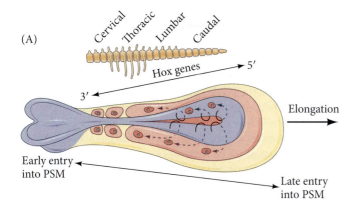

(A)

After D. Noordermeer et al. 2014, *eLife* 3: e02557

(B) Transition to chromatin states

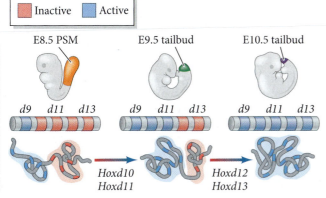

FIGURE 17.8 The spatiotemporal collinearity of Hox gene expression in the presomitic mesoderm (PSM) is correlated with chromatin remodeling. (A) Illustration showing the successive migration of cells into the PSM as the tailbud elongates, which correlates with the onset of progressively more 5′ Hox gene expression and the development of different vertebral identities. (B) Changes in chromatin structure permit progressive access for differential Hox gene expression over the course of PSM elongation. Colored PSM and tailbuds represent tissues used to analyze *Hoxd* chromatin structure at the different embryonic stages (E8.5–E10.5) (B after D. Noordermeer et al. 2014. *eLife* 3: e02557.)

anterior Hox genes will ingress early, whereas cells that express more posterior Hox genes will ingress later (Iimura and Pourquié 2006; Denans et al. 2015). During this period of paraxial mesoderm development, the axis is elongating as new progenitor cells enter the PSM from the caudal progenitor zone and occupy sequentially more posterior positions over time. Thus, more-anterior PSM cells will express more 3′ Hox genes, while the later-incorporated PSM cells will be located in the posterior and express more 5′ Hox genes (across paralogues)—all of which results in the final identity of somites along the anterior-posterior axis (see Figure 17.8A; reviewed in Casaca et al. 2014).

This temporal activation of Hox genes has been called the Hox clock (Duboule and Morata 1994). How is such a linear activation of Hox genes carried out on the cellular level? Research has shown that the Hox genes change from a tightly packed to an unpacked architecture[1] in a chronological order that matches the order of their expression in the PSM: first the 3′ Hox genes (*Hoxd4*) show signs of unpacked structure, followed by the *Hoxd8-9* clusters, then the *Hoxd10*, and finally the more 5′ *Hoxd11-12* (**FIGURE 17.8B**; Montavon and Duboule 2013; Noordermeer et al. 2014). Moreover, it appears that once a cell has adopted its position in the PSM, it has also adopted its particular chromatin state for its Hox genes, and all its daughter cells retain a fixed memory of that state.

Somitogenesis

How does the presomitic mesoderm become partitioned into the correct number of appropriately sized and bilaterally symmetrical somites? As the mesenchymal cells of the PSM mature, they become organized into "whorls" of cells that constitute the somite precursors and are sometimes called **somitomeres** (Meier 1979). These somite precursors undergo cell organizational changes such that the outer cells adhere together to form an epithelium while the inner cells remain mesenchymal. The first somites

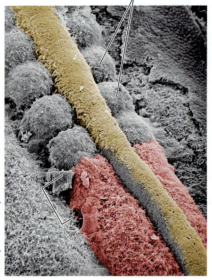

Somites

Courtesy of K. W. Tosney

FIGURE 17.9 Neural tube and somites seen by scanning electron microscopy. When the surface ectoderm is peeled away, well-formed somites are revealed, along with paraxial mesoderm (red) that has not yet separated into distinct somites. A rounding of the paraxial mesoderm into a somitomere (area inside bracket) is seen at the lower left, and neural crest cells can be seen migrating ventrally from the roof of the neural tube (yellow).

appear just posterior to the region of the otic vesicle, and new somites "bud off" from the rostral end of the PSM at regular intervals (**FIGURE 17.9**). The formation of somites is called **somitogenesis**, and this process involves the periodic creation of an epithelial fissure by the mesenchymal cells of the PSM. These medial-to-lateral partitions establish an epithelial boundary between the posterior half of the next anterior somite and the anteriormost extent of the PSM. Therefore, whole somites with anterior and posterior borders are not formed simultaneously; rather, one boundary at a time is created at regular intervals. When a new boundary is formed, it creates the posterior half of a somite (thus completing a fully formed somite) as well as establishes the anterior half of the next somite to be formed.

To accurately describe the somites that have formed and the position of somitomeres in the presomitic mesoderm, a numbering scheme was established using Roman numerals (Pourquié and Tam 2001). The most recently formed somite is always position I, and each additionally older somite is numbered counting up: II, III, IV, and so on. Moving posteriorly into regions where somites have not yet formed, positions of future somites are numbered counting backward, 0, –I, –II, –III, and so on. Somitomere 0 (zero) shares a boundary with somite I and is always the next somite to be formed.

Because individual embryos in any species can develop at slightly different rates (as when chick embryos are incubated at slightly different temperatures), the number of somites formed is usually the best indicator of how far development has proceeded. The number of somites in an adult individual is species-specific. Chicks have about 50 pairs of somites, mice have 65 pairs (many of them in the tail), zebrafish have 33 pairs, and humans generally have between 38 and 45 pairs (Müller and O'Rahilly 1986). Some snakes have as many as 500 pairs of somites!

Axis elongation: A caudal progenitor zone and tissue-to-tissue forces

Compared with most other vertebrates, the snake clearly has a longer anterior-posterior axis relative to its other axes; thus, one important factor influencing somitogenesis might be the process by which this axis is elongated. Previously, we alluded to the origin of cells that make up the PSM, from which cells in the anterior paraxial mesoderm arise during gastrulation by ingression through the primitive streak in amniotes or via convergence on the midline in fish and amphibians. As we discussed previously in this chapter, however, the posteriormost region of the tail contains a population of multipotent progenitor cells that have the potential to contribute to both the neural tube (*Sox2*-expressing) and the paraxial mesoderm (*Tbx6*-expressing); hence, these cells are called **neuromesoderm progenitors**, or **NMPs** (Tzouanacou et al. 2009). The nascent paraxial mesoderm cells are released from this pool and become positioned in the tail end of the rods of the presomitic mesoderm.

Although the mechanisms vary somewhat among species, the three most significant factors driving axis elongation from the caudal progenitor zone are *cell proliferation, cell migration,*[2] and *intertissue adhesion*. To exemplify the contributions of these three mechanisms, we will examine their roles in the zebrafish tailbud. The zebrafish tailbud can be divided into four regions that reflect different cell behaviors: the dorsal medial zone, the progenitor zone, the maturation zone, and a region occupied by the emerging PSM (**FIGURE 17.10A**). By using a nuclear-localized transgenic reporter, the directional movements of cells within the tailbud have been traced (Lawton et al. 2013). This analysis demonstrated that the bipotential **neuromesodermal** stem cells reside in the dorsal medial zone (see Figure 17.10A), which is positioned just dorsal to the neural tube and the axial and paraxial mesoderm within the tailbud (Martin and Kimelman 2012). These NMP cells first rapidly move posteriorly by **collective migration**[3] into the progenitor zone (tip of the tailbud; see Figure 17.10A). In the progenitor zone, cell velocities slow due to a reduction in "coherence" and concomitant cell mixing. The authors of this study appropriately compare this effect to the flow of traffic. Automobiles all moving in the same direction can achieve high speeds, but when vehicles change direction, swerve in and out of lanes, or even turn around entirely, it triggers a dramatic reduction in speed in the group of vehicles. In the context of the movement of NMP cells, it is thought that this change in community behavior might enable these cells to both change directions and begin to

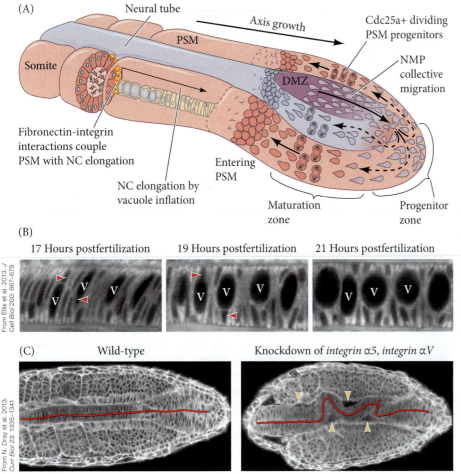

(A)

Neural tube

Axis growth

PSM

Cdc25a+ dividing PSM progenitors

Somite

NMP collective migration

DMZ

Fibronectin-integrin interactions couple PSM with NC elongation

Entering PSM

NC elongation by vacuole inflation

Maturation zone

Progenitor zone

From Ellis et al. 2013. J Cell Biol 200: 667–679

(B)

17 Hours postfertilization   19 Hours postfertilization   21 Hours postfertilization

From N. Dray et al. 2013. Curr Biol 23: 1335–1341

(C)   Wild-type   Knockdown of *integrin α5, integrin αV*

FIGURE 17.10   A model of axis elongation in zebrafish. (A) Bipotential neuromesodermal progenitor (NMP) cells from the dorsal medial zone (DMZ) collectively migrate to the progenitor zone, where they diverge either toward a neural lineage and into the neural tube or bilaterally migrate toward the maturation zones and on into the PSM. Progenitor cells transiently express Cdc25a in the maturation zone to trigger a round of division. Chordamesoderm cells inflate vacuoles that exert pressure on surrounding tissues, which results in a posterior extension of the notochord (NC). Fibronectin-integrin interactions tether the PSM to the notochord, resulting in a posterior pulling of the PSM during notochord elongation. These three processes—cell migration, cell division, and PSM-NC-coupled tissue shifting—together drive axis elongation. (B) High magnification of notochord cells in the zebrafish embryo shows vacuoles (V) filling over time, which contributes to notochord extension (red arrowheads point out nuclei). (C) Double knockdown of *integrin α5* and *integrin αV* by morpholino prevents adhesion of notochord cells with the extracellular matrix; as a result, the notochord buckles as it attempts to elongate (red line and arrowheads).

synchronize their developmental trajectories. These progenitor cells turn anteriorly to migrate bilaterally into the maturation zones on either side of the posteriormost axial mesoderm and finally into the PSM region (see Figure 17.10A).

When NMP cells migrate through the maturation zone, as the term *maturation* implies, they begin to express mesodermal markers (*Msgn1* and *Tbx6* genes). However, they also transiently express Cdc25a, which promotes *one* cycle of division in these cells before they move into the PSM and differentiate (Bouldin et al. 2014). At least in the zebrafish tailbud, although cells are proliferating, cell *migration* appears to be the more significant contributor to axis elongation (reviewed in McMillen and Holley 2015).

FURTHER DEVELOPMENT

LEVERAGING AN EXTRACELLULAR PUSH   In addition to cell migration and proliferation, intertissue adhesive forces also contribute to the extension of the anterior-posterior axis. The principal tissues at play are the paraxial mesoderm and the adjacent notochord. As maturing NMP cells move into the PSM, a fibronectin matrix becomes progressively deposited on the surface of the PSM and the interfaces between the paraxial mesoderm (somites and PSM) and the notochord (see Figure 17.10A). During this period of axis elongation, chordamesodermal cells are undergoing significant cellular changes, which result in a stiffening and directed extension of the notochord (Ellis et al. 2013a). Specifically, chordamesodermal cells use endosomal trafficking to inflate large vacuoles (yes, that is very cool), which leads to an increase in cell size that exerts pressure on the surrounding tissues (**FIGURE 17.10B**). Chordamesodermal cells also secrete a sheath of extracellular matrix components (collagen, laminin) that surrounds the notochord and resists expansion from the internal pressure of the cells. As a result of this architecture and inflating chordamesoderm, the notochord elongates in the direction of least resistance—toward the tail (see Figure 17.10A; Ellis et al. 2013b). At least in zebrafish, the paraxial mesoderm essentially "hitches a ride" on the notochord by using the fibronectin matrix and integrin receptors to mechanically couple the posterior extension of the PSM with the elongation of the notochord (**FIGURE 17.10C**; Dray et al. 2013; McMillen and Holley 2015).

Once the PSM is growing, how does it get chopped up into repeated segments? A critical insight into the process was revealed when *Xenopus* and mouse embryos were experimentally reduced in size: the resulting number of somites generated was normal, but each somite was smaller than normal (Tam 1981). This result suggested that a regulatory mechanism controls the number of somites independent of the size of the segmenting tissue. We can therefore refine our questions to ask, What mediates the epithelialization of PSM cells to physically create a boundary and consequently a somite? What mechanism(s) define the position and timing of this boundary formation?

HOW SOMITES TRANSITION FROM A MESENCHYMAL TO AN EPITHELIAL ARCH-ITECTURE Somite architecture is built of epithelial blocks, although the PSM only supplies mesenchymal cells. Therefore, the embryo must transform the mesenchymal cells into epithelial cells in a **mesenchymal-epithelial transition (MET)**. This process involves the upregulation of *Mesodermal posterior* (*Mesp*), which codes for a transcription factor that regulates the onset of the MET. As a somite is forming, *Mesp* expression quickly becomes restricted to the anterior half of the somite (**FIGURE 17.11A**). A principal function of Mesp is to upregulate *Eph* in the anterior portion of somitomeres (**FIGURE 17.11B,C**). Eph activity at the presumptive anterior border of a somitomere (S – I) triggers the upregulation of its own ligands, the ephrins, in the opposing posterior half of the more anterior somitomere (S0; see Figure 17.11B,C), which is a pattern that is sequentially repeated over the course of somitogenesis (**FIGURE 17.12**; Watanabe and Takahashi 2010; Fagotto et al. 2014; Cayuso et al. 2015; Liang et al. 2015).

We saw in Chapter 15 that the Eph tyrosine kinase receptors and their ephrin ligands are able to elicit cell-to-cell repulsion between the posterior region of a somite and migrating neural crest cells. Similarly, separation of the somite from the anterior end of the presomitic mesoderm occurs at the border between ephrin- and Eph-expressing cells (see Figure 17.11C; Durbin et al. 1998). Interfering with this signaling (by injecting embryos with mRNA encoding dominant negative Ephs) leads to the formation of abnormal somite boundaries. Moreover, in *fused somites* (*tbx6*) zebrafish mutants, *Eph A4* is lost, and *ephrin B2* is ubiquitously expressed throughout the paraxial

Images A and B from L. Durbin et al. 2000. *Development* 127: 1703–1713

(A) *Mesp-a*

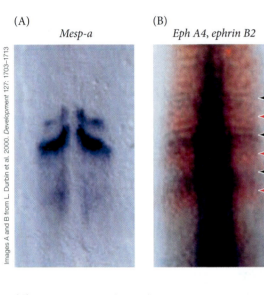

(B) *Eph A4, ephrin B2*

(C) Paraxial mesoderm   NC

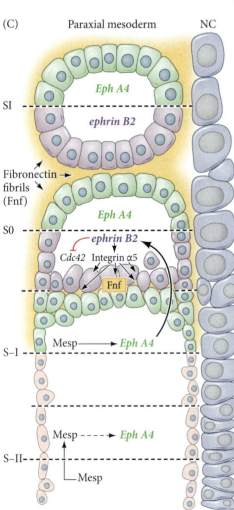

SI

Fibronectin fibrils (Fnf)

S0

Eph A4

ephrin B2

Cdc42 Integrin α5

Fnf

S–I

Mesp ⟶ *Eph A4*

S–II

Mesp ⤏ *Eph A4*

Mesp

FIGURE 17.11 Eph-ephrin signaling regulates epithelialization during somite boundary formation. Expression of *Mesodermal posterior-a* (*Mesp-a*; dark purple) (A) and of *Eph A4* (black arrows) and *ephrin B2* (red arrows) (B) in the paraxial mesoderm of zebrafish embryos (dorsal views). (C) Model of Mesp and Eph-ephrin signaling fostering mesenchymal-epithelial transitions that define the apposing cells of a somite boundary. Mesp-a becomes restricted to the anterior half of the S–I somitomere, which upregulates *Eph A4* within this domain. In turn, *Eph A4* upregulates its binding partner *ephrin B2* in the cells of the presumptive posterior S0 somitomere, which triggers epithelialization and formation of a boundary. Fissure formation is facilitated by repression of *Cdc42* and activation of integrin α5-fibronectin interactions downstream of *ephrin B2*. NC, notochord.

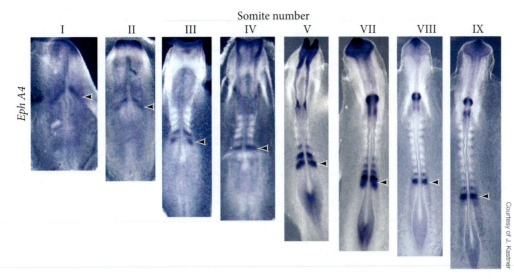

Somite number

I   II   III   IV   V   VII   VIII   IX

Eph A4

FIGURE 17.12 Ephrin and its receptor constitute a possible fissure site for somite formation. In situ hybridization shows that *Eph A4* (dark blue; arrowheads) is expressed as new somites form in the chick embryo.

mesoderm, and consequently no somite boundaries are formed (Barrios et al. 2003). The Eph A4–ephrin B2 signaling leads to epithelialization immediately after somite fission by regulating two downstream factors, Rho GTPases and integrin-fibronectin interactions. (See **Further Development 17.2, From Cell Surface Interactions to Cytoskeletal Rearrangements**, online.)

How a somite forms: The clock-wavefront model

Somites appear on both sides of the embryo at exactly the same time. Even when isolated from the rest of the body, presomitic mesoderm will segment at the appropriate time and in the correct direction (Palmeirim et al. 1997). The current predominant model to explain synchronized somite formation is the clock-wavefront model proposed by Cooke and Zeeman (1976; reviewed by Hubaud and Pourquié 2014). In this model, two converging systems interact to regulate (1) where a boundary will be capable of forming (the wavefront) and (2) when epithelial boundary formation should occur (the clock).

The wavefront, more aptly called the **determination front**, is established by a caudal^{HIGH}-to-rostral^{LOW} gradient of FGF activity within the PSM, which is inverse to the rostral^{HIGH}-to-caudal^{LOW} gradient of retinoic acid (**FIGURE 17.13**). FGF signaling maintains PSM cells within an immature state; therefore, cells that become positioned at lower threshold concentrations of FGF activity will become competent to form a boundary. However, only those cells that are both competent and receiving the temporal instructions to form a boundary will actually undergo epithelialization (*activation of the Mesp-to-Eph cascade*).

The instructions for when to create a segmentation furrow are largely controlled by the clock-like oscillating signals of the Notch pathway. Each oscillation of Notch-Delta organizes groups of presomitic cells that will then segment together at the appropriate threshold of FGF signaling (see Maroto et al. 2012). In the chick embryo, a new somite is formed about every 90 minutes. In mouse embryos, this time frame is more variable but is closer to every 2 hours, whereas in zebrafish, a somite is formed approximately every 30 minutes (Tam 1981; Kimmel et al. 1995).

WHERE A SOMITE BOUNDARY FORMS: THE DETERMINATION FRONT Recall the maturation of a newly born NMP cell from the tailbud; as it enters the PSM, the axis continues to grow, and this cell will eventually find itself part of a somite. Interestingly, the periodicity at which a group of cells contributes to a newly formed somite generally occurs at the same distance from the tailbud, although some exceptions exist. This finding suggests that the posterior extension of the tailbud strongly influences the location of boundary

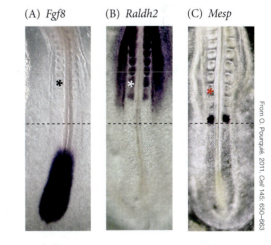

(A) *Fgf8*   (B) *Raldh2*   (C) *Mesp*

FIGURE 17.13 Somites form at the junction of retinoic acid (anterior) and FGF (posterior) domains. Asterisks show the last formed somite. The dashed line approximates the boundary region where somites are being determined. (A) *Fgf8* expression (purple) in the posterior part of the embryo. (B) RNA for *Raldh2* (retinoic acid–synthesizing enzyme) in the central part of the embryo. (C) *Mesp* stain shows the region where the somite formation will occur later.

formation, and this is indeed the case. Earlier we discussed the role of the opposing anterior-to-posterior gradients of RA and Fgf8/Wnt3a signaling in mediating NMP cell specification (see Figure 17.6). This robust morphogenetic mechanism is similarly used to influence the maturation of PSM cells to become competent to form boundaries and, as such, is called the *wavefront*, *wave*, or *determination front* (see Figure 17.13B). We will use *determination front* throughout the rest of this chapter to reduce potential confusion with discussions of oscillating waves of gene expression covered later in this section.

Elegant somite inversions were performed in the chick embryo to identify the location of the determination front in the PSM (**FIGURE 17.14A**; Dubrulle et al. 2001). As we already know, an anterior-posterior prepattern develops in the somitomeres of the PSM before the somites form. In the chick inversion experiments, inverting somitomere 0 resulted in an unchanged, committed gene expression pattern, as illustrated by the retention of a reversed posterior-to-anterior pattern in this inverted tissue (this somitomere region was "determined"). However, similar inversions at somitomeres –III and –VI caused variable patterning changes ("labial") and a complete reassignment of polarity ("undetermined"), respectively (**FIGURE 17.14B–D**). From this work, the determination front was identified to reside at somitomere –IV.

This determination front is at the trailing edge of an Fgf8 gradient originating from the tailbud and node of the primitive streak. Most interesting is how the Fgf8 gradient is shaped in the PSM. *Fgf8* is transcribed only in the tailbud and *not* in the PSM (**FIGURE 17.14E,F**); therefore, as the tailbud grows caudally, so does the source of cells actively transcribing *Fgf8*. One mechanism playing a major role in shaping the Fgf8 gradient is RNA decay (Dubrulle and Pourquié 2004). The amount of *Fgf8* transcript in a PSM cell will steadily decline over time due to RNA decay, which will thereby create a caudal-to-rostral gradient of Fgf8 activity (**FIGURE 17.14G**). In this way, a gradient of Fgf8 provides different concentration thresholds across the PSM. In addition, the absence of *Fgf* transcription in the PSM is maintained by the increasing concentration of the repressive retinoic acid from the somites and anterior PSM. What is the cellular outcome of these opposing morphogens?

These and other findings (Dubrulle et al. 2001) suggest that the Fgf8 morphogen *is* the molecular determination front for epithelialization, and at somitomere –IV the threshold drops low enough to permit those cells at that axial location to be competent to form a boundary. More specifically, the cells at the Fgf8 determination front become competent to respond to the "molecular clock," whose alarm goes off when a boundary should be created.

WHEN A SOMITE BOUNDARY FORMS: THE CLOCK   Scientists use the analogy of a clock to describe the controls behind the periodicity of boundary (and somite) formation. What would a molecular clock look like in an embryo? What duration constitutes a "time period" for this clock? In the context of a cell, a clock can simply be the regular fluctuation of a protein's activity on and off, provided this change in activity is in fact repeated and rhythmic in nature. In the context of a tissue, however, this fluctuating protein timepiece needs to somehow be communicated across a field of cells. Thus, one model for a molecular clock of somitogenesis would posit that the activity of a protein that regulates the mesenchymal-epithelial transitions (METs) at a somite boundary might become functional in one cell of the PSM and as part of its function relay this event to its neighboring cells through cell-to-cell interactions until its activity is cyclically inhibited. In this way, each cell across a tissue could experience "on" and "off" states of protein activity, or *the ticking of a clock*.

One of the key "clock" components that maintains the pace of somitogenesis is the Notch signaling pathway (see Wahi et al. 2014). When a small group of cells from a region constituting the posterior border at the presumptive somite boundary is transplanted into a region of presomitic mesoderm that would not ordinarily be part of the boundary area, a new boundary is created. The

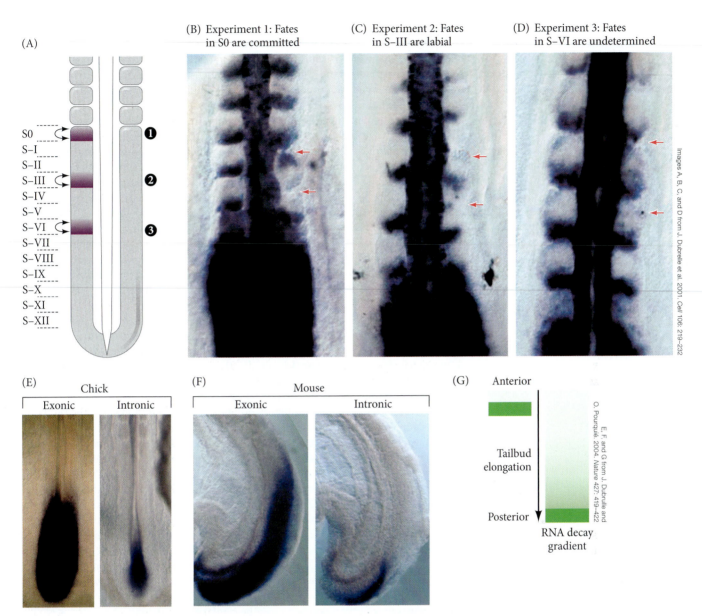

Images A, B, C, and D from J. Dubrulle et al. 2001, *Cell* 106: 219–232

E. F. and G from J. Dubrulle and O. Pourquié. 2004. *Nature* 427: 419–422

FIGURE 17.14 A posterior *Fgf8* gradient establishes the determination front. (A) Schematic of a series of somitomere tissue reversals in which presomitic tissue at three different axial locations (1, 2, and 3) was flipped along the anterior-posterior axis. (B–D) *c-delta1* gene expression in chick embryos that had somitomere reversal at the axial level denoted in the schematic. For each example, the control side is on the left and the experimental side is on the right. Flipping somitomere S0 (B) showed complete commitment to maintain the positional expression of *c-delta1*; thus, it was already determined. In contrast, flipping somitomeres at positions S–III (C) and S–VI (D) showed, respectively, disorganized expression and normal posterior expression of *c-delta1*. These data suggested that cells become determined to form boundary locations at S–IV. Red arrows indicate points of surgical inversion. (E,F) *Fgf8* expression in the chick and mouse. Exonic and intronic in situ probes to reveal, respectively, any cell with *Fgf8* mRNA or nuclear RNA (pre-mRNA). Careful examination of these results demonstrates two important properties of the *Fgf8* gradient. The first is that the *Fgf8* gene is actively transcribed only in the tailbud (intronic probe). (G) The second property is that a gradient of *Fgf8* in the presomitic mesoderm is established by a mechanism of RNA decays. Green bar represents the posterior movement of cells actively transcribing *Fgf8*, and the trailing gradient is made by RNA decay over time in cells no longer transcribing *Fgf8*.

transplanted boundary cells instruct the cells anterior to them to epithelialize and separate. Nonboundary cells will not induce border formation when transplanted to a non-border area. However, these cells can acquire boundary-forming ability if an activated Notch protein is electroporated into them, which demonstrates that Notch signaling can induce the METs that underlie somite formation (Sato et al. 2002). (See **Further Development 17.3, Notch Signaling and Somite Formation**, online.)

The molecular clock of somitogenesis directs the timing of the formation of the somite boundaries. If Notch activity represents the mechanism of this timekeeper, then it must show an on-off rhythmicity that is also transferred across cells. The endogenous level of Notch activity in the mouse PSM has been visualized and shown to oscillate in a segmentally defined pattern that correlates with boundary formation (Morimoto et al. 2005; Aulehla et al. 2008). Like a wave of gene expression across the PSM, cells experience Notch upregulation and then downregulation from the caudal to rostral extents of the PSM. This wave of Notch expression crashes on somite 0, where a somite boundary forms at the interface between the Notch-expressing and Notch-nonexpressing areas.

In addition, Notch signaling provides a mechanism to transfer this signal from cell to cell across the PSM. As we discussed in Chapter 4, full-length Notch forms a transmembrane protein that binds to its receptor, Delta, in adjacent cells. Delta is also a transmembrane protein, and initial upregulation and presentation of Notch will trigger a concomitant upregulation of Delta in adjacent cells, which in turn will reinforce Notch in the other cells surrounding it. This is a pattern-generating mechanism by Notch-Delta signaling known as **lateral inhibition**. This receptor-to-receptor pairing provides a mechanism for signal transfer throughout the PSM; however, this mechanism predicts that the expression of Notch and Delta should display a mosaic pattern over the entire PSM, yet it doesn't. Like Notch, Delta also displays oscillatory expression patterns with a posterior-to-anterior progression across the PSM—a key feature of the clock mechanism. How is this possible?

Although there are interspecies differences in exactly which gene products oscillate, in all vertebrate species the on-off ticking of the clock involves a negative feedback loop of the Notch signaling pathway (Krol et al. 2011; Eckalbar et al. 2012). Thus, in all vertebrates at least one of the Notch target genes having dynamic oscillating expression in the presomitic mesoderm is also able to *inhibit* the gene for Notch, which establishes a negative feedback loop. These inhibitory proteins are unstable, and when the inhibitor is degraded, Notch becomes active again. Such feedback creates a cycle (the "clock") in which the gene for Notch is turned on and off by the absence or presence of a protein that Notch itself induces. These on-off oscillations could provide the molecular basis for the periodicity of somite segmentation (Holley and Nüsslein-Volhard 2000; Jiang et al. 2000; Dale et al. 2003). Such oscillating Notch targets include *Hairy1*, *Hairy/Enhancer of split-related proteins* (*Her*), and *Lunatic fringe*, all of which are activated by Notch, expressed in a similar oscillating pattern throughout the PSM from the tailbud to the last formed somite, and negatively feed back to repress Notch signaling (Chipman and Akam 2008; Pueyo et al. 2008). For instance, the *Hairy1* gene was the first Notch target found to show a rhythmic pattern of expression (**FIGURE 17.15**). The *Hairy1* gene is expressed first in a broad domain at the caudal end of the presomitic mesoderm. This expression domain moves anteriorly while narrowing until it reaches the rostral end of the PSM, at which time a new wave of expression begins in the caudal end. Aulehla and colleagues (2008) observed the waves of transcriptional activity of *Lunatic fringe* in live mice embryos, which

FIGURE 17.15 Somite formation correlates with wavelike expression of the *Hairy1* gene in the chick. (A) In the posterior portion of a chick embryo, somite SI has just budded off the presomitic mesoderm. Expression of the *Hairy1* gene (purple) is seen in the posterior half of this somite as well as in the posterior portion of the presomitic mesoderm and in a thin band that will form the posterior half of the next somite (S0). (B) A posterior fissure (small arrow) begins to separate the new somite from the presomitic mesoderm. The posteriormost region of *Hairy1* expression shifts anteriorly. (C) The newly formed somite, now referred to as SI, retains the expression of *Hairy1* in its posterior half. Again, the posteriormost region of *Hairy1* expression in the PSM continues to shift anteriorly as well as to shorten. The former SI somite, now called SII, undergoes differentiation. (D) Creation of this newly formed somite SI is complete, and a new cycle of *Hairy1* expression begins again. In the chick, formation of each somite and the wave of *Hairy1* expression through the PSM take about 90 minutes. (After I. Palmeirim et al. 1997. *Cell* 91(5): P639–P648.)

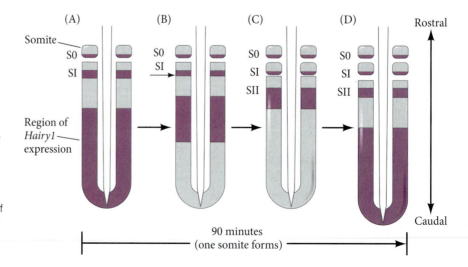

(A) Mouse

Wild-type

Lfng^{–/–}

Dll3^{–/–}

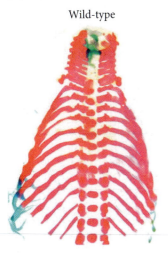

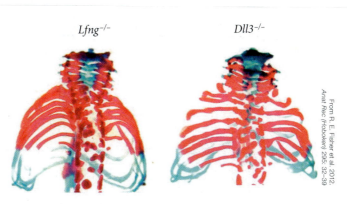

From R. E. Fisher et al. 2012, *Anat Rec (Hoboken)* 295: 32–39

(B) Human

Lfng C-to-A missense
mutation (inactive enzyme)

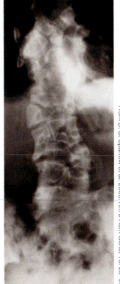

From D. B. Sparrow et al. 2006, *Am J Hum Genet* 78: 28–37

FIGURE 17.16 Notch-Delta signaling is essential for proper somito-genesis in mice and humans. In the mouse, loss of either the Notch target *Lunatic fringe* (*Lfng*) or its binding partner *Distal-less-3* (*Dll3*) results in severe vertebral malformations (A), which are particularly similar to the phenotype caused by known mutations in the human *LUNATIC FRINGE* (B).

mimic the pattern seen with *Hairy1*. The time it takes for a wave of expression to cross the presomitic mesoderm is 90 minutes in the chick, which is—not coincidentally—the time it takes to form a pair of somites in this species. This dynamic expression is not due to cell movement but to cells turning the gene on and off in different regions of the tissue through loops of negative feedback (Johnston et al. 1997; Palmeirim et al. 1997; Jouve et al. 2000, 2002; Dale et al. 2003). It should not be surprising, therefore, that loss of function of Notch or its downstream cycling target genes in mice and humans leads to major segmentation defects of the vertebrae, such as the spinal deformities in scoliosis and spondylocostal dysostosis (**FIGURE 17.16**; Zhang et al. 2002; Sparrow et al. 2006).

TERMINATING THE NOTCH CLOCK WITH EPITHELIALIZATION As we discussed earlier, Mesp is a global regulator of the Eph-ephrin cascade that triggers MET and boundary formation. Mesp is activated by Notch, and as a transcription factor, Mesp contributes to the suppression of Notch (Morimoto et al. 2005). This cycle of alternating activation and suppression causes Mesp expression to oscillate in time and space as well. Mesp is initially expressed in a somite-wide domain; it is then repressed in the posterior half of this domain but maintained in the anterior half, where it in turn represses Notch activity. Wherever Mesp expression is maintained, that group of cells becomes the most anterior in the next somite; Eph A4 is then induced, and the boundary forms immediate-ly anterior to those cells (see Figure 17.11C; Saga et al. 1997). In the posterior half of the prospective somite, where Mesp is not expressed, Notch activity induces the expression of the transcription factor Uncx4.1, which contributes to the specification of the somite's posterior identity (Takahashi et al. 2000; Saga 2007). In this way, the somite boundary is determined, and the somite is given anterior-posterior polarity at the same time.

BEING AT THE RIGHT PLACE AT THE RIGHT TIME: CONNECTING THE CLOCK AND THE DETERMINATION FRONT More posteriorly positioned PSM cells experiencing waves of Notch signaling do not prematurely epithelialize because these cells are not competent to sufficiently respond to Notch signaling due to the influence of FGFs. As long as the presomitic mesenchyme is in a region of relatively high Fgf8 concentration, the clock will not function. At least in zebrafish, this lack of function appears to be due to the repression of Delta, the major ligand of Notch. The binding of Fgf8 to its receptor enables the expression of the Her13.2 protein, which is necessary to inhibit transcrip-tion of Delta (see Dequéant and Pourquié 2008). FGF signals are needed to get cells to

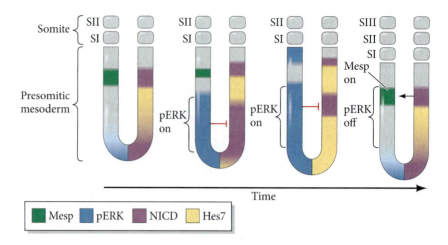

migrate anteriorly out of the tailbud, but as long as FGFs activate ERK transcription factors, the cells remain unresponsive to Notch ligands. It has been proposed that Fgf8 synthesis also cycles, but at a different frequency than that of the Notch ligands (Niwa et al. 2011; Pourquié 2011). Thus, through a combination of a concentration gradient of Fgf8 and its own unique pattern of cycling (probably by synthesizing its own inhibitors and by inhibition by retinoic acid), FGF signaling is downregulated in certain areas of the paraxial mesoderm, and cells in these regions become progressively more competent to respond to Notch signals (**FIGURE 17.17**; for a review, see Hubaud and Pourquié 2014). FGFs therefore establish the *placement* (i.e., determination front) of cells that are competent to respond to oscillating Notch signals (the clock), which can induce the METs that result in somite formation.[4] (See **Further Development 17.4, How Many Somites to Form? The Rate of Oscillations versus the Rate of Axis Elongation**, online.)

Linking the clock-wavefront to Hox-mediated axial identity and the end of somitogenesis

Somitogenesis cannot continue indefinitely; rather, it must end and reach this end with the appropriate anterior-posterior identities laid down. As previously mentioned, Hox genes play pioneer regulatory roles in the specification of axial identity from head to tail with spatial and temporal collinearity. How are the clock and the determination front connected to the role of Hox genes?

If Fgf8 protein levels are manipulated to create extra (albeit smaller) somites, the appropriate Hox gene expression will be activated in the appropriately numbered somite, even if it is in a different position along the anterior-posterior axis, which suggests that the determination front (FGF gradient) primarily influences somite size as opposed to Hox gene expression. Mutations that affect the autonomous segmentation clock, however, do affect activation of the appropriate Hox genes (Dubrulle et al. 2001; Zákány et al. 2001). The regulation of Hox genes by the segmentation clock presumably allows coordination between the formation and the specification of the new segments. How the mechanisms of somitogenesis feed into Hox gene regulation is not completely known. In *Xenopus*, research has shown that *XDelta2*, an oscillating gene and receptor to Notch, can upregulate at least three different Hox paralogue groups and initiate a positive feedback loop with these Hox proteins (Peres et al. 2006). Based on this association, it is tempting to speculate that the segmentation clock might directly trigger the timed activation of Hox genes; it is unknown, however, whether this influence occurs in a collinear fashion or involves chromatin modifications as described above for Hox gene expression.

Once Hox gene expression along the trunk is initiated, it will feed back on the determination front to terminate axis elongation and end somitogenesis. Specifically, the collinear activation of the expression of Hox genes closer to the 5′ end of the clusters results in a progressively greater repression of Wnt signaling in the tailbud (**FIGURE 17.18**; Denans et al. 2015). Recall that Wnt3a is secreted from the tailbud and displayed in a caudal-to-rostral gradient similar to Fgf8. Wnt3a functions to promote the migration of neuromesoderm progenitors into the PSM, and in doing so fuels PSM

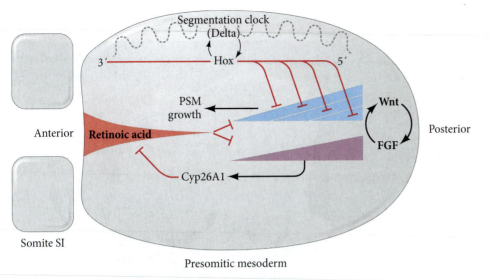

FIGURE 17.18 Model of the regulatory mechanisms governing somitogenesis. The molecular segmentation clock through Notch-Delta dictates the order of Hox gene expression, which functions in part to repress Wnt signaling and, indirectly, Fgf8 expression. Thus, the levels of the posterior-originating Wnt/Fgf8 determination front are influenced by Hox genes as well as by the development of anterior structures through retinoic acid signaling. Retinoic acid inhibits Fgf8 and Wnt3a, whereas Fgf8 is capable of repressing retinoic acid through Cyp26A1. Via this balance of signaling, anterior somites form before posterior somites. (© Michael Barresi.)

growth and axis elongation (Dunty et al. 2008). Thus, as new cells move into the PSM over time and start to express increasingly more 5′ Hox genes with temporal collinearity, Wnt signaling is progressively inhibited, and tailbud growth slows. In amniotes, the segmentation clock does not significantly alter its rate during this time; therefore, the rate of somite formation will outpace tailbud growth and exhaust the PSM, leading to the end of somitogenesis (Denans et al. 2015). Moreover, Hox gene-mediated slowing of tailbud growth is reinforced by the indirect repression of FGF signaling in two ways. First, the encroaching reach of retinoic acid from the somites will inhibit FGF expression with greater effectiveness; second, Wnt and FGF signaling will function in a positive feedback loop, maintaining each other's expression in the tailbud (Aulehla et al. 2003; Young et al. 2009; Naiche et al. 2011). Due to the progressive repression of Wnt signaling, FGF will indirectly also become downregulated over time. That, in turn, will lead to the downregulation of Cyp26A1, providing further redundancy for the hyperactivation of retinoic acid (Iulianella et al. 1999). Taken together, this provides a model in which the temporal collinear activation of 5′ Hox genes puts the brakes on axis elongation through the direct inhibition of Wnt signaling, which indirectly halts the determination front and exhausts the PSM (Denans et al. 2015).

Sclerotome Development

As a somite matures, it becomes divided into two major compartments, the sclerotome and the dermomyotome. These compartments are found in all vertebrates, and similar structures are found in the cephalochordate *Amphioxus* (lancelets), one of our closest invertebrate relatives, indicating that they are ancient embryonic structures (Devoto et al. 2006; Mansfield et al. 2015). How the sclerotome and dermomyotome develop is a complex story involving epithelial-mesenchymal transitions (EMTs) and signaling cascades. Here we turn to the development of the sclerotome, with a discussion of dermomyotome formation to follow.

Shortly after somite formation, the outer epithelial cells and inner core of mesenchymal cells begin to show signs of differentiation (**FIGURE 17.19A,E**). The first visible indicator occurs in the ventromedial portion of the somite, where an EMT occurs to form the sclerotome (**FIGURE 17.19B**). This EMT is important to establish a migratory population of cells capable of moving into position around the midline axial structures and building the vertebral column. Just prior to the transition, sclerotome progenitor cells express the transcription factor Pax1, which is required for their transition into a mesenchyme and subsequent differentiation into cartilage (Smith and Tuan 1996). In this transition, the epithelial cells lose N-cadherin expression and become motile mesenchyme (**FIGURE 17.19C,D,F**; Sosic et al. 1997). Sclerotome cells also express inhibitors of the muscle-forming transcription factors—myogenic regulatory factors, or MRFs—which we will discuss soon (Chen et al. 1996).

(A) 2-Day embryo

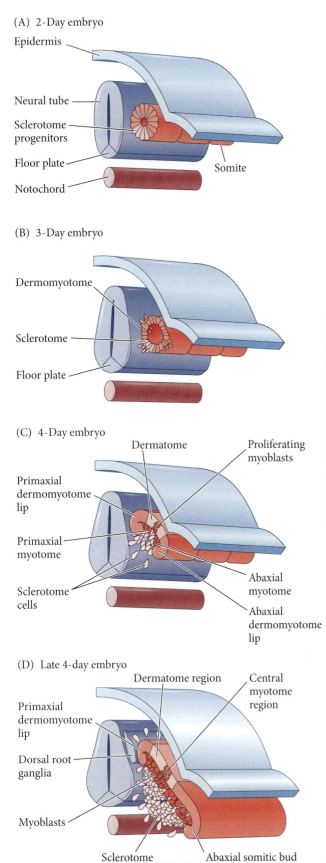

Epidermis

Neural tube

Sclerotome progenitors

Floor plate

Notochord

Somite

(B) 3-Day embryo

Dermomyotome

Sclerotome

Floor plate

(C) 4-Day embryo

Dermatome

Proliferating myoblasts

Primaxial dermomyotome lip

Primaxial myotome

Sclerotome cells

Abaxial myotome

Abaxial dermomyotome lip

(D) Late 4-day embryo

Dermatome region

Central myotome region

Primaxial dermomyotome lip

Dorsal root ganglia

Myoblasts

Sclerotome

Abaxial somitic bud

(E)

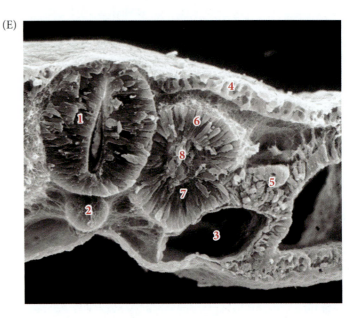

FIGURE 17.19 Transverse section through the trunk of a chick embryo on days 2–4. (A) In the 2-day somite, the sclerotome cells can be distinguished from the rest of the somite. (B) On day 3, the sclerotome cells lose their adhesion to one another and migrate toward the neural tube. (C) On day 4, the remaining cells divide. The medial cells form a primaxial myotome beneath the dermomyotome, and the lateral cells form an abaxial myotome. (D) A layer of muscle cell precursors (the myotome) forms beneath the epithelial dermomyotome. (E,F) Scanning electron micrographs correspond to (A) and (D), respectively; 1, neural tube; 2, notochord; 3, dorsal aorta; 4, surface ectoderm; 5, intermediate mesoderm; 6, dorsal half of somite; 7, ventral half of somite; 8, somitocoel/arthrotome; 9, central sclerotome; 10, ventral sclerotome; 11, lateral sclerotome; 12, dorsal sclerotome; 13, dermomyotome. (A,B after J. Langman. 1981. *Medical Embryology*, 4th Ed. Williams & Wilkins, Baltimore; C,D after C. P. Ordahl. 1993. In *Molecular Basis of Morphogenesis*, M. Bernfield [Ed.], pp. 165–170. Wiley-Liss, New York.)

(F)

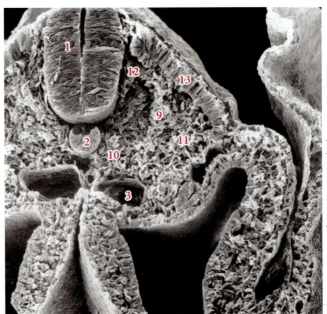

Images E and F from B. Christ et al. 2004. *Anat Embryol (Berl)* 208: 333–350, courtesy of H. J. Jacob and B. Christ.

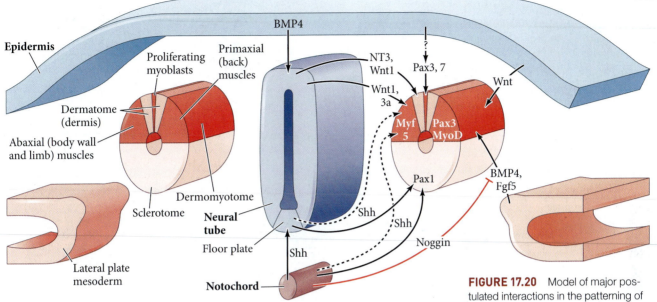

FIGURE 17.20 Model of major postulated interactions in the patterning of the somite. The sclerotome is white; dermomyotome regions are red and pink. A combination of Wnts (probably Wnt1 and Wnt3a) is induced by BMP4 in the dorsal neural tube. These Wnt proteins, in combination with low concentrations of Sonic hedgehog from the notochord and floor plate, induce the primaxial myotome, which synthesizes the myogenic transcription factor Myf5. High concentrations of Shh from the notochord and neural tube floor plate induce Pax1 expression in those cells fated to become the sclerotome. Certain concentrations of neurotrophin 3 (NT3) from the dorsal neural tube appear to specify the dermatome, while Wnt proteins from the epidermis, in conjunction with BMP4 and Fgf5 from the lateral plate mesoderm, are thought to induce the primaxial myotome. The proliferating myoblasts are characterized by Pax3 and Pax7 and are induced by Wnts in the epidermis. (After G. Cossu et al. 1996. *Trends Genet* 12: 218–223.)

As mentioned near the start of this chapter, the mesenchymal cells that make up the sclerotome can be subdivided into several regions (see Figure 17.2E). Although most sclerotome cells become precursors of the vertebral and rib cartilage, the dorsal sclerotome forms the syndetome, giving rise to tendons, and the medial sclerotome cells closest to the neural tube generate the meninges (coverings) of the spinal cord as well as give rise to blood vessels that will provide the spinal cord with nutrients and oxygen (Halata et al. 1990; Nimmagadda et al. 2007). The cells in the center of the somite (which remain mesenchymal) also contribute to the sclerotome, becoming the vertebral joints, the cartilaginous discs between the vertebrae (intervertebral discs), and the portions of the ribs closest to the vertebrae (Mittapalli et al. 2005; Christ et al. 2007; Scaal 2015). This region of the somite has been called the arthrotome.

Like the proverbial piece of real estate, the destiny of a particular region of the somite depends on three things: location, location, location. As shown in **FIGURE 17.20**, the locations of the somitic regions place them close to different signaling centers, such as the notochord and floor plate (sources of Sonic hedgehog and Noggin), the neural tube (source of Wnts and BMPs), and the surface epithelium (also a source of Wnts and BMPs). Sclerotome precursors reside in the ventromedial portion of the somite and therefore are in closest proximity to the notochord. These cells are induced to become the sclerotome by notochord-derived paracrine factors, especially Sonic hedgehog (Fan and Tessier-Lavigne 1994; Johnson et al. 1994). If portions of the chick notochord are transplanted next to other regions of the somite, those regions will also become sclerotome cells. The notochord and somites also secrete Noggin and Gremlin, two BMP antagonists. The absence of BMPs is critical in permitting Sonic hedgehog to induce cartilage expression, and if either of these inhibitors is deficient, the sclerotome fails to form, and the chicks lack normal vertebrae.

Vertebrae formation

Sonic hedgehog is required for specification of sclerotome fates, but what directs sclerotome cell migration toward and around the notochord and neural tube to form the vertebrae? The notochord appears to induce its surrounding mesenchyme cells to secrete epimorphin. Epimorphin then attracts sclerotome cells to the region around the notochord and neural tube, where they begin to condense and differentiate into cartilage

(A)

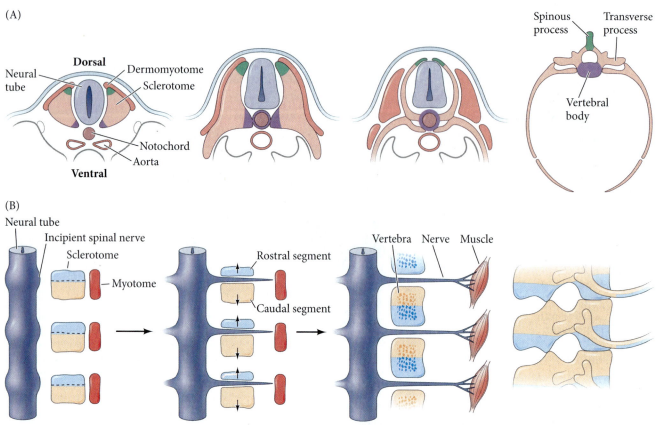

(B)

FIGURE 17.21 Resegmentation of the sclerotome to form vertebrae. (A) Illustration of the sequential development of sclerotome into the vertebrae. (B) Each sclerotome splits into an anterior and a posterior segment. As the spinal neurons grow outward to innervate the muscles from the myotome, the anterior segment of each sclerotome combines with the posterior segment of the next anterior sclerotome to form a vertebral rudiment. (A after B. Christ et al. 2000. *Anat Embryol (Berl)* 202: 179–194; B after W. J. Larsen. 1998. *Essentials of Human Embryology*. Churchill Livingstone: New York; H. Aoyama and K. Asamoto. 2000. *Mech Dev* 99: 71–82.)

(**FIGURE 17.21A**). In addition, the more dorsal migration of sclerotome cells over the top of the neural tube to form the spinous process of the vertebra appears to be induced by the secretion of platelet-derived growth factor (PDGF) from the sclerotome cells immediately below them. The migrating cells are able to respond to these PDGF signals by expressing TGF-β type II receptors (Wang and Serra 2012).

Before the sclerotome-derived cells form a vertebra, each sclerotome must split into an anterior and a posterior segment (**FIGURE 17.21B**). As motor neurons from the neural tube grow laterally to innervate the newly forming muscles, the anterior segment of each sclerotome recombines with the posterior segment of the next anterior sclerotome to form the vertebral rudiment in a process known as **resegmentation** (Remak 1850). This division of neighboring somitic contributions was confirmed through quail-chick chimeras in which the anterior or posterior portion of quail somites was transplanted into a chick somite at the identical location. Quail-specific antigens are easily identified on cells, enabling the differentiated structures to be traced back to the donor tissue (Aoyama and Asamoto 2000; Huang et al. 2000). Although these experiments in the chick supported distinct resegmentation of sclerotome halves, in zebrafish there may be more mixing of contributions from both halves of the sclerotome to vertebrae (Morin-Kensicki et al. 2002). This resegmentation involving the sclerotome but not the myotome enables the muscles to coordinate the movement of the skeleton and permits the body to move laterally, a movement that is reminiscent of the strategy used by insects when constructing segments out of parasegments (see Chapter 9). The bending and twisting movements of the spine are permitted by the intervertebral (synovial) joints that form from the arthrotome region of the sclerotome. Removal of sclerotome cells from the arthrotome leads to the failure of synovial joints to form and to the fusion of adjacent vertebrae (Mittapalli et al. 2005).

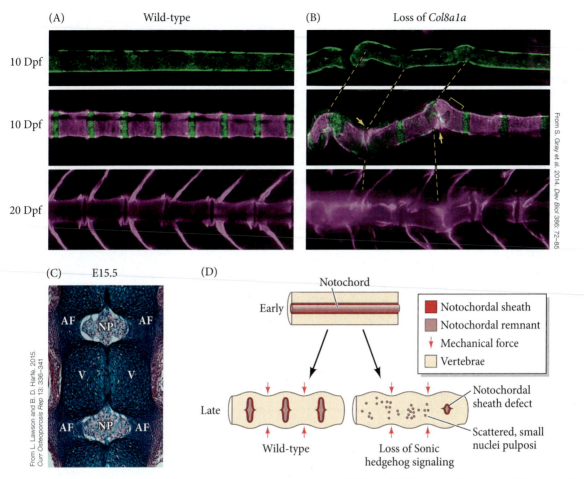

From S. Gray et al. 2014. *Dev Biol* 386: 72–85

From L. Lawson and B. D. Harfe. 2015. *Curr Osteoporosis Rep* 13: 336–341

FIGURE 17.22 Development of the spinal column and intervertebral discs. (A) Collagen 8a1a is normally expressed throughout the spinal column, as revealed by this *Col2a1* transgenic zebrafish reporter (green). Dpf, days postfertilization. (B) Loss of *Col8a1a* clearly results in a failure to properly form a straight spine and the presence of fused vertebrae, as visualized with alizarin red staining (magenta) for bone. (C) A vertebra with associated nucleus pulposus (NP) in an E15.5 mouse. V, vertebra; AF, annulus fibrosis. (D) A model of how the notochordal sheath functions to maintain small portions of the notochord as they develop into the nuclei pulposi. Loss of proper Sonic hedgehog signaling (in *smoothened* mutants) results in varying degrees of notochordal sheath reduction and consequently a failure to form nuclei pulposi. (D after K. S. Choi and B. D. Harfe. 2011. *Proc Natl Acad Sci USA* 108: 9484–9489.)

FURTHER DEVELOPMENT

THE NOTOCHORD SUPPORTS VERTEBRAL MORPHOGENESIS AND BECOMES PART OF THE INTERVERTEBRAL DISC The notochord, with its secretion of Sonic hedgehog, is critical for sclerotome development. As we discussed earlier, the notochord is also important for elongation of the axis, which has ramifications for spine morphogenesis. In zebrafish, improper vacuole filling of notochordal cells results in buckling of the notochord and subsequent vertebral fusions and associated spine defects (Ellis et al. 2013). Additional evidence that proper notochord formation is necessary for proper spine formation comes from experiments that destroy the integrity of the extracellular matrix sheath surrounding the notochord. When one of the collagens in this sheath is inhibited from forming in zebrafish embryos, the notochord bends, which leads to irregular bone deposition, vertebral fusion, and curvature of the spine akin to scoliosis in humans (**FIGURE 17.22A,B**; Gray et al. 2014).

What happens to the notochord in the adult? A common misconception is that after the notochord has provided inductions and axial support, it completely degenerates. There is some truth to this idea in that some of the notochordal cells appear to die by apoptosis once the vertebrae have formed, likely signaled by mechanical forces. It is interesting, however, that those same tensile forces from invading vertebrae also segment the notochord into smaller units, which are retained and develop into the **nuclei pulposi** (see Figure 17.2F; Aszódi et

? Developing Questions

Some species do not develop nuclei pulposi; the chick is one such species. What becomes of notochord cells in these animals? Do they all degenerate, or do some contribute to other structures of the spine? In addition, although mechanical forces seem to contribute to the development of nuclei pulposi, do any molecular cues also guide their formation? Consider the array of guidance cues expressed in midline structures (Eph-ephrins, netrins, slits); could they be instructing the coalescence of notochord cells into segmented foci? Amazingly, despite the "central" role the notochord plays in development, fundamental questions about the development of its own cell lineage remain.

al. 1998; Choi et al. 2008; Guehring et al. 2009; McCann et al. 2011; Risbud and Shapiro 2011; reviewed in Chan et al. 2014, Lawson and Harfe 2015). Most important is that the notochordal origin of nuclei pulposi has been demonstrated through lineage tracing experiments in the mouse (Choi et al. 2008; McCann et al. 2011). The nuclei pulposi form a gel-like mass in the center of the intervertebral discs, which are surrounded by the annulus fibrosus, a connective tissue derived from sclerotome (**FIGURE 17.22C**). Those are the spinal discs that "slip" in certain back injuries.

Little is known about the mechanisms regulating intervertebral disc formation, but it appears that the notochordal sheath is essential to the development of the nuclei pulposi (Choi and Harfe 2011; Choi et al. 2012). In experiments that cause a weakened extracellular sheath to form, pressure from the forming vertebral bodies disperses the notochord cells, and the nuclei pulposi fail to form (**FIGURE 17.22D**; Choi and Harfe 2011).

Tendon formation: The syndetome

The most dorsal part of the sclerotome will become the fourth compartment of the somite, the syndetome. The tendon-forming cells of the syndetome can be visualized by their expression of the *Scleraxis* gene (**FIGURE 17.23**; Schweitzer et al. 2001; Brent et al. 2003). Because there is no obvious morphological distinction between the sclerotome and syndetome cells (they are both mesenchymal), our knowledge of this somitic compartment had to wait until we had molecular markers (Pax1 for the sclerotome, Scleraxis for the syndetome) that could distinguish between them and allow researchers to follow the cells' fates.

Because the tendons connect muscles to bones, it is not surprising that the syndetome (Greek *syn*, "connected") is derived from the most dorsal portion of the sclerotome—that is, it is derived from sclerotome cells adjacent to the muscle-forming myotome (**FIGURE 17.24A**). The syndetome is made from the myotome's secretion of Fgf8 onto the immediately subjacent row of sclerotome cells (Brent et al. 2003; Brent and Tabin 2004). Other transcription factors limit the expression of Scleraxis to the anterior and posterior portions of the syndetome, causing two stripes of Scleraxis expression (**FIGURE 17.24B**). Meanwhile, the developing cartilage cells, under the influence of Sonic hedgehog from the notochord and floor plate, synthesize the Sox5 and Sox6 transcription factors that block *Scleraxis* transcription while activating the cartilage-promoting factor Sox9 (Yamashita et al. 2012). In this way, the cartilage protects itself

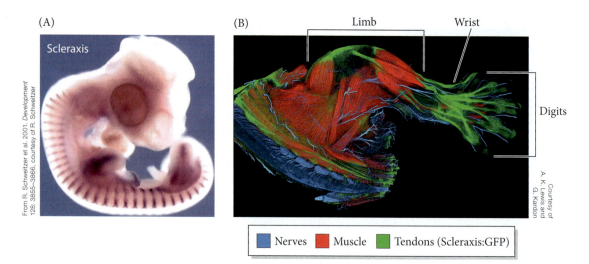

(A)

Scleraxis

From R. Schweitzer et al. 2001. *Development* 128: 3855–3866, courtesy of R. Schweitzer

(B)   Limb   Wrist

Digits

Courtesy of A. K. Lewis and G. Kardon

■ Nerves ■ Muscle ■ Tendons (Scleraxis:GFP)

FIGURE 17.23 Scleraxis is expressed in the progenitors of the tendons. (A) In situ hybridization showing the pattern of *Scleraxis* expression in the developing chick embryo. (B) Limb, wrist, and digits of a newborn mouse, showing Scleraxis (fused to GFP) in the tendons (green) connecting muscles (stained red with antibodies to myosin). The neurons have been stained blue by antibodies to neurofilament proteins.

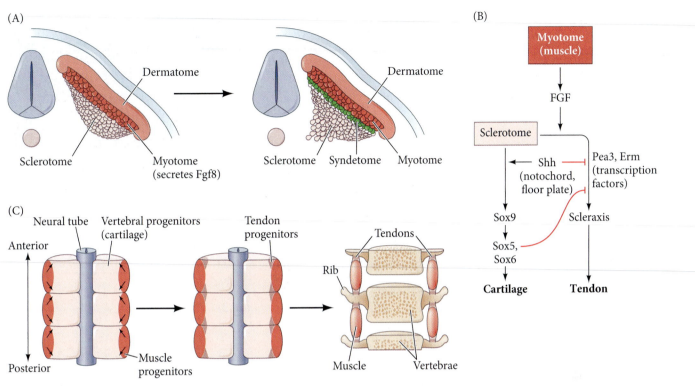

FIGURE 17.24 Induction of Scleraxis in the chick sclerotome by Fgf8 from the myotome. (A) The dermatome, myotome, and sclerotome are established before the tendon precursors are specified. Tendon precursors (syndetome) are specified in the dorsalmost tier of sclerotome cells by Fgf8 received from the myotome. (B) Pathway by which Fgf8 signals from the muscle precursor cells induce the subjacent sclerotome cells to become tendons. (C) Syndetome cells migrate (small arrows) along the developing vertebrae. They differentiate into tendons that connect the ribs to the intercostal muscles beloved by devotees of spareribs. (A,C after A. E. Brent et al. 2003. *Cell* 113: 235–248.)

from any spread of the Fgf8 signal. The tendons then associate with the muscles directly above them and with the skeleton (including the ribs) on either side of them (**FIGURE 17.24C**; Brent et al. 2005). (See **Further Development 17.5, Formation of the Dorsal Aorta**, online.)

Dermomyotome Development

The dermomyotome occupies the dorsolateral half of the somite, and in contrast to the complete epithelial-mesenchymal transition (EMT) exhibited by the sclerotome, it maintains much of its epithelial structure. Through a variety of analyses, including fate mapping with chick-quail chimeras, the dermomyotome can be subdivided into three functionally distinct regions: the dermatome, the myotome, and migratory myoblasts (see Figure 17.2; Ordahl and Le Douarin 1992; Brand-Saberi et al. 1996; Kato and Aoyama 1998). The cells in the two lateralmost portions of this epithelium are called the dorsomedial and ventrolateral lips (closest to and farthest from the neural tube, respectively), which together function as progenitor zones that generate the myotome for the formation of skeletal muscle cells of the body and limbs. Muscle precursor cells—**myoblasts**—from the dorsomedial and ventrolateral lips will migrate beneath the dermomyotome to produce the myotome (see Figure 17.24A). Those myoblasts in the myotome closest to the neural tube form the centrally located **primaxial muscles**, which include the intercostal musculature between the ribs and the deep muscles of the back; those myoblasts farthest from the neural tube produce the **abaxial muscles** of the body wall, limbs, and tongue.[5] The dermatome is located in the centralmost region of the dermomyotome and will form back dermis and several other derivatives. The boundary between the primaxial and abaxial muscles and between the somite-derived and lateral plate-derived dermis is called the **lateral**

FIGURE 17.25 Primaxial and abaxial domains of vertebrate mesoderm. (A) Mesoderm (red) differentiation in an early-stage chick embryo. (B) Day-9 chick embryo in which *Prox1* gene expression is revealed by a dark stain. *Prox1* is expressed in the abaxial region of the chick trunk. The boundary between the stained and unstained regions is the lateral somitic frontier (dotted line). (C) Day-13 chick; regionalization of the mesoderm is apparent. (A,C after B. B. Winslow et al. 2007. *Dev Dyn* 236: 2371–2381.)

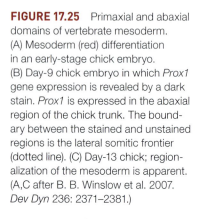

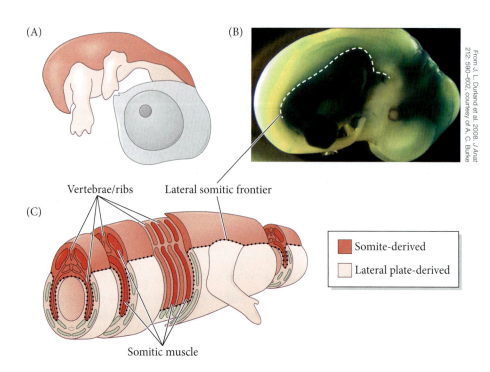

(A)

(B)

From J. L. Durland et al. 2008. *J Anat* 212: 590–602, courtesy of A. C. Burke

Vertebrae/ribs   Lateral somitic frontier

(C)

☐ Somite-derived
☐ Lateral plate-derived

Somitic muscle

somitic frontier (**FIGURE 17.25**; Christ and Ordahl 1995; Burke and Nowicki 2003; Nowicki et al. 2003). Various transcription factors distinguish the primaxial and abaxial muscles.

The dermis of the ventral portion of the body is derived from the lateral plate, and the dermis of the head and neck comes, at least in part, from the cranial neural crest. However, in the trunk, the dermatome's major product is the precursors of the dermis of the back. In addition, recent studies have shown that this central region of the dermomyotome also gives rise to a population of muscle cells (Gros et al. 2005; Relaix et al. 2005). Therefore, some researchers (Christ and Ordahl 1995; Christ et al. 2007) prefer to retain the term *dermomyotome* (or *central dermomyotome*) for this epithelial region. Soon, this part of the somite also undergoes an EMT. FGF signals from the myotome activate transcription of the *Snail2* gene in the central dermomyotome cells, and the Snail2 protein is a well-known regulator of EMT (see Figure 15.9; Delfini et al. 2009). During EMT, the mitotic spindles of the epithelial cells are realigned so that cell division takes place along the dorsal-ventral axis. The ventral daughter cell joins the other myoblasts from the myotomes, while the other daughter cell locates dorsally, becoming a precursor of the dermis. Reminiscent of sclerotome EMT progression, the N-cadherin holding these cells together is downregulated, and the two daughter cells go their separate ways, with the remaining N-cadherin found only on those cells entering the myotome (Ben-Yair and Kalcheim 2005).

The muscle precursor cells that delaminate from the epithelial plate to join the primary myotome cells remain undifferentiated, and they proliferate rapidly to account for most of the myoblast cells. While most of these progenitor cells differentiate to form muscles, some remain undifferentiated and surround the mature muscle cells. These undifferentiated cells become skeletal muscle stem cells called **satellite cells**, and they are responsible for postnatal muscle growth and muscle repair.

Determination of the central dermomyotome

The central dermomyotome generates muscle precursors as well as the dermal cells that constitute the dermis of the dorsal skin. The dermis of the ventral and lateral sides of the body is derived from the lateral plate mesoderm that forms the body wall. The maintenance of the central dermomyotome depends on Wnt6 from the epidermis (Christ et al. 2007), and its EMT appears to be regulated by neurotrophin 3 (NT3) and Wnt1, two factors secreted by the neural tube (see Figure 17.20). Antibodies that block

NT3 activity prevent the conversion of epithelial dermatome into the loose dermal mesenchyme that migrates beneath the epidermis (Brill et al. 1995). Removing or rotating the neural tube prevents this dermis from forming (Takahashi et al. 1992; Olivera-Martinez et al. 2002). The Wnt signals from the epidermis promote the differentiation of the dorsally migrating central dermomyotome cells into dermis (Atit et al. 2006).

Muscle precursor cells and dermal cells are not the only derivatives of the central dermomyotome, however. Atit and colleagues (2006) have shown that **brown adipose cells** ("brown fat") are also somite-derived and appear to come from the central dermomyotome. Brown fat plays an active role in energy utilization by burning fat to produce heat (unlike the better-known white adipose tissue, or "white fat," which stores fat). Tseng and colleagues (2008) have found that skeletal muscle and brown fat cells share the same somitic precursor that originally expresses myogenic regulatory factors. In brown fat precursor cells, the transcription factor PRDM16 is induced (probably by BMP7); PRDM16 appears to be critical for the conversion of myoblasts to brown fat cells because it activates a battery of genes that are specific for the fat-burning metabolism of brown adipocytes (Kajimura et al. 2009).

Determination of the myotome

All the skeletal musculature in the vertebrate body with the exception of the head muscles comes from the dermomyotome of the somite. The myotome forms from the lateral edges, or "lips," of the dermomyotome and folds to form a layer between the more peripheral dermomyotome and the more medial sclerotome. The major transcription factors associated with (and causing) muscle development are the **myogenic regulatory factors** (**MRFs**, sometimes called the myogenic bHLH proteins). This family of transcription factors includes MyoD, Myf5, myogenin, and MRF4 (**FIGURE 17.26**). Each member of this family can activate the genes of the other family members, leading to positive feedback regulation so powerful that the activation of an MRF in nearly any cell in the body converts that cell into muscle.[6]

The MRFs bind to and activate genes that are critical for muscle function. For instance, the MyoD protein appears to directly activate the muscle-specific creatine phosphokinase gene by binding to the DNA immediately upstream from it (Lassar et al. 1989). There are also two MyoD-binding sites on the DNA adjacent to the genes encoding a subunit of the chicken muscle acetylcholine receptor (Piette et al. 1990). MyoD also directly activates its own gene. Therefore, once the *MyoD* gene is activated, its protein product binds to the DNA immediately upstream of *MyoD* and keeps this gene

FIGURE 17.26 Differential gene expression in myotome. (A) The primaxial myotome is thought to be specified by a combination of Wnts (probably Wnt1 and Wnt3a) from the dorsal neural tube and low concentrations of Sonic hedgehog from the floor plate of the neural tube. Pax3 in somitic cells allows expression of Myf5 in response to paracrine factors, which allows the cells to synthesize the myogenic transcription factor Myf5. In combination with a Six protein and Mef2, Myf5 activates the genes responsible for activating myogenin and MRF4. (B) BMP4 is inhibited by Noggin produced by cells that migrate specifically to the lips of the somite. In the absence of BMP4, Wnt proteins from the epidermis are thought to induce the abaxial myotome. (After V. G. Punch et al. 2009. *Wiley Interdiscip Rev Syst Biol Med* 1: 128–140.)

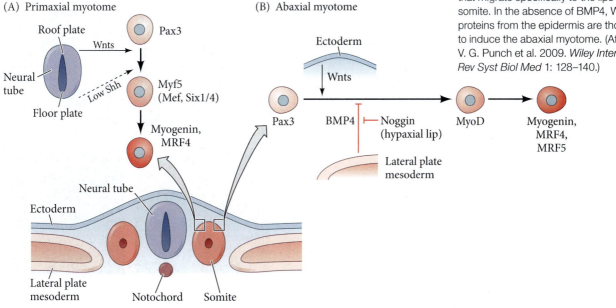

active. Many of the MRFs are active only if they associate with a muscle-specific cofactor from the Mef2 (myocyte enhancer factor-2) family of proteins. MyoD can activate the *Mef2* gene and thereby regulate the differential timing of muscle gene expression.

As discussed earlier, the myotome is induced in the somite at two different places by at least two distinct signals (see Punch et al. 2009). Studies using transplantation and knockout mice indicate that the *primaxial* myoblasts from the medial portion of the somite are induced by factors from the neural tube—probably Wnt1 and Wnt3a from the dorsal region and low levels of Sonic hedgehog from the floor plate of the neural tube (see Figure 17.20; Münsterberg et al. 1995; Stern et al. 1995; Borycki et al. 2000). These factors induce the Pax3-containing cells of the somite to activate the *Myf5* gene in the primaxial myotome. Myf5 (in concert with Mef2 and either Six1 or Six4) activates the *Myogenin* and *MRF4* genes, whose proteins activate the muscle-specific gene regulatory network (see Figure 17.26A; Buckingham et al. 2006). The cells of the primaxial myotome appear to be originally confined by the laminin extracellular matrix that outlines the dermomyotome and myotome. As the myoblasts mature, however, this matrix dissolves, and the primaxial myoblasts migrate along fibronectin cables. Eventually, they align, fuse, and elongate to become the deep muscles of the back, connecting to the developing vertebrae and ribs (Deries et al. 2010, 2012).

The abaxial myoblasts that form the limb and ventral body wall musculature arise from the lateral edge of the somite. Two conditions appear to be necessary to produce these muscle precursors: (1) the presence of Wnt signals and (2) the absence of BMPs (see Figure 17.26B; Marcelle et al. 1997; Reshef et al. 1998). Wnt proteins (especially Wnt7a) are made in the epidermis (see Figure 17.22; Cossu et al. 1996a; Pourquié et al. 1996; Dietrich et al. 1998), but the BMP4 made by the adjacent lateral plate mesoderm would normally prevent muscles from forming.

What, then, is inhibiting BMP activity? Several studies on chick embryos have found that the dorsomedial and ventrolateral lips of the dermomyotome have attached at their tips a population of cells that secrete the BMP inhibitor Noggin (Gerhart et al. 2006, 2011). These Noggin-secreting cells arise in the blastocyst, become part of the epiblast, and distinguish themselves by expressing the mRNA for MyoD but not translating this mRNA into protein. These particular cells migrate to become paraxial mesoderm, specifically sorting out to the dorsomedial and ventrolateral lips of the dermomyotome. There they synthesize and secrete Noggin, thus promoting differentiation of myoblasts. If Noggin-secreting cells are removed from the epiblast, there is a decrease in the skeletal musculature throughout the body, and the ventral body wall is so weak that the heart and abdominal organs often are herniated through it (**FIGURE 17.27**). This

FIGURE 17.27 Ablating Noggin-secreting epiblast cells results in severe muscle defects. Noggin-secreting epiblast cells were ablated in stage-2 chick embryos using antibodies against G8. (A) The control embryo has normal morphology (upper photograph) and abundant staining of myosin (lower photograph, red) in the muscles. (B) Embryos whose Noggin-secreting epiblast cells were ablated have severe eye defects, severely reduced somatic musculature, and herniation of abdominal organs through the thin abdominal wall (upper photograph, arrow). Severely reduced musculature (sparse myosin in lower photograph) is characteristic of these embryos.

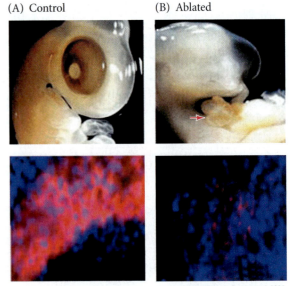

(A) Control   (B) Ablated

All images from J. Gerhart et al. 2006.
J Cell Biol 175: 283–292, courtesy of
J. Gerhart and M. George-Weinstein

defect can be prevented by implanting Noggin-releasing beads into somites lacking Noggin-secreting cells. Once BMP is inhibited, Wnt7 can induce MyoD in the competent dermomyotome cells, which activates the battery of MRF proteins that generate the muscle precursor cells. (See **Further Development 17.6, An Emerging Model of Neural Crest-Regulated Myogenesis, Further Development 17.7, Osteogenesis: The Development of Bones, Further Development 17.8, Paracrine Factors, Their Receptors, and Human Bone Growth,** and **Further Development 17.9, Maturation of Muscle**, all online.)

Closing Thoughts on the Opening Photo

The formation of segments, or somites, is a highly regulated process that determines "what, when, where, and how many" somites an organism makes. This beautiful image of a garter snake embryo stained with alcian blue was produced by Anne C. Burke and illustrates the grand nature of somitogenesis. Segments of paraxial mesoderm are carved up into sequential blocks, through the orchestration of an Fgf8 determination front, a Notch-Delta molecular clock, and an Eph-ephrin mediated boundary formation.

Photograph courtesy of Anne C. Burke

Endnotes

[1] How does one examine chromatin states? The Duboule lab used a new technique called circular chromosome conformation capture (or 4C-seq) to observe the three-dimensional genomic organization of Hox gene clusters and identify which Hox genes are in tightly packed or loosely packed chromatin states during paraxial mesoderm development.

[2] In amniotes, cell migration from the caudal progenitor zone has been interpreted to be more a result of tissue deformation as opposed to individual cell migration (see Bénazéraf et al. 2010; Bénazéraf and Pourquié 2013).

[3] *Collective migration* is when self-propelled migrating cells exert directionally coordinated forces on one another, as opposed to individually migrating cells with loose contacts or the movement of a group of cells by tissue pushing due to proliferation or intercalation. Other cells, such as neural crest and even metastasizing cancer cells, have been suggested to use collective migration.

[4] Pourquié (2011) has noted that this situation seems a lot like the one that exists in the limb bud (see Chapter 19), wherein a pool of

newly formed cells that has been maintained by FGFs in a relatively undifferentiated and migratory state becomes differentiated into periodic elements (limb cartilage) by interacting gradients of FGFs and retinoic acid.

[5] As used here, the terms *primaxial* and *abaxial* designate the muscles from the medial and lateral portions of the somite, respectively. The terms *epaxial* and *hypaxial* are commonly used, but these terms are derived from secondary modifications of the adult anatomy (the hypaxial muscles being innervated by the ventral regions of the spinal cord) rather than from the somitic myotome lineages (see Nowicki et al. 2003).

[6] A general rule of development, as in the U.S. Constitution, is that powerful entities must be powerfully regulated. As a result of their power to convert any cell into muscle, the MRFs are among the most powerfully controlled entities of the genome. They are controlled at several points in transcription as well as in processing, translation, and posttranslational modification (see Sartorelli and Juan 2011; Ling et al. 2012).

Snapshot Summary
Paraxial Mesoderm

1. The paraxial mesoderm forms blocks of tissue called somites. Somites give rise to three major divisions: the sclerotome, the myotome, and the central dermomyotome.

2. The spatiotemporal expression of 3′-to-5′ Hox genes along the paraxial mesoderm correlates with a progressive loosening of chromatin structures via epigenetic regulation as well as with the timing of ingression into the paraxial mesoderm along the anterior-posterior axis. Posterior gradient signals of

FGFs and Wnts maintain neuromesoderm progenitor (NMP) cells in the progenitor state, whereas opposing gradients of retinoic acid promote differentiation of these cells. These antagonistic signals establish where a new somitic boundary will form in the presomitic mesoderm.

3. Cyclic activation of Notch-Delta signaling throughout the presomitic mesoderm establishes the timing of segment formation, and Eph receptor systems are involved

(Continued)

Snapshot Summary *(continued)*

in the physical formation of boundaries. Moreover, Rho GTPases and integrin-fibronectin interactions also appear to be important in causing presomitic mesodermal cells to become epithelial.

4. The sclerotome forms the vertebral cartilage. In thoracic vertebrae, the sclerotome cells also form the ribs. The intervertebral joints as well as the meninges and dorsal aorta cells also come from the sclerotome.

5. The primaxial myotome forms the back musculature. The abaxial myotome forms the muscles of the body wall, limbs, and tongue.

6. The central dermomyotome forms the dermis of the back as well as precursors of muscle and brown fat cells.

7. The somite regions are specified by paracrine factors secreted by neighboring tissues. The sclerotome is specified to a large degree by Sonic hedgehog, which is secreted by the notochord and floor plate cells. The two myotome regions are specified by different factors, and in both instances, myogenic regulatory factors (MRFs) are induced in the cells that will become muscles.

8. The major lineages that form the skeleton are the somites (axial skeleton), the lateral plate mesoderm (appendages), and the neural crest and head mesoderm (skull and face).

9. Tendons are formed through the conversion of the dorsalmost layer of sclerotome cells into syndetome cells by FGFs secreted by the myotome.

Go to oup.com/he/barresi12xe for Further Developments, Scientists Speak interviews, Watch Development videos, Dev Tutorials, and complete bibliographic information for all literature cited in this chapter.

Intermediate and Lateral Plate Mesoderm

Heart, Blood, and Kidneys

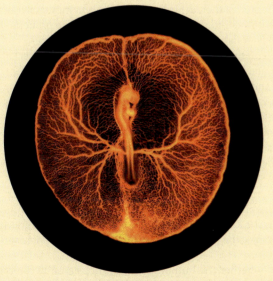

How is the heart formed, and how does it connect to the arteries and veins?

© Vincent Pasque

WHILE THE AXIAL AND PARAXIAL MESODERM form the notochord and somites of the dorsum, the intermediate and lateral plate mesoderm extend around the sides and front of the body. The **intermediate mesoderm** forms the urogenital system, consisting of the kidneys, the gonads, and their associated ducts. The outer (cortical) portion of the adrenal gland also derives from this region. Farthest away from the notochord, the **lateral plate mesoderm** gives rise to the heart, blood vessels, and blood cells of the circulatory system, as well as to the lining of the body cavities. It gives rise to the pelvic and limb skeleton (but not the limb muscles, which are somitic in origin). Lateral plate mesoderm also helps form a series of extraembryonic membranes that are important for transporting nutrients to the embryo (**FIGURE 18.1**).

The four subdivisions of the mesoderm (axial, paraxial, intermediate, and lateral plate) are thought to be specified along the mediolateral (center-to-side) axes by increasing amounts of BMPs (Pourquié et al. 1996; Tonegawa et al. 1997). The more lateral mesoderm of the chick embryo expresses higher levels of BMP4 than do the midline areas, and one can change the identity of the mesodermal tissue by altering BMP expression. While it is not known how this patterning is accomplished, it is thought that the different BMP concentrations may cause differential expression of the Forkhead (Fox) family of transcription factors. The *Foxf1* gene is transcribed in those regions that will become the lateral plate and extraembryonic mesoderm, whereas *Foxc1* and *Foxc2* are expressed in the paraxial mesoderm that will form the somites (Wilm et al. 2004). If *Foxc1* and *Foxc2* are both deleted from the mouse genome, the paraxial mesoderm is respecified as intermediate mesoderm and initiates expression of the *Pax2* gene, which encodes a major transcription factor of the intermediate mesoderm (see Figure 18.1B).

In this chapter, we will focus on those organs that make and circulate the blood. The blood cells are made by the lateral plate mesoderm, as is the heart and most of the blood vessels that circulate the blood. The kidney, from the intermediate mesoderm, filters wastes out of the blood, and also has a major influence on blood pressure, composition, and volume.

Intermediate Mesoderm: The Kidney

Physiologist and philosopher of science Homer Smith noted in 1953 that "our kidneys constitute the major foundation of our philosophical freedom. Only because they work the way they do has it become possible for us to have bone, muscles, glands, and brains." While this

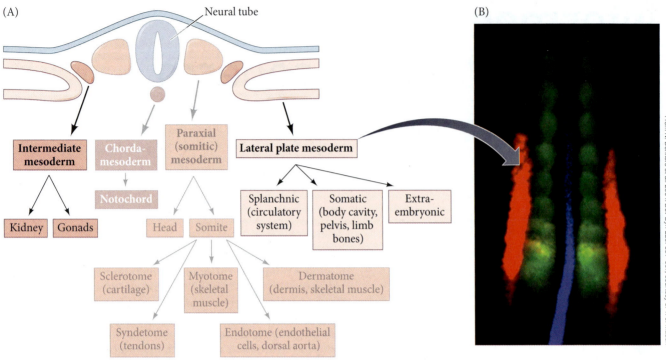

(A)

Neural tube

(B)

From N. Denkers et al. 2004. *Dev Dyn* 229: 661–667, courtesy of T. J. Mauch

FIGURE 18.1   Major lineages of the amniote mesoderm. (A) Schematic of the mesodermal compartments of the amniote embryo. (B) Staining for the medial mesodermal compartments in the trunk of a 12-somite chick embryo (~33 hours). In situ hybridization was performed with probes binding to *Chordin* mRNA (blue) in the notochord, *Paraxis* mRNA (green) in the somites, and *Pax2* mRNA (red) in the intermediate mesoderm.

statement may smack of hyperbole, the human kidney is a remarkably intricate organ whose importance cannot be overestimated. Its functional unit, the **nephron**, contains more than 10,000 cells and at least 12 different cell types, each cell type having a specific function and being located in a particular place in relation to the others along the length of the nephron.

Mammalian kidney development progresses through three major stages. The first two stages are transient; only the third and final stage persists as a functional kidney. Early in development (day 22 in humans; day 8 in mice), the **pronephric duct** arises in the intermediate mesoderm ventrolateral to the anterior somites. The cells of this duct migrate caudally, and the anterior region of the duct induces the adjacent mesenchyme to form the **pronephros**, or tubules of the initial kidney (**FIGURE 18.2A**). The pronephric tubules form functioning kidneys in fish and in amphibian larvae, but they are not believed to be active in amniotes. In mammals, the pronephric tubules and the anterior portion of the pronephric duct degenerate, but the more caudal portions of the pronephric duct and its derivatives persist and serve as the central component of the excretory system throughout development (Toivonen 1945; Saxén 1987). This remaining duct is often referred to as the **nephric**, or **Wolffian**, **duct**.

As the pronephric tubules degenerate, the middle portion of the nephric duct induces a new set of kidney tubules in the adjacent mesenchyme. This set of tubules constitutes the **mesonephros**, sometimes called the mesonephric kidney (**FIGURE 18.2B**; Sainio and Raatikainen-Ahokas 1999). In some mammalian species, the mesonephros functions briefly in urine filtration, but in mice and rats, it does not function as a working kidney. In humans, about 30 mesonephric tubules form, beginning around day 25. As more tubules are induced caudally, the anterior mesonephric tubules begin to regress through apoptosis (interestingly, in mice the anterior tubules remain while the posterior ones regress; **FIGURE 18.2C,D**).

Although it remains unknown whether the human mesonephros actually filters blood and makes urine, it definitely provides important developmental functions during its brief existence. First, it is one of the main sources of the hematopoietic stem cells necessary for blood cell development (Medvinsky and Dzierzak 1996; Wintour et al. 1996). Second, in male mammals, some of the mesonephric tubules persist to become the tubes that transport the sperm from the testes to the urethra (the epididymis and vas deferens; see Chapter 6).

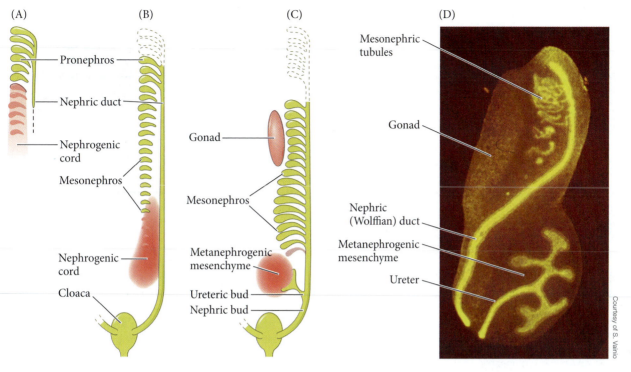

FIGURE 18.2 General scheme of development in the vertebrate kidney. (A) The original tubules, constituting the pronephros, are induced from the nephrogenic mesenchyme by the pronephric duct as it migrates caudally. (B) As the pronephros degenerates, the mesonephric tubules form. (C) The final mammalian kidney, the metanephros, is induced by the ureteric bud, which branches from the nephric duct. (D) Intermediate mesoderm of a 13-day mouse embryo showing initiation of the metanephric kidney (bottom) while the mesonephros is still apparent. The duct tissue is stained with a fluorescent antibody to a cytokeratin found in the pronephric duct and its derivatives. (A–C after L. Saxén. 1987. *Organogenesis of the Kidney*. Cambridge University Press: Cambridge, UK.)

The permanent kidney of amniotes, the **metanephros**, originates through a complex set of interactions between epithelial and mesenchymal components of the intermediate mesoderm (reviewed in Costantini and Kopan 2010; McMahon 2016). In the first steps, the kidney-forming **metanephric mesenchyme** (also called the metanephrogenic mesenchyme) becomes committed in the posterior regions of the intermediate mesoderm, where it induces the formation of a branch from each of the paired nephric ducts. These epithelial branches are called the **ureteric buds**. These buds eventually grow out from the nephric duct to become the collecting ducts and ureters that take the urine to the bladder. When the ureteric buds emerge from the nephric duct, they enter the metanephric mesenchyme. The ureteric buds induce this mesenchymal tissue to condense around them and to differentiate into the nephrons of the mammalian kidney. As this mesenchyme begins to differentiate, it tells the ureteric bud to branch and grow. These reciprocal inductions form the kidneys.

Specification of the Intermediate Mesoderm: Pax2, Pax8, and Lim1

The intermediate mesoderm of the chick embryo acquires its ability to form kidneys through its interactions with the paraxial mesoderm. While its bias to become intermediate mesoderm is probably established through a BMP gradient, specification appears to become stabilized through signals from the paraxial mesoderm. Mauch and her colleagues (2000) showed that signals from the paraxial mesoderm induced primitive kidney formation in the intermediate mesoderm of the chick embryo. They cut developing embryos such that the intermediate mesoderm could not contact the paraxial mesoderm on one side of the body. That side of the body (where contact with the paraxial mesoderm was abolished) did not form kidneys, but the undisturbed side was able to form kidneys (**FIGURE 18.3A,B**). Thus, paraxial mesoderm appears to be both necessary and sufficient for inducing kidney-forming ability in the intermediate mesoderm. In support of this, paraxial mesoderm can even induce lateral plate mesoderm to generate pronephric tubules when co-cultured together. No other cell type can accomplish this.

These interactions induce the expression of a set of homeodomain transcription factors—including Lim1 (sometimes called Lhx1), Pax2, and Pax8—that cause the intermediate mesoderm to form the kidney (**FIGURE 18.3C**; Karavanov et al. 1998; Kobayashi

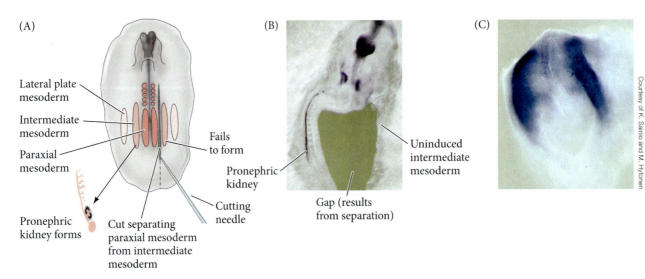

(A) Lateral plate mesoderm

Intermediate mesoderm

Paraxial mesoderm

Pronephric kidney forms

Cut separating paraxial mesoderm from intermediate mesoderm

Fails to form

Pronephric kidney

Cutting needle

(B)

Gap (results from separation)

Uninduced intermediate mesoderm

(C)

Courtesy of K. Sainio and M. Hytonen

FIGURE 18.3 Signals from the paraxial mesoderm induce pronephros formation in the intermediate mesoderm of the chick embryo. (A) The paraxial mesoderm was surgically separated from the intermediate mesoderm on the right side of the body. (B) As a result, a pronephric kidney (Pax2-staining duct) developed only on the left side. (C) Lim1 expression in an 8-day mouse embryo, showing the prospective intermediate mesoderm. (A,B after T. J. Mauch et al. 2000. *Dev Biol* 220: 62–75.)

et al. 2005; Cirio et al. 2011). In the chick embryo, Pax2 and Lim1 are expressed in the intermediate mesoderm, starting at the level of the sixth somite (i.e., only in the trunk, not in the head). If Pax2 is experimentally induced in the presomitic mesoderm, it converts that paraxial mesoderm into intermediate mesoderm, causing it to express Lim1 and form kidneys (Mauch et al. 2000; Suetsugu et al. 2005). Similarly, in mouse embryos with knockouts of both the *Pax2* and *Pax8* genes, the mesenchymal-epithelial transition necessary to form the nephric duct fails, the cells undergo apoptosis, and no kidney structures form (Bouchard et al. 2002). Moreover, in the mouse, Lim1 and Pax2 appear to induce one another. (See **Further Development 18.1, Hox Makes Cells Competent to Express Lim and Make a Kidney**, online.)

Reciprocal Interactions of Developing Kidney Tissues

The kidney of amniotes, the type of kidney you also possess, forms from two distinct progenitor cell populations derived from the intermediate mesenchyme—the ureteric bud and the metanephric mesenchyme. The ureteric bud gives rise to all of the cell types that compose the mature collecting ducts and the ureter, while the metanephric mesenchyme gives rise to all of the cell types that compose the mature nephron, as well as to some vascular and stromal derivatives. These two groups of cells—the ureteric bud and the metanephric mesenchyme—interact and reciprocally induce each other to form the kidney (**FIGURE 18.4**). The metanephric mesenchyme causes the ureteric bud to elongate and branch. The tips of these branches induce the loose mesenchyme cells to form pretubular renal aggregates. Each aggregated nodule proliferates and differentiates into the intricate structure of a renal nephron. Each pretubular aggregate first undergoes a mesenchymal-epithelial transition, becoming a polarized renal vesicle. Subsequently, this renal vesicle elongates into a comma (C) shape and then forms a characteristic S-shaped tube. Soon afterward, the cells of this epithelial structure begin to differentiate into regionally specific cell types, including the Bowman's capsule cells, the podocytes, and the cells of the proximal and distal renal tubules. While this transformation is happening, the cells of the S-shaped tubule closest to the ureteric bud break down the basal lamina of the ureteric bud epithelium and migrate into the duct region. This creates an open connection between the ureteric bud and the newly formed nephron tubule, allowing material to pass from one into the other (Bard et al. 2001; Kao et al. 2012). These mesenchyme-derived tubules form the mature nephrons of the functioning kidney, while the branched ureteric bud gives rise to the collecting ducts and to the ureter, which drains the urine from the kidney.

Clifford Grobstein (1955, 1956) documented this reciprocal induction in vitro. He separated the ureteric bud from the metanephric mesenchyme and cultured them either individually or together. In the absence of mesenchyme, the ureteric bud does

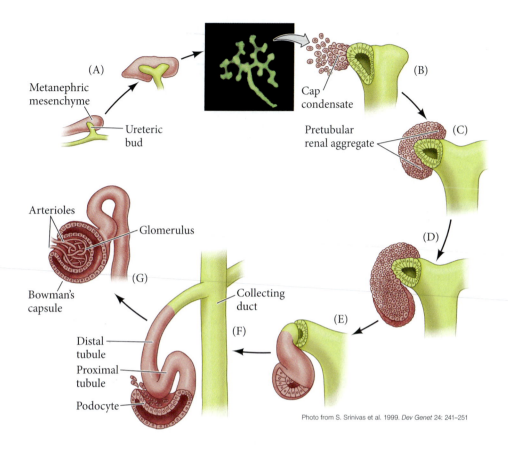

(A) Metanephric mesenchyme

Ureteric bud

Cap condensate

Pretubular renal aggregate

(B)

(C)

(D)

(E)

(F)

(G)

Arterioles

Glomerulus

Bowman's capsule

Collecting duct

Distal tubule

Proximal tubule

Podocyte

Photo from S. Srinivas et al. 1999. *Dev Genet* 24: 241–251

FIGURE 18.4 Reciprocal induction in the development of the mammalian kidney. (A) As the ureteric bud enters the metanephric mesenchyme, the mesenchyme induces the bud to branch. (B–G) At the tips of the branches, the epithelium induces the mesenchyme to aggregate and cavitate to form the renal tubules and glomeruli (where the blood from the arteriole is filtered). When the mesenchyme has condensed into an epithelium, it digests the basal lamina of the ureteric bud cells that induced it and connects to the ureteric bud epithelium. A portion of the aggregated mesenchyme (the pretubular condensate) becomes the nephron (renal tubules and Bowman's capsule), while the ureteric bud becomes the collecting duct for the urine. (After L. Saxén. 1984. In *Modern Biological Experimentation*, ed. C. Chagas, pp. 155–163. Pontificia Academia Scientiarum. Città del Vaticano; H. Sariola. 2002. *Curr Opin Nephrol Hypertens* 11: 17–21.)

not branch. In the absence of the ureteric bud, the mesenchyme soon dies. But when they are placed together, the ureteric bud grows and branches, and nephrons form throughout the mesenchyme. This has been confirmed by experiments using GFP-labeled proteins to monitor cell division and branching (**FIGURE 18.5**; Srinivas et al. 1999).

Mechanisms of reciprocal induction

The induction of the metanephros can be viewed as a dialogue between the ureteric bud and the metanephric mesenchyme. As the dialogue continues, both tissues are altered. We will eavesdrop on this dialogue, which has become a model for organogenesis (Costantini 2012; Krause et al. 2015a). Many of the paracrine factors causing the mutual induction of the kidney nephron and its collecting ducts have been identified, and there is a possibility (Krause et al. 2015b) that these proteins are packaged as exosomes whose contents would be concentrated on the neighboring cells.

STEP 1: FORMATION OF METANEPHRIC MESENCHYME AND THE URETERIC BUD

The metanephric mesenchyme and the ureteric bud are more alike than they appear.

0.5 mm

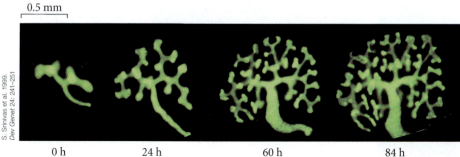

S. Srinivas et al. 1999. *Dev Genet* 24: 241–251

0 h   24 h   60 h   84 h

FIGURE 18.5 Kidney branching observed in vitro. (A) A kidney rudiment from an 11.5-day mouse embryo was placed into culture. This transgenic mouse had a *GFP* gene fused to a *Hoxb7* promoter, so it expressed green fluorescent protein in the nephric (Wolffian) duct and in the ureteric buds. Since GFP can be photographed in living tissues, the kidney could be followed as it developed.

FIGURE 18.6 Creating organoids of kidneys from induced pluripotent stem cells. (A) Schematic mechanism of generating the ureteric bud and the metanephric mesoderm from the posterior mesodermal precursor cells. Those precursor cells migrating from the posterior mesoderm early during gastrulation leave the area of Wnt and proceed toward areas of more FGFs and retinoic acid (RA). These become the progenitors of the ureteric bud epithelium. Those migrating later, after remaining longer in the Wnt-dominated area, become the progenitors of the metanephric mesenchyme. (B) RA signaling in the late primitive streak stage. An RA-degrading enzyme, Cyp26, is both regulated by Wnt and expressed in the presomitic mesoderm region and shields posterior mesodermal precursor (PMP) cells from RA signaling. (C) Immunofluorescence microscope analysis of a kidney organoid formed from human iPSCs that were exposed sequentially to Wnt signal promoters and FGF signals. The insert is a higher-magnification view of a nephron segmented into four compartments, including the collecting duct (green), distal (yellow) and proximal (blue) tubules, and the glomerulus (red). (After M. Takasato et al. 2015. *Nature* 526: 564–568.)

Both come from the intermediate mesoderm, and both are generated through the actions of Wnt and FGF signaling pathways. The ureteric epithelium comes from early migrating intermediate mesoderm, which is exposed to Wnt signals for only a short time and then is exposed for a longer time to the posterior signals Fgf9 and retinoic acid. The cells that become the metanephric mesenchyme migrate through the primitive streak later and are thus exposed to Wnt signals for a longer period of time. They then experience FGF and retinoic acid signals (**FIGURE 18.6A,B**; Takasato et al. 2015), which induce a set of transcription factors that enable the metanephric mesenchyme to respond to the ureteric bud. Only the metanephric mesenchyme has the competence to respond to the ureteric bud and form kidney tubules (Saxén 1970; Sariola et al. 1982).

FURTHER DEVELOPMENT

THE KIDNEY ORGANOID When human induced pluripotent stem cells (iPSCs) cells are cultivated sequentially in activators of the Wnt and FGF pathways, they become either ureteric epithelium or metanephric mesenchyme, depending on their length of exposure to each factor, just as in the embryo. Even more remarkably, when these cell types are cultured together, organoids resembling the kidney are generated (**FIGURE 18.6C**; Takasato and Little 2015). While these organoids do not have the intricate nephron structure, the major cell types of the nephron and collecting ducts are formed.

STEP 2: THE METANEPHRIC MESENCHYME INDUCES OUTGROWTH OF THE URETERIC BUD The stage is now set for the secretion of paracrine factors that can induce the ureteric buds to emerge. **Glial-derived neurotrophic factor** (**GDNF**) secreted

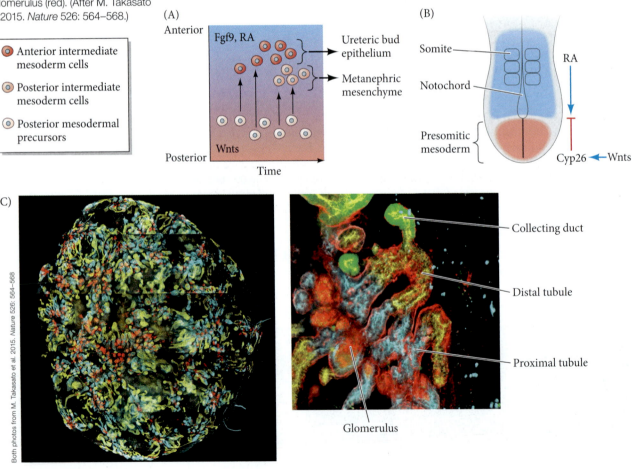

Both photos from M. Takasato et al. 2015. *Nature* 526: 564–568

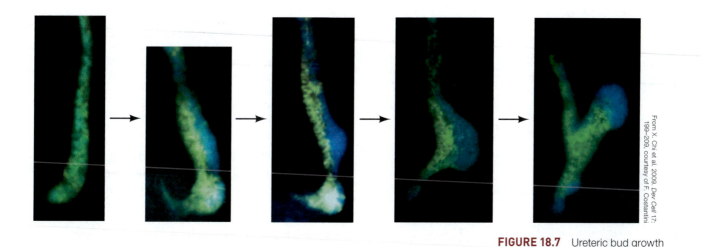

From X. Chi et al. 2009. *Dev Cell* 17: 199–209, courtesy of F. Costantini

from the metanephric mesenchyme causes the outgrowth of the ureteric bud only from those nephric duct cells expressing the GDNF receptor Ret (**FIGURE 18.7**). (See **Further Development 18.2, The Metanephric Mesenchyme Secretes GDNF to Induce and Direct the Ureteric Bud**, online.)

FIGURE 18.7 Ureteric bud growth is dependent on GDNF and its receptors. When mice are constructed from Ret-deficient cells (green) and Ret-expressing cells (blue), the cells expressing Ret migrate to form the tips of the ureteric bud.

STEP 3: THE URETERIC BUD PREVENTS MESENCHYMAL APOPTOSIS The third signal in kidney development is sent from the ureteric bud to the metanephric mesenchyme. The factors secreted from the ureteric bud include Fgf2, Fgf9, and BMP7. If left uninduced by the ureteric bud, the mesenchyme cells undergo apoptosis (Grobstein 1955; Koseki et al. 1992). However, if induced by the ureteric bud, the mesenchyme cells are rescued from the precipice of death and are converted into proliferating stem cells (Bard and Ross 1991; Bard et al. 1996).

STEP 4: THE MESENCHYME INDUCES THE BRANCHING OF THE URETERIC BUD GDNF and Wnts from the mesenchyme, as well as other paracrine factors (FGFs and BMPs), have all been implicated in directing the branching of the ureteric bud, probably working as "pushes" and "pulls" on cell division and on the extracellular matrix (Ritvos et al. 1995; Miyazaki et al. 2000; Lin et al. 2001; Majumdar et al. 2003). GDNF not only induces the initial ureteric bud from the nephric duct but can also induce secondary buds from the ureteric bud once the bud enters the mesenchyme (**FIGURE 18.8**; Sainio et al. 1997; Shakya et al. 2005; Chi et al. 2008). GDNF also induces Wnt11 synthesis in the responsive cells at the tip of the bud (see Figure 18.9A), and Wnt11 reciprocates by regulating GDNF levels (Majumdar et al. 2003; Kuure et al. 2007). The cooperation between the GDNF/Ret pathway and the Wnt pathway appears to coordinate the balance between branching and metanephric mesenchyme proliferation such that continued kidney development is ensured. In this way, two groups of stem cells are maintained: the **ureteric bud tip cells** and the **mesenchyme cap cells** (Mugford et al. 2009; Barak et al. 2012).

(A)

(B)

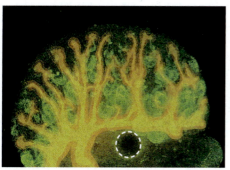
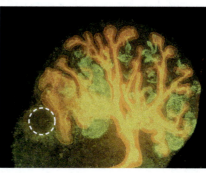

From K. Sainio et al. 1997. *Development* 124: 4077–4087, courtesy of K. Sainio

FIGURE 18.8 Effect of GDNF on branching of the ureteric epithelium. The ureteric bud and its branches are stained orange (with antibodies to cytokeratin 18), while the nephrons are stained green (with antibodies to nephron brush border antigens). (A) A 13-day embryonic mouse kidney cultured for 2 days with a control bead (circle) has a normal branching pattern. (B) A similar kidney cultured for 2 days with a GDNF-soaked bead shows a distorted pattern, as new branches are induced in the vicinity of the bead.

(A)

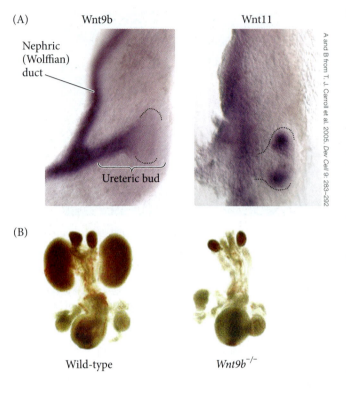

Wnt9b

Wnt11

Nephric (Wolffian) duct

Ureteric bud

A and B from T. J. Carroll et al. 2005, *Dev Cell 9*: 283–292

(B)

Wild-type

Wnt9b⁻/⁻

FIGURE 18.9 Wnts are critical for kidney development. (A) In the 11-day mouse kidney, Wnt9b is found on the stalk of the ureteric bud, while Wnt11 is found at the tips. Wnt9b induces the metanephric mesenchyme to condense; Wnt11 will partition the metanephric mesoderm to induce branching of the ureteric bud. Borders of the bud are indicated by a dotted line. (B) A wild-type 18.5-day male mouse (left) has normal kidneys, adrenal glands, and ureters. In a *Wnt9b*-deficient mouse (right), the kidneys are absent.

STEP 5: WNT SIGNALS CONVERT THE AGGREGATED MESENCHYME CELLS INTO A NEPHRON Wnt9b and Wnt6 from the ureteric bud are critical for transforming the metanephric mesenchyme cells into tubular epithelium. These paracrine signals induce yet another Wnt in the mesenchyme, Wnt4, which acts in an autocrine manner to complete the transition from mesenchymal mass to epithelium (**FIGURE 18.9**; Stark et al. 1994; Kispert et al. 1998; Itäranta et al. 2002).

The epithelium hollows out to form the renal vesicle, which immediately becomes polarized in a proximal (near the ureteric bud) to distal direction. A combination of signaling factors (especially Notch proteins) is critical for differential gene expression along the length of the new epithelium. As the epithelium changes shape to form the C- and S-shaped tubules, the regions of the nephron become specified (Georgas et al. 2009). The mechanism by which the nephron connects to the ureteric bud remains elusive.

STEP 6: INSERTING THE URETER INTO THE BLADDER The branching epithelium becomes the collecting system of the kidney. This epithelium collects the filtered urine from the nephron and secretes antidiuretic hormone for the resorption of water (a process that, not so incidentally, makes life on land possible). The original stalk of the ureteric bud, situated above the first branch point, becomes the ureter, the tube that carries urine into the bladder. The junction between the ureter and the bladder is extremely important, and hydronephrosis, a birth defect leading to abnormalities of renal filtration, occurs when this junction is not properly placed and urine cannot enter the bladder.

The ureter is made into a watertight connecting duct by the condensation of mesenchymal cells around it. These mesenchymal cells become smooth muscle cells capable of wavelike contractions (peristalsis) that allow the urine to move into the bladder. These cells also secrete BMP4 (Cebrian et al. 2004), which causes differentiation of this region of the ureteric bud into the ureter. BMP inhibitors protect the region of the ureteric bud that forms the collecting ducts from this differentiation.

To complete kidney development, the ureter must establish connections with the endodermally derived bladder. The bladder develops from a portion of the **cloaca**,[1] which will become the waste receptacle for both the intestine and the kidney (**FIGURE 18.10A,B**). In mammals, the cloaca becomes divided by a septum into the urogenital sinus and the rectum. Part of the urogenital sinus becomes the bladder, while another part becomes the urethra (which will carry the urine out of the body).

The ureteric bud grows toward the bladder through an ephrin-mediated pathway (Weiss et al. 2014). Once at the bladder, the urogenital sinus cells of the bladder wrap themselves around both the ureter and the nephric duct. Then the nephric duct migrates ventrally, opening into the urethra rather than into the bladder (**FIGURE 18.10C–F**). Expansion of the bladder then moves the ureter to its final position at the neck of the bladder (Batourina et al. 2002; Mendelsohn 2009). In females, the entire nephric duct degenerates, while the Müllerian duct opens into the vagina (see Chapter 6). In males, the nephric duct also forms the sperm outflow track, so males expel sperm and urine through the same opening.

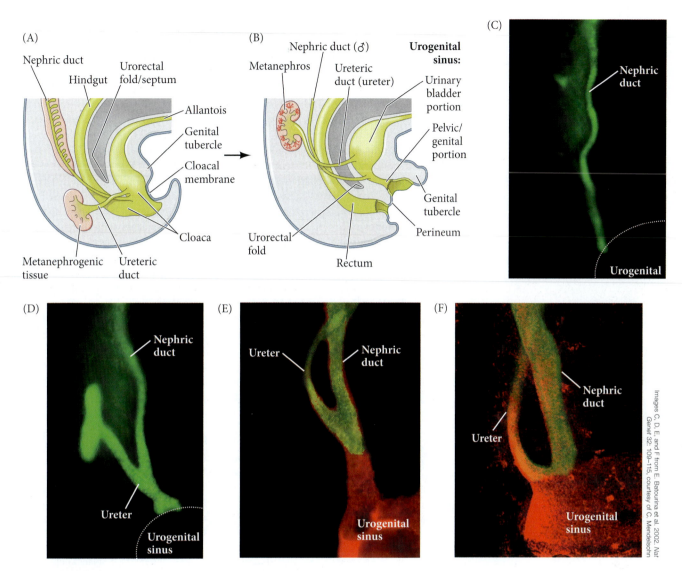

FIGURE 18.10 Development of the bladder and its connection to the kidney via the ureter. (A) The cloaca originates as an endodermal collecting area that opens into the allantois. (B) The urorectal septum divides the cloaca into the future rectum and the urogenital sinus. The bladder forms from the anterior portion of the sinus, and the urethra develops from the posterior region of the sinus. The space between the rectal opening and the urinary opening is the perineum. (C–F) Insertion of the ureter into the embryonic mouse bladder. (C) Day-10 mouse urogenital tract. The nephric duct is stained with GFP fused to a *Hoxb7* promoter. (D) Urogenital tract from a day-11 embryo, after ureteric bud outgrowth. (E) Whole mount urogenital tract from a day-12 embryo. The ducts are stained green and the urogenital sinus red. (F) The ureter separates from the nephric duct and forms a separate opening into the bladder. (A,B after L. R. Cochard. 2002. *Netter's Atlas of Human Embryology*. MediMedia USA: Peterboro, NJ.)

Images C, D, E, and F from E. Batourina et al. 2002. *Nat Genet* 32: 109–115, courtesy of C. Mendelsohn

Thus, the blood-filtering kidneys emerge from the mutual induction of two parts of the intermediate mesoderm, the ureteric bud and the metanephric mesenchyme. Now we can focus more laterally, on the lateral plate mesoderm, and discern the genesis of the heart, vessels, and blood.

Lateral Plate Mesoderm: Heart and Circulatory System

Consisting of a heart, blood cells, and an intricate system of blood vessels, the circulatory system is the vertebrate embryo's first functional unit, providing nourishment to the developing organism. Few events in biology are as thought-provoking and accessible as watching the heart beating in a 2-day chick embryo, pumping the first blood cells into vessels that have not even formed valves yet. In 1651, amid the chaos of the

English civil wars, William Harvey, physician to the king, was comforted by viewing the heart as the undisputed ruler of the body, through whose divinely ordained powers the lawful growth of the organism was assured. Later embryologists saw the heart as more of a servant than a ruler, the chamberlain of the household who assured that nutrients reached the apically located brain and the peripherally located muscles. In either metaphor, the heart and its circulation (which Harvey discovered) were seen to be critical for development. As Harvey argued persuasively in 1651, the chick embryo must form its own blood without any help from the hen, and this blood is crucial in embryonic growth. How this happened was a mystery to him.

The heart and circulatory system that were so intriguing to Harvey arise from the vertebrate embryo's lateral plate mesoderm. The lateral plate mesoderm resides on the lateral side of each of the two bands of intermediate mesoderm (see Figure 18.1). Each of these lateral plates splits horizontally into two layers. The dorsal layer is the **somatic (parietal) mesoderm**, which underlies the ectoderm and, together with the ectoderm, forms the **somatopleure**. The ventral layer is the **splanchnic (visceral) mesoderm**, which overlies the endoderm and, together with the endoderm, forms the **splanchnopleure** (**FIGURE 18.11A**). The space between these two layers becomes the body cavity—the **coelom**—which stretches from the future neck region to the posterior of the body. (See **Further Development 18.3, Coelom Formation**, online.)

During later development, the right- and left-side coeloms fuse, and folds of tissue extend from the somatic mesoderm, dividing the coelom into separate cavities. In

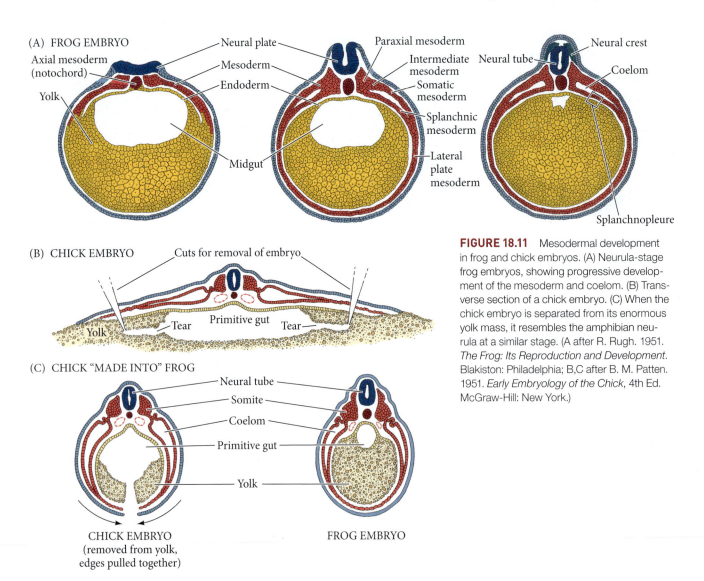

FIGURE 18.11 Mesodermal development in frog and chick embryos. (A) Neurula-stage frog embryos, showing progressive development of the mesoderm and coelom. (B) Transverse section of a chick embryo. (C) When the chick embryo is separated from its enormous yolk mass, it resembles the amphibian neurula at a similar stage. (A after R. Rugh. 1951. *The Frog: Its Reproduction and Development.* Blakiston: Philadelphia; B,C after B. M. Patten. 1951. *Early Embryology of the Chick*, 4th Ed. McGraw-Hill: New York.)

mammals, these mesodermal folds subdivide the coelom into the **pleural**, **pericardial**, and **peritoneal cavities**, enveloping the thorax, heart, and abdomen, respectively. The mechanism for creating the linings of these body cavities from the lateral plate mesoderm has changed little throughout vertebrate evolution, and the development of the amniote mesoderm can be compared with similar stages of frog embryos (**FIGURE 18.11B,C**). The development of the circulatory system provides excellent examples of induction, specification, cell migration, organ formation, and the role of stem cells in both embryonic development and adult tissue regeneration.

Heart Development

The circulatory system is the first working unit in the developing embryo, and the heart is the first functional organ. Like other organs, the heart arises through the specification of precursor cells, the migration of these precursor cells to the organ-forming region, the specification of cell types through signaling interactions within and between tissues, and the coordination of morphogenesis, growth, and cell differentiation.

A minimalist heart

Both the chick and the mammalian heart are complex, rather baroque structures. However, the amniote heart evolved from far simpler pumps (Stolfi et al. 2010). The four-chambered masterpiece that is the mammalian heart is a developmental elaboration of the single-chambered tunicate heart that forms from about two dozen cells. In tunicates (noted to be one of the closest invertebrate relatives to the vertebrates; see Chapter 10), cardiac precursor cells form bilateral cell clusters that migrate anteriorly and ventrally along the endoderm and fuse at the ventral midline (**FIGURE 18.12**; Davidson et al. 2005). The few cells that form the tunicate heart appear to have the same basic pattern of transcription factors that we see in the chick and mouse heart lineages. (See **Further Development 18.4, Molecular Mechanisms of Heart Development in the Tunicate *Ciona*,** online.)

Formation of the heart fields

While tunicate embryos develop rapidly and from a small number of cells, the vertebrate heart arises from two regions of splanchnic mesoderm—one on each side of the body—that interact with adjacent tissue to become specified for heart development.

In the early amniote gastrula, the heart progenitor cells (about 50 of them in mice) are located in two small patches, one on each side of the epiblast, close to the rostral portion of the primitive streak. These cells migrate together through the streak and form two groups of lateral plate mesoderm cells, positioned anteriorly at the level of the node (Tam et al. 1997; Colas et al. 2000). The general specification of a **heart field**, also known as **cardiogenic mesoderm**, has already started during this cellular migration. Labeling experiments by Stalberg and DeHann (1969) and Abu-Issa and Kirby (2008) have shown that the progenitor cells of the heart field migrate such that the medial-lateral (center-to-side) arrangement of these early cells will become the anterior-posterior (rostral-caudal) axis of a linear **heart tube**.

FIGURE 18.12 Heart development in the tunicate *Ciona*. (A) In the tailbud-stage embryo, transgenic Mesp-GFP glows in regions where Mesp is activated by Tbx6 in the B8.9 and B8.10 blastomeres. (B) At a slightly later stage, the heart precursors migrate into the head region. (C) Ventrolateral view in which both left and right heart and muscle precursors can be observed. Cell divisions are forming both the heart (left) and the anterior muscles.

(A)

Anterior pharyngeal muscle precursors

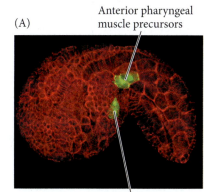

Heart precursors

(B)

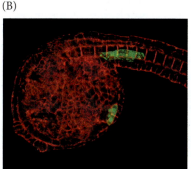

(C)

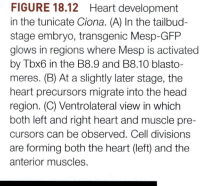
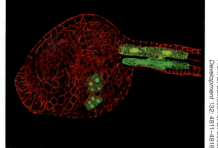

(A)

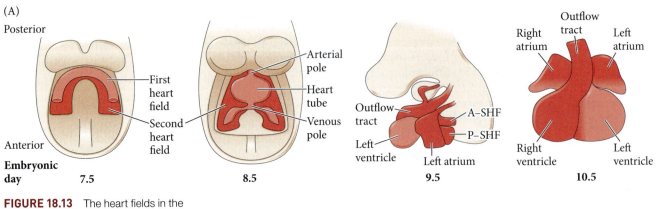

Embryonic day

7.5   8.5   9.5   10.5

FIGURE 18.13 The heart fields in the mouse embryo. (A) On embryonic day 7.5, the heart fields from each side of the body have joined into a common cardiac crescent that contains the first and second heart fields. The first heart field contributes primarily to the left ventricle. By day 10.5, the second heart field contributes to the other three chambers—the right ventricle and the left and right atria—as well as to the outflow tract that originally includes the aorta and the pulmonary artery. Anterior and posterior second heart fields are shown (A-SHF and P-SHF). (B) A possible lineage tree showing the cooperation of first and second heart fields in forming the heart and also showing the mixture of heart, lung, and pulmonary blood vessel cells existing in the second heart field. The dashed line indicates that the exact location of the separation of lung, pulmonary, facial muscle, and heart cell progenitors is not known. Some of the transcription factors associated with these cardiopharyngeal mesoderm (CPM) progenitor cells are listed beneath them. (A after R. G. Kelly. 2012. *Curr Top Dev Biol* 100: 33–65; B after R. Diogo et al. 2015. *Nature* 520: 466–473 and T. Peng et al. 2013. *Nature* 500: 589–592.)

(B)

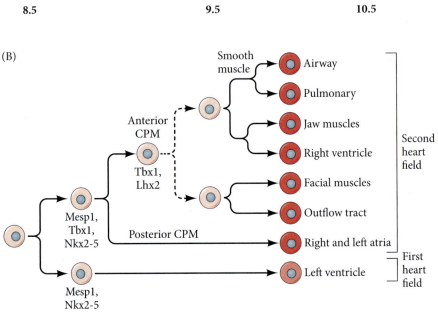

The vertebrate heart field is divided into at least two regions (**FIGURE 18.13**). The first heart field appears to form the scaffold of the developing heart. The progenitor cells of the first field fuse at the midline to form the primary heart tube that gives rise to the muscular regions of the left and right ventricles (de la Cruz and Sanchez-Gomez 1998). However, these cells have limited proliferative ability and therefore will generate only the major portion of the left ventricle of the adult heart (i.e., the chamber that pumps blood into the aorta). The progenitors of the second heart field add cells to both anterior and posterior ends of the heart tube (Meilhac et al. 2015). On the posterior end, these cells will produce the atria (one in fish, two in all other vertebrates) and contribute to the inlet part of the heart. On the anterior end, the second heart field in amniotes will generate the right ventricle (and a portion of the single ventricle in fish) as well as the outflow region (the conus arteriosus and the truncus arteriosus), which becomes the base of the aorta and pulmonary arteries (de la Cruz et al. 1989; Kelly 2012; Liu and Stainier 2012). It is only through the process of looping that the atria are brought anterior to the ventricles to form the adult four-chambered heart.

The second heart field is a remarkable group of cells because it contains not only the progenitor cells for the heart, but also the cells that will generate the facial muscles, the pulmonary artery and vein, and the lung mesenchyme (Lescroart et al. 2010, 2015; Peng et al. 2013). Therefore, in a remarkable way, the heart precursor cells are coordinated with the development of the face and lungs. The common precursor of pharyngeal and cardiac mesoderm is thought to be derived from a similar group of pharynx and heart progenitor cells that are found in certain deuterostome invertebrates such as tunicates (Diogo et al. 2015).

All the cell types of the heart—the **cardiomyocytes** that form the muscular layers, the **endocardium** that forms the internal layer, the **endocardial cushions** of the valves, the **epicardium** that forms the coronary blood vessels that feed the heart, and the **Purkinje fibers**[2] that coordinate the heartbeat—are generated from these heart fields[3] (Mikawa 1999; van Wijk et al. 2009). Moreover, as we will discuss later, it appears that each progenitor cell is capable of becoming any of the differentiated heart cell types. The cardiac precursor cells will be supplemented by cardiac neural crest cells; the latter help make the outflow tract and the septum that separates the aorta from the pulmonary trunk (see Figure 15.21; Porras and Brown 2008).

Specification of the cardiogenic mesoderm

The cardiogenic mesoderm cells are specified by their interactions with the pharyngeal endoderm and notochord. The heart does not form if this anterior endoderm is removed, and posterior endoderm cannot induce heart cells to form (Nascone and Mercola 1995; Schultheiss et al. 1995). BMPs (especially BMP2) from the anterior endoderm promote both heart and blood development. Endodermal BMPs also induce Fgf8 synthesis in the endoderm directly beneath the cardiogenic mesoderm, and Fgf8 appears to be critical for the expression of cardiac proteins (Alsan and Schultheiss 2002).

Inhibitory signals prevent heart structures from forming where they should not be made. First, the notochord secretes Noggin and Chordin, blocking BMP signaling in the center of the embryo, and specific Noggin-secreting cells in the myotome prevent heart cell specification of the somites. Second, Wnt proteins from the neural tube, especially Wnt3a and Wnt8, inhibit heart formation but promote blood formation. The anterior endoderm, moreover, produces Wnt inhibitors, which prevent Wnts from binding to their receptors. In this way, cardiac precursor cells are specified in those places where BMPs (from the lateral mesoderm and endoderm) and Wnt antagonists (from the anterior endoderm) coincide (**FIGURE 18.14**; Marvin et al. 2001; Schneider and Mercola 2001; Tzahor and Lassar 2001; Gerhart et al. 2011). In the absence of Wnt signals, BMPs activate **Nkx2-5** and **Mesp1**, two genes that are critical in the regulatory network that specifies the heart cells. (See **Further Development 18.5, A Gene Regulatory Network at the Heart of Cardiogenic Mesoderm Specification**, online.)

Migration of the cardiac precursor cells

As the presumptive heart cells move anteriorly between the ectoderm and endoderm toward the middle of the embryo, they remain in close contact with the endoderm surface (Linask and Lash 1986). In the chick, the directionality of this migration appears to be provided by the foregut endoderm. If the cardiac region endoderm is rotated with respect to the rest of the embryo, migration of the cardiogenic mesoderm cells is reversed. It is thought that the endodermal component responsible for this movement is an anterior-to-posterior concentration gradient of fibronectin. Antibodies against fibronectin stop the migration, whereas antibodies against other extracellular matrix components do not (Linask and Lash 1988).

FIGURE 18.14 Model of inductive interactions involving the BMP and Wnt pathways that form the boundaries of the cardiogenic mesoderm. Wnt signals from the neural tube instruct lateral plate mesoderm (LPM) to become precursors of the blood and blood vessels. In the anterior portion of the body, Wnt inhibitors (Dickkopf [Dkk], Crescent, Cerberus) from the pharyngeal endoderm prevent Wnt from functioning, allowing later signals (BMP, Fgf8) to convert lateral plate mesoderm into cardiogenic mesoderm. BMP signals will also be important for the differentiation of hemangiogenic (blood, blood vessel) mesoderm. In the center of the embryo, Noggin and Chordin signals from the notochord block BMPs. Thus, the cardiac and blood-forming fields do not form in the center of the embryo.

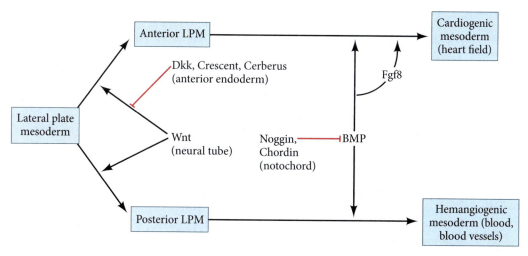

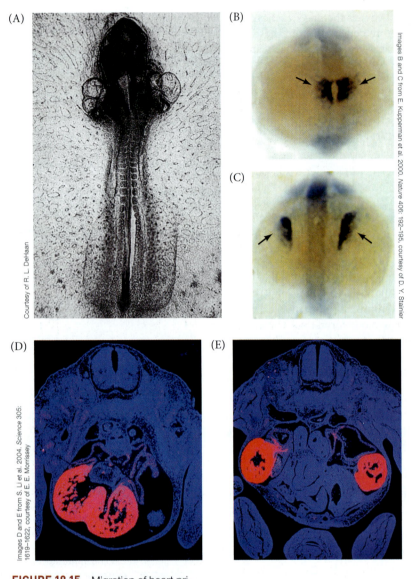

(A)

Courtesy of R. L. DeHaan

(B)

(C)

Images B and C from E. Kupperman et al. 2000. *Nature* 406: 192–195, courtesy of D. Y. Stainier

(D)

(E)

Images D and E from S. Li et al. 2004. *Science* 305: 1619–1622, courtesy of E. E. Morrisey.

FIGURE 18.15 Migration of heart primordia. (A) Cardia bifida (two hearts) in a chick embryo, induced by surgically cutting the ventral midline, thereby preventing the two heart primordia from fusing. (B) Wild-type zebrafish and (C) *miles apart* mutant, stained with probes for the cardiac myosin light chain. There is a lack of migration in the *miles apart* mutant. (D) Mouse heart stained with antisense RNA probe to ventricular myosin shows fusion of the heart primordia in a wild-type 13.5 day embryo. (E) Cardia bifida in a *Foxp4*-deficient mouse embryo. Interestingly, each of these hearts has ventricles and atria, and they both loop and form all four chambers with normal left-right asymmetry.

This movement produces two populations of migrating cardiac precursor cells, one on the right side of the embryo and another on the left. Each side has its own first and second heart fields, and each of these populations starts to form its own heart tube. In the chick, the fields are brought together around the seven-somite stage, when the foregut is formed by the inward folding of the splanchnopleure. This movement places the two cardiac tubes together (Varner and Taber 2012). The two endocardial tubes lie within the common tube for a short time, but eventually these two tubes also fuse. The bilateral origin of the heart can be demonstrated by surgically preventing the merger of the lateral plate mesoderm (Gräper 1907; DeHaan 1959). This manipulation results in a condition called **cardia bifida**, in which two separate hearts form, one on each side of the body (**FIGURE 18.15**). Thus, endoderm specifies heart progenitors, gives directionality to their migration, and mechanically pulls the two heart fields together. (See **Further Development 18.6, Bilateral Origin of the Heart**, online.)

As the cells of the first heart field migrate along the endoderm to form the heart tube, the cells of the second heart field remain in contact with the pharyngeal endoderm. Here they are kept in a state of proliferation by a combination of paracrine factors (probably Sonic hedgehog, Fgf8, and Wnts; Chen et al. 2007; Lin et al. 2007). The cells of the second heart field can be distinguished by their expression of the Islet1 transcription factor. These cells also begin to synthesize and secrete Fgf8, which acts in an autocrine manner to stimulate the cells to migrate and add themselves to the anterior and posterior portions of the heart tube formed by the first heart field progenitor cells (Park et al. 2008). The anterior region of the second heart field contributes to the right ventricle and the outflow tract, whereas the posterior region generates the atria (Zaffran et al. 2004; Verzi et al. 2005; Galli et al. 2008).

As the second heart field precursor cells migrate, the posterior region becomes exposed to increasingly higher concentrations of retinoic acid (RA) produced by the posterior mesoderm. RA is critical in specifying these posterior precursor cells to become the inflow, or "venous," portions of the heart—the sinus venosus and atria. Originally, these fates are not fixed, as transplantation or rotation experiments show that these precursor cells can regulate and differentiate in accordance with a new environment. But once the posterior cardiac precursors enter the realm of active RA synthesis, they express the gene for retinaldehyde dehydrogenase-2; they can then produce their own RA, and their posterior fate becomes committed (**FIGURE 18.16**; Simões-Costa et al. 2005).

As in kidney development, retinoic acid regulates the expression of Hox genes (especially *Hoxa1*, *Hoxb1*, and *Hoxa3*), which appear to promote different regional identities in the second heart field precursors (Bertrand et al. 2011). In mice, the outflow tract region, as well as the cardiac neural crest cells that enter this region of the second heart field, display differential Hox gene expression based on their exposure to RA (Diman et al. 2011). This ability of RA to specify and commit heart precursor cells

(A) Chick, stage 8

(B) Mouse, 8 days

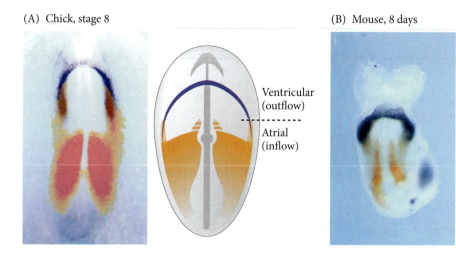

Ventricular
(outflow)

Atrial
(inflow)

From M. Simões-Costa et al. 2005. *Dev Biol* 277: 1–15, courtesy of J. Xavier-Neto

FIGURE 18.16 Double in situ hybridization for the expression of *Raldh2* (orange), which encodes the retinoic acid-synthesizing enzyme retinaldehyde dehydrogenase-2, and *Tbx5* (purple), a marker for the early heart fields. In the developmental stages seen here, the heart precursor cells are exposed to progressively increasing amounts of retinoic acid. (A) Chick, stage 8 (26–29 hours). (B) Mouse, 8 days.

to become atria explains its teratogenic effects on heart development, wherein exposure of vertebrate embryos to RA can cause expansion of atrial tissues at the expense of ventricular tissues (Stainier and Fishman 1992; Hochgreb et al. 2003). (See **Further Development 18.7, Fusion of the Heart and the First Heartbeats**, online.)

Initial heart cell differentiation

One of the most important discoveries of cardiac development was the demonstration that the different cells of the heart—the ventricular myocytes, the atrial myocytes, the smooth muscles that generate the venous and arterial vasculature, the endothelial lining of the heart and valves, and the epicardium that forms an envelope for the heart—are all derived from the same progenitor cell type (Kattman et al. 2006; Moretti et al. 2006; Wu et al. 2006). Indeed, there appears to be an early multipotent progenitor cell population that bears responsibility for forming the entire circulatory system. Under one set of influences, its descendants become **hemangioblasts**, those cells that form blood vessels and blood cells; under the conditions in the heart fields, its descendants form **multipotent cardiac precursor cells** (**FIGURE 18.17**; Linask 2003; Anton et al. 2007). Once particular pioneer transcription factors (Nkx2-5, Mesp1, Gata4, and Tbx5) are activated within these different descendants, self-sustaining gene regulatory networks differentiate the lineages of the heart. (See **Further Development 18.8, A Self-Sustaining Gene Regulatory Network Differentiates the Heart**, online.)

FIGURE 18.17 Model for early cardiovascular lineages. The splanchnic mesoderm gives rise to two lineages, both of which have Flk1 (a VEGF receptor) on their cell membranes. The earlier population gives rise to the hemangioblasts (precursors to blood cells and blood vessels), whereas the later population gives rise to the cardiac (heart) precursor cells. This latter population in turn gives rise to a variety of cell types whose relationships are still obscure; however, all the cell types of the heart can be traced back to the cardiac precursor cells. (After R. Anton. 2007. *BioEssays* 29: 422–426 and D. M. DeLaughter et al. 2011. *Birth Defects Res A: Clin Mol Teratol* 91: 511–525.)

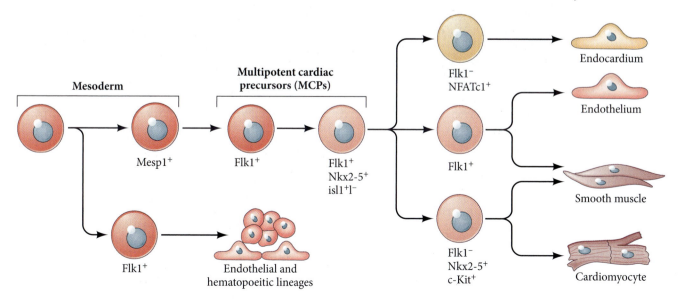

Mesoderm

Multipotent cardiac precursors (MCPs)

Mesp1+

Flk1+

Flk1+
Nkx2-5+
isl1+l−

Flk1−
NFATc1+

Flk1+

Flk1−
Nkx2-5+
c-Kit+

Flk1+

Endothelial and hematopoeitic lineages

Endocardium

Endothelium

Smooth muscle

Cardiomyocyte

Looping of the heart

In 3-day chick embryos and 4-week human embryos, the heart is a two-chambered tube, with an atrium to receive blood and a ventricle to pump blood out. (In the chick embryo, the unaided eye can see the remarkable cycle of blood entering the lower chamber and being pumped out through the aorta.) Looping of the heart converts the original anterior-posterior polarity of the heart tube into the right-left polarity seen in the adult organism. When this looping is complete, the portion of the heart tube destined to become the atria lies anterior to the portion that will become the ventricles (**FIGURE 18.18**).

This critically important process begins with the anterior part of the heart specifying the direction of the bend. Looping begins immediately after the onset of rhythmic heart contractions and the initiation of blood flow; pressure from the blood flow helps drive looping to completion (Hove et al. 2003; Groenendijk et al. 2005). As the

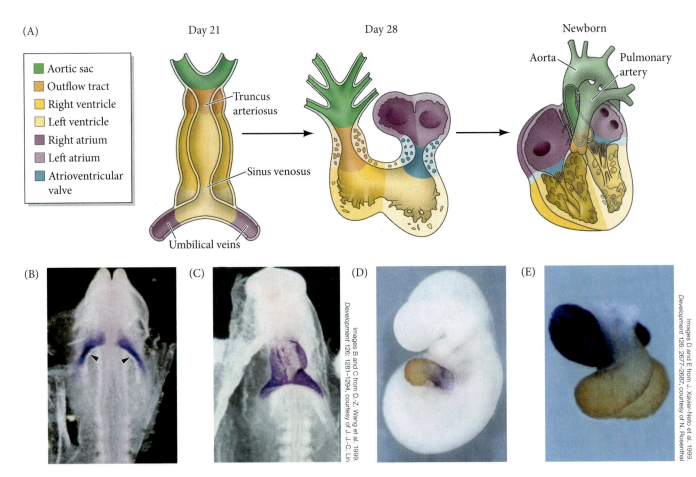

FIGURE 18.18 Cardiac looping and chamber formation. (A) Schematic diagram of cardiac morphogenesis in humans. On day 21, the heart is a single-chambered tube. Regional specification of the tube is shown by the different colors. By day 28, cardiac looping has occurred, placing the presumptive atria anterior to the presumptive ventricles. In the newborn, the valves and chambers establish circulatory routes such that the left ventricle pumps into the aorta and the right ventricle pumps into the pulmonary artery to the lungs. (B,C) *Xin* expression in the fusion of left and right heart primordia of a chick. The cells fated to form the myocardium are stained for the *Xin* message, whose protein product is essential for heart-tube looping. (B) Stage-9 chick neurula, in which Xin protein (purple) is seen in the two symmetrical heart-forming fields (arrowheads). (C) Stage-10 chick embryo showing fusion of the two heart-forming regions prior to looping. (D,E) Specification of the atria and ventricles occurs even before heart looping. The atria and ventricles of the mouse embryo express different types of myosin proteins; here, atrial myosin stains blue and ventricular myosin stains orange. (D) In the tubular heart (prior to looping), the two myosins (and their respective stains) overlap at the atrioventricular channel joining the future regions of the heart. (E) After looping, the blue stain is seen in the definitive atria and inflow tract, while the orange stain is seen in the ventricles. The unstained region above the ventricles is the truncus arteriosus. Derived primarily from the neural crest, the truncus arteriosus becomes separated into the aorta and pulmonary arteries. (A after D. Srivastava and E. N. Olson. 2000. *Nature* 407: 221–226.)

Within figure (A):

Day 21 · Day 28 · Newborn

Legend:
- Aortic sac
- Outflow tract
- Right ventricle
- Left ventricle
- Right atrium
- Left atrium
- Atrioventricular valve

Truncus arteriosus · Sinus venosus · Umbilical veins · Aorta · Pulmonary artery

(B) (C) Images B and C from D.-Z. Wang et al. 1999. *Development* 126: 1281–1294, courtesy of J. J.-C. Lin

(D) (E) Images D and E from J. Xavier-Neto et al. 1999. *Development* 126: 2677–2687, courtesy of N. Rosenthal

bending of the heart tube deepens, an increasing volume of blood enters the heart. The volume differences are thought to be transmitted to the cells through the extracellular matrix and the cytoskeleton (Linask et al. 2005; Garita et al. 2011). Precise chamber alignment is needed for the correct signaling for the formation of the heart valves and the ventricular and atrial septa, and to allow for the heart to become connected to the embryonic vasculature that has been developing concomitantly within the embryo.

FURTHER DEVELOPMENT

Forming heart valves

As the heart is looping, changes in the endocardium start forming the valves. The beginning of heart valve development is the formation of endocardial cushions in the canal between the atrium and the ventricle and in the outflow tract of the looping heart tube (Armstrong and Bischoff 2004). Cushion development is initiated by the myocardium's signaling to endocardial cells to express the *Twist* gene. Twist protein is a transcription factor that initiates epithelial-mesenchymal transition and cell migration. And these endocardial cells leave the endocardium and migrate to form the endocardial cushions (Barnett and Desgrosellier 2003; Shelton and Yutzey 2008). Twist also activates the gene for Tbx20, and together Twist and Tbx20 activate the proteins that cause the proliferation and strengthening of the valves. (See **Further Development 18.9, Changing Heart Anatomy at Birth**, online.)

Blood Vessel Formation

Although the heart is the first functional organ of the body, it does not begin to pump until the vascular system of the embryo has established its first circulatory loops. Rather than sprouting from the heart, the blood vessels form independently, linking up to the heart soon afterward. Everyone's circulatory system is different, since the genome cannot encode the intricate series of connections between the arteries and veins. Indeed, chance plays a major role in establishing the microanatomy of the circulatory system. However, all circulatory systems in a given species look very much alike because the development of the circulatory system is severely constrained by physiological, evolutionary, and physical parameters.

Vasculogenesis: The initial formation of blood vessels

The development of blood vessels occurs by two temporally separate processes: **vasculogenesis** and **angiogenesis** (**FIGURE 18.19**). During vasculogenesis, a network of blood vessels is created de novo from the lateral plate mesoderm. During angiogenesis, this primary network is remodeled and pruned into a distinct capillary bed, arteries, and veins.

   In the first phase of vasculogenesis, a combination of BMP, Wnt, and Notch signals activates the Etv2 transcription factor in lateral plate mesoderm cells leaving the primitive streak in the posterior of the embryo, converting them into hemangioblasts.[4] Labeling zebrafish embryos with fluorescent probes to make single-cell fate maps confirms that hemangioblasts are the common progenitor for both the hematopoietic (blood cell) and endothelial (blood vessel) lineages in zebrafish (Paik and Zon 2010). This population of bipotent progenitor cells is found only in the ventral portion of the aorta, the region that had been known to produce these two cell types. The pathway permitting such aortic cells to differentiate into hemangioblasts appears to be induced by the *Cdx4* gene, while the determination of whether the hemangioblast becomes a blood cell precursor or a blood vessel precursor is regulated by the Notch signaling pathway. Notch signaling increases the conversion of hemangioblasts into blood cell precursors, whereas reduced amounts of Notch cause hemangioblasts to become endothelial (Vogeli et al. 2006; Hart et al. 2007; Lee et al. 2009). Notch signaling activates the expression of the Runx1 transcription factor, which, as we will shortly see, appears to be conserved throughout vertebrates in inducing the conversion of endothelial cells to blood stem cells (Burns et al. 2005, 2009). (See **Further Development 18.10, Constraints on the Formation of Blood Vessels**, online.)

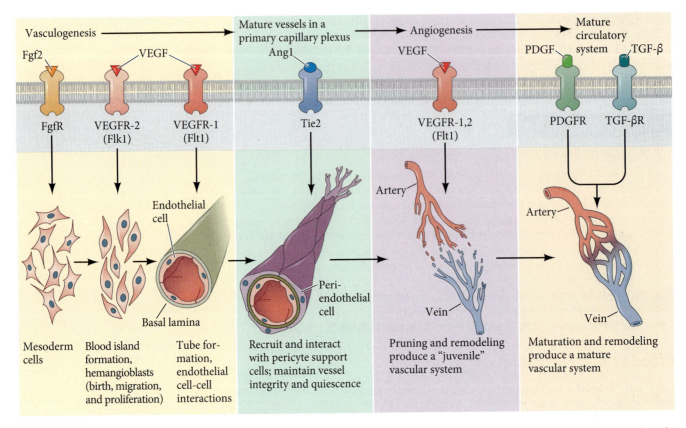

FIGURE 18.19 Vasculogenesis and angiogenesis. Vasculogenesis involves the formation of blood islands containing hemangioblasts and the construction of capillary networks from them (left panel). Angiogenesis involves the formation of new blood vessels by remodeling and building on older ones. Angiogenesis finishes the circulatory connections begun by vasculogenesis. The major paracrine factors involved in each step are shown at the top of the diagram, and their receptors (on the vessel-forming cells) are shown beneath them. (After D. Hanahan. 1997. *Science* 277: 48–50 and W. Risau. 1997. *Nature* 386: 671–674.)

SITES OF VASCULOGENESIS   In amniotes, formation of the primary vascular networks occurs in two distinct and independent regions. First, **extraembryonic vasculogenesis** occurs in the **blood islands** of the yolk sac. These are the blood islands formed by the hemangioblasts, and they give rise to the early vasculature needed to feed the embryo and also to a red blood cell population that functions in the early embryo (**FIGURE 18.20A**). Mouse studies (Frame et al. 2016) suggest that definitive (adult) blood stem cells also emerge in these yolk sac blood islands. Second, **intraembryonic vasculogenesis** forms the dorsal aorta, and vessels from this large vessel connect with capillary networks that form from mesodermal cells within each organ. The first scaffold of the dorsal aorta came from cells of the somite that migrate ventrally (see Further Development 17.5, online).

FURTHER DEVELOPMENT

DEVELOPMENT OF THE BLOOD ISLAND   The aggregation of endothelium-forming cells in the yolk sac is a critical step in amniote development, for the blood islands that line the yolk sac produce the veins that bring nutrients to the embryo and transport gases to and from the sites of respiratory exchange (**FIGURE 18.20B**). In birds, these vessels are called the **vitelline veins**; in mammals, they are called the **omphalomesenteric veins** or, more usually, **umbilical veins**. In the chick, blood islands are first seen in the area opaca, when the primitive streak is at its fullest extent (Pardanaud et al. 1987). They form cords of hemangioblasts, which

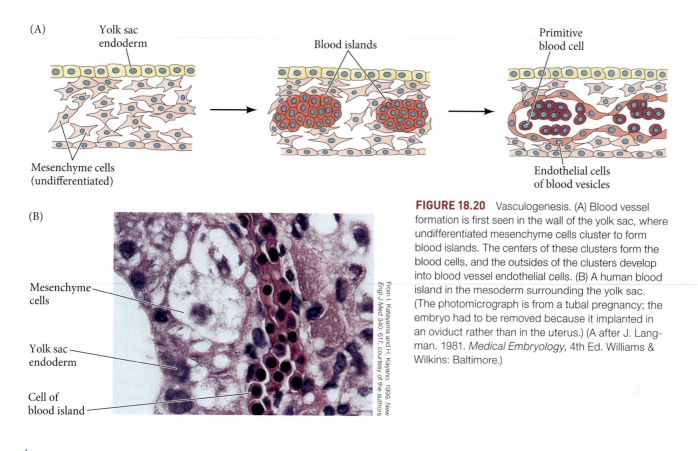

(A)

Yolk sac endoderm

Blood islands

Primitive blood cell

Mesenchyme cells (undifferentiated)

Endothelial cells of blood vesicles

(B)

Mesenchyme cells

Yolk sac endoderm

Cell of blood island

From I. Katayama and H. Kayano. 1999. *New Engl J Med* 340: 617, courtesy of the authors

FIGURE 18.20 Vasculogenesis. (A) Blood vessel formation is first seen in the wall of the yolk sac, where undifferentiated mesenchyme cells cluster to form blood islands. The centers of these clusters form the blood cells, and the outsides of the clusters develop into blood vessel endothelial cells. (B) A human blood island in the mesoderm surrounding the yolk sac. (The photomicrograph is from a tubal pregnancy; the embryo had to be removed because it implanted in an oviduct rather than in the uterus.) (A after J. Langman. 1981. *Medical Embryology*, 4th Ed. Williams & Wilkins: Baltimore.)

soon become hollowed out and become the flat endothelial cells lining the vessels (while the central cells give rise to blood cells). As the blood islands grow, they eventually merge to form the capillary network draining into the two vitelline veins, which bring food and blood cells to the newly formed heart.

GROWTH FACTORS AND VASCULOGENESIS Three growth factors are critically responsible for initiating vasculogenesis (see Figure 18.19). One of these, **basic fibroblast growth factor** (**Fgf2**), is required for the generation of hemangioblasts from the splanchnic mesoderm. When cells from quail blastodiscs are dissociated in culture, they do not form blood islands or endothelial cells. However, when these cells are cultured with Fgf2 protein, blood islands emerge and form endothelial cells (Flamme and Risau 1992). Fgf2 is synthesized in the chick embryonic chorioallantoic membrane and is responsible for the vascularization of this tissue (Ribatti et al. 1995).

The second family of proteins involved in vasculogenesis is the **vascular endothelial growth factors** (**VEGFs**). This family includes several VEGFs, as well as placental growth factor (PlGF), which direct the expansive growth of blood vessels in the placenta. Each VEGF appears to enable the differentiation of the angioblasts (blood vessel progenitor cells) and their multiplication to form endothelial tubes. The most important VEGF in normal development, VEGF-A, is secreted by the mesenchymal cells near the blood islands, and hemangioblasts and angioblasts have receptors for this VEGF (Millauer et al. 1993). If mouse embryos lack the genes encoding either VEGF-A or its major receptor (the Flk1 receptor tyrosine kinase), yolk sac blood islands fail to appear and vasculogenesis fails to take place (**FIGURE 18.21A**; Ferrara et al. 1996). Mice lacking genes for the Flk1 receptor protein have blood islands and differentiated endothelial cells, but these cells are not organized into blood vessels (Fong et al. 1995; Shalaby et al. 1995). VEGF-A is also important in forming blood vessels to the developing bone and kidney. (See **Further Development 18.11, VEGF and Your Green Tea Diet**, online.)

A third set of proteins, the **angiopoietins**, mediate the interaction between the endothelial cells and the **pericytes**—smooth muscle-like cells that the endothelial cells recruit to cover them. Mutations of either the angiopoietins or their receptor protein,

(A)

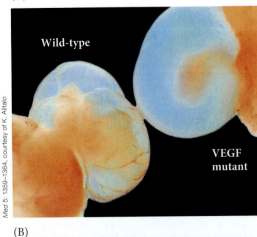

Wild-type

VEGF mutant

(B)

FIGURE 18.21 VEGF and its receptors in mouse embryos. (A) Yolk sacs of a wild-type mouse and a littermate heterozygous for a loss-of-function mutation of VEGF-A. The mutant embryo lacks blood vessels in its yolk sac and dies. (B) In a 9.5-day mouse embryo, VEGFR-3 (red), a VEGF receptor found on tip cells, is found at the angiogenic front of the capillaries (stained green).

Tie2, lead to malformed blood vessels deficient in the smooth muscles that normally surround them (Davis et al. 1996; Suri et al. 1996; Vikkula et al. 1996; Moyon et al. 2001). (See **Further Development 18.12, Arterial, Venous, and Lymphatic Vessels**, online.)

Angiogenesis: Sprouting of blood vessels and remodeling of vascular beds

After an initial phase of vasculogenesis, angiogenesis begins. By this process, the primary capillary networks are remodeled and veins and arteries are made (see Figure 18.19). The critical factor for angiogenesis is VEGF-A (Adams and Alitalo 2007). In many cases, VEGF-A secreted by an organ will induce the migration of endothelial cells from existing blood vessels into that organ, causing the endothelial cells to form capillary networks there. Other factors, including hypoxia (low oxygen levels), can also induce the secretion of VEGF-A and thus induce blood vessel formation.

During angiogenesis, some endothelial cells in the existing blood vessel respond to the VEGF signal and begin "sprouting" to form a new vessel. These are known as the **tip cells**, and they differ from the other vessel cells. (If all the endothelial cells responded equally, the original blood vessel would fall apart.) The tip cells express the Notch ligand Delta-like-4 (Dll4) on their cell surfaces. Dll4 activates Notch signaling in the adjacent cells, preventing them from responding to VEGF-A (Noguera-Troise et al. 2006; Ridgway et al. 2006; Hellström et al. 2007). If Dll4 expression is experimentally reduced, tip cells form along a large portion of the blood vessel in response to VEGF-A.

The tip cells produce filopodia that are densely packed with VEGFR-2 (VEGF receptor-2) on their cell surfaces. They also express another VEGF receptor, VEGFR-3, and blocking VEGFR-3 greatly suppresses sprouting (**FIGURE 18.21B**; Tammela et al. 2008). These receptors enable the tip cell to extend toward the source of VEGF, and when the cell divides, the division is along the gradient of VEGFs. Indeed, the filopodia of the tip cells act just like the filopodia of neural crest cells and neural growth cones, and they respond to similar cues (Carmeliet and Tessier-Lavigne 2005; Eichmann et al. 2005). Semaphorins, netrins, neuropilins, and split proteins all have roles in directing the sprouting tip cells to the source of VEGF. (See **Further Development 18.13, Anti-Angiogenesis in Normal and Abnormal Development**, online.)

Hematopoiesis: Stem Cells and Long-Lived Progenitor Cells

Each day we lose and replace about 300 *billion* blood cells. As blood cells are destroyed in the spleen, their replacements come from populations of stem cells. As we described in Chapter 5, a stem cell is capable of extensive proliferation, creating both more stem cells (self-renewal) and differentiated cell progeny (see Figures 5.1 and 5.3). In the case of **hematopoiesis**—the generation of blood cells—stem cells divide to produce (1) more stem cells and (2) progenitor cells that can respond to the environment around them to differentiate into about a dozen mature blood cell types (Notta et al. 2016). The critical stem cell in hematopoiesis is the **pluripotent hematopoietic stem cell**, or simply the **hematopoietic stem cell (HSC)**, which is capable of producing all the blood cells and lymphocytes of the body. The HSC can achieve this by generating a series of intermediate progenitor cells whose potency is restricted to certain lineages.

Sites of hematopoiesis

In the early 1900s, numerous investigators (looking at many different vertebrate species, including mongooses, bats, and humans) observed the emergence of blood cells from the ventral endothelium of the aorta (Adamo and Garcia-Cardeña 2012). In the 1960s, however, experiments on mice concluded that all hematopoietic stem cells are derived from cells originating in the extraembryonic blood islands surrounding the yolk sac. The aortic hematopoiesis was thought of as an intermediate stop that the stem cells made on their way to the spleen and bone marrow (the sites of adult hematopoiesis in mice).

However, in 1975, Françoise Dieterlen-Lièvre transplanted early chick yolk sacs onto 2-day (pre-circulation) quail embryos. Chick and quail blood cells can be readily distinguished under the microscope, and the chimeric animal survives. Dieterlen-Lièvre's analysis indicated that all the blood cells of the late quail embryo originated from the quail host and not from the transplanted chick yolk sac. Moreover, hematopoietic activity within the embryo was restricted to one major site: the ventral portion of the aorta (Dieterlen-Lièvre and Martin 1981). The grafting of splanchopleure from this **aorta-gonad-mesonephros (AGM) region** from one genetically variant mouse to another confirmed that in mammals, too, definitive hematopoiesis takes place from inside the embryo (Godin et al. 1993; Medvinsky et al. 1993). Soon afterward, hematopoietic stem cells were identified in clusters of cells that were observed on the ventral region of the 10.5-day embryonic mouse aorta (Cumano et al. 1996; Medvinsky and Dziermak 1996).

While there is evidence that some yolk sac hematopoietic stem cells persist in the adult mouse (see Samokhvalov et al. 2007; Frame et al. 2016), it is generally thought that the yolk sac hematopoietic stem cells in mammals produce blood cells that allow oxygen to be transported to the early embryo, but that nearly all the stem cells found in the adult are those from the AGM region that have migrated to the bone marrow (Jaffredo et al. 2010).

In 2009, several laboratories proposed a new mechanism for blood cell production. This new hypothesis was based on the discovery of a new cell type in the AGM region, the **hemogenic endothelial cell**.[5] The sclerotome produces hemangioblasts that migrate to the dorsal aorta and replace most of the primary dorsal aorta cells (see Further Development 17.5, online). Before their replacement, the remaining primary, lateral plate-derived endothelial cells of the dorsal aorta (now in the ventral area of the blood vessel) give rise to blood-forming stem cells. These blood vessel-derived hematopoietic stem cells are the critical source of adult blood stem cells (see Chapter 5). By analyzing the types of cells made by the blood vessel endothelium, researchers were able to isolate the hemogenic endothelial cells and showed that they produce the hematopoietic stem cells that migrate to the liver and bone marrow (Eilken et al. 2009; Lancrin et al. 2009). Furthermore, the transition from endothelial cell to hematopoietic stem cell was mediated by the activation of the Runx1 transcription factor (**FIGURE 18.22**). In mice lacking the *Runx1* gene, blood stem cells failed to form in the yolk sac, umbilical arteries, dorsal aorta, and placental vessels (Chen et al. 2009; Tober et al. 2016).

The *Runx1* gene appears to be regulated by a complex and dynamic circuitry. Moreover, Runx1 protein expression is not initiated until after the heart starts beating. If cardiac mutations prevent fluid flow through the aorta, Runx1 is not expressed. Rather, shear forces (i.e., friction) from the fluid flow are required to activate the *Runx1* gene in the ventral endothelium of the dorsal aorta (Adamo et al. 2009; North et al. 2009).[6] The shear forces appear to elevate levels of nitric oxide (NO) in the endothelium. NO, in turn, activates (perhaps through cGMP) *Runx1* and other genes known to be critical for blood cell formation. The transition from hemogenic endothelial cell to HSC does not appear to be caused by an asymmetrical cell division. Rather, there is a rearrangement of cytoskeleton and tight junctions that resemble an epithelial-mesenchymal transition, such as those seen in the sclerotome or dermamyotome (Yue et al. 2012).

In non-amniote vertebrates, the splanchnopleure is also the source of the hematopoietic stem cells, and BMPs are crucial in inducing the blood-forming cells in all vertebrates studied. In *Xenopus*, the ventral mesoderm forms a large blood island that is the first site of hematopoiesis. Ectopic BMP2 and BMP4 can induce blood cell and blood vessel formation in *Xenopus*, and interference with BMP signaling prevents blood formation (Maéno et al. 1994; Hemmati-Brivanlou and Thomsen 1995). In the zebrafish, both yolk sac and aortic hematopoiesis are seen. As in the mammalian embryo, the second wave of hematopoiesis is from the aorta. Hematopoietic stem cells can be seen arising from the ventral aortic endothelium (Bertrand et al. 2010; Kissa and Herbomel 2010), and the same genetic pathways leading to Runx1 expression (including the BMPs) regulate this second, definitive wave of hematopoiesis (Mullins et al. 1996; Paik and Zon 2010).

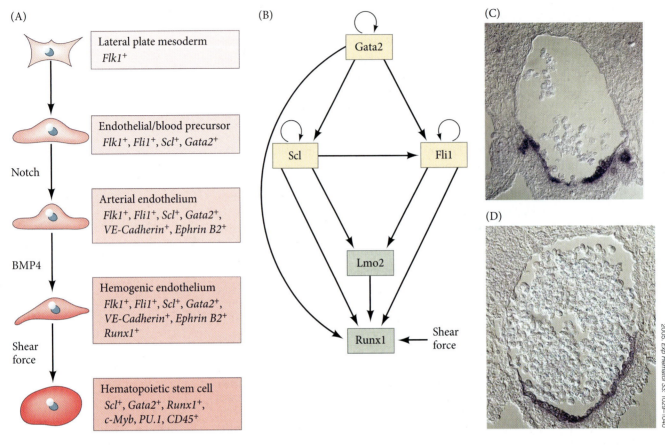

(A)

Lateral plate mesoderm
Flk1⁺

Notch

Endothelial/blood precursor
Flk1⁺, *Fli1*⁺, *Scl*⁺, *Gata2*⁺

Arterial endothelium
Flk1⁺, *Fli1*⁺, *Scl*⁺, *Gata2*⁺,
VE-Cadherin⁺, *Ephrin B2*⁺

BMP4

Hemogenic endothelium
Flk1⁺, *Fli1*⁺, *Scl*⁺, *Gata2*⁺,
VE-Cadherin⁺, *Ephrin B2*⁺
Runx1⁺

Shear
force

Hematopoietic stem cell
Scl⁺, *Gata2*⁺, *Runx1*⁺,
c-Myb, *PU.1*, *CD45*⁺

(B) Gata2 / Scl / Fli1 / Lmo2 / Runx1 ← Shear force

(C) (D)

Images C and D from T. Jaffredo et al. 2005. *Exp Hematol* 33: 1029–1040

FIGURE 18.22 Pathway for hematopoietic stem cell formation. (A) In the developing mouse, hematopoietic stem cells arise from the hemogenic endothelium of the aorta. Runx1 is critical for this conversion of endothelial cells into blood stem cells. The lateral plate mesodermal heritage of the hemogenic endothelium is seen, as is the juxtacrine and paracrine factors that brought it to its fate. The transcription factors associated with each stage are shown at the right. (B) A simplified view of the factors activating *Runx1* in the gene regulatory network establishing the hematopoietic stem cell in the mouse. The Gata2, Fli1, and Scl transcription factors bind at adjacent sites on a single enhancer 23 base pairs downstream from the *Runx1* transcription initiation site. Scl is critical for preventing blood and vascular cells from becoming heart muscle. The mechanism by which shear force is mechanotransduced to help activate *Runx1* remains unknown. (C) Runx1 expression (purple) in the stage-19 chick embryo; Runx1-expressing cells have become part of the blood vessel. (D) Runx1-expressing cells at stage 21 in the chick embryo. Hematopoietic colonies are visible. (A after G. M. Swiers et al. 2010. *Int J Dev Biol* 54: 1151–1163; B after J. E. Pimanda and B. Göttgens. 2010. *Int J Dev Biol* 54: 1201–1211.)

The bone marrow HSC niche

The hematopoietic stem cells of the aorta generate HSCs that come to reside first in the liver and then in the bone marrow (Coskun and Hirschi 2010). In humans, the aorta generates blood cells around days 27 through 40 (Tavian and Péault 2005). The bone marrow HSC is a remarkable cell, in that it is the common precursor of red blood cells (erythrocytes), white blood cells (granulocytes, neutrophils, and platelets), monocytes (macrophages and osteoclasts), and lymphocytes. When transplanted into inbred, irradiated mice (that are genetically identical to the donor cells and whose own stem cells have been eliminated by radiation), HSCs can repopulate the mouse with all the blood and lymphoid cell types. It is estimated that only about 1 in every 10,000 blood cells is a pluripotent HSC (Berardi et al. 1995). In humans, "bone marrow transplants" are used to transfer healthy HSCs into people whose lymphocytes, red blood cells, or white blood cells have been wiped out by disease, drugs, or radiation. In recent years, more than 50,000 such transplantations have been performed annually (Gratwohl et al. 2010).

The maintenance of the HSC depends on the stem cell niche, and especially on the ability of the HSC to receive the paracrine factor **stem cell factor**, or **SCF**. SCF binds to the Kit receptor protein. (This binding is critical for sperm and pigment stem cells as well as HSCs.) Because it was important to determine which cells of the stem cell niche were supplying SCF, Ding and colleagues (2012) constructed genetically recombined mice in which the gene for SCF was replaced by the gene for green fluorescent protein in all or in selected cell types. When all the cell types of the niche expressed GFP instead of SCF, the HSCs died. When the researchers deleted SCF production only in certain cell types, they found that replacing SCF with GFP in blood cells, bone cells, or marrow mesenchymal cells did not

(A)

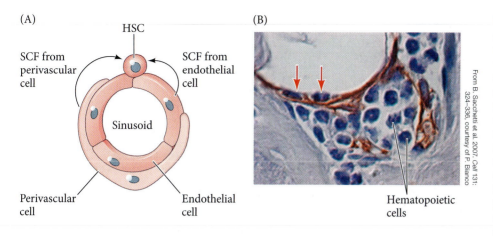

(B)

FIGURE 18.23 The home for HSCs appears to be a niche where stem cell factor (SCF) can be made by the perivascular (subendothelial) cells as well as by the endothelial cells of the bone marrow sinusoids. (A) Simplified diagram of the sinusoid with its endothelial cells and a surrounding perivascular cell. (B) Development of a stem cell niche when human perivascular cells (stained brown with antibodies to the subendothelial cell marker CD146) implanted into a mouse. At 8 weeks, processes from perivascular cells establish contacts with hematopoietic stem cells (as in human bone marrow). The red arrows show hematopoietic cells between endothelial and perivascular cells. (A after I. A. Shestopalov and L. I. Zon. 2012. *Nature* 481: 453–455.)

block HSC maintenance. However, when they got rid of SCF expression in either the endothelial cells or the perivascular cells surrounding the endothelial cells, many fewer HSCs survived. And when SCF synthesis was turned off in both of these cell types (but not in the others), all the HSCs perished. It appears, then, that the SCF needed for HSC survival is made primarily by the perivascular cells, with some contribution from the endothelial cells (**FIGURE 18.23**).

SCF is not the only paracrine factor that HSCs require; there are numerous others, and these probably render the HSCs competent to respond to the paracrine and juxtacrine factors that will direct cell differentiation (Morrison and Scadden 2014). Stem cell niches often contain long-term, quiescent HSCs that are used to generate progenitor cells on a continual basis, as well as shorter-acting HSCs that can respond to immediate physiological needs (see Figure 18.24). Wnts that activate the noncanonical pathways are secreted by niche osteoblasts to maintain the quiescent HSCs, whereas the canonical Wnt pathway may be critical for inducing them to become the rapidly proliferating HSCs (Reya et al. 2003; Sugimura et al. 2012). It is probable that in adult mammals, the maintenance of the billions of blood cells is not dependent on a small number of hematopoietic stem cells but on the steady-state production of numerous long-lived progenitor cells that are specified for either a single lineage or multiple lineages (Sun et al. 2014).

FURTHER DEVELOPMENT

Hematopoietic inductive microenvironments

The major models of blood cell production envision blood differentiation as progressing down a series of less potent precursor cells. At the top are the multipotent hematopoietic stem cells (HSCs) which can give rise to more restricted multipotent stem cells (such as the common myeloid progenitor cells, CMPs), and finally to lineage-committed progenitor cells. Endocrine, paracrine, and juxtacrine factors are thought to push blood cell differentiation down one path or another (**FIGURE 18.24**).

One of the major endocrine factors (i.e., hormones) is **erythropoietin**, which appears to cause the CMP to make more megakaryocyte/erythroid progenitor cells (MEPs) and biases the MEPs to make more erythrocytes (Lu et al. 2008; Klimchenko et al. 2009). The paracrine factors involved in blood cell and lymphocyte formation are the **cytokines**. Cytokines can be made by several cell types, but they are collected and concentrated by the extracellular matrix of the stromal (mesenchymal) cells at the sites of hematopoiesis (Hunt et al. 1987; Whitlock et al. 1987). For instance, granulocyte-macrophage colony-stimulating factor (GM-CSF) and the multilineage growth factor interleukin 3 (IL3) both bind to the heparan sulfate glycosaminoglycan of the bone marrow stroma (Gordon et al. 1987; Roberts et al. 1998). The extracellular matrix is then able to present these paracrine factors to the stem cells in concentrations high enough

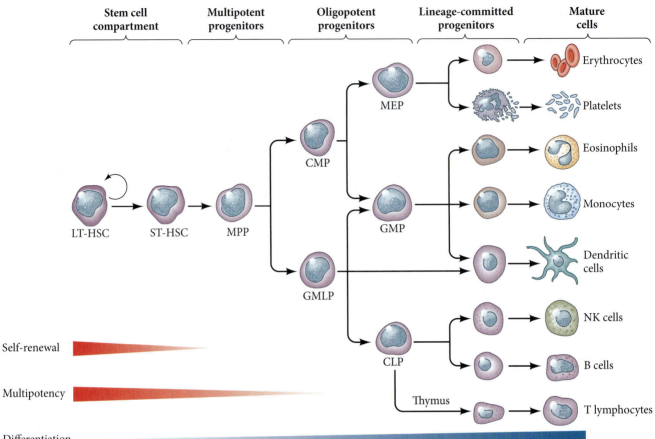

FIGURE 18.24 Hierarchy of hematopoietic lineages. At the top of the hierarchy are the long-term hematopoietic stem cells (LT-HSCs), which give rise to short-term HSCs (ST-HSCs) that retain limited self-renewing capabilities (see Chapter 5). Rapidly dividing multipotent progenitors (MPPs) still possess the potential to generate either myeloid (red blood cell types) or lymphoid lineages (white blood cell types), beyond which differentiation becomes increasingly restricted. Progeny of the MPP include the common myeloid progenitors (CMPs) and granulocyte-macrophage-lymphocyte progenitors (GMLPs). Further differentiation takes place, producing common lymphoid progenitors (CLPs), granulocyte-macrophage progenitors (GMPs), and megakaryocyte-erythrocyte progenitors (MEPs). These progenitors will further differentiate into the various red and white blood cell types. (After S. M. Cullen et al. 2014. *Curr Top Dev Biol* 107: 39–75.)

to bind to their respective receptors. At different stages of maturation, the stem cells become competent to respond to different factors.

The developmental path taken by a descendant of a pluripotent HSC depends on which growth factors it meets, and is therefore determined by the stromal cells. Wolf and Trentin (1968) demonstrated that short-range interactions between stromal cells and stem cells determine the developmental fates of the stem cells' progeny. These investigators placed plugs of bone marrow in a spleen and then injected stem cells into it. Those CMPs that came to reside in the spleen formed colonies that were predominantly erythroid, whereas those that came to reside in the bone marrow formed colonies that were predominantly granulocytic. Colonies that straddled the borders of the two tissue types were predominantly erythroid in the spleen and granulocytic in the marrow. Such regions of determination are referred to as **hematopoietic inductive microenvironments** (**HIMs**). As expected, the HIMs induce different sets of transcription factors in these cells, and these transcription factors specify the fate of the particular cells (see Kluger et al. 2004).

The transcription factors of the HIM may act by pulling the equilibrium of the stem cells' transcription network in different directions (Krumsiek et al. 2011; Wontakal et al. 2012). By decomposing the interactions into negative feedback loops and feedforward loops (of both activation and repression), there appear to be only four stable configurations that this network can have. Such stable configurations are called "attractor states" in systems theory, and these attractor states correspond to four cell types. Moreover, certain mutations will make certain attractor states impossible, and these are the mutations that block the differentiation of certain cell types.

This scheme of blood cell differentiation going through cell types of progressively diminished potency may not work at all stages of life. Notta and colleagues (2016) used single-cell transcriptomes to show that the midlevel stem cells (such as the CMP) were not present during later stages of human development. It is possible that later in life, blood cell production follows immediately from the HSC.

Coda

We ask a lot of our circulatory system. We require a precise flow of blood through the valves each second of our lives; we demand fine-tuned coordination between our brain, heart, bone marrow, and hormones such that the cardiac muscle contractions adapt to our physiological needs; and we demand that the production of our blood cells—cells made by precursors that formed in our embryo—be so precise that we get neither cancer nor anemia. Given all this, it is not surprising that blood cell differentiation, heart development, kidney development, and blood vessel formation are now among the most important fields of study in medical science. Congenital heart defects are among the most prevalent types of birth defects, and cardiovascular disease is the most common cause of death in industrialized nations. The questions of cardiogenesis, kidney formation, angiogenesis, and hematopoiesis that engaged Aristotle and Harvey still excite major research programs.

Closing Thoughts on the Opening Photo

The heart and vascular system of the chick have been studied for centuries (see Figure 1.3). This modern image, a fluorescence micrograph, depicts the 2-day chick embryo. It was compiled about 45 hours after the egg was laid, at the point where the heart begins to beat. The vascular system was revealed by injecting fluorescent beads into the circulatory system. The heart is the first functioning organ, yet as seen here, most of the circulation goes to the extraembryonic region, bringing in nutrients from the yolk and exchanging gases. One of the first events in the formation of the heart is the establishing of polarity. The atria receive the blood, while the ventricles pump it.

© Vincent Pasque

Endnotes

[1] The term *cloaca* is Latin for "sewer"—a bad (yet funny) joke on the part of early European anatomists.

[2] Note that these specialized myocardial nerve fibers are not the same thing as the Purkinje *neurons* of the cerebellum mentioned in Chapter 13. Both were named for the nineteenth-century Czech anatomist and histologist Jan Purkinje.

[3] A third heart field, extending more posteriorly, may exist (Bressan et al. 2013). In chick embryos, this third heart field includes those cells that generate the pacemaker myocytes that stimulate the rhythmic contractions of the heart muscles.

[4] The prefixes *hem-* and *hemato-* refer to blood (as in hemoglobin). Similarly, the prefix *angio-* refers to blood vessels. The suffix *-blast* denotes a rapidly dividing cell, usually a stem cell. The suffixes *-poiesis* and *-poietic* refer to generation or formation (*poeisis* is also the root of the word *poetry*). Thus, hematopoietic stem cells are those cells that generate the different types of blood cells. The Latin suffix *-genesis* (as in *angiogenesis*) means the same as the Greek *-poiesis*. The

names of zebrafish hematopoietic mutants can be very poetic. Most are named after wines, and one of the genes producing a bloodless phenotype is named *vlad tepes*, after the historic Vlad Dracula.

[5] The relationship of the hemogenic endothelial cell and the hemangioblast is controversial. In general, it is thought that hemangioblasts generate the hemogenic endothelial cells (see Ueno and Weissman 2010) and that the hemangioblast is a precursor to the hemogenic endothelium.

[6] Mechanotransduction of biophysical forces such as shear stress from blood flow is a major player in cardiovascular development (see Linask and Watanabe 2015). Recall that it is also required for normal heart development (Mironov et al. 2005) and for the correct patterning of blood vessels (Lucitti et al. 2007; Yashiro et al. 2007). It is also needed for the fragmentation of the platelet precursor cell—the megakaryocyte—into platelets. The megakaryocyte in the bone marrow inserts small processes into the blood vessels surrounding the stem cell niche, and the shear force there fragments these processes into platelets (Junt et al. 2007).

Snapshot Summary

Intermediate and Lateral Plate Mesoderm

1. The intermediate mesoderm forms the kidneys, adrenal glands, and gonads. It is specified through interactions with the paraxial mesoderm that require Pax2, Pax8, and Lim1.

2. The metanephric kidney of mammals is formed by reciprocal interactions of the metanephric mesenchyme and a branch of the nephric duct called the ureteric bud. The ureteric bud and the metanephric mesenchyme are specified depending on the length of time their progenitor cells are exposed to Wnt and FGF signals.

3. The metanephric mesenchyme secretes GDNF, which induces formation of the ureteric bud.

4. The ureteric bud secretes Fgf2, Fgf9, and BMP7 to prevent apoptosis in the metanephric mesenchyme. Without these factors, this kidney-forming mesenchyme dies.

5. The ureteric bud secretes Wnt9b and Wnt6, which induce the competent metanephric mesenchyme to form epithelial tubules.

6. The lateral plate mesoderm splits into two layers. The dorsal layer is the somatic (parietal) mesoderm, which underlies the ectoderm and forms the somatopleure. The ventral layer is the splanchnic (visceral) mesoderm, which overlies the endoderm and forms the splanchnopleure.

7. The space between the two layers of lateral plate mesoderm forms the body cavity, or coelom.

8. The vertebrate heart arises from splanchnic mesoderm on both sides of the body. This region of cells is called the heart field, or cardiogenic mesoderm. The cardiogenic mesoderm is specified by BMPs in the absence of Wnt signals.

9. The Nkx2-5, Mesp1, and Gata transcription factors are important in committing the cardiogenic mesoderm to become heart cells. These cardiac precursor cells migrate from the sides to the midline of the embryo, in the neck region.

10. There are two major heart fields on each side of the body. Each heart field has two regions: The first heart field forms the scaffold of the heart tube and will form the left ventricle. The rest of the heart is made largely by the second heart field.

11. A cardiac precursor cell can form each of the major lineages of the heart. The cardiogenic mesoderm forms the endocardium (which is continuous with the blood vessels) and the myocardium (the muscular component of the heart).

12. The endocardial tubes form separately and then fuse. The looping of the heart transforms the original anterior-posterior polarity of the heart tube into a right-left polarity.

13. Retinoic acid is important in determining the anterior-posterior polarity of the heart and kidneys.

14. Tbx transcription factors are critical for specifying the heart chambers.

15. Blood vessels are constructed by two processes: vasculogenesis and angiogenesis. Vasculogenesis involves the condensing of splanchnic mesoderm cells to form blood islands. The outer cells of these islands become endothelial (blood vessel) cells. Angiogenesis involves the remodeling of existing blood vessels.

16. Numerous paracrine factors are essential in blood vessel formation. Fgf2 is needed for specifying the hemangioblasts. VEGF-A is essential for the differentiation of the angioblasts. Angiopoietins allow the smooth muscle cells (and smooth muscle-like pericytes) to cover the vessels.

17. The pluripotent hematopoietic stem cell (HSC) generates other pluripotent stem cells, as well as lineage-restricted stem cells. It gives rise to both blood cells and lymphocytes.

18. In vertebrates, HSCs are thought to originate from hemogenic endothelial cells that characterize the blood islands, the dorsal aorta, and the placental vessels. The generation of multipotent HSCs (definitive HSCs) appears to be derived from the ventral portion of the aorta.

19. The common myeloid progenitor (CMP) is a blood stem cell that can generate the more committed stem cells for the different blood lineages. Hematopoietic inductive microenvironments (HIMs) determine the blood cell differentiation.

20. HSCs depend on stem cell factor (SCF), which is provided to them primarily by the perivascular cells of the sinusoids contained in the stem cell niche.

Go to oup.com/he/barresi12xe for Further Developments, Scientists Speak interviews, Watch Development videos, Dev Tutorials, and complete bibliographic information for all literature cited in this chapter.

Development of the Tetrapod Limb

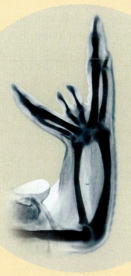

How many fingers am I holding up?

After L. S. Honig and D. Summerbell. 1985.
J Embryol Exp Morphol 87: 163–174

CONSIDER YOUR LIMB. It has fingers or toes at one end, a humerus or femur at the other. You won't find anyone with fingers in the middle of their arm. Also consider the subtle but obvious differences between your hands and your feet. If your fingers were replaced by toes, you would certainly know it. Despite these differences, the bones of your feet are similar to the bones of your hand. It's easy to see that they share a common pattern. And finally, consider that both your hands are remarkably similar in size, as are both your feet. These commonplace phenomena present fascinating questions to the developmental biologist. How is it that vertebrates have four limbs and not six or eight? How is it that the little finger develops at one edge of the limb and the thumb at the other? How does the forelimb grow differently than the hindlimb? How can limb size be so precisely regulated? Is there a conserved set of developmental mechanisms that can explain why our hands have five digits, a chick's wing three digits, and a horse's hoof one?

Limb Anatomy

As the name denotes, **tetrapods** are four-limbed vertebrates (amphibians, reptiles, birds, and mammals). The bones of any tetrapod limb—be it arm or leg, wing or flipper—consist of a proximal **stylopod** (humerus/femur) adjacent to the body wall, a **zeugopod** (radius-ulna/tibia-fibula) in the middle region, and a distal **autopod** (carpals-fingers/tarsals-toes) (**FIGURE 19.1**).[1] Fingers and toes can be referred to as phalanges or, more generally, digits. The positional information needed to construct a limb has to function in a three-dimensional[2] coordinate system:

- The first dimension is the *proximal-distal axis* ("close-far"; that is, shoulder-to-finger or hip-to-toe). The bones of the limb are formed by endochondral ossification. They are initially cartilaginous, but eventually most of the cartilage is replaced by bone. Somehow the limb cells make differently shaped parts when making the stylopod than when making the autopod.

- The second dimension is the *anterior-posterior axis* (thumb-to-pinkie). Our little fingers or toes mark the posterior end, and our thumbs or big toes are at the anterior end. In humans, it is obvious that each hand develops as a mirror image of the other. One can imagine other arrangements—such as the thumb developing on the left side of both hands—but these patterns do not occur.

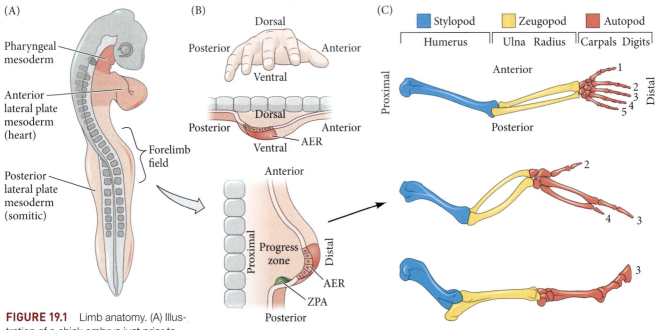

FIGURE 19.1 Limb anatomy. (A) Illustration of a chick embryo just prior to limb growth showing three important mesodermal cell types as well as the emergence of the limb field. (B) Axis orientation and anatomy of the limb bud. AER, apical ectodermal ridge; ZPA, zone of polarizing activity. (C) Skeletal pattern of the human arm, chick wing, and horse forelimb. (According to convention, the digits of the chick wing are numbered 2, 3, and 4. The cartilage condensations forming the digits appear similar to those forming digits 2, 3, and 4 of mice and humans; however, new evidence suggests that the correct designation may be 1, 2, and 3.) (A after M. Tanaka. 2013. *Dev Growth Differ* 55: 149–163; B after M. Logan. 2003. *Development* 130: 6401–6410.)

- Finally, limbs have a *dorsal-ventral axis*: our palms (ventral) are readily distinguishable from our knuckles (dorsal).

The Limb Bud

The first visible sign of limb development is the formation of bilateral bulges called **limb buds** at the presumptive forelimb and hindlimb locations (**FIGURE 19.2A**). Fate-mapping studies on salamanders, pioneered by Ross Granville Harrison's laboratory (see Harrison 1918, 1969), showed that the center of this disc of cells in the somatic region of the lateral plate mesoderm normally gives rise to the limb itself. Adjacent to it are the cells that will form the peribrachial (around the limb) flank tissue and the shoulder girdle. However, if all these cells are extirpated from the embryo, a limb will still form (albeit somewhat later) from an additional ring of cells that surrounds this area but would not normally form a limb. If this surrounding ring of cells is included in the extirpated tissue, no limb will develop. This larger region, representing all the cells in the area capable of forming a limb on their own, is the **limb field**.

The cells that make up the limb bud are derived from the posterior lateral plate mesoderm, adjacent somites, and the bud's overlying ectoderm. Lateral plate mesenchyme cells migrate within the limb fields to form the limb *skeletal* precursor cells, while mesenchymal cells from the somites at the same level migrate in to establish the limb *muscle* precursor cells (**FIGURE 19.2B,C**). This accumulating heterogeneous population of mesenchymal cells proliferates under the ectodermal tissue, creating the limb bud.

Even the early limb bud possesses its own organization such that the main direction of growth occurs along the proximal-to-distal axis (somites-to-ectoderm), with lesser growth occurring along the dorsal-to-ventral and anterior-to-posterior axes (see Figure 19.1B). The limb bud is further regionalized into three functionally distinct domains:

1. The highly proliferative mesenchyme that fuels limb bud growth is known as the **progress zone** (**PZ**) mesenchyme (also called the undifferentiated zone).

2. The cells found within the most posterior region of the progress zone constitute the **zone of polarizing activity** (**ZPA**), since it patterns cell fates along the anterior-posterior axis.

3. The **apical ectodermal ridge** (**AER**) is a thickening of the ectoderm at the apex of the developing limb bud (**FIGURE 19.2D**).

(A)

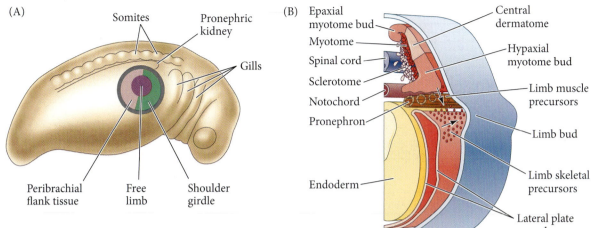

Somites

Pronephric kidney

Gills

Peribrachial flank tissue

Free limb

Shoulder girdle

(B)

Epaxial myotome bud

Myotome

Spinal cord

Sclerotome

Notochord

Pronephron

Endoderm

Central dermatome

Hypaxial myotome bud

Limb muscle precursors

Limb bud

Limb skeletal precursors

Lateral plate mesoderm

(C)

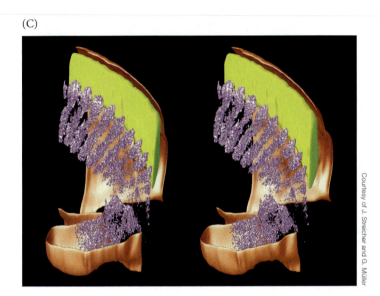

Courtesy of J. Streicher and G. Müller

(D)

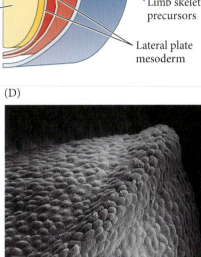

Courtesy of K. W. Tosney

FIGURE 19.2 The limb bud. (A) Prospective forelimb field of the salamander *Ambystoma maculatum*. The central area contains cells destined to form the limb per se (the free limb). The cells surrounding the free limb give rise to the peribrachial flank tissue and the shoulder girdle. The ring of cells outside these regions usually is not included in the limb, but can form a limb if the more central tissues are extirpated. (B) Emergence of the limb bud. Proliferation of mesenchymal cells (arrows) from the somatic region of the lateral plate mesoderm causes the limb bud in the amphibian embryo to bulge outward. These cells generate the skeletal elements of the limb. Contributions of myoblasts from the lateral myotome provide the limb's musculature. (C) Entry of myoblasts (purple) into the limb bud. This computer stereogram was created from sections of an in situ hybridization to the *Myf5* mRNA found in developing muscle cells. If you can cross your eyes (or try focusing "past" the page, looking through it to your toes), the three-dimensionality of the stereogram will become apparent. (D) Scanning electron micrograph of an early chick forelimb bud, with the apical ectodermal ridge in the foreground. (A after D. L. Stocum and J. F. Fallon. 1982. *J Embryol Exp Morphol* 69: 7–36.)

Hox Gene Specification of Limb Skeleton Identity

Homeobox transcription factors, or Hox genes, play an essential role in specifying whether a particular mesenchymal cell will become stylopod, zeugopod, or autopod. Understanding their role has given researchers substantial new insight into the development and evolution of the vertebrate limb.

From proximal to distal: Hox genes in the limb

Based on the expression patterns of *Hoxa* and *Hoxd* gene complexes in the limb buds of mice, Mario Capecchi's laboratory (Davis et al. 1995) proposed a model for how these Hox genes specify the identity of a limb region (**FIGURE 19.3A,B**). Here, *Hox9* and *Hox10* paralogues specify the stylopod, *Hox11* paralogues specify the zeugopod, and *Hox12* and *Hox13* paralogues specify the autopod. This scenario has been confirmed by numerous experiments. For instance, when all six alleles of the three *Hox10* paralogues (*Hox10aaccdd*) were knocked out in mouse embryos, the hindlimbs of the resulting mice lacked a femur and patella, which would normally form in the stylopod. The forelimbs had humeruses, however, because the *Hox9* paralogues are expressed in the forelimb stylopod but not in the hindlimb stylopod (Wellik and Capecchi 2003). When all six alleles of the three *Hox11* paralogues were knocked out, the resulting hindlimbs had femurs but neither tibias

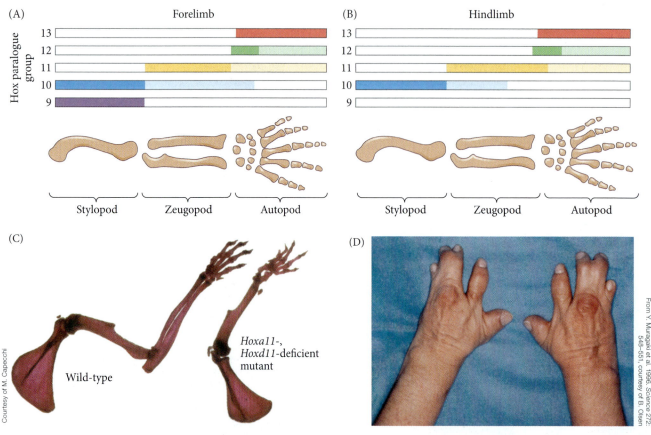

(A) Forelimb
(B) Hindlimb

Stylopod Zeugopod Autopod

(C) Wild-type *Hoxa11-, Hoxd11*-deficient mutant

Courtesy of M. Capecchi

(D)

From Y. Muragaki et al. 1996, *Science* 272: 548–551, courtesy of B. Olsen

FIGURE 19.3 Deletion of limb bone elements by the deletion of paralogous Hox genes. (A) 5′ Hox gene patterning of the forelimb. *Hox9* and *Hox10* paralogues specify the humerus (stylopod). *Hox10* paralogues are expressed to a lesser extent in the radius and ulna (zeugopod). *Hox11* paralogues are chiefly responsible for patterning the zeugopod. *Hox12* and *Hox13* paralogues function in the autopod, with *Hox12* paralogues functioning primarily in the wrist and to a lesser extent in the digits. (B) A similar but somewhat differing pattern is seen in the hindlimb. (C) Forelimb of a wild-type mouse (left) and of a double-mutant mouse that lacks functional *Hoxa11* and *Hoxd11* genes (right). The ulna and radius are severely reduced or absent in the mutant. (D) Human polysyndactyly ("many fingers joined together") syndrome results from a homozygous mutation at the *HOXD13* loci. This syndrome includes malformations of the urogenital system, which also expresses *HOXD13*. (A,B after D. M. Wellik and M. R. Capecchi. 2003. *Science* 301: 363–367.)

nor fibulas (and the forelimbs lacked the ulnas and radii). Thus, the *Hox11* knockout got rid of the zeugopods (**FIGURE 19.3C**). Similarly, knocking out all the paralogous *Hoxa13* and *Hoxd13* loci resulted in loss of the autopod (Fromental-Ramain et al. 1996). Humans homozygous for a *HOXD13* mutation show abnormalities of the hands and feet wherein the digits fuse (**FIGURE 19.3D**), and humans with homozygous mutant alleles of *HOXA13* also have deformities of their autopods (Muragaki et al. 1996; Mortlock and Innis 1997). In both mice and humans, the autopod (the most distal portion of the limb) is affected by the loss of function of the most 5′ Hox genes.

FURTHER DEVELOPMENT

From fins to fingers: Hox genes and limb evolution

How did the vertebrate appendage evolve into the limbs we find so useful today? The fossil record points to an important transition in forelimb morphology from the pectoral fins of ray-finned fish to the digited limbs of the tetrapod, a transition that provided the opportunity for aquatic life to explore terrestrial habitats. Understanding the evolutionary history of the tetrapod limb can help us analyze the developmental mechanisms that are essential for today's limb morphology. The discovery of the Devonian fossil *Tiktaalik roseae*, a "fish with fingers," highlights the importance of joint development in limb evolution. Fish fins, including those of some of the most primitive species, develop using the same three Hox gene expression phases as tetrapods use to form their limbs (Davis et al. 2007; Ahn and Ho 2008). The independent modification of fin bones into limb bones may have been made possible by the joints. The joints of *Tiktaalik*'s pectoral fins are very similar to those of amphibians and indicate that *Tiktaalik* had mobile wrists and a substrate-supported stance in which the elbow and shoulder could flex (**FIGURE 19.4**; Shubin et al. 2006; Shubin 2008). In addition, the presence of wristlike structures and the loss of dermal scales in these regions suggest that

FIGURE 19.4 Limb evolution. (A) *Tiktaalik roseae*, a fish with wrists and fingers, lived in shallow waters about 375 million years ago. This reconstruction shows *Tiktaalik*'s fishlike gills, fins, scales, and (lack of) neck. The external nostrils on its snout, however, indicate that it could breathe air. (B) Fossilized *Tiktaalik* bones reveal the beginnings of digits, wrists, elbows, and shoulders and suggest that this amphibian-like fish could propel itself on stream bottoms and perhaps live on land for short durations. The joints of the fin included a ball-and-socket joint in the shoulder and a planar joint that allowed the wrist to bend. Other joints allowed the animal to perch on its substrate. (C) Resistant contact with a substrate would have allowed flexion at the proximal joints (shoulder and elbow) and extension at the distal ones (wrist and digits). (B,C after N. H. Shubin et al. 2006. *Nature* 440: 764–771.)

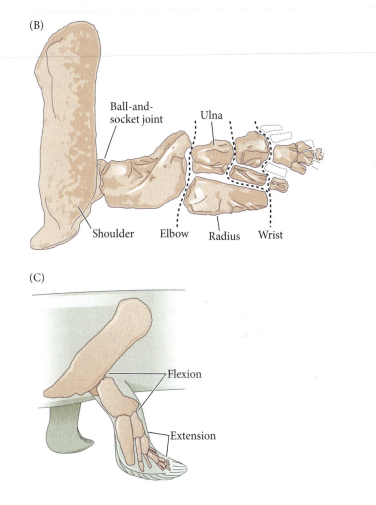

this Devonian fish was able to propel itself on moist substrates. Thus, *Tiktaalik* is thought to be a transition between fish and amphibians—a "fishapod" (as one of its discoverers, Neil Shubin, called it) "capable of doing push-ups."

What types of molecular and morphological changes occurred along the different branches that led to the ray-finned fish on the one hand and terrestrial tetrapods on the other? In those fish that are most closely related to tetrapods (lobe-finned fish such as coelacanths and lungfish), the more proximal bones of the pectoral fin are homologous to the stylopod segment of tetrapod forelimbs and are similarly responsible for articulation around the pectoral girdle, or shoulder. However, the fin of ray-finned fish diverged in form, and this is most evident in the more distal elements and in particular the autopod (digits). Ray-finned fish lack an endoskeleton associated with the autopod, whereas ancestral fish in the Sarcopterygian clade (lobbed-fin fish) display expanded endochondral skeletons in their fins (as in *Tiktaalik*). Thus, adaptation targeting the developmental mechanisms governing the more distal limb skeleton was the primary basis for limb evolution. (See **Further Development 19.1, Homology between the Limb Buds of Fish and Tetrapods**, online.)

We hope this brief examination of the evolution of the tetrapod limb, from fish fins to human hands, has illuminated the importance of Hox gene regulation during limb development. Hox genes are critical for specifying fates along each axis of the limb, and their expression is under the influence of signals emanating from the flank (proximal) and AER (distal), among other regions. What are these signals, and how do they function to (1) determine where limbs form, (2) promote limb bud outgrowth and patterning, and (3) specify fates along the anteroposterior and dorsoventral axes?

Determining What Kind of Limb to Form and Where to Put It

Because the limbs, unlike the heart or brain, are not essential for embryonic or fetal life, one can experimentally remove or transplant parts of the developing limb, or create limb-specific mutants, without interfering with the vital processes of the organism. Such experiments have shown that certain basic "morphogenetic rules" for forming a limb appear to be the same in all tetrapods. Grafted pieces of reptile or mammalian limb buds can direct the formation of chick limbs, and regions taken from frog limb buds can direct the patterning of salamander limbs (Fallon and Crosby 1977; Sessions et al. 1989; Hinchliffe 1991). Moreover, *regenerating* salamander limbs appear to follow many of the same rules as developing limbs (see Chapter 22; Muneoka and Bryant 1982). What are these morphogenetic rules?

Specifying the limb fields

Limbs do not form just anywhere along the body axis; rather, they are generated at discrete positions. Early fate-mapping and transplantation studies in the chick demonstrated that there are two specific regions, or fields, of somitic and lateral plate mesoderm that are determined to form limbs long before any visible signs of wings or legs emerge. The mesodermal cells that give rise to a vertebrate limb have been identified by (1) removing certain groups of cells and observing that a limb does not develop in their absence ("lose it"; see Detwiler 1918; Harrison 1918), (2) transplanting groups of cells to a new location and observing that they form a limb in this new place ("move it"; see Hertwig 1925), and (3) marking groups of cells with dyes or radioactive precursors and observing that their descendants partake in limb development ("find it"; Rosenquist 1971).

Vertebrates have no more than four limb buds per embryo, and limb buds are always paired opposite each other with respect to the midline. Although the limbs of different vertebrates differ with respect to the somite level at which they arise, their position is constant with respect to the level of Hox gene expression along the anterior-posterior axis (see Chapter 12). For instance, in fish (in which the pectoral and pelvic fins correspond to the anterior and posterior limbs, respectively), amphibians, birds, and mammals, the forelimb buds are found at the most anterior expression region of *Hoxc6*, the position of the first thoracic vertebra (Oliver et al. 1988; Molven et al. 1990; Burke et al. 1995).[3] It is probable that positional information from the Hox gene expression domains causes the paraxial mesoderm in the limb-forming regions to be different from all other paraxial mesoderm. Transplantation experiments in which paraxial mesoderm (somites) from different locations is placed adjacent to the flank lateral plate shows that the paraxial mesoderm from limb-forming regions promotes limb bud formation, whereas the paraxial mesoderm from limbless flank actively represses limb formation (**FIGURE 19.5**; Noro et al. 2011).

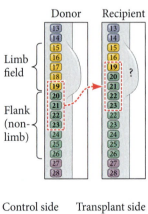

FIGURE 19.5 Transplantation of different regions of the presomitic mesoderm (PSM) to the limb field causes changes in limb size. (A) Transplanting PSM from the level of the presumptive forelimb to the flank (region between the forelimb and hindlimb) results in a larger forelimb bud (area between arrowheads). (B) Transplantation of PSM from the flank region to the level of the presumptive forelimb results in a smaller forelimb bud. (After M. Noro et al. 2011. *Dev Dyn* 240: 1639–1649.)

(A)
Transplantation of somites from limb to flank level

(B)
Transplantation of somites from flank to limb level

Control side   Transplant side

Control side   Transplant side

24 hrs

36 hrs

Results: Larger limb bud

Results: Smaller limb bud

After M. Noro et al. 2011. *Dev Dyn* 240:1639–1649

FIGURE 19.6 Multilimbed Pacific tree frog (*Hyla regilla*), the result of infestation of the tadpole-stage developing limb buds by trematode cysts. The parasitic cysts apparently split the developing limb buds in several places, resulting in extra limbs. In this adult frog's skeleton, the cartilage is stained blue, and the bones are stained red.

Courtesy of S. Sessions

FURTHER DEVELOPMENT

The limb bud potential

When it first forms, the limb field has the ability to regulate for lost or added parts. In the tailbud stage of the spotted salamander (*Ambystoma maculatum*), any half of the limb disc is able to generate an entire limb when grafted to a new site (Harrison 1918). This potency can also be shown by splitting the limb disc vertically into two or more segments and placing thin barriers between the segments to prevent their reunion. When this is done, each segment develops into a full limb. Thus, like an early sea urchin embryo, the limb field represents a "harmonious equipotent system" wherein a cell can be instructed to form any part of the limb. The plasticity of the limb bud has been highlighted by a remarkable experiment of nature. In numerous ponds in the United States, multilegged frogs and salamanders have been found (**FIGURE 19.6**). The presence of these extra appendages has been linked to the infestation of the larval abdomen by parasitic trematode worms, which apparently split the developing tadpole limb buds, and the limb bud fragments develop as multiple limbs (Sessions and Ruth 1990; Sessions et al. 1999).

Induction of the early limb bud

Differential Hox gene expression along the anterior-posterior axis in the trunk sets up a prepattern of tissue identities that includes the limb field location, but what mechanisms are then triggered to initiate limb bud formation? The process can be divided into four stages: (1) making mesoderm permissive for limb formation; (2) specifying forelimb and hindlimb; (3) inducing epithelial-mesenchymal transitions; and (4) establishing two positive feedback loops for limb bud formation.

1. MAKING MESODERM PERMISSIVE FOR FORELIMB FORMATION BY RETINOIC ACID In Chapter 17, we described the antagonistic relationship of retinoic acid (RA) and Fgf8 during somitogenesis. Recall that Fgf8 is expressed by the caudal (posterior) progenitor zone, which is located just posterior to the forelimb field (and present in a gradient that is high posteriorly along the presomitic mesoderm). RA, by contrast, is generated more anteriorly in somites and anterior presomitic mesoderm. Relevant for forelimb development, Fgf8 is also expressed just anterior to the forelimb field, in the heart lateral plate mesoderm (**FIGURE 19.7**). Investigations into forelimb development in chick and mouse, as

(A) RARE

Wild-type   *rdh10*⁻/⁻

Heart   Neural tube

(B) Fgf8

Caudal progenitor zone

(C) Tbx5

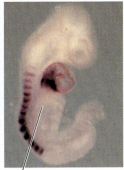

Forelimb field

From T. J. Cunningham et al. 2013. *Cell Rep* 3: 1503–1511/CC BY-NC-ND 3.0

FIGURE 19.7 Antagonism between retinoic acid (RA) and Fgf8 determines the pattern of Tbx5 expression in the mouse forelimb field. As seen in control subjects on the left, reporter expression from the RA regulatory element (RARE) falls directly between Fgf8 expression in the heart (anterior) and throughout the caudal progenitor zone. (A) RARE reporter expression is almost completely repressed by the loss of the enzyme retinaldehyde dehydrogenase 10 (*rdh10* mutant, right), the exception being minimal expression in the neural tube. (B) In contrast, loss of *Rdh10* results in an expansion of Fgf8 expression into the axial region typical occupied by RA. (C) Lack of RA signaling through the loss of *Rdh10* also leads to a reduction of Tbx5 expression in the forelimb field.

well as the pectoral fin of zebrafish, suggest that the anterior and posterior expression of Fgf8 functions to inhibit forelimb bud initiation (Tanaka 2013; Cunningham and Duester 2015). For instance, gain of Fgf8 function, either by its direct application or by constitutive activation of FGF receptors (FGFRs), results in a loss of the forelimb field (marked by loss of expression of *Tbx5*; see Figure 19.7C) and truncated limbs (Marques et al. 2008; Cunningham et al. 2013).

In contrast, RA is present throughout the somitic regions of the trunk adjacent to the forelimb field, where it is needed to repress trunk Fgf8 expression and thus promote forelimb bud initiation. Targeted loss of RA synthesis (by mutation or pharmacological inhibition) in the vertebrate forelimb results in an expansion of Fgf8 expression toward the presumptive forelimb field, a reduction in *Tbx5* expression, and the failure to form forelimb buds—all of which are consistent with a gain of Fgf8 signaling. Because RA functions as a transcription factor ligand and has been shown to directly repress *Fgf8* gene transcription (Kumar and Duester 2014), the current model of forelimb bud initiation begins with RA restricting Fgf8 expression from the presumptive limb region. In the absence of Fgf8, the lateral plate mesoderm is permissive for forelimb bud formation and development (**FIGURE 19.8A**; Tanaka 2013; Cunningham and Duester 2015).

Although RA-Fgf8 antagonism plays a role during forelimb initiation and the early stages of somitogenesis, RA is largely dispensable for hindlimb development. Loss of RA synthesis in mice does not affect hindlimb bud formation, and overall hindlimb size and patterning are normal (Cunningham et al. 2013). Thus, it is currently unknown what signaling mechanisms underlie hindlimb bud initiation.

2. FORELIMB AND HINDLIMB SPECIFICATION BY TBX5 AND ISLET1 Early specification of forelimb and hindlimb identity begins in the limb fields prior to bud formation through the expression of particular transcription factors (Agarwal et al. 2003; Grandel and Brand 2011). In mice, the gene encoding Tbx5 is transcribed in limb fields of the forelimbs, while the genes encoding Islet1, Tbx4, and Pitx1 are expressed in presumptive hindlimbs (Chapman et al. 1996; Gibson-Brown et al. 1996; Takeuchi et al. 1999;

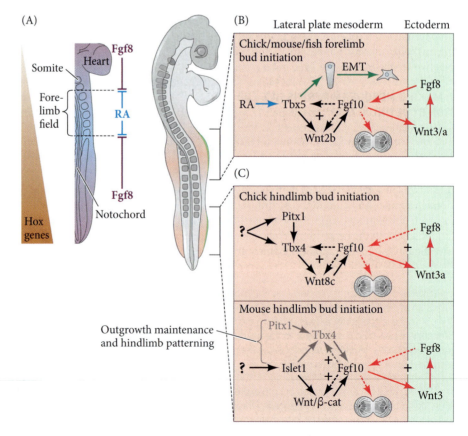

FIGURE 19.8 Model for forelimb field initiation. (A) Initially, axial-level Hox genes regulate Fgf8 and retinoic acid expression, which function as antagonistic signals that induce Tbx5 expression for initiation of the forelimb field. (B) Model of the positive feedback loops between signaling factors that promote forelimb development across species. (C) Many of the essential factors patterning hindlimb development between chick and mouse are the same and also generate positive feedback loops of signaling. Some of the initiation factors for hindlimb induction do differ, however; namely, Islet1 is required for the mouse but not the chick hindlimb. (A after T. J. Cunningham and G. Duester. 2015. *Nat Rev Mol Cell Biol* 16: 110–123.)

Kawakami et al. 2011).[4] Downstream of the regulatory function of these transcription factors is Fgf10, the primary inducer for limb bud formation through the initiation of cell shape changes and proliferation for outgrowth, as discussed below (**FIGURE 19.8B,C**).

Several laboratories (e.g., Logan et al. 1998; Ohuchi et al. 1998; Rodriguez-Esteban et al. 1999; Takeuchi et al. 1999) have provided evidence that Tbx4 and Tbx5 are critical in the specification of hindlimbs and forelimbs, respectively. Before limb formation, there are normally regions of Tbx4 expression in the posterior portion of the lateral plate mesoderm (including the region that will form the hindlimbs) and regions of Tbx5 expression in the anterior portion (including the region that will form the forelimbs). When Fgf10-secreting beads were used to induce an ectopic limb between the chick hindlimb and forelimb buds (**FIGURE 19.9A,B**), the type of limb produced was determined by which Tbx protein was expressed. Limb buds induced by placing FGF beads close to the hindlimb (opposite somite 25) expressed *Tbx4* and became hindlimbs. Limb buds induced close to the forelimb (opposite somite 17) expressed *Tbx5* and developed as forelimbs (wings). Limb buds induced in the center of the flank tissue expressed *Tbx5* in the anterior portion of the limb and *Tbx4* in the posterior portion; these limbs developed as chimeric structures, with the anterior resembling a forelimb and the posterior resembling a hindlimb (**FIGURE 19.9C–E**). Moreover, when a chick embryo was made to express *Tbx4* throughout the flank tissue (by infecting the tissue with a virus that expressed *Tbx4*), limbs induced in the anterior region of the flank often became legs instead of wings (**FIGURE 19.9F,G**).

In further support of Tbx5 being the critical factor for the initiation and specification of the forelimb limb bud, loss of the *Tbx5* gene in chicks,

FIGURE 19.9 Fgf10 expression and action in the developing chick limb. (A) Fgf10 becomes expressed in the lateral plate mesoderm in precisely those positions (arrows) where limbs normally form. (B) When transgenic cells that secrete Fgf10 are placed in the flanks of a chick embryo, the Fgf10 can cause the formation of an ectopic limb (arrow). (C) Limb type in the chick is specified by Tbx4 and Tbx5. In situ hybridizations show that during normal chick development, Tbx5 (blue) is found in the anterior lateral plate mesoderm, whereas Tbx4 (red) is found in the posterior lateral plate mesoderm. Tbx5-containing limb buds produce wings, whereas Tbx4-containing limb buds generate legs. If a new limb bud is induced with an FGF-secreting bead, the type of limb formed depends on which *Tbx* gene is expressed in the limb bud. If placed between the regions of *Tbx4* and *Tbx5* expression, the bead will induce the expression of *Tbx4* posteriorly and *Tbx5* anteriorly. The resulting limb bud will also express *Tbx5* anteriorly and *Tbx4* posteriorly and will generate a chimeric limb. (D) Expression of *Tbx5* in the forelimb (w, wing) buds and in the anterior portion of a limb bud induced by an FGF-secreting bead (red arrowhead). Staining for *Mrf4* mRNA marks the somite positions. (E) Expression of *Tbx4* in the hindlimb (le, leg) buds and in the posterior portion of an FGF-induced limb bud (red arrowhead). (F) A chimeric limb (red arrow) induced by an FGF bead. (G) At a later stage of development, the chimeric limb contains anterior wing structures (feathers) and posterior leg structures (scales). (C after H. Ohuchi and S. Noji. 1999. *Cell Tissue Res* 296: 45–56.)

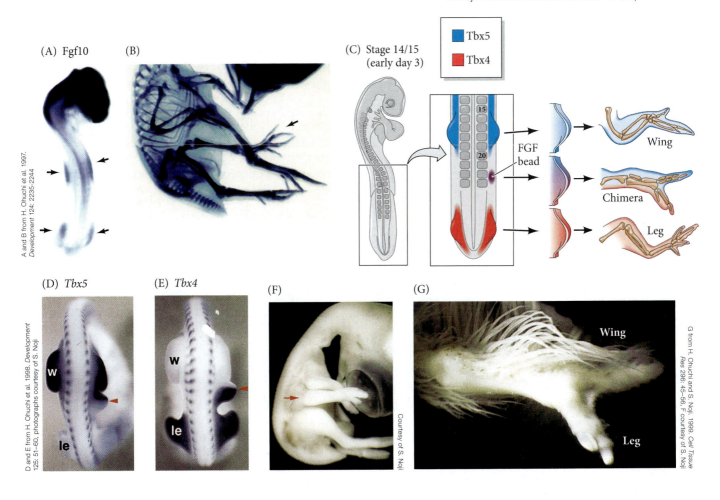

A and B from H. Ohuchi et al. 1997. *Development* 124: 2235-2244

(A) Fgf10

(B)

(C) Stage 14/15 (early day 3)

■ Tbx5
■ Tbx4

FGF bead

Wing

Chimera

Leg

(D) *Tbx5*

(E) *Tbx4*

(F)

(G)

D and E from H. Ohuchi et al. 1998. *Development* 125: 51–60, photographs courtesy of S. Noji

w

le

w

le

Courtesy of S. Noji

Wing

Leg

G from H. Ohuchi and S. Noji. 1999. *Cell Tissue Res* 296: 45–56, F courtesy of S. Noji

mice, and fish results in a complete failure of forelimb formation that includes even the most proximal shoulder girdle structure (Garrity et al. 2002; Agarwal et al. 2003; Rallis et al. 2003). However, the role of Tbx4 in hindlimb specification may differ between chicks and mice. In chicks, loss of Tbx4 function in the hindlimb field completely inhibits leg initiation and growth (Takeuchi et al. 2003); in mice, hindlimb bud growth and initial patterning appear normal when *Tbx4* is knocked out (Naiche and Papaioannou 2003), although leg development is arrested prematurely. This finding suggests that in mice, Tbx4 normally plays more of a role in maintaining outgrowth of the hindlimb than in its initial formation.

More recent investigations have revealed two additional transcription factors involved in the initiation of the hindlimb: Pitx1 and Islet1. Indeed, misexpression of *Pitx1* in the mouse forelimb causes its muscles, bones, and tendons to develop into ones that look like those of a hindlimb (Minguillon et al. 2005; DeLaurier et al. 2006; Ouimette et al. 2010); *Tbx4* expressed in the mouse forelimb will not have this effect. Additionally, Pitx1 protein activates hindlimb-specific genes in the forelimb, including *Hoxc10* and *Tbx4*. Interestingly, a mutation in the human *PITX1* gene that causes a haploinsufficiency in Pitx1 protein results in a bilateral "club foot" phenotype (Alvarado et al. 2011). These results indicate that Pitx1 is sufficient for hindlimb specification; however, the hindlimb is neither completely lost nor severely mispatterned in *Pitx1*-null mice, although some hindlimb structures are malformed (Duboc and Logan 2011). This observation suggests that yet another factor may be involved. Islet1, a homeodomain transcription factor, is transiently expressed in the hindlimb field before *Fgf10* expression and leg bud formation in mice (Yang et al. 2006). When *Islet1* is inactivated specifically in the lateral plate mesoderm, hindlimbs do not form, which is consistent with a role in hindlimb initiation (Itou et al. 2012). Transcriptional regulation of *Islet1* and of *Pitx1* are independent of each other's function, as are their roles in hindlimb development. Despite both genes being documented to similarly upregulate *Fgf10* and *Tbx4*, Islet1 functions to induce hindlimb bud initiation (see Figure 19.8C, black arrows), whereas Pitx1 plays a role in hindlimb patterning (see Figure 19.8C, gray arrows).

3. INDUCTION OF EPITHELIAL-MESENCHYMAL TRANSITIONS BY TBX5 Prior to limb bud formation, the lateral plate mesoderm of the somatopleure displays characteristics of a pseudostratified epithelium with apical-basal polarity (**FIGURE 19.10**). This tissue architecture is perplexing, since these cells contribute to the progress zone of the limb bud, which is made up of mesenchymal cells. Research shows that the epithelial cells making up the mesoderm of the early somatopleure undergo an epithelial-mesenchymal transition (EMT) specifically in the limb fields before any signs of such cell behavior are observed in the mesoderm of flank regions (Gros and Tabin 2014). Lineage tracing of the somatopleural mesoderm reveals a visible change from epithelial to mesenchymal morphology over the course of 24 hours. At least in the case of the forelimb,

FIGURE 19.10 Epithelial-mesenchymal transitions of the epithelial mesoderm of the somatopleure during limb bud formation. (A) The mesoderm (lateral plate mesoderm) of the early somatopleure is epithelial. (B–D) Over a 24-hour period, this mesoderm (labeled with GFP) undergoes an epithelial-mesenchymal transition in the areas of the limb fields.

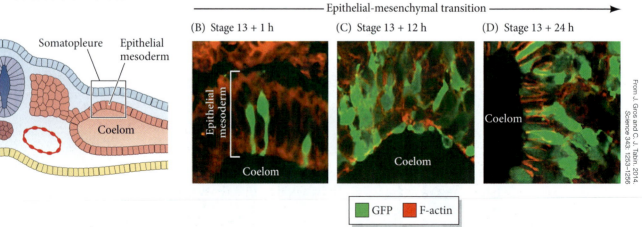

Epithelial-mesenchymal transition

(A)

Somatopleure / Epithelial mesoderm

Coelom

(B) Stage 13 + 1 h

Epithelial mesoderm

Coelom

(C) Stage 13 + 12 h

Coelom

(D) Stage 13 + 24 h

Coelom

GFP   F-actin

Tbx5 knockout mice show a significant loss of limb bud mesenchyme, suggesting that Tbx5 is a major regulator of EMT in the forelimb field (see Figure 19.8B, green arrows). It is unknown whether Islet1, Fgf10, or other factors (Tbx4, Pitx1) are similarly required for EMT in the hindlimb.

4. ESTABLISHMENT OF TWO POSITIVE FEEDBACK LOOPS FOR LIMB BUD FORMATION BY FGF-WNT   Through the upregulation of *Tbx5* in the forelimb and *Islet1* in the hindlimb, the mesenchyme cells commit toward limb bud development and secrete the paracrine factor Fgf10. Fgf10 provides the signal to initiate and propagate the limb-forming interactions between ectoderm and mesoderm, and these signaling interactions directly promote limb bud formation and growth.

Formation of the limb is arguably one of the most remarkable morphological events in embryonic development. Limb development is a prolonged process of outgrowth, and Fgf10 possesses the morphogenetic power to induce limb formation. Remember that a bead containing Fgf10 placed ectopically beneath the flank ectoderm can induce extra limbs to form (see Figure 19.9B,C; Ohuchi et al. 1997; Sekine et al. 1999). Growth of the limb bud is then maintained through a *positive feedback loop*, whereby Wnt/β-catenin and the transcription factors it initiates perpetuate Fgf10 signaling (see Figure 19.8B,C, black dashed arrows). Fgf10 does not just maintain a limb-promoting signaling loop, but also directly induces formation of a new signaling tissue—the AER.

Fgf10 secreted by the limb field mesenchyme induces the overlying ectoderm to form the apical ectodermal ridge (see Figure 19.2D; Xu et al. 1998; Yonei-Tamura et al. 1999). The AER runs along the distal margin of the limb bud and will become a major signaling center for the developing limbs (Saunders 1948; Kieny 1960; Saunders and Reuss 1974; Fernandez-Teran and Ros 2008). Fgf10 is capable of inducing the AER in the competent ectoderm between the dorsal and ventral sides of the embryo, the boundary of which is required for AER formation (Carrington and Fallon 1988; Laufer et al. 1997a; Rodriguez-Esteban et al. 1997; Tanaka et al. 1997).

Fgf10 stimulates Wnt3 (Wnt3a in chicks; Wnt3 in humans and mice) in the prospective limb bud surface ectoderm. The Wnt protein acts through the canonical β-catenin pathway to induce Fgf8 expression in this same ectoderm (Fernandez-Teran and Ros 2008). This causes the surface ectoderm to elongate and physically become the AER.

One of the main functions of the AER is to tell the mesenchyme cells directly beneath it to continue making Fgf10. In this way, a second positive feedback loop is created wherein mesodermal Fgf10 tells the surface ectoderm to continue to make Fgf8, and the surface ectoderm continues to tell the underlying mesoderm to make Fgf10 (see Figure 19.8B,C, red arrows; Mahmood et al. 1995; Crossley et al. 1996; Vogel et al. 1996; Ohuchi et al. 1997; Kawakami et al. 2001). The continued expression of FGFs maintains mitosis in the mesenchyme beneath the AER, which fuels the outgrowth of the limb.

Outgrowth: Generating the Proximal-Distal Axis of the Limb

The apical ectodermal ridge

The apical ectodermal ridge is a multipurpose signaling center that will influence patterning along all axes of limb development (**FIGURE 19.11A,B**). The diverse roles of the AER include (1) maintaining the mesenchyme beneath it in a plastic, proliferating state that enables the linear (proximal-distal, or shoulder-finger) growth of the limb; (2) maintaining the expression of those molecules that generate the anterior-posterior (thumb-pinkie) axis; and (3) interacting with the proteins specifying the anterior-posterior and dorsal-ventral (knuckle-palm) axes so that each cell is given instructions on how to differentiate (see Figure 19.1).

The proximal-distal growth and differentiation of the limb bud are made possible by a series of interactions between the AER and the limb bud mesenchyme directly (200 μm) beneath it. As mentioned earlier, this mesenchyme is called the progress zone (PZ) mesenchyme because its proliferation extends the limb bud (Harrison 1918; Saunders

<aside>
(?) Developing Questions

Autonomous or nonautonomous? Perhaps that should be the question regarding hindlimb bud formation. Retinoic acid antagonism of Fgf8 is an important nonautonomous mechanism required to induce forelimb development, but this "battle of the paracrine factors" does not play out for induction of the hindlimb field. Is there enough prepatterning from *Hox* and *Islet1* gene expression to support an autonomous mechanism of hindlimb bud induction? Moreover, how important is the fourth dimension—that of time—in influencing hindlimb development? How might you experimentally approach these questions?
</aside>

FIGURE 19.11 Manipulation of the apical ectodermal ridge (AER). (A) In the normal 3-day chick embryo, Fgf8 (dark purple) is expressed in the AER of both forelimb and hindlimb buds. (B) Expression of *Fgf8* RNA in the AER, the source of mitotic signals to the underlying mesoderm. (C) Summary of experiments demonstrating the effect of the AER on the underlying mesenchyme. (C after N. K. Wessells. 1973. *Tissue Interactions in Development: An Addison-Wesley Module in Biology, no 9.* Addison-Wesley Longman: Boston.)

(A)   (B)

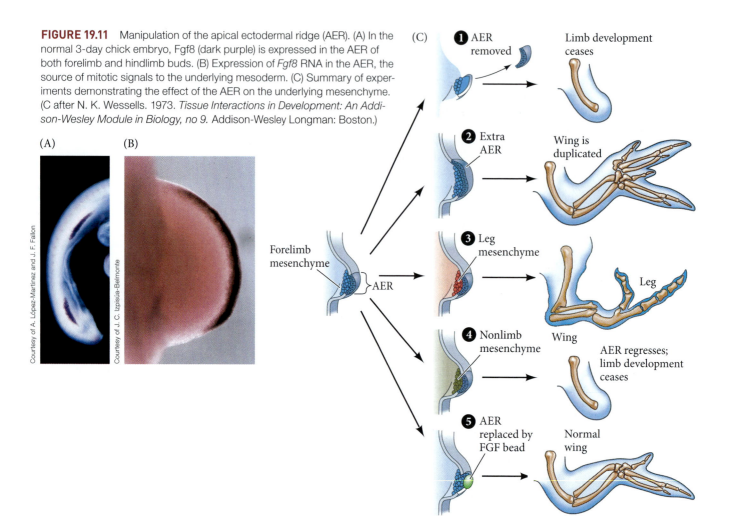

1948; Tabin and Wolpert 2007). These interactions were demonstrated by the results of several experiments on chick embryos (**FIGURE 19.11C**):

- If the AER is removed at any time during limb development, further development of distal limb skeletal elements ceases.

- If an extra AER is grafted onto an existing limb bud, supernumerary structures are formed, usually toward the distal end of the limb.

- If leg mesenchyme is placed directly beneath the wing AER, distal hindlimb structures (toes) develop at the end of the limb. (If this mesenchyme is placed farther from the AER, however, the hindlimb [leg] mesenchyme becomes integrated into wing structures.)

- If limb mesenchyme is replaced by nonlimb mesenchyme beneath the AER, the AER regresses and limb development ceases.

Thus, although the mesenchyme cells induce and sustain the AER and determine the type of limb to be formed, the AER is responsible for the sustained outgrowth and development of the limb (Zwilling 1955; Saunders et al. 1957; Saunders 1972; Krabbenhoft and Fallon 1989). The AER keeps the mesenchyme cells directly beneath it in a state of mitotic proliferation and prevents them from forming cartilage (see ten Berge et al. 2008).

Fgf8 is the major active factor in the AER, and Fgf8-secreting beads can substitute for the AER functions in inducing limb growth (see Figure 19.11C, panel 5). There are other FGFs made by the AER, including Fgf4, Fgf9, and Fgf17 (Lewandoski et al. 2000; Boulet et al. 2004). Loss of any one of these FGFs causes only mild to no defects in skeletal pattern, though, suggesting that significant redundancy exists within this family for limb patterning. However, genetic removal of multiple FGF genes demonstrated increasingly severe and specific skeletal malformations with each additional FGF gene removed, which supports the idea that AER-derived FGFs exhibit some control over patterning (see Figure 19.8B,C, red arrows; Mariani et al. 2008). (See **Further Development 19.2, Induction of and by the AER**, online.)

Specifying the limb mesoderm: Determining the proximal-distal polarity

THE ROLE OF THE AER   In 1948, John Saunders made a simple and profound observation: if the AER is removed from an early-stage wing bud, only a humerus forms. If the AER is removed slightly later, humerus, radius, and ulna form (Saunders 1948; Iten 1982; Rowe et al. 1982). Explaining how this happens has not been easy. First it had to be determined whether the positional information for proximal-distal polarity resided in the AER or in the progress zone mesenchyme. Through a series of reciprocal transplantations, this specificity was found to reside in the mesenchyme. If the AER had provided the positional information—somehow instructing the undifferentiated mesoderm beneath it as to what structures to make—then older AERs combined with younger mesoderm should have produced limbs with deletions in the middle, whereas younger AERs combined with older mesoderm should have produced duplications of structures. This was not found to be the case; rather, normal limbs formed in both experiments (Rubin and Saunders 1972). But when the entire progress zone (including both the mesoderm and the AER) from an early embryo was placed on the limb bud of a later-stage embryo, new proximal structures were produced beyond those already present (**FIGURE 19.12A**). Conversely, when old progress zones were added to young limb buds, distal structures developed such that digits were seen to emerge from the humerus without an intervening ulna and radius (**FIGURE 19.12B**; Summerbell and Lewis 1975). These experiments demonstrated that the mesenchyme specifies the skeletal identities along the proximal-distal axis, which beckons the next question: How?

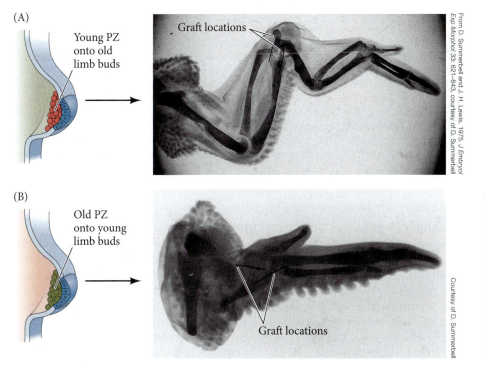

(A) Young PZ onto old limb buds

Graft locations

From D. Summerbell and J. H. Lewis. 1975. *J Embryol Exp Morphol* 33: 621–643, courtesy of D. Summerbell

(B) Old PZ onto young limb buds

Graft locations

Courtesy of D. Summerbell

FIGURE 19.12   Control of proximal-distal specification of the limb is correlated with the age of the progress zone (PZ) mesenchyme. (A) An extra set of ulna and radius formed when an early wing-bud progress zone was transplanted to a late wing bud that had already formed ulna and radius. (B) Lack of intermediate structures are seen when a late wing-bud progress zone was transplanted to an early wing bud.

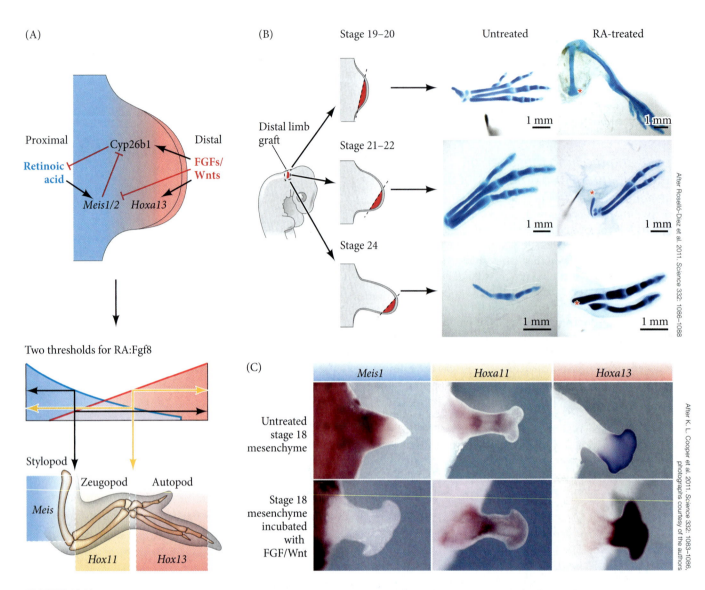

FIGURE 19.13 (A) Model for limb patterning, whereby the proximal-distal axis is generated by opposing gradients of retinoic acid (RA) (blue shading) from the proximal flank and of FGFs and Wnts (pink shading) from the distal AER. (B) Grafting procedure in chick embryos showing transplantation of limb bud tips to the head region of another chick embryo. Bud tips were either untreated or treated by inserting a bead soaked in RA (asterisks mark bead location). Results showed that RA proximalizes the bones forming from the transplanted mesenchyme. Untreated limb bud tips (red) generated specific limb cartilage depending on age. However, when the bud tip was treated with 1 mg/ml RA, the skeleton that formed became more proximal. (C) Treatment with FGFs and Wnts changes the expression pattern of specific proximodistal transcription factors in transplanted mesenchyme (dark staining). *Meis1* is specific for the stylopod, *Hoxa11* is specific for the zeugopod, and *Hoxa13* is specific for the autopod. Mesenchyme from the earliest limb bud (stage 18) will form all three types of cartilage. When the mesenchyme is first incubated in Fgf8 and Wnt3a, however, the autopod transcription factor (*Hoxa13*) is greatly expressed, whereas the stylopod marker (*Meis1*) is drastically reduced. The addition of RA in culture is required to maintain competence to express *Meis1* proximally (not shown). (A after S. Mackem and M. Lewandoski. 2011. *Science* 332: 1038–1039; B after Roselló-Diez et al. 2011. *Science* 332: 1086–1088; C after K. L. Cooper et al. 2011. *Science* 332: 1083–1086.)

GRADIENT MODELS OF LIMB PATTERNING In 2010, evidence on chick limb patterning converged on a model of two opposing gradients: a gradient of FGFs and Wnts from the distal AER, and a second gradient of retinoic acid from the proximal flank tissue (**FIGURE 19.13A**). Such a two-gradient explanation had been proposed earlier (see Maden 1985; Crawford and Stocum 1988a,b; Mercader et al. 2000), but actual evidence for the model came in 2011 from mesenchyme transplantation experiments (Cooper et al. 2011; Roselló-Díez et al. 2011).

Researchers took undifferentiated limb bud mesenchyme cells and "repacked" them into the ectodermal hull of a young limb bud. As expected, the age of the mesenchyme determined the type of bones formed. However, the type of bone formed became more proximal (in the stylopod direction) if young limb bud mesenchyme had been treated with RA in the presence of Wnt and FGF, and it became more distal (toward the autopod) if the mesenchyme had been treated with only FGFs and Wnts (**FIGURE 19.13B,C**). Moreover, if the actions of FGFs were inhibited, the bones became more proximal; and if RA synthesis was inhibited, the bones became more distal. Thus, there appears to be a balance between the proximalizing of the bones by RA from the flank and the distalizing of bones by the FGFs and Wnts of the AER. The opposing gradients may accomplish this balance by laying down a segmental pattern of different transcription factors in the mesenchyme. Such opposing gradients are probably a common mechanism for cell specification, as we've already seen in the early *Drosophila* embryo (see Chapter 9). (**See Further Development 19.3, Mechanistic Support for the Dual Gradient Model of Limb Patterning,** and **Further Development 19.4, Alternative Views on the Dual Gradient Model: Can a Single Gradient Do the Job?**, both online.)

Turing's model: A reaction-diffusion mechanism of proximal-distal limb development

Genes and proteins don't produce a skeleton. Cells do. The cell types of the stylopod and the autopod are identical; it is only how they are arranged in space that differs. Amazingly, dissociated limb mesenchyme placed in culture is capable of self-organizing, expressing 5′ Hox genes, and forming limblike structures with rods and nodules of cartilage (Ros et al. 1994), which raises the foundational question, How do these cells "know" to organize appropriately? Applied to the embryonic limb, why is only one cartilage element formed in the stylopod while two are formed in the zeugopod and several in the autopod? How do the gradients surrounding these cells tell them how to create different parts of the skeleton in different places? Why are the fingers and toes always at the distal end of the limb? The answers may come from a model that involves the diffusion of two or more negatively interacting signals. This is known as the reaction-diffusion mechanism for developmental patterning.

THE REACTION-DIFFUSION MODEL   The **reaction-diffusion mechanism** is a mathematical model formulated by Alan Turing (1952) to explain how complex chemical patterns can be generated out of substances that are initially homogeneously distributed. Turing was the British mathematician and computer scientist who broke the German "Enigma" code during World War II, as recounted in the 2014 film *The Imitation Game*. Two years before his death, Turing provided biologists with a basic mathematical model for explaining how patterns can be self-organized. Although some scientists began applying his model in the 1970s to pattern chondrogenesis in the limb (Newman and Frisch 1979), it was not until quite recently, with the accumulation of experimental evidence, that the model gained broad acceptance.

The uniqueness of Turing's model lies in the "reaction" portion of his mechanism. There is no dependence on molecular prepatterns; rather, interactions between two molecules can spontaneously produce a nonuniform pattern (reviewed in an approachable manner in Kondo and Miura 2010). Turing realized that generation of such patterns would not occur in the presence of just a single diffusible morphogen, but that it *could* be achieved by two homogeneously distributed substances (which we will call morphogen *A*, for "activator," and morphogen *I*, for "inhibitor") *if the rates of production of each substance depended on the other* (**FIGURE 19.14A,B**).

The Turing model provides a framework for a system of "local autoactivation-lateral inhibition" (LALI) to generate stable patterns that could be used to drive developmental change (Meinhardt 2008). (Other "Turing-type" reaction-diffusion systems are also used by cells, with similar results.) In Turing's model, morphogen *A* promotes the production of more morphogen *A* (autoactivation) as well as production of morphogen *I*. Morphogen *I*, however, inhibits the production of morphogen *A* (lateral inhibition). Turing's mathematical analysis shows that if *I* diffuses more readily than *A*, sharp waves of concentration differences will be generated for morphogen *A* (**FIGURE 19.14C**).

FIGURE 19.14 Reaction-diffusion (Turing) mechanism of pattern generation. (A) The Turing model is based on the interaction of two factors, one that is both autoactivating and able to activate its own inhibitor. These interactions can lead to self-generating patterns of alternative cell fates, which may resemble the stripes of a flag or more labyrinth-like patterns. (B) Generation of periodic spatial heterogeneity can occur spontaneously when two reactants, I and A, are mixed together under the conditions that I inhibits A, A catalyzes production of both I and A, and I diffuses faster than A (Time 1). Time 2 illustrates the conditions of the reaction-diffusion mechanism yielding a peak of A and a lower peak of I at the same place. (C) The distribution of the reactants is initially random, and their concentrations fluctuate over a given average. As A increases locally, it produces more I, which diffuses to inhibit more peaks of A from forming in the vicinity of its production. The result is a series of A peaks ("standing waves") at regular intervals. (D,E) Computer simulations of the limb elements that would result from a Turing mechanism of self-generation. (D) Cross-sectional view of the activator morphogen TGF-β at successive stages of chick limb development (increasing time is shown from bottom to top). The concentration of TGF-β is indicated by color (low = green; high = red). (E) Three-dimensional view of the cells undergoing condensation into bone (gray) as predicted by this computer simulation. Note that the number of "bones" in each region of the limb correlates with the number of TGF-β concentration peaks over developmental time, as shown in (D). (A,B from S. Kondo and T. Miura. 2010. *Science* 329: 1616–1620, courtesy of S. Miyazawa.)

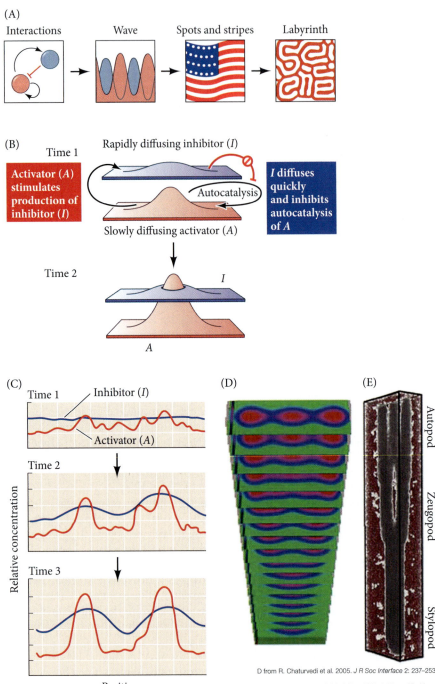

D from R. Chaturvedi et al. 2005. *J R Soc Interface* 2: 237–253

E from Y. T. Zhang et al. 2013. *Math Biosci* 243: 1–17, modified from T. M. Cickovski. 2005. *IEEE/ACM Trans Comput Biol Bioinform* 2: 273–288

The diffusion of the interacting signals may initially be random, yet due to the activator-inhibitor dynamics of this LALI-type Turing model, there will be alternating areas of high and low concentrations of a morphogen, which can produce different cell fates. When the activating morphogen is above a certain threshold level, a cell (or group of cells) can be instructed to differentiate in a certain way.

Turing's model has produced fascinating results when applied to limb development (**FIGURE 19.14D,E**). It appears that reaction-diffusion dynamics can tell us how the limb bud acquires its proximal-distal polarity as well as how the number of digits is regulated at the distal tip of the limb (Turing and digits will be covered later in this chapter). The reaction-diffusion system has been proposed to be sufficient for establishing patterns of precartilage and noncartilage tissues (Zhu et al. 2010).

"TURING" ALONG THE LIMB   Stuart Newman's laboratory has demonstrated that a reaction-diffusion mechanism can pattern limb mesenchyme, and that size and shape matter (Hentschel et al. 2004; Chaturvedi et al. 2005; Newman and Bhat 2007; Zhu et al. 2010; reviewed in Zhang et al. 2013). To mathematically model limb chondrogenesis within a Turing framework, the key parameters need to be identified. During chondrogenesis along the proximal-distal axis, the AER is seen as dividing the limb into two domains: the *inhibitory domain* (also called the *apical zone*), the most distal mesenchyme subjacent to the AER, in which precartilage condensation is repressed; and the *active zone*, which lies proximally adjacent to the inhibitory domain and is the morphogenetically active domain where cartilage-forming condensations coalesce. A third domain, the "frozen zone" well beyond the influence of the AER, contains the formed cartilage primordia of the skeleton at proximal regions of the developing limb. It is within the active zone of the limb mesenchyme that the Turing parameters apply (**FIGURE 19.15**). The active and frozen zones are further defined (both in tissue and in differential equations of the math model) by their unique expression of FgfR2 and FgfR3, respectively (Szebenyi et al. 1995; Hentschel et al. 2004).

The limb mesenchyme cells within the active zone synthesize *activators* of cartilage nodule formation. These activators include TGF-β, BMPs, Activins, and certain carbohydrate-binding proteins called galectins. Galectins can induce the formation of certain cell adhesion molecules and extracellular matrix proteins, such as fibronectin, that cause cells to aggregate together to form the cartilaginous skeleton. These same cells, however, also synthesize *inhibitors* of aggregation, such as Noggin and inhibitory galectins. As a result, what were once cartilage-forming aggregates inhibit the areas surrounding them from forming such aggregates (see Figure 19.15, lower panel).

At different sizes of the limb, different numbers of precartilaginous condensations can form. First, a single condensation can fit (humerus), then two (ulna and radius), then several (wrist, digits). In this reaction-diffusion hypothesis, the aggregations of precartilage mesenchyme actively recruit more cells from the surrounding area and laterally inhibit the formation of other foci of condensation. The number of these condensations, then, depends on the geometry of the active zone and the strength of the lateral inhibition. Once formed, the aggregates of mesenchyme interact with one another not only to recruit more cells but also to express the transcription factors (Sox9) and extracellular matrix (collagen 2) characteristic of cartilage (Lorda-Diez et al. 2011).

According to the model, waves of synthesis and inhibition would form the original pattern of the limb. By placing such constraints as geometry, diffusibility, and the rates of synthesis and degradation of each activator and inhibitor, Zhu and colleagues have been able to model the types of skeleton formed as the limb bud grows. First, the computer model accurately mimicked the normal patterning of the limb (**FIGURE 19.16A**). Next, it simulated the aberrant skeletons formed as a result of manipulations (**FIGURE 19.16B**) and mutations (**FIGURE 19.16C**). Altering the geometries could also yield the patterns seen in fossil limbs (**FIGURE 19.16D**).

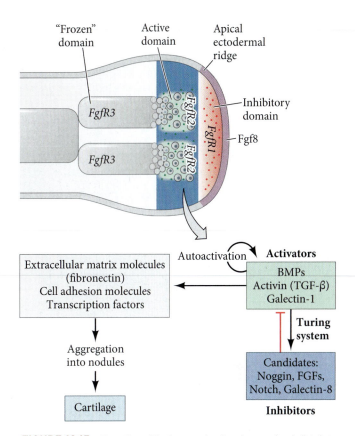

FIGURE 19.15   Reaction-diffusion mechanism for proximal-distal limb specification. In the inhibitory domain immediately outside the AER, cells are kept dividing by FGFs and Wnts and are prevented from forming cartilage. Behind this area, in the active domain, cartilaginous nodules actively form according to a reaction-diffusion mechanism. Here each cell secretes and can respond to activating paracrine factors of the TGF-β superfamily (TGF-β, BMPs, Activin) and cell adhesion factors such as galectin-1. These factors stimulate their own synthesis as well as that of the extracellular matrix and cell adhesion proteins that promote aggregation. The activating cells also stimulate the synthesis of diffusible inhibitors of aggregation (including Noggin and galectin-8), preventing cell adhesion in neighboring regions. The places where nodules can form are governed by the geometry of the limb bud (i.e., the geometry decides how many "waves" of activator will be allowed). In the "frozen" domain, the aggregated nodules can now differentiate into cartilage, thus "freezing" the configuration. (After J. Zhu et al. 2010. *PLOS ONE* 5: e10892.)

(A)

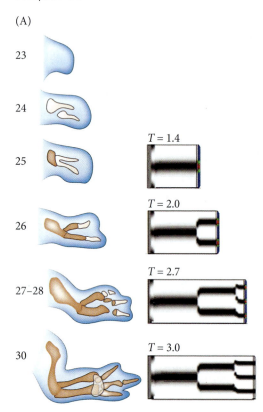

FIGURE 19.16 Computer simulations of limb development can be made after providing experimentally derived parameters of limb bud geometry and of rates of diffusion, synthesis, and degradation of activators and inhibitors. (A) This model predicts the sequential formation of the single stylopod, the double zeugopod, and the multiple cartilaginous rods of the autopod. Stages of chick embryos are on the left. T (time) is in arbitrary units for comparing the computer-generated stages. (B,C) The chick model also mimics the observed patterns of skeletal aberration caused by (B) experimental manipulations and (C) mutations. (D) Changing the parameters slightly will also generate the observed limb skeletons found in fossils, such as the fishlike aquatic reptile *Brachypterygius* (whose forelimb had a paddlelike shape) and *Sauripterus*, one of the first land-dwelling reptiles. (After J. Zhu et al. 2010. *PLOS ONE* 5: e10892.)

19.16A, after J. Zhu et al. 2010. *PLOS ONE* 5: e10892/CC 4.0, based on S. A. Newman and H. L. Frisch. 1979. *Science* 205: 662–668. 19.16B, after J. Zhu et al. 2010. *PLOS ONE* 5: e10892/CC BY 4.0; J. W. Saunders. 1948. *J Exp Zool* 108: 363–403. 19.16C, after J. Zhu et al. 2010. *PLOS ONE* 5: e10892/CC BY 4.0; Y. Litingtung et al. 2002. *Nature* 418: 979–983. 19.16D, after J. Zhu et al. 2010. *PLOS ONE* 5: e10892/CC BY 4.0; A. M. Kirton. 1983. *A Review of British Upper Jurassic Ichthyosaurs*. Unpublished Ph.D. dissertation. University of Newcastle-upon-Tyne: UK; N. Shubin et al. 2009. *Nature* 457: 818–823, based on illustration by K. Monoyios

Specifying the Anterior-Posterior Axis

The specification of the anterior-posterior axis of the limb is the earliest restriction in limb bud cell potency from the pluripotent condition. In the chick, this axis is specified shortly before a limb bud is recognizable.

Sonic hedgehog defines a zone of polarizing activity

Viktor Hamburger (1938) showed that as early as the 16-somite stage of a chick embryo, prospective wing mesoderm transplanted to the flank area develops into a limb with

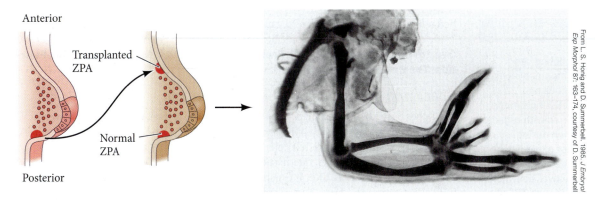

From L. S. Honig and D. Summerbell. 1985. *J Embryol Exp Morphol* 87: 163–174, courtesy of D. Summerbell

FIGURE 19.17   When a ZPA is grafted to anterior limb bud mesoderm, duplicated digits emerge as a mirror image of the normal digits. (After L. S. Honig and D. Summerbell. 1985. *J Embryol Exp Morphol* 87: 163–174.)

the anterior-posterior and dorsal-ventral polarities of the donor graft, not those of the host tissue. Several later experiments (Saunders and Gasseling 1968; Tickle et al. 1975) suggested that the anterior-posterior axis is specified by a small block of mesodermal tissue near the posterior junction of the young limb bud and the body wall. When tissue from this region is taken from a young limb bud and transplanted to a position on the anterior side of another limb bud, the number of digits on the resulting wing is doubled (**FIGURE 19.17**). Moreover, the structures of the extra set of digits are mirror images of the normally produced structures. Polarity is maintained, but the information is now coming from both an anterior and a posterior direction. Thus, this region of the mesoderm has been called the zone of polarizing activity, or ZPA.

The search for the molecule(s) that confer polarizing activity on the ZPA became one of the most intensive quests in developmental biology. In 1993, Riddle and colleagues showed by in situ hybridization that *Sonic hedgehog* (*Shh*), a vertebrate homologue of the *Drosophila hedgehog* gene, was expressed specifically in that region of the limb bud known to be the ZPA (**FIGURE 19.18A**). As evidence that this association between the ZPA and *Sonic hedgehog* was more than just a correlation, Riddle and colleagues (1993) demonstrated that the secretion of Shh protein is sufficient for polarizing activity. They transfected embryonic chick fibroblasts (which normally would never synthesize Shh) with a viral vector containing the *Shh* gene (**FIGURE 19.18B**). The gene became expressed, translated, and secreted in these fibroblasts, which were then inserted under the anterior ectoderm of an early chick limb bud. Mirror-image digit duplications like those induced by ZPA transplants were the result. Moreover, beads containing Sonic

FIGURE 19.18   Sonic hedgehog protein is expressed in the ZPA. (A) In situ hybridization showing the sites of Sonic hedgehog expression (arrows) in the posterior mesoderm of the chick limb buds. These are precisely the regions that transplantation experiments defined as the ZPA. (B) Shh is sufficient to serve the function of the ZPA. When Shh is ectopically produced on the anterior margin of a limb bud of a chick embryo (by grafting recombinant cells expressing Shh), the resulting limbs exhibit a mirror-image digit duplication. (B after R. D. Riddle et al. 1993. *Cell* 75: 1401–1416, based on J. W. Saunders and M. Gasseling. 1968. In *Epithelial-Mesenchymal Interactions: 18th Hahnemann Symposium*. Raul Fleischmajer, Rupert E. Billingham (Eds.), pp. 78–97. Williams & Wilkins: Baltimore.)

(A)

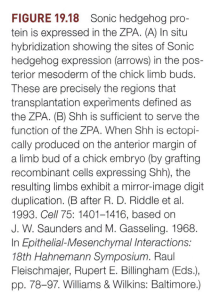

Courtesy of R. D. Riddle

(B) Stage 19–23 embryo

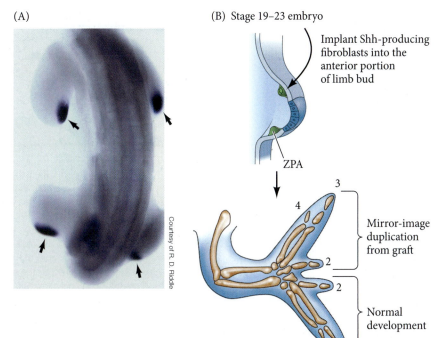

Implant Shh-producing fibroblasts into the anterior portion of limb bud

ZPA

3
4

Mirror-image duplication from graft

2

2

Normal development

4
3

(?) Developing Questions

Mathematical modeling can point the developmental biologist toward new questions and experiments, and Turing's model has certainly done so with regard to pattern formation during organogenesis. For example, which factors in the active zone are the main "reactive" activators and inhibitors? Although TGF-β is a well-supported candidate for the chondrogenic activator, there has been little experimental data to characterize the potential inhibitors that mathematical modeling is predicting. Another parameter to evaluate is cell movements.

hedgehog protein were shown to cause the same duplications (López-Martínez et al. 1995; Yang et al. 1997). Thus, Sonic hedgehog appears to be the active agent of the ZPA. (See **Further Development 19.5, From Humans to Cats, a Natural Gain of Shh Function: The Extra Toes Mutation**, online.)

Specifying digit identity by Sonic hedgehog

How does Sonic hedgehog specify the identities of the digits? When scientists were able to perform fine-scale fate-mapping experiments on the Shh-secreting cells of the ZPA, they were surprised to find that cells that expressed Shh at any time did not undergo apoptosis (programmed cell death; see Chapter 15; Further Development 15.24, online) in the way the AER does after it finishes its job. Rather, the descendants of Shh-secreting cells become the bone and muscle of the posterior limb (Ahn and Joyner 2004; Harfe et al. 2004). Indeed, digits 5 and 4 (and part of digit 3) of the mouse hindlimb are formed from the descendants of Shh-secreting cells (**FIGURE 19.19**).

It seems that specification of the digits is primarily dependent on the amount of time the *Shh* gene is expressed and only a little bit on the concentration of Shh protein that other cells receive (see Tabin and McMahon 2008). The difference between digits 4 and 5 is that the cells of the more posterior digit 5 express *Shh* longer and are exposed to Shh (in an autocrine manner) for a longer time. Digit 3 is made up of some cells that secreted Shh for a shorter period than those of digit 4, and they also depend on Shh diffusion from the ZPA (indicated by digit 4 being lost when Shh is modified such that it cannot diffuse away from cells). Digit 2 is dependent entirely on Shh diffusion for its specification, and digit 1 is specified independently of Shh. Indeed, in a naturally occurring chick mutant that lacks Shh expression in the limb, the only digit that forms is digit 1. Furthermore, when the genes for Shh and Gli3 are conditionally knocked out in the mouse limb, the resulting limbs have numerous digits, but the digits have no obvious specificity (Litingtung et al. 2002; Ros et al. 2003; Scherz et al. 2007). Vargas and Fallon (2005) propose that digit 1 is specified by *Hoxd13* in the absence of *Hoxd12*. Forced expression of *Hoxd12* throughout the digit primordia leads to the transformation of digit 1 into a more posterior digit (Knezevic et al. 1997).

By using conditional knockouts of the mouse *Shh* gene (i.e., researchers could stop Shh expression at different times during mouse development), Zhu and Mackem (2011) found that Sonic hedgehog works by two temporally distinct mechanisms. The first

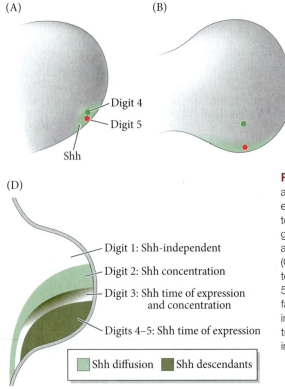

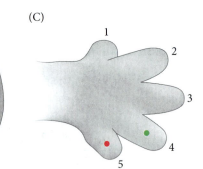

FIGURE 19.19 The descendants of Shh-secreting cells form digits 4 and 5 and contribute to the specification of digits 2 and 3 in the mouse limb. (A) In the early mouse hindlimb bud, the progenitors of digit 4 (green dot) and the progenitors of digit 5 (red dot) are both in the ZPA and express Sonic hedgehog (light green shading). (B) At later stages of limb development, the cells forming digit 5 are still expressing Shh in the ZPA, but the cells that form digit 4 no longer do. (C) When the digits form, the cells in digit 5 will have seen high levels of Shh protein for a longer time than the cells in digit 4. (D) Schematic by which digits 4 and 5 are specified by the amount of time they were exposed to Shh in an autocrine fashion; digit 3 is specified by the amount of time the cells were exposed to Shh in both an autocrine and a paracrine fashion. Digit 2 is specified by the concentration of Shh its cells received by paracrine diffusion, and digit 1 is specified independently of Shh. (After B. D. Harfe et al. 2004. *Cell* 118: 517–528.)

phase involves the specification of digit identity (from the posterior pinky to the anterior thumb). In this phase, Shh acts as a morphogen, with the digit identities being specified first by the concentration of Shh in that region of the limb bud, and then by the duration of exposure to Shh. In the second phase, Shh works as a mitogen to stimulate the proliferation and expansion of the limb bud mesenchyme, thus helping shape the limb bud. (See **Further Development 19.6, The Mechanism by which Sonic Hedgehog Establishes Digital Identity**, online.)

Sonic hedgehog and FGFs: Another positive feedback loop

When the limb bud is relatively small, an initial positive feedback loop is established between Fgf10 produced in the mesoderm and Fgf8 produced in the ectoderm, promoting limb outgrowth (**FIGURE 19.20A**). As the limb bud grows, the ZPA is established, and another regulatory loop is created (**FIGURE 19.20B**). BMPs in the mesoderm would downregulate FGFs in the AER were it not for the Shh-dependent expression of a BMP inhibitor, Gremlin (Niswander et al. 1994; Zúñiga et al. 1999; Scherz et al. 2004; Vokes et al. 2008). Sonic hedgehog in the ZPA activates Gremlin, which inhibits BMPs, thus

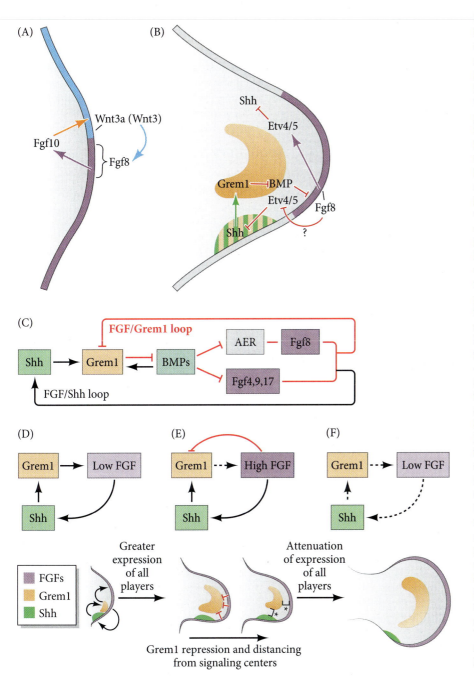

FIGURE 19.20 Early interactions between the AER and limb bud mesenchyme. (A) In the limb bud, Fgf10 from mesenchyme generated by the lateral plate mesoderm activates a Wnt (Wnt3a in chicks; Wnt3 in mice and humans) in the ectoderm. Wnt activates the β-catenin pathway, which induces synthesis of Fgf8 in the region near the AER. Fgf8 activates Fgf10, causing a positive feedback loop. (B) As the limb bud grows, Sonic hedgehog (Shh) in the posterior mesenchyme creates a new signaling center that induces posterior-anterior polarity, and it also activates Gremlin (Grem1) to prevent mesenchymal BMPs from blocking FGF synthesis in the AER. Moreover, Fgf8 operates in part by differentially regulating *Etv4/5* (genes within the E-twenty-six superfamily of transcription factors) along the anterior-posterior axis of the limb bud, which in turn reinforces a gradient of Shh expression from the posterior. (C) Two feedback loops link the AER and ZPA. In the positive feedback loop (black arrow, below), FGFs 4, 9, and 17 from the AER activate Shh, stabilizing the ZPA. In the reciprocal inhibitory loop (red; above), Shh from the ZPA activates Gremlin (Grem1), which blocks BMPs, thus preventing BMP-mediated inactivation of FGFs in the AER. (D) The feedback loops create a mutual accelerated synthesis of Shh (ZPA) and FGFs (AER). (E) As FGF concentration climbs, it eventually reaches a threshold where it inhibits Gremlin, thus allowing the BMPs to begin repressing the AER FGFs. As more cells multiply in the area not expressing Gremlin (brackets and asterisks), the Gremlin signal near the AER is too weak to prevent the BMPs from repressing the FGFs. (F) At that point, the AER disappears, removing the signals that stabilize the ZPA. The ZPA then also disappears. (A,B after M. Fernandez-Teran and M. A. Ros. 2008. *Int J Dev Biol* 52: 857–871; C after J. M. Verheyden and X. Sun. 2008. *Nature* 454: 638–641.)

promoting the maintenance of FGF expression and continued limb bud outgrowth. FGF in turn inhibits repressors of Shh to complete the positive feedback loop. As with most multigene pathways, however, this interactivity is more complicated than presented here.

Depending on the levels of FGFs in the apical ectodermal ridge, the zone of polarizing activity can be either activated or shut down; two feedback loops have been demonstrated (**FIGURE 19.20C**; Verheyden and Sun 2008; Bénazet et al. 2009). At first, relatively low levels of AER FGFs activate Shh and keep the ZPA functioning. The FGF signals appear to inhibit the proteins Etv4 and Etv5, which are repressors of *Sonic hedgehog* transcription (Mao et al. 2009; Zhang et al. 2009). Thus, the AER and ZPA mutually support each other through the positive loop of Sonic hedgehog and FGFs (Todt and Fallon 1987; Laufer et al. 1994; Niswander et al. 1994). In the more anterior region of the limb bud, Fgf8 positively regulates Etv4/5, which in turn represses Shh in this region, further reinforcing the posterior-to-anterior gradient of Shh from the ZPA (Mao et al. 2009).

As a result of Shh stimulation through FGF signaling, levels of Gremlin (a powerful BMP antagonist) become high, and the positive FGF/Shh loop sustains limb growth (**FIGURE 19.20D**). As long as the Gremlin signal can diffuse to the AER, FGFs will be made and the AER maintained. However, as FGF levels consequently also rise, a negative feedback loop to *block* Gremlin expression in the distal mesenchyme is triggered (**FIGURE 19.20E**). This repression of Gremlin synthesis paired with the progressive expansion of the limb bud creates increased distance between Gremlin and the signaling centers (AER and ZPA) in the distalmost mesenchyme. At that time, the BMPs abrogate FGF synthesis, the AER collapses, and the ZPA (with no FGFs to support it) is terminated. The embryonic phase of limb development ends (**FIGURE 19.20F**).

Hox genes are part of the regulatory network specifying digit identity

As mentioned early in this chapter, Hox genes are critical for specifying fates along each axis of the limb, and their expression—especially that of the *Hoxd* cluster—functions in two phases (Zakany et al. 2004; Tarchini and Duboule 2006; see also Abbasi 2011). The first phase is important for the specification of the stylopod and zeugopod, as discussed earlier (**FIGURE 19.21**). The later phase of *Hoxd* expression helps specify the autopod. There are two major "early" *cis*-regulatory regions involved, composed of numerous enhancers that work together to activate the *Hoxd* genes in a specific temporal and spatial array—the ELCR and POST regulatory regions.

The major early regulatory, or *early limb control regulatory* (ELCR), region activates transcription in a time-dependent manner: the closer the gene is to the ELCR region, the earlier it is activated. The second early regulatory region, POST ("posterior restriction"), imposes spatial restrictions on the expression of the 5' *Hoxd* genes (*Hoxd10–13*) such that the genes closest to this region have the most restricted expression domains, starting from the posterior margin of the limb bud (see Figure 19.21A,B).

This situation creates a pattern of nested Hoxd proteins essential for activating the long-range enhancer (ZRS enhancer) of the *Sonic hedgehog* gene, thereby activating Shh expression in the posterior limb bud mesoderm and forming the ZPA (Tarchini et al. 2006; Galli et al. 2010). In addition, the ZPA is also influenced by the presence of Hoxb8 in the flank mesenchyme, which appears to help define the posterior boundary of the forelimb bud (see Figure 19.21B). If *Hoxb8* is expressed ectopically in the anterior compartment of the mouse forelimb bud, a ZPA will also form there (Charite et al. 1994; Hornstein et al. 2005).

The ZPA now feeds back to alter the *Hoxd* gene expression patterns. Sonic hedgehog expressed from the posterior margin activates a second set of enhancers called the global control region, or GCR (see Figure 19.21C,D; Spitz et al. 2003; Montavon et al. 2011). The Hox genes closest to the GCR are expressed most broadly. This expression inverts the original pattern of *Hoxd10–13* expression such that *Hoxd13* is expressed at the highest level and extends most anteriorly. *Hoxd12*, *Hoxd11*, and *Hoxd10* are expressed in slightly narrower domains, so that the most anterior digit (e.g., the thumb) expresses *Hoxd13* but no other Hox gene (see Figure 19.21B, last diagram in panel; Montavon et al. 2008). Thus, the first phase of *Hoxd* gene expression helps specify the ZPA, while

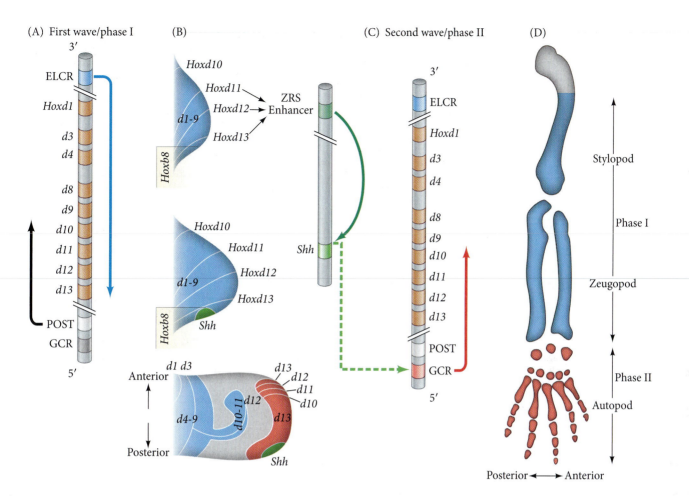

FIGURE 19.21 Changes in *Hoxd* gene expression regulate patterning of the tetrapod limb in two independent phases. (A) The first phase of *Hoxd* expression is initiated as the limb bud forms. The early limb control regulatory (ELCR) element activates those genes closest to it earlier than those genes away from it, whereas the POST regulatory element acts negatively to restrict the anterior expression of these genes in the opposite direction. (B) This results in expression domains such that *Hoxd13* expression is confined to the most posterior region, whereas *Hoxd12* is allowed to expand more anteriorly. The 5' *Hoxd* genes activate the long-range Shh enhancer (ZRS), thereby creating the ZPA in the posterior limb mesoderm. White lines in the limb bud show the boundaries of gene expression. (C) In the second phase, Shh activates the GCR regulatory locus, which inverts the *Hoxd* expression pattern such that *Hoxd13* is located more anteriorly than the other *Hoxd* genes (B, bottom red). (D) Skeletal elements specified by the early (blue) and later (red) phases. (After A. A. Abbasi. 2011. *Dev Dyn* 240: 1005–1016.)

in the second phase of *Hoxd* expression the ZPA instructs the expression patterns, and these patterns define the identities of the digits. In support of this model, transplantation of either the ZPA or other Shh-secreting cells to the anterior margin of the limb bud at this stage leads to the formation of mirror-image patterns of *Hoxd* expression and results in mirror-image digits (Izpisúa-Belmonte et al. 1991; Nohno et al. 1991; Riddle et al. 1993). (See **Further Development 19.7, A Turing Model for Self-Organizing Digit Skeletogenesis**, online.)

Generating the Dorsal-Ventral Axis

The third axis of the limb distinguishes the dorsal half of the limb (knuckles, nails, claws) from the ventral half (pads, soles). In 1974, MacCabe and colleagues demonstrated that the dorsal-ventral polarity of the limb bud is determined by the ectoderm encasing it. If the ectoderm is rotated 180° with respect to the limb bud mesenchyme, the dorsal-ventral axis is partially reversed—that is, the distal elements (digits) are "upside-down"—which suggests that the late specification of the dorsal-ventral axis of the limb is regulated by its ectodermal component(s).

One molecule that appears to be particularly important in specifying dorsal-ventral polarity is Wnt7a. The *Wnt7a* gene is expressed in the dorsal (but not the ventral) ectoderm of chick and mouse limb buds (**FIGURE 19.22A**; Dealy et al. 1993; Parr et al. 1993). When Parr and McMahon (1995) knocked out the *Wnt7a* gene, the resulting mouse embryos had ventral footpads on both surfaces of their paws, showing that Wnt7a is needed for the dorsal patterning of the limb.

Wnt7a is the first known dorsal-ventral axis gene expressed in limb development. It induces activation of the *Lmx1b* (also known as *Lim1*) gene in the dorsal mesenchyme. *Lmx1b* encodes a transcription factor that appears to be essential for specifying dorsal cell fates in the limb. If Lmx1b protein is expressed in the ventral mesenchyme cells, those cells develop a dorsal phenotype (Riddle et al. 1995; Vogel et al. 1995; Altabef and Tickle 2002). Human and mouse *lmx1b* mutants also reveal this gene's importance for specifying dorsal limb fates. *Lmx1b* knockouts in mice produce a syndrome in which the dorsal limb phenotype is lacking and those cells have taken on ventral fates, exhibiting footpads, ventral tendons, and sesamoids (all ventral-specific structures; **FIGURE 19.22B,C**). Similarly in humans, loss-of-function mutations in the *LMX1B* gene result in nail-patella syndrome (no nails on the digits, no kneecaps), in which the dorsal sides of the limbs have been ventralized (Chen et al. 1998; Dreyer et al. 1998). The Lim1 protein probably specifies the cells to differentiate in a dorsal manner, which is critical, as we saw in Chapter 15, for the innervation of motor neurons (whose growth cones recognize inhibitory factors made differentially in the dorsal and ventral compartments of the limb bud). Conversely, the transcription factor Engrailed-1 marks the ventral ectoderm of the limb bud and is induced by BMPs in the underlying mesoderm (**FIGURE 19.23**). If BMPs are knocked out in the early limb bud, Engrailed-1 is not expressed, and Wnt7a is expressed in both dorsal and ventral ectoderm. The result is a malformed limb that is dorsal on both sides (Ahn et al. 2001; Pizette et al. 2001).

(A) Gene expression

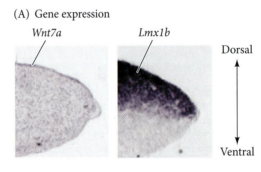

(B) Wild-type

(C) *lmx1b* mutant

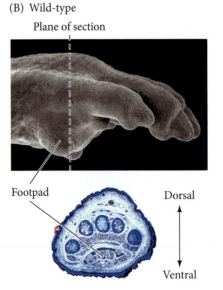

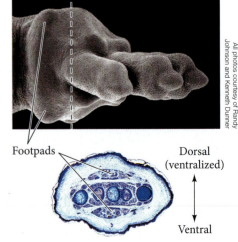

FIGURE 19.22 Lmx1b-dependent dorsal-ventral patterning by Wnt7a. (A) *Wnt7a* and *Lmx1b* are both expressed in the dorsal limb bud. *Wnt7a* is restricted to the epidermis, however, whereas *Lmx1b* is present throughout the dorsal mesenchyme. (B,C) Loss of *Lmx1b* ventralizes the forelimb, as evidenced by the presence of footpads on both sides of the paw in the mutant phenotype.

All photos courtesy of Randy Johnson and Kenneth Dunner

The dorsal-ventral axis is also coordinated with the other two axes. Indeed, the *Wnt7a*-deficient mice described earlier lacked not only dorsal limb structures but also posterior digits, suggesting that Wnt7a is also needed for the anterior-posterior axis (Parr and McMahon 1995). Yang and Niswander (1995) made a similar set of observations in chick embryos. These investigators removed the dorsal ectoderm from developing limbs and found that the result was the loss of posterior skeletal elements from the limbs. The reason these limbs lacked posterior digits was that *Shh* expression was greatly reduced. Viral-induced expression of *Wnt7a* was able to substitute for the dorsal ectoderm signal and restore *Shh* expression and posterior phenotypes. These findings showed that Sonic hedgehog synthesis is stimulated by the combination of Fgf4 and Wnt7a proteins. Conversely, overactive Wnt signaling in the ventral ectoderm causes an overgrowth of the AER and extra digits, indicating that the proximal-distal patterning is not independent of dorsal-ventral patterning either (Loomis et al. 1998; Adamska et al. 2004).

Thus, at the end of limb patterning, BMPs are responsible for simultaneously shutting down the AER, indirectly shutting down the ZPA, and inhibiting the Wnt7a signal along the dorsal-ventral axis (Pizette et al. 2001). The BMP signal eliminates growth and patterning along all three axes. When exogenous BMP is applied to the AER, the elongated epithelium of the AER reverts to a cuboidal epithelium and ceases to produce FGFs; and when BMPs are inhibited by Noggin, the AER continues to persist days after it would normally have regressed (Gañan et al. 1998; Pizette and Niswander 1999).

Cell Death and the Formation of Digits and Joints

Apoptosis—programmed cell death—plays a role in sculpting the tetrapod limb. Indeed, cell death is essential if our joints are to form and if our fingers are to become separated (Zaleske 1985; Zuzarte-Luis and Hurle 2005). The death (or lack of death) of specific cells in the vertebrate limb is genetically programmed and has been selected for over the course of evolution.

Sculpting the autopod

The difference between a chicken's foot and the webbed foot of a duck is the presence or absence of cell death between the digits (**FIGURE 19.24**). Saunders and colleagues have shown that after a certain stage, chick cells between the digit cartilage are destined to die, and will do so even if transplanted to another region of the embryo or placed in culture (Saunders et al. 1962; Saunders and Fallon 1966). Before that time, however,

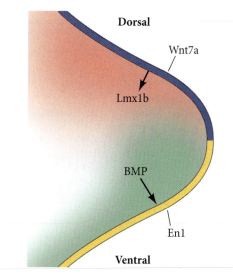

FIGURE 19.23 Model of dorsal-to-ventral patterning in the limb bud by Wnt and BMP signaling. Wnt7a induces dorsal cell fates of the limb bud through Lmx1b, whereas BMP signaling functions through Engrailed-1 (En1) to regulate ventral limb patterning. (After K. Ahn et al. 2001. *Development* 128: 4449–4461.)

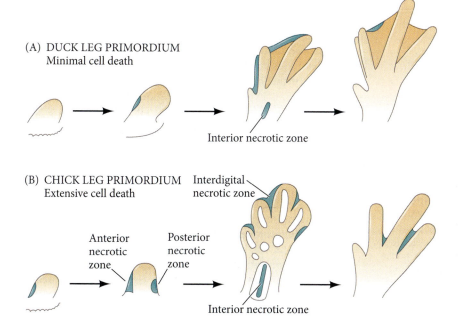

(A) DUCK LEG PRIMORDIUM
Minimal cell death

Interior necrotic zone

(B) CHICK LEG PRIMORDIUM
Extensive cell death

Interdigital necrotic zone

Anterior necrotic zone

Posterior necrotic zone

Interior necrotic zone

FIGURE 19.24 Patterns of cell death in leg primordia of duck (A) and chick (B) embryos. Blue shading indicates areas of cell death. In the duck, the regions of cell death are very small, whereas there are extensive regions of cell death in the interdigital tissue of the chick leg. (After J. W. Saunders and J. F. Fallon. 1966. In *Major Problems in Developmental Biology*, M. Locke [Ed.], pp. 289–314. Academic Press: New York and London.)

transplantation to a duck limb will save them. Between the time when the cell's death is determined and when death actually takes place, levels of DNA, RNA, and protein synthesis in the cell decrease dramatically (Pollak and Fallon 1976).

In addition to the **interdigital necrotic zone**, three other regions of the limb are "sculpted" by cell death. The ulna and radius are separated from each other by an **interior necrotic zone**, and the **anterior** and **posterior necrotic zones** further shape the end of the limb (see Figure 19.24B; Saunders and Fallon 1966). Although these zones are referred to as *necrotic*, this term is a holdover from the days when no distinction was made between necrotic (pathologic or traumatic) cell death and apoptotic cell death. These cells die by apoptosis, and the death of the interdigital tissue is associated with the fragmentation of their DNA (Mori et al. 1995).

The signal for apoptosis in the autopod is provided by the BMP proteins, whose expression, interestingly, is dependent on upregulated synthesis of interdigital RA (Cunningham and Duester 2015). BMP2, BMP4, and BMP7 are each expressed in the interdigital mesenchyme, and blocking BMP signaling (by infecting progress zone cells with retroviruses carrying dominant negative BMP receptors) prevents interdigital apoptosis (Yokouchi et al. 1996; Zou and Niswander 1996; Abara-Buis et al. 2011). Because these BMPs are expressed throughout the progress zone mesenchyme, it is thought that cell death would be the default state unless there were active suppression of the BMPs. This suppression may come from the Noggin protein, which is made in the developing cartilage of the digits and in the perichondrial cells surrounding it (Capdevila and Johnson 1998; Merino et al. 1998). If Noggin is expressed throughout the limb bud, no apoptosis is seen.

Forming the joints

The function first ascribed to BMPs was the formation, not the destruction, of bone and cartilage tissue. In the developing limb, BMPs induce the mesenchymal cells—depending on the stage of development—either to undergo apoptosis or to become cartilage-producing chondrocytes. The same BMPs can induce death or differentiation, depending on the responding cell's history. This **context dependency** of signal action is a critical concept in developmental biology. It is also critical for the formation of joints. Macias and colleagues (1997) have shown that during early limb bud stages (before cartilage condensation), beads secreting BMP2 or BMP7 cause apoptosis. Two days later, the same beads cause the limb bud cells to form cartilage.

In the normally developing limb, BMPs use both of these properties to form joints. Several BMPs are made in the perichondrial cells surrounding the condensing chondrocytes and promote further cartilage formation (**FIGURE 19.25A**). Another BMP, GDF5, is expressed at the regions between the bones, where joints will form, and appears to be critical for joint formation (**FIGURE 19.25B**; Macias et al. 1997; Brunet et al. 1998). Mouse mutations of the gene for GDF5 produce brachypodia, a condition characterized by the lack of limb joints (Storm and Kingsley 1999). In mice homozygous for loss-of-function of the BMP antagonist Noggin, no joints form. Rather, BMP7 in these *Noggin*-defective embryos appears to recruit nearly all the surrounding mesenchyme into the digits (**FIGURE 19.25C**). (See **Further Development 19.8, Joints Need Blood Vessels, Wnt Proteins, and Working Muscles to Form, Further Development 19.9, Continued Limb Growth: Epiphyseal Plates, Further Development 19.10, Fibroblast Growth Factor Receptors: Dwarfism,** and **Further Development 19.11, Growth Hormone and Estrogen Receptors**, all online.)

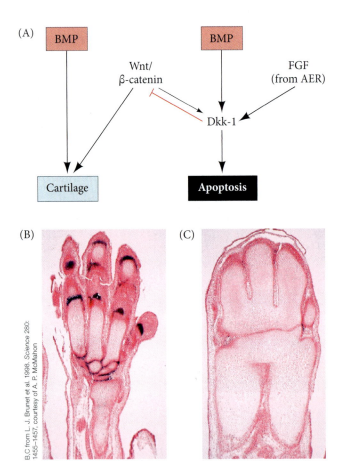

B,C from L. J. Brunet et al. 1998. *Science* 280: 1455–1457, courtesy of A. P. McMahon

FIGURE 19.25 Possible involvement of BMPs in stabilizing cartilage and apoptosis. (A) Model for the dual role of BMP signals in limb mesodermal cells. BMP can be received in the presence of FGFs (to produce apoptosis) or Wnts (to induce bone). When FGFs from the AER are present, Dickkopf (Dkk) is activated. This protein mediates apoptosis and at the same time inhibits Wnt from aiding in skeleton formation. (B,C) Effects of Noggin. (B) A 16.5-day autopod from a wild-type mouse, showing *GDF5* expression (dark blue) at the joints. (C) A 16.5-day *Noggin*-deficient mutant mouse autopod, showing neither joints nor *GDF5* expression. Presumably, in the absence of Noggin, BMP7 was able to convert nearly all the mesenchyme into cartilage. (A after L. Grotewold and U. Rüther. 2002. *EMBO J* 21: 966–975.)

Evolution by Altering Limb Signaling Centers

Charles Darwin wrote in *On the Origin of Species*, "What can be more curious than that the hand of a man, formed for grasping, that of a mole for digging, the leg of a horse, the paddle of the porpoise, and the wing of the bat should all be constructed on the same pattern and should include similar bones, and in the same relative positions?" Darwin recognized that the differences between horse legs, porpoise flippers, and human hands are underlain by a similar pattern of bone formation. He proposed that these bones evolved from a common ancestor, although he did not know how. C. H. Waddington pointed out that such evolution was predicated on changing the development of the limb in ways that could be selected. In other words, he proposed that changes in development caused the arrival of new variations. These variations could then be tested by natural selection. We have now discovered several ways by which "tinkering" with limb signaling molecules can generate new limb morphologies. In these ways, limb evolution can be caused by developmental changes.[5] (See **Further Development 19.12, Dinosaurs and Chicken Fingers**, online.)

WEB-FOOTED FRIENDS   We can start with the sculpting of the autopod. The regulation of BMPs is critical in creating the webbed feet of ducks (Laufer et al. 1997b; Merino et al. 1999). The interdigital regions of duck feet exhibit the same pattern of BMP expression as the webbing of chick feet. However, whereas the interdigital regions of the chick feet appear to undergo BMP-mediated apoptosis, developing duck feet synthesize the BMP inhibitor Gremlin and block this regional cell death (**FIGURE 19.26**). Moreover, the webbing of chick feet can be preserved if Gremlin-soaked beads are placed in the interdigital regions. Thus, the evolution of web-footed birds probably involved the inhibition of BMP-mediated apoptosis in the interdigital regions. In Chapter 24, we will find that the bat embryo uses a similar mechanism to acquire its wings.

TINKERING WITH THE SIGNALING CENTERS: MAKING WHALES   Numerous transition fossils attest to the evolution of modern cetaceans (whales, dolphins, porpoises) from hoofed land mammals (Gingrich et al. 1994; Thewissen et al. 2007, 2009). Numerous changes in the anatomy were made, but few are as striking as the conversion of a forelimb into a flipper and the elimination of the hindlimb altogether. These events were accomplished by modifying the signaling centers of the ancestral cetacean limb

FIGURE 19.26   Autopods of chick (upper row) and duck (lower row) are shown at similar stages. Both show BMP4 expression (dark blue) in the interdigital webbing; BMP4 induces apoptosis. The duck foot (but not the chicken foot) expresses the BMP4-inhibitory protein Gremlin (dark brown; arrows) in the interdigital webbing. Thus, the chicken foot undergoes interdigital apoptosis (as seen by neutral red dye accumulation in the dying cells), but the duck foot does not.

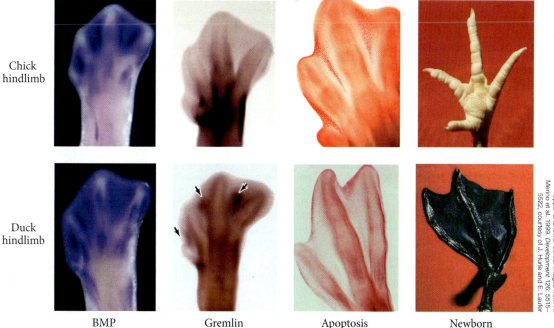

Chick hindlimb

Duck hindlimb

BMP   Gremlin   Apoptosis   Newborn

Upper and lower far left and far right images courtesy of J. Hurle and E. Laufer. Upper and lower center images from R. Merino et al. 1999. *Development* 126: 5515–5522, courtesy of J. Hurle and E. Laufer.

From L. N. Cooper 2009, PhD thesis, Kent State University; courtesy of L. N. Cooper and the Thewissen laboratory. Clearing and staining completed by Dr. Sirpa Nummela

Remnants of pelvic girdle

Hyperphalangy in digits 2 and 3

FIGURE 19.27 An approximately 110-day embryo of a pantropical spotted dolphin (*Stenella attenuata*), stained to show bones (red) and cartilage (blue). Hyperphalangy (extremely long fingers) is seen in the forelimb (correlating with continued *Fgf8* expression in the AER), and a rudimentary hindlimb is seen (correlating with the reduction of AER signaling following the elimination of Shh from the ZPA).

buds in three ways. First, the FGF signaling of the forelimb AER was preserved for a much longer duration, which caused the formation of longer fingers by the continual addition of phalanges. Second, interdigital apoptosis was prevented by blocking BMP activity in a manner similar to that described above for duck feet. Third, the Sonic hedgehog signal from the hindlimb ZPA ceased early in development. Once the ZPA signal diminished, the AER could not be sustained, and the hindlimb ceased to develop (Thewissen et al. 2006). **FIGURE 19.27** shows the elongated flipper phalanges and truncated hindlimb of an embryonic dolphin. Thus, despite creationists claiming that there is no way whales could have evolved from land mammals (see Gish 1985), in fact the combination of developmental biology and paleontology explains the phenomenon extremely well.

Closing Thoughts on the Opening Photo

Perhaps the more appropriate question for this image might have been "*Which fingers am I holding up?*" The skeletal elements in this chick wing reveal a mirror-image duplication of the digits, which we now know was due to the misexpression of Sonic hedgehog from the anterior side of the limb bud. Gradients of signaling factors along the major axes play essential roles in establishing the correct number and pattern of structures in the arm and hand. Just as important for limb development is the underlying gene regulatory network that Hox genes regulate as well as the self-organizing interactions between the cells that ultimately build the tissues of the limb.

After L. S. Honig and D. Summerbell. 1985.
J Embryol Exp Morphol 87: 163–174

Endnotes

[1] These terms can be difficult to remember, but knowing their word origins can help. Stylo = like a pillar; zeugo = joining; auto = self; pod = foot.

[2] Actually, it is a four-dimensional system, in which time is the fourth axis. Developmental biologists get used to seeing nature in four dimensions.

[3] Interestingly, Hox gene expression in at least some snakes (such as *Python*) creates a pattern in which each somite is specified to become a thoracic (ribbed) vertebra. The patterns of Hox gene expression associated with limb-forming regions in the snake are not seen (see Chapter 17; Cohn and Tickle 1999).

[4] *Tbx* stands for "T-box," a specific DNA-binding domain. The *T* (*Brachyury*) gene and its relatives have a sequence that encodes this domain. We discussed *Tbx5* in the context of heart ventricle development in Chapter 18.

[5] Earlier in this chapter, we noted that developmental biologists get used to thinking in four dimensions. Evolutionary developmental biologists have to think in *five* dimensions: the three standard dimensions of space, the dimension of developmental time (hours or days), and the dimension of evolutionary time (millions of years).

Snapshot Summary
Development of the Tetrapod Limb

19

1. The positions where limbs emerge from the body axis depend on Hox gene expression.

2. The proximal-distal axis of the developing limb is initiated by the induction of the ectoderm at the dorsal-ventral boundary by Fgf10 from the mesenchyme. This induction forms the apical ectodermal ridge (AER). The AER secretes Fgf8, which keeps the underlying mesenchyme proliferative and undifferentiated. This area of mesenchyme is called the progress zone.

3. Tbx5 induces the forelimb, whereas Tbx4 (chick) and Islet1 (mouse) induce hindlimb identity.

4. Two opposing gradients—one of FGFs and Wnts from the AER, the other of retinoic acid from the flank—pattern the proximal-distal axis of the chick limb.

5. As the limb grows outward, the stylopod forms first, then the zeugopod, and last the autopod. Each phase of limb development is characterized by a specific pattern of Hox gene expression.

6. Turing-type models suggest that a reaction-diffusion mechanism can explain the constant pattern of stylopod-zeugopod-autopod seen in tetrapod limbs.

7. The anterior-posterior axis is defined by the expression of Sonic hedgehog in the zone of polarizing activity, a region in the posterior mesoderm of the limb bud. If ZPA tissue (or Shh-secreting cells or beads) is placed in the anterior margin of a limb bud, a second, mirror-image pattern of Hox gene expression occurs, along with a corresponding mirror-image duplication of the digits.

8. The ZPA is maintained by the interaction of FGFs from the AER with mesenchyme made competent to express Sonic hedgehog by its expression of particular Hox genes. Sonic hedgehog acts in turn, probably in an indirect manner, to change the expression of the Hox genes in the limb bud.

9. Sonic hedgehog specifies digits in at least two ways. It works through BMP inhibition in the interdigital mesenchyme, and it also regulates the proliferation of digit cartilage.

10. The dorsal-ventral axis is formed in part by the expression of Wnt7a in the dorsal portion of the limb ectoderm. Wnt7a also maintains the expression level of Sonic hedgehog in the ZPA and of Fgf4 in the posterior AER. Fgf4 and Shh reciprocally maintain each other's expression.

11. Levels of FGFs in the AER can either support or inhibit the production of Shh by the ZPA. As the limb bud grows and more FGFs are produced in the AER, Shh expression is inhibited. This in turn causes the lowering of FGF levels, and eventually proximal-distal outgrowth ceases.

12. Cell death in the limb is mediated by BMPs and is necessary for the formation of digits and joints. Differences between the unwebbed chicken foot and the webbed duck foot can be explained by differences in the expression of Gremlin, a protein that antagonizes BMPs.

13. By modifying paracrine factor secretion, different limb morphologies can form, initiating the development of webbed feet, flippers, or hands. By eliminating the synthesis of certain paracrine factors, limbs can be prevented from forming (as in whales).

14. BMPs are involved both in inducing apoptosis and in differentiating the mesenchymal cells into cartilage. The regulation of BMP effects by Noggin and Gremlin proteins is critical in forming the joints between the bones of the limb and in regulating proximal-distal outgrowth.

Go to oup.com/he/barresi12xe for Further Developments, Scientists Speak interviews, Watch Development videos, Dev Tutorials, and complete bibliographic information for all literature cited in this chapter.

The Endoderm
Tubes and Organs for Digestion and Respiration

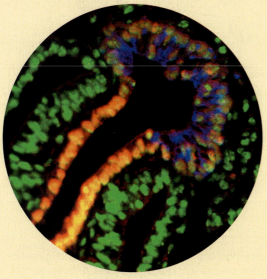

How do some gut cells become pancreas cells while the neighboring gut cells become liver or intestine?

Courtesy of Ken S. Zaret

THE ENDODERM FORMS THE EPITHELIUM OF THE GUT AND RESPIRATORY TUBES of the *adult* amniote, where it is essential for the exchange of gases and foods. In the *embryonic* amniote, whose food and oxygen come from the mother via the placenta, the endoderm's first major function is to induce the formation of several mesodermal organs. As we have seen in earlier chapters, the endoderm is critical for instructing the formation of the notochord, heart, blood vessels, and even the mesodermal germ layer. The endoderm's second embryonic function is to construct the linings of two tubes within the vertebrate body. The **digestive tube** extends the length of the body, and buds from the digestive tube form the liver, gallbladder, and pancreas. The **respiratory tube** forms as an outgrowth of the digestive tube and eventually bifurcates into the two lungs. The region of the digestive tube anterior to the point where the respiratory tube branches off is the **pharynx**. A third embryonic function is to form the epithelium of several glands. Epithelial outpockets of the pharynx give rise to the tonsils and to the thyroid, thymus, and parathyroid glands.

The endoderm arises from two sources. The main source is the set of cells that enters the interior of the embryo through the primitive streak during gastrulation. This is often called the **definitive endoderm**. It replaces the **visceral endoderm** that is primarily forming the yolk sac. However, not all the visceral endoderm is removed. Live-cell imaging studies using fluorescent markers show that definitive endoderm doesn't replace visceral endoderm as a sheet (**FIGURE 20.1**; Kwon et al. 2008; Viotti et al. 2014a). Rather, individual definitive endoderm cells can be seen intercalating into the visceral endoderm layer. The descendants of these epiblast cells remain in the embryonic region while most (but not all) of the original visceral endoderm becomes extraembryonic. The Sox17 transcription factor marks the endoderm in many species, and the *Sox17* gene appears to be activated in some cells as they leave the primitive streak. In mutants lacking *Sox17*, the definitive endoderm does not form (Viotti et al. 2014b).

The presence of Sox17 protein makes the definitive endoderm cells different from the mesodermal cells that also travel through the primitive streak; the mesodermal cells express the gene for the Brachyury transcription factor, and Brachyury appears to be critical for the development of the mesoderm. Whether *Sox17* or *Brachyury* is expressed appears to depend on the concentration of Nodal secreted from the visceral endoderm. High levels of Nodal induce *Sox17*, while BMPs and FGFs act against Nodal and specify the migrating cells to become mesoderm (**FIGURE 20.2**; Vincent et al. 2003; Dunn et al. 2004).

Primitive streak

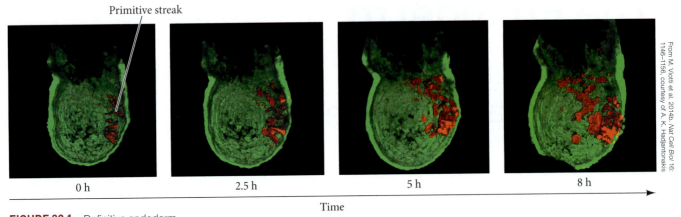

0 h   2.5 h   5 h   8 h

Time

From M. Viotti et al. 2014b. *Nat Cell Biol* 16: 1146–1156, courtesy of A. K. Hadjantonakis

FIGURE 20.1 Definitive endoderm cells from mouse epiblast replace the cells of the visceral endoderm. Here, the visceral endoderm of the embryo has been genetically marked with green fluorescent protein and can be seen enveloping the epiblast. Epiblast cells were randomly labeled with red dye and filmed as they migrated through the primitive streak and replaced the visceral endoderm cells.

The definitive endoderm of the gut is then defined by three regions, each with a distinct embryonic origin and delineated by its location along the anterior-posterior axis (Gordillo et al. 2015). As should be all too apparent by now, the A-P axis of vertebrates is specified by gradients of Wnts, FGFs, and BMPs, each of which is at the highest concentration posteriorly (see Figure 20.2). The endoderm near the head will form the anterior foregut cells, which will generate the precursors of the lung and thyroid glands. The endoderm in the posterior becomes a collection of midgut-hindgut precursor cells and will form the intestinal progenitor cells. The region between them—in the area of

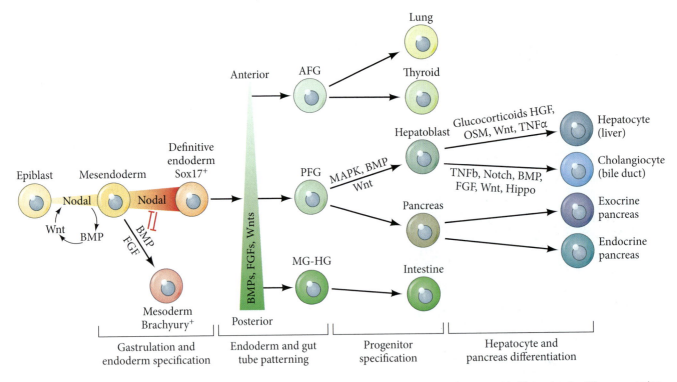

FIGURE 20.2 Signals mediating endoderm progenitor fates. Nodal signaling activates epiblast cells to assume a mesendodermal fate and migrate through the primitive streak. Those cells exposed to high Nodal concentrations then tend to become definitive endoderm cells. Those receiving BMPs and FGFs tend to become mesodermal. The later fate of the definitive endoderm depends to a large degree on where the cell resides along the anterior-posterior axis. Those in the posterior region are exposed to high levels of Wnts, BMPs, and FGFs and become midgut-hindgut (MG-HG) cells that generate the intestines. Cells exposed to somewhat lower levels of these paracrine factors give rise to the posterior foregut (PFG) cells, the precursors of the liver (hepatocytes and cholangiocytes) and pancreas (which will divide into exocrine and endocrine progenitors). Those definitive endoderm cells seeing very low levels of these paracrine factors generate the anterior foregut (AFG) cells that become the precursors of the lung and thyroid gland. Many lineages were left out for simplification. (After M. Gordillo et al. 2015. *Development* 142: 2094–2108.)

moderate BMPs, FGFs, and Wnts—becomes the posterior foregut precursors. These are the cells that give rise to the epithelium of the pancreas and the liver.

The gut cells initially form a flat sheet beneath the embryo (in the chick or human) or around the embryo (in the mouse). From the flat sheet of definitive endoderm, these cells form a tube. Mammalian gut tube development begins at two sites that migrate toward each other and fuse in the center (Lawson et al. 1986; Franklin et al. 2008). In the foregut, cells from the lateral portions of the anterior endoderm move ventrally to form the tube of the **anterior intestinal portal** (**AIP**); the **caudal intestinal portal** (**CIP**) forms from the posterior endoderm. The AIP and CIP migrate toward each other and come together to form the midgut (**FIGURE 20.3**).

The openings of the anterior and posterior ends of the gut tube are unique in that they are the only regions of the embryo where endoderm meets ectoderm. At first, the oral end is blocked by a region of endodermal cells that join to the mouth ectoderm (the **stomodeum**) at the **oral plate**. Eventually (at about 22 days in human embryos), the oral plate breaks, creating the oral opening of the digestive tube. The opening itself is lined by ectodermal cells. This arrangement creates an interesting situation, because the oral plate ectoderm is in contact with the brain ectoderm, which has curved around toward the ventral portion of the embryo. These two ectodermal regions interact with each other, with the roof of the oral region forming Rathke's pouch and becoming the glandular portion of the pituitary gland. The neural tissue on the floor of the diencephalon gives rise to the infundibulum, which becomes the neural portion of the pituitary. Thus, the pituitary gland has a dual origin, which is reflected in its adult functions. There is a similar meeting of endoderm and ectoderm at the anus; this is called the **anorectal junction**.

The Pharynx

The anterior endodermal portion of the digestive and respiratory tubes begins in the pharynx. Using a reporter gene (the above-mentioned *Sox17*) that becomes activated only in the endoderm, Rothova and colleagues (2012) found that there is a dividing line between ectoderm and endoderm in the mammalian mouth. In mammals, the epithelial layers of the teeth and the major salivary glands are from the ectoderm. The epithelia of the anterior taste buds are ectodermal (generated by cranial placodes; see Chapter 16), but the epithelia of the posterior taste buds, as well as some of the posterior salivary and mucus glands, are derived from the endoderm.

The embryonic pharynx in mammals contains four pairs of endoderm-derived **pharyngeal pouches**. Between these pouches are four **pharyngeal arches** (**FIGURE 20.4**). The first pair of pharyngeal pouches become the auditory cavities of the middle ear and the associated eustachian tubes. The second pair of pouches give rise to the walls of the tonsils. The thymus is derived from the third pair of pharyngeal pouches; the thymus will direct the differentiation of T lymphocytes during later stages of development. One pair of parathyroid glands is also derived from the third pair of pharyngeal pouches, while the other pair is derived from the fourth pair of pouches. In addition to these paired pouches, a small, central diverticulum is formed between the second pharyngeal pouches on the floor of the pharynx. This pocket of endoderm and mesenchyme will bud off from the pharynx and migrate down the neck to become the thyroid gland. The respiratory tube sprouts from the pharyngeal floor (between the fourth pair of pharyngeal pouches) to form the lungs, as we will soon see.

Many streams of cranial neural crest cells migrate into the pouches that are forming the thyroid, parathyroid, and thymus glands. Sonic hedgehog from the endoderm appears to act as a survival factor, preventing apoptosis of the neural crest cells (Moore-Scott and Manley 2005). In addition, genetic analysis combined with transplantation studies of zebrafish has shown that FGFs (mainly Fgf3 and Fgf8) from the ectoderm and mesoderm are important not only for the migration and survival of neural crest cells but also for the formation of the pouches themselves. Mice deficient in both *Fgf8* and *Fgf3* genes lack all the pharyngeal pouches, even when endoderm is present. Instead of migrating laterally and ventrally to form pouches, the endoderm remains in the anterior pharynx and does not spread out (Crump et al. 2004).

(?) Developing Questions

The gut tube is not symmetrical. It rotates in specific directions, placing the stomach near the heart and the appendix on the right-hand side. What molecular and cellular events cause gut looping?

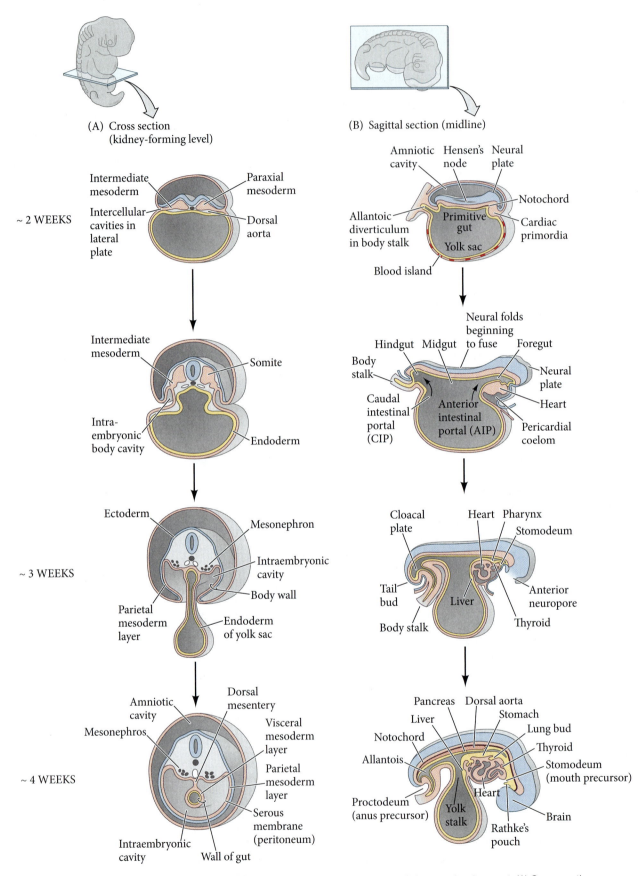

(A) Cross section
(kidney-forming level)

(B) Sagittal section (midline)

~ 2 WEEKS

Intermediate
mesoderm

Intercellular
cavities in
lateral
plate

Paraxial
mesoderm

Dorsal
aorta

Amniotic
cavity

Hensen's
node

Neural
plate

Notochord

Allantoic
diverticulum
in body stalk

Primitive
gut

Yolk sac

Cardiac
primordia

Blood island

Intermediate
mesoderm

Somite

Intra-
embryonic
body cavity

Endoderm

Neural folds
beginning
to fuse

Hindgut   Midgut

Body
stalk

Caudal
intestinal
portal
(CIP)

Anterior
intestinal
portal (AIP)

Foregut

Neural
plate

Heart

Pericardial
coelom

~ 3 WEEKS

Ectoderm

Parietal
mesoderm
layer

Mesonephron

Intraembryonic
cavity

Body wall

Endoderm
of yolk sac

Cloacal
plate

Tail
bud

Body stalk

Heart   Pharynx

Stomodeum

Liver

Anterior
neuropore

Thyroid

~ 4 WEEKS

Amniotic
cavity

Mesonephros

Intraembryonic
cavity

Dorsal
mesentery

Visceral
mesoderm
layer

Parietal
mesoderm
layer

Serous
membrane
(peritoneum)

Wall of gut

Pancreas

Liver

Notochord

Allantois

Proctodeum
(anus precursor)

Yolk
stalk

Dorsal aorta

Stomach

Lung bud

Thyroid

Stomodeum
(mouth precursor)

Heart

Brain

Rathke's
pouch

FIGURE 20.3 Endodermal folding during early human development. (A) Cross sections through the kidney-forming region. (B) Sagittal sections through the embryo's midline. (After T. W. Sadler. 2009. *Langman's Medical Embryology*. Lippincott, Williams & Wilkins: Hagerstown, MD.)

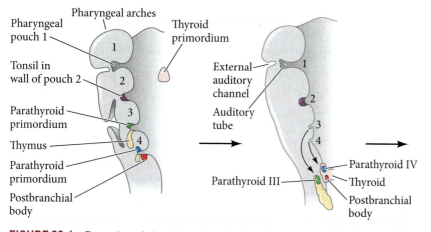

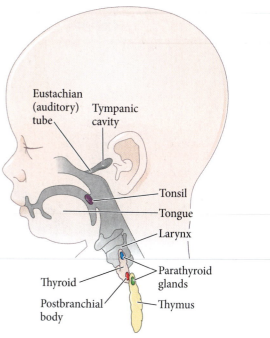

FIGURE 20.4 Formation of glandular primordia from the pharyngeal pouches. The pouches are symmetrical; only the right side is shown. The end of each of the first pharyngeal pouches becomes the tympanic cavity of the middle ear and the eustachian tube. The second pouches receive aggregates of lymphoid tissue and become the tonsils. The dorsal portion of the third pharyngeal pouches forms part of the parathyroid gland, while the ventral portion forms the thymus. Both migrate caudally and meet with the tissue from the fourth pharyngeal pouches to form the rest of the parathyroid and the postbranchial body. The thyroid, which had originated in the midline of the pharynx, also migrates caudally into the neck region. (After B. M. Carlson. 2014. *Human Embryology and Developmental Biology*, Fifth Edition, pp. 294–334. Elsevier/Saunders: Philadelphia.)

The Digestive Tube and Its Derivatives

During formation of the endodermal tube, mesenchyme cells from the splanchnic portion of the lateral plate mesoderm wrap around the endoderm (see Figure 20.3). The endodermal cells generate only the lining of the digestive tube and its glands, while mesenchyme cells from the splanchnic mesoderm surround the tube and will provide the future connective tissue and the smooth muscles that generate peristaltic contraction. Posterior to the pharynx, the digestive tube constricts to form the esophagus, which is followed in sequence by the stomach, small intestine, and large intestine.

As **FIGURE 20.5A** shows, the stomach develops as a dilated region of the gut close to the pharynx. The intestines develop more caudally, and the connection between the intestine and yolk sac is eventually severed. The intestine originally ends in the endodermal cloaca, but after the cloaca separates into the bladder and rectal regions (see Chapter 18), the intestine joins with the rectum. At the caudal end of the rectum, a depression forms where the endoderm meets the overlying ectoderm, and a thin **cloacal membrane** separates the two tissues. When the cloacal membrane eventually ruptures, the resulting opening becomes the anus.

Specification of the gut tissue

The production of endoderm is one of the first decisions made by the embryo, and the transcription factor Sox17 is critical in this specification. In amphibian embryos, endoderm is specified autonomously by the presence of Sox17. Dominant negative forms of Sox17 (having repressive instead of activator subunits) block endoderm formation in the vegetal blastomeres of amphibians, while the overexpression of the wild-type form expands the endodermal domain (Hudson et al. 1997; Henry and Melton 1998). Mice and zebrafish deficient in *Sox17* have defective gut endoderm, and when *Sox17* is expressed experimentally in embryonic stem cells, these stem cells produce endodermal derivatives (Kanai-Azuma et al. 2002; Takayama et al. 2011).

Although Sox17 helps specify the digestive tube, it does not give the tube its remarkable polarity. The digestive tube proceeds from the pharynx to the anus, differentiating along the way into the esophagus, stomach, duodenum, and intestines, and putting out branches that become (among other things) the epithelia of the thyroid, thymus,

(A)

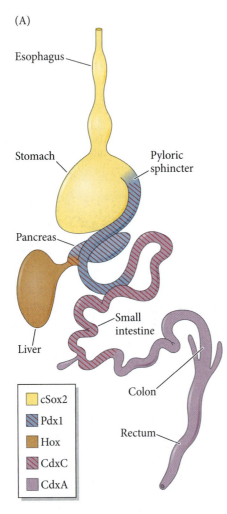

Esophagus

Stomach

Pyloric sphincter

Pancreas

Small intestine

Liver

Colon

Rectum

	cSox2
	Pdx1
	Hox
	CdxC
	CdxA

(B)

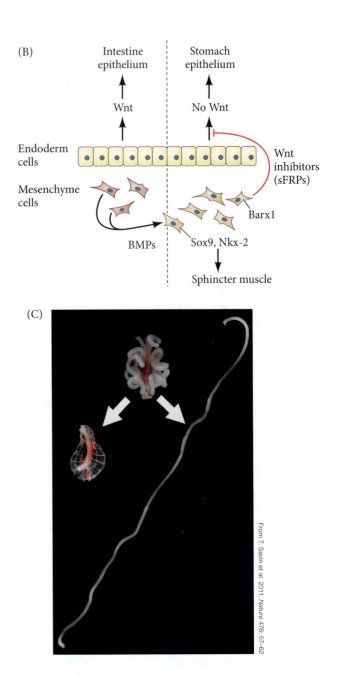

Intestine epithelium

Stomach epithelium

Wnt

No Wnt

Endoderm cells

Wnt inhibitors (sFRPs)

Mesenchyme cells

Barx1

BMPs

Sox9, Nkx-2

Sphincter muscle

(C)

From T. Savin et al. 2011. *Nature* 476: 57–62

FIGURE 20.5 Regional specification of the gut endoderm and splanchnic mesoderm through reciprocal interactions. (A) Regional transcription factors of the (mature) chick gut endoderm. (CdxA and C are the avian homologues of Cdx1 and 2.) These factors are seen prior to interactions with the mesoderm, but they are not stabilized. (B) Possible mechanism by which mesenchymal cells may induce endoderm to become either intestine or stomach, depending on the region. (C) Surgically separating the chick endodermal gut tube from the dorsal mesentery at embryonic day 12 causes the mesentery to shrivel up and the gut tube to straighten. The original gut-mesentery association (top) holds the digestive tube in place. When the two parts are separated on embryonic day 20, the mesentery (left) shrivels up, while the gut tube (right) straightens. (A after A. Grapin-Botton et al. 2001. *Genes Dev* 15: 444–454; B after B. M. Kim et al. 2005. *Dev Cell* 8: 611–622; C after T. Savin et al. 2011. *Nature* 476: 57–62.)

pancreas, and liver. What tells the endodermal tube to become these particular tissues at particular places? Why do we never see a mouth opening directly into a stomach? The endoderm and the splanchnic lateral plate mesoderm undergo a complicated set of interactions, and the signals for generating the different gut tissues appear to be conserved throughout the vertebrate classes (see Wallace and Pack 2003). One possible model for the polarity of the gut tube starts off with the specification of the pharynx and then follows with the specification of the remainder of the gut tube. Studies on chick embryos using beads containing either retinoic acid (RA) or inhibitors of its synthesis (Bayha et al. 2009) show that the pharynx can develop only in areas containing little or no RA, whereas the RA gradient patterns the pharyngeal arch endoderm in a graded manner. This is probably accomplished by activating and repressing particular sets of transcription factor genes.

The second phase of gut specification is thought to involve signals from the splanchnic mesoderm-derived mesenchyme surrounding the endodermal tube. As the digestive tube encounters different mesenchymes, mesenchyme cells instruct the endoderm to differentiate into esophagus, stomach, small intestine, and colon (Okada 1960; Gumpel-Pinot et al. 1978; Fukumachi and Takayama 1980; Kedinger et al. 1990).

FURTHER DEVELOPMENT

WNT AND MOLECULAR PATHWAYS REGULATING GUT TUBE DEVELOPMENT

Wnt signals are thought to be especially important in specifying regions of the gut tube. The initial ("default") specification of the entire gut tube is thought to be anterior (i.e., stomach/esophagus). However, graded Wnt signaling from the posterior mesoderm (instructed by RA and FGF gradients) provides a signal that induces in the gut endoderm the posteriorizing transcription factors Cdx1 and Cdx2 (see Figure 20.5A), as well as the paracrine factor Indian hedgehog. At high concentrations, the Cdx transcription factors induce formation of the large intestine, whereas at lower concentrations they induce formation of the small intestine. Indeed, when β-catenin is artificially expressed in the *foregut* tissue, the *Cdx2* gene is activated and the anterior endoderm tissue is transformed into the more posterior, intestinal type of tissue (Sherwood et al. 2011; Stringer et al. 2012).

The molecular pathways by which Wnt signals from the mesenchyme influence the gut tube are just becoming known (**FIGURE 20.5B**). Cdx2, for instance, suppresses genes such as *Hhex* and thereby prevents the stomach, liver, and pancreas from forming in the posterior (Bossard and Zaret 2000; McLin et al. 2007). In the anterior regions of the gut tube (which form the thymus, pancreas, stomach, and liver), Wnt signaling is blocked. In the stomach-forming domain, the mesenchyme lining the gut tube expresses the transcription factor Barx1, which activates production of two Frzb-like Wnt antagonists (the proteins sFRP1 and sFRP2) that block Wnt signaling in the vicinity of the stomach but not around the intestine. (Indeed, *Barx1*-deficient mice do not develop stomachs and express intestinal markers in that tissue; Kim et al. 2005.)

Wnt-based polarity may be transient and may need further interactions between the endoderm and the surrounding mesenchymes. These interactions may involve Shh being made by the endoderm and secreted at different concentrations at different sites, which functions to induce a nested pattern of differential Hox gene expression in the mesoderm surrounding the gut tube (Roberts et al. 1995, 1998). Evidence for the importance of Hox genes already exists for the hindgut region (Roberts et al. 1995; Yokouchi et al. 1995), where retroviral-mediated misexpression of Hox genes in the mesoderm alters the differentiation of the adjacent endoderm (Roberts et al. 1998). The Hox genes are thought to specify the mesoderm so that it can further interact with the endodermal tube and more finely specify its regions.

Once the boundaries of the transcription factors are established, differentiation can begin. The regional differentiation of the mesoderm (into smooth muscle types) and the regional differentiation of the endoderm (into different functional units, such as the stomach, duodenum, and small intestine) are synchronized. For instance, in certain regions the intestinal mesenchyme instructs more anterior mesoderm to become the smooth muscles of the pyloric sphincter rather than the non-sphincter smooth muscles of the stomach or intestine (see Figure 20.5B; Theodosiou and Tabin 2005).

The interaction between the splanchnic mesoderm and the endoderm continues long after the specification stage of development. One derivative of the splanchnic mesoderm is the **dorsal mesentery**, a fibrous membrane that connects the gut to the body wall. The looping of the intestinal tube is driven by a combination of growth intrinsic to the endoderm coupled with the connection of that tube to the dorsal mesentery (Savin et al. 2011). If the connection is severed, the mesentery shrinks and the gut becomes a long, thin tube with no folding (**FIGURE 20.5C**).

Interactions of mesoderm and endoderm are also important for the formation of villi in the gut. The differentiation of smooth muscle in the mesoderm constricts the underlying growing endoderm and mesenchyme, thus creating compressive stresses that cause the endoderm to buckle, which ultimately leads to the formation of villi (Shyer et al. 2013). This buckling localizes the intestinal stem cells at the base of the villi. Originally, all gut tube cells have the potential to become stem cells, but the buckling allows certain tissues to interact more readily than others and causes inhibitory paracrine factors (most importantly BMP4) to restrict stem cell formation to

? Developing Questions

One of the major human birth defects is pyloric stenosis, a condition in which the muscles of the pyloric sphincter thicken and prevent food from entering the intestine. How do these muscles arise and function, and what might cause the defect?

those regions farthest away from the tip of the villi (Shyer et al. 2015). (See **Further Development 20.1, Stem Cells of Endodermal Origin**, online.)

Accessory organs: The liver, pancreas, and gallbladder

Endoderm forms the lining of three accessory organs—the liver, pancreas, and gallbladder—that develop immediately caudal to the stomach. The **hepatic diverticulum** buds off endoderm and extends out from the foregut into the surrounding mesenchyme. The endoderm comprising this bud comes from two populations of cells: a lateral group that exclusively forms liver cells, and ventral-medial endoderm cells that form several midgut regions, including the liver (Tremblay and Zaret 2005). The mesenchyme induces this endoderm to proliferate, branch, and form the glandular epithelium of the liver. A portion of the hepatic diverticulum (the region closest to the digestive tube) continues to function as the drainage duct of the liver, and a branch from this duct produces the gallbladder (**FIGURE 20.6**). The pancreas develops from the fusion of distinct dorsal and ventral diverticula. As they grow, they come closer together and eventually fuse. In humans, only the ventral duct survives to carry digestive enzymes into the intestine. In other species (such as the dog), both dorsal and ventral ducts empty into the intestine.

The posterior foregut endoderm contains progenitor cells that can give rise to the pancreas, liver, and gallbladder. There is an intimate relationship between the splanchnic lateral plate mesoderm and the foregut endoderm. Just as the foregut endoderm is critical in specifying the cardiogenic mesoderm, the mesoderm, especially the blood-vessel endothelial cells, induce the endodermal tube to produce the liver primordium and the pancreatic rudiments.

The chromatin of the multipotent stem cells of the ventral foregut endoderm may be primed for their differential activation. The genes involved in forming the liver progenitor cells are silenced in a different manner than those genes involved in forming the pancreatic progenitor cells. Thus, a single signal may be able to de-repress an entire battery of specification genes (Xu et al. 2011; Zaret 2016).

LIVER FORMATION   The expression of liver-specific genes (such as those for α-fetoprotein and albumin) can occur in any region of the gut tube that is exposed to cardiogenic mesoderm. However, this induction can occur only if the notochord is removed. If the notochord is placed adjacent to the portion of the endoderm normally induced by cardiogenic mesoderm to become liver, the endoderm will not form liver (hepatic) tissue. Therefore, the developing heart appears to induce the liver to form, while the presence of the notochord inhibits liver formation (**FIGURE 20.7**). This induction is probably due to FGFs secreted by the developing heart and endothelial cells (Le Douarin 1975; Gualdi et al. 1996; Jung et al. 1999; Matsumoto et al. 2001). BMP (and possibly Wnt) signals from the lateral plate mesoderm are also needed for liver formation (Zhang et al. 2004; Ober et al. 2006). Thus, the heart and endothelial

FIGURE 20.6   Pancreatic development in humans. (A) At 30 days, the ventral pancreatic bud is close to the liver primordium. (B) By 35 days, it begins migrating posteriorly and (C) comes into contact with the dorsal pancreatic bud during the sixth week of development. (D) In most individuals, the dorsal pancreatic bud loses its duct into the duodenum; however, in about 10% of the population, the dual duct system persists. (After T. W. Sadler. 2018. *Langman's Medical Embryology.* Wolters Kluwer Health: Philadelphia.)

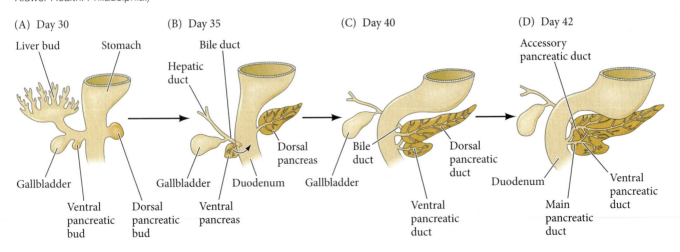

(A) Day 30 — Liver bud, Stomach, Gallbladder, Ventral pancreatic bud, Dorsal pancreatic bud

(B) Day 35 — Hepatic duct, Bile duct, Gallbladder, Duodenum, Ventral pancreas, Dorsal pancreas

(C) Day 40 — Bile duct, Gallbladder, Dorsal pancreatic duct, Ventral pancreatic duct

(D) Day 42 — Accessory pancreatic duct, Duodenum, Main pancreatic duct, Ventral pancreatic duct

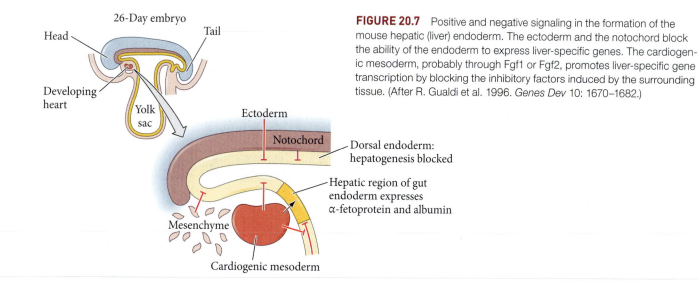

26-Day embryo

Head

Tail

Developing heart

Yolk sac

Ectoderm

Notochord

Dorsal endoderm: hepatogenesis blocked

Hepatic region of gut endoderm expresses α-fetoprotein and albumin

Mesenchyme

Cardiogenic mesoderm

FIGURE 20.7 Positive and negative signaling in the formation of the mouse hepatic (liver) endoderm. The ectoderm and the notochord block the ability of the endoderm to express liver-specific genes. The cardiogenic mesoderm, probably through Fgf1 or Fgf2, promotes liver-specific gene transcription by blocking the inhibitory factors induced by the surrounding tissue. (After R. Gualdi et al. 1996. *Genes Dev* 10: 1670–1682.)

cells have a developmental function in addition to their circulatory roles: they help induce the formation of the liver bud by secreting paracrine factors.

But in order to respond to the FGF signal, the endoderm has to become competent. This competence is given to the foregut endoderm by the Forkhead transcription factors (Foxa1 and Foxa2), which open the chromatin surrounding the liver-specific genes. Mouse embryos lacking the expression of these Forkhead transcription factors in their endoderm fail to produce a liver (Lee et al. 2005; Hirai et al. 2010; Parviz et al. 2003). Once the signal is given, other Forkhead transcription factors, such as HNF4α, become critical for the morphological and biochemical differentiation of the hepatic bud into liver tissue (Parviz et al. 2003). (See **Further Development 20.2, Differentiation of the Cells of the Liver**, online.)

PANCREAS FORMATION   The formation of the pancreas may be the flip side of liver formation. Whereas the heart cells promote and the notochord prevents liver formation, the notochord may actively promote pancreas formation and the heart cells block it. It seems this particular region of the digestive tube has the ability to become either pancreas or liver. One set of conditions (presence of heart, absence of notochord) induces the liver, while the opposite conditions (presence of notochord, absence of heart) cause the pancreas to form.

The notochord activates pancreas development by repressing *Shh* expression in the endoderm (Apelqvist et al. 1997; Hebrok et al. 1998). (This was a surprising finding, since the notochord is a source of Shh protein and an inducer of further *Shh* gene expression in ectodermal tissues.) Sonic hedgehog is expressed throughout the gut endoderm, *except* in the region that will form the pancreas. The notochord in this region secretes Fgf2 and activin, which are able to downregulate *Shh* expression. If *Shh* is experimentally expressed in this region, the tissue reverts to being intestinal (Jonnson et al. 1994; Ahlgren et al. 1996; Offield et al. 1996).

The lack of Shh in the pancreas-forming region of the gut seems to enable this region to respond to signals coming from the blood vessel endothelium. Indeed, pancreatic development is initiated at precisely those three locations where the foregut endoderm contacts the endothelium of the major blood vessels. It is at those points—where the endodermal tube meets the aorta and the vitelline veins—that the transcription factors Pdx1 and Ptf1a are expressed (**FIGURE 20.8A–C**; Lammert et al. 2001; Yoshitomi and Zaret 2004). If the blood vessels are removed from this area, the *Pdx1*- and *Ptf1a*-expressing regions fail to form and the pancreatic endoderm fails to bud. If more blood vessels form in this area, more of the endodermal tube becomes pancreatic tissues. (See **Further Development 20.3, Insulin-Secreting Pancreatic Cells**, and **Further Development 20.4, A Model for a Hierarchical Dichotomous System for Determining Cell Types**, both online.)

FIGURE 20.8 Induction of *Pdx1* gene expression in the gut epithelium. (A) In the chick embryo, *Pdx1* (purple) is expressed in the gut tube and is induced by contact with the aorta and vitelline veins. The regions of *Pdx1* gene expression create the dorsal and ventral rudiments of the pancreas. (B) In the mouse embryo, only the right vitelline vein survives, and it contacts the gut endothelium. *Pdx1* gene expression is seen only on this side, and only one ventral pancreatic bud emerges. (C) In situ hybridization of *Pdx1* mRNA in a section through the region of contact between the blood vessels and the gut tube of a mouse embryo. The regions of *Pdx1* expression show as deep blue. (D) Blood vessels (stained red) direct islets (stained green with antibodies to insulin) of chick embryo to differentiate. The nuclei are stained deep blue. (A–C after E. Lammert et al. 2001. *Science* 294: 564–567, courtesy of D. Melton.)

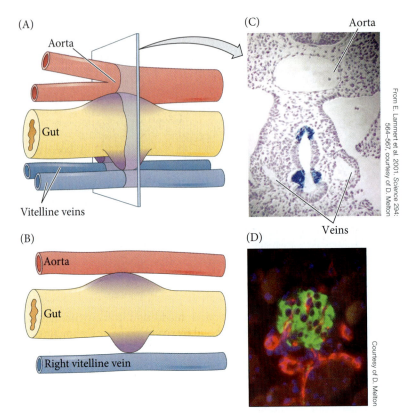

From E. Lammert et al. 2001. *Science* 294: 564–567, courtesy of D. Melton

Courtesy of D. Melton

FIGURE 20.9 Production of functional insulin-secreting human β cells from adult cells. The adult skin cell is converted into an induced pluripotent stem cell (iPSC) by the transcription factors mentioned in Chapter 5. The iPSC can become almost any cell in the embryo. To make the cell into a pancreatic β cell, researchers sequentially mimicked the conditions seen in the embryo. This meant providing it certain paracrine factors and paracrine factor inhibitors. The iPSC first became a cell type having the transcription factor pattern of primitive endoderm. Then it became sequentially, like a foregut cell, a pancreatic cell, a pancreatic endocrine cell, and finally (after some intermediary steps not shown here), a pancreatic β cell. These β cells could be transferred into mice, where they were able to regulate glucose levels and cure the mouse model of diabetes. (After F. W. Pagliuca et al. 2014. *Cell* 159: 428–439; A. Rezania et al. 2014. *Nat Biotechnol* 32: 1121–1133.)

GENERATING FUNCTIONAL PANCREATIC β CELLS   One of the most important potential medical applications of developmental biology is the conversion or replacement of missing or damaged cells with new functional cells. In Chapter 5, we discussed induced pluripotent stem cells (iPSCs). Human skin cells can be transformed into pluripotent stem cells by activating certain transcription factors that revert the adult cell to a condition similar to—perhaps identical with—the embryonic stem cells of the inner cell mass. Paracrine factors added at the right times and in the right amounts can replicate in vitro the conditions of the embryo and cause an iPSC to differentiate into a particular cell type. In 2014, two laboratories found the right sequence of conditions to induce the formation of functional, insulin-secreting pancreatic β cells (Pagliuca et al. 2014; Rezania et al. 2014). These β cells were able to cure diabetes in mice (**FIGURE 20.9**). (See **Further Development 20.5, Defining the Factors for β Cell Differentiation by iPSC**, online.)

THE GALLBLADDER   The origin of the gallbladder is not well characterized. In fact, fate mapping of the endoderm that gives rise to mouse gallbladder was accomplished

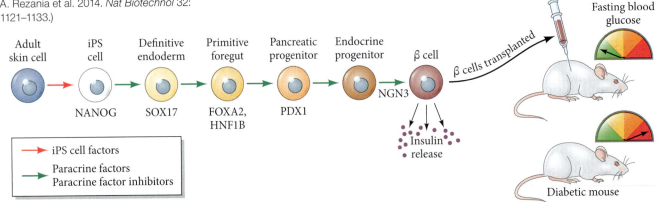

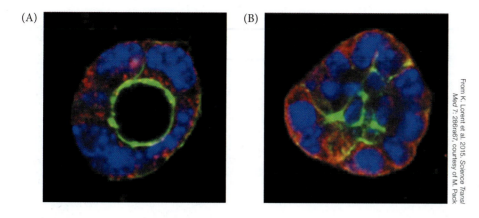

(A)   (B)

From K. Lorent et al. 2015. *Science Transl Med* 7: 286ra67, courtesy of M. Pack

FIGURE 20.10 Biliary atresia caused by plant compound. (A) Gallbladder cells in 3D culture form spheres with open lumen, resembling normal bile ducts. (B) In mice treated with the teratogenic compound biliatresone from pigweed, the polarity of the cells is altered and the lumen is occluded.

only in 2015. This showed that most of the gallbladder progenitors were located in the lateralmost region of the foregut endoderm, at the level corresponding to the junction of the first and second somites (Uemura et al. 2015).

One interesting finding about the development of the gallbladder is that some mammals (including some human infants) are born with the disease biliary atresia, where the bile ducts of the gallbladder are blocked. No one knows how this happens. However, an outbreak of biliary atresia in Australian livestock focused attention on this disease. An international group of scientists concluded that prolonged drought had resulted in expanded use of the pigweed plant as fodder, and that this plant contained a teratogenic toxin, biliatresone, that specifically interfered with gallbladder development (**FIGURE 20.10**). By screening thousands of compounds in zebrafish (which can be done relatively cheaply compared to other organisms), the scientists found that this particular compound occluded the developing bile ducts. This study showed that an environmental compound could cause this disease, and biologists are now searching for this compound in other plants.

The Respiratory Tube

Although they have no role in digestion, the lungs are in fact a derivative of the digestive tube. In the center of the pharyngeal floor, between the fourth pair of pharyngeal pouches, the **laryngotracheal groove** extends ventrally (**FIGURE 20.11A–C**). This groove then bifurcates into the branches that form the paired bronchi and lungs. The laryngotracheal endoderm becomes the lining of the trachea, the two bronchi, and the air sacs (alveoli) of the lungs. Sometimes this separation is not complete and a baby is born with a connection between the gut tube and the respiratory tube. This digestive and respiratory condition is called a **tracheal-esophageal fistula** and must be surgically repaired so that the baby can breathe and swallow properly.

The splitting of the trachea from the esophagus is another example of the interactions between the endoderm and specific mesenchyme. At this later point in development, the difference is between dorsal and ventral regions of the body. Wnt signals from the mesenchyme cause β-catenin to accumulate in the region of the gut tube that will become the lungs and trachea. Without these signals, the separation of the gut tube from the tracheal tube and the trachea's development into the lungs fails to happen (Goss et al. 2009). Conversely, extra lungs can form if β-catenin is expressed ectopically in the gut tube (Harris-Johnson et al. 2009).

The dorsal portion of the respiratory tube remains in contact with mesenchyme that contains the Barx1 transcription factor and is producing the Wnt-blocking soluble Frizzled-related proteins (sFRPs; similar to Frzb, see Chapter 11). sFRPs can bind to Wnts and keep them from reaching their cell membrane receptors, thus blocking Wnt activity and helping to specify the epithelium of the esophagus. The ventral portion of the respiratory tube, however, comes into contact with a mesenchyme that does not produce sFRPs. The Wnt signals, which were blocked earlier, here convert the tube into the ciliated respiratory epithelium of the trachea (**FIGURE 20.11D**; Woo et al. 2011).

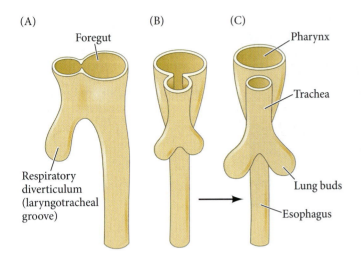

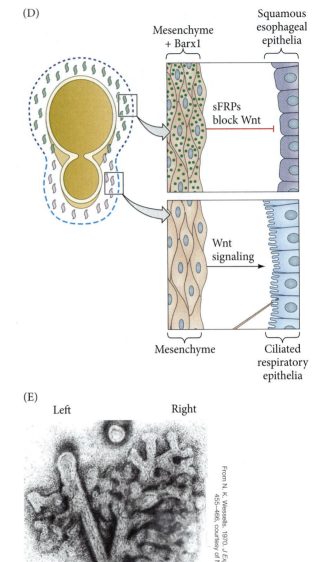

FIGURE 20.11 Partitioning of the foregut into the esophagus and respiratory diverticulum during the third and fourth weeks of human gestation. (A,B) Lateral and ventral views, end of week 3. (C) Ventral view, week 4. (D) Model for the roles of Wnt signaling and *Barx1*-expressing mesenchyme in differentiating the esophagus and the trachea. The absence of Barx1 protein and presence of Wnt signaling result in the expression of *Nkx2-1* and the differentiation of respiratory epithelium. When Barx1 is present, Wnt signaling is blocked and the Sox2 transcription factor is transcribed, helping to make the region esophageal. (E) After embryonic mouse lung epithelium had branched into two bronchi, the entire rudiment was excised and cultured. The right bronchus was left untouched while the tip of the left bronchus was covered with tracheal mesenchyme. The tip of the right bronchus formed the branches characteristic of the lung, whereas hardly any branching occurred in the left bronchus. (A–C after T. W. Sadler. 2018. *Langman's Medical Embryology.* Wolters Kluwer Health: Philadelphia; D after J. Woo et al. 2011. *PLOS One* 6: e22493.)

Epithelial-mesenchymal interactions and the biomechanics of branching in the lungs

As in the digestive tube, the regional specificity of the mesenchyme determines the differentiation of the developing respiratory tube. In the developing mammal, respiratory epithelium grows straight in the region of the neck to form the trachea. After entering the thorax, it branches, forming the two bronchi and then the lungs. The respiratory epithelium of an embryonic mouse can be isolated soon after it has split into two bronchi, and the two sides can be treated differently. **FIGURE 20.11E** shows the result when the right bronchial epithelium was allowed to retain its lung mesenchyme while the left bronchus was surrounded with tracheal mesenchyme (Wessells 1970). The right bronchus proliferated and branched under the influence of the lung mesenchyme, whereas the left bronchus continued to grow in an unbranched manner. Moreover, the differentiation of the respiratory epithelia into trachea cells or lung cells depends on the mesenchyme it encounters (Shannon et al. 1998).

You might think that the elaborate three-dimensional branching pattern of the mammalian lung is one of random design (**FIGURE 20.12A**). To the surprise of many, the

(?) Developing Questions

When a baby is born prematurely, its lung cells are often not differentiated. What can physicians do to accelerate lung development?

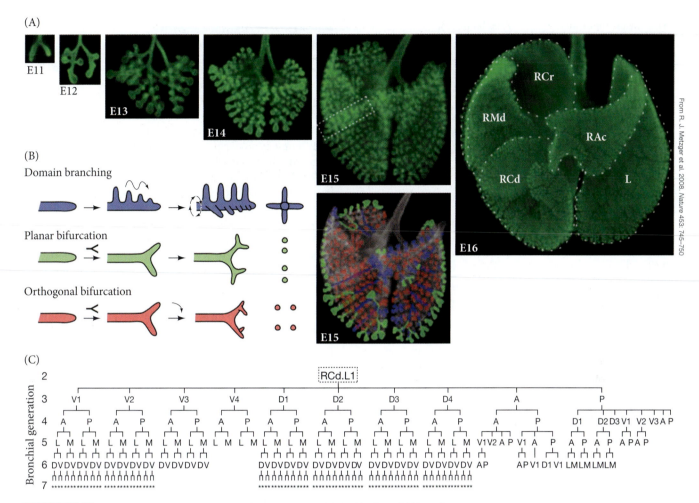

FIGURE 20.12 Morphogenesis of branches in the mouse bronchial epithelium. (A) Ventral views of the mouse lung at different embryonic days (E) visualized with antibody labeling to the E-Cadherin protein. Scale bar 500 μm. Each lobe at E16 is outlined, right cranial (RCr), right middle (RMd), accessory (RAc), right caudal (RCd), and left (L) lobes. (B) Illustration of the three different modes of bronchial epithelial branching. Lobes that display these different modes are pseudocolored in the E15 image of the mouse lung to the right. (C) The lineage tree for each branch of the right caudal (RCd), lateral 1 (L1) lobe (boxed in A). The orientation of each branch relative to its founding bronchi has been mapped, whether it's anterior (A), dorsal (D), lateral (L), medial (M), posterior (P), ventral (V), or varying (*) to the parental branch is indicated. (B,C after R. J. Metzger et al. 2008. *Nature* 453: 745–750.)

treelike branching pattern of the lung is highly stereotypical and follows three modes of differing geometric behaviors: domain branching, planar bifurcation, and orthogonal bifurcation (Metzger et al. 2008). Much like the teeth of a spiraling comb, *domain branching* forms rows, or domains, of successive outgrowths (lobes) along the proximal to distal axis of a parental epithelial tube (**FIGURE 20.12B**). At consistent locations, new rows of branches grow from a circumferentially shifted position about the dorsal, ventral, medial, or lateral aspects of the bronchus (the spiraling comb). Further branching can occur by *planar bifurcation*, such that the tips of each lobe of a given branch split at their midline, with each repeated branch point occurring along the same anterior to posterior plane. Later-forming branches similarly bifurcate at their tips, but in most a 90-degree rotation occurs between each branching event to create an alternating pattern of perpendicularly positioned branch points, a pattern known as *orthogonal bifurcation*. Remarkably, these three different modes are utilized with such precision during lung morphogenesis that the lineage of each alveolus can be traced back to the originating parental bronchus (**FIGURE 20.12C**; Metzger et al. 2008).

(A)

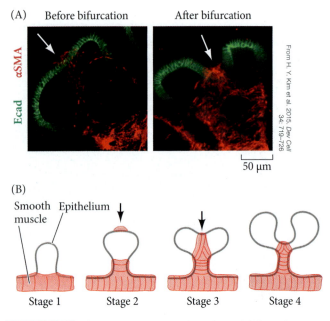

Before bifurcation   After bifurcation

αSMA

Ecad

50 μm

From H. Y. Kim et al. 2015. *Dev Cell* 34: 719–726

(B)

Smooth Epithelium
muscle

Stage 1   Stage 2   Stage 3   Stage 4

FIGURE 20.13 Smooth muscle may force branch bifurcation of the mouse lung. (A) Localized smooth muscle differentiation (αSMA, red) at the presumptive cleft of the lung epithelium (green, E-cadherin labeling) as well as after branch bifurcation (arrowheads). (B) Illustration of the time course of smooth muscle and lung epithelium interactions during branch bifurcation. (B after H. Y. Kim et al. 2015. *Dev Cell* 34: 719–726.)

FURTHER DEVELOPMENT

Different forces can make the same branch

"When you come to a fork in the road, take it." This comic quote by the late Yogi Berra may actually have some wisdom when it comes to lung development. It seems that the mouse and chick have decided to take very different approaches to forcing a fork in the road. In the mammalian lung, differentiation of smooth muscle is required by the distal tip mesenchyme of an outgrowing branch. This smooth muscle prefigures the emergence of a cleft within the branch, which soon connects with the surrounding smooth muscle cells around the proximal regions of the branch (**FIGURE 20.13**; Spurlin and Nelson 2017; Kim et al. 2015). Pharmacological disruption of this differentiation prevents terminal branching, suggesting that contraction of smooth muscle localized to the branch tip promotes branch bifurcation in the mouse. In the avian airway, however, branch points are built off of a parental branch by triggering localized apical constriction, which forces a new bud to form (**FIGURE 20.14**; Kim et al. 2013; Spurlin and Nelson 2017). Importantly, for both mouse and chick lung epithelium, bud outgrowth and branching develop in response to the focal expression of FGFs by the adjacent mesenchyme (see Figure 20.14).

To read about how lung maturation may induce labor, see **Further Development 20.6, Immune System Regulation of the Final Steps in Lung Development**, online.

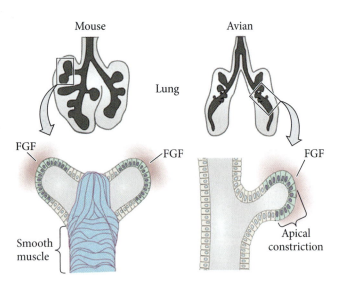

Mouse   Avian

Lung

FGF   FGF   FGF

Smooth
muscle

Apical
constriction

FIGURE 20.14 The mouse and avian airways form branches via different mechanisms. The mouse uses smooth muscle contraction to create a bifurcation in a lung branch bud, while the chick uses apical constriction to induce new branch points. Mesenchyme signaling by FGF family members at the sites of outgrowth is required for both biomechanical processes of branch morphogenesis in the mouse and avian lung. (After J. W. Spurlin and C. M. Nelson. 2017. *Philos Trans R Soc Lond B Biol Sci* 372: 20150527.)

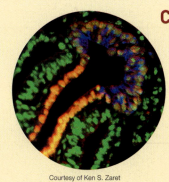

Closing Thoughts on the Opening Photo

The budding of the liver rudiment can be seen here in this 9-day mouse embryo. The nuclei are stained green, and the gut precursor cells are stained orange with an antibody to the FoxA2 transcription factor. The blue-staining cells are hepatoblasts and will become liver; these cells express a homeodomain-containing transcription factor, Hex, that changes the structure of the epithelial cells and enables them to proliferate into the mesenchyme.

Courtesy of Ken S. Zaret

Snapshot Summary
The Endoderm

1. The Sox17 transcription factor is crucial for the specification of the endoderm. In vertebrates, the endoderm constructs the digestive (gut) tube and the respiratory tube.

2. The gut tube is divided into three regions by the gradient of Wnts, BMPs, and FGFs along its anterior-posterior axis. The endoderm in the posterior becomes a collection of midgut-hindgut precursor cells forming the intestines. The endoderm near the head will form the anterior foregut cells, which will generate the precursors of the lung and thyroid glands. The region between them becomes the posterior foregut precursor cells, giving rise to the pancreas, the gallbladder, and the liver.

3. Four pairs of pharyngeal pouches become the endodermal lining of the eustachian tubes, the tonsils, the thymus, and the parathyroid glands. The thyroid also forms in this region of endoderm.

4. Gut tissues form by reciprocal interactions between the endoderm and the mesoderm. Wnt signals from the mesoderm and Sonic hedgehog from the endoderm appear to play a role in inducing a nested pattern of Hox gene expression in the mesoderm surrounding the gut. The regionalized mesoderm then instructs the endodermal tube to become the different organs of the digestive tract.

5. The two major cell types of the liver are the hepatocytes, which regulate body metabolism, and the cholangiocytes, which line the ducts.

6. The endoderm helps specify the splanchnic mesoderm; the splanchnic mesoderm, especially the heart and the blood vessels, helps specify the endoderm.

7. The pancreas forms in a region of endoderm that lacks *Shh* expression. The Pdx1 and Ptf1a transcription factors are expressed in this region.

8. By mimicking the conditions of embryonic development, human iPSCs can be transformed into precursors of pancreatic β cells and generate cells that secrete insulin.

9. The respiratory tube is derived as an outpocketing of the digestive tube. The regional specificity of the mesenchyme it meets determines whether the tube remains straight (as in the trachea) or branches (as in the bronchi and alveoli).

10. Three modes of branching morphogenesis exist in the developing lung: domain branching, planar bifurcation, and orthogonal bifurcation. Branching development in the respiratory bud is controlled by biomechanical forces generated by smooth muscle contraction (in mouse) or by apical constriction (in chick).

Go to oup.com/he/barresi12xe for Further Developments, Scientists Speak interviews, Watch Development videos, Dev Tutorials, and complete bibliographic information for all literature cited in this chapter.

Metamorphosis
The Hormonal Reactivation of Development

21

How might the larval foods help the survival of the adult form?

FEW EVENTS IN ANIMAL DEVELOPMENT are as spectacular as **metamorphosis**, the hormonal reactivation of developmental phenomena that gives the animal a new form. Animals (including humans) whose young are essentially smaller, less sexually mature versions of the adult are referred to as **direct developers**. Most animal species, however, are **indirect developers**, whose life cycle includes a larval stage with characteristics very different from those of the adult organism, which emerges only after a period of metamorphosis.

Metamorphosis is both a developmental and an ecological transition. Very often, larval forms are specialized for some function, such as growth or dispersal, whereas the adult is specialized for reproduction. *Cecropia* moths, for example, hatch from eggs and develop as wingless juveniles—caterpillars—for several months. After metamorphosis, the adult insects spend only a day or so as fully developed winged moths and must mate quickly before they die. The adult moths never eat, and in fact have no mouthparts during this brief reproductive phase of the life cycle. Metamorphosis is initiated by specific hormones that reactivate developmental processes throughout the entire organism, changing it morphologically, physiologically, and behaviorally to prepare itself for a new mode of existence. Ecologically, metamorphosis is associated with changes of habitat, food, and behaviors (Jacobs et al. 2006).

Among indirect developers, there are two major types of larvae. Larvae that represent dramatically different body plans than the adult form and that are morphologically distinct from the adult are called **primary larvae**. Sea urchin larvae, for instance, are bilaterally symmetrical organisms that float among and collect food in the plankton of the open ocean. The sea urchin *adult*, on the other hand, is pentameral (i.e., has fivefold symmetry) and feeds by scraping algae from rocks on the seafloor. There is no trace of the adult form in the body plan of the larva.[1]

Secondary larvae are found among those animals whose larvae and adults possess the same basic body plan. Thus, despite the obvious differences between the caterpillar and the butterfly, these two life stages retain the same major body axes and develop by deleting and modifying old parts while adding new structures into a preexisting framework. Similarly, the frog tadpole, although specialized for an aquatic environment, is a secondary larva, organized on the same pattern as the adult will be (Jagersten 1972; Raff and Raff 2009).

Metamorphosis is one of the most striking of developmental phenomena, and the extensive *morphological* changes undergone by some species have fascinated developmental anatomists for centuries

(Merian 1705; Swammerdam 1737). But we know only an outline of the *molecular* bases of metamorphosis, and only for a handful of species.

Amphibian Metamorphosis

Amphibians are actually named for their ability to undergo metamorphosis, their appellation coming from the Greek *amphi* ("double") and *bios* ("life"). Amphibian metamorphosis is associated with morphological changes that prepare an aquatic organism for a primarily terrestrial existence. In **urodeles** (salamanders), these changes include the resorption of the tail fin, the destruction of the external gills, and a change in skin structure. In **anurans** (frogs and toads), the changes are more dramatic, with almost every organ subject to modification (**TABLE 21.1**; see also Figure 1.7). The changes in amphibian metamorphosis are initiated by thyroid hormones such as **thyroxine (T_4)** and **tri-iodothyronine (T_3)** that travel through the blood to reach all the organs of the larva. When the larval organs encounter these thyroid hormones, they can respond in any of four ways: growth (as in the hindlimbs of the frog), death (as in the tail of the frog), remodeling (as in the frog's intestine), and respecification (as in the liver enzymes of the frog).

Morphological changes associated with amphibian metamorphosis

GROWTH OF NEW STRUCTURES The hormone tri-iodothyronine induces certain adult-specific organs to form. As seen in Chapter 1, the limbs of the adult frog emerge from specific sites on the metamorphosing tadpole. Similarly, in the eye, eyelids and the nictitating membranes (the so-called "third eyelid" in frogs) both emerge. Moreover, T_3 induces the proliferation and differentiation of new neurons to serve these organs. As the limbs grow out from the body axis, new neurons proliferate and differentiate in the spinal cord. These neurons send axons to the newly formed limb musculature (Marsh-Armstrong et al. 2004). Blocking T_3 activity prevents these neurons from forming and causes paralysis of the limbs.

One readily observed consequence of anuran metamorphosis is the movement of the eyes to the front of the head from their originally lateral position (**FIGURE 21.1A,B**).[2] The lateral eyes of the tadpole are typical of preyed-upon herbivores, whereas the frontally located eyes of the frog befit its more predatory lifestyle. To catch its prey, the frog needs

System	Larva	Adult
Locomotory	Aquatic; tail fins	Terrestrial; tailless tetrapod
Respiratory	Gills, skin, lungs; larval hemoglobins	Skin, lungs; adult hemoglobins
Circulatory	Aortic arches; aorta; anterior, posterior, and common jugular veins	Carotid arch; systemic arch; cardinal veins
Nutritional	Herbivorous; long spiral gut; intestinal symbionts; small mouth, horny jaws, labial teeth	Carnivorous; short gut; proteases, large mouth with long tongue
Nervous	Lack of nictitating membrane; porphyropsin, lateral line system, Mauthner neurons	Development of ocular muscles, nictitating membrane, tympanic membrane; rhodopsin; lateral line system lost, Mauthner neurons degenerate
Excretory	Largely ammonia, some urea (ammonotelic)	Largely urea; high activity of enzymes of ornithine-urea cycle (ureotelic)
Integument	Thin, bilayered epidermis with thin dermis; no mucous or granular glands	Stratified squamous epidermis with adult keratins; well-developed dermis contains mucous and granular glands secreting antimicrobial peptides

Source: Data from C. Turner and J. T. Bagnara. 1976. *General Endocrinology*. Saunders: Philadelphia; D. S. Reilly et al. 1994. *Dev Biol* 162: 123–133.

to see in three dimensions. That is, it has to acquire a *binocular field of vision*, where inputs from both eyes converge in the brain (see Figure 15.34B). In the tadpole, the right eye innervates the left side of the brain, and vice versa; there are no ipsilateral (same-side) projections of the retinal neurons. During metamorphosis, however, ipsilateral pathways emerge alongside the contralateral (opposite-side) pathways, enabling input from both eyes to reach the same area of the brain (Currie and Cowan 1974; Hoskins and Grobstein 1985a).

In *Xenopus*, these new pathways result not from the remodeling of existing neurons but from the formation of new neurons that differentiate in response to thyroid hormones (Hoskins and Grobstein 1985a,b). The ability of these axons to project ipsilaterally results from the induction of ephrin B in the optic chiasm by the thyroid hormones (Nakagawa et al. 2000). Ephrin B is also found in the optic chiasm of mammals (which have ipsilateral projections throughout life) but not in the chiasm of fish and birds (which have only contralateral projections). As we saw in Chapter 15, ephrins can repel certain neurons, causing them to project in one direction rather than another (**FIGURE 21.1C,D**).

CELL DEATH DURING METAMORPHOSIS The hormone T_3 also induces certain larval-specific structures to die. T_3 causes the degeneration of the paddlelike tail and the oxygen-procuring gills that were important for larval (but not adult) movement and respiration. While it is obvious that the tadpole's tail muscles and skin die, is this death murder or induced suicide? In other words, is T_3 telling the cells to kill themselves, or is T_3 telling something else to kill the cells? Recent evidence suggests that the first part of tail resorption is caused by suicide, but the last remnants of the tadpole tail must be killed off by other means. When tadpole muscle cells were injected with a dominant negative T_3 receptor (and therefore could not respond to T_3), the muscle cells survived, indicating that T_3 told them to kill themselves by apoptosis (Nakajima and Yaoita 2003; Nakajima et al. 2005). This was confirmed by the demonstration that the death of the tadpole muscle cells is prevented by blocking the activity of the apoptosis-inducing enzyme caspase-9 (Rowe et al. 2005). However, later in metamorphosis, the tail muscles are eaten by macrophages, perhaps because the extracellular matrix that supported the muscle cells has been digested by proteases.

Death also comes to the tadpole's red blood cells. During metamorphosis, tadpole hemoglobin is replaced by adult hemoglobin, which binds oxygen more slowly and releases it more rapidly (McCutcheon 1936; Riggs 1951). The red blood cells carrying the tadpole hemoglobin have a different shape than the adult red blood cells, and these larval red blood cells are specifically digested by macrophages in the liver and spleen after the adult red blood cells are made (Hasebe et al. 1999).

REMODELING DURING METAMORPHOSIS Among frogs and toads, certain larval structures are remodeled for adult needs. The larval intestine, with its numerous coils for digesting plant material, is converted into a shorter intestine for a carnivorous diet. Schrieber and his colleagues (2005) have demonstrated that the new cells of the adult intestine are derived from functioning cells of the larval intestine (instead of there being a subpopulation of stem cells that give rise to the adult intestine). As the extracellular matrix of the old intestine dissolves, most of the intestinal epithelial cells die. Those that survive appear to dedifferentiate and become intestinal stem cells (Stolow and Shi 1995; Ishizuya-Oka et al. 2001; Fu et al. 2005; Hasabe et al. 2013).

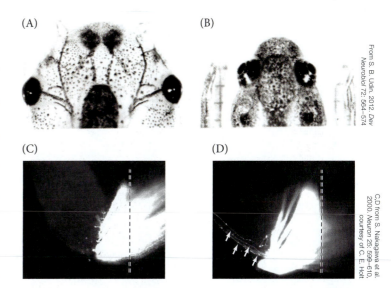

(A) (B) (C) (D)

From S. B. Udin, 2012. *Dev Neurobiol* 72: 564–574

C,D from S. Nakagawa et al. 2000. *Neuron* 25: 599–610, courtesy of C. E. Holt

FIGURE 21.1 Eye migration and associated neuronal changes during metamorphosis of the *Xenopus laevis* tadpole. (A) The eyes of the tadpole are laterally placed, so there is relatively little binocular field of vision. (B) The eyes migrate dorsally and rostrally during metamorphosis, creating a large binocular field for the adult frog. (C,D) Retinal projections of metamorphosing tadpole. The dye DiI was placed on a cut stump of the optic nerve to label the retinal projection. (C) In early and middle stages of metamorphosis, axons project across the midline (dashed line) from one side of the brain to the other. (D) In late metamorphosis, ephrin B is produced in the optic chiasm as certain neurons (arrows) are formed that project ipsilaterally.

(A) Tadpole

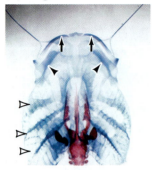

(B) Early metamorphosis

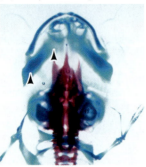

(C) Late metamorphosis

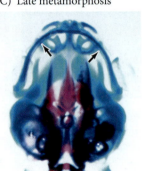

(D) Froglet

From D. L. Berry et al. 1998. *Dev Biol* 203: 24–35, courtesy of D. D. Brown

FIGURE 21.2 Changes in the *Xenopus* skull during metamorphosis. Whole mounts were stained with alcian blue to stain cartilage and alizarin red to stain bone. (A) Prior to metamorphosis, the pharyngeal (branchial) arch cartilage (open arrowheads) is prominent, Meckel's cartilage (arrows) is at the tip of the head, and the ceratohyal cartilage (arrowheads) is relatively wide and anteriorly placed. (B–D) As metamorphosis ensues, the pharyngeal arch cartilage disappears, Meckel's cartilage elongates, the mandible (lower jawbone) forms around Meckel's cartilage, and the ceratohyal cartilage narrows and becomes more posteriorly located.

Much of the nervous system is remodeled as neurons grow and innervate new targets. While some neurons (like those in the optic pathway) emerge, other larval neurons, such as certain motor neurons in the tadpole jaw, switch their allegiances from larval muscle to newly formed adult muscle (Alley and Barnes 1983). Still others, such as the cells innervating the tongue muscle (a newly formed muscle not present in the larva), lie dormant during the tadpole stage and form their first synapses during metamorphosis (Grobstein 1987). The lateral line system of the tadpole (which allows the tadpole to sense water movement and helps it hear) degenerates, and the ears undergo further differentiation (see Fritzsch et al. 1988). The middle ear develops, as does the tympanic membrane characteristic of frog and toad outer ears.[3] Thus, the anuran nervous system undergoes enormous restructuring as some neurons die, others are born, and others change their specificity.

The shape of the anuran skull also changes significantly as practically every structural component of the head is remodeled (Trueb and Hanken 1992; Berry et al. 1998). The most obvious change is that new bone is being made. The tadpole skull is primarily neural crest-derived cartilage; the adult skull is primarily neural crest-derived bone (**FIGURE 21.2**; Gross and Hanken 2005). As the lower jaw of the adult forms, Meckel's cartilage elongates to nearly double its original length, and dermal bone forms around it. While Meckel's cartilage is growing, the gills and pharyngeal arch cartilage (which were necessary for aquatic respiration in the tadpole) degenerate. Other cartilage is extensively remodeled. Thus, as in the nervous system, some skeletal elements proliferate, some die, and some are remodeled. The mechanisms by which one hormone signals differential effects in different and often adjacent tissues remain unknown. (See **Further Development 21.1: Biochemical Respecification in the Liver**, online.)

Hormonal control of amphibian metamorphosis

The control of metamorphosis by thyroid hormones was first demonstrated in 1912 by J. F. Gudernatsch, who discovered that tadpoles metamorphosed prematurely when fed powdered horse thyroid glands. In a complementary study, Bennet Allen (1916) found that when he removed or destroyed the thyroid rudiment of early tadpoles, the larvae never metamorphosed but instead grew into giant tadpoles. Subsequent studies showed that the sequential steps of anuran metamorphosis are regulated by increasing amounts of thyroid hormone (see Saxén et al. 1957; Kollros 1961; Hanken and Hall 1988). Some events (such as the development of limbs) occur early, when the concentration of thyroid hormones is low; other events (such as the resorption of the tail and remodeling of the intestine) occur later, after the hormones reach higher concentrations. These observations gave rise to a **threshold model**, wherein the different events of metamorphosis are triggered by different concentrations of thyroid hormones. Although the threshold model remains useful, molecular studies have shown that the timing of the events of amphibian metamorphosis is more complex than just increasing hormone concentrations.

The metamorphic changes of frog development are brought about by (1) the secretion of the hormone thyroxine (T_4) into the blood by the thyroid gland; (2) the conversion of T_4 into the more active hormone, tri-iodothyronine (T_3) by the target

FIGURE 21.3 Metabolism of thyroxine (T$_4$) and tri-iodothyronine (T$_3$). T$_4$ serves as a prohormone. It is converted in the peripheral tissues to the active hormone T$_3$ by deiodinase II. T$_3$ can be inactivated by deiodinase III, which converts T$_3$ into di-iodothyronine, which is not thought to induce metamorphosis.

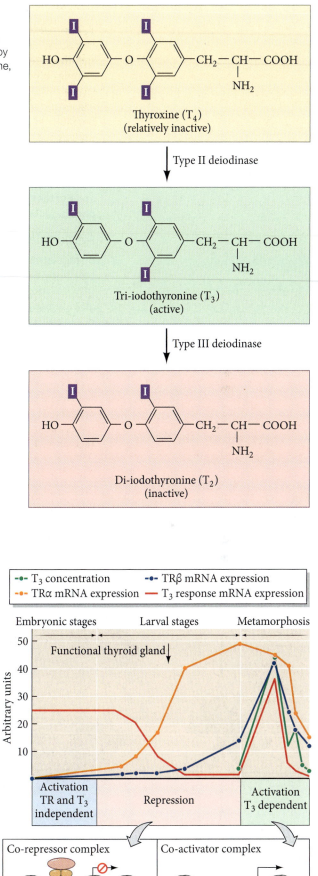

tissues; and (3) the degradation of T$_3$ in the target tissues (**FIGURE 21.3**). Once inside the cell, T$_3$ binds to the nuclear **thyroid hormone receptors** (**TRs**) with much higher affinity than does T$_4$ and causes these transcription factors to become transcriptional activators of gene expression. Thus, the levels of both T$_3$ and TRs in the target tissues are essential for producing the metamorphic response in each tissue (Kistler et al. 1977; Robinson et al. 1977; Becker et al. 1997).

The concentration of T$_3$ in each tissue is regulated by the concentration of T$_4$ in the blood and by two critical intracellular enzymes that remove iodine atoms from T$_4$ and T$_3$. **Type II deiodinase** removes an iodine atom from the outer ring of the precursor hormone (T$_4$) to convert it into the more active hormone T$_3$. **Type III deiodinase** removes an iodine atom from the inner ring of T$_3$ to convert it into an inactive compound (T$_2$) that will eventually be metabolized to tyrosine (Becker et al. 1997). Tadpoles that are genetically modified to overexpress type III deiodinase in their target tissues never complete metamorphosis (Huang et al. 1999); therefore, the regulation of metamorphosis involves tissue-specific regulation of the form of the hormone that binds most effectively to its receptor.

Thyroid hormone receptors are nuclear proteins, and there are two major types. In *Xenopus*, **thyroid hormone receptor α** (**TRα**) is widely distributed throughout all tissues and is present even before the organism has a thyroid gland. Yet, in an example of a positive feedback loop, the gene encoding **thyroid hormone receptor β** (**TRβ**) is itself directly activated by TRβ bound to thyroid hormone. TRβ levels are very low before the advent of metamorphosis; as the levels of thyroid hormone increase during metamorphosis, so do intracellular levels of TRβ (**FIGURE 21.4**; Yaoita and Brown 1990; Eliceiri and Brown 1994). As we will see, this positive regulation of hormone receptor gene expression by its own gene product is a common feature of metamorphosis across animal taxa.

The TRs do not work alone, but form dimers with the retinoid receptor RXR. TR-RXR dimers bind thyroid hormones and can then upregulate transcription (Mangelsdorf and Evans 1995; Wong and Shi 1995; Wolffe and Shi 1999). Importantly, the TR-RXR receptor complex is physically associated with appropriate promoters and enhancers even before it binds

FIGURE 21.4 The hormonal control of *Xenopus* metamorphosis. During premetamorphosis, thyroid hormone titers are low, and unliganded TRα binds to the chromatin and fixes transcriptional repressors that stabilize nucleosomes. During metamorphic climax, blood levels of thyroid hormones increase, and the TRα binds thyroxine. This causes the exchange of transcriptional repressors for transcriptional activators. The nucleosomes disperse, and the T$_3$-sensitive genes are activated. One of these genes encodes TRβ, which further accelerates the metamorphic responses. Eventually, feedback inhibition lowers the amount of circulating thyroid hormones, and metamorphosis ends. (After A. Grimaldi. 2013. *Biochim Biophys Acta* 1830: 3882–3892.)

T_3 (Grimaldi et al. 2012). In its unbound state, TR-RXR is a transcriptional *repressor*, recruiting histone deacetylases and other co-repressor proteins to its target genes and stabilizing repressive nucleosomes around the promoter. However, when the TR-RXR complex binds T_3, the repressors leave the complex and are replaced by co-activators such as histone acetyltransferase. These co-activators cause the dispersal of the nucleosomes and the activation of those same genes previously inhibited (Sachs et al. 2001; Buchholz et al. 2003; Grimaldi et al. 2013). Thus, the TRs have a dual function: when unliganded, they repress gene expression, preventing early metamorphosis; but when bound to T_3, they activate the expression of these same genes (see Figure 21.4). (See **Further Development 21.2, From Pre-Metamorphosis to Metamorphic Climax, Further Development 21.3, Differential Tissue Responses to Thyroid Hormones,** and **Further Development 21.4, Variations on the Themes of Amphibian Metamorphosis**, all online.)

Regionally specific developmental programs

By regulating the amount of T_3 and TRs in their cells, the different regions of the tadpole body can respond to thyroid hormones at different times. The type of response (proliferation, apoptosis, differentiation, migration) is determined by other factors already present in the different tissues. The same stimulus causes some tissues to degenerate while stimulating others to develop and differentiate, as exemplified by the process of tail degeneration: thyroid hormone instructs the limb bud muscles to grow (they die without thyroxine) while instructing the tail muscles to undergo apoptosis (Cai et al. 2007).

The resorption of the tail structures is relatively rapid, since the bony skeleton does not extend to the tail (Wassersug 1989). After apoptosis has taken place, macrophages collect in the tail region and digest the debris with their enzymes (especially collagenases and metalloproteinases), and the tail becomes a large sac of proteolytic enzymes (Kaltenbach et al. 1979; Oofusa and Yoshizato 1991; Patterson et al. 1995).[4] The tail epidermis acts differently than the head or trunk epidermis. During **metamorphic climax**, the larval skin is instructed to undergo apoptosis. The tadpole head and body are able to generate new epidermis from epithelial stem cells. The tail epidermis, however, lacks these stem cells and fails to generate new skin (Suzuki et al. 2002).

Organ-specific responses to thyroid hormones have been dramatically demonstrated by transplanting a tail tip to the trunk region and placing an eye cup in the tail (Schwind 1933; Geigy 1941). Tail-tip tissue placed in the trunk is not protected from degeneration, but the eye cup retains its integrity even when it lies within the degenerating tail (**FIGURE 21.5**). Thus, the way a tissue responds to the thyroid hormone is inherent in the tissue itself; it is not dependent on its position within the larva.

The metamorphosis of tadpoles into frogs is one of the most rapid and accessible examples of development, obvious even to the eyes of children. Yet it still presents an enormous set of enigmas. As Don Brown and Liquan Cai (Cai et al. 2007) have asked, "What will encourage the modern generation of scientists to study the wonderful biological problems presented by amphibian metamorphosis?" Recent work has shown the importance of metamorphosis for studying regeneration, and it is also a critical area in which development and ecology have a marked impact on each other.

(A)

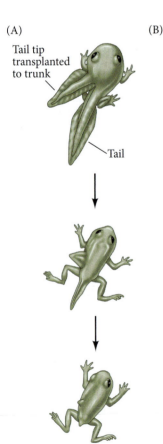

Tail tip transplanted to trunk

Tail

(B)

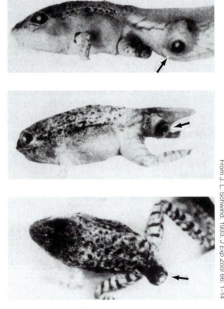

From J. L. Schwind. 1933. *J Exp Zool* 66: 1–14

FIGURE 21.5 Regional specificity during frog metamorphosis. (A) Tail tips regress even when transplanted into the trunk. (B) Eye cups remain intact even when transplanted into the regressing tail. (A, after R. Geigy. 1941. *Rev Suisse Zool* 48: 483–494.)

Metamorphosis in Insects

Insects are the most speciose of Earth's animals, and the diversity of their life cycles makes science fiction pale by comparison. There are three major patterns of insect development. A few insects, such as springtails, have no larval stage and undergo direct, or **ametabolous**, development (**FIGURE 21.6A**). Immediately after hatching, ametabolous insects have a **pronymph** stage bearing the structures that enabled it to get out of the egg. But after this transitory stage, the insect looks like a small adult; it grows larger after each molt with a new cuticle, but is unchanged in form (Truman and Riddiford 1999).

Other insects, notably grasshoppers and bugs, undergo a gradual, or **hemimetabolous**, metamorphosis (**FIGURE 21.6B**). After spending a very brief period of time as a pronymph (whose cuticle is often shed as the insect hatches), the insect looks like an immature adult and is called a **nymph**. The rudiments of the wings, genital organs, and other adult structures are present and become progressively more mature with each molt. At the final molt, the emerging insect is a winged and sexually mature adult, or **imago**.

In the **holometabolous** development of insects such as flies, beetles, moths, and butterflies, there is no pronymph stage (**FIGURE 21.6C**). The juvenile form that hatches from the egg is called a **larva**. The larva (a caterpillar, grub, or maggot) undergoes a series of molts as it becomes larger. The stages between these larval molts are called **instars**. The number of larval molts before becoming an adult is characteristic of a species, although environmental factors can increase or decrease the number. The larval instars grow in a stepwise fashion, each instar being larger than the previous one. Finally, there is a

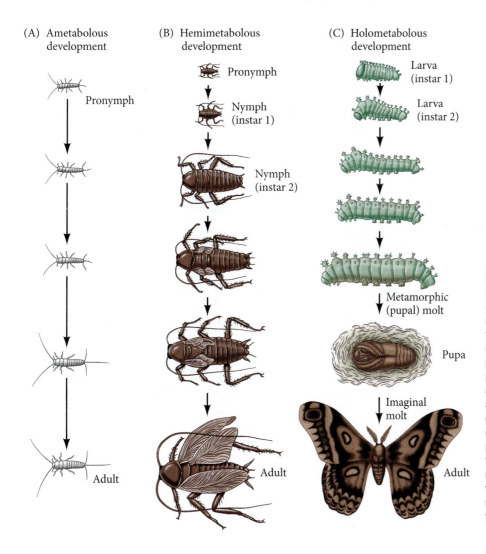

(A) Ametabolous development

Pronymph

Adult

(B) Hemimetabolous development

Pronymph

Nymph (instar 1)

Nymph (instar 2)

Adult

(C) Holometabolous development

Larva (instar 1)

Larva (instar 2)

Metamorphic (pupal) molt

Pupa

Imaginal molt

Adult

FIGURE 21.6 Modes of insect development. Molts are represented as arrows. (A) Ametabolous (direct) development in a silverfish. After a brief pronymph stage, the insect looks like a small adult. (B) Hemimetabolous (gradual) metamorphosis in a cockroach. After a very brief pronymph phase, the insect becomes a nymph. After each molt, the next nymphal instar looks more like an adult, gradually growing wings and genital organs. (C) Holometabolous (complete) metamorphosis in a moth. After hatching as a larva, the insect undergoes successive larval molts until a metamorphic molt causes it to enter the pupal stage. Then an imaginal molt turns it into an adult that ecloses from the pupal case with a new cuticle.

dramatic and sudden transformation between the larval and adult stages: after the final instar, the larva undergoes a **metamorphic molt** to become a **pupa**. The pupa does not feed, and its energy must come from those foods it ingested as a larva. During pupation, adult structures form and replace the larval structures. Eventually, an **imaginal molt** forms the adult (imago) cuticle beneath the pupal cuticle, and then the adult later emerges from the pupal case at adult eclosion. While the larva is said to hatch from an egg, the imago is said to eclose from the pupa. Carroll Williams (1959) characterized holometabolous metamorphosis as the switch between foraging and reproduction: "The earth-bound early stages built enormous digestive tracts and hauled them around on caterpillar treads. Later in the life-history these assets could be liquidated and reinvested in the construction of an entirely new organism—a flying-machine devoted to sex."

Imaginal discs

In holometabolous insects, the transformation from juvenile into adult occurs within the pupal cuticle. Most of the larval body is systematically destroyed by programmed cell death, while new adult organs develop from relatively undifferentiated nests of **imaginal cells**. Thus, within any larva there are two distinct populations of cells: the larval cells, which are used for the functions of the juvenile insect; and thousands of imaginal cells, which lie within the larva in clusters, awaiting the signal to differentiate.

There are three main types of imaginal cells (**FIGURE 21.7**):

1. The cells of **imaginal discs** will form the cuticular structures of the adult, including the wings, legs, antennae, eyes, head, thorax, and genitalia.

2. **Histoblasts** (tissue-forming cells) are imaginal cells that will form the adult abdomen.

3. Clusters of imaginal cells within each organ will proliferate to form the adult organ as the larval organ degenerates.

In the newly hatched larva, the imaginal discs are visible as local thickenings of the epidermis. Each disc in the early *Drosophila* larva has about 10–50 cells, and there are 19 such discs in these flies. The epidermis of the head, thorax, and limbs comes from

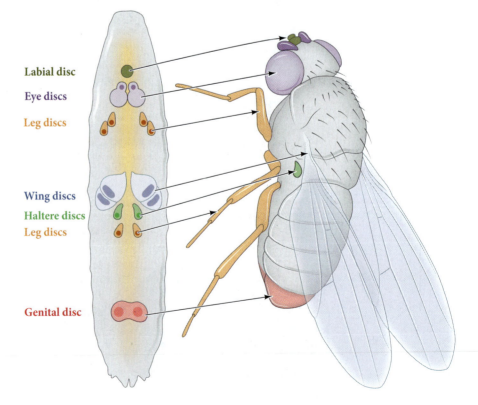

Labial disc

Eye discs

Leg discs

Wing discs

Haltere discs

Leg discs

Genital disc

FIGURE 21.7 Locations and developmental fates of imaginal discs and imaginal tissues in the third instar larva (left) of *Drosophila melanogaster*. (After J. S. Jaszczak and A. Halme. 2016. *Curr Opin Genet Dev* 40: 87–94.)

(A)

A and B from J. W. Fristrom et al. 1977.
Am Zool 17: 671–684.; courtesy of D. Fristrom

FIGURE 21.8 Imaginal disc elongation. Scanning electron micrograph of *Drosophila* third instar leg disc (A) before and (B) after elongation.

(B)

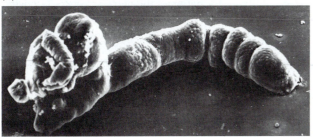

nine bilateral pairs of discs, whereas the epidermis of the genitalia is derived from a single disc at the midline.

Whereas most larval cells have a very limited mitotic capacity, imaginal discs divide rapidly at specific characteristic times. As their cells proliferate, the discs form a tubular epithelium that folds in on itself in a compact spiral (**FIGURE 21.8A**). At metamorphosis, these cells proliferate even further as they differentiate and elongate (**FIGURE 21.8B**). The fate map and elongation sequence of one of the six *Drosophila* leg discs is shown in **FIGURE 21.9**. At the end of the third instar, just before pupation, the leg disc is an epithelial sac connected by a thin stalk to the larval epidermis. On one side of the sac, the epithelium is coiled into a series of concentric folds "reminiscent of a Danish pastry" (Kalm et al. 1995). As pupation begins, the cells at the center of the disc telescope

FIGURE 21.9 Sequence of leg imaginal disc development in *Drosophila*. Specification of the disc type occurs within the embryo. Proliferation of the disc cells and specification as to the type of leg cell that each will produce are accomplished in the larval stages. Elongation of the disc takes place in the early pupal ("prepupa") stage, and differentiation of the leg tissues occurs while the insect is a pupa. T_1, basitarsus; T_{2-5}, tarsal segments 2–5. (After D. Fristrom and J. W. Fristrom. 1975. *Dev Biol* 43: 1–23; L. von Kalm et al. 1995. *BioEssays* 17: 693–702.)

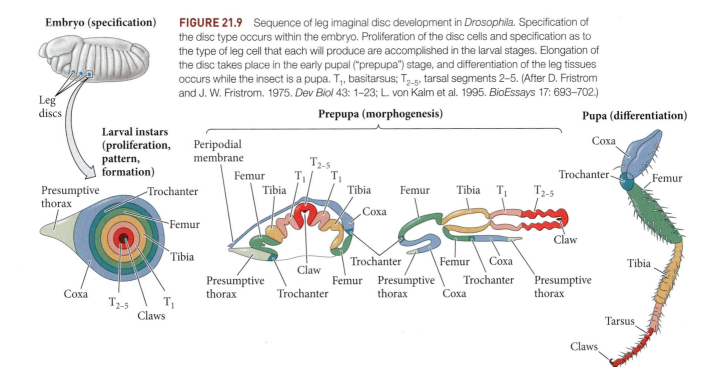

out to become the most distal portions of the leg—the claws and the tarsus. The outer cells become the proximal structures—the coxa and the adjoining epidermis (Schubiger 1968). After differentiating, the cells of the appendages and epidermis secrete a cuticle appropriate for each specific region. Although the disc is composed primarily of epidermal cells, a small number of **adepithelial cells** migrate into the disc early in development. During the pupal stage, these cells give rise to the muscles and nerves that serve the legs. (See **Further Development 21.5, Specification and Proliferation**, online.)

EVERSION AND DIFFERENTIATION The mature leg disc in the third instar of *Drosophila* does not look anything like the adult structure. It is determined but not yet differentiated; its differentiation requires a signal, in the form of a set of pulses of the "molting" hormone 20-hydroxyecdysone (20E; see Figure 21.10A). The first pulse, occurring in the late larval stages, initiates formation of the pupa, arrests cell division in the disc, and initiates the cell shape changes that drive the **eversion** of the leg. The cells of early third instar discs are tightly arranged along the proximal-distal axis. When the hormonal signal to differentiate is given, the cells change their shape and the leg is everted, the central cells of the disc becoming the most distal (claw) cells of the limb (Condic et al. 1991; Taylor and Adler 2008). The leg structures will differentiate within the pupa, so that by the time the adult fly ecloses they will be fully formed and functional: in fact, the adult uses its legs in its final escape from the pupal case. (See **Further Development 21.6, Insect Metamorphosis**, and **Further Development 21.7, Parasitoid Wasp Development**, both online.)

Hormonal control of insect metamorphosis

Although the details of insect metamorphosis differ among species, the general pattern of hormonal action is very similar. Like amphibian metamorphosis, the metamorphosis of insects is regulated by systemic hormonal signals, which are controlled by neurohormones from the brain (for reviews, see Gilbert and Goodman 1981; Riddiford 1996). Insect molting and metamorphosis are controlled by two effector hormones: the steroid 20-hydroxyecdysone (20E) and the lipid juvenile hormone (JH) (**FIGURE 21.10A**). 20E initiates and coordinates each molt (whether larva-to-larva, larva-to-pupa, or pupa-to-adult) and regulates the changes in gene expression that occur during metamorphosis. High levels of JH prevent the ecdysone-induced changes in gene expression that are necessary for metamorphosis. Thus, the presence of JH during a larval molt ensures that the result of that molt is another larval instar, not a pupa or an adult. When the concentration of JH becomes low enough, the 20E-induced molt produces a pupa instead of a larva. When 20E acts in the absence of JH, the imaginal discs differentiate and the molt gives rise to an adult (**FIGURE 21.10B**).

The molting process is initiated in the brain, by the release of **prothoracicotropic hormone** (**PTTH**) in response to neural, hormonal, or environmental signals (see Figure 21.10B). This peptide hormone stimulates the production of **ecdysone** by the **prothoracic gland** by activating the RTK (receptor tyrosine kinase) pathway in those cells (Rewitz et al. 2009; Ou et al. 2011). Ecdysone is then modified in peripheral tissues to become the active molting hormone 20E. Each molt is initiated by one or more pulses of 20E. For a larval molt, the first pulse produces a small rise in the 20E concentration in the larval hemolymph (blood) and elicits a change in cellular commitment in the epidermis. A second, larger pulse of 20E initiates the differentiation events associated with molting. These pulses of 20E commit and stimulate the epidermal cells to synthesize enzymes that digest the old cuticle and synthesize a new one.

⟨?⟩ Developing Questions

How might diseases such as malaria be controlled by altering insect metamorphosis?

FURTHER DEVELOPMENT

TITRATING JUVENILE HORMONE FOR EACH TRANSITION Larval-to-larval molts are produced when there are large circulating titers of JH. Juvenile hormone is secreted by the **corpora allata**. The secretory cells of the corpora allata are active during larval molts but inactive during the metamorphic molt and the imaginal molt. In the last larval instar, the level of JH drops through two mechanisms: the medial nerve from the brain to the corpora allata inhibits these glands from

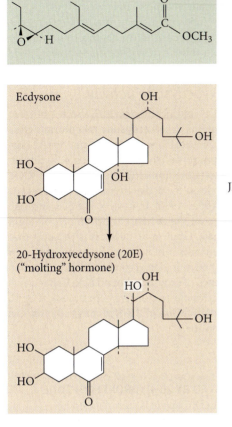

(A) Juvenile hormone (JH)

Ecdysone

20-Hydroxyecdysone (20E) ("molting" hormone)

FIGURE 21.10 Regulation of insect metamorphosis. (A) Structures of juvenile hormone (JH), ecdysone, and the active molting hormone 20-hydroxyecdysone (20E). (B) General pathway of insect metamorphosis. 20E and JH together cause molts that form the next larval instar. When the concentration of JH becomes low enough, the 20E-induced molt produces a pupa instead of a larva. When 20E acts in the absence of JH, the imaginal discs differentiate and the molt gives rise to an adult (imago). (A after L. I. Gilbert and W. Goodman. 1981. In *Metamorphosis: A Problem in Developmental Biology*, L. I. Gilbert and E. Frieden [Eds.], pp. 139–176. Plenum: New York; B after L. I. Gilbert et al. 1980. *Recent Prog Horm Res* 36: 401.)

(B) Brain — Neurosecretory cells of prothoracic gland

Juvenile hormone

Corpora allata

Prothoracicotropic hormone (PTTH)

JH receptor (JHR)

Prothoracic gland

JH-JHR

"Differentiating signal"

"Molting signal"

Ecdysone

20E

Cuticle

Chromosomes
RNA (L)
Protein synthesis
Larval structures

Chromosomes
RNA (P)
Protein synthesis
Pupal structures

Chromosomes
RNA (A)
Protein synthesis
Adult structures

Larva

Pupa

Adult

producing JH, and there is a simultaneous increase in the body's ability to degrade existing JH (Safranek and Williams 1989). This triggers the release of PTTH from the brain (Nijhout and Williams 1974; Rountree and Bollenbacher 1986). PTTH, in turn, stimulates the prothoracic gland to secrete a small amount of ecdysone. The resulting pulse of 20E, in the absence of high levels of JH, commits the epidermal cells to pupal development. Larva-specific mRNAs are not replaced, and new mRNAs are synthesized whose protein products inhibit the transcription of the larval messages.

There are two major pulses of 20E during *Drosophila* metamorphosis. A pulse in the third instar larva triggers the "prepupal" morphogenesis of the leg and wing imaginal discs, as well as the death of the larval hindgut. The larva stops eating and migrates to find a site to begin pupation. A second pulse 10–12 hours later tells the prepupa to become a pupa. The head inverts and the salivary glands degenerate (Riddiford 1982; Nijhout 1994). Hence, the first pulse of 20E during the last larval instar inactivates larva-specific genes and initiates the morphogenesis of imaginal disc structures. The second pulse transcribes pupa-specific genes and initiates the molt (Nijhout 1994). At the imaginal molt, when 20E acts in the absence of juvenile hormone, the imaginal discs fully differentiate and the molt gives rise to an adult.

The molecular biology of 20-hydroxyecdysone activity

ECDYSONE RECEPTORS Like amphibian thyroid hormones, 20E cannot bind to DNA by itself. It must first bind to nuclear proteins called **ecdysone receptors** (**EcRs**). EcRs are evolutionarily related to, and almost identical in structure to, the thyroid hormone receptors of amphibians. An EcR protein forms an active molecule by dimerizing with an Ultraspiracle (Usp) protein. Usp is the homologue of amphibian RXR, which we learned earlier dimerizes with TR to form the active thyroid hormone receptor (Koelle et al. 1991; Yao et al. 1992; Thomas et al. 1993). In insects, the EcR and Usp proteins attach to the DNA and then dimerize on the enhancer or promoter element of the ecdysone-responsive genes (Szamborska-Gbur et al. 2014). It is thought that in the absence of hormone-bound EcR, Usp recruits inhibitors of the transcription of ecdysone-responsive genes (Tsai et al. 1999). This inhibition is converted into activation when ecdysone binds to its receptor. The presence of ecdysone-bound EcR-Usp recruits histone methyltransferases that *activate* the ecdysone-responsive genes (Sedkov et al. 2003). (See **Further Development 21.8, Identification of 20-Hydroxyecdysone as a Metamorphic Transcriptional Regulator**, online.)

FURTHER DEVELOPMENT

THE DEVELOPMENTAL CASCADES INITIATED BY 20-HYDROXYECDYSONE
A simplified framework of metamorphosis in *Drosophila* is summarized in **FIGURE 21.11**. First, the "molting" hormone 20-hydroxyecdysone (20E) binds to the EcR/Usp receptor complex, which activates the "early response genes," including *E74* and *E75* (the puffs in Figure 2 in Further Development 21.8, online), as well as *Broad* and the *EcR* gene itself. The transcription factors encoded by

FIGURE 21.11 20-hydroxyecdysone initiates developmental cascades. (A) Schematic of the major gene expression cascade in *Drosophila* metamorphosis. When 20E binds to the EcR/Usp receptor complex, it activates the early response genes, including *E74*, *E75*, and *Broad*. Their products activate the "late genes." The activated EcR/Usp complex also activates a series of genes whose products are transcription factors and which activate the *βFTZ-F1* gene. The βFTZ-F1 protein modifies the chromatin so that the next 20E pulse activates a different set of late genes. The products of these genes also inhibit the early-expressed genes, including those for the EcR receptor. (After K. King-Jones et al. 2005. *Cell* 121: 773–784.)

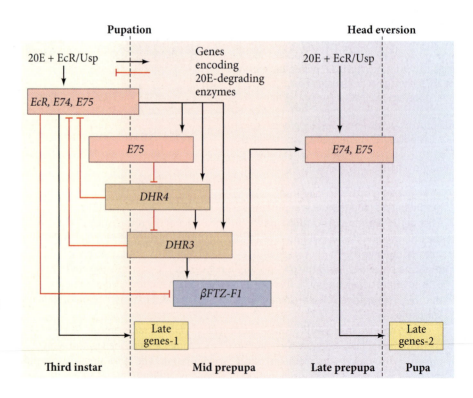

these genes activate a second series of genes, such as *E75*, *DHR4*, and *DHR3*. The products of these genes are transcription factors that work together to form the pupa. Second, the products of the second-wave genes shut off the early response genes so that they do not interfere with this second burst of 20E. Third, 20E activates the genes whose products inactivate and degrade ecdysone itself. In this way, the nucleus is cleared of the hormone so that it can respond to a second pulse. Moreover, 20E usually inhibits the gene encoding βFTZ-F1. Now this transcription factor can be synthesized, and it enables a new set of genes to respond to the second burst of 20E (Rewitz et al. 2009). Moreover, DHR4 coordinates growth and behavior in the larva. It allows the larva to stop feeding once it reaches a certain weight and to begin searching for a place to glue itself to and form a pupa (Urness and Thummel 1995; Crossgrove et al. 1996; King-Jones et al. 2005). (See **Further Development 21.9, Understanding the Different Effects of 20E**, and **Further Development 21.10, Precocenes and Synthetic JH**, both online.)

Determination of the wing imaginal discs

When ecdysone signaling is unaffected by juvenile hormone, it activates the growth and differentiation of imaginal discs that have already been determined. For example, the largest of *Drosophila*'s imaginal discs is that of the wing, containing some 60,000 cells. (In contrast, the leg and haltere discs each contain about 10,000 cells; Fristrom 1972.) The wing discs are distinguished from the other imaginal discs by the expression of the *vestigial* gene (Kim et al. 1996). When this gene is expressed in any other imaginal disc, wing tissue emerges.

ANTERIOR AND POSTERIOR COMPARTMENTS The axes of the wing are specified by gene expression patterns that divide the embryo into discrete but interacting compartments (**FIGURE 21.12A**; Meinhardt 1980; Causo et al. 1993; Tabata et al. 1995). The anterior-posterior axis of the wing begins to be specified during the first instar larva. Here, the expression of the *engrailed* gene distinguishes the posterior compartment of the wing from the anterior compartment. The Engrailed transcription factor is expressed only in the posterior compartment, and in those cells, it activates the gene for the paracrine factor Hedgehog. In a complex manner, the diffusion of Hedgehog activates the gene encoding the BMP homologues Decapentaplegic (Dpp) and Glass-bottom boat (Gbb) in a narrow stripe of cells in the anterior region of the wing disc (Ho et al. 2005).

These BMPs establish a gradient of BMP signaling activity (Matsuda and Shimmi 2012), which can be measured by the phosphorylation of the Mad transcription factor (a Smad protein), because BMPs activate Mad. Dpp is a short-range paracrine factor, whereas Gbb exhibits a much longer range of diffusion to create a gradient (**FIGURE 21.12B**; Bangi and Wharton 2006). This signaling gradient regulates the amount of cell proliferation in the wing regions and also specifies cell fates (Rogulja and Irvine

FIGURE 21.12 Compartmentalization and anterior-posterior patterning in the wing imaginal disc. (A) In the first instar larva, the anterior-posterior axis has been formed and can be recognized by the expression of the *engrailed* gene in the posterior compartment. Engrailed, a transcription factor, activates the *hedgehog* gene. Hedgehog acts as a short-range paracrine factor to activate *decapentaplegic* (*dpp*) in the anterior cells adjacent to the posterior compartment, where Dpp and a related protein, Glass-bottom boat (Gbb), act over a longer range. (B) Dpp and Gbb proteins create a concentration gradient of BMP-like signaling, measured by the phosphorylation of Mad (pMad). High concentrations of Dpp plus Gbb near the source activate both the *spalt* (*sal*) and *optomotor blind* (*omb*) genes. Lower concentrations (near the periphery) activate *omb* but not *sal*. When Dpp plus Gbb levels drop below a certain threshold, *brinker* (*brk*) is no longer repressed. L2–L5 mark the longitudinal wing veins, with L2 being the most anterior. (A after L. Wolpert et al. 1998. *Principles of Development*. Oxford University Press: Oxford; B after E. Bangi and K. Wharton. 2006. *Dev Biol* 295: 178–193.)

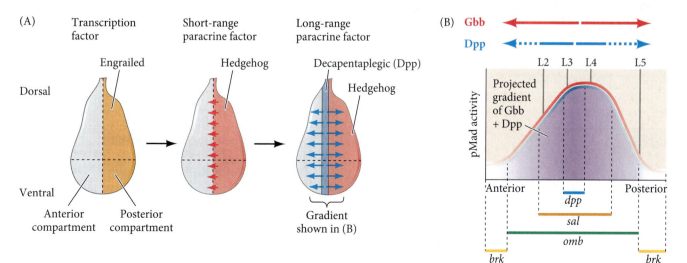

2005; Hamaratoglu et al. 2014). Several transcription factor genes respond differently to activated Mad. At high levels, the *spalt* (*sal*) and *optomotor blind* (*omb*) genes are activated, whereas at low levels (where Gbb provides the primary signal), only *omb* is activated. Below a particular level of phosphorylated Mad activity, the *brinker* (*brk*) gene is no longer inhibited; thus, *brk* is expressed outside the signaling domain. Specific cell fates of the wing are specified in response to the action of these transcription factors. (For example, the fifth longitudinal vein of the wing is formed at the border of *optomotor blind* and *brinker*; see Figure 21.12B). Although experimental evidence shows that Dpp regulates wing growth, the mechanisms by which it does this remain unknown (Hamaratoglu et al. 2014; Hariharan 2015).

DORSAL-VENTRAL AND PROXIMAL-DISTAL AXES The dorsal-ventral axis of the wing is formed at the second instar stage by the expression of the *apterous* gene in the prospective dorsal cells of the wing disc (Blair 1993; Diaz-Benjumea and Cohen 1993). Here, the upper layer of the wing is distinguished from the lower layer of the wing blade (Bryant 1970; Garcia-Bellido et al. 1973). The *vestigial* gene remains active in the ventral portion of the wing disc (**FIGURE 21.13A**). The dorsal portion of the wing synthesizes transmembrane proteins that prevent the intermixing of the dorsal and ventral cells (Milán et al. 2005). At the boundary between the dorsal and ventral compartments, the Apterous and Vestigial transcription factors interact to activate the gene encoding the Wnt paracrine factor Wingless (**FIGURE 21.13B**). Neumann and Cohen (1996) showed that Wingless protein acts as a growth factor to promote the cell proliferation that extends the wing. Wingless also helps establish the proximal-distal axis of the wing: high levels of Wingless activate the *Distal-less* gene, which specifies the most distal regions of the wing (Neumann and Cohen 1996, 1997; Zecca et al. 1996). This occurs in the central region of the disc and "telescopes" outward as the distal margin of the wing blade (**FIGURE 21.13C**). Thus, a battery of paracrine factors patterns the wing disc, giving each cell an identity along the dorsal-ventral, proximal-distal, and anterior-posterior axes. In metamorphosis, we see a reprising of the developmental phenomena that generated the larva itself. (See **Further Development 21.11, Homologous Specification**, and **Further Development 21.12, Metamorphosis of the Pluteus Larva**, both online.)

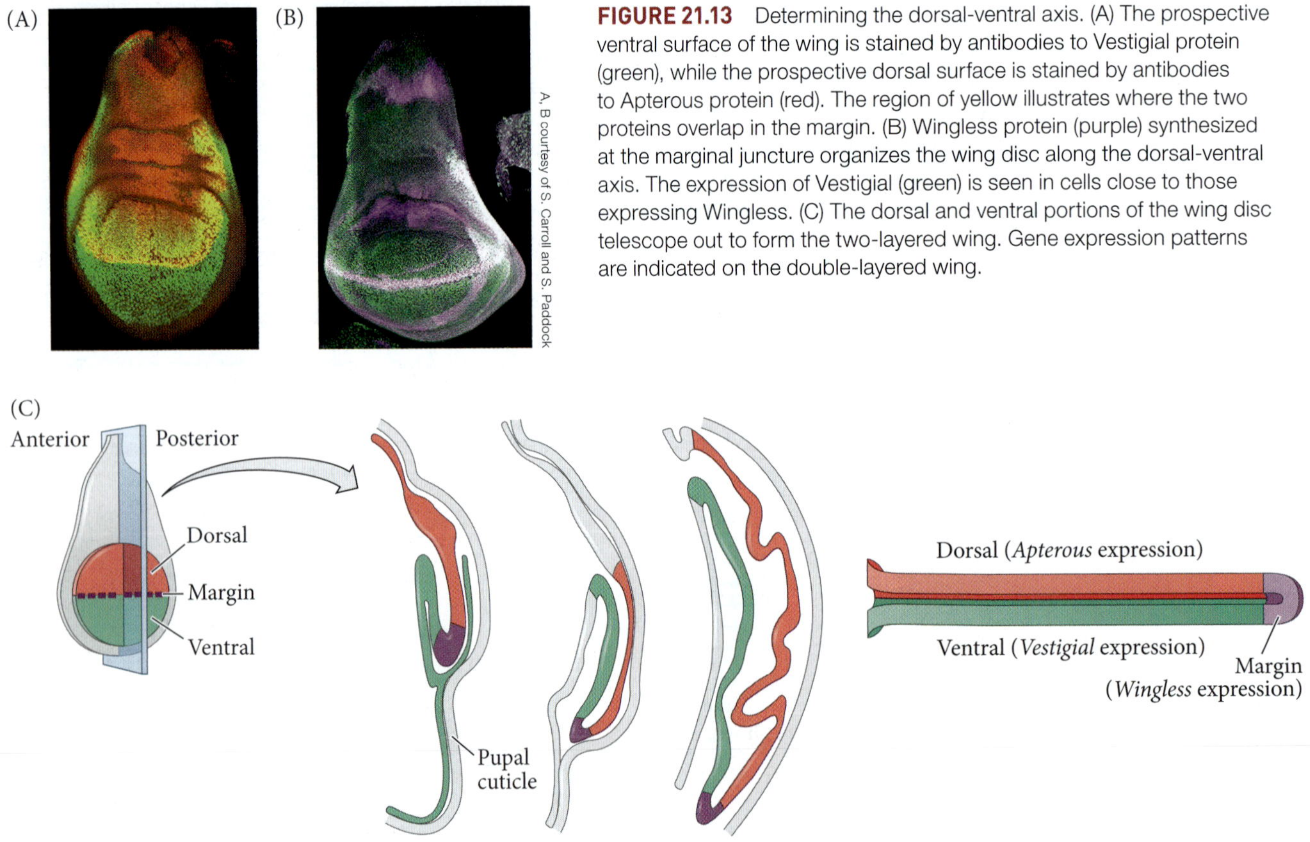

(A)

(B)

A, B courtesy of S. Carroll and S. Paddock

FIGURE 21.13 Determining the dorsal-ventral axis. (A) The prospective ventral surface of the wing is stained by antibodies to Vestigial protein (green), while the prospective dorsal surface is stained by antibodies to Apterous protein (red). The region of yellow illustrates where the two proteins overlap in the margin. (B) Wingless protein (purple) synthesized at the marginal juncture organizes the wing disc along the dorsal-ventral axis. The expression of Vestigial (green) is seen in cells close to those expressing Wingless. (C) The dorsal and ventral portions of the wing disc telescope out to form the two-layered wing. Gene expression patterns are indicated on the double-layered wing.

(C)

Anterior · Posterior

Dorsal
Margin
Ventral

Pupal cuticle

Dorsal (*Apterous* expression)
Ventral (*Vestigial* expression)
Margin (*Wingless* expression)

Closing Thoughts on the Opening Photo

© The Natural History Museum/Alamy Stock Photo

As Alfred Lord Tennyson (1886) intuited, "The old order changeth, yielding place to the new." Metamorphosis separates an individual into two distinct life cycle stages, with different anatomy, different physiology, and different ecological niches. The insect life cycle was discovered and documented in the early eighteenth century by Maria Sibylla Merian, an artist who, among other things, painted the butterflies of Surinam in South America. This portion of a lithograph by Merian (1705) shows the larval, pupal, and adult forms of *Morpho deidamia*. The caterpillar is eating the leaves of the Barbados cherry tree; the pupa of this species resembles that tree's leaves. Merian also noticed that the larvae of different species need different plants than does the adult butterfly. In many instances, the larval food plant contains noxious chemicals that the adult absorbs. Monarch butterfly caterpillars, for instance, obtain toxic alkaloids from plants; these toxins render the metamorphosed adult very unpalatable to birds (and the birds learn not to eat a monarch butterfly).

Endnotes

[1] Although there is controversy on the subject, larvae probably evolved after the adult form had been established. In other words, animals evolved through direct development, and larval forms came about as specializations for feeding or dispersal during the early part of the life cycle (Jenner 2000; Rouse 2000; Raff and Raff 2009). Even so, the biphasic life cycle may be a trait characteristic of metazoans (see Degnan and Degnan 2010).

[2] One of the most spectacular movements of eyes during metamorphosis occurs in flatfish such as flounder. Originally, a flounder's eyes, like the lateral eyes of other fish species, are on opposite sides of its face. However, during metamorphosis, one of the eyes migrates across the head to meet the eye on the other side (Hashimoto et al. 2002; Bao et al. 2005). This allows the flatfish to dwell on the ocean bottom, looking upward.

[3] Tadpoles experience a brief period of deafness as the neurons change targets; see Boatright-Horowitz and Simmons 1997.

[4] Interestingly, the degeneration of the human tail, which takes place during week 4 of gestation, resembles the resorption of the tadpole tail (see Fallon and Simandl 1978).

(21) Snapshot Summary
Metamorphosis

1. Amphibian metamorphosis includes both morphological and biochemical changes. Some structures are remodeled, some are replaced, and some new structures are formed.

2. The hormone responsible for amphibian metamorphosis is tri-iodothyronine (T_3). The synthesis of T_3 from thyroxine (T_4) and the degradation of T_3 by deiodinases can regulate metamorphosis in different tissues. T_3 binds to thyroid hormone receptors and acts predominantly at the transcriptional level.

3. Many changes during amphibian metamorphosis are regionally specific. The tail muscles degenerate; the trunk muscles persist. An eye will persist even if transplanted into a degenerating tail.

4. Metamorphic change in amphibians can be brought about by cell death, cell differentiation, or by cell-type switching.

5. The specific timing of metamorphic events can be orchestrated by the different events occurring at different levels of thyroid hormones.

6. Animals with direct development do not have a larval stage. Primary larvae (such as those of sea urchins) specify their body axes differently than the adult, whereas secondary larvae (such as those of insects and amphibians) have body axes that are the same as adults of the species.

7. Ametabolous insects undergo direct development. Hemimetabolous insects pass through nymph stages wherein the immature organism is usually a smaller version of the adult.

8. In holometabolous insects, there is a dramatic metamorphosis from larva to pupa to sexually mature adult. In the stages between larval molts, the larva is called an instar. After the last instar, the larva undergoes a metamorphic molt to become a pupa. The pupa undergoes an imaginal molt to become an adult.

9. During the pupal stage, the imaginal discs and histoblasts grow and differentiate to produce the structures of the adult body.

10. The anterior-posterior, dorsal-ventral, and proximal-distal axes are sequentially specified by interactions between different compartments in the imaginal discs. The disc "telescopes out" during development, its central regions becoming distal.

11. Molting is caused by the hormone 20-hydroxyecdysone (20E). In the presence of high levels of juvenile hormone, the molt gives rise to another larval instar. In low concentrations of juvenile hormone, the molt produces a pupa; if no juvenile hormone is present, the molt is an imaginal molt.

(Continued)

Snapshot Summary (*continued*)

12. The ecdysone receptors are almost identical in structure to the thyroid hormone receptors of amphibians and are evolutionarily related to these receptors.

13. The ecdysone receptor gene that can form at least three different proteins. The types of ecdysone receptors in a cell may influence the response of that cell to 20E. The ecdysone receptors bind to DNA to activate or repress transcription.

Go to oup.com/he/barresi12xe for Further Developments, Scientists Speak interviews, Watch Development videos, Dev Tutorials, and complete bibliographic information for all literature cited in this chapter.

Regeneration
The Development of Rebuilding

22

Four heads better than one?

Courtesy of Junji Morokuma and Michael Levin

DEVELOPMENT NEVER CEASES. Throughout life, we continuously generate new blood cells, epidermal cells, and digestive tract epithelium from stem cells. A more obvious recurrence of embryonic-like development is the rebuilding of whole tissues following injury or loss through the process of regeneration. Whether it's Ponce de Leon's Fountain of Youth or Marvel's superhero Deadpool, regeneration is a process that has captivated the imagination of writers, artists, and Hollywood alike. It's not all science fiction. Regeneration has also caught the wonder of scientists who have made great strides in dissecting the developmental mechanisms underlying the ability of some species to exhibit a fantastic potential for regeneration. Some adult salamanders, for instance, can regrow limbs and tails after these appendages have been amputated (a process surprisingly similar to the depiction of Deadpool's "baby hand").[1] It is difficult to behold the phenomenon of limb regeneration in salamanders without wondering why we humans cannot grow back our arms and legs. What gives these animals an ability we so sorely lack? Experimental biology was born of the efforts of eighteenth-century naturalists to answer this question (see Morgan 1901). The regeneration experiments of Abraham Tremblay (using hydra, a cnidarian), René Antoine Ferchault de Réaumur (crustaceans), and Lazzaro Spallanzani (salamanders) set the standard for experimental research and for the intelligent discussion of one's data (see Dinsmore 1991). More than two centuries later, we are beginning to find answers to the great questions of regeneration, and at some point we may be able to alter the human body so as to permit our own limbs to regenerate.

Defining the Problem of Regeneration

"I'd give my right arm to know the secret of regeneration." This quote from Oscar E. Schotté (quoted in Goss 1991) captures the fascination science has had with the remarkable ability of some organisms to rebuild themselves. **Regeneration** is the reactivation of developmental mechanisms in postembryonic life to restore missing or damaged tissues. The potential benefit of harnessing the powers of regeneration in humans would mean that severed limbs could be restored, diseased organs could be removed and then regrown, and nerve cells altered by age, disease, or trauma could once again function normally. Before modern medicine can succeed in coaxing human bone or neural tissue to regenerate, we must first understand how regeneration occurs in those species that routinely exhibit this ability.

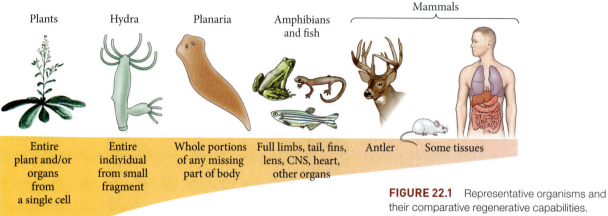

FIGURE 22.1 Representative organisms and their comparative regenerative capabilities. (© Michael F. Barresi.)

1. Morphological memory map

2. Acknowledgment of difference

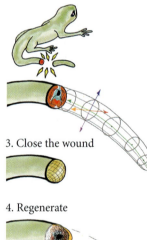

3. Close the wound

4. Regenerate

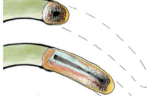

5. Stop regenerating

FIGURE 22.2
Conceptualized steps of regeneration.
(© Michael F. Barresi.)

Our knowledge of the roles of paracrine factors in organ formation and our ability to clone the genes that produce those factors have propelled what Susan Bryant (1999) has called "a regeneration renaissance." *Renaissance* literally means "rebirth," and because regeneration can involve a return to the embryonic state, the term is apt in many ways.

Although some form of regeneration takes place in nearly all species, several organisms have emerged as particularly fruitful models for the study of regeneration (**FIGURE 22.1**). Some of the most incredible regenerative abilities are seen in the plant kingdom, in which de novo organ generation is possible even from a single cell. Not far behind in this totipotent ability are hydra and planarians, whose ability for near-total regeneration is unmatched among animals. They are able to regenerate complete organs following amputation, or even complete individuals from very small fragments. Certain salamanders are unique among tetrapods in being able to regenerate whole limbs, and frog larvae are often used to study the regeneration of the tail and the lens of the eye. Zebrafish have recently proved advantageous for investigating the mechanisms of central nervous system, retina, heart, liver, and fin regeneration. And although mammals are unable to rebuild whole appendages, individual tissues and organs do possess variable regenerative capabilities; most notable are the antlers of deer. How is any of this possible? Trying to unravel the mysteries of regeneration is a daunting task.

For any organism, we can conceptualize several steps necessary for the regeneration of an injured structure (**FIGURE 22.2**):

1. Prior to any injury, the cells and tissues must have an "idea" of their own identity within the organism, their shape, and even their position in relation to the other cells of the body: a "morphological memory map."

2. After an injury, the cells and tissues of the organism need to recognize that something has changed and that an exact replacement is to be made.

3. The organism needs to rapidly respond by closing the wound.

4. The true regenerative response commences. Implementing some of the same mechanisms used during embryonic development, a regenerating structure initiates cell proliferation, growth of the tissues, and a re-patterning of the cells so that they differentiate into the structure that was lost.

5. Regeneration must end, having formed the correct size and shape of the lost structure, which is both integrated with and in proportion to the rest of the body.

Can you imagine the mechanisms involved in these steps? Regeneration is difficult, and yet some organisms are masters at it. Regeneration requires a cell-based system of awareness of the whole organism, interactions with wound-induced immune responses, and a dramatic resurgence of developmental morphogenesis.

MODES OF REGENERATION   Despite the many differences in regenerative potential across species, each organism discussed in this chapter exemplifies one or more of the four modes of regeneration (**FIGURE 22.3**):

1. *Stem cell–mediated regeneration.* Stem cells allow an organism to regrow certain organs or tissues that have been lost; examples include the regrowth of hair shafts from follicular stem cells in the hair bulge and the continual replacement of blood cells from the hematopoietic stem cells in the bone marrow.

2. *Epimorphosis.* In some species, adult structures can undergo *de*differentiation to form a relatively undifferentiated mass of cells (a blastema) that then redifferentiates during morphogenesis of the new structure. Such regeneration is characteristic of regenerating amphibian limbs.

3. *Morphallaxis.* Here, regeneration occurs through the re-patterning of existing tissues. There is often little new growth but rather cell death and a change in cell type (i.e., transdifferentiation into a different cell fate). This results in a rescaling of the whole organism as well as regeneration of the missing part. This type of regeneration is best represented by the hydra.

4. *Compensatory regeneration.* Here, the differentiated cells divide but maintain their differentiated functions. The new cells do not come from stem cells, nor do they come from the dedifferentiation of adult cells. Each cell produces cells similar to itself; no mass of undifferentiated tissue forms. This type of regeneration is characteristic of the mammalian liver.

FIGURE 22.3   Four different modes of regeneration.

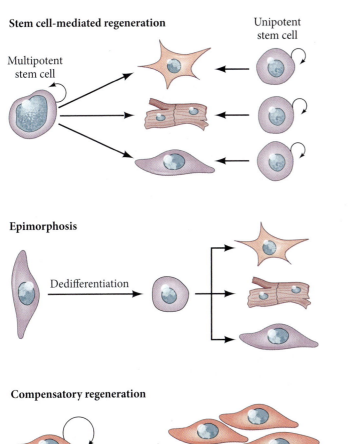

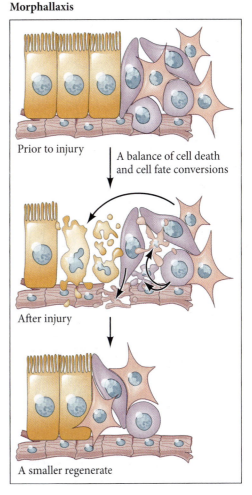

Regeneration, a Recapitulation of Embryonic Development?

The creation and use of stem cells, the processes of cell proliferation and differentiation, and tissue morphogenesis are the core mechanisms behind all modes of postembryonic regeneration touched on above; they are also the basic mechanisms behind embryogenesis. So is regeneration just a recapitulation of embryonic development (**FIGURE 22.4**)? As with most things in developmental biology, the answer invariably is yes and no.

Regeneration certainly falls victim to the old saying "If it ain't broke, don't fix it." If the mechanisms employed during embryogenesis were sufficient to build all the cells and tissues of the entire organism in the first place, is there really a need to reinvent these mechanisms to accomplish the same task later in life, or are there constraints imposed by the postembryonic environment that necessitate novel solutions? The answer to this FAQ is yes, regeneration co-opts the mechanisms of embryogenesis whenever possible, *and* yes, regeneration requires context-specific adaptations.

There are four principal differences between regeneration and embryogenesis that prevent regeneration from being a strict recapitulation of embryonic development. These involve the immune response, reprogramming, integration, and termination.

1. *Immune response.* The most significant difference between regeneration and embryonic development is that regeneration is a response to an injury. Often this means the presence of trauma, necrotic cell death, escaping fluids, and exposed, unprotected tissues. The immune response includes the initial attention to closing the wound and in many cases (species-dependent) the deployment of phagocytic cells to clean up the injury site (see Figure 22.4; 1a,1b).

2. *Induced reprogramming.* Following the injury response, cells (resident or from afar) need to become activated in order to take on an immature, embryonic-like state before they can take advantage of the developmental programs used to rebuild tissues. Regeneration may require novel processes to reprogram adult cells to become embryonic-like (see Figure 22.4; 2a,2b).

3. *System integration.* Regeneration of adult tissues involves the formation of new cells that become organized and integrated into existing differentiated tissues. This would include the proper integration of blood vessels and nerve connections, for instance (see Figure 22.4; 3a,3b).

4. *Size recognition and termination.* Regeneration happens in the context of neighboring adult tissues that are still functioning and possibly still growing. This suggests that the regenerating tissues need to employ specific mechanisms of communication to recognize their spatial relationship with the surrounding tissues and with the organism as a whole. These mechanisms must also ensure that regrowth is calibrated to the appropriate scale and is terminated once that size is achieved (see Figure 22.4; 4).

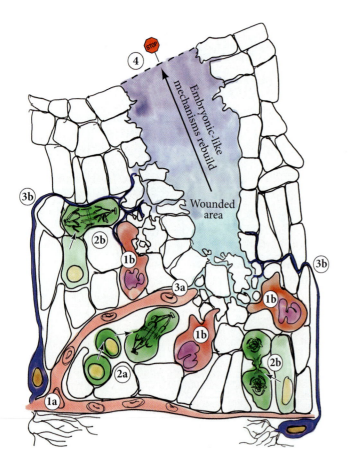

FIGURE 22.4 Regeneration novelties. Wounds induce an inflammatory response from the circulatory system (1a) that includes macrophage invasion (1b). Small black arrows indicate direction of cell lineage. Regeneration requires the recruitment of stem cells through the activation of resident multipotent stem cells (2a) and/or the reprogramming of differentiated cells into progenitor cells (2b). There must be mechanisms to integrate newly regenerating cells with existing tissues, such as blood vessels (3a) and neurons (3b). Once tissue repair is underway, regeneration must possess some unique way to properly scale the size of the repair and stop the process when the correct proportions are achieved (4). © Michael F. Barresi.

Researchers in the field of regeneration are grappling with unraveling how these processes that are unique to regeneration work and how the mechanisms of embryonic development are used to fuel the regenerative process. As you proceed through this chapter, think about how the specific details of regeneration fit into these two conceptual bins: (1) redeployment of embryonic mechanisms and (2) regeneration-specific responses.

An Evolutionary Perspective on Regeneration

Why can't humans regrow a limb following amputation, like the salamander, or repair an injured brain or heart, like the zebrafish? Such abilities seem to be reserved for comic-book characters or for mostly aquatic animals that evolved earlier than humans. The ability to regenerate from injury seems to have clear adaptive benefits. If this is the case and some of our metazoan ancestors possessed super-regenerative abilities, then did the ancestors that led to humans and other mammals lose this trait over the course of evolution? Perhaps considering the evolutionary history of regeneration can help us better understand why some organisms today can robustly regenerate while others exhibit only the limited capacities for tissue repair.

Two prevailing thoughts about the evolution of regeneration have emerged (**FIGURE 22.5**). As suggested above, there could have been an ancient ancestor with expansive regenerative capacities that were selectively retained and lost over evolutionary history (a macroevolutionary effect). Alternatively, regenerative abilities could have evolved independently in different species (a microevolutionary effect), which suggests that divergent lineages converged on the innovation of regeneration independently (that is, they exhibited **convergent evolution**). These two ideas are not mutually exclusive; a combination of both evolutionary effects would posit that an ancient ancestor passed on its regenerative functions to some but not all descendants, while unrelated lineages simultaneously evolved their own mechanisms of regeneration.

Although researchers have not yet determined whether a sole regenerating common ancestor existed or whether regeneration is a product of convergent evolution, there are representative regenerating species across plant and animal phyla. Plants have been shown to be capable of regenerating the entire plant from individual cells in culture—these cells are therefore totipotent. More important, plants possess a diversity of regenerative abilities, such as complete body regeneration by green algae, the re-creation of meristems by some bryophytes (e.g., liverworts), and, depending on the injury, complete regeneration of full shoot or root organs by seed-bearing plants. Thus, much of the plant kingdom has maintained expansive regenerative abilities over the course of evolution. This makes sense from a natural selection perspective. Plants, in general, cannot move—they cannot run away from a predator or seek shelter when wounded. It therefore makes sense that plants would have evolved extreme plasticity in their ability to heal, repair, and replace tissues.

Compared with that of plants, the regenerative trait in metazoans is more sporadically dispersed across the tree of life. If the ability to regenerate was derived from a common ancestor, then the modern species most directly related to that ancestor could be expected to retain similar abilities to regenerate. As noted above, planarian flatworms are among the most well studied and impressive whole-body regenerators. The flatworm-like clade Acoelomorpha (acoels), which has comparable powers of regeneration, is a 550-million-year-old sister group to all other bilaterally symmetrical animals (see Figure 22.5A, yellow). Just like the planarians, today's acoels are capable of total regeneration following amputation of an entire half of the body, and they do so through stem cells called neoblasts, just as do planarians (De Mulder et al. 2009; Raz et al. 2017). Even more impressive is that this regenerative process depends on Wnt signaling, as does regeneration throughout bilaterians (Srivastava et al. 2014). These data strongly suggest that the underlying molecular and cellular mechanisms of regeneration were likely present in the common ancestor to all bilaterians.

If regeneration was in fact a trait of the urbilaterian ancestor, then how far back could the origin of this repair behavior go? As noted in Chapter 1 (see Figures 1.20 and 1.25), the poriferans (sponges) may represent the most basal of all metazoan organisms. Interestingly, the regenerative abilities of sponges were discovered more

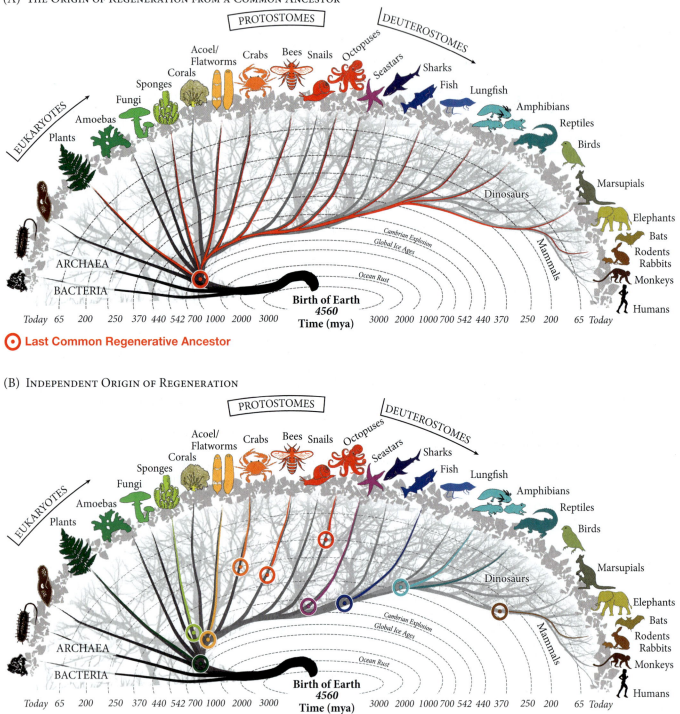

FIGURE 22.5 The evolution of regeneration. Schematic of two possible origins of regeneration that are not mutually exclusive. (A) Origin of regeneration from a single common ancestor. This ancestor is theoretically positioned at the origin of multicellularity on the tree of life (follow red lines from common ancestor to derived lineages). Acoels and sponges show regenerative abilities, suggesting their ancestor might represent a common ancestor to all regenerative metazoan life. (B) Alternatively, regeneration may have evolved numerous times through convergent evolution. © Michael F. Barresi.

than a century ago (Wilson 1907); however, their full capacities for regeneration and the underlying cellular behaviors are only now starting to be discovered (Adamska 2018; Funayama 2018). Remarkably, some sponges (the four classes Homoscleromorpha, Demospongia, Calcarea, and Hexactinellida) exhibit the greatest plasticity for complete body regeneration of any animals studied. Not only can the adults of some species fully regenerate their body tissue following large ablations, but some are even capable of regenerating a completely new individual from dissociated cells! That's right, here is an animal whose regenerative abilities seem to match the totipotency of a plant. Dissociated cells of at least two species (*Spongilla lacustris* and *Haliclona* cf. *permollis*) were capable of forming aggregates, sorting the cells appropriately, removing debris, adhering to a substrate, and completing cell differentiation into functionally organized sponge tissues. Importantly, this totipotent feature is not widespread across sponge species, but clearly all sponges possess expansive regenerative abilities (Eerkes-Medrano et al. 2015). Recent findings support a model in which at least two resident cell types, choanocytes and archeocytes, function as pluripotent stem cells to fuel much of the regeneration (**FIGURE 22.6**). Additionally, these same cell types along with other epithelial cells have been shown to undergo a process of transdifferentiation to directly replace lost tissues (Borisenko et al. 2015; Ereskovsky et al. 2015, 2017), and in at least one species, *Halisarca dujardini*, a blastema-like structure is created at the site of injury. In this species, choanocytes respond to an injury by undergoing an epithelial-mesenchymal transition and migrating to the wound site. There, they mix with similarly undifferentiated cells derived from neighboring archeocytes and form a proliferative mass that then serves to regenerate the injured tissue (Borisenko et al. 2015). This type of cellular response is highly suggestive of a blastema-based mechanism of regeneration, which is seen in many different animal regenerators across phyla. Thus, today's descendants from perhaps the most basal metazoan, the sponge, demonstrate three of the observed modes of regeneration: stem cell-mediated regeneration, morphallaxis (or transdifferentiation), and epimorphosis (blastema formation).

WHY ARE SO MANY ANIMALS UNABLE TO REGENERATE? This is the most common question in the field of regeneration. When in doubt, blame natural selection. Differences in regenerative capacity may be related to the array of behavioral options available to motile animals for escaping predators or the life-threatening outcomes associated with blood loss. You will learn later in this chapter that although scar formation is amazingly efficient at repairing a wound, it also inhibits the regenerative process. One hypothesis concerning why an animal might need to produce a scar that inhibits regeneration is that animal survival is in some cases enhanced by rapid scarring rather than by the slower process of regeneration. Consider this: when facing a life-threatening injury, isn't it more important to prevent immediate death from blood loss than to slowly regenerate a lost limb? Regenerative medicine researchers are currently trying to decipher the mechanisms that control whether an injury is healed by scarring or by a regenerative response, in the hope of finding a way to induce pro-regenerative interventions.

(?) Developing Questions

If there is a single common ancestor of all regenerating organisms, then how far back do you think regeneration goes? We mentioned the possibility that it could go back as far as the ancestors of acoels or perhaps sponges, but could the origins lie at the beginning of multicellularity, or even in some way relate to the behaviors of single eukaryotic cells? The unicellular ciliate genus *Stentor* is capable of regenerating! Could our unicellular ancestor regenerate?

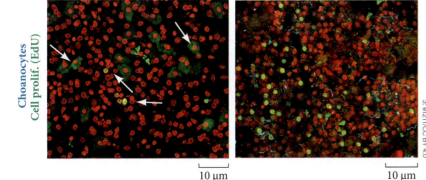

(A) Control sponge surface (B) 12 h post injury

Choanocytes
Cell prolif. (EdU)

10 μm   10 μm

From I. E. Borisenko et al. 2015, *PeerJ* 3: e1211/CC BY 4.0

FIGURE 22.6 Sponge choanocytes proliferate in response to injury. Proliferating cells in S phase of the cell cycle are labeled with EdU (green); choanocytes are stained with anti-tubulin (blue); red fluorescence shows all nuclei. (A) Uninjured control. (B) Recruitment and increased proliferation of choanocytes at the wound surface.

Although we will explore later in this chapter the role the extracellular matrix plays in establishing the balance between scarring versus regenerative healing, our hypothesis is not by itself a satisfying answer to the larger question of why regenerative abilities might have been reduced over the course of evolution. With exceptions such as deer regenerating antlers, an intriguing fact is that the best regenerators happen to live in a predominantly aqueous environment. What role, if any, does water play in the regenerative process? There are likely many evolutionary trade-offs that contributed to reductions in regenerative ability; thus, a single answer seems improbable. However, addressing these questions is propelling the study of regeneration to new heights. Students today should feel particularly lucky to be part of this regenerative renaissance, as so many of these foundational questions can for the first time begin to be addressed. To productively explore these questions, we first need to learn about the current understanding of the mechanisms driving different regenerative processes.

Regenerative Mechanics

We have already described diverse modes of regeneration among organisms, from whole body repair to organ- or tissue-restricted healing. What are the developmental mechanisms underlying these different regenerative capacities among plants and animals, and do they share any common characteristics? The answer is that they share several characteristics. Stem cells, cell cycle control, morphogens, and the composition of the extracellular matrix are the key recurring features.

Plant Regeneration
A totipotent way of regenerating

If you have ever mowed a lawn, then you know plants possess a relentless ability to regrow, again and again and again. Even more astounding is how whole plants can be regenerated from isolated parts; place the end of a severed bunch of celery in a bowl of water and it will sprout new roots and shoots (**FIGURE 22.7**). It was fortunate for astronaut/botanist Mark Watney in *The Martian* that pieces of potato could be put in the soil to regenerate completely new plants (see the 2015 film adaptation of Andy Weir's book). There exists a great diversity of regenerative strategies in plants, many identified by the agricultural industry. Here we will focus on the primary modes of plant regeneration in nature, which provide striking similarities to the mechanisms of animal regeneration, despite their independent evolution. These modes include a period of cellular reprogramming, the creation of totipotent cells, and the subsequent derivation of plant organs that are patterned based on positional information.

REGENERATION BY A SINGLE CELL FOR A SINGLE CELL The *Acetabularia* alga is colloquially called the "mermaid's wine glass" due to its shape, a cuplike structure standing about 6 cm high on a central stalk that is often affixed by its basal rhizoids to a rock at sea (**FIGURE 22.8**). Remarkably, each *Acetabularia* plant is a single cell! That's right. It has one cell membrane spanning the entire organism, and for most of its life, an *Acetabularia* alga has only one nucleus restricted to the basal end of the cell. What is even more remarkable is how this "wine glass" can regenerate an entire new "cup" from its stalk.

In the 1940s, before the central dogma of molecular biology (DNA to RNA to protein) was known, Joachim Hämmerling discovered that if the basal portion of an *Acetabularia* cell (still containing the nucleus) was severed, it was always capable of repeatedly regenerating the cup. The more apical the cut, the more apically restricted the regeneration (Mandoli 1998). Hämmerling's experiments set the stage for the discovery not only of mRNA, but of the importance of positional information along the apical-basal axis for proper regenerative patterning of the organism—whether a single cell or, as you will soon learn, a multicellular plant or animal.

FIGURE 22.7 Cut, provide water and sunlight, and watch regeneration happen in a plant. Author Michael Barresi cut a complete bunch of celery near its basal end and immersed the roots in tap water. Within a week, new shoots emerged from the central whorl.

REGENERATION BY A SINGLE CELL FOR A WHOLE PLANT

The evolution of multicellular embryophytes began with the formation of multicellular colonies of green algae. Keeping this evolutionary origin in mind is important when considering plant regeneration. It should thus not be surprising that a common green alga such as *Bryopsis plumosa* can regenerate intact cells from subprotoplasts (clumps of cytoplasm without a cell membrane or wall) and whole algal sheets from just single (albeit multinucleated) cells (**FIGURE 22.9**; Kim et al. 2001). Land plants, however, have evolved their own, more complex totipotent powers of regeneration. In the right culture conditions, single adult plant cells or explants from specific differentiated tissues can undergo reprogramming to form a totipotent **callus** (an unorganized mass of largely unspecified cells), which can be further directed by hormones to develop into shoots and/or roots. Some of the first examples of this ability in plants came from studies of carrot and tobacco (Steward et al. 1958; Vasil and Hildebrandt 1965; see also Birnbaum and Sánchez Alvarado 2008).

The mechanisms underlying **de novo regeneration** (i.e., regeneration from small portions of tissue in a culture environment) of plant organs rely heavily on some of the same hormone signaling pathways used in embryogenesis, especially those of auxin (see Figure 4.30) and cytokinin. Supersaturating concentrations of auxin are necessary for establishing the callus. Subsequently, combinations of auxin and cytokinin in specific concentration ratios can be used to control the type of tissue that develops from the callus. A high auxin:cytokinin ratio promotes root development, whereas the opposite (high cytokinin:auxin) promotes shoot development (Skoog and Miller 1957; see also Su and Zhang 2014). Thus, the core developmental mechanisms that originally built the plant embryo are very much recapitulated during regeneration in the adult plant. (See **Further Development 22.1, A "PLETHORA" of Controls**, online.)

A plant's meri-aculous healing abilities

As you by now know, plants are indeterminant growers. Their stubborn, lifelong continuation of development is enabled by the persistence of meristematic tissues. Plant meristems are miraculous in the sense that they can create all of

FIGURE 22.8 The *Acetabularia* alga, or "mermaid's wine glass." Six separate individuals are shown in this photograph. This plant can regenerate the upper half of its body from the lower half. Why is that amazing? Because an entire plant is a single cell!

Photo courtesy of Alphonse Ralph Cavaliere, Gettysburg University

(A) *Bryopsis plumera*

Photo by Courtney Janiak

(B)   (C)   (D)

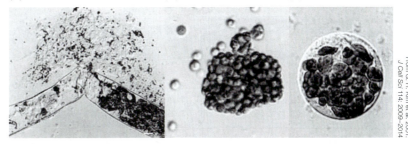

Regeneration

From G. H. Kim et al. 2001. J Cell Sci 114: 2009–2014

FIGURE 22.9 Regeneration in *Bryopsis plumosa*. (A) This green alga is a single huge, multinucleate cell. (B–D) Intact protoplasts can regenerate from clumps of cytoplasm. (B) A broken cell spilling its contents into the surrounding medium. (C) A clump of organelles. (D) A clump of organelles that has regenerated a primary envelope, just 20 minutes after wounding. Later, this subprotoplast will regenerate a true cell membrane.

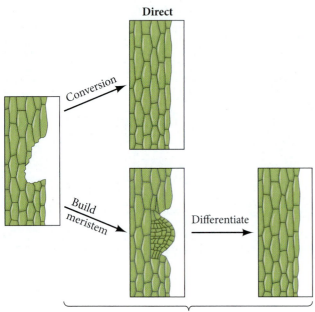

Direct

Conversion

Build meristem → Differentiate

Indirect

FIGURE 22.10   Direct and indirect ways for a plant to regenerate. Note, this is a simplistic representation of the options following an injury in a plant. The indirect pathway often forms a callus first, which then leads to a meristem that can give rise to roots or shoots.

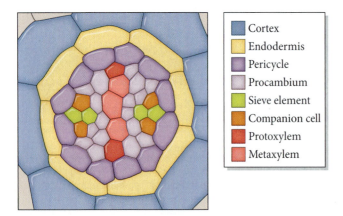

Color	Tissue
	Cortex
	Endodermis
	Pericycle
	Procambium
	Sieve element
	Companion cell
	Protoxylem
	Metaxylem

FIGURE 22.11   Pericycle and procambium are adult stem cells in the plant. Cross sectional illustration of the root of *Arabidopsis thaliana*. (After Miyashima et al. 2013. *EMBO J* 32: 178–193.)

a plant's organ types from a self-perpetuating undifferentiated mass of stem cells. Therefore, when a plant is injured, it can take advantage of the existing shoot and root apical meristematic tissues to produce new cells for regeneration. However, what if an injury also results in the loss of one of a plant's meristems, perhaps chewed off by an herbivore or even laser ablated by a researcher? Remember, plant cells cannot migrate throughout the plant as mesenchymal cells can in animals. Therefore, the cells in the wounded area have two options (**FIGURE 22.10**):

1. They can regenerate lost tissue in an *indirect* manner, by *first developing into a new meristem* that can then give rise to the appropriate tissue types.

2. Alternatively, they can take a *direct* approach to regeneration by *transdifferentiating* (changing directly) into the lost cell types.

But is every cell in a plant capable of responding in these ways to an injury? It has recently been demonstrated that not all cells are competent to form a totipotent callus. Instead, resident **adult stem cells** become activated following injury to initiate either a direct or an indirect regenerative response. More specifically, in *A. thaliana* roots, **pericycle** and **procambium** cells are known to serve as a plant's adult stem cells (**FIGURE 22.11**; Sugimoto et al. 2010; Liu et al. 2014; Jouannet et al. 2015; reviewed in Kareem et al. 2016a,b). The procambium plays a critical meristematic role during embryonic development of vascular plants, as do the pericycle cells situated between the endodermis and the stele (vascular tissues) in the root. Postembryonically, adult stem cells are considered *partially* differentiated. For instance, although adult stem cells in the pericycle express genes specific for their tissue type and region, they also retain the expression of genes associated with the initiation of lateral root development. When roots branch, adult stem cells in the pericycle form a new root apical meristem, which then generates a new lateral root. It appears that an adult stem cell's ability to initiate this program for lateral root development—in which the cell re-enters the cell cycle and produces new multipotent cells—means that the cell also retains the potential to produce callus that can differentiate into root and shoot tissues.

FROM INJURY TO TRANSCRIPTION FACTORS TO PATTERNING FACTORS   Injuries to a plant trigger an immediate disruption in the distribution of the hormone auxin, which leads to a cascade of events supporting new meristem development and subsequent regeneration. For instance, targeted ablation of the root apical meristem's quiescent center (QC) causes a change in the auxin responsiveness of cells near the disrupted region, which is soon followed by upregulation of genes required to regenerate a new QC. The genes for the PLETHORA (PLT) family of transcription factors, which also have a role in root apical meristem development are the first genes to be induced near the injury site. PLT transcription factors then activate expression of *SHORTROOT* (*SHR*) and subsequent expression of *SCARECROW* (*SCR*) (**FIGURE 22.12**). Importantly, expression of this network of genes causes redistribution

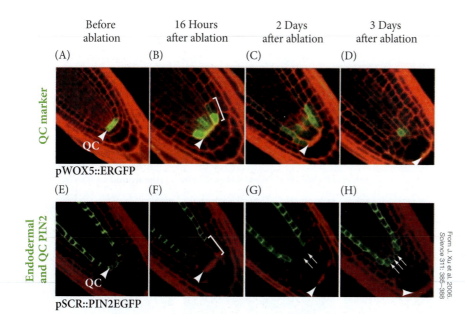

From J. Xu et al. 2006. *Science* 311: 385–388

FIGURE 22.12 An *Arabidopsis thaliana* root tip showing regeneration of the quiescent center (QC) of the root apical meristem over time. (A–D) Arrowheads show the location of the ablated QC. The QC is visualized by GFP expression driven by the *pWOX5* promoter (green). The bracket in (B) indicates expanded expression of the QC marker. (E–H) Visualization of PIN2 localization in endodermal cells. Expression of a PIN2:GFP fusion is driven by the *pSCR* promoter (green). The bracket in (F) denotes a loss of PIN2 expression, while arrows in (G) and (H) indicate the renewal of PIN2 expression in the cells corresponding to the newly regenerated QC.

of PIN transporters, such that auxin flows are reoriented along the apical-basal and lateral axes in ways that support not only QC formation but appropriate organ regeneration.

The PLT-led gene network is also required for the initiation of a new stem cell niche needed to promote normal lateral root development, thereby exemplifying how an embryogenic program can be redeployed during postembryonic development. Moreover, it was recently discovered that PLT and SCR interact together with a plant-specific PCNA (proliferating cell nuclear antigen) transcription factor (Shimotohno et al. 2018).[2] PCNA has long been directly implicated in the control of the cell cycle and cell proliferation, from yeast and plants to humans (Strazalka and Ziemienowicz 2011; Li 2015). Therefore, PLT interaction with PCNA provides a mechanism by which differentiated cells could re-enter the cell cycle to initiate a regenerative response—a mechanism that might exemplify deep homology between plant and animal regeneration pathways. (See **Further Development 22.2, A Lateral Root Way to be Call(o)us**, online.)

THE CANALIZATION OF VASCULAR REGENERATION As vascular tissues, xylem and phloem are some of the most critical structures needing rapid repair following injury. Indeed, a cut through the vascular tissue of a plant's stem triggers the regeneration of new connections between the vascular conduits above and below the cut. Interestingly, removal of shoot and leaf structures above the cut prevents regeneration of xylem and phloem. If, however, an exogenous source of auxin is applied to the ablated stem or leaf branch, then vascular regeneration is restored (**FIGURE 22.13**). These results suggest that an apical source of auxin alone is sufficient to initiate properly oriented vascular reconnections. Closer observation of the regenerated connections reveals a consistent developmental progression. Vascular differentiation occurs first at the site of auxin application and then proceeds basally in a single-cell file until it contacts the existing vasculature. This line of prevascular cells channels (canalizes) auxin flow toward more basal, as yet undifferentiated, cells in the same line, thus progressing the regeneration of xylem. This pattern of regeneration has led to the **canalization model** for vascular development, in which the vasculature apical to a healing wound or developing tissue serves as the auxin source, and the regenerating (or developing) row of increasingly basal differentiating vascular cells serves as the auxin sink (Bennett et al. 2014). Thus, much like the rush of water that can carve new paths for a river, auxin signaling can carve a new canal for the literal return of water flow in a plant.

The canalization model would suggest that auxin signaling is capable of establishing a self-organizing pattern that leads to correct vascular positioning. When auxin reaches a cell, it causes auxin transporters to be asymmetrically relocalized to the basal end of that cell, directing auxin efflux downward to the next cell in line. As more cells

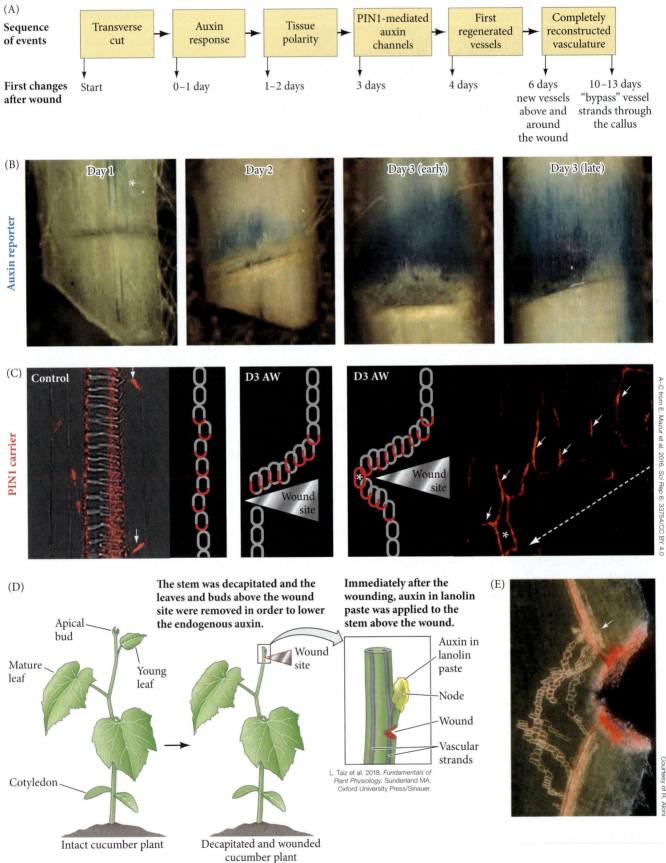

(A)

Sequence of events

Transverse cut → Auxin response → Tissue polarity → PIN1-mediated auxin channels → First regenerated vessels → Completely reconstructed vasculature

First changes after wound

Start | 0–1 day | 1–2 days | 3 days | 4 days | 6 days new vessels above and around the wound | 10–13 days "bypass" vessel strands through the callus

(B) **Auxin reporter**

Day 1 | Day 2 | Day 3 (early) | Day 3 (late)

(C) **PIN1 carrier**

Control | D3 AW | D3 AW | Wound site

(D)

Apical bud
Young leaf
Mature leaf
Cotyledon

Intact cucumber plant

The stem was decapitated and the leaves and buds above the wound site were removed in order to lower the endogenous auxin.

Decapitated and wounded cucumber plant

Immediately after the wounding, auxin in lanolin paste was applied to the stem above the wound.

Wound site

Auxin in lanolin paste
Node
Wound
Vascular strands

L. Taiz et al. 2018. *Fundamentals of Plant Physiology.* Sunderland MA: Oxford University Press/Sinauer.

(E)

are added to the line, the downward movement of auxin is amplified and narrowed.
The net result is the formation of a canal of vascular differentiation. As shown in
Figure 22.13, this mechanism begins with a rapid auxin response above and around
the wound. This is followed by the redistribution of new PIN1 auxin efflux transporters to the basal and lateral sides of differentiating cells, to direct auxin flow around
the wound. Last, new vessel strands circumvent the wound and connect to existing
vascular tissues (Mazur et al. 2016). (See **Further Development 22.3, Epigenetic
Control of Plant Regeneration**, online.)

Whole Body Animal Regeneration

Hydra: Stem cell-mediated regeneration, morphallaxis, and epimorphosis

"We shall never be destroyed! Cut off a limb, and two more shall take its place! …Hail Hydra!"
(*Strange Tails*, Marvel Comics). Like the fictional terrorist organization, the *real* hydras,[3]
freshwater cnidarians, are a superpower when it comes to their regenerative abilities.
Most hydras are tiny—about 0.5 cm long. A hydra has a tubular, radially symmetrical
body, with a "head" at its distal end and a "foot" at its proximal end. The "foot," or
basal disc, enables the animal to stick to rocks or the undersides of pond plants. The
"head" consists of a conical **hypostome** region that contains the mouth and a ring of
tentacles (which catch food) beneath it. Hydras are diploblastic animals, having only
ectoderm and endoderm (**FIGURE 22.14A**). Their two epithelial layers are referred to
as **myoepithelia** because they possess characteristics of both epithelial and muscle
cells. Although hydras lack a true mesoderm, they do contain secretory cells, gametes,
stinging cells (nematocytes), and neurons that are not part of the two epithelial layers
(**FIGURE 22.14B**; Li et al. 2015). Hydras can reproduce sexually, but they do so only
under adverse conditions (such as crowding or cold temperatures). They usually multiply asexually, by budding off a new individual. The buds form about two-thirds of the
way down the animal's body axis.

ROUTINE CELL REPLACEMENT BY THREE TYPES OF STEM CELLS   A hydra's body is
not particularly stable. In humans and flies, for instance, a skin cell in the body's trunk
is not expected to migrate and eventually be sloughed off from the face or foot—but that
is exactly what happens in hydra. The cells of the body column are constantly undergoing mitosis and are eventually displaced to the extremities of the column, from which
they are shed (**FIGURE 22.14C**; Campbell 1967a,b). Thus, each cell plays several roles,
depending on how old it is, and the signals specifying cell fate must be active all the
time. In a sense, a hydra's body is always regenerating.

This cellular replacement is generated from three cell types. Endodermal and
ectodermal cells are unipotent progenitor cells that divide continuously, producing
more lineage-restricted epithelia. The third cell type is a multipotent **interstitial stem
cell** found within the ectodermal layer (see Figure 22.14B). This stem cell generates
neurons, secretory cells, nematocytes, and gametes. The most significant cell proliferation by each of these three types of stem cells occurs within the central region
of the body, after which displaced myoepithelia and migrating interstitial progeny
move to and differentiate at the apical and basal extremities (Buzgariu et al. 2015).

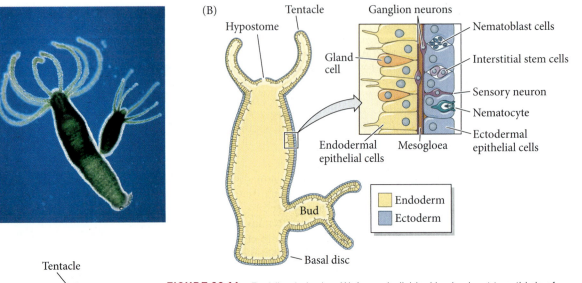

(A)

© Biophoto Associates/Science Source

(B)

Hypostome
Tentacle
Ganglion neurons
Gland cell
Nematoblast cells
Interstitial stem cells
Sensory neuron
Nematocyte
Ectodermal epithelial cells
Endodermal epithelial cells
Mesogloea
Bud
Basal disc

Endoderm
Ectoderm

(C)

Tentacle
20 Days
4 Days
Little or no cell movement
2 Days
8 Days
20 Days
Foot (little or no cell movement)

FIGURE 22.14 Budding in hydra. (A) A new individual buds about two-thirds of the way down the side of an adult hydra. (B) Schematic of the myoepithelium, with its unipotent endodermal and ectodermal cells and its multipotent interstitial stem cells. (C) Cell movements in *Hydra* were traced by following the migration of labeled tissues. The arrows indicate the starting and leaving positions of the labeled cells. The bracket indicates regions in which no net cell movement took place. Cell division takes place throughout the body column except at the tentacles and foot. (B after Q. Li et al. 2015. *J Genet Genomics* 42: 57–70; C after R. D. Campbell. 1967. *J Morphol* 121: 19–28.)

Compared with the myoepithelial stem cells (endoderm and ectoderm), interstitial stem cells are paused in G2 phase of the cell cycle for a longer period and cycle at a faster rate (Buzgariu et al. 2014), suggesting that the interstitial stem cells are poised to immediately respond to a need for cell replacement through rapid proliferation. These three cell types are all that are needed to form a hydra, and if hydra cells are separated and reaggregated, a new hydra will form (Gierer et al. 1972; Technau 2000; Bode 2011).

THE HEAD ACTIVATOR Experimental embryology—indeed, experimental biology—can be said to have started with Tremblay's studies of hydra regeneration. In 1741, Tremblay reported that "the story of the Phoenix who is reborn from his own ashes, fabulous as it is, offers nothing more marvelous than the discovery of which we are going to speak." He found that when he cut a hydra into as many as 40 pieces, "there are reborn as many complete animals similar to the first." Each piece would regenerate a head at its original apical end and a foot at its original basal end.

Every portion of the hydra's body column along the apical-basal axis is potentially able to form both a head and a foot. The animal's polarity, however, is coordinated by a series of morphogenetic gradients that permit the head to form only at one place and the basal disc to form only at another. Evidence for such gradients was first obtained from grafting experiments begun by Ethel Browne in the early 1900s. When hypostome tissue from one hydra is transplanted into the middle of another hydra, the transplanted tissue forms a new apical-basal axis, with the hypostome extending outward (**FIGURE 22.15A**). When a basal disc is grafted to the middle of a host hydra, a new axis also forms, but with the opposite polarity, extending a basal disc (**FIGURE 22.15B**). When tissues from both ends are transplanted simultaneously into the middle of a host, either no new axis is formed or the new axis has little polarity (**FIGURE 22.15C**; Browne 1909; Newman 1974). These experiments have been interpreted to indicate the existence of a **head activation gradient** (highest at the hypostome) and a **foot activation gradient** (highest at the basal disc). The head activation gradient can be measured by implanting rings

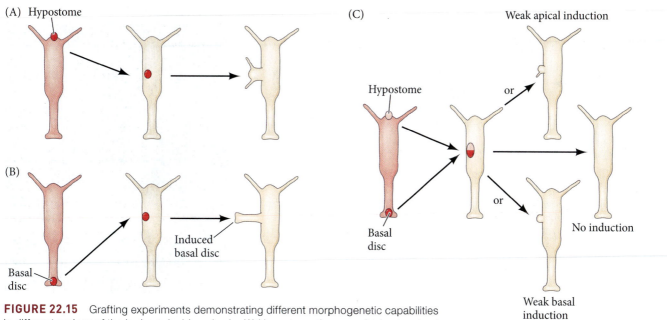

FIGURE 22.15 Grafting experiments demonstrating different morphogenetic capabilities in different regions of the hydra apical-basal axis. (A) Hypostome tissue grafted onto a host trunk induces a secondary axis with an extended hypostome. (B) Basal disc tissue grafted onto a host trunk induces a secondary axis with an extended basal disc. (C) If hypostome and basal disc tissues are transplanted together, only weak (if any) inductions are seen. (After S. A. Newman. 1974. *J Embryol Exp Morphol* 31: 541–555.)

of tissue from various levels of a donor hydra into a particular region of the host trunk (Wilby and Webster 1970; Herlands and Bode 1974; MacWilliams 1983b). The higher the level of head activator in the donor tissue, the greater the percentage of implants that will induce the formation of new heads. The head activation factor is concentrated in the hypostome and decreases linearly toward the basal disc.

THE HYPOSTOME AS ORGANIZER   Ethel Browne (1909; see also Lenhoff 1991) noted that the hypostome acted as an "organizer" of the hydra. This notion has been confirmed by Broun and Bode (2002), who demonstrated that (1) when transplanted, the hypostome can induce host tissue to form a second body axis; (2) the hypostome produces the head activation signal; (3) the hypostome is the only "self-differentiating" region of the hydra; and (4) the hypostome also produces a "head inhibition signal" that suppresses the formation of new organizing centers. (See **Further Development 22.4, The Organizing Properties of the Hypostome**, and **Further Development 22.5, Ethel Browne and the Organizer**, both online.)

A GRADIENT OF WNT3 IS THE INDUCER   The major head inducer of the hypostome organizer is a set of Wnt proteins acting through the canonical β-catenin pathway (Hobmayer et al. 2000; Broun et al. 2005; Lengfeld et al. 2009; see also Bode 2009). These Wnt proteins are seen in the apical end of the early bud, defining the hypostome region as the bud elongates (**FIGURE 22.16A**). If the Wnt signaling inhibitor GSK3 is itself inhibited throughout the body axis, ectopic tentacles form at all levels, and each piece of the trunk has the ability to stimulate the outgrowth of new buds. Similarly, transgenic hydra made to globally misexpress the downstream Wnt effector β-catenin form ectopic buds all along the body axis and even on top of newly formed ectopic buds (**FIGURE 22.16B**; Gee et al. 2010). When the hypostome is brought into contact with the trunk of an adult hydra, it induces expression of the *Brachyury* gene in a Wnt-dependent manner—just as vertebrate organizers do—even though hydras lack mesoderm (Broun et al. 1999; Broun and Bode 2002). These results strongly indicate that Wnt proteins (in particular, Wnt3) function as the head organizer during normal hydra development, but do they function similarly during regeneration? (See **Further Development 22.6, Morphallaxis and Epimorphosis in Hydra**, and **Further Development 22.7, The Head Inhibition Gradients**, both online.)

(A) *Wnt3* mRNA expression

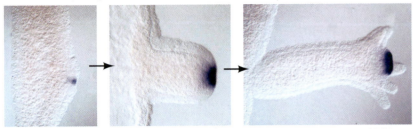

From B. Hobmayer et al. 2000. *Nature* 407: 186–189, courtesy of T. W. Holstein and B. Hobmayer

FIGURE 22.16 Wnt/β-catenin signaling during hydra budding. (A) *Wnt3* mRNA expression (purple) in the hypostome during early bud (left), midstage bud (center), and a bud with early tentacles (right). (B) Transgenic hydra made to misexpress β-catenin (the downstream Wnt effector) have numerous ectopic buds (including buds formed on top of other buds, such as the example marked with an arrow).

(B) Misexpression of β-catenin

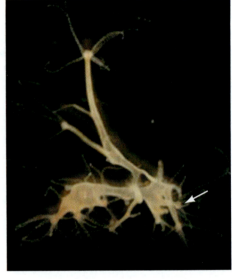

From L. Gee et al. 2010. *Dev Biol* 340: 116–124

? Developing Questions

Does the gradient model for planarian regeneration completely cut it? The simplest slice that cuts this flatworm into anterior and posterior halves is the most puzzling. This single cut results in cells on the anterior side of the cut making a tail, and on the posterior side making a head—totally different anatomical structures from cells that were direct neighbors before the cut! So local positional information (which would have been nearly identical in both sets of cells) may not be sufficient to explain how the change in fates needed for this regeneration is orchestrated.

Stem cell-mediated regeneration in flatworms

Planarian flatworms can reproduce asexually by binary fission, during which they split themselves in half, separating their posterior end from their anterior end, and each segment regenerates the lost parts. During regeneration, each piece re-creates all the appropriate cell types that make up the planarian, such as photoreceptors, nervous system, epithelium, muscle, intestines, pharynx, and gonads (see Roberts-Galbraith and Newmark 2015). Only recently has it been shown that the cells capable of this regeneration are the same pluripotent stem cells that repair and replace body parts. It has been known since the 1700s that when planarians are cut in half, just as occurs in asexual reproduction, the head half will regenerate a tail from the wound site while the tail half will regenerate a head (**FIGURE 22.17A,B**; Pallas 1766). Not until 1905, however, did Thomas Hunt Morgan and C. M. Child realize that such polarity indicated an important principle of development (see Sunderland 2010).[4] Morgan pointed out that if both the head and the tail were cut off a flatworm, thus trisecting the animal, the medial segment would regenerate a head from the former anterior end and a tail from the former posterior end, but never the reverse (**FIGURE 22.17C**). Furthermore, if the medial segment were sufficiently thin, the regenerating portions would be abnormal (**FIGURE 22.17D**). Both Morgan (1905) and Child (1905) postulated a gradient of anterior-producing materials concentrated in the head region. The middle segment would be told what to regenerate at both ends by the concentration gradient of these materials. If the piece were too narrow, however, the gradient would not be sensed within the segment.

THE BLASTEMA AND ADULT PLURIPOTENT STEM CELLS What cells form the new head or tail in planarians? For decades it was believed that the old cells *dedifferentiated* at the cut ends of the planarian to form a **regeneration blastema**, a collection of relatively undifferentiated cells that would be organized into new structures by paracrine factors located at the wound surface (see Baguñà 2012). In 2011, however, a series of experiments by Wagner and colleagues provided substantial evidence that dedifferentiation does *not* occur. Rather, the regeneration blastema forms from pluripotent stem cells called **clonogenic neoblasts** (**cNeoblasts**), a set of pluripotent cells in flatworms that serve as stem cells to replace the aging cells of the adult body (**FIGURE 22.18A**; Newmark and Sánchez Alvarado 2000; Pellettieri and Sánchez Alvarado 2007; reviewed in Adler and Sánchez Alvarado 2015; Zhu and Pearson 2016; Reddien 2018).

Clonogenic neoblasts can migrate to a wound site and regenerate the tissue. Wagner and colleagues were able to show that if planarians were irradiated at a dosage that

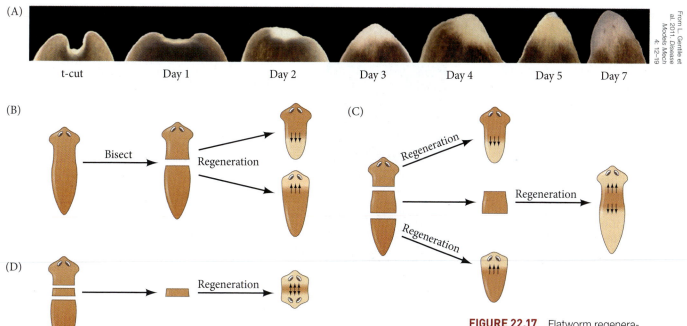

From L. Gentile et al. 2011. *Disease Models Mech* 4: 12–19

(A) t-cut   Day 1   Day 2   Day 3   Day 4   Day 5   Day 7

(B) Bisect   Regeneration

(C) Regeneration   Regeneration   Regeneration

(D) Regeneration

FIGURE 22.17 Flatworm regeneration and its limits. (A) Time course of a planarian flatworm regenerating a new head following head amputation. (B) If the planarian is cut in half, the anterior portion of the lower half regenerates a head while the posterior of the upper half regenerates a tail. The same tissue can generate a head (if it is at the anterior portion of the tail piece) or a tail (if it is at the posterior portion of the head piece). (C) If a flatworm is cut into three pieces, the middle piece will regenerate a head from its anterior end and a tail from its posterior end. (D) If the middle slice is too narrow, there is no discernible morphogen gradient within it, and regeneration is abnormal. (B–D after R. J. Goss. 1969. *Principles of Regeneration*, pp. 56–73. Academic Press: New York.)

destroyed nearly all neoblasts (dividing cells are killed more readily by radiation than nondividing cells, which is the basis for irradiating cancer sites), there would be some individual animals in which a single cNeoblast survived. From this neoblast, dividing progenitor cells formed, ultimately producing cell types of all germ layer origins. This single-cell response demonstrated that neoblasts are pluripotent cells residing in the adult body, capable of regenerating all tissues of the planarian (**FIGURE 22.18B**).

If cNeoblasts are essential for regeneration, their total loss should prevent regeneration. Next the researchers irradiated planarians so that all dividing cells were destroyed (**FIGURE 22.18C**). These planarians died because of failed tissue replacement. However, transplantation of a single cNeoblast into such an irradiated flatworm could, in some cases, restore all the cells of the organism. Not only did the flatworm survive, but it split into more planarians, and all cells of these new planarians had the same genotype as the single donor neoblast. These results conclusively demonstrated that the production of new cells during regeneration in the flatworm is from adult pluripotent stem cells (Wagner et al. 2011). (See **Further Development 22.8, cNeoblast Specialization**, online.)

HEAD-TO-TAIL POLARITY   As we saw with hydra, Wnt signaling appears to play a major role in establishing a polarity of differential cell fates. In hydra, this polarity is positively regulated by Wnt/β-catenin signaling along the apical-basal axis. In planarians, Wnt/β-catenin functions to establish anterior-posterior polarity of the regenerating flatworm; here, though, Wnts functioning through β-catenin promote tail development while repressing head regeneration (Gurley et al. 2008; Petersen and Reddien 2008, 2011). In fact, *Wnt* expression is excluded from the head, and functional proteins are presumed to be present in a tail-to-head gradient. Thus, the role that Wnt signaling plays in axis patterning is thematically similar in hydra and planarians, but its effects are different.

Several labs took a comparative approach to understanding the control over polarity during head regeneration in planarians. When planarians of the species *Procotyla fluviatilis* and *Dendrocoelum lacteum* are decapitated, they are unable to regenerate their heads. Researchers saw this distinction as an opportunity to identify the genes that are essential for the regeneration capabilities of planarian species such as *Dugesia japonica*. Comparing the transcriptomes of anterior-facing blastema cells from regeneration-competent

FIGURE 22.18 Cell production during planarian regeneration is accomplished by a pluripotent stem cell population of neoblasts. (A) Neoblasts in the planarian flatworm *Schmidtea mediterranea*, labeled with an RNA probe to soxP-2. Each pluripotent neoblast generates a colony of neoblast cells (red; nuclei are stained blue) in the flatworm. These clonogenic neoblast cells produce the differentiating cells of the regenerating flatworm. Neoblasts are scattered throughout the body posterior to the eyes (although they are not present in the centrally located pharynx). (B) Irradiation with 1750 rad kills almost all neoblasts. If even one survives, a single clonogenic neoblast can divide to generate a colony of dividing cells that will ultimately produce the differentiated cells of the organs. (C) Irradiation with 6000 rad eliminates all dividing cells. Transplanting a single clonogenic neoblast from a donor strain (red) not only results in the production of all the cell types in the organism but also restores the organism's capacity for regeneration. (B,C after E. Tanaka and P. W. Reddien. 2011. *Dev Cell* 21: 172–185.)

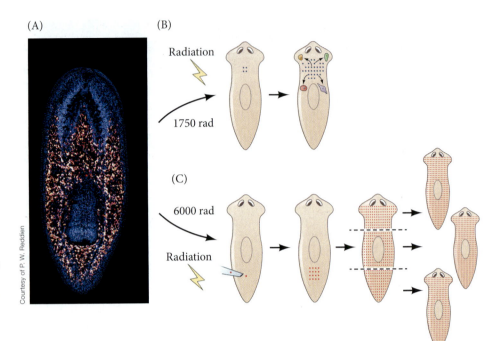

Courtesy of P. W. Reddien

species with those from regeneration-deficient species revealed an upregulation of genes indicative of highly active Wnt/β-catenin signaling only in the blastemas of regeneration-deficient planarians (Sikes and Newmark 2013). Most remarkable is that inhibition of Wnt/β-catenin signaling in regeneration-deficient species yields fully functional regenerated heads (**FIGURE 22.19**; Liu et al. 2013; Sikes and Newmark 2013; Umesono et al. 2013). These results demonstrate the inhibitory function of Wnt signaling on head specification during regeneration. In regenerating planarians, β-catenin is activated (via Wnts) in the posterior-facing blastema, which generates a tail. As in vertebrate development, an anterior polarity in planarian regeneration is dependent on repression of Wnt signaling, which prevents β-catenin accumulation and allows head formation. If β-catenin is eliminated from the posterior (tail-forming) blastema by RNA interference, that blastema will form a head (**FIGURE 22.20**). Indeed, when RNAi completely eliminates β-catenin from non-regenerating flatworms, the entire organism becomes a head, with normal-size eyes all around the periphery (Gurley et al. 2008; Iglesias et al. 2008) (See **Further Development 22.9, Notum Heads Wnt off at the Slice**, online.).

A MORPHOLOGICAL MEMORY MAP THAT FLEXES ITS PCG MUSCLES

There seem to be two possible ways in which an organism can rebuild lost parts: it can either create new populations of cells that restore the missing parts to make a full-size individual, or it can reassign identities to cells that are already there, reducing the "cellular real estate" occupied by each anatomical region to produce a smaller regenerated individual. Planarians apparently use both

From S. Y. Liu et al. 2013. *Nature* 500: 81–84

4 days   10 days   21 days

FIGURE 22.19 Restoration of head regeneration in *Dendrocoelum lacteum* by knockdown of β-catenin. This regeneration-incompetent flatworm cannot regrow a head following amputation (top row). If, however, the gene for β-catenin is knocked down with RNA interference (bottom row), this species is capable of regenerating its head following amputation over a 21-day period.

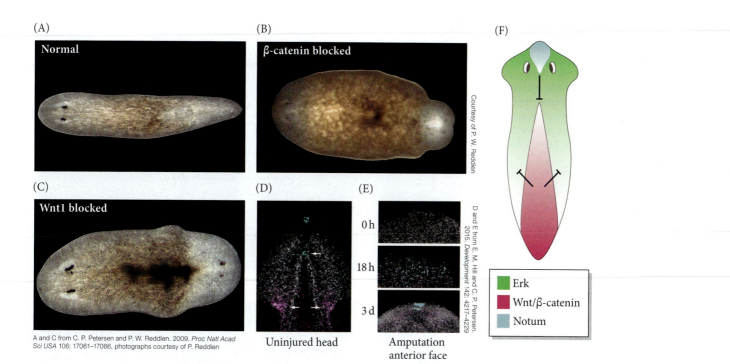

FIGURE 22.20   Polarity in planarian regeneration. Normally, Wnts are produced in the posterior blastema, and the result is a tail (A). If the Wnt pathway is blocked by using RNA interference against either β-catenin (B) or Wnt1 (C) mRNA, then the posterior blastema regenerates a head, resulting in a worm with heads at both ends. (D,E) Expression of *wnt11-6* (magenta) and *notum* (blue) in the planarian head (D) and in the anterior-facing blastema (E). *Chat* is a gene marker for nervous system cells (D, gray), and Hoechst stain labels all nuclei (E, gray). (F) Proposed model of anterior-posterior polarity. Wnt inhibits the anteriorly expressed head inducer Erk to promote tail, while Wnt is restricted from the most anterior head regions by Notum.

approaches. As discussed above, planarians do form a characteristic blastema that fuels new cell generation at the wound site. But planarians also transform existing tissues by changing cell identities to create an accurately proportioned yet smaller regenerate. This may sound crazy, since it must require a global re-patterning of cell identities across all axes. How could this happen? Is there a morphological map that provides the cells with the necessary positional information for this restructuring?

These questions will challenge researchers for a long time, but recent results suggest that a map of cell fate may in fact exist in the flatworm, and that it is related to the expression of embryonic patterning genes that confer fates on cells relative to the cells' positions. These "positional control genes," or PCGs, are expressed throughout planarian life (**FIGURE 22.21A**; Scimone et al. 2016; Reddien 2018). Their expression across all axes is linked to the organization of muscle fibers (**FIGURE 22.21B–D**; Witchley et al. 2013). The muscle fibers are organized in three different layers—circular, diagonal, and longitudinal—that occupy increasingly inner positions in the worm (**FIGURE 22.22A**). Importantly, loss of muscle fibers from different layers causes different defects in regeneration (**FIGURE 22.22B**). Knockdown of *myoD* reduces the number of longitudinal fibers and also abolishes regenerative capacity, whereas loss of *nkx1-1* reduces the number of circular fibers and results in a bifurcated midline and duplicated head following an anterior amputation (Scimone et al. 2017).

A MODEL FOR HOW INJURY CHANGES US   If the pattern of muscle-mediated PCG expression in planarians represents a morphological memory map of positional information, then how is this map used to change cell fates for the regeneration of the missing parts? We already know that antagonistic anterior-to-posterior gradients involving canonical Wnt signaling play a major role in establishing the

FIGURE 22.21 Positional control genes (PCGs) in planarians. (A) Fluorescent in situ hybridization of select PCGs across all axes of the adult planarian. (B–D) PCGs are expressed primarily in differentiated muscle. Some regions of the planarian (B, white box, expanded in C) show near-perfect co-localization of PCGs (magenta) with muscle (green). For example, *wntP-2* is seen expressed in the extended nucleus of this isolated muscle fiber (D).

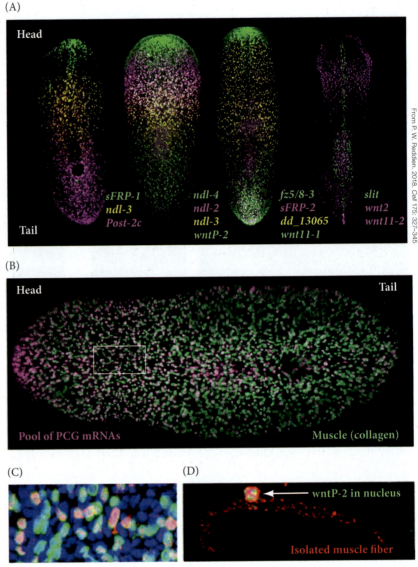

anterior-posterior axis. Wnt signaling promotes tail development in a regenerate while repressing head formation, and Notum, normally expressed in the head, inhibits Wnt expression, allowing a head to form in a regenerate (Birkholz et al. 2018; Reddien 2018; Rink 2018). Furthermore, these two antagonistic gradients are reestablished in an anterior-facing blastema forming at the cut surface of a tail piece, with Notum expressed in the most anterior region and Wnt expression being highest more posteriorly (see Figure 22.20E). In fact, there are some 200 genes that are "generically" induced by wounding, yet evidence suggests that the induction of both *notum* and *wnt1* in opposing anterior and posterior positions, respectively, functions to reset the pattern of PCGs to foster normal regeneration (Wenemoser et al. 2012; Wurtzel et al. 2015). Moreover, this antagonistic relationship may be reinforced through an autoregulatory function of Wnt/β-catenin signaling, which serves to maintain its own level of expression in coordination with a similar but antagonistic autoregulatory head-patterning system (likely including Notum) (Stückemann et al. 2017). The overall model, therefore, is that following injury, induction of these two opposing self-organizing gradient systems effects the re-assignment of cell identities across the anterior-posterior axis of the regenerating flatworm. This in turn resets the expression of PCGs,

(A)

(B)

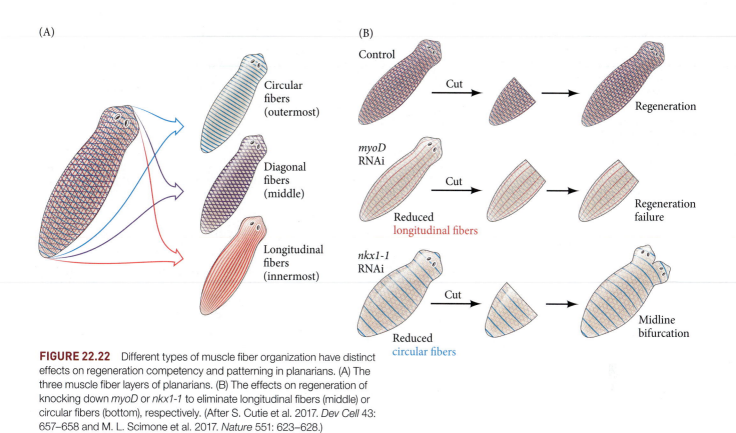

FIGURE 22.22 Different types of muscle fiber organization have distinct effects on regeneration competency and patterning in planarians. (A) The three muscle fiber layers of planarians. (B) The effects on regeneration of knocking down *myoD* or *nkx1-1* to eliminate longitudinal fibers (middle) or circular fibers (bottom), respectively. (After S. Cutie et al. 2017. *Dev Cell* 43: 657–658 and M. L. Scimone et al. 2017. *Nature* 551: 623–628.)

thereby programming the developing cNeoblasts underlying the muscles to produce a smaller but proportionally correct flatworm (**FIGURE 22.23**).

FURTHER DEVELOPMENT

AN ALTERNATIVE MORPHOLOGICAL MEMORY MAP: "IT'S ELECTRIC," BIOELECTRIC In addition to considering the roles played by PCGs and the underlying gene regulatory networks in controlling cell identity, we might also consider the influence that the physiological states of cells may have on regeneration. The differences in the endogenous electrical properties of cells across an individual are now being proposed as the basis for a second map of positional information, known as the bioelectric pattern (McLaughlin and Levin 2018).

Bioelectric signaling refers to communication between cells based on changes in the transmembrane voltage potential (differences in the electric charge across a cell's membrane) that is known to affect cellular behavior. Could the electrical states of cells also affect regeneration? Consider what happens to a tissue during an injury. The cell-cell junctions and even cell membranes can be seriously compromised, resulting in a chaotic flood of ion movement. Changes in the distribution of charged ions in or outside the cell will rapidly alter its membrane potential (the voltage across its membrane, Vmem) (**FIGURE 22.24A**). Thus, increases or decreases in Vmem (hyperpolarization or depolarization, respectively) will occur at injury sites. In addition, differences in bioelectric properties among cells establish visible patterns across whole tissues and organisms. For instance, a Vmem gradient exists in a planarian from the most depolarized cells in the head to the most hyperpolarized cells in the tail (see Figure 22.24A). Pharmacological or molecular manipulation of ion channels or gap junctions to induce widespread depolarization following amputation of the planarian head and tail will cause the midbody piece to regenerate two

(?) Developing Questions

What is it about incurring a wound that triggers the induction of *wnt* or *notum*? That is, what is the actual wound "signal"?

FIGURE 22.23 Overall model for planarian regeneration. Injury-induced wound signaling induces a re-patterning of the anterior-to-posterior expression of *notum* and *wnt1* at the cut surfaces. This new morphogen pattern reinforces PCG expression programs that lead to cNeoblast specialization and subsequent differentiation. (After P. W. Reddien. 2018. *Cell* 175: 327–345.)

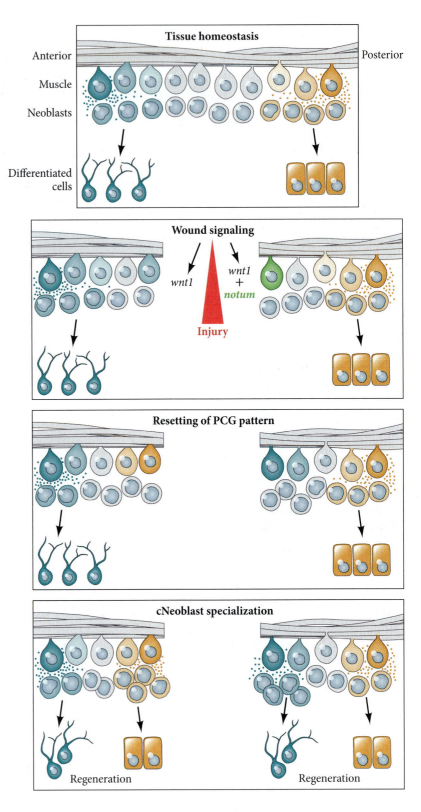

oppositely facing heads; if instead widespread hyperpolarization is induced, no head will form (**FIGURE 22.24B,C**; Beane et al. 2011).

How could differences in membrane voltage influence whole tissue reconstruction? We just learned how the expression of positional control genes affects neoblast activation and specialization, and that one critical player in controlling anterior-posterior regeneration is Wnt/β-catenin signaling. Although double

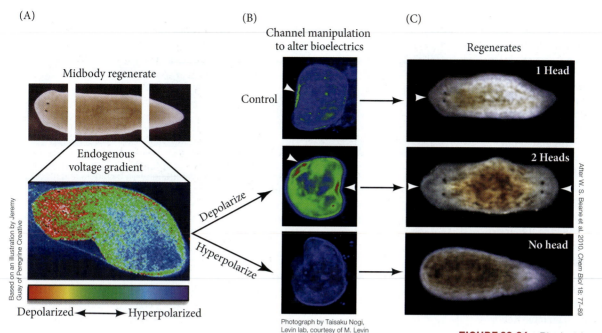

Based on an illustration by Jeremy Guay of Peregrine Creative

Photograph by Taisaku Nogi, Levin lab, courtesy of M. Levin

After W. S. Beane et al. 2010. *Chem Biol* 18: 77–89

FIGURE 22.24 Bioelectric regulation of planarian regeneration. (A) An amputated midbody trunk immediately displays a gradient of membrane potential differences across its anterior-posterior axis. (B) Pharmacological or molecular manipulation of ion transport can establish depolarized or hyperpolarized states, as visualized with DiBAC, a vital voltage dye indicator. (C) Depolarization causes an ectopic head to form in the posterior, while hyperpolarization blocks any head from forming.

heads form on a midbody slice of a planarian as a result of a loss of Wnt/β-catenin signaling, a similarly treated midbody slice can be rescued if treated with a hyperpolarizing drug (**FIGURE 22.25**; Beane et al. 2011). Therefore, bioelectric states may function through many of the known PCGs in planarians. This implies that the morphogenetic map is highly dynamic, capable of translating physiological effects into genetic programs that affect form (a type of "physiological epigenetics") (Levin et al. 2017, 2018).

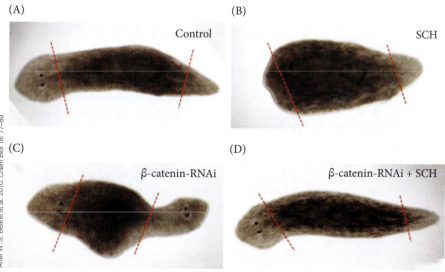

After W. S. Beane et al. 2010. *Chem Biol* 18: 77–89

FIGURE 22.25 Bioelectric interactions with Wnt/β-catenin signaling. Dotted red lines indicate the positions of the original cuts to produce a midsection. Tissue outside these lines has been regenerated. (A) Control. (B) Treatment with SCH-28080 (SCH), an inhibitor of the proton/potassium exchanger (H,K-ATPase) that is known to set up the voltage gradient in planarians, results in a hyperpolarized condition leading to no head formation. (C) In contrast, knockdown of β-catenin produces two axial heads. (D) Remarkably, SCH-28080 treatment reverses this affect; it prevents ectopic head development during posterior regeneration.

Developing Questions

Fun fact! We told you that planarian midbody slices briefly exposed to depolarization will regenerate biaxial heads. What we did not mention is that if those two heads are again amputated but in the absence of any bioelectric modifiers, then the midbody piece, surprisingly, will still regenerate double heads! These animals are permanently two-headed. This amputation can be repeated over and over again with the same result. Since splitting in two and regenerating is a normal pattern of reproduction in flatworms, this two-headed morphology can be considered an inherited change to the animal's target morphology. This suggests that a brief change in bioelectrics leads to a permanent rewriting of the target morphology. How can this alternative epigenetic mechanism of pattern formation be transmitted from one generation to the next?

(A) Control 45 days post-amputation

(B) Cocktail-treated 45 days post-amputation

FIGURE 22.26 Restoration of postmetamorphic frog regeneration by bioelectric modulation. (A) Following limb amputation of a control froglet, no regeneration occurs. (B) After 24 hours of ionophore exposure ("cocktail treatment") to alter the bioelectric state of the amputated limb, blastema formation and regrowth of the entire limb are rescued (distalmost elements, including toes, are seen; arrowheads).

Further evidence for a mechanistic role of bioelectric signaling comes from comparative studies of transcriptomes in *Xenopus* embryos, *axolotl* regenerates, and differentiating human mesenchymal cells exposed to states of depolarization. The results have revealed conserved transcriptional networks associated with this depolarized bioelectric state (Pai et al. 2015). Remarkably, modulation of bioelectric states can even restore regenerative ability in the limbs of postmetamorphic frogs (**FIGURE 22.26**; Tseng and Levin 2013). Moreover, depolarizing human bone-forming cells and fat cells partially reprograms them into mesenchymal stem cells, some of which are able to re-form both bone-forming cells and fat cells (Sundelacruz et al. 2013). The field of bioelectrics in development and regeneration is still immature, but its expected applications in regenerative medicine have "exciting potential."

Tissue-Restricted Animal Regeneration
Salamanders: Epimorphic limb regeneration

When an adult salamander limb is amputated, the remaining limb cells reconstruct a new limb, complete with all its differentiated cells arranged in the proper order. Remarkably, the limb regenerates only the missing structures and no more. For example, when the limb is amputated at the wrist, the salamander forms a new wrist and foot, but not a new elbow. In some way, the salamander limb "knows" where the proximal-distal axis has been severed and is able to regenerate from that point on (**FIGURE 22.27**).

Limb regeneration in salamanders is powered by the formation of the regeneration blastema at the distalmost end of the amputated limb (**FIGURE 22.28**). As is similar in planarians, the blastema is an aggregation of relatively undifferentiated cells. The stages of limb regeneration in the salamander are as follows (**FIGURE 22.29**; Hass and Whited 2017):

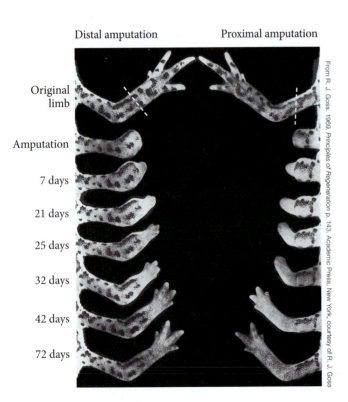

Distal amputation   Proximal amputation

Original limb

Amputation

7 days

21 days

25 days

32 days

42 days

72 days

FIGURE 22.27 Regeneration of a salamander forelimb. The amputation shown on the left was made below the elbow; the amputation shown on the right cut through the humerus. In both instances, the correct positional information was respecified, and a normal limb was regenerated within 72 days.

(A)

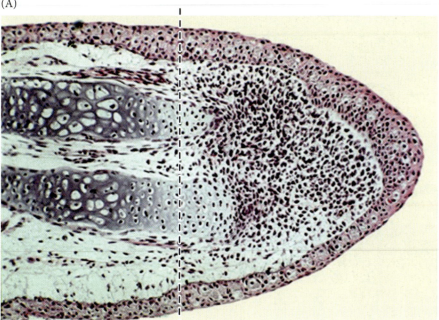

A and B after A. Simon and E. M. Tanaka. 2013. *Wiley Interdiscip Rev Dev Biol* 2: 291–300

(B)

Nerve

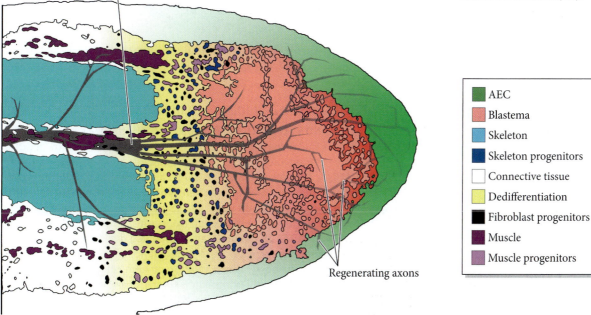

Regenerating axons

🟩	AEC
🟥	Blastema
🟦	Skeleton
🟦	Skeleton progenitors
⬜	Connective tissue
🟨	Dedifferentiation
⬛	Fibroblast progenitors
🟪	Muscle
🟪	Muscle progenitors

FIGURE 22.28 Anatomy of the limb blastema. (A) Longitudinal section of a regenerating newt limb following amputation (dashed line) and stained with hematoxylin/eosin. (B) Artistic representation of the different cell and tissue components in the amputated limb. An outer epidermal thickening called the apical epidermal cap (AEC; green) covers the wound. Proximal to the cut plane lie the preexisting differentiated tissues of muscle (purple), skeleton (light blue), nerve (dark gray), and connective tissue (white). Once the wound is covered, the cells from the distal tip of the existing tissues undergo dedifferentiation (yellow zone), which produces specified, lineage-restricted progenitors for each tissue type (muscle progenitors, light purple; skeleton progenitors, indigo blue; fibroblast progenitors, black; regenerating axons, gray graded from proximal to distal). These progenitor cells form a mass of proliferative cells directly beneath the AEC that constitutes the blastema (red).

1. Blood and immune cells flood the amputated area, and a blood clot quickly forms.

2. Wounding triggers an activation of stem/progenitor cell proliferation.

3. Epidermal cells along the edge of the cut migrate over the wound to form the **wound epidermis**.

4. Through cell proliferation and continued migration, the wound epidermis thickens into the **apical epidermal cap** (**AEC**).

5. Signals from the AEC to the amassing population of progenitor cells underneath it foster development of the regeneration blastema.

6. Continued proliferation and progressive differentiation of the blastema fuel the outgrowth of the limb regenerate.

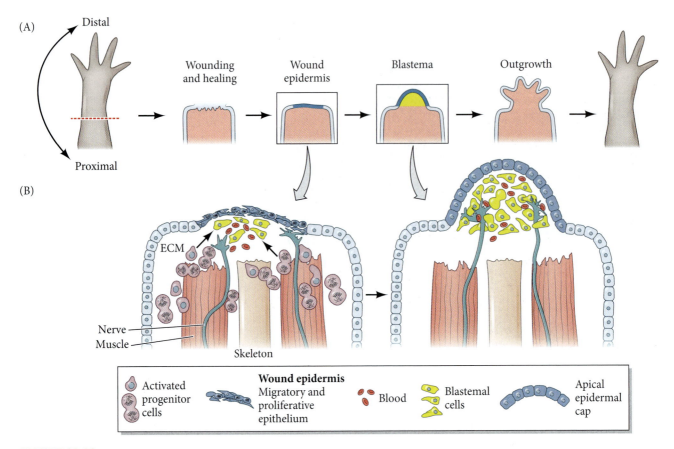

FIGURE 22.29 Stages of limb regeneration in the salamander. (After J. L. Whited and C. J. Tabin. 2009. *J Biol* 8: 5. doi:10.1186/jbiol105; B. J. Haas and J. L. Whited. 2017. *Trends Genet* 33: 553–565.)

(?) Developing Questions

How does a localized wound trigger a global activation of stem cell proliferation across the entire body, the outcome of which is a localized regeneration blastema?

Defining the cells of the regeneration blastema

DEDIFFERENTIATION AND STEM CELL ACTIVATION   Historically, the salamander limb blastema has served as the primary example of epimorphic regeneration by dedifferentiation of the tissue at the wound, which then proliferates and redifferentiates into the new limb parts (see Brockes and Kumar 2002; Gardiner et al. 2002; Simon and Tanaka 2013). Bone, dermis, and cartilage just beneath the site of amputation contribute to the regeneration blastema, as do satellite cells from nearby muscles (Morrison et al. 2006). More recent studies using the axolotl (Mexican salamander) have revealed that stem cells throughout the salamander may be activated during regeneration and at least in part contribute to the formation and maintenance of the blastema (Payzin-Dogru and Whited 2018). Curiously, this activation of stem cell proliferation was a systemic response to a local injury; the contralateral limbs, heart, liver, and spinal cord all showed increased proliferation (Johnson et al. 2018).

The current work on amphibian regeneration is requiring that we abandon an earlier held idea that there is a distinct difference between planarian stem cell-mediated repair and the dedifferentiated responses in the salamander. This idea is now being modified because of our new understanding that there are actually more similarities than differences in these two regenerating systems (Nacu and Tanaka 2011).

FATES RESTRICTED   When a salamander limb is amputated, a blood clot forms. Within 6–12 hours, epidermal cells from the remaining stump migrate to cover the wound surface, forming the wound epidermis. Unlike in wound healing in mammals, no scar forms, and the dermis does not move with the epidermis to cover the site of amputation. The nerves innervating the limb degenerate for a short distance proximal to the plane of amputation (see Chernoff and Stocum 1995).

During the next 4 days, the extracellular matrices (ECMs) of the tissues beneath the wound epidermis are degraded by proteases, liberating single cells that undergo dramatic dedifferentiation: bone cells, cartilage cells, fibroblasts, and myocytes all lose their differentiated characteristics. Genes that are expressed in differentiated tissues (such as the *mrf4* and *myf5* genes expressed in muscle cells) are downregulated, while there is a dramatic increase in the expression of genes, such as *msx1*, that are associated with the proliferating progress zone mesenchyme of the embryonic limb (Simon et al. 1995). These dedifferentiated cells, paired with activated stem cells, migrate under the wound epidermis to give rise to the regeneration blastema underlying the apical epidermal cap. These cells are the ones that will continue to proliferate and that will eventually redifferentiate to form the new structures of the limb (Butler 1935). The AEC acts similarly to the apical ectodermal ridge (AER) during normal limb development (see Chapter 19; Han et al. 2001).

One of the major questions of regeneration is whether the cells contributing to the blastema retain a "memory" of what they had been. In other words, do new muscles arise from old muscle cells that dedifferentiated, or can any cell of the blastema become a muscle cell? Kragl and colleagues (2009) found that the blastema is not a collection of homogeneous, fully dedifferentiated cells. Rather, in the regenerating limbs of the axolotl, muscle cells arise only from old muscle cells, dermal cells arise only from old dermal cells, and cartilage arises only from old cartilage or old dermal cells (**FIGURE 22.30**). Thus, the blastema is also not a collection of unspecified multipotent progenitor cells. Rather, the cells retain their specification, and the blastema is a heterogeneous assortment of *restricted* progenitor cells.

THE REQUIREMENT FOR NERVES AND THE APICAL EPIDERMAL CAP The growth of the regeneration blastema depends on the presence of both the apical epidermal cap and nerves. If, following an amputation, the AEC is prevented from forming by immediately pulling the dorsal and ventral skin over the wound and suturing it closed, then all regeneration is halted—no blastema forms, no outgrowth occurs. Interestingly, this is the protocol physicians practice when someone has an amputated digit or limb. In fact, if a child's amputated digit is not closed in this way and is instead left open and kept sterile,

FIGURE 22.30 Blastema cells retain their specification even when they dedifferentiate. (A,B) Schematic representation of the procedure wherein a particular tissue (in this case, cartilage) is transplanted from a salamander expressing a *GFP* transgene into a wild-type salamander limb. Later, the limb is amputated through the region of the limb containing GFP expression, and a blastema is formed containing GFP-expressing cells that had been cartilage precursors. The regenerated limb is then studied to see if GFP is found only in the regenerated cartilage tissues or in other tissues. The dashed lines in (B) mark the position of the amputation. (C) Longitudinal section of a regenerated limb 30 days after amputation. Muscle cells are stained red; nuclei are stained blue. The majority of GFP-expressing cells (green) were found in regenerated cartilage; no GFP was seen in the muscle. (After M. Kragl et al. 2009. *Nature* 460: 60–65, courtesy of E. Tanaka.)

(A)

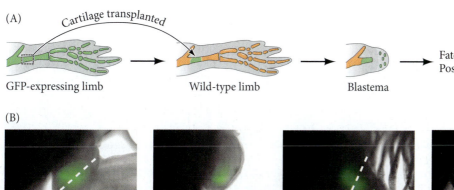

Cartilage transplanted

GFP-expressing limb → Wild-type limb → Blastema → Fate? Positional identity?

(B)

 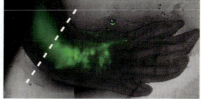

Graft   Amputation   Blastema   Regenerated limb

(C)   Regenerated limb

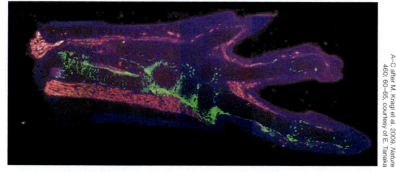

A–C after M. Kragl et al. 2009. *Nature* 460: 60–65, courtesy of E. Tanaka

it will regenerate. Therefore, allowing time for the wound epidermis to form an AEC is important, whether you are a human or a salamander.

The AEC stimulates the growth of the blastema by secreting Fgf8 (just as the apical ectodermal ridge does in normal limb development), but the effect of the AEC is only possible if nerves are present (Mullen et al. 1996). Both sensory and motor axons innervate the blastema such that sensory axons make direct contact with the AEC and motor axons terminate in the blastema mesenchyme (see Figure 22.28B). Singer and others demonstrated that a minimum number of nerve fibers of either sensory or motor type must be present for regeneration to take place (Todd 1823; Singer 1946, 1952, 1954; Singer and Craven 1948; Sidman and Singer 1960). Moreover, these nerve fibers are necessary for the proliferation and outgrowth of the blastema (**FIGURE 22.31**; Farkas et al. 2016; Farkas

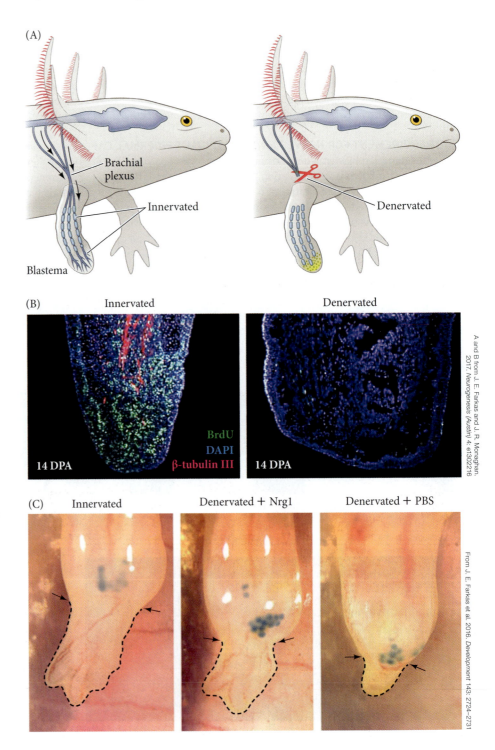

FIGURE 22.31 The requirement for nerves and sufficiency of Neuregulin for axolotl limb regeneration. (A) Illustration of the experimental procedure for a denervated limb. (B) As seen in the control regenerate (left), the blastema is normally highly innervated (indicated by labeling with β-tubulin III, a neural marker; red) and densely populated with proliferating cells (BrdU, green). Denervation (right) nearly abolishes all cell proliferation. DPA, days post-amputation. (C) Beads coated with Neuregulin-1 (Nrg1; blue) implanted into a denervated amputated limb are sufficient to rescue limb regeneration to the point of digit formation as compared to a control bead soaked in phosphate buffered saline (PBS).

A and B from J. E. Farkas and J. R. Monaghan. 2017. *Neurogenesis (Austin)* 4: e1302216

From J. E. Farkas et al. 2016. *Development* 143: 2724–2731

and Monaghan 2017). If the limb is first denervated and then amputated, no regeneration will occur. If a wound is made in the epidermis of an intact proximal limb and a nerve is diverted to the wound area, an ectopic blastema-like bud will form, but not a fully regenerated limb. These two experiments suggest that the nerve is both necessary and sufficient for blastema formation.[5] To induce a complete ectopic limb, not only does a nerve need to be diverted to the wound site, but an epidermal graft from the opposite side of the limb (from a posterior to an anterior location) needs to be placed near the wound (**FIGURE 22.32A–D**; Endo et al. 2004). These results suggest that during normal limb regeneration, the regenerating nerves deliver important signals to the AEC. They also suggest, however, that signals from nerves are not alone sufficient for ectopic limb growth; for that growth, positional cues from an epidermis that are different from the positional cues at the wound site itself are also needed (see Yin and Poss 2008; McCusker and Gardiner 2011, 2014). What could these regeneration-promoting signals from the nerve and AEC signals be? (See **Further Development 22.10, A Recapitulation of Limb Development**, and **Further Development 22.11, Newt Anterior Gradient Protein**, both online.).

Luring the mechanisms of regeneration from zebrafish organs

So far in this chapter, we have discussed such diverse mechanisms of regeneration as morphallaxis in hydra, the planarian's deployment of pluripotent stem cells, and the elegant epimorphosis exhibited by the salamander's blastema. Next we will expand our understanding of these regenerative processes by looking at organ regeneration in zebrafish (*Danio rerio*). The zebrafish has increasingly been employed to study organ regeneration due to its regenerative competencies and its genetic and technical advantages. Most notable has been its use in studying the molecular regulation of regeneration in the fin, heart, central nervous system, eye, liver, pancreas, kidney, bone, and sensory hair cells of the inner ear and lateral line system (reviewed in Shi et al. 2015; Zhong et al. 2016). Here we highlight some of the insights gleaned from investigating zebrafish fin and heart regeneration.

WNT UPON A FIN The zebrafish caudal fin fans out along the dorsal-ventral axis with 16 to 18 segmented bony rays separated from one another by inter-ray tissue. The regenerative capacity of the fin is so robust and easily accessible that clipping caudal fins for molecular analysis is a routine procedure in most zebrafish research labs. The fin ray is made primarily of bone, but it also includes a diverse set of other cell types, including fibroblasts, blood vessels, nerves, and pigment cells. Like the salamander limb, upon amputation the zebrafish fin rays first close the wound with epidermal cells that form an apical epidermal cap, and most tissue types undergo

FIGURE 22.32 Induction of ectopic limbs in salamanders. (A) Schematic showing experiment in which a nerve is diverted to a wound site in the limb epidermis (gray square) and an epidermal skin graft from the posterior portion of the contralateral limb (blue square) is grafted next to the wound site. (B,C) Results of this experiment show that a limb blastema is induced (B), which develops into a full limb (C; arrow). (D) The regenerated accessory limb (arrow) is correctly patterned, as seen by the alcian blue staining of cartilaginous elements. (E) Accessory limbs can be induced solely through the application of beads coated with BMP2 (or BMP7) and Fgf2/8 to a wound location in the limb.

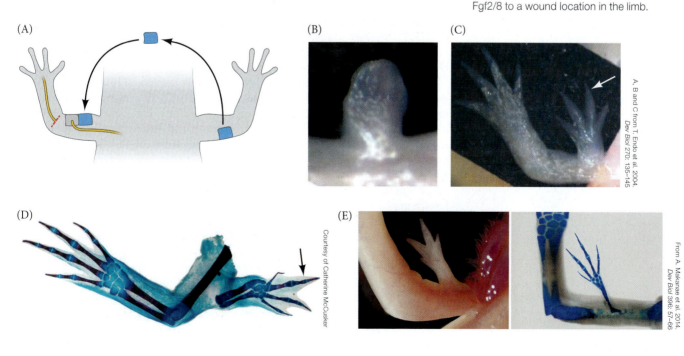

(A) (B) (C) (D) (E)

A, B and C from T. Endo et al. 2004, *Dev Biol* 270: 135–145

Courtesy of Catherine McCusker

From A. Makanae et al. 2014, *Dev Biol* 396: 57–66

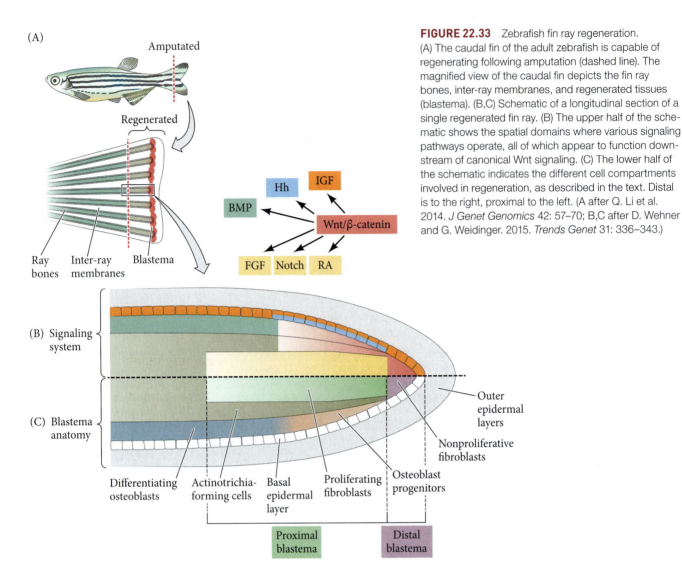

FIGURE 22.33 Zebrafish fin ray regeneration. (A) The caudal fin of the adult zebrafish is capable of regenerating following amputation (dashed line). The magnified view of the caudal fin depicts the fin ray bones, inter-ray membranes, and regenerated tissues (blastema). (B,C) Schematic of a longitudinal section of a single regenerated fin ray. (B) The upper half of the schematic shows the spatial domains where various signaling pathways operate, all of which appear to function downstream of canonical Wnt signaling. (C) The lower half of the schematic indicates the different cell compartments involved in regeneration, as described in the text. Distal is to the right, proximal to the left. (A after Q. Li et al. 2014. *J Genet Genomics* 42: 57–70; B,C after D. Wehner and G. Weidinger. 2015. *Trends Genet* 31: 336–343.)

dedifferentiation, proliferation, and distal migration to establish a blastema (Knopf et al. 2011; Stewart and Stankunas 2012). Some contributions may occur from as yet unidentified resident stem/progenitor cells capable of generating osteoblasts in the absence of any committed bone cells (Singh et al. 2012).

The regenerated fin can be divided into four basic sections (**FIGURE 22.33**): (1) the *distal blastema*, made up of nonproliferative fibroblasts; (2) the *proliferative proximal blastema*, which constitutes the major portion of the dedifferentiated mesenchyme; (3) the *differentiating proximal blastema*, which functions to add differentiated cells to existing and newly formed tissues during outgrowth; and (4) the laterally located *epidermal layers*, which serve as complex signaling centers throughout regeneration. Despite the seemingly simple distal outgrowth that typifies fin regeneration, the molecular regulation of this process is highly complex and involves all major signaling pathways known to play roles during embryonic development. We will focus on just one here, the Wnt/β-catenin signaling pathway, which appears to be the first "domino" that begins to reveal the regenerative pattern of fin rays (comprehensively reviewed in Wehner and Weidinger 2015). The Wnt/β-catenin pathway is active in the distal blastema and the lateralmost regions of the proximal proliferative blastema constituting progenitors of osteoblasts and actinotrichia (fibrous elements of the fin; see Figure 22.33). Loss and gain of function of Wnt/β-catenin signaling results in the decrease and increase of blastema proliferation and regeneration rate, respectively (Kawakami et al. 2006;

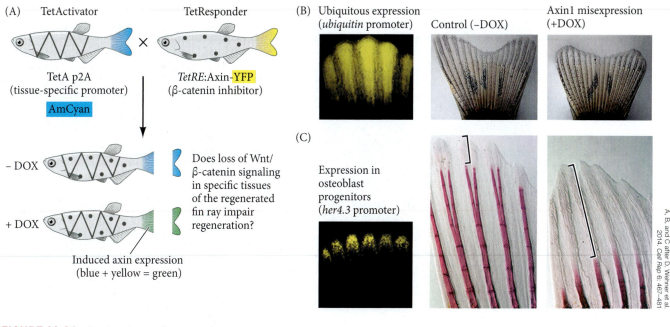

FIGURE 22.34 Testing the spatiotemporal requirements of Wnt/β-catenin signaling during fin regeneration. (A) Experimental design using the Tet/On system for inducible gene expression. One transgenic fish possesses a tissue-specific promoter (*ubiquitin* or *her4.3*), driving the expression of tetracycline and cyan fluorescent protein. A different transgenic fish has a transgene with the *tet* promoter driving Axin1-YFP expression; functional transcription of this transgene will occur only in combination with doxycycline (DOX). Crossing the two fish generates double transgenic individuals (represented by dots and stripes) that enable the spatial (tissue-specific promoter) and temporal (DOX) expression of Axin1.

(B,C) Misexpression of Axin1 disrupts regeneration. The photos at left show the pattern of Axin1 misexpression (yellow) driven by the *ubiquitin* (B) or *her4.3* (C) promoter. The photos at right show the tail fin 12 days after amputation, with (+DOX) or without (–DOX) Axin1 misexpression. (B) When Wnt/β-catenin is inhibited by misexpressing Axin1 throughout the fin (*ubiquitin* promoter), there is a reduced rate of fin regeneration. (C) When Axin1 is misexpressed only in the osteoblast progenitor cells of the fin (*her4.3* promoter), ossification of the fin ray bones during regeneration is severely impaired (red stain, brackets). (After D. Wehner et al. 2014. *Cell Rep* 6: 467–481.)

Stoick-Cooper et al. 2007; Huang et al. 2009; Wehner et al. 2014). The Weidinger Lab demonstrated that misexpression of the β-catenin inhibitor Axin1 throughout the fin blastema or just in the lateral progenitor zones resulted in significantly reduced fin rays as well as a failure of fin rays to ossify (**FIGURE 22.34**; Wehner et al. 2014). It appears, however, that Wnt/β-catenin functions indirectly through the modulation of other mitogenic regulators, such as Hedgehog proteins (Sonic hedgehog and Indian hedgehog), Fgf8, retinoic acid, and insulin-like growth factor.

EPIMORPHOSIS, COMPENSATION, AND TRANSDIFFERENTIATION: THE HEART OF THE MATTER The zebrafish has a relatively simple tubular heart. Venous blood enters the sinus venosus, passes into the single atrium, is pumped into the single ventricle, and leaves the heart through the bulbus arteriosis. Several different injury models—involving surgical removal of pieces of the myocardium, cryo-injury, and genetically induced tissue-specific ablation—have been adopted to study regeneration of heart tissues in zebrafish (reviewed in Shi et al. 2015; Mokalled and Poss 2018). The zebrafish heart retains the ability to regenerate throughout the life of the fish, which is in part due to the sustained mitotic capacity of the cardiac myocytes (muscle cells) that constitute a majority of the heart tissue (Poss et al. 2002). It is clear that a major contribution to regeneration of the adult heart comes directly from preexisting cardiac myocytes. Indeed, use of the "Zebrabow" transgenic system (see Chapter 2) for tracing the lineage of cells under the control of heart-specific promoters has demonstrated that previously differentiated cardiac myocytes give rise to clones of regenerated cells (**FIGURE 22.35**; Gupta and Poss 2012). These studies and the work described in the Further Development sections below demonstrate that

(A)

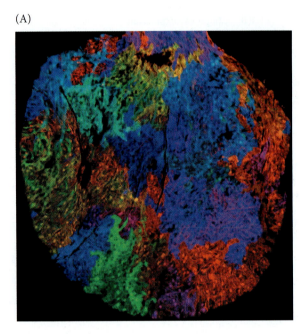

(B) Control (uninjured)

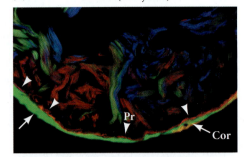

14 days post-ablation

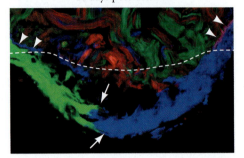

60 days post-ablation

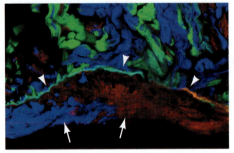

FIGURE 22.35 Preexisting cardiomyocytes contribute to ventricular regeneration in zebrafish. (A) Double-transgenic zebrafish were used to produce multicolor clonal labeling only in cardiomyocytes, as controlled by the *cmlc2* promoter for Cre expression. The "ER" in CreER denotes its estrogen-responsive control, which enables researchers to use the drug tamoxifen to induce recombination at any time they desire. This image is of a 6-week-postfertilization heart ventricle following recombination at 4 days postfertilization. Patches of distinct colors can be seen, which indicates that the heart is derived from only several dozen cardiac progenitors. (B) These preexisting clonally labeled cardiac myocytes are seen contributing to a majority of the regenerated ventricular tissue. Arrowheads and arrows indicate the primordial (Pr) and cortical layers (Cor), respectively.

the zebrafish uses dedifferentiation, transdifferentiation, blastema formation, and compensatory proliferation to regenerate the heart. The stages of zebrafish heart regeneration are summarized below (**FIGURE 22.36**; González-Rosa et al. 2017):

1. **The early responses (see Figure 22.36C):**

   a. Injury by amputation, freezing, or other means induces an inflammatory response required for regeneration to proceed. Macrophages, neutrophils, and other cells invade the site of injury.

   b. Localized apoptosis occurs near the edges of the injury.

   c. The normally tightly adherent endocardial cells ball up and start to express both embryonic and cytokine genes.

2. **The regeneration scaffold (see Figure 22.36D):**

   a. Both the endocardial cells and epicardial cells proliferate and migrate over the inner and outer surfaces of the injured tissue.

   b. The myofibroblasts and extracellular matrix components increase within the injured area, creating fibrotic, scarlike tissue.

3. **The cardiac blastema (see Figure 22.36E):**

   a. The myocardium (trabecular and cortical) at the edge of the injury undergoes significant proliferation.

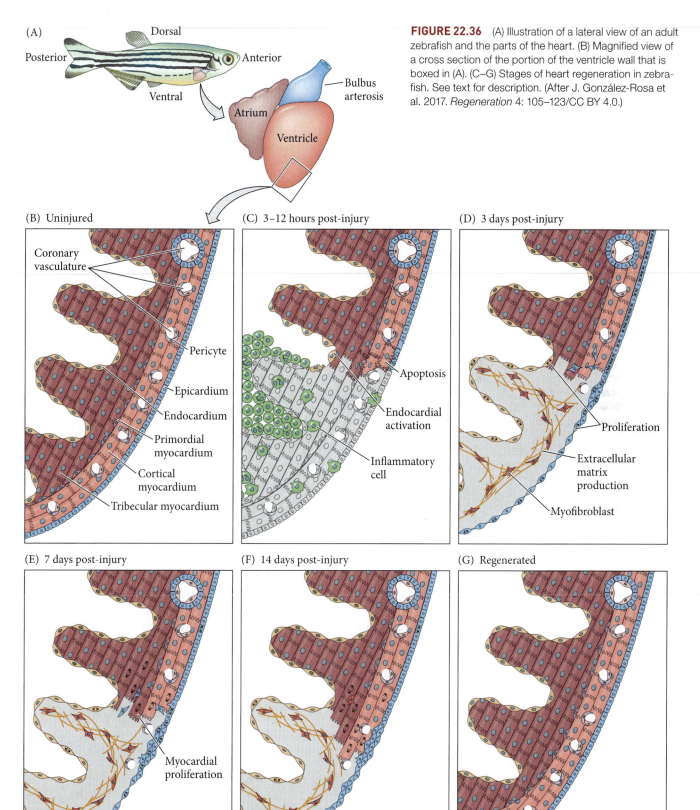

FIGURE 22.36 (A) Illustration of a lateral view of an adult zebrafish and the parts of the heart. (B) Magnified view of a cross section of the portion of the ventricle wall that is boxed in (A). (C–G) Stages of heart regeneration in zebrafish. See text for description. (After J. González-Rosa et al. 2017. *Regeneration* 4: 105–123/CC BY 4.0.)

(A)

Dorsal
Posterior
Anterior
Ventral
Bulbus arterosis
Atrium
Ventricle

(B) Uninjured

Coronary vasculature
Pericyte
Epicardium
Endocardium
Primordial myocardium
Cortical myocardium
Tribecular myocardium

(C) 3–12 hours post-injury

Apoptosis
Endocardial activation
Inflammatory cell

(D) 3 days post-injury

Proliferation
Extracellular matrix production
Myofibroblast

(E) 7 days post-injury

Myocardial proliferation

(F) 14 days post-injury

(G) Regenerated

Thickened cortical myocardium

4. **Compensatory proliferation and revascularization (see Figure 22.36F):**
   a. New coronary vessels rapidly form. This revascularization is necessary for complete regeneration.
   b. Proliferation of cardiomyocytes increases throughout the heart.
   c. Fibrotic tissue is progressively lost over the course of regeneration.

5. **Integration (see Figure 22.36G):**
   a. The fibrotic scar is completely removed and replaced with a slightly thicker myocardium than that of an uninjured heart.
   b. The regenerated myocardium becomes functionally integrated into the rest of the heart.

FURTHER DEVELOPMENT

WE'LL COMPENSATE FOR YOUR ABSENCE   Work from many labs has shown that heart regeneration in the adult zebrafish is carried out primarily through the dedifferentiation of preexisting cardiomyocytes (epimorphosis), establishment of a blastema at the wound site through local proliferation and migration, and finally, redifferentiation of blastema cells to repair the heart (Curado and Stainier 2006; Lepilina et al. 2006; Kikuchi et al. 2010; Zhang et al. 2013). In addition, it has been shown that healthy ventricular tissue far from the acute injury also responds by increasing proliferation (hyperplasia), which is a *compensatory* mechanism of regeneration (Poss et al. 2002; Sallin et al. 2015). Compensatory regeneration is often accompanied by hypertrophic growth (i.e., increase in cell size), but this has yet to be shown to occur in zebrafish heart regeneration.

FURTHER DEVELOPMENT

WE'LL CHANGE TO FILL THE VOID   The developing zebrafish heart also appears to be capable of morphallaxis or transdifferentiation during regeneration. Researchers working with zebrafish larvae induced severe injury to the ventricular tissue of the larval heart by causing apoptosis of the ventricular cardiomyocytes (Zhang et al. 2013). They did so by targeting the expression of nitroreductase (NTR) in ventricular cardiomyocytes using the *ventricular myosin heavy chain* (*vmhc*) promoter and inducing cell death by administering an NTR-reactive cytotoxic prodrug. This procedure severely ablated ventricular tissue in the larval heart. What happened next was remarkable. Neighboring differentiated *atrial* cardiomyocytes responded to the injury by migrating into the damaged ventricular tissue and upregulating ventricle-specific genes such as *vmhc* (**FIGURE 22.37A**). Months later, fate mapping of these migrating atrial cardiomyocytes revealed that they remained in the ventricular wall, contributing to a fully regenerated and functioning ventricle and heart (**FIGURE 22.37B**). Zhang and colleagues went on to show that Notch-Delta signaling is highly upregulated in the atrial myocardium and is required for this atrial-mediated repair of ventricular damage (**FIGURE 22.37C**). Pharmacological inhibition of Notch signaling with DAPT exposure during ablation of ventricle tissue severely impairs heart regeneration (**FIGURE 22.37D**; Zhang et al. 2013). These results suggest that at least during larval development, cardiac myocytes are capable of undergoing transdifferentiation to support regeneration of the heart. Thus, it appears that the cells of the zebrafish heart employ multiple mechanisms to power its regeneration: blastema formation through epimorphosis, compensatory proliferation, and Notch-mediated transdifferentiation. (See **Further Development 22.12, Immune Cells to the Regeneration!**, online.)

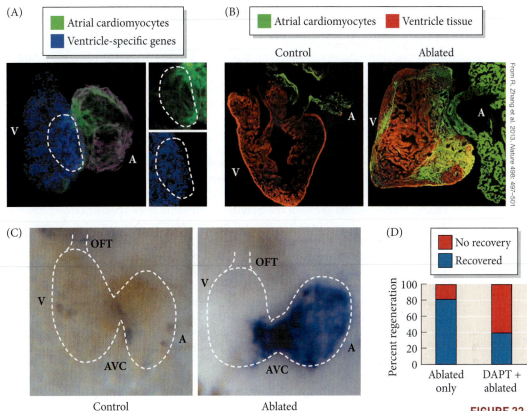

From R. Zhang et al. 2013. *Nature* 498: 497–501

Regeneration in Mammals

Although mammals do not have the same level of regenerative abilities as the other vertebrates covered in this chapter, they can regenerate certain structures. We cover below a few instances of regeneration in mammals: the well-known compensatory regeneration of the liver, the phenomenon of youthful digit-tip regeneration, a surprising instance of heart regeneration in neonatal mice, and the super abilities of the spiny mouse to regrow skin.

Compensatory regeneration in the mammalian liver

According to Greek mythology, Prometheus's punishment for bringing the gift of fire to humans was to be chained to a rock and have an eagle tear out and eat a portion of his liver each day. His liver then regenerated each night, providing a continuous food supply for the eagle and eternal punishment for Prometheus. Today the standard assay for liver regeneration is a partial hepatectomy, whereby specific lobes of the liver are removed (after anesthesia is administered, unlike in the story of Prometheus), leaving the other hepatic lobes intact. Although the removed lobe does not grow back, the remaining lobes enlarge to compensate for the loss of the missing tissue (Higgins and Anderson 1931). The amount of liver regenerated is equivalent to the amount of liver removed. Such **compensatory regeneration**—the division of differentiated cells to recover the structure and function of an injured organ—has been demonstrated in the zebrafish heart as described above and in the mammalian liver.

The human liver regenerates by the proliferation of existing tissue. Surprisingly, the regenerating liver cells do not fully dedifferentiate when they re-enter the cell cycle. No regeneration blastema is formed. Rather, mammalian liver regeneration appears to have two other lines of defense, the first of which consists of normal, mature, adult hepatocytes. These mature cells, which are usually not dividing, are instructed to rejoin the cell cycle and proliferate until they have compensated for the missing part. The second line of defense, discussed below, is a population of hepatic progenitor cells that

FIGURE 22.37 Transdifferentiation of atrial cardiomyocytes into ventricular cardiomyocytes during larval heart regeneration in the zebrafish. (A) After researchers induced apoptosis of ventricular cardiomyocytes, fate mapping of differentiated atrial cardiomyocytes (green) shows migration from the atrium to the wound area where ventricle tissue has been ablated. At the transition point during this migration, the atrial cells start expressing ventricular markers, suggesting transdifferentiation. (B) Twelve months after ablation of ventricular tissue, infiltrating atrial cardiomyocytes (green) fully differentiate into the ventricular tissue (red) and contribute to a functional adult heart. (C) Notch-Delta signaling is required for the successful regenerative contributions from atrial cardiomyocytes. Twenty-four hours after ventricular cell ablation (right photograph), *deltaD* along with other related genes (blue stain) are highly upregulated in the atrial cardiomyocytes, particularly in those migrating toward the ventricular tissue. (D) Pharmacological inhibition of Notch signaling with DAPT exposure during ablation of ventricle tissue severely impairs heart regeneration. A, atrium; V, ventricle; AVC, atrioventricular canal; OFT, outflow tract. (D after R. Zhang et al. 2013. *Nature* 498: 497–501.)

are normally quiescent but that are activated when the injury is severe and adult hepatocytes cannot regenerate well due to senescence, alcohol abuse, or disease.

In normal liver regeneration, the five types of liver cells—hepatocytes, duct cells, fat-storing (Ito) cells, endothelial cells, and Kupffer macrophages—all begin dividing to produce more of themselves. Each type retains its cellular identity, and the liver retains its ability to synthesize the liver-specific enzymes necessary for glucose regulation, toxin degradation, bile synthesis, albumin production, and other hepatic functions even as it regenerates itself (Michalopoulos and DeFrances 1997).

There are probably several redundant pathways that initiate liver cell proliferation and regeneration (**FIGURE 22.38**; Riehle et al. 2011). Global gene profiling indicates that the end result of these pathways is to downregulate (but not totally suppress) the genes involved in the differentiated functions of liver cells while activating those genes committing the cell to mitosis (White et al. 2005). The removal or injury of the liver is sensed through the bloodstream: some liver-specific factors are lost while others (such as bile acids and gut lipopolysaccharides) increase. These lipopolysaccharides activate some non-hepatocytes to secrete paracrine factors that allow the remaining hepatocytes to re-enter the cell cycle. The Kupffer cell secretes interleukin 6 (IL6) and tumor necrosis factor-α (which are usually involved in activating the adult immune system), and the stellate cells secrete the paracrine factors **hepatocyte growth factor** (**HGF**, or **scatter factor**) and TGF-β. The specialized blood vessels of the liver also produce HGF as well as Wnt2 (Ding et al. 2010). The trauma of partial hepatectomy may activate metalloproteinases that digest the extracellular matrix and permit the hepatocytes to separate and proliferate. These enzymes also may cleave HGF to its active form (Mars et al. 1995). Together, the factors produced by the endothelial cells, Kupffer cells, and stellate cells allow the hepatocytes to divide by preventing apoptosis, activating cyclins D and E, and repressing cyclin inhibitors such as p27 (see Taub 2004).

The liver stops growing when it reaches the appropriate size; the mechanism for how this is achieved is not yet known. One clue, though, comes from parabiosis experiments, in which the circulatory systems of two rats are surgically joined together. Partial hepatectomy in one parabiosed rat causes the other rat's liver to enlarge (Moolten and Bucher 1967). Therefore, some factor, or factors, in the blood appear to be establishing the size of the liver. Huang and colleagues (2006) have proposed that these factors are bile acids that are secreted by the liver and positively regulate hepatocyte growth. Partial hepatectomy stimulates the release of bile acids into the blood. These bile acids are received by the hepatocytes and activate the Fxr transcription factor, which promotes cell division. Mice without functional Fxr protein cannot regenerate their livers. Therefore, bile acids (a relatively small percentage of the products secreted by the liver) appear to regulate the size of the liver, keeping it at a particular volume of cells. The molecular mechanisms by

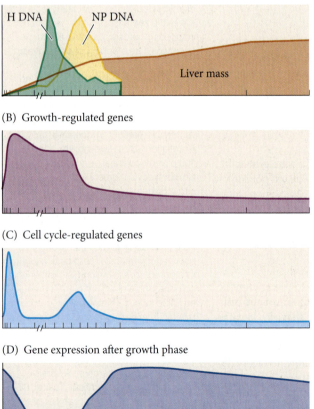

(A) DNA synthesis

H DNA   NP DNA

Liver mass

(B) Growth-regulated genes

(C) Cell cycle-regulated genes

(D) Gene expression after growth phase

02 4 8   24 36 48 60 75   168   216
Hours post-hepatectomy

FIGURE 22.38 Correlation of changes in gene expression with increases in liver mass following partial hepatectomy in mammals. (A) Initial peaks in DNA synthesis are seen in both hepatocytes (H DNA; green) and thereafter in nonparenchymal cells (NP DNA; yellow). This burst in DNA synthesis corresponds with the upregulation of growth-regulated (B) and cell cycle-regulated (C) gene expression, both of which taper off as the liver mass (brown shading in A) reaches its normal volume. (D) Overall gene expression remains high after the growth phase, reflecting the functionality of the regenerated liver tissue. (After R. Taub. 2004. *Nat Rev Mol Cell Biol* 5: 836–847, based on R. Taub. 2003. In *Hepatology: A Textbook of Liver Disease*, 4th ed., D. Z. Zakim & W. B. Boyer [Eds.], Saunders: Philadelphia.)

which these factors interact and by which the liver is first told to begin regenerating and then to stop regenerating after reaching the appropriate size remain to be discovered. (See **Further Development 22.13, Oval Cells and Liver Regeneration,** online.)

Young at heart

Mammals, humans included, are capable of regenerating the tips of their digits provided the organism is young enough. As in a regenerating salamander limb, a blastema composed of progenitor cells forms at the tip of the digit, and the new epidermis is regenerated from ectoderm-restricted progenitor cells while new bone comes from osteoblast progenitor cells (Fernando et al. 2011; Lehoczky et al. 2011; Rinkevich et al. 2011). This regenerative capacity seen in the young is not restricted to just the digits, however. Heart tissue can also regenerate in mice, but only within the first week of neonatal life. After that, the ability is lost. This age-related loss of regeneration is suggested to be associated with the generally widespread temporal withdrawal of cells from the cell cycle, as seen in differentiating cardiomyocytes (Porrello et al. 2011). It is known, though, that cardiomyocytes in adult mammals (as in adult zebrafish) will respond to a heart attack by re-entering the cell cycle, presumably contributing to injury repair (Senyo et al. 2013).

Perhaps regeneration in mammals is not much different from that in a fish or salamander? It has recently been shown that the neonatal regenerative ability of the heart in mice is dependent on cardiac neural inputs. Does this sound similar to the requirement for innervation during salamander limb regeneration? (See Figure 22.31 and Figure 2 in Further Development 22.11, online.) Similarly, in the neonatal mouse, mechanical denervation or chemical inhibition of cholinergic nerve function prevented both myocardial cell proliferation and regeneration of heart tissue. Moreover, Neuregulin-1 (as in salamander limb regeneration) is upregulated in the epicardium of injured neonatal mouse hearts, and this expression is lost if the heart is denervated. Misexpression of Neuregulin-1 (along with Nerve Growth Factor) in the injured denervated neonatal mouse heart can rescue its regenerative ability (Mahmoud et al. 2015). This dependence on innervation as well as the role of Neuregulin-1 signaling is also conserved during zebrafish heart regeneration (see Figures 1 and 2 in Further Development 22.12, online). Furthermore, it appears that Neuregulin signaling can also function as a mitogenic signal in the uninjured heart, causing myocardial hyperplasia (increase in cell number) and leading to abnormally large hearts (Gemberling et al. 2015). These results identify Neuregulin signaling as a conserved mechanism for neural-controlled regeneration; however, its mitogenic powers suggest that significant regulatory controls over this system must be in place to "switch" it on during injury and off upon completion of regeneration.

The spiny mouse, at the tipping point between scar and regeneration

While in Kenya conducting some ecological fieldwork, Dr. Ashley Seifert overheard that there was a mouse that could essentially "jump out of its skin." As creepy as that picture might seem, the African spiny mouse (*Acomys cahirinus*) can in fact shed large pieces of full-thickness skin (epidermis and dermis), presumably as an evolved predator defense mechanism (**FIGURE 22.39A**). Fortunately for the mouse, its ability to shed skin is paired with an ability to grow the skin back without any detrimental scarring (Seifert et al. 2012; Seifert and Muneoka 2018). Seifert and colleagues have demonstrated this by cutting 4-mm holes in the mouse's outer ear (pinna), a technique that is often used to tag experimental mice. Watching the repair of these holes, the researchers verified that the spiny mouse could regenerate the appropriate epidermal and dermal tissues of the ear without scarring (**FIGURE 22.39B–E**), whereas non-regenerative species, including

(A)

(B)

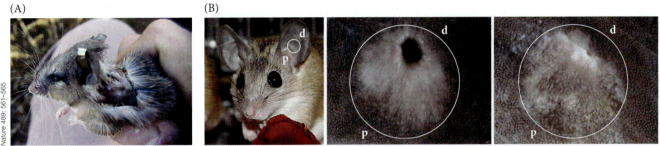

(C) Hole / Distal / Proximal

(D) D30 — Blastema — p ... d

(E) D40 — p ... d

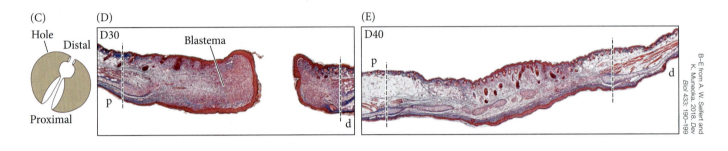

(F)

Control: D10, D20, D34

Macrophage depletion: D10, D20, D34, D70

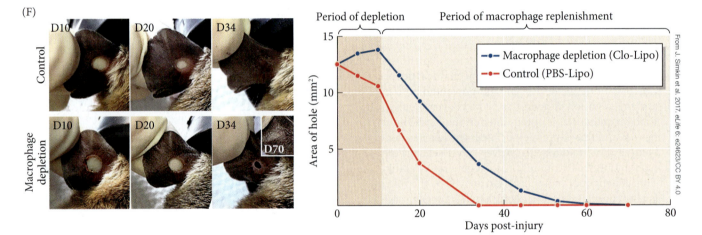

Period of depletion | Period of macrophage replenishment

Area of hole (mm²) vs. Days post-injury

— Macrophage depletion (Clo-Lipo)
— Control (PBS-Lipo)

FIGURE 22.39 The spiny mouse regenerates instead of scarring. (A) The spiny mouse has extremely fragile skin that can tear completely off, down to the muscle. This missing skin did regenerate back! (B) The 4-mm ear pinna hole punch is a tractable model for looking at complex tissue regeneration in mammals. The middle and right images show a pinna hole 30 and 40 days, respectively, post-injury. (C) Regeneration occurs preferentially along the proximal half of the ear hole, where the most significant blastema forms.

(D) 30 Days post-injury. (E) Regeneration is complete by 40 days post-injury, with all tissue types present. (F) Macrophage depletion with clodronate liposome (Clo-Lipo) injections represses ear hole regeneration in the spiny mouse (images at left, days 10, 20, 34, and 70) until this depletion is stopped. As macrophages return over time, so do regenerative abilities (right). Dotted lines in D and E mark the boundaries of injury. p, proximal; d, distal; PBS-Lipo, phosphate buffered saline control.

the mouse used as a model organism (*Mus musculus*) and the noted "healer" strain of mouse (Murphy Roths Large (MRL/MpJ)), could only partially heal the holes with fibrotic scars (Gawriluk et al. 2016).

In the beginning of this chapter, we noted that one of the primary differences between the processes of regeneration and the mechanisms involved in embryonic development was the active involvement of the immune system during regeneration (see Figure 22.4). Macrophages are well known to function as local mediators of inflammatory and restorative responses for both fibrotic scar and regenerative outcomes. It was demonstrated that macrophages were required for blastema formation and successful regeneration of the salamander limb and zebrafish fin (Godwin et al. 2013; Petrie et al. 2014). Similarly, depletion of macrophages in the spiny mouse impairs regeneration of the hole in the ear pinna (**FIGURE 22.39F**). Curiously, Simkin and colleagues also showed that during normal regeneration in the spiny mouse ear hole, macrophages were at the border of but not so much within the blastema (Simkin et al. 2017). So how could these macrophages be functioning to influence regenerative behaviors?

One factor to consider while pondering this question is that not all macrophages are alike. Macrophages can shift their "phenotype" to the type of macrophages needed, which generally fluctuates between being pro-inflammatory or pro-restorative (that is, M1 or M2 macrophages respectively). These different types of macrophages are known to express and secrete important factors that can either promote scar formation by upregulating the formation of collagens, for example, or repress fibrotic conditions by upregulating matrix metalloproteinases designed to break down collagen and other extracellular matrix components. One leading hypothesis suggests that in regeneration-competent vertebrates, the initial involvement of macrophages fosters a pro-regenerative environment that permits blastema formation. The limited regenerative abilities in other vertebrates, such as humans, however, may be due to an initial macrophage response that induces inflammation and scar formation. Not surprisingly, gaining an understanding of the role of macrophages and the composition of the extracellular matrix is now a major focus of regeneration research and the development of therapies in the field of regenerative medicine (see Londono and Badylak 2015; Costa et al. 2017).

Courtesy of Junji Morokuma and Michael Levin

Closing Thoughts on the Opening Photo

After reading this chapter, you should be able to identify this image as a planarian with extra heads. This image was produced by the Levin lab after researchers manipulated the bioelectric state of this worm during regeneration. The idea of dual epigenetic and genetic driven maps for positional identity represents arguably the most exciting and challenging problem in the field of regenerative development. Understanding how the cells and tissues of an individual organism can interpret injuries and orchestrate the proper scaling of regeneration is nothing short of marvelous.

Endnotes

[1] In the 2016 film *Deadpool*, Wade Wilson regenerates his hand following its complete amputation. It first forms a "babylike" or small hand that then grows larger. The axolotl, or Mexican salamander, also first regenerates a much smaller but well-patterned arm and hand that then subsequently grow into the correct size perfectly proportional to the animal's body size. However, based on the 2018 *Deadpool 2* production, do you think Wade's regenerative powers are more like those of the salamander or perhaps of *hydra*? (And no, we don't mean Captain America's archenemy here.)

[2] Specifically, this transcription factor is TCP20, a Class I member of the teosinte-branched cycloidea PCNA (TCP) proteins that was shown to interact directly with PLT and SCR (Shimotohno et al. 2018).

[3] *Hydra* is both the genus name and the common name for these animals. For simplicity, we will use the common (unitalicized) form in this discussion.

[4] Before 1910, "fly lab" maestro Morgan was well known for his research into flatworm regeneration. Indeed, it was only in 1900 that Morgan first mentioned *Drosophila*—as food for his flatworms! He was even able to "stain" the flatworms' digestive tubes by feeding them pigmented *Drosophila* eyes. Later, when he founded modern genetics, Morgan renounced flatworms as a model for heredity in favor of *Drosophila* (see Mittman and Fausto-Sterling 1992).

[5] Although these experiments demonstrate the dependence on the nerve, it may very well be a developed dependence. The 1970 work by Singer and colleagues showed that if you remove the nerve during development and then conduct an aneurogenic limb experiment with this animal, regeneration still occurs. The limb does *not* actually need the nerve to regenerate if it never had it to begin with. This phenomenon is known as nerve addiction. How this occurs is unknown.

Snapshot Summary
Regeneration

1. There are four modes of regeneration. In stem-cell mediated regeneration (characteristic of planarians), new cells are routinely produced to replace the ones that die. In epimorphosis (seen in regenerating salamander limbs and fish fins), tissues form a regeneration blastema, divide, and redifferentiate into the new structure. In morphallaxis (characteristic of hydra), there is a re-patterning of existing tissue with little or no growth. In compensatory regeneration (as in the mammalian liver), cells divide but retain their differentiated state.

2. Regeneration has been seen in sponges, which may be related to our most basal ancestors. However, whether all regenerative abilities today are derived from a basal

3. ancestor or whether the tree of life is filled with examples of the convergent evolution of regeneration remains to be determined.

4. Plants can follow both indirect and direct paths for totipotent regeneration through the transdifferentiation of mature cells or the establishment of de novo meristems.

5. Hydra appear to have a head activation gradient and a foot activation gradient.

6. In planarian flatworms, regeneration occurs by forming a regeneration blastema produced by pluripotent clonogenic neoblasts. Gradients of positional control genes such as *Wnt* appear to direct the anterior-posterior differentiation of these cells in a pattern regulated by the head-expressed Wnt inhibitor Notum.

7. In the regenerating limb blastemas of amphibians, cells do not become multipotent. Rather, cells retain their specifications, with cartilage arising from preexisting cartilage, neurons coming from preexisting neurons, and muscles coming either from preexisting muscle cells or from muscle stem cells.

7. Mitogens such as Neuregulin-1 are provided by the innervating nerves and are capable of inducing the regeneration of limbs even in the absence of nerves. Salamander limb regeneration appears to use the same pattern-formation system as the developing limb.

8. Multiple modes of regeneration have been discovered to operate in zebrafish. Distal regeneration of the zebrafish fin occurs largely through the dedifferentiation of existing cell types followed by the active proliferation of a blastema-like outgrowth. Zebrafish heart tissue also employs an initial mode of epimorphosis followed by a compensatory regenerative period of proliferation and transdifferentiation.

9. In the mammalian liver, no regenerating blastema is formed, and the liver regenerates the same volume as it lost. Each cell appears to generate its own cell type. A reserve population of multipotent progenitor cells divides when these tissues cannot regenerate the missing portions.

10. The African spiny mouse has revealed the importance of macrophages in promoting a regenerative environment that prevents fibrotic scar formation.

Go to oup.com/he/barresi12xe for Further Developments, Scientists Speak interviews, Watch Development videos, Dev Tutorials, and complete bibliographic information for all literature cited in this chapter.

Development in Health and Disease

Birth Defects, Endocrine Disruptors, and Cancer

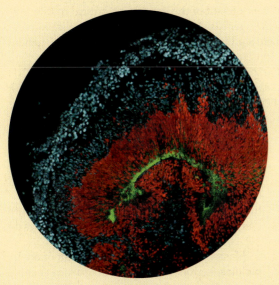

A "brain" made by stem cells shows layering similar to that of normal human brains. How might a virus like Zika get access to developing neurons?

From Nowakowski et al. 2016. *Cell Stem Cell* 18: 591–596

"THE AMAZING THING ABOUT MAMMALIAN DEVELOPMENT," says British medical geneticist Veronica van Heyningen (2000), "is not that it sometimes goes wrong, but that it ever succeeds." It is indeed amazing that any of us is here, because relatively few human conceptions develop successfully to birth. Fertility studies (Mantzouratou and Delhanty 2011; Chavez et al. 2012) suggest that only 20–50% of human cleavage-stage embryos successfully implant in the uterus. Many of these embryos have chromosomal anomalies that are expressed so early that the embryo dies before the woman realizes she has conceived. Of the embryos that *do* implant successfully, studies from the 1980s suggest that only about 40% survive to term (Edmonds et al. 1982; Boué et al. 1985). Further studies (Winter 1996; Epstein 2008) estimate that some 2.5% of babies who do come to term have a recognizable birth defect.

Although the body has remarkable back-up pathways and redundancies that permit a great deal of flexibility, abnormal phenotypes can still emerge. There are three major pathways to abnormal development:

1. *Genetic mechanisms.* Mutations in genes or changes in the number of chromosomes can alter development.
2. *Environmental mechanisms.* Agents (usually chemicals) from outside the body cause deleterious phenotypic changes by inhibiting or enhancing developmental signals.
3. *Stochastic (random) events.* Chance plays a role in determining the phenotype, and some developmental anomalies are just "bad luck."

Most of this chapter will deal with genetic and environmental effects.[1] We will start, however, by briefly examining the role of random events.

The Role of Chance

Although physicians and researchers often parse developmental anomalies into those caused by internal (genetic) means versus those caused by external (environmental) agents, more consideration is now being given to the role of stochastic factors—randomness—in developmental defects (Molenaar et al. 1993; Holliday 2005; Smith 2011). Even an embryo with wild-type genes and a favorable environment may develop an abnormal phenotype as the result of "bad luck." Developmental outcomes are probabilistic rather than predetermined (Wright 1920; Gottlieb 2003; Kilfoil et al. 2009). Consider, for example, X-chromosome inactivation in females (see Chapter 6).

If a woman carries one normal and one mutant allele for an X-linked blood clotting factor, statistically we would expect the wild-type allele to be inactivated in about 50% of her cells. If the wild-type allele is inactivated in 50% of the liver cells that produce clotting factor, the woman is phenotypically normal. But what would happen if, just by chance, 95% of the wild-type X chromosomes were inactivated in those liver cells? Only 5% of her X chromosomes would express the wild-type allele, and she would have an abnormality. Indeed, there have been cases of female identical twins where, in one twin, chance resulted in the inactivation of a large percentage of her X chromosomes carrying the normal clotting factor allele; that twin had severe hemophilia (inability of the blood to clot). The other twin, with a lower percentage of her normal X chromosomes inactivated, was not affected (Tiberio 1994; Valleix et al. 2002).

Such variability is not limited to genes on the X chromosome. Measurements of gene expression in individual cells show that protein synthesis is a stochastic process, with random fluctuations in both transcription and translation leading to variations in the levels of proteins produced at any given time (Raj and van Oudenaarden 2008; Stockholm et al. 2010). Cell specification, developmental signaling, and cell migration are thought to be influenced by chance fluctuations in the amounts of transcription factors, paracrine factors, and receptors produced at a particular moment. Thus, genetically identical animals raised in precisely the same environments can have vastly different phenotypes (Gilbert and Jorgensen 1998; Vogt et al. 2008; Ruvinsky 2009; Zhou et al. 2013). Recent studies have shown that in many insects, the absence of particular symbiotic bacteria also causes birth defects (Kikuchi et al. 2018).

Genetic Errors of Human Development

Those abnormalities caused by intrinsic genetic events may result from mutations, aneuploidies (improper chromosome number), or chromosomal translocations (Opitz 1987).

The developmental nature of human syndromes

Human birth defects, which range from life-threatening to relatively benign, are often linked into **syndromes** (Greek, "running together"), with several abnormalities associated with each other. Genetically based syndromes are caused either by (1) a chromosomal event (such as an aneuploidy) in which several genes are deleted or added, or by (2) **pleiotropy**—the production of several effects by a single gene or pair of genes (see Grüneberg 1938; Hadorn 1955). Syndromes are said to have *mosaic pleiotropy* when the effects are produced independently as a result of the gene being critical in different parts of the body. For instance, the *KIT* gene is expressed in blood stem cells, pigment stem cells, and germ stem cells, where it promotes their proliferation. When this gene is defective, the resulting syndrome consists of anemia (lack of red blood cells), albinism (lack of pigment cells), and sterility (lack of germ cells).

Syndromes are said to have *relational pleiotropy* when a defective gene in one part of the embryo causes a defect in another part, even though the gene is not expressed in the second tissue. For example, failure of *MITF* expression in the pigmented retina prevents this structure from fully differentiating. This failure of pigmented retina growth in turn causes a malformation of the choroid fissure of the eye, resulting in the drainage of vitreous humor. Without this fluid, the eye fails to enlarge (hence microphthalmia, "small eye"). The lenses and corneas are therefore smaller, even though they themselves do not express *MITF*.

Mosaic syndromes can be the result of **aneuploidies**—errors in the number of particular chromosomes. Even an extra copy of the tiny chromosome 21 disrupts numerous developmental functions. This **trisomy 21** causes a set of anomalies—among them facial muscle changes, heart and gut abnormalities, and cognitive problems—collectively known as **Down syndrome** (**FIGURE 23.1**).

FIGURE 23.1 Down syndrome. (A) Down syndrome, caused by a third copy of chromosome 21, is characterized by a particular facial pattern, cognitive deficiencies, the absence of a nasal bone, and often heart and gastrointestinal defects. (B) The nuclear staining results shown here detect chromosome number using fluorescently labeled probes that bind to DNA on chromosomes 21 (pink) and 13 (blue). This person has Down syndrome (trisomy 21) but has the normal two copies of chromosome 13.

(A)

© Denys Kuvaiev/Alamy Stock Photo

(B)

Courtesy of Abbott Laboratories

Certain genes on chromosome 21 encode transcription factors and regulatory microR-NAs, and the extra copy of chromosome 21 probably causes an overproduction of these regulatory factors (Bras et al. 2018). Such overproduction results in the misregulation of genes necessary for heart, muscle, and nerve formation (Chang and Min 2009; Korbel et al. 2009; Antonarakis et al. 2017). One such regulatory microRNA encoded on chromosome 21, miRNA-155, is found throughout the developing human fetus. This miRNA downregulates translation of the messages for certain transcription factors necessary for normal neural and heart development and is highly elevated in the brains and hearts of people with Down syndrome (Elton et al. 2010; Wang et al. 2013).

Genetic and phenotypic heterogeneity

In pleiotropy, the same gene can produce different effects in different tissues. However, the opposite phenomenon is an equally important feature of genetic syndromes: mutations in different genes can produce the same phenotype. If several genes are part of the same signal transduction pathway, a mutation in any of them often produces a similar phenotypic result. This production of similar phenotypes by mutations in different genes is called **genetic heterogeneity**. The syndrome of sterility, anemia, and albinism caused by the absence of Kit protein (discussed above) can also be caused by the absence of its paracrine ligand, stem cell factor (SCF). Another example is cyclopia (see Figure 4.23B), a phenotype that can be produced either by mutations in the gene for Sonic hedgehog, *or* by mutations in the genes activated by Shh, *or* in the genes controlling cholesterol synthesis (since cholesterol is essential for Shh signaling).

Not only can different mutations produce the same phenotype, but the same mutation can produce a different phenotype in different individuals, a phenomenon known as **phenotypic heterogeneity** (Wolf 1995, 1997; Nijhout and Paulsen 1997). Phenotypic heterogeneity comes about because genes are not autonomous agents. Rather, they interact with other genes and gene products, becoming integrated into complex pathways and networks. Bellus and colleagues (1996) analyzed the phenotypes derived from the same mutation in the *FGFR3* gene in 10 unrelated families. These phenotypes ranged from relatively mild anomalies to potentially lethal malformations. Defects in the gene producing myosin-7 usually cause the degeneration of ear and eye neurons, leading to a syndrome of blindness and deafness. However, the Nobel Prize-winning molecular biologist James Watson has these mutant genes and showed no defect in his hearing or sight (Green and Annas 2008). (See **Further Development 23.1, Preimplantation Genetics**, online.)

Teratogenesis: Environmental Assaults on Animal Development

The summer of 1962 brought two portentous events. The first was the publication of the landmark book *Silent Spring* by Rachel Carson, in which she documented that the pesticide DDT was destroying bird eggs and preventing reproduction in several species. Her work is credited with spurring the modern environmental movement. The second event was the discovery that thalidomide, a sedative used to help manage pregnancies, could cause limb and ear abnormalities in the human fetus (Lenz 1962; see Chapter 1).[2] These two revelations showed that the embryo was vulnerable to environmental agents. Indeed, Rachel Carson made the connection, commenting, "It is all of a piece, thalidomide and pesticides. They represent our willingness to rush ahead and use something without knowing what the results will be" (Carson 1962).

Exogenous agents that cause birth defects are called **teratogens** (**TABLE 23.1**). Most teratogens produce their effects during certain critical time frames. Human development is usually divided into an **embryonic period** (to the end of week 8) and a **fetal period** (the remaining time in utero). Most organ systems form during the embryonic period; the fetal period is generally one of growth and modeling. Thus, maximum fetal susceptibility to teratogens is between weeks 3 and 8 (**FIGURE 23.2**). The nervous system, however, is constantly forming and remains susceptible throughout development. Prior to week 3, exposure does not usually produce congenital anomalies because a teratogen encountered at this time either damages most or all of the cells of an embryo, resulting in its death, or kills only a few cells, allowing the embryo to fully recover.

TABLE 23.1 Some agents thought to cause disruptions in human fetal development[a]

DRUGS AND CHEMICALS	
Alcohol	Valproic acid
Aminoglycosides (gentamicin)	Warfarin
Aminopterin	**IONIZING RADIATION (X-RAYS)**
Antithyroid agents (PTU)	**HYPERTHERMIA (FEVER)**
Bromine	**INFECTIOUS MICROORGANISMS**
Cortisone	Coxsackievirus
Diethylstilbesterol (DES)	Cytomegalovirus
Diphenylhydantoin	Herpes simplex
Heroin	Parvovirus
Lead	Rubella (German measles)
Methylmercury	*Toxoplasma gondii* (toxoplasmosis)
Penicillamine	*Treponema pallidum* (syphilis)
Retinoic acid (isotretinoin, Accutane)	Zika virus
Streptomycin	**METABOLIC CONDITIONS IN THE MOTHER**
Tetracycline	Autoimmune disease (including Rh incompatibility)
Thalidomide	Diabetes
Trimethadione	Dietary deficiencies, malnutrition
	Phenylketonuria

[a]This list includes known and possible teratogenic agents and is not exhaustive.

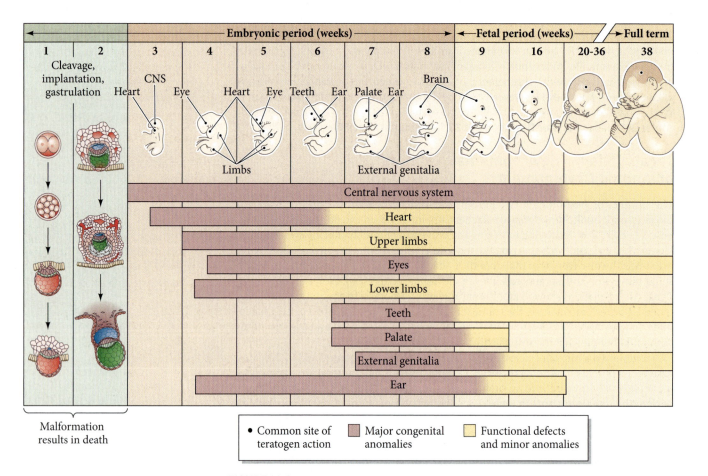

FIGURE 23.2 Weeks of gestation and sensitivity of embryonic organs to teratogens. (After K. L. Moore and T. V. N. Persaud. 1993. *The Developing Human: Clinically Oriented Embryology*, 5th ed., p. 156. W. B. Saunders: Philadelphia.)

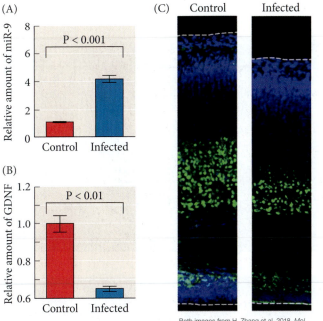

(A)

(B)

(C)

Control   Infected

Both images from H. Zhang et al. 2018. *Mol Neurobiol* doi: 10.1007/s12035-018-1358-4

FIGURE 23.3 Possible mechanism for Zika virus-induced cell death. Fetal mouse brains were infected with Zika virus and monitored days later. (A) Zika infection upregulates the miR-9 microRNA, which is associated with cell death and decreased neuron formation. (B) Subsequent down-regulation of GDNF, a neural growth factor that is a target of miR-9. (C) Decrease in the cortical neurons in the Zika virus-infected brains. Blue fluorescence is from DAPI, which stains DNA, and green fluorescence is from Tbr1, a marker of newborn neurons. Error bars represent standard error of the mean. (After H. Zhang et al. 2018. *Mol Neurobiol.* https://doi.org/10.1007/s12035-018-1358-4.)

The largest class of teratogens includes drugs and chemicals. Viruses, radiation, high body temperature, and metabolic conditions in the mother can also act as teratogens. Some chemicals found naturally in the environment can cause birth defects. For example, jervine and cyclopamine are products of the plant *Veratrum californicum* that block Sonic hedgehog signaling and lead to cyclopia (see Figure 4.23B). Nicotine, a natural product concentrated in tobacco smoke, is associated with impaired lung and brain development (Dwyer et al. 2008; Maritz and Harding 2011).

Some viruses can cause congenital anomalies. The mosquito-borne Zika virus has been implicated in microcephaly, a birth defect characterized by a small brain and head (Centers for Disease Control and Prevention 2016; Mlakar et al. 2016). Evidence indicates that in pregnant women, Zika virus directly infects the neural progenitor cells of the fetal cerebral cortex, resulting in the death of these cells, which would result in the smaller brain and head size of the newborn (Tang et al. 2016; Merfeld et al. 2017). One mechanism for this is through Zika virus altering microRNA levels, leading to the activation of natural cell death pathways and reduced stem cell division (**FIGURE 23.3**; Bhagat et al. 2018; Zhang et al. 2018).

Although different agents are teratogenic in different organisms (see Gilbert and Epel 2015), laboratory animals have been used to screen compounds that have a high probability of being hazardous. *Xenopus* and zebrafish, as we saw in Chapters 11 and 12, undergo early development using the same basic paracrine factors and transcription factor pathways that we use. These model organisms have been especially important in identifying teratogenic molecules in the environment. Studies on zebrafish, for instance, found that water-soluble components from the 2010 *Deepwater Horizon* oil spill in the Gulf of Mexico caused numerous developmental anomalies traceable to neural crest cell migration (**FIGURE 23.4**; de Soysa et al. 2012).

Alcohol as a teratogen

In terms of the frequency of its effects and its cost to society, the most devastating teratogen is undoubtedly alcohol (ethanol). Babies born with **fetal alcohol syndrome** (**FAS**) are characterized by small head size, an indistinct philtrum (the pair of ridges that run between the nose and mouth above the center of the upper lip), a narrow vermillion border on the upper lip, and a low nose bridge (Lemoine et al. 1968; Jones and Smith 1973).

> **(?) Developing Questions**
>
> In the United States, pregnant women are warned not to drink water from lakes that lie near abandoned mines. Do you think this warning is warranted? Why?

FIGURE 23.4 Water-soluble crude oil components from the *Deepwater Horizon* oil spill are teratogenic in zebrafish. Compared with normal zebrafish of the same age, zebrafish embryos exposed to oil spill components produced larvae with severe developmental anomalies, including reduction in the size of head, gill, and thoracic cartilages (blue staining) associated with cranial neural crest cell migration.

Normal

Affected

Both images from T. Y. de Soysa et al. 2012. *BMC Biol* 10: 40/CC BY 2.0

(A)

(B)

(C)

(D)

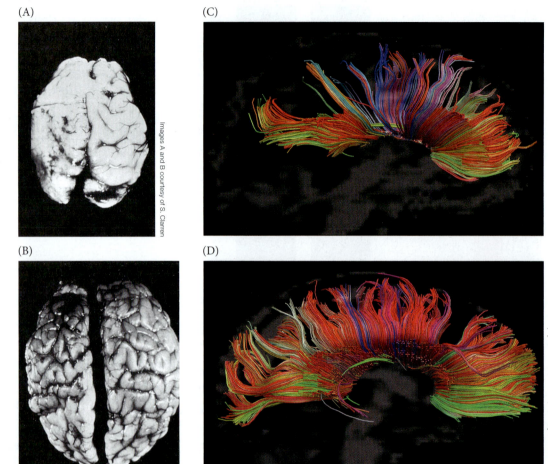

Images A and B courtesy of S. Clarren

Images C and D from J. R. Wozniak and R. L. Muetzel, 2011. *Neuropsychol Rev* 21: 133–147, courtesy of the authors

FIGURE 23.5 Effects of alcohol on fetal brains. (A,B) Comparison of a brain from an infant with fetal alcohol syndrome (A) with a brain from a normal infant of the same age (B). The brain from the infant with FAS is smaller, and the pattern of convolutions is obscured by glial cells that have migrated over the top of the brain. (C,D) Regionally specific abnormalities of the corpus callosum seen by diffusion tensor imaging of myelinated neurons. The difference in fiber tracks in a child with FASD (C) compared with those of a same-age unaffected child (D) suggests that there are significant abnormalities in neurons that would normally project through the posterior regions of the brain into the cortex of the parietal and temporal lobes.

The brains of such children may be dramatically smaller than normal and often show poor development, the results of deficiencies of neuronal and glial migration (**FIGURE 23.5A,B**; Clarren 1986). FAS is the most prevalent type of congenital intellectual disability syndrome, occurring in approximately 1 out of every 650 children born in the United States (May and Gossage 2001). Although the IQs of children with FAS vary substantially, the mean is about 68 (Streissguth and LaDue 1987). Most adults and adolescents with FAS cannot handle money and have difficulty learning from past experiences (see Dorris 1989; Kulp and Kulp 2000).

Fetal alcohol syndrome represents only a portion of a range of defects caused by prenatal alcohol exposure. The term **fetal alcohol spectrum disorder** (**FASD**) has been coined to encompass all of the alcohol-induced malformations and functional deficits that occur. In many FASD children, behavioral abnormalities exist without any gross physical changes in head size or notable reductions in IQ (NCBDD 2009). However, recent techniques that can identify neural tracts in the brain have found subtle abnormalities that correlate with altered mental processing speed and executive functioning (such as planning, memorizing, and retaining information) (**FIGURE 23.5C,D**; Wozniak and Muetzel 2011).

As with other teratogens, the amount and timing of fetal exposure to alcohol, as well as the genetic background of the fetus, contribute to the developmental outcome. Variability in the mother's ability to metabolize alcohol also may account for some outcome differences (Warren and Li 2005). While FASD is most strongly associated with high levels of alcohol consumption, the results of laboratory animal studies suggest that even a single episode of consuming the equivalent of two alcoholic drinks during pregnancy may lead to loss of fetal brain cells (**FIGURE 23.6**; Sulik 2005). ("One drink" is defined as 12 oz. of beer, 5 oz. of wine, or 1.5 oz. of hard liquor.) It is important to note that alcohol can cause permanent damage to a fetus at a time before most women even realize they are pregnant.

FURTHER DEVELOPMENT

HOW ALCOHOL AFFECTS DEVELOPMENT: LESSONS FROM THE MOUSE   When mice are exposed to alcohol at the time of gastrulation, ethanol induces defects of the face and brain that are comparable to those in humans with FAS (**FIGURE 23.6**; Sulik 2005). As in human fetuses, the nose and upper lip of the ethanol-exposed pups are poorly developed, and nervous system problems involve failure to close the neural tube and incomplete development of the forebrain (see Chapters 13 and 14). This mouse model of FAS can be used to study the ways by which ethanol causes its effects on the embryo.

It appears that ethanol works on several processes, and can interfere with cell migration, proliferation, adhesion, and survival. Hoffman and Kulyk (1999) showed that instead of migrating and dividing, neural crest cells of alcohol-exposed fetuses prematurely initiate their differentiation into facial cartilage. Among the numerous genes that are misregulated following maternal alcohol exposure in mice are several that encode proteins involved in cell movement and neuron outgrowth (Green et al. 2007). In later-stage mouse embryos exposed to ethanol, the death of neural crest-derived cells is seen as early as 12 hours following the exposure. When the time of alcohol exposure corresponds to the third and fourth weeks of human development, cells that should form the median portion of the forebrain, upper midface, and cranial nerves are killed. This has been confirmed in early chick embryos, in which transient ethanol exposure at environmentally relevant doses (about 25 mM) decimates migrating cranial neural crest cells, causing cell death throughout the head region (Flentke et al. 2011).

Normal

Alcohol-exposed

Images A and C from K. K. Sulik et al. 1981. *Science* 214: 936–938, courtesy of K. Sulik

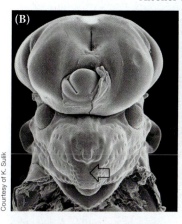

Courtesy of K. Sulik

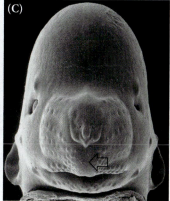

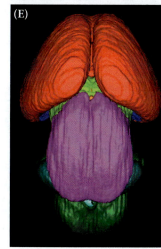

Images D and E from E. A. Godin et al. 2010. *Alcohol Clin Exp Res* 34: 98–111

FIGURE 23.6   Alcohol-induced craniofacial and brain abnormalities in mice. (A–C) Normal (A) and abnormal (B,C) day-14 embryonic mice. In (B), the anterior neural tube failed to close, resulting in exencephaly, a condition in which the brain tissue is exposed to the exterior. Later in development, the exposed brain tissue will erode away, resulting in anencephaly. (B,C) Prenatal alcohol exposure can also affect facial development, resulting in a small nose and an abnormal upper lip (open arrow). These facial features are present in fetal alcohol syndrome. (D,E) Three-dimensional reconstructions prepared from magnetic resonance images of the brains of normal (D) and alcohol-exposed (E) 17-day embryonic mice. In the alcohol-exposed specimen, the olfactory bulbs (pink) are absent and the cerebral hemispheres (red) are abnormally united in the midline. Light green, diencephalon; magenta, mesencephalon; teal, cerebellum; dark green, pons and medulla.

One reason for this cell death in mouse embryos is that alcohol treatment results in the generation of superoxide radicals that can damage cell membranes (**FIGURE 23.7A–C**; Davis et al. 1990; Kotch et al. 1995; Sulik 2005). In model systems, antioxidants have been effective in reducing both the cell death and the malformations caused by alcohol (Chen et al. 2004).

Abnormal signaling may also underlie excessive cell death. In alcohol-exposed embryos, expression of Sonic hedgehog (which is important in establishing the facial midline structures) is downregulated. While the mechanism for this downregulation remains incompletely understood, the finding that Shh-secreting cells placed into the head mesenchyme can prevent the alcohol-induced death of cranial neural crest cells highlights the importance of the Shh pathway as a target for alcohol's teratogenesis (Ahlgren et al. 2002; Chrisman et al. 2004).

Another mechanism that may be involved in alcohol's teratogenesis is its interference with the ability of the cell adhesion molecule L1 to hold cells together and signal to each other (Littner et al. 2013). This interference with morphogenesis occurs at the protein level, not at the gene level. Ramanathan and colleagues (1996) have shown that at levels as low as 7 mM—an alcohol concentration produced in the blood or brain with a single drink—alcohol can block the adhesive function of the L1 protein in vitro (**FIGURE 23.7D**). Moreover, mutations in the human L1 gene cause a syndrome of intellectual disability and malformations similar to that seen in severe FAS cases. L1 and the loss of neurons can lead to abnormal excitation in the neural circuits. Such changes in neuron firing can lead

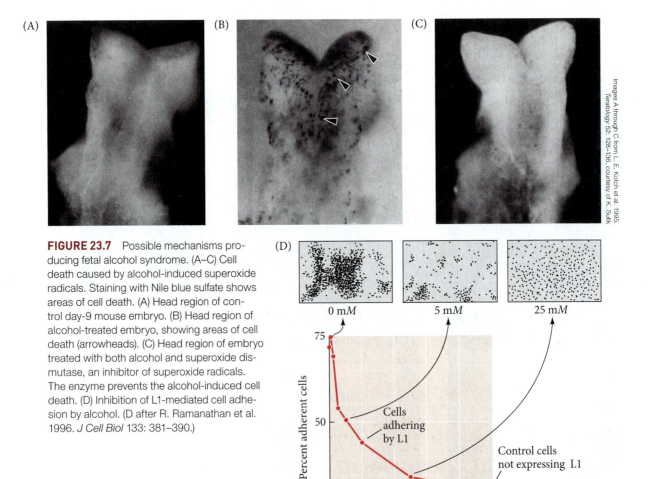

FIGURE 23.7 Possible mechanisms producing fetal alcohol syndrome. (A–C) Cell death caused by alcohol-induced superoxide radicals. Staining with Nile blue sulfate shows areas of cell death. (A) Head region of control day-9 mouse embryo. (B) Head region of alcohol-treated embryo, showing areas of cell death (arrowheads). (C) Head region of embryo treated with both alcohol and superoxide dismutase, an inhibitor of superoxide radicals. The enzyme prevents the alcohol-induced cell death. (D) Inhibition of L1-mediated cell adhesion by alcohol. (D after R. Ramanathan et al. 1996. *J Cell Biol* 133: 381–390.)

Images A through C from L. E. Kotch et al. 1995. *Teratology* 52: 128–136, courtesy of K. Sulik

to behavioral deficits and the death of more neurons (Granato and Dering 2018). Thus, alcohol can cross the placenta, enter the fetus, and block several critical functions in brain and facial development.

Retinoic acid as a teratogen

In some instances, even a compound involved in normal development can have deleterious effects if it is present in large enough amounts or at particular times. As we have seen in earlier chapters, retinoic acid (RA) is a vitamin A derivative that is important in specifying the anterior-posterior axis and in forming the jaws and heart of the mammalian embryo. As Piersma and colleagues (2017) have written, "The region-specific homeostasis of RA in the embryo is in many ways the driving force determining developmental cell proliferation versus differentiation. As a consequence, RA concentrations are carefully controlled in time and space in the developing embryo."

In its pharmaceutical form, 13-*cis*-retinoic acid (also called isotretinoin and sold under the trademark Accutane) has been useful in treating severe cystic acne and has been available for this purpose since 1982. The deleterious effects of administering large amounts of RA (or its vitamin A precursor) to pregnant animals had been known since the 1950s (Cohlan 1953; Giroud and Martinet 1959; Kochhar et al. 1984). However, about 160,000 women of childbearing age (15–45 years) have taken isotretinoin since it was introduced, and some have used it during pregnancy. Isotretinoin-containing drugs now carry a strong warning against their use by pregnant women. In the United States, retinoic acid exposure is a critical public health concern because there is significant overlap between the population using acne medicine and the population of women of childbearing age—and because an estimated 50% of pregnancies in the U.S. are unplanned (Finer and Zolna 2011).

Lammer and co-workers (1985) studied a group of women who inadvertently exposed themselves to RA and who elected to remain pregnant. Of their 59 fetuses, 26 were born without any noticeable anomalies, 12 aborted spontaneously, and 21 were born with obvious anomalies. The affected infants had a characteristic pattern of anomalies, including absent or defective ears, absent or small jaws, cleft palate, aortic arch abnormalities, thymus deficiencies, and abnormalities of the central nervous system. These anomalies are largely due to the failure of cranial neural crest cells to migrate into the pharyngeal arches of the face to form the jaw and ears (Moroni et al. 1994; Studer et al. 1994). Radioactively labeled RA binds to the cranial neural crest cells and arrests both their proliferation and their migration (Johnston et al. 1985; Goulding and Pratt 1986). The teratogenic period during which cranial neural crest cells are affected occurs on days 20–35 in humans (days 8–10 in mice).

Retinoic acid probably disrupts these cells in several ways. One mechanism is that excess RA activates the negative feedback pathway that usually ensures the proper amount of this compound. Transient large increases in RA thus activate the synthesis of RA-degrading enzymes, causing a long-lasting *decrease* of RA. It is this deficiency in RA that results in the malformations (Lee et al. 2012). This explains why high amounts of retinoic acid produce phenotypes similar to those seen in deficiencies of retinoic acid.

Interference with RA signaling may be a wider public health concern for another reason. Glyphosate-based herbicides (such as Roundup) have been reported to upregulate the activity of endogenous RA (Paganelli et al. 2010). When *Xenopus* embryos were incubated in solutions containing ecologically relevant concentrations of these herbicides, RA-responsive reporter gene activation was dramatically altered, and the embryos exhibited cranial neural crest defects and facial disorders similar to those seen in RA teratogenesis (**FIGURE 23.8**).

Glyphosate is the most widely used (and profitable) herbicide in North America, where over 180 tons of it have been applied. It acts by blocking a plant enzyme that is critical for the synthesis of certain amino acids. One of the powerful abilities of genetic engineering has been to manufacture wide-spectrum herbicides such as Roundup and then breed crop plants that are resistant to this herbicide. This means that if you spray a large area, all the weeds will be killed and the only plants remaining will be those that are glyphosate-resistant. This, unfortunately, has led to the demise of certain weeds (such as milkweed) that are the major food sources for insects such

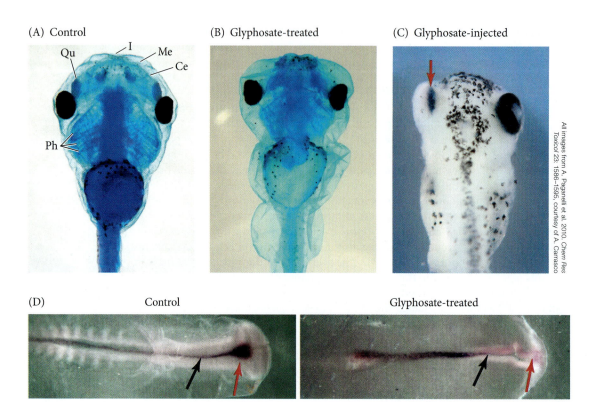

(A) Control

(B) Glyphosate-treated

(C) Glyphosate-injected

(D) Control   Glyphosate-treated

All images from A. Paganelli et al. 2010. *Chem Res Toxicol* 23: 1586–1595, courtesy of A. Carrasco

FIGURE 23.8 Glyphosate herbicide teratogenicity. (A) *Xenopus* tadpole raised under control conditions, stained with alcian blue to show facial cartilages. Ph, pharyngeal; Ce, ceratohyal; I, infrarostral; Me, meckel; Qu, quadrate. (B) *Xenopus* tadpole raised in environmentally relevant concentrations of glyphosate and similarly stained. Its pharyngeal arches and midline facial cartilage (cranial neural crest derivatives) failed to develop properly. (C) If an embryo is injected such that only one side (arrow) is exposed to glyphosate, that side shows cranial neural crest anomalies. (D) Control chick embryos show *sonic hedgehog* gene expression in the notochord (black arrow) and prechordal mesoderm (red arrow). Chick embryos grown in glyphosate show a severe reduction of *sonic hedgehog* expression in the prechordal (craniofacial) mesoderm.

> **(?) Developing Questions**
>
> In the northern United States, certain ponds contain a high proportion of frogs that have six or more limbs. What are the possible causes of these malformations?

as the monarch butterfly (Pelton et al. 2018). It has also led to increased amounts of glyphosate in our food supply. As we will see in the next chapter, normal animal development relies on having a healthy microbiome—the population of microbes in and on our bodies. These microbes use the same enzyme that glyphosate inhibits, and they can also be destroyed by this "herbicide." Thus, a new area of teratology has recently been initiated, where developmental defects are caused by the ability of a chemical to destroy the bacteria that give signals for normal development (Aitbali et al. 2018). (See **Further Development 23.2, The Developmental Origins of Adult Human Disease**, online.)

Endocrine Disruptors: The Embryonic Origins of Adult Disease

A specialized area of teratogenesis involves the misregulation of the endocrine system during development. Endocrine disruptors are exogenous (coming from outside the body) chemicals that disrupt development by interfering with the normal functions of hormones (Colborn et al. 1993, 1997). The phenotypic changes produced by endocrine disruptors are not the obvious anatomical birth defects produced by classic teratogens. Rather, the anatomical alterations induced by endocrine disruptors are often seen only microscopically; the major changes are physiological. The effects of endocrine disruptors often become manifest later in adult life and may persist for generations after exposure to the disruptor. A recent study (Balalian et al. 2019) found that prenatal and early childhood exposure to a certain endocrine disrupter prevented the normal functioning of the limb muscles 11 years later.

Endocrine disruptors can interfere with hormone function in many ways:

- They can mimic the effect of a natural hormone, such as diethylstilbestrol (DES), which binds to the estrogen receptor and mimics estradiol, which is active in building the female reproductive tract.

- They can act as antagonists and inhibit the binding of a hormone to its receptor or block the synthesis of a hormone. Vinclozolin, used to prevent mold growth on grapes and berries, binds to the testosterone receptor and prevents differentiation of male organs and behaviors (Kelce et al. 1994; Hotchkiss et al. 2003).

- They can affect the synthesis, elimination, or transportation of a hormone in the body. The herbicide atrazine, for example, elevates the synthesis of estrogen and can convert testes into ovaries in frogs.

- Some endocrine disruptors can "prime" the organism to be more sensitive to hormones later in life. Bisphenol A exposure during fetal development, for example, makes breast tissue more responsive to steroid hormones during puberty.

It was long thought that there were only a few teratogenic agents, and they were only harmful to fetuses exposed to high doses of the chemical. We now recognize that endocrine disruptors are everywhere in our technological society, and that low-dose exposure to endocrine disruptors in utero can be sufficient to produce significant disabilities later in life. Endocrine disruptors include chemicals in the materials that line baby bottles and the brightly colored plastic containers from which we drink our water; chemicals used in cosmetics, sunblocks, and hair dyes; and chemicals that prevent clothing from being highly flammable. When so many chemicals are involved, we are exposed to not just one but multiple endocrine disruptors, simultaneously and continuously. In many cases, more damage is done at a very low dose (such as 25 ng/kg/day) of an endocrine disruptor than at higher doses (see Myers et al. 2009; Belcher et al. 2012; Vandenberg et al. 2012). This can be due to higher concentrations of the hormone mimic activating negative feedback processes (as we discussed concerning retinoic acid) or to the chemical binding to different receptors at low doses and thereby initiating different pathways (Speroni et al. 2017; Villar-Pazos et al. 2017; Acevedo et al. 2018). There are numerous endocrine disruptors (see Gilbert and Epel 2015; Kabir et al. 2015). We encounter many of these substances on a daily basis, and we are exposed to them prenatally. (See **Further Development 23.3, DDT as an Endocrine Disruptor**, online.)

Diethylstilbestrol (DES)

One of the first endocrine disruptors to be identified was the potent environmental estrogen **diethylstilbestrol**, or **DES**. This drug was thought to ease pregnancy and prevent miscarriages, and it is estimated that in the United States, more than 1 million pregnant women and their fetuses were exposed to DES between 1947 and 1971. (This is probably a small fraction of exposures worldwide.) Although research from the 1950s showed that in fact DES had no beneficial effects on pregnancy, it continued to be prescribed until the FDA banned it in 1971. The ban was imposed when a specific type of tumor (clear-cell adenocarcinoma) was discovered in the reproductive tracts of some of the women whose mothers had taken DES during pregnancy.

DES interferes with sexual and gonadal development by causing cell type changes in the female reproductive tract (the derivatives of the Müllerian duct, which forms the upper portion of the vagina, cervix, uterus, and oviducts; see Figure 6.7) (**FIGURE 23.9**). In many cases, DES destroys the boundary between the oviduct and the uterus (the uterotubal junction), resulting in infertility, low fertility, and a high risk for other reproductive health problems (Robboy et al. 1982; Newbold et al. 1983; Hoover et al. 2011).

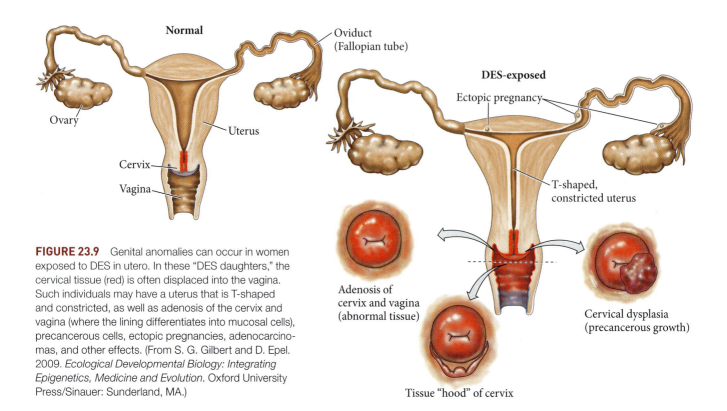

FIGURE 23.9 Genital anomalies can occur in women exposed to DES in utero. In these "DES daughters," the cervical tissue (red) is often displaced into the vagina. Such individuals may have a uterus that is T-shaped and constricted, as well as adenosis of the cervix and vagina (where the lining differentiates into mucosal cells), precancerous cells, ectopic pregnancies, adenocarcinomas, and other effects. (From S. G. Gilbert and D. Epel. 2009. *Ecological Developmental Biology: Integrating Epigenetics, Medicine and Evolution*. Oxford University Press/Sinauer: Sunderland, MA.)

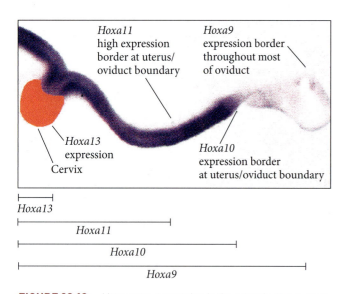

FIGURE 23.10 *Hoxa* gene expression in the reproductive system of a normal 16.5-day embryonic female mouse. A whole mount in situ hybridization of the *Hoxa13* probe is shown (red) along with a probe for *Hoxa10* (purple). *Hoxa9* expression extends throughout the uterus and through much of the presumptive oviduct. *Hoxa10* expression has a sharp anterior border at the transition between the presumptive uterus and the oviduct. *Hoxa11* has the same anterior border as *Hoxa10*, but its expression diminishes closer to the cervix. *Hoxa13* expression is found only in the cervix and upper vagina. (After L. Ma et al. 1998. *Dev Biol* 197: 141–154.)

Symptoms similar to human DES syndrome occur in mice exposed to DES in utero, allowing the mechanisms of this endocrine disruptor to be uncovered. Normally the regions of the female reproductive tract are specified by the *Hoxa* genes (**FIGURE 23.10**). The effects of DES on the female mouse reproductive tract appear to be the result of altered *Hoxa10* expression in the Müllerian duct (Ma et al. 1998). When DES was injected under the skin of pregnant mice, the DES almost completely repressed the expression of *Hoxa10* in the Müllerian duct of the developing fetuses (**FIGURE 23.11**). This repression was most pronounced in the stroma (mesenchyme) of the duct, where experimental embryologists had localized the effect of DES (Boutin and Cunha 1997). In addition, in *Hoxa10* knockout mice (Benson et al. 1996; Ma et al. 1998), there is a transformation of part of the uterus into oviduct tissue, as well as abnormalities of the uterotubal junction.

The Wnt proteins provide a link between Hox gene expression and uterine morphology. The Hox and Wnt proteins are both involved in the specification and morphogenesis of the reproductive tissues (**FIGURE 23.12**). The reproductive tracts of DES-exposed female mice resemble those of *Wnt7a* knockout mice. Hox genes and the Wnt genes communicate to keep each other activated; however, DES, acting through the estrogen receptor, represses the *Wnt7a* gene. This prevents maintenance of the Hox gene expression pattern, and prevents activation of another Wnt gene, *Wnt5a*, which encodes a protein necessary for cell proliferation (Miller et al. 1998; Carta and Sassoon 2004). (See **Further Development 23.4, DES as an Obesogen**, online.)

The effect of DES on fertility is a complex story of public policy, medicine, and developmental biology (Bell 1986;

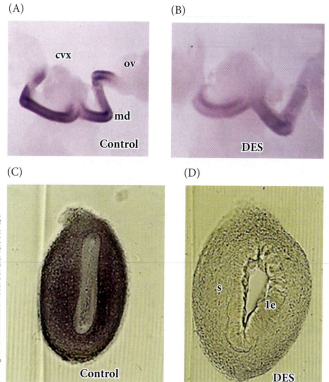

(A) cvx / ov / md / Control

(B) DES

(C) Control

(D) s / le / DES

FIGURE 23.11 In situ hybridization of a *Hoxa10* probe shows that DES exposure represses *Hoxa10*. (A) Normal 16.5-day embryonic female mice show *Hoxa10* expression from the boundary of the cervix through the uterus primordium and most of the oviduct (cvx, cervix; md, Müllerian duct; ov, ovary). (B) In mice exposed prenatally to DES, this expression is severely repressed. (C) In control female mice at 5 days after birth (when reproductive tissues are still forming), a section through the uterus shows abundant expression of *Hoxa10* in the uterine mesenchyme. (D) In female mice that are given high doses of DES 5 days after birth, *Hoxa10* expression in the mesenchyme is almost completely suppressed (le, luminal epithelium; s, stroma).

Palmlund 1996), and endocrine disruption by estrogenic compounds is ongoing. The Chapel Hill Consensus of 2007 (stemming from a conference sponsored by the Environmental Protection Agency and the National Institute of Environmental Health Sciences) claimed that some of the major constituents of plastics were estrogenic compounds, and that they were present in doses large enough to have profound effects on human sexual development (vom Saal et al. 2007). The most important of these compounds is bisphenol A.

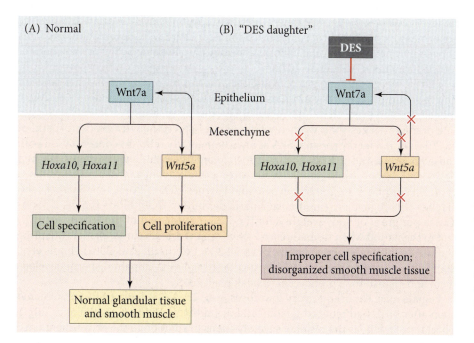

FIGURE 23.12 Misregulation of Müllerian duct morphogenesis by DES. (A) During normal morphogenesis, the *Hoxa10* and *Hoxa11* genes in the mesenchyme are activated and maintained by Wnt7a from the epithelium. Wnt7a also induces *Wnt5a* in the mesenchyme, and Wnt5a protein both maintains *Wnt7a* expression and causes mesenchymal cell proliferation. Together, these factors specify and order the morphogenesis of the uterus. (B) DES, acting through the estrogen receptor, blocks *Wnt7a* expression. Proper activation of the Hox genes and *Wnt5a* in the mesenchyme does not occur, leading to a radically altered morphology of the female genitalia. (After J. Kitajewsky and D. Sassoon. 2000. *BioEssays* 22: 902–910.)

Bisphenol A (BPA)

In the early years of hormone research, the steroid hormones were very difficult to isolate, so chemists manufactured synthetic analogues that would accomplish the same tasks. Bisphenol A, one of these analogues, was first synthesized as an estrogenic compound in the 1930s. Later, polymer chemists realized that BPA could be used in plastic production, and today it is one of the top 50 chemicals produced worldwide. Four corporations in the United States make almost 2 billion pounds of it each year for use in the resin lining most cans, as well as the polycarbonate plastic in baby bottles, children's toys, and water bottles. It is also used in dental sealant and (strange as it sounds) in cash register receipts. In its modified form, tetrabromo-bisphenol A, it is the major flame retardant on fabrics.

Human exposure comes primarily from BPA that has leached from food containers (von Goetz et al. 2010). Babies and infants acquire BPA through polycarbonate bottles; teenagers and adults get most of their BPA by consuming canned food that has been stored in containers lined with BPA-containing resins. Since 95% of urine samples taken from people in the U.S. and Japan have measurable BPA levels (Calafat et al. 2005), and since BPA can inhibit normal primate brain development at concentrations lower than what the U.S. Environmental Protection Agency (EPA) considers safe (Lernath et al. 2008), health concerns have been raised over the roles that BPA might play in causing reproductive failure, cancer, and behavioral anomalies. (See **Further Development 23.5, BPA and Altered Behavior**, online.)

BPA AND REPRODUCTIVE HEALTH   BPA does not remain fixed in plastic forever (Krishnan et al. 1993; vom Saal 2000; Howdeshell et al. 2003). If you let water sit in an old polycarbonate rat cage at room temperature for a week, you can measure about 300 µg per liter of BPA in the water. That is a biologically active amount—a concentration that will reverse the sex of a male frog and cause weight changes in the uterus of a young mouse. Such leached BPA also can cause chromosome anomalies. When a laboratory technician mistakenly rinsed some polycarbonate cages in an alkaline detergent, the female mice housed in these cages were found to have meiotic abnormalities in 40% of their oocytes (the normal level of such abnormalities is about 1.5%). When BPA was administered to pregnant mice under controlled circumstances, Hunt and her colleagues (2003) showed that short, low-dose exposure to BPA was sufficient to cause meiotic defects in maturing mouse oocytes. This effect was also seen in primates. Exposure of fetal female monkeys to low doses of BPA (at levels comparable to that found in human serum) caused ovarian and meiotic abnormalities similar to those observed in mice. There were several abnormalities of ovarian function, including abnormal meiotic chromosome behavior and aberrant follicle formation (Hunt et al. 2012).

BPA crosses the human placenta and accumulates in concentrations that can alter development in laboratory animals (Ikezuki et al. 2002; Schönfelder et al. 2002). Indeed, women exposed to high levels of BPA during pregnancy had an 83% higher rate of miscarriages than women who had not been so heavily exposed (Lathi et al. 2014). In model organisms, BPA at environmentally relevant concentrations can cause abnormalities in fetal gonads, prostate enlargement, low sperm counts, and behavioral changes when these fetuses become adults (vom Saal et al. 1998; Palanza et al. 2002; Kubo et al. 2003; vom Saal and Hughes 2005). When vom Saal and colleagues (1997) gave pregnant mice 2 parts per billion (ppb) BPA—that is, 2 nanograms per gram of body weight—for the 7 days at the end of pregnancy (equivalent to the period when human reproductive organs are developing), male offspring showed an increase in prostate size of about 30% (Wetherill et al. 2002; Timms et al. 2005). In organ cultures of human testicular tissue, BPA impaired sperm development and testosterone production (Eladak et al. 2018).

Female mice exposed to very low doses (e.g., 25 ng/kg/day) of BPA in utero and soon after birth had reduced fertility and fecundity as adults (Cabaton et al. 2011). This lower fertility is the result of several actions in addition to the above-mentioned effects on developing eggs. First, BPA and other endocrine disruptors have been found to prevent the sex-specific maturation of those parts of the mouse brain regulating ovulation (Rubin and de Sauvage 2006; Gore et al. 2011). Second, female mice exposed in utero to low doses of BPA (2000 times lower than the dosage considered safe by the U.S. government) had alterations in the organization of their uterus, vagina, breast tissue, and ovaries, as well as altered estrous cycles as adults (Howdeshell and vom Saal 2000; Markey et al. 2003; Acevedo et al. 2018). Third, female mice exposed to short exposures of "safe" levels of BPA had deficiencies in their placentas, leading to poor fetal growth (Müller et al. 2018). And fourth, BPA has been found to alter the gamete-specific methylation pattern of imprinted genes in mouse embryos and placentas (Susiarjo et al. 2013).

Removing BPA from the environment is important but is proving to be difficult. Even BPS and BPF, the BPA analogues introduced by the chemical industry as "safe" alternatives, have been shown to be endocrine disruptors with estrogenic effects (Rochester and Bolden 2015; Le Fol et al. 2017). (See **Further Development 23.6, Testicular Dysgenesis**, and **Further Development 23.7, BPA and Cancer Susceptibility**, both online.)

Atrazine: Endocrine disruption through hormone synthesis

The enzyme aromatase can convert testosterone into estrogen, and this estrogen is able to induce female sex determination in many vertebrates. In turtles, for instance, estrogen downregulates testis-forming genes and upregulates the genes producing ovaries (Valenzuela et al. 2013; Bieser and Wibbels 2014). BPA and other estrogenic endocrine disruptors can also reverse the sex of turtles raised at "male" temperatures (Jandegian et al. 2015). This and other studies (e.g., see Bergeron et al. 1994, 1999) have important consequences for conservation efforts to protect endangered species (including turtles, amphibians, and crocodilians) in which hormones can effect changes in primary sex determination.

The survival of some amphibian species may be at risk from herbicides that promote estrogens at the expense of testosterone, severely depleting the number, function, and fertility of the males. One such case involves the development of hermaphroditic and demasculinized frogs after exposure to extremely low doses of the weed killer atrazine, one of the most widely used herbicides in the world, and one that is found in streams and ponds throughout the United States (**FIGURE 23.13**). Atrazine induces aromatase, which, as mentioned earlier, converts testosterone into estrogen. Hayes and colleagues (2002a) found that exposing tadpoles to atrazine concentrations as low as 0.1 ppb produced gonadal and other sexual anomalies in male frogs. At 0.1 ppb and higher, many male tadpoles developed ovaries in addition to testes. At 1 ppb atrazine, the vocal sacs (which a male frog must have in order to signal and obtain a potential mate) failed to develop properly. Similar experiments in outdoor environments more similar to natural conditions (Langlois et al. 2010) also showed that male leopard frogs (*Rana pipiens*) had been transformed into females by atrazine.

FIGURE 23.13 Demasculinization of frogs by low amounts of atrazine. (A) Testis of a frog from a natural site having 0.5 parts per billion (ppb) atrazine. The testis contains three lobules that are developing both sperm and an oocyte. (B) Two testes of a frog from a natural site containing 0.8 ppb atrazine. These organs show severe testicular dysgenesis, which characterized 28% of the frogs found at that site. (C) Effect of a 46-day exposure to 25 ppb atrazine on testosterone levels in the blood plasma of sexually mature male *Xenopus*. Levels in control males were some 10-fold higher than in control females; atrazine-treated males had plasma testosterone levels at or below those of control females. (C after T. B. Hayes et al. 2002. *Proc Natl Acad Sci USA* 99: 5476–5480. © 2002 National Academy of Sciences, U.S.A.)

(A)

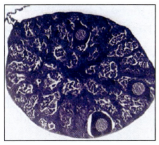

Images A and B from T. B. Hayes et al. 2003. *Envir Health Perspec* 111: 568–575, photographs courtesy of T. Hayes

(B)

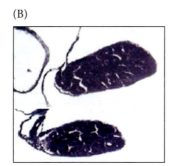

(C)

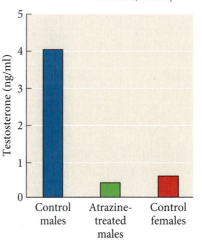

In laboratory experiments, the testosterone levels of adult male frogs were reduced by 90% (to levels of control females) when they were exposed, as sexually mature adults, to 25 ppb atrazine (Hayes et al. 2002a). This is an ecologically relevant dose, since the allowable amount of atrazine in U.S. drinking water is 3 ppb, and atrazine levels can reach 224 ppb in streams of the midwestern United States (Battaglin et al. 2000; Barbash et al. 2001). Even at doses as low as 2.5 ppb, the sexual behavior of the male frogs was severely diminished. Matings became relatively rare, and in 10% of the cases, males exposed to atrazine became functional egg-laying females (Hayes et al. 2010).

In a field study, Hayes and his colleagues collected leopard frogs and water samples from eight sites across the central United States (Hayes et al. 2002b, 2003). They sent the water samples to two separate laboratories to determine their atrazine levels, and they coded the frog specimens so that the technicians dissecting the gonads did not know which site the animals came from. The results showed that the water from all but one site contained atrazine—and this was the only site from which the frogs had no gonadal abnormalities. At concentrations as low as 0.1 ppb, leopard frogs displayed testicular dysgenesis (stunted growth of the testes) or conversion to ovaries. In many examples, oocytes were found in the testes (see Figure 23.13).

Atrazine's ability to feminize male gonads has been seen in all classes of vertebrates. Indeed, low sperm count, poor semen quality, and decreased fertility have been observed in men who are routinely exposed to atrazine (Swan et al. 2003; Hayes et al. 2011). Concern over atrazine's apparent ability to disrupt sex hormones in both wildlife and humans has resulted in bans on the use of this herbicide by France, Germany, Italy, Norway, Sweden, and Switzerland (Dalton 2002). However, drug companies in the United States have lobbied successfully to keep atrazine in the American markets (see Blumenstyk 2003; Aviv 2014).

Fracking: A potential new source of endocrine disruption

U.S. government regulations do not cover the compounds added to the environment by the hydraulic fracturing ("fracking") procedures used to extract methane (natural gas) from shale. A total of 632 chemicals have been identified as being used in this procedure. About 25% of them are known to cause tumors, and more than 35% of them are known to affect the endocrine system (Colborn et al. 2011). It is estimated that about 50% of the fluid used in fracking returns to the surface (Department of Energy 2009). Water samples taken from both standing water and groundwater at fracking sites contained estrogenic compounds, antiestrogenic compounds, and anti-androgenic (anti-testosterone) compounds (Kassotis et al. 2014; Webb et al. 2014). Some compounds in the water activated enhancers of estrogen-responsive genes, and others prevented the activation of testosterone-responsive genes.

Prenatal exposure to purified mixtures of the chemicals found in fracking water (at concentrations found in drinking water near fracking sites) caused reproductive deficiencies in female mice and altered pituitary hormone levels in both sexes (Kassotis et al. 2015, 2016). One of the sites where the water was tested had been a ranch, but ranching had to be discontinued because the animals were no longer producing offspring. A recent study concerning fracking in rural Colorado documented an increased incidence of congenital heart disease in children born in families residing close to the fracking wells (McKenzie et al. 2014). Thus, fracking appears to be a source of chemicals that can impede fertility and brain development, and the methane produced is a greenhouse gas that accelerates global warming.

Transgenerational Inheritance of Developmental Disorders

A lifetime of chopping wood will not give your offspring bulging biceps, nor would the loss of your arms in an accident cause your offspring to be born armless. This is because the environmental agents—exercise or trauma in these cases—do not cause mutations in the DNA. In order to be transmitted, mutations must not only be somatic, but also enter the germ line. Thus, genetic mutations acquired in skin cells that are overexposed to sunlight will not be transmitted. However, one of the most surprising results of contemporary developmental genetics has been the discovery that certain environmentally induced phenotypes *can* be transmitted from generation to generation.

DNA methylation seems to be a mechanism that can circumvent the mutational block to the transmission of acquired traits (see Chapter 3).

Certain agents can cause the same alterations of DNA methylation throughout the body, and these alterations in methylation can be transmitted by the sperm and egg. Jablonka and Raz (2009) have documented dozens of cases where different "epialleles"—DNA containing different methylation patterns—can be stably transmitted from generation to generation. In mammals, epiallelic inheritance was first documented by studies of the endocrine disruptor vinclozolin, a fungicide widely used on grapes. When injected into pregnant rats during particular days of gestation, vinclozolin caused testicular dysgenesis in the male offspring. The testes started forming normally, but as the mouse got older, its testes degenerated and quit producing sperm. What's more interesting is that the male mice fathered by mice with this induced testicular dysgenesis (often by artificial means) also exhibit testicular dysgenesis. So do their male offspring and the subsequent generation's male offspring (Anway et al. 2005; Anway and Skinner 2006; Guerrero-Bosagna et al. 2010). Thus, when a pregnant rat is given vinclozolin, even her great-grandsons are affected (**FIGURE 23.14**).

The mechanism for this inheritance in rats appears to be DNA methylation (Skinner 2016). The promoters of more than 100 genes in the Sertoli cells (see Chapter 6) have their methylation patterns changed by vinclozolin, and altered promoter methylation can be seen in the sperm DNA for at least three subsequent generations (Guerrero-Bosagna et al. 2010; Stouder and Paolini-Giacobino 2010). These genes include those whose products are necessary for cell proliferation, G proteins, ion channels, and receptors. It is important to note that by the third (F_3) generation, there could have been no

(A)

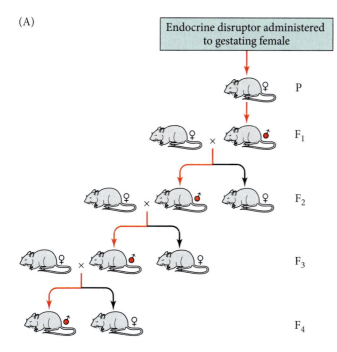

FIGURE 23.14 Epigenetic transmission of endocrine disruption. (A) Transmission of testicular dysgenesis syndrome (red circles) is shown through four generations of mice. The only mice exposed in utero were the F_1 generation. (B,C) Cross section of the seminiferous tubules from the testes of (B) a control male rat and (C) a male rat whose grandfather was born from a mother injected with vinclozolin. The arrow in (B) shows the tails of normal sperm. The arrow in (C) shows the lack of germ cells in the much smaller tubule of the rat descended from the vinclozolin-injected female; this rat was infertile under normal conditions. (A after M. D. Anway and M. K. Skinner 2006. *Endocrinology* 147: S43–S49.)

(B)

(C)

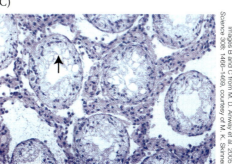

Images B and C from M. D. Anway et al. 2005.
Science 308: 1466–1469, courtesy of M. K. Skinner

direct exposure to vinclozolin. The first fetus is inside the treated mother; and that fetus has germ cells (of the F_2 generation) inside itself. But even though the offspring of the F_3 and F_4 generations have never been exposed to vinclozolin, their phenotype is changed by the initial injection to their great-grandmother.

Similar studies have indicated that other endocrine disruptors—DES, bisphenol A, and PCBs—also have transgenerational effects (Skinner et al. 2010; Walker and Gore 2011; Gillette et al. 2018). Indeed, behavioral changes induced by BPA in mice may last at least four generations, and DNA methylation differences can be observed in mouse brains for generations after the pregnant mouse is given BPA (Wolstenhome et al. 2012; Drobná et al. 2018). The public health ramifications of this type of inheritance are just beginning to be explored.

Cancer as a Disease of Development

In addition to teratogenesis and endocrine disruption, tumor formation is also recognized as a disease of abnormal development (e.g., see Virchow 1858; Stevens and Little 1953; Auerbach 1961; Pierce et al. 1978). A *New York Times* headline pithily summarized the relationship between cancer and embryos: "A Tumor, the Embryo's Evil Twin." **Carcinogenesis** (cancer formation) is more than just genetic changes in the cells giving rise to the tumor (see Hanahan and Weinberg 2000; Versteeg 2014; Jamshidi et al. 2017). Rather, cancer can be seen as a disease of cell-cell communication, wherein the cancer cells are permitted to express many processes that were useful to embryonic cells. When the processes are restarted, the restrictions that had prevented these adult cells from dividing, migrating, attaching to a new tissue, implanting into the tissue, and recruiting new blood vessels from the new tissue become abrogated. Indeed, the formation and migration of cancer cells have recently been compared to the development and implantation of the blastocyst into the uterus (Mor et al. 2017; Vento-Tormo et al. 2018).

It was once thought that carcinogenesis and metastasis could come about only by the proliferation of a cell that had acquired mutations enabling it to become "autonomous." This **somatic mutation theory** characterized cancer by intracellular mechanisms that enable a cell to become independent of its environment. Cancer was caused by the proliferation of a "renegade cell." But this turns out to be only part of the explanation. In the late twentieth century, biologists returned to the earlier idea of cancer being a defect of communication between tissues. This alternative view became known as the **tissue organization field theory** (**TOFT**), which saw cancer as "development gone awry" (Sonnenschein and Soto 1999, 2016).

According to the TOFT, carcinogens disrupt the reciprocal communication between the mesenchymal and epithelial tissues of an organ. Mutation is one way of doing this, but not the major way. Moreover, the TOFT adopted the premise that proliferation and motility were constitutive properties of cells. Cells usually divide and move. Stasis has to be enforced by cell-cell communication. When this reciprocal communication is disturbed, the constraints imposed by the tissue over the cells in its field of influence are relaxed. As a consequence, the cells are free to express their capacity to proliferate and to move, thus forming tumors, invading tissues, and traveling to distant locations where they form metastases.

Carcinogenesis is being recast as a stepwise progression of conditions that depends on reciprocal interactions between incipient cancer cells and the supporting cells of their tissue environment. The progressive alteration of cell-cell interactions leads to aberrant tissue architecture, and possibly to the formation of niches that generate cancer cells. Indeed, cancer cells appear to proceed by recapitulating steps of normal development, including the formation of a niche in which to proliferate. Thus, both cancer and congenital anomalies can be seen as diseases of tissue organization, differentiation, and intercellular communication. As we will see, they are often caused by defects in the same pathways. There are many reasons to view malignancy and metastasis in terms of development, three of which we will discuss here:

1. Context-dependent tumor formation
2. Defects in cell-cell communication as the initiator of cancers
3. Cancer stem cells

CONTEXT-DEPENDENT TUMORS Many tumor cells have normal genomes, and whether or not these tumors become malignant (migratory and invasive) depends on their environment (Pierce et al. 1974; Mack et al. 2014). The most remarkable of these cases is the **teratocarcinoma**, a tumor of germ cells or stem cells (Illmensee and Mintz 1976; Stewart and Mintz 1981). Teratocarcinomas are malignant growths of cells that resemble the inner cell mass of the mammalian blastocyst, and they can kill the organism. However, if a teratocarcinoma cell is placed on the inner cell mass of a mouse blastocyst, it will integrate into the blastocyst, lose its malignancy, and divide normally. Its cellular progeny can become part of numerous embryonic organs. Should its progeny form part of the germ line, the sperm or egg cells formed from the tumor cell will transmit the tumor genome to the next generation. Thus, whether the cell becomes a tumor or becomes part of the embryo can depend on its surrounding cells.

It is possible that the stem cell environment suppresses tumor formation by its secretion of inhibitors of the paracrine pathways. For instance, many tumor cells, such as melanomas, secrete the paracrine factor Nodal. This aids their proliferation and also helps supply them with blood vessels. When placed in an environment of embryonic stem cells (which secrete Nodal *inhibitors*), aggressive melanoma tumors (which are derived from neural crest cells) become normal pigment cells (Hendrix et al. 2007; Postovit et al. 2008). Remarkably, such malignant melanoma cells, when transplanted into early chick embryos, downregulate their Nodal expression and migrate as nonmalignant cells along the neural crest cell pathways (**FIGURE 23.15**; Kasemeier-Kulesa et al. 2008).

DEFECTS IN CELL-CELL COMMUNICATION According to the TOFT, tissue interactions are required to prevent cells from dividing, leading to the premise that cancer can be caused by miscommunication between

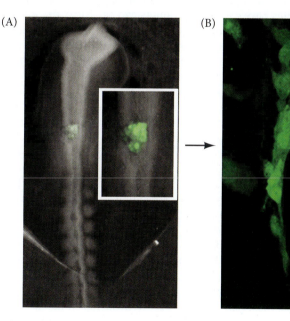

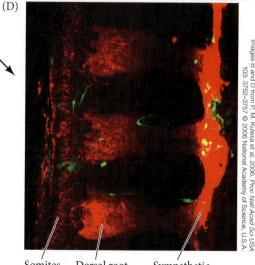

FIGURE 23.15 When aggressively metastatic human melanoma cells are injected into a 2-day chick embryo dorsal neural tube (A), they form normal migratory chains (B) and follow the neural crest migration roots to integrate into facial cartilage (C) and sympathetic ganglia (D). There, they form nonmalignant melanocytes.

FIGURE 23.16 Evidence that the stroma regulates the production of epithelial (parenchymal) tumors. (A) Schematic drawing of the experimental protocol. The mammary gland tissue contains both epithelium and stroma (mesenchymal cells). The two groups of cells can be isolated and then recombined. One can add a cancer-forming substance (carcinogen) to the epithelium and not the stroma, or to the stroma but not the epithelium. Then one can combine them so that the carcinogen has been experienced by the epithelium (but not the stroma), by the stroma (but not the epithelium), by both the stroma and the epithelium, or by neither. (B) Results when the cancer-causing mutagen N-nitrosomethylurea (NMU) or just the control vehicle (VEH) was applied to either the stroma or the epithelium and transplanted back into the rat mammary gland. On the horizontal axis, the label to the left of each slash indicates the treatment applied to the epithelium and the label to the right indicates that applied to the stroma. Only animals whose stroma was treated with NMU developed tumors, regardless of whether the epithelium was exposed to NMU or not. NMU-treated, intact animals (positive controls) developed tumors, but none of the rats receiving control solutions (negative controls) had tumors. (B after M. V. Maffini et al. 2004. *J Cell Sci* 117: 1495–1502.)

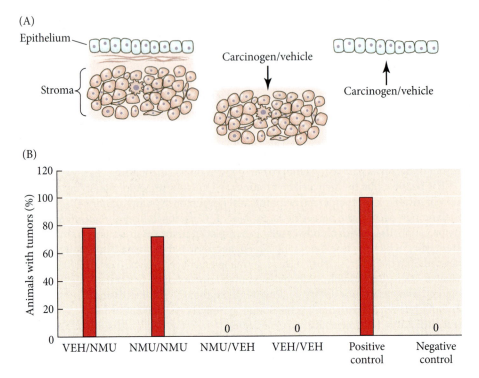

cells. Thus, tumors can arise through defects in tissue architecture, and the surroundings of a cell are critical in determining malignancy (Sonnenschein and Soto 1999, 2000; Bissell et al. 2002). Studies have shown that tumors can be caused by altering the structure of the tissue, and that these tumors can be suppressed by restoring an appropriate tissue environment (Coleman et al. 1997; Weaver et al. 1997; Booth et al. 2011). In particular, although 80% of human tumors arise from epithelial cells, these cells do not always appear to be the site of the cancer-causing lesion. Rather, epithelial cell cancers are often caused by defects in the mesenchymal stromal cells that surround and sustain the epithelia. When Maffini and colleagues (2004) recombined normal and carcinogen-treated epithelia and mesenchyme in rat mammary glands, tumorous growth of mammary epithelial cells occurred not in carcinogen-treated epithelia but only in epithelia placed in combination with mammary mesenchyme that had been exposed to the carcinogen. Thus, the carcinogen caused defects in the mesenchymal stroma of the mammary gland, and the treated mammary stroma somehow could no longer provide the epithelial cells with restrictions on dividing and moving (**FIGURE 23.16**). These findings have led to a new appreciation of the ways in which stromal cells can regulate cancer initiation in the adjacent epithelium (see Wagner 2016).

This brings us to the next notion: that tumors can occur through disruptions of paracrine signaling between cells. Rubin and de Sauvage (2006) concluded that "several key signaling pathways, such as Hedgehog, Notch, Wnt and BMP/TGF-β/Activin, are involved in most processes essential to the proper development of an embryo. It is also becoming increasingly clear that these pathways can have a crucial role in tumorigenesis when reactivated in adult tissues through sporadic mutations or other mechanisms." We saw this before, in the discussion of Nodal secretion by melanoma cells. These findings demonstrate the importance of stromal tissue just mentioned. Many tumors, for instance, secrete the paracrine factor Sonic hedgehog (Shh), which can act in one of two ways. First, it can act in an autocrine fashion, stimulating the cells that produce it to grow (**FIGURE 23.17A,B**; Rubin and de Sauvage 2006; Zhao et al. 2009); in such cases, as in certain medulloblastomas and leukemias, Shh pathway inhibitors can stop the cancer growth. Second, the Shh produced by the tumor may act not on the tumor but on the stromal meserchyme cells around it, causing the stromal cells to produce factors that support tumor growth (**FIGURE 23.17C**). If the Shh pathway is blocked, the tumor regresses (Yauch et al. 2008, 2009; Tian et al. 2009). Cyclopamine, a teratogen

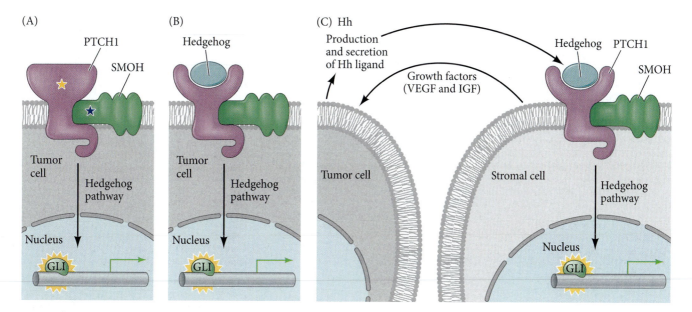

FIGURE 23.17   Mechanisms by which the Hedgehog (Hh) pathway (see Figure 4.22) can lead to cancer. (A) When Shh is a mitogen (as it is for cerebellar granule neuron progenitor cells or hematopoietic stem cells), loss-of-function mutations in the Hh ligand Patched (PTCH1; yellow star) or gain-of-function mutations in the Patched inhibitor Smoothened (SMOH; blue star) activate the Hedgehog pathway, even in the absence of Shh or another Hedgehog protein. (B) In the autocrine model, tumor cells both produce and respond to the Hh ligand. (C) In the paracrine model, tumor cells produce and secrete the Hh ligand, and the surrounding stromal cells receive the Hh protein. The stromal cells respond by producing growth factors, such as VEGF or IGF, that support tumor growth or survival. GLI is the transcription factor activated by the Hh pathway. (After L. L. Rubin and F. J. de Sauvage. 2006. *Nat Rev Drug Discov* 5: 1026–1033.)

that blocks Shh signaling, can prevent certain of these tumors from growing (Berman et al. 2002, 2003; Thayer et al. 2003; Song et al. 2011).

Thus, the same chemicals that can cause teratogenesis by blocking a pathway in embryonic development may be useful in blocking the activation of cancer stem cells. Even the classic teratogen thalidomide is being "rehabilitated" for use in the fight against cancer.

THE CANCER STEM CELL HYPOTHESIS   In 1971, Pierce and Johnson reported that "malignant tissue, like normal tissue, maintains itself by proliferation and differentiation of its stem cells." That same year, Pierce and Wallace demonstrated the presence of stem cells—cells that would cause the same cancer when transplanted into other animals—in rat tumors. The similarities between normal stem cells and cancer stem cells was highlighted when lineage tracing revealed that the stem cells of intestinal adenomas (the precursor of intestinal cancer) are Lgr5[+] cells and have the same relationship to the Paneth cells (see Chapter 5) as do normal intestinal stem cells (Schepers et al. 2013).

In numerous cancers, a rapidly dividing cancer stem cell (CSC) population gives rise to more cancer stem cells as well as to populations of relatively slowly dividing differentiated cells (Lapidot et al. 1994; Chen et al. 2012; Driessens et al. 2012; Schepers et al. 2012). Indeed, when tumor cells are transplanted from one animal to another, only the CSCs can give rise to new heterogeneous tumors (Gupta et al. 2009; Singh and Settleman 2010).

The origins of CSCs remain uncertain and may be different for different tumor types. Some researchers feel that the CSCs come from either normal adult stem cells or progenitor (transit-amplifying) cells. This appears to be the case in colon cancer, in which a normal intestinal stem cell (responsible for the daily renewal of all the epithelial cell types in the intestine) seems to become a cancer stem cell (de Sousa e Melo et al. 2017). In other cases, CSCs appear to form when an epithelial cell starts (but does

FIGURE 23.18 Tumor cells can construct their own niches. (A) Using a cell-lineage-tracing approach, lung tumor cells were found to divide to produce two cell types: a tumor cell and, unexpectedly, a support cell that is not a tumor cell but that forms part of the tumor niche. (B) The support cell (purple), derived from a tumor cell, can synthesize and secrete a functional Wnt protein. When the Wnt paracrine factor binds to a Wnt receptor on a tumor cell (green), it activates the cell's Wnt pathway, which drives tumor cell proliferation. (After M. Huch and E. L. Rawlins. 2017. *Nature* 545: 292–293.)

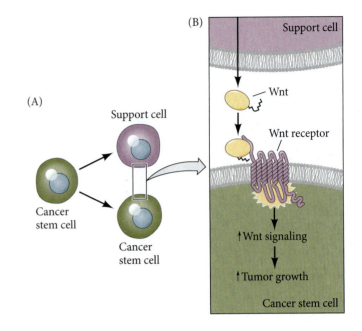

not complete) epithelial-mesenchymal transition (EMT). These EMTs were seen in the development of neural crest cells and somites (see Chapters 15 and 17). A cell that becomes an epithelial tumor (a carcinoma) may go through a series of changes from being an epithelial cell tightly joined to its neighbors to being a mesenchyme-like cell that can travel to other sites and begin new tumors (Pastushenko et al. 2018). These changes could be induced by alterations in the stromal cells supporting the cell.

Interestingly, CSCs may be able to create their own niches. It is the more differentiated cells in the tumor rather than the CSCs that become the niches that secrete the paracrine and juxtracrine factors (such as Wnts and Notch) needed for further CSC growth (**FIGURE 23.18**; Lim et al. 2017; Tammela et al. 2017). In some tumors (such as aggressive glioblastomas), these niche cells can even differentiate into blood vessel endothelial cells, thereby creating the tumor's own vasculature (El Hallani et al. 2010; Ricci-Vitiani et al. 2010; Wang et al. 2010).

Development-based therapies for cancer

Cancer is not so much the result of a cell gone bad as it is of cell relationships gone awry. Cancers are often diseases of developmental signaling, and several types of cancer cells can be normalized when placed back into regions of embryos that express certain paracrine factors or their inhibitors. This developmental view of cancer allows us to explore new avenues for cancer treatment. One such mode of treatment, **differentiation therapy**, was considered possible as long as 40 years ago but was not feasible at the time.

In 1978, Pierce and his colleagues noted that cancer cells were in many ways reversions to embryonic cells, and they hypothesized that cancer cells should revert to normalcy if they were made to differentiate. Also in 1978, Sachs discovered that certain leukemias could be controlled by making the leukemic cells differentiate rather than proliferate. One of these leukemias, acute promyelocytic leukemia (APL), is caused by a somatic recombination creating a "new" transcription factor, one of whose subunits is a retinoic acid receptor. This receptor, even in the absence of retinoic acid, binds to the RA binding sites in DNA, where it represses RA-responsive genes as well as creates a larger condensed chromatin structure (Nowak et al. 2009). Expression of this "new" transcription factor in neutrophil progenitors causes the cell to become malignant (Miller et al. 1992; Grignani et al. 1998). Treatment of APL patients with all-*trans* retinoic acid results in remission in more than 90% of cases because the additional RA is able to effect the differentiation of the leukemic cells into normal neutrophils (Hansen et al. 2000; Fontana and Rishi 2002). Retinoic acid therapy also restores normal differentiation to nasopharyngeal carcinomas (**FIGURE 23.19**; Yan et al. 2014; Yan and Liu 2016).

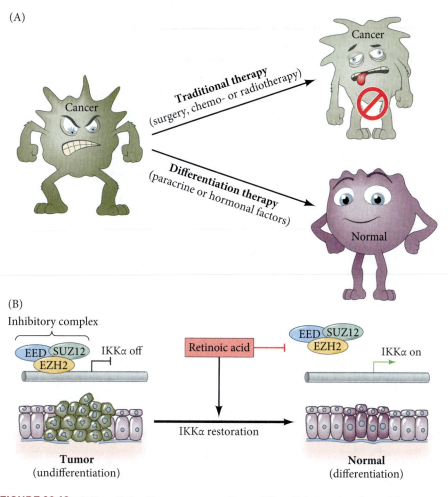

FIGURE 23.19 Differentiation therapy. (A) While traditional cancer treatments (surgery, chemotherapy, and radiation) aim to kill tumor cells, differentiation therapy is based on the concept that a tumor is a developmental disorder and can be induced to become a normal "adult" cell. (B) Retinoic acid can induce malignant nasopharyngeal carcinoma cells to become normal adult cells. Nasopharyngeal carcinoma cells are like undifferentiated neck cells, and the enzyme IKKα is epigenetically suppressed by a complex containing enhancer of zeste homologue 2 (EZH2) (left). Retinoic acid treatment prevents the formation of that inhibitory complex and restores IKKα expression. The tumor cells differentiate into normal cells (right). (After M. Yan and Q. Liu. 2016. *Chin J Cancer* 35: 3.)

Coda

Developmental biology is increasingly important in modern medicine. Preventive medicine, public health, and conservation biology demand that we learn more about the mechanisms by which industrial chemicals and drugs can damage embryos. The ability to effectively and inexpensively assay compounds for potential harm is critical. Developmental biology also provides new ways of understanding carcinogenesis and new approaches to preventing and curing cancers. And finally, developmental biology is providing the explanations for how mutated genes and aneuploidies cause their aberrant phenotypes.

It is critical to realize that the agents we put into the environment, the cosmetics we put on our skin, and the substances we eat and drink can reach developing embryos, fetuses, and larvae. Developing organisms have different physiologies as they construct, rather than merely sustain, their phenotypes, and chemicals that appear harmless to adults may disrupt the development of embryos. It takes a community to raise an embryo.

Closing Thoughts on the Opening Photo

In this cross section of a stem cell-derived cerebral organoid (a "mini-brain in a dish"), the radial glial cells that function as neural stem cells are stained red and the mature neurons are blue. The green dye binds to the AXL receptor proteins at the base of the neural stem cells. These membrane-bound receptor proteins are not found on most other cells, but they are found in high density on neural stem cells during the second trimester of human pregnancy, when they are involved in cell division. Zika virus is thought to attach to these AXL proteins to get into the cells (Nowakowski et al. 2016; Persaud et al. 2018).

From Nowakowski et al. 2016. *Cell Stem Cell* 18: 591–596

Endnotes

[1] This chapter focuses on human health and provides general information about a variety of medical topics. It is not intended to provide medical advice for specific persons or disorders. For a more thorough account of development and disease, please see Gilbert and Epel 2015.

[2] It took over 50 years to uncover the mechanism of thalidomide's teratogenicity. Thalidomide promotes the degradation of the SALL4 transcription factor that is used in the formation of limbs and ears. The loss-of-function mutation of SALL4 in humans produces the same types of congenital anomalies (Donovan et al. 2018).

Snapshot Summary
Development in Health and Disease

1. Developmental anomalies due to genetic errors and environmental influences result in a relatively low rate of survival of all human conceptions.

2. Chance plays a role in developmental outcomes. There is large variation in the amounts of transcription and translation, such that at different times, cells are making more or fewer developmentally important proteins.

3. Pleiotropy occurs when several different effects are produced by a single gene. In mosaic pleiotropy, each effect is caused independently by the expression of the same gene in different tissues. In relational pleiotropy, abnormal gene expression in one tissue influences other tissues, even though those other tissues do not express that gene.

4. Genetic heterogeneity occurs when mutations in more than one gene can produce the same phenotype. Phenotypic heterogeneity arises when the same gene can produce different defects (or differing severities of the same defect) in different individuals.

5. Teratogenic agents include chemicals such as alcohol and retinoic acid, as well as heavy metals, certain pathogens, and ionizing radiation. These agents adversely affect normal development and can result in malformations and functional deficits.

   - There may be multiple effects of alcohol on cells and tissues that result in a syndrome of cognitive and physical abnormalities.

   - The compound retinoic acid is active in development, and too much or too little of it can cause congenital anomalies.

6. Endocrine disruptors can bind to or block hormone receptors or block the synthesis, transport, or excretion of hormones. Presently, bisphenol A and other endocrine disruptive compounds are being considered as possible agents of reproductive failure, cancer, and behavioral anomalies.

   - Environmental estrogens can cause reproductive system anomalies by suppressing Hox gene expression and Wnt pathways.

   - In some instances, endocrine disruptors methylate DNA, and these patterns of methylation can be inherited from one generation to the next.

7. Cancer can be seen as a disease of altered development. Cancers metastasize in manners similar to embryonic cell movement, and some tumors revert to nonmalignancy when placed in environments that support normal morphogenesis and curtail excessive cell proliferation.

   - Cancers can arise from errors in cell-cell communication. These errors include alterations of paracrine factor synthesis.

   - In many instances, tumors have a rapidly dividing cancer stem cell population, which produces more cancer stem cells as well as more quiescent and differentiated cells. These differentiated cells can act as a niche for the cancer stem cells.

Go to oup.com/he/barresi12xe for Further Developments, Scientists Speak interviews, Watch Development videos, Dev Tutorials, and complete bibliographic information for all literature cited in this chapter.

Development and Evolution

Developmental Mechanisms of Evolutionary Change

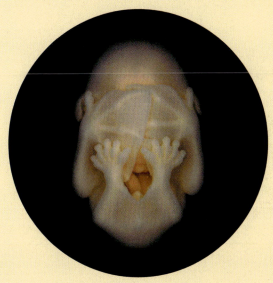

What changes in development might be needed for the evolution of a nonflying mammal into a bat?

Photograph courtesy of R. R. Behringer

WHILE HE WAS WRITING ON THE ORIGIN OF SPECIES, Charles Darwin consulted his friend Thomas Huxley concerning the origins of variation. In his response, Huxley noted that many differences between organisms could be traced to differences in their development, and that these differences are "the result not so much of the development of new parts as of the modification of parts already existing and common to both the divergent types" (Huxley 1857).

Huxley's response expresses a major tenet of **evolutionary developmental biology**, a relatively new science that views evolution as the result of changes in development. If development is the change of gene expression and cell position over time, then evolution is the change of development over time. This new field—colloquially referred to as **evo-devo**—is producing a new model of evolution that integrates developmental biology, paleontology, and population genetics to explain and characterize the diversity of life (Raff 1996; Hall 1999; Arthur 2004; Carroll et al. 2005; Kirschner and Gerhart 2005). In other words, evolutionary developmental biology links genetics with evolution through the agencies of development. As Thomas Huxley's grandson, Julian Huxley, observed in 1942, "A study of the effects of genes during development is as essential for an understanding of evolution as are the study of mutation and that of selection." Contemporary evolutionary developmental biology is analyzing how changes in development can create the diverse variation that natural selection can act on. Rather than concentrating on the "survival of the fittest," evolutionary developmental biology gives us new insights into the "arrival of the fittest" (Carroll et al. 2005; Gilbert and Epel 2015). (See **Further Development 24.1, Relating Evolution to Development in the Nineteenth Century**, online.)

The Developmental Genetic Model of Evolutionary Change

Preconditions for Evolution: The Developmental Structure of the Genome

If natural selection can operate only on existing variants, where does all that variation come from? If, as Darwin (1868) and Huxley concluded, variation arose from changes in development, then how could the development of an embryo change when development is so finely tuned and complex? How could such change occur without destroying the entire organism?[1] The matter remained a mystery until evolutionary developmental biologists demonstrated that large morphological changes could arise during development because of two conditions that underlie the development of all multicellular organisms: **modularity** and **molecular parsimony**. (See **Further Development 24.2, "Intelligent Design" and Evolutionary Developmental Biology**, online.).

Modularity: Divergence through dissociation

One of the most important discoveries of evolutionary developmental biology is that not only are anatomical units modular (such that one part of the body can develop differently than the others), but the DNA regions that form the enhancers of genes are also modular. This genetic modularity means that there can be multiple enhancers for each gene and that each enhancer region can have binding sites for multiple transcription factors. The modularity of enhancer elements allows particular sets of genes to be activated together and permits a particular gene to become expressed in several discrete places. Thus, if by mutation a particular gene loses or gains a modular enhancer element, the organism containing that particular set of enhancers will express that gene in different places or at different times than organisms retaining the original allele. This mutability can result in the development of different anatomical and physiological morphologies (Sucena and Stern 2000; Shapiro et al. 2004), and major morphological changes can proceed through a mutation in a DNA regulatory region. Thus, the modularity of enhancers can be critical in providing selectable variation. Indeed, mutations affecting enhancer sequences are now thought to be the most important cause of morphological divergence between groups of animals (Carroll 2008; Stern and Orgogozo 2008).

PITX1 AND STICKLEBACK EVOLUTION   The importance of enhancer modularity has been dramatically demonstrated by the analysis of evolution in threespine stickleback fish (*Gasterosteus aculeatus*). Freshwater sticklebacks evolved from marine sticklebacks about 12,000 years ago, when marine populations colonized the newly formed freshwater lakes at the end of the last ice age. Marine sticklebacks (**FIGURE 24.1A**) have pelvic spines that serve as protection against predation, lacerating the mouths of predatory fish that try to eat the sticklebacks. (Indeed, the scientific name of the fish translates as "bony stomach, with spines.") Freshwater sticklebacks do not have pelvic spines (**FIGURE 24.1B**). This may be because the freshwater fish lack the piscine predators that the marine fish face, and instead must deal with invertebrate predators

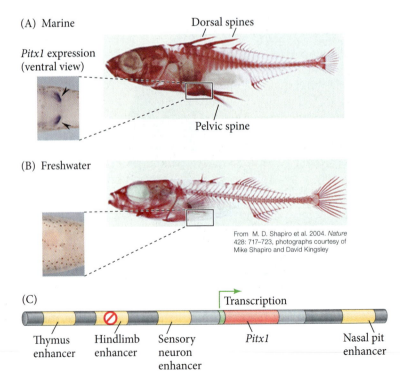

FIGURE 24.1 Evolution through enhancer modularity during development. Loss of *Pitx1* gene expression in the pelvic region of freshwater populations of the threespine stickleback (*Gasterosteus aculeatus*). Bony plates and pelvic spines characterize marine populations of this species (A). In freshwater populations (B), the pelvic spines are absent, as is much of the bony armor. In magnified ventral views of embryos (inset photos), in situ hybridization reveals *Pitx1* expression (purple) in the pelvic area (as well as in sensory neurons, thymic cells, and nasal regions) of the marine population. The staining in the pelvic region is absent in freshwater populations, although it is still seen in the other areas. The arrowheads point to *Pitx1* expression in the ventral region that forms the pelvic spines of the marine populations. (C) Model for the evolution of pelvic spine loss. Four enhancer regions are postulated to reside near the *Pitx1* coding region. These enhancers direct the expression of this gene in the thymus, pelvic spines, sensory neurons, and nasal pit, respectively. In freshwater populations of threespine sticklebacks, the pelvic spine (hindlimb) enhancer module has been mutated, and the *Pitx1* gene fails to function there. (After M. D. Shapiro et al. 2004. *Nature* 428: 717–723.)

(A) Marine

Pitx1 expression (ventral view)

Dorsal spines

Pelvic spine

(B) Freshwater

From M. D. Shapiro et al. 2004. *Nature* 428: 717–723, photographs courtesy of Mike Shapiro and David Kingsley

(C)

Transcription

Thymus enhancer

Hindlimb enhancer

Sensory neuron enhancer

Pitx1

Nasal pit enhancer

that can easily capture them by grasping onto such spines. Thus, a pelvis without spines was selected in freshwater populations of this species.

To determine which genes might be involved in these pelvic differences, researchers mated individuals from marine (spined) and freshwater (spineless) stickleback populations. The resulting offspring were bred to each other and produced numerous progeny, some of which had pelvic spines and some of which didn't. Using molecular markers to identify specific regions of the parental chromosomes, Shapiro and co-workers (2004) found that the major gene for pelvic spine development mapped to the distal end of chromosome 7. After testing numerous candidate genes in this region (e.g., genes known to be active in the pelvic and hindlimb structures of mice), they found that the gene encoding the transcription factor Pitx1 was located here.

When Shapiro and colleagues compared the amino acid sequences of the Pitx1 proteins of marine and freshwater sticklebacks, there were no differences. However, there was a critically important difference when they compared the expression patterns of the *Pitx1* gene. In both populations, *Pitx1* was expressed in the precursors of the thymus, nose, and sensory neurons. In the marine populations, *Pitx1* was also expressed in the pelvic region. But in the freshwater populations, the pelvic expression of *Pitx1* was absent or severely reduced (**FIGURE 24.1C**). Since the coding region of *Pitx1* was not mutated and the difference between the freshwater and marine populations involved the expression pattern of this gene, it was reasonable to conclude that the enhancer region allowing expression of *Pitx1* in the pelvic area (i.e., the pelvic spine enhancer) does not function in the freshwater populations.

This conclusion was confirmed when high-resolution genetic mapping showed that the DNA of the "hindlimb" enhancer of *Pitx1* differed between sticklebacks with pelvic spines and those without pelvic spines (Chan et al. 2010).[2] When this 2.5-kb DNA fragment from marine (spined) fish was fused to a gene for green fluorescent protein and inserted into fertilized freshwater stickleback eggs, GFP was expressed in the pelvis. Moreover, when this same fragment taken from marine sticklebacks was placed next to the *Pitx1*-coding sequence of freshwater (spine-deficient) fish and then injected into fertilized eggs of the spine-deficient fish, pelvic spines formed in the freshwater fish. Once it is translated, Pitx1 functions by binding to an enhancer of the Tbx4 gene, activating that gene to promote hindlimb development (Logan and Tabin 1999; see Chapter 19).

RECRUITMENT   Modularity allows the recruitment (or "co-option") of entire suites of characters into new places. In Chapter 10, we discussed the recruitment of the skeleton-forming genes (the skeletogenic subroutine) into the developmental repertoire of the sea urchin micromeres. In most echinoderm groups, the skeletogenic genes are activated only in the adult and are used to form the hard exoskeletal plates. However, in sea urchins (and not in any other echinoderm group) this set of genes has come under the control of the micromere double-negative gate because of changes in the enhancer of one of these genes. Thus, the skeleton is made by larval mesenchymal cells (Gao and Davidson 2008).

Another example of recruitment is seen in insects, in the wing structure that defines the beetles. Beetles are one of the most successful animal groups on the planet, accounting for roughly 20% of extant animal species (Hunt et al. 2007). They differ from other insects in forming an elytron, a forewing encased in a hard exoskeleton. This makes them the "living jewels" so beloved of naturalists (**FIGURE 24.2**).[3] In beetles, as in *Drosophila*, the *Apterous* gene is expressed in the dorsal compartment of the wing imaginal discs, and the Apterous transcription factor organizes

FIGURE 24.2   Elytra are the hardened forewings that characterize Coleoptera, the beetles. Elytra are formed through the recruitment of the genetic module for exoskeleton development into the module for dorsal forewing development. (A) The elytra of a "ladybug" beetle. Its forewings are ornamented with exoskeleton, and its hindwings are extended. (B) These "living jewels" from the Oxford Museum of Natural History illustrate some of the diversity of beetle elytra.

(A)

© F1online digitale Bildagentur GmbH/Alamy

(B)

© Jochen Tack/Alamy

the tissue to differentiate dorsal wing structures. However, in beetles (and in no other known insect), Apterous protein also activates the exoskeleton genes in the forewing while repressing them in the hindwing (Tomoyasu et al. 2009). Thus, a new type of wing emerges from the recruitment of one module (the subroutine of exoskeletal development) into another (the subroutine of dorsal forewing development). (See **Further Development 24.3, Correlated Progression**, online.)

Molecular parsimony: Gene duplication and divergence

The second precondition for macroevolution through developmental change is molecular parsimony, sometimes called the "small toolkit." In other words, although development differs enormously from lineage to lineage, development within all lineages uses the same types of molecules. The transcription factors, paracrine factors, adhesion molecules, and signal transduction cascades are remarkably similar from one phylum to another. Many of the signal transduction cascades appear to have evolved in single-celled protists and to have been recruited later for metazoan activities (Booth and King 2016; Sébé-Pedrós et al. 2016). The evolution of animals appears to have involved the evolution of enhancers and the evolution of new pathways, especially those using BMPs and Wnts (Paps and Holland 2018; Sébé-Pedrós et al. 2018). This establishment was rapid, and the development of jellyfish and flatworms uses the same major kit of transcription factors and paracrine factors as does that of flies and vertebrates (Finnerty et al. 2004; Carroll et al. 2005; Putnam et al. 2007; Ryan et al. 2007; Hejnol et al. 2009).

THE SMALL TOOLKIT Certain transcription factors and paracrine factors (such as those of the BMP, Hox, and Pax groups) are found in all animal phyla. In fact, some "toolkit genes" appear to play the same *roles* in multiple animal lineages. The BMPs appear to be used throughout the animal kingdom to specify the dorsal-ventral axis (**FIGURE 24.3A**); the Wnt and Hox genes are used to specify the anterior-posterior axis in all the bilaterians (**FIGURE 24.3B**); and the *Pax6* gene appears to be involved in specifying light-sensing organs, irrespective of whether the eye is that of a mollusk (e.g., squid), an insect, or a primate (**FIGURE 24.3C**).[4] Similarly, homologues of *Otx* specify head formation in both vertebrates and invertebrates; and though insect and vertebrate hearts are very different, both are formed using *tinman/Nkx2-5* (see Erwin 1999). Certain microRNAs appear to be found in all animals, and these appear to play the same or very similar developmental roles in whatever phylum they are found (Christodoulou et al. 2010). These include miRNA-124, which is found in the central nervous systems of protostomes and deuterostomes; miRNA-12, which is found in guts throughout the animal kingdom; and miRNA-92, which helps specify ciliated locomotor cells in deuterostome and protostome larvae. Discovering that the same set of transcription factors and microRNAs causes the specification of the same types of cells throughout the animal kingdom is a powerful argument for the protostomes and deuterostomes being derived from a common ancestor that used these factors in similar ways to specify its organs (Davidson and Erwin 2010).

In some instances, homologous pathways made of homologous components are used for the same function in both protostomes and deuterostomes. This has been called **deep homology** (Shubin et al. 1997, 2009). Conserved similarities in both the pathway and its function over millions of years of phylogenetic divergence are considered to be evidence of deep homology between these modules (Shubin et al. 1997). One example is the Chordin/BMP4 interaction discussed in Chapter 11. In both vertebrates and invertebrates, Chordin/Short-gastrulation (Sog) inhibits the lateralizing effects of BMP4/Decapentaplegic (Dpp), thereby allowing the ectoderm protected by Chordin/Sog to become the neurogenic ectoderm.[5] These reactions are so similar that *Drosophila* Dpp protein can induce ventral fates in *Xenopus* and can substitute for Sog (**FIGURE 24.4**; Holley et al. 1995).

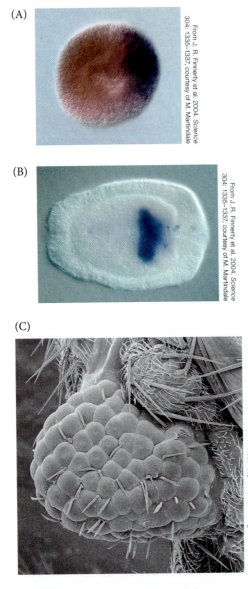

(A)

From J. R. Finnerty et al. 2004. *Science* 304: 1335–1337, courtesy of M. Martindale

(B)

From J. R. Finnerty et al. 2004. *Science* 304: 1335–1337, courtesy of M. Martindale

(C)

From G. Halder et al. 1995. *Science* 267: 1788–1792, courtesy of W. J. Gehring and G. Halder

FIGURE 24.3 Evidence of the evolutionary conservation of regulatory genes. (A) The cnidarian homologue of the vertebrate *Bmp4* and *Drosophila Decapentaplegic* genes is expressed asymmetrically at the edge of the blastopore in the embryo of the sea anemone *Nematostella*. This gene represents an ancestral form of the protostome and deuterostome forms of the gene. (B) The Hox gene *Anthox6*, a cnidarian member of the paralogue 1 group of Hox genes, is expressed at the blastopore side of the larval sea anemone. (C) The *Pax6* gene for eye development is an example of a gene ancestral to both protostomes and deuterostomes. The micrograph shows ommatidia of the compound insect eye emerging in the leg of a fruit fly (a protostome) in which mouse (deuterostome) *Pax6* cDNA was expressed in the leg disc.

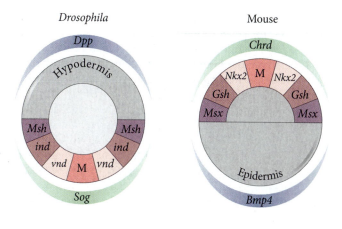

FIGURE 24.4 The same set of instructions forms the nervous systems of both protostomes and deuterostomes. In the fruit fly (a protostome), the TGF-β superfamily member *Dpp* (*Decapentaplegic*) is expressed dorsally and is opposed by *Short-gastrulation* (*Sog*) ventrally. In the mouse (a deuterostome), the TGF-β superfamily member *Bmp4* is expressed ventrally and is countered dorsally by *Chordin* (*Chrd*), a homologue of *Short-gastrulation*. The highest concentration of Chordin/Sog becomes the midline (M). The midline is dorsal in vertebrates and ventral in insects, and the concentration gradient of the TGF-β superfamily protein (BMP4 or Dpp) activates genes specifying the regions of the nervous system in the same order in both groups: *vnd/Nkx2*, followed by *ind/Gsh*, and finally *Msh/Msx*. These genes have been seen to be expressed in a similar fashion in cnidarians. (After H. Reichert and A. Simeone. 2001. *Philos Trans R Soc Lond B* 356: 1533–1544.)

DUPLICATION AND DIVERGENCE   One theme that resounds through studies of paracrine and transcription factors is that these proteins (and the genes that encode them) come in families (Holland et al. 2017). How do gene families come into existence? The answer is through duplication of an original gene and the subsequent independent mutation of the original duplicates (**FIGURE 24.5**). This creates a family of genes that are related by common descent (and which are often still adjacent to each other). This scenario of **duplication and divergence** is seen in the Hox genes, the globin genes, the collagen genes, the *Distal-less* genes, and in many paracrine factor families (e.g., the Wnt genes). Each member of such a gene family is homologous to the others (i.e., their sequence similarities are due to descent from a common ancestor and are not the result of convergence for a particular function), and they are called **paralogues**. Susumu Ohno (1970), one of the founders of the gene family concept, likened gene duplication to a sneaky criminal circumventing surveillance. While the "police force" of natural selection makes certain that there is a "good" gene properly performing its function, that gene's duplicate, unencumbered by the constraints of selection, can mutate and undertake new functions.

Such "subfunctionalization" has since been shown to be the case in many gene families, including the Hox genes. Hox genes represent an especially complex and important case of duplication and divergence. We find that (1) there are related Hox genes in

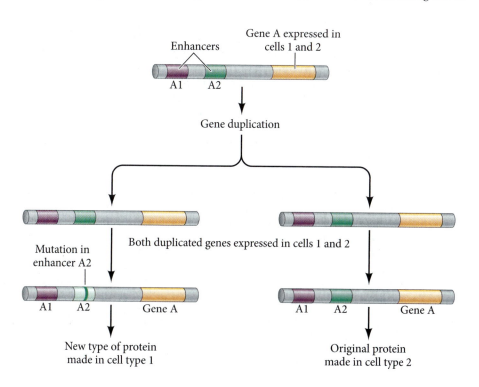

FIGURE 24.5 Duplication and divergence. Duplication of a gene that is expressed in several different cell types may be followed by mutations in the duplicated genes. This can lead to a subdivision of the gene's original function, such that each of the duplicated genes is expressed in a different cell type. In the hypothetical case described here, a mutation in one of the duplicated gene enhancers leads to a new pattern of gene expression and a different functional protein in cell type 1.

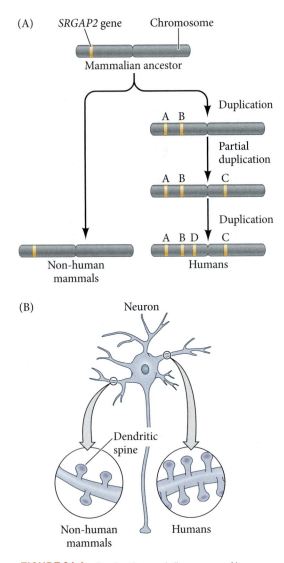

(A) *SRGAP2* gene   Chromosome

Mammalian ancestor

Duplication

A B

Partial duplication

A B   C

Duplication

A B D   C

Non-human mammals   Humans

(B)   Neuron

Dendritic spine

Non-human mammals   Humans

FIGURE 24.6 Duplication and divergence of human *SRGAP2*. (A) The *SRGAP2* gene is found as a single copy in the genomes of all mammals except humans. In the lineage giving rise to humans, duplication events gave rise to four similar versions of the gene, designated A–D. (B) The "ancestral" gene, *SRGAP2A*, with minor contributions from *SRGAP2B* and *D*, enables the maturation of dendritic spines (protuberances) on the surfaces of neurons. *SGRAP2C* is a partial duplication, and its product inhibits *SRGAP2A*, slowing dendritic spine maturation and promoting neuronal migration. This partial duplication may have allowed for the evolution of longer maturation time and greater flexibility in the human brain. (After D. H. Geschwind and G. Konopka. 2012. *Nature* 486: 481–482.)

each animal group (such as *Deformed, Ultrabithorax,* and *Antennapedia* in *Drosophila,* or the 39 Hox genes in mammals); and (2) there are several clusters of Hox genes in vertebrates (the 39 mammalian Hox genes, for example, are clustered on four different chromosomes). The similarity of all the Hox genes is best explained by descent from a common ancestral gene. This would mean that in *Drosophila,* the *Deformed, Ultrabithorax,* and *Antennapedia* genes all emerged as duplications of an original gene. The sequence patterns of these three genes (especially in the homeodomain region) are extremely well conserved. Such tandem gene duplications are thought to be the result of errors in DNA replication, and such errors are not uncommon. Once replicated, the gene copies can diverge by random mutations in their coding sequences and enhancers, developing different expression patterns and new functions (Lynch and Conery 2000; Damen 2002; Locascio et al. 2002).

Moreover, each *Drosophila* Hox gene has a homologue (and sometimes several) in vertebrates. In some cases, the homologies go very deep and can also be seen in the gene's functions. Not only is the vertebrate *Hoxb4* gene similar in sequence to its *Drosophila* homologue, *Deformed* (*Dfd*), but human *HOXB4* can perform the functions of *Dfd* when introduced into *Dfd*-deficient *Drosophila* embryos (Malicki et al. 1992). As mentioned in Chapter 12, the Hox genes of insects and humans are not just homologous—they occur in the same order on their respective chromosomes. Their expression patterns are also remarkably similar: the more 3′ Hox genes have more anterior expression boundaries (see Figure 12.20).[6] Thus, these genes are homologous between species (as opposed to members of a gene family being homologous within a species). Genes that are homologous between species are called **orthologues**.

One of the most important gene duplication events in human evolution may have been the duplication of *SRGAP2,* a gene that may have enabled the expansion of the human cerebral cortex. The protein encoded by this gene is expressed in the mammalian brain cortex and appears to *slow down* cell division and decrease the length and density of dendritic processes. However, humans differ from all other animals (including chimpanzees) by having duplicated this gene twice. Moreover, the second duplication event was not complete, so one of the newly formed genes is only a partial duplicate. This partial gene produces a truncated SRGAP2 protein, SRGAP2C, which is also made in the cerebral cortex and which *inhibits* the activity of normal SRGAP2 made from the complete genes. As a result, cell division in the cerebral cortex continues for longer periods of time, and the dendrites are larger and have more connections (**FIGURE 24.6**; Charrier et al. 2012; Dennis et al. 2012). Based on genomic evidence, these gene duplication events are calculated to have taken place about 2.4 million years ago. This would be about the time of *Australopithecus,* the increase in primate brain size, and the first known use of tools (Tyler-Smith and Xue 2012).

Mechanisms of Evolutionary Change

In 1975, Mary-Claire King and Allan Wilson published a paper titled "Evolution at Two Levels in Humans and Chimpanzees." This study showed that despite the large anatomical differences between chimpanzees and humans, their protein-encoding DNA was almost identical. The differences were to be found in the regulatory genes that acted during development:

> The organismal differences between chimpanzees and humans would … result chiefly from genetic changes in a few regulatory systems, while amino acid substitutions in general would rarely be a key factor in major adaptive shifts.

In other words, the allelic substitutions of the genes that encode protein sequences—which seem to be pretty much the same for chimpanzees and humans—were not seen as being important for large-scale evolution. This has been confirmed by subsequent genomic studies (Deline et al. 2018). The important differences are where, when, and how much the genes are activated. In 1977, the idea that changes in gene regulation during development is critical for evolution was extended by François Jacob, the Nobel laureate who helped establish the operon model of gene regulation. First, Jacob said, evolution works with what it has: it combines existing parts in new ways rather than creating new parts. Second, he predicted that such "tinkering" would be most likely to occur in those genes that construct the embryo, not in the genes that function in adults (Jacob 1977).

Wallace Arthur (2004) catalogued four ways in which Jacob's "tinkering" can take place at the level of gene expression to generate phenotypic variation available for natural selection:

1. Heterotopy (change in location)
2. Heterochrony (change in time)
3. Heterometry (change in amount)
4. Heterotypy (change in kind)

These changes can only be accomplished if the gene expression patterns are modular—that is, if they are controlled by different enhancer elements. The modularity of development allows one part of the organism to change without necessarily affecting the other parts.[7]

Heterotopy

One important way of creating new structures is to alter the *location* where a transcription factor or paracrine factor is expressed. This spatial alteration of gene expression is called **heterotopy** (Greek, "different place"). Heterotopy allows different cells to take on a new identity (as sea urchin micromeres did when they recruited the genes for the skeleton formation; see Chapter 10) or to activate or inhibit a paracrine factor-mediated process in a new area of the body (as when Gremlin inhibits BMP-mediated apoptosis in the webbing between digits; see Figure 19.26). There are many other examples, some of which we describe next.

HOW THE BAT GOT ITS WINGS AND THE TURTLE GOT ITS SHELL   In Chapter 1, we mentioned that wings evolved in the bat by changes in the development of the forelimb, such that the cells in the interdigital webbing did not die. It turns out that the bat retains its forelimb webbing in a manner very similar to how the duck embryo retains its hindlimb webbing—by blocking the BMPs that would otherwise cause the interdigital cells to undergo apoptosis. Both Gremlin and FGF signaling appear to block BMP functions in the bat wing. Unlike other mammals, bats express Fgf8 in their interdigital webbing, and this protein is critical for maintaining the cells there. If FGF signaling is inhibited (by drugs such as SU5402), BMPs can induce apoptosis of the forelimb webbing, just as in other mammals (Laufer et al. 1997; Weatherbee et al. 2006). The Fgf8 in the webbing also appears to be responsible for providing the mitotic signal that extends the digits of the bat, thereby expanding its wing (Hockman et al. 2008; Sears 2008).

The formation of the turtle shell also uses BMPs and FGFs, but in different ways. What distinguishes turtles from other vertebrates is their ribs—they migrate laterally into the dermis instead of forming a rib cage (**FIGURE 24.7**). Certain regions of the turtle dermis attract rib precursor cells, and these dermal regions differ from those of other vertebrates because they synthesize Fgf10. Fgf10 seems to attract the ribs, since the ribs do not enter the dermis if the Fgf10 signal is blocked (Burke 1989; Cebra-Thomas et al. 2005). Once inside the dermis, the rib cells do what rib cells are expected to do—they undergo endochondral ossification, wherein the cartilage cells are replaced by bone. To do this, BMPs are made. But the rib is embedded in dermis, and the dermal cells can also respond to the BMPs by becoming bone (Cebra-Thomas et al. 2005; Rice et al. 2015). In this way, each of the newly positioned ribs instructs the dermis around it to become bone, and thus the turtle gets its shell. These conclusions about turtle

(A)

(B)

(C)

(D)

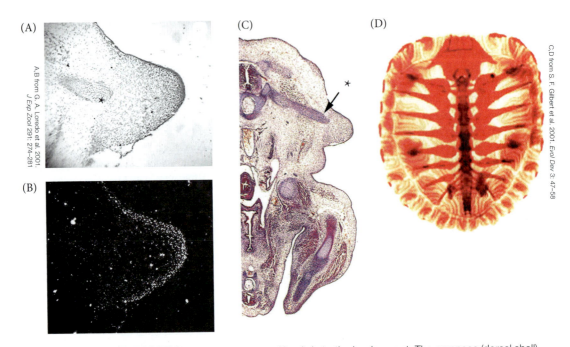

FIGURE 24.7 Heterotopy on several levels in turtle development. The carapace (dorsal shell) of the turtle is formed through sequential layers of heterotopies. *Fgf10* expression in certain regions of the dermis impels rib precursor cells to migrate laterally into the dermis instead of forming a rib cage. (A,B) Cross section of early turtle embryo as the rib enters the dermis (A, brightfield; B, autoradiograph staining for *Fgf10*). (C) Half cross section of a slightly later turtle embryo, showing a rib (arrow) extending from the vertebra into the region of the dermis that will expand to form the shell. (D) Hatchling turtle stained with alizarin red to show bones. Bones can be seen in the dermis around the ribs that entered into it. Heterotopies include *Fgf10* expression, rib placement, and bone location. Asterisks in (A) and (C) point to migrating rib cells.

development have facilitated new paleontological theories of the turtle's evolutionary origins (Nagashima et al. 2009; Lyson et al. 2013).

THE TWO LIPS OF TULIPS   The enormous diversity of angiosperm flowers came into being largely by homeotic mutations (Thiessen 2010; Moyroud and Glover 2017). This diversity includes the petals of the rose, which appear to have come from stamens (Ronse de Craene 2003; Dubois et al. 2010), and the petals of tulips, which appear to have come from sepals (**FIGURE 24.8**; Kanno et al. 2003). As discussed in Chapter 6, petal identity is generated by the expression of class A, B, and E floral organ identity genes, while sepals are produced when just class A and E genes are expressed. In tulips, class A and B genes are seen in the first two whorls, converting the sepals into petals. When *Arabidopsis thaliana* embryos were experimentally induced to express class B floral organ identity genes in the first whorl, the plants formed petal-like structures rather than sepals (Krizek and Meyerowitz 1996).

Heterochrony

Heterochrony (Greek, "different time") is a shift in the relative order or timing of two developmental processes. Heterochrony can be seen at any level of development, from gene regulation to adult animal behaviors (West-Eberhard 2003). In heterochrony, one module changes its time of expression or growth rate relative to the other modules of the embryo.

Heterochronies are quite common in vertebrate evolution (McNamara 2012). We have already discussed the extended growth of the human brain. Another example is found in marsupials, whose jaws and forelimbs develop at a faster rate than do those of placental mammals, allowing the marsupial newborn to climb into the maternal pouch and suckle (Smith 2003; Sears 2004). Birds are thought to have arisen, in part, through the heterochronic growth of dinosaur skulls and long bones (Bhullar et al. 2012; McNamara and

(A)

(B)

Alberto Salguero/CC BY-SA 3.0

Bernd Haynold/CC BY-SA 3.0

FIGURE 24.8 (A) Floral organ identities are specified according to the classic gene boundaries of the ABCDE model in *Arabidopsis thaliana*. (B) In tulips, the boundaries have changed such that class B genes are also expressed in the first whorl, causing petal like-sepals ("tepals") to form. (A,B diagrams after E. Moyroud and B. Glover. 2017. *Curr Biol* 27: R941–R951.)

Long 2012). The enormous number of vertebrae and ribs formed in embryonic snakes (more than 500 in some species) is likewise due to heterochrony. Snake-specific chromatin rearrangements permit the expression of the Oct4 transcription factor in the somitic mesoderm for far longer than in other vertebrates (Aires et al. 2016). Moreover, the segmentation reactions cycle nearly four times faster relative to tissue growth in snake embryos than they do in related vertebrate embryos (Gomez et al. 2008). Similarly, the elongated fingers in the dolphin flipper appear to be the result of the prolonged heterochronic expression of *Fgf8*, which as we saw in Chapter 19 encodes a major paracrine factor for limb outgrowth (Richardson and Oelschläger 2002; Cooper 2010).

Heterochronies are responsible for the myriad flower shapes, even within closely related groups (Li and Johnston 2000; Buendia-Monreal and Gillmor 2018). For instance, the flowers of the derived hummingbird-pollinated species of *Delphinium* differ from the more basal bee-pollinated species due to an overall decreased growth rate, coupled with an extended growth period for the reward-providing petals (Guerrant 1982). Similarly, differences in floral development in the family Solanaceae (potatoes, nightshades, petunias, and peppers) are often brought about by heterochronic shifts in petal production (Kostyun and Moyle 2017). (See **Further Development 24.4, Allometry**, online.)

Heterometry

Heterometry is a change in the *amount* of a gene product or structure. We mentioned heterometric changes in Chapter 16 when we discussed the evolution of the blind Mexican cavefish (see Figure 16.10). In this example, the overproduction of Sonic hedgehog protein (Shh) in the midline prechordal plate downregulates the *Pax6* gene, preventing eye formation. But overexpression of *Shh* has other consequences as well. Not only does it cause degeneration of the eyes, but it also causes the jaw size and number of taste buds to increase (Franz-Odendaal and Hall 2006; Yamamoto et al. 2009). Since cavefish live in complete darkness, the expansion of their jaw size and gustatory sense at the expense of sight can be selected.

Heterometry can also be seen in flower organ specification. In some angiosperm species that are dioecious (having separate male and female plants) the downregulation of either the class B or C floral organ identity genes can determine the sex of

Bmp4 expression

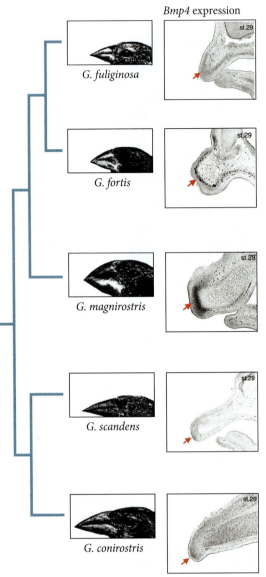

G. fuliginosa

G. fortis

G. magnirostris

G. scandens

G. conirostris

From A. Abzhanov et al. 2004. *Science* 305: 1462–1465

FIGURE 24.9 Correlation between beak shape and the expression of *Bmp4* in five species of Darwin's finches. In the genus *Geospiza*, the ground finches (represented by *G. fuliginosa*, *G. fortis*, and *G. magnirostris*) diverged from the cactus finches (represented by *G. scandens* and *G. conirostris*). The differences in beak morphology correlate with heterochronic and heterometric changes in *Bmp4* expression in the beak. BMP4 (red arrow) is expressed earlier and at higher levels in the seed-crushing ground finches. The photographs of the embryonic beaks were taken at the same stage (stage 29) of development. This gene expression difference provides one explanation for the role of natural selection on these birds. (After A. Abzhanov et al. 2004. *Science* 305: 1462–1465.)

the flower. For instance, in male spinach plants, class B floral organ identity genes are expressed in high amounts in the third whorl, giving rise to anthers. These class B genes are only weakly expressed in female plants, which have no anthers (Pfent et al. 2005). Thus, in some plant species, sex may be determined by the expression levels of floral organ identity genes.

DARWIN'S FINCHES　One of the best-known examples of heterometry involves Darwin's celebrated finches, a set of 15 closely related birds collected by Charles Darwin and his shipmates during their visit to the Galápagos and Cocos islands in 1835. These birds helped Darwin frame his evolutionary theory of descent with modification, and they still serve as one of the best examples of adaptive radiation and natural selection (see Weiner 1994; Grant and Grant 2008). Systematists have shown that these finch species evolved in a particular manner, with a major speciation event being the split between the cactus finches and the ground finches. The ground finches evolved deep, broad beaks that enable them to crack open seeds, whereas the cactus finches evolved narrow, pointed beaks that allow them to probe cactus flowers and fruits for insects and flower parts. Schneider and Helms (2003) had shown that species differences in the beak pattern were caused by changes in the growth of the neural crest-derived mesenchyme of the frontonasal process (i.e., those cells that form the facial bones), and Abzhanov and his colleagues (2004) found a remarkable correlation between the beak shape of the finches and the timing and amount of *Bmp4* expression (**FIGURE 24.9**). No other paracrine factor showed such differences. *Bmp4* expression in ground finches started earlier and was much greater than *Bmp4* expression in cactus finches. In all cases, the *Bmp4* expression pattern correlated with the breadth and depth of the beak.

The importance of these expression differences was confirmed experimentally by changing the *Bmp4* expression pattern in chick embryos to mimic the heterometric and heterochronic changes in the ground finches (Abzhanov et al. 2004; Wu et al. 2004). When *Bmp4* expression was enhanced in the frontonasal process mesenchyme, the chick developed a broad beak reminiscent of the beaks of the ground finches. Conversely, when BMP signaling was inhibited in this region (by Noggin, a BMP inhibitor), the beak lost depth and width.[8]

But the story goes on. Gene chip technology showed that the level of expression of another gene, *Calmodulin*, is 15-fold higher in embryonic beaks of sharp-beaked cactus finches than in blunt-beaked ground finches. Moreover, when Calmodulin was upregulated in the embryonic chicken beak, the chick beak too became long and pointed. BMP4 and Calmodulin, therefore, represent two targets for natural selection, one regulating the breadth and depth of the beak, and the other regulating length. Together they explain the shape variations of Darwin's finches (Abzhanov et al. 2006; Campàs et al. 2011).

SNAKE LIMBS　How did snakes lose their limbs? Embryonic pythons actually make the first parts of their legs, forming a transient limb. But the region creating the limb (the AER; see Chapter 19) is unable to be sustained due to weak production of Sonic hedgehog (Shh), which in turn is due to a deletion in the distal enhancer for the *Shh* gene. This deleted region contained binding sites for important transcription factors, including Hox proteins. The snake enhancer, therefore, cannot respond to Hox proteins because the Hox proteins can't bind to it. If the snake enhancer replaces the normal enhancer in embryonic mice, the mice develop severely truncated limbs (**FIGURE 24.10A**; Kvon et al. 2016; Leal and Cohn 2016, 2018). The normal phenotype is restored if the missing transcription factor-binding binding site is added to the snake enhancer (**FIGURE 24.10B**).

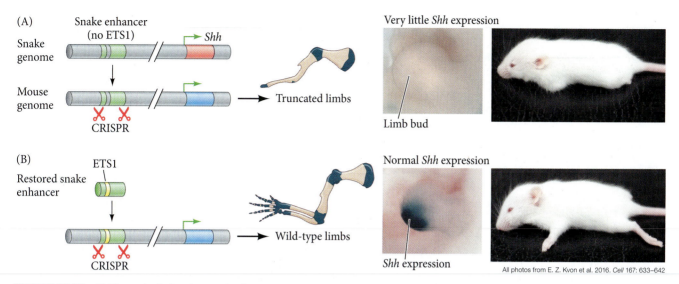

FIGURE 24.10 (A) The snake limb enhancer for *Shh* is missing an element (ETS1) that is crucial for directing limb development. When the snake *Shh* enhancer replaces the mouse *Shh* enhancer in embryonic mice, very little Shh (dark blue) is seen in the limb buds, and the mouse develops a truncated limb. (B) If the missing ETS1 site from the mouse *Shh* enhancer (or other reptilian *Shh* enhancers) is added to the snake enhancer, it functions normally, directing mouse *Shh* expression in the limb buds and making full-length limbs. (After E. Z. Kvon et al. 2016. *Cell* 167: 633–642.)

Heterotypy

In heterochrony, heterotopy, and heterometry, mutations affect the regulatory regions of the gene. The changes of **heterotypy** affect the actual coding region of the gene, and thus can change the functional properties of the protein being synthesized. Changes in the coding sequence of transcription factors can have profound consequences in animal and plant evolution (Wang et al. 2005).

WHY INSECTS HAVE SIX LEGS   Insects have only six legs while most other arthropod groups (think of spiders, millipedes, centipedes, lobsters, and shrimp) have many more. How is it that the insects form legs only in their three thoracic segments and have no legs on their abdominal segments? The answer seems to be found in the relationship between Ultrabithorax (Ubx) protein and the *Distal-less* gene. In most of the arthropod groups, Ubx does not inhibit *Distal-less*. However, in the insect lineage, a mutation occurred that replaces a group of nucleotides in the *Ubx* gene with a sequence encoding a stretch of alanine residues (**FIGURE 24.11**; Galant and Carroll 2002; Ronshaugen et al. 2002). This polyalanine region represses *Distal-less* transcription in the abdominal segments.

FIGURE 24.11 Changes in Ubx protein associated with the insect clade in the evolution of arthropods. Of all arthropods, only the insects have Ubx protein that is able to repress *Distal-less* gene expression and thereby inhibit abdominal legs. This ability to repress *Distal-less* is due to a mutation that is seen only in the insect *Ubx* gene. (After R. Galant and S. B. Carroll. 2002. *Nature* 415: 910–913, and M. Ronshaugen et al. 2002. *Nature* 415: 914–917.)

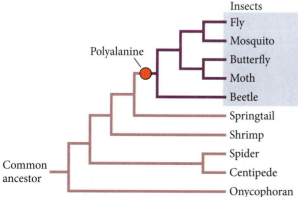

When a brine shrimp *Ubx* gene is experimentally modified to encode the insect polyalanine region, the shrimp embryo represses the *Distal-less* gene. The ability of Ubx to inhibit *Distal-less* thus appears to be the result of a gain-of-function mutation that characterizes the insect lineage.

MAIZE: WHY CORN IS SO EASY TO EAT   Corn (*Zea mays*) is an organism whose propagation throughout the world may have been effected through a mutation of a transcription factor (Wang et al. 2005). Many steps are postulated to have occurred between the ancestral teosinte plant and modern corn, and one of the most economically important steps was the liberation of the kernel from its hard protective envelope (the glume). This change exposed the kernels on the cob, making them more readily harvested for human consumption. This critical event in corn evolution and domestication is controlled by a single gene—*teosinte glume architecture* (*tga1*)—that encodes a transcription factor active in the development of the inflorescence (the "ear") of the plant. A mutation in *tga1* results in a single amino acid substitution, from lysine in teosinte to asparagine in corn. This change appears to cause the Tga1 protein to be degraded faster, preventing the completion of the glume surface that would normally cover the kernel. When corn *tga1* is placed into teosinte, the teosinte kernels become corn-like; and when teosinte *tga1* is expressed in corn, the kernels become teosinte-like.

Developmental Constraints on Evolution

There are only about three dozen major animal lineages, and they encompass all the different body plans seen in the animal kingdom. One can easily envision other body plans by imagining animals that do not exist; science fiction writers do it all the time. So why don't we see more body plans among the living animals? To answer this, we have to consider the constraints imposed on evolution. This notion of constraint is used differently by different groups of scientists. While many population biologists see constraints as limiting "ideal" adaptations (such as constraints on optimal foraging), developmental biologists see constraints as limiting the possibility of certain phenotypes even existing (see Amundson 1994, 2005).

Physical constraints

The laws of diffusion, hydraulics, and physical support are immutable and will permit only certain physical phenotypes to arise. For example, blood cannot circulate to a rotating organ; thus, a vertebrate on wheeled appendages (of the sort that Dorothy saw in Oz) cannot exist, and this entire evolutionary avenue is closed off. Similarly, structural parameters and fluid dynamics would prohibit the existence of 6-foot-tall mosquitoes or 25-foot-long leeches.

Morphogenetic constraints

Bateson (1894) and Alberch (1989) noted that when organisms depart from their normal development, they do so in only a limited number of ways. For instance, although there have been many modifications of the vertebrate limb over 300 million years, some modifications are never seen (such as a middle digit shorter than its surrounding digits, or a zeugopod more proximal than the stylopod; see Chapter 19) (Holder 1983; Wake and Larson 1987). These observations suggest a limb construction scheme that follows certain rules (Oster et al. 1988; Newman and Müller 2005).

   One of the major sources of morphogenetic constraints lies in the limited ways that differentiated patterns can arise from homogeneity. Chief among these patterning mechanisms is the **reaction-diffusion mechanism**. This mechanism for developmental patterning, formulated by Alan Turing (1952), is a way of generating complex chemical patterns out of substances that are initially homogeneously distributed. As Turing explained, this patterning could be achieved by two homogeneously distributed substances, if the rates of production of each substance depended on the other and their diffusion rates differed. The patterns this can produce are stable and could be used to drive developmental change. We saw how incredibly useful this was in explaining the development and possible evolutionary trajectories of the vertebrate limb in Chapter

19 (see Figure 19.15). Turing's model has been used to explain the formation of the limbs and digits of tetrapods (see pp. 531–534 and Further Development 19.7, online), as well as the stripes of zebras and angelfish, and the formation of tooth cusps. (See **Further Development 24.5, How Do Zebras (and Angelfish) Get Their Stripes?** and **Further Development 24.6, How Do the Correct Number of Cusps Form in a Tooth?**, both online.)

Pleiotropic constraints and redundancy

Pleiotropy, the ability of a gene to play different roles in different cells, is the "opposite" of modularity, involving the connections between parts rather than their independence (Lonfat et al. 2014; Hu et al. 2017). Pleiotropies may underlie the constraints seen in mammalian development. Galis speculates that mammals have only seven cervical vertebrae (whereas birds may have dozens) because the Hox genes that specify these vertebrae have become linked to stem cell proliferation in mammals (Galis 1999; Galis and Metz 2001; Abramovich et al. 2005; Schiedlmeier et al. 2007). Thus, changes in Hox gene expression that might facilitate evolutionary changes in the skeleton might also *mis*regulate cell proliferation and lead to cancers. Galis supports this speculation with epidemiological evidence showing that changes in skeletal morphology correlate with childhood cancer. The intraembryonic selection against having more or fewer than seven cervical ribs appears to be remarkably strong. At least 78% of human embryos with an extra anterior rib (i.e., six cervical vertebrae) die before birth, and 83% die by their first birthday. These deaths appear to be caused by multiple congenital anomalies or cancers (**FIGURE 24.12**); Galis et al. 2006).

Other constraints involve redundancies. In these cases, two or more different genes can produce the same effect; if one gene misfunctions, the other can fulfill the job. In insects and mammals, multiple enhancers provide robust patterns of gene expression, enabling the phenotype to be stable (Osterwalder et al. 2018).

(A)

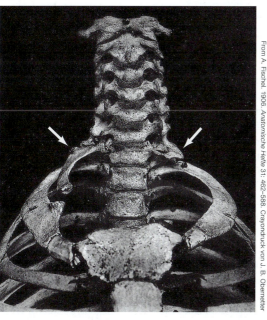

From A. Fischel, 1906. *Anatomische Hefte* 31: 462–588. Crayondruck von J. B. Obernetter

(B)

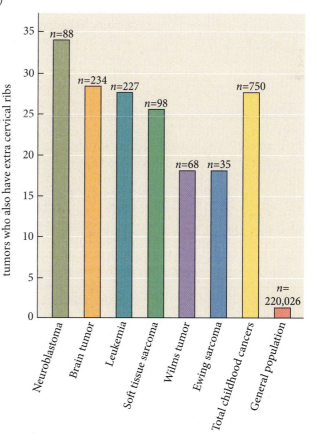

FIGURE 24.12 (A) A skeleton with extra cervical ribs (arrows). (B) Nearly 80% of fetuses with extra cervical ribs die before birth. Those surviving often develop cancers very early in life. This indicates strong selection against changes in the number of mammalian cervical ribs. (B after R. Schumacher et al. 1992. *Eur J Pediatrics* 151: 432–434.)

Ecological Evolutionary Developmental Biology

An organism's development is not scripted by its genes alone. Developmental instructions are also given by the outside environment, especially by symbiotic microbes. The study of the relationship between developing organisms and their environment has given rise to the field of ecological developmental biology. If evolutionary changes are generated by heritable changes in development, and development involves interactions with environmental factors, then alterations in the environmental interactions during development may affect evolution. As Armin Moczek (2015) has written, "To develop is to interact with the environment. To evolve is to alter these interactions in a heritable manner." This thinking has given rise to a subset of evolutionary developmental biology called **ecological evolutionary developmental biology**, or **eco-evo-devo** (Abouheif et al. 2014; Gilbert and Epel 2015; Sultan 2015, 2017).[9] Here we discuss three major aspects of eco-evo-devo: developmental plasticity, the inheritance of epialleles, and developmental symbiosis.

Plasticity-First Evolution

In the early 1900s, some evolutionary biologists speculated that the environment could select one of a variety of environmentally induced phenotypes and that this phenotype would then become genetically "fixed"—that is, normative for the species. One of the most important of these plasticity-first hypotheses involves the concept of **genetic assimilation**, defined as the process by which a phenotypic character initially produced only in response to some environmental influence becomes, through a process of selection, taken over by the genotype so that the phenotype forms even in the absence of the environmental influences that gave rise to it (King and Stanfield 1985). The idea of genetic assimilation was introduced independently by Waddington (1942, 1953, 1961) and Schmalhausen (1949) to explain the remarkable outcomes of artificial selection experiments in which an environmentally induced phenotype became expressed even in the *absence* of the external stimulus that was initially necessary to induce it.

Genetic assimilation in the laboratory

Genetic assimilation is readily demonstrated in the laboratory. Suzuki and Nijhout (2006) have shown genetic assimilation in the larvae of the tobacco hornworm moth (*Manduca sexta*; **FIGURE 24.13**). By judicious selection protocols, Suzuki and Nijhout were able to breed lines in which the environmentally induced phenotype (larval color) was selected for and was eventually produced without the environmental agent (temperature shock). The underlying genetic differences concerned the ability of heat stress to raise juvenile hormone titers in the larvae. Therefore, at least in the laboratory, genetic assimilation can be shown to work.

One of the most famous laboratory demonstrations of genetic assimilation was C. H. Waddington's experiments using laboratory strains of *Drosophila* that had a particular reaction norm in their response to ether. Embryos exposed to ether at a particular stage developed a phenotype similar to that of the *bithorax* mutation and had four wings instead of two. The flies' halteres—balancing structures on the third thoracic segment—were transformed into wings (see Chapter 9). Generation after generation was exposed to ether, and individuals showing the four-winged state were selectively bred each time. After 20 generations, the mated *Drosophila* produced the mutant phenotype even when no ether was applied (**FIGURE 24.14**; Waddington 1953, 1956).

In 1996, Gibson and Hogness repeated Waddington's *bithorax* experiments and obtained similar results. Moreover, they found four distinct alleles of the *Ultrabithorax* (*Ubx*) gene existing in the population prior to the ether exposure. *Ubx* is the homeotic gene whose loss-of-function mutations are responsible for the genetically inherited four-winged fly phenotype (see Figure 9.25). Gibson concluded that "In our experiment, we show that differences in the *Ubx* gene are the cause of these morphological changes," but they did not know what the ether was doing to elicit such changes.

It appears that the answer involves a set of proteins called **heat shock proteins**, such as Hsp90. Hsp90 is a molecular "chaperone" that is required to keep many proteins

(A)

Courtesy of Fred Nijhout

FIGURE 24.13 Effect of selection on temperature-mediated larval color change in the black mutant of the moth *Manduca sexta*. (A) The two color morphs of *Manduca sexta* larvae. (B) Changes in the coloration of heat-shocked larvae in response to selection. One group was selected for increased greenness upon heat treatment (polyphenic; green line), with the "greenest" larvae being bred for the next generation. Another group was selected for decreased color change (i.e., remaining black) upon heat treatment (monophenic; red line). The remaining larvae were not selected (blue line). The color score (0 for completely black, 4 for completely green) indicates the relative amount of colored regions in the larvae. The mono-phenic line lost its plasticity after the seventh generation. (C) Reaction norm for generation-13 flies reared at constant temperatures between 20°C and 33°C, and heat-shocked at 42°C. Note the steep polyphenism at about 28°C. (After Y. Suzuki and H. F. Nijhout. 2006. *Science* 311: 650–652.)

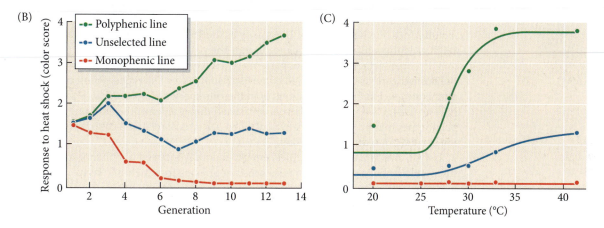

in their proper three-dimensional conformation. There are numerous genetic muta-tions that are usually not expressed as mutant phenotypes because Hsp90 can correct the small changes in mutant protein structure. However, when heat shock or ether is applied, there are numerous other proteins that need the attention of Hsp90, and there is not enough Hsp90 to go around. Thus, the variation that has existed but has not

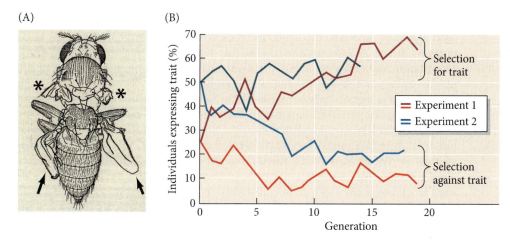

FIGURE 24.14 Phenocopy of the *bithorax* mutation. (A) A *bithorax* (four-winged) phenotype produced after treatment of the embryo with ether. The forewings have been removed to show the aberrant metathorax. Asterisks mark forewing stumps; arrows point to the extra pair of wings. This particular individual is actually from the "assimilated" stock that produced this phenotype without being exposed to ether. (B) Selection experiments for or against the *bithorax*-like response to ether treatment. Two experiments are shown (red and blue lines). In both cases, one group was selected for the trait and the other group was selected against the trait. (After C. H. Waddington. 1956. *Evolution* 10: 1–13.)

been expressed then becomes expressed. In this way, Hsp90 appears to be a "capacitor" for evolutionary change, allowing genetic changes to accumulate until environmental stress releases them to reveal their effects on phenotype. When the environment changes and these combinations are expressed, most of them are predicted to be neutral or detrimental. But some might be beneficial and selected in the new environment. Continued selection would enable the genetic fixation of the adaptive physiological response (Rutherford and Lindquist 1998; Queitsch et al. 2002). Hsp90 was found to buffer genetic variation in plants and vertebrates (Sangster et al. 2008; Rohner et al. 2014). Sangster and colleagues conclude that "Hsp90 appears to fulfill Waddington's concept of a developmental buffer or molecular canalization mechanism; one which can lead to the assimilation of novel traits on a large scale."

Genetic assimilation in natural environments

We know of several instances in which it appears that phenotypic variation due to developmental plasticity was later fixed by genes. The first involves pigment variations in butterflies (Hiyama et al. 2012). As early as the 1890s (Standfuss 1896; Goldschmidt 1938), scientists used heat shock to disrupt the pattern of butterfly wing pigmentation. In some instances, the color patterns that develop after temperature shock mimic the normal genetically controlled patterns of subspecies (ecotypes) living at different temperatures. Further observations of numerous species of butterflies (Shapiro 1976; Nijhout 1984; Otaki et al. 2010) have confirmed that temperature variation can induce phenotypes that mimic genetically controlled patterns of related races or species existing in colder or warmer conditions.

In fish, amphibians, and reptiles, environmentally induced jaw phenotypes appear to be able to be assimilated. There are frog species in which an environmentally induced carnivorous phenotype in one species is a genetically transmitted phenotype in a close relative (Levis and Pfennig 2018; Levis et al. 2018). Cichlid fish also display remarkably different jaw structures, depending on what they eat. In the frogs (as in the butterflies), this genetic assimilation is thought to be due to newly acquired regulation of the hormones involved in metamorphosis. The larger cichlid jaws appear to have been environmentally induced by physical stress activating the Wnt pathway to make bone. Now, however, some species activate the Wnt pathway without needing the environmental cue (Parsons et al. 2014; Hu and Albertson 2017).

Another dramatic case of genetic assimilation concerns the tiger snake (*Notechis scutatus*), which has a head structure that can be altered by diet. The tiger snake can develop a bigger head to ingest bigger prey. This plasticity is seen when the snake's diet includes both large and small mice. However, on some islands the diet contains only large mice, and here the snakes are born with large heads, and there is no plasticity. As Aubret and Shine (2009) claim, this is "clear empirical evidence of genetic assimilation, with the elaboration of an adaptive trait shifting from phenotypically plastic expression through to canalization within a few thousand years." Thus, an *environmentally* induced phenotype might become the standard *genetically* induced phenotype in one part of the range of that organism.

There are at least two important evolutionary advantages to the fixation of environmentally induced phenotypes (West-Eberhard 1989, 2003). First, *the phenotype is not random.* The environment elicited the novel phenotype, and the phenotype has already been tested by natural selection. This would eliminate a long period of testing phenotypes derived by random mutations. As Garson and colleagues (2003) note, although mutation is random, developmental parameters may account for some of the directionality in morphological evolution.

Second, *the phenotype already exists in a large portion of the population.* One of the problems of explaining new phenotypes is that the bearers of such phenotypes are "monsters" compared with the wild-type. How would such mutations, perhaps present in only one individual or one family, become established and eventually take over a population? The developmental model solves this problem: this phenotype has been around for a long while, and the capacity to express it is widespread in the population; it merely needs to be genetically stabilized by modifier genes that already exist in the population.

Given these two strong advantages, the genetic assimilation of morphs originally produced through developmental plasticity may contribute significantly to the origin of new species. Ecologist Mary Jane West-Eberhard (2005) has noted that "contrary to popular belief, environmentally initiated novelties may have greater evolutionary potential than mutationally induced ones."

Selectable Epigenetic Variation

Changes in development provide the raw material of variation. But we have seen in this chapter and earlier in the book that developmental signals can come from the environment as well as from the nuclei and cytoplasm. Although these ideas sound "Lamarckian," they have nothing to do with what Lamarck proposed, which was that phenotypes acquired by use or disuse could be transmitted to the germ line. The giraffe reaching higher in the trees for leaves does not make the neck of its offspring longer, children of weight lifters don't inherit their parents' physiques, and accident victims who have lost limbs can rest assured that their children will be born with normal arms and legs. In these examples, the DNA of the germ cells has not been altered by the environmentally induced variation, and the trait will not be transmitted from one generation to the next.

But what if an environmental agent were to cause changes not only in the somatic DNA but also in the germline DNA? Then the effect might be able to be transmitted from one generation to the next. This could be done not by genetic changes to the chromatin, but by epigenetic changes.

While alleles are variants of DNA sequences that can be transmitted from one generation to the next, **epialleles** are variants of chromatin structure that can be inherited between generations. In most known cases, epialleles are differences in DNA methylation patterning that are able to affect the germ line and thereby be transmitted to offspring. The asymmetrical *peloria* variant of the toadflax plant (*Linaria vulgaris*; **FIGURE 24.15**) was first described by Linnaeus in 1742 as a stably inherited form. In 1999, Coen showed that this variant was due not to a distinctive allele, but rather to a stable epiallele. Instead of carrying a mutation in the *CYCLOIDEA* gene, the *peloria* form of this gene was hypermethylated. It does not matter to the developing system whether a gene has been inactivated by a mutation or by altered chromatin configuration (Cubas and Coen 1999). The effect is the same.

There are dozens of examples of epiallelic inheritance (Jablonka and Raz 2009; Gilbert and Epel 2015). These include the following:

(A)

Courtesy of R. Grant-Downton

(B)

Courtesy of R. Grant-Downton

FIGURE 24.15 Epigenetic forms of toadflax. (A) Typical *Linaria*, with a relatively unmethylated *cycloidea* gene. (B) The *cycloidea* gene of the *peloria* variant is relatively heavily methylated. The epialleles that create the different phenotypes of this species are stably inherited.

- **Diet-induced DNA methylation**. In the *viable-Agouti* phenotype in mice, methylation differences affect coat color and obesity. When a pregnant female is fed a diet high in methyl donors, the specific methylation pattern at the *Agouti* locus is transmitted not only to the progeny developing in utero, but also to the progeny of those mice and to their progeny (Jirtle and Skinner 2007). Similarly, enzymatic and metabolic phenotypes are established in utero by protein-restricted diets in rats when protein restriction during a grandmother rat's pregnancy leads to a specific methylation pattern in her pups and grandpups (Burdge et al. 2007; Lillycrop and Burdge 2015).

- **Endocrine disruptor-induced DNA methylation**. The endocrine disruptors vinclozolin (a fungicide used on crops), methoxychlor (an insecticide), and bisphenol A (used in plastics) have the ability to alter DNA methylation patterns in the germ line, thereby causing developmental anomalies and predispositions to diseases in the grandpups of mice exposed to these chemicals in utero (see Figure 23.14; Anway et al. 2005, 2006a,b; Newbold et al. 2006; Gillette et al. 2018; Mennigen et al. 2018).

- **Behavior-induced DNA methylation and microRNA differences**. Stress-resistant behavior of rats was shown to be due to methylation patterns, induced by maternal care, in the glucocorticoid receptor genes. Meaney (2001) found that rats that received extensive maternal care had less stress-induced anxiety and, if female, developed into mothers that gave their offspring similar levels of maternal care. Recent evidence (Gapp et al. 2014; Rodgers et al. 2015) suggests that environmentally induced behaviors can be transferred by sperm microRNAs.

- **Light-induced DNA methylation differences**. *Polygonum* buckwheat stem and leaf morphology are influenced by the light conditions experienced by an individual's parents. These altered phenotypes are transmitted by epialleles having different methylation patterns (Baker et al. 2018).

Thus, epigenetic DNA variation enables environmental agents to affect heritable phenotypes in ways other than DNA mutations.

Evolution and Developmental Symbiosis

One important aspect of development involves **sympoiesis**, developmental interactions with an expected population of symbionts. There are three major ways that evolution can occur through changes in sympoiesis.

First, symbionts can be a source of selectable variation. Symbionts can be transmitted either directly or indirectly from parent to offspring through the mother and thus constitute a separate source of inheritance (Funkhauser and Bordenstein 2013; Roughgarden et al. 2018). The pea aphid (*Acyrthosiphon pisum*), for example, has numerous species of symbionts living within its cells. One species of symbiotic bacterium, *Buchnera aphidicola*, can provide the aphid with either higher fecundity or greater heat tolerance, depending on which allele of a heat shock protein the bacterium produces. Another symbiotic bacterium, a species of *Rickettsiella*, contains alleles that can alter the aphid's color (**FIGURE 24.16**). A third bacterial symbiont, *Hamiltonella defensa*, can (if it is the appropriate strain) provide proteins that defend the host aphid against parasitoid wasps (Dunbar et al. 2007; Oliver et al. 2009; Tsuchida et al. 2010). All three of these symbionts are thought to be transmitted by the mother into the cytoplasm of the egg or embryo, and certain genetically different *Buchnera* strains have been transmitted for millions of generations. Thus, symbiont alleles can alter development to produce different selectable phenotypes.

Second, symbiont-host relationships can be a source of reproductive isolation. Reproductive isolation, in which biological or geographic barriers prevent members of the same species from producing fertile offspring, is necessary for the formation of new species, and there are many ways to accomplish it. Both prezygotic (before

(?) Developing Questions

Recent evidence suggests that symbionts may be pivotal in evolutionary events. How might bacteria be involved with reproductive isolation and with the origins of multicellular animals?

(A) Without *Rickettsiella*

4 days old   8 days old   12 days old

(B) With *Rickettsiella*

4 days old   8 days old   12 days old

All images from S. F. Gilbert and D. Epel. 2015. *Ecological Developmental Biology*, 2nd Edition. Oxford University Press/Sinauer: Sunderland, MA, photographs courtesy of T. Tsuchida.

FIGURE 24.16   The color of adult pea aphids depends on whether or not their cells contain *Rickettsiella* bacterial symbionts. (A) Without *Rickettsiella*, red aphid newborns become red adults. (B) With *Rickettsiella*, red aphid newborns become green adults.

fertilization) and postzygotic (after fertilization) modes of reproductive isolation have been attributed to bacterial symbionts. As an example of prezygotic reproductive isolation, bacteria in the food eaten by *Drosophila* larvae influence the flies' mating preferences as adults. The bacteria change the cuticle pheromones that guide sexual choice (Sharon et al. 2010). Postzygotic reproductive isolation can be brought about by cytoplasmic incompatibility. Bacteria in the egg cytoplasm of one group of wasps would not allow the egg to develop if it was fertilized by wasps that did not have that symbiont (Brucker and Bordenstein 2013).

Third, Lynn Margulis and Dorion Sagan (1993a; 2003) speculated that symbionts were behind many, if not all, of the major transitions of evolution. "Evolution," they wrote, "is the acquiring of new genomes." The process by which new organisms form by acquiring symbionts is called **symbiogenesis** (Merezhkowsky 1909; Margulis 1993b), and processes such as herbivory (Vermeij 2004), multicellularity, and pregnancy are thought to have originated through symbioses.

The evolution of multicellularity

Genomic analyses agree that animals probably arose from a group of protists very much like today's choanoflagellates. Choanoflagellates are single-celled, and their name comes from their resemblance to the choanocytes (collar cells) of sponges. In filtered seawater, one such protist, *Salpingoeca rosetta*, proliferates asexually, forming more single-celled protists. However, when cultured in media containing the bacterium *Algoriphagus machipongonensis*, the cells do not separate. Rather, they form sheets or rosettes, whereby the cells are connected by an extracellular matrix and

FIGURE 24.17 Two morphologies of the choanoflagellate protist *Salpingoeca rosetta*. (A) Single-celled form. (B) Colonial form, with multiple cells linked by an extracellular matrix. The bacterium *Algoriphagus machipongonensis*, often found with *S. rosetta*, can convert the organism from dividing into individual cells to forming multicellular "rosettes."

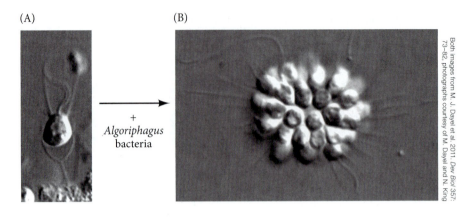

(A)   (B)

+
Algoriphagus
bacteria

Both images from M. J. Dayel et al. 2011. *Dev Biol* 357: 73–82, photographs courtesy of M. Dayel and N. King

cytoplasmic bridges (**FIGURE 24.17**; Dayel et al. 2011; Brunet and King 2017). This rosette formation may make feeding more efficient (Roper et al. 2013). The sphingolipids in the cell envelope of the bacteria are able to effect this transition, and the bacteria are found naturally with colonial forms of this choanoflagellate species. Thus, multicellularity may have arisen by developmental changes induced by neighboring bacteria.

The evolution of placental mammals

A different type of symbiont—retroviruses—may have been critical in the formation of the uterus and the placenta, two of the key structures distinguishing placental mammals (and allowing development to occur inside the mother). Proteins derived from retroviruses enable the mammalian female reproductive tract to form and function (Dupressoir et al. 2011; Lynch et al. 2011.) The activation of the uterine prolactin gene by progesterone is made possible by an enhancer that originated on a retrovirus. This enhancer is seen only in placental mammals and not in prototherians (egg-laying monotremes such as the platypus) or marsupials. In addition, the evolution and functioning of the placenta also appear to have been mediated, at least in part, by proteins that arise from retroviral genes that have integrated into the mammalian genomes (Villarreal and Ryan 2011). (See **Further Development 24.7, Transposable Elements and the Origin of Pregnancy**, online.)

Coda

In the late 1800s, experimental embryology separated itself from evolutionary biology to mature on its own. However, one of the pioneers, Wilhelm Roux, promised that once it did mature, embryology would return to evolutionary biology with powerful mechanisms to help explain how evolution takes place. Evolution is a theory of change, and population genetics can identify and quantify the dynamics of such change. However, Roux realized that evolutionary biology needed a theory of body construction that would provide a means by which a specific mutation becomes manifest as a selectable phenotype. The creativity of evolution is two-fold. Development is the creativity of an artist, while natural selection is the creativity of a curator.

When confronted with the question of how the arthropod body plan arose, Hughes and Kaufman (2002) begin their study thus:

> To answer this question by invoking natural selection is correct—but insufficient.
> The fangs of a centipede … and the claws of a lobster accord these organisms
> a fitness advantage. However, the crux of the mystery is this: From what
> developmental genetic changes did these novelties arise in the first place?

This is exactly the question for which modern developmental biology has been able to provide some answers, and it continues to do so.

Closing Thoughts on the Opening Photo

How does something new enter into the world? Evolutionary changes in anatomy take place through changes in development. Bats provide us with excellent examples of characteristic anatomical features that can be linked to changes in the expression of developmental regulatory genes. In the evolution of a flying mammal, small changes in the expression of developmental regulatory genes have driven large changes in forelimb morphology. Developmental biologists have identified molecular changes that are critically important in retaining the bat forelimb webbing (such as the expression of BMP inhibitors and FGFs in the webbing), elongating the fingers, and reducing the ulna (see Sears 2008; Behringer et al. 2009). This embryo of the fruit bat *Carollia perspicillata* shows forelimb webbing and extension of its digits.

Photograph courtesy of R. R. Behringer

Endnotes

[1] Circumventing this problem was critical for defeating an argument made by those who dismiss evolution as the source of biodiversity. Most proponents of intelligent design allow microevolution (which is confined to changes within a species) but deny the macroevolutionary changes (between species and forming species) explained in this chapter.

[2] Interestingly, the loss of the pelvic spines in several stickleback populations appears to have been the result of independent losses of this *Pitx1* expression domain. This finding suggests that if the loss of *Pitx1* expression in the pelvis occurs, this trait can be readily selected (Colosimo et al. 2004). Here we see that by combining the approaches of population genetics and developmental genetics, one can determine the mechanisms by which evolution can occur.

[3] Both Darwin and Wallace were avid beetle collectors, but it was the geneticist J. B. S. Haldane whose remark may best reflect the prominence of these insects. When asked by a cleric what the study of nature could tell us about God, Haldane is said to have replied, "He has an inordinate fondness for beetles."

[4] This doesn't mean that the eye is the only thing that is specified by Pax6, or that *Pax6* hasn't become regulated by different proteins in different phyla (Lynch and Wagner 2011).

[5] In addition to this central inhibitory reaction, there are other reactions that add to the deep homology of the instructions for forming the protostome and deuterostome nerve cord. The proteins involved in the diffusion and stability of BMPs and Chordin also are conserved between insects and vertebrates (Larrain et al. 2001).

[6] Hox genes appear to have been important in axis specification even before bilaterian phyla evolved (He et al. 2018; Technau and Genikhovich 2018). However, their proteins should not be thought of as directly specifying body segments. Rather, they are intermediaries, carrying out regional-specific instructions from other, non-Hox processes, such as FGF gradients in vertebrates and gap genes in flies. Different combinations of Hox genes enable cell proliferation and cell adhesion in tissue-specific manners (Gawne et al. 2018; Mallo 2018).

[7] This chapter concentrates on transcriptional-level changes that can generate new morphological forms, but morphological changes can be instigated at these levels as well. Abzhanov and Kaufman (1999), for instance, have shown that posttranscriptional regulation of the *Sex combs reduced* gene is critical in converting legs into maxillipeds in the terrestrial crustacean sowbug *Porcellio scaber*.

[8] Note the important principle: Paracrine factors and transcription factors perform different functions in different cell types, depending on other proteins. BMP can cause apoptosis in limb webbing, mitosis in facial mesenchyme, bone formation in ribs, and epidermal specification in the ectoderm.

[9] Ecological evolutionary developmental biology (eco-evo-devo) and ecological developmental biology (eco-devo), are overlapping but distinct disciplines. The former uses eco-devo principles to explain evolution. The latter looks at the relationship between development and environment but doesn't necessarily put that look into an evolutionary context.

Snapshot Summary
Development and Evolution

1. Evolution is the result of inherited changes in development. Modifications of embryonic or larval development can create new phenotypes that can then be selected.

2. Homology means that similarity between organisms or genes can be traced to descent from a common ancestor. In some instances, certain genes specify the same traits throughout the animal phyla.

3. The modularity of development allows parts of the embryo to change without affecting other parts. This modularity of development is due in large part to the modularity of enhancers.

4. New gene transcription can be caused by modifying existing DNA elements to become enhancers, by mutating the DNA sequences bound by transcription factors to eliminate an enhancer, or by adding an enhancer sequence or mutating an existing one.

5. Recruitment (co-option) of existing genes and pathways for new functions is a fundamental mechanism for creating new phenotypes. Such instances include the formation of beetle elytra and the production of the larval skeleton of sea urchins.

(Continued)

Snapshot Summary   *(continued)*

6. Like structures and genes, signal transduction pathways can be homologous, with homologous proteins organized in homologous ways. These pathways can be used for different developmental processes both in different organisms and within the same organism.

7. The formation of new cell types may result from duplicated genes whose regulation has diverged. The Hox genes and many other gene families started as single genes that were duplicated.

8. Evolution can occur through the "tinkering" of existing genes. The ways of effecting evolutionary change through development at the level of gene expression are: change in location (heterotopy), change in timing (heterochrony), change in amount (heterometry), and change in kind (heterotypy).

9. Heterotopy appears to account for the evolution of wings in bats, shells in turtles, and the petal-like sepals in tulips.

10. Heterochrony has been important in limb formation throughout the animal kingdom and in floral development in angiosperm plants.

11. Heterometry can account for the loss of limbs in snakes and for specification of sex organs in some angiosperm plants.

12. Heterometry combined with heterochrony can account for the beak phenotypes in Darwin's finches and for the size of the human brain.

13. Heterotypy can change the functional properties of proteins and can have significant developmental effects. The constraint on insect anatomy of having only six legs is one example; the evolution of the corn kernel is another.

14. Developmental constraints prevent certain phenotypes from arising. Such constraints may be physical (no rotating limbs), morphogenetic (no middle digit shorter than its neighbors), pleiotropic (more cervical vertebrae in birds than in mammals) or redundancies that stabilize the phenotype.

15. Genetic assimilation—wherein a phenotypic character initially induced by the environment becomes, through a process of selection, produced by the genotype in all permissive environments—has been well documented in the laboratory.

16. Epigenetic inheritance systems include epialleles, wherein inherited patterns of DNA methylation can regulate gene expression. A heavily methylated gene can be as nonfunctional as a genetically mutant allele.

17. Symbiotic organisms are often needed for development to occur, and variants of these organisms may cause different modes of development. Moreover, symbiotic microbes may have facilitated major evolutionary events, such as the origins of multicellularity and of the uterus and placenta in placental mammals.

18. Evolutionary developmental biology is able to show how small genetic or epigenetic changes can generate large phenotypic changes and enable the production of new anatomical structures.

Go to oup.com/he/barresi12xe for Further Developments, Scientists Speak interviews, Watch Development videos, Dev Tutorials, and complete bibliographic information for all literature cited in this chapter.

Appendix

A Quick Guide to Finding and Comprehending Research Articles in Developmental Biology

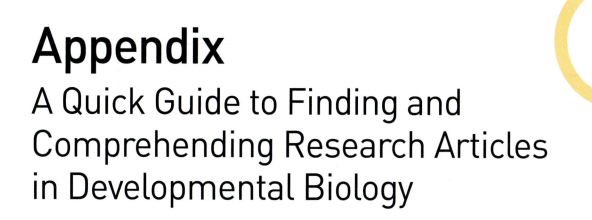

Research for Research

If you ask a scientist why they get up in the morning to do this work (or go to bed in the morning as some experiments may require), a common response is for the thrill of discovering new knowledge. But what knowledge is considered new? In order to determine the next best questions to investigate, scientists must first understand what the currently accepted ideas in the field may be—an understanding most certainly requiring analysis of the field's scientific literature.

Finding the research

We present here a suggested process for conducting research in the scientific literature (**FIGURE A.1**). When beginning this process consider establishing a literal or virtual

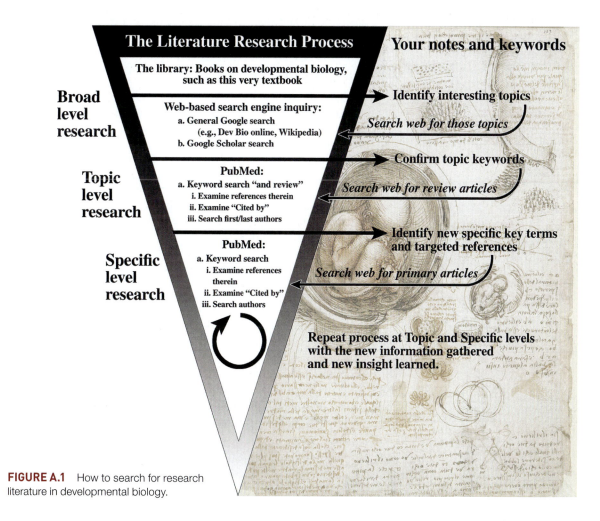

The Literature Research Process

Your notes and keywords

The library: Books on developmental biology, such as this very textbook

Broad level research

Web-based search engine inquiry:
a. General Google search
(e.g., Dev Bio online, Wikipedia)
b. Google Scholar search

Identify interesting topics

Search web for those topics

Topic level research

PubMed:
a. Keyword search "and review"
    i. Examine references therein
    ii. Examine "Cited by"
    iii. Search first/last authors

Confirm topic keywords

Search web for review articles

Specific level research

PubMed:
a. Keyword search
    i. Examine references therein
    ii. Examine "Cited by"
    iii. Search authors

Identify new specific key terms and targeted references

Search web for primary articles

Repeat process at Topic and Specific levels with the new information gathered and new insight learned.

FIGURE A.1 How to search for research literature in developmental biology.

notebook designated to serve as a recorded history of your journey. It is particularly helpful to establish some guiding questions that can be used to evaluate the utility of a given reference source.

Are you interested in figuring out why our arms are different from our legs? Start with reading the chapter on limb development in this very textbook. Visit a library for additional books on this broad topic. During your exploration of these resources, note the aspects of this topic that pique your curiosity, and, most importantly, identify subtopics within your broad interest, along with specific "keywords."

Once you are prepared with some foundational understanding of your topic and a starting bank of keywords, use your favorite search engine (e.g., Google) and enter some of your keywords to see what kinds of websites are discovered. Most certainly you may find the online resources of this text book (which you should take advantage of) as well as websites like Wikipedia. The goal at this level of your literature investigation is to continue to build a record of pertinent topics that could increase your comprehension of the subject and to guide your research toward more specific subtopics.

Navigating the PubMed database

PubMed is the primary database we recommend for searching directly for research articles in developmental biology (**FIGURE A.2A**; https://www.ncbi.nlm.nih.gov/pubmed). As you march deeper into the literature research process, we recommend first collecting relevant "review" articles on your topic (see Figure A.1, Topic level research). A review article is a synopsis of a particular topic within a field. These articles are written in a more accessible manner and provide an overarching summary of a given subtopic. A review article frames the problem within the broader scientific field, and also often identifies some of the most immediate questions that researchers should contemplate trying to answer. Most importantly, statements made in reviews are backed up with citations to supporting papers in the primary literature. These citations are fantastic clues to help you find important papers you should contemplate reading yourself. With those references noted, return to PubMed and search for them directly (**FIGURE A.2B**).

Getting a PDF of an article

Next, you need to obtain the electronic full text or PDF versions of the articles you want to read. Not all articles are freely accessible. If you are affiliated with a university or the like, then your institution may have subscriptions to certain journals, in which case access to the PDF versions can be granted through your library's resources. Fortunately, some journals are open access, and all "PubMed Central" articles are free to download. The full text can be accessed via the PubMed page of your selected article (see Figure A.2B), by clicking "Full text links" at the upper right. This will bring you to the journal's website and the article's access point (**FIGURE A.2C**). The article access pages of different journals all look different, but there will be a clickable link to download the PDF somewhere on this page.

Defining the Anatomy of a Research Paper

Once you have found an article of interest, the next step is to extract and analyze the findings of the study it describes. What were the authors' research questions and hypotheses? What approaches did the authors use to test their hypotheses, and are their conclusions actually supported by their stated findings?

The typical research article is divided into the following sections: Title, Authors, Abstract and/or Summary, Introduction, Materials and Methods, Results, Discussion, Acknowledgements, and References. To learn about these sections and how to read a research article, and to find more extensive tips on finding relevant articles, please consult the online version of the Appendix on the Companion Website.

(A)

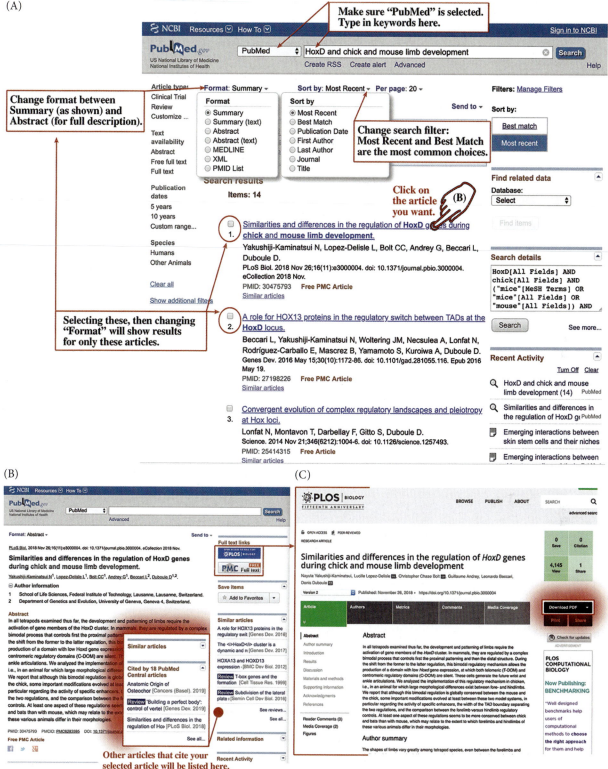

FIGURE A.2 Navigating PubMed. (A) The PubMed homepage provides an opportunity to search the PubMed database with keywords. Several dropdown menus enable you to refine your search criteria and display your results in different formats. (B) When you click on an article in your results, PubMed sends you to a page with both the abstract and a potential mechanism to acquire a PDF of the article. This page also lists similar articles and sometimes articles that have cited your chosen article. (C) Clicking "Full text links" will bring you to a page where the PDF can be downloaded. If an article is not freely available, then you may have to look for the specific article of interest through an institution's library page to take advantage of potential subscriptions.

Glossary

A

Abaxial muscles   Muscles derived from the lateral portions of the myotome.

Acetylation   See **Histone acetylation**.

Achondroplasia   Condition wherein chondrocytes stop proliferating earlier than usual, resulting in short limbs (achondroplasic dwarfism). Often caused by mutations that activate the *FgfR3* gene prematurely.

Acron   Anterior region of the body of an arthropod (including insects); in front of the mouth and includes the brain.

Acrosomal process   A fingerlike process in the sperm head extended by the polymerization of actin filaments during the early stages of fertilization in sea urchins and many other species. It contains surface molecules for species-specific recognition between sperm and egg.

Acrosome (acrosomal vesicle)   Cap-like organelle that, together with the sperm nucleus, forms the sperm head. Contains proteolytic enzymes that can digest the extracellular coats surrounding the egg, allowing the sperm to gain access to the egg cell membrane, to which it fuses.

Acrosome reaction   The Ca^{2+}-dependent fusion of the acrosome with the sperm cell membrane, resulting in exocytosis and release of proteolytic enzymes that allow the sperm to penetrate the egg extracellular matrix and fertilize the egg.

Actinomyosin contractions   Contractile forces within a cell caused by myosin attaching to and moving along filamentous actin. Examples: contractions of muscle cells; apical constrictions of neural plate cells at the hinge points.

Actinopterygian fishes   Ray-finned fish; includes teleosts.

Activins   Members of the TGF-β superfamily of proteins; with Nodal, important in specifying the different regions of the mesoderm and for distinguishing the left and right body axes of vertebrates.

Adepithelial cells   Cells that migrate into the imaginal discs early in the development of the holometabolous insect larva; these cells give rise to muscles and nerves in the pupal stage.

Adhesion   Attachment between cells or between a cell and its extracellular substrate. The latter provides a surface for migrating cells to travel along.

Adult pluripotent stem cells   Stem cells in an adult organism capable of regenerating all cell types of the adult. Example: neoblasts of planarian flatworms.

Adult stem cell niche   Niche that houses adult stem cells and regulates stem cell self-renewal, survival, and differentiation of the progeny that leave the niche.

Adult stem cells   Stem cells found in the tissues of organs after the organ has matured. Adult stem cells are usually involved in replacing and repairing tissues of that particular organ, forming a subset of cell types. Compare with **Adult pluripotent stem cells**; **Embryonic stem cells**.

Afferent   Carrying to, as in neurons that carry information to the central nervous system (spinal cord and brain), from sensory receptor cells (e.g., sound waves from the ear, light signals from the retina, touch sensations from the skin); or vessels that carry fluid (e.g., blood) to a structure.

Aging   The time-related deterioration of the physiological functions necessary for survival and reproduction.

Allantois   In amniote species, extra-embryonic membrane that stores urinary wastes and helps mediate gas exchange. It is derived from splanchnopleure at the caudal end of the primitive streak. In mammals, the size of the allantois depends on how well nitrogenous wastes can be removed by the chorionic placenta. In reptiles and birds, the allantois becomes a large sac, as there is no other way to keep the toxic by-products of metabolism away from the developing embryo.

Allometry   Developmental changes that occur when different parts of an organism grow at different rates.

Alternation of generations   In plants, a life cycle in which a haploid multicellular gamete-producing stage (the gametophyte) alternates with a diploid multicellular spore-producing stage (the sporophyte).

Alternative pre-mRNA splicing   A means of producing multiple different proteins from a single gene by splicing together different sets of exons to generate different types of mRNAs.

Amacrine neurons   Neurons of the vertebrate neural retina that lack large axons. Most are inhibitory. See also **Neural retina**.

Ametabolous   A pattern of insect development in which there is no larval stage and the insect undergoes direct development to a small adult form following a transitory pronymph stage.

Amnion   "Water sac." A membrane enclosing and protecting the embryo and its surrounding amniotic fluid. Derived from the two layers of the somatopleure: ectoderm that supplies epithelial cells, and mesoderm that generates the connective tissue.

Amniote egg   Egg that develops extra-embryonic membranes (the amnion, chorion, allantois, and yolk sac) that provide nourishment and other environmental needs to the developing embryo. Characteristic of the amniote vertebrates: the reptiles and birds, in which the egg typically develops in a shell outside the mother's body; and the mammals, where the egg has become modified to develop inside the mother.

Amniotes   The groups of vertebrates in which the embryo develops an amnion (water sac) that surrounds the body of the embryo. Includes reptiles, birds, and mammals. Compare with **Anamniote**.

Amniotic fluid   A secretion that serves as a "shock absorber" for the developing embryo while preventing it from drying out.

Ampulla   Latin, "flask." The segment of the mammalian oviduct, distal to the uterus and near the ovary, where fertilization takes place.

Anagen   The growth phase of a hair follicle, during which the hair grows in length.

Analogous   Structures and/or their respective components whose similarity arises from their performing a similar function rather than their arising from a common ancestor (e.g., the wing of a butterfly vs. the wing of a bird). Compare with **Homologous**.

Anamniotes The fish and amphibians; i.e., the vertebrate groups that do not form an amnion during embryonic development. Compare with **Amniotes**.

Anchor cell (AC) The cell connecting the overlying gonad to the vulval precursor cells in *C. elegans*. If the anchor cell is destroyed, the VPCs will not form a vulva, but instead become part of the hypodermis.

Androgen insensitivity syndrome Intersex condition in which an XY individual has a mutation in the gene encoding the androgen receptor protein that binds testosterone. This results in a female external phenotype, lack of a uterus and oviducts and presence of abdominal testes.

Androgen Masculinizing substance, usually a steroid hormone such as testosterone.

Anencephaly A congenital defect (almost always lethal) resulting from failure to close the anterior neuropore. The forebrain remains in contact with the amniotic fluid and subsequently degenerates, so the vault of the skull fails to form.

Aneuploidy Condition in which one or more chromosome(s) is either lacking or present in multiple copies.

Aneurogenic Devoid of any neural innervation.

Angioblasts From *angio*, blood vessel; and *blast*, a rapidly dividing cell (usually a stem cell). The progenitor cells of blood vessels.

Angiogenesis Process by which the primary network of blood vessels created by vasculogenesis is remodeled and pruned into a distinct capillary bed, arteries, and veins.

Angiopoietins Paracrine factors that mediate the interaction between endothelial cells and pericytes.

Animal cap In amphibians, the roof of the blastocoel (in the animal hemisphere).

Animal hemisphere The upper half of an egg containing the animal pole. In the amphibian embryo, the cells in the animal hemisphere having little yolk divide rapidly and become actively mobile ("animated").

Animal pole The pole of the egg or embryo where the concentration of yolk is relatively low; opposite end of the egg from the vegetal pole.

Anoikis Rapid apoptosis that occurs when epithelial cells lose their attachment to the extracellular matrix.

Anorectal junction The meeting of endoderm and ectoderm at the anus in vertebrate embryos.

Antennapedia complex A region of *Drosophila* chromosome 3 containing the homeotic genes *labial* (*lab*), *Antennapedia* (*Antp*), *sex combs reduced* (*scr*), *deformed* (*dfd*), and *proboscipedia* (*pb*), which specify head and thoracic segment identities.

Anterior heart field Cells of the heart field forming the outflow tract (conus and truncus arteriosus, right ventricle).

Anterior intestinal portal (AIP) The posterior opening of the developing foregut region of the primitive gut tube; it opens into the future midgut region which is contiguous with the yolk sac at this stage.

Anterior necrotic zone A zone of programmed cell death on the anterior side of the developing tetrapod limb that helps shape the limb.

Anterior neuropore See **Neuropore**.

Anterior visceral endoderm (AVE) Mammalian equivalent to the chick hypoblast and similar to the head portion of the amphibian organizer, it creates an anterior region by secreting antagonists of Nodal.

Anterior-posterior (anteroposterior or AP) axis The body axis defining the head versus the tail (or mouth versus anus). When referring to the limb, this refers to the thumb (anterior)-pinkie (posterior) axis.

Anther A pollen-producing organ on the stamen (male portion) of a flower.

Anti-Müllerian hormone (AMH) TGF-β family paracrine factor secreted by the embryonic testes that induces apoptosis of the epithelium and destruction of the basal lamina of the Müllerian duct, preventing formation of the uterus and oviducts. Also known as anti-Müllerian factor, or AMF. Sometimes called Müllerian-inhibiting factor (MIF).

Anticlinal divisions In plants, cell divisions in which the new cell walls are laid down perpendicular to the surface of the plant. Compare with **Periclinal divisions**.

Anurans Frogs and toads. Compare with **Urodeles**.

Aorta-gonad-mesonephros region (AGM) A mesenchymal area in the lateral plate splanchnopleure near the ventral aorta that produces the hematopoietic stem cells.

Aortic arches These begin as symmetrically arranged, paired vessels that develop within the paired pharyngeal arches and link the ascending/ventral and descending/dorsal paired aortae. Some of the aortic arches degenerate.

Aperture An opening. In plants, this is an area on the walls of a pollen grain where the wall is thinner or softer so that the expanding pollen tube can exit.

Apical At the apex, or top. Example: The apical layer of your epidermis is the outermost layer, facing the external environment.

Apical constriction Constriction of the apical end of a cell, caused by localized contraction of actinomyosin complexes at the apical border.

Apical ectodermal fold (AEF) The ectoderm overlying the mesenchyme of the developing fish fin that promotes fin ray development; derived from the original apical ectodermal ridge, which becomes the AEF in ray-finned fish after the proximal patterning of the stylopod.

Apical ectodermal ridge (AER) A ridge along the distal margin of the limb bud that will become a major signaling center for the developing limb. Its roles include (1) maintaining the mesenchyme beneath it in a plastic, proliferating state that enables the linear (proximal-distal) growth of the limb; (2) maintaining the expression of those molecules that generate the anterior-posterior axis; and (3) interacting with the proteins specifying the anterior-posterior and dorsal-ventral axes so that each cell is given instructions on how to differentiate.

Apical epidermal cap (AEC) Forms in the wound epidermis of an amputated salamander limb and acts similarly to the apical ectodermal ridge during normal limb development.

Apicobasal axis Apical-to-basal axis.

Apomixis Asexual reproduction in plants, where an embryo forms from cells that did not divide meiotically.

Apoptosis Programmed cell death. Apoptosis is an active process that prunes unneeded structures (e.g., frog tails, male mammary tissue), controls the number of cells in particular tissues, and sculpts complex organs (e.g., palate, retina, digits, and heart). See also **Anoikis**; **Necrotic zones**.

Aqueous humor Nourishing fluid that bathes the lens of the vertebrate eye and supplies pressure needed to stabilize the curvature of the eye.

Archenteron The primitive gut of an embryo. In the sea urchin, it is formed by invagination of the vegetal plate into the blastocoel.

Area opaca The peripheral ring of avian blastoderm cells that have not shed their deep cells.

Area pellucida A 1-cell-thick area in the center of the avian blastoderm (following shedding of most of the deep cells) that forms most of the actual embryo.

Aromatase Enzyme that converts testosterone to estradiol (a form of estrogen). Excess aromatase in the environment is linked to herbicides and other chemicals and is believed to contribute to reproductive disorders (demasculinization and feminization, particularly in male amphibians).

Arthrotome Mesenchymal cells in the center of the somite that contribute to the sclerotome, becoming the vertebral joints, the intervertebral discs, and those portions of the ribs closest to the vertebrae.

Astrocytes See **Astroglial cells**.

Astroglial cells (astrocytes) A diverse class of star (astro) shaped glial cells that carry out an array of functions, including establishing the blood-brain barrier, responding to inflammation in the CNS, and supporting synapse homeostasis and neural transmission.

Autocrine interaction The same cells that secrete paracrine factors also respond to them.

Autonomous specification A mode of cell commitment in which the blastomere inherits a determinant, usually a set of transcription factors from the egg cytoplasm, and these transcription factors regulate gene expression to direct the cell into a particular path of development.

Autophagy Intracellular system that removes and replaces damaged organelles and senescent cells.

Autopod The distal bones of a vertebrate limb: carpals and metacarpals (forelimb), tarsals and metatarsals (hindlimb), and phalanges ("digits"; the fingers and toes).

Autotrophy The process found in organisms such as plants and bacteria of synthesizing complex organic compounds using inorganic substances and light or chemical energy.

Auxin A family of plant hormones that regulate the growth and shape of the embryo as well as that of organs during adult stages. The most common is indole-3-acetic acid (IAA).

Axial protocadherin A type of protocadherin expressed in the presumptive notochord cells that allow them to separate from the paraxial (somite-forming) mesoderm to form the notochord. Found in amphibian embryos. See **Protocadherin**.

Axis determination The developmental process whereby the major body axes of an animal or plant are specified.

Axon Thin extension of the nerve cell body. Transmits signals (action potentials) to targets in the central and peripheral nervous systems. Axonal migration is crucial to development of the vertebrate nervous system.

Axoneme The portion of a cilium or flagellum consisting of two central microtubules surrounded by a row of 9 doublet microtubules. The motor protein dynein attached to the doublet microtubules provides the force for ciliary and flagellar function.

B

Bacteroid A special state of a bacterium, which has become differentiated as part of a symbiotic complex with legume roots.

Basal At the base, or bottom. Example: The basal layer of epidermal cells is the lowest layer, which sits on the basal lamina.

Basal disc The "foot" of a hydra; enables the animal to stick to rocks or the undersides of pond plants.

Basal lamina Specialized, closely knit sheets of extracellular matrix that underlie epithelia, composed largely of laminin and type IV collagen. Epithelial cells adhere to the basal lamina in part via binding between integrins and laminin. Sometimes called the basement membrane.

Basal layer (stratum germinativum) The inner layer of both the embryonic and adult epidermis. This layer contains epidermal stem cells attached to a basement membrane.

Basal transcription factors Transcription factors that specifically bind to the CpG-rich sites, forming a "saddle" that can recruit RNA polymerase II and position it for transcription.

Basic fibroblast growth factor (Fgf2) One of three growth factors required for the generation of hemangioblasts from the splanchnic mesoderm. See also **Angiopoietins**; **Vascular endothelial growth factors** (**VEGFs**).

Bergmann glia Type of glial cell; extends a thin process throughout the developing neuroepithelium of the cerebellum.

bHLH proteins The basic helix-loop-helix family of transcription factors, including such proteins as scleraxis, the MRFs (MyoD, Myf5, and myogenin), and c-Myc.

Bicoid Anterior morphogen critical for establishing anterior-posterior polarity in the *Drosophila* embryo. Functions as a transcription factor to activate anterior-specific gap genes and as a translational repressor to suppress posterior-specific gap genes.

Bilaminar germ disc An amniote embryo prior to gastrulation; consists of epiblast and hypoblast layers.

Bilateral holoblastic cleavage Cleavage pattern, found primarily in tunicates, in which the first cleavage plane establishes the right-left axis of symmetry in the embryo and each successive division orients itself to this plane of symmetry. Thus the half-embryo formed on one side of the first cleavage plane is the mirror image of the other side.

Bilaterians (triploblasts) Those animals characterized by bilaterian body symmetry and the presence of three germ layers (endoderm, ectoderm, and mesoderm). Includes all animal groups except the sponges, cnidarians, ctenophores, and placozoans.

Bindin A 30,500-Da protein on the acrosomal process of sea urchin sperm that mediates the species-specific recognition between the sperm and the egg vitelline envelope during fertilization.

Bindin receptors Species-specific receptors on the vitelline envelope of sea urchin eggs that bind to the bindin on the acrosomal process of sperm during fertilization.

Biofilm Mats of microorganisms, such as bacteria, that generate an extracellular matrix. These regulate larval settlement of many marine invertebrate species.

Bipolar interneurons Neurons in the neural retina, positioned between the photoreceptors (rods and cones) and ganglion cells for transmitting signals from the photoreceptors to the ganglion cells.

Bipotential (indifferent) gonad Common precursor tissue derived from the genital ridge in mammals, from which the male and female gonads diverge.

Bisphenol A (BPA) Synthetic estrogenic chemical compound used in plastics and flame retardants. BPA has been associated with meiotic defects, reproductive abnormalities, and precancerous conditions in rodents.

Bithorax complex The region of *Drosophila* chromosome 3 containing the homeotic gene *Ultrabithorax* (*Ubx*), which is required for the identity of the third thoracic segment, and the *abdominal A* (*abdA*) and *Abdominal B* (*AbdB*) genes, which are responsible for the segmental identities of the abdominal segments.

Bivalent The four chromatids of a homologous pair of chromosomes and their synaptonemal complex during prophase of the first meiotic division. Also called a tetrad.

Blastema A group of undifferentiated progenitor cells that form in some organisms at the site of an amputation; the blastema is able to grow and differentiate to replace the amputated tissue.

Blastocoel A fluid-filled cavity of the blastula stage of an embryo.

Blastocyst A mammalian blastula. The blastocoel is expanded and the inner cell mass is positioned on one side of the ring of trophoblast cells.

Blastoderm The layer of cells formed during cleavage at the animal pole in telolecithal eggs, as in fish, reptiles, and birds. Because the yolk concentrated in the vegetal region of the egg impedes cleavage, only the small amount of yolk-free cytoplasm at the animal pole is able to divide in these eggs. During development, the blastoderm spreads around the yolk as it forms the embryo.

Blastodisc Small region at the animal pole of the telolecithal eggs of fish, birds, and reptiles, containing the yolk-free cytoplasm where cleavage can occur and that gives rise to the embryo. Following cleavage, the blastodisc becomes the blastoderm.

Blastomere A cleavage-stage cell resulting from mitosis.

Blastopore The invagination point where gastrulation begins. In deuterostomes, this marks the site of the anus. In protostomes, this marks the site of the mouth.

Blastula Early-stage embryo consisting of a sphere of cells surrounding an inner fluid-filled cavity, the blastocoel.

Blood islands Aggregations of hemangioblasts in the splanchnic mesoderm. It is generally thought that the inner cells of these blood islands become blood progenitor cells, while the outer cells become angioblasts.

BMP family See **Bone morphogenetic proteins**.

BMP4 A protein in the BMP family, used extensively in neural development; for example, BMP4 is produced by target organs innervated by the trigeminal nerve and causes differential growth and differentiation of the neurons. Also involved in bone differentiation. See **Bone morphogenetic proteins**.

Bone marrow-derived stem cells (BMDCs) See **Mesenchymal stem cells**.

Bone morphogenetic proteins (BMPs) Members of the TGF-β superfamily of proteins. Originally identified by their ability to induce bone formation, BMPs are extremely multifunctional, having been found to regulate cell division, apoptosis, cell migration, and differentiation.

Bottle cells Invaginating cells during amphibian gastrulation, in which the main body of each cell is displaced toward the inside of the embryo while maintaining contact with the outside surface by way of a slender neck.

Brain-derived neurotrophic factor (BDNF) A paracrine factor that regulates neural activity and appears to be critical for synapse formation by inducing local translation of neural messages in the dendrites. BDNF is required for the survival of a particular subset of neurons in the striatum (a region of the brain involved in movement).

Brainbow Genetic method used to trigger the expression of different combinations and amounts of different fluorescent proteins within cells, labeling them with a seeming "rainbow" of possible colors that can be used to identify each individual cell in a tissue, organ, or whole embryo.

Branchial arches See **Pharyngeal arches**.

Brown adipose cells (brown fat) Adipose cells derived from the central dermomyotome. Brown fat cells produce heat, as opposed to white adipose cells that store lipids. Brown fat cells contain numerous mitochondria and dissipate energy as heat instead of synthesizing ATP.

Bulge A region of the hair follicle that is a niche for adult stem cells.

C

Cadherins **C**alcium-**d**ependent ad**he**sion molecules. Transmembrane proteins that interact with other cadherins on adjacent cells and are critical for establishing and maintaining intercellular connections, spatial segregation of cell types, and the organization of animal form.

Cajal-Retzius cells Reelin-secreting cells in the neocortex just under the pial surface. Reelin directs the migration of newly born neurons toward the pial surface.

Callus In plants, a growing mass of unorganized and undifferentiated cells that cover a wound; these cells can be induced to form a plant meristem and develop into shoots and/or roots.

Calorie restriction Dietary restriction as a means of extending mammalian longevity (at the expense of fertility).

Cambrian explosion The rapid diversification of life in the Cambrian period (approximately 541 million years ago) during which most major animal groups appeared, including those with species living today.

Canalization model In plants, a model for explaining the polarized transport of auxin along the apical to basal axis during vascular development and regeneration. In this model, the vasculature apical to the developing tissue or healing wound serves as the auxin source, and the row of increasingly basal developing or regenerating vascular cells serve as the auxin sink.

Canalization See **Robustness**.

Cancer stem cell hypothesis The hypothesis that the malignant part of a tumor is either an adult stem cell that has escaped the control of its niche or a more differentiated cell that has regained stem cell properties.

Cap sequence   See **Transcription initiation site**.

Capacitation   The set of physiological changes by which mammalian sperm become capable of fertilizing an egg.

Carcinogenesis   The initiation of cancer whereby a non-cancerous cell is transformed into a cancerous cell.

Cardia bifida   A condition in which two separate hearts form, resulting from manipulation of the embryo or genetic defects that prevent fusion of the two endocardial tubes.

Cardiac neural crest   Subregion of the cranial neural crest that extends from the otic (ear) placodes to the third somites. Cardiac neural crest cells develop into melanocytes, neurons, cartilage, and connective tissue. Cardiac neural crest also contributes to the muscular-connective tissue wall of the large arteries (the "outflow tracts") of the heart, as well as contributing to the septum that separates pulmonary circulation from the aorta.

Cardiogenic mesoderm   See **Heart fields**.

Cardiomyocytes   Cardiac cells derived from heart field tissue that form the muscular layers of the heart and its inflow and outflow tracts.

Carpel   The female reproductive organ of an angiosperm flower, containing stigma, an ovary, and often a style. Sometimes called a **Pistil**.

Catagen   Regression phase of the hair follicle regeneration cycle.

Catenins   A complex of proteins that anchor cadherins inside the cell. The cadherin-catenin complex forms the classic adherens junctions that help hold epithelial cells together and, by binding to the actin (microfilament) cytoskeleton of the cell, integrate the epithelial cells into a mechanical unit. One of them, β-catenin, can also be a transcription factor.

β-Catenin   A protein that can act as an anchor for cadherins or as a transcription factor (induced by the Wnt pathway). It is important in the specification of germ layers throughout the animal phyla.

CatSper   Cation (usually, Ca^{2+}) channels that seem to be specific to sperm. They are critical for sperm cell locomotion and guidance.

Caudal intestinal portal (CIP)   The anterior opening of the developing hindgut region of the primitive gut tube; it opens into the future midgut region, which is contiguous with the yolk sac at this stage.

Caudal progenitor zone   A region in the tailbud of vertebrate embryos that is made up of multipotent neuromesoderm progenitor cells. See also **Neuromesoderm progenitors**.

Caudal   Referring to the tail.

Cavitation   In mammalian embryos, a process whereby the trophoblast cells secrete fluid into the morula to create a blastocoel. The membranes of trophoblast cells pump sodium ions (Na^+) into the central cavity, drawing in water osmotically and thus creating and enlarging the blastocoel.

Cell adhesion molecules   Adhesion molecules that hold cells together. The major group of these is the cadherins. See also **Cadherins**.

Cell lineage   The series of cell types starting from an undifferentiated, pluripotent stem cell through stages of increasing differentiation to the terminally differentiated cell type.

Cell wall   An extracellular layer outside the cell membrane and surrounding the cells of plants, fungi, and many prokaryotes. In land plants, cell walls are rigid, containing cellulose, and other polymers; they inhibit cell movement and restrict the planes of cell division.

Cellular blastoderm   Stage of *Drosophila* development in which all the cells are arranged in a single-layered jacket around the yolky core of the egg.

Cellularization   The process of creating individual cells out of a multinucleated cell (a syncytium) by separating the nuclei by cell membranes. Example: During early development in the *Drosophila* embryo, the syncytial blastoderm is turned into a cellular blastoderm when cell membranes grow inward between the peripheral nuclei, separating off individual cells from the inner yolky cytoplasm.

Central dogma   The explanation of the transfer of information coded in DNA to make proteins: DNA is transcribed into RNA, which is then translated into proteins.

Central nervous system (CNS)   The brain and spinal cord of vertebrates.

Centrolecithal   Type of egg, such as those of insects, that has yolk in the center and undergoes superficial cleavage.

Centromere   A region of a chromosome where sister chromatids are attached to each other by the kinetochore.

Centrosome-attracting body (CAB)   Cellular structure that, in some invertebrate blastomeres, positions the centrosomes asymmetrically and recruits particular mRNAs so that the resulting daughter cells are different sizes and have different properties.

Cephalic furrow   A transverse furrow formed during gastrulation in *Drosophila* that separates the future head region (procephalon) from the germ band, which will form the thorax and abdomen.

Cephalic neural crest   See **Cranial neural crest**.

Cephalic   Referring to the head.

Charophytic algae   A group of freshwater green algae, an ancient member of which is thought to have been the common ancestor of all terrestrial plants (embryophytes).

Chemoaffinity hypothesis   Hypothesis put forth by Sperry in 1965 suggesting that nerve cells in the brain acquire individual chemical tags that distinguish them from one another and that these guide the assembly and organization of the neural circuits in the brain.

Chemoattractant   A biochemical that causes cells to move toward it.

Chemotaxis   Movement of a cell down a chemical gradient, such as sperm following a chemical (chemoattractant) secreted by the egg.

Chiasmata   Points of attachment between homologous chromosomes during meiosis that are thought to represent regions where crossing-over is occurring.

Chimera   An organism consisting of a mixture of cells from two individuals.

Chimeric embryo   Embryo made from tissues of more than one genetic source.

ChIP-Seq   **Ch**romatin **i**mmuno**p**recipitation **seq**uencing. A lab protocol used to identify the precise DNA sequences bound by particular transcription factors or nucleosomes containing specific modified histones.

Choanoblastaea   Considered to be the ancestor of all metazoans (multicellular animals); thought to have been a sphere of choanocytes that lived in the open ocean.

Choanocyte   Found in sponges, it is a type of cell that contains a central flagellum surrounded by a collar of microvilli; choanocytes power the unidirectional flow of water through the sponge and also function as multipotent stem cells during sponge regeneration.

Choanoflagellates A group of unicellular and colonial free-living eukaryotes. The cells have the same basic structure as choanocytes of sponges, with a central flagellum surrounded by a collar of microvilli. Ancient choanoflagellates are considered to have been the common ancestor of all metazoans. See **Choanoblastaea**.

Chondrocyte-like osteoblasts Cranial neural crest cells undergoing early stages of intramembranous ossification. These cells downregulate Runx2 and begin expressing the *osteopontin* gene, giving them a phenotype similar to a developing chondrocyte.

Chondrocytes Cartilage cells.

Chondrogenesis Formation of cartilage, in which chondrocytes differentiate from condensed mesenchyme.

Chordamesoderm Axial mesoderm in a chordate embryo that produces the notochord.

Chordate An animal that has, at some stage of its life cycle, a notochord and a dorsal nerve cord or neural tube.

Chordin A paracrine factor with organizer activity. Chordin binds directly to BMP4 and BMP2 and prevents their complexing with their receptors, thus inducing dorsal ectoderm to form neural tissue.

Chorioallantoic membrane Forms in some amniote species, such as birds, by fusion of the mesodermal layer of the allantoic membrane with the mesodermal layer of the chorion. This extremely vascular envelope is crucial for bird development and is responsible for transporting calcium from the eggshell into the embryo for bone production.

Chorion An extraembryonic membrane essential for gas exchange in amniote embryos. It is generated from the extraembryonic somatopleure. The chorion adheres to the shell in birds and reptiles, allowing the exchange of gases between the egg and the environment. It forms the embryonic/fetal portion of the placenta in mammals.

Chorionic villus sampling Taking a sample from the placenta at 8–10 weeks of gestation to grow fetal cells to be analyzed for the presence or absence of certain chromosomes, genes, or enzymes.

Chromatid Half of a mitotic prophase chromosome, which consists of duplicate "sister" chromatids that

are attached to each other by the kinetochore.

Chromatin The complex of DNA and protein in which eukaryotic genes are contained.

Chromosome diminution The fragmentation of chromosomes just prior to cell division, resulting in cells in which only a portion of the original chromosome survives. Chromosome diminution occurs during cleavage in *Parascaris aequorum* in the cells that will generate the somatic cells while the future germ cells are protected from this phenomenon and maintain an intact genome.

Chromosomes Structures in a cell that contain genetic material in the form of deoxyribonucleic acid (DNA). In eukaryotes, they are contained within the nucleus and consist of DNA and proteins; they become condensed prior to cell division.

Ciliary body A vascular structure at the junction between the neural retina and the iris that secretes the aqueous humor.

Cis-regulatory elements Regulatory elements (promoters and enhancers) that reside on the same stretch of DNA as the gene they regulate.

Cleavage A series of rapid mitotic cell divisions following fertilization in many early embryos; cleavage divides the embryo without increasing its mass.

Cleavage furrow A groove formed in the cell membrane in a dividing cell due to tightening of the microfilamentous ring.

Cleft of Brachet In amphibian gastrulation, it is a region of extracellular matrix that separates the ectoderm from the involuting mesendoderm.

Cloaca Latin, "sewer." An endodermally lined chamber at the caudal end of the embryo that will become the receptacle for waste from the intestine and the kidneys and products from the gonads. Amphibians, reptiles, and birds retain this organ and use it to void gametes and both liquid and solid wastes. In mammals, the cloaca becomes divided by a septum into the urogenital sinus and the rectum.

Cloacal membrane At the caudal end of the hindgut formed by closely apposed endoderm and ectoderm; future site of the anus.

Cloning See **Somatic cell nuclear transfer**.

Clonogenic neoblasts (cNeoblasts) Pluripotent stem cells in flatworms

that migrate to a wound site and regenerate the tissue; form the regeneration blastema of flatworms.

cNeoblasts See **Clonogenic neoblasts**.

Cochlea In amniotes, the portion of the inner ear involved in hearing. The mammalian cochlea is derived from the otic placode. It is spiral in shape and contains the organ of Corti, the sensory structure for hearing.

Cochleovestibular ganglion The ganglion adjacent to the otic vesicle. It forms the major neural connection between the brain and the inner ear structures.

Coelom Space between the somatic mesoderm and splanchnic mesoderm that becomes the body cavity. In mammals, the coelom becomes subdivided into the pleural, pericardial, and peritoneal cavities, enveloping the thorax, heart, and abdomen, respectively.

Coherence Scientific evidence that fits into a system of other findings and is therefore more readily accepted.

Cohesin proteins Protein rings that encircle the sister chromatids during meiosis, provide a scaffold for the assembly of the meiotic recombination complex, resist the pulling forces of the spindle microtubules, and thereby keep the sister chromatids attached during the first meiotic division and promote pairing of homologous chromosomes, allowing recombination.

Collective cell migration The movement of a sheet of cells, wherein the cells at the leading edge of the sheet in part provide the locomotory force, using pseudopodia to pull the rest of the cells forward. Cells behind the leading edge, being surrounded by other cells, are prevented from forming locomotory pseudopodia due to contact inhibition of cell movement. However, during chemoattractive responses the cells at the rear of the cluster can form actomyosin arrays that squeeze these cells forward.

Collective migration Migration of self-propelled cells exerting directionally coordinated forces upon one another, as opposed to individually migrating cells or the movement of a group of cells caused by tissue pushing due to proliferation or intercalation.

Colonial theory The theory, originally proposed by Ernst Haeckel in 1874, that multicellular organisms arose through the symbiosis of unicellular organisms of the same species.

Columns of Terni Groups of preganglionic autonomic motor neurons in the thoracic region that are among the motor neurons whose axons are migrating away from the spinal cord during vertebrate development.

Combinatorial association In developmental genetics, the principle that enhancers contain regions of DNA that bind transcription factors, and it is this combination of transcription factors that activates the gene.

Commensalism A symbiotic relationship that is beneficial to one partner and neither beneficial nor harmful to the other partner.

Commissureless (Comm) An endosomal protein in *Drosophila*, expressed in axons prior to their crossing the midline; functions to route Robo proteins to the lysosome instead of permitting their expression in the cell membrane.

Commitment Describes a state in which a cell's developmental fate has become restricted even though it is not yet displaying overt changes in cellular biochemistry and function.

Committed stem cells Includes multipotent and unipotent stem cells that have the potential to become any of a relatively few cell types (multipotent) or only one cell type (unipotent).

Compaction A unique feature of mammalian cleavage, mediated by the cell adhesion molecule E-cadherin. The cells in the early (around eight-cell) embryo change their adhesive properties and become tightly attached to each other.

Comparative embryology Study of how anatomy changes during the development of different organisms.

Compensatory regeneration Form of regeneration in which the differentiated cells divide but maintain their differentiated functions (e.g., mammalian liver).

Competence The ability of cells or tissues to respond to a specific inductive signal.

Conditional specification The ability of cells to achieve their respective fates by interactions with other cells. What a cell becomes is in large measure specified by paracrine factors secreted by its neighbors.

Cones Color-sensitive photoreceptor cells in the neural retina. See **Neural retina**.

Congenital adrenal hyperplasia A condition causing female pseudo-hermaphroditism due to the presence of excess testosterone.

Congenital defect Any defect that an individual is born with. Congenital defects can be hereditary. or they can have an environmental cause (e.g., exposure to teratogenic plants, drugs, chemicals, radiation, etc.). They can also be idiopathic (i.e., cause is unknown).

Conjoined twins Monozygotic twins that share some part of their bodies; they may even share a vital organ, such as a heart or liver.

Consensus sequence When referring to an intron, these are located at the 5′ and 3′ ends of the introns that signal the "splice sites" of the intron.

Conserved dopamine neurotrophic factor (CDNF) A neurotrophin that enhances the survival of the midbrain dopaminergic neurons. See also **Neurotrophin**.

Contact inhibition of locomotion The mechanism whereby cells are prohibited from forming locomotory pseudopodia along contact surfaces with other cells. These interactions with the cell membranes of other cells prevent "backward" migration over other cells and result in "forward" migration of the leading edge of cells.

Context dependency The meaning or role of an individual component of a system (such as a transcription factor) is dependent on its context. For example, in the formation of tetrapod limb joints, the same BMPs can induce either cell death or cell differentiation, depending on the stage of the responding cell.

Conus arteriosus Cardiac outflow tract; along with the truncus arteriosus will become the base of the aorta and pulmonary arteries.

Convergent evolution The independent evolution of similar traits among organisms not closely related but adapting to a similar environment.

Convergent extension A phenomenon wherein cells intercalate to narrow the tissue and at the same time move it forward. Mechanism used for elongation of the archenteron in the sea urchin embryo, notochord of the tunicate embryo, and involuting mesoderm of the amphibian. This movement is reminiscent of traffic on a highway when several lanes must merge to form a single lane.

Coordinated gene expression The simultaneous expression of many different genes in a specific cell type. Its basis is often a single transcription factor (e.g., Pax6) that is crucial to several different enhancer sequences; the different enhancers are differentially "primed," and the binding of the same factor to all of them activates all the genes at once.

Cornified layer (stratum corneum) The outer layer of the epidermis, consisting of keratinocytes that are now dead, flattened sacs of keratin protein with their nuclei pushed to one edge of the cell. These cells are continually shed throughout life and are replaced by new cells.

Corona radiata The innermost layer of cumulus cells around a mammalian egg, immediately adjacent to the zona pellucida.

Corpora allata Insect glands that secrete juvenile hormone (JH) during larval molts.

Correlative evidence Evidence based on the association of events. The "find it" of "find it, lose it, move it." See also **Gain-of-function evidence**; **Loss-of-function evidence**.

Cortex An outer structure (in contrast with medulla, an inner structure).

Cortical cytoplasm A thin layer of gel-like cytoplasm lying immediately beneath the cell membrane of a cell. In an egg, the cortex contains high concentrations of globular actin molecules that will polymerize to form microfilaments and microvilli during fertilization.

Cortical granule reaction The basis of the **slow block to polyspermy** in many animal species, including sea urchins and most mammals. A mechanical block to polyspermy that in sea urchins becomes complete about a minute after successful sperm-egg fusion, in which enzymes from the egg's cortical granules contribute to the formation of a fertilization envelope that blocks further sperm entry.

Cortical granules Membrane-bound, Golgi-derived structures located in the egg cortex; contain enzymes and other components. The exocytosis of these granules at fertilization is homologous to the exocytosis of the acrosome in sperm in the acrosome reaction.

Cortical plate The layer of cells in the developing cerebrum of mammals formed by neurons in the ventricular zone migrating outward along radial glial processes to a position near the outer surface of the brain, where they will set up the six layers of the neocortex.

Cotyledons The embryonic leaves contained within the seed of a plant, which provide nutrients that support embryogenesis and germination of the seedling. During germination, they emerge from the seed before true leaves are formed.

CpG islands Regions of DNA rich in the CpG sequence: a cytosine and a guanosine connected by a normal phosphate bond. Promoters often contain such islands, and transcription is often initiated nearby, possibly because they bind the basal transcription factors that recruit RNA polymerase II.

Cranial (cephalic) neural crest Neural crest cells in the future head region that migrate to produce the craniofacial mesenchyme, which differentiates into the cartilage, bone, cranial neurons, glia, and connective tissues of the face. These cells also enter the pharyngeal arches and pouches to give rise to thymic cells, the odontoblasts of the tooth primordia, and the bones of the middle ear and jaw.

Cranial sensory placodes Ectodermal thickenings that form in the cranial region of the vertebrate embryo; includes the olfactory (nasal), otic (ear), and lens (eye) placodes, and placodes that give rise to sensory neurons of various cranial nerves. Also called cranial ectodermal placodes.

Craniorachischisis The failure of the entire neural tube to close.

Cranium The vertebrate skull, composed of the neurocranium (skull vault and base) and the viscerocranium (jaws and other pharyngeal arch derivatives).

Cre-lox A site-specific recombinase technology that allows control over the spatial and temporal pattern of a gene knockout or gene misexpression.

CRISPR **C**lustered **R**egularly **I**nterspaced **S**hort **P**alindromic **R**epeat. A stretch of DNA in prokaryotes that when transcribed into RNA serve as guides for recognizing segments of viral DNA. Used in association with Cas9 (CRISPR-associated enzyme 9) in a method for gene editing that is relatively fast and inexpensive.

Crossing over The exchange of genetic material during meiosis, whereby genes from one chromatid are exchanged with homologous genes from another.

Crown cells Cells neighboring the nodal cells, critical to setting up the left-right axis in the mammalian embryo. Crown cells each have a single immobile cilium that senses the left-to-right movement of fluids caused by the motile cilia on node cells. This sets up a cascade of events within the crown cells that serve to maintain Nodal expression on the left side, where it can activate *Pitx1* genes, which determine the left- and right-sidedness.

Crypt A deep tubular recess or pit. Example: intestinal crypts between the intestinal villi.

Crystallins Transparent, lens-specific proteins.

Cumulus A layer of cells surrounding the mammalian egg, made up of ovarian follicular (granulosa) cells that nurture the egg until it is released from the ovary. The innermost layer of cumulus cells, the corona radiate, is released with the egg at ovulation.

Cutaneous appendages Species-specific epidermal modifications that include hairs, scales, scutes, feathers, hooves, claws, and horns.

CXCR4 The receptor for stromal-derived factor 1 (SDF1). See also **Stromal-derived factor 1**.

Cyclic adenosine 3′,5′-monophosphate (cAMP) An important component of several intracellular signaling cascades and the soluble chemotactic substance that directs the aggregation of the myxamoebae of *Dictyostelium* to form a grex.

Cyclin B The larger subunit of mitosis-promoting factor, shows the cyclical behavior that is key to mitotic regulation, accumulating during S and being degraded after the cells have reached M. Cyclin B regulates the small subunit of MPF, the cyclin-dependent kinase.

Cyclin-dependent kinase Small subunit of MPF, activates mitosis by phosphorylating several target proteins, including histones, the nuclear envelope lamin proteins, and the regulatory subunit of cytoplasmic myosin, resulting in chromatin condensation, nuclear envelope depolymerization, and the organization of the mitotic spindle. Requires cyclin B to function.

Cyclooxygenase-2 (COX2) An enzyme that generates prostaglandins from the fatty acid arachidonic acid.

Cyclopia Congenital defect characterized by a single eye, caused by mutations in genes that encode either Sonic hedgehog or the enzymes that synthesize cholesterol and can be induced by certain chemicals that interfere with the cholesterol biosynthetic enzymes.

Cystoblasts/cystocytes Derived from the asymmetric division of the germline stem cells of *Drosophila*, a cystoblast undergoes four mitotic divisions with incomplete cytokinesis to form a cluster of 16 cystocytes (one ovum and 15 nurse cells) interconnected by ring canals.

Cytokines Paracrine factors important in cell signaling and the immune response. During blood formation, they are collected and concentrated by the extracellular matrix of the stromal (mesenchymal) cells at the sites of hematopoiesis and are involved in blood cell and lymphocyte formation.

Cytokinesis The division of the cell cytoplasm into two daughter cells. The mechanical agent of cytokinesis is a contractile ring of microfilaments made of actin and the motor protein myosin. Each daughter cell receives one of the nuclei produced by nuclear division (karyokinesis).

Cytonemes Specialized filopodial projections that extend out from a cell (sometimes more than 100 µm) to make contact with another cell producing a paracrine factor. Paracrine factors may be delivered to target cells by attaching to receptors on the tips of cytonemes and traveling down the length of the cytonemes to the body of the target cells. A cytoneme can also extend from a cell producing a paracrine factor to make contact with a target cell.

Cytoplasmic bridges Continuity between adjacent cells that results from incomplete cytokinesis, e.g., during gametogenesis.

Cytoplasmic determinants Factors within the egg cytoplasm that determine cell fate; these are molecules, often transcription factors, that regulate gene expression. In autonomous specification, these are apportioned to different blastomeres of the early embryo. See **Cytoplasmic determination factors**.

Cytoplasmic determination factors Factors found in the cytoplasm of a cell that determine cell fate. Example: gradients of different cytoplasmic determination factors that dictate cell fate along the anterior-posterior axis are found in the syncytial blastoderm of the *Drosophila* embryo.

Cytoplasmic polyadenylation-element-binding-protein (CPEB)   Protein that binds to mRNA in the 3' UTR and helps to control translation. When phosphorylated, it allows for the elongation of the polyadenine (polyA) tail on the mRNA.

Cytotrophoblast   Mammalian extraembryonic epithelium composed of the original trophoblast cells, it adheres to the endometrium through adhesion molecules and, in species with invasive placentation such as the mouse and human, secretes proteolytic enzymes that enable the cytotrophoblast to enter the uterine wall and remodel the uterine blood vessels so that the maternal blood bathes fetal blood vessels.

D

Dauer larva   A metabolically dormant larval stage in *C. elegans*. See also **Diapause**.

De novo regeneration   Regeneration of a structure anew (de novo) from cells that are reprogrammed to form the structure rather than from differentiated cells of that structure. For example, in certain plants *de novo* regeneration of whole organs is possible from single cells.

Decidua   The maternal portion of the placenta, made from the endometrium of the uterus.

Decussate phyllotaxis   In plants, the arrangement of lateral organs, such as leaves, in which successive opposite pairs of organs are offset by 180 degrees.

Deep cells   A population of cells in the zebrafish blastula between the enveloping layer (EVL) and the yolk syncytial layer (YSL) that give rise to the embryo proper.

Deep homology   Signal transduction pathways composed of homologous proteins arranged in a homologous manner that are used for the same function in both protostomes and deuterostomes.

Definitive endoderm   The endoderm that enters the interior of the amniote embryo through the primitive streak during gastrulation, replacing the visceral endoderm, which is primarily forming the yolk sac and allantois along with the splanchnic lateral plate mesoderm.

Delamination   The splitting of one cellular sheet into two more or less parallel sheets.

Delta protein   Cell surface ligand for Notch; participates in juxtacrine interaction and activation of the Notch pathway.

Dendrites   The fine, branching extensions (dendritic arbor) emanating from neurons; dendrites pick up electric impulses from other cells.

Dendritic arbor   Extensive branching found in the dendrites of some neurons, such as Purkinje neurons.

Dental lamina   A broad epidermal thickening in the jaw that later resolves into separate placodes, which together with the underlying mesenchyme form teeth.

Dermal bone   Bone that forms in the dermis of the skin, such as most of the bones of the skull and face. They can be derived from head mesoderm or cranial neural crest-derived mesenchymal cells.

Dermal papilla   A component of mesenchymal-epithelial induction during hair formation; a small node formed by dermal fibroblasts beneath the epidermal hair germ that stimulates proliferation of the overlying epidermal basal stem cells, which will give rise to the hair shaft.

Dermal tissue   In animals, the tissue (the dermis) that underlies the epidermis; together they make up the skin. In plants, the tissue that makes up the outer layer (the epidermis) of the plant; epidermal cells and guard cells that surround stomata are examples of cell types found in dermal tissue.

Dermatome   The central portion of the dermomyotome that produces the precursors of the dermis of the back and a population of muscle cells.

Dermomyotome   Dorsolateral portion of the somite that contains skeletal muscle progenitor cells (including those that migrate into the limbs) and the cells that generate the dermis of the back.

Descent with modification   Darwin's theory to explain unity of type by descent from a common ancestor and to explain adaptations to the conditions of particular environments by natural selection.

Determination front   Equivalent to the "wavefront" of the "clock-wavefront" model for somite formation; where boundaries of somites form, determined by a caudal^{HIGH}-to-rostral^{LOW} gradient of FGF in the presomitic mesoderm.

Determination   The stage of commitment following specification; the determined stage, assumed irreversible, is when a cell or tissue is capable of differentiating autonomously even when placed into a non-neutral environment.

Determined   Committed to a given fate. A cell is determined if it maintains its developmental maturation toward its fate even when placed in a new environment. See also **Determination**.

Deuterostomes   In the deuterostome animal groups (including echinoderms, tunicates, cephalochordates, and vertebrates), during embryonic development the first opening (i.e., the blastopore) becomes the anus while the second opening becomes the mouth (hence, *deutero stoma*, "mouth second"). Compare with **Protostomes**.

Development   The process of progressive and continuous change that generates a complex multicellular organism from a single cell. Development occurs throughout embryogenesis, maturation to the adult form, and continues into senescence.

Developmental biology   The discipline that studies embryonic and other developmental processes, such as replacement of old cells by new, regeneration, metamorphosis, aging, and development of disease states such as cancer.

Developmental constraints   In evolution, the limitation of the number and forms of possible phenotypes that can be created by the interactions that are possible among molecules and between modules in the developing organism.

Developmental plasticity   The ability of an embryo or larva to react to an environmental input with a change in form, state, movement, or rate of activity (i.e., phenotypic change).

Dextral coiling   Right-coiling. In a snail, having its coils open to the right of its shell. See also **Sinistral coiling**.

Diacylglycerol (DAG)   Second messenger generated in the IP$_3$ pathway from membrane phospholipid phosphatidylinositol 4,5-bisphosphate (PIP2), along with IP$_3$. DAG activates protein kinase C, which in turn activates a protein that exchanges sodium ions for hydrogen ions, raising the pH within the cell.

Diakinesis   Greek, "moving apart." In the first meiotic division, the stage that marks the end of prophase I, when the nuclear envelope breaks down and chromosomes migrate to the metaphase plate.

Diapause   A metabolically dormant, nonfeeding stage of an organism during which development and aging are suspended; can occur at the embryonic, larval, pupal, or adult stage.

Dickkopf   German, "thick head," "stubborn." A protein that interacts directly with the Wnt receptors, preventing Wnt signaling.

Dictyate resting stage   The prolonged diplotene stage of the first meiotic division in mammalian primary oocytes. They remain in this stage until just prior to ovulation, when they complete meiosis I and are ovulated as secondary oocytes.

Diencephalon   The caudal subdivision of the prosencephalon that will form the optic vesicles, retinas, pineal gland, and the thalamic and hypothalamic brain regions, which receive neural input from the retina.

Diethylstilbestrol (DES)   A potent environmental estrogen. DES administration to pregnant women interferes with sexual and gonadal development in their female offspring resulting in infertility, subfertility, ectopic pregnancies, adenocarcinomas, and other effects.

Differential adhesion hypothesis   A model explaining patterns of cell sorting based on thermodynamic principles. Cells interact so as to form an aggregate with the smallest interfacial free energy and therefore, the most thermodynamically stable pattern.

Differential gene expression   A basic principle of developmental genetics: In spite of the fact that all the cells of an individual body contain the same genome, the specific proteins expressed by the different cell types are widely diverse. Differential gene expression, differential nRNA processing, differential mRNA translation, and differential protein modification all work to allow the extensive differentiation of cell types.

Differential RNA processing   The splicing of mRNA precursors into messages that specify different proteins by using different combinations of potential exons.

Differentiation   The process by which an unspecialized cell becomes specialized into one of the many cell types that make up the body.

Differentiation therapy   Treatments for cancer that use transcription factors

and other molecules to "normalize" cancers—that is, to cause cancerous cells to revert to differentiation rather than continued proliferation.

Digestive tube   The primitive gut of the embryo, which extends the length of the body from the pharynx to the cloaca. Buds from the digestive tube form the thyroid, thymus, and parathyroid glands, lungs, liver, gallbladder, and pancreas.

5α-Dihydrotestosterone (DHT)   A steroid hormone derived from testosterone by the action of the enzyme 5α-ketosteroid reductase 2. DHT is required for masculinization of the male urethra, prostate, penis, and scrotum.

Diploblasts   "Two-layer" animals; they possess endoderm and ectoderm but most species lack true mesoderm. Includes the ctenophores (comb jellies) and cnidarians (jellyfish, corals, hydra, sea anemones). Compare with **Bilaterians**.

Diplotene   Greek, "double threads." In the first meiotic division, the fourth and last stage of prophase I, when the synaptonemal complex breaks down and the two homologous chromosomes start to separate but remain attached at points of chiasmata where crossing-over is occurring. Follows pachytene stage.

Direct development   Embryogenesis characterized by the lack of a larval stage, where the embryo proceeds to construct a small adult.

Discoidal cleavage   Meroblastic cleavage pattern for telolecithal eggs, in which the cell divisions occur only in the small blastodisc, as in birds, reptiles, and fish.

Discoidal meroblastic cleavage   See **Discoidal cleavage**.

Disruption   Abnormality or congenital defect caused by exogenous agents (teratogens) such as plants, chemicals, viruses, radiation, or hyperthermia.

Dissociation   The ability of one module to develop differently from other modules.

Distal tip cell   A single nondividing cell located at the end of each gonad in *C. elegans* that maintains the nearest germ cells in mitosis by inhibiting their going into meiosis.

Dizygotic twins   "Two eggs." Twins that result from two separate but approximately simultaneous fertilization events. Genetically such "fraternal" (Latin *frater*, "brother") twins are full

siblings. Compare with **Monozygotic twins**.

Dmrt1   Protein that in birds, frogs, and fish appears to activate *Sox9*, the central male-determining gene in vertebrates. Dmrt1 is also needed for maintaining testicular structures in mammals.

DNA methylation   A method of controlling the level of gene transcription in vertebrates by the enzymatic methylation of the promoters of inactive genes. Certain cytosine residues that are followed by guanosine residues are methylated and the resulting methylcytosine stabilizes nucleosomes and prevents transcription factors from binding. Important in X chromosome inactivation and DNA imprinting.

DNA-binding domain   Transcription factor domain that recognizes a particular DNA sequence.

Dormancy   In seed plants, a prolonged period of quiescence that a seed can undergo prior to germination.

Dorsal blastopore lip   Location of the involuting marginal zone cells of amphibian gastrulation. Migrating marginal cells sequentially become the dorsal lip of the blastopore, turn inward and travel along the inner surface of the outer animal cap cells (i.e., the blastocoel roof).

Dorsal closure   A process that brings together the two sides of the epidermis of the *Drosophila* embryo at the dorsal surface.

Dorsal mesentery   A derivative of the splanchnic mesoderm, this fibrous membrane connects the endoderm to the body wall. Involved in the looping of the developing intestines.

Dorsal root ganglia (DRG)   Sensory spinal ganglia derived from the trunk neural crest that migrate along the ventral pathway and stay in the sclerotome. Sensory neurons of the DRG connect centrally with neurons in the dorsal horn of the spinal cord.

Dorsal-ventral (dorsoventral or DV) axis   The plane defining the back (dorsum) versus the belly (ventrum). When referring to the limb, this axis refers to the knuckles (dorsal) and palms (ventral).

Dorsolateral hinge points (DLHPs)   In the formation of the avian and mammalian neural tube, two hinge regions in the lateral sides of the neural plate that bend the two sides of the plate inward toward each other after the medial hinge point (MHP) has bent the plate along its midline.

Dorsolateral pathway Pathway taken by trunk neural crest cells traveling dorsolaterally beneath the ectoderm to become melanocytes.

Dosage compensation Equalization of expression of X chromosome-encoded gene products in male and female cells. Can be achieved by (1) doubling the transcription rate of the male X chromosomes (*Drosophila*), (2) partially repressing both X chromosomes (*C. elegans*), or (3) inactivating one X chromosome in each female cell (mammals).

Double fertilization In angiosperms, a process wherein one sperm nucleus combines with the egg nucleus to produce a zygote, while the other sperm combines with somatic cells to produce the triploid endosperm.

Double-negative gate A mechanism whereby a repressor locks the genes of specification, and these genes can be unlocked by the repressor of that repressor. (In other words, activation by the repression of a repressor.)

Doublesex (*Dsx*) A *Drosophila* gene active in both males and females, but whose RNA transcript is spliced in a sex-specific manner to produce sex-specific transcription factors: the female-specific transcription factor activates female-specific genes and inhibits male development; the male-specific transcription factor inhibits female traits and promotes male traits.

Down syndrome Syndrome caused by having an extra copy of chromosome 21 in humans; includes anomalies such as facial muscle changes, heart and gut abnormalities, and cognitive problems.

Ductus arteriosus A vessel that forms from left aortic arch VI, serves as a shunt between the embryonic/fetal pulmonary artery and the descending aorta in mammals. It normally closes at birth (if not, a pathological condition results called patent ductus arteriosus).

Duplication and divergence Tandem gene duplications resulting from replication errors. Once replicated, the gene copies can diverge by random mutations developing different expression patterns and new functions.

Dynein A motor protein that travels along microtubules. It is an ATPase, an enzyme that hydrolyzes ATP, converting the released chemical energy into mechanical energy. In cilia and flagella, dynein is attached to the axoneme microtubules that provides the force for propulsion by allowing the active sliding of the outer doublet microtubules, causing the flagellum or cilium to bend.

Dysgenesis Greek, "bad beginning." Refers to defective development.

E

20E See **20-Hydroxyecdysone**.

E-cadherin A type of cadherin expressed in epithelial tissues as well as all early mammalian embryonic cells (the E stands for epithelial). See **Cadherins**.

Early allocation and progenitor expansion model An alternative to the progress zone model of proximal-distal specification of the limb, wherein the cells of the entire early limb bud are already specified; subsequent cell divisions simply expand these cell populations.

Ecdysone Insect steroid hormone, secreted by the prothoracic glands, that is modified in peripheral tissues to become the active molting hormone 20-hydroxyecdysone. Crucial to insect metamorphosis.

Ecdysone receptor (EcR) Nuclear protein that binds to ecdysone in insects; when bound it forms an active complex with another protein that binds to DNA, inducing transcription of ecdysone-responsive genes. Evolutionarily related to, and almost identical in structure to, thyroid hormone receptors.

Ecdysozoans One of the two major protostome groups; characterized by exoskeletons that periodically molt. The arthropods (including insects and crustaceans) and the nematodes (roundworms, including the model organism *C. elegans*) are two prominent groups. See also **Lophotrochozoans**.

Ecological evolutionary developmental biology (eco-evo-devo) The science studying the ways by which developmental changes initiated by the environment can affect evolution. It deals primarily with the evolutionary aspects of developmental symbiosis, developmental plasticity, and niche construction.

Ecomorph See **Morph**.

Ectoderm Greek *ektos*, "outside." The cells that remain on either the outside (amphibian) or dorsal (avian, mammalian) surface of the embryo following gastrulation. Of the three germ layers, the ectoderm forms the nervous system from the neural tube and neural crest; also generates the epidermis covering the embryo.

Ectodermal appendage placodes Thickenings of the epidermal ectoderm involved in forming non-sensory structures such as hair, teeth, feathers, mammary and sweat glands.

Ectodermal appendages Structures that form from specific regions of epidermal ectoderm and underlying mesenchyme through a series of interactive inductions; includes hairs, scales, scutes (such as the coverings of turtle shells), teeth, sweat glands, mammary glands, and feathers.

Ectodermal placodes Thickenings of the surface ectoderm in embryos that become the rudiment of numerous organs. Includes cranial placodes, ectodermal appendage placodes.

Ectodysplasin (EDA) cascade A gene cascade specific for cutaneous appendage formation. Vertebrates with dysfunctional EDA proteins exhibit a syndrome called anhidrotic ectodermal dysplasia characterized by absent or malformed cutaneous appendages (hair, teeth, and sweat glands).

Efferent Carried away from. Often used in reference to neurons that carry information away from the central nervous system (brain and spinal cord) to be acted on by the peripheral nervous system (muscles), or a vessel that carries fluid away from a structure. Compare with **Afferent**.

Efferent ducts Ducts that link the rete testis to the Wolffian duct, formed from remodeled tubules of the mesonephric kidney.

Efflux transport A process found whereby transporter proteins move a substance from the inside to the outside of a cell. Example: PIN proteins in plants use efflux transport to move auxin from inside to outside the cell.

Egg chamber An ovariole or egg tube (over a dozen per ovary) in which the *Drosophila* oocyte will develop, containing 15 interconnected nurse cells and a single oocyte.

Egg jelly A glycoprotein meshwork outside the vitelline envelope in many species, most commonly it is used to attract and/or to activate sperm.

Embryo A developing organism prior to birth or hatching. In humans, the term embryo generally refers to the early stages of development, starting with the fertilized egg until the end of organogenesis (first 8 weeks of gestation). After this, the developing human is called a fetus until its birth.

Embryo sac   The female gametophyte of an angiosperm. Found inside the ovule, it consists of eight or fewer cells formed by the division of the haploid nucleus of the megaspore cell. See also **Megaspores**.

Embryogenesis   The stages of development between fertilization and hatching (or birth).

Embryology   The study of animal development from fertilization to hatching or birth.

Embryonic axis   Any of the positional axes in an embryo; includes anterior-posterior (head-tail), dorsal-ventral (back-belly), and right-left.

Embryonic epiblast   In mammals, the epiblast cells that contribute to the embryo proper that split off from the epiblast cells that line the amniotic cavity.

Embryonic germ cells (EGCs)   Pluripotent embryonic cells with characteristics of the inner cell mass derived from PGCs that have been treated particular paracrine factors to maintain cell proliferation.

Embryonic period   In human development, the first 8 weeks in utero prior to the fetal period; the time during which most organ systems form.

Embryonic shield   A localized thickening on the future dorsal side of the fish embryo; functionally equivalent to the dorsal blastopore lip of amphibians.

Embryonic stem cells (ESCs)   Pluripotent stem cells derived from cultures of the mammalian inner cell mass blastomeres that are capable of generating all the cell types of the body.

Embryophytes   Land plants. These are named embryophytes because they all undergo embryogenesis.

Emergent properties   See **Level-specific properties and emergence**.

EMT   See **Epithelial-mesenchymal transition**.

Enamel knot   The signaling center for tooth development, a group of cells induced in the epithelium by the neural crest-derived mesenchyme that secretes paracrine factors that pattern the cusp of the tooth.

Endfoot   A swollen membrane extension at the basal end of the neuroepithelial or radial glial cell. Radial glial endfeet wrap around blood vessels in the brain to establish the blood-brain barrier.

Endoblast   See **Secondary hypoblast**.

Endocardial cushions   Tissue in the developing vertebrate heart derived from the endocardium. It forms the septa that divide the atrioventricular area of the originally tubular heart into left and right atria and ventricles in amniotes; in amphibians, it separates the two atria (the ventricle remains undivided; in fish, all chambers remain undivided). The endocardial cushions also form the atrioventricular valves.

Endocardium   The internal lining of the heart chambers, derived from the heart fields.

Endochondral ossification   Bone formation in which mesodermal mesenchyme becomes cartilage and the cartilage is replaced by bone. It characterizes the bones of the trunk and limbs.

Endocrine disruptors   Hormonally active compounds in the environment (e.g., DES; BPS; aromatase) that can have major detrimental effects on development, particularly of the gonads. Many endocrine disruptors are also obesogens (cause increased production of fat cells and fat accumulation).

Endocrine factors   Hormones that travel through the blood to their target cells and tissues to exert their effects.

Endoderm   Greek *endon*, "within." The innermost germ layer; forms the epithelial lining of the respiratory tract, the gastrointestinal tract, and the accessory organs (e.g., liver, pancreas) of the digestive tract. In the amphibian embryo, the yolk-containing cells of the vegetal hemisphere become endoderm. In amniote embryos, the endoderm is the most ventral of the three germ layers, and also forms the epithelium of the yolk sack and allantois.

Endomesoderm   The combination of endodermal and mesodermal cells.

Endometrium   The epithelial lining of the uterus.

Endosome   A membrane-bound vesicle that is internalized by a cell through endocytosis. Internalizing ligand-receptor complexes in endosomes is a common mechanism in paracrine signaling.

Endosperm   A triploid seed tissue found in angiosperms, providing nutrition for the developing embryo.

Endosteal osteoblasts   Osteoblasts that line the bone marrow and are responsible for providing the niche that attracts hematopoietic stem cells (HSCs), prevents apoptosis, and keeps the HSCs in a state of plasticity.

Endosymbiosis   Greek, "living within." Describes the situation in which one cell lives inside another cell or one organism lives within another.

Endothelins   Small peptides secreted by blood vessels that have a role in vasoconstriction and can direct the migration of certain neural crest cells as well as the extension of certain sympathetic axons that have endothelin receptors, e.g. targeting of neurons from the superior cervical ganglia to the carotid artery.

Endothelium   The single-layer sheet of epithelial cells lining of the blood vessels.

Energids   In *Drosophila*, the nuclei at the periphery of the syncytial blastoderm and their associated cytoplasmic islands of cytoskeletal proteins.

Enhancer   A DNA sequence that controls the efficiency and rate of transcription from a specific promoter. Enhancers bind specific transcription factors that activate the gene by (1) recruiting enzymes (such as histone acetyltransferases) that break up the nucleosomes in the area or (2) stabilizing the transcription initiation complex.

Enhancer modularity   The principle that having multiple enhancers allows a protein to be expressed in several different tissues while not being expressed at all in others, according to the combination of transcription factor proteins the enhancers bind.

Enteric ganglia   See **Parasympathetic ganglia**.

Enterocoely   The embryonic process of forming the coelom by extending mesodermal pouches from the gut. Typical of most deuterostomes. See also **Schizocoely**.

Enveloping layer (EVL)   A cell population in the zebrafish embryo at the mid-blastula transition made up of the most superficial cells from the blastoderm, which form an epithelial sheet a single cell layer thick. The EVL is an extraembryonic protective covering that is sloughed off during later development.

Environmental integration Describes the influence of cues from the environment surrounding the embryo, fetus, or larva on their development.

Ependyma Epithelial lining of the spinal cord canal and the ventricles of the brain.

Ependymal cells Epithelial cells that line the ventricles of the brain and canal of the spinal cord; they secrete cerebrospinal fluid.

Eph receptors Receptor for ephrin ligands.

Ephrins Juxtacrine ligands. Binding between an ephrin ligand on one cell and an Eph receptor on an adjacent cell results in signals being sent to both cells. These signals are often those of either attraction or repulsion, and ephrins are often seen directing cell migration and defining where cell boundaries are to form. As well as directing neural crest cell migration, ephrins and Eph receptors function in the formation of blood vessels, neurons, and somites.

Epialleles Variants of chromatin structure that can be inherited between generations. In most known cases, epialleles are differences in DNA methylation patterning that are able to affect the germ line and thereby be transmitted to offspring.

Epiblast The outer layer of the thickened margin of the epibolizing blastoderm in the gastrulating fish embryo or the upper layer of the bilaminar gastrulating embryonic disc in amniotes (reptiles, birds and mammals. The epiblast contains ectoderm precursors in fish and all three germ layer precursors of the embryo proper (plus the amnion) in amniotes. It also forms the avian chorion and allantois.

Epiboly The movement of epithelial sheets (usually of ectodermal cells) that spread as a unit (rather than individually) to enclose the deeper layers of the embryo. Epiboly can occur by the cells dividing, by the cells changing their shape, or by several layers of cells intercalating into fewer layers. Often, all three mechanisms are used.

Epibranchial placodes A subgroup of cranial placodes that form in the pharyngeal region of the vertebrate embryo. Gives rise to the sensory neurons of three cranial nerves: facial (VII), glossopharyngeal (IX), and vagus (X).

Epicardium The outer surface of the heart that forms the coronary blood vessels that feed the heart, derived from the heart fields.

Epidermal placodes The thickenings of epidermal ectoderm associated with ectodermal appendages. See **Ectodermal appendage placodes**.

Epidermis Outer layer of the skin, derived from ectoderm.

Epididymis Derived from the Wolffian duct, the tube adjacent to the testis that links the efferent tubules to the ductus deferens.

Epigenesis The view supported by Aristotle and William Harvey that the organs of the embryo are formed de novo ("from scratch") at each generation.

Epigenetics The study of mechanisms that act on the phenotype without changing the nucleotide sequence of the DNA. Specifically, these changes work "outside the gene" (i.e., epigenetically) by altering gene *expression* rather than by altering the gene sequence as mutation does. Epigenetic changes can sometimes be transmitted to future generations, a phenomenon referred to as **epigenetic inheritance**.

Epimorphic regeneration See **Epimorphosis**.

Epimorphin A multifunctional protein: in the membranes of mesenchymal cells, it directs epithelial morphogenesis; when expressed by sclerotome cells, it acts to attract further prechondrogenic sclerotome cells to the region of the notochord and neural tube, where the cells form vertebrae.

Epimorphosis Form of regeneration observed when adult structures undergo *de*differentiation to form a relatively undifferentiated mass of cells that then redifferentiates to form the new structure (e.g., amphibian limb regeneration).

Epiphyseal plates Cartilaginous growth zones at the proximal and distal ends of the long bones that allow continued bone growth.

Episomal vectors Vehicles for gene delivery usually derived from viruses that do not insert themselves into host DNA.

Epithelial cells Cells of an epithelium. See **Epithelium**

Epithelial-mesenchymal interactions Induction involving interactions of sheets of epithelial cells with adjacent mesenchymal cells. Properties of these interactions include regional specificity (when placed together, the same epithelium develops different structures according to the region from which the mesenchyme was taken), genetic specificity (the genome of the epithelium limits its ability to respond to signals from the mesenchyme, i.e., the response is species-specific).

Epithelial-mesenchymal transition (EMT) An orderly series of events whereby epithelial cells are transformed into mesenchymal cells. In this transition, a polarized stationary epithelial cell, which normally interacts with basement membrane through its basal surface, becomes a migratory mesenchymal cell that can invade tissues and form organs in new places. See also **Mesenchymal-to-epithelial transition**.

Epithelium Epithelial cells tightly linked together on a basement membrane to form a sheet or tube with little extracellular matrix. Plural, epithelia.

Equatorial region The junction between the inner acrosomal membrane and the sperm cell membrane in mammals. It is exposed by the acrosomal reaction and is where membrane fusion between sperm and egg begins.

Equivalence group In the development of *C. elegans*, the group of six vulval precursor cells, each of which is competent to become induced by the anchor cell.

Erythroblast Cell that matures from the proerythroblast and synthesizes enormous amounts of hemoglobin.

Erythrocyte The mature red blood cell that enters the circulation where it delivers oxygen to the tissues. It is incapable of division, RNA synthesis, or protein synthesis. Amphibians, fish, and birds retain the functionless nucleus; mammals extrude it from the cell.

Erythroid progenitor cell A committed stem cell that can form only red blood cells.

Erythropoietin A hormone that acts on erythroid progenitor cells to produce proerythroblasts, which will generate red blood cells.

ESCs See **Embryonic stem cells**.

Estrogen A group of steroid hormones (including **estradiol**) needed for complete postnatal development of the Müllerian ducts (in females) and the Wolffian ducts (in males). Necessary for fertility in both sexes.

Estrus   Greek *oistros*, "frenzy." The estrogen-dominated stage of the ovarian cycle in female non-human mammals that are spontaneous or periodic ovulators, characterized by the display of behaviors consistent with receptivity to mating. Also called "heat."

Euchromatin   The comparatively open state of chromatin that contains most of the organism's genes, most of which are capable of being transcribed. Compare with **Heterochromatin**.

Eukaryotic initiation factor-4 (eIF4E)   A protein that is important for the initiation of translation. Binds to the 5′ cap of mRNAs and contributes to the protein complex that mediates RNA unwinding; brings the 3′ end of the message next to the 5′ end, allowing the mRNA to bind to and be recognized by the ribosome. Interacts with eukaryotic initiation factor-4G (eIF4G), a scaffold protein that allows the mRNA to bind to the ribosome.

Eukaryotic organisms   Organisms whose cells contain membrane-bound organelles, including a nucleus with chromosomes that undergo mitosis. Can be single-celled or multicellular.

Evolutionary developmental biology ("evo-devo")   A model of evolution that integrates developmental genetics and population genetics to explain and the origin of biodiversity.

Exine   The outer coating of a pollen grain or spore. It is extremely resistant to decay.

Exon   In a gene, the region or regions of DNA that encode the protein. Compare with **Intron**.

Exosomes   A type of membrane-bound extracellular vesicles that are released from cells and which may contain proteins and RNAs that can influence the development of other cells that receive them.

Extant   Still in existence, such as extant species.

Extended evolutionary synthesis   A model of evolution emphasizing developmental plasticity, epigenetic inheritance, niche constriction, and the reciprocal interactions between the organism and its environment.

External granular layer   A germinal zone of cerebellar neuroblasts that migrate from the germinal neuroepithelium to the outer surface of the developing cerebellum.

External YSL (eYSL)   A region of the yolk syncytial layer (YSL) in fish embryos. The eYSL forms from yolk syncytial nuclei that move further vegetally, staying ahead of the blastoderm margin as the blastoderm expands to surround the yolk cell. See **Yolk syncytial layer**.

Extracellular matrix (ECM)   Macromolecules secreted by cells into their immediate environment, forming a region of noncellular material in the interstices between the cells. Extracellular matrices are made up of collagen, proteoglycans, and a variety of specialized glycoprotein molecules such as fibronectin and laminin.

Extraembryonic endoderm   Formed by delamination of the hypoblast cells from the avian epiblast or mammalian inner cell mass to line the yolk sac.

Extraembryonic vasculogenesis   The formation of blood islands in the yolk sac (i.e., outside the embryo).

Eye field   Region in the anterior portion of the neural tube that will develop into the neural and pigmented retinas.

F

Fasciculation   In neural development, the process of one axon adhering to and using another axon for growth.

Fast block to polyspermy   Mechanism by which additional sperm are prevented from fusing with a fertilized sea urchin egg by changing the electric potential to a more positive level. Has not been demonstrated in mammals.

Fate map   Diagrams based on having followed cell lineages from specific regions of the embryo in order to "map" larval or adult structures onto the region of the embryo from which they arose. The superimposition of a map of "what is to be" onto a structure that has yet to develop into these organs.

Female pronucleus   The haploid nucleus of the egg.

Fertilization cone   An extension from the surface of the egg where the egg and sperm have fused during fertilization. Caused by polymerization of actin, it provides a connection that widens the cytoplasmic bridge between egg and sperm, allowing the sperm nucleus and proximal centriole to enter the egg.

Fertilization envelope   Forms from the vitelline envelope of the sea urchin egg following cortical granule release. Glycosaminoglycans released by the cortical granules absorb water to expand the space between the cell membrane and fertilization envelope.

Fertilization   Fusion of male and female gametes followed by fusion of the haploid gamete nuclei to restore the full complement of chromosomes characteristic of the species and initiation in the egg cytoplasm of those reactions that permit development to proceed.

Fetal alcohol syndrome (FAS)   Condition of babies born to alcoholic mothers, characterized by small head size, specific facial features, and small brain that often shows defects in neuronal and glial migration. FAS is the most prevalent congenital mental retardation syndrome. Another term, **fetal alcohol spectrum disorder (FASD)** has been coined to encompass the less visible behavioral effects on children exposed prenatally to alcohol.

Fetal period   In human development, the period following the embryonic period, from the end of 8 weeks to birth; the period after organ systems have mostly formed and generally growth and modeling is occurring.

α-Fetoprotein   In mammals, a protein that binds and inactivates fetal estrogen, but not testosterone, in both male and female fetuses and shown in rodents to be critical for normal sexual differentiation of the brain.

Fetus   The stage in mammalian development between the embryonic stage and birth, characterized by growth and modeling. In humans, from the ninth week of gestation to birth.

Fibroblast growth factor 9 (Fgf9)   A growth factor involved in testes development in mammals, stimulating proliferation and differentiation in Sertoli cells and their maintenance of *Sox9* expression. Suppresses Wnt4 signaling, which would otherwise direct ovarian development. Also involved in metanephric kidney development, promoting development of a population of stem cell for nephrons. See also **Fibroblast growth factors**.

Fibroblast growth factor receptors (FGFRs)   A set of receptor tyrosine kinases that are activated by FGFs, resulting in activation of the dormant kinase and phosphorylation of certain proteins (including other FGF receptors) within the responding cell.

Fibroblast growth factors (FGFs) A family of paracrine factors that regulate cell proliferation and differentiation.

Fibronectin A very large (460 kDa) glycoprotein dimer synthesized by numerous cell types and secreted into the extracellular matrix. Functions as a general adhesive molecule, linking cells to one another and to other substrates such as collagen and proteoglycans, and provides a substrate for cell migration.

Filopodia Long, thin processes containing microfilaments; cells can move by extending, attaching, and then contracting filopodia. Produced e.g. by migrating mesenchyme cells in sea urchin embryos, growth cones for nerve outgrowth, tip cells in blood vessel formation.

Fin bud Bud of tissue in fish embryos that gives rise to a fin; homologous to the limb bud of tetrapods.

First polar body The smaller cell produced when a primary oocyte goes through its first meiotic division, producing one large cell, the secondary oocyte, which retains most of the cytoplasm, and a tiny cell, the first polar body, which is ultimately lost. Both cells are haploid.

Flagellum A long, motile extension of a cell containing a central axoneme of microtubules in a 9+2 arrangement (9 outer doublets and 2 central singlets). Its whipping action ("flagellation") functions for propulsion, as in the tail of a sperm.

Floor plate Ventral region of the neural tube important in establishing dorsal-ventral polarity. Induced to form by Sonic hedgehog secreted from the adjacent notochord. It becomes a secondary signaling center that also secretes Sonic hedgehog, establishing a gradient that is highest ventrally.

Floral MADS-box genes Genes of flowering plants that code for MADS-box transcription factors involved in floral development. The MADS-box is a conserved sequence that codes for a specific DNA-binding domain.

Floral meristem identity genes A set of genes in angiosperms whose expression starts flower development, activating the floral organ identity genes to switch meristem cells from a proliferative to a reproductive fate. See also **Floral organ identity genes**.

Floral organ identity genes Angiosperm genes (sets A, B, C, D, and E) that determine the locations and fates of floral meristem cells.

Fluorescent dye Compounds, such as fluorescein and green fluorescent protein (GFP), that emit bright light at a specific wavelength when excited with ultraviolet light.

Focal adhesions Where the cell membrane adheres to the extracellular matrix in migrating cells, mediated by connections between actin, integrin and the extracellular matrix.

Follicle A small group of cells around a cavity. E.g., mammalian ovarian follicle, composed of a single ovum surrounded by granulosa cells and thecal cells; hair follicle, feather follicle, where a hair or feather is produced.

Follicle-stimulating hormone (FSH) A peptide hormone secreted by the mammalian pituitary that promotes ovarian follicle development and spermatogenesis.

Follicular stem cells Multipotent adult stem cells that reside in the bulge niche of the hair follicle. They give rise to the hair shaft and sheath.

Follistatin A paracrine factor with organizer activity, an inhibitor of both activin and BMPs, causes ectoderm to become neural tissue.

Foot activation gradient A gradient, highest at the basal disc, that appears to be present in *Hydra* that permits the basal disc to form only in one place.

Foramen ovale In the fetal mammalian heart, an opening in the septum separating the right and left atria.

Forkhead transcription factors Transcription factors (e.g., Fox proteins, HNF4α) that are especially important in the endoderm that will form liver, where they help activate the regulatory regions surrounding liver-specific genes.

Forward genetics Genetic technique of exposing an organism to an agent that causes random mutations and screening for particular phenotypes. Compare with **Reverse genetics**.

Frizzled Transmembrane receptor for Wnt family of paracrine factors.

Frontonasal process Cranial prominence formed by neural crest cells from the midbrain and rhombomeres 1 and 2 of the hindbrain that forms the forehead, the middle of the nose, and the primary palate.

Fruit In flowering plants, a ripened and mature seed-containing ovary or ovaries.

G

G protein A protein that binds GTP and is activated or inactivated by GTP modifying enzymes (such as GTPases). They play important roles in the RTK pathway and in cytoskeletal maintenance.

Gain-of-function evidence A strong type of evidence, wherein the initiation of the first event causes the second event to happen even in instances where or when neither event usually occurs. The "move it" of "Find it, lose it, move it." See also **Correlative evidence**; **Loss-of-function evidence**.

Gamete A specialized reproductive cell through which sexually reproducing parents pass chromosomes to their offspring; an egg or a sperm.

Gametogenesis The production of gametes.

Gametophytic Referring to the gametophyte, the haploid stage of the alternating life cycle in plants and algae that produces gametes (eggs and sperm); the sexual phase. Compare with **Sporophytic**.

Ganglia Clusters of neuronal cell bodies whose axons form a nerve. Singular, ganglion.

Gap genes *Drosophila* zygotic genes expressed in broad (about three segments wide), partially overlapping domains. Gap mutants lacked large regions of the body (several contiguous segments).

Gastrula A stage of the embryo following gastrulation that contains the three germ layers that will interact to generate the organs of the body.

Gastrulation A process involving movement of the blastomeres of the embryo relative to one another resulting in the formation of the three germ layers of the embryo.

GDNF See **Glial-derived neurotrophic factor**.

GEF See **GTP exchange factor**; **RTK pathway**.

Gene regulatory networks (GRNs) Patterns generated by the interactions among transcription factors and their enhancers that help define the course that development follows.

Generative cell In angiosperms, the cell of a pollen grain that divides to produce the two male gamete nuclei.

Genetic assimilation The process by which a phenotypic character initially produced only in response to some environmental influence becomes, through a process of selection, taken over by the genotype so that it is formed even in the absence of the environmental influence that had first been necessary.

Genetic heterogeneity The production of similar phenotypes by mutations in different genes.

Geniculate placodes A pair of epibranchial placodes in vertebrates; they give rise to the sensory components of the paired 7th cranial nerve (the facial nerve) and innervate tastebuds, tonsils and ear lobes.

Genital disc Region of the *Drosophila* larva that will generate male or female genitalia. Male and female genitalia are derived from separate cell populations of the genital disc, as induced by paracrine factors.

Genital ridge A thickening of the splanchnic mesoderm and of the underlying intermediate mesodermal mesenchyme on the medial edge of the mesonephros; it forms the testis or ovary. Also called the germinal ridge. See also **Germinal epithelium**.

Genital tubercle A structure cranial to the cloacal membrane during the indifferent stage of differentiation of the mammalian external genitalia. It will form either the clitoris in the female fetus or the penis in the male.

Genome The complete DNA sequence of an individual organism.

Genomic equivalence The theory that every cell of an organism has the same genome as every other cell.

Genomic imprinting A phenomenon in mammals whereby only the sperm-derived or only the egg-derived allele of the gene is expressed, sometimes due to inactivation of one allele by DNA methylation during spermatogenesis or oogenesis.

Germ band A collection of cells along the ventral midline of the *Drosophila* embryo that forms during gastrulation by convergence and extension that includes all the cells that will form the trunk of the embryo and the thorax and abdomen of the adult.

Germ cell lineage The cells that form the gametes, including the primordial germ cells, the developing sperm and eggs, and the mature gametes.

Germ cells A group of cells set aside for reproductive function; germ cells become the cells of the gonads (ovary and testis) that undergo meiotic cell divisions to generate the gametes. Compare with **Somatic cells**.

Germ layer One of the three layers of the embryo, ectoderm, mesoderm, and endoderm, in triploblastic organisms, or of the two layers, ectoderm and endoderm, in diploblastic organisms, generated by the process of gastrulation, that will form all of the tissues of the body except for the germ cells.

Germ line The line of cells that become germ cells, separate from the somatic cells, found in many animals, including insects, roundworms, and vertebrates. Specification of the germ line can occur autonomously from determinants found in cytoplasmic regions of the egg, or can occur later through induction by neighboring cells.

Germ plasm Cytoplasmic determinants (mRNA and proteins) in the eggs of some species, including frogs, nematodes, and flies, that autonomously specify the primordial germ cells.

Germ plasm theory The first testable model of cell specification, proposed by Weismann in 1888, in which each cell of the embryo would develop autonomously. Instead of dividing equally, the chromosomes were hypothesized to divide in such a way that different chromosomal determinants entered different cells. Only the nuclei in those cells destined to become germ cells (gametes) were postulated to contain all the different types of determinants. The nuclei of all other cells would have only a subset of the original determinants.

Germ ring A thickened ring of cells in the margin of the deep cells that appears in a fish embryo once the blastoderm has covered about half of the yolk cell. Composed of a superficial layer, the epiblast, and an inner layer, the hypoblast.

Germ stem cell (GSC) In female *Drosophila*, the stem cell that gives rise to the oocyte.

Germarium In female *Drosophila*, niche in the anterior region of an ovariole containing germ stem cells and several somatic cell types.

Germinal crescent A region in the anterior portion of the avian and reptilian blastoderm area pellucida containing the hypoblasts displaced by migrating endodermal cells. It contains the primordial germ cells (precursors of the germ cells), which later migrate through the blood vessels to the gonads.

Germinal epithelium Epithelium of the bipotential gonad, derived from splanchnic mesoderm, that will form the somatic (i.e., non-germ cell) component of the gonads.

Germinal neuroepithelium A layer of rapidly dividing neural stem cells one cell layer thick that constitute the original neural tube.

Germinal ridge See **Genital ridge**.

Germinal vesicle breakdown (GVBD) Disintegration of the primary oocyte nuclear membrane (germinal vesicle) upon resumption of meiosis during oogenesis.

Germline stem cells In *Drosophila*, pole cell (primordial germ cells) derivatives that divide asymmetrically to produce another stem cell and a differentiated daughter cell called a cystoblast, which in turn produces a single ovum and 15 nurse cells.

GFP See **Green fluorescent protein**.

Glia Supportive cells of the central nervous system, derived from the neural tube, and of the peripheral nervous system, derived from neural crest.

Glial guidance A mechanism important for positioning young neurons in the developing mammalian brain (e.g., the granule neuron precursors travel on the long processes of the Bergmann glia in the cerebellum).

Glial-derived neurotrophic factor (GDNF) A paracrine factor that binds to the Ret receptor tyrosine kinase. It is produced by the gut mesenchyme that attracts vagal and sacral neural crest cells, and it is produced by the metanephrogenic mesenchyme to induce the formation and branching of the ureteric buds.

Glochidium The larva of some freshwater bivalve molluscs, such as uniod clams; has a shell resembling a tiny bear trap, used to attach to the gills or fins of fish. It feeds on the fish's body fluids until it drops off to metamorphose into an adult clam.

Glycogen synthase kinase 3 (GSK3) Targets β-catenin for destruction.

Glycosaminoglycans (GAGs) Complex acidic polysaccharides consisting of unbranched chains assembled from many repeats of a two-sugar unit. The carbohydrate component of proteoglycans.

Gonadal sex determination See **Primary sex determination**.

Gonadotropin-releasing hormone (GRH; GnRH) Peptide hormone released from the hypothalamus that stimulates the pituitary to release the gonadotropins follicle-stimulating hormone and luteinizing hormone, which are required for mammalian gametogenesis and steroidogenesis.

Gonialblast   In male *Drosophila*, a committed progenitor cell that divides to become the precursors of the sperm.

Gonocytes   Mammalian primordial germ cells (PGCs) that have arrived at the genital ridge of a male embryo and have become incorporated into the sex cords.

Granule cells   Derived from neuroblasts of the external granule layer of the developing cerebellum. Granule neurons migrate back toward the ventricular (ependymal) zone, where they produce a region called the internal granule layer.

Granulosa cells   Cortical epithelial cells of the fetal ovary, granulosa cells surround individual germ cells that will become the ova and will form, with thecal cells, the follicles that envelop the germ cells and secrete steroid hormones. The number of granulosa cells increase and form concentric layers around the oocyte as the oocyte matures prior to ovulation.

Gray crescent   A band of inner gray cytoplasm that appears following a rotation of the cortical cytoplasm with respect to the internal cytoplasm in the marginal region of the 1-cell amphibian embryo. Gastrulation starts in this location.

Gray matter   Regions of the brain and spinal cord rich in neuronal cell bodies. Compare with **White matter**.

Green fluorescent protein (GFP)   A protein that occurs naturally in certain jellyfish. It emits bright green fluorescence when exposed to ultraviolet light. The *GFP* gene is widely used as a transgenic label for cells in developmental and other research, since cells that express GFP are easily identified by a bright green glow.

GRNs   See **Gene regulatory networks**.

Ground tissue   In plants, all tissue that is not dermal or vascular; functions primarily in storage, support, and photosynthesis, and includes filler tissue called parenchyma and the more supportive collenchyma and sclerenchyma.

Growth and differentiation factors (GDFs)   See **Paracrine factors**.

Growth cone   The motile tip of a neuronal axon; leads nerve outgrowth.

Growth factor   A secreted protein that binds to a receptor and initiates signals to promote cell division and growth.

Growth plate closure   Causes the cessation of bone growth at the end of puberty. High levels of estrogen induce apoptosis in the hypertrophic chondrocytes and stimulate the invasion of bone-forming osteoblasts into the growth plate.

GTP exchange factor (GEF)   In the RTK (receptor tyrosine kinase) pathway, this factor exchanges a phosphate that transforms a bound GDP on a G protein into a bound GTP, activating the G protein. See also **RTK pathway**.

GTPase-activating protein (GAP)   The protein that enables the G protein Ras to quickly return to its inactive state. This is done by hydrolyzing Ras's bound GTP back to GDP. Ras becomes activated through the RTK pathway.

Gurken   A protein coded for by the *gurken* gene. The *gurken* message is synthesized in the nurse cells of the *Drosophila* ovary and transported into the oocyte, where it is translated into the protein along an anterior-posterior gradient. It signals follicle cells closest to the oocyte nucleus to posteriorize; part of the process that will set up the anterior-posterior axis of the egg and future embryo.

Gynandromorph   Greek *gynos*, "female"; *andros*, "male." An animal in which some body parts are male and others are female. Compare with **Hermaphrodite**.

Gyrencephalic   Having numerous folds in the cerebral cortex, as in humans and cetaceans. Compare with **Lissencephalic**.

H

Hair cells   Sensory receptors that transform the movement of fluid into electrical signals. In the inner ear, they are in the organ of Corti of the cochlea for hearing and the semicircular canals for balance. They are also in the lateral line organs of fish and amphibians for detecting movement and changes in pressure in the water.

Hair follicle stem cells   See **Follicular stem cells**.

Halteres   A pair of balancers on the third thoracic segment of two-winged flies, such as *Drosophila*.

Haptotaxis   Directional migration of cells on a substrate, up a gradient of adhesiveness.

Hatched blastula   Free-swimming sea urchin embryo, after the cells of the animal hemisphere synthesize and secrete a hatching enzyme that digests the fertilization envelope.

Head activation gradient   A morphogenetic gradient in *Hydra* that is highest at the hypostome and permits the head to form.

Head mesoderm   Mesoderm located anterior to the trunk mesoderm, consisting of the unsegmented paraxial mesoderm and prechordal mesoderm. This region provides the head mesenchyme that forms much of the connective tissues and musculature of the head.

Head process   In avian embryos, the anterior portion of the chordamesoderm that passes through Hensen's node and migrates anteriorly, ahead of the notochordal mesoderm, to come to lie underneath cells that will form forebrain and midbrain.

Heart fields (cardiogenic mesoderm)   In vertebrates, two regions of splanchnic mesoderm, one on each side of the body, that become specified for heart development. In amniotes, the cardiac cells of the heart field migrate through the primitive streak during gastrulation such that the medial-lateral arrangement of these early cells will become the anterior-posterior (rostral-caudal) axis of the developing heart tube.

Heart tube   Linear (anterior-to-posterior) structure formed at the midline of the heart fields; will become the atria, ventricles, and the base of the aorta and pulmonary arteries.

Heat shock proteins   Intercellular proteins induced by stress, they help other proteins to fold correctly and maintain their functions.

Hedgehog   A family of paracrine factors used by the embryo to induce particular cell types and to create boundaries between tissues. Hedgehog proteins must become complexed with a molecule of cholesterol in order to function. Vertebrates have at least three homologues of the *Drosophila hedgehog* gene: *sonic hedgehog* (*shh*), *desert hedgehog* (*dhh*), and *indian hedgehog* (*ihh*).

Hedgehog pathway   Proteins activated by the binding of a Hedgehog protein to the Patched receptor. When Hedgehog binds to Patched, the

Patched protein's shape is altered such that it no longer inhibits Smoothened. Smoothened acts to release the Ci protein from the microtubules and to prevent its being cleaved. The intact Ci protein can now enter the nucleus, where it acts as a transcriptional *activator* of the same genes it once repressed.

Hemangioblasts   Rapidly dividing cells, usually stem cells, that form blood vessels and blood cells.

Hematopoiesis   The generation of blood cells.

Hematopoietic inductive microenvironments (HIMs)   Cell regions that induce different sets of transcription factors in multipotent hematopoietic stem cells; these transcription factors specify the developmental path taken by descendants of those cells.

Hematopoietic stem cell (HSC)   A pluripotent stem cell type that generates a series of intermediate progenitor cells whose potency is restricted to certain blood cell lineages. These lineages are then capable of producing all the blood cells and lymphocytes of the body.

Hemimetabolous   A form of insect metamorphosis that includes pronymph, nymph, and imago (adult) stages.

Hemogenic endothelial cell   Primary endothelial cells of the dorsal aorta, in the ventral area, derived from the lateral plate. They give rise to the hematopoietic stem cells (HSCs) that migrate to the liver and bone marrow and become the adult hematopoietic stem cells.

Hensen's node   In avian embryos, a regional thickening of cells at the anterior end of the primitive streak. The center of Hensen's node contains a funnel-shaped depression (sometimes called the primitive pit) through which cells can enter the embryo to form the notochord and prechordal plate. Hensen's node is the functional equivalent of the dorsal lip of the amphibian blastopore (i.e., the organizer) and the fish embryonic shield. Also known as the **primitive knot**.

Heparan sulfate proteoglycans (HSPGs)   One of the most widespread proteoglycans in the extracellular matrix, it can bind many members of different paracrine families and present them in high concentrations to their receptors.

Hepatectomy   Surgical removal of part of the liver.

Hepatic diverticulum   The liver precursor, a bud of endoderm that extends out from the foregut into the surrounding mesenchyme.

Hepatocyte growth factor (HGF)   Paracrine factor secreted by the stellate cells of the liver that allows the hepatocytes to re-enter the cell cycle during compensatory regeneration. Also called scatter factor.

Hermaphrodite   An individual in which both ovarian and testicular tissues exists, having either ovotestes (gonads containing both ovarian and testicular tissue) or an ovary on one side and a testis on the other. Compare with **Gynandromorph**.

Heterochromatin   Chromatin that remains condensed throughout most of the cell cycle and replicates later than most of the other chromatin. Usually transcriptionally inactive. Compare with **Euchromatin**.

Heterochronic parabiosis   Surgically joining the circulatory systems of two animals of different ages; has been used to study the effects of aging on stem cells in mice.

Heterochrony   Greek, "different time." A shift in the relative timing of two developmental processes as a mechanism to generate phenotypic variation available for natural selection. One module changes its time of expression or growth rate relative to the other modules of the embryo.

Heterometry   Greek, "different measure." A change in the amount of a gene product as a mechanism to generate phenotypic variation available for natural selection.

Heterophilic binding   Binding between different molecules, as when a receptor in the membrane of one cell binds to a different type of receptor in the membrane of another cell.

Heterotopy   Greek, "different place." The spatial alteration of gene expression (e.g., transcription factors or paracrine factors) as a mechanism to generate phenotypic variation available for natural selection.

Heterotypy   Greek, "different kind." The alteration of the actual coding region of the gene, changing the functional properties of the protein being synthesized, as a mechanism to generate phenotypic variation available for natural selection.

High CpG-content promoters (HCPs)   Promoters with many CpG islands; these promoters often regulate developmental genes required for the

construction of the organism; their default state is "on." See also **CpG islands**.

Histoblast nests   Clusters of imaginal cells that will form the adult abdomen in holometabolous insects.

Histoblasts   Imaginal cells that will form the adult abdomen of a holometabolous insect; they are carried around in the larval stages, growing in number by cell division, until the pupal stage when they begin their differentiation. See **Histoblast nests**.

Histone acetylation   The addition of negatively charged acetyl groups to histones, which neutralizes the basic charge of lysine and loosens the histones, and thus activates transcription.

Histone acetyltransferases   Enzymes that place acetyl groups on histones (especially on lysines in histones H3 and H4). Acetyltransferases destabilize the nucleosomes so that they come apart easily, thus facilitating transcription.

Histone deacetylases   Enzymes that remove acetyl groups from histones, stabilizing the nucleosomes and preventing transcription.

Histone methylation   The addition of methyl groups to histones. Can either activate or further repress transcription, depending on the amino acid that is methylated and the presence of other methyl or acetyl groups in the vicinity.

Histone methyltransferases   Enzymes that add methyl groups to histones and either activate or repress transcription.

Histone   Positively charged proteins that are the major protein component of chromatin. See also **Nucleosome**.

Holobiont   Term for the composite organism of a host and its persistent symbionts.

Holoblastic cleavage   Greek *holos*, "complete." Refers to a cell division (cleavage) pattern in the embryo in which the entire egg is divided into smaller cells, as it is in echinoderms, amphibians and mammals.

Holometabolous   The type of insect metamorphosis found in flies, beetles, moths, and butterflies. There is no pronymph stage. The insect hatches as a larva (a caterpillar, grub, or maggot) and progresses through instar stages as it gets bigger between larval molts, a metamorphic molt to become a pupa, an imaginal molt and finally the emergence (eclose) of the adult (imago).

Homeobox   A 180-base pair DNA sequence that characterizes genes that code for homeodomain proteins, including Hox genes.

Homeorhesis   How the organism stabilizes its different cell lineages while it is still constructing itself.

Homeostasis   Maintenance of a stable physiological state by means of feedback responses.

Homeotic complex (Hom-C)   The region of *Drosophila* chromosome 3 containing both the Antennapedia complex and the bithorax complex.

Homeotic mutants   Result from mutations of homeotic selector genes, in which one structure is replaced by another (as where an antenna is replaced by a leg).

Homeotic selector genes   A class of *Drosophila* genes regulated by the protein products of the gap, pair-rule, and segment polarity genes whose transcription determines the developmental fate of each segment.

Homeotic transformations   The replacement of one structure by another during development due to a homeotic mutation. See **Homeotic mutants**.

Homing   The ability of a cell to migrate and find its tissue specific destination.

Homodimer   Two identical protein molecules bound together.

Homologous   Structures and/or their respective components whose similarity arises from their being derived from a common ancestral structure. For example, the wing of a bird and the forelimb of a human. Compare with **Analogous**.

Homologue   (1) One of a pair (or larger set) of chromosomes with the same overall genetic composition. For example, diploid organisms have two copies (homologues) of each chromosome, one inherited from each parent. (2) Evolutionary features in different species that are similar by reason of descent from a common ancestor.

Homophilic binding   Binding between like molecules, as when a receptor in the membrane of one cell binds to the same type of receptor in the cell membrane of another cell.

Horizontal neurons   Neurons in the neural retina that transmit electrical impulses in the plane of the retina; help to integrate sensory signals coming from many photoreceptor cells.

Horizontal transmission   When a host that is born free of symbionts but subsequently becomes infected, either by its environment or by other members of the species. Can also refer to the transfer of genes from one organism to another without involving reproduction, as can occur in bacteria. Compare with **Vertical transmission**.

Host   The larger organism in a symbiotic relationship in which one of the organisms involved is much larger than the other, and the smaller organism may live on the surface or inside the body of the larger. Also refers to the organism receiving a graft from a donor in a tissue transplant.

Hox genes   Large family of related genes that dictate (at least in part) regional identity in the embryo, particularly along the anterior-posterior axis. Hox genes encode transcription factors that regulate the expression of other genes. All known mammalian genomes contain four copies of the Hox complex per haploid set, located on four different chromosomes (*Hoxa* through *Hoxd* in the mouse, *HOXA* through *HOXD* in humans). The mammalian Hox/HOX genes are numbered from 1 to 13, starting from that end of each complex that is expressed most anteriorly.

Hu proteins   RNA binding proteins that stabilize mRNAs involved in neuronal development, preventing them from being quickly degraded. Examples: HuA, HuB, HuC, and HuD.

Hub   A regulatory microenvironment in *Drosophila* testes where the stem cells for sperm reside.

Hyaline layer   A coating around the sea urchin egg formed by the cortical granule protein hyalin. The hyaline layer provides support for the blastomeres during cleavage.

Hydatidiform mole   A human tumor which resembles placental tissue, arise when a haploid sperm fertilizes an egg in which the female pronucleus is absent and the entire genome is derived from the sperm, which precludes normal development and is cited as evidence for genomic imprinting.

20-Hydroxyecdysone (20E)   An insect hormone, the active form of ecdysone, that initiates and coordinates each molt, regulates the changes in gene expression that occur during metamorphosis, and signals imaginal disc differentiation.

Hyperactivation   The increased and more forceful motility displayed by capacitated sperm of some mammalian species. Hyperactivation has been proposed to help detach capacitated sperm from the oviductal epithelium, allow sperm to travel more effectively through viscous oviductal fluids, and facilitate penetration of the extracellular matrix of the cumulus cells.

Hypermorphosis   The extension of development beyond its ancestral state; it is an evolutionary mechanism whereby the total developmental time is extended without altering the rate of development. Example: In humans, the extension of the fetal rate of brain growth beyond birth.

Hypertrophic chondrocytes   Formed during the fourth phase of endochondral ossification, when the chondrocytes, under the influence of the transcription factor Runx2, stop dividing and increase their volume dramatically.

Hypoblast islands (primary hypoblast)   Derived from area pellucida cells of the avian blastoderm that migrate individually into the subgerminal cavity to form individual disconnected clusters containing 5–20 cells each. Does not contribute to the embryo proper.

Hypoblast   The inner layer of the thickened margin of the epibolizing blastoderm in the gastrulating fish embryo or the lower layer of the bilaminar embryonic blastoderm in birds and mammals. The hypoblast in fish (but not in birds and mammals) contains the precursors of both the endoderm and mesoderm. In birds and mammals, it contains precursors to the extraembryonic endoderm of the yolk sac.

Hypomorphic mutations   Mutations that reduce gene function, as opposed to a "null" mutation that results in the loss of a protein's function.

Hypostome   A conical region of the "head" of a hydra that contains the mouth.

I

Imaginal cells   Cells carried around in the larva of the holometabolous insect that will form the structures of the adult. During the larval stages, these cells increase in number, but do not differentiate until the pupal stage; include imaginal discs, histoblasts, and clusters of imaginal cells within each larval organ.

Imaginal discs   Clusters of relatively undifferentiated cells set aside to produce adult structures. Imaginal discs will form the cuticular structures of the adult, including the wings, legs, antennae, halters, eyes, head, thorax, and genitalia in holometabolous insects.

Imaginal molt   Final molt in a holometabolous insect when the adult (imago) cuticle forms beneath the pupal cuticle, and the adult later emerges from the pupal case at adult eclosion.

Imaginal rudiment   Develops from the left coelomic sac of the pluteus larva and will form many of the structures of the adult sea urchin.

Imago   A winged and sexually mature adult insect.

In situ   Latin, "on site." In its natural position or environment.

In situ probe   Complementary DNA or RNA used to localize a specific DNA or RNA sequence in a tissue.

Indel   An insertion or deletion of DNA bases.

Indeterminate growth   Growth that is not halted, as opposed to determinate growth, which stops once a structure reaches a genetically predetermined size. Unlike animals, plants can have indeterminate growth.

Indifferent gonad   See **Bipotential gonad**.

Indirect developers   Animals for which embryonic development includes a larval stage with characteristics very different from those of the adult organism, which emerges only after a period of metamorphosis.

Indole-3-acetic acid (IAA)   The most common type of auxin. See **Auxin**.

Induced pluripotent stem (iPS) cells   Adult cells that have been converted to cells with the pluripotency of embryonic stem cells. Usually accomplished by the activation of certain transcription factors.

Inducer   Tissue that produces a signal (or signals) that induces a cellular behavior in some other tissue.

Induction   The process by which one cell population influences the development of neighboring cells via interactions at close range.

Inductive signals   Signals produced by inducers; often these signals are secreted proteins called paracrine factors. See **Inducer** and **Paracrine factor**.

Inflorescence meristem   The meristem that develops from the shoot apical meristem when the plant begins flowering; the inflorescence meristem produces the floral meristems, which form the carpels, stamens, petals and sepals of each flower. **See Meristem.**

Ingression   Migration of individual cells from the surface layer into the interior of the embryo. The cells become mesenchymal (i.e., they separate from one another) and migrate independently.

Initial cells   Totipotent stem cells generated by the early plant embryo; there are two clusters of these: the shoot apical meristem and the root apical meristem.

Inner cell mass (ICM)   A small group of internal cells within a mammalian blastocyst that will eventually develop into the embryo proper and its associated yolk sac, allantois, and amnion.

Inositol 1,4,5-trisphosphate (IP$_3$)   A second messenger generated by the phospholipase C enzyme that releases intracellular Ca^{2+} stores. Important in the initiation of both cortical granule release and sea urchin development.

Inside-out gradient of development   The developmental process in the neocortex and the rest of the brain in which the neurons with the earliest birthdays form the layer closest to the ventricle and subsequent neurons travel greater distances to form the more superficial layers.

Instar   The stages between larval molts in holometabolous insects. During these stages, the larva (caterpillar, grub, or maggot) feeds and grows larger between each molt, until the end of the final instar stage, when the larva is transformed into a pupa.

Instructive interaction   A mode of inductive interaction in which a signal from the inducing cell is necessary for initiating new gene expression in the responding cell.

Insulator   DNA sequence that limits the range within which an enhancer can activate a given gene's expression (thereby "insulating" a promoter from being activated by another gene's enhancers).

Insulin signaling pathway   Pathway involving a receptor for insulin and insulin-like proteins; may be an important component of genetically limited life spans, wherein downregulation of the pathway can correspond to an increased lifespan.

Insulin-like growth factors (IGFs)   Growth factors that initiate an FGF-like signal transduction cascade that interferes with the signal transduction pathways of both BMPs and Wnts. IGFs are required for the formation of the anterior neural tube, including the brain and sensory placodes of amphibians.

Integration   A principle of the theoretical systems approach: How the parts are put together, and how they interact to form the whole.

Integrins   A family of receptor proteins, named for the fact that they *integrate* extracellular and intracellular scaffolds, allowing them to work together. On the extracellular side, integrins bind to sequences found in several adhesive proteins in extracellular matrix, including fibronectin, vitronectin (in the basal lamina of the eye), and laminin. On the cytoplasmic side, integrins bind to talin and α-actinin, two proteins that connect to actin microfilaments. This dual binding enables the cell to move by using myosin to contract the actin microfilaments against the fixed extracellular matrix.

Integuments   An outer protective layer such as the skin. In plants, the seed coat comes from the integuments of the ovule.

Interdigital necrotic zone   A zone of programmed cell death in the developing tetrapod limb that separates the digits from one another; when the cells in this zone do not die, webbing between the digits remains, as in the duck foot.

Interior necrotic zone   A zone of programmed cell death in the developing tetrapod limb that separates the radius from the ulna.

Interkinesis   The brief period between the end of meiosis I and the beginning of meiosis II.

Interkinetic nuclear migration   The movement of nuclei within certain cells as they go through the cell cycle; seen in the germinal neuroepithelium in which nuclei translocate from the basal end to the apical end of the cells near the ventricular surface, where they undergo mitosis, after which they slowly migrate basally again.

Intermediate mesoderm   Mesoderm immediately lateral to the paraxial mesoderm. It forms the outer (cortical) portion of the adrenal gland and the urogenital system, consisting of the kidneys, gonads, and their associated ducts.

Intermediate progenitor cells (IP cells)   Neuron precursor cells of the subventricular zone; derived from radial glial cells.

Intermediate spermatogonia The first committed stem cell type of the mammalian testis, they are committed to becoming spermatozoa.

Intermediate zone See **Mantle zone**.

Internal granular layer A layer in the cerebellum that is formed by the migration of granule cells from the external granular layer back toward the ventricular zone.

Internal YSL (iYSL) A region of the yolk syncytial layer (YSL) in fish embryos. The iYSL forms from yolk syncytial nuclei that move under the blastoderm as it expands to surround the yolk cell. See **Yolk syncytial layer**.

Intersex A condition in which male and female traits are observed in the same individual.

Interstitial stem cell A type of stem cell found within the ectodermal layer of *Hydra* that generates neurons, secretory cells, nematocytes, and gametes.

Intine The inner wall of a pollen grain, composed mostly of cellulose.

Intraembryonic vasculogenesis The formation of blood vessels during embryonic organogenesis. Compare with **Extraembryonic vasculogenesis**.

Intramembranous bone Bone formed by intramembranous ossification.

Intramembranous ossification Formation of bone directly in mesenchyme without a cartilaginous precursor. There are three main types of intramembranous bone: sesamoid bone and periosteal bone, which come from mesoderm, and dermal bone which originate from cranial neural crest-derived mesenchymal cells.

Introns Non-protein-coding regions of DNA within a gene. Compare with **Exon**.

Invagination The infolding of a region of cells, much like the indenting of a soft rubber ball when it is poked.

Invariant cell lineage When each cell of an embryo gives rise to the same number and type of cells in every embryo of that species, as seen in the embryos of the roundworm *Caenorhabditis elegans*.

Involuting marginal zone (IMZ) Cells that involute during *Xenopus* gastrulation, includes precursors of the pharyngeal endoderm, head mesoderm, notochord, somites, and heart, kidney, and ventral mesoderm.

Involution Inturning or inward movement of an expanding outer layer so that it spreads over the internal surface of the remaining external cells.

Ionophore A compound that allows the diffusion of ions such as Ca^{2+} across lipid membranes, permitting them to traverse otherwise impermeable barriers.

Ipsilateral An anatomical term meaning on the same side of the body.

Iris A pigmented ring of muscular tissue in the eye that controls the size of the pupil and determines eye color.

Isolecithal Greek, "equal yolk." Describes eggs with sparse, equally distributed yolk particles, as in sea urchins, mammals, and snails.

Isthmus The narrow segment of the mammalian oviduct adjacent to the uterus.

Izumo A protein found on the equatorial region of mature mammalian sperm that binds to Juno, on the cell membrane of the oocyte. These proteins help stabilize sperm-egg binding.

J

Jagged protein Ligand for Notch, participates in juxtacrine interaction and activation of the Notch pathway.

JAK Janus kinase proteins. Linked to FGF receptors in the JAK-STAT cascade.

JAK-STAT cascade A pathway activated by paracrine factors binding to receptors that span the cell membrane and are linked on the cytoplasmic side to JAK (**Ja**nus **k**inase) proteins. The binding of ligand to the receptor phosphorylates the STAT (**s**ignal **t**ransducers and **a**ctivators of **t**ranscription) family of transcription factors.

Junctional neurulation Formation of the neural tube in the transition zone between the primary neural tube (which lies anterior to the hindlimbs) and the secondary neural tube (which extends posteriorly from the sacral region in mammals or is just in the tail region in fish and amphibians).

Juno A protein found anchored to the mammalian oocyte cell membrane that is critical in binding to Izumo, a protein on mammalian sperm.

Juvenile hormone (JH) A lipid hormone in insects that prevents the ecdysone-induced changes in gene expression that are necessary for metamorphosis. Thus, its presence during a molt ensures that the result of that molt is another larval instar, not a pupa or an adult.

Juxtacrine interactions When cell membrane proteins on one cell surface interact with receptor proteins on adjacent (juxtaposed) cell surfaces.

Juxtacrine signaling Signaling between cells that are juxtaposed, i.e., in direct contact with one another.

K

Kairomones Chemicals that are released by a predator and can induce defenses in its prey.

Karyokinesis The mitotic division of the cell's nucleus. The mechanical agent of karyokinesis is the mitotic spindle.

Keratinocytes Differentiated epidermal cells that are bound tightly together and produce a water-impermeable seal of lipid and protein.

Koller's sickle See **Primitive streak**.

Kupffer's vesicle Transient fluid-filled organ housing the cilia that control left-right asymmetry in zebrafish.

L

Labioscrotal folds Folds surrounding the cloacal membrane in the indifferent stage of differentiation of mammalian external genitalia. They will form the labia majora in the female and the scrotum in the male. Also called urethral folds or genital swellings.

lacZ **gene** The *E. coli* gene for β-galactosidase; commonly used as a reporter gene.

Lamellipodia Broad locomotory pseudopods containing actin networks; found on migrating cells, also found on growth cones of neurons.

Laminae Layers. In the brain, neurons are organized into laminae and clusters (nuclei).

Laminin A large glycoprotein and major component of the basal lamina, plays a role in assembling the extracellular matrix, promoting cell adhesion and growth, changing cell shape, and permitting cell migration.

Lampbrush chromosomes Chromosomes in an amphibian primary oocyte during the diplotene stage of the first meiotic prophase that stretch out large loops of DNA, representing sites of upregulated RNA synthesis.

Lanugo The first hairs of human embryos, usually shed before birth.

Large micromeres A tier of cells produced by the fifth cleavage in the sea urchin embryo when the micromeres divide. Become the primary mesenchyme cells, which form the skeletal spicules of the larva.

Larva The sexually immature stage of an organism, often of significantly different appearance than the adult and frequently the stage that lives the longest and is used for feeding or dispersal.

Larval settlement Ability of marine larvae to suspend development until they experience a particular environmental cue for settlement.

Laryngotracheal groove An outpouching of endodermal epithelium in the center of the pharyngeal floor, between the fourth pair of pharyngeal pouches, that extends ventrally. The laryngotracheal groove then bifurcates into the branches that form the paired bronchi and lungs.

Last eukaryotic common ancestor (LECA) The hypothetical ancestor to all plants, animals and fungi, thought to have been a unicellular protist that had flagella and mitochondria.

Last universal common ancestor (LUCA) The ancient common ancestor to all life on Earth.

Lateral inhibition The inhibition of a cell by the activity of a neighboring cell.

Lateral line placodes Paired ectodermal placodes that form as posterior cranial placodes in amphibians and fish. They generate neuromasts, which contain mechanosensing hair cells and their innervating neurons; function to detect the flow of water, weak bioelectric fields, and changes in pressure. See **Hair cells**.

Lateral plate mesoderm Mesodermal sheet lateral to the intermediate mesoderm. Gives rise to appendicular bones, connective tissues of the limb buds, circulatory system (heart, blood vessels, and blood cells), muscles and connective tissues of the digestive and respiratory tracts, and lining of coelom and its derivatives. It also helps form a series of extraembryonic membranes that are important for transporting nutrients to the embryo.

Lateral somitic frontier The boundary between the primaxial and abaxial muscles and between the somite-derived and lateral plate-derived dermis.

Leader sequence See **5′ Untranslated region**.

Leghemoglobin An oxygen-carrying protein, similar to hemoglobin, found in the nitrogen-fixing root nodules of legumes. It is a plant protein whose genes are induced by symbiotic bacteria. This protein protects the nitrogenase enzyme from being inactivated by oxygen.

Lens placode Paired epidermal thickenings induced by the underlying optic cups to invaginate to form the lens vesicles, which differentiate into the adult transparent eye lenses that allow light to impinge on the retinas.

Lens vesicle The vesicle that forms from the lens placode. It differentiates into the lens. It also induces the overlying ectoderm to become the cornea, and induces the inner side of the optic cup to differentiate into the neural retina.

Leptotene Greek, "thin thread." In the first meiotic division, the first stage of prophase I, when the chromatin is stretched out thinly such that one cannot identify individual chromosomes. DNA replication has already occurred, and each chromosome consists of two parallel chromatids.

Level-specific properties and emergence A principle of the theoretical systems approach: The properties of a system at any given level of organization cannot be totally explained by those of levels "below" it.

Leydig cells Testis cells derived from the interstitial mesenchyme cells surrounding the testis cords that make the testosterone required for secondary sex determination and, in the adult, required to support spermatogenesis.

Life expectancy The length of time an average individual of a given species can expect to live; it is characteristic of populations, not of species.

Ligand A molecule secreted by a cell that elicits a response in another cell by binding to a receptor on that cell.

Lim genes Genes coding for transcription factors that are structurally related to proteins encoded by Hox genes.

Limb bud A circular bulge that will form the future limb. The limb bud is formed by the proliferation of mesenchyme cells from the somatic layer of the limb field lateral plate mesoderm (the limb *skeletal* precursor cells) and from the somites (the limb *muscle* precursor cells).

Limb field An area of the embryo containing all of the cells capable of forming a limb.

Lineage tracing Tracking the development of cell maturation over time. Embryonic cells can be labeled and tracked to see what they become in the larva or adult organism; this allows the development of a fate map. See **Fate map**.

Lineage-restricted stem cells Stem cells derived from multipotent stem cells, and which can now generate only a particular cell type or set of cell types.

Lissencephalic Having a cerebral cortex that lacks folds, as in mice. Compare with **Gyrencephalic**.

Long noncoding RNAs (lncRNAs) Transcriptional regulators that inactivate genes on one of the two chromosomes of a diploid organism. For example, Xist is a lncRNA involved in the inactivation of genes on the second X chromosome of females. Some lncRNAs appear to be specific for either the maternal or paternal copy of a gene.

Lophotrochozoans One of two major protostome groups, many of which are characterized spiral cleavage and by the larval form known as the trochophore. A diverse group that includes the annelids (segmented worms such as earthworms), molluscs (e.g., snails), and flatworms (e.g., *Planaria*). See also **Ecdysozoans**.

Loss-of-function evidence The absence of the postulated cause is associated with the absence of the postulated effect. The "lose it" of "find it, lose it, move it." See also **Correlative evidence**; **Gain-of-function evidence**.

Low CpG-content promoters (LCPs) These promoters are usually found in the genes whose products characterize mature, fully differentiated cells. The CpG sites are usually methylated and their default state is "off," although they can be activated by specific transcription factors. See also **CpG islands**.

Lumen The hollow space within any tubular or globular structure or organ.

Luteinizing hormone (LH) A hormone secreted by the mammalian pituitary that stimulates the production of steroid hormones, such as estrogen from the ovarian follicle cells and testosterone from the testicular Leydig cells. A surge in LH levels causes the primary oocyte to complete meiosis I and prepares the follicle for ovulation.

Lymphatic vasculature The vessels of the circulatory system that transport lymph (as opposed to the blood vessels of the circulatory system).

M

Macromeres Larger cells generated by asymmetrical cleavage, e.g., the four large cells generated by the fourth cleavage when the vegetal tier of the sea urchin embryo undergoes an unequal equatorial cleavage.

MADS-box transcription factors A family of proteins that all share a conserved motif in their DNA-binding domains; they are found in diverse groups of eukaryotes.

Male pronucleus The haploid nucleus of the sperm.

Malformation Abnormalities caused by genetic events such as gene mutations, chromosomal aneuploidies, and translocations.

Mammalian gut-associated lymphoid tissue (GALT) Lymphoid tissue that mediates mucosal immunity and oral immune tolerance, allowing mammals to eat food without creating an immune response to it. Intestinal microbes are critical for the maturation of GALT.

Mantle zone (intermediate zone) Second layer of the developing spinal cord and medulla that forms around the original neural tube. Because it contains neuronal cell bodies and has a grayish appearance grossly, it will form the gray matter.

Marginal zone (1) The third and outer zone of the developing spinal cord and medulla composed of a cell-poor region composed of axons extending from neurons residing in the mantle zone. Will form the white matter as glial cells cover the axons with myelin sheaths, which have a whitish appearance. (2) In amphibian gastrula, where gastrulation begins, the region surrounding the equator of the blastula, where the animal and vegetal hemispheres meet. (3) In bird and reptile gastrulae (= marginal belt), a thin layer of cells between the area pellucida and the area opaca, important in determining cell fate during early development.

Maskin Protein in amphibian oocytes that creates a repressive loop structure in messenger RNA, preventing its translation. It creates the loop by binding to two other proteins, cytoplasmic polyadenylation-element-binding protein (CPEB) and eIF4E

factor, which are bound to opposite ends of the mRNA.

Master regulator Transcription factors that can control cell differentiation by (1) being expressed when the specification of a cell type begins, (2) regulating the expression of genes specific to that cell type, and (3) being able to redirect a cell's fate to this cell type.

Maternal contributions The stored mRNAs and proteins within the cytoplasm of the egg, produced from the maternal genome during the primary oocyte stage. See also **Maternal message**.

Maternal effect An effect occurring during embryonic development that is controlled by gene products that were stored in the egg while it was in the ovary; these gene products were made using the maternal genome through gene transcription within the egg itself, prior to its undergoing meiosis, or within nurse cells that transport gene products into the egg. See **Maternal message**.

Maternal effect genes Genes belonging to the maternal genome that are used to make messenger RNAs or proteins that are localized to different regions of an egg and affect embryonic development, as seen in the *Drosophila* egg. See **Maternal effect**.

Maternal message Messenger RNA that is either made in the egg and stored in the egg's cytoplasm while the egg is a primary oocyte or is made in nurse cells within the ovary and transported to the egg's cytoplasm. At the primary oocyte stage, the egg within the ovary is still diploid, as are any surrounding nurse cells. The mRNA, therefore, is being made from the maternal genome.

Maternal-to-zygote transition The embryonic stage when maternally provided mRNAs are degraded and control of development is handed over to the zygote's own genome; often occurs in the mid-blastula stage. Seen in many different animal groups.

Maximum lifespan Maximum number of years an individual of a given species has been known to survive and is characteristic of that species.

Mechanical anisotropy Having a difference in a mechanical property, such as stretchability, along different axes.

Medial hinge point (MHP) In birds and mammals, formed by the cells at the midline of the neural plate. MHP cells become anchored to the notochord

beneath them and form a hinge, which forms a furrow at the dorsal midline and helps bend the neural plate as it forms a neural tube.

Mediator A large, multimeric complex of nearly 30 protein subunits that in many genes is the link that connects RNA polymerase II (bound to the promoter) to an enhancer sequence, thus forming a pre-initiation complex at the promoter.

Medullary cord Forms by condensation of mesenchyme cells and then mesenchymal-to-epithelial transition in the caudal region of the avian embryo during the process of secondary neurulation. It will then cavitate to form the caudal section of the neural tube.

Megasporangium The structure in which megaspores form.

Megaspores Haploid cells derived from the diploid megaspore mother cell. At least of one these four cells produces the female gametophyte that produces the egg cell.

Meiosis A unique division process that in animals occurs only in germ cells, to reduce the number of chromosomes to a haploid complement. All other cells divide by mitosis. Meiosis differs from mitosis in that (1) meiotic cells undergo two cell divisions without an intervening period of DNA replication, and (2) homologous chromosomes (each consisting of two sister chromatids joined at a kinetochore) pair together and recombine genetic material.

Melanoblasts Pigment progenitor cells.

Melanocyte stem cells Adult stem cells derived from trunk neural crest cells that form melanoblasts and come to reside in the bulge niche of the hair or feather follicle and which give rise to the pigment of the skin, hair, and feathers.

Melanocytes Cells containing the pigment melanin. Derived from neural crest cells that undergo extensive migration to all regions of the epidermis.

Meltrins Set of metalloproteinases involved in cell fusion events, such as the fusion of myoblasts to form a myofiber and of macrophages to form osteoclasts. See also **Metalloproteinases**.

Meridional When pertaining to cell division, a cleavage that is transverse through the equatorial plane.

Meristem Tissue in plants containing undifferentiated and actively dividing cells. This is where production of new plant tissue occurs. Different types of meristems give rise to different structures of the plant. The two main meristems are the shoot and root apical meristems.

Meroblastic cleavage Greek *meros*, "part." Refers to the cell division (cleavage) pattern in zygotes containing large amounts of yolk, wherein only a portion of the cytoplasm is cleaved. The cleavage furrow does not penetrate the yolky portion of the cytoplasm because the yolk platelets impede membrane formation there. Only part of the egg is destined to become the embryo, while the other portion—the yolk—serves as nutrition for the embryo, as in insects, fish, reptiles, and birds.

Meroistic oogenesis Type of oogenesis found in certain insects (including *Drosophila* and moths), in which cytoplasmic connections remain between the cells produced by the oogonium.

Mesencephalon The midbrain, the middle vesicle of the developing vertebrate brain; major derivatives include optic tectum and tegmentum. Its lumen becomes the cerebral aqueduct.

Mesenchymal cells Unconnected or loosely connected cells that can act as independent migratory units. These are in contrast to epithelial cells.

Mesenchymal stem cells (MSCs) Also called bone marrow-derived stem cells, or BMDCs. Multipotent stem cells that originate in the bone marrow, MSCs are able to give rise to numerous bone, cartilage, muscle, and fat lineages.

Mesenchymal-to-epithelial transition (MET) Transformation of mesenchymal cells into epithelial cells. Occurs, for example, during somite formation, when somites form from presomitic mesoderm. See also **Epithelial-mesenchymal transition**.

Mesenchyme cap cells A population of multipotent stem cells, derived from metanephric kidney mesenchyme, that cover the tips of the ureteric bud branches and can form all the cell types of the nephron.

Mesenchyme Loosely organized embryonic connective tissue consisting of scattered fibroblast-like and sometimes migratory mesenchymal cells separated by large amounts of extracellular matrix.

Mesendodermal Pertaining to mesendoderm, which develops into mesoderm and endoderm. Synonym for endomesodermal.

Mesentoblast In snail embryos, the 4d blastomere whose progeny give rise to most of the mesodermal (heart, kidney and muscles) and endodermal (gut tube) structures.

Mesoderm Greek *mesos*, "between." The middle of the three embryonic germ layers, lying between the ectoderm and the endoderm. The mesoderm gives rise to muscles and skeleton, connective tissue, the urogenital system (kidneys, gonads, and ducts), blood and blood vessels, and most of the heart.

Mesodermal mantle The cells that involute through the ventral and lateral blastopore lips during amphibian gastrulation and will form the heart, kidneys, bones, and parts of several other organs.

Mesomeres The eight cells generated in the sea urchin embryo by the fourth cleavage when the four cells of the animal tier divide meridionally into eight blastomeres, each with the same volume.

Mesonephric duct See **Wolffian duct**.

Mesonephros The second kidney of the amniote embryo, induced in the adjacent mesenchyme by the middle portion of the nephric (Wolffian) duct. It functions briefly in urine filtration in some mammalian species and mesonephric tubules form the tubes that transport the sperm from the testes to the urethra (the epididymis and vas deferens). Forms the adult kidney of anamniotes (fish and amphibians).

Mesp1 A gene for a transcription factor that is critical in the regulatory network that specifies heart cells. These networks are conserved across vertebrates. See also **Nkx2-5**.

Messenger RNA (mRNA) RNA that codes for a protein and leaves the nucleus after being processed from nuclear RNA in a manner that excises non-coding domains and protects the ends of the strand.

MET See **Mesenchymal-to-epithelial transition**.

Metalloproteinases Matrix metalloproteinases (MMP). Enzymes that digest extracellular matrices and are important in many types of tissue remodeling in disease and development, including metastasis, branching morphogenesis of epithelial organs,

placental detachment at birth, and arthritis.

Metamorphic climax When the major metamorphic changes, such as tail and gill resorption, and intestinal remodeling, occur in the amphibian. The concentration of T_4 rises dramatically and TRβ levels peak.

Metamorphic molt Pupal molt; in holometabolous insects, the molt at the end of the final instar stage, when the larva becomes a pupa.

Metamorphosis Changing from one form to another, such as the transformation of an insect larva to a sexually mature adult or a tadpole to a frog.

Metanephric mesenchyme An area of mesenchyme, derived from posterior regions of the intermediate mesoderm, involved in mesenchymal-epithelial interactions that generate the metanephric kidney and will form the secretory nephrons. Also called metanephrogenic mesenchyme.

Metanephros/metanephric kidney The third kidney of the amniote embryo and the permanent kidney of amniotes.

Metaphase plate A structure present during mitosis or meiosis in which the chromosomes are attached via their kinetochores to the microtubule spindle and are lined up between the two poles of the cell. If the metaphase plate forms midway between the two poles the division will be symmetrical; if it is closer to one pole, it will be asymmetrical, producing one larger cell and one smaller.

Metastasis The invasion of cancerous cells into other tissues.

Metazoa Animals.

Metencephalon The anterior subdivision of the rhombencephalon; gives rise to the cerebellum (coordinates movements, posture, and balance) and pons (fiber tracts for communication between brain regions).

Methylation See **Histone methylation**.

5-Methylcytosine A "fifth" base in DNA, made enzymatically after DNA is replicated by converting cytosine to 5-methycytosine—only cytosines followed by a guanosine can be converted. In mammals, about 5% of cytosines in DNA are converted to 5-methylcytosine.

Microfilaments Long cables of polymerized actin and a major component of the cytoskeleton. In combination with myosin, forms contractile forces necessary for cytokinesis; formed during fertilization in the egg's cortex to extend microvilli; attached indirectly by other molecules to transmembrane adhesion molecules, such as cadherins and integrins.

Microglia   Small glial cells of the central nervous system that carry out an immune function by engulfing dying and dysfunctional neurons and glia.

Micromeres   Small cells created by asymmetrical cleavage, e.g., four small cells generated by the fourth cleavage at the vegetal pole when the vegetal tier of the sea urchin embryo undergoes an unequal equatorial cleavage.

Micropyle   The only place where *Drosophila* sperm can enter the egg, at the future dorsal anterior region of the embryo, a tunnel in the chorion (eggshell) that allows sperm to pass through it one at a time.

MicroRNA (miRNA)   Small (about 22 nucleotide) RNAs complementary to a portion of a particular mRNA that regulates translation of a specific message. MicroRNAs usually bind to the 3′ UTR of mRNAs and inhibit their translation.

Microspikes   Essential for neuronal pathfinding, microfilament-containing pointed filopodia of the growth cone that elongate and contract to allow axonal migration. Microspikes also sample the microenvironment and send signals back to the soma.

Microsporangia   The places within the anther where the microspores are produced.

Microspores   In plants, the haploid cell that develops into the pollen tube, the male gametophyte.

Microtubule organizing centers (MTOCs)   Eukaryotic cell structure that polymerizes tubulin into microtubules. These are critical in making cilia, flagella, and the spindles of the meiotic and mitotic spindle apparatus.

Microvilli   Small microfilament-containing projections that extend from the surface of cells; e.g., on the egg surface during fertilization where they may aid sperm entry into the cell.

Mid-blastula transition (MBT)   The transition from the early rapid biphasic (only M and S phases) mitoses of the embryo to a stage characterized by (1) mitoses that include the "gap" stages (G1 and G2) of the cell cycle, (2) loss of synchronicity of cell division, and (3) transcription of new (zygotic) mRNAs needed for gastrulation and cell specification.

Midpiece   Section of sperm flagellum near the head that contains rings of mitochondria that provide the ATP needed to fuel the dynein ATPases and support sperm motility.

miRNA   See **MicroRNA**.

MITF   **Mi**crophthalmia-associated **t**ranscription **f**actor. A transcription factor necessary for melanoblast specification and pigment production. Its name comes from the fact that a mutation in the gene for this transcription factor causes small eyes (microphthalmia) in mice.

Mitosis-promoting factor (MPF)   Consists of cyclin B and a cyclin-dependent kinase (CDK), required to initiate entry into the mitotic (M) phase of the cell cycle in both meiosis and mitosis.

Model systems   Species that are easily studied in the laboratory and have special properties that allow their mechanisms of development to be readily observed (e.g., sea urchins, *Drosophila*, *C. elegans*, zebrafish, and mouse).

Modularity   A principle of the theoretical systems approach. The organism develops as a system of discrete and interacting modules.

Module   A discrete unit of growth, characterized by more internal than external integration.

Molecular parsimony   The principle that development in all lineages uses the same types of molecules (the "small toolkit"). The "toolkit" includes transcription factors, paracrine factors, adhesion molecules, and signal transduction cascades that are remarkably similar from one phylum to another.

Monospermy   Only one sperm enters the egg, and a haploid sperm nucleus and a haploid egg nucleus combine to form the diploid nucleus of the fertilized egg (zygote), thus restoring the chromosome number appropriate for the species.

Monozygotic twins   Greek, "one-egg." Genetically "identical" twins; form when the cells of a single early-cleavage embryo become dissociated from one another, either by the separation of early blastomeres or by the separation of the inner cell mass into two regions within the same blastocyst. Compare with **Dizygotic twins**.

Morph   One of several different potential phenotypes that result from environmental conditions. Also called an ecomorph.

Morphallactic regeneration   See **Morphallaxis**.

Morphallaxis   Type of regeneration that occurs through the repatterning of existing tissues with little new growth (e.g., *Hydra*).

Morphogenesis   The organization of the cells of the body into functional structures via coordinated cell growth, cell migration, and cell death.

Morphogenetic determinants   Transcription factors or their mRNAs that will influence the cell's development.

Morphogens   Greek, "form-givers." Diffusible biochemical molecules that can determine the fate of a cell by their concentrations, in that cells exposed to high levels of a morphogen will activate different genes than those exposed to lower levels.

Morpholino   An antisense oligonucleotide against an mRNA; used to experimentally inhibit protein expression.

Morula   Latin, "mulberry." Vertebrate embryo of 16–64 cells; precedes the blastula or blastocyst stage. Mammalian morula occurs at the 16-cell stage, consists of a small group of internal cells (that will form the inner cell mass) surrounded by a larger group of external cells (that will form the trophoblast).

Mosaic embryos   Embryos in which most of the cells are determined by autonomous specification, with each cell receiving its instructions independently and without cell-cell interaction.

Mosaic pleiotropy   A gene is independently expressed in several tissues. Each tissue needs the gene product and develops abnormally in its absence.

mRNA cytoplasmic localization   The spatial regulation of mRNA translation, mediated by (1) diffusion and local anchoring, (2) localized protection, and (3) active transport along the cytoskeleton.

Müller glial cells   Cells of the neural retina that support and maintain the neurons therein.

Müllerian duct (paramesonephric duct)   Duct running lateral to the mesonephric (Wolffian) duct in both male and female mammalian embryos. These ducts regress in the male fetus, but form the oviducts, uterus, cervix, and upper part of the vagina in the female fetus. Compare with **Wolffian duct**.

Müllerian-inhibiting factor (MIF)   See **Anti-Müllerian hormone**.

Multicellular eukaryotic organism   A eukaryotic organism with multiple cells that remain together as a functional whole; subsequent generations form the same coherent individuals composed of multiple cells. (Includes plants, fungi, and animals.)

Multicellularity   Consisting of multiple cells.

Multipotent   Refers to the ability of a stem cell to generate different cell types with restricted specificity for the tissue in which they reside. Example: Most adult stem cells in organs of animals are multipotent.

Multipotent cardiac progenitor cells   Progenitor cells of the heart field that form cardiomyocytes, endocardium, epicardium, and the Purkinje fibers of the heart.

Multipotent stem cells   Adult stem cells whose commitment is limited to a relatively small subset of all the possible cells of the body.

Mutualism   A form of symbiosis in which the relationship benefits both partners.

Mycorrhizae   Fungi that form symbiotic relationships with plants to extend their roots. While the plant supplies the fungus with sugars, the fungus absorbs water and mineral nutrients from the soil.

Myelencephalon   The posterior subdivision of the rhombencephalon; becomes the medulla oblongata.

Myelin sheath   Modified oligodendrocyte (in CNS) or Schwann cell (in peripheral NS) plasma membrane that surrounds nerve cell axons, providing insulation that confines and speeds electrical impulses transmitted along axons.

Myoblast   Muscle precursor cell.

Myocardium   Heart muscle.

Myoepithelia   Epithelia whose cells possess characteristics of both epithelial and muscle cells, e.g., the two epithelial layers of *Hydra*.

Myofiber   Muscle cell that is multinucleate and forms from the fusion of myoblasts; skeletal muscle cell.

Myogenic regulatory factors (MRFs)   Basic helix-loop-helix transcription factors (such as MyoD, Myf5 and myogenin) that are critical regulators of muscle development.

Myogenin   A myogenic regulatory factor that regulates several genes involved in the differentiation and repair of skeletal muscle cells. See also **Myogenic regulatory factors**.

Myostatin   Greek, "muscle stopper." A member of the TGF-β family, it negatively regulates muscle development. Genetic defects in the gene or its negative regulatory miRNA cause huge muscles to develop in some mammals, including humans.

Myotome   Portion of the somite that gives rise to all skeletal muscles of the vertebrate body except for those in the head. The myotome has two components: the primaxial component, closest to the neural tube, which forms the musculature of the back and rib cage, and the abaxial component, away from the neural tube, which forms the muscles of the limbs and ventral body wall.

N

N-cadherin   A type of cadherin that is highly expressed on cells of the developing central nervous system (the N stands for Neural). May play roles in mediating neural signals. See also **Cadherins**.

NAD⁺ kinase   Activated during the early response of the sea urchin egg to the sperm, converts NAD^+ to $NADP^+$, which can be used as a coenzyme for lipid biosynthesis and may be important in the construction of the many new cell membranes required during cleavage. $NADP^+$ is also used to make NAADP.

Naïve   Unaffected, lacking experience.

Naïve ESC   An embryonic stem cell (ESC) that is the most immature, undifferentiated ESC with the greatest potential for pluripotency. Compare with **Primed ESC**. Also see **Embryonic stem cells**.

Naïve pluripotent state   The most immature, undifferentiated state of embryonic stem cells with the greatest potential for pluripotency.

Nanos   Protein critical for the establishment of anterior-posterior polarity of the *Drosophila* embryo. *Nanos* mRNA is tethered to the posterior end of the oocyte and is translated after ovulation and fertilization. Nanos protein diffuses from the posterior end, while Bicoid diffuses from the anterior end, setting up two opposing gradients that establish the anterior-posterior polarity of the embryo.

Necrosis   Pathologic cell death caused by factors such as inflammation of toxic injury. Compare with **Apoptosis**.

Necrotic zones   Regions of the tetrapod limb "sculpted" by apoptotic (programmed) cell death; the term "necrotic" zone is a holdover from a time when no distinction was made between necrosis and apoptosis. The four necrotic regions are interdigital, anterior, posterior, and interior.

Negative feedback loop   A process in which the product of the process inhibits an earlier step in the process.

Neocortex   A layer of gray matter in the cerebrum that is a distinguishing feature of the mammalian brain; it stratifies into six layers of neuronal cell bodies, each with different functional properties.

Neoteny   Retention of the juvenile body form throughout life while the germ cells and reproductive system mature (e.g., the Mexican axolotl). See also **Progenesis**.

Nephric duct   See **Wolffian duct**.

Nephron   Functional unit of the kidney.

Nerve growth factor (NGF)   Neurotrophin involved primarily in the growth of nerve cells. Released from potential target tissues, it works at short ranges as either a chemotactic factor or chemorepulsive factor for axonal guidance. Also important in the selective survival of different subsets of neurons.

Netrins   Paracrine factors found in a gradient that guide axonal growth cones. They are important in commissural axon migration and retinal axon migration. Netrin-1 is secreted by the floor plate; netrin-2 is secreted by the lower region of the spinal cord.

Neural crest   A transient band of cells, arising from the lateral edges of the neural plate, that joins the neural tube to the epidermis. It gives rise to a cell population—the neural crest cells—that detach during formation of the neural tube and migrate to form a variety of cell types and structures, including sensory neurons, enteric neurons, glia, pigment cells, and (in the head) bone and cartilage.

Neural crest cells   See **Neural crest**.

Neural crest effectors   Transcription factors (e.g., MITF and Rho GTPase) activated by neural crest specifiers that give the neural crest cells their migratory properties and some of their differentiated properties.

Neural crest specifiers   A set of transcription factors (e.g., FoxD3, Sox9, Id,

Twist, and Snail) induced by the neural plate border-specifying transcription factors, that specify the cells that are to become the neural crest.

Neural folds Thickened edges of the neural plate that move upward during neurulation and migrate toward the midline and eventually fuse to form the neural tube.

Neural groove U-shaped groove that forms in the center of the neural plate during primary neurulation.

Neural keel A band of neural precursor cells that are brought into the dorsal midline during convergence and extension movements in the epiblast of the fish embryo. It extends over the axial and paraxial mesoderm and eventually forms a rod of tissue that separates from the epidermal ectoderm and develops a slit-like lumen to become the neural tube.

Neural plate border specifiers A set of transcription factors (e.g., Distal-less-5, Pax3, and Pax7), induced by the neural plate inductive signals, that collectively confer upon the border region the ability to form neural crest and dorsal neural tube cell types. Induce expression of neural crest specifiers.

Neural plate border The border between the neural plate and the epidermis.

Neural plate inductive signals Paracrine factors (e.g., BMPs, Wnts, FGFs and Notch) that interact to specify the boundaries between neural and non-neural ectoderm during gastrulation. In amphibians, inductive signals secreted by the notochord are sufficient to specify neural plate; in chick, signals secreted by the ventral ectoderm and paraxial mesoderm specify the boundaries.

Neural plate The region of the dorsal ectoderm that is specified to be neural ectoderm. It later folds upward to become the neural tube.

Neural restrictive silencer element (NRSE) A regulatory DNA sequence found in several mouse genes that prevents a promoter's activation in any tissue except neurons, limiting the expression of these genes to the nervous system.

Neural restrictive silencer factor (NRSF) A zinc finger transcription factor that binds the NRSE and is expressed in every cell that is *not* a mature neuron.

Neural retina Derived from the inner layer of the optic cup, composed of a layered array of cells that include the light- and color-sensitive photoreceptor cells (rods and cones), the cell bodies of the ganglion cells, bipolar interneurons that transmit electric stimuli from the rods and cones to the ganglion cells, Müller glial cells that maintain its integrity, amacrine neurons (which lack large axons), and horizontal neurons that transmit electric impulses in the plane of the retina.

Neural stem cells (NSCs) Stem cells of the central nervous system capable of neurogenesis throughout life. In vertebrates, NSCs retain much of the characteristics of their embryonic progenitor cell, the radial glial cell.

Neural tube The embryonic precursor to the central nervous system (brain and spinal cord).

Neuroblast An immature dividing precursor cell that can differentiate into the cells of the nervous system.

Neurocranium The vault and base of the skull.

Neuromesodermal progenitors (NMPs) In vertebrate embryos, a population of multipotent progenitor cells with the potential to contribute to both the neural tube and paraxial (somitic) mesoderm, found in the posteriormost region (caudal end) of the embryo.

Neuromesodermal Giving rise to both neural and somitic cell types. See **Neuromesodermal progenitors (NMPs)**.

Neurons Nerve cells; cells specialized for the conduction and transmission of information via electrical and chemical signals.

Neuropore The two open ends (anterior neuropore and posterior neuropore) of the neural tube that later close.

Neurotransmitters Molecules (e.g., acetylcholine, GABA, serotonin) secreted at the ends of axons. These molecules cross the synaptic cleft and are received by the adjacent neuron, thus relaying the neural signal. See also **Synapse**.

Neurotrophin/neurotropin Neuro*trophic* (Greek, "nourishing") refers to a factor's ability to keep the neuron alive, usually by supplying growth factors. Neuro*tropic* (Latin, "turning") refers to a substance that attracts or repulses neurons. Because many factors have both properties, both terms are used; in the recent literature, neurotrophin appears to be preferred. See also **Nerve growth factor**;

Brain-derived neurotrophic factor; Conserved dopamine neurotrophic factor (CDNF); Neurotrophins 3 and 4/5.

Neurotrophins 3 and 4/5 (NT3, NT4/5) Neurotrophin 3 attracts sensory neurons from the dorsal root ganglia; NT4/5 attracts facial motor neurons and cerebellar granule cells.

Neurula Refers to an embryo during neurulation (i.e., while the neural tube is forming).

Neurulation Process of folding of the neural plate and closing of the cranial and caudal neuropores to form the neural tube.

Newt anterior gradient protein (nAG) Factor released by neurons in the blastema of a regenerating salamander limb that is thought to be the nerve-derived factor necessary for proliferation of the blastema cells.

Nieuwkoop center The dorsalmost vegetal blastomeres of the amphibian blastula, formed as a consequence of the cortical rotation initiated by the sperm entry; an important signaling center on the dorsal side of the embryo. One of its main functions is to induce the Organizer.

Nkx2-5 A gene that codes for a transcription factor that is critical in the regulatory network that specifies the heart cells. These networks are conserved across vertebrates. See also **Mesp1**.

NMPs See **Neuromesoderm progenitors**.

Nodal A paracrine factor and member of the TGF-β family involved in establishing left-right asymmetry in vertebrates and invertebrates.

Node The mammalian homologue of Hensen's node.

Nodose placodes A pair of epibranchial placodes in vertebrates; they give rise to the sensory components of the paired 10th cranial nerve (the vagus nerve) that innervates many organs of the body, such as the heart, lungs, and gastrointestinal tract.

Noggin A BMP antagonist (i.e., blocks BMP signaling).

Non-skeletogenic mesenchyme Formed from the veg2 layer of the 60-cell sea urchin embryo, it generates pigment cells, immunocytes, and muscle cells. Also called secondary mesenchyme.

Noninvoluting marginal zone (NIMZ) Region of cells on the exterior of the gastrulating amphibian embryo that do not involute. They expand by epiboly along with the animal cap cells to cover the entire embryo, eventually forming the surface ectoderm.

Notch protein Transmembrane protein that is a receptor for Delta, Jagged, or Serrate, participants in juxtacrine interactions. Ligand binding causes Notch to undergo a conformational change that enables a part of its cytoplasmic domain to be cut off by the presenilin-1 protease. The cleaved portion enters the nucleus and binds to a dormant transcription factor of the CSL family. When bound to the Notch protein, the CSL transcription factors activate their target genes.

Notochord A transient mesodermal rod in the most dorsal portion of the embryo that plays an important role in inducing and patterning the nervous system. Characteristic feature of chordates.

Nuclear RNA (nRNA) The original transcription product. Sometimes called heterogeneous nuclear RNA (hnRNA) or pre-messenger RNA (pre-mRNA); contains the cap sequence, the 5′ UTR, exons, introns, and the 3′ UTR.

Nuclear RNA selection Means of controlling gene expression by processing specific subsets of the nRNA population into mRNA in different types of cells.

Nuclei pulposi A gel-like mass in the center of the intervertebral discs derived from notochordal cells.

Nucleosome The basic unit of chromatin structure, composed of an octamer of histone proteins (two molecules each of histones H2A, H2B, H3, and H4) wrapped with two loops containing approximately 147 base pairs of DNA.

Nucleus (1) The membrane-enclosed organelle housing the eukaryotic chromosomes. (2) An organized cluster of the cell bodies of neurons in the brain with specific functions and connections.

Nurse cells Cells that provides nourishment to a developing egg. In *Drosophila* ovarioles, fifteen interconnected nurse cells generate mRNAs and proteins that are transported to a single developing oocyte.

Nymph Insect larval stage that resembles an immature adult of the species. Becomes progressively more mature through a series of molts.

O

Obesogens Substances that increase the production and accumulation of fat and adipose (fat) cells in the body. Several endocrine disruptors, including DES and BPA, have been shown to be obesogens.

Obligate mutualism Symbiosis in which the species involved are interdependent with one another to such an extent that neither partner could survive without the other.

Olfactory placodes Paired epidermal thickenings that form the nasal epithelium (smell receptors) as well as the ganglia for the olfactory nerves.

Oligodendrocytes A type of glial cell within the central nervous system that wrap themselves around axons to produce a myelin sheath. Also called oligodendroglia.

Omphalomesenteric (umbilical) veins The veins that form from yolk sac blood islands. These veins bring nutrients to the mammalian embryo and transport gases to and from sites of respiratory exchange with the mother.

Oncogenes Regulatory genes that promote cell division, reduce cell adhesion, and prevent cell death. Can promote tumor formation and metastasis. Proto-oncogenes are the normal version of these genes, which, when overexpressed or misexpressed through mutations or inappropriate methylations, are called oncogenes and can result in cancer.

Oocyte A developing egg. A **primary oocyte** is in a stage of growth, has not gone through meiosis, and has a diploid nucleus. A **secondary oocyte** has completed its first meiotic division but not the second, and is haploid.

Oogenesis The development of the egg (ovum), including meiotic divisions and maturation.

Oogonia In animals, the female germ line cells that divide mitotically to become oocytes. Singular, oogonium.

Optic chiasm In vertebrates, the part of the lower side of the brain where the two optic nerves cross the midline, forming an x-shaped structure that then extends and innervates target cells on the contralateral side of the brain. (*Chiasm* comes from *Chi*, Greek for the letter *X*.)

Optic cups Double-walled chambers formed by the invagination of the optic vesicles.

Optic nerve Cranial nerve (CN II) that forms from axons of the neural retina that grow back to the brain by traveling down the optic stalk.

Optic vesicle Extend from the diencephalon and activate the head ectoderm's latent lens-forming ability.

Oral plate A region where the ectoderm of the stomodeum meets the endoderm of the primitive gut. It later breaks open to form the oral opening.

Organ of Corti The receptor organ in the cochlea of the inner ear. It contains hair cells that reside in a fluid-filled chamber. Pressure waves from the movement of fluid in this chamber are transformed by the hair cells into action potentials that are sent to the brain by the auditory nerve and interpreted as sound.

Organization/activation hypothesis The theory that sex hormones act during the fetal or neonatal stage of a mammal's life to organize the nervous system in a sex-specific manner, and that during adult life, the same hormones may have transitory motivational (or "activational") effects.

Organizer In amphibians, the dorsal lip cells of the blastopore and their derivatives (notochord and head endomesoderm). Functionally equivalent to Hensen's node in chick, the node in mammals, and the shield in fish. Organizer action establishes the basic body plan of the early embryo. Also known as the Spemann Organizer or (more correctly) the Spemann-Mangold organizer.

Organogenesis Interactions between, and rearrangement of, cells of the three germ layers to produce tissues and organs.

Organoids Rudimentary organs, usually the size of a pea, grown in culture from pluripotent stem cells.

Orthologues Genes from different species that are similar in DNA sequence because those genes were inherited from a common ancestor. Compare with **Paralogues**.

Oskar A protein involved in setting up the anterior-posterior axis of the *Drosophila* egg and future embryo by binding *nanos* mRNA in the posterior region of the egg, which will establish the posterior end of the future embryo.

Ossification See **Osteogenesis**.

Osteoblast A committed bone precursor cell.

Osteoclasts Multinucleated cells derived from a blood cell lineage that enter the bone through the blood vessels and destroy bone tissue during remodeling.

Osteocytes Bone cells. Derived from osteoblasts that become embedded in the calcified osteoid matrix.

Osteogenesis Bone formation; the transformation of mesenchyme into bone tissue through a progression from osteoclast to osteoblast to osteocyte. See **Endochondral ossification**; **Intramembranous ossification**.

Osteoid matrix A collagen-proteoglycan secreted by osteoblasts that is able to bind calcium.

Otic cup The structure formed during the morphogenesis of the inner ear when the otic placode invaginates to the point of forming the shape of a cup. This stage of inner ear development comes after the otic pit stage. Once the edges of the otic cup come together and fuse, the structure is called the otic vesicle (otocyst).

Otic pit The structure formed during the morphogenesis of the inner ear when the otic placode starts to invaginate, creating an indentation.

Otic placodes Paired epidermal thickenings that invaginate to form the inner ear labyrinth, whose neurons form the acoustic ganglia that enable us to hear.

Outer radial glia (oRG) Progenitor cells that reside in the subventricular zone of the cerebrum and give rise to intermediate progenitor (IP) cells.

Outflow tract In the developing heart, made up of the conus arteriosus and truncus arteriosus; becomes the base of the aorta and the pulmonary arteries.

Oval cells A population of progenitor cells in the liver that divide and form new hepatocytes and bile duct cells when hepatocytes themselves are unable to regenerate the liver sufficiently.

Ovariole The *Drosophila* egg chamber.

Ovary The structure that produces an ovum, the female gamete. In mammals, there is a pair of ovaries in the abdomen, which propel eggs into the oviducts. In angiosperms, the ovary is part of the carpel that contains the ovule(s).

Ovastacin A protease released by the cortical granules of mammalian eggs after fertilization to digest ZP2 and thereby prevent further sperm from entering the egg.

Oviparity Young hatch from eggs ejected by the mother, as in birds, amphibians, and most invertebrates.

Ovoviviparity Young hatch from eggs held within the mother's body where they continue to develop for a period of time, as in certain reptiles and sharks. Compare with **Viviparity**.

Ovulation Release of the egg from the ovary.

Ovules In angiosperms, the structure comprising the megasporangium and the integument, which, after fertilization, develops into a seed.

Ovum The mature egg (at the stage of meiosis at which it is fertilized). Plural, **ova**.

P

P-cadherin A type of cadherin found predominantly on the placenta, where it helps the placenta stick to the uterus (the P stands for placenta). See also **Cadherins**.

P-granules The germ plasm in *C. elegans*. Isolated to a single germline precursor cell (P4 blastomere) early in cleavage.

p53 A transcription factor that can stop the cell cycle, cause cellular senescence in rapidly dividing cells, instruct the initiation of apoptosis, and activate DNA repair enzymes. One of the most important regulators of cell division.

Pachytene Greek, "thick thread." In the first meiotic division, the third stage of prophase I during which the chromatids thicken and shorten and can be seen by light microscopy as individual chromatids. Crossing over occurs during this stage.

Pair-rule genes *Drosophila* zygotic genes, regulated by gap gene proteins. Pair-rule genes are each expressed in seven stripes that divide the embryo into transverse bands perpendicular to the anterior-posterior axis. Pair-rule mutants lack portions of every other segment.

PAL-1 A maternally expressed transcription factor in the oocyte of the nematode *C. elegans* that is required for the differentiation of the P1 lineage of cells. P1 is one of the cells of the two-cell embryo.

PAR proteins Found in the cytoplasm of oocytes of the nematode *C. elegans*; involved in determining the anterior-posterior axis of the embryo following fertilization.

Paracrine factor A secreted, diffusible protein that provides a signal that interacts with and changes the cellular behavior of neighboring cells and tissues.

Paracrine interaction An interaction whereby proteins synthesized by one cell diffuse over a distance to induce changes in neighboring cells.

Paracrine signaling Signaling between cells that occurs across long distances through the secretion of **paracrine factors** into the extracellular matrix.

Paralogues Genes that are similar in sequence because they are the result of gene duplication events in an ancestral species. Compare with **Orthologues**.

Parasegment A "transegmental" unit in *Drosophila* that includes the posterior compartment of one segment and the anterior compartment of the immediately posterior segment; appears to be the fundamental unit of embryonic gene expression.

Parasitism Type of symbiosis in which one partner benefits at the expense of the other.

Parasympathetic (enteric) ganglia Ganglia of the parasympathetic ("rest and digest") nervous system derived from vagal and sacral neural crest cells.

Paraxial (somitic) mesoderm Thick bands of embryonic mesoderm immediately adjacent to the neural tube and notochord. In the trunk, paraxial mesoderm gives rise to somites, in the head it (along with the neural crest) gives rise to the skeleton, connective tissues and musculature of the face and skull.

Paraxial protocadherin Adhesion protein expressed specifically in the paraxial (somite-forming) mesoderm during amphibian gastrulation; essential for convergent extension.

Parietal endoderm Cells of the primitive endoderm that contact the trophoblast of the mammalian embryo. See **Primitive endoderm**.

Parietal mesoderm See **Somatic mesoderm**.

Parthenogenesis Greek, "virgin birth." When an ovum is activated in the absence of sperm. Normal development can proceed in many invertebrates and some vertebrates.

Pathway selection The first step in the specification of axonal connection, wherein the axons travel along a route that leads them to a particular region of the embryo.

Pattern formation The set of processes by which embryonic cells form ordered spatial arrangements of differentiated tissues.

Peri-implantation In placental mammals, the embryonic period from when the blastocyst is free in the uterus through its first interactions with the uterine endometrium.

Pericardial cavity The division of the coelom that surrounds the heart. Compare with **Peritoneal cavity**; **Pleural cavity**.

Perichondrium Connective tissue that surrounds most cartilage, except at joints.

Periclinal divisions In plants, cell divisions in which the new cell walls are laid down parallel to the surface of the plant. Compare with **Anticlinal divisions**.

Pericycle A layer of cells in the roots of plants, located between the endodermis and vascular tissue. It contains adult stem cells that can form new root apical meristems to grow lateral roots.

Pericytes Smooth-musclelike cells recruited to cover endothelial cells during vasculogenesis.

Periderm A temporary epidermis-like covering in the embryo that is shed once the inner layer differentiates to form a true epidermis.

Periosteal bone Bone that adds thickness to long bones and is derived from mesoderm via intramembranous ossification.

Periosteum A fibrous sheath containing connective tissue, capillaries, and bone progenitor cells and that covers the developing and adult bone.

Peripheral nervous system All the nerves and neurons lying outside the CNS (central nervous system; brain and spinal cord).

Peritoneal cavity The division of the coelom that encloses the abdominal organs. Compare with **Pericardial cavity**; **Pleural cavity**.

Permissive interaction Inductive interaction in which the responding tissue has already been specified, and needs only an environment that allows the expression of these traits.

Petals In an angiosperm flower, a non-sexual, non-photosynthetic, modified leaf. These are frequently brightly colored and can attract pollinating insects to the flower.

Petrosal placodes A pair of epibranchial placodes in vertebrates; they give rise to the sensory components of the paired 9th cranial nerve (the glossopharyngeal nerve) that innervates the tongue, carotid sinus, and carotid body, among other structures.

Pharyngeal arches Paired bars of mesenchymal tissue (derived from paraxial mesoderm, lateral plate mesoderm, and neural crest cells), covered by endoderm internally and ectoderm externally. Found near the pharynx of the vertebrate embryo, the arches form gill supports in fish, and many skeletal and connective tissue structures in the face, jaw, mouth, and larynx in other vertebrates. Also called branchial arches.

Pharyngeal clefts Clefts (invaginations) of external ectoderm that separate the pharyngeal arches. In amniotes, there are four pharyngeal clefts in the early embryo, but only the first becomes a structure (the external auditory meatus).

Pharyngeal pouches Inside the pharynx, these are where the pharyngeal epithelium (endoderm) pushes out laterally to form pairs of pouches between the pharyngeal arches. These give rise to the auditory tube, wall of the tonsil, thymus gland, parathyroids and thyroid.

Pharyngula Term often applied to the late neurula stage of vertebrate embryos.

Pharynx The region of the digestive tube anterior to the point at which the respiratory tube branches off.

Phenotypic heterogeneity Refers to the same mutation producing different phenotypes in different individuals.

Phenotypic plasticity The ability of an organism to react to an environmental input with a change in form, state, movement, or rate of activity.

Pheromones Vaporized chemicals emitted by an individual that results in communication with another individual. Pheromones are recognized by the vomeronasal organ of many mammalian species and play a major role in sexual behavior.

Phloem In vascular plants, the conduits that carry sugars produced by photosynthesis, along with other metabolites, from sources to sinks—primarily from the leaves to the non-photosynthetic parts of the plant.

Phosphatidylinositol 4,5-bisphosphate (PIP$_2$) A membrane phospholipid that during the IP$_3$ pathway is split by the enzyme phospholipase C (PLC) to yield two active compounds: IP$_3$ and diacylglycerol (DAG). The IP$_3$ pathway is activated during fertilization, launching the slow block to polyspermy and activating the egg to start developing.

Phospholipase C (PLC) Enzyme in the IP$_3$ pathway that splits membrane phospholipid phosphatidylinositol 4,5-bisphosphate (PIP$_2$) to yield IP$_3$ and diacylglycerol (DAG).

Phragmoplast A structure found in plants during cytokinesis that forms between the two daughter nuclei. Made up of cellulose-filled fusing vesicles, microtubules, microfilaments, and endoplasmic reticulum, it builds the cell wall between the two daughter cells.

Phyllotaxis The arrangement of leaves along a plant stem.

Phylotypic stage The stage that typifies a phylum, such as the late neurula or pharyngula of vertebrates, and which appears to be relatively invariant and to constrain its evolution.

Phytohormone A hormone, such as auxin, found in plants.

Pial surface The outer surface of the brain; "pial" refers to its being next to the pia mater, one of the meninges of the brain.

PIE-1 A maternally expressed transcription factor in the oocyte of the nematode *C. elegans* that is necessary for germline cell fate.

Pigmented epithelium Another term for pigmented retina. See **Pigmented retina**.

Pigmented retina The melanin-containing layer of the vertebrate eye that lies behind the neural retina. It forms from the outer layer of the optic cup. The black melanin pigment absorbs light coming through the neural retina, preventing it from bouncing back through the neural retina, which would distort the image perceived. Also referred to as **pigmented epithelium**.

Pioneer nerve fibers Axons that go ahead of other axons and serve as guides for them.

Pioneer transcription factors Transcription factors (e.g., Fox A1 and Pax7) that can penetrate repressed chromatin and bind to their enhancer

DNA sequences, a step critical to establishing certain cell lineages.

Pistil See **Carpel**.

Piwi One of the proteins, along with Tudor, Vasa, and Nanos, expressed in germ cells to suppress gene expression.

Placenta The organ in placental mammals that serves as the interface between fetal and maternal circulations and has endocrine, immune, nutritive and respiratory functions. It consists of a maternal portion (the uterine endometrium, or decidua, which is modified during pregnancy) and a fetal component (the chorion).

Placodes An area of ectodermal thickening. These include the cranial placodes (e.g., the olfactory, lens, and otic placodes); and the epidermal placodes of cutaneous appendages such as hair and feathers, which are formed via inductive interactions between the dermal mesenchyme and the ectodermal epithelium.

Plasmodesmata (singular, Plasmodesma) Cytoplasmic channels that form between adjacent plant cells, allowing for direct transport of substances between the cells.

Plastids Organelles found in plant cells that perform many functions, including photosynthesis. Example: Chloroplasts.

PLC See **Phospholipase C**.

PLCζ (Phospholipase C zeta) A soluble form of phospholipase C found in the head of mammalian sperm that is released during gamete fusion in fertilization. It sets off the IP_3 pathway in the egg that results in Ca^{2+} release and activation of the egg.

Pleiotropy The production of several effects by one gene or pair of genes.

Pleural cavity The division of the coelom that surrounds the lungs. Compare with **Pericardial cavity**; **Peritoneal cavity**.

Pluripotent hematopoietic stem cell See **Hematopoietic stem cell (HSC)**.

Pluripotent Latin, "capable of many things." A single pluripotent stem cell has the ability to give rise to different types of cells that develop from the three germ layers (mesoderm, endoderm, ectoderm) from which all the cells of the body arise. The cells of the mammalian inner cell mass (ICM) are pluripotent, as are embryonic stem cells. Each of these cells can generate any cell type in the body, but because the distinction between ICM and

trophoblast has been established, it is thought that ICM cells are not able to form the trophoblast. Germ cells and germ cell tumors (such as teratocarcinomas) can also form pluripotent stem cells. Compare with **Totipotent**.

Pluteus larva Type of larva found in sea urchins and brittle stars; a planktonic larva that is bilaterally symmetrical, ciliated, and has long arms supported by skeletal spicules.

Polar body The smaller cell, containing hardly any cytoplasm, generated during the asymmetrical meiotic division of the oocyte. The first polar body is haploid and results from the first meiotic division and the secondary polar body is also haploid and results from the second meiotic division.

Polar granule component (PGC) A protein important for germ line specification and localized to *Drosophila* polar granules. PGC inhibits transcription of somatic cell-determining genes by preventing the phosphorylation of RNA polymerase II.

Polar granules Particles containing factors important for germ line specification that are localized to the pole plasm and pole cells of *Drosophila*.

Polar lobe An anucleate bulb of cytoplasm extruded immediately before first cleavage, and sometimes before the second cleavage, in certain spirally cleaving embryos (mostly in the mollusc and annelid phyla). It contains the determinants for the proper cleavage rhythm and the cleavage orientation of the D blastomere.

Polarization The first stage of cell migration, wherein a cell defines its front and its back ends, directed by diffusing signals (such as a chemotactic protein) or by signals from the extracellular matrix. These signals will reorganize the cytoskeleton such that the front part of the cell will form lamellipodia (or filopodia) with newly polymerized actin.

Pole cells About five nuclei in the *Drosophila* embryo that reach the surface of the posterior pole during the ninth division cycle and become enclosed by cell membranes. The pole cells give rise to the gametes of the adult.

Pole plasm Cytoplasm at the posterior pole of the *Drosophila* oocyte that contains the determinants for producing the abdomen and the germ cells.

Pollination The process by which pollen grains are transferred from the male anther of the flower to the

female stigma of that flower or another flower.

PolyA tail A series of adenine (A) residues that are added by enzymes to the 3′ terminus of the mRNA transcript in the nucleus. The polyA tail confers stability on the mRNA, allows the mRNA to exit the nucleus, and permits the mRNA to be translated into protein.

Polyadenylation The insertion of a "tail" of some 200–300 adenylate residues on the RNA transcript, about 20 bases downstream of the AAUAAA sequence. This polyA tail (1) confers stability on the mRNA, (2) allows the mRNA to exit the nucleus, and (3) permits the mRNA to be translated into protein.

Polycomb proteins Family of proteins that bind to condensed nucleosomes, keeping the genes in an inactive state.

Polydactyly The presence of extra (supernumerary) digits, such as the dew claw on Great Pyrenees dogs.

Polyphenism A type of phenotypic plasticity, refers to discontinuous ("either/or") phenotypes elicited by the environment. Compare with **Reaction norm**.

Polyspermy The entrance of more than one sperm during fertilization resulting in aneuploidy (abnormal chromosome number) and either death or abnormal development. An exception, called physiological polyspermy, occurs in some organisms such as *Drosophila* and birds, where multiple sperm enter the egg but only one sperm pronucleus fuses with the egg pronucleus.

Polytene chromosomes Chromosomes in the larval cells of *Drosophila* (but not the imaginal cells that give rise to the adult) in which the DNA undergoes many rounds of replication without separation, forming large "puffs" that are easily visible and indicate active gene transcription.

Population asymmetry Mode of maintaining homeostasis in a population of stem cells in which some of the cells are more prone to producing differentiated progeny, while others divide to maintain the stem cell pool.

Posterior marginal zone (PMZ) The end of the chick blastoderm where primitive streak formation begins and acts as the equivalent of the amphibian Nieuwkoop center. The cells of the PMZ initiate gastrulation and prevent other regions of the margin from forming their own primitive streaks.

Posterior necrotic zone A zone of programmed cell death on the posterior side of the developing tetrapod limb that helps shape the limb.

Posterior neuropore See **Neuropore**.

Posterior progenitor zone A region in the tailbud of vertebrate embryos that is made up of multipotent neuromesoderm progenitor cells. Also referred to as **Caudal progenitor zone**.

Postsynaptic cell The target cell that receives chemical neurotransmitters from a presynaptic neuron, causing depolarization or hyperpolarization of the target cell's membrane.

Posttranslational regulation Modifications that determine whether the translated protein will be active. These modifications can include cleaving an inhibitory peptide sequence; sequestration and targeting to specific cell regions; assembly with other proteins to form a functional unit; binding an ion (such as Ca^{2+}); or modification by the covalent addition of a phosphate or acetate group.

Potency In referring to stem cells, the power to produce different types of differentiated cells.

Pre-granulosa cells The cells of the primordial follicle in the ovary that develop most closely to the germ cells. These become the granulosa cells of the follicle.

Pre-initiation complex The complex of RNA polymerase II at the promoter with transcription factors on the enhancer, as brought together by the Mediator molecules. See also **Mediator**.

Pre-metamorphosis The first stage in amphibian metamorphosis; the thyroid gland has begun to mature and is secreting low levels of T_4 and very low levels of T_3. The TRα receptor is present, but the TRβ receptor is not.

Prechordal plate See **Prechordal plate mesoderm**.

Prechordal plate mesoderm Precursor of the head mesoderm. The mesoderm cells that move inward during gastrulation ahead of the chordamesoderm.

Precursor cells (precursors) Widely used term to denote any ancestral cell type (stem or progenitor cells) of a particular lineage (e.g., neuronal precursors; blood cell precursors).

Predator-induced polyphenism The ability to modulate development in the presence of predators in order to express a more defensive phenotype.

Preeclampsia Medical condition of pregnant women characterized by hypertension, poor renal filtration, and fetal distress. A leading cause of premature birth and both fetal and maternal deaths. **Pluripotent hematopoietic stem cell** See **Hematopoietic stem cell (HSC)**.

Preformationism The view, supported by the early microscopist Marcello Malpighi, that the organs of the embryo are already present, in miniature form, within the egg (or sperm). A corollary, *embôitment* (encapsulation), stated that the next generation already existed in a prefigured state within the germ cells of the first prefigured generation, thus ensuring that the species would remain constant.

Preimplantation genetics Testing for genetic diseases using blastomeres from embryos produced by in vitro fertilization before implanting the embryo in the uterus.

Prenatal diagnosis The use of chorionic villus sampling or amniocentesis to diagnose many genetic diseases before a baby is born.

Presomitic mesoderm (PSM) The mesoderm that will form the somites. Also known as the segmental plate.

Presynaptic neuron Neuron that transmits chemical neurotransmitters to a target cell, causing the depolarization or hyperpolarization of the target cell's membrane.

Primary capillary plexus A network of capillaries formed by endothelial cells during vasculogenesis.

Primary cilium A single, non-motile cilium found on most cells; lacks a central pair of microtubules and is involved in part of the hedgehog signaling pathway by transporting signaling molecules on its microtubules using motor proteins.

Primary embryonic induction The process whereby the dorsal axis and central nervous system forms through interactions with the underlying mesoderm, derived from the dorsal lip of the blastopore in amphibian embryos.

Primary hypoblast During gastrulation in the chick embryo, a cellular layer that forms from cells that first delaminate from the anterior epiblast to form islands of disconnected cells and then migrate to form the primary hypoblast layer below the anterior epiblast.

Primary larvae Larvae that represent dramatically different body plans than the adult form and that are morphologically distinct from the adult; the plutei of sea urchins are such larvae. Compare with **Secondary larvae**.

Primary mesenchyme See **Skeletogenic mesenchyme**.

Primary neurulation The process that forms the anterior portion of the neural tube. The cells surrounding the neural plate direct the neural plate cells to proliferate, invaginate, and pinch off from the surface to form a hollow tube.

Primary oocytes A developing egg that has passed through the oogonial stage and is in a stage of growth prior to any meiotic division. Contains a large nucleus called a germinal vesicle. At this stage, mRNA (maternal mRNA) is being made and stored in the egg. In mammals, a primary oocyte is arrested in first meiotic prophase until just prior to ovulation, when the first meiotic division is completed and the egg becomes a secondary oocyte. The second meiotic division is then arrested and is not completed until after fertilization.

Primary sex determination (or gonadal sex determination) The determination of the gonads to form either the egg-forming ovaries or sperm-forming testes. Primary sex determination is chromosomal and is not usually influenced by the environment in mammals, but can be affected by the environment in other vertebrates.

Primary spermatocytes Derived from mitotic division of the type B spermatogonia, these are the cells that first go through a period of growth and then enter meiosis.

Primaxial muscles The intercostal musculature between the ribs and the deep muscles of the back, formed from those myoblasts in the myotome closest to the neural tube.

Primed Prepared; in the context of embryonic stem cells, ready for differentiation.

Primed ESC An embryonic stem cell (ESC) that was cultured from an inner cell mass cell that already had some maturation toward the epiblast lineage. Compare with **Naïve ESC**. Also see **Embryonic stem cells**.

Primed pluripotent state The state of an embryonic stem cell that has undergone some maturation toward the epiblast lineage.

Primitive endoderm The layer of endoderm cells created during early mammalian development when the inner cell mass splits into two layers. The lower layer, in contact with the blastocoel, is the primitive endoderm, and is homologous to the hypoblast of the avian embryo. It will form the inner lining of the yolk sac and will be used for positioning the site of gastrulation, regulating the movements of cells in the epiblast, and promoting the maturation of blood cells. It is an extraembryonic layer that does not provide cells to the body of the embryo.

Primitive groove A depression that forms within the primitive streak that serves as an opening through which migrating cells pass into the deep layers of the embryo.

Primitive knot/pit See **Hensen's node**.

Primitive pit During chick gastrulation, a funnel-shaped depression at the center of Hensen's node through which cells migrate to form the notochord and prechordal plate. See **Hensen's node**.

Primitive streak The first morphological sign of gastrulation in amniotes, it first arises from a local thickening of the epiblast at the posterior edge of the area pellucida, called Koller's sickle. Homologous to the amphibian blastopore.

Primordial germ cells (PGCs) Gamete progenitor cells, which typically arise elsewhere and migrate into the developing gonads.

Proacrosin The inactive form of a mammalian sperm proteinase that is stored in the acrosome and released during the acrosomal reaction and helps the sperm move through the zona pellucida of the egg.

Procambium In plants, a layer of stem cells that produces the vascular tissue; it can also give rise to the pericycle in the roots.

Proembryo During development in seed plants, the stage created by the asymmetrical first cleavage of the zygote. At this two-cell stage, the smaller apical cell will give rise to all parts of the plant proper, except the tip of the root; the larger basal cell will generate the root apex and the suspensor, which connects that embryo to the nutrients in the seed.

Proerythroblast A red blood cell precursor.

Progamic phase The events of pollen development from pollination to fertilization; the period of pollen tube growth through the female pistil.

Progenesis Condition in which the gonads and germ cells develop at a faster rate than the rest of the body, becoming sexually mature while the rest of the body is still in a juvenile phase. Compare with **Neoteny**.

Progenitor An ancestor in the direct lineage; a predecessor or precursor.

Progenitor cells Relatively undifferentiated cells that have the capacity to divide a few times before differentiating and, unlike stem cells, are not capable of unlimited self-renewal. They are sometimes called **transit amplifying cells** because they divide while migrating.

Progerias Premature aging syndromes; in humans and mice, appear to be caused by mutations that prevent the functioning of DNA repair enzymes.

Progesterone A steroid hormone important in the maintenance of pregnancy in mammals. Progesterone secreted from the cumulus cells may act as a chemotactic factor for sperm.

Programmed cell death See **Apoptosis**.

Progress zone (PZ) Highly proliferative limb bud mesenchyme directly beneath the apical ectodermal ridge (AER). The proximal-distal growth and differentiation of the limb bud are made possible by a series of interactions between the AER and the progress zone. Also called the undifferentiated zone.

Progress zone model Model for specification of proximal-distal specification of the limb that postulates that each mesoderm cell is specified by the amount of time it spends dividing in the progress zone. The longer a cell spends in the progress zone, the more mitoses it achieves and the more distal its specification becomes.

Prometamorphosis The second stage in amphibian metamorphosis, during which the thyroid matures and secretes more thyroid hormones.

Promoter Region of a gene containing the DNA sequence to which RNA polymerase II binds to initiate transcription. See also **CpG islands**; **Enhancer**.

Pronephric duct Arises in the intermediate mesoderm, migrates caudally, and induces the adjacent mesenchyme to form the pronephros, or tubules of the initial kidney of the embryo. The pronephric tubules form functioning kidneys in fish and in amphibian larvae but are not believed to be active in amniotes. As

the duct continues growing downward it induces the mesonephric mesenchyme to form tubules, at which point it is called the mesonephric duct. Also called Wolffian duct and nephric duct.

Pronephros The first region of kidney mesenchyme to form kidney tubules in vertebrates. The pronephros is a functioning kidney in fish and amphibian larvae, but is not believed to be active in amniotes, and degenerates after other regions of the kidney develop.

Pronuclei The male and female haploid nuclei within a fertilized egg that fuse to form the diploid nucleus of the zygote.

Pronymph The stage immediately after hatching in ametabolous insects, when the organism bears the structures that enabled it to get out of the egg; after this stage, the insect looks like a small adult.

Prosencephalon The forebrain; the most anterior vesicle of the developing vertebrate brain. Will form two secondary brain vesicles: the telencephalon and the diencephalon.

Protamines Basic proteins, tightly compacted through disulfide bonds, that package the DNA of the sperm nucleus.

Protein-protein interaction domain A domain of a transcription factor that enables it to interact with other proteins on the enhancer or promoter.

Proteoglycans Large extracellular matrix molecules consisting of core proteins (such as syndecan) with covalently attached glycosaminoglycan polysaccharide side chains. Two of the most widespread are heparan sulfate proteoglycan and chondroitin sulfate proteoglycan.

Proteome The number and type of proteins encoded by the genome.

Prothoracic gland In insects, a gland that secretes ecdysone, a molting hormone; production of ecdysone is stimulated by the prothoracicotropic hormone.

Prothoracicotropic hormone (PTTH) A peptide hormone that initiates the molting process in insects when it is released by neurosecretory cells in the brain in response to neural, hormonal, or environmental signals. PTTH stimulates the production of ecdysone by the prothoracic gland.

Protocadherins A class of cadherins that lack the attachment to the actin skeleton through catenins. They are an important means of keeping migrating epithelia together, and they are important in separating the notochord from surrounding mesoderm during its formation.

Protostomes Greek, "mouth first." Animals that form their mouth regions from the blastopore, such as molluscs. Compare with **Deuterostomes**.

Proximal-distal axis The close-far axis, e.g., shoulder-finger or hip-toe (in relation to the body's center).

Pseudohermaphroditism Intersex conditions in which the secondary sex characteristics differ from what would be expected from the gonadal sex. Male pseudohermaphroditism (e.g., androgen insensitivity syndrome) describes conditions wherein the gonadal sex is male and the secondary sex characteristics are female, while female pseudohermaphroditism describes the reverse situation (e.g., congenital adrenal hyperplasia).

Pupa A non-feeding stage of a holometabolous insect following the last instar when the organism is going through metamorphosis, being transformed from a larva into an adult (imago).

Purkinje fibers Modified heart muscle cells in the inner walls of the ventricles, specialized for rapid conduction of the contractile signal. Essential for synchronizing the contractions of the ventricles in amniotes.

Purkinje neurons Large, multibranched neurons that are the major cell type of the cerebellum.

Q

Quiescence A period of inactivity or dormancy; usually affiliated with a period of stem cell behavior.

R

R-cadherin A type of cadherin critical in forming the retina (the R stands for retina). See **Cadherins**.

R-spondin1 (Rspo1) Small, soluble protein that upregulates the Wnt pathway and is critical for ovary formation in mammals.

RA See **Retinoic acid**.

Radial glial cells (radial glia) Neural progenitor cells found in the ventricular zone (VZ) of the developing brain. At each division, they generate another VZ cell and a more committed cell type that leaves the VZ to differentiate.

Radial holoblastic cleavage Cleavage pattern in echinoderms. The cleavage planes, which divide the egg completely into separate cells (holoblastic), are parallel or perpendicular to the animal-vegetal axis of the egg.

Radial intercalation In fish embryos, the movement of deep epiblast cells into the more superficial epiblast layer, helping to power epiboly during gastrulation.

Random epigenetic drift The hypothesis that the chance accumulation of inappropriate epigenetic methylation due to errors made by the DNA methylating and demethylating enzymes could be the critical factor in aging and cancers.

Ras A G-protein in the RTK pathway. Mutations in the *RAS* gene account for a large proportion of cancerous human tumors.

Rathke's pouch An outpocketing of the ectoderm in the roof of the oral region that forms the glandular portion of the pituitary gland in vertebrates. It meets the infundibulum, an outpocketing of the floor of the diencephalon, which forms the neural portion of the pituitary gland.

Reaction norm A type of phenotypic plasticity in which the genome encodes the potential for a continuous range of potential phenotypes; the environment the individual encounters determines which of the potential phenotypes develops. Compare with **Polyphenism**.

Reaction-diffusion mechanism Model for developmental patterning, especially that of the limb, wherein two homogeneously distributed substances (an activator, substance *A*, that activates itself as well as forming its own, faster-diffusing inhibitor, substance *I*) interact to produce stable complex patterns during morphogenesis. According to this model, set forth in the early 1950s by mathematician Alan Turing, the patterns generated by this reaction-diffusion mechanism represent regional differences in the concentrations of the two substances.

Reactive oxygen species (ROS) Metabolic by-products that can damage cell membranes and proteins and destroy DNA. ROS are generated by mitochondria due to insufficient reduction of oxygen atoms and include superoxide ions, hydroxyl ("free") radicals, and hydrogen peroxide.

Receptor A protein that functions to bind a ligand. See also **Ligand**.

Receptor tyrosine kinase (RTK) A receptor that spans the cell membrane and has an extracellular region, a transmembrane region, and a cytoplasmic region. Ligand (paracrine factor) binding to the extracellular domain causes a conformational change in the receptor's cytoplasmic domains, activating kinase activity that uses ATP to phosphorylate specific tyrosine residues of particular proteins.

Reciprocal inductions A common sequential feature of induction: One tissue induces another, and that tissue then acts back on the original inducing tissue and induces it, thus the inducer becomes the induced.

Reelin An extracellular matrix protein found in the developing cerebellum and cerebrum. In the cerebellum it permits neurons to bind to glial cells as neurons migrate and form layers; in the cerebrum it directs migration of neurons toward the pial surface.

Regeneration blastema A collection of relatively undifferentiated cells that are organized into new structures by paracrine factors located at the cut surface. The collection of cells may be derived from differentiated tissue near the site of amputation that dedifferentiate, go through a period of mitosis, and then redifferentiate into the lost structures, as in the regenerating salamander limb, or may be from pluripotent stem cells that migrate to the cut surface, as in flatworm regeneration.

Regeneration The ability to reform body structure or organ that has been damaged or destroyed by trauma or disease.

Regenerative medicine The therapeutic use of stem cells to correct genetic pathologies (e.g., sickle-cell anemia) or repair damaged organs.

Regulation The ability to respecify cells so that the removal of cells destined to become a particular structure can be compensated for by other cells producing that structure. This is seen when an entire embryo is produced by cells that would have contributed only certain parts to the original embryo. It is also seen in the ability of two or more early embryos to form one chimeric individual rather than twins, triplets, or a multiheaded individual.

Relational pleiotropy The action of a gene in one part of the embryo that affects other parts, not by being expressed in these other parts but by having initiated a cascade of events that affect these other parts.

Reporter gene A gene with a product that is readily identifiable and not usually made in the cells of interest. Can be fused to regulatory elements from a gene of interest, inserted into embryos, and then monitored for reporter gene expression. If the sequence contains an enhancer, the reporter gene should become active at particular times and places. The genes for green fluorescent protein (*GFP*) and β-galactosidase (*lacZ*) are common examples.

Repressor A DNA- or RNA-binding regulatory element that actively represses the transcription of a particular gene.

Resact A 14-amino-acid peptide that has been isolated from the egg jelly of the sea urchin *Arbacia punctulata* that acts as a chemotactic factor and sperm-activating peptide for sperm of the same species, i.e., it is species-specific and is thereby a mechanism to ensure that fertilization is also species-specific. See also **Sperm-activating peptide**.

Resegmentation Occurs during formation of the vertebrae from sclerotomes; the rostral segment of each sclerotome recombines with the caudal segment of the next anterior sclerotome to form the vertebral rudiment and this enables the muscles of the vertebral column derived from the myotomes to coordinate the movement of the skeleton, permitting the body to move laterally.

Respiratory tube The future respiratory tract, which forms as an epithelial outpocketing of the pharynx, and eventually bifurcates into the two lungs.

Responder During induction, the tissue being induced. Cells of the responding tissue must have receptors for the inducing molecules and be competent to respond to the inducer.

Resting membrane potential The membrane potential (membrane voltage) normally maintained by a cell, determined by the concentration of ions on either side of the membrane. Generally this is −70mV, where the inside of the cell is negatively charged with respect to the exterior.

Rete testis A network of thin canals that convey sperm from the seminiferous tubules to the efferent ducts.

Reticulocyte Cell derived from the mammalian erythroblast that has expelled its nucleus. Although reticulocytes, lacking a nucleus, can no longer synthesize globin mRNA, they can translate existing messages into globins. A reticulocyte differentiates into a mature red blood cell (erythrocyte), in which even translation of mRNA doesn't take place.

Retina See **Neural retina**.

Retinal ganglion cells (RGCs) Neurons in the retina of the eye whose axons are guided to the optic tectum of the brain. Guidance cues come from netrin, slit, semaphorin, and ephrin families of molecules.

Retinal homeobox (Rx) A transcription factor coded for by the *Rx* gene. Produced in the eye field and helps specify the retina.

Retinoic acid (RA) A derivative of vitamin A and morphogen involved in anterior-posterior axis formation. Cells receiving high levels of RA express posterior genes.

Retinoic acid-4-hydroxylase An enzyme that degrades retinoic acid.

Retinotectal projection The map of retinal connections to the optic tectum. Point-for-point correspondence between the cells of the retina and the cells of the tectum that enables the animal to see an unbroken image.

Reverse development The transformation of a mature stage of an organism to a more juvenile stage of its life cycle. Seen in certain hydrozoan species where the sexually mature adult-stage medusa is able to revert to the polyp stage.

Reverse genetics Genetic technique of knocking out or knocking down the expression of a gene in an organism and then studying the phenotype that results. Compare with **Forward genetics**.

Rheotaxis A form of movement where a cell or animal turns to face a current of gas or liquid. Sperm traveling through the female reproductive tract are thought to have positive rheotaxis.

Rho GTPases A family of molecules including RhoA, Rac1, and Cdc42 that convert soluble actin into fibrous actin cables that anchor at the cadherins. These help mediate cell migration by lamellipodia and filopodia and the

cadherin-dependent remodeling of the cytoskeleton.

Rhombencephalon The hindbrain, the most caudal vesicle of the developing vertebrate brain; will form two secondary brain vesicles, the metencephalon and myelencephalon.

Rhombomeres Periodic swellings that divide the rhombencephalon into smaller compartments, each with a different fate and different associated nerve ganglia.

Right-left axis Specification of the two lateral sides of the body.

Ring canals The cytoplasmic interconnections between the cystocytes that become the ovum and nurse cells in an ovariole of *Drosophila*.

RNA interference (RNAi) Process by which miRNAs inhibit expression of specific genes by degrading their mRNAs.

RNA polymerase II An enzyme that binds to a promoter on DNA and, when activated, catalyzes the transcription of an RNA template from the DNA.

RNA-induced silencing complex (RISC) A complex containing several proteins and a microRNA, which can then bind to the 3′ UTR of messages and inhibit their translation.

RNA-Seq (RNA sequencing) Using next-generation sequencing technology to sequence and quantify the RNA present in a biological sample.

Robo proteins See **Roundabout proteins**.

Robustness (canalization) The ability of an organism to develop the same phenotype despite perturbations from the environment or from mutations. It is a function of interactions within and between developmental modules.

Rods Photoreceptors in the neural retina of the vertebrate eye that are more sensitive to low light than cones. They contain only one light-sensitive pigment and therefore do not transmit information about color.

Roof plate Dorsal region of the neural tube important in the establishment of dorsal-ventral polarity. The adjacent epidermis induces expression of BMP4 in the roof plate cells, which in turn induces a cascade of TGF-β proteins in adjacent cells of the neural tube.

Root apical meristem (RAM) In plants, the meristem at the tip of a growing root. See **Meristem**.

Rosettes Pinwheel-like structures, such as the structures made up of small clusters of neural stem cells surrounded by ciliated ependymal cells found in the V-SVZ of the mammalian cerebrum.

Rostral-caudal Latin, "beak-tail." An anterior-posterior positional axis; often used when referring to vertebrate embryos or brains.

Rotational cleavage The cleavage pattern for mammalian and nematode embryos. In mammals, the first cleavage is a normal meridional division while in the second cleavage, one of the two blastomeres divides meridionally and the other divides equatorially. In *C. elegans*, each asymmetrical division produces one founder cell that produces differentiated descendants; and one stem cell. The stem cell lineage always undergoes meridional division to produce (1) an anterior founder cell and (2) a posterior cell that will continue the stem cell lineage.

Roundabout proteins (Robo) Proteins that are receptors for slit proteins, involved in controlling the crossing of the midline of commissural axons.

Royalactin Protein that induces a honeybee larva to become a queen. Fed to the larva by worker bees, the protein binds to EGF receptors in the larva fat body and stimulates the production of juvenile hormone, which elevates the levels of yolk proteins necessary for egg production.

RTK pathway The receptor tyrosine kinase (RTK) is dimerized by ligand, which causes autophosphorylation of the receptor. An adaptor protein recognizes the phosphorylated tyrosines on the RTK and activates an intermediate protein, GEF, which activates the Ras G protein by allowing the phosphorylation of the GDP-bound Ras. At the same time, the GAP protein stimulates the hydrolysis of this phosphate bond, returning Ras to its inactive state. The active Ras activates the Raf protein kinase C (PKC), which in turn phosphorylates a series of kinases. Eventually, an activated kinase alters gene expression in the nucleus of the responding cell by phosphorylating certain transcription factors (which can then enter the nucleus to change the types of genes transcribed) and certain translation factors (which alter the level of protein synthesis). In many cases, this pathway is reinforced by the release of Ca^{2+}.

S

Sacral neural crest Neural crest cells that lie posterior to the trunk neural crest and along with the vagal neural crest generate the parasympathetic (enteric) ganglia of the gut that are required for peristaltic movement in the bowels.

Sarcopterygian fishes Lobe-finned fish, including coelacanths and lungfish. Tetrapods evolved from sarcopterygian ancestors.

Satellite cells Populations of muscle stem cells and progenitor cells that reside alongside adult muscle fibers and can respond to injury or exercise by proliferating into myogenic cells that fuse and form new muscle fibers.

Scatter factor See **Hepatocyte growth factor**.

Schizocoely The embryonic process of forming the coelom by hollowing out a previously solid cord of mesodermal cells. Typical of protostomes. See also **Enterocoely**.

Schwann cell Type of glial cell of the peripheral nervous system that generates a myelin sheath, allowing rapid transmission of electrical signals along an axon.

Sclerotomes Blocks of mesodermal cells in the ventromedial half of each somite that will differentiate into the vertebrae, intervertebral discs (except for the nuclei pulposi) and ribs, in addition to the meninges of the spinal cord and the blood vessels that serve the spinal cord. They are also critical in patterning the neural crest and motor neurons.

Sebaceous gland Glands that are associated with hair follicles and produce an oily substance, **sebum**, that serves to lubricate the hair and skin.

Secondary hypoblast Underlies the epiblast in the bilaminar avian blastoderm. A sheet of cells derived from deep yolky cells at the posterior margin of the blastoderm that migrates anteriorly, displacing the hypoblast islands (primary hypoblast). Hypoblast cells do not contribute to the avian embryo proper, but instead form portions of the external membranes, especially the yolk sac, and provide chemical signals that specify the migration of epiblast cells. Also called endoblast.

Secondary larvae Larvae that possess the same basic body plan as the adult; caterpillars and tadpoles are examples. Compare with **Primary larvae**.

Secondary mesenchyme See **Nonskeletal mesenchyme**.

Secondary neurulation The process that forms the posterior portion of the neural tube by the coalescence of mesenchyme cells into a solid cord that subsequently forms cavities that coalesce to create a hollow tube.

Secondary oocyte The haploid oocyte following the first meiotic division (this division also generates the first polar body).

Secondary sex determination Developmental events, directed by hormones produced by the gonads that affect the phenotype outside the gonads. This includes the male or female duct systems and external genitalia, and, in many species, sex-specific body size, vocal cartilage, and musculature.

Secondary spermatocytes A pair of haploid cells derived from the first meiotic division of a primary spermatocyte, which then complete the second division of meiosis to generate the four haploid spermatids.

Seed The embryonic plant, a ripened ovule, enclosed by a protective coat.

Seed coat The outer protective jacket of the seed. The seed coat forms from the two integumental layers of the ovule.

Segment polarity genes *Drosophila* zygotic genes, activated by the proteins encoded by the pair-rule genes, whose mRNA and protein products divide the embryo into segment-sized units, establishing the periodicity of the embryo. Segment polarity mutants showed defects (deletions, duplications, polarity reversals) in every segment.

Segmental plate A synonym for presomitic mesoderm, the mesoderm that will form the somites.

Segmentation genes Genes whose products divide the early *Drosophila* embryo into a repeating series of segmental primordia along the anterior-posterior axis. Include gap genes, pair-rule genes, and segment polarity genes.

Selective affinity Principle that explains why disaggregated cells reaggregate to reflect their embryonic positions. Specifically, the inner surface of the ectoderm has a positive affinity for mesodermal cells and a negative affinity for the endoderm, while the mesoderm has positive affinities for both ectodermal and endodermal cells.

Self-renewal   The ability of a cell to divide and produce a replica of itself.

Semaphorins   Extracellular matrix proteins that repel migrating neural crest cells and axonal growth cones.

Seminiferous tubules   In male mammals, form in the gonad from the testis cords. They contain Sertoli cells (nurse cells) and spermatogonia (sperm stem cells).

Senescence   The physiological deterioration that characterizes old age.

Sensory placodes   In vertebrates, the ectodermal placodes that contribute to the sense organs, forming the olfactory epithelium, inner ear, lens of the eye, and cranial sensory ganglia, as well as lateral line organs in amphibians and fish.

Sepals   Outer structures of the flower, usually protective and photosynthetic, that surround the inner, fertile, portions of the flower.

Septum   A partition that divides a chamber, such as the atrial septa that split the developing atrium into left and right atria. Plural, septa.

Sertoli cells   Large secretory support cells in the seminiferous tubules of the testes involved in spermatogenesis in the adult through their role in nourishing and maintaining the developing sperm cells. They secrete AMH in the fetus and provide a niche for the incoming germ cells. They are derived from somatic cells, which are in turn derived from the genital ridge epithelium.

Sesamoid bone   Small bones at joints that form as a result of mechanical stress (such as the patella). They are derived from mesoderm via intramembranous ossification.

Sex-lethal (Sxl)   An autosomal gene in *Drosophila* involved in sex determination. It codes for a splicing factor that initiates a cascade of RNA processing events, which eventually lead to male-specific and female-specific transcription factors, the Doublesex proteins. See ***Doublesex***.

Shh   See **Sonic hedgehog**.

Shield   See **Embryonic shield**.

Shoot apical meristem (SAM)   In plants, the meristem at the tip of a growing shoot that is the source of stem cells for all plant organs above ground, such as leaves and flowers. See **Meristem**.

Signal transduction cascades   Pathways of response whereby paracrine factors bind to a receptor that initiates a series of enzymatic reactions within the cell that in turn have often several responses as their end point, such as the regulation of transcription factors (such that different genes are expressed in the cells reacting to these paracrine factors) and/or the regulation of the cytoskeleton (such that the cells responding to the paracrine factors alter their shape or are permitted to migrate).

Silencer   A DNA regulatory element that binds transcription factors that actively repress the transcription of a particular gene.

Single stem cell asymmetry   A mode of stem cell division in which two types of cells are produced at each division, a stem cell and a developmentally committed cell.

Sinistral coiling   Left-coiling. In a snail, having its coils open to the left of its shells. See also **Dextral coiling**.

Sinus venosus   The posterior region of the developing heart, where the two major vitelline veins bringing blood to the heart fuse. Inflow tract to the atrial area of the heart.

Sinusoidal endothelial cells   Cells that line the large blood channels (sinusoids) of the liver and critical to liver function. Also provides paracrine factors needed for division of hepatoblast stem cells during liver regeneration. Long considered mesodermal in origin, they now are known to be derived at least in part by specialized endodermal cells.

Sirtuin genes   Encode histone deacetylation (chromatin-silencing) enzymes that guard the genome, preventing genes from being expressed at the wrong times and places, and may help repair chromosomal breaks. They may be important defenses against premature aging.

Sister chromatids   Each of a pair of newly replicated chromatids. They have the same DNA sequence and are joined by a centromere.

Skeletogenic mesenchyme   Also called primary mesenchyme, formed from the first tier of micromeres (the large micromeres) of the 60-cell sea urchin embryo. They ingress, moving into the blastocoel, and form the larval skeleton.

SKN-1   A maternally expressed transcription factor in the oocyte of the nematode *C. elegans* that controls the fate of the EMS cell, one of the cells of the 4-cell stage that marks the ventral region of the developing embryo.

Slit proteins   Proteins of the extracellular matrix that are chemorepulsive; involved in inhibiting migration of neural crest cells and in controlling growth of commissural axons.

Slow block to polyspermy   See **Cortical granule reaction**.

Smad family   Transcription factors activated by members of the TGF-β superfamily that function in the SMAD pathway. See also **SMAD pathway**.

SMAD pathway   The pathway activated by members of the TGF-β superfamily. The TGF-β ligand binds to a type II TGF-β receptor, which allows that receptor to bind to a type I TGF-β receptor. Once the two receptors are in close contact, the type II receptor phosphorylates a serine or threonine on the type I receptor, thereby activating it. The activated type I receptor can now phosphorylate the Smad proteins. Smads 1 and 5 are activated by the BMP family of TGF-β factors, while the receptors binding activin, Nodal, and the TGF-β family phosphorylate Smads 2 and 3. These phosphorylated Smads bind to Smad4 and form the transcription factor complex that will enter the nucleus.

Small micromeres   A cluster of cells produced by the fifth cleavage at the vegetal pole in the sea urchin embryo when the micromeres divide.

Solenoids   Structures, created from tightly wound nucleosomes stabilized by histone H1, that inhibit transcription of genes by preventing transcription factors and RNA polymerases from gaining access to the genes.

Soma   Greek, "body." Can refer to the cell body (particularly of neurons) or to the cells that form an organism's body (as distinct from the germ cells).

Somatic (parietal) mesoderm   Derived from lateral mesoderm closest to the ectoderm (dorsal) and separated from other components of lateral mesoderm (splanchnic, near endoderm, ventral) by the intraembryonic coelom. Together with the overlying ectoderm, the somatic mesoderm comprises the somatopleure, which will form the body wall. The somatic mesoderm also forms part of the lining of the coelom. Not to be confused with somitic (paraxial) mesoderm.

Somatic cell nuclear transfer (SCNT)   Less accurately known as "cloning," the procedure by which a cell nucleus is transferred into an activated enucleated egg and directs the development of a complete organism with the same genome as the donor cell.

Somatic cells   Cells that make up the body—i.e., all cells in the organism that are not germ cells. Compare with **Germ cells**.

Somatic mutation theory (SMT)   A hypothesis for cancer initiation, positing that carcinogenesis is a cellular phenomenon, caused by mutations in otherwise normal cells, which instruct the cell to proliferate.

Somatopleure   Made up of somatic lateral plate mesoderm and overlying ectoderm.

Somites   Segmental blocks of mesoderm formed from paraxial mesoderm adjacent to the notochord (the axial mesoderm). Each contain major compartments: the sclerotome, which forms the axial skeleton (vertebrae and ribs), and the dermomyotome, which goes on to form dermatome and myotome. The dermatome forms the dermis of the back; the myotome forms musculature of the back, rib cage, and ventral body. Additional muscle progenitors detach from the lateral edge of the dermomyotome and migrate into the limbs to form the muscles of the fore- and hindlimbs.

Somitic mesoderm   See **Paraxial mesoderm**. Not to be confused with **somatic mesoderm**.

Somitogenesis   The process of segmentation of the paraxial mesoderm to form somites, beginning cranially and extending caudally. Its components are (1) periodicity, (2) fissure formation (to separate the somites), (3) epithelialization, (4) specification, and (5) differentiation.

Somitomeres   Early pre-somites, consisting of paraxial mesoderm cells organized into whorls of cells.

Sonic hedgehog (Shh)   The major hedgehog family paracrine factor. Shh has distinct functions in different tissues of the embryo. For example, it is secreted by the notochord inducing the ventral region of the neural tube to form the floor plate. It is also involved in the establishment of left-right asymmetry, primitive gut tube differentiation, proper feather formation in birds, differentiation of the sclerotome, and patterning the anterior-posterior axis of limb buds.

Sox9   An autosomal gene involved in several developmental processes, most notably bone formation. In the genital ridge of mammals, it induces

testis formation, and XX humans with an extra copy of *SOX9* develop as males.

Specification   The first stage of commitment of cell or tissue fate during which the cell or tissue is capable of differentiating autonomously (i.e., by itself) when placed in an environment that is neutral with respect to the developmental pathway. At the stage of specification, cell commitment is still capable of being reversed.

Specified   The stage during development when a cell is capable of differentiating autonomously when placed in a neutral environment, such as a petri dish. A specified cell's commitment to cell identity is still labile, however, and can be altered if the cell is transplanted to a population of differently specified cells. **The specified stage precedes determination. Compare with Determination.**

Spemann's Organizer   See **Organizer**.

Sperm head   Consists of the nucleus, acrosome, and minimal cytoplasm.

Sperm-activating peptides (SAPs)   Small chemotactic peptides found in the jelly of echinoderm eggs. They diffuse away from the egg jelly in seawater and are species specific, only attracting sperm of the same species. Resact, found in the sea urchin *Arbacia punctulata*, is an example.

Spermatids   Haploid sperm cells, the stage following the second meiotic division. In mammals, spermatids are still connected to one another by cytoplasmic bridges, allowing for diffusion of gene products across the cytoplasmic bridges.

Spermatogenesis   The production of sperm.

Spermatogonia   Sperm stem cells. When a spermatogonium stops undergoing mitosis, it becomes a primary spermatocyte and increases in size prior to meiosis.

Spermatogonial stem cell population   In mammals, the group of stem cells that will form the germ cell lineage leading up to the sperm. They divide mitotically, but not completely, forming clusters of spermatogonia.

Spermatozoa   The male gamete or mature sperm cell.

Spermiogenesis   The differentiation of the mature spermatozoa from the haploid round spermatid.

Spina bifida   A congenital defect resulting from incomplete closure of the

spine around the spinal cord, usually in the lower back. There are differing degrees of severity, the most severe being when the neural folds also fail to close.

Spiral phyllotaxis   In plants, the arrangement of lateral organs, such as leaves and flowers in the inflorescence, in which successive organs emerge sequentially around the apex in accordance with Fibonacci's "golden angle" of 137.5 degrees between each organ.

Spiral holoblastic cleavage   Characteristic of several animal groups, including annelid worms, some flatworms, and most molluscs. Cleavage is at oblique angles to the animal-vegetal axis, forming a "spiral" arrangement of daughter blastomeres. The cells touch one another at more places than do those of radially cleaving embryos, assuming the most thermodynamically stable packing orientation.

Splanchnic (visceral) mesoderm   Also called the visceral mesoderm and splanchnic lateral plate mesoderm; derived from lateral mesoderm closest to the endoderm (ventral) and separated from other component of lateral mesoderm (somatic, near ectoderm, dorsal) by the intraembryonic coelom. Together with the underlying endoderm, it forms the splanchnopleure. The splanchnic mesoderm will form the heart, capillaries, gonads, the visceral peritoneum and serous membranes that cover the organs, the mesenteries, and blood cells.

Splanchnopleure   Made up of splanchnic lateral plate mesoderm and underlying endoderm. See **Splanchnic mesoderm**.

Spliceosome   A complex made up of small nuclear RNAs (snRNAs) and splicing factors, that binds to splice sites and mediates the splicing of nRNA.

Splicing   The cutting, rearranging, and ligating back together of the mRNA precursor into separate messages that specify different proteins by using different combinations of potential exons. **See Differential RNA processing.**

Splicing enhancer   A *cis*-acting sequence on nRNA that promotes the assembly of spliceosomes at RNA cleavage sites.

Splicing factors   Proteins that bind to splice sites or to the areas adjacent to them.

Splicing isoforms Different proteins encoded by the same gene and generated by alternative splicing.

Splicing silencer A *cis*-acting sequence on nRNA that acts to exclude exons from an mRNA sequence.

Sporophytic Referring to the sporophyte, the diploid growth stage in the alternating life cycle of plants and algae. Compare with **Gametophytic**.

Src family kinases (SFK) Family of enzymes that phosphorylate tyrosine residues; involved in many signaling events, including the responses of growth cones to chemoattractants.

Sry **S**ex-determining **r**egion of the **Y** chromosome. The *Sry* gene encodes the mammalian testis-determining factor. It is probably active for only a few hours in the genital ridge, during which time it synthesizes the Sry transcription factor, whose primary role is to activate the *Sox9* gene required for testis formation.

Stamen The male organ of a flower. It usually comprises a stalk (filament) and a pollen-producing anther.

STAT **S**ignal **t**ransducers and **a**ctivators of **t**ranscription. A family of transcription factors, part of the JAK-STAT pathway. Important in the regulation of human fetal bone growth.

Stem cell A relatively undifferentiated cell from the embryo, fetus, or adult that, that divides and when it does so, produces (1) one cell that retains its undifferentiated character and remains in the stem cell niche; and (2) a second cell that leaves the niche and can undergo one or more paths of differentiation. See also **Adult stem cell**; **Embryonic stem cell**.

Stem cell factor (SCF) Paracrine factor important for maintaining certain stem cells, including hematopoietic, sperm, and pigment stem cells. Binds to the Kit receptor protein.

Stem cell mediated regeneration Process by which stem cells allow an organism to regrow certain organs or tissues (e.g., hair, blood cells) that have been lost.

Stem cell niche An environment (regulatory microenvironment) that provides a milieu of extracellular matrices and paracrine factors that allows cells residing within it to remain relatively undifferentiated. Regulates stem cell proliferation and differentiation.

Stereoblastulae Blastulae that have no blastocoel, e.g., blastulae produced by spiral cleavage.

Steroidogenic factor 1 (*Sf1*) A transcription factor that in mammals is necessary for creating the bipotential gonad. It declines in the developing ovary but remains at high levels in the developing testis, masculinizing both Leydig and Sertoli cells.

Stigma The surface of a carpel, usually at the peak of the style, that receives pollen.

Stochastic Pertaining to a random process that provides a set of random variables that can be analyzed statistically, but not necessarily predicted.

Stomata (singular, stoma) In plants, the pores in the epidermis of leaves and other organs that allow for gas exchange. Each pore is bordered by two guard cells that control the size of the pore, opening and closing the stoma in response to environmental conditions.

Stomodeum An ectoderm-lined invagination in the oral region of the embryo that meets the endoderm of the closed gut tube to form the oral plate.

Stratum corneum The outermost layer of the epidermis in the skin of tetrapods (amphibians, reptiles including birds, and mammals). It is a protective layer consisting of cornified cells, which are dead cells filled with keratin protein that are shed and replaced throughout the life of the organism.

Stratum germinativum See **Basal layer**.

Stromal derived factor 1 (SDF1) A chemoattractant. SDF1 is secreted, for example, by ectodermal placodes, thereby attracting cranial neural crest cells toward the placode.

Style A stalk, often elongated, between the stigma that receives pollen and the ovary, where the ovule is located.

Stylopod The proximal bones of a vertebrate limb, adjacent to the body wall; either the humerus (forelimb) or the femur (hindlimb).

Subgerminal cavity A space between the blastoderm and the yolk of avian eggs which is created when the blastoderm cells absorb water from the albumen ("egg white") and secrete fluid between themselves and the yolk.

Subgranular zone (SGZ) A region of the hippocampus in the cerebrum that contains neural stem cells, allowing for adult neurogenesis in this region.

Subventricular zone A region in the vertebrate cerebrum that is formed as progenitor cells migrate away from the ventricular zone.

Sulcus limitans A longitudinal groove that divides the developing spinal cord and medulla into dorsal (receives sensory input) and ventral (initiates motor functions) halves.

Superficial cleavage The divisions of the cytoplasm of centrolecithal zygotes that occur only in the rim of cytoplasm around the periphery of the cell due to the presence of a large amount of centrally-located yolk, as in insects.

Surfactant A secretion of specific proteins and phospholipids such as sphingomyelin and lecithin produced by the type II alveolar cells of the lungs very late in gestation. The surfactant enables the alveolar cells to touch one another without sticking together.

Suspensor A plant structure within the germinating seed that connects the plant embryo to the nutrients within the seed. It develops from the basal cell of the proembryo. See **Proembryo**.

Symbiogenesis A hypothesis for the origin of eukaryotic cells, wherein the first eukaryotic cells emerged from the fusion of prokaryotic organisms, one forming the nucleus, the other forming the mitochondrion.

Symbiont The smaller organism in a symbiotic relationship in which the other organism is much larger and serves as the host, while the smaller organism may live on the surface or inside the body of the larger.

Symbiosis Greek, "living together." Refers to any close association between organisms of different species.

Sympoiesis The phenomena of development through the interactions of multiple species, wherein symbionts provide developmental signals needed by the host, and the host often reciprocates in facilitating symbiont reproduction.

Synapse Junction at which a neuron contacts its target cell (which can be another neuron or another type of cell) and information in the form of neurotransmitter molecules (e.g., acetylcholine, GABA, serotonin) is exchanged across the synaptic cleft between the two cells.

Synapsis The highly specific parallel alignment (pairing) of homologous chromosomes during the first meiotic division.

Synaptic cleft   The small cleft that separates the axon of a signaling neuron from the dendrite or soma of its target cell.

Synaptonemal complex   The proteinaceous ribbon that forms during synapsis between homologous chromosomes, holding them together. A ladderlike structure with a central element and two lateral bars that are associated with the homologous chromosomes. See **Synapsis**.

Syncytial blastoderm   Describes the *Drosophila* embryo during cleavage when nuclei have divided, but no cell membranes have yet formed to separate the nuclei into individual cells.

Syncytial specification   The interactions of nuclei and transcription factors, which eventually result in cell specification, that take place in a common cytoplasm, as in the early *Drosophila* embryo.

Syncytiotrophoblast   A population of cells from the mammalian trophoblast that undergoes mitosis without cytokinesis resulting in multinucleate cells. The syncytiotrophoblast tissue is thought to further the progression of the embryo into the uterine wall by digesting uterine tissue.

Syncytium   Many nuclei residing in a common cytoplasm, results either from karyokinesis without cytokinesis or from cell fusion.

Syndetome   Greek *syn*, "connected." Derived from the most dorsal sclerotome cells, which express the *scleraxis* gene and generate the tendons.

Syndrome   Greek, "happening together." Several malformations or pathologies that occur concurrently. Genetically based syndromes are caused either by (1) a chromosomal event (such as trisomy 21, or Down syndrome) where several genes are deleted or added, or (2) by one gene having many effects.

Systems theory   In development, refers to an approach that views the organism as coming together through the interactions of its component processes. Although the emphasis applied to each varies, the theoretical systems approach can be characterized by six principles: (1) context-dependent properties; (2) level-specific properties and emergence; (3) heterogeneous causation; (4) integration; (5) modularity and robustness; and (6) homeorhesis (stability while undergoing change).

T

T-box (Tbx)   A specific DNA-binding domain found in certain transcription factors, including the *T* (*Brachyury*) gene, *Tbx4* and *Tbx5*. Tbx4 and Tbx5 help specify hindlimbs and forelimbs, respectively.

Target selection   The second step in the specification of axonal connection, wherein the axons, once they reach the correct area, recognize and bind to a set of cells with which they may form stable connections.

Telencephalon   The anterior subdivision of the prosencephalon; will eventually form the cerebral hemispheres.

Telogen   The resting phase of the hair follicle regeneration cycle.

Telolecithal   Describes the eggs of birds and fish which have only one small area at the animal pole of the egg that is free of yolk.

Telomerase   Enzyme complex that can extend the telomeres to their full length and maintains telomere integrity.

Telomeres   Repeated DNA sequences at the ends of chromosomes that provide a protective cap to the chromosomes.

Telson   A tail-like structure; the posterior most segment of certain arthropods. Seen in insect larvae such as *Drosophila*.

Temporal colinearity   The mechanism that controls the timing of Hox gene activation, which occurs anteriorly first and progressively more posteriorly; sets up spatial colinearity of Hox gene expression relative to their 3'-to-5' genomic organization.

Teratocarcinoma   A tumor derived from malignant primordial germ cells and containing an undifferentiated stem cell population (embryonal carcinoma, or EC cells) that has biochemical and developmental properties similar to those of the inner cell mass. EC cells can differentiate into a wide variety of tissues, including gut and respiratory epithelia, muscle, nerve, cartilage, and bone.

Teratogens   Greek, "monster-formers." Exogenous agents that cause disruptions in development resulting in teratogenesis, the formation of congenital defects. Teratology is the study of birth defects and of how environmental agents disrupt normal development.

Terminal end bulbs   The ends of the extensive branches of ducts in the mammary glands of mammals. Under the influence of estrogens at puberty, the ducts grow by the elongation of these buds.

Testis cords   Loops in the medullary (central) region of the developing testis formed by the developing Sertoli cells and the incoming germ cells. Will become the seminiferous tubules and site of spermatogenesis.

Testis-determining factor   A protein encoded by the *Sry* gene on the mammalian Y chromosome that organizes the gonad into a testis rather than an ovary.

Testosterone   A steroid hormone that is androgenic. In mammals, it is secreted by the fetal testes and masculinizes the fetus, stimulating the formation of the penis, male duct system, scrotum, and other portions of the male anatomy, as well as inhibiting development of the breast primordia.

Tetrad   See **Bivalent**.

Tetrapods   Latin, "four feet." Includes the vertebrates amphibians, reptiles, birds, and mammals. Evolved from lobe-finned fish (sarcopterygian) ancestors.

TGF-β family   **T**ransforming **g**rowth **f**actor-β. A family of growth factors within the TGF-β superfamily.

TGF-β superfamily   More than 30 structurally related members of a group of paracrine factors. The proteins encoded by TGF-β superfamily genes are processed such that the carboxy-terminal region contains the mature peptide. These peptides are dimerized into homodimers (with themselves) or heterodimers (with other TGF-β peptides) and are secreted from the cell. The TGF-β superfamily includes the TGF-β family, activin family, bone morphogenetic proteins (BMPs), Vg1 family, and other proteins, including glial-derived neurotrophic factor (GDNF; necessary for kidney and enteric neuron differentiation) and anti-Müllerian hormone (AMH; involved in mammalian sex determination).

Thecal cells   Steroid hormone-secreting cells of the mammalian ovary that, together with the granulosa cells, form the follicles surrounding the germ cells. They differentiate from mesenchyme cells of the ovary.

Thermotaxis   Migration that is directed by a gradient of temperature, either up or down the gradient.

Threshold model   A model of development wherein biological events are triggered when a specific concentration of a morphogen or hormone is reached.

Thyroid hormone receptors (TRs)   Nuclear receptors that bind the thyroid hormones tri-iodothyronine (T_3), as well as thyroxine (T_4). Once bound to the hormone, the TR becomes a transcriptional activator of gene expression. There are several different TR types, including TRα and TRβ.

Thyroxine (T_4)   Thyroid hormone containing four iodine molecules; is converted to the more active T_3 form through removal of one iodine molecule. Increases basal metabolic rate in cells. Initiates metamorphosis in amphibians.

Tip cells   Certain endothelial cells that can respond to vascular endothelial growth factor (VEGF) and begin "sprouting" to form a new vessel during angiogenesis. See also **Ureteric bud tip cells.**

Tissue engineering   A regenerative medicine approach whereby a scaffold is generated from material that resembles extracellular matrix or decellularized extracellular matrix from a donor, is seeded with stem cells, and is used to replace an organ or part of an organ.

Tissue organization field theory (TOFT)   A hypothesis for cancer initiation, positing that carcinogenesis is a tissue-based phenomenon and is caused by agents that interfere with the cell-cell communication that prevents cells from proliferating.

Torpedo   The receptor protein for Gurken. When expressed in terminal follicle cells in a *Drosophila* egg chamber, it binds to Gurken produced by the egg, which signals these follicle cells to differentiate into posterior follicle cells and synthesize a molecule that activates protein kinase A in the egg; part of the process that sets up the anterior-posterior axis of the egg and future embryo.

Totipotent   Latin, "capable of all." Describes the potency of certain stem cells to form all structures of an organism, such as the earliest mammalian blastomeres (up to the 8-cell stage), which can form both

trophoblast cells and the embryo precursor cells. Compare with **Pluripotent.**

Tracheal-esophageal fistula   An abnormal connection between the gut tube (esophagus) and respiratory tube (trachea) that can occur in babies when the separation of these two tubes by the laryngotracheal groove during embryonic development is not complete. It is a condition that must be surgically repaired so the baby can breathe and swallow properly. See **Laryngotracheal groove.**

Trans-activating domain   The transcription factor domain that activates or suppresses the transcription of the gene whose promoter or enhancer it has bound, usually by enabling the transcription factor to interact with the proteins involved in binding RNA polymerase or with enzymes that modify histones.

trans-regulatory elements   Soluble molecules whose genes are located elsewhere in the genome and which bind to the *cis*-regulatory elements. They are usually transcription factors or microRNAs.

Transcription elongation complex (TEC)   A complex of several transcription factors that breaks the connection between RNA polymerase II and the Mediator complex, allowing transcription (which has been initiated) to proceed.

Transcription elongation suppressor   A repressive transcription factor that functions to prevent the transcription elongation complex from associating with RNA polymerase II, pausing transcription.

Transcription factor   A protein that binds to DNA with precise sequence recognition for specific promoters, enhancers, or silencers.

Transcription factor domains   The three major domains are a DNA-binding domain, a trans-activating domain and a protein-protein interaction domain.

Transcription initiation site   DNA sequence of a gene that codes for the addition of a modified nucleotide "cap" at the 5′ end of the RNA soon after it is transcribed. Also called the cap sequence.

Transcription termination sequence   DNA sequence of a gene where transcription is terminated. Transcription continues for about 1000 nucleotides beyond the AATAAA site of the 3′

untranslated region of the gene before being terminated.

Transcription   The process of copying DNA into RNA.

Transcription-associated factors (TAFs)   Proteins that stabilize RNA polymerase on the promoter of a gene and enable it to initiate transcription.

Transcriptional co-regulators   Proteins, recruited by transcription factors, that make modifications in chromatin structure, which either enhance or repress transcription of specific genes.

Transcriptome   Total messenger RNAs (mRNAs) expressed by genes in an organism or a specific type of tissue or cell.

Transdifferentiation   The transformation of one cell type into another.

Transforming growth factor   See **TGF-β superfamily.**

Transgene   Exogenous DNA or gene introduced through experimental manipulation into a cell's genome.

Transit amplifying cells   See **Progenitor cells.**

Transition zone   In neural tube development in vertebrates, the zone between the region that undergoes primary neurulation and the region that undergoes secondary neurulation. The size of this zone varies among different species. See also **Primary neurulation** and **Secondary neurulation.**

Translation initiation site   The ATG codon (becomes AUG in mRNA), which signals the beginning of the first exon (protein-coding region) of a gene.

Translation termination codon   Sequence in a gene, TAA, TAG, or TGA, which is transcribed as a codon in the mRNA—when a ribosome encounters this codon, the ribosome dissociates and the protein is released.

Translation   The process in which the codons of a messenger RNA are translated into the amino acid sequence of a polypeptide chain.

Trefoil stage   A stage in certain spirally cleaving embryos, wherein a particularly large polar lobe is extruded at first cleavage, giving the appearance of a third cell forming before the polar lobe is reabsorbed back into the CD blastomere.

Tri-iodothyronine (T_3)   The more active form of thyroid hormone, produced through the removal of an iodine molecular from thyroxine (T_4). See **Thyroxine (T_4)**.

Trigeminal placode In vertebrates, a pair of intermediate cranial placodes that are subdivided into the ophthalmic and maxomandibular placodes and that generate the distal neurons of the paired trigeminal ganglions, the sensory ganglions of the paired 5th cranial nerve (the trigeminal nerve).

Triploblasts See **Bilaterians**.

Trisomy 21 Condition (in humans) of having three copies of chromosome 2 (an example of aneuploidy). Causes Down syndrome.

Trithorax Family of proteins that are recruited to retain the memory of the transcriptional state of regions of DNA as the cell goes through mitosis; keeps active genes active.

Trophectoderm cells In the mammalian embryo, the outer layer of cells of the blastocyst that surround the inner cell mass and blastocoel; develop into the embryonic side of the placenta.

Trophoblast The external cells of the early mammalian embryo (i.e., the morula and the blastocyst) that will bind to the uterus. Trophoblast cells form the chorion (the embryonic portion of the placenta). Also called trophectoderm.

Truncus arteriosus Cardiac outflow tract precursor that along with the conus arteriosus will form the base of the aorta and pulmonary artery.

Trunk neural crest Neural crest cells migrating from this region become the dorsal root ganglia containing the sensory neurons, sympathetic ganglia, adrenal medulla, the nerve clusters surrounding the aorta, and Schwann cells if they migrate along a ventral pathway, and they generate melanocytes of the dorsum and belly if they migrate along a dorsolateral pathway.

Tube cell Also called the vegetative cell, it is one of two cells produced by the division of the microspore nucleus in angiosperm pollen grains. It engulfs the generative cell, to give rise to the pollen tube.

Tubulin A dimeric protein that polymerizes to form microtubules. Microtubules are a major component of the cytoskeleton; they are found in centrioles and basal bodies; they also form the mitotic spindle and axoneme of cilia and flagella.

Tudor One of the proteins, along with Piwi, Vasa, and Nanos, expressed in germ cells to suppress gene expression. Also involved in anterior-posterior polarity in the *Drosophila* embryo by localizing Nanos, a posterior morphogen.

Tumor angiogenesis factors Factors secreted by microtumors; these factors (including VEGFs, Fgf2, placenta-like growth factor, and others) stimulate mitosis in endothelial cells and direct the cell differentiation into blood vessels in the direction of the tumor.

Tumor suppressor genes Regulatory genes whose gene products protect against a cell progressing towards cancer. Gene products may inhibit cell division or increase the adhesion between cells; they can also induce apoptosis of rapidly dividing cells. Cancer can result from either mutations or inappropriate methylations that inactivate tumor suppressor genes.

Tunica albuginea In mammals, a thick, whitish capsule of extracellular matrix that encases the testis.

"Turing-type" model See **Reaction-diffusion model**.

Type A spermatogonia In mammals, sperm stem cells that undergo mitosis and maintain the population of Type A spermatogonia while also generating Type B spermatogonia.

Type B cells A type of neural stem cell found in the rosettes of the V-SVZ of the cerebrum; fuel the generation of specific types of neurons in the olfactory bulb and striatum.

Type B spermatogonia In mammals, precursors of the spermatocytes and the last cells of the line that undergo mitosis. They divide once to generate the primary spermatocytes.

Type II deiodinase Intracellular enzyme that removes an iodine atom from the outer ring of thyroxine (T_4), converting it into the more active T_3 hormone.

Type III deiodinase Intracellular enzyme that removes an iodine atom from the inner ring of T_3 to convert it into the inactive compound T_2, which will eventually be metabolized to tyrosine.

Type IV collagen A type of collagen that forms a fine meshwork; found in the basal lamina, an extracellular matrix that lies underneath epithelia.

U

Umbilical cord Connecting cord derived from the allantois that brings the embryonic blood circulation to the uterine vessels of the mother in placental mammals.

Umbilical veins See **Omphalomesenteric veins** and **Vitelline veins**.

Undifferentiated zone See **Progress zone**.

Unipotent stem cells Stem cells that generate only one cell type, such as the spermatogonia of the mammalian testes that only generate sperm.

Unsegmented mesoderm Bands of paraxial mesoderm prior to their segmentation into somites.

3′ Untranslated region (3′ UTR) A region of a eukaryotic gene and RNA following the translation termination codon that, although transcribed, is not translated into protein. It includes the region needed for insertion of the polyA tail on the transcript that allows the transcript to exit the nucleus.

5′ Untranslated region (5′ UTR) Also called a leader sequence or leader RNA; a region of a eukaryotic gene or RNA. In a gene, it is a sequence of base pairs between the transcription initiation and translation initiation sites; in an RNA, it is its 5′ end. These are not translated into protein, but can determine the rate at which translation is initiated.

Ureteric bud tip cells A population of stem cells that form at the tips of the ureteric bud branches during metanephric kidney formation.

Ureteric buds In amniotes, paired epithelial branches induced by the metanephrogenic mesenchyme to branch from each of the paired nephric ducts. Ureteric buds will form the collecting ducts, renal pelvis, and ureters that take the urine to the bladder.

Urodeles Amphibian group that includes the salamanders. Compare with **Anurans**.

Urogenital sinus In mammals, the region of the cloaca that is separated from the rectum by the urogenital septum. The bladder forms from the anterior portion of the sinus, and the urethra develops from the posterior region. In females, also forms Skene's glands; in males it also forms the prostate gland.

Uterine cycle   A component of the menstrual cycle, the function of the uterine cycle is to provide the appropriate environment for the developing blastocyst.

V

Vagal neural crest   Neural crest cells from the neck region, which overlaps the cranial/trunk crest boundary. Together with the sacral neural crest, generates the parasympathetic (enteric) ganglia of the gut, which are required for peristaltic movement of the bowels.

Vas (ductus) deferens   Derived from the Wolffian duct, the tube through which sperm pass from the epididymis to the urethra.

Vasa   One of the proteins, along with Tudor, Piwi, and Nanos, expressed in germ cells to suppress gene expression. Also involved in anterior-posterior polarity in the *Drosophila* embryo by localizing Nanos, a posterior morphogen.

Vascular endothelial growth factors (VEGFs)   A family of proteins involved in vasculogenesis that includes several VEGFs, as well as placental growth factor. Each VEGF appears to enable the differentiation of the angioblasts and their multiplication to form endothelial tubes. Also critical for angiogenesis.

Vascular tissue   The conducting tissue in vascular plants that transports fluids and nutrients; its major components are the xylem and phloem. See **Xylem** and **Phloem**.

Vasculogenesis   The de novo creation of a network of blood vessels from the lateral plate mesoderm. See also **Extraembryonic vasculogenesis**.

Vegetal hemisphere   The bottom portion of an ovum, where yolk is more concentrated. The yolk can be an impediment to cleavage, as in the amphibian embryo, causing the yolk-filled cells to divide more slowly and undergo less movement during embryogenesis.

Vegetal plate   Area of thickened cells at the vegetal pole of the sea urchin blastula.

Vegetal pole   The yolk containing end of the egg or embryo, opposite the animal pole.

Vegetal rotation   During frog gastrulation, internal cell rearrangements place the prospective pharyngeal endoderm cells adjacent to the blastocoel and immediately above the involuting mesoderm.

VEGF   See **Vascular endothelial growth factors**.

VegT pathway   Involved in dorsal-ventral polarity and specification of the organizer cells in the amphibian embryo. The VegT pathway activates the expression of Nodal-related paracrine factors in the cells of the vegetal hemisphere of the embryo, which in turn activate the Smad2 transcription factor in the mesodermal cells above them, activating genes that give these cells their "organizer" properties.

Vellus   Short and silky hair of the fetus and neonate that remains on many parts of the human body that are usually considered hairless, such as the forehead and eyelids. In other areas of the body, vellus hair gives way to longer and thicker "terminal" hair.

Ventral (ventrolateral) pathway   Migration pathway of trunk neural crest cells that travel ventrally through the anterior of the sclerotome and contribute to the sympathetic and parasympathetic ganglia, adrenomedullary cells, and dorsal root ganglia.

Ventral furrow   Invagination of the prospective mesoderm, about 1000 cells constituting the ventral midline of the embryo, at the onset of gastrulation in *Drosophila*.

Ventricular (ependymal) cells   Cells derived from the neuroepithelium that line the ventricles of the brain and secrete cerebrospinal fluid.

Ventricular radial glia (vRG)   Progenitor cells that reside in the ventricular zone. They give rise to neurons, outer radial glia (oRG), and intermediate progenitor (IP) cells. See also **Ventricular zone**.

Ventricular zone (VZ)   Inner layer of the developing spinal cord and brain. Forms from the germinal neuroepithelium of the original neural tube and contains neural progenitor cells that are a source of neurons and glial cells. Will form the ependyma.

Ventricular-subventricular zone (V-SVZ)   Region of the cerebrum that contains neural stem cells and is capable of neurogenesis in the adult.

Vertical transmission   In referring to symbiosis, the transfer of symbionts from one generation to the next through the germ cells, usually the eggs.

Vg1   A family of proteins that is part of the TGF-β superfamily. Important in specifying mesoderm in amphibian embryos. See also **TGF-β superfamily**.

Visceral endoderm   A region of the primitive endoderm where the cells contact the epiblast in the mammalian embryo. See **Primitive endoderm**.

Visceral mesoderm   See **Splanchnic mesoderm**.

Viscerocranium   The jaws and other skeletal elements derived from the pharyngeal arches.

Vital dyes   Stains used to label living cells without killing them. When applied to embryos, vital dyes have been used to follow cell migration during development and generate fate maps of specific regions of the embryo.

Vitelline envelope   In invertebrates, the extracellular matrix that forms a fibrous mat around the egg outside the cell membrane and is often involved in sperm-egg recognition and is essential for the species-specific binding of sperm. The vitelline envelope contains several different glycoproteins. It is supplemented by extensions of membrane glycoproteins from the cell membrane and by proteinaceous "posts" that adhere the vitelline envelope to the membrane.

Vitelline veins   The veins, continuous with the endocardium, that carry nutrients from the yolk sac into the sinus venosus of the developing vertebrate heart. In birds, these veins form from yolk sac blood islands, and bring nutrients to the embryo and transport gases to and from the sites of respiratory exchange. In mammals they are called omphalomesenteric veins or umbilical veins.

Vitellogenesis   The formation of yolk proteins, which are deposited in the primary oocyte.

Viviparity   Young are nourished in and born from the mother's body rather than hatched from an egg, as in placental mammals. Compare with **Oviparity**.

Vulval precursor cells (VPCs)   Six cells in the larval stage of *C. elegans* that will form the vulva via inductive signals.

W

White matter   The axonal (as opposed to neuronal) region of the brain and spinal cord. Name derives from the fact that myelin sheaths give the axons a whitish appearance. Compare with **Gray matter**.

Whole mount in situ hybridization　A technique designed for staining specific DNA or RNA sequences within intact tissues and whole embryos. This allows researchers to look at entire embryos or their organs without sectioning them. By using dye-labeled RNA probes that target mRNAs expressed by specific genes, researchers can use the technique to show regions of expression of these genes.

Wholist organicism　Philosophical notion stating that the properties of the whole cannot be predicted solely from the properties of its component parts, and that the properties of the parts are informed by their relationship to the whole. It was very influential in the construction of developmental biology.

Whorled phyllotaxis　In plants, the arrangement of lateral organs in which a set of organs (3 or more) emerges simultaneously in a ring around the stem or apex.

Whorls　In plants, a set of leaves, sepals, petals, or branches that have a whorled phyllotaxis. See **Whorled**.

Wnt pathways　Signal transduction cascades initiated by the binding of a Wnt protein to its receptor Frizzled on the cell membrane. This binding can initiate any of number of different pathways ("canonical" and "noncanonical") to activate Wnt-responsive genes in the nucleus.

Wnt4　A protein in the Wnt family; in mammals it is involved in primary sex determination, kidney development, and the timing of meiosis. It is expressed in the bipotential gonads, but becomes undetectable in XY gonads becoming testes; it is maintained in XX gonads becoming ovaries. See also **Wnts**.

Wnt7a　A Wnt protein especially important in specifying dorsal-ventral polarity in the tetrapod limb; expressed in the dorsal, but not ventral, ectoderm of limb buds. If expression in this region is eliminated, both dorsal and ventral sides of the limb form structures appropriate for the ventral surface, such as ventral footpads on both surfaces of a paw. See also **Wnts**.

Wnts　A gene family of cysteine-rich glycoprotein paracrine factors. Their name is a fusion of the name of the *Drosophila* segment polarity gene *wingless* with the name of one of its vertebrate homologues, *integrated*. Wnt proteins are critical in establishing the polarity of insect and vertebrate limbs, promoting the proliferation of stem cells, and in several steps of urogenital system development.

Wolffian (nephric) duct　In vertebrates, the duct of the developing excretory system that grows down alongside the mesonephric mesoderm and induces it to form kidney tubules. In amniotes, it later degenerates in females, but in males, becomes the epididymis and vas deferens.

Wound epidermis　In salamander limb regeneration, the epidermal cells that migrate over the stump amputation to cover the wound surface immediately following amputation; later thickens to form the apical ectodermal cap.

X

X chromosome inactivation　In mammals, the irreversible conversion of the chromatin of one X chromosome in each female (XX) cell into highly condensed heterochromatin—a Barr body—thus preventing excess transcription of genes on the X chromosome. See also **Dosage compensation**.

Xylem　In vascular plants, the conduits for bringing water and nutrients upward through the plant.

Y

Yellow crescent　Region of the tunicate zygote cytoplasm extending from the vegetal pole to the equator that forms after fertilization by the migration of cytoplasm containing yellow lipid inclusions; will become mesoderm. Contains the mRNA for transcription factors that will specify the muscles.

Yolk cell　The cell containing the yolk in a fish embryo, once the yolk-free cytoplasm at the animal pole of the egg divides to form individual cells above the yolky cytoplasm. Initially, all the cells maintain a connection with the underlying yolk cell.

Yolk plug　The large endodermal cells that remain exposed on the vegetal surface surrounded by the blastopore of the amphibian gastrulating embryo.

Yolk sac　The first extraembryonic membrane to form, derived from splanchnopleure that grows over the yolk to enclose it. The yolk sac mediates nutrition in developing birds and reptiles. It is connected to the midgut by the yolk duct (vitelline duct), so that the walls of the yolk sac and the walls of the gut are continuous.

Yolk syncytial layer (YSL)　A cell population in the zebrafish cleavage stage embryo formed at the ninth or tenth cell cycle, when the cells at the vegetal edge of the blastoderm fuse with the underlying yolk cell, producing a ring of nuclei in the part of the yolk cell cytoplasm that sits just beneath the blastoderm. Important for directing some of the cell movements of gastrulation.

Z

Zebrabow　Transgenic zebrafish used to trigger the expression of different combinations and amounts of different fluorescent proteins within cells, labeling them with a seeming "rainbow" of possible colors that can be used to identify each individual cell in a tissue, organ, or whole embryo.

Zeugopod　The middle bones of the vertebrate limb; the radius and ulna (forelimb) or tibia and fibula (hindlimb).

Zona pellucida　Glycoprotein coat (extracellular matrix) around the mammalian egg, synthesized and secreted by the growing oocyte.

Zona proteins 1, 2, and 3 (ZP1, ZP2, ZP3)　The three major glycoproteins found in the zona pellucida of the mammalian egg; the human zona pellucida also contains ZP4. Involved in binding sperm in a relatively, but not absolutely, species-specific manner.

Zone of polarizing activity (ZPA)　A small block of mesodermal tissue in the very posterior of the limb bud progress zone. Specifies the anterior-posterior axis of the developing limb through the action of the paracrine factor Sonic hedgehog.

Zygote　A fertilized egg with a diploid chromosomal complement in its zygote nucleus generated by fusion of the haploid male and female pronuclei.

Zygotene　Greek, "yoked threads." In the first meiotic division, it is the second stage of prophase I, when homologous chromosomes pair side by side; follows leptotene.

Index

A

A cells in ventricular-subventricular zone, 138, 139, 156n6
Abaxial dermomyotome lip, *480*
Abaxial muscles, *481*, 485, 486, 488, 489n5
Abaxial myoblasts, 488
Abaxial myotome, *480*, 487
ABCDE model, 67–69, *184*, *185–186*
AbdA (Abdominal A) in *Drosophila*, 268, 354
AbdB (Abdominal B) in *Drosophila*, 268, 354
Abdomen development in *Drosophila*, *253*, *254*, *256*, *257*, *259*, *260*, *261*, 268
Abdominal A (AbdA) gene in *Drosophila*, 268, 354
Abdominal B (AbdB) gene in *Drosophila*, 268, 354
Abnormal development. *See* Developmental abnormalities
Accutane, 627
Acetabularia, *586*, *587*
Acetylation, *55*, 62–63
Acetylcholine, *437*
O-Acetyltransferase Porcupine, 108
Acidic FGF (Fgf1), 102, *555*
Acila castrensis (nutclam), *226*, *227*
Acoelomorpha (acoels), regeneration of, 583
Acomys cahirinus (African spiny mouse), skin regeneration in, 615–616, *617*
Acorns, *33*
Acron in *Drosophila*, *257*, *259*, 260
Acrosin, 216
Acrosomal process, 194, *199*, 200, 201, *202*
Acrosome (acrosomal vesicle), 194, *195*, 198, *199*, 200
    exocytosis of, 200, 205, 214
Acrosome reaction
    in external fertilization, *199*, 200, *201*
    in internal fertilization, 212, 214
    species-specific, 200, *201*
Actin
    in acrosome reaction, *200*
    in amphibian gastrulation, *304*
    in axon growth cones, *422*, 423
    cadherins and catenins binding to, 91
    in *Caenorhabditis elegans*, 241
    in cell division, *46*, *47*, 197
    in *Drosophila*, 48, 250, 251, 252, *270*
    in eggs, 197, 201
    in epithelial-mesenchymal transition, 94, 95
    and Hedgehog, 107
    and integrins, 94
    in neuron migration in neocortex development, 390, *391*
    in oogenesis, 183
    polymerization of, 201

in primary neurulation, *368*
in sea urchins, 287
in snails, 229
in sperm, 194
in sperm–egg fusion, 215
treadmilling of, 423
in tunicates, 290
and Wnt signaling pathway, 111
in zebrafish, 325
α-Actinin, 94
Actinomyosin, 369
    in axon guidance, *423*
    in neural crest migration, *417*, 418
Active zone in reaction-diffusion mechanism, 533
Activin, 101, 111
    in amphibians, *313*, 317
    in birds, in left-right axis formation, 344
    in cancer, 638
    in embryonic stem cell differentiation, 148, *149*
    functions of, 112
    gradient of, 100, *101*
    in limb development and reaction-diffusion mechanism, 533
    in neural tube specification, 377, *378*
    in organoid formation, *153*
    in pancreas development, 555
    in sea urchins, 282
Actomyosin in zebrafish, 327
Acute promyelocytic leukemia, 640
Acyrthosiphon pisum (pea aphid), 660, *661*
Adaptations, 27, *32*, 652
    of butterfly larvae, 36–37n10
Adenohypophyseal placode, 442, *443*
Adenoma, intestinal, 639
Adenosine diphosphate (ADP), *102*
Adenosine monophosphate, cyclic, 213
Adenosine triphosphatase (ATPase), 194
Adenosine triphosphate (ATP), 102, 194, *213*
Adepithelial cells, 572
Adherens junctions, 91, 95, 135
Adhesion of cells. *See* Cell adhesion
Adipocytes, 146, 458, *459*
Adipose tissue, 487
ADMP in amphibians, 315
Adolescence, 396–397. *See also* Puberty
Adrenal gland
    congenital hyperplasia of, 168
    as intermediate mesoderm derivative, 462, 491
    as neural crest derivative, *400*, 401, 410
Adult stem cells, 129, 135–147
    in aging, 129
    in cancer, 639–640
    clinical trials on, 156
    in *Drosophila* ovary, *135*, 136–137

ectodermal appendage, 457–458
embryonic stem cells compared to, 131
environmental influences on, 140–141
in flatworms, 595
hematopoietic, 129, 143–145, 508, 511, 512–513
intestinal, 142–143
in laboratory studies, 147
lineage of, *128*
mesenchymal, 146–147
multipotent, 135
neural, 137–141, *386*
niches of, 135–145
in plant regeneration, 588
self-renewal of, 129
Aepyornis, 221n2
Aequorin, 205
African clawed frog, 6, *296*. *See also* *Xenopus laevis*
African spiny mouse (*Acomys cahirinus*), skin regeneration in, 615–616, *617*
AGAMOUS and *AGAMOUS* (AG and AG), *184*, *185*, *186*
AGAMOUS1, *68*, 69
Aging, 156
    adult stem cells in, 129
    chromosome changes in, 183
    growth differentiation factor 11 in, 141, 157n9
    oocytes in, 183
    ovarian germ cells in, *182*
Agouti gene, 660
Agrin, *437*
Air sac primordium, 118, *119*
Albinism in *KIT* mutation, 620, 621
Albumen, *344*
Albumin, 84, 554, *555*
    in bird eggs, 337
    in sperm capacitation, *213*
Alcohol
    as teratogen, 35, 623–627
    zebrafish research on, 324
Algae, 30, *32*, *33*, 36
    regeneration of, 583, 586, 587
Algoriphagus machipongonensis, 661–662
Alisphenoid bone, 418
Alkaline phosphatase, 79
Allantois, *175*, 335, *499*
    in birds, 338
    in mammals, 348, *352*
Allen, Bennet, 566
Alligators, 173, 457
Allocortex, 398n2
Alpha-melanocyte-stimulating hormone, *403*
Alternation of generations, *32*, 34, 186
Alternative pre-mRNA splicing, *71*, 71–72

Alveoli
  of lungs, 557, 559
  of mammary glands, 457
Alx1 and *Alx1* in sea urchins, *281, 282,*
    285
Alzheimer disease, 150, 152
Ambystoma jeffersonianum (Jefferson
    salamander), 221n3
Ambystoma maculatum (spotted
    salamander), *519, 523*
Ambystoma mexicanum (Mexican
    salamander), *2, 6*
  pharyngeal arches in, *18*
  regeneration in, 20, 135, 602, 604, 605,
    606, 617n1
Ambystoma plantineum (silvery
    salamander), 221n3
Ameloblasts, 457
American Paint Horses, *414*
Ametabolous development, 569
gamma-Aminobutyric acid, *138,* 140
Amnion, *175, 335, 336*
  in anamniotic vertebrates, 295
  in birds, 338
  in mammals, 348, 351, *352*
    in twins, 358
Amnioserosa, 252, *253, 271*
Amniotes, 335
  cell migration from caudal progenitor
    zone in, 489n2
  cranial sensory placodes in, *442, 443,*
    444
  egg in, 335, *336*
  endoderm functions in, 547
  heart development in, 501, 502
  kidney development in, 494
  left-right axis formation in, 356
  mesoderm lineages in, *492*
  neural tube closure in, 370–371
  phylogenetic relationships of, *336*
  primary neurulation in, 367
  somitogenesis in, 479
  vasculogenesis in, 508
Amniotic cavity, *336, 349,* 351, *352, 550*
Amniotic fluid, 3, 351
Amoebas, in tree of life, *27, 584*
Amot (angiomotin), *135*
Amphibians, 295–322
  anterior-posterior axis in, 306–307, 309
    primary embryonic induction of, 308
    regional specificity of neural
      induction along, 318–321
  atrazine affecting, 633
  axis formation in, 306–322
    head inducer in, 320–321
    induction of neural ectoderm and
      dorsal mesoderm in, 314–315
    molecular mechanisms of, 308–318
    organizer formation in, 309–314
    organizer functions in, 314
    progressive determination in,
      306–307
    regional specificity of neural
      induction in, 318–321

axon outgrowth in, 420–421
BMP in, 314–318, 320–321, 333n3
bottle cells in, 299, 300–302, 303
β-catenin in, *304,* 307, 310–312, 313, *314*
  in neural ectoderm formation, 316, 317
  in trunk patterning, 321
cell recombination assays, 89–90
cleavage in, *15,* 297–298, 310, 312, *313,*
  346, 359, 360
collective cell migration in, 303–304
convergent extension in, 299, 304–305,
  316, 408
cortical rotation of cytoplasm in, 296,
  297, 311, 312, *316*
cranial placodes in, 443
differential cell affinity in, *89, 90*
dorsal blastopore lip in, 300–303
dorsal mesoderm induction in, 314–315
dorsal-ventral axis in, 296–297, 306–
  308, 309–318, 321
  and BMP, 314–318
  specification of, *310*
early development of, 295–322
endoderm formation in, 551
evolutionary transition of fish to,
  520–521
eye development in, 96–97, 451
  in metamorphosis, 564–565, 568
fertilization in, *8, 9, 10,* 296–297, 307, 310
fibronectin in, 303–304
frogs. *See* Frogs
gastrulation in, *16,* 296–297, 299–305,
  307, *359, 360*
  and BMP, 315, *316*
  epiboly in, 299, 300, 303
  in frogs, *8, 9, 10, 297,* 298, 299–305,
    360
  involution and cell migration in, 299,
    302–304
  and left-right axis specification, 322
  and lens induction, 97
  mediolateral intercalation in,
    304–305
  mid-blastula transition in
    preparation for, 298–299
  and organizer functions, 314, 315
  radial intercalation in, 300, 304, 305
  vegetal rotation in, 299, 300–302
genetic assimilation in, 658
germ layer specification in, 306
gray crescent in, 296, 297, 300, 307
jaw phenotypes in, 658
kidney development in, 303, 305, 309, 492
lateral line placodes in, 442
left-right axis in, 306, 309, 322
life cycle of, 8–11
limb development in, *519, 522*
  in metamorphosis, 564, 566
  in regeneration, 522, 602–607
metamorphosis in, *8,* 9–11, 563, 564–568
  climax in, *11, 567,* 568
  hormonal control of, 11, 564, 565,
    566–568
  morphological changes in, 564–566

regionally specific developmental
  programs in, 568
mid-blastula transition in, 298–299,
  306, 312
neural crest in, 408, 409, 410
neural ectoderm and dorsal mesoderm
  induction in, 314–317
neural tube in, 9, *10,* 309, 314, 318,
  333n4, 372
  primary embryonic induction of,
    307–308
  region-specific induction of, 318
Nieuwkoop center in, 309–310, 313, 314
oocytes of, 196
organizer in, 307–314
  definition of, 307
  fish organizer compared to, 333n5
  formation of, 309–314
  functions of, 314–321
  in neural ectoderm and dorsal
    mesoderm induction, 314–317
  and Nieuwkoop center, 309, 314
  in primary embryonic induction,
    307–308
  in regional specificity of neural
    induction along anterior-posterior
    axis, 318–321
paraxial mesoderm in, 470
primary embryonic induction in,
  307–308
reaggregation of cells in, *89*
regeneration of, *580,* 602
  in salamanders, *2, 4, 6,* 20, 522, 579,
    580, 583, 602–607, 617n1
regional specificity of neural induction
  in, 318–321
salamanders. *See* Salamanders
Spemann and Mangold research on,
  307–308, 309
sperm entry point in, 296, 297, 298, 300,
  307
  and organizer formation, 309, 310, 312
temporal specificity of neural induction
  in, *319*
in tree of life, *27, 584*
trunk patterning in, 321
vital dye staining of, *22*
Amphimixis, 193
Amphioxus, 26, 27, 225, 226
  neurulation in, 366
  pharyngeal arches in, 246n2
  sclerotome and dermomyotome in, 479
Amplexus, *10*
Ampulla, *345*
Amygdala, 397
Amyotrophic lateral sclerosis, 152
Anagen, 458, *459*
Analogous structures, 28, 224
Anamniotes, 295
  amphibians as, 295. *See also*
    Amphibians
  cleavage in, 346
  fish as, 295. *See also* Fish
  lateral line placodes in, 442

Anaphase, 178–179
  in *Drosophila,* 251
  in mice, 183
  in snails, 236
Anchor cells, 122–123
Androgen, 167, 168. *See also* Testosterone
Androgen insensitivity syndrome, 167, 168
Anemia
  Fanconi, 150
  in *KIT* mutation, 620, 621
  sickle cell, 151, 151
Anemones, 197
Anencephaly, 370, 372, 625
Aneuploidy, 183, 620–621
Ang1 in vasculogenesis, 508
Angelfish, 655
Angier, Natalie, 65
Angioblasts, 509
Angiogenesis, 507, 508, 510, 515n4
  and anti-angiogenesis, 510
Angiomotin (Amot), 135
Angiopoietin, 509–510
Angiopoietin-1 and hematopoietic stem
      cells, 144, 145
Angiosperms
  ABCDE model of, 67–69, 184, 185–186
  alternation of generations in, 186
  apomixis in, 220
  *Arabidopsis thaliana. See Arabidopsis
      thaliana*
  fertilization in, 12, 187, 188, 217–220
    double, 187, 189, 198, 220
  floral meristem identity genes in, 184,
      185, 186
  floral organ identity genes in, 67–69,
      184, 185–186, 650, 651–652
  gametogenesis in, 12, 13, 186–189
  germ cells in, 186
  heterochrony of, 651
  heterometry of, 651–652
  heterotopy of, 650, 651
  leaf-to-flower transition in, 12, 67–69,
      184
  life cycle of, 7–8, 11–13, 187
  pollination in, 217–218
  polyspermy prevention in, 220
  reproductive phase in, 12
  sex determination in, 184–186, 651–652
  timing of flowering in, 71, 184
  vegetative phase in, 13
Animal cap
  in amphibians, 298, 304, 309, 317
  in sea urchins, 279
  in snails, 237
  in zebrafish, 328, 330, 331, 332
Animal hemisphere
  in amphibians, 298
  in sea urchins, 279
  in zebrafish, 323
Animal pole, 14, 43
  in amphibians, 298, 301, 305
  in birds, 336, 337
  in sea urchins, 276, 278, 279
  in snails, 227, 228, 238

  in tunicates, 291
  in zebrafish, 325, 327, 328, 329, 330, 331
Animals. *See also specific animals.*
  bilateral symmetry in, 29–30
  cell behavior in embryo, 19
  cleavage patterns in, 13–14
  as direct developers, 563
  gametogenesis in, 174–183
  gastrulation in, 14–16, 224
  germ layers in, 17–18
  as indirect developers, 563
  life cycle in, 6–7, 563
  metamorphosis in, 2, 563–568
  phylogenetic relationships of, 225
  regeneration in, 2, 591–617
  in tree of life, 27, 225
Aniridia, 97, 98
Anisotropy, 232, 233, 235
Annelids, 224, 225, 226
Annulus fibrosus, 483
Anoikos, 142
Anorectal junction, 549
Antennae, 123
  of *Drosophila,* 269, 273
Antennapedia and *Antennapedia* in
      Drosophila, 268, 269, 353, 648
Anterior cranial placodes, 442, 443
Anterior foregut cells, 548
Anterior intestinal portal, 549, 550
Anterior necrotic zone, 541, 542
Anterior visceral endoderm, 352, 353
Anterior-posterior axis, 17, 223
  in amphibians, 306–307, 309
    primary embryonic induction of, 308
    regional specificity of neural
        induction along, 318–321
  in birds, 338, 339, 342, 344
  in *Caenorhabditis elegans,* 240, 241, 243,
      244
  in central nervous system, 375–376
  definition of, 16
  in *Drosophila,* 48, 249, 255–261, 272,
      353, 354
    in wing development, 575–576
  in flatworm regeneration, 597–598, 600
  Hox genes in, 361, 376, 467–469, 478–
      479, 646
    in limb development, 522, 523
    in mammals, 353–356
  in limb development, 517, 518, 522,
      534–539
    Hox genes in, 522, 523
  in mammals, 352–356
  neural tube closure in, 371
  in otocyst, 448
  in paraxial mesoderm, 465–469
    clock-wavefront model of, 478–479
    elongation of, 470–473
  planar cell polarity pathway in, 111
  in sponges, 30
  in zebrafish, 331
Antheridium, 33
Anthers, 186, 187, 188, 652
Anti-angiogenesis, 510

Anticlinal divisions, 133, 135
Antidiuretic hormone, 498
Anti-Müllerian hormone, 111
  in androgen insensitivity syndrome, 167
  in primary sex determination, 161, 165
  in secondary sex determination, 166–167
Antioxidants, 626
Antipodal cells, 188, 189
Antisense RNA, 74, 78
  and DIG-labeled probe, 53, 78–79
Antler regeneration, 580, 586
Anurans
  frogs. *See* Frogs
  metamorphosis of, 564–568
  toads, 564, 565, 566
Anus, 549, 551
  in amphibians, 303, 305
  in *Caenorhabditis elegans,* 240
  in deuterostomes, 276
  in sea urchins, 286, 287
  in snails, 227
Aorta, 410, 416, 420, 463, 550
  and cardiac neural crest, 419, 420
  and heart field formation, 502, 503
  hematopoiesis in, 510–511, 512
  and looping of heart, 506
  and pancreas development, 555, 556
  sclerotome in development of, 463, 464
  and vasculogenesis, 508
Aorta-gonad-mesonephros region, 511
Aortic arch, 419, 420
AP1 and *AP1* (APETALA1 and
      APETALA1), 184, 185, 186
AP2-2 (APETALA2-2), 68, 69
AP3 and *AP3* (APETALA3 and
      APETALA3), 67, 185, 186
APC protein, 110, 422
APC/C protein, 179
Apertures in pollen grains, 218–219
Apes, brain growth and development in,
      392–393, 396
APETALA1 and *APETALA1* (AP1 and
      AP1), 184, 185, 186
APETALA2-2 (AP2-2), 68, 69
APETALA3 and *APETALA3* (AP3 and
      AP3), 67, 185, 186
Aphids, 660, 661
Apical-basal axis
  in central nervous system, 137
  in cerebral organoid formation, 154, 155
  in hydra regeneration, 592–593
  in mice, 135
  in plants, 223
    asymmetrical cell division in, 46–47,
        114
    auxin in, 113–114, 115
    in regeneration, 586
  in sea urchins, 277, 284
Apical cells, 13, 46–47
Apical constriction, 369
  in *Caenorhabditis elegans,* 243
Apical ectodermal ridge, 518, 521, 527–531
  in cetaceans, 544
  and dorsal-ventral axis, 541

FGFs in, 537–538
functions of, 527
in gradient model, *530*
induction of, 527
in limb mesoderm specification, 529–531
and mesenchyme interactions, 527–528, *537*
in reaction-diffusion mechanism, 533, *534*
Apical epidermal cap
in salamander limb regeneration, 603, 605–607
in zebrafish fin regeneration, 607
Apical zone in reaction-diffusion mechanism, 533
aPKC protein, *352*
Apodemus sylvaticus (wood mouse), *212*
Apomixis, 220
Apoptosis
in amphibian metamorphosis, 565, 568
in anoikis, 142
Bcl-x in, 72
BMP in, 112, 542, 543, 649
in *Caenorhabditis elegans*, 245
in cetaceans, 544
in hair regeneration, 458
in injury response, 610, *611*
in kidney development, 492, 494, 497
in limb development, 536, 541–542, 543
in morphogenesis, 19
in neurons, 436
in notochord, 461, 483
in polar body, 183
Apterous and *apterous* in *Drosophila*, 576, 645–646
APX-1 in *Caenorhabditis elegans*, 243
Arabidopsis thaliana
ABCDE model of, *184*, 185–186
alternative pre-mRNA splicing in, 71
apical-basal axis in, 13, *46*, 114, *115*
asymmetrical division in, *46*, 114
auxin in, 114, *115*
double fertilization in, 198
floral organ identity genes in, *67*, 69, *185*, 650, *651*
floral structure in, *185*
flowering signal in, 184
heterotopy in, 650, *651*
life cycle of, 11–13
meristem of, *12*, 13, 114, 132–134, 588
as model organism, *6*, 11, 231
phyllotaxis of, 231–235
pollen germination in, 218, *219*
regeneration of, 588, *589*, *590*
spiral patterns in, 226, 231
Arbacia lixula, 201
Arbacia punctulata, 199, *199*
Archaea in tree of life, *27*, 584
Archaeopteryx, *26*, 29
Archenteron
in amphibians, 300, *301*, 304, 305, *308*, 318, *319*, 322
in sea urchins, *279*, 283, *285*, 286–288
in snails, 237

Archeocytes, 585
Area opaca in birds, *337*, 338, *339*, *340*, 342, *343*, 508
Area pellucida in birds, *337*, 338, *339*, 340, 342, *343*
ARF (AUXIN RESPONSE FACTOR), 114–116, 234
Arginine-glycine-aspartate sequence, 94
Argonaute family proteins, 75–76
ARHGAP11A, 395
ARHGAP11B, 395–396, 397
Aristotle, 2–3, 335
Armadillo and *armadillo* in *Drosophila*, 260, 266
Arrector pili muscles, 458
Arrow worms, *276*
Arthropods, 224
body plan in, 662
crustacean, 25–26
limb development in, 653
in tree of life, *225*
Arthrotome, *463*, *464*, 482
Arthur, Wallace, 649
Artificial selection, 28, 656
Ascidians, 288. *See also* Tunicates
Asperger syndrome, 152
ASPM, 396
Assimilation, genetic, 656–659
Astroglia, 431
Astyanax mexicanus (Mexican cave fish), 450, *451*, 651
Asymmetrical cell division
and apical-basal axis in plants, 46–47, 114
of radial glial cells, 387, 388, 392
of stem cells, 127–128, 129, 131, 386–387, 453
in tunicates, 289
ATG sequence, *56*, 57
Atrazine, 629, 633–634
Atria, cardiac
differentiation of, 505
and heart fields, 502, 504
and looping of heart, 506, 507
retinoic acid affecting, 505
specification of, *506*
in zebrafish, 609, 612, *613*
Attractor states, 514
Auditory cortex, 386
Auditory tube, *551*
Autism spectrum disorder, 152, 157n21
Autoactivation in Turing model, 531–532
Autocrine interactions, 100, 267
Autologous structures, 153, 157n23
Autonomous specification, 41–43, 45, 292n6
in *Caenorhabditis elegans*, 242
cytoplasmic determinants of, 41–43
definition of, 41
in frogs, 44, 45
in plants, 46–47
in sea urchins, 45, 278, 282
in snails, 41, 226, 235, 237
in tunicates, 41–43, 44, 289, 290

Autopod, 517, *518*, 541–542
apoptosis in sculpting of, 541–542, 543
gradient model on, *530*, 531
Hox gene specification of, 519, 520, *530*, 538, *539*
reaction-diffusion model on, 531, *532*, *534*
Autotrophy, 33
Auxin, 47, 113–116
in apical-basal polarity, 113–114, *115*
discovery of, 114
efflux transport of, 114, *115*, 232–235
functions of, 113
and phyllotaxis of plants, 232–235
and regeneration in plants, 587, 588–591
signaling pathway, 114–116
synthesis and distribution of, 114, *115*
Auxin repressor protein, 114–116
AUXIN RESPONSE FACTOR (ARF), 114–116, 234
AUX-Rep (auxin repressor protein), 114–116
Avian development. *See* Birds
Axial mesoderm, 461, 491, *500*
Axin in Wnt signaling pathway, 110
Axin1 in zebrafish fin regeneration, 609
Axis determination, 16, 223
in amphibians, 306–322
in birds, 338, 342–345, *355*
in *Caenorhabditis elegans*, 240, 241–242
in *Drosophila melanogaster*, 223, 247–274
in mammals, 351–356, 361n1
in otocyst, 447–448
in sea urchins, 286
in snails, 235–237
species differences in, 223
in tunicates, 290
in zebrafish, 330–333
Axis elongation in somitogenesis, 470–473, 479
AXL receptor proteins, 642
Axolotl salamanders, *2*, *6*
pharyngeal arches in, *18*
regeneration in, 20, 135, 602, 604, 605, *606*, 617n1
Axonemes, 194, *195*
Axons, 420–436
address selection, 424, 432, 436
in central nervous system, 439n12
chemoaffinity hypothesis on, 434
of commissural neurons, 424, 428–429, *430*
and enteric neural crest cell migration, 412–413
formation of, 420
growth cones of, 421–423. *See also* Growth cones
guidance of, 78, 423–430
in marginal zone, 382
of motor neurons, 424–427, 436, *437*
in optic system, 424, 431, *432*
outgrowth of, 420–421
pathway selection, 424

in peripheral nervous system, 439n12
as pioneer nerve fibers, 423–424
of Purkinje neurons, 384, 421
of retinal ganglion, 431, *432*, 433–435
in salamander regeneration, *603, 606*
specificity of connections, 423–424, 431
synapse formation, 421, 436–438
target selection, 424, 432–435, 436

B

B lymphocytes, *130, 514*
B type stem cells in V-SVZ, 137–141, 156n6
Bacteria
   symbiotic, 32–33, 620, 660–662
   in tree of life, *27, 584*
bag of marbles (bam) in *Drosophila, 136,* 137
Balance function of ear, 441, 442, 444, 446, *447*
Baleen whales, 26
bam (bag of marbles) in *Drosophila, 136,* 137
Bardet-Biedl syndrome, 118
Barnacles, 25–26
Barr bodies, *169*
Barx1 and *Barx1, 552,* 553, 557, *558*
Basal bodies, 356
Basal cell carcinoma, 108
Basal cell nevus syndrome, 108
Basal cells, 13, 46–47
Basal disc in hydra, 591, 592, *593*
Basal epidermis, 453
Basal lamina, *93, 94,* 453
   in birds, 340
   and epibranchial placode, *446*
   in epithelial-mesenchymal transition, 94, 95, 340
   in mammals, *352*
   in neural crest migration, 410
   in sea urchins, 284, 285
   of ureteric bud, 494, *495*
Basal layer of epidermis, 453, 454
Basal phyla, 224, *225*
Basal transcription factors, 58
Basement membrane, *93, 95, 353,* 453
Basic fibroblast growth factor, 102. *See also* Fgf2
Basic helix-loop-helix (bHLH), *66,* 487
Basic leucine zipper (bZip), *66*
Basilar papilla, 444, 447
Basket Tree, *2*
Basophils, *130*
Bats
   hematopoiesis in, 510
   as pollinators, 217
   in tree of life, *27, 584*
   wings of, 28, *29, 543,* 649, 663
BCIP (5-bromo-4-chloro-3-indoyl-phosphate), *79*
Bcl-2 in mammals, 345
Bcl-X and *Bcl-x, 71,* 72
BDNF (brain-derived neurotrophic factor), 150, 432
Beak patterns, heterometry in, 652
Bees, *27, 218, 584*

Beetles, 569, 645–646, 663n3
Behavior-induced DNA methylation, 660
Bergmann glia, *384*
Beta cells of pancreas, 556
β-globin gene, 56, 58
bFGF (basic fibroblast growth factor). *See* Fgf2
bHLH (basic helix-loop-helix), *66,* 487
Bicarbonate in sperm capacitation, *213*
Bicoid and *bicoid* in *Drosophila,* 262, 264, *265,* 273n2
   and anterior-posterior axis, *255, 256*–260
   and mRNA, 73, 77
Bilaminar germ disc, 351
Bilateral cleavage
   holoblastic, 14, *15,* 288–289
   meroblastic, *15*
Bilateral symmetry, 29–30, *32*
   of triploblasts, 224
   of tunicates, *288*
Bilaterians, 29–30, 224–226
   anterior-posterior axis in, 646
   regeneration in, 583
   in tree of life, *225*
Bile acids, 614
Bile ducts, *548, 554,* 557
Biliary atresia, 557
Biliatresone, 557
Bin3 in *Drosophila,* 258
Bindin, *200, 201, 202, 204, 207*
Bindin receptors, 204, *207*
Biodiversity, evolution as source of, 663n1
Bioelectric pattern in flatworms, 599–602, 617
Biology
   central dogma of, 52–53, 586
   developmental. *See* Developmental biology
Bipotential gonads, 160, 161, *162, 163,* 167
Birds, 335, 336–345
   as amniotes, 335
   axis specification in, 338, 342–345, *355*
   basilar papilla in, 444
   beak patterns in, 652
   chickens. *See* Chick development
   cleavage in, 14, *15,* 335, 336–338, 359, *360*
   cochlea in, 447
   collapsins in, 426
   ear development in, 342, 459
   early development in, 335, 336–345
   eggs of, 196, 336–338
   fertilization in, 336
   gastrulation in, *16,* 338–342, *343,* 348, 359, *360, 463*
   germ layers in, 338
   gonads in, 164
   as gynandromorphs, 169, 189
   heterochrony in, 650
   heterometry in, 652
   limb development in, 522, 541–542, 543
   lung development in, 560
   neural crest in, *367, 368,* 402, 410, 419

cardiac, *420*
   migration of, 22, *23,* 401, 439n2
   neural tube in, *342, 366, 367, 374, 378, 382, 480, 485*
   closure of, 370, 371
   neurulation in, 366, 367, 368, 374, 463
   phylogenetic relationships of, *296, 336*
   as pollinators, 217, *218*
   presomitic mesoderm in, *467, 468*
   primitive streak in, 338–340, *341, 343,* 359
   sex determination in, 159, 169
   temporal collinearity of Hox genes in, 468–469
   in tree of life, *27, 584*
   vasculogenesis in, 508–509
   vertebrae in, 655
   vitelline veins in, 508–509
   web-footed, 541–542, *543*
   wings of, 28
Birdwing butterfly (*Ornithoptera croesus*), *170*
Birth defects, 34, 35, 619, 620–621
   in kidney, 498
   in neural tube, 372–374, 380
   pyloric stenosis in, 553
   in teratogen exposure, 35–36, 621–628
Bisphenol A (BPA), 629, 632–633
   DNA methylation from, 660
   in plastic products, 631, 632–633
   transgenerational effects of, 636
Bithorax and *bithorax* in *Drosophila,* 268, 353, 656, *657*
Bivalent configuration, 177, *178*
Bladder, 551
   formation of, 166, 167
   ureter insertion into, 498, *499*
Blastema, 20, 585
   in African spiny mouse, *616,* 617
   in flatworms, 594, 595–596, 597, 598
   macrophages in formation of, 616
   in salamanders, 20, 602–607
   in young mammals, 615
   in zebrafish, 608–609, 610, 612
Blastocoel
   in amphibians, 298, 318
      in axis formation, *308*
      in gastrulation, 300, 301, 302, 303, 304, *305*
   in birds, 339, 341
   in *Caenorhabditis elegans,* 243
   definition of, 36n4
   in mammals, 348, *349*
      in gastrulation, 351
      in mice, 134, *346*
      in peri-implantation period, 348
      in twins, *358*
   in sea urchins, 277, 278, *279, 283, 284, 285,* 287, 288
   in tunicates, 289
Blastocyst, *345, 346, 347,* 348, 361n4
   definition of, 36n4
   inner cell mass of, 131, 134–135, 148, 358
   in peri-implantation period, 348, *349, 350*

species differences in, 156n1
in twins, 358
Blastoderm
  in birds, 3, *340, 341*, 342
    in axis specification, 342, 343, *344*
    cleavage of, 336–338
    in gastrulation, 338, 340, 342
  cellular, *47, 48*, 250–252
  cranial sensory placodes originating
    in, 441
  in *Drosophila*, 47–48, 100, *171, 249,*
    250–252, 255, 258, 263, 270
  syncytial. *See* Syncytial blastoderm
  in zebrafish, 325, 326–327, 328
Blastodisc
  in birds, 336, *337*, 338, *344*
  in mammals, *349*
  in zebrafish, 324–325
Blastomeres, 13–14
  in amphibians, 298, *298*, 313, 551
  in *Caenorhabditis elegans*, 242–243
  definition of, 7, 36n4
  in *Drosophila*, 249
  in mammals, 13–14, 350
    in cleavage, 346, 347
    in twins, 358
  in sea urchins, 275, 276, 277, 278–279, 288
    conditional specification of, 44–45
  in snails, 41, 226–227
    and axis determination, 235, 236–237
    in cleavage, 226–227, 228, 230
    D-quadrant, 230, 236–237
  in tunicates, *288*, 290, 291
    autonomous specification of, 42–43,
      290
    in cleavage, 289
    and fate maps, 20–21, *43*
  in zebrafish, 324, 325, 331
Blastopore
  in amphibians, *8, 9, 10*, 298
    formation of, *302*
    in gastrulation, 300–303
  in *Caenorhabditis elegans*, 245
  definition of, 36n4
  in sea urchins, *283, 286, 287*
  in snails, 227, *228*, 237, *238*
  in zebrafish, *323*, 330–331
Blastopore lip
  in amphibians, 300–303, 304, 305
  dorsal. *See* Dorsal blastopore lip
  in fish, 330–331
  lateral, *301, 303, 305*, 331
  in snails, 237
Blastula, 16
  in amphibians, *10*, 53, 298, 299, 300,
    314, 317
  cell fate determination, *40*, 44
  definition of, 7, 36n4
  in fish, *326*
    in zebrafish, *73, 76*, 119
  in sea urchins, 277–278, *279*, 283, 286,
    292n1
  stereoblastulae, 226
BLBP (brain lipid binding protein), *138*

Blebbistatin, *147*
Blind Mexican cavefish, 450, *451*, 651
Blood
  generation of cells in, 510–515
  and hematopoietic stem cells. *See*
    Hematopoietic stem cells
  regeneration of, 581
Blood flow
  and heart development, 506–507, 515n6
  and hematopoiesis, 511, 515n6
  shear stress in, 511, *512*, 515n6
Blood islands, *367*, 508–509, 511
Blood vessel development, 491, 507–510
  angiogenesis in, 507, *508*, 510, 515n4
  cardiac neural crest in, 401
  cell differentiation in, 505
  endoderm in, 547
  heart field formation in, 502, 503
  hemangiogenic mesoderm in, *503*
  lateral plate mesoderm in, 491, 499,
    507–510
  pharyngeal arches in, *416*
  in regeneration, 582, 612
  sclerotomes in, 464
  vasculogenesis in, 507–510
  in zebrafish, *323*, 612
Blood–brain barrier, 141
BMP (bone morphogenetic protein), 101,
    111, 112
  and adult neural stem cells in V-SVZ,
    138, 140
  in amphibians, 314–318, 320–321, 333n3
  in apoptosis, 112, 542, 543, 649
  in bat wing development, 649, 663
  in birds, 344–345, 542, 543
  in cancer, 638
  in cerebellum development, 384
  in cetaceans, 544
  and Chordin interactions, 646, *647*,
    663n5
  in cranial placode induction, 443, 445
  diffusion of, 117
  in digestive tube development, *552*
  in dorsal-ventral axis, 314–318, 646
    in limb development, 540, 541
    of neural tube, 377, 378
  in *Drosophila*, 112, *136*, 137, 318, 575
  and ectoderm derivatives, *364*, 365
  in ectodermal appendage development,
    456
  in endoderm specification, 548–549
  in epidermis specification, 365, 453
  evolution of pathway, 646
  in eye development, 449, 451, *452*
  in facial features, 419
  functions of, 112, 663n8
  in hair follicle regeneration, 458, *459*
  in heart development, 503
  in hematopoiesis, 511
  inhibitors of. *See* BMP inhibitors
  in kidney development, 497
  in limb development, 537, 538
    in apoptosis, 542, 543
    context dependency of actions, 542

  in dorsal-ventral axis, 540, 541
  and reaction-diffusion mechanism,
    533
  in liver development, 554
  in mesoderm specification, 465, 491,
    493, 503, *548*
  in myotome determination, 488–489
  in neural crest delamination, 406–407
  in neural crest specification, 404, *405*
  in neural plate folding, *369*, 370
  and primordial germ cell migration,
    175, 176
  in sclerotome development, 481
  in secondary neurulation, 374
  and small toolkit in evolution, 646
  in spermatogenesis, 181
  in tooth development, 455
  in turtle shell development, 649
  in vasculogenesis, 507
  in zebrafish, 330, 331–332, *608*
BMP inhibitors, 112, 140, 361, 488–489
  in amphibians, 314–318, *321*
  in cranial sensory placode induction, *443*
  in finch beak shape, 652
  in hair follicle regeneration, *459*
  in joint formation, 542
  in kidney development, 498
  in neural crest development, *405*
BMP1, 112
BMP2
  in amphibians, 315, 316
  in cardiogenic mesoderm specification,
    503
  in hair follicle regeneration, *459*
  in hematopoiesis, 511
  in limb development, 542
  in salamander limb regeneration, *607*
BMP3 in facial features, 419
BMP4 and Bmp4, 112
  and adult intestinal stem cell niche, *142*
  in amphibians, *314*, 315, 316, 318
  in birds, 344, 465, 652
  and Chordin, 646, *647*, 663n5
  deep homology of, 646, *647*
  in dorsal-ventral specification of neural
    tube, 377, *378*
  in embryonic stem cell differentiation,
    148, *149*
  in eye development, 97, *452*
  functions of, 112
  in gut development, 553
  in hair regeneration, *459*
  in hematopoiesis, 511, *512*
  in kidney development, 498
  in limb development, 542, *543*
  in mesoderm specification, 465, 491
  in myotome determination, *487*, 488
  in organoid formation, *153*
  and primordial germ cells, 175
  in sclerotome development, *481*
  in tooth development, 455, *455*
Bmp6 in hair growth, 458, *459*
BMP7 and Bmp7
  in amphibians, 315

and brown fat, 487
in dorsal-ventral specification of neural tube, 377, *378*
in eye development, *452*
functions of, 112
in kidney development, 497
in limb development, 542
in salamander limb regeneration, *607*
in tooth development, *456*
BMP11 (growth differentiation factor 11), *138*, 141, 157n9
Boc (Brother of Cdo) protein, 428
in axon guidance, 428, *430*, 431
in Hedgehog pathway, *106*, 107, 428
Body plan
evolution of, 654, 662
in primary larvae, 563
in secondary larvae, 563
Boettcher's cells, *447*
Bone development. *See* Skeleton development
Bone marrow
hematopoietic stem cells in, *130*, 143–145, 157n12, 511, 512–513, 581
mesenchymal stem cells in, 146
transplantation of, 143, 157n12, 512
Bone morphogenetic protein. *See* BMP
Bonobos, brain development in, *395*, 396
Border cells, 447
Bottle cells in amphibians, 299, 300–302, 303, 305
Boveri, Theodor, 203, 238
Bowman's capsule, 494, *495*
BPA. *See* Bisphenol A
Brachydanio rerio. See Zebrafish
Brachypodia, 542
Brachypterygius, 534
Brachyury and *Brachyury*, 544n4
in amphibians, 303, 306
and *goosecoid* expression, *101*
in hydra regeneration, 593
in mammals, 353, 361
in mesoderm specification, 465, 547
in notochord formation, 292n7
in stem cell differentiation, *149*
in tunicates, 290, 291, 292
Brain-derived neurotrophic factor (BDNF), 150, 432
Brain growth and development, 363, 381–398
in adolescence, 396–397
adult organization specified in embryonic brain, 386
alcohol affecting, 624, 625, 626–627
in amphibians, 313, 314, 318, *320*, 321
anterior-posterior axis in, 375–376
axon guidance in, 417, 431
in birds, 341, 342, *375, 376*
BPA affecting, 632, 633, 636
in *Caenorhabditis elegans*, 242
Cajal-Retzius cells in, 390–391, 396
cell division in, 386–387, 392
cell migration in, 384, 389–391, 392, 396
cerebral folds in, 393–394, 395, 397, 398n3

in cerebral organoid, 153, 154–155
dorsal-ventral axis in, 376–377
in Down syndrome, 621
ectoderm in, 363, *364*, 549
fate mapping of, *24*
fracking chemicals affecting, 634
functional regions in, 386
genes in, 395–396
in humans, 381, 392–397
in infants, growth rate of fetal neurons in, 392–393
interkinetic nuclear migration in, 386–387
layers in, 382, 389–391, 396
in mammals, 381, 385, 392–397
mechanisms regulating, 386–392
in microcephaly, 155
myelination in, 397
neocortex in, 385, 389–392
neural stem cells in, *120*, 137–141, 386–387
neuroanatomy in, 381–386
neurogenesis in, 387–388, 392
neuron differentiation in, *129*
Notch signaling in, 139, 156n7, 392
number of neurons in, 381, 421
primary cilium in, *120*, 392
radial glial cells in, 386, 389–390, 392, 394, 395
retinal ganglion axons in, 431
signaling mechanisms in, 390–392
SRGAP2 in, 648
subgranular zone in, 137
synapses in, 393
transcription factors in, *59*
in tunicates, 290, 291
ventricular-subventricular zone in, 137–141
weight of brain in, 393
in zebrafish, *323*, 387
Zika virus affecting, 623
Brain lipid binding protein (BLBP), *138*
Brain organoid, 153, 154–155
Branchial arches, *18*, 439, *566*. *See also* Pharyngeal arches
Brand, Michael, 118, 156n5
Breast cancer, 638
Breast development. *See* Mammary gland development
Brenner, Sydney, 238
Bridges, Calvin, *248*
Brine shrimp, 654
brinker (brk), 575, 576
Broad in *Drosophila* metamorphosis, 574
5-Bromo-4-chloro-3-indoyl-phosphate (BCIP), *79*
Bronchi, 557, 558–559
Brooks, William Keith, 39
Brother of Cdo (Boc) protein, 428
in axon guidance, 428, *430*, 431
in Hedgehog pathway, *106*, 107, 428
Brown adipose cells, 487
Browne, Ethel, 592, 593
Bryant, Susan, 580

Bryophytes, 583
Bryopsis plumosa, 587
Buchnera aphidicola, 660
Buckwheat, 660
Budding in hydra, 591, *592, 594*
Bulge region in hair follicles, 458, *459*
Burke, Anne C., 489
Butterflies, 563, 577
adaptations of, 36–37n10
as gynandromorphs, *170*
holometabolous development of, 569
monarch, 36–37n10, 577, 628
pigment variations in, 658
buttonhead in *Drosophila, 260*
bZip (basic leucine zipper), *66*

C
C cells in ventricular-subventricular zone, *138, 139, 140, 141*, 156n6
Cactus finches, 652
Cactus protein in *Drosophila*, 270–271
Cadherin, 30, 91–92
in amphibian gastrulation, 304, 305
in cell sorting, 92, 392
in epithelial-mesenchymal transition, 94, 95
in neocortex development, 390, *391*, 392
in neural crest delamination, 407–408
in sea urchins, 285
E-Cadherin, *92, 95*
in *Caenorhabditis elegans*, 245
in *Drosophila, 136*, 137
in lung development, *559, 560*
in mammals, 135, 346, *350*, 351, *352*
in neural crest delamination, 407–408
in neural tube formation, 372
in zebrafish, 327, 392
N-Cadherin, *92*
in epithelial-mesenchymal transition, 486
in motor neuron guidance, 426
in neocortex development, 390, *391*, 392
in neural crest delamination, 407–408
in neural crest migration, 409, *417*
in neural tube formation, 372
in retinal ganglion axon guidance, 431
in sclerotome development, 479
in synapse formation, 436
P-Cadherin, 92
R-Cadherin, 431
VE-Cadherin, *512*
Caenorhabditis elegans, 226, 238–245
antisense RNA in, 74
autonomous specification in, 242
axis determination in, 240, 241–242
blastomere identity in, 242–243
cell fusion in, 245
cell lineage in, 238, *239*, 240, 243, *244*
cleavage in, 240, 241, 242
conditional specification in, 242–243
fertilization in, 239–240, 241
gastrulation in, 243–245
genome of, 238, 239
hermaphroditism in, 121, 238, 239, *240*, 245

LIN-14 and *lin-14* in, 74
as model organism, 5, 6
proteoglycans in, 93
vulva in, 121–123, *240*, 245
Cajal-Retzius cells, 390–391, 396
Calbindin, *386*
Calcarea, 585
Calcium channels, 212
Calcium ions
  in acrosome reaction, 214
  in amphibian gastrulation, 305
  in angiosperm fertilization, 220
  in cortical granule reaction, 205–206, 207, 208
  in egg activation, 182, 206–209, 249
  and G protein, *208*
  in pollen tube elongation, *219*, 220
  in sperm hyperactivation, 212
  in sperm rheotaxis, 212
  in Wnt signaling pathway, *110*, 111
  in zebrafish, 325
Calcium-dependent adhesion molecules, 91. *See also* Cadherin
Callose, 219
Callus in plant regeneration, 587, 588
Calmodulin and *Calmodulin*, 652
Calretinin, *386*
Cambrian explosion, 33
Camerarius, Rudolph, 190n7
Canalization model of plant regeneration, 589–591
Cancer, 636–640
  cell-to-cell communication in, 636, 637–639
  in childhood, in cervical vertebrae anomalies, 655
  collective cell migration in, 489n3
  context-dependent, 637
  development-based therapy in, 640
  in diethylstilbestrol exposure, 629, *630*
  epithelial cell, 638, 640
  epithelial-mesenchymal transition in, 95, 284, 640
  from fracking chemicals, 634
  Hedgehog in, 108
  HeLa cell line in study of, 147
  hematopoietic stem cells in, 145
  in mammary gland, 638
  metastasis of, 636
  somatic mutation theory of, 636
  stem cells in, 637, 639–640
  syncytium in, 49n4
  tissue organization field theory of, 636, 637
Canonical Wnt/β-catenin pathway, 109–111
Cap cells, *136*, 137
Cap sequence, 57
Capacitation of sperm, 212, 213, 214
Capecchi, Mario, 519
CAR cells and hematopoietic stem cells, *144*, 145
Carcinogenesis, 636
  somatic mutation theory of, 636

stem cell hypothesis on, 639–640
tissue organization field theory of, 636, 637
Cardia bifida, 504
Cardiac myocytes
  in mice, neonatal, 615
  in zebrafish, 609–612, *613*, 615
Cardiac neural crest, 401, 419–420, 503
Cardiogenic mesoderm, 501–503, 554, *555*
Cardiomyocytes, *149*, 502
Cardiopharyngeal mesoderm, *502*
Carollia perspicillata (fruit bat), 663
Carotid artery, *416*, 432
Carotid body, 420, 442
Carotid sinus, 442
Carpal bones, 517, *518*
Carpels, 185, 188, *189*
  in *Arabidopsis thaliana*, 12, 13, *185*
  and floral organ identity genes, 68–69, *185*, 186
  in pollen tube elongation and navigation, 219, 220
Carrot, 587
Carson, Rachel, 621
Cartesian coordinate model, 272
Cartilage development, 536
  mesenchymal stem cells in, 146
  neural crest cells in, 401, 402, *403*, 415
  reaction-diffusion mechanism in, 533
  in salamander regeneration, 605
  and tendon formation, 484–485
  Turing model on, 531, 533
Cas9 (CRISPR associated enzyme 9), 82–83
  in *Caenorhabditis elegans*, 239
  in zebrafish, 82, 324
Casein in JAK-STAT pathway, 104
Caspase3, *438*
Caspase-9 in amphibian metamorphosis, 565
Catagen, 458, *459*
β-Catenin
  and adult stem cells in V-SVZ, *139*
  in amphibians, *304*, 307, 310–312, *313*, *314*
    in neural ectoderm formation, 316, 317
    in trunk patterning, 321
  in *Caenorhabditis elegans*, 242
  in digestive tube development, 553
  in *Drosophila*, *266*
  in ectodermal appendage development, 456, *457*
  in flatworm regeneration, 595–596, *597*, 598, 600–601
  in hydra regeneration, 593, *594*
  in limb development, *524*, 527, *537*
  in lung development, 557
  in mammals, in primary sex determination, 160, *161*, *163*, 164, 165
  in sea urchins, 280–281, *285*, 292n3
  in snails, 237
  in tooth regeneration, 457

in tunicates, 290, *291*
and Wnt, 109–111, 456, *524*, 527
in zebrafish, 331–332, 333n6
  in fin regeneration, 608–609
Catenins, 109–111
  and cadherins, 91, *92*
  in epithelial-mesenchymal transition, *95*
Caterpillars, 577
  metamorphosis of, 563, 570
  of Monarch butterflies, 577
Cats, X-chromosome inactivation in, *169*
CatSper calcium channels, 212, 214
Caudal and *caudal* in *Drosophila*, 73, *256*, 258–259, 260, 262, *265*
Caudal intestinal portal, 549, *550*
Caudal lateral epiblast, 341, *342*, 379
Caudal progenitor zone, 466, *467*, 469, 470–473
  cell migration from, 470, 489n2
  in forelimb formation, 523
Caudal vertebrae, 355
Cavefish, 450, *451*, 651
Cavitation, 348
  in secondary neurulation, 374
  in zebrafish, 380n1
CBCC (crypt base columnar cell), 142
CBF1 in Notch signaling, *121*
CBFα1 and mesenchymal stem cell differentiation, *147*
CD45 in hematopoietic stem cell formation, *512*
Cdc25a, 471
Cdc42, *423*, 472
Cdk1 in egg activation, 210
Cdk5 regulatory subunit-associated protein 2 (CDK5RAP2), 155, 157n24
Cdx transcription factors, 553
Cdx1 in digestive tube development, 553
Cdx2 and *Cdx2*, 134, *135*
  in digestive tube development, 553
  in mammals, 347, *350*
Cdx4 in vasculogenesis, 507
CdxA, *552*
CdxC, *552*
Cecropia, 563
Celery, 586
Cell adhesion, 87, 88–92, 95
  in axis elongation, 470, 471
  cadherins in, 30, 91–92, 120
  differential adhesion in, 90–91
  differential affinity in, 89–90
  differential interfacial tension in, 93
  equilibrium in, 91
  extracellular matrix in, 93
  fibronectin in, 93
  heterotypic aggregates in, 92
  homotypic aggregates in, 92
  in inner cell mass, *135*
  laminin in, 94
  in motor neuron guidance, 426
  of neuromesodermal progenitors, 470, 471
  in organoids, 153
  surface tension in, 91, *92*

Cell adhesion molecules
    alcohol affecting, 626
    in axon guidance, 423, 426
    cadherins as, 30, 91–92, 120
    juxtacrine signaling of, 120
    in limb development and reaction-
        diffusion mechanism, 533
    in mammals, 346
    in neural tube closure, 372
    in organoid formation, *153*
    in ventricular-subventricular zone, *138,*
        139
Cell affinity, differential, 89–90
Cell cycle, 14, 387
    in amphibians, 298
    in hair follicles, 458, *459*
    in hydra, 592
    in mammals, 345
Cell death
    in alcohol exposure, 625–626
    in amphibian metamorphosis, 565
    in apoptosis, 19. *See also* Apoptosis
Cell differentiation, 4, 39–49, 95
    autonomous specification in, 41–43
    commitment in, 39–41, 128
    conditional specification in, 44–45
    definition of, 4, 39
    in dorsolateral pathway of neural crest,
        414
    gene expression in, 51–85
    genomic equivalence in, 51, 53–54
    in heart development, 505
    juxtacrine signaling in, 120–123
    lineage tracing studies of, 41
    in organoids, 153
    reprogramming of, 67
    syncytial specification in, 47–48
Cell division, 19, 95
    actin in, *46,* 47, 197
    asymmetrical, 46, 386–387. *See also*
        Asymmetrical cell division
    in birds, 340
    in brain development, 386–387, 392
    in *Drosophila,* 250, *251,* 252
    in epidermis, 453
    in hair follicles, 458
    in mammals, 345–347
    in meiosis. *See* Meiosis
    in mitosis. *See* Mitosis
    in neural crest, 402
    in plants, 13, 34, 46, *46*
    in primary neurulation, 368
    of radial glia, 387, 388
    of stem cells, 127–128, 131, 386–387, 388
    symmetrical, 387
    in zebrafish, 324–327, *387*
Cell fate determination, 39, 40, 41
    autonomous specification in, 41–43
    in *Caenorhabditis elegans,* 242–243, *243*
    conditional specification in, 44–45
    definition of, 40
    in *Drosophila,* 260
    homeotic regulation of, 69
    lateral inhibition in, 123

morphogen gradients in, 100
    in organ of Corti, 448
    in otocyst, 448
    in paraxial mesoderm, 466
    in plants, 46–47
    reprogramming of, 67
    in sea urchins, 275, 277, 278
    in snails, 237
    syncytial specification in, 47–48
Cell fate mapping. *See* Fate mapping
Cell fate maturation, 40–41
Cell identity, 39–49, 120–123
Cell lineage, 41, 42–43
    in *Caenorhabditis elegans,* 238, *239,* 240,
        243, *244*
    and gene regulatory networks, 70
    of germ cells, 159, 174, 175
    invariant, 238
    of neural crest, 402, *403*
    of radial glia, 388
    of stem cells, 128, 143–145, 387, 388
Cell membrane, 19
    adhesive properties of, 91
    and cell-to-cell communication, 87, 88
    of egg, *196,* 197–198
        fusion with sperm cell membrane,
            200, 201–202, *203,* 214–215
        membrane potential of, 203–204, *210*
        in polyspermy prevention, 203–205
    focal protrusions as signaling sources,
        117–119
    Notch protein in, 120
    primary cilium of, 118, *120*
    of sperm, 194, *195,* 201–202
        equatorial region of, 215
        fusion with egg cell membrane, *200,*
            201–202, *203,* 214–215
Cell migration, 16, 95
    alcohol exposure affecting, 625
    in amphibians, 299, *301,* 303–304, 307
    of axon growth cones, 421–423
    in birds, 338, 340–341, 342
    in brain development, 384, 389–391,
        392, 396
    from caudal progenitor zone, 470, 489n2
    collective. *See* Collective migration
    extracellular matrix in, 93, 94
    fibronectin in, 94
    in heart development, 501, 503–505
    of hematopoietic stem cells, 511
    laminin in, 94
    in mammals, 351, 353
    in morphogenesis, 19
    of myoblasts, 485, 488
    of neural crest, 405–418. *See also* Neural
        crest, migration of
    of neuromesodermal progenitors,
        470–471
    in organogenesis, 7
    into paraxial mesoderm, 468
    of primordial germ cells, 175, *175*
    radial glial cells as scaffold in, 382, 385,
        386, 389–390
    Reelin regulation of, 390–391

in sclerotome development, 479
    in sea urchins, 284, 285
    thermodynamic model of, 90–91
    traffic movement analogy of, 405–406,
        421
    types of movements in, 16
    in vertebrae formation, 481–482
    in zebrafish, 327–328, 330
Cell position
    in conditional specification, 44–45, 47
    and regeneration, 586
    in syncytial specification, 48
Cell proliferation
    in axis elongation, 470, 471
    in regeneration, 585
Cell recombination assays, 89–90
Cell shape changes, 19, 87, 94
Cell signaling, 39, 87–125. *See also* Cell-to-
        cell communication
Cell sorting
    in brain development, 392
    cadherins in, 92, 392
    differential adhesion in, 90–91
    differential affinity in, 89–90
    differential interfacial tension in, 93
    equilibrium in, 91
    hierarchy of, 90, *91*
    in organoids, 153
Cell specification, 39–49
    autonomous, 41–43. *See also*
        Autonomous specification
    in birds, 340
    conditional, 41, 44–45. *See also*
        Conditional specification
    definition of, 39, 40
    levels of commitment in, 39–41
    lineage tracing of, 41, 42–43
    in paraxial mesoderm, 465–466
    in plants, 46–47
    syncytial, 41, 47–48, 249, *257*
    in zebrafish, 328
Cell-to-cell communication, 87–125
    of adult neural stem cells in V-SVZ, 139
    bioelectric, in flatworms, 599–602
    cadherins in, 91–92
    in cancer, 636, 637–639
    in conditional specification, 44, 45
    differential adhesion in, 90–91
    differential affinity in, 89–90
    extracellular matrix in, 87, *88,* 93–94
    induction and competence in, 95–99
    juxtacrine signaling in, 88–92, 100,
        120–123
    paracrine signaling in, 88, 100–119
    in plants, 114
    signal transduction in, 88, 101–102
    thermodynamic model of, 90–91
Cell wall remodeling in mechanical stress,
        232
Cellular blastoderm, *47,* 48, 250–252
Cellularization, *47,* 48
    in birds, *337*
    in *Drosophila, 171, 249, 251,* 252, *257,* 263
    in rat heart, 99

Cellulose, and phyllotaxis of plants, 232, *233, 235*
Central canal of spinal cord, *383*
Central cell, 188, 189, 217
Central dermomyotome, 486–487
Central dogma of biology, 52–53, 586
Central nervous system, 365–379
    adult neural stem cells in, 137–141
    in amphibians, 316
    anatomy of, 381–386
    anterior-posterior axis in, 375–376
    auditory pathway in, 444
    brain in, 381–398. *See also* Brain growth and development
    dorsal-ventral axis in, 376–379
    as ectoderm derivative, 363, *364*
    neural plate as precursor of, 365
    neuroepithelial cells in, 381–382
    neurulation of, 365–374
    patterning of, 375–379
    primary embryonic induction of, 307–308
    spinal cord in. *See* Spinal cord
    teratogens affecting, *622*
    tracts in, 428, *429*, 439n12
Central zone stem cells in plants, *132, 133*
Centrioles
    of egg, 202
    of radial glia, 392
    of sperm, 194, *195*, 202, 210, 211, 217
        in amphibians, 296, *297*
        in polyspermy, 202, *203*
        in tunicates, 289
Centrolecithal cleavage, 14, *15*
Centromeres, 190n6
Centrosome-attracting body, 289, 292n5
Centrosomes, 210, *217*, 221
    in *Caenorhabditis elegans*, 241
    in *Drosophila*, *251*
    in polyspermy, *203*
    in snails, 235, *236*
    in tunicates, 289
Cephalic furrow in *Drosophila*, 252, *253*
Cephalic neural crest, 401. *See also* Cranial neural crest
Cephalochordates, *276*
Ceratohyal cartilage in amphibian metamorphosis, *566*
Cerberus and *cerberus*
    in amphibians, *302*, 312, 320, *321*
    in birds, 344–345, 361
    in cranial sensory placode induction, 443
    in heart development, *503*
    in mammals, *352, 353*, 356, 361
Cerebellum, *375, 376, 383*
    glia scaffold in development of, 389–390
    organization of, 384
    Purkinje neurons in, *383, 384, 421*
Cerebral hemispheres, 376
Cerebral organoid, 153, 154–155, 642
Cerebrospinal fluid, 381
    and adult neural stem cells, 137, *138*, 140
    and radial glial stem cells, 392

Cerebrum, 154, *375, 383*
    in adolescence, 397
    adult neural stem cells in, 137–141
    folds of, 393–394, *395*, 397, 398n3
    functional regions of, 386
    neocortex of, 385, 389–392
    neurogenesis in, 388
    organization of, *383, 384*–386
    protomap in development of, 386
    radial glia in, 386, 389–390, 394, 395
    signaling mechanisms in development of, 390–392
Cervical cancer, 147
Cervical ganglia, 432
Cervical vertebrae, 355, 461, 467, *468*, 655
    and childhood cancer, 655
Cervix, uterine, 166
    diethylstilbestrol exposure affecting, *629, 630, 631*
Cetaceans, 543–544
CF transmembrane conductance regulator, 151
Chalaza, *337*
Chance and random events
    abnormal development in, 619–620
    mutations in, 658
Chara braunii, 33, 36
Chara zeylanica, 33
Charophytic algae, *32, 33*
Chase and run model of cranial neural crest, 416–418
Chat in flatworm regeneration, *597*
Chemical gradients in sperm migration, 213
Chemicals, developmental abnormalities from, *622*, 623, 629
    atrazine, 629, 633–634
    bisphenol A, 631, 632–633, 636, 660
    in fracking procedures, 634
    glyphosate, 627–628
    vinclozolin, 629, 635–636, 660
Chemoaffinity hypothesis, 434
Chemoattraction
    in axon guidance, 422, 425, 426, 432, *435*
    in neural crest migration, *417*, 418
    of sperm to egg, 198, 199
Chemorepulsion
    in axon guidance, 422, 424–429, *430*, 431, 432, 434, *435*
    in neural crest migration, 417
Chemotrophins, 432–433
Chiasmata in meiosis, 178
Chick development
    alcohol affecting, 625
    axis specification in, *342*, 344–345, *355*
    axon guidance in, 428
    beak patterns in, 652
    blastoderm in, 3, *339*
    brain in, *375, 376*
    cell fate in, 338
    cleavage in, *337, 360*
    cranial placodes in, *445*
    dermomyotome in, *463, 480, 486, 488*
    digestive tube in, 552

    ear in, 459
    endoderm in, 549
    epithelial-mesenchymal transition in, 95
    fate maps of, *24, 340*
    Fgf8 in, *103*
    fibronectin in, 94
    filopodial protrusions in, *119*
    gastrulation in, *341, 343, 360, 463*
    genetic labeling experiments on, 22, *23*
    germ layers in, 17
    glyphosate affecting, *628*
    Hamburger-Hamilton stages in, 20
    Harvey on, 3
    heart in, 499, 500, 501, 503, 504, *505*, 506, 515, 515n3
    hematopoiesis in, 511, *512*
    historical studies of, 335
    Hox genes in, 355, 468–469
    kidneys in, 493–494
    lack of teeth in, 455
    limbs in, *518, 521, 522, 523, 524*, 525–526
        and apical ectodermal ridge, *528*
        and apoptosis, 541–542, 543
        and digit identity, 536
        dorsal-ventral axis in, 540, 541
        gradient model of, 530
        reaction-diffusion model of, *534*
        and zone of polarizing activity, 534–535
    lungs in, 560
    Malpighi on, 3
    melanoma in, 637
    mesoderm in, 465, 491, *492*, 493–494, *500*
        paraxial, 468–469
    as model system, 6, 335
    motor neurons in, 424, *425, 427*
    neural crest in, 22, *23*, 401, 402, 410, 439n2
    neural tube in, 3, *378, 382, 480*
        closure of, 371
    neurulation in, 366, 367, 368, 371, *463*
        secondary, 374
    optic vesicle in, *103*
    pancreas in, *556*
    primitive streak in, *340, 341, 359, 463*
    retinal ganglion axons in, 434
    rhombomeres in, 376
    sclerotome in, *480*, 481, 482, *485*
    somitogenesis in, 470, 473, 474, *475, 476*
    Sonic hedgehog in, *107*, 108, *119*
    tendon formation in, *484, 485*
    transgenic DNA analysis of, 23, *24*
    vasculogenesis in, 508, 509, 515
    vertebrae formation in, 355, 482, 484
    von Baer's laws on, 24
    webbed feet in, 541–542, 543
    wings in, *525, 528*, 544
Chick-quail chimeras, 22, *23*, 482, 485, 511
Child, C. M., 594
Chimeras, 22–23, *24*
    chick-quail, 22, *23*, 482, 485, 511
    mouse, 82, 148

Chimpanzees
  brain development in, *393, 395, 395, 396*
  evolution of, 648–649
ChiP-Seq (chromatin
      immunoprecipitation-sequencing),
      79–80
Chironomus tentans, 53
Choanoblastaea, 30–31
Choanocytes, 30, 585, 661
Choanoflagellates, 30, *31, 32,* 661–662
Cholangiocytes, *548*
Cholesterol
  and Hedgehog, 105, 106, 107, 621
  in neural tube formation, 372, *373*
  and sperm capacitation, 212, *213*
Chondrocytes, *405,* 542
Chondrogenesis, 536. *See also* Cartilage
      development
  mesenchymal stem cells in, 146
  Turing model on, 531, *533*
Chondroitin sulfate, 93, 411, *432*
Chordamesoderm, 461, *462*
  in amphibians, 302, *303,* 307–308
  in axis elongation, 471–472
  in birds, 341
  in mammals, *352*
  in primary embryonic induction, 307–308
  specification of, 465
  in tunicates, 290
  in zebrafish, 329
Chordates, 224
  notochord of, 26–27, *32,* 226, *276, 288, 296*
  phylogenetic tree of, *296*
  in tree of life, *225*
  tunicates as, 288, 292
Chordin and *chordin, 71*
  in amphibians, 312, *313,* 315, 316, *317,*
      318, 320, 321
  and BMP, 112, 315, 316, *646, 647,* 663n5
  in chick development, *492*
  in *Drosophila,* 318
  in heart development, 503
  in mammals, 353
  in zebrafish, 330, 331
Chorion, 335, *336*
  in birds, 338
  in mammals, 347, *349,* 358
Chromatids, 177, 178, 179
Chromatin, 55–56
  in amphibians, 298–299
  circular chromosome conformation
      capture in analysis of, 489n1
  decondensation of, 210
  and enhancer-promoter interactions, 58
  epigenetic modification of, 62–65, 131
  euchromatic, 56, 62
  in fusion of genetic material in
      fertilization, 217
  heterochromatic, 56, 62
  in Hox genes, 459
  immunoprecipitation-sequencing
      (ChiP-Seq), 79–80
  of liver-specific genes, 555
  in meiosis, 177, 179

  in multipotent stem cells, 554
  poised, 65
  in sea urchins, 275
  structure of, 55–56, *459*
Chromosome 21 trisomy, 151, 620–621
Chromosomes
  abnormalities of, 34, 35, 619–621, 632
  age-related changes in, 183
  chromatin in. *See* Chromatin
  in fusion of genetic material in
      fertilization, 210–211, 216–217
  in meiosis, 159, 176–179, 182
  in polyspermy, 202, *203*
  polytene, 53
  in sex determination, 159, 160–173
    in *Drosophila,* 159, 169–173
    in mammals, 159, 160–169
  triploid, 202
Ci and *ci* (Cubitus interruptus and *cubitus
      interruptus*), 106, 107, *260, 266*
Cichlid fish, 658
Cilia
  in amphibians, 322
  in mammals, 356, *357*
  primary cilium compared to, 118, *120*
  respiratory, 221n1
  in sea urchins, 277
  in zebrafish, 332–333
Ciliary ganglion, 416
Ciliary muscle, 416
Ciliopathies, 118
Cilium, primary, 118, *120, 123, 138*
Ciona, 291, 366
    C. intestinalis, 6, 49n1, *276*
    C. savignyi, 42
Circadian rhythms
  in flowering signals, 184
  in hematopoietic stem cells, 145
Circular chromosome conformation
      capture (4C-seq), 489n1
Circulatory system development, 499–515
  and adult neural stem cells in V-SVZ,
      140–141
  in amphibian metamorphosis, *564*
  blood vessels in, 507–510. *See also* Blood
      vessel development
  heart in, 499–507. *See also* Heart
      development
  hematopoiesis in, 510–515
  lateral plate mesoderm in, 462, 491, *492,*
      499–515
Circumpharyngeal crest, 419
Cirrhosis of liver, 150
Cis-regulatory elements, 57, 58, 61, 69,
      85n4
CLASP protein, *422*
Claudius' cells, *447*
CLAVATA1 (CLV1), *133,* 134
CLAVATA2 (CLV2), *133,* 134
CLAVATA3 (CLV3), *133,* 134
Clavicle, 416
Claws of *Drosophila, 571, 572*

Cleavage
  in amphibians, *15,* 297–298, 310, 312,
      313, 346, 359, *360*
    in frogs, *8, 9, 10,* 297–298, *360*
  in anamniotes, 346
  in animal life cycle, 7
  Aristotle on, 2–3
  autonomous specification in, 41
  axis determination in, 223
  in birds, 14, *15,* 335, 336–338, 359, *360*
  in *Caenorhabditis elegans,* 240, 241, 242
  conditional specification in, 44
  definition of, 7, 223
  in *Drosophila,* 14, 47, 249, 250–252
  holoblastic. *See* Holoblastic cleavage
  in lophotrochozoans, 224
  in mammals, 13–14, *15,* 335, 345–347,
      359, *360,* 619, *622*
  meridional. *See* Meridional cleavage
  meroblastic. *See* Meroblastic cleavage
  patterns of, 2–3, 13–14
  radial, 14, *15,* 276–277, 297–298
  in reptiles, 14, *15,* 335
  rotational, 14, *15,* 240, 346
  in sea urchins, 14, 276–277, 278
  in seed plants, 13
  in snails, 14, 226–230
  spiral, 14, *15,* 224, 226–230
  superficial, 14, *15,* 250
  syncytial specification in, 47
  in tunicates, 288–290
  in zebrafish, *323, 324–327,* 359, *360*
Cleavage Under Targets and Release
      Using Nuclease (CUT&RUN), 79, 80
Cleft of Brachet, 302, *303, 304, 305*
Clitoris, 166, *167*
Cloaca, *167, 493, 498, 499,* 515n1, 551
Cloacal membrane, 551
Clock-wavefront model, 473–479
Cloning
  disease development in, 85n1
  of frogs, 53, 67
  of mammals, 53–54, *54*
  reprogramming of cell identity in, 67
Clonogenic neoblasts (cNeoblasts) in
      flatworms, 594–595, *596,* 599, *600*
Club foot, 526
Clustered regularly interspaced short
      palindromic repeats (CRISPR),
      82–83
CLV1 (CLAVATA1), *133,* 134
CLV2 (CLAVATA2), *133,* 134
CLV3 (CLAVATA3), *133,* 134
Clypeaster japonicus, 211
Clypeolabrum in *Drosophila,* 265
Cnidarians, 29–30, *32,* 174, *197*
  as diploblasts, 224
  hydra as, 591. *See also* Hydra
  mesoderm in, 224
  in tree of life, *225*
CO and *CO* (CONSTANS and
      CONSTANS), 184
Coccygeal vertebrae, 355
Cochlea, 444, 446–448, 459, 460n1

Cochlear duct, 447
Cochleovestibular ganglion, *445, 446*
Cockroach development, *569*
Coelacanths, 521
Coelom, 224, *463*, 500–501
    in birds, *367*
    extraembryonic, *367, 463*
    intraembryonic, *463*
    and mesoderm development, *500*
    pericardial, *367, 550*
    in sea urchins, 278
Cohesin proteins in meiosis, 178–179, 183
Col2a1, 483
Col8a1a, 483
Coleoptera, *645*
Collagen
    in axis elongation, 471
    in chondrogenesis, 533
    in *Drosophila,* 137
    in extracellular matrix, 93, 94, 533
    in gene duplication and divergence, 647
    in intervertebral disc development, 483
    in mammals, *352*
    in mesenchymal stem cell
        differentiation, 147
    in scar formation, 617
    type IV, 94
Collagen 2, 533
Collagen2a1, 483
Collagen 8a1a, *483*
Collagenase in amphibian
        metamorphosis, 568
Collapsins, 426–427
Collar cells, 30, 661
Collecting ducts, 494, 495, *496*
Collective migration
    in amphibians, 303–304
    definition of, 489n3
    of neural crest, 409, 416, 439, 489n3
    of neuromesodermal progenitors,
        470–471
Colon, 552
    cancer of, 639
Colonial theory on multicellularity, 30
Comb jellies, 224
Commissural neurons, 424, 428–429, *430*
Commissureless (Comm) protein, *430*
Commitment to cell identity, 39–41, 128
Common myeloid progenitors, 513–515
Comparative anatomy, 2–3
Comparative embryology, 2–6
Compensatory regeneration, 581
    in mammalian liver, 613–615
    in zebrafish, 612
Competence, 96
    and induction, 95–99
Complement 3a in neural crest migration,
        409
Complementary DNA (cDNA), 81, 83,
        315, 316
Complete cleavage. *See* Holoblastic
        cleavage
Conditional mutagenesis, 83–84
Conditional specification, 41, 44–45, 292n6

in *Caenorhabditis elegans,* 242–243
    cell position in, 44–45, 47
    in plants, 47
    in sea urchins, 44–45, 275, 278–279, 282
    in tunicates, 290, 291
Confetti mice, *143,* 402, *403,* 439n5
Congenital disorders, 35–36, 621
    adrenal hyperplasia in, 168
    of cervical ribs and vertebrae, 655
    of heart, in fracking chemical exposure,
        634
    of neural tube, 372–374, 380
    in teratogen exposure, 621–628
Conjoined twins, 333, 358–359, 361n6
Conklin, Edwin Grant, 20–21, 42, 49n2
Connective tissue
    as neural crest derivative, *400,* 401, 415
    as somite derivative, 462
Consensus sequences, 72
CONSTANS and *CONSTANS* (CO and
        CO), 184
Contact inhibition in neural crest
        migration, *407,* 408–409, 416, 418
Context dependency
    of BMP actions, 542
    of tumors, 637
Conus arteriosus, 502
Convergent evolution, 583
Convergent extension
    in amphibians, 299, 304–305, *316,* 408
    in birds, 340
    in sea urchins, 288
    in tunicates, *291*
    in zebrafish, 329
Corals, *27, 584*
Coriander, *33*
Cork oak acorn, *33*
Corn (*Zea mays*), 654
Cornea, *59, 60,* 401
Cornified layer of epidermis, 453
Corona radiata, 198
Corpora allata, 572–573
Cortex of eggs, 197–198
    granules in, *197,* 198, 204
        reaction in polyspermy prevention,
            203, 204–206, 207, 208, 215
    rotation in amphibians, 296, *297, 311,*
        312, 316
Cortical granule reaction
    calcium ions in, 205–206, 207, 208
    in external fertilization, 203, 204–206,
        207
    in internal fertilization, 215
Cortical granule serine protease, 204, *205*
Cortical plate, *383, 388, 389, 390, 394*
Cortical rotation of cytoplasm in
        amphibians, 296, *297, 311,* 312, *316*
Cos2 protein, *106*
Cotyledons, *12, 13, 132, 187*
    and auxin signaling, 114, *115*
Cows, 212, *213*
Coxa in *Drosophila, 571,* 572
CpG islands, 58, 63–64
Crabs, *27, 584*

Cranial nerve development, 376
    alcohol affecting, 625
    cranial placodes in, 442, *446*
    neural crest in, *401,* 416, 439n4
    pharyngeal arch in, *416*
Cranial neural crest, 401, 414–419, 439
    chase and run model of, 416–418
    and cranial sensory placodes, 441–442
    glyphosate affecting, *628*
    Hox genes affecting, 401–402
    migration of, 409, 549
    and pharyngeal arches, 401, 439n1
    potency of, 401–402
    retinoic acid affecting, 627
    in skull development, 418–419
Cranial sensory ganglia, 441, 442
Cranial sensory placodes, 404, 441–452
    anterior, 442, *443*
    in ear development, 404, 441, 442,
        444–448
    in eye development, 404, 441, 442,
        448–452
    induction of, 443–445
    intermediate, 442, *443*
    posterior, 442, *443*
Craniopagus, *359*
Craniorachischisis, *370,* 372
Cranium, 414
    in amphibian metamorphosis, 566
    neural crest in development of, 418–419
CRE gene, *391*
Cre-lox technique, 83–84, 439n9
Cre-recombinase, 84
Creatine phosphokinase, 487
Crepidula, 237, 238
Crescent
    gray, in amphibians, 296, 297, 300, 307
    in heart development, *503*
    yellow, in tunicates, 42, 289, 290
Crickets, 175
CRISPR/Cas9 genome editing, 82–83,
        239, 324
Crocodiles, *18*
    atrazine affecting, 633
    temperature-dependent sex
        determination in, 173, 190n5
Crossing-over phase in meiosis, 178
Crown cells, 356
Crustacean arthropods, 25–26
Crypt, intestinal, 142–143
Crypt base columnar cell, 142
Crystallin and *crystallin* in lens
        development, 61, 97
CSL transcription factors, 121
Ctenophores, *32,* 224, *225,* 246n1
Cubitus interruptus and *cubitus
        interruptus, 106,* 107, *260, 266*
Cucumber plants, *590*
Culture, stem cells in, 129, 134, 146–147,
        148–150
    and organoids, 153–155
Cumulus in mammalian eggs, 198, 211,
        213, 214

CUT&RUN (Cleavage Under Targets and Release Using Nuclease), 79, 80
Cuvier, Georges, 26
CXCL12 and *CXCL12*, *144*, 145
CXCR4 in neural crest migration, *417*, 418
Cyanobacteria, 32–33
Cyclin
  in egg activation, 210
  in liver regeneration, 614
  in meiosis, 178, 216
Cyclin B, 178, 210
Cyclin D, 614
Cyclin dependent kinase (Cdk), 155, 157n24, 210
Cyclin E, 614
CYCLOIDEA, 659
Cyclopamine, 107, 108, 623, 638–639
Cyclopia, 107, 450, 621, 623
Cyp26 in kidney development, *496*
Cyp26A1 in somitogenesis, 479
Cyp26b, 466
Cyp26b1 and meiosis timing, 176, *177*
Cystic fibrosis, 151
Cystoblasts, *136*, 137
Cytokines, 113, 513
Cytokinesis, 131
  in angiosperms, 189
  in brain development, 392
  in oogenic meiosis, 183
  phragmoplast mode of, 34, 47
Cytokinin, 47, 587
Cytonemes, 117–118, *119*
Cytoplasm
  of eggs, 195–197. *See also* Egg, cytoplasm of
  of sperm, 194, 195, *195*
Cytoplasmic determinants, 196
  in autonomous specification, 41–43, 235, 242
  in *Caenorhabditis elegans,* 242
  in *Drosophila,* 48, 249
  in snails, 228–230
  in stem cell regulation, 131
  in zebrafish, 325
Cytosine, 64–65
Cytotrophoblast, 100, *349*

D

D blastomeres, 230, 236–237, *237*
d4EHP in *Drosophila,* 259
D114, and adult intestinal stem cell niche, *142*
Dab1 and *Dab1* (Disabled-1 and *Disabled-1*), 390–391
Dachshund dogs, 19, 28
Dalmatian dogs, *414*
Dandelions, 220
Danio rerio. See Zebrafish
Darwin, Charles, 113, 543, 643, 652
  on artificial selection, 28
  on barnacles, 25–26
  and evolutionary embryology, 24, 25–26, 27

on finches, 652
  on tunicates, 292
Darwin, Erasmus, 159, 190n7
Dauerblastula, 279
Davidson, Eric, 69–70, 279
DCC in axon guidance, 428, *430, 432*
DCX (double cortin), *138*
DDT, 621
DDX17, 396
De novo regeneration, 587
De-adhesion in sea urchins, 285
Decapentaplegic and *decapentaplegic* (Dpp and *dpp*), 112, 646
  in cytoneme-mediated signaling, 118, *119*
  in *Drosophila, 271, 272,* 318, 575–576, *647*
  in snails, *236*
Decellularized heart, 99
Decidua, 348
Decussate pattern, 231, *231*
Dedifferentiation, *66, 67*
Deep homology, 646, 663n5
Deep sequencing, 80–81, *81*
Deepwater Horizon oil spill, 623
Deer antler regeneration, 586
Definitive endoderm, 547, 548–549
deformed gene in *Drosophila, 268,* 354, 648
Deiodinase II, 567
Deiodinase III, 567
Deiters' cells, 447
Delamination, 16
  of muscle precursor cells, 486
  of neural crest, *400,* 406–408
  sensory ganglia generated in, 445–446
Delphinium, 651
Delta protein, 120–123, 282, 476
  in birds, *342, 475*
  in *Caenorhabditis elegans,* 121, 122–123
  in eye development, 451, *452*
  in sea urchins, *281,* 282
  in somitogenesis, 473, *475, 476, 477, 478, 479*
  in zebrafish heart regeneration, 612, *613*
Delta1 in birds, *342, 475*
Delta1 in somitogenesis, *475*
deltaD in zebrafish heart regeneration, *613*
Delta-like-4 (Dll4) in angiogenesis, 510
Demospongia, 585
Dendrites of Purkinje neurons, 384, *421*
Dendritic arbors, 384
Dendritic cells, *514*
Dendrocoelum lacteum, 595, 596
Dental lamina, 454, 457
Denticles in *Drosophila, 267*
Dentin, 457
Dermal papilla, 458, *459*
Dermal tissue in angiosperms, 13
Dermatogen stage in plant development, *12, 132,* 133
Dermatome, *463, 464,* 485–489
Dermis development, 453, 458
  dermatome in, *464,* 486
  dermomyotome in, *464,* 486–487

and ectodermal appendages, 454, 455
  lateral plate mesoderm in, 486
  neural crest in, 486
  somites in, 462
Dermomyotome, *410, 464,* 479, *481,* 485–489
  abaxial, *480*
  central, 486–487
  in chick development, *463, 480, 486,* 488
  FGF secretion of, 425
  primaxial, *480*
DES (diethylstilbestrol), 629–631, 636
The Descent of Man (Darwin), 26
Descent with modification, 27, 652
Desert hedgehog and *desert hedgehog,* 105
DeSimone, Doug, 303, 304
Determination front in somitogenesis, 473–479
Deuterostomes, 224–226, 275–293, *276*
  heart development in, 502
  sea urchins as, 275–288. *See also* Sea urchins
  in tree of life, *27, 225,* 584
  tunicates as, 288–291. *See also* Tunicates
Development, definition of, 1
Developmental abnormalities, 619–642
  cancer in, 636–640
  disruptors in, 35, *622,* 628–634
  environmental factors in, 35–36, 619, 621–634
  genetic factors in, 35, 619–621
  random stochastic factors in, 619–620
  transgenerational inheritance of, 634–636
Developmental biology, 2–6
  definition of, 36n1
  ecological, 663n9
  ecological evolutionary, 656–662, 663n9
  evolutionary, 643–646
  introduction to, 1–37
  model systems in, 5–6
  research methods in, 20–23
Developmental genetics
  and evolutionary change, 643–655
  research tools in, 78–84
Developmental symbiosis, 660–662
Dextral coiling, 228–230
DHR3 in *Drosophila* metamorphosis, *574, 575*
DHR4 in *Drosophila* metamorphosis, *574, 575*
Diabetes
  induced pluripotent stem cells in, 151
  neural tube formation in, 372, *373*
  pancreatic β-cells in, 556
  regenerative medicine in, 150
  transcription factors in, 67
Diacylglycerol in egg activation, *207, 210*
Diakinesis, 178
Diazepam-binding inhibitor protein, 140
Dicer RNase, 75
Dickkopf and *dickkopf* (Dkk and *dkk*), 109
  in amphibians, *306, 320,* 321
  in heart development, *503*

in limb development, *542*
in mammals, 353
in zebrafish, 332
Dicranoweisia cirrata, 33
Dictyate resting stage in oogenesis, 182
Diencephalon, *375*, 376, 386, *450*, 549
optic vesicle formation in, 448
Dieterlen-Lièvre, Françoise, 511
Diethylstilbestrol (DES), 629–631, 636
Diet-induced DNA methylation, 660
Differential adhesion hypothesis, 90–91
Differential gene expression, 51–85. *See also* Gene expression
Differential interfacial tension hypothesis, 93
Differentiation of cells. *See* Cell differentiation
Differentiation therapy in cancer, 640, *641*
Diffusion
of mRNA, 76, *77*
of paracrine factors, 117
Digestive tube, 547–557
in accessory organ formation, 554–557
lung origin in, 557
Digit development
anatomy in, 517, *518*
apoptosis in, 541–542
dorsal-ventral axis in, 539
in *Drosophila, 571, 572*
evolution of, 520–521
constraints on, 654–655
Hox genes in, 536, 538–539
necrotic zones in, *541*, 542
in polysyndactyly, *520*
reaction-diffusion model of, 531–533
and regeneration in young mammals, 615
Sonic hedgehog in, 536–537
Turing model of, 531–533
zone of polarizing activity in, 535
Digoxigenin, *53*, 78–79
Dihydrotestosterone and sex determination in mammals, *161*
Di-iodothyronine in amphibian metamorphosis, *567*
Dinosaurs, 650
Dioecious plants, 186, 651–652
Diploblasts, 30, 224
Diplotene stage in meiosis, 178, 182
Dipterans, 269, 273n6
Direct developers, 563
Disabled-1 and *Disabled-1* (Dab1 and *Dab1*), 390–391
Discoidal cleavage, 14, *15*
in birds, 336–337
meroblastic, 325, 336–337
in zebrafish, 325
Disheveled (Dsh), 110, 111
in amphibian organizer formation, 310, *311*, 312, *313*
in sea urchins, 280
Dispatched protein, *105*
Disruptors in development, 35, *622*, 628–636

Distal-less and *Distal-less* (Dll and *Dll*), 576
duplication and divergence of, 647
in somitogenesis, *477*
Ubx inhibition of, 653–654
Distal-less-3 (Dll3), 477
Divergence, 644–648
and modularity, 644–646
and molecular parsimony, 646–648
Dix3 in eye development, *452*
Dizygotic twins, 357, 358
Dkk and *dkk. See* Dickkopf and *dickkopf*
Dll and *Dll. See* Distal-less and *Distal-less*
Dll4 (Delta-like-4) in angiogenesis, 510
Dlx2 in neural crest migration, 417
Dlx5/6 in neural crest specification, 404, *405*
Dmrt1 and *Dmrt1*, 164, 165, 173, 174
DNA
in brain development, 396
cap sequence, 57
central dogma of biology on, 52, *52*
complementary (cDNA), 81, 83, 315, 316
CpG islands of, 58, 63–64
exon and intron regions of, 56, *56*
in fusion of genetic material in fertilization, 210–211, 216–217
and gene anatomy, 55–61
in genomic equivalence, 53–54
in meiosis, 177
methylation of. *See* DNA methylation
mitochondrial, 215
in oocyte, 216–217
repair of, 82, 83
in sperm, 194, 216
synthesis of, 209–210, 211, 216–217, 387
in liver regeneration, *614*
transcription of, 51, 52, 62–70
in transgenic chimeras, 22–23
DNA methylation, 63–65
in BPA exposure, 633
diet-induced, 660
from endocrine disruptors, 660
and genomic imprinting, 65
inheritance of, 64–65, 634–636, 659–660
mechanisms of, 64
and neural tube defects, 373
at promoters, 63–64, *64*
from vinclozolin, 635–636, 660
DNA methyltransferase-1 (Dnmt1), 64–65
DNA methyltransferase-3 (Dnmt3), 64, *65*
Dogs
artificial selection of, 28
cell division patterns in, 19
pancreas development in, 554
variable melanoblast migration in, 414
Dolphins, 398n3, 543, 544, 651
Domain branching, 559
Dopamine, 150
Dormancy, 8, 34
Dorsal blastopore lip, 300–303, *305*, 309
in fish, 330
and head induction, 320

and organizer functions, 314, 316, 317, *321*
and primary embryonic induction, 307
Dorsal closure in *Drosophila, 252*
Dorsal ectoderm
in amphibians, 314
in *Drosophila, 271*
Dorsal mesentery, 553
Dorsal mesoderm
in amphibians, 309, *321*
in axis formation, 308
convergent extension of, 304–305
induction of, 314–315
and organizer formation, 313, *313*
and organizer functions, 314
in zebrafish, 330
Dorsal nerve cord in tunicates, 288
Dorsal protein and *dorsal* gene in *Drosophila*, 270–271, *272*
Dorsal root, *383*, 401
Dorsal root ganglia, 401
axons from, 426–427, 432
as neural crest derivative, 401, 402, *403*, 410
Dorsalin, 377, *378*
Dorsal-ventral axis, *17*, 223
in amphibians, 296–297, 306–307, 309–318, 321
and BMP, 314–318
specification of, *310*
in birds, 338
BMP in, 314–318, 377, 378, 646
in limb development, 540, 541
in *Caenorhabditis elegans*, 240, 241–242, *243*, 244
definition of, 16
in *Drosophila melanogaster*, 249, 269–271, 272
in wing development, 576
in limb development, 518, 539–541
in mammals, 356
in nervous system development, 363, 376–379
in otocyst, 447–448
in zebrafish, *323*, 330–332, 607
Dorsolateral hinge points, *368*, 369, 370, 371
Dorsolateral pathway in neural crest migration, 401, 410, 411, 413–414
Double cortin (DCX), *138*
Double fertilization, *187*, 189, 198, 220
Double-negative gates in sea urchins, 281–282, 285, 286
Doublesex and *doublesex* (Dsx and *dsx*) in *Drosophila, 171*, 172–173
Double-stranded RNA (dsRNA), 75, 82
Down syndrome, 151, 620–621
Dpp and *dpp* (decapentaplegic and *decapentaplegic*), 112, 646
in cytoneme-mediated signaling, 118, *119*
in *Drosophila, 271*, 272, 318, 575–576, 647
in snails, *236*
DR5 and *DR5*, 114, *115*
Driesch, Hans, 44–45, 49n3, 275, 292n6
Driever, Wolfgang, 257

Drosha RNase, 75
Drosophila hedgehog, 105, 124n5, 535
Drosophila melanogaster, 247–274
    air sac primordium in, 118, *119*
    anterior-posterior axis in, 48, 249, 255–
        261, 272, 353, 354
        in wing development, 575–576
    axis specification in, 223, 247–274
    axon guidance in, 428, *429, 430*
    bacterial symbionts of, 661
    BMP in, 112, *136,* 137, 318, 575
    Cartesian coordinate model of, 272
    cell specification in, 531
    cleavage in, 14, 47, 249, 250–252
    CRISPR/Cas9 genome editing in, 82
    cytoneme-mediated morphogen
        signaling in, 118, *119*
    cytoplasmic determinants in, 48, 249
    cytoplasmic polarity in, 256–259
    dorsal-ventral axis in, 249, 269–271, 272
        in wing development, 576
    early development of, 249–254
    fertilization in, 249–250
    forward genetics mutagenesis screens
        on, 81, 254
    gastrulation in, *249,* 252–254, 271
    genetic assimilation studies of, 656–658
    genomic equivalence in, 53
    germ cells in, 174
    germ stem cells in, 136–137
    as gynandromorph, 169–170
    homeotic selector genes in, 268–269
    Hox genes in, 353, 354, 469, 648
    imaginal discs in, 83, 118, *119,* 254,
        273n1, 570–572, 645–646
        leg, *570,* 571–572, *573,* 575
        wing, 118, *119, 570,* 573, 575–576,
            645–646
    jaw development in, 83
    leg eversion in, 572
    life cycle of, *249*
    maternal effect genes in, 229, 252, 255,
        256, 258, 262, 270
    maternal gradients in, 256–259
    metamorphosis in, *249,* 570–576
    mid-blastula transition in, 252
    mitosis rate in, 14
    as model organism, 6, 247, *248*
    Morgan studies of, 617n4
    morphogen gradients in, 100
    mRNA in, 73, 76–77
    nervous system development in, 121
    Notch pathway in, 121, 272
    paracrine factors in, 101, *117*
    parasegments in, 261–262, 267, 273
    polytene chromosomes in, 53, 247, *248*
    proteoglycans in, 93
    proximal-distal axis in, 273, 576
    retinal neurons in, 433
    salivary glands in, 272
    segmentation of, 53, 108, *249,* 255–261,
        273n4
        in gastrulation, 252, *253,* 254
        segmentation genes in, 260–267

    sex determination in, 159, 169–173
    signal transduction pathways in, 104
    stripped patterns in, 263–264, 267
    syncytial specification in, 47–48, 249,
        257
    terminal gene group in, 260
    toggle switches in development of, *262,*
        263
    translational regulation in, 73
    wing development in, *268, 269,* 273,
        645–646, 656, *657*
    Wnt in, 108, 265, *266,* 318
Drugs, developmental abnormalities
    from, *622, 623,* 629
    diethylstilbestrol, 629–631, 636
    retinoic acid, 627
    thalidomide, 35–36, 621, 642n2
Dscam and DSCAM, 72, 428, 433
Dsh (Disheveled), 110, 111
    in amphibian organizer formation, 310,
        311, 312, *313*
    in sea urchins, 280
Dsx and *dsx* (Doublesex and *doublesex*) in
    Drosophila, 171, 172–173
Ducks, webbed feet of, 541–542, *543*
Duct cells in liver regeneration, 614
Ductus arteriosus, *416*
Dugesia japonica, 595
Duodenum, 551, *554*
Duplication and divergence, 647–648
Dye marking, 21–22
Dynein, *77, 194, 213,* 221n1, 332, 356

E

E cells in ventricular-subventricular zone,
    138, *138*
E74 in *Drosophila* metamorphosis, 574
E75 in *Drosophila* metamorphosis, 574,
    575
Ear development, *18,* 459, *622*
    in amphibian metamorphosis, 566,
        577n3
    anatomy in, 447
    in birds, 342, 459
    blood vessel formation in, *414*
    cranial placodes in, 404, 441, 442,
        444–448
    neural crest in, 401, 404, *415,* 416, *416*
    pharyngeal arches and pouches in, *416,*
        549, *551*
    in phenotypic heterogeneity, 621
    retinoic acid affecting, 627
    thalidomide affecting, 35, 621, 642n2
    transcription factors in, 65
    in zebrafish, *323*
Ear drum, 444
Early limb control regulatory region
    (ELCR), 538, *539*
Ecdysone in insect metamorphosis, 572,
    573, 574–575
Ecdysone receptors in insect
    metamorphosis, 574
Ecdysozoans, 224, *225*
echidna hedgehog, 124n5

Echinoderms, 224, *276*
    cleavage in, *15,* 346
    sea urchins as, 275–288. *See also* Sea
        urchins
    in tree of life, *225*
Echinometra lucunter, 201
Ecological developmental biology, 663n9
Ecological evolutionary developmental
    biology, 656–662
    compared to ecological developmental
        biology, 663n9
    developmental symbiosis in, 660–662
    plasticity-first evolution in, 656–659
    selectable epigenetic variation in,
        659–660
EcR and *EcR* in insect metamorphosis, 574
Ectoderm
    in amphibians, *8, 9,* 298, 305
        and BMP, 314–318
        epiboly of, 300, *302*
        in eye development, 96–97, *98, 99,* 451
        in gastrulation, 299, 300, *301,* 302,
            303, *305*
        neural, 298, 314–317
        and organizer functions, 314–317, *321*
        in primary embryonic induction, 307,
            308
        radial intercalation of, 300
    in bilateral symmetry, 29–30
    in birds, 340, 341, 342, *367, 368*
    brain, 549
    cell recombination assays, 89
    definition of, 7, 36n6
    derivatives of, 18, 363, *364, 365,* 399,
        441, 549
    differential affinity of, 89
    in diploblasts, 224
    dorsal, *271,* 314
    in *Drosophila,* 252, *253,* 254, *271, 272*
    and embryonic stem cells, *149*
    and endoderm at digestive tube, 549
    epidermis formed from, 363, *364, 365,*
        441, *443,* 453
    in eye development, 96–97, *98, 99,* 448,
        451, *452*
    in gastrulation, 16
        in amphibians, 299, 300, *301,* 302,
            303, *305*
    in hydra, 591, *592*
    in limb development, 539
        apical ectodermal ridge in. *See* Apical
            ectodermal ridge
    in mammals, *349, 352,* 353
    neural. *See* Neural ectoderm
    in neural tube closure, 371, *372*
    in neural tube dorsal-ventral
        patterning, 377
    and neurulation, 365
        primary, *367, 368, 369,* 374
        secondary, 374
    non-neural, *405,* 441, *444,* 448, 451
    oral plate, 549
    pharyngeal, in heart development, 419
    placodes in, 404, 441

in sea urchins, 278, 279, 283, *285, 286*
in snails, 230
in triploblasts, 224
in tunicates, *288, 289, 290*
in zebrafish, *323, 327, 328, 330*
Ectodermal appendages, 454–458, *459*
signaling pathways in the development of, 456
stem cells of, 457–458, *459*
Ectomesoderm, *228*
Edn (endothelin), *403, 414, 432*
Edn1 (endothelin-1), *403*
Edn3 (endothelin-3), *403*
EDNRB2 and *Ednrb2*, 414
Efflux transport of auxin, 114, *115*, 232–235
EGF (epidermal growth factor), 113, 140
EGFR (epidermal growth factor receptor), *138*, 140, *453*
Egg, 193, *199*
and acrosome reaction, 200
activation of, 193, 206–210
in angiosperms, 220
calcium ions in, 182, 206–209, 249
in *Drosophila*, 249–250
in external fertilization, 206–210
G protein in, *208*
in internal fertilization, 216
protein and DNA synthesis in, 209–210, 211
of amniotes, 335, *336*
of amphibians, 296, 297, *298*, 306, 307, *311*
cortical rotation in, 296, *297, 311*, 312, *316*
of frogs, 9, *10*, 11, 14
anatomy of, 195–198
of angiosperms, 12, 188, 189, 217–220
of birds, 196, 335, 336–338, 342
of *Caenorhabditis elegans*, 239–240, 241, 242
cell membrane of, *196, 197*–198
fusion with sperm cell membrane, *200, 201*–202, *203, 214*–215
membrane potential of, 203–204, *210*
in polyspermy prevention, 203–205
sodium ion channels in, 203–204
cytoplasm of, 194, 195–197
in amphibians, 296, *297, 311, 312, 316*
bacteria in, 661
in *Caenorhabditis elegans*, 241, 242
calcium ions in, 205–206
cleavage patterns, 14, *15*
determinants in. *See* Cytoplasmic determinants
in *Drosophila*, 48, 249, 250
localization of mRNA in, 76–77
in meiosis, 182, 183
morphogenetic factors in, 196
protective chemicals in, 196
in sea urchins, 279, 280, 282
transcription factors in, 69
in tunicates, 289, *289*
volume and composition of, 196, 221n2
in zebrafish, 324, 325, 326

definition of, 221n2
DNA methylation in, 65, 635
of *Drosophila*, 48, 249–250
external fertilization of, 198–211
fusion of genetic material with sperm, 210–211
and germ layers, 17–18
internal fertilization of, 211–217
of mammals, 198, 211–217, 335, 345–347
maturation of, 196, *197*
and oogenesis, 174, 176, 182–183. *See also* Oogenesis
and polyspermy prevention, 202–206
of sea urchins, 196, *197*, 198–211, 279, 280, 282
size of, 19, 345
species variations in, 221n2
and sperm recognition, 193, 194, 197, 198–201, 214
structure of, 193–194, *195*–198
of tunicates, 289
of zebrafish, 324–325, 326, 330
Egg chamber in *Drosophila*, 255
Egg jelly, 9, 197, 199
and acrosome reaction, 200, *201*
sperm recognition of, 201
Eggan, Kevin, 152
Eichele, Gregor, 381
eIF4E (eukaryotic initiation factor 4E), 72
Elbow, evolution of, *521*
ELCR (early limb control regulatory) region, 538, *539*
Elephants, *27*, 457, *584*
Elytra, *645*
Embryo. *See also specific organisms.*
in animal life cycle, 7
cell behavior in, 19
cell specification in, 39–49
chimeric, 22, *23*
cleavage in, 13–14. *See also* Cleavage
definition of, 1, 6
development of, compared to regeneration, 582–583, 616
differential gene expression in, 51–85
gastrulation in, 14–16. *See also* Gastrulation
germ layers in, 17–18
in model systems, 5
mosaic, 45
phylotypic stage in, 24
in plant life cycle, 8, *12, 187*
regulative, 45
research methods on, 20–23
Embryo sac, 188, 189, *219*, 220
Embryoid bodies, 154, *154*
Embryology, 1
comparative, 2–6
evolutionary, 24–34
medical, 34–36
von Baer's laws on, 24–25, 27
Embryonic germ cells, 148
Embryonic shield in zebrafish, 329, 330, 332

Embryonic stem cells, *66*, 131–135, 156n2
adult stem cells compared to, 131
clinical trials on, 156
in Cre-lox technique, 84
differentiation of, 148–149
human, 147–150
in laboratory, 147–150
naïve, 148, 150, 157n18
organoids derived from, 153
pluripotent, 129, 131–135, 147–150, 347
primed, 148, 150, 157n18
and regenerative medicine, 150
sources of, 148
Embryophytes, 31, *32*, 34
empty spiracles in *Drosophila*, 260, 354
Emx in mammals, 354
Enamel knot, 454, 456
END-1 in *Caenorhabditis elegans*, 243
End3, 414
Endfoot, 381, *382*
Endoblast in birds, 338, *339*
Endocardial cushions, 502–503, 507
Endocardium, 502, 507
in zebrafish, 610, *611*
Endocrine disruptors, 628–636
atrazine as, 629, 633–634
bisphenol A as, 629, 632–633, 660
diethylstilbestrol as, 629–631, 636
DNA methylation from, 660
fracking chemicals as, 634
transgenerational inheritance of disorders from, 634–636
vinclozolin as, 629, 635–636, 660
Endocrine system
disruptors affecting, 628–634
hormones in. *See* Hormones
pancreas in, *548*
Endoderm, 547–561
in amphibians, 301–302, *305, 316*, 320, 551
in axis formation, 308
in frogs, 8, 9, *10, 301*
in gastrulation, 299, 300, 302, 303, 305
in lens induction, 97
and Nieuwkoop center, 309
and organizer formation, 313
and organizer functions, 314, *321*
in primary embryonic induction, *308*
specification of, 299, 306
in bilateral symmetry, 29–30
in birds, *339*, 340–342, *343, 367*
in *Caenorhabditis elegans*, 242, *243*, 244
cell recombination assays, 89
cell sorting in, *90*
in cranial placode induction, 443
definition of, 7, 36n6
definitive, 547, *548*
derivatives of, 18, 547, 549
in digestive tube development, 547–557
in diplobasts, 224
in *Drosophila*, 252, *253*
in eye development, 97, 451
functions of, 547

in gastrulation, 16
in heart development, 419, 503, 504
in hydra, 591, 592
laryngotracheal, 557
in mammals, 348–351, *352*
    in humans, 549, *550*
parietal, 351
pharyngeal. *See* Pharyngeal endoderm
and primary neurulation, *367*
in respiratory tube development, 547, 557–560
in sea urchins, 278, 279, 281, 282, 283, *286*, 287, 292n3
in snails, 228, 237, *238*
sources of, 547
in tunicates, *288*, 289, 290, 291
visceral, 351, 352, 353, 547
in zebrafish, *327*, 328, 330, *331*
Endolymphatic duct, 447
Endomesoderm
in amphibians, *301*, 302, 307, 320, *321*
in birds, 342
in sea urchins, 278, 281
Endoplasmic reticulum
calcium ions released from, 205, 206, 207
and cortical granule reaction, 205
hedgehog in, *105*
in tunicates, 289
Wnt in, 108, 111
Endosperm, *12*, 13, *187*, 188, 189, 217
in double fertilization, 220
Endosteal hematopoietic stem cell niche, 143, 144, 145
Endosymbiosis, 32–33, 620, 660–662
Endothelial cells
in angiogenesis, 510
in brain, 381
and hematopoietic stem cells, 143, 511, *512, 513*, 515n5
hemogenic, 511, *512*, 515n5
in liver regeneration, 614
in vasculogenesis, 507, *508*, 509
Endothelial precursor cells, 464
Endothelin, *403*, 414, 432
Endothelin receptor, 414
Endothelin-1, *403*
Endothelin-3, *403*, 414
Endotome, 464
Energids, 250, *250*
Engrailed and *engrailed* in *Drosophila*, 255, 260, 261, 265–267, *268*
in Cartesian coordinate model, *272*
in wing development, 575
Engrailed-1 in limb development, 540, *541*
Enhancer of zeste homologue 2 (EZH2), *641*
Enhancers, 57, 58–61, 65
activation of, *59*, 60
brain-specific, *59*
evolution of, 644–645, 646
identification of, 60–61
limb-specific, *59*
modularity of, *59*, 60, 61, 644–645

pancreas-specific, *59*
and promoter interactions, 58–59
and reporter genes, 60
and transcription factors, 58–60, 61, 65, 66–67
Enteric ganglia, 401, 412
Enteric neural crest, 412–413
Enterocoely, 226
Enveloping layer in zebrafish, *323, 326*, 327, 328
Environmental influences
in anterior-posterior axis formation, 353
in conditional specification, 44
disruptors in, 35, 628–636
on frogs, 9, 658
in genetic assimilation, 656–659
on neural stem cells, 140–141
in neural tube defects, 372–374, 380
on plants, 9, 133, 184
question of, 5
in sex determination, 159, 164–165, 173–174
    atrazine in, 633–634
temperature in, 164–165, 173–174, 656–658
teratogens in, 35–36, 621–628
Eomesodermin and *eomesodermin*, 306, 327
Eosinophils, *130, 514*
Epaxial muscles, 489n5
Ependyma, 382, 384
Ependymal cells, 382
Ephrin
in amphibian metamorphosis, 565
in axon guidance, 423, 425, 426, 431, *432, 433*, 434–435
in hematopoietic stem cell formation, *512*
in juxtacrine signaling, 120
in kidney development, 498
in neural crest migration, 411, 412, 414, 417, 418
in somitogenesis, 472–473, 477, 489
Ephrin A, 434–435
Ephrin A2, 434, *435*
Ephrin A5, 425, 434, *435*
Ephrin B, 565
Ephrin B2, *432*, 472, 473, *512*
Ephrin receptors, 120
in axon guidance, 426, 431, *432, 433*, 434
EphA4, 425, 472, 473, *473*, 477
EphB1, *432*
EphB2, 414
in juxtacrine signaling, 120
in melanoblast migration, 414
in mesenchymal-epithelial transition, 472
in motor neuron guidance, 425
in neural crest migration, 418
in somitogenesis, 472, 473, *473*, 477
Epialleles, 635, 659–660
Epiblast, 374, 379, 547, *548*
in birds, *337, 338, 339*, 340, 341, 359, 488
and embryonic stem cells, *149*

in epithelial-mesenchymal transition, *95*
in mammals, 134, *135*, 156n1, *349, 352*
    in anterior-posterior axis formation, 352, 353
    in gastrulation, 348, 350, 351
    in peri-implantation period, *350*
Noggin-secreting, 488
in zebrafish, *327, 328, 329*, 330
Epiboly, 16
in amphibians, 299, 300, *302, 303*
in birds, 342
in *Caenorhabditis elegans*, 245
in snails, *238*
in zebrafish, *323, 326*, 327–328, *329*, 330
Epibranchial ganglia and nerves, 446
Epibranchial placodes, 442, *443*, 444–448
induction of, 444–445
sensory ganglia generated from, 446
Epicardium, 503, 505
in mice, 615
in zebrafish, 610, *611*
Epidermal growth factor, 113, 140
Epidermal growth factor receptor, *138*, 140, *453*
Epidermal placodes, 454
Epidermal stem cells, 453
Epidermis, 441, 453–459
in amphibians, 306, 307
    and BMP, 314, 315, 317, *320*
    formation of, *320*
    in gastrulation, 300
    in metamorphosis, 568
    and organizer functions, 314, 315, 317
    in primary embryonic induction, *308*
    wound epidermis, 603, 604, 605, 606
in birds, 340, *368*
cell recombination assays, 89
cell sorting in, 89–90, *90*
derivatives of, *364*
as ectoderm derivative, 363, *364*, 365, 441, *443*, 453
and ectodermal appendages, 454
functions of, 363
interfollicular, 458
loss and replacement of, 453–454, 460n2
and neural crest, *400, 404, 410*
in neural tube formation, 372
in neurulation, *366, 368, 369*, 372
origin of, 453–454
regeneration of, 453–454
selective affinity of, 89
specification of, 365, 453
wound epidermis in salamanders, 603, 604, 605, 606
in zebrafish, 330
Epididymis, 167, 492
Epigenetics, 17, 62–65
chromatin modification and gene expression in, 62–65
definition of, 62
DNA methylation in, 63–65, 635–636, 659–660
and evolution, 659–660
in flatworm regeneration, 601, 602, 617

in fusion of pronuclei in fertilization, 216–217
histone modification in, 62–63, 131
in neural tube defects, 372, 373–374, 380
in plant regeneration, 591
in stem cell regulation, 131, 137
in vinclozolin exposure and testicular dysgenesis, 635–636
Epimorphin in vertebrae formation, 481
Epimorphosis, 581, 585, 591
in salamanders, 602–603, 604, 607
in zebrafish, 609–612
Epithalamus, 375
Epithelial cell cancers, 638, 640
Epithelial stem cells in amphibian metamorphosis, 568
Epithelial-mesenchymal interactions in lung development, 558–559
Epithelial-mesenchymal transition, 19, 94–95
in birds, 340
in cancer, 95, 284, 640
in limb development, 526–527
in mammals, 351, 352
in neural crest, 95, 399, 400, 406–408
in sclerotome development, 479, 485
in sea urchins, 283–286
in somites, 464, 486
Epithelium
adhesion to laminin, 94
and basal lamina, 93, 94
and cadherins, 91
collective migration of, 409
definition of, 19
delamination of, 446
dental, 455
and extracellular matrix, 93
in gastrulation, 16, 30
in lung development, 558–559, 560
mesenchyme transition to. See Mesenchymal-epithelial transition
in morphogenesis, 19, 87
olfactory, 404, 441
oral, 441
in recombination experiments, 455
respiratory, 558
retinal pigmented, 448, 449
transition to mesenchyme. See Epithelial-mesenchymal transition
Equatorial region of sperm cell membrane, 215, 215
Equivalence group, 122
Erk and ERK
in flatworm regeneration, 597
in RTK pathway, 103
in somitogenesis, 478
Erlandson, Axel, 2
Erythrocytes. See Red blood cells
Erythropoietin, 113, 513
Escherichia coli, 60, 73, 82
Esophagus, 551, 552, 552, 557, 558
Estrogen
in androgen insensitivity syndrome, 167
and atrazine, 629, 633

and bisphenol A, 632
and diethylstilbestrol, 629, 631
and fracking chemicals, 634
in mammary gland development, 457
in primary sex determination, 161, 163
in secondary sex determination, 166
ET and ET in eye development, 449
Ethanol
as teratogen, 35, 623–627
zebrafish research on, 324
Ethical issues in research, 148
ETS and Ets
in RTK signal transduction pathway, 103
in sea urchins, 281
ETS1 and Ets1
in limb development, 653
neural crest potency, 402
in sea urchins, 282
Etv2 in vasculogenesis, 507
Etv4/5 and Etv4/5 in limb development, 537, 538
Euchromatin, 56, 62, 299
Eukaryotes
gene anatomy in, 55, 56–57
and last eukaryotic common ancestor, 31–33, 36, 223–224, 225
meiosis in, 176
multicellular, 223–224
in tree of life, 27, 584
Eukaryotic genes, 55, 56–57
Eukaryotic initiation factor 4E (eIF4E), 72
Eustachian tube, 549, 551
Even-skipped and even-skipped in Drosophila, 260, 263, 264, 265, 266, 267
Evolution, 5, 24–34, 643–664
adaptations in, 27, 32
artificial selection in, 28, 656
biodiversity in, 663n1
biphasic life cycle in, 577n1
brain development in, 393, 395, 396
Cambrian explosion in, 33
constraints on, 654–655
convergent, 583
descent with modification in, 27, 652
developmental genetic model of, 643–655
early development strategies in, 223–226
and ecological evolutionary developmental biology, 656–662
epigenetic variation in, 659–660
genetic assimilation in, 656–659
heterochrony in, 650–651
heterotopy in, 649–650
limb development in, 520–521, 543–544
modularity in, 644–646
molecular parsimony in, 644, 646–648
multicellularity in, 30–31, 223–224, 661–662
natural selection in. See Natural selection
pharyngeal arches in, 18
placental mammals in, 662
plasticity-first, 656–659
preconditions for, 643–648
recruitment in, 645–646

regeneration in, 583–586
segmentation of body plan in, 461
small toolkit in, 646
sweat glands in, 458
symbiosis in, 660–662
transitional morphological states in, 29
tree of life in, 27, 27, 225, 584
tunicates in, 292
Evolutionary developmental biology, 643–646
developmental genetic model of evolutionary change in, 643–655
ecological, 656–662
modularity in, 644–646
molecular parsimony in, 644, 646–648
Evolutionary embryology, 24–34
developmental relatedness in, 27–31
homologous and analogous structures in, 28
land plants in, 31–34
Exencephaly, 370, 372, 625
Exocrine pancreas, 548
Exocytosis, 198
of acrosome, 200, 205, 214
in cortical granule reaction, 205, 205, 207, 207
Exons, 56, 56, 57, 59
definition of, 56, 85n2
in pre-mRNA processing, 71–72
splicing of, 58, 71, 72
Exonucleases, 57
External fertilization, 9, 198–211
acrosome reaction in, 199, 200, 201
attraction of sperm in, 198–199
cell membrane fusion in, 199, 200, 201–202
early responses in, 208, 209
egg activation in, 206–210
in frogs, 9, 10
genetic fusion in, 210–211
late responses in, 209–210
polyspermy prevention in, 202–206, 208, 210
protein and DNA synthesis in, 209–210
recognition events in, 198–201
in sea urchins, 198–211
summary of events in, 199
External granular layer, 383, 384
External yolk syncytial layer, 326, 327, 328
Extracellular envelope of eggs, 197–198
Extracellular matrix, 87
and cell migration during morphogenesis, 19
and cell-to-cell communication, 87, 88, 93–94
collagen-based, 32
composition of, 93–94
definition of, 93
evolution of, 32
and motor neuron guidance, 426
in neurulation, 366
permissive interactions in, 99
receptors for, 94
in regeneration, 617

of liver, 614
   in salamanders, 604–605
   in zebrafish, 610, *611*
   in sea urchins, 284, *285*, 287
Extraembryonic ectoderm in mammals, 353
Extraembryonic endoderm in mammals, *349*
Extraembryonic membranes, 338
   lateral plate mesoderm in development of, 462, 491, *492*
   in mammals, 347
   in twins, *358*
Extraembryonic mesoderm, *351*
   in birds, 340, *340*, 367
   in mammals, *349*
Extraembryonic vasculogenesis, 508
exuperantia in *Drosophila*, *256*
Eya1 in eye development, *452*
Eya1/2 in cranial placode induction, *443*, 444
Eya2 in neural crest specification, *404*
Eye development, 95–97, 448–451, *622*
   in amphibians, 96–97, *98*, 451
      in metamorphosis, 564–565, 568
   cell signaling in, 121
   cranial sensory placodes in, 404, 441, 442, 448–452
   cyclopia in, 107, 450, 621, 623
   in *Drosophila*, 269, 450, *570*
   enhancers in, *59*, 61
   eye field formation in, 449–450
   Fgf8 and *Fgf8* in, 97, 102, *103*, 451, *452*
   in fish
      in metamorphosis, 577n2
      in Mexican cavefish, 450, *451*, 651
      in zebrafish, *323*, 450
   heterometry in, 651
   induction of, 95–97, 102, *103*, 451, *452*
   of lens. *See* Lens development
   lens placode in, 449
   microphthalmia in, 620
   neural crest in, 401, 404, 448
   Notch in, 121
   optic vesicle in, 376
   Pax6 and *Pax6* in, 83, 97, *98*, 448, 449–450, 451, *452*, 646
   in phenotypic heterogeneity, 621
   of retina. *See* Retina development
   in vertebrates, 95–97, 448–451
Eye fields, 449–450
eyeless in *Drosophila*, 269
EZH2 (enhancer of zeste homologue 2), *641*

F

Face development
   alcohol affecting, 623, 625, 626
   in cyclopia, 107, 450, 621, 623
   in Down syndrome, 620
   neural crest in, *400*, 401, 414–416, 418–419, 439
   Sonic hedgehog in, 107
Facial bones, *415*
Facial muscles, 502

Facial nerve, 376, 416
FAK (focal adhesion kinase), 423
Fallopian tubes. *See* Oviducts
Fanconi anemia, 150
Fast block to polyspermy, 203–204, 208, *210*
Fat
   brown, 487
   mesenchymal stem cells in formation of, 146
   subcutaneous, 453, 458, *459*
   white, 487
Fate mapping
   in autonomous specification, 41–43
   of birds, *24*, 340
   in conditional specification, 44
   in direct observation, 20–21
   of *Drosophila*, 271
   with fluorescent dyes, *22*
   in lineage tracing experiments, 41, 42–43
   of neural crest, 402, 410, 439n5
   of sea urchins, 278–279
   with transgenic DNA, *24*
   of tunicates, 20–21, 41–43, 289–290
   of zebrafish, 22, 49, 324, 327
FD and *FD* (FLOWERING LOCUS D and *FLOWERING LOCUS D*), 184
Feathers, 441, 454
Femur, 517, 519, *571*
Ferchault de Réaumur, René Antoine, 579
Ferns, 186
Ferrets, brain development in, *389*, *394*
Fertility
   in androgen insensitivity syndrome, 167
   bisphenol A affecting, 632–633
   diethylstilbestrol affecting, 629, 630–631
   fracking chemicals affecting, 634
   and genetic mechanisms of gonadal sex determination, 163
   in tubulin mutations, 183
Fertilization, 193–221
   acrosome reaction in, *199*, *200*, *201*, 212, 214
   activation of egg in, 206–210
   in amphibians, *8*, *9*, *10*, 296–297, 307, 310
      sperm entry point in, 296, 297, 298, 300, 307, 309, 310, 312
   in angiosperms, 12, *187*, 188, 217–220
      double, *187*, 189
   in animal life cycle, 7
   in birds, 336
   in *Caenorhabditis elegans*, 239–240, 241, *241*
   cell membrane fusion in, 201–202, 214–215
   definition of, 7, 193
   in *Drosophila melanogaster*, 249–250
   external, 198–211. *See also* External fertilization
   functions of, 193
   and gamete structure, 193–198

genetic fusion in, 210–211, 216–217
historical descriptions of, 194
internal, 211–217. *See also* Internal fertilization
   in mammals, 182, 211–217, 345
      in twins, 357–359
   and meiosis in secondary oocytes, 182
   monospermy in, 202
   in plant life cycle, 7, 8
   polyspermy prevention in, 202–206, *210*, 215–216, 220
   in sea urchins, 194, *196*, 198–211
   sex determination at time of, 159
   in tunicates, 194, *289*
   in zebrafish, 322, 325
Fertilization cone, 201, *202*
Fertilization envelope, 194, 204–205, 209
   in sea urchins, 277, 278, 283
Fetal alcohol spectrum disorder, 624
Fetal alcohol syndrome, 623–624
   mouse model of, 625–627
Fetal neuronal growth rate, 392–393
Fetal period, 621
   gonad development in, 161, 163, 167, 168
   teratogen exposure in, 621, *622*
α-Fetoprotein, 554, *555*
FGF (fibroblast growth factor), 101, 102–104
   acidic (Fgf1), 102, *555*
   in amphibians, 321
   basic, 102, *149*. *See also* Fgf2
   in bat wing development, 649, 663
   in birds, 341, *342*
   in cranial placode induction, 443, 444, 445
   in cytoneme-mediated signaling, 118, *119*
   diffusion of, 117
   in digestive tube development, 553
   in ectodermal appendage development, 456
   in embryonic stem cell differentiation, 148, *149*
   in endoderm specification, 547, 548–549
   in epithelial-mesenchymal transition, 486
   in eye development, 450, 451, *452*
   in hair follicle regeneration, 458, *459*
   and JAK-STAT pathway, 104
   keratinocyte growth factor (Fgf7), 102, 453, *459*
   in kidney development, 496, 497
   in limb development, *525*
      and apical ectodermal ridge interactions, 528–529
      and apoptosis, *542*
      in cetaceans, 544
      in dorsal-ventral axis, 541
      in feedback loop with Sonic hedgehog, 537–538
      in gradient model, 530–531
      in limb bud, 489n4, 527
      and reaction-diffusion mechanism, *533*

in liver development, 554, 555
in lung development, 560, *560*
in mammals, in anterior-posterior axis formation, 353
and mesenchymal stem cells, 146
in mesoderm specification, *467, 548*
in motor neuron guidance, 425
in neural crest migration, 419, 549
in neural crest specification, 404, *405*
and neural progenitors in caudal spinal cord, *379*
in otocyst axis determination, 447
and radial glial stem cell fate, 392
and RTK pathway, 102–104
in sea urchins, *285*, 286
in somitogenesis, 473, 474, 477–479
in tunicates, 291
in turtle shell development, 649
in zebrafish, *331, 332*
    in fin regeneration, *608*
Fgf1 (acidic FGF), 102, *555*
Fgf2, 102
in cerebral organoid formation, *154*
in embryonic stem cell differentiation, *149*
in kidney development, 497
in liver development, *555*
in pancreas development, 555
in salamander limb regeneration, *607*
in vasculogenesis, *508,* 509
Fgf3 and *Fgf3*
in neural crest migration, 549
in pharyngeal pouch formation, 549
and phenotypic heterogeneity, 621
Fgf4 and *Fgf4*
in dog breeds, 28
in limb development, 528, *537*, 541
in organoid formation, *153*
in tooth development, *456*
Fgf5 and *Fgf5*
in dog breeds, 28
in sclerotome development, *481*
Fgf7 (keratinocyte growth factor), 102, *453, 459*
Fgf8 and *Fgf8*
in bat wing development, 649
in birds, *103, 342,* 344, 419
in dolphins, 651
in ectodermal appendage development, 456
in eye development, 97, 102, *103,* 451, *452*
in heart development, 503, 504
in limb development, 527
    and apical ectodermal ridge interactions, 538
    in dolphins, *544*
    in feedback loop with Sonic hedgehog, 537, 538
    of forelimb, 523–524, 527, *544*
    in gradient model, *530*
    of hindlimb, 527
    and reaction-diffusion mechanism, *533*

in neural crest migration, 419, 549
and neuromesodermal progenitors, 379
in paraxial mesoderm specification, 466, *467*
in pharyngeal pouch formation, 549
and receptor tyrosine kinase pathway, 102–104
in salamander limb regeneration, 606, *607*
in somitogenesis, *473,* 474, *475,* 477–478, *479,* 489
in tendon formation, 484, 485
in tooth development, 455
Fgf9
in gametogenesis, 176
in kidney development, 496, 497
in limb development, 528, *537*
in organoid formation, *153*
and primary sex determination in mammals, *163,* 165
Fgf10 and *Fgf10*
in ectodermal appendage development, 456
in epidermis regeneration, *453*
in hair follicle regeneration, *459*
in limb development, *524, 525,* 526, 527
    in apical ectodermal ridge induction, 527
    in feedback loop with Sonic hedgehog, 537
    in limb bud formation, 527
    in turtle shell development, 649, *650*
Fgf17 in limb development, 528, *537*
Fgf18 in hair growth, 458, *459*
Fgf20 in ectodermal appendage development, 456
FGFR (fibroblast growth factor receptor), *71,* 102
in limb development, 524
in reaction-diffusion mechanism, 533
in vasculogenesis, *508*
FgfR2, 533
FgfR3, 533
Fibroblast, *66,* 453
dermal, 458
differentiated and dedifferentiated, *66, 67*
in zebrafish fin regeneration, *608*
Fibroblast growth factor. *See* FGF
Fibroblast growth factor receptor. *See* FGFR
Fibronectin
in amphibian gastrulation, 303–304, *305*
in axis elongation, 471, *472*
in birds, 342
in cardiac precursor cell migration, 503
in extracellular matrix, 93–94
fibrils of, 93
in limb development and reaction-diffusion mechanism, 533
in myoblast migration, 488
in neural crest migration, 411

in neuron migration in neocortex development, 390, *391*
receptors for, 94
in somitogenesis, 471, 472, 473
Fibula, 517, 520
Field mouse (*Peromyscus maniculatus*), 212
Figs, 190n7
Filopodia
in amphibians, *305*
of axon growth cones, 421, *422,* 434
in birds, 342
cytoneme projections of, 117–118, *119*
in neural tube closure, 371
in neuron migration in neocortex development, 390, *391*
in sea urchins, 285, *287,* 288
of tip cells, 510
Fimbriae, *345*
Finches, heterometry in, 652
Fingers, 517. *See also* Digit development
Fins, 522, 524
and limb evolution, 520–521
of zebrafish, 524
    regeneration of, 607–609
Fire, Andrew, 75
Fish, 295
blastopore lip of, 330–331
cleavage in, 14, *15,* 359, *360*
eye development in, 450
    in metamorphosis, 577n2
    in Mexican cavefish, 450, *451,* 651
    in zebrafish, *323,* 450
fins of, 520–521, 522, 524
    regeneration in, 607–609
gastrulation in, 359, *360*
genetic assimilation in, 658
jaw phenotypes in, 658
kidney development in, *323,* 492
lateral line placodes in, 442
metamorphosis in flatfish, 577n2
Mexican cavefish, 450, *451,* 651
neural crest in, 402, 410, 417, 439, 439n6
neural tube in, 372, 378
neurulation in, 366
paraxial mesoderm in, 470
pharyngeal arches in, *18,* 246n2
phylogenetic tree of, *296*
regeneration in, 457
    in zebrafish, 135, 580, 583, 607–612
retinal ganglion axon guidance in, 431, 433
sex determination in, 164
stickleback, 644–645, 663n2
Tiktaalik roseae, 26, 29, 520–521
in tree of life, *27, 584*
von Baer's laws on, 24
zebrafish. *See* Zebrafish
Fish organizer, 326, 333n5
5′ untranslated region (5′ UTR), *56, 57,* 72, 73, 85n3
Flagella of sperm, 194, *195,* 210, 212, 221n1
Flatfish, metamorphosis in, 577n2

Flatworms, 224, 646
bioelectric pattern in, 599–602, 617
blastema in, 594
cleavage in, 226
polarity in, 595–596, 597
positional control genes in, 596–601
regeneration of, 6, 580, 594–602, 607, 617
epigenetics in, 601, 602, 617
evolution of, 583
gradient model of, 594
in tree of life, 27, 225, 584
Fli1 in hematopoietic stem cell formation, 512
Flies
chromosomal sex determination in, 159
fruitfly. See Drosophila melanogaster
holometabolous development of, 569
Flk1and Flk1
in heart development, 505
in hematopoietic stem cell formation, 512
in vasculogenesis, 508, 509
FLM (FLOWERING LOCUS M), 71
Floor plate, 377, 378, 379, 383, 481
in birds, 342, 480
and myotome determination, 487, 488
Floral meristem identity genes, 184, 185, 186
Floral organ identity genes, 67–69, 184, 185–186
heterometry of, 651–652
heterotopy of, 650, 651
transcriptional regulation of, 67–69
Flounders, metamorphosis in, 577n2
FLOWERING LOCUS D and FLOWERING LOCUS D (FD and FD), 184
FLOWERING LOCUS M (FLM), 71
FLOWERING LOCUS T and FLOWERING LOCUS T (FT and FT), 184
Flowering plants. See Angiosperms
Flowers
ABCDE model of, 67–69, 184, 185–186
components of, 68, 185
floral organ identity genes of, 67–69, 184, 185–186, 650, 651–652
and gametogenesis, 12, 13, 186–189
heterochrony of, 651
heterometry of, 651–652
heterotopy of, 650, 651
perfect or bisexual, 184, 185, 187
and sex determination, 184–186
timing of, 71, 184
transition from vegetative phase to, 12, 67–69, 184
whorled arrangement of, 68–69, 185, 186, 231, 232, 650, 651
Floxed genes, 84
Flt1 in vasculogenesis, 508
Fluorescence activated cell sorting, 81
Fluorescent dyes, 21–22, 22
Focal adhesion kinase (FAK), 423

Focal adhesions in axon guidance, 423
Focal membrane protrusions, 117–119
Folic acid in neural tube formation, 372–374, 380
Follicles
hair. See Hair follicle
ovarian, 163
Follistatin
in amphibians, 315, 320, 321
in gonadal sex determination mammals, 163
in zebrafish, 330
Follistatin in gonadal sex determination, 163
Foot activation gradient in hydra, 592
Forebrain, 375, 376
alcohol affecting development of, 625
in amphibians, 10, 307, 314, 318, 321
in anencephaly, 372
in birds, 341, 375
eye field formation in, 449, 451
in mammals, 353, 354
and neural crest, 401
and neural tube closure, 371
Foregut, 549, 553
and accessory organ development, 554, 555, 558
in amphibians, 313
in birds, 340, 341, 343, 367
in humans, 357
multipotent stem cells in, 554
neural crest cells in, 412
in sea urchins, 286, 287
Forehead, 415
Forelimb development
in bats, 28, 29, 663
in cetaceans, 543, 544
epithelial-mesenchymal transition in, 526–527
evolution of, 520, 521, 543, 544
heterochrony in, 650
homologous and analogous, 28, 28
Hox genes in, 519, 520
limb bud in, 518, 522
in mice, 28, 29, 524, 526
retinoic acid in, 523–524, 527
in salamander regeneration, 602
Tbx5 in, 524–525, 527
Forkhead proteins, 491
Formin and formin in snail coiling, 229
Forward genetics, 81, 254
Fox proteins in mesoderm specification, 491
Fox12 in gonadal sex determination, 163
Foxa and foxa in sea urchins, 70, 282
FoxA1 and Foxa1 in liver development, 67, 555
FoxA2 and Foxa2
in liver development, 555, 561
in mammals, 361
and pancreatic β cells, 556
Foxc1 in mesoderm specification, 491
Foxc2 in mesoderm specification, 491
FoxD in tunicates, 291

FoxD3 and Foxd3
in neural crest delamination, 407
in neural crest migration, 417
in neural crest specification, 404, 405, 414
foxD4 in amphibians, 315, 317
FoxE3 in eye development, 452
Foxf1 in mesoderm specification, 491
Foxn2/3 in sea urchins, 285
Foxp1, 377, 438
Foxp4 in heart development, 504
Fracking chemicals, 634
Fragile-X syndrome, 152
Fraternal twins, 357, 358
Frizzled
in Drosophila, 266
in synapse formation, 436, 438
in Wnt signaling pathway, 109, 110, 111
Frizzled-3 and Frizzled-3, 436, 438
Frizzled receptor, 109
Frizzled-related protein (Frzb) in amphibians, 312, 320, 321
Frogs
African clawed. See Xenopus laevis
anterior-posterior axis in, 307
atrazine affecting development of, 629, 633–634
autonomous and conditional specification in, 44, 45
axon outgrowth in, 420–421
BPA affecting development of, 632
β-catenin in, 310, 312
cleavage in, 8, 9, 10, 297–298, 360
cloning experiments with, 53, 67
cortical cytoplasm rotation in, 297
cranial neural crest in, 417
dorsal-ventral axis in, 310
fertilization in, 8, 9, 10
gametogenesis in, 8, 9, 11
gastrulation in, 8, 9, 10, 297, 299–305, 360
mid-blastula transition in preparation for, 298
genetic assimilation in, 658
genomic equivalence in, 53
germ layer specification in, 306
gray crescent in, 297
jaw phenotypes in, 658
left-right axis specification in, 322
lens induction in, 97, 98
life cycle in, 8–11
limb development in, 521–522, 523
in metamorphosis, 564, 566
in regeneration, 602
mesoderm development in, 500, 501
metamorphosis in, 8, 9–11, 563, 564–568, 658
morphogen gradients in, 100, 101
neural tube in, 333n4, 372
organogenesis in, 8, 9, 10
phylogenetic tree of, 296
polyspermy prevention in, 204
primary embryonic induction in, 307, 308

primary sex determination in, 164
regeneration in, 602
retinal ganglion axon guidance in, 431, 433–434
third eyelid in, 564
translational regulation in, 73
transplantation experiments in, 53, 295
Frontal bone, 418, *419*
Frontal lobe, 397
Frontonasal process, 415, 416, 419
Frozen zone in reaction-diffusion mechanism, 533
Fruit, 188, 189
Fruit bat *(Carollia perspicillata),* 663
Fruitfly. *See Drosophila melanogaster*
Frzb in amphibians, 312, *320,* 321
FT and *FT* (FLOWERING LOCUS T and *FLOWERING LOCUS T),* 184
βFTZ-F1 and β*FTZ-F1* in *Drosophila* metamorphosis, *574, 575*
Fundulus, 326
Fungi in tree of life, *27, 584*
Fungicides, vinclozolin as, 629, 635–636, 660
Fura-2, 205
Fused and *fused* in *Drosophila, 106,* 107, *260*
fused somites, 472
Fushi tarazu and *fushi tarazu,* in *Drosophila, 255, 260, 261, 263,* 265, *266,* 267
Fusogenic proteins, 202
Futile cycle mutants, *73*
Fxr in liver regeneration, 614

G
G protein in egg activation, *208*
GABA (gamma-aminobutyric acid), *138,* 140
Gaiano, Nicholas, 387–388
GAL4 and *GAL4,* 83
β-Galactosidase, *59,* 60–61, 418, *419,* 439n9
Galectins in limb development, 533
Gallbladder, 547, 554, 556–557
Gallus gallus, 6, 335, *336. See also* Chick development
Gametes, 193
in animal life cycle, 7
definition of, 7, 159
in external fertilization, 198–211
and gametogenesis. *See* Gametogenesis
germ cell lineage of, 159
of hydra, 591
in internal fertilization, 211–217
in plant life cycle, 7, 8
structure of, 193–198
Gametogenesis
in angiosperms, 12, 13, 186–189
in animals, 174–183
definition of, 7
in frogs, *8, 9,* 11
induced pluripotent stem cells in, 152
in mammals, 174, 175, 176, 180–183
meiosis in, 176–179

in oogenesis, 182–183, 195–196, *197, 198,* 216, *217,* 221n2
in spermatogenesis, 180, 181, *181*
oogenesis in. *See* Oogenesis
primordial germ cells in, 174–176, 180, 182
spermatogenesis in. *See* Spermatogenesis
Gametophytic stage, 187–188, 189
and alternation of generations, *32,* 34
definition of, 7
Gamma-aminobutyric acid (GABA), *138,* 140
Ganglia, 376, 401, 439n3
cervical, 432
cochleovestibular, *445,* 446
dorsal root, 401, 402, *403,* 410, 426–427, 432
enteric, 401, 412
parasympathetic, 401
retinal. *See* Retinal ganglion
sensory, 441, *442,* 445–446
sympathetic, 402, *403*
trigeminal, 416, 442
Gap genes in *Drosophila, 255–256, 260,* 261, 262–263
GAP protein in RTK signal transduction pathway, *103*
Garter snake, 489
Gasterosteus aculeatus (threespine stickleback fish), 644–645
Gastrula, 7
amphibian, 307, *308, 309, 317, 318, 319,* 321
tunicate, *291*
zebrafish, *329*
Gastrulation, 7, 14–16, 224
in amphibians. *See* Amphibians, gastrulation in
axis determination in, 16, 223
in birds, *16,* 338–342, 343, 348, 359, *360, 463*
in *Caenorhabditis elegans,* 243–245
definition of, 7, 223
in *Drosophila, 249,* 252–254, 271
in mammals, *16,* 348–351, 359, *360,* 361n1
in conjoined twins, 358
teratogen sensitivity at time of, *622*
origins of, 30
in protostomes, 224
in reptiles, 348
in sea urchins, 279, 283–288
in snails, 226, *228,* 237, *238*
in tunicates, *16,* 289
in vertebrates, 448, *452*
in zebrafish, *323,* 327–330, *360*
Gata1, *58*
Gata2 and *Gata2* in hematopoietic stem cell formation, *512*
Gata4 and *Gata4*
in heart development, 505
in primary sex determination, *161, 163,* 164

Gata6 in mammals
in gastrulation, 351
in peri-implantation period, *350*
in trophoblast-inner cell mass determination, 347
Gbb (Glass-bottom boat) in *Drosophila,* 575–576
GBP (GSK3-binding protein) in amphibian organizer formation, 310, *311,* 312
Gbx2 in neural crest specification, 404, *405*
Gcm transcription factor, 282
GDF5 and *GDF5,* 542
GDF11, *138,* 141, 157n9
GDNF (glial-derived neurotrophic factor), 111, *403,* 412
in kidney development, 496–497
in spermatogenesis, 181
in Zika virus infection, *623*
GEF protein, *103*
Gene editing, CRISPR/Cas9 technique, 82–83, 239, 324
Gene expression, 51–85
anatomy in, 55–61
and cell differentiation, 51–85
central dogma in, 52–53
definition of, 51
epigenetics in, 62–65
gene regulatory networks in, 69–70
and genomic equivalence, 51, 53–54
heterochrony in, 649, 650–651
heterometry in, 649, 651–652
heterotopy in, 649–650
integrins in, 94
messenger RNA translation in, 52, 72–77
morphogen gradients in, 100
of multiple genes, 66
phenotypic variation in, 649
posttranslational protein modification in, 52, 77–78
pre-messenger RNA processing in, 51, 70–72
regulation of, 51–52, 62–78
research tools on, 78–84
transcription in, 51, 52, 62–70
Gene knockdowns, 81–82
Gene knockouts, 81–82, 84
Gene regulatory elements
enhancers, 57, 58–61
promoters, 56–57, 58
silencers, 57, 58, 61
summary of, 61
transcription factors, 65–69
Gene regulatory networks, 69–70, 80, 95, 292n6
in cranial placode induction, 444
in flatworm regeneration, 599
in heart development, 505
in hematopoietic stem cell formation, *512*
muscle-specific, 488
in neural crest specification, 404, *405*
in sea urchins, 69–70, 279–282, 284, 285

Gene therapy, combined with induced pluripotent stem cells, 150–151
Generation time in model systems, 5
Generative cells, 188
Genes
    anatomy of, 55–61
    chromatin composition of, 55–56
    consensus sequences in, 72
    duplication and divergence of, 647–648
    expression of. See Gene expression
    function tests, 81–84
    heterotypy in, 649, 653–654
    orthologues, 648
    pleiotropy of, 655
Genetic assimilation, 656–659
Genetic labeling, 22, 23
Genetic malformations, 35
Genetic research, 78–84
    Drosophila in, 81, 82, 247, 248, 254, 256
    on gene expression, 78–81
    on gene function, 81–84
    in situ hybridization in, 78–79
    labeling in, 22, 23
    reporter genes in, 60
    sequencing technologies in, 79–81
    transgenic DNA chimeras in, 22–23, 24
Genetic syndromes, 35, 620–621
Genetics
    in birth defects, 34–35, 619, 620–621
    and epigenetics. See Epigenetics
    and evolutionary change, 643–655
    forward, 81
    heterogeneity in, 621
    modularity in, 644–646
    in primary sex determination, 163–166
    reverse, 81–82
    in secondary sex determination, 167–169
    in transgenerational inheritance, 634–636
Geniculate placode, 442, 446
Genital disc in Drosophila, 570, 571
Genital ridges, 161, 162, 163, 165, 175
Genital tubercle, 166, 167
Genome
    of Caenorhabditis elegans, 238, 239
    CRISPR/Cas9 genome editing, 82–83
    definition of, 7, 53
    developmental structure of, 643–648
    equivalence of, 51, 53–54
    of frogs, 9
    in mammals, 346
        in humans, 238, 396
    mitochondrial, 210
    in model systems, 5–6
    in somatic cell nucleus, 51
    zygotic, 14
Genomic equivalence, 51, 53–54, 67
Genomic imprinting, 65, 85n5
Genotypes, and snail coiling, 229
Geospiza, 652
Germ band in Drosophila, 252, 253, 262
Germ cells
    in angiosperms, 186
    in animal life cycle, 7

in Caenorhabditis elegans, 244, 245
    definition of, 7
    embryonic, 148
    in frogs, 11
    in gametogenesis, 174–175
    lineage of, 159, 174
    in mammals, 161, 163
    meiosis of, 11
    migration of, 19
    in ovary, 182
    primordial. See Primordial germ cells
    teratocarcinoma of, 637
Germ layers, 36n6
    in amphibians, 298, 306
    in bilateral symmetry, 29–30
    in birds, 338
    cell recombination assays, 89
    definition of, 7
    in diploblasts, 224
    in early organogenesis, 17–18
    ectoderm. See Ectoderm
    endoderm. See Endoderm
    in gastrulation, 16
    in mammals, 349
    mesoderm. See Mesoderm
    in triploblasts, 224–226
    in zebrafish, 327–330, 331
Germ line, 159, 174, 175
Germ ring in zebrafish, 327, 328
Germ stem cells, 136, 136–137
Germ wall margin in birds, 339
German shepherd dogs, 19
Germarium, 137
Germinal crescent, 341
Germinal neuroepithelium, 382, 386–387
GFAP (glial fibrillary acidic protein), 138, 139
GFP (green fluorescent protein), 23, 24, 36n7, 60, 67, 84
    and auxin, 114, 115
Giant and giant in Drosophila, 260, 262, 263, 264, 265
Gibberellin, 184
Gill arches, 18
Gills, 439
    loss of, in metamorphosis, 564, 565, 566
Giraffes, 461, 659
Glass-bottom boat (Gbb) in Drosophila, 575–576
Gli proteins in Hedgehog pathway, 106, 107, 118, 428, 639
Gli3 and Gli3 in limb development, 534, 536
Glial cells, 381
    Bergmann, 384
    in fetal alcohol syndrome, 624
    as neural crest derivative, 400, 401, 402, 403, 410, 411, 415
    radial. See Radial glial cells
    in retinal ganglion axon guidance, 431
    as scaffold in brain development, 389–390
    in ventricular-subventricular zone, 140

Glial fibrillary acidic protein (GFAP), 138, 139
Glial-derived neurotrophic factor (GDNF), 111, 403, 412
    in kidney development, 496–497
    in spermatogenesis, 181
    in Zika virus infection, 623
Glioblastoma, 640
Globin genes, 3, 56, 58, 647
Globular stage in plant development, 12, 13, 132
Glomerulus, 495, 496
Glossopharyngeal nerve, 376, 416
GLP-1 protein in Caenorhabditis elegans, 243
Glutathione, 209
Glycogen synthase kinase 3 (GSK3), 110
    in amphibian organizer formation, 310, 311, 312, 313
    in hydra, 593
    in sea urchins, 280
Glycogen synthase kinase 3 inhibitor (GSK3i), 148
Glycoproteins
    in extracellular matrix, 93–94
    in zona pellucida, 214
Glycosaminoglycans
    in cortical granules, 198, 205, 205
    in extracellular matrix, 93
    in polyspermy prevention, 198, 205, 205
Glyphosate, 627–628
Glypicans, 108, 109
gmnn in amphibians, 317
Goats, 346
Golden hamster, 214
Goldschmidt, Richard, 238
Golgi apparatus in sperm, 194, 195
Gonad development, 160–169, 493
    atrazine affecting, 633, 634
    in Caenorhabditis elegans, 121–122
    fibronectin in, 94
    intermediate mesoderm in, 462, 491, 492
    primordial germ cells in, 175
    in pseudohermaphroditism, 168
Gonadal rudiments, 161–166, 174
Gonadal sex determination, 161–166, 169
Gonadotropin-releasing hormone, 442
Gonocytes, 180
gooseberry in Drosophila, 260
Goosecoid and goosecoid
    in amphibians, 314
        Activin-induced expression of, 100, 101
    in gastrulation, 302, 303, 315
    in organizer formation, 312, 313
    in zebrafish, 332
Gorillas, brain development in, 393
Gorse (Ulex europeaus), 218
Gradient models of tetrapod limb patterning, 530–531
Grainyhead and Grainyhead, 372
Granular layers
    of cerebellum, 383, 384
    of epidermis, 453

Granule cells, 381, *383*
  in cerebellum, *383*, 384
  in cerebral cortex, *385*, *386*
Granulocyte-macrophage colony-stimulating factor, 513
Granulocyte-macrophage lymphocyte progenitors, *514*
Granulocyte-macrophage progenitors, *514*
Granulocytes, and hematopoietic stem cells, 512
Granulosa cells, 163
Grasshoppers, 569
Gray crescent in amphibians, 296, 297, 300, 307
Gray matter, 382, 385
Green algae, 583, 587
Green fluorescent protein (GFP), 23, *24*, 36n7, 60, *67*, 84
  and auxin, 114, *115*
Green sea urchin, *276*
Gremlin protein
  in limb development, 537, 538, 543, 649
  in sclerotome development, 481
Gremlin1 in limb development, *537*
gRNA (guide RNA), 82, 83
Ground finches, 652
Ground tissue in angiosperms, 13
Growth, question of, 4
Growth cones
  cell adhesion mechanisms, 426
  collapse of, 426–427, 434
  of commissural neurons, 428–429
  as locomotory apparatus, 421
  microspikes of, 421, *422*
  of motor neurons, 436, *437*
  repulsion of, 426–427
  of retinal ganglion axons, 431, *432*, 434
Growth differentiation factor 5 (GDF5 and *GDF5*), 542
Growth differentiation factor 11 (GDF11), *138*, 141, 157n9
Gsh gene, *647*
GSK3 (glycogen synthase kinase 3), 110
  in amphibian organizer formation, 310, *311*, 312, *313*
  in hydra, 593
  in sea urchins, *280*
GSK3-binding protein (GBP) in amphibian organizer formation, 310, *311*, 312
GSK3i (glycogen synthase kinase 3 inhibitor), 148
Guanine monophosphate, cyclic (cGMP), 511
Guanosine, methylated, 57
Gudernatsch, J. F., 566
Guide RNA (gRNA), 82, 83
Guinea pigs, 212
Gurdon, John, 53
Gurken and *gurken* in *Drosophila*, 269–270
Gut development
  in amphibians, 300, 322
    in metamorphosis, 564, 565
  in *Caenorhabditis elegans*, 242, *244*, 245

digestive tube in, 547–557
looping in, 549, 553
in mammals, 549
neural crest cells in, 401, 412–413
in sea urchins, 279, *286*, 287, 288
in snails, 237
tissue specification in, 551–554
villi in, 553–554
in zebrafish, *323*
Gymnosperms, 220
Gynandromorphs, 169–170, 189
Gyrencephaly, 393–394
Gyri, cerebral, 393–394, 395, 397

H

Hair cells, sensory, 444, 447
Hair development, 454, 455, 456
  neural crest in, 410, 439n4
  non-sensory placodes in, 441
  and regeneration, 458, *459*, 581
  signaling pathways in, 456
  stem cells in, 458, *459*
Hair follicle
  bulge region in, 458, *459*
  development of, 146, 454, 455
  melanocytes in, 453
  regeneration of, 146, 458
  stem cells of, 146, 458, *459*
Hair follicle stem cells, 146, 458, *459*
Hair placode, 454, 455
Hair shaft, 458, *459*
hairy in *Drosophila*, *260*, 263
Hairy1 in somitogenesis, 476–477
Hairy/Enhancer of split-related proteins, 476, 477
Haldane, J.B.S., 663n3
Haliclona cf. *permollis*, 585
Halisarca dujardini, 585
Halteres, 269, 273, *570*, 575
Hamburger, Viktor, 20, 534
Hamburger-Hamilton stages in chick development, 20
Hamiltonella defensa, 660
Hämmerling, Joachim, 586
Hamsters, *214*
Haploid cells, 176, 178
  in angiosperm life cycle, 187, 188, *189*
  in oogenesis, 183
  in spermatogenesis, 180, 181–182
HAR1 (human accelerated region-1), 396, 397
Harrison, Ross Granville, 309, 420–421, 518
Harvey, William, 3, 36n2, 500
Hatched blastula of sea urchins, 278, 283
Hawkmoths, *422*
HbA, 151
HbS, 151
Head activation gradient in hydra, 592–593
Head activator in hydra, 592–593
Head development
  alcohol affecting, 624, 626
  in amphibians, 303, *305*, 307, 318–321
    in metamorphosis, 566

in birds, *340*, 341
cranial sensory placodes in, 441–452
in *Drosophila*, 252, 256, 257–259, 260, 354
  and axis formation, *254*
  in gastrulation, *253*
  homeotic selector genes in, 268, 269
  parasegments in, *261*
of face. *See* Face development
in flatworm regeneration, 595–596, 598, 599–601, 602, 617
in mammals, 354, 361
mesoderm in, 462
neural crest in, *400*, 414–416, 418–419
pharyngeal arches in, *18*
of skull, 414, 418–419, 566
in snakes, 658
in zebrafish, *327*, 330, 331
Zika virus affecting, 623
Head ectoderm in eye development, *97*, 448
Head endoderm in cranial placode induction, 443
Head mesoderm, 461, 462
  in amphibians, *302*, 303, *305*, 314, 318
  in birds, 341
  in cranial placode induction, 443
  in skull development, 418
Head process, 341, *343*, 353
Hearing, 441, 442, 444, 446
  auditory cortex in, 386
  cochlea in, 444, 446–448
  otic placode in, 442
Heart development, 491, 499–507, 550, 622
  in amphibians, *298*, 303, 305, 322
  in birds, 340
  and blood vessel development, 507
  cardiac neural crest in, 419–420, 503
  cardiogenic mesoderm in, 501–503, 554
  cell differentiation in, 505
  cell migration in, 90
  in Down syndrome, 620, 621
  endoderm in, 547
  fibronectin in, 94
  fracking chemicals affecting, 634
  heart fields in, 501–503, 504
  in Holt-Oram syndrome, 35
  lateral plate mesoderm in, 491, 499–507
  in left-right axis, 322, 343, 356, *357*
  and liver development, 554–555
  looping in, 506–507
  in mammals, 356, *357*, 501
    and regeneration in neonates, 615
  precursor cells in, 503–505
  recellularization in, 99
  in tunicates, 290, 291, 501, 502
  valves in, 507
  in zebrafish, *323*, 332, *504*
    in regeneration, 609–612, 615
Heart fields, 501–503, 504
Heart stage in plant development, *12*, 13, *132*
Heart tube, 501, 502, 504, 506

Heat shock proteins (Hsp), 660
in androgen insensitivity syndrome, *168*
in *Drosophila*, 77, 656–658
HEC1 and *HEC1*, 133
Hedgehog and *hedgehog*, 101, 105–108
in cancer, 108, 638, *639*
in ciliopathies, 118
in *Drosophila*, 105, 107, *119*, 124n5
as segment polarity gene, *260*, 265, *266*, 267
signaling pathway of, *106*
in wing development, 575
functions of, 107
as morphogens, 105
naming of, 125n5
and primary cilium, 118, *120*
processing and secretion of, 105–106
signaling pathway of, 106–107, 108, 118
Sonic. *See* Sonic hedgehog and *sonic hedgehog*
in zebrafish fin regeneration, *608*, 609
HeLa cells, 147, 157n17
Hemangioblasts, 505, 507
differentiation of, 507
generation of, 509
and hemogenic endothelial cells, 515n5
in vasculogenesis, *508*
Hemangiogenic mesoderm, *503*
Hematopoiesis, 510–515
inductive microenvironments in, 513–515
sites of, 510–511
Hematopoietic inductive microenvironments, 513–515
Hematopoietic stem cell niche, 143–145, 157n13, 512–513
Hematopoietic stem cells, 129, 130, 143–145, 508, 510–515, 515n4
in bone marrow, *130*, 143–145, 157n12, 511, 512–513, 581
in endosteal niche, 143, 144, 145
in inductive microenvironments, 513–515
long-term, *514*
mesonephros as source of, 492
multipotent, 130, 513
pathway in formation of, *512*
in perivascular niche, 143, 144, 145, 157n13, 513
short-term, *514*
in umbilical cord, 157n16
Hemichordates, *225*, *276*
Hemimetabolous development, 569
Hemmati-Brivanlou, Ali, 317
Hemogenic endothelial cells, 511, *512*, 515n5
Hemoglobin, 77
in amphibian metamorphosis, 565
genes for, 51, 52, 56, 151
in sickle cell anemia, 151
steps in production of, *56*
Hensen, Viktor, 361n5, 420

Hensen's node, 374, *550*
in birds, *343*, 359, 420
in gastrulation, 338, 340, 341, 342
and left-right axis specification, *344*
and primary neurulation, *367*
in mice, 361n5
Heparan sulfate proteoglycans
functions in extracellular matrix, 93, 117
in germ stem cell development, 137
and Hedgehog secretion, *105*
in hematopoiesis, 513
in paracrine factor diffusion, 117
in RTK pathway, *103*
and Wnt secretion, 108
Hepatectomy, partial, compensatory regeneration in, 613–615
Hepatic diverticulum, 554
Hepatic duct, 554
Hepatic progenitor cells, *548*
in liver regeneration, 613–614
Hepatoblasts, *548*, 561
Hepatocyte growth factor (HGF), 113, 614
Hepatocytes, *149*, *548*
in liver regeneration, 613, 614
Her protein and *Her* gene in somitogenesis, 476, *477*
Herbicides
atrazine, 629, 633–634
glyphosate, 627–628
Hermaphroditism, 166, 190n2
in atrazine exposure, 633
in *Caenorhabditis elegans*, 121, 238, 239, 240, 245
and pseudohermaphroditism, 168
Hertwig, Oscar, 221
Hes gene and Notch activity, 140
Hes protein and neural stem stems in V-SVZ, *138*
Hes7 in somitogenesis, *478*
HesC and *HesC* in sea urchins, 281–282
Heterochromatin, 56, 62
Heterochrony, 649, 650–651
Heterogeneity, genetic, 621
Heterogeneous nuclear RNA, 52, 57
Heterometry, 649, 651–652
Heterophilic binding, 88, *88*
Heterotopy, 649–650
Heterotypy, 649, 653–654
Hex transcription factor, 561
Hexactinellida, 585
HGF (hepatocyte growth factor), 113, 614
Hhex
in amphibians, 302, 313
in digestive tube development, 553
High CpG-content promoters, 63–64
Hindbrain, *375*, 376
in amphibians, 307, 314, 318
anterior-posterior patterning of, 376, 415
in birds, 341, 342, *375*, 376
in mammals, 353
and neural crest, *401*, 415, 416
and neural tube closure, *370*, 371
segmentation of, 376, 415

Hindgut, *550*
in *Drosophila* metamorphosis, 573
Hox genes in development of, 553
neural crest cells in, 412
in sea urchins, *286*, 287
Hindlimb development
in amphibian metamorphosis, 564
apical ectodermal ridge in, 528
in cetaceans, 543, 544
digit specification in, 536
Hox genes in, 519, *520*
Islet1 in, 524, 526, 527
limb bub in, 518, 527
Sonic hedgehog in, 536
specification in, 524–526
Hinge points, 369–370
dorsolateral, *368*, 369, 370, 371
medial, *368*, 369, 370, 377
Hippo signaling, *135*
Hippocampus, *375*
Hirschsprung disease, 412
His, Wilhelm, 420
Histoblasts, 570
Histone, 55
acetylation of, 62–63
enzymes modifying, 65–66
epigenetic modification of, 62–63, 131
methylation of, 62–63, 64
neural tube defects in modifications of, *373*, 374
in sea urchin fertilization, 210
Histone acetyltransferase, 62, 65, 568
Histone deacetylase, 62, 568
Histone methyltransferase, 62
in insect metamorphosis, 574
and Polycomb proteins, 63
and Trithorax proteins, 65–66, 67
HLH-1 and *hlh-1* in *Caenorhabditis elegans*, 242, *243*
HNF1B and pancreatic β cells, 556
HNF4α and Hnf4α in liver development, 83–84, 555
Hodgkin, Jonathan, 239
Holoblastic cleavage, 2, 14, *15*
bilateral, 14, *15*, 288–289
in *Caenorhabditis elegans*, 240
in mammals, 335
radial, *15*
in amphibians, 297–298
in sea urchins, 276–277
rotational, *15*
spiral, 14, *15*
in snails, 226–230
in tunicates, 288–289
Holometabolous development, 569–570
metamorphosis in, 569–576
Holtfreter, Johannes, 87, 89, 90, 91, 314
Holt-Oram syndrome, 35
Holtzer, Howard, 99
Holub, Miroslav, 351
Homeodomain transcription factors, *66*
Homeotic complex (Hom-C) in *Drosophila*, 268, 353, 354
Homeotic mutants, 268–269, 273n5

Homeotic selector genes in *Drosophila*, 256, 268–269
Homeotic transformations in floral organs, 69
Homo sapiens, *6*, 375, 380n3. *See also* Humans
Homology, 28, 30, 224
    deep, 646, 663n5
Homology directed repair, *82*, 83
Homophilic binding, 88
Homoscleromorphs, 30, 31, 585
homothorax in *Drosophila*, 269
Horizontal plane, *17*
Hormones. *See also specific hormones.*
    in amphibian metamorphosis, 11, 564, 566–568
    endocrine disruptors affecting, 628–634
    in insect metamorphosis, 572–575
    paracrine factors compared to, 96
    in plants, *32*, 33, 47
        in regeneration, 587, 588–591
    in sex determination, 166–169
    in sperm migration, 213
Horses, *414, 518*
Hox clock, 469
Hox code hypothesis, 353–356
Hox genes
    in amphibians, *321*
    in anterior-posterior axis, 361, 376, 467–469, 478–479, 646
        in limb development, 522, 523
        in mammals, 353–356
    chromatin structure in, 459
    circular chromosome conformation capture in analysis of, 489n1
    and comparative anatomy, 355
    diethylstilbestrol exposure affecting, 630, *631*
    in digestive tube development, *552*, 553
    in *Drosophila*, 353, 354, 469, 648
    duplication and divergence of, 647–648
    in ectodermal appendage development, 455
    evolution of, 646, 663n6
    in heart development, 504
    as homeodomain transcription factors, *66*
    in limb development, 519–521, 522, 527, 544
        in anterior-posterior axis, 522, 523
        in digit specification, 536, 538–539
        of forelimb, 524
        in gradient model, *530*
        in limb bud induction, 523
        in reaction-diffusion model, 531
    in mammals, 353–356
    in motor neuron guidance, 424
    and neural cell differentiation, 414
    and neural crest potency, 401–402
    orthologues of, 648
    paralogues of, 354–355, 519–520
    and paraxial mesoderm, 467–469, 478–479, 522

    regional-specific instructions of, 663n6
    ribosomal Rpl38 protein and, 74
    in skeleton development, 74, 401–402
    spatiotemporal collinearity of, 467–469, 478–479
    temporal activation of, 468–469
    in vertebrae development, 74, 355–356, 467, 655
Hox5 and *Hox5*, 355
Hox6 and *Hox6*, 355
Hox9 and *Hox9*, 355, 519, *520*
Hox10 and *Hox10*, 355, *356*, 519, *520*
Hox11, 355–356, 519, 520
hox11/13b, 70
Hox12, 519, *520*
Hox13, 519, *520*
Hoxa and HoxA
    in anterior-posterior axis, 353, 354, 355, 467, *468*
    diethylstilbestrol affecting, 630, *631*
    in ectodermal appendage development, 455
    in heart development, 504
    in limb development, 519, 520, *530*
Hoxb and HoxB
    in anterior-posterior axis, 354, 356, 467, *468*
    in heart development, 504
    in kidney development, *499*
    orthologues of, 648
Hoxc and HoxC
    in anterior-posterior axis, 354, 355
    in limb development, 522
Hoxd and HoxD, 469
    in anterior-posterior axis, 354, 355
    in limb development, 519, 520, 536, 538–539
Hsp (heat shock proteins), 660
    in androgen insensitivity syndrome, *168*
    in *Drosophila*, 77, 656–658
Hsp83 and *hsp83* in *Drosophila*, 77
Hsp90 in *Drosophila*, 656–658
Hu proteins, 72–73, 85n8
huckebein in *Drosophila*, *260*, 262
human accelerated region-1 (HAR1), 396, 397
Humans
    anterior-posterior axis formation in, 352
    birth defects in, 34–36, 619, 620–621
    brain growth and development in, *375*, 381, 392–397
    cervical vertebrae in, 461
    cleavage in, 346, 619, *622*
    developmental abnormalities in, 34–36, 619–642
    digestive tube in, 549
    embryonic period of development, 1, 24, 621, *622*
    endoderm in, 549, *550*
    environmental factors affecting development of, 619, 621–634
    epidermis in, 453–454, 460n2

    evolution of, 28, 648–649
    eye development in, 97, *98*, 450
        cyclopia in, 107, 621
        microphthalmia in, 620
    fertilization in, *345*
    fetal period of development, 621, *622*
    gallbladder in, 557
    gastrulation in, *351*
    genetic factors affecting development of, 34–35, 619–621
    genome of, 238, 396
    gonad development in, 161, 162, 163, 190n1
    hair growth in, 458
    health and disease in, 619–642
    heart development in, 506
    hematopoiesis in, 510
    kidney development in, *357*, 492, *550*
    left-right asymmetry in, 16, *357*
    limb development in, 520, 540
    liver regeneration in, 613
    meiosis in, 178
    mesenchymal stem cells in, 146
    microRNAs in, 74
    mouse model of, 336
    neural crest in, 415, *415, 420*, 439
    neural tube closure in, *370, 371*
        defects of, *370*, 372–374
    nucleosomes in, 55
    number of cell types in, 36n3
    oogenesis in, 182
    optic nerve in, 431
    organoids in study of, 153–155
    pancreas development in, *357*, 554
    respiratory tube in, *558*
    signal transduction pathways in, 104
    skeletal anatomy in, *518*
    somitogenesis in, 470, 477
    Sonic Hedgehog in, 108
    sperm of, 180–181, 212
    spinal cord in, *383*
    stem cell model of development and disease in, 147–157
    stochastic factors affecting development in, 619–620
    sweat glands in, 458
    tail degeneration in, 577n4
    teratogen exposure in, 35–36, 621–628
    tissue and germ layer formation in early embryo, *349*
    in tree of life, 27, *584*
    twin, 357–359
    vasculogenesis in, *509*
    vertebrae formation in, 483
    Wnt in, 108
    X-chromosome inactivation in, 169
    zygote size of, 345
Humerus, 517, *518, 520*, 529, 533
Hunchback and *hunchback* in *Drosophila*, 255, 258, 259–260, 262, 263, 264, *265*
Huxley, Julian, 643
Huxley, Thomas, 643
Hyalin, 205

Hyaline layer in sea urchins, 283–284, 287
Hybridization, in situ, 53, *59*, 78–79
Hydra, *26*, 617n3
    budding in, 591, *592, 594*
    cell migration and death in, 157n11
    cell replacement in, 591–592
    as diploblasts, 224
    head activator in, 592–593
    hypostome as organizer in, 593
    as model organism, *6*
    regeneration in, 6, 135, 579, 580, 591–593, 607
    reproduction of, 591
    totipotent stem cells in, 128
Hydrogen peroxide in cortical granule reaction, *205*
Hydroid cells, *32, 33*
Hydronephrosis, 498
20-Hydroxyecdysone in insect metamorphosis, 572–575
5-Hydroxytryptamine, *138*
Hyla regilla (Pacific tree frog), *523*
Hyoid bone, *416*
Hyoid cartilage, *415, 416*
Hypaxial muscles, 489n5
"Hyperion" redwood sequoia, *2*
Hypermorphosis, 393
Hypoblast, 361
    in birds, *337, 339*
        in gastrulation, 338, 340, 341, 342
        primary, 338
        secondary, 338, 340
    in epithelial-mesenchymal transition, *95*
    in mammals, 348–351
    in zebrafish, *327, 328, 329, 330*
Hypocotyl, 114, *115*
Hypodermis in *Caenorhabditis elegans*, *239, 243*, 245
Hypomorphic mutations, 150
Hypophysis, *12, 132*
Hypostome in hydra, 591, *592*, 593
Hypothalamus, *375, 450*

I

IAA (indole-3-acetic acid), 114, *116. See also* Auxin
Identical twins, 357–358, *359*
IGF. *See* Insulin-like growth factor
Ihh and *ihh* (Indian hedgehog and *indian hedgehog*), 105, 553, 609
IKKα in nasopharyngeal carcinoma, *641*
IL. *See* Interleukins
Ilyanassa snails, 226, *227, 236, 237*
Imaginal cells, 570
Imaginal discs, 83, 254, 273n1, 570–572, 573, 574
    wing, 118, *119, 570*, 573, 575–576, 645–646
Imaginal molt, *569*, 570, 574
Imago, 569, 570
Immortalized tumor cells, 147
Immune system in regeneration, 582
Immunoprecipitation sequencing, 79–80

Implantation, *345*, 348
    and peri-implantation period, 348, *350*
    teratogen sensitivity at time of, *622*
Incus, *415*, 416
ind gene, *647*
Indels, 82, *82*
Indeterminate growth, 1–2, 8, *32*, 133, 587–588
Indian hedgehog and *indian hedgehog* (Ihh and *ihh*), 105, 553, 609
Indirect developers, 563
Indole-3-acetic acid (IAA), 114, *116. See also* Auxin
Induced pluripotent stem cells (iPSCs), *66, 67*, 150–156, 556
    applications to human development and disease, 150–153
    formation of insulin-producing β-cells from, 556
    medical uses of, 150–152, 156, 157n20
    organoids derived from, 153, 496
    patient-specific, 150–151
Inducers, 96, 100
    of amphibian lens, 97
    paracrine factors as, 96, 100–119
    signal transduction cascade in response to, 101–102
Induction, 95–99
    and competence, 95–99
    of cranial placodes, 443–445
    definition of, 96
    discovery of, 18
    of eye development, 95–97, 102, *103*, 451, *452*
    instructive and permissive interactions in, 99
    juxtacrine signaling in, 121–123
    of kidney development, 100, 495–499
    of limb buds, 523–527
    of neural crest cells, 404
    paracrine signaling in, 96, 100–119, 121–123
    primary embryonic, 307–308
Infertility. *See* Fertility
Inflorescence meristem, 132, *231, 232, 234*
Infundibulum, *345*, 549
Ingression in gastrulation, 16
    in sea urchins, 283–286
Inheritance
    of DNA methylation patterns, 64–65
    genomic imprinting in, 65
    of histone methylation patterns, 63
    Mendelian, 65
Inhibitory domain in reaction-diffusion mechanism, 533
Initial cells in plants, 131–132, 133
Injury
    epithelial-mesenchymal transition in, 95
    regeneration in, 582, 583. *See also* Regeneration
    scar formation in, 585, 586
Inner cell mass, 347–348, *349, 350*
    and Hippo signaling, *135*
    in mice, 134–135, *346*, 348, 350

pluripotent cells in, 131, 134–135, 148, 347
    in twins, 358
Inner ear, 441, 459
    anatomy of, 447
    and axis determination in otocyst, 447, 448
    blood vessel formation in, *414*
    and neural crest, 404
    and otic placode, 442, 444, 446
    sensory epithelium of, 442
Inositol 1,4,5-trisphosphate (IP$_3$), 206–208, *210*, 216
Insects, 224
    ametabolous development of, 569
    cleavage patterns in, 14, *15*
    dipteran, 269, 273n6
    Drosophila. See Drosophila melanogaster
    environmental influences on development of, 5
    as gynandromorphs, 169–170
    hemimetabolous development of, 569
    heterotypy and number of legs in, 653–654
    holometabolous development of, 569–570
    life cycle of, 569
    limb development in, 653–654
    metamorphosis in, 563, 569–576
        in *Drosophila*, *249*, 570–576
        hormonal control of, 572–575
        20-hydroxyecdysone in, 572–575
        imaginal discs in, 570–572, 575–576
    as pollinators, 217, *218*
    recruitment in, 645–646
    sex determination in, 159, 169
    symbiotic bacteria and birth defects in, 620
    syncytial specification in, 47
In situ hybridization, 53, *59*, 78–79
Instars, 569
    in *Drosophila*, 571, 572, 573–574, 576
Instructive interactions, 99, 124n2
Insulin, 51, 77, 556
Insulin-like growth factor (IGF)
    in amphibians, *320*, 321
    in cancer, *639*
    in epidermis regeneration, *453*
    and radial glial stem cell fate, 392
    in zebrafish fin regeneration, *608*, 609
Integrins, 94, *453*
    in axis elongation, *471, 472*
    in motor neuron guidance, 426
    in neocortex development, 390
    in somitogenesis, *471, 472*, 473
Integuments, 188, *189*
Intelligent design, 663n1
Interdigital necrotic zone, *541, 542*
Interior necrotic zone, *541, 542*
Interkinesis, 178
    nuclear migration in, 386–387
Interleukins, 113
    IL1-β, *138*
    IL3, 513
    IL6, 614

Intermediate cranial placodes, 442, *443*
Intermediate mesoderm, 461, 462, 491–499, 500
    derivatives of, 462, *462*, 491
    in gonad development, 462, 491, *492*
    in kidney development, 462, 491–499
    specification of, 465, 491, 493–494
intermediate neuroblast defective in *Drosophila*, 271, *272*
Intermediate progenitor cells, 388
Intermediate zone (mantle zone), 382, *383*
    in cerebellum, *383*
    in cerebral cortex, *383*, 385, *388*, *394*
    in spinal cord or medulla, *383*
Internal fertilization, 198, 211–217
    acrosome reaction in, 214
    cell membrane fusion in, 214–215
    egg activation in, 216
    gamete translocation in, 211–212
    genetic fusion in, 216–217
    polyspermy prevention in, 215–216
    recognition at zona pellucida in, 214
    sperm capacitation in, 212
    sperm hyperactivation in, 212–213
    thermal and chemical gradients in, 213
Internal granular layer, *383*, 384
Internal yolk syncytial layer, 326, *327*, 328
International Society for Stem Cell Research, 156
Interneurons, 377, *378*
Interphase, 48, *236*, *251*
Intersex conditions, 168, 190n2, 190n3
Interstitial stem cells in hydra, 591–592
Intervertebral discs, 462, *463*, 464, 483–484
Intervertebral joints, 482
Intestinal organoid, 153
Intestinal stem cells, 142–143, 639
    in amphibian metamorphosis, 565
    clonogenic nature of, *143*
Intestines, 551, 552, 553
    adenoma of, 639
    in amphibian metamorphosis, 564, 565, 566
    in *Caenorhabditis elegans*, *239*, *240*, *242*, *244*, 245
    looping of, 549, 553
    progenitor cells, 548
    in snails, 237
    villi of, 142, 553–554
Intine, 188
Intraembryonic vasculogenesis, 508
Intramembranous bones, 418
Introns, 56, 57, *59*
    in alternative pre-mRNA splicing, *71*, 72
    enhancers in, 58, 60
    noncoding information in, 56, 57
    removal of, *56*, *57*, 70, 72
Invagination, 16
    in amphibians, 299, 300–302
    in *Caenorhabditis elegans*, 243
    in *Drosophila*, 251–252, 259, 271
    of optic vesicle, 448–449

    of otic placode, 445
    in sea urchins, 278–279, 283, 286–288
Invariant cell lineage, 238
Invertebrates, 226
    cleavage patterns in, 14
    evolutionary embryology of, 26, 27
    tunicates as, 288, 292
Involucrin, *453*
Involuting marginal zone in amphibians, 302, 303, 304, 305
Involution, 16, 302
    in amphibians, 299, 302–304, *305*
    in zebrafish, 328, *329*
IP_3 (inositol 1,4,5-trisphosphate), 206–208, *210*, 216
Iris, 97, *98*
Islet1 and *Islet1* (Isl1 and *Isl1*)
    in heart development, 504, *505*
    in limb development, 524, 526, 527
    in motor neuron guidance, 424, *425*, *438*
Islet2 and *Islet2* (Isl2 and *Isl2*)
    in motor neuron guidance, 424, *425*, *438*
    in retinal ganglion axon guidance, 431
Isocortex, 398n2
Isolecithal cleavage, 14, *15*
Isotretinoin, 627
Isthmus, *345*
Itasaki, Nobue, 459
Ito cells in liver regeneration, 614
Izumo protein, 215, *215*

J

Jacob, François, 649
Jacobson, Marcus, 433
Jaffe, Laurinda, 203
Jagged, 120, *121*
Jagged1 (Jag1)
    in epidermis regeneration, *453*
    and hematopoietic stem cells, 145
    and neural stem cells in V-SVZ, *138*, 140
Jagged2 in epidermis regeneration, *453*
JAK (Janus kinase), 104, *110*
Jak2, *104*
JAK-STAT pathway, 104
Janus kinase (JAK), 104, *110*
Jaw development
    in amphibian metamorphosis, 566
    in *Drosophila*, 83
    GAL4-UAS system in study of, 83
    genetic assimilation in, 658
    heterochrony in, 650
    heterometry in, 651
    neural crest in, 401, 402, 415, 416, 439
    pharyngeal arch in, 18, *416*
    retinoic acid affecting, 627
Jefferson salamander (*Ambystoma jeffersonianum*), 221n3
Jellyfish, 224, 646
Jervine, 107, *107*, *450*, 623
Joint development, 541–542
    evolution of, 520–521
    intervertebral, 482

Junctional neurulation, 366
Juno protein, 215, *215*, 216
Just, Ernest E., 87
Juvenile hormone in insect metamorphosis, 572–573, 574, 575
Juxtacrine signaling, 88–92, 100
    cadherins in, 91–92, 120
    in *Caenorhabditis elegans* vulva induction, 121–123
    cell adhesion in, 91–92, 120
    for cell identity, 120–123
    in conditional specification, 44, 45
    differential cell affinity in, 89–90
    Eph receptors and ligands in, 120
    Notch proteins in, 120–121
    in stem cell regulation, 130
    thermodynamic model of, 90–91

K

Kallmann syndrome, 442
Kartagener triad, 221n1
Karyokinesis, 250
KDM6B and *Kdm6b* in temperature-dependent sex determination, 174
Keller, Ray, 305
Keratin, 304, *305*, 453, 455
Keratinocyte growth factor (Fgf7), 102, *453*, *459*
Keratinocytes, 453, 458
Kidney development, *167*, *462*, 491–499
    in amphibians, 303, 305, 309, 492
    in birds, 340, *340*
    in humans, 357, 492, *550*
    induction of, 100, 495–499
    intermediate mesoderm in, 462, 491–499
    in mammals, 492, *493*, 498
    paraxial mesoderm in, 493
    reciprocal interactions in, 494–499
    stages of, 492–493
    in zebrafish, *323*
Kidney organoid, 153, 496
Kidney tubules, 492, *493*, 494, *495*, 496
Kierkegaard, Søren, 49
Kinesin, 77, 310, *311*
Kinetochores, 190n6
    in aneuploidy, 183
    in meiosis, 178, 179
King, Mary-Claire, 648
Kit and *Kit*, 414, 620, 621
    in heart development, *505*
    and hematopoietic stem cells, 512
Klein, Allon, 49
Klf4 and *Klf4*, 66, 67, 150, *151*
Klimt, Gustav, 361n4
Knirps and *knirps* in *Drosophila*, *260*, 262, 263, *265*
Knockdown techniques, 81–82, 150
Knockout techniques, 81–82, 84, 150
    in *Caenorhabditis elegans*, 239
Koepfli, J. B., 114
Koller's sickle, 338, *339*, *340*
Kornberg, Thomas, 118
Kowalevsky, Alexander, 26, 27, 292

Krüppel and *Krüppel,* in *Drosophila,* 255, 260, 261, 262, 263, 264, 265
Kugelkern protein, 251
Kupffer macrophages in liver regeneration, 614
Kupffer's vesicle in zebrafish, 323, 332, 333n6

L

L1 protein and *L1* gene, 61, 626
Labia majora, 166, 167
Labial disc in *Drosophila,* 570
labial gene in *Drosophila,* 268, 354
Labioscrotal folds, 166, 167
Lacks, Henrietta, 147, 157n17
Lactation, 104, 457
LacZ and *lacZ,* 59, 60, 61, 123, 143, 264
Ladybug beetles, 645
LAG-2 and *lag-2,* 123
Lamellipodia
    in amphibian gastrulation, 303–304, 305
    in axon growth cones, 422
    in neural crest cell migration, 405
Lamina
    basal. *See* Basal lamina
    in brain, 382, 389, 390
    dental, 454, 457
    reticular, 93
Laminin
    in axis elongation, 471
    in extracellular matrix, 93, 94, 488
    in mammals, 350
    in mesenchymal stem cell differentiation, 146
    in motor neuron guidance, 426
    in neural crest migration, 411
    receptors for, 94
    in retinal ganglion axon guidance, 431
    in sea urchins, 284, 285
    in synapse formation, 436, 437
Laminin5, 453
Lancelet (*Amphioxus*), 26, 27, 225, 226
    neurulation in, 366
    pharyngeal arches in, 246n2
    sclerotome and dermomyotome in, 479
Land plants, developmental history of, 31–34
Language learning, 397
Large intestine, 551, 553
Larvae
    adaptations of, 36–37n10
    in animal life cycle, 7, 563
    and evolution of biphasic life cycle, 577n1
    of holometabolous insects, 569–570
        imaginal discs of, 570–572
        molts of, 569, 570
    metamorphosis of, 563
    primary, 563
    secondary, 563
    of sponges, 30
    taxonomic classification based on, 25–26

Laryngeal cartilage, 415, 416
Laryngotracheal endoderm, 557
Laryngotracheal groove, 557
Last common regenerative ancestor, 584
Last eukaryotic common ancestor, 31–33, 36, 223–224, 225
Last universal common ancestor, 27
Lateral axis, 17
Lateral blastopore lip
    in amphibians, 301, 303, 305
    in zebrafish, 331
Lateral ectoderm in *Drosophila,* 271
Lateral geniculate nucleus, 386, 431
Lateral inhibition, 123, 476
    in reaction-diffusion model, 531–532, 533
Lateral line placodes, 442
Lateral motor column, 424, 438
Lateral nasal process, 415
Lateral plate mesenchyme, 494
    derivatives of, 462, 492
    in limb development, 518
Lateral plate mesoderm, 461, 462, 463, 481, 491, 499–515
    in birds, 340
    in blood vessel formation, 499, 507–510
    derivatives of, 462, 486, 491
    in dermis development, 486
    in heart development, 499–507
    in limb development, 518, 519, 522, 523
    in mammals, 356
    specification of, 465
    splanchnic, 552
Lateral somitic frontier, 485–486
Lateral ventricles, neural stem cells in, 137–141
Ldb1, 58
Le Douarin, Nicole, 402
Leader sequence (5′ UTR), 57
LEAFY (LFY), 184, 186
LEF in Wnt signaling pathway, 110
Left-handed (sinistral) coiling, 228–230, 245
Left-right axis, 16, 223, 361
    in amphibians, 306–307, 309, 322
    in birds, 338, 343–345
    in *Caenorhabditis elegans,* 240, 242
    in mammals, 16, 353, 356, 357
    in zebrafish, 331, 332–333
Lefty and *lefty* in zebrafish, 331
Lefty1
    in birds, 345
    in mammals, 353, 361
Leg eversion in *Drosophila,* 572
Leg imaginal discs in *Drosophila,* 570, 571–572, 573, 575
Lens development, 448
    enhancers in, 59, 60, 61, 66
    Fgf8 in, 97, 102, 103
    induction of, 96–97, 98, 102, 103, 451, 452
    placode in, 404, 442, 443, 448, 449, 451, 452
    in vertebrates, 96–97, 98, 448, 451, 452

Lens placode, 442, 443, 448, 449, 451, 452
Leopard frog (*Rana pipiens*), 8–11
    atrazine affecting, 633, 634
    gray crescent of, 297
Leptoid cells, 33
Leptotene stage, 177, 178
Leukemia, 638, 640
Leukemia inhibitory factor (LIF), 148
Lewis, Edward, 254
Leydig cells, 161, 163, 167, 180
Lfng (Lunatic fringe) in somitogenesis, 476–477
LFY (LEAFY), 184, 186
Lgr in Wnt signaling pathway, 110
Lgr5, 142, 143, 639
Lhx in eye development, 450
Lhx1 and *Lhx1,* 361
    in kidney development, 493–494
    in limb development, 540
    in motor neuron guidance, 425
Lhx2 and *Lhx2*
    in cerebral cortex formation, 385
    in eye development, 449
    in heart development, 502
Lhx3 and *Lhx3,* 377
    in motor neuron guidance, 425
    in tunicates, 290
Lhx9 and *Lhx9* in primary sex determination, 161, 163, 164
LIF (leukemia inhibitory factor), 148
Life cycles, 6–13
    of angiosperms, 7–8, 11–13, 187
    of animals, 6–7, 563
    Aristotle on, 2–3
    of frogs, 8–11
    gametogenesis in, 176–179
    of insects, 249, 569
Ligands
    in cell-to-cell signaling, 88
    in signal transduction pathways, 101, 102
Light
    and DNA methylation, 660
    and photoperiods, 9, 184
Lillie, Frank, 206
Lily pollen tubes, 219
Lim protein, 66
    in motor neuron guidance, 424–425
    in retinal ganglion axon guidance, 431
Lim1
    in kidney development, 493–494
    in limb development, 540
    in motor neuron guidance, 425
Limb buds, 489n4, 518, 519, 522, 544
    and apical ectodermal ridge, 527–528, 537
    dorsal-ventral axis in, 539–541
    feedback loops in formation of, 527, 537–538
    forelimb, 524
    Hedgehog in, 105–106, 653
    of hindlimb, 518, 527
    induction of, 523–527
    motor neuron guidance to, 425

potential of, 523
Turing model on development of, 531–533
and zone of polarizing activity, 535
Limb development, 517–545
in amphibian metamorphosis, 564, 566
anatomy in, 517–518
anterior-posterior axis in, 517, 518, 522, 534–539
Hox genes in, 522, 523
apical ectodermal ridge in, 518, 527–531
apoptosis in, 536, 541–542
autopod, 517, 518, 541–542
computer simulation of, 533, 534
cytoneme-mediated signaling in, 118
dorsal-ventral axis in, 518, 539–541
in Drosophila
leg eversion in, 572
leg imaginal discs in, 570, 571–572, 573, 575
proximal-distal axis in, 273, 576
wing imaginal discs in, 118, 119, 570, 573, 575–576, 645–646
evolution of, 520–521, 543–544
constraints on, 654–655
heterometry in, 652, 653
heterotypy in, 653–654
of forelimb. See Forelimb development
gradient models of, 530–531
Hedgehog in, 105–106
of hindlimb. See Hindlimb development
homologous and analogous structures in, 28
Hox genes in. See Hox genes, in limb development
in insects, 653–654
joints in, 520, 542
limb buds in. See Limb buds
limb fields in. See Limb fields
mesoderm in, 523–524, 529–531
polarity in, 108
proximal-distal axis in, 517, 518, 527–533
in Drosophila, 273, 576
in salamander regeneration, 602
reaction-diffusion mechanism in, 531–533, 654–655
in regeneration, 20, 522, 579
in salamanders, 2, 4, 20, 522, 579, 580, 583, 602–607, 617n1
in salamanders, 518, 519, 522, 523
morphogenetic rules in, 522
in regeneration, 2, 4, 20, 522, 579, 580, 583, 602–607, 617n1
in snakes, loss of, 652, 653
teratogens affecting, 622
in tetrapods, 517–545
thalidomide affecting, 35–36, 621, 642n2
transcription factors in, 59
transplantation experiments on, 518, 521–523, 530–531, 534–535
Turing model of, 531–533, 654–655
of wings. See Wing development

Wnt in, 108, 524, 533, 537, 542
in gradient model, 530–531
in limb bud formation, 527
zone of polarizing activity in, 534–536
Limb fields, 518, 523
epithelial-mesenchymal transition in, 526–527
forelimb, 523, 524
specification of, 522
LIN-3 and lin-3, 122
lin-4, 74–75
LIN-12 and lin-12, 122–123
LIN-14 and lin-14, 74, 75
Linaria vulgaris (toadflax), 659
Lineage tracing, 41, 42–43. See also Cell lineage
Lineage-committed progenitors, 514
Lipids
fusion-promoting, 211
and Hedgehog, 105, 106, 107
and NAD+ kinase, 209
in tunicate egg cytoplasm, 289
and Wnt, 108, 109
Lipopolysaccharides in liver regeneration, 614
Lissencephaly, 393–394, 395
Lithium chloride treatment of sea urchin embryos, 280, 281
Liver development, 549, 550, 554–555, 561
in amphibian metamorphosis, 564
Cre-lox technique in study of, 83–84
digestive tube in, 547, 552
FoxA1 and Foxa1 in l, 67, 555
hematopoietic stem cells in, 511, 512
HNF4α and Hnf4α in, 83–84, 555
in left-right axis, 343, 356, 357
in mammals
in left-right axis, 356, 357
in regeneration, 581, 613–615
and pancreas development, 555
progenitor cells in, 548
in regeneration, 613–614
size regulation in, 614–615
Wnt signaling in, 553, 554
in zebrafish, 323
Liver disease, 150, 152
Liverwort regeneration, 583
Lizards, 24, 447
Lmx1b and Lmx1b in limb development, 540, 541
Lobbed-fin fish, 521
Local autoactivation-lateral inhibition in Turing model, 531–532
Loeb, Jacques, 206
Looping
of heart, 506–507
of intestines, 549, 553
Lophotrochozoans, 224, 225
Lou Gehrig's disease, 152
Low CpG-content promoters, 64, 64
loxP sequences, 84, 84
LRP5/6, 109, 110
LRP6 and Lrp6, 119, 266
Lumbar vertebrae, 355, 356, 467

Lumens in secondary neurulation, 374
Lunatic fringe (Lfng) in somitogenesis, 476–477
Lung bud, 550
Lung cancer, 640
Lung development, 548
branching in, 558–559, 560
digestive tube in, 557
epithelial-mesenchymal interactions in, 558–559
in heart field formation, 502
in humans, 357, 393
respiratory tube in, 547, 549
Lungfish, 27, 521, 584
Luteinizing hormone, 182
Lymnaea (Radix) snails, 228, 229, 245
Lymphocytes
B cells, 130, 514
and hematopoietic stem cells, 512, 513, 514
T cells, 130, 420, 514, 549
Lymphoid progenitor cells, 130
Lysine, 62, 63, 66
Lytechinus pictus, 287, 288
Lytechinus variegatus, 276
acrosome reaction in, 201
cleavage in, 276–277, 277
gastrulation in, 283
invagination of vegetal plate in, 286

M

Macho and macho in tunicates, 43, 44, 290
Macho-1 and macho-1 in tunicates, 290
Macroclemys temminckii (snapping turtle), 173
Macroevolution, 646, 663n1
Macroglia, 388
Macromeres
in amphibians, 298
in sea urchins, 276–277, 279, 280, 281, 283
in snails, 227, 229, 237, 238
D-quadrant, 236, 237
Macrophages, 130, 144, 616–617
in African spiny mouse, 616, 617
in amphibian metamorphosis, 565, 568
and hematopoietic stem cells, 512
in injury response, 610, 616
Kupffer, in liver regeneration, 614
M1 and M2 types of, 617
Macular degeneration, 151
Mad and MAD in Drosophila, 136, 575–576
MADS-box transcription factors, 66
in angiosperms, 67, 68, 69, 185
L-Maf, 61
in lens induction, 97, 102, 103, 452
Malleus, 415, 416
Malpighi, Marcello, 3
Mammals, 335, 345–359
adult neural stem stems in, 120, 137
as amniotes, 335
anterior-posterior axis in, 352–356
in central nervous system, 375–376
Hox genes affecting, 353–356
uterus shape affecting, 353

axis formation in, 351–356, 361n1
blastocyst of, 156n1. *See also* Blastocyst
brain in, 385, 395
cavitation in, 348
cleavage in, 13–14, *15*, 335, 345–347, *359*, *360*, 619, *622*
cloning of, 53–54
collapsins in, 426
cortical granule reaction in, 205
dorsal-ventral axis in, 356
ear in, *18*, 460n1
early development in, 335, 345–359
ectodermal appendages in, 454
eggs of, 198, 211–217, 335
    activation of, *207*, 216
    size of, 345
epidermis regeneration in, *453*
fertilization in, 211–217, 345
    and meiosis in secondary oocytes, 182
    in twins, 357–359
gametogenesis in, 174, 175, 176, 180–183
gastrulation in, *16*, 348–351, *360*, 361n1
    in conjoined twins, 358
    teratogen sensitivity at time of, *622*
genomic equivalence in, 53–54
genomic imprinting in, 65
gut tube in, 549
health and disease in, 619–642
heart development in, 356, *357*, 501
    and regeneration in neonates, 615
hematopoiesis in, 511, 513
human. *See* Humans
inner cell mass in, 131, 134–135, *346*, 347–348
kidney development in, 492, *493*, 498
left-right axis in, 353, 356, *357*
limb development in, 521, 522
liver development in, 356, *357*
    in regeneration, 581, 613–615
lung development in, 558–559
mammary glands in, 457
mice. *See* Mice
motor neuron synapse in, *437*
mouth of, 549
muscle development in, 436
neural crest in, 410, *415*
neural tube in, 333n4
    closure of, 370, 371, *371*
neurulation in, 366
omphalomesenteric veins in, 508
paracrine factors in, 101
peri-implantation period in, 348, *350*
pharynx in, 549
phylogenetic relationships of, *296*, 336
placental, 14, 345
    evolution of, 662
    phylogenetic relationships of, *336*
    sex determination in, 159, 160–169
pleiotropy in, 655
pluripotent stem cells in, 128, 131, 347
polyspermy prevention in, 204, 205, 215–216

primordial germ cells in, 174, 175–176
regeneration in, 580, 583, 613–617
    compensatory, 581, 613–615
    of liver, 581, 613–615
retinal ganglion axon growth in, 431
sex chromosomes in, 159, 160–169
sex determination in, 159, 160–169
skin in, 453
Sonic Hedgehog in, 108
sperm in, 160, 180–182, 194, *195*, 198, 211–217, 345
sweat glands in, 458
tooth development in, 456, 457
twin, 357–359
umbilical veins in, 508
Mammary ducts, 457
Mammary gland cancer, 638
Mammary gland development, 454, 455, 456
    bisphenol A affecting, 629
    BPA affecting, 633
    non-sensory placodes in, 441
    in puberty, 457, 629
    signaling pathways in, 104, 456
    stem cells in, 457
Mammary placodes, 454, 456
Mammary ridges, 454
Mandible, *416*
    in amphibian metamorphosis, *566*
    in *Drosophila*, 261, 265
Mandibular process, *455*
Manduca sexta, 73, *422*, 656, *657*
Mangold, Hilde, 307–308, 309, 333n1
Mangold, Otto, 318
Mantle zone (intermediate zone), 382, *383*
    in cerebellum, *383*
    in cerebral cortex, *383*, 385, *388*, *394*
    in spinal cord or medulla, *383*
MAP kinase, *210*, 216
MAPK/ErK pathway inhibitor (MEK*i*), 148
Marginal zone of brain, 382, *383*
Marginal zone of embryo
    in amphibians, 300, *301*, 302
    in birds, *337*, 338, 340
Margulis, Lynn, 661
Marsupials, *27*, 156n1, *584*, 650
Maskin, 210
Maternal contribution, 73
Maternal effect genes
    in *Drosophila*, 229, 252, 255, *258*, 262, 270
    in snail coiling, 228–229, 245
    in zebrafish, 325
Maternal messages in *Drosophila*, 252
Maternal-to-zygotic transition in *Drosophila*, 252
Matrix metalloproteinases, 617
    MMP2 in polyspermy prevention, 216
Maturation zone, neuroprogenitor cell migration through, 471
Maxilla, *416*
    in *Drosophila*, 261, 265
Maxillary process, 419, *455*

Maxillomandibular placode, 442
Maxillomandibular process, 415
Mechanical anisotropy, 232, *233*, 235
Mechanical stress, and phyllotaxis, 232, *233*, 235
Mechanoreceptor neurons, *427*
Mechanotransduction, 515n6
Meckel's cartilage, *415*
    in amphibian metamorphosis, 566, *566*
MeCP2 and *MeCP2*, 64, 152
MED-1 in *Caenorhabditis elegans*, 242, *243*
MED-2 in *Caenorhabditis elegans*, 242, *243*
Medial geniculate nucleus, 386
Medial hinge point, 368, 369, 370, 377
Medial motor column, 424
Mediator complex, *59*
Medical embryology, 34–36
Mediolateral intercalation in amphibian gastrulation, 304–305
Medulla oblongata, *375*, 376, 382, *383*
Medulla of adrenal gland, *400*, 401
Medullary cord, 374, 380n2
Medulloblastoma, 638
Mef2 and *Mef2*, 487, 488
Megagametophytes, *187*
Megakaryocyte-erythrocyte progenitors, 513, *514*
Megakaryocytes, 515n6
Megason, Sean, 49
Megasporangium, *187*, *188*, *189*
Megaspores, *187*, *188*, *189*, *219*
Megasporocytes, *187*, *189*
Meiosis, 159, 176–179
    in amphibians, 11
    in angiosperms, *12*, 187, *188*, 189
    BPA affecting, 632
    in *Caenorhabditis elegans*, *240*
    definition of, 159
    in *Drosophila*, 249
    genomic variation in, 159, 178
    in mammals, 182, 345
        and sex determination, 160, 163
    in oogenesis, 182–183, *195*–196, *197*, *198*, 216, *217*, 221n2
    in spermatogenesis, 180, 181
    spindles in, *54*, *178*, 179, 183
    stages of, 177–179
    timing of, 176, *177*
    in tunicates, 289
Meis transcription factors, *59*
Meis1 in limb development, *530*
MEK, *103*
MEK*i* (MAPK/ErK pathway inhibitor), 148
Melanin, 449, 451
Melanoblasts, 413–414
Melanocyte stem cells, 458
Melanocytes, 22, 401, 402, *403*, *405*, 410, 415, 453
α-Melanocyte-stimulating hormone, *403*
Melanoma, 637
Mello, Craig, 75
Melton, Douglas, 317

Membrane potential of egg cell membrane, *210*
in polyspermy prevention, 203–204
Mendelian inheritance, 65
Merian, Maria Sibylla, 577
Meridional cleavage, 227
in amphibians, 297
in *Caenorhabditis elegans,* 240
in mammals, 346
in sea urchins, 276, 277
in tunicates, 288
in zebrafish, 325
Meristem, 114, 132–134
of *Arabidopsis thaliana, 12,* 13, 114, 132–134
in regeneration, 588
and floral meristem identity genes, 184, *185, 186*
and gametogenesis, 186
germ cells in, 186
in indeterminate growth, 8, 587–588
inflorescence, 132, *231, 232, 234*
patterning genes, 132–133
in regeneration, 583, 587–588
root. *See* Root apical meristem
shoot. *See* Shoot apical meristem
and spiral patterns, *231,* 232
stem cells in, 8, 232, 588
Mermaids wine glass, 586, *587*
Meroblastic cleavage, 3, 14, *15*
in birds, 335, 336–337
discoidal, 325, 336–337
in reptiles, 335
in zebrafish, 324–325
Mesencephalon, *375, 376, 450*
Mesenchymal stem cells, 146–147, 457
and hematopoietic stem cells, *144,* 145
Mesenchymal-epithelial transition
in kidney development, 494
in secondary neurulation, 374
in somitogenesis, 472, 475, 477, 478
Mesenchyme
in amphibians, *301*
definition of, 19
dental, 455
and ectodermal appendages, 454, 455, 456
lateral plate, *462, 492, 494, 518*
in limb development, 518
and apical ectodermal ridge interactions, 527–528, *537*
and mesoderm specification in proximal-distal axis, 529
and reaction-diffusion mechanism, 533
in liver development, 554, *555*
in lung development, 558–559, 560
metanephric, 493, 494–499
migration of, 94, 95, 351, 456
collective, 409
in morphogenesis, 19, 87
nephrogenic, *493*
neural crest, *367*
non-skeletogenic or secondary, 278

in recombination experiments, 455
in sea urchins, 278, 279–286, 287
in secondary neurulation, 374
skeletogenic, 278, 279–286, 287
somitomeres in, 469
transition of epithelial cells to. *See* Epithelial-mesenchymal transition
in tunicates, *288,* 291
Mesenchyme cap cells, 497
Mesendoderm, 30, *153*
in amphibians, 303–304
in birds, 342
in snails, *238*
in zebrafish, *327, 329, 331*
Mesentery, dorsal, 553
Mesentoblast, 228, 237
Mesocortex, 398n2
Mesoderm
in amphibians, 298, 299, 309, 314–315, *321*
in axis formation, 307, 308
convergent extension of, 304–305
in gastrulation, 299, 300, *301,* 302, 303, 304–305
head, *302,* 303, *305,* 314, 318
in lens induction, 97
and Nieuwkoop center, 309
and organizer formation, 313
and organizer functions, 314
in primary embryonic induction, 307–308
specification of, 306
vegetal induction of, *314*
axial, 461, 491
in bilateral symmetry, 29–30
in birds, 340–342, *367*
in *Caenorhabditis elegans, 243,* 245
cardiogenic, 501–503, 554, *555*
cardiopharyngeal, *502*
cell recombination assays, 89–90
cell sorting in, 89–90
convergent extension of, 304–305, 408
in cranial placode induction, 443
definition of, 7, 36n6
derivatives of, 18, *462*
differential affinity of cells in, 89
in diploblasts, 224
dorsal. *See* Dorsal mesoderm
in *Drosophila,* 252, *253,* 262, 271
and embryonic stem cells, *149*
in eye development, 97, *452*
in gastrulation, 16
in gonad development, 462, 491, *492*
head. *See* Head mesoderm
hemangiogenic, *503*
intermediate, 491–499. *See also* Intermediate mesoderm
in kidney development, 462, 491–499
lateral plate, 499–515. *See also* Lateral plate mesoderm
in limb development, 523–524, 529–531
in mammals, *349,* 351, *352,* 353
paraxial, 461–490. *See also* Paraxial mesoderm

pharyngeal, 442, *518*
prechordal, 313, 320, 340, *352,* 462
presomitic. *See* Presomitic mesoderm
in sea urchins, 278, 281, 283, 288
in secondary neurulation, 374
in skull development, 418
in snails, 228
somatic, 500
specification of, *331,* 529–531, 547
in amphibians, 306
of intermediate mesoderm, 465, 491, 493–494
splanchnic. *See* Splanchnic mesoderm
in triploblasts, 224
trunk, 461–462
in tunicates, 289, 290
ventral, 309, 313, 315, 316, 330
in zebrafish, *323, 327,* 328, 329, 330, *331*
Mesodermal mantle, 305
Mesodermal posterior and *mesodermal posterior* (Mesp and *Mesp*)
in heart development, *501, 502,* 503, 505
in somitogenesis, 472, 473, 477, *478*
Mesogenin 1 and *Mesogenin 1,* 466, 471
Mesomeres in sea urchins, 276–277, *279, 280,* 283
Mesonephric duct, *162*
Mesonephros, *167,* 492, *493,* 550
retinoic acid secretion from, 176, *177*
Mesothorax in *Drosophila, 254*
Mesp and *Mesp* (*Mesodermal posterior*)
in heart development, *501, 502,* 503, 505
in somitogenesis, 472, 473, 477, *478*
Messenger RNA (mRNA), 14
active transport along cytoskeleton, 77
in amphibians, 306, 315–318
β-globin gene, *56*
in brain development, 396
and centrosome-attracting body, 289, 292n5
cytoplasmic localization of, 76–77
differential longevity of, 72–73
diffusion and local anchoring of, 76, 77
in *Drosophila,* 252, 255, 256–258, *259,* 263, 267, 270
in egg activation, 210
in egg cytoplasm, 196
gene transcription into pre-messenger RNA, 51, 62–70
Macho, 43, 44
maternal, 73, 217
processing of pre-mRNA into, 51, 52, 70–72
in snails, 229, 235, *236*
stored oocyte, 14, 73
Sxl in, 170
translation of, 52, 72–77, 210
Metabolism
in pregnancy, disorders of, *622,* 623
in sea urchin egg, activation of, 206–210
Metalloproteinases, 95, 617
in amphibian metamorphosis, 568
in liver regeneration, 614
in polyspermy prevention, 216

Metamorphic climax in amphibians, *11, 567*, 568
Metamorphic molt, *569,* 570
Metamorphosis, 563–578
   in amphibians, 564–568. *See also* Amphibians, metamorphosis in
   in animal life cycle, 7
   definition of, 2
   in insects, 563, 569–576. *See also* Insects, metamorphosis in
   model systems on, 6
   in sea urchins, 278, 288, 563
   in sponges, 30
   in tunicates, 288, 290
Metanephric mesenchyme, 493, 494
   and ureteric bud interactions, 494–499
Metanephros, 493, 495
Metaphase, 178, *179,* 182, *183, 197*
   in *Caenorhabditis elegans,* 241
   in mammals, 216, 221n2
   in mice, *183*
   in sea urchins, *203*
   in sheep cloning, *54*
   in snails, *236*
Metaphase plate, 178, *217, 241*
Metastasis, 95, 636
   collective cell migration in, 489n3
   of melanoma, *637*
Metathorax in *Drosophila, 254*
Metaxylem, *588*
Metazoans, 224, 246n1
   branches of, 224
   diploblastic, 224
   regeneration in, 583
   triploblastic, 224–226
Metencephlalon, *375, 450*
Methoxychlor, 660
Methyl CpG binding protein-2 (MeCP2), 64, 152
Methylation, *55*
   DNA, 63–65. *See also* DNA methylation
   histone, 62–63, 64
Mexican cavefish, 450, *451,* 651
Mexican salamander (*Ambystoma mexicanum*), *2,* 6
   pharyngeal arches in, *18*
   regeneration in, *2,* 20, 135, 602, 604, 605, *606,* 617n1
Mice
   adult stem cells in, 141, 142, 156n6
   alcohol affecting development of, 625–627
   axis formation in, 352, 353, 354, 355–356
   axon guidance in, *430*
   β-globin gene in, *58*
   BPA affecting development of, 632, 633, 636, 660
   brain development in, *391, 393, 394,* 395
      alcohol affecting, 625
      cerebral cortex in, 385, 386
      neocortex in, 392
   cavitation in, 348
   cleavage in, 346, 360

confetti model, *143,* 402, *403,* 439n5
Cre-lox technique in study of, 83–84
CRISPR/Cas9 genome editing in study of, 83
diabetes in, 556
diethylstilbestrol exposure affecting, 630, *631*
digestive tube development in, 551, 553
DNA methylation in, 660
embryonic stem cells in, 148, 150
endoderm in, *548, 549,* 551
   anterior visceral, 353
enhancers in, 60–61
epiblast in, 352
epidermis in, 453
eye development in, *449,* 450
fertilization in, 198, *212,* 215
gallbladder development in, 556–557
gastrulation in, 351, *360,* 625
gene knockouts in study of, 82
genomic imprinting in, 85n5
gonads in, 162, 164, *165,* 190n1
growth differentiation factor 11 in, 141
gut tissue specification in, 551
heart development in, 501, *502, 504, 505, 506*
   and regeneration in neonates, 615
Hedgehog in, 105–106, 107
hematopoiesis in, 510–511, 512, *513*
Hensen's node in, 361n5
Hox genes in, *74,* 354, 355–356, *468*
   in limb development, 519–520
Hu proteins in, 85n8
hypoblast in, 361
induced pluripotent stem cells in, 151
inner cell mass in, 134–135, *346,* 348, 350
kidney development in, *108,* 492, 494, *495, 497, 498, 499*
limb development in, 105–106, 526, *653*
   and apoptosis, *542*
   of digits, 28, *29,* 536
   dorsal-ventral axis in, 540, 541
   epithelial-mesenchymal transition in, 527
   of forelimb, 28, *29,* 523, *524,* 527
   of hindlimb, *524,* 536
   Hox genes in, 519–520
liver development in, *555,* 561
lung development in, 558, *559,* 560
mammary glands in, 454, 455, 456
meiosis timing in, *177*
melanoblast migration in, *414*
mesoderm specification in, 465, *467,* 491, 494
messenger RNA translation in, 74
as model organism, 6, 336, 345, 616
motor neurons in, *425, 436, 437, 438*
muscles in, 60–61
Myf5 and *Myf5* in, 60–61
myotome determination in, 488
neural crest in, 402
   cardiac, *420*
   cranial, 415, 418, *419*

   enteric, *413*
   lineage tracing of, *403*
   trunk, 411, *414*
neural restrictive silencer element in, 61
neural tube in, *370, 371, 372, 373,* 374
neurulation in, 366
notochord in, 351, *352*
odd-skipped related 1 gene in, 53
oogenesis in, *183, 198*
pancreas development in, 556
peri-implantation period in, *350*
polyspermy prevention in, 215
presomitic mesoderm in, *468*
primitive streak in, 361
primordial germ cell migration in, *175*
proteoglycans in, 93
Purkinje neurons in, *421*
radial glial cells in, 395
Reelin regulation of neuronal migration in, *391*
regeneration in
   of heart, 615
   of skin, 615–616
reporter genes in, *60*
reprogramming of cell identity in, 67
respiratory tube development in, *558*
sex determination in, 160
silencers in, *61*
somitogenesis in, 470, 472, 473, 475, 477
sperm in, 181, *195, 212,* 345
tendon formation in, *484*
tooth development in, 455, 457
vasculogenesis in, 508, 509
vertebrae in, *74,* 355, 461, *468,* 483
Wnt in, *108*
X-chromosome inactivation in, 169
Zika virus infection in, *623*
microcephalin-1, 396
microcephalin-5, 396
Microcephaly, 155, 623
Microevolution, 663n1
Microfilaments, 197
Microgametophyes, *187*
Micromeres
   in amphibians, 298
   in sea urchins, 278, 279, 645
      apical constriction of, 285
      cell-specifying signals from, 282
      in cleavage, 276–277
      in epithelial-mesenchymal transition, 283–286
      induction ability, *279,* 292n3
      specification of, 45, 278, 280–281
   in snails, 227, 230, 237, *238*
Microphthalmia, 620
Microphthalmia-associated transcription factor, 65, *66,* 85n6, 414, 439n7, 451, 620
Micropyle, 219
   in *Drosophila,* 249–250, *257*
   in plants, 188, *189,* 219, 220
MicroRNA (miRNA), 74–76
   in cleavage, 345
   in Down syndrome, 621

and maternal-to-zygotic transition, 76, 76
and RNA interference, 75–76
and small toolkit of evolution, 646
stress affecting, 660
Microspikes of axon growth cones, 421, 422
Microsporangia, 187, 188
Microspores, 187
Microsporocytes, 187, 188
Microtubule organizing centers, 183, 211
Microtubules
    in amphibians
        in cortical rotation of cytoplasm, 296, 297
        in organizer formation, 310, 311, 312, 313
    in axon growth cones, 421, 422, 423
    in *Caenorhabditis elegans*, 241
    in *Drosophila*, 48, 250, 251, 252, 255
    in fusion of genetic material in fertilization, 211, 217
    in meiosis, 178, 179, 183
    in plants
        in asymmetrical cell division, 46, 47
        in phyllotaxis, 232, 233, 235
    in sea urchins, 221
    in snails, 229
    in sperm, 194, 195
    in tunicates, 289
    in zebrafish, 325, 326
Microvilli of eggs, 197, 202, 205
Mid-blastula transition
    in amphibians, 298–299, 306, 312
    in *Drosophila*, 252
    in zebrafish, 325–326
Midbrain, 375, 376
    in amphibians, 314
    in birds, 341, 375
    in mammals, 354
    and neural crest, 401, 416
    and neural tube closure, 371
Midgut, 549, 550
    in sea urchins, 286, 287
Midgut-hindgut cells, 548
Midpiece of sperm, 194, 195
Midsagittal plane, 17
miles apart zebrafish mutation, 504
Milk lines, 454
Milk production, 104, 457
Milk teeth, 457
Milkweed, 627–628
MITF and *MITF*, 65, 66, 85n6, 414, 439n7, 451, 620
Mitochondria
    DNA in, 215
    genome in, 210
    maternal, 215
    in sperm, 194, 195, 210, 215
Mitosis, 1, 163
    in angiosperms, 187, 188, 189, 219
    in *Caenorhabditis elegans*, 240
    cleavage in, 7, 14
    in *Drosophila*, 250, 251, 571

in hydra, 591
in limb development, 528
in multicellular eukaryotic organisms, 223
of neural stem cells, 387
rate of, 7
in snails, 236
in somites, 464
in sperm stem cells, 180
in zebrafish, 325, 609
Mitosis-promoting factor, 210
Mitotic spindles, 14, 155, 157n24, 178, 183, 194
    in brain development, 387, 392
    in epithelial-mesenchymal transition, 486
    in mammals, 217
    in sea urchins, 202, 210
    in snails, 227, 228, 236
    in zebrafish, 325
MMP (matrix metalloproteinase), 617
    MMP2 in polyspermy prevention, 216
Model systems and organisms, 5–6. *See also specific organisms.*
    Arabidopsis thaliana, 6, 11, 231
    for axis determination, 223
    Caenorhabditis elegans, 6
    chickens, 6, 335
    Drosophila melanogaster, 6, 247, 248
    mice, 6, 336, 345, 616
    regeneration in, 6, 580
    sea squirts, 6
    sea urchins, 6, 276
    snails, 226
    for teratogenesis, 623
    tunicates, 276
    Xenopus laevis, 6, 623
    zebrafish, 6, 296, 322–324, 623
Modularity, 644–646
    of enhancers, 59, 60, 61, 644–645
    recruitment in, 645–646
    of silencers, 61
Mole embryos, 26
Molecular parsimony, 644, 646–648
Mollusks, 224
    autonomous specification in, 235
    axis determination in, 235–237
    cleavage in, 15, 226–230
    gastrulation in, 237
    snails. *See* Snails
    in tree of life, 225
Molts, 569, 570
    in *Drosophila*, 572
    hormone control of, 572–575
    imaginal, 569, 570, 574
    metamorphic, 569, 570
MOM-2 in *Caenorhabditis elegans*, 243
MOM-5 in *Caenorhabditis elegans*, 243
Monarch butterflies, 36–37n10, 577, 628
Mongoose, 510
Monkeys, 27, 584, 632
Monocytes, 130, 512, 514
MONOPTEROS (MP), 234, 235
Monospermy, 202

Monotremes, 335, 336
Monozygotic twins, 357–358, 359
Morgan, Thomas Hunt, 247, 248, 594, 617n4
Morgan Lilian, 248
Morphallaxis, 581, 585, 591, 607
    in zebrafish, 612
Morpho deidamia, 577
Morphogenesis, 4, 87–125
    in animals, 19
    cadherins in, 91–92
    and constraints on evolution, 654–655
    definition of, 4, 87
    differential cell affinity in, 89–90
    of eye, 448–451
    juxtacrine signaling in, 88–92
    of otic placode, 444, 445
    in plants, 19
Morphogenetic determinants, 124n3
Morphogens
    auxin as, 113–116
    compared to morphogenetic determinants, 124n3
    definition of, 100, 124n3
    in *Drosophila*, 255, 256, 257–258, 260, 270
    and filopodial cytonemes, 118, 119
    gradients of, 100–101
    Hedgehog proteins as, 105
    in motor neuron guidance, 425
    paracrine factors as, 100–101
    in Turing model of reaction-diffusion, 531–533
Morpholinos, 78, 81–82, 316
Morula, 135, 298, 345, 346, 347, 348
Mosaic embryos, 45
Mosaic pleiotropy, 620
Moss, 186
Moths
    genetic assimilation studies of, 656, 657
    holometabolous development of, 569
    metamorphosis of, 563
    silkworm, 7
Motor neurons, 376, 383, 411, 424–427
    activity-dependent survival of, 436
    in amyotrophic lateral sclerosis, 152
    muscle cell targets of, 424–427
    regenerative medicine in injuries of, 150
    in salamander limb regeneration, 606
    and Sonic hedgehog, 105, 377, 378
    synapse formation, 436, 437, 438
Motor root, 383
Mouse development. *See* Mice
Mouth, 549, 550
    of *Caenorhabditis elegans*, 244
    of deuterostomes, 226
    of *Drosophila*, 260
    of mammals, 549
    of protostomes, 224, 226, 276
    of sea urchins, 288
    of snails, 227, 237
    teratogens affecting development of, 622
MP (MONOPTEROS), 234, 235

MRF (myogenic regulatory factor), 487–489, 489n6
  in myotome determination, 487–489
  in salamander regeneration, 605
  in sclerotome development, 479, 481
MRF4 and *mrf4*, 487, 488, 605
MRF5, 487
Msgn1 and *Msgn1*, 466, 471
msh (muscle segment homeobox) in *Drosophila*, 271, 272, 647
MSL in *Drosophila*, 248
msl2 in *Drosophila*, 171
Msx in mice, 647
Msx1 and *msx1*
  in neural crest specification, 404, 405
  in salamander regeneration, 605
Mud snail, 226, 227
Mules, 54
Müller, Johannes, 25
Müllerian ducts, 161, 166, 167, 498
  in androgen insensitivity syndrome, 167
  diethylstilbestrol affecting, 629, 630, 631
Mullins, Mary, 331
Multicellularity
  evolution of, 30–31, 223–224, 661–662
  of sea urchins, 211
Multipotent cardiac precursor cells, 505
Multipotent neural crest cells, 402, 403
Multipotent stem cells, 128, 129, 135
  of foregut, 554
  hematopoietic, 130, 513, 514
  mesenchymal, 146
  neuroepithelial, 382
Muotri, Alysson, 152, 157n22
Mus musculus, 6, 336, 616. See also Mice
Muscle development
  in amphibians, 309
  in *Caenorhabditis elegans*, 239, 242, 243, 244, 245
  dermomyotome in, 464
  in *Drosophila*, 262, 572
  in flatworm regeneration, 597, 598, 599
  in limbs, 518, 519, 536
  Macho and *macho* in, 43, 44, 290
  motor neurons in, 424–427, 436, 437
  Myf5 and *Myf5* in, 60–61, 487, 488
  myoblasts in, 485
  MyoD and *MyoD* in, 66, 487–488, 489
  myogenic regulatory factors in, 487–489
  myotome in, 464, 480, 482, 485, 487–489
  neural crest cells in, 401, 402, 415, 416, 418
  Pax7 in, 67
  pharyngeal arch in, 416
  in salamander regeneration, 603, 604, 605
  satellite cells in, 486
  in sea urchins, 278
  of skeletal muscle. See Skeletal muscle
  of smooth muscle, 505, 551, 553, 560
  somites in, 462

synapse formation in, 436
  and tendon development, 485
  in tunicates, 43, 288, 290, 291
  in zebrafish heart regeneration, 609–612
muscle segment homeobox (msh) in *Drosophila*, 271, 272, 647
Muscular dystrophy, 462
Mutations, 619, 620–621
  carcinogenesis in, 636
  conditional, 83–84
  Cre-lox technique in study of, 83–84, 84
  CRISPR/Cas9 genome editing in study of, 82, 82–83
  duplication and divergence in, 647–648
  in forward genetics, 81, 254
  in gene knockout experiments, 82
  hypomorphic, 150
  modularity in, 644–646
c-Myb in hematopoietic stem cell formation, 512
c-Myc and *c-Myc*, 66, 67, 150, 151
Myelencephalon, 375, 450
Myelin, 382, 397
Myelination, 397
Myeloid progenitor cells, 130
Myeloproliferative disease, 145
Myf5 and *Myf5*, 60–61, 487
  in myotome determination, 487, 488
  in salamander regeneration, 605
  in sclerotome development, 481
Myoblasts, 480, 481, 485, 486
  abaxial, 488
  and brown fat, 487
  in limb bud, 519
  migration of, 485, 488
  primaxial, 488
Myocardium, 506, 507
  in mice, 615
  in zebrafish, 610, 611, 612
Myocyte enhancer factor-2 (Mef2), 487, 488
Myocytes
  cardiac, 609–612, 613, 615
  differentiation of, 505
MyoD and *MyoD*, 66, 487–488
  in flatworm regeneration, 597, 599
  in mesenchymal stem cell differentiation, 147
  in myotome determination, 487–488, 489
Myoepithelia in hydra, 591, 592
Myofibroblasts in zebrafish heart regeneration, 610, 611
Myogenic bHLH proteins, 487
Myogenic regulatory factor (MRF), 487–489, 489n6
  in myotome determination, 487–489
  in salamander regeneration, 605
  in sclerotome development, 479, 481
Myogenin and *Myogenin*, 487, 488
Myoplasm in tunicates, 290
Myosin
  in heart development, 506
  in oogenesis, 183
  in tunicates, 290

Myosin-7, 621
Myotome, 463, 464, 482, 485–489
  abaxial, 480
  in chick development, 480, 485
  determination of, 487–489
  in limb development, 519
  primaxial, 480, 481

N

NAD⁺ kinase in egg activation, 209, 210
NADPH, 209
Nail-patella syndrome, 540
Naïve embryonic stem cells, 148, 150, 157n18
Nakamura, Osamu, 309
Nanog, 347
  in gastrulation, 351
  and induced pluripotent stem cells, 150, 556
  and pancreatic β cells, 556
  in peri-implantation period, 350
  and pluripotency of embryonic stem cells, 148
  and pluripotency of inner cell mass, 134, 347
  in trophoblast-inner cell mass determination, 347
Nanos and *nanos*
  and cytoplasmic localization of mRNA, 76, 77
  in *Drosophila*, 73, 77, 256, 257, 258, 259, 260
  in gametogenesis, 174, 175, 177
  in snails, 237
Nanos2 and *nanos2*, 177
Nasal bone, 418
Nasopharyngeal carcinoma, 641
Nasrat in *Drosophila*, 256
Natriuretic peptide precursor, 403
Natural selection, 643, 647, 649
  and environmentally induced phenotypes, 658
  in finches, 652
  and limb development, 543
  and regeneration, 583, 585
NCAM, 432
Necrotic zones in digit development, 541, 542
Nematocytes in hydra, 591, 592
Nematodes, 224
  Caenorhabditis elegans, 238–245. See also *Caenorhabditis elegans*
  cleavage in, 346
  signal transduction pathways in, 104
  syncytia in, 49n4
  in tree of life, 225
Neoblasts, 583
  clonogenic, in flatworms, 594–595, 596, 599, 600
Neocortex development, 385, 398n2
  Cajal-Retzius cells in, 390–391, 396
  folding in, 394
  neurogenesis in, 388
  radial glia in, 389–390, 394
  signaling mechanisms in, 390–392

Neomycin cassettes, 82
Nephric bud, *493*
Nephric (Wolffian) ducts, 166, 167
  in bladder development, 498, *499*
  in gonad development, *161, 162*
  in kidney development, 492, *493*, 494, *495*
Nephrogenic cord, *493*
Nephrogenic mesenchyme, *493*
Nephrons, 492, 494, 495, 498
Nerve addiction, 617n5
Nerve growth factor (NGF), 432, 615
Nervous system development
  adult neural stem cells in, 137–141
  in amphibians, 303, 307–308, 314–321
    in metamorphosis, *564, 566*
  anterior-posterior axis in, 375–376
  axonal pathways in, 420–436
  in birds, 341
  in *Caenorhabditis elegans,* 239, 242, *244*
  of central nervous system. *See* Central nervous system
  of cranial nerves. *See* Cranial nerve development
  cranial sensory placodes in, 441–442
  dorsal-ventral axis in, 363, 376–379
  in *Drosophila*, 254, 262, *271, 272,* 572
  neural crest in, 363, *400,* 401
  neural tube in, 363–380
  neurulation in, 365–374
  Notch in, 121
  of peripheral nervous system. *See* Peripheral nervous system
  pioneer nerve fibers in, 423–424, 439n10
  primary embryonic induction in, 307–308
  in regeneration, 582
    in salamanders, *603, 604,* 605–607
  Sonic hedgehog in, 107–108
  synapse formation in, 436–438
  teratogens affecting, 621, *622*
    alcohol, 624, 625, 626–627
  in tunicates, 290
  in zebrafish, *323*
Netrin
  in angiogenesis, 510
  in axon guidance, 423, 426, 428, *429, 430,* 431, *432*
  as paracrine factor, 113
Netrin-1 and *netrin-1,* 428, *429*
Netrin-2, 428, *429*
Neural crest, 363, 399–420, *442*
  alcohol affecting, 625
  in amphibians, 408, 409, 410
  in birds, *367, 368,* 402, 410, 419
    cardiac, *420*
    migration of, 22, *23,* 401, 439n2
  cardiac, 401, 419–420, 503
  cranial, 414–419. *See also* Cranial neural crest
  and cranial placodes, *442, 443,* 446
  definition of, 363

delamination of, *400,* 406–408
derivatives of, *364, 399, 400,* 401–402, 410
  head skeleton, 418–419
as ectoderm derivative, 363, *364*
enteric, 412–413
epithelial-mesenchymal transition in, 95, 399, *400,* 406–408
in eye development, 401, 404, 448
in ganglia, 376
and Hox genes, 401–402
induction of, 404
lineage tracing of, 402, *403*
melanocytes originating in, 22, 401
and mesoderm development, *500*
in mice. *See* Mice, neural crest in
migration of, 18, 399, *400,* 405–418, *470,* 549
  alcohol affecting, 625
  in birds, 22, *23,* 401, 439n2
  chase and run model of, 416–418
  collective, 409, 416, 439, 489n3
  contact inhibition in, *407,* 408–409, 416, 418
  directional dispersal in, 412
  dorsolateral pathway, 401, *403,* 410, 411, 413–414
  regions of, 401
  retinoic acid affecting, 627
  traffic movement analogy of, 405–406, 421
  transgenic techniques in study of, 23
  ventral pathway, 401, 410–413
  in zebrafish, *409,* 549, 623
multipotency of, 402, *403*
and primary neurulation, *367, 368,* 369
regionalization of, 401–402
sacral, 401, 412
specification of, *400,* 404, 410
transplantation experiments with, 22, 402, 410
trunk, 410–414. *See also* Trunk neural crest
vagal, 401, 412
Neural crest specifiers, 404, *405*
Neural ectoderm, 374, *405,* 441, 444
  in amphibians, 298, 314–317
  in birds, *340, 341*
  in *Drosophila, 271, 272*
  in neural tube closure, *371*
  in primary neurulation, 374
  in tunicates, *288,* 290
  in zebrafish, 330
Neural folds, *444*
  in birds, *340,* 367, *368, 375*
  cell density in, 369
  convergence of, *368,* 369
  and neural crest, *400*
  in neural tube closure, *370, 371*
Neural groove, *343, 367, 369, 372*
Neural keel in zebrafish, 330, 380n1
Neural plate, 365–374
  in amphibians, *89,* 298, 307, 333n2, *372*
  in birds, 340, 341, 367, *368, 375*

cell sorting in, 89, *90*
in cranial placode induction, *442, 443*
definition of, 363
derivatives of, *364*
as ectoderm derivative, 363, *364,* 365
in eye development, 449, *452*
and neural crest, *400,* 404, *405*
neuroepithelial cells in, 381
optic vesicle region of, 451
and placode specification, *442, 443*
in primary neurulation, 365–374
in zebrafish, 380n1
Neural plate border specifiers, 404, *405*
Neural progenitor cells, 386, 388, *445, 446*
Neural restrictive silencer element, 61
Neural restrictive silencer factor, 61
Neural retina, *103,* 431, 448, 449, 451, *452. See also* Retina development
Neural stem cells, 137–141
  adult, 137–141, *386*
  astroglial characteristics of, 156n5
  in brain development, 386–387
  cell lineage studies of, 387, 388
  in cerebral organoid, 155
  division of, 386–387
  environmental influences on, 140–141
  in germinal neuroepithelium, 382
  interkinetic nuclear migration of, 386–387
  neuroepithelial cells as, 382
  in neurogenesis, 388
  and primary cilium, *120*
  radial glial cells as, 382
  in ventricular-subventricular zone, 137–141
  in zebrafish, 156n5
Neural tube, 363–380
  in amphibians, *9, 10,* 309, 314, 318, 333n4, *372*
    primary embryonic induction of, 307–308
    region-specific induction of, 318
  anterior-posterior axis in, 375, 376
  in birds, *342, 366, 367, 374, 378, 382, 480, 485*
    closure of, 370, 371
  closure of, *368, 369,* 370–374, 376
    alcohol exposure affecting, 625
  defects in, 372–375, 380
  definition of, 363
  derivatives of, *364*
  differentiation of, 381, *383*
  dorsal-ventral axis in, 376–379
  as ectoderm derivative, 363, *364*
  enhancers, *59, 60*
  epithelial-mesenchymal transition in, *95*
  floor plate of, 377, *378, 480, 481*
  folic acid affecting formation of, 372–374, 380
  germinal neruoepithelium in, 382
  in mammals, 333n4, *354*
  and mesoderm, 461, 465, 470, *500*
  and myotome determination, *487,* 488

and neural crest, 399, *400,* 408, *410, 420*
neuroepithelial cells in, 381, 382
primary embryonic induction of, 307–308
in primary neurulation, 365–374
progenitor cells in, 379
region-specific induction of, 318
roof plate of, 377, *378*
in sclerotome development, *480*
in secondary neurulation, 365–366, 374
and somites, *470*
and Sonic hedgehog, 107–108
and Sox2, 465
in tunicates, 289, 290
and vertebrae formation, 481–482
in zebrafish, 330, 333n4, 372, 378, *471*
Neuregulin, *403*
in mammalian heart regeneration, 615
in salamander limb regeneration, *606,* 615
Neuregulin-1, *606, 615*
Neuroblasts
delamination of, 446
in *Drosophila,* 254, 272
epibranchial, 446
Neurocranium, 418
Neuroectoderm
in cerebral organoid formation, *154*
in zebrafish, *327*
Neuroepithelial cells, 381–382, *383*
in cerebral organoid formation, 154, 155
division of, 386–387
germinal, 382, 386–387
in neurogenesis, 388
Neurofilament protein, *438*
Neurogenesis, 387–388
adult neural stem cells in, 137–141
in brain development, 387–388, 392
neurogenin in amphibians, 315
Neuromesodermal progenitors, 379, *467,* 470–471, 473–474
Neuromesodermal stem cells, 470
Neuromuscular junction, *437*
Neuronal precursor cells, *129*
Neurons, 381
activity-dependent survival of, 436
alcohol affecting development of, 624, 625, 626–627
in amphibian metamorphosis, 565, 566
apoptosis of, 436
axon guidance of, 78
in *Caenorhabditis elegans,* 239, 242, *244*
in cerebral cortex, 385–386, 394
commissural, 424, 428–429, *430*
deep layer, *388*
differentiation of, *129*
glia as scaffold for, 389–390
and Hu proteins, 72–73, 85n8
human fetal, postnatal growth rate of, 392–393
in hydra, 591, *592*
migration of, 389–391
motor. *See* Motor neurons
in neocortex development, 390

as neural crest derivative, *400,* 401, 402, *403, 405,* 410, 411, 415
in neurogenesis, 388
number in human brain, 381, 421
pioneer, 439n10
postsynaptic, 436
precursor cells of, 72–73, *389,* 390
presynaptic, 436
and protein degradation, 78
Purkinje, 381
Reelin regulation of migration, 390–391
rhombomeres producing, 376
sensory. *See* Sensory neurons
specificity of connections, 424
spinal, 376
synapse of, 436–438. *See also* Synapse
upper layer, *388*
Neuropilin
in angiogenesis, 510
in axon guidance, 424, 426
in neural crest migration, 412
Neuropilin-2, 412, 424
Neuropores, anterior and posterior, *367, 370,* 371
Neurotransmitters, 436, *437*
Neurotrophin (NT), 113, *403,* 439n14
in axon guidance, *427,* 432
as chemotactic protein, 432
as paracrine factor, 113
Neurotrophin 3 (NT3)
in axon guidance, *427,* 432
in dermomyotome determination, 486–487
in sclerotome development, *481*
Neurotrophin 4/5 in axon guidance, 432
Neurula, 9, *10, 291,* 365
cranial placode precursors in, *442, 443*
Neurulation, 365–374
in birds, *340, 463*
junctional, 366
primary, 365–374, 379
secondary, 365–366, 374, 379
transition zone in, 366
Neutrophils, *130,* 512, 610
Newman, Stuart, 533
Newts, *21,* 298, 307, 314, *319*
regeneration in, *603*
Next-generation sequencing technology, 80–81
NGF (nerve growth factor), 432, 615
NGN3 and pancreatic β cells, *556*
Niches of stem cells, 130, 131, 135–145
bone marrow, 512–513
spermatogonia in, 180
Nicotine, 623
Nieuwkoop, Pieter, 309
Nieuwkoop center, 309–310, 313, 314
Nilsson, Lisa, 397
Nitric oxide in endothelium, 511
Nitroblue tetrazolium chloride, *79*
Nitroreductase, 612
NK cells, *514*
nkx1-1 in flatworm regeneration, 597, *599*
Nkx2 and *Nkx2, 552, 647*

Nkx2-1, 558
Nkx2-5 and *Nkx2-5, 502, 503,* 505, *646*
Nkx6-1, *377, 379*
Nlcam in eye development, 451
no tail gene in zebrafish, *328*
Nodal and *nodal,* 101, 111, 359, 361
in amphibians, 361
in germ layer specification, 306
in left-right axis formation, 322
in tail development, 333n3
in birds, 359, 361
in left-right axis formation, 343, *344,* 345
in conjoined twins, 359
in endoderm specification, 547, *548*
functions of, 112
in mammals, 361
in anterior-posterior axis formation, 353
in left-right axis formation, 356
in melanoma, 637, 638
in snail coiling, 229–230
in zebrafish, 330, 331, *332,* 333, 361
Nodal-related proteins in amphibians, 312–314, 320
Node, in mammals
in anterior-posterior axis formation, 352, 353
formation of, 351
in left-right axis formation, 356
Nodose placode, 442, *446*
Noggin and *noggin*
in amphibians, 312, *313,* 315, 316, *320, 321*
as BMP inhibitor, 112, 140, 315, 316, 542
in eye development, 449
in finch beak shape, 652
in heart development, 503
in limb development, 533, 541, 542
in mammals, 353
in mesoderm specification, 465, 503
in myotome determination, *487,* 488–489
in neural crest specification, 404
in neural plate folding, *369,* 370
and neural stem cells in V-SVZ, *138,* 140
in sclerotome development, 481
in zebrafish, 330, 332
Noncanonical Wnt pathways, *110,* 111
Noncoding sequences, 52, 56, 57–61
in brain development, 396
Non-homologous end joining, 82
Non-skeletogenic mesenchyme in sea urchins, 278, *286,* 287, 288
Noradrenaline, *144,* 145
Norrin, 315
Nose development, *415,* 416
alcohol affecting, 623, 625
cranial sensory placodes in, 404, 441, 442
in mammals, 353
in zebrafish, *323*

Notch
    in angiogenesis, 510
    in brain development, 139, 156n7, 392
    in *Caenorhabditis elegans,* 121, 122–123
    in cancer, 638, 640
    in *Drosophila,* 121, 272
    in epidermis differentiation, 453
    in hematopoietic stem cell formation, *512*
    as juxtacrine factor, 120–121
    in kidney development, 498
    in nervous system development, 121
    in otic and epibranchial induction, 445
    in pattern formation, 120–121
    in sea urchins, 282
    signaling pathway, 120–123, 282
        lateral inhibition in, 476
    in somitogenesis, 473, 474–478, *479,* 489
    and stem cells in V-SVZ, *138,* 139, 140, 156n6, 156n7
    in vasculogenesis, 507
    in zebrafish regeneration, *608, 612, 613*
Notch intracellular domain, 139, 140–141, *142*
Notechis scutatus (tiger snake), 658
Notochord, 36n5, 36n9, 226, *410*
    in amphibians, 9, 298, *301,* 303, *305,* 307, 308
        and BMP inhibitors, 316, 317
        and dorsal mesoderm induction, 333n2
        and hindbrain structures, 318
        and Nieuwkoop center, 309
        and organizer formation, 313
        and organizer functions, 314, 316, 317
        and regional specificity of neural induction, 318
    in axis elongation, 471–472
    in birds, 338, 340, 341, 342, *375, 463,* 484
        in primary neurulation, *367, 368*
        in secondary neurulation, *374*
    and cardiogenic mesoderm specification, 503, 554
    cell sorting in, 89–90
    chordamesoderm in formation of, 461–462
    in chordates, 26–27, *32,* 226, *276,* 288, *296*
    definition of, 27, 36n9
    in dorsal-ventral specification of neural tube, 377, *378*
    endoderm in formation of, 547
    and evolutionary embryology, 26–27
    and eye field development, *450*
    and liver development, 554, *555*
    in mammals, 351, *352,* 353
    in neural plate folding, *369,* 370
    and neurulation, *366, 367*
    in pancreas development, 555
    and paraxial mesoderm, 461
    in primary embryonic induction, 307, 308

    in sclerotome development, *480,* 481
    in tunicates, 26, 27, 36n9, 288, 289, 290, 291, 292
    and vertebrae formation, 481–484
    in zebrafish, *323,* 329, 330, 331, *471,* 483
Notum and *Notum,* 109
    in flatworm regeneration, *597, 598, 600*
NT. *See* Neurotrophin
Nuclear envelope, 65
    degeneration of, 210
    in meiosis, 177, 178, 183, *183*
Nuclear RNA, heterogeneous, 52, 57
Nucleosomes, 55
Nucleus, in brain, 382, 390, 398n1
Nucleus, cellular, 398n1
    in cloning experiments, 53–54, 67
    in eggs, 196, *197,* 210–211, 216–217
    genomic equivalence of, 51, 53–54, 67
    in interphase, 48
    migration of, 386–387
    position and fate of, 45
    in sperm, 194, *195,* 210–211, 216–217
    in syncytial specification, 47–48
    triploid, 202
Nucleus pulposus, 462, *463,* 483–484
Numb and NUMB, 140, 392
Nurse cells in *Drosophila,* 255, 258
Nüsslein-Volhard, Christiane, 81, 108, 254, 255, 256, 257
Nutclam (*Acila castrensis*), 226, *227*
Nymph stage, 569

O
Oak acorns, *33*
Observation of living embryos, 20–21
OCT1, *104*
Oct3/4 and *Oct3/4, 66,* 67
Oct4 and *Oct4, 350*
    and embryonic stem cell pluripotency, 148, 347
    and induced pluripotent stem cells, 150
    and inner cell mass pluripotency, 134, *135,* 347
    in mammals, 347
    in snakes, 651
Octopus, *27, 584*
odd-paired in *Drosophila,* 260
Odd-skipped and *odd-skipped* in *Drosophila,* 53, 78, *79,* 260, 267
Odd-skipped related 1 gene, 53
Odontoblasts, 401, 457
Oil spills, 623
Olfactory epithelium, 404, 441
Olfactory lobes, *375*
Olfactory nerve, 442
Olfactory placode, 442, *443*
Olfactory process, *450*
Oligodendrocytes, 382
Oligopotent progenitors, *514*
omb (optomotor blind) in *Drosophila, 575,* 576
Omphalomesenteric veins, 508
On the Generation of Animals (Aristotle), 2

On the Generation of Living Creatures (Harvey), 3
On the Origin of Species (Darwin), 24, 543
Oocytes
    age-related changes in, 183
    atrazine affecting, 634
    attraction of sperm to, 199
    BPA affecting, 632
    of *Caenorhabditis elegans,* 241
    and chromosomal sex determination, 159
    cytoplasm of, 195–196
    of *Drosophila,* 255
        dorsal-ventral patterning in, 269–270
        fertilization and activation of, 249–250
        and germ stem cells, 136, 137
    generated from induced pluripotent stem cells, 152
    of mammals, 163, 182–183, 211–212, 345
    maturation of, 182
    messenger RNA in, 14, 73
    primary, 182, 183, *197*
    of sea urchins, *280*
    secondary, 183
    in sheep cloning, *54*
    translational regulation in, 73
    translocation to oviduct, 211
    of tunicates, *289*
Oogenesis, 174, 176
    DNA methylation in, 65
    in *Drosophila,* 255, 258
    in mammals, 182–183
    meiosis in, 182–183, 195–196, *197, 198,* 216, *217,* 221n2
Oogonia, *33,* 182
    of *Drosophila,* 255
Openbrain in neural tube formation, 372
Ophthalmic placode, 442
Opium poppy, *218*
Optic chiasm, 431, *432, 433*
    in amphibian metamorphosis, 565
Optic cup, 431, 448, 449, 450, 451, *452*
Optic cup organoid, 153
Optic disc, 431, *432*
Optic nerve, 431, *432, 433,* 449
Optic tectum, 431, *432,* 433–435
Optic tract, 433
Optic vesicle, *375,* 376, *450, 452*
    in chick development, *103*
    and induction of eye development, 96, 97, *98,* 102, *103,* 451
    in vertebrates, 96, 97, 448–449, 451
optomotor blind (omb) in *Drosophila, 575,* 576
Oral plate, 549
Orangutans, *393, 395*
Organ of Corti, 444, 447, 448
Organizer
    in amphibians, 307–314. *See also* Amphibians, organizer in
    in birds, 342–345
    in fish, 326, 333n5

Organogenesis
  in animal life cycle, 7
  definition of, 7
  in frogs, *8, 9, 10*
  germ layers in, 17–18
  in humans, organoids in study of,
    153–155
  in plant life cycle, 8
Organoids, 153–155
  cerebral, 153, 154–155, 642
  intestinal, 153
  kidney, 153, 496
  optic cup, 153
Ornithoptera croesus (birdwing butterfly),
    170
orthodenticle in *Drosophila, 260,* 354
Orthogonal bifurcation, 559
Orthologues, 648
Oskar and *oskar* in *Drosophila, 77, 256, 258*
Osteoblasts
  in digit regeneration in young
    mammals, 615
  in fin regeneration in zebrafish, 608,
    609
  and hematopoietic stem cells, 143, *144,*
    145, 513
  and mesenchymal stem cells, *146*
Osteoclasts, *144,* 512
Otic cup, 446, 447
Otic-epibranchial development,
    444–448
Otic pit, 445
Otic placode, *370,* 442, *443,* 444–448, 459
  induction of, 444–445, *446*
  morphogenesis of, *444,* 445, *446*
  specification of, *443*
Otic vesicle, 445, 446, 459
  axis determination in, 447–448
  closure of, 445
  saccular part of, *447*
  utricular part of, *447*
Otocyst, *445,* 446, 447–448
Otx proteins
  in sea urchins, 281
  and small toolkit of evolution, 646
Otx2 and *Otx2,* 361
  in eye development, *97,* 449, *452*
  in mammals, 354
Outer ear, 444
Outer radial glia, 388, 394, 395
Ovarian follicles, 163
Ovaries, 160, 182–183, 211, *345*
  age-related changes in, *182,* 183
  in angiosperms, 12, 185, *187,* 188, *189*
  atrazine affecting, 633, 634
  BPA affecting, 632, 633
  in *Caenorhabditis elegans,* 121, *240*
  in frogs, 11
  germ stem cells in development of,
    136–137
  in hermaphroditism, 166
  and oogenesis, 176, 182–183
  and primary sex determination, 161,
    162, 163–164

and secondary sex determination, 166,
    167
  temperature affecting development of,
    174
  Wnt in development of, *108,* 160
Ovastacin, 215
Oviducts, 345, 346, 348
  in *Caenorhabditis elegans,* 240
  capacitation of sperm in, 212
  diethylstilbestrol exposure affecting,
    629, 630, *631*
  differentiation of, 166
  hyperactivation of sperm in, 212–213
  translocation of gametes to, 211–212
Oviparity, 2
Ovoperoxidase, *205*
Ovotestes, 166
Ovothiols, 209
Ovoviviparity, 2
Ovulation, *345*
Ovules, *187,* 188–189
  in ABCDE model, *185, 186*
  and fertilization, 217, 218, 219
  and pollen tube navigation, 219
Ovum, 183. *See also* Egg
Oxygen, and hematopoietic stem cells,
    144, 145
Oyster toadfish, 27, 37n11

P

P5.p cells, 122
P6.p cells, 122
P7.p cells, 122
p27 in liver regeneration, 614
p300 in Notch signaling, *121*
Pachygryria, 394
Pachytene stage in meiosis, 178
Pacific tree frog (*Hyla regilla*), 523
Paired and *paired* in *Drosophila, 260,* 267
Pair-rule genes in *Drosophila, 255,* 256,
    260, 261, 263–265, *266*
PAL-1 protein in *Caenorhabditis elegans,*
    242, *243*
Palmitic acid, *105,* 108
Palmitoleic acid, 108, *109*
Pancreas development, 547, *548,* 549, *550,*
    554, 555–556
  β-cells in, 556
  digestive tube in, 547, 552
  enhancers in, *59, 60, 61*
  in humans, *357,* 554
  and liver development, 555
  progenitor specification in, *548*
  Wnt signaling in, 553
  in zebrafish, *323*
Pancreatic duct, *554*
Pander, Christian, 17, 18
Paneth cells, 142, 143, 639
pangolin in *Drosophila,* 260
PAR and *par*
  in brain development, 392
  in *Caenorhabditis elegans,* 241, 242
  in mice, 135

  in sea urchins, 277
  in tunicates, 289
PAR-1 in *Caenorhabditis elegans,* 241
Par-3 in brain development, 392
PAR-3 in *Caenorhabditis elegans,* 241
PAR-6 in *Caenorhabditis elegans,* 241
Parabiosis
  heterochronic, 141
  isochronic, *141*
Paracrine factors, 88, 100–119
  in amphibians, 306, *320*
  in *Caenorhabditis elegans* vulva
    induction, 121–123
  cell biology of, 117–119
  in conditional specification, 44, 45
  in cranial placode induction, 443–444
  definition of, 96
  diffusion of, 117
  duplication and divergence of, 647
  endocrine factors compared to, 96
  in epithelial-mesenchymal transition,
    94, 95
  in extracellular matrix, 93
  in eye development, 96, 97
  fibroblast growth factor, 101, 102–104
  in gene regulatory network, 69
  Hedgehog family, 101, 104–108
  in hematopoiesis, 513
  in heterotopy, 649–650
  as inducers, 96, 100–119
  in mesenchymal stem cell
    differentiation, 146
  as morphogens, 100–101
  in oogenesis, 182
  in secondary sex determination, 166
  signal transduction cascades, 101–102
  in stem cell regulation, 130
  TGF-β superfamily, 101, 111–116
  Wnt family, 101, 108–111
Paralogues, 647
  of Hox genes, 354–355
Parapagus, *359*
Parasegments in *Drosophila,* 261–262, 267,
    273
Parasite infestation, extra limb
    development in, 523
Parasitoid wasps, 660, 661
Parasympathetic nervous system, 401
Parathyroid gland, 416, 419, 439n1, 547,
    549, *551*
Paraxial mesoderm, 374, 461–490, 491
  in amphibians, *500*
  anterior-posterior axis of, 465–469
    in clock-wavefront model, 478–479
    elongation of, 470–473
  in birds, *340, 341, 342*
  derivatives of, 462
  dermomyotome development, 479,
    485–489
  and Hox genes, 467–469, 478–479, 522
  and intermediate mesoderm
    specification, 493
  in kidney development, 493, *494*
  in limb development, 522

and neural progenitors in caudal spinal cord, *379*
in neurulation, *366*
sclerotome development, 479–485
in somitogenesis, 469–479, 489
specification of, 465–466, 491
Paraxis, *492*
Parietal bone, 418, *419*
Parietal endoderm, 351
Parietal mesoderm, 500, *550*
Parkinson disease, 150, 152
Partitioning defective proteins. *See* PAR and *par*
Partitioning proteins, 135
Patched and *patched*
in cancer, 108, *639*
in cytoneme-mediated signaling, 118, *119*
in *Drosophila*, *260*, *266*
and Hedgehog, 106–107, 108, 118, 639
and primary cilium, 118
Patella, and Hox genes in mice, 519
Patella snail, 41
Pattern formation, 4, 39–49
Notch pathway in, 120–121
spiral. *See* Spiral patterns
Turning model of, 531–533
Pax proteins, *66*
and small toolkit in evolution, 646
Pax1 in sclerotome development, 479, *481*
Pax2 and *Pax2*
in eye development, *451*
in intermediate mesoderm, 491, *492*, 493–494
in kidney development, 493–494
in ventral retinal cells, 434
Pax3 and *Pax3*
in myotome determination, *487*, 488
in neural crest specification, 404, *405*
in neural tube formation, 372, *377*
in sclerotome development, *481*
Pax6 and *Pax6*, 59, 61, 663n4
in blastema transplant experiments, 20
enhancers, *59*, *60*, *61*, 66
in eye development, 83, 97, 98, 448, 449–450, *451*, *452*, 646
in Mexican cavefish, *451*, 651
and neural progenitors in caudal spinal cord, *379*
in neural tube formation, *377*, 378
targeted expression of, 83
Pax7, 65–66
in neural crest specification, 404, *405*
and neural progenitors in caudal spinal cord, *379*
in neural tube formation, 377, *377*, *378*
as pioneer transcription factor, 67
in sclerotome development, *481*
Pax8 and *Pax8*
in intermediate mesoderm specification, 493–494
in kidney development, 493–494
Pbx1 transcription factor, *59*
PCBs (polychlorinated biphenols), 636

PCNA (proliferating cell nuclear antigen), 589, 617n2
PDGF. *See* Platelet-derived growth factor
PDGFR (platelet-derived growth factor receptor), *350*, 508
Pdx1 and *Pdx1*
in digestive tube development, *552*
in pancreas development, 555, *556*
Pea aphid (*Acyrthosiphon pisum*), 660, *661*
Pectin, 218
Pectoral fins, 520–521, 524
Pectoral girdle, 521
Pelle protein kinase in *Drosophila*, 271
Penis, 167
Peribrachial flank tissue, 518, *519*
Pericardial cavity, 501
Periclinal divisions, 133, 135
Pericycle, 588
Pericytes, 509
Periderm, 453
Periglomerular cells, *386*
Peri-implantation period, 348, *350*
Perineum, *499*
Peripheral nervous system, 439n12
cranial nerves in. *See* Cranial nerve development
as ectoderm derivative, 363, *364*
as neural crest derivative, *400*, 411, 412
neural crest origin of, 363, *364*
Peritoneal cavity, 501
Perivascular hematopoietic stem cell niche, 143, 144, 145, 157n13, 513
Permissive interactions, 99, 124n2
Peromyscus maniculatus (field mouse), *212*
Pesticides as teratogens, 621
Petals, 185, 650, 651
and floral organ identity genes, 68–69, *185*, *186*
Petrosal placode, 442, *446*
pH in activated egg, 209, 210
PHA-4 and *pha-4* in *Caenorhabditis elegans*, *242*, *243*
Phalangeal cells in ear development, 447
Phalangeal process in ear development, 447
Phalanges, 517. *See also* Digit development
Pharyngeal arches, 18, 226, 246n2, 549, *551*
in amphibian metamorphosis, 566
and cardiac neural crest, 419, *420*
in chick development, *103*
and cranial neural crest, 401, 415–416, 439, 439n1
derivatives of, *416*
glyphosate affecting, *628*
retinoic acid affecting, 552
Pharyngeal ectoderm in heart development, 419
Pharyngeal endoderm
in amphibians, 298, 301–302, *305*, *316*, 320
in axis formation, 308
and organizer formation, 313
and organizer functions, 314

in birds, 341, 342, *343*
and epibranchial placode, 446
in heart development, 419, 503, 504
Pharyngeal mesoderm, 442, *518*
Pharyngeal pouches, 439n1, 549, *551*, 557
Pharyngula, *360*
Pharynx, 547, 549, *550*, *551*, *558*
in *Caenorhabditis elegans*, *239*, *240*, *242*, *243*, *244*, *245*
in mammals, 549
specification of, 552
Phenotypes
environmentally induced, 656–659
transgenerational inheritance of, 634–636
and floral organ identity genes, *68*
in genetic heterogeneity, 621
in plasticity-first evolution, 656–659
sexual, 166–169
and snail coiling, 229
stochastic factors affecting, 619–620
Phenotypic heterogeneity, 621
Philtrum of upper lip, *415*
Phloem, 13, 19
in canalization model of regeneration, 589–591
Phocomelia, 35, *35*
Phosphatidylinositol bisphosphate (PIP$_2$)
in egg activation, *207*
Phospholipase C (PLC)
in egg activation, *207*, 208, 210, 216
in Wnt signaling pathway, *110*, 111
Phosphorylation, 102, *103*, 104, 110, 112, *113*
in *Caenorhabditis elegans*, 241
in sperm capacitation, *213*
Photoperiods, 9, 184
Photoreceptors, 449, 450
Photosynthesis, 33
Phragmoplast, *32*, *34*, *46*, 47
Phyllotaxis, 231–235
auxin in, 232–235
mechanical stress in, 232, *233*, 235
Phylogenetic relationships, *225*, *296*, *336*
of model organisms, 6
Phylotypic stage, 24, 36n8
Phylum, 36n8
Phytohormones, *32*, 33, 47
PI and *PI* (PISTILLATA and *PISTILLATA*), 68, 184, *185*
Pia mater, 381
Pial surface, 381, *382*, 390
PIE-1 in *Caenorhabditis elegans*, 242
Piebaldism, *414*
Pigment cells
melanocytes as, 22, 401, 402, *403*, *405*, 410, 415, 453
as neural crest derivatives, *400*, 401, 402, 410, 413–414, 415
in retina development, 90, 448, 449, 451
Pigmentation
in amphibian cytoplasm, *297*
of hair, 458

of skin, 453, 458
in tunicates, 289
Pigmented epithelium, retinal, 448, 449, 451
Pigweed, teratogenic effects of, 557
Pillar cells, 447
PIN proteins in auxin distribution, 114, *115*
and phyllotaxis, 232–235
and regeneration, 589, *590,* 591
PIN1 and *PIN1,* 114, *115*
in phyllotaxis, 233, 234, 235
in regeneration, *590,* 591
PIN7, *115*
Pineal gland, *449*
Pinwheel structures in V-SVZ, 138, 139
Pioneer nerve fibers, 423–424, 439n10
Pioneer transcription factors, 66–67, 69, *70*
PIP$_2$ in egg activation, *207*
PISTILLATA and *PISTILLATA* (PI and *PI),* 68, 184, *185*
PISTILLATA2, 68
Pistils, 185
Pituitary gland, 9, 442, 549
and luteinizing hormone, 182
and prolactin, *104*
Pitx1 and *Pitx1*
in humans (*PITX1),* 526
in limb development, 524, 526, 527
in snail coiling, 230, *230*
in stickleback fish, 644–645, 663n2
Pitx2 and *Pitx2,* 361
in amphibians, 322
in birds, in left-right axis formation, 343, *344,* 345
in humans (*PITX2),* 361n3
in mammals, in left-right axis formation, 356
in mice, 361n3
in zebrafish, 332
Piwi proteins in gametogenesis, 174
PKA (protein kinase A), *106,* 118
in acrosome reaction, 214
in sperm capacitation, *213*
PKC (protein kinase C), *103*
in egg activation, *210*
in inner cell mass development, 135
PKC-3 in *Caenorhabditis elegans, 241*
Pkd2 protein, 356
Placenta, 335, *336,* 347, 348
blood vessels in, 509
BPA affecting, 633
cytotrophoblast cells, 100
evolution of, 662
and retroviruses, 662
Placental growth factor (PlGF), 509
Placental mammals, 14, 345
evolution of, 662
phylogenetic relationships of, *336*
sex determination in, 159, 160–169
Placodes, 404, 441–452
cranial, 404, 441–452
epidermal, 454
hair, 454, 455

mammary, 454, 456
in neural crest migration, *417,* 418
non-sensory, 441
Placozoans, *225*
Planar bifurcation, 559
Planar cell polarity pathway, *110,* 111
in amphibians, 305
in mammals, 356
in neural crest migration, 409
in zebrafish, 329
Planarian flatworms, 580, 583, 594–602. *See also* Flatworms
Plants
apical-basal axis in, 223
asymmetrical cell division in, 46–47, 114
auxin in, 113–114, *115*
in regeneration, 586
autonomous and conditional specification in, 46–47
auxin in, 113–116
axis determination in, 223
cell wall remodeling in mechanical stress, 232
developmental history of, 31–34
embryogenesis in, 13, 31
flowering. *See* Angiosperms
gametogenesis in, 12, 13, 186–189
indeterminate growth of, 1–2, 8, *32,* 133, 587–588
life cycle of, 7–8
light-induced DNA methylation in, 660
meristem of, 132–134, 587–588. *See also* Meristem
phyllotaxis of, 231–235
regeneration of, 1–2, 580, 586–591
canalization model of, 589–591
direct and indirect methods in, 588
evolution of, 583, 585
sex determination in, 184–186
single cell, 586
spiral patterns in, 226, 231–235, 245
totipotent cells in, 131–132, 133–134, 585, 586–591
in tree of life, 27, 225, 584
vascular tissue of, 13, 19
in canalization model of regeneration, 589–591
Plastic products, BPA in, 631, 632–633, 660
Plasticity-first evolution, 656–659
Plastids, *32,* 33
Platelet-derived growth factor (PDGF)
in angiogenesis, *508*
in cranial placode induction, *443*
in hair follicle regeneration, 458, *459*
and mesenchymal stem cells, 146
in vertebrae formation, 482
Platelet-derived growth factor receptor (PDGFR), *350, 508*
Platelets, *130,* 512, *514,* 515n6
Platyhelminths, 226
PLC (phospholipase C)
in egg activation, *207,* 208, *210,* 216
in Wnt signaling pathway, *110,* 111

PLC$_\zeta$ (PLC-zeta), 216
Pleiotropy, 620, 621, 655
PLETHORA and *PLETHORA,* 132, 133, 617n2
in plant regeneration, 588–589
PLT2 and *PLT2, 132,* 133
Pleural cavity, 501
PLT and *PLT,* 132, 133, 617n2
in plant regeneration, 588–589
PLT2 and *PLT2, 132,* 133
Pluriblast, 156n1
Pluripotent cells, 6, 128–129
embryonic stem cells, 129, 131–135, 147–150, 347
in flatworm regeneration, 594–595, *596*
hematopoietic stem cells, 510, 512, 513
induced, 150–156. *See also* Induced pluripotent stem cells
in inner cell mass, 131, 134–135, 347
in laboratory, 147–150
in mammals, 347
mesenchymal, 146
organoids derived from, 153–155
in regenerative medicine, 150
self-organizing ability of, 154
sources of, 148
Pluteus larvae, 44, *45*
sea urchin, *279, 283,* 288
Pmar1 and *Pmar1* in sea urchins, 281–282
Pmar1-HesC double-negative gate, 281
Podocytes, 494, *495*
Polar bodies, 189, *197*
first, 183
in insects, 221n4
in mammalian eggs, *198, 217*
second, 183
Polar nuclei, *187, 189, 219*
Polarizing activity zone in limb development, 534–536
Pole cells in *Drosophila,* 250, 252, *253, 257*
polehole in *Drosophila, 256*
Pollen grains, 187–188
apertures in, 218–219
and fertilization in angiosperms, 217–219
stamen production of, 12
and stigma interactions, 217–218, 219
structure of, 188
Pollen tubes, *187,* 188, 189, 217, 218–220
Pollination, 190n7, 217–218, *218*
and heterochrony of flowers, 651
PolyA tail, 57
and differential mRNA longevity, 72
and RNA interference, 76
and selective inhibition of translation, 73
Polyadenylation, 57
Polycarbonate plastics, BPA in, 632–633
Polychlorinated biphenols (PCBs), 636
Polycomb proteins, 63, 67
Polygonum buckwheat, 660
Polymerase chain reaction technique, 81
Polyneuronal innervation, 436

Polyspermy prevention, 198
  in angiosperms, 220
  cortical granule reaction in, 203,
    204–206
  in *Drosophila*, 250
  in external fertilization, 202–206
  fast block in, 203–204, 208, *210*
  in internal fertilization, 215–216
  slow block in, 203, 204–205, 208, *210*,
    215–216
  zinc spark in, 216
Polysyndactyly, *520*
Polytene chromosomes, 53, 247, *248*
Polytubey, 220
Pons, *375*, 376
POP-1 and *pop-1* in *Caenorhabditis elegans*,
    242, 243
Population asymmetry, 128
Porcellio scaber, 663n7
Porcupine gene, 108, 124n8
Poriferans, 583. *See also* Sponges
Porpoises, 543
Positional control genes in flatworms,
    596–601
Postbranchial body, *551*
Posterior cranial placodes, 442, *443*
Posterior foregut cells, *548, 549*
Posterior marginal zone in birds, 338, *339*,
    342–343, 361
Posterior necrotic zone, *541, 542*
Posterior restriction regulatory region,
    538, *539*
Postsynaptic cells, 436
Posttranslational protein modification, 52,
    52, 77–78
Potatoes, regeneration of, 586
Potency
  of cardiac precursor cells, 505
  of neural crest cells, 401–402, *403*
  of pluripotent cells. *See* Pluripotent cells
  of stem cells, 128–130
    multipotent. *See* Multipotent stem
      cells
  of totipotent cells. *See* Totipotent cells
POU transcription factor family, *66*
The Power of Movement in Plants (Darwin),
    113
PRDM16 transcription factor, and brown
    fat, 487
Prechordal mesoderm, 313, 320, 462
  in birds, 340
  in mammals, *352*
Prechordal plate, 303, 314, 330, 448, *450*
Prechordal plate mesoderm, 341
Precursor cells, 130
  cardiac, 503–505
  endothelial, 464
  migration of, 390
  muscle, 486
  neuronal, *129*
  Sertoli cell, 166
  uterine, 122, 123
  vulval, 121–122
Preformationism, 17

Pregnancy
  mammary gland growth in, 457
  maternal age in, *183*
  metabolic disorders in, *622, 623*
  teratogen exposure in, 35–36, 621–634
  uterus in, 166
Pre-granulosa cells, 163
Pre-messenger RNA (pre-mRNA), 57
  β-globin gene, *56*
  central dogma of biology on, 52
  gene transcription into, 51, 62–70
  noncoding information in, 52
  processing of, 51, 52, 70–72
  splicing of, 71–72
  Sxl in, 170, 171
  transformer in, 172
Pre-placodal region, *442, 443, 444*
  in eye development, 448, *452*
Presenilin-1 protease, 121
Presomitic mesoderm, 461, 464, 467, *468*
  axis elongation, 470–473
  in birds, *342*
  clock-wavefront model of, 473–479
  determination front in, 474
  epithelialization of, 472
  in forelimb formation, 523
  maturation of, 474
  origin of, 470
  in somitogenesis, 469–479
  spatiotemporal collinearity of Hox
    genes in, 467–469
  specification of, 465–466
  transplantation experiments with, 467,
    468, 522
Presynaptic neurons, 436
Prickle protein, *291*
Primary cilium, 118, *120*, 123, *138*
  of radial glia, 392
Primary embryonic induction in
    amphibians, 307–308
Primary hypoblast, 338
Primary larvae, 563
Primary mesenchyme, 278
Primary neurulation, 365–374, 379
Primary oocytes, *197*
Primary sex determination, 161–166, 169
  atrazine affecting, 633–634
Primates, brain development in, 392–393,
    395, 396
Primaxial dermomyotome lip, *480*
Primaxial muscles, *481, 485, 486, 488, 489*n5
Primaxial myoblasts, 488
Primaxial myotome, *480, 487*, 488
Primed embryonic stem cells, 148, 150,
    157n18
Primitive endoderm, 348–351, 352
Primitive groove, *340, 351*
  in birds, 338
Primitive knot, 338
Primitive pit, 338, *367*
Primitive streak
  in amniotes, *442*
  in birds, 338–340, 341, *342*, 343, 359,
    375, 463

  in axis formation, 343, *344*
  elongation of, 340
  formation of, 338–340, 342
  in primary neurulation, *367*
  regression of, 341, 342, *343*
  in mammals, *349, 352*, 361
    in anterior-posterior axis formation,
      352, 353
    in gastrulation, 351
Primordial germ cells, 134, *135*
  in gametogenesis, 174–176, 180, 182
  migration of, 175
  pluripotent embryonic stem cells from,
    148, 152
  in zebrafish, *323*
proboscipedia in *Drosophila*, 268, 354
Procambium, 588
Procephalon in *Drosophila*, 265
Procollagen, *71*
Procotyla fluviatilis, 595
Proctodeum, *550*
Proembryo, 46–47
Progamic phase in angiosperms, 217
Progenitor cells, *128*, 129–130, 386, 441
  of breast, 457
  cardiac, 501, 502, 504
  in caudal region of spinal cord, *379*
  in cerebellum, 384
  in heart development, 501, 502, 504
  hematopoietic, *514*
  hepatic, *548*, 613–614
  induced pluripotent stem cells derived
    from, 151
  intermediate, 388
  intestinal, 548
  lymphoid, 130
  mesenchymal stem cells as, 146
  myeloid, *130*
  neural, 386, 388, *445, 446*
  in neural tube, 379
  in neurogenesis, 388
  neuromesodermal, 379, *467*, 470–471,
    473–474
  in plant meristem, 133
  preneural, *379*
  proneural, *379*
Progesterone, 163
  and acrosome reaction in sperm, 214
  and chemical gradients in sperm
    migration, 213
  in mammary gland development, 457
  and retroviruses, 662
Progress zone in limb development, 518,
    526
  and apical ectodermal ridge
    interactions, 527–528
  BMP in, 542
  and mesoderm specification in
    proximal-distal axis, 529
  in salamander regeneration, 605
Prokaryotic genes, 55, 82
Prolactin, *104*
  in mammary gland development, 457
  and retroviruses, 662

Proliferating cell nuclear antigen (PCNA), 589, 617n2
Promiscuity of females, 212
Promoters, 56–57, 58, *59*, 85n3
  chromatin immunoprecipitation-sequencing in analysis of, 79–80
  CpG-content and DNA methylation, 63–64
  and enhancer interactions, 58–59, 60
Promyelocytic leukemia, acute, 640
Pronephric duct, 492, *493*
Pronephros, 492, *493, 494,* 519
Pronucleus
  of egg, 196, *197,* 211, 216–217, 221
    in *Caenorhabditis elegans,* 241
    in *Drosophila,* 250
    in tunicates, *289*
  fusion in fertilization, 7, 211, 216–217
  of sperm, 211, 216–217, 221
    in *Caenorhabditis elegans,* 241
    in *Drosophila,* 250
    in tunicates, *289*
Pronymph stage, 569
Prophase, 177–178, *179,* 182, 183
  in mice, *183*
  in snails, *236*
Prosencephalon, *375,* 376
Prostaglandin PGD_2, 458
Prostate gland, 167, 632
Proteasomes, 78
Protein
  in cell-to-cell signaling, 88
  central dogma on synthesis of, 52–53
  conformational changes in, 88
  degradation of, 78
  in egg cytoplasm, 196
  posttranslational modification of, 52, 77–78
  synthesis in activated egg, 209–210, 211
Protein kinase A (PKA), *106,* 107
  in acrosome reaction, 214
  in sperm capacitation, *213*
Protein kinase C (PKC), *103*
  in egg activation, *210*
  in inner cell mass development, 135
Protein kinase C-3 (PKC-3) in *Caenorhabditis elegans,* 241
Proteoglycans, 93
  heparan sulfate. *See* Heparan sulfate proteoglycans
Prothoracic gland in insect metamorphosis, 572, 573
Prothoracicotropic hormone in insect metamorphosis, 572, 573
Prothorax in *Drosophila,* 254
Protists, 661
Protostomes, 224, *276. See also specific organisms.*
  Caenorhabditis elegans, 226, 238–245
  Drosophila melanogaster, 247–274
  snails, 226–237
  in tree of life, *27, 225,* 584
Protoxylem, *588*
Prox1 gene, *486*

Proximal-distal axis in limb development, 517, 518, 527–533
  in *Drosophila,* 273, 576
  in salamander regeneration, 602
  Turing model of, 531–533
Pruning, synaptic, 397
Pseudohermaphroditism, 168
Ptc2 in eye development, *451*
Pteridium aquilinum, 33
Ptf1a and *Ptff1a* in pancreas development, 555
PTTH (prothoracicotropic hormone) in insect metamorphosis, 572, 573
PU.1 in hematopoietic stem cell formation, *512*
Puberty
  brain development in, 396–397
  gonad development in, 162, 166
  mammary gland growth in, 457, 629
  meiosis timing in, 176
  oogenesis in, 182
Pulmonary artery, 416
  and cardiac neural crest, 419, *420*
  and heart field formation, 502, *502, 503*
  and looping of heart, *506*
Pulmonary vein, 502
Pulp cells, 457
Pumilio and *pumilio* in *Drosophila, 256, 258*
Pupa, 570
  of *Drosophila,* 571–572, 573–574
Purkinje, Jan, 515n2
Purkinje fibers, 503, 515n2
Purkinje neurons, 381, *383,* 515n2
  in cerebellum, *383, 384, 421*
  dendrites of, *384, 421*
Pygopagus, *359*
Pyloric sphincter, *552,* 553
Pyloric stenosis, 553
Pyramidal cells, *385*
Pythons, 652

Q

Quail
  cardiac neural crest in, *420*
  and chick chimeras, 22, *23,* 482, 485
  hematopoiesis in, 511
  vasculogenesis in, 509
Quiescence period, 34
Quiescent center, *12, 132,* 588–589

R

R5LE mRNA, *236*
Rabbits, *27,* 584
Rac, *110*
  in amphibian gastrulation, 303, *305*
Rac1
  in amphibian gastrulation, *305*
  in axon guidance, *423*
  in neural crest migration, *417*
Radial cleavage, 14, *15*
  in amphibians, 297–298
  in sea urchins, 276–277

Radial glial cells, 137, 382, 386, 388
  cell lineage studies of, 388
  in cerebral cortex, 386, 394, 395
  in cerebral organoid, *154,* 155, 642
  division of, 387, 388, 392
  functions of, 382
  interkinetic nuclear migration of, 387
  in neocortex, 389–390
  in neurogenesis, 388
  outer, 388, 394, 395
  as scaffold, 382, 385, 386, 389–390, 394
  as stem cells, 382, 386
  ventricular, 388, 392, 395
Radial intercalation
  in amphibians, 300, 304, 305
  in zebrafish, 327
Radiation exposure as teratogen, *622,* 623
Radius (bone), 517, *518,* 520, 529, 533
Radix snails, 228, *229,* 245
Raldh2 and *Raldh2, 467*
  in birds, *342*
  in heart development, *505*
  in somitogenesis, *473*
Ramón y Cajal, Santiago, 363, 420, 428
Rana pipiens (leopard frog), 8–11
  atrazine affecting, 633, 634
  gray crescent of, *297*
Random events
  abnormal development in, 619–620
  mutations in, 658
RARβ in birds, *342*
Ras, in RTK pathway, *103*
Ras G protein in RTK pathway, *103*
Rathke, Heinrich, 17, 18
Rathke's pouch, 442, 549, *550*
Rats
  adult neurogenesis in, 137
  carcinogenesis in, *638*
  commissural axons in, *429*
  dorsal root ganglia in, 432
  epiallelic inheritance in, 635–636
  eye development in, 97, *98*
  heart recellularization in, 99
  kidney development in, 492
  liver regeneration in, 614
  motor neuron injury treatment in, 150
  stress and DNA methylation in, 660
Rawles, Mary, 22, 410
Ray-finned fish, 521
Rdh10 in forelimb formation, *523*
Reaction-diffusion mechanism in limb development, 531–533, 654–655
Reaggregation of cells, 89
Recellularization of heart, 99
Receptor tyrosine kinase (RTK), 102
  in amphibians, 321
  in *Caenorhabditis elegans* vulva induction, 121, 122
  and fibroblast growth factors, 102–104, 321
  in insect metamorphosis, 572
Receptors. *See also specific receptors.*
  definition of, 88

for extracellular matrix molecules, 94
in signal transduction pathways, 102
Recombination technology in Cre-lox technique, 83–84
Recruitment, 645–646
Rectum, 498, *499*, 551, *552*
in *Caenorhabditis elegans*, 240
Red blood cells, 1
in amphibian metamorphosis, 565
differentiation of, 39–40
globin genes in, 3, 64
and hematopoietic stem cells, *130*, 512, 513, *514*
hemoglobin in, 51, 56
immature, ribosomes in, 73
Red-eared slider turtle (*Trachemys scripta elegans*), 173–174
Redwood sequoias, *2*
Reelin and *Reelin*, 390–391, 396
Regeneration, 4, 579–618
blastema in, 20, 585
in flatworms, 594, 595–596, 597, 598
in salamanders, 20, 602–607
in young mammals, 615
in zebrafish, 608–609, 610, 612
compensatory, 581
in mammalian liver, 581, 613–615
in zebrafish, 612
de novo, 587
definition of, 579
embryonic development compared to, 582–583, 616
of epidermis, 453–454
epimorphosis in, 581, 585, 591, 607
in salamanders, 602–603, 604
evolution of, 583–586
experimental research on, 579
in flatworms, 6, *580*, 594–602, 607
of hair, 458, *459*, 581
in hydra, 6, 135, 579, *580*, 591–593, 607
immune response in, 582
in mammals, 580, 583, 613–617
compensatory, 581, 613–615
of liver, 581, 613–615
in model organisms, 6, 580
modes of, 581
morphallaxis in, 581, 585, 591, 607
in zebrafish, 612
permissive interactions in, 99
in plants, 1–2, 580, 586–591
canalization model of, 589–591
direct and indirect methods in, 588
evolution of, 583, 585
and recellularization of rat heart, 99, *99*
reprogramming in, 582, 583
in salamanders. *See* Salamanders, regeneration in
size recognition and termination in, 582
stem cells in. *See* Stem cells, in regeneration
steps in, 580
system integration in, 582
of teeth, 456, 457

in zebrafish, 135, 580, 583, 607–612
of fin, 607–609
of heart, 609–612, 615
Regenerative medicine, 150, 156, 585
Regional specificity of neural induction in amphibians, 318–321
Regulative embryos, 45
Regulatory elements, noncoding, 57–61
Relational pleiotropy, 620
Remodeling of plant cell walls in mechanical stress, 232
Renal tubules, 496
pronephric and mesonephric, 492, *493*
proximal and distal, 494, *495*, 496
Reporter genes, *59*, 60
Repressors (silencers), 57, 58, 61, 65
Reproduction, 4, 159
endocrine disruptors affecting, 629–634
atrazine, 633–634
bisphenol A, 632–633
diethylstilbestrol, 629–631
fracking chemicals, 634
female promiscuity in, 212
fertility in. *See* Fertility
fertilization in. *See* Fertilization
meiosis and genomic variation in, 159
as plant development phase, 12, 13
transition from vegetative phase to, *12*, 67–69, 184
symbionts affecting, 660–661
Reproductive isolation, 660–661
Reprogramming in regeneration, 582, 583
Reptiles, *32*
as amniotes, 335
cleavage in, 14, *15*, 335
environmental sex determination in, 173–174
gastrulation in, 348
jaw phenotypes in, 658
limb development in, 521
neural tube closure in, 370
phylogenetic relationships of, *296, 336*
regeneration in, 457
in tree of life, *27, 584*
Resact, 199
Resegmentation of sclerotome, 482
Respiratory system
in amphibian metamorphosis, *564*
lungs in. *See* Lung development
Respiratory tube, 547, 549, 557–560
Responders, 96, 100
Resting potential of egg cell membrane, in polyspermy prevention, 203–204
Ret protein, 412, 497
Rete testis, 162, *162*
Reticular lamina, *93*
Retina development, 90, *103*, 431, 448–451
in amphibian metamorphosis, 565
axon guidance in, 431, *432*, 433–435
cell migration in, 90
enhancers in, *59*, 60
eye field in, 449–450
induction of, 96, 451, *452*

pigmented cells in, 90, 448, 449, 451
in vertebrates, 96, 448, 449–451
Retinal ganglion, 431, *432*, 449
chemoaffinity hypothesis on, 434
number in optic nerve, 431
target selection by, 433–435
Retinal homeobox in eye development, 449–450, 451
Retinal pigmented epithelium, 448, 449, 451
Retinaldehyde dehydrogenase-2, 504, *505, 523*
Retinoic acid
in amphibians, 321
in anterior-posterior axis formation, 321, 353, 449, 627
in birds, 342
in cranial placode induction, 443
in differentiation therapy for cancer, 640, *641*
in digestive tube development, 552, 553
and glyphosate interactions, 627
in heart development, 504–505, 627
in kidney development, 496
in leukemia, 640
in limb development, 527
in apoptosis, 542
in forelimb formation, 523–524, 527
in gradient model, 530–531
in hindlimb formation, 527
in mammals, 353, 627
and meiosis timing, 176, *177*
and neuromesodermal progenitors, 379
in oogenesis, 182
in otocyst axis determination, *447*, 448
in paraxial mesoderm specification, 466, *467*
in somitogenesis, 473, 474, 479
as teratogen, 35, 505, 627
in zebrafish, 324
in fin regeneration, *608*, 609
Retinoic acid receptor RARβ in birds, *342*
Retinoic acid regulatory element (RARE), *523*
Retinoid receptor RXR in amphibian metamorphosis, 567–568
Retinotectal projection, 433–435
Retroviruses, 662
Rett syndrome, 152
REV and *REV*, *132*, 133
Reverse genetics, 81–82
Reverse transcriptase, 81
REVOLUTA and *REVOLUTA*, *132*, 133
Rheotaxis, 212
Rhizoids, *33*
Rho GTPases, *110*, 111
in axon guidance, 423
Cdc42, *423, 472*
in epithelialization of somites, 473
Rac1, *305*, 417, 423
RhoA, *110*, 409, 423
Rhombencephalon, *375*, 376
Rhombomeres, 376
and cranial neural crest migration, 415, 416, 417

Rib meristem, *132, 133*
Ribosomes
   in egg cytoplasm, 196
   selective activation of mRNA
      translation, 73–74
Ribs, 461, 462, 467
   cervical, 655
   sclerotomes in development of, *463, 464*
   and tendon development, *485*
   in turtle shell development, 649, *650*
Rickettsiella, 660, *661*
Rieger's syndrome, 361n3
Right-handed (dextral) coiling, 228–230
RNA
   antisense, 74
   double-stranded, 75, 82
   guide, 82, 83
   heterogeneous nuclear, 52, 57
   in human brain development, 396
   messenger. *See* Messenger RNA
   microRNA. *See* MicroRNA
   noncoding information in, 52, 56, 57–61
      in brain development, 396
   pre-messenger. *See* Pre-messenger
      RNA
   sex-specific slicing in *Drosophila*, 171,
      172
   short-interfering, *75*
   transcription of DNA in, 51, 52, 62–70
RNA interference (RNAi), 75, 81–82, 150
   in flatworms, 596, *597*
RNA polymerase II, 56–57, 58
   β-globin gene, *56*
   and DNA methylation at promoters, 64
   and enhancer activation, 60
   and transcription, *52*, 58, *59*, 60, 66
RNA-induced silencing complex (RISC),
   75, 75–76
RNA-Seq, 80–81
Robo (roundabout) protein, 428–429, *430*,
   431, *432*, 439n11
Robo1, 428–429, *430*
Robo2, 428–429, *430*, 431, *432*
Robo3, 428, *430*
Rodents
   epiallelic inheritance in, 635–636
   mice. *See* Mice
   rats. *See* Rats
   tooth regeneration in, 457
   in tree of life, *27*, 584
Roof plate, 377, *378, 379*, 383
Root apical meristem, *32*, 132
   in *Arabidopsis thaliana, 12, 13*, 114, 132,
      133, 588, *589*
   and auxin, 114, *115*, 588
   quiescent center of, 588–589
   in regeneration, 588
   totipotent stem cells in, 133
Roots
   basal cells in development of, 13
   regeneration of, 588
Ror, in Wnt signaling pathway, *110*, 111
Rosa26-LacZ reporter, *143*
Roses, floral organ identity genes in, 69

Rosette formation
   in sponges, 661–662
   in V-SVZ, 138, *139*
Rostand, Jean, 295
Rostral-caudal axis, 368, 379
Rotational cleavage, 14, *15*
   in *Caenorhabditis elegans*, 240
   in mammals, 346
Roundabout (Robo) proteins, 428–429,
   430, 431, *432*, 439n11
Roundup, 627–628
Roundworm, 5, 6. *See also Caenorhabditis
   elegans*
Roux, Wilhelm, 44, 45, 662
RpI38 protein, 74
R-spondin1 (Rspo1) in primary sex
   determination, *161, 163, 164*
RTK. *See* Receptor tyrosine kinase
Runt and *runt* in *Drosophila*, 171, 260, 263,
   267
Runx1 and *Runx1*
   in hematopoiesis, 511, *512*
   in vasculogenesis, 507
Rx and *Rx* in eye development, 449–450,
   451
RXR in amphibian metamorphosis,
   567–568
Ryk in Wnt signaling pathway, *110*, 111
Ryk receptors in Wnt signaling pathway,
   110

S

S locus, 218
Sacral neural crest, 401, 412
Sacral vertebrae, 355–356, 366
SACY in sperm capacitation, *213*
Sagan, Dorion, 661
St. John's wort, 152
Saint-Hilaire, Geoffroy, 318
sal (spalt) in *Drosophila*, 575, *576*
Salamanders
   cleavage in, 297–298
   fertilization in, 221n3
   limb development in, 518, *519*, 522, 523
      morphogenetic rules in, 522
      in regeneration, 2, 4, 20, 522, 579,
         580, 583, 602–607, 617n1
   metamorphosis in, 564
   as model organism, 6
   regeneration in, 2, 4, 579, 580, 583,
      602–607, 617n1
      blastema in, 20, 602–607
      evolution of, 583
      as model organism, 6
      morphogenetic rules in, 522
   regional specificity of neural induction
      in, 318, *319*
   transplantation experiments with, 20,
      295, 318, *319*
Salivary glands, 272, 549, 574
SALL4 transcription factor, 642n2
Salpingoeca rosetta, 661–662
Sarcopterygians, 521
Sartorius muscle, *425*

Satellite cells, 486
Saunders, John, 529
Sauripterus, 534
Scales, as ectodermal appendage, 454
Scar formation, 585, 586, 617
SCARECROW (SCR), 588–589, 617n2
Scatter factor, 614
SCF. *See* Stem cell factor
SCH-28080 treatment, flatworm
   regeneration in, *601*
Schizocoely, 224, 226
Schlopp, Steffen, 118
Schotté, Oscar E., 579
Schultz, Jack, 247
Schwann, Theodor, 420
Schwann cells, 420, 436, *437*
   as neural crest derivative, *400*, 402, *403*,
      410
Scl and *Scl* in hematopoietic stem cell
   formation, *512*
Scleraxis and *Scleraxis* in tendon
   formation, 484–485
Sclerotomes, 439n6, *463, 464*, 479–485
   in axon guidance, 426
   dorsomedial, *463, 464*
   hemangioblasts from, 511
   in neural crest migration, *410*, 411, 412
   resegmentation of, 482
   in tendon formation, 484–485
   ventral-posterior, *463, 464*
   ventromedial, *463, 464*
   in vertebrae formation, *463, 464*,
      481–484
Scoliosis, 477, 483
SCR (SCARECROW), 588–589, 617n2
Scrotum, 167, *357*
Scute protein in *Drosophila*, 171
Scutes of turtles, 454, 456
SDF1 (stromal-derived factor-1), *144*, 417,
   418
Sea squirts, 6, 226, 288
   fate mapping of, 41
   notochord in, 26
Sea urchins, 275–288
   acrosome reaction in, *199, 200, 201*
   archenteron invagination in, 286–288
   autonomous specification in, 45, 278,
      282
   blastula in, 277–278, 292n1
   cleavage in, 14, 276–277, 278
   conditional specification in, 44–45, 275,
      278–279, 282
   convergent extension in, 288
   early development of, 275–282
   eggs in, 196, *197*, 198–211
   epithelial-mesenchymal transition in,
      283–286
   fate maps and blastomere
      determination in, 278–279
   fertilization envelope in, 204–205
   fertilization in, 194, *196*, 198–211, 221
   gastrulation in, *16*, 279, 283–288
   gene regulatory networks in, 69–70,
      279–282, 284, 285

limb field in, 523
metamorphosis in, 278, 288, 563
micromeres in. *See* Micromeres, in sea urchins
as model organism, 6, *276*
polyspermy prevention in, 202–206
recruitment in, 645
skeletogenic mesenchyme in, 278, 279–286, *287*
ingression of, 283–286
specification of, 279–282
sperm in, 196, 198–211
stored oocyte mRNA in, 73
vegetal cell specification in, 282
Seal limb, *28*
Seastars, *27,* 226, *584*
Sebaceous glands, 458
Secondary hypoblast, 338, *339,* 340
Secondary larvae, 563
Secondary mesenchyme, 278
Secondary neurulation, 365–366, 374, 379
Secondary sex determination, 161, 166–169
Secreted Frizzled-related protein (sFRP), 109, *552,* 553, 557
Securin, 216
Seed coat, *12, 187,* 188, 189
Seeds, 188
dispersal of, 34
dormancy of, 8, 34
evolution of, *32,* 34
in life cycle of flowering plants, 8, *12, 13, 187*
and regeneration, 583
SEEDSTICK (STK), *185*
Segment polarity genes in *Drosophila, 255, 256, 260,* 261, 265–267
Segmental plate, *340,* 461
Segmentation
of *Drosophila,* 53, 108, *249,* 255–261, 273n4
in gastrulation, 252, *253,* 254
segmentation genes in, 260–267
evolution of, 461
of sclerotome in vertebrae formation, 482
in somitogenesis, 469–470, 489
Segmentation genes in *Drosophila,* 260–267
Selective affinity, 89–90
Self-renewal of stem cells, 127, 129, 387
Semaphorin
in angiogenesis, 510
in axon guidance, 423, 424, *425,* 426–427, 431, *432*
in neural crest migration, 411, 412, 417
as paracrine factor, 113
Semaphorin-3, 426–427
Semaphorin-3A, *432*
Semaphorin-3F, 411, 412, 424, *425*
Semicircular canals, 444, 446, 447, 459
Seminiferous tubules, 161, 162, *165,* 180
Sensory ganglia, 441, *442,* 445–446

Sensory neurons, 376, *377,* 382, *383, 425*
cranial placodes forming, 445
ganglia of, 441, *442,* 445–446
in head, 441–442
as neural crest derivative, 401, 402, 410
in salamander limb regeneration, 606
Sensory placodes, 404, 441
cranial, 404, 441–452. *See also* Cranial sensory placodes
Sensory root (dorsal), *383,* 401
SEPALLATA (SEP), *185*
Sepals, *189*
in ABCDE model, 185, 186
and floral organ identity genes, 68–69, *185,* 186
in tulips, 650, *651*
Separase in meiosis, *179*
Sequencing techniques
ChiP-Seq, 79–80
next-generation, 80–81
RNA-Seq, 80–81
Serotonin, *138,* 140
Serrate protein in Notch pathway, 120, *121*
Sertoli cell precursors, 166
Sertoli cells, 161–162, 163, *165,* 166
anti-Müllerian hormone of, 167
Desert hedgehog in, 105
retinoic acid synthesis in, 176
in spermatogenesis, 180, 181
Set1, and germ stem cell self-renewal, 137
Sex chromosomes
in birds, 159
in *Drosophila,* 159, 169–173
in mammals, 159, 160–169
sex combs reduced gene, 268, 272, 663n7
Sex determination, 159–191
in angiosperms, 184–186, 651–652
atrazine affecting, 633–634
chromosomal, 159, 160–173
in *Drosophila,* 159, 169–173
in mammals, 159, 160–169
in *Drosophila,* 159, 169–173
environmental, 159, 164–165, 173–174
in atrazine exposure, 633–634
temperature-dependent, 164–165, 173–174
gametogenesis in, 174–183
gonadal, 161–166, 169
hormones in, 166–169
in mammals, 159, 160–169
primary, 161–166, 169
atrazine affecting, 633–634
secondary, 161, 166–169
Sex lethal protein and gene (Sxl and *Sxl*), 170–172, 173
Sexual phenotype, 166–169
Sexual reproduction, 159
Sf1 and *Sf1* in primary sex determination, 161, 163, 164
SFK (Src family kinase)
in axon guidance, 430
in egg activation, *207,* 208
sFRP (secreted Frizzled-related protein), 109, *552,* 553, 557

sFRP1, 553
sFRP2, 553
Sgo2 in meiosis, *179*
Sharks, *27, 584*
SHATTERPROOF (SHP), *185*
Shear stress in blood flow, 511, *512,* 515n6
Sheep
brain development in, *394*
cloning of, 54
cyclopia in, *107*
Shells
of bird eggs, *337*
of turtles, 454, 649–650
Shoot apical meristem, *32,* 67, 132
in *Arabidopsis thaliana, 12,* 13, 114, 132–134
and auxin, 114
in leaf-to-flower transition, 184
and phyllotaxis, *231, 233*
in regeneration, 588
totipotency in, 133–134
Short gastrulation and *short gastrulation* (Sog and *Sog*), 646
in *Drosophila, 254, 272,* 318, *647*
Short-interfering RNA, 75
SHORTROOT (SHR), 588
Shoulder, 518, *519,* 521, 526
SHP (SHATTERPROOF), *185*
SHR (SHORTROOT), 588
Shrimp, 25, 654
Siamois and *siamois* in amphibians, 312, 313, 315, 317
Sickle-cell anemia, 151
Sieve element, *588*
Signal transducers and activators of transcription (STAT), 104
Signal transduction pathways, 101–102
cadherins in, 91
in *Caenorhabditis elegans* vulva induction, 121–123
fibroblast growth factors in, 102–104
Hedgehog in, 106–107
JAK-STAT, 104
protein conformational changes in, 88
RTK, 102–104
Signaling
cell-to-cell communication in, 87. *See also* Cell-to-cell communication
juxtacrine, 88–92. *See also* Juxtacrine signaling
paracrine, 88, 101–102. *See also* Paracrine factors
Silencers, 57, 58, 61, 65
Silent Spring (Carson), 621
Silkworm moths, 7
Silverfish, ametabolous development of, *569*
Silvery salamander (*Ambystoma plantineum*), 221n3
Single stem cell asymmetry, 128, *128*
Sinistral coiling, 228–230, 245
Sinus venosus, 504
Sinusoidal microvessels, 145, 157n14
SisA in *Drosophila, 171*

Six proteins in myotome determination, *487*, 488

Six1 and *Six1*
  in cranial placode induction, *443*, 444
  in myotome determination, *487*, 488
  in neural crest specification, *404*

Six3 and *Six3* in eye development, *449, 450, 452*

Six4 and *Six4*
  in cranial placode induction, *443*, 444
  in myotome determination, *487*, 488
  in neural crest specification, *404*

Skeletal muscle
  myotome in formation of, 485, 487–489
  satellite cells in, 486
  syncytia in, 49n4

Skeletogenic mesenchyme in sea urchins, 278, 279–286, 287
  ingression of, 283–286
  specification of, 279–282

Skeleton development
  in amphibian metamorphosis, 566
  BMP in, 112
  Hox genes in, 74, 401–402
  Indian hedgehog in, 105
  JAK-STAT pathway in, 104
  lateral plate mesoderm in, 462, 491, *492*
  in limbs, 518, *519,* 529, 531, 536
  neural crest in, 401–402, *403,* 415, 416, 418–419
  in salamander regeneration, *603,* 604
  in sea urchins, 278, 285–286, 645
  somites in, 462
  in young mammals during digit regeneration, 615

Skene's glands, 166, *167*

Skin, 453
  African spiny mouse regeneration of, 615–616, *617*
  in amphibian metamorphosis, *564,* 568
  loss and replacement of, 453–454, 460n2
  pigmentation of, 453

Skin cancer, Sonic Hedgehog in, 108

Skloot, Rebecca, 157n17

SKN-1 in *Caenorhabditis elegans, 242, 243*

Skull, 414
  in amphibian metamorphosis, 566
  neural crest in development of, 418–419

Slimb protein, *106*

Slit
  in axon guidance, 423, 426, 428–429, *430,* 431, *432*
  in *Drosophila,* 428
  in neural crest migration, 412
  as paracrine factor, 113

Sloppy-paired and *sloppy-paired* in *Drosophila, 260, 266,* 267

Slow block to polyspermy, *210*
  in external fertilization, 203, 204–205, 208
  in internal fertilization, 215–216

Smad, 112, *113*

Smad and SMAD, 112, *113*

in amphibians, 306, 312–313, 315
  naming of, 124n10
  in neural crest specification, 404
  in zebrafish, *331*

Smad1, 112, *113*

Smad2, 112, *113*
  in amphibians, 306, 312–313, 315

Smad3, 112, *113*

Smad4, 112, *113*

Smad5, 112, *113*

Small intestine, 551, 552, 553

Small toolkit, 646

Smith, Homer, 491

Smooth muscle
  cardiovascular, 505
  digestive, 551, 553
  respiratory, 560

Smoothened and *smoothened, 266*
  in cancer, 108, *639*
  and Hedgehog, *106,* 107, 108, 118, *120,* 429, *639*
  and primary cilium, 118, *120*
  in vertebrae development, *483*

Snail protein and *snail* gene
  in birds, *344,* 345
  in *Drosophila,* 272
  in epithelial-mesenchymal transition, 486
  in neural crest delamination, 407
  in neural crest specification, 404, *405*
  in sea urchins, 295
  in tunicates, 290

Snail2 and *Snail2,* 407, 486

Snails, 226–230, 235–237
  autonomous specification in, 41, 226, 235, 237
  axis determination in, 235–237
  cleavage in, 14, 226–230
  gastrulation in, 226, *228,* 237, *238*
  as model organism, 226
  spiral patterns in, 226–230, 245
  in tree of life, *27, 584*

Snakes, 26, 489
  anterior-posterior axis in, 470
  genetic assimilation in, 658
  heterochrony in, 651
  heterometry and loss of limbs in, 652, *653*
  somitogenesis in, 470
  vertebrae in, 461

Snapping turtle (*Macroclemys temminckii*), 173

Sodium ions
  in cavitation in mammals, 348
  in fast block to polyspermy, 203–204, 208

Sodium pump, *207,* 348

Sog and *Sog* (short gastrulation), 646
  in *Drosophila, 254, 272,* 318, *647*

Solanaceae, 651

Solenoids, 55

Somatic cells, 7
  in *Caenorhabditis elegans, 238, 239,* 245
  in cloning experiments, 53–54, 67

genes in nucleus of, 51, 53–54
  in gonad development, 163

Somatic mesoderm, 500

Somatic mutation theory, 636

Somatopleure, 500, 526

Somites, 461, 462
  in amphibians, 9, *298, 307, 308*
  anterior-posterior specification by Hox genes, 467–469, 478–479
  and axon guidance, 426
  in birds, 340, *342, 343, 367*
  cell types of, 464–465
  clock-wavefront model on, 473–479
  definition of, 461
  derivatives of, 462, 464–465
  dermomyotome development, 479, 485–489
  differentiation of, *463*
  epithelial-mesenchymal transition in, 464, 486
  eye field formation in, 449
  formation of, 469–479
  in mammals, *354*
  mesenchymal-epithelial transition in, 473–474
  in neural crest migration, 410, 411, 412
  in neural tube closure, *370*
  numbering of, 470
  sclerotome development, 479–485
  and somitogenesis, 469–479
  in zebrafish, *323,* 330, 470–471

Somitic mesoderm, 462

Somitogenesis, 469–479, 489
  axis elongation in, 470–473, 479
  caudal progenitor zone in, 470–473
  clock-wavefront model of, 473–479
  determination front in, 473–479
  mesenchymal-epithelial transition in, 472, 475, 477, 478
  species differences in, 470

Somitomeres, 469–470, 474, *475*

Songbirds, 137

Sonic hedgehog and *sonic hedgehog,* 105, 124n5
  and adult neural stem cells in V-SVZ, *138,* 140, 157n8
  alcohol affecting, 626
  in axon guidance, 428, *429, 430,* 431
  in birds, *342,* 344–345, *628*
  in cancer, 638–639
  in cerebellum, 384
  in cetaceans, 544
  in cranial placode induction, 443
  and cytoneme-mediated signaling, 118, *119*
  in digestive tube development, 553
  in eye development, 107, 450, *451,* 621, 651
  in cyclopia, 107, 450, 623
  in Mexican cavefish, 450, *451,* 651
  functions of, 105, 107–108
  glyphosate affecting, *628*
  in heart development, 504
  heterometry of, 652, *653*

in limb development, 534–539, 544, *653*
  in cetaceans, 544
  and digit identity, 536–537
  in dorsal-ventral axis, 541
  in feedback loop with FGFs, 537–538
  in mice, 105, *653*
  and zone of polarizing activity, 534–536
in myotome determination, 487, 488
in neural crest migration, 549
in neural plate folding, *369*
and neural progenitors in caudal spinal cord, *379*
in neural tube development, 372, 377, *378, 379*
in otocyst axis determination, 447–448
in pancreas development, 555
and radial glial stem cell fate, 392
in sclerotome development, 481, 483
in snakes, 652, *653*
in tooth development, *456*
in vertebrae development, *483*
in zebrafish fin regeneration, 609
Sowbugs, 663n7
Sox transcription factors, 66, 97, 365
Sox1, 365
Sox2 and *Sox2*, 61, 66, 67, *350*
  in cerebral organoid formation, *154*
  in digestive tube development, *552*
  and embryonic stem cell pluripotency, 148, 347
  in eye development, 450
  and induced pluripotent stem cells, 150
  and inner cell mass pluripotency, 134, 347
  in neural crest delamination, 407
  neural plate expression of, 365
  and neural progenitors in V-SVZ, *141*
  in neural tube development, 470
  in paraxial mesoderm specification, 465, 466, 467, 470
  in respiratory tube development, *558*
  in secondary neurulation, 374
Sox3 and *sox3*
  in amphibians, 315
  neural plate expression of, 365
Sox5 in tendon formation, 484, *485*
Sox6 in tendon formation, 484, *485*
sox8 and neural crest potency, 402
Sox9 and *Sox9*
  in digestive tube development, *552*
  in neural crest specification, 404, *405*
  in primary sex determination, 161, *163*, 164–166
  in reaction-diffusion model of limb development, 533
  in tendon formation, 484, *485*
Sox10
  and epibranchial placode, *446*
  in neural crest specification, 404, *405*
sox11 in amphibians, 317
Sox17 and *Sox17*, 549
  in amphibian germ layer specification, 306

in endoderm specification, 551
in gut tissue specification, 551
in mammals, *350*
and pancreatic β cells, *556*
in visceral and definitive endoderm, 547, *548*
SoxC in sea urchins, 282
Spallanzani, Lazzaro, 579
spalt (sal) in *Drosophila, 575, 576*
Spatiotemporal collinearity of Hox genes, 467–469, 478–479
Spätzle protein in *Drosophila,* 271
Spemann, Hans, 307–308, 309, 451
Sperm, 193, *199*
  acrosome reaction in, 200, 212, 214
  in aggregates or train, 212
  of amphibians, 9, 11, 296, *311*
    entry point of, 296, 297, 298, 300, 307, 309, 310, 312
  anatomy of, 194, *195*
  of angiosperms, 12, *187,* 188, 189, 217–220
  atrazine affecting, 634
  binding to extracellular matrix of egg, 198
  binding to zona pellucida, 214
  BPA affecting, 632
  of *Caenorhabditis elegans,* 239–240, 241
  capacitation of, 212, 213, 214
  cell membrane of, 194, *195,* 201–202
    fusion with egg cell membrane, *200,* 201–202, *203,* 214–215
  centrioles in, 194, *195,* 202, 210, 211, 217
    in amphibians, 296, *297*
    in polyspermy, 202, *203*
    in tunicates, 289
  and chromosomal sex determination, 159
  cytoplasm of, 194, 195
  DNA methylation in, 635
  of *Drosophila,* 249–250
  in external fertilization, 198–211
  formation of, 174, 176, 180–182, 194
  gene methylation in, 65
  genetic fusion with egg, 210–211
  hyperactivation of, 212–213
  in internal fertilization, 211–217
  of mammals, 160, 180–182, 194, *195,* 198, 211–217, 345
  maturation of, *180,* 194, 212, *213*
  and monospermy, 202
  motility of, 194, 212
  outflow tract, 498
  and polyspermy prevention, 202–206, 215–216, 220
  progesterone response of, 213, 214
  recognition between egg and, 193, 194, 197, 198–201, 214
    species-specific, 198–199
  rheotaxis of, 212
  of sea urchins, 196, 198–211
  sex chromosomes in, 160
  size changes, 19

stem cells in production of, 152, 162, 176, 180, 181, 182
structure of, 193–194
thermotaxis of, 213
translocation to oviduct, 212
of tunicates, 289
Sperm head, 194, *195*
Sperm train, 212
Sperm-activating peptides, 199
Spermatheca, 240
Spermatids, 180, 181–182
Spermatocytes, 180, *181*
  primary, 181
  secondary, 181
Spermatogenesis, 174, 176, 195
  Desert hedgehog in, 105
  DNA methylation in, 65
  in mammals, 180–182
Spermatogonia, 130, 162, 163, 180, 181, 182
Spermatozoa, 182
Spermiogenesis, 180, *181,* 182
Spina bifida, *370,* 372
Spinach plants, 186, 652
Spinal cord
  in amphibians, 307, 309, 318
  anterior-posterior axis in, 376
  in birds, 341, 342, *375*
  columns of Terni in, 424
  commissural neurons from, 428–429, *430*
  differentiation of neural tube into, 381, *383*
  dorsal-ventral axis in, 376–377, *378*
  as ectoderm derivative, 363, *364*
  in humans, *375*
  injury of, 151
  motor neurons from, 424–427, 436
  neural crest cells, *411*
  in neural tube closure, 371
  neuroanatomy of, 382, *383*
  progenitors in caudal region of, 379
Spinal nerves, *383,* 482
Spindles
  meiotic, *54, 178,* 179, 183
  mitotic. *See* Mitotic spindles
Spine
  dorsomedial sclerotome in development of, *463,* 464
  intervertebral discs in, 462, *463,* 464, 483–484
  intervertebral joints in, 482
  vertebrae in. *See* Vertebrae
Spinous layer of epidermis, *453*
Spinous process, *482*
Spiny mouse, skin regeneration in, 615–616, 617
Spiral cleavage
  holoblastic, 14, *15,* 226–230
  in lophotrochozoans, 224
Spiral patterns, 223
  dextral coiling in, 229–230
  in plants, 226, 231–235, 245

sinistral coiling in, 228–230, 245
in snails, 226–230, 245
Spiralia, 224
Splanchnic mesoderm, *463*, 500
and accessory organ development, 554
in digestive tube development, 551, 552–553
in heart development, 501
in vasculogenesis, 509
Splanchnopleure, 500, 504, 511
Spleen, 343, 356, *357*, 511
Spliceosomes, 72
Splicing, 58, 71–72
isoforms in, 71, 72, 85n8
reverse genetics in study of, 81
sex-specific, in *Drosophila*, 171, 172
Splicing factors, 72
Split proteins in angiogenesis, 510
Spondylocostal dysostosis, 477
Sponges, *32*, 246n1
choanocytes in, 30, 661–662
gastrulation in, 30
larva of, *26*
pumping action of, *26*
regeneration of, 583–585
in tree of life, *27, 225, 584*
Spongilla lacustris, 585
Sporophytic stage, 8, 13, 186, *187,* 188
and alternation of generations, *32, 34*
definition of, 7
Spotted dolphin (*Stenella attenuata*), 544
Spotted salamander (*Ambystoma maculatum*), *519, 523*
Springtails, ametabolous development of, 569
SPTLC1, 396
Squamosal bone, 418
Src family kinase (SFK)
in axon guidance, 430
in egg activation, *207, 208*
SRGAP2 and *SRGAP2*, 648
Sry and *Sry,* 66
in androgen insensitivity syndrome, 167
and primary sex determination, 161, *163,* 164–166
Stamens, 187, 188, 650
in ABCDE model, 185, 186
in *Arabidopsis thaliana*, 12, 13, *185*
and floral organ identity genes, 68–69, *185,* 186
Stapedial artery, *416*
Stapes, *415,* 416
STAT (signal transducers and activators of transcription), 104
Stat5, *104*
Statocysts, 227
Steinberg, Malcom, 90
Stellate cells in liver regeneration, 614
Stem cell factor, 113
and auxin signaling in plants, *116*
in bone marrow hematopoietic stem cell niche, 512–513
and primordial germ cell migration, 175

Stem cell therapy, 143, 150, 156, 157n12
Stem cells, 127–158
adult, 129, 135–147. *See also* Adult stem cells
in amphibian metamorphosis, 565, 568
in *Caenorhabditis elegans,* 240, 245
in cancer, 637, 639–640
committed, 128
in culture, 129, 134, 146–147, 148–150
definition of, 4
differentiation potential of, *129*
division of, 127–128, 129, 131, 386–387, 388
ectodermal appendage, 457–458, *459*
embryonic, 131–135. *See also* Embryonic stem cells
epidermal, 453
in flatworms, 594–602
future research on, 156
in germinal neuroepithelium, 382
hair follicle, 146, 458, *459*
hematopoietic. *See* Hematopoietic stem cells
in human model of development and disease, 147–157
in hydra, 591–592
and indeterminate growth of plants, 8
in inner cell mass, 131, 134–135
intestinal, 142–143, 565, 639
medical uses of, 143, 150–152, 156, 157n12
melanocyte, 458
in meristem, 8, 232, 588
mesenchymal, *144,* 145, 146–147, 457
in mesoderm specification, 466
multipotent. *See* Multipotent stem cells
neural. *See* Neural stem cells
neuroepithelial cells as, 382
neuromesodermal, 470
niches of, 130, 131, 135–145
bone marrow, 512–513
spermatogonia in, 180
pluripotent, 6, *66,* 128–129, 131–135
induced. *See* Induced pluripotent stem cells
radial glia as, 382, 386
in regeneration, 4, 581, 582, 585, 607
in flatworms, 594–602
in hydra, 591–592
medical uses of, 150, 156
in plants, 588
in salamanders, 604, 605
regulation of, 130–131
self-renewal of, 127, 129, 387
of skeletal muscle, 486
spermatogonial, 162, 163, 176, 180, 181, 182
teratocarcinoma of, 637
totipotent, 128, *129,* 131
undifferentiated, 127
unipotent, 130
Stenella attenuata (spotted dolphin), *544*
Stentor, 585
Stereoblastula, 36n4, 226, 237

Stickleback fish, 644–645, 663n2
Stigma, 185, *187,* 188, *189,* 217
and pollen interactions, 217–218, 219
and pollen tube navigation, 219
STK (SEEDSTICK), *185*
Stochastic factors in abnormal development, 619–620
Stöhr, Philipp, 458
Stomach development, *357, 550,* 551, 552, *554*
Wnt signaling in, 553
Stomodeum, 549, *550*
Stra8 and meiosis timing, 176, *177*
Stratum corneum, 453
Stratum germinativum, 453
Stress, DNA methylation in, 660
Stripped patterns in *Drosophila*, 263–264, 267
Stromal cells, 143, *144,* 145
Stromal-derived factor-1 (SDF1), *144, 417,* 418
Strongylocentrotus, 284
S. franciscanus, 201, *202*
S. purpuratus, 6, 201, *202, 278*
Sturtevant, Alfred, 245
Styela partita
bilateral symmetry of, *288*
cytoplasmic rearrangement in, *289*
fate map of, 20–21, 42–43, *289*
Styles, 185, *187,* 188, *189, 218*
and pollen tube navigation, 219
Styloid process, *415*
Stylopod, 517, *518*
evolution of, 521
gradient model on, *530,* 531
Hox gene specification of, 519, *520, 530, 539*
reaction-diffusion model on, 531, *532, 534*
Subcutaneous fat, 453, 458
Subgerminal cavity, 337
Subgranular zone, 137
Subventricular zone, *383, 385,* 388, *389, 394*
adult neural stem cell niche of, 137–141
SuFu (Suppressor of Fused) protein, *106,* 107, 118
Sulcus, cerebral, 393–394, *395*
Sulcus limitans, 382, *383*
Superficial cleavage, 14, *15,* 250
Supernumerary nipples, 454
Supernumerary teeth, 456
Superoxide radicals, 626
Suppressor of Fused (SuFu) protein, *106,* 107, 118
Surface tension in cell adhesion, 91, *92*
Survival of the fittest, 643
Suspensor, 13, 46, *46*
swallow gene in *Drosophila*, 256
Sweat glands, 441, 454, 458
Swim and *swim,* 117
Sxl and *Sxl* (sex lethal protein and gene), 170–172, 173
Sycamore trees, *2*
SYCP1 in meiosis, *179*

Symbiogenesis, 661
Symbiosis, 660–662
    bacteria in, 32–33, 620, 660–662
Sympathetic nervous system, 401
    ganglia in, 402, 403
    and hematopoietic stem cells, 144, 145
Sympoiesis, 660–662
Synapse, 421, 436–438
    definition of, 436
    in infants, 393
    pruning of, 397
    of Purkinje neurons, 384
Synapsis in meiosis, 177
Synaptic cleft, 436
Synaptic vesicles, 436
Synaptonemal complex in meiosis, 177,
        178, 179
Syncytial blastoderm, 47–48
    and anterior-posterior axis, 255, 258,
        273n3
    cellularization in, 47, 48, 251, 252
    cleavage in, 47, 250
    differential RNA splicing in, 171
    and dorsal-ventral axis, 270
    and mid-blastula transition, 252
    Sex-lethal expression in, 171
Syncytial cables in sea urchins, 283, 286,
        287
Syncytial specification, 41, 47–48, 249, 257
Syncytiotrophoblast, 349, 351
Syncytium, 47, 49n4
    in Caenorhabditis elegans, 245
    in Drosophila, 47–48, 249, 250, 257, 263
Syndetome, 463, 464, 484–485
Synergids, 188, 189, 220
Syngamy, 193
Synovial joints, 482

T

T lymphocytes, 130, 420, 514, 549
T3 (tri-iodothyronine) in amphibian
        metamorphosis, 564, 565, 566–568
T4 (thyroxine) in amphibian
        metamorphosis, 564, 566–567, 568
TAA codon, 57
TAA1 (TRYPTOPHAN
        AMINOTRANSFERASE OF
        ARABIDOPSIS), 114, 115, 116
TAA1-related (TAR) enzymes, 115
Tadpoles
    atrazine affecting, 633
    metamorphosis in, 8, 9–11, 563,
        564–568
    organogenesis in, 9, 10
    parasite infestation affecting limb
        development in, 523
    as secondary larvae, 563
TAG codon, 57
Tail development
    in amphibians, 318, 333n3
        and anterior-posterior axis, 320–321
        loss of, in metamorphosis, 564, 565,
            566, 568
    axis elongation in, 470–471, 472

    in Drosophila, 257
    in flatworm regeneration, 595–596, 597,
        598
    in mammals, 361
        in humans, and degeneration, 577n4
        progenitor cells in, 470, 471, 473–474
        and secondary neurulation, 366
        somitogenesis in, 473–474, 478–479
    in sperm, 195
    in zebrafish, 330, 331, 471
Tailless and tailless in Drosophila, 260, 262,
        265
Takahaski, Kazutoshi, 150
Talin, 94
Talpid in limb development, 534
Tamoxifen, 84
Tarsal bones, 517
Tarsus in Drosophila, 571, 572
Taste buds, 442, 549, 651
TATA-binding protein, 57
TATA-box, 57
Taylor, Doris, 99
Taz, 135
TBP in JAK-STAT pathway, 104
Tbr and Tbr in sea urchins, 281, 282
Tbx and Tbx, 525, 544n4
    in Caenorhabditis elegans, 242, 243
Tbx1 in heart development, 502
Tbx4 and Tbx4 in limb development,
        524–526, 527, 645
Tbx5 and Tbx5, 35
    in dorsal retinal cells, 434
    in heart development, 505
    in limb development, 523, 524–525, 527
        and epithelial-mesenchymal
            transition, 526–527
        in limb bud formation, 527
Tbx6 and Tbx6
    in heart development, 501
    in paraxial mesoderm, 465–466, 467,
        470, 471, 472
    in presomitic mesoderm, 465–466, 467
    in secondary neurulation, 374
    in somitogenesis, 472
    in tunicates, 290
Tbx16 in mesoderm specification, 465
Tbx20 in heart development, 507
TBX-35 and tbx-35 in Caenorhabditis
        elegans, 242, 243
Tbxt and tbxt, 303. See also Brachyury and
        Brachyury
TCF
    in sea urchins, 281
    in Wnt signaling pathway, 110
Tcf3 in amphibians, 312, 313
TEAD, 135
Teenage brains, 396–397
Teeth. See Tooth development
Tel gene in sea urchins, 282
Telencephalon, 375, 376, 450
Teleost fish, 380n1
Telogen, 458, 459
Telolecithal cleavage, 14, 15
Telolecithal eggs, 324, 336

Telomeres, 178
Telophase, 178, 183
Telson in Drosophila, 259–260
Temperature
    and genetic assimilation, 656–658
    and sex determination
        in angiosperms, 184
        in animals, 164–165, 173–174, 190n5
    and thermal gradient in sperm
        migration, 213
Temporal bone, 416
Temporal collinearity of Hox genes, 468–469
Temporal specificity of neural induction
        in amphibians, 319
Tendon development, 464, 484–485
Tennyson, Alfred, 577
Tentacles of hydra, 592
Teosinte, 654
Teosinte-branched cycloidea, 617n2
Teosinte glume architecture 1 (Tga1 and
        tga1), 654
Teratocarcinoma, 147, 637
Teratogens, 35–36, 621–628
    alcohol as, 35, 623–627
    biliary atresia from, 557
    cyclopamine as, 107, 623, 638–639
    cyclopia from, 107, 621, 623
    definition of, 124n6
    endocrine disruptors as, 628–634
    glyphosate as, 627–628
    retinoic acid as, 35, 505, 627
    thalidomide as, 35–36, 621, 642n2
Terminal end buds, 457
Testes, 160
    in androgen insensitivity syndrome,
        167
    atrazine affecting, 633, 634
    bisphenol A affecting, 632
    in Drosophila, 137
    in hermaphroditism, 166
    in primary sex determination, 161–166
    and secondary sex determination, 166,
        167
    and spermatogenesis, 176, 181, 194
    temperature affecting development of,
        174
    vinclozolin affecting, 635–636
Testis cords, 162
Testosterone
    in androgen insensitivity syndrome,
        167, 168
    and atrazine, 633, 634
    BPA affecting production of, 632
    and fracking chemicals, 634
    in primary sex determination, 161, 163
    in secondary sex determination, 166,
        167
Tetrad configuration, 177
Tetrapods, 246n2, 296
    limb development in, 517–545. See also
        Limb development
Tfap2 and Tfap2
    in neural crest specification, 404, 405
    in neural tube formation, 372

tfap2b and neural crest potency, 402

TFIIB, 66

TGA codon, 57

Tga1 and *tga1* (teosinte glume architecture 1), 654

TGF (transforming growth factor), 124n9

Tgfα in epidermis regeneration, *453*

TGF-β
and adult hematopoietic stem cells, 145
in amphibians, *306, 313, 314*
in angiogenesis, *508*
in cancer, 638
in *Drosophila* germ stem cell development, *136,* 137
in ectodermal appendage development, 456
in limb development, *532, 533, 536*
in liver regeneration, 614
and mesenchymal stem cells, 146
in neural crest development, *405*
in neural tube specification, 377–378, 379
in regenerative medicine, 150
in sea urchins, 282
in zebrafish, 330

TGF-β receptors, 112, *113*
in angiogenesis, *508*
in vertebrae formation, 482

TGF-β superfamily, 100, 101, 111–116

Tgf-β2 in hair follicle regeneration, *459*

Thalamus, *375,* 386

Thalidomide, 35–36, 621, 642n2
in cancer therapy, 639

Thecal cells, 163

Thermal gradient in sperm migration, 213

Thermodynamic model of cell interactions, 90–91

Thermotaxis, 213

Thimann, Kenneth, 114

Thomas, E. Donnall, 157n12

Thompson, J. V., 25

Thoracic vertebrae, 355, *356,* 366, 461, 467, *468,* 522

Thoracopagus, *359*

Thorax development in *Drosophila, 253, 254,* 257, 258, 259, 260
in homeotic mutants, 268–269
homeotic selector genes in, 268, 269
in prepupa stage, *571*

3' untranslated region (3' UTR), *56,* 57, 72, 85n3
and axis determination in snails, 235, *236*
and cytoplasmic localization of mRNA, 76, 77
and differential mRNA longevity, 72
in *Drosophila,* 258, 259
and microRNAs, 74, 75, 76
and RNA-induced silencing complex, 75
and stored oocyte mRNA, 73

Threespine stickleback fish (*Gasterosteus aculeatus*), 644–645

Threshold model of amphibian metamorphosis, 566

Thrombopoietin, *144,* 145

Thymidine, radioactive, 387

Thymus gland development
digestive tube in, 551
neural crest in, 549
pharyngeal arches and pouches in, 401, 416, 419–420, 439n1, 547, 549, *551*
Wnt signaling in, 553

Thyroid gland development, 548
in amphibian metamorphosis, 11
digestive tube in, 551
endoderm in, *548, 550*
neural crest in, 549
pharyngeal arches and pouches in, 416, 419, 439n1, 547, 549, *551*
progenitor specification in, *548*

Thyroid hormone receptors in amphibian metamorphosis, 567–568

Thyroid hormones in amphibian metamorphosis, 564, 565, 566–568

Thyroxine (T₄) in amphibian metamorphosis, 564, 566–567, 568

Tibia, 517, 519, *571*

Tie2 in vasculogenesis, *508,* 510

Tiger snakes, 658

tiggywinkle hedgehog, 124n5

Tiktaalik roseae, 26, 29, 520–521

tinman gene, 646

Tip cells in angiogenesis, 510

Tissue organization field theory, 636, 637

Toadflax (*Linaria vulgaris*), 659

Toads, metamorphosis of, 564, 565, 566

Tobacco, 587

Tobacco hornworm moth (*Manduca sexta*), 73, 656, *657*

Toes, 517. *See also* Digit development

Toll protein in *Drosophila,* 271

Toll receptor in *Drosophila,* 271

Tongue, 442
in amphibian metamorphosis, 566

Tonsils, 442, 547, 549, *551*

Tooth development, 454, 455, 456, 549
ectoderm in, 454, 549
epithelium in, 441, 455, 549
lack of, in chickens, 455
in mammals, 456, 457
milk teeth in, 457
neural crest in, 401, 415, 439n4
non-sensory placodes in, 441
odontoblasts in, 401
reaction-diffusion mechanism in, 655
and regeneration, 456, 457
signaling pathways in, 456
stem cells in, 146, 152, 457
teratogens affecting, *622*

Tooth Fairy Project, 152, 157n22

TOPLESS, 115, *116*

Torenia fournieri, 220

Torpedo and *torpedo* in *Drosophila,* 269–270

Torpedo stage in angiosperm embryogenesis, *12,* 13

torso gene in *Drosophila,* 260

torsolike gene in *Drosophila,* 256

Totipotent cells, 128, *129,* 131, 217
in mammals, 347
in plants, 131–132, 133–134, 585, 586–591
in regeneration, 585, 586–591

Townes, P. L., 89

Tra and *tra* in *Drosophila, 171,* 172–173, 190n4

Trachea, *357,* 557, 558

Tracheal rings, *415*

Tracheal-esophageal fistula, 557

Trachemys scripta elegans (red-eared slider turtle), 173–174

Tracts in central nervous system, 439n12

Transcriptase, reverse, 81

Transcription, 51, 62–70
β-globin gene, *56*
cap sequence in, 57
central dogma of biology on, 52
chromatin immunoprecipitation-sequencing in analysis of, 79–80
initiation of, 57, 58
in meiosis, 178
microRNAs in, 74–76
nucleosomes affecting, 55
regulation of, 51, *58,* 62–70
co-regulators in, 58
enhancers in, 58–61, 65, 66–67
epigenetic modification of chromatin in, 62–65
of floral organ identity genes, 67–69
promoters in, 56–57, 58, 64
silencers in, 58, 65
of stem cells, 131
termination sequence, 57

Transcription factors, 48, 58, 61, 62, 65–69
basal, 58
brain-specific, *59*
chromatin immunoprecipitation-sequencing in analysis of, 79–80
definition of, 58
duplication and divergence of, 647
and enhancers, 58–61, 65, 66–67
and gene regulatory networks, 69–70
in heterotopy, 649–650
inducing pluripotent stem cells, 150
limb-expressed, *59*
maternal, 69, *70*
in meristem, 132–133
nucleosomes affecting, 55
pioneer, 66–67, 69, *70*
and pluripotency of inner cell mass, 134
and promoters, 56–57, 58, 64
and silencers, 58, 65
in stem cell regulation, 131
synthesis of, 63

Transcription termination sequence, 57

Transcriptome, 80, 81

Transcriptomics, 81

Transfer RNA (tRNA) in egg cytoplasm, 196

Transformer and *transformer* (Tra and *tra*) in *Drosophila, 171, 172–173*
Transforming growth factor (TGF), 124n9
 Tgfα in epidermis regeneration, *453*
 TGF-β. *See* TGF-β
Transgenerational inheritance of developmental disorders, 634–636
Transgenic DNA chimeras, 22–23, *24*
Transit-amplifying cells, 129
 in adult intestinal stem cell niche, *142,* 143
Transition zone in neurulation, 366
Translation, 52, 72–77
 β-globin gene, *56*
 central dogma of biology on, 52, *52*
 differential mRNA longevity in, 72–73
 and differential posttranslational protein modification, 52, 77–78
 and 5′ untranslated region, *56,* 57
 inhibitors of, 210
 initiation of, *56, 57,* 72
 microRNAs in, 74–76
 ribosomal selectivity in, 73–74
 RNA interference in, 75
 selective activation of, 73–74
 selective inhibition of, 73
 termination of, 57
 and 3′ untranslated region, *57,* 72
Translation initiation site, *56, 57*
Translation termination codon, 57
Translocation of gametes to oviduct, 211–212
Transplantation experiments
 with amphibians, 20, 53, 295, 307–308, 318, *319,* 451, 568
 cell fate in, 40, 44
 conditional specification in, 44
 with flatworms, 595
 with hydra, 592, *593*
 induced pluripotent stem cells in, 151
 limb development in, 518, 521–523, 530–531, 534–535
 with neural crest, 22, 402, 410
 with optic vesicle, 451
 with presomitic mesoderm, 467, *468, 522*
 regional specificity of neural induction in, 318, *319*
 of Spemann and Mangold, 307–308
 transgenic DNA chimeras in, 23–24
 with zebrafish, 330, *330*
Transplantation of bone marrow, 143, 157n12, 512
Trans-regulatory elements, 85n4
Transverse process, *482*
Tree of life, 27, *225, 584*
Trematode infestation, extra limb development in, 523
Tremblay, Abraham, 579, 592
Trigeminal ganglion, 416, 442
Trigeminal nerve, 376, 416, 442
Trigeminal placode, 442, *443*
Tri-iodothyronine (T₃) in amphibian metamorphosis, 564, 565, 566–568

Triploblasts, 224–226
Triploid nucleus, 202
Trisomy syndromes
 Down syndrome (trisomy 21), 151, 620–621
 maternal age in, *183*
Trithorax proteins, 63, 65–66, 67
Triturus, 319
 T. cristatus, 307, *308*
 T. taeniatus, 307, *308*
tRNA (transfer RNA) in egg cytoplasm, 196
Trochoblasts, 41
Trochophore, 224
Trochus snail, *227*
Trophectoderm, 347, *350*
 and embryonic stem cells, *149*
 and inner cell mass, 134, 135
Trophoblast in mammals, *346,* 347, *349*
 in cavitation, 348
 in gastrulation, 348, 351
 in twins, 358
TRPC6 and *TRPC6* in autism, 152
Truncus arteriosus, *420,* 502, *506*
trunk in *Drosophila, 256*
Trunk mesoderm, 461–462
Trunk neural crest, 401, 410–414
 Hox genes affecting, 401–402
 migration pathways, 401, 410–414
 dorsolateral, 401, 410, 411, 413–414
 ventral, 401, 410–413
 potency of, 401–402
Trunk region
 in amphibians, 321
 in zebrafish, *327,* 331
Trypsin, 90, 124n1
TRYPTOPHAN AMINOTRANSFERASE OF ARABIDOPSIS (TAA1), 114, *115, 116*
Tube cells, 188
Tube protein in *Drosophila,* 271
Tubules in kidney, 496
 pronephric and mesonephric, 492, *493*
 proximal and distal, 494, *495, 496*
Tubulin
 and mesenchymal stem cell differentiation, *147*
 in oocytes, 183, 217
 polymerization into microtubules, 423
 in snail coiling, *229*
 in sperm flagellum, 194
 in zebrafish, 328
Tudor and *tudor, 174, 256*
Tulips, 650, *651*
Tumor cells, immortalized, 147
Tumor necrosis factor-α in liver regeneration, 614
Tunica albuginea, *162, 163*
Tunicates, 226, *276,* 288–291
 autonomous specification in, 41–43, 44, 289, 290
 bilateral symmetry in, *288*
 as chordates, 288, 292
 cleavage in, 288–290

conditional specification in, 290, 291
early development in, 288–291
as evolutionary link, 292
fate mapping of, 20–21, 41–43, 289–290
fertilization in, 194, *289*
gastrulation in, *16,* 289
heart development in, 290, 291, 501, 502
metamorphosis in, 288, 290
as model organism, *276*
muscle development in, 43, *288,* 290, 291
notochord in, 26, 27, 36n9, 288, 289, 290, 291, 292
pharyngeal arches in, 246n2
phylogenetic tree of, *296*
pigmentation in, 289
yellow crescent in, 42, 289, 290
Tunnel of Corti, 447
Turing, Alan, 531, 654
Turing model, 531–533, 654–655
Turtles
 atrazine affecting development of, 633
 environmental sex determination in, 5, 173–174, 190n5, 633
 gonad development in, 164
 scute of, 456
 shells of, 649–650
Twin protein and *twin* gene in amphibians
 in neural ectoderm induction, 315, 317
 in organizer formation, 312, 313, *313*
Twins, 357–359
 conjoined, 333, 358–359, 361n6
 X chromosome inactivation in, 620
Twist and *Twist, 271,* 407, 507
Tympanic cavity, *551*
Tympanic membrane, 444
Type A spermatogonia, 181
Type B cells in ventricular-subventricular zone, 137–141
Type B spermatogonia, 181
Type II deiodinase, 567
Type III deiodinase, 567
Tyrosine kinase, 102
 in amphibians, 321
 in egg activation, *207*
 in sperm capacitation, *213*
Tyrosine kinase receptor. *See* Receptor tyrosine kinase

U

Ubiquitin, *116*
Ubx and *Ubx* (Ultrabithorax and *Ultrabithorax*)
 in *Drosophila,* 268, 269, 354, 648, 656
 heterotypy of, 653–654
Udx1, *205,* 209
Ulex europeaus (gorse), *218*
Ulna, 517, *518,* 520, 529, 533
Ultrabithorax and *Ultrabithorax* (Ubx and *Ubx*)
 in *Drosophila,* 268, 269, 354, 648, 656
 heterotypy of, 653–654
Ultraspiracle (Usp), 574
Umbilical cord, stem cells in, 146, 157n16

Umbilical veins, 508
Uncx4.1 in somitogenesis, 477
Undifferentiated stem cells, 127
Unequal radial holoblastic cleavage in amphibians, 297–298
Unipotent stem cells, 130
Unpaired protein in *Drosophila, 171*
Untranslated regions
    3' UTR, *56, 57. See also* 3' untranslated region
    5' UTR, *56, 57, 72, 73, 85n3*
Ureter, *167, 493, 494, 498, 499*
Ureteric bud, *493, 494*
    basal lamina of, 494, *495*
    and metanephric mesenchyme interactions, 494–499
Ureteric bud tip cells, 497
Urethra, 166, 167, 498, *499*
Uridine triphosphate nucleosides, *78, 79*
Urochordates, *225, 226, 276*
Urodeles
    metamorphosis in, 564
    salamanders as, 564. *See also* Salamanders
Urogenital development
    intermediate mesoderm in, 491–499
    kidneys in, 491–499. *See also* Kidney development
    Wnt in, 108
Urogenital sinus, 166, 167, 498, *499*
Usp (Ultraspiracle), 574
Uterus, *345, 346, 347, 348, 351*
    in anterior-posterior axis formation, 353
    BPA affecting development of, 633
    in *Caenorhabditis elegans, 240,* 245
    contraction in translocation of sperm, 212
    diethylstilbestrol exposure affecting, 629, 630, *631*
    differentiation of, 166, *167*
    implantation in, 348
    and retroviruses, 662

V

V3 interneurons, 377
Vagal neural crest, 401, 412
Vagina, 166, *167*
    BPA affecting development of, 633
    diethylstilbestrol exposure affecting, 629, *630*
    sperm deposition in, 212
Vagus nerve, *416, 442*
valois in *Drosophila, 256*
Valves, cardiac, 507
Vas deferens, 167, 492
Vasa and *vasa, 174–175, 256*
Vascular endothelial growth factor (VEGF)
    in cancer, *639*
    in embryonic stem cell differentiation, *149*
    in sea urchins, 286
    in vasculogenesis and angiogenesis, *508, 509, 510*

Vascular endothelial growth factor receptor (VEGFR), *508,* 510
Vascular tissue of plants, 13, 19
    in canalization model of regeneration, 589–591
Vasculogenesis, 507–510
    extraembryonic, 508
    growth factors in, 509–510
    intraembryonic, 508
    sites of, 508
VCAM1 adhesion molecule, *138, 139*
Vegetal cells
    in amphibians, 298, 305, 307
        functions of, 306
        in mesoderm induction, *314*
        and Nieuwkoop center, 309–310
        in organizer formation, 312–314
        rotation in gastrulation, 299, 300–302
        specification as endoderm, 299
    in sea urchins, *280, 281, 282,* 284, 286–287
    in tunicates, 290
Vegetal plate in sea urchins, 277, 284, 286–287
Vegetal pole, 14, *43*
    in amphibians, *298, 300, 301*
    in sea urchins, 276, *277, 278, 279*
    in snails, 227, *228, 237*
    in tunicates, *288,* 289
    in zebrafish, *323, 327, 329,* 330
Vegetal rotation in amphibian gastrulation, 299, 300–302
Vegetative cell in angiosperms, 218
Vegetative phase in plant development, *12,* 13, 133
    and transition to reproductive phase, *12,* 67–69, 184
VEGF (vascular endothelial growth factor)
    in cancer, *639*
    in embryonic stem cell differentiation, *149*
    in sea urchins, 286
    in vasculogenesis and angiogenesis, *508, 509,* 510
VEGF-A, 509, 510
VEGFR (vascular endothelial growth factor receptor), *508,* 510
VEGFR-1, *508*
VEGFR-2, *508,* 510
VEGFR-3, 510
VegT and *VegT* in amphibians, 299, 306, *314*
Vena cava, 357
Ventral axis, *17*
Ventral blastopore lip
    in amphibians, *301, 303,* 305
    in zebrafish, 330–331
Ventral ectoderm in *Drosophila, 271*
Ventral furrow in *Drosophila, 252, 253*
Ventral mesoderm
    in amphibians, 309, 313, 315, 316
    in zebrafish, 330

ventral nervous system defective gene in *Drosophila,* 271
Ventral pathway in neural crest migration, 401, 410–413
Ventral root, *383*
Ventral uterine precursor cell, 122, 123
Ventricles, cardiac
    differentiation of, 505
    and heart fields, 502, *502*
    and looping of heart, 506, 507
    retinoic acid affecting, 505
    specification of, *506*
    in zebrafish, 609, 612, *613*
Ventricular cells, 382
ventricular myosin heavy chain (vmhc), 612
Ventricular radial glia, 388, 392, 395
Ventricular zone, 382, *383*
    in cerebellum, *383,* 384
    in cerebral cortex, *383, 386, 388, 389, 394*
    neurogenesis in, 388, 392
    in spinal cord or medulla, *383*
Ventricular-subventricular zone (V-SVZ), 137–141
    adult neural stem cell niche of, 137–141
    promoting differentiation in, 139–140
    rosette formation in, 138, 139
Veratrum californicum, 107, 623
Vernalization, *184*
Versican, 417
Vertebrae
    cervical, 355, 461, 467, *468,* 655
    coccygeal, 355
    comparative anatomy of, 355
    epithelial-mesenchymal transition in, 95
    and Hox genes, 74, 355–356, 467, 655
    lumbar, 355, 356, 467
    sacral, 355–356, 366
    sclerotomes in development of, *463, 464,* 481–484
    somites in development of, 462, 467
    thoracic, 355, *356, 366,* 461, 467, *468,* 522
Vertebrates, 224, *276*
    anamniotic, 295
    anterior-posterior axis in, 318
    axonal pathways in, 420–436
    brain growth and development in, 381–398
    cleavage in, 14
    epidermis in, 453
    evolutionary embryology of, 24–25, 26–27
    eye development in, 95–97, 448–451
    heart development in, 502
    Hedgehog pathway in, 105, *106,* 107
    hematopoiesis in, 511
    limb development in, 522, 524
    mesoderm in, 461
    neural crest in, 399–420
    neural tube in, 363–380
    neurulation in, 365–374
    Nodal in, 112

pharyngeal arches in, 18
phylogenetic relationships of, *296, 336*
phylotypic stage in development of, 24
planar cell polarity pathway in, 111
pre-mRNA splicing in, 71
in tree of life, *225*
von Baer's laws on, 24–25
Wnt signaling pathways in, 111
Vestibular system, *447*
    cochleovestibular ganglion in, 446
    otic placode in, 442
    semicircular canals in, 446
Vestibulocochlear nerve, 376, 442
Vestigial and *vestigial* in *Drosophila*, 575, 576
Vg1 and *Vg1*, 101, 111
    in amphibians, 306, *313, 314*
    in birds, 361
Viceroy butterfly, 36–37n10
Villi, intestinal, 553–554
    and adult intestinal stem cell niche, 142
Vinclozolin, 629, 635–636, 660
Vinculin, *94*
Viruses
    retroviruses, 662
    teratogenic effects of, *622,* 623
    Zika virus, 623, 642
Visceral endoderm, 351, 352, 353, 547
Visceral mesoderm, 500, *550. See also*
        Splanchnic mesoderm
Viscerocranium, 418
Vismodegib, 108
Visual cortex, 386
Vital dyes, 21
Vitelline envelope
    in acrosome reaction, 200
    in birds, 342
    in *Caenorhabditis elegans*, 242
    in external fertilization, *199,* 204
    and fertilization envelope, 204–205
    functions of, 197
    in sea urchins, *196, 197*
    in sperm–egg recognition, 197, 214
Vitelline membrane in birds, *337*
Vitelline veins, 508–509, 555, *556*
Viviparity, 2
vlad tepes, 515n4
vmhc (ventricular myosin heavy chain), 612
vnd gene, *647*
von Baer, Karl Ernst, 17, 24, 194
von Baer's laws, 24–25, 27
von Kölliker, Albert, 194
V-SVZ. *See* Ventricular-subventricular
        zone
Vsx2 in eye development, 451
Vulva in *Caenorhabditis elegans*, 121–123,
        240, 245
Vulval precursor cells, 121–122

W

Waddington, C. H., 656
Wagner, Dan, 49
Wasps, parasitoid, 660, 661

Water supply
    atrazine in, 633–634
    bisphenol A, 632–633
    fracking chemicals in, 634
Watermelon seeds, 220
Watson, James, 621
Wavefront, in clock-wavefront model of
        somitogenesis, 473–479
Webbing
    of bat wings, 649, 663
    of bird feet, 541–542, 543
Weismann, August, 44, 45
West-Eberhard, Mary Jane, 659
Whales, 26, 398n3, 543–544
White adipose tissue, 487
White blood cells, 512
White matter, 382, *383, 389,* 397
Whittaker, J. R., 43
Whole mount in situ hybridization, 78–79
Whorls, 68, 185, 186, *231,* 232
    and heterochrony, 650, *651*
    and heterometry, 652
Wieschaus, Eric, 108, 254, 255
Williams, Carroll, 570
Wilmut, Ian, 53–54
Wilson, A. C., 396
Wilson, Allan, 648
Wilson, Edmund B., 13, 45, 279
Wing development
    apical ectodermal ridge in, 528
    in bats, 28, *29,* 543, 649, 663
    in butterflies, pigment variations in,
        658
    in chickens, *525, 528,* 534–535, 544
    in *Drosophila,* 268, 269, 273, 645–646,
        656, *657*
        imaginal discs in, 118, *119, 570, 573,*
            575–576, 645–646
    homologous and analogous, 28, *28*
    progress zone in, *529*
    recruitment in, 645–646
    wing bud in, *529*
Wing imaginal disc in *Drosophila,* 118,
        119, 570, 573, 575–576, 645–646
Wingless and *wingless* in *Drosophila,* 108,
        260, 265, *266,* 267, *272,* 576
    diffusion of, *117*
Wnt and *Wnt,* 101, 108–111
    absence of, 109–110
    in amphibians, 305, *306,* 310, *311,* 312,
        313, *317*
        in anterior-posterior axis, 321
        gradient of, 321
        and neural induction along anterior-
            posterior axis, 318–321
        in organizer formation, 313
        in tail development, 333n3
        in trunk patterning, 321
    in axis determination, 243, 646
    in birds, *342,* 361
    in *Caenorhabditis elegans,* 243
    and calcium, *110,* 111
    in cancer, 638, 640
    and β-catenin, 109–111, 456, *524,* 527

in cranial placode induction, 443,
    444–445
in dermomyotome determination,
    486–487
diethylstilbestrol exposure affecting,
    630, *631*
diffusion of, 117
in digestive tube development, *552,* 553
in *Drosophila,* 108, 265, *266,* 318
in ectodermal appendage development,
    456
in embryonic stem cell differentiation,
    148, *149*
in endoderm specification, 548–549
in eye development, 449, 451, *452*
in facial features, 419
functions of, 108
in heart development, 503, 504
and hematopoietic stem cells, 145, 513
in jaw development, 658
in kidney development, *108,* 496, 497,
    498
in limb development, 108, *524, 537*
    and apoptosis, *542*
    in gradient model, 530–531
    in limb bud formation, 527
    and reaction-diffusion mechanism,
        533
in liver development, 553, 554
in lung development, 557
in mammals, 353
    in primary sex determination, 160,
        161, *163*
in myotome determination, *487,* 488
in neural crest delamination, 406, *407*
in neural crest migration, 409
in neural crest specification, 404, *405*
and neural progenitors in caudal spinal
        cord, *379*
and neuromesodermal progenitors, 379
in otocyst axis determination, 447
in paraxial mesoderm specification, *467*
in planar cell polarity, *110,* 111
in regeneration, 583
    in flatworms, 595–596, 597–598,
        600–601
    in hair follicles, 458, *459*
    in hydra, 593, *594*
    in teeth, 457
    in zebrafish, 608–609
in respiratory tube development, *558*
in sclerotome development, 481
in sea urchins, *280*
secretion of, 108–109
signaling compounds, 124n7
signaling pathways, 108–111, 359
    evolution of, 646
in somitogenesis, 478–479
in spermatogenesis, 181
in vasculogenesis, 507
in zebrafish, 111, 332, 608–609
Wnt inhibitors
    in cranial sensory placode induction,
        443

in heart development, 503
in neural crest development, *405*
Wnt inhibitory factor (Wif), 109
Wnt1 and *Wnt1*
  in dermomyotome determination, 486
  in flatworm regeneration, *597, 598, 600*
  in myotome determination, *487, 488*
  in sclerotome development, *481*
Wnt2, 614
Wnt2b, *142*
Wnt3 and *Wnt3*
  in hydra regeneration, 593, *594*
  in limb development, 527, *537*
  tectum gradient of, 435
Wnt3a, *109*, 353
  and adult intestinal stem cell niche, *142*
  in heart development, 503
  in limb development, *524*, 527, *537*
  in myotome determination, *487, 488*
  in organoid formation, *153*
  in paraxial mesoderm specification, *467*
  in sclerotome development, *481*
  in somitogenesis, 474, 478, *479*
Wnt4 and *Wnt4*
  in kidney development, *108, 498*
  in oogenesis, 176
  in primary sex determination in
    mammals, *161, 163,* 164
Wnt5, 111
Wnt5a and *Wnt5a*, 111, 630, *631*
Wnt6, 486, 498
Wnt7, 489
Wnt7a and *Wnt7a*
  diethylstilbestrol affecting, 630, *631*
  in limb development, 540–541
  in myotome determination, 488
Wnt8 and *Wnt8*
  in amphibians, 303, 321
  in heart development, 503
Wnt8a, 118, *119*, 332
Wnt8c, *342, 524*
Wnt9b, 498
Wnt11 and *Wnt11*, 111
  in amphibian organizer formation, 310,
    311, 312, *313*
  in kidney development, 497, *498*
Wnt11-6 in flatworm regeneration, *597*
wntP-2 in flatworm regeneration, *598*
Wolffian (nephric) ducts, 166, 167
  in bladder development, 498, *499*
  in gonad development, *161, 162*
  in kidney development, 492, 493, 494,
    495
Wolpert, Lewis, 14
Wood mouse (*Apodemus sylvaticus*), 212
Wound epidermis in salamanders, 603,
  604, 605, 606
Wound healing
  epithelial-mesenchymal transition in,
    95
  regeneration in, 582, 583. *See also*
    Regeneration
  scar formation in, 585, 586
WOX2, 133

Wright, Frank Lloyd, 283
Wrists, 533
  evolution of, 520–521
  in salamander regeneration, 602
Wt1 and *Wt1*, 72
  in primary sex determination, *161, 163,
    163,* 164
WUSCHEL and *WUSCHEL* (WUS and
  WUS), *132,* 133, 134

X

X chromosomes, 3
  in *Drosophila*, 169–173, *248*
  inactivation of, 168–169, 619–620
  in mammals, 159, 160–169
  stochastic factors in disorders of,
    619–620
Xbra and *Xbra*
  Activin gradient affecting expression
    of, 100, *101*
  in amphibian gastrulation, *302*
Xenopus laevis
  Activin gradient in, 100, *101*
  anterior-posterior axis in, 307, 320, 321
  axis formation in, 223, *311*
  BMP and neural ectoderm induction in,
    315–317, *318*
  β-catenin in, 310, 312
  cleavage in, *298, 360*
  dorsal-ventral axis in, *310*
  early development of, *10*
  extracellular matrix in, *93*
  eye development in, 96, 97, 98, 99, *449*
    in metamorphosis, 565
  fibronectin in, *93*
  gastrulation in, *93,* 299–305, *360*
  glyphosate affecting development of,
    628
  head development in, 320
  hematopoiesis in, 511
  left-right axis specification in, 322
  metamorphosis in, 565, *566, 567*
  mid-blastula transition in, 298
  as model organism, 6, 623
  neural crest in, 408, 409, *417*
  neural tube formation in, 372
  neurulation in, 366, *372*
  organizer functions in, *321*
  phylogenetic tree of, *296*
  primary embryonic induction in, 307,
    308
  regeneration in, 602
  retinoic acid affecting development of,
    627
  retinotectal projection in, 433
  somitogenesis in, 472, 478
  teratogenesis in, 623, 627, *628*
  thyroid hormone receptors in, 567
  trunk patterning in, 321
  vegetal induction of mesoderm in, *314*
Xenopus nodal-related and *Xenopus nodal-
  related* (Xnr and *Xnr), 313, 314, 322*
Xin protein in heart development, *506*

Xlim1 protein in amphibian organizer
  formation, 312
Xnr and *Xnr* (*Xenopus nodal-related*), in
  amphibians, 313, *314, 322*
Xnr1 and *Xnr1, 322*
XO genotype
  in *Drosophila,* 169, 170
  in mammals, 160, 169
Xrx1 and *Xrx1* in eye development, *449*
Xwnt8 protein, *314, 320*
XX chromosomes, 3
  in *Drosophila,* 170, 171
  in mammals, 159, 160, 163, 164, *165*
XXY chromosomes, 160, 190n1
XY chromosomes, 3, 159
  in *Drosophila,* 170, 171
  in mammals, 160, 161, 163, 164, *165*
    and androgen insensitivity
      syndrome, 167, *168*
Xylem, 13, 19
  in canalization model of regeneration,
    589–591

Y

Y chromosomes, 190n1
  in mammals, 159, 160–169
  Sex-determining Region of, 164–166
Yamanaka, Shinya, 67, 150
Yamanaka factors, *66*
Yap, *135*
Yellow crescent in tunicates, 42, 289, 290
Yolk, 14, *15,* 196
  in bird eggs, 196, 336–337, 338, 342
    in axis specification, *344*
  in zebrafish eggs, 324, 325, 326
Yolk cell in zebrafish, 325, 326, 327, 328,
  329, 330
Yolk plug in amphibians, *302,* 303
Yolk sac, 335, *336*
  in birds, 338, *341*
  blood islands of, 508–509, 511
  hematopoiesis in, 511
  in mammals, 348, *349,* 351
    in twins, *358*
  vasculogenesis in, 508–509
Yolk syncytial layer in zebrafish, *323, 326,*
  327–328, 332
  external, 326, 327, 328
  internal, 326, 327, 328
YUCCA (YUC), 114, *115, 116*

Z

Z1.ppp cell, 122, *123*
Z4.aaa cell, 122, *123*
Zea mays (corn), 654
Zeb-2 and *Zeb-2* in neural crest
  delamination, 407
Zebrafish, *296,* 322–333
  anterior-posterior axis in, *327,* 331
  axis elongation in, 470–471, 472
  brain development in, *323,* 387
  cavitation in, 380n1
  cleavage in, *323,* 324–327, *359, 360*
  conjoined twinning in, 333

cranial neural crest in, 439
CRISPR/Cas9 genome editing in, 82, 324
dorsal-ventral axis in, *323*, 330–332, 607
early development of, 322–333
embryogenesis in, *323*
embryonic shield in, 329, *330*, 332
eye development in, *323*, 450
fate mapping of, *22*, *49*, 324, 327
fertilization in, 322, 325
fins of, 524
    regeneration in, 607–609
forward genetics mutagenesis screens on, 81
Futile cycle mutant, *73*
gallbladder development in, 557
gastrulation in, *323*, 327–330, *360*
germ layers in, 327–330, *331*
green fluorescent protein in, *60*
gut tissue specification in, 551
heart development in, *323*, 332, *504*
    in regeneration, 609–612, 615
hematopoiesis in, 511
hypoblast in, *327*, 328, 329, 330
interkinetic nuclear migration in, 387
left-right axis in, 331, 332–333
maternal contributions in, *73*
medullary cord in, 380n2
microRNAs and maternal-to-zygotic transition in, *76*
as model organism, 5, 6, *296*, 322–324, 623
mutagenesis screens in, 324
neural crest migration in, *409*, 549, 623
neural keel in, 330, 380n1
neural stem cells in, 156n5
neural tube in, 330, 333n4, 372, 378, *471*
paraxial mesoderm in, 472

phylogenetic tree of, *296*
presomitic mesoderm in, 465
radial glia in, 387
reaction-diffusion mechanism in, 655
regeneration in, 135, 580, 583, 607–612
    of fin, 607–609
    of heart, 609–612, 615
retinotectal projection in, *433*
sclerotome resegmentation in, 482
somitogenesis in, 470, 472, 473, 477
stored oocyte mRNA in, *73*
tail development in, 330, 331, 471
teratogenesis in, 623
in tree of life, *225*
vasculogenesis in, 507
vertebrae formation in, 482, 483
Wnt in, 111, *119*, 332, 608–609
Zeugopod, 517, *518*
    gradient model on, *530*
    Hox gene specification of, 519, *520*, *530*, 539
    reaction-diffusion model on, *532*, *534*
Zic1 in neural crest specification, 404, *405*
Zic2 and *zic2*
    in amphibians, *317*
    in retinal ganglion axon guidance, *432*
ZicL in tunicates, *291*
Zika virus, 623, 642
Zinc
    in neural tube formation, 372, *373*, 380
    in polyspermy prevention, 216
Zinc-finger transcription factor family, *66*
Zinc spark in polyspermy prevention, 216
Zona pellucida, 198, 212, *345*
    binding of sperm to, 214
    in polyspermy prevention, 215, 216
    in sperm–egg recognition, 214
Zona proteins, 214
    in polyspermy prevention, 215

Zone of polarizing activity, 518
    in cetaceans, 544
    and digit identity, 536
    and dorsal-ventral axis, 541
    and Hox genes, 538–539
    Sonic hedgehog in, 534–536, 537, 538
ZW chromosomes in birds, 159, 189
Zygote, 4
    of amphibians, 9, 296, 298
    of angiosperms, 12, 13, *46*, *187*, 188, 220
    in animal life cycle, 7, 14
    of *Arabidopsis thaliana*, *12*, *46*, 126, *132*
    in asymmetrical cell division, 46–47
    of *Caenorhabditis elegans*, 240, 241, 242, 243
    chromatin in, 85n1
    cleavage of, 7, 14, 134, 223
    and CRISPR/Cas9 genome editing, *82*
    definition of, 1
    in double fertilization, *187*, 220
    of *Drosophila*, 250, 252, 259
    and embryonic germ cell layers, *17*
    formation in fertilization, 1, 193
    fusion of genetic material in fertilization, 210–211, 216–217
    and genomic equivalence, 53
    homozygotes and heterozygotes, 81
    of mammals, 216–217, 345, 346
        of mice, 134, 164
        in twins, 357–359
    mitosis of, 1, 7
    in plant life cycle, 12, 13
    of sea urchins, 202, 210, 211
    totipotency of, *129*
    of tunicates, 42
    and X-chromosome inactivation, *169*
    of zebrafish, 326
Zygotene stage, 177, *178*
ZZ chromosomes in birds, 159, 189